1st International Conference on Science and Technology for Sustainable Industry (ICSTSI 2020)

IOP Conference Series: Materials Science and Engineering
Volume 980

Banjarbaru, Indonesia
6 – 7 August 2020

Editor:

Rodiansono

ISBN: 978-1-7138-4794-6
ISSN: 1757-8981

Printed from e-media with permission by:

Curran Associates, Inc.
57 Morehouse Lane
Red Hook, NY 12571

Some format issues inherent in the e-media version may also appear in this print version.

Institute of Physics
Dirac House, Temple Back
Bristol BS1 6BE UK

Phone: 44 1 17 929 7481
Fax: 44 1 17 920 0979

techtracking@iop.org

Additional copies of this publication are available from:

Curran Associates, Inc.
57 Morehouse Lane
Red Hook, NY 12571 USA
Phone: 845-758-0400
Fax: 845-758-2633
Email: curran@proceedings.com
Web: www.proceedings.com

TABLE OF CONTENTS

IOP Conf. Series: Materials Science and Engineering **980** (2020) 011001 doi:10.1088/1757-899X/980/1/011001

PREFACE

The 1st **International Conference on Science and Technology for Sustainable Industry** (ICSTSI) is organized as a collaboration between The Institution of Research and Standardization of Industry Banjarbaru (Balai Riset dan Standardisasi Industri Banjarbaru) and The Chemistry Department of Lambung Mangkurat University. The theme of this conference is *"Emerging Science and Technology as a Solution for Global Challenge on Research and Technology Based on Sustainable Resources"*. Sustainability remains a crucial issue regarding how the industrial sectors use technology to explore and utilize resources in the era of industrial revolution 4.0. This revolution has started with the applications of advanced manufacturing and digital information technologies. The world is undergoing a fundamental transformation, which is changing the way we live, work and think. The fourth industrial revolution presents a future filled with exciting opportunities and new challenges and the potential to meet specific requirements during the COVID-19 crisis. Industry 4.0 technologies have the capability of providing better digital solutions for our daily lives during this crisis.

The main goal of this conference is to bring together scientists, academics, and practitioners who work in various fields to exchange their ideas by presenting their research findings. This two-day of conference are certainly the moment of scientific forum that can be used as a means of knowledge exchange to connect participants (in any discipline) with each other, so they can discuss their work, learning from one another, and achieving improvements in results.

In the initial plan, this conference should have been held in Banjarbaru, a city in the province of South Kalimantan, Indonesia on August 6th-7th 2020. Unfortunately, due to the extraordinary global health issue with the coronavirus disease (Covid-19) that causes the large-scale social restrictions in the framework of the acceleration of Covid-19 handling, ICSTSI was held virtually while maintaining its original schedule. There were about 250 attendees that participated in the conference meeting, and 154 of them were presenting their papers in the plenary sessions. The meeting was done by virtual instead of being postponed because virtual meetings have become the new normal under COVID-19. The virtual meeting was held by Zoom meeting video-conference method. This virtual conference could be used as a tool to speed up existing discussions and timelines among scientists. Each presenter delivered the paper for approximately 7 minutes presentation and followed by 3 minutes discussions and Q&A sessions. The event was appreciated by the participants because it was well-organized and being one of the pioneers in the new normal video-conference trends. The YouTube channel of the opening ceremony ICSTSI 2020 could be found here https://s.id/u1Kmu.

The plenary session in the first day consisted of the presentation by keynote speakers and poster presentations. The parallel session in the second day contained of invited speakers and oral presentations, divided into 6 different topics: material and applied chemistry; wood and non-wood forest products technology; food, cosmetics, and medicine; analysis and method validation industrial, process optimalization technology; biorefinery, bioenergy, and renewable energy and biotechnology; and waste treatment and environmental management. The keynote and invited speakers who had been presented on this conference are listed below.

Keynote Speakers:
Dr. Ir. Doddy Rahadi, MT (Ministry of Industry, Indonesia)
Prof. Dr. Shinso Yokota (Utsunomiya University, Japan)
Assoc. Prof. Dr. Azlan Kamari (Universiti Pendidikan Sultan Idris, Malaysia)
Prof Dr. Ir. Umar Santoso, M. Sc. (Faculty of Agricultural Technology Gadjah Mada University, Indonesia)
Mukhlis Bahrainy (The CEO of Pachira Group)
Prof. Dr. Is Fatimah (The Islamic University of Indonesia)

Invited Speakers:
Prof. Rodiansono, Ph.D (Lambung Mangkurat University)
Prof. Gustan Pari (Forest Products Research and Development Center)
Dr. Nazarni Rahmi (The Institution of Research and Standardization of Industry Banjarbaru, Ministry of Industry)
Ir. Lies Indriati (Center for Pulp and Paper, Ministry of Industry)
Dr. Lisman Suryanegara (Center for Research of Biomaterial, Indonesian Institute of Science)
Dr. Aris Mukimin (Center for Industrial Pollution Prevention Technology, Ministry of Industry)

The committee has received 154 papers from both oral and poster presentations, with the participants spread from many countries like Indonesia, Japan and Malaysia. The participants of conference came from various research institutions and universities such as Indonesian Rubber Research Institute, Palembang Institute for Industrial Research and Standardization; Center for Craft and Batik; Bandar Lampung Institute for Industrial Research and Standardization; Sembawa Research Centre; Department of Metallurgical and Materials Engineering; University of Indonesia; Department of Chemical Engineering; Politeknik Negeri Sriwijaya, Indonesia; Department of Agroindustrial Technology, Bogor Agricultural University; Department of Agroindustrial Technology, Lambung Mangkurat University; Department of Chemical Engineering, Engineering Faculty, University of Lambung Mangkurat; Departement of Chemistry, Gadjah Mada University; Wetland-Based Materials Research Center, Lambung Mangkurat University; State Islamic University of Sunan Gunung Djati, Bandung; Center for Ceramic, Bandung; Department of Chemistry, Universitas Palangka Raya; Institution of Research and Standardization of Industry – Padang; Environmental Technic Study Program of Lambung Mangkurat University; Department of Chemistry, Universiti Pendidikan Sultan Idris; Department of Leather Processing Technology, Politeknik ATK Yogyakarta; Faculty of Animal Science, Gadjah Mada University; Banjarbaru Enviroment and Forestry Research Development Institute; Research Division for Natural Product Technology (BPTBA), Indonesian Institute of Sciences (LIPI); Department of Pharmaceutical Chemistry, Gadjah Mada University; Institute of Research and Standardization of Industry, Banjarbaru; Indonesian Swampland Agricultural Research Institute, Banjarbaru; Research Division for Natural Products Technology-Indonesian Institute of Sciences; Faculty of Agriculture, Kyushu University; Faculty of Agriculture, Pattimura University; Accounting, Pamulang University; Center of energy resources technology and chemical industry, BPPT; Chemical Engineering,

Pamulang University. These papers were reviewed by scientific committee and 85 full papers were accepted to be published in IOP Proceeding Conference Series. We would like to thank the participants who have submitted their papers. We are grateful to IOP Publishing as our publication media partner. We also highly appreciate the sponsors who financially supported this event especially to the Industrial Research and Development Council Ministry of Industry, Indonesia.

LIST OF COMMITTEE

Steering Committee

- Head of The Institution of Research and Standardization of Industry Banjarbaru
- Head of Research and Development of Industry, Ministry of Industry
- Head of Agro Industry Research Center, Ministry of Industry
- Head of Department of Chemistry, Faculty of Mathematics and Natural Sciences, Lambung Mangkurat University
- Dean of Faculty of Mathematics and Natural Sciences, Lambung Mangkurat University
- Chairman of Indonesia Association of Food Technologist

Organizing Committee:

Person in Charge	:	Ratri Yuli Lestari, S.Hut., M.Env.
Chair	:	Dr. Nazarni Rahmi, S.TP., M.Si
Vice Chair	:	Sunardi, S.Si., M.Sc., Ph.D.
Secretariat	:	Utami Irawati, S.Si., M.E.S., Ph.D; Rais Salim, S.Hut.; Faiza Elisa Hasfianti,S.Hut., Fitri Yuliati, S.Si.
Treasurer	:	Endang Gembirawati, A.Md, Maria Dewi Astuti, S.Si., M.Si.
Event and protocoler	:	Hamlan Ihsan, S.Si; Desi Mustika Amaliyah, S.T., M.M.; M. Listianto Raharjo, S.T.; Drs. Taufiqur Rohman, M.Si.
Plenary session	:	Rodiansono, Ph.D.; Dr. Evy Setiawati, S.Si., M.T.; I Dewa Gede P. P., S.Si.
Equipments and transportation	:	Isa Anshari; Saibatul Hamdi, S.Hut, M.P.; Budi Tri Cahyana, S.T.; Ahmad Budi Junaidi, S.Si., M.Sc.; Dr. Rahmat Yunus, M.Si.
Food and Beverages	:	Rufida, A.Md.; Nadra Khairiah, S.Si, Dahlena Ariyani, S.Si., M.Si.; Kholifatu Rosyidah, S.Si., M.Si.
Publication and Documentation	:	Dwi Harsono, S.Hut.; M. Sobirin; Vemby Danang N.; Dwi Rasy Mujiyanti, S.Si., M.Si.; Noer Komari, S.Si., M.Kes.

Scientific Committee:

Editor in Chief:

- Prof. Rodiansono, Ph.D. (Chemistry Department, Lambung Mangkurat University)

Scientific Committee Members:

- Prof. Ris. Dr. Muhammad Hanafi (RC Chemistry-Indonesian Institute of Sciences LIPI)
- Prof. Dr. Enos Tangke Arung, S.Hut, M.P. (Faculty of Forestry, Mulawarman University, Samarinda)
- Prof. Dr. Is Fatimah (Department of Chemistry, Islamic University of Indonesia)

IOP Conf. Series: Materials Science and Engineering **980** (2020) 011001 doi:10.1088/1757-899X/980/1/011001

- Prof. Iriyanti Fatyasari Nata, S.T., M.T., PhD. (Faculty of Engineering, Lambung Mangkurat University)
- Prof. Rudianto Amirta, S.Hut, M.P. (Faculty of Forestry-Mulawarman University, Samarinda)
- Prof. Dr. Ir. Subyakto, M.Sc. (Center for Research of Biomaterial, Indonesian Institute of Science)
- Prof. Ris. Dr. Ir. Gustan Pari, M.S. (Forest Products Research and Development Center, Ministry of Forestry, Indonesia)
- Dr. Evy Setiawati, S.Si., M.T. (The Institution of Research and Standardization of Industry Banjarbaru)
- Dr. Ragil Widyorini, S.T.M.T. (Faculty of Forestry, Gadjah Mada University)
- Ir. Lies Indriati (Center for Pulp and Paper (BBPK), Ministry of Industry Indonesia)
- Dr. Dede Heri Yuli Yanto (Center for Research of Biomaterial, Indonesian Institute of Sciences-LIPI)
- Sunardi, S.Si., M.Sc., Ph.D. (Chemistry Department, Lambung Mangkurat University)
- Dr. Lisman Suryanegara, S.Si., M.Agr. (Center for Research of Biomaterial, Indonesian Institute of Sciences-LIPI)
- Muthia Elma, ST., M.Sc., PhD. (Chemical Engineering Study Program Engineering Faculty)
- Agung Nugroho, S.TP., M.Sc., Ph.D. (Department of Agro-Industrial Technology, Faculty of Agriculture, Lambung Mangkurat University)
- Dr. Hendra Wijaya, S.Si., M.Si. (Center for Agro-Based Industry-BBIA, Ministry of Industry Indonesia)
- Dr. Nazarni Rahmi, S.TP., M.Si. (The Institution of Research and Standardization of Industry Banjarbaru)
- Dr. Wiwin Tyas Istikowati, S.Hut., M.Sc. (Faculty of Forestry, Lambung Mangkurat University)
- Dr. Ihak Sumardi, S.Hut, M.Si., PhD. (Bandung Institute of Technology)
- Utami Irawati, Ph.D. (Chemistry Department, Lambung Mangkurat University)
- Adi Setyo Purnomo, S.Si., M.Sc., Ph.D. (Department of Chemistry, Faculty of Science and Data Analytics, Institut Teknologi Sepuluh Nopember, Surabaya, Indonesia)
- Dr. Maherawati, S.TP., M.P. (Faculty of Agriculture, Tanjung Pura University, Pontianak, Indonesia)
- Dr. Andriati Ningrum, S.TP, M.Sc. (Faculty of Agriculture Technology, Gadjah Mada University, Yogyakarta, Indonesia)

IOP Conf. Series: Materials Science and Engineering **980** (2020) 011001 doi:10.1088/1757-899X/980/1/011001

CONFERENCE DOCUMENTATION

Figure 1. Conference guests and participants.

Figure 2. Conference opening ceremony.

IOP Conf. Series: Materials Science and Engineering **980** (2020) 011001 doi:10.1088/1757-899X/980/1/011001

Figure 3. Conference invited speaker.

Figure 4. Conference committee.

Figure 5. Behind the scenes.

IOP Conf. Series: Materials Science and Engineering **980** (2020) 011002 doi:10.1088/1757-899X/980/1/011002

Peer review declaration

All papers published in this volume of IOP Conference Series: Materials Science and Engineering have been peer reviewed through processes administered by the Editors. Reviews were conducted by expert referees to the professional and scientific standards expected of a proceedings journal published by IOP Publishing:

- **Type of peer review: Double-blind**
  All papers came through a basic review which includes checking of initial technical criteria (paper scope and similarity checks). The initial technical criteria were checked by the editorial team. Any paper that out of scope was immediately rejected. The accepted papers received were continued to a peer-review process (submission structure, English language usage, and manuscript content) by two professional experts in a related subject area. The identities of both the authors and reviewers were kept hidden. After completing the peer review process, the process was continued by proofreading and layout.

- **Conference submission management system:**
  Google forms that connected to email of 1st International Conference on Science and Technology for Sustainable Industry

- **Number of submissions received:** 154 papers

- **Number of submissions sent for review:** 120 papers

- **Number of submissions accepted:** 85 papers

- **Acceptance Rate (Number of Submissions Accepted / Number of Submissions Received X 100):** 55.19%

- **Average number of reviews per paper:** two reviews

- **Total number of reviewers involved:** 23 reviewers

- **Any additional info on review process:**
  The reviewers were asked to answer a number of questions related to the quality of the content (reviewed form). Thus, the review processes were not personal and formalized. The review process for each paper took more than one month. The decision to accept or reject was based on the suggestions of each reviewer. If differences of opinion occurred between two reviewers, the editorial team pointed out one additional reviewer to review. The decision was based on majority votes based on all comments.

- **Contact person for queries:** seminar.brsbb@gmail.com

IOP Conf. Series: Materials Science and Engineering **980** (2020) 012001 doi:10.1088/1757-899X/980/1/012001

The effect of natural rubber on physical and mechanical properties of rubber seal for LPG tube valve

H Handayani*, A Ramadhan, A Cifriadi, N A Kinasih, A F Falaah and D R Maspanger
Indonesian Rubber Research Institute, Salak Street No. 1, Bogor, 16128, Indonesia

Email: hani.handayani@puslitkaret.co.id

Abstract. The price of natural rubber (NR) continues to decline. In early January 2020 the price of Standard Indonesian Rubber (SIR) 20 was around 1.83 US$/kg and in May 2020 dropped to 1.09 US$/kg. Under this condition, NR downstream program becomes a necessity to increase domestic consumption of NR so that the sustainable rubber industry can be achieved. One prospective product and can be substituted using NR is the rubber seal for Liquified Petroleum Gas (LPG) tube valve. Based on the rubber seal quality requirements in Standar Nasional Indonesia (SNI) 7655: 2010, the rubber raw material should use rubber which is resistant to n-pentane. Nitrile Butadiene Rubber (NBR) synthetic rubber is suitable, meanwhile NR is not resistant to n-pentane so it is not suitable to produce rubber seal for LPG tube valve. Hence, this research studied the effect of NR on physical and mechanical properties of rubber seal compared to NBR compound. The results showed that NR improved elastic properties such as tensile strength, elongation at break and compression set, but decreased the resistance to n-pentane. Mixing NBR and NR with appropriate compositions can produce a better rubber seal for LPG tube valve.

1. Introduction

Liquified Petroleum Gas (LPG) tube valve must be equipped with rubber seal to prevent gas leakage during usage or gas filling and to strengthen the regulator position. The product quality of the rubber seal in Indonesia must meet the national standard of SNI 7655:2010. This standard has been enforced compulsory referring to the Indonesian Ministry of Industry regulation No: 67 / M-IND / PER / 6/2012.

Rubber seal product quality requires raw material which is resistant to *n*-pentane according to SNI 7655: 2010. One kind of rubber that is resistant to *n*-pentane is Nitrile Butadiene Rubber (NBR) which is imported from abroad with the price in the domestic market of 3.50-5.00 USD/kg. The presence of nitrile groups in NBR structure makes it polar rubber so that it is more resistant to non-polar solvents and oils, while it has low mechanical and dynamical properties [1,2,3]. Natural rubber (NR) has excellent mechanical and dynamical properties [4], but it has low chemical resistance [5]. As a result, it cannot be used for making rubber seal of LPG tube valve. To increase the resistance of natural rubber to oil or gas, modification is required. Modification of natural rubber is needed to enhance the mechanical properties of NBR, to improve NR resistance to *n*-pentane, and primarily to increase domestic consumption of NR.

Modification of natural rubber can be done both chemically and physically. Chemical modification is done through an epoxidation reaction to produce oil-resistant natural rubber that is called epoxidized natural rubber, but it requires a relatively high cost [6]. Meanwhile, physical modification is done through mixing (blending) natural rubber with synthetic rubber NBR [7,8,9], and it requires lower costs than using modified natural rubber. Thus, this method was chosen to make formulas of rubber seal compounds.

A study on vulcanized rubber of NR/NBR mixtures (41% acrylonitrile) in different compositions with variations in the amount of carbon black has been done by [10]. The results showed that the compositions of the NR/NBR mixtures (40/60, 60/40 and 80/20) had a modulus of 100% which were greater than the modulus of 100% NR and NBR. Meanwhile, the tensile strength decreased with more NBR content in the mixture. The tensile strength of the mixtures was smaller than NR and NBR especially at low hardness. This indicated that NR and NBR were not compatible each other. The compression set of the NR/NBR mixtures increased with the smaller NR content in the mixtures. Compression continued to improve with the increased hardness of vulcanized rubber. The best compression sets were obtained from the compositions of 20/80 and 40/60 NR/NBR mixtures. The swelling properties using ASTM oil no.3 showed that the composition of the 20/80 NR/NBR mixture was best followed by the 40/60 composition. Baker concluded that NBR (41% acrylonitrile) could be substituted with a mixture of 20/80 NR/NBR.

Another study that investigated the properties of NR/NBR mixture without fillers by using dicumyl peroxide as a vulcanizing agent has been conducted by [11]. The result showed that the tensile strength trend with varied NR/NBR mixture compositions was different from those shown by [10]. The swelling properties were getting better with more NBR in the mixtures.

Several studies also have shown that the NR/NBR mixture with compatibilizer or homogenizer could increase the mechanical properties [8,12] and the swelling behavior in ASTM oil No. 1 and No. 2 [9,13], toluene [13]. Mixing natural rubber with NBR rubber can be done by using solvents, by mechanical or using rubber powder. The most common method used in the industry is by mechanical in the banbury or open mill. In this research, the natural rubber was mixed with NBR in an open mill. However, there was little information about the effect of natural rubber on physical and mechanical properties of rubber seal for LPG tube valve. Therefore, this study aims to study the effect of natural rubber on physical and mechanical properties of rubber seal compounds, such as hardness, tensile strength, elongation at break, compression set, swelling properties and resistance to ozone.

2. Materials and Methods

2.1. Materials

The materials used in this study include Natural Rubber (NR), *Standard Indonesia Rubber* (SIR) 20 obtained from local suppliers, Nitrile Butadiene Rubber (NBR) Krynac 3345 from Arlanxeo, and technical quality of rubber compounding chemicals from local suppliers. The rubber compounding chemicals include: sulfur as the vulcanization agent, stearic acid and zinc oxide as the activator, Si-69 (bis-triethoxy silyl propyl-tetrasulfide) as the coupling agent, red pigment as the color pigment, silica zeosil 175 MP as the filler, PEG (polyethylen glycol) as the dispersing agent, Antilux 654A as the antiozonant, MMB (mercaptomethyl benzimidazole) and TMQ (trimethyl dihydroquinoline) as the antioxidant, DOP (dioctyl phthalate) as the plasticizer for NBR, paraffinic oil as the plasticizer for NR, CBS (cyclohexyl benzothiazole sulfenamide) and TMTD (tetramethyl thiuram disulfide) as the accelerator, and lastly PVI (N-cyclohexylthio phthalimide) as the retardant.

2.2. Rubber seal compounding

Natural rubber and NBR were blended with other ingredients in the two-roll open mill according to the compound formulas. The addition of chemicals used a unit of phr (per hundred rubbers). This was used for calculating formulas where other chemicals were calculated as parts per 100 weight of rubber. Compounds were formulated with variations in 5 types of elastomer as seen in table 1. The compounds were then cured in a press machine at 150 °C to form the vulcanized rubber. The physical properties were then tested for hardness, tensile strength, elongation at break, and compression set at room temperature.

Table 1. The compounding formulas with various kinds of elastomer.

Compounding ingredients	Compounding Formulas (per hundred rubber, phr)				
	A (0% NR)	B (10% NR)	C (20% NR)	D (30% NR)	E (100% NR)
NBR 3345	100	90	80	70	-
NR	-	10	20	30	100
Silica 175 MP	40	40	40	40	40
Silane Si-69	2	2	2	2	2
PEG	2	2	2	2	2
Color pigment	1	1	1	1	2
Antilux 654A	3.5	3.5	3.5	3.5	3.5
MMB	2	2	2	2	2
TMQ	1.5	1.5	1.5	1.5	1.5
Paraffinic oil	-	-	-	-	8
DOP	8	8	8	8	-
Zinc Oxide	2	2	2	2	2
Stearic Acid	0.5	0.5	0.5	0.5	0.5
CBS	2.4	2.4	2.4	2.4	2.4
TMTD	3	3	3	3	3
PVI	0.5	0.5	0.5	0.5	0.5
Sulphur	0.4	0.4	0.4	0.4	0.4

2.3. Curing characteristics

Before curing, the mixtures of natural rubber and NBR were tested for the curing characteristics using MDR 2000 rheometer. The rheometer test was aimed at obtaining optimum vulcanization time and other maturation characteristics, such as maximum torque, minimum torque and scorch time.

2.4. Physical and mechanical testing

The vulcanized rubber was prepared in the form of test samples on the basis of standard methods. To test the hardness, the standard method was based on ASTM D 2240-15. Besides, the standard method for testing the tensile was based on ASTM D 412-16 and for elongation at break was based on SNI ISO 37:2015. The standard method for testing the compression set was according to ISO 815-1:2014 and for testing the swelling volume in *n*-pentane was according to ISO 1817:2015. The results of these physical properties were compared to the physical properties of rubber seal of LPG tube valve in SNI 7655:2010 [14].

3. Results and discussion

3.1. Curing characteristics

The curing characteristics, expressed in terms of maximum torque (S'_{max}) value, minimum torque (S'_{min}) value, delta torque ($\Delta S' = S'_{max} - S'_{min}$), optimum cure time (t90) and scorch time (ts2) of rubber compounds are shown in table 2. From the table, it can be seen that the compounds which used natural rubber at minimum 20% NR (compounds C, D and E) obtained higher maximum torque and delta torque values than full NBR compound (compound A). It means that more cross-linking was formed in the compounds using minimum 20% NR. Besides that, all compounds with natural rubber got lower optimum cure time and scorch time than that of full NBR compound, meaning that the compounds which used NR cured faster than full NBR compound. Faster cure or vulcanization time is needed by the industry to reduce cost production. From this curing characteristics aspect, it can be concluded that the substitution of minimum 20% NBR by natural rubber is better in NBR-NR compounding formulation.

Table 2. Curing characteristics of rubber seal compounds with variations in elastomer.

Curing Characteristics	Compounding Formulation				
	A (0% NR)	B (10% NR)	C (20% NR)	D (30% NR)	E (100% NR)
ΔS' (Kg-cm)	20.15	17.42	22.76	21.56	21.91
S'_{max} (Kg-cm)	21.36	18.88	24.64	23.29	23.13
S'_{min} (Kg-cm)	1.21	1.46	1.88	1.73	1.22
T_{90} (min)	18.23	15.08	16.24	17.59	15.34
TS_2 (min)	4.08	3.22	2.26	3.28	2.34

3.2. Physical and mechanical properties

The physical and mechanical properties of rubber seal compounds for LPG tube valve can be seen in table 3. It shows that the hardness of compounds A-E were slightly greater than the requirement of SNI 7655: 2010. This might be caused by the excessive amount of filler added which increased the hardness. In order to reduce the hardness value to fulfill the standard requirement, either the filler dose has to be reduced or the plasticizer dose has to be increased. On the other hand, the values of elasticity such as tensile strength and elongation at break of all compounds fulfilled the standard requirements. This was because of the use of natural rubber in the compounds that improved the elasticity. Besides that, the values for resistance of ozone obtained from all the compounds also could meet the standard requirements. At the same time, the values of swelling properties in *n*-pentane from compounds A-D met the standard requirements, while compound E which used 100% natural rubber was not resistant to n-pentane because of the polarity of natural rubber. Furthermore, the compression set values from all of the compounds did not meet the standard requirements. Thus, reformulation is still required to improve the compression set properties of the rubber seal for LPG tube valve.

Table 3. Physical properties of rubber seal compounds with variations in elastomer.

Physical Properties	SNI 7655:2010 Requirement	Compounding Formulation				
		A (0% NR)	B (10% NR)	C (20% NR)	D (30% NR)	E (100% NR)
Hardness (Shore A)	60 ± 5	67	67	67	67	68
Tensile strength (MPa)	Min. 10	12.6	12.6	13.5	14.0	23.4
Elongation at break (%)	Min. 300	340	350	400	420	610
Compression set, temp. 27±2 °C (%)	Max. 10	12.77	11.43	11.51	11.21	10.19
Swelling volume in *n*-pentane at RT, 168 h	Max. +35	13.24	20.29	22.21	31.64	69.74
Swelling volume in n-pentane at 70 °C, 96 h	Max. -12	-4.30	-7.6	-7.61	-8.34	-3.26
Resistance to ozone	No crack	No crack	No crack	No crack	No crack	No crack

4. Conclusion

From this research, it can be concluded that the use of natural rubber as the substitution of NBR in rubber seal compounds improved the elasticity properties of the rubber seal for LPG tube valve. The usage of minimum 20% NR showed better curing characteristics than that of the full NBR compound. Meanwhile, in terms of physical and mechanical properties, the use of 30% natural rubber resulted in the best properties and meet most of the requirements of SNI 7655: 2010, except for the compression set. In conclusion, the natural rubber as the substitution for NBR in making the rubber seal for LPG tube valve can be used at a range of 20%-30%, but reformulation is still necessary to improve the compression set properties.

Acknowledgments

This work was carried out with the financial support of the Indonesia Endowment Fund for Education or *Lembaga Pengelola Dana Pendidikan* (LPDP), Ministry of Finance of the Republic of Indonesia through the Productive and Innovative Research or *Riset Inovatif Produktif* (RISPRO) grant with the contract number 57/LPDP/2019. Hani Handayani, Arief Ramadhan and Adi Cifriadi as the main contributors of this paper are also thankful to the other members as their co-authors.

References

[1] Thabet A M, Abouel-Kasim A, Bayoumi M R and El-Sebaie M G 2009 Effect of carbon black loading in the swelling and compression set behaviour of SBR and NBR rubber compounds *Mater. Design* **30(5)** 1561-68
[2] Wang H M, Lu X R and Wang S J 2014 The swelling properties of nitrile rubber with different acrylonitrile contents in cyclohexane medium research *Adv. Mat. Res.* **936** 1942-47
[3] Choi S S and Ho H S 2009 Influence of the swelling temperature and acrylonitrile content of NBR on the water swelling behaviors of silica-filled NBR vulcanizates *J. Ind. Eng. Chem.* **15(2)** 167-70
[4] Ismail H, Majid R A and Taib R M 2014 Effects of dynamic vulcanization on tensile, morphological, and swelling properties of poly(vinyl chloride) (PVC)/ epoxidized natural rubber (ENR)/(Kenaf core powder) composites *J. Vinyl Addit. Techn.* **22(3)** 1-7
[5] Yasin K A, Ansarifar A, Hameed S and Wang L 2011 A new method for crosslinking and reinforcing acrylonitrile-butadiene rubber using a silanized silica nanofiller *Polym. Advan. Technol.* **22(2)** 215-24
[6] Gelling I R 1991 Epoxidised natural rubber *J. Nat. Rubb. Res.* **6(3)** 184-205
[7] Lewan M V 1998 *NR/NBR blends – basic problems and solutions in blend of natural rubber* (London: Chapman & Hall)
[8] Kongsin K and Lewan M V 1998 *Improving the morphology and properties of NR/NBR blends with polychloroprene as the compatibilizing agent in blend of natural rubber* (London: Chapman & Hall)
[9] Karnika de Silva, K G and Lewan M V 1998 Improving phase morphology and properties of NR/NBR blends with compatibilizer *J. Rubb. Res. Institute of Sri Lanka* **81** 38-50
[10] Baker C, Hallam W G and Smith, I F 1974 *NR Tech.* **5(2)** 29
[11] Ahmad M and Wheelans M A 1984 *NR Tech.* **15(4)** 78
[12] Go J H and Ha C S 1996 Rheology and properties of EPDM/BR blends with or without a homogenizing agent or a coupling agent *J. Appl. Polym. Sci.* **62(3)** 509-21
[13] Edirisinghe D and Freakley P K 2003 Effect of varied carbon black distribution on the morphology and properties of blends of natural and nitrile rubber *J. Rubb. Res. Institute of Sri Lanka* **86** 58-79
[14] Standar Nasional Indonesia (SNI) 7655:2010 about *Rubber seal of LPG Tube Valve* (Jakarta: Badan Standardisasi Nasional (BSN))

IOP Conf. Series: Materials Science and Engineering **980** (2020) 012002 doi:10.1088/1757-899X/980/1/012002

Thermal aging and chemical resistance evaluation of carbon black filled natural rubber blending: effect of the composition of acrylo nitrile and styrene butadiene rubber

T Susanto[1*], Rahmaniar[1], Farida[2], D W Lestari[2] and K Abdullah[3]

[1] Palembang Institute for Industrial Research and Standardization, Ministry of Industry, Indonesia
[2] Center for Craft and Batik, Ministry of Industry, Indonesia
[3] Bandar Lampung Institute for Industrial Research and Standardization, Ministry of Industry, Indonesia

Email: 3trisusanto87@gmail.com

Abstract. Natural rubber (NR) has excellent tensile and abrasion properties, but its thermal and chemical resistance needs to be improved. The present study investigates the effect of acrylonitrile-butadiene rubber (NBR) and styrene-butadiene rubber (SBR) composition in NR compounding filled carbon black (CB). The effect of chemical and heat resistance was examined by varying the addition of SBR/NBR (phr/phr) as follows 40/0; 35/5; 30/0; 25/15; and 20/20 into 60 phr (part hundred rubber) NR filled by 50 phr CB respectively. Compared with NR composites, SBR/NBR addition improved the hardness and density slightly. While tensile strength, modulus, abrasion resistance, and compression set decreased significantly. The volume, hardness, and physical properties were measured before and after 28 days of immersion in acidic, alkalis, and salty solutions. It was found that the higher loading SBR/NBR enhances chemical resistance. The hydrochloric and sulphuric acid immersion influenced the physical changes more vigorously. The thermal aging evaluation was observed by measuring the hardness and tensile strength at hot air circulation of 70°C for 24; 48 and 72 hours. The higher SBR/NBR loading could minimize the changes in hardness and tensile strength. To sum up, SBR/NBR addition improves the chemical and heat resistance on ternary blends.

1. Introduction

The demand for rubber and derivative rubber products for transportation, industrial machinery, mining applications, oil gas exploration, Etc has increased. Thus, many new nano-composites as popular advanced engineering materials have been developed to supply that demand. Due to the high price and less efficient nano-composites of polymer and fillers such as silica, carbon black, clay, and montmorillonite, new polymeric materials have been focused on mixing binary or ternary blends [1,2]. This method reduces production costs with low investment and a highly efficient process. The combination of some polymers may produce new polymeric materials with desired physical, mechanical, tensile, and other properties. Generally, the blending of rubber-rubber is preferable for commercial industrial production rather than rubber-plastic blends. Binary or ternary rubber blends have a high processing ability and a fine dispersion due to the crystalline structure and phase of each component are nearly similar [3,4]. However, the specified desired properties of the blends depend on the mixture ratio, polarity of polymers, composition of each polymer, miscibility, and the homogeneity of the polymers [5,6].

Natural Rubber (NR) has been the most well-known base polymer for polymeric materials for some advantages including good compression set, abrasion resistance, and tensile properties [7,8]. In addition, the availability of NR is higher than synthetic rubber (petroleum-based polymer). Thus, the production cost of NR-based products could be more profitable and efficient. Based on the chemical structure of NR that has double bond carbon (cis, 1-4 isoprene), some negative aspects may affect the physical and mechanical properties of NR [9]. For instance, the resistance of NR to oil, gas, ozone, oxygen, solvent, acidic solution, base solution, heating, and other chemicals is relatively lower compared to other synthetic rubber [10]. Therefore some efforts have been researched like chemical modification (epoxidation, chlorination, hydrogenation, grafting, de-hydrogenation, Etc) [11]. Due to the high price and complex technology of chemical modification, blending NR with other synthetic rubber was chosen to design polymeric material based natural rubber [1].

Since massive demand for coal as energy globally has risen, some coal exploration as open-pit mining has been developed in many areas. The issue of coal derivative products as new alternative energy becomes a trigger to increase the total coal production. Consequently, the demand for rubber products related to coal machinery, conveyor belt, idler roller belt, bucket elevator, and other open-pit mining equipment has doubled [12]. Rubber products for this application need special requirements particularly high chemical resistance, for instance, from acid water used during transporting the coal from the open-pit until the processing unit [12,13]. In addition, the products also shall have excellent physic-mechanical and thermal aging properties. Therefore, this research focuses on designing the ternary natural rubber blends that could comply with the rubber products' requirement standard.

Previous studies reported that ternary blends of NR/BR/SBR of 25/25/50 phr reinforced SiO_2 nanoparticles successfully improved the processing ability and mechanical properties by adding an optimum number of SiO_2 [11]. Burgaz, Gencoglu, and Goksuzoglu [10] found that optimum carbon black filled in ternary blends of NR/BR and NR/BR/SBR improve compression set, tensile strength and stress relaxation, and thermo-mechanical properties. Besides, Abdelsalam, Araby, El-Sabbagh, Abdelmoneim, and Hassan [5] stated that optimum carbon black filler in ternary blends of 30/30/40 phr of NR/SBR/NBR could increase the compatibility between CB and the polymer matrix. Commercial carbon black has been used as a reinforcing filler for polymer matrix since it is very fine dispersed and contributes massive enhancement on tensile and physic mechanical properties for organic polymers [5,8,14]. Based on those findings, carbon black was chosen as a reinforcing filler for ternary blends in this research.

To improve the chemical resistance of natural rubber-based polymeric materials, some additives were added such as nano alumina to against hydrochloric acid aqueous and sodium hydroxide; barium sulfate against hydrochloric acid aqueous; ZnO against acidic and base solution 15. Bach, Vu, and Vu [11] used carbon fillers, graphite, shungite filler, and silicon carbide to enhance rubber composites' chemical resistance in the oil pump impellers. The acrylonitrile-butadiene rubber (NBR) has been used widely in rubber composites for oil gas application since it has a good oil abrasion and chemical resistance [3,16,18]. Styrene-butadiene rubber is also commonly mixed with organic polymer because it has better abrasion resistance, crack initiation resistance, high filler loading capacity, and lower heat build-up behavior [2,4,19].

Previous research reported that binary blending of SBR and NR results in rubber that has excellent tensile strength, elastic modulus, and wear resistance with the good processing-ability and cure rate index [8,11]; the binary blending of NBR and SBR also could improve some mechanical and oil resistance [2,4]. However, research on ternary blends of NR, SBR, and NBR with predominantly NR filled carbon black is very limited. There is still a lack of information of processing-ability in which

IOP Conf. Series: Materials Science and Engineering **980** (2020) 012002 doi:10.1088/1757-899X/980/1/012002

NR is more dominant than SBR/NBR filled carbon black using homogenizer UB 4000. The scientific findings on the NBR addition effect against acidic and base solutions on NR/SBR polymer are very limited. To the best of our knowledge, no systematic work has been reported about the effect of SBR/NBR addition into dominated NR-filled carbon black polymeric materials on physico-mechanical thermal aging and chemical resistance particularly the resistance to the acidic and base solution of ternary natural rubber blends. The result of this research would be valuable in formulating natural rubber products for open pit mining application such as conveyor belt, idler roller, rubber pad, seal, and other rubber spare parts.

2. Experimental method

2.1. Materials

Natural rubber Standard Indonesian Rubber (SIR-10) was bought from Felda Indorubber crumb rubber factory in South Sumatera, Indonesia. The specifications of SIR 10 are as follows Volatile Matter 0,003%; PRI (Plasticity Retention Index) 68% and Mooney viscosity (ML1+4 of 82 at 102°C). Styrene-Butadiene Rubber (SBR) Bayer M Ltd Type 1502 has specifications as follows density of 0,945; styrene content 23% weight; Mooney viscosity (ML1+4)of 54 at 102°C. Acrylonitrile Butadiene Rubber (NBR) Bayer M Ltd has specifications as follows acrylonitrile content 32%; density of 1,17; Mooney viscosity (ML1+4 of 47 at 102°C).

Filler: carbon black high abrasion furnace as N330 that has 28-36 nm particle size and 70-99 m^2/g surface area. Main accelerator TMTD and secondary DPG. Compatibilizer used in this ternary blend was Ultra-blend 4000 (UB4000). Plasticizers and softeners: aromatic oil, Stearic Acid, paraffin wax (Oleo-chemicals Industry, PT Sumi Asih). Activators: ZnO (NC 105 Global Chemical Co, Ltd). Other standard chemicals for rubber compounding include age resister, sulfur curative agent, coumarone resin, 6PPD (Brataco Chemical), and other chemicals.

2.2. Rubber Compounding

The formulation of rubber mastication was listed in table 1. NR was masticated firstly for about 3-5 minutes, then SBR and NBR were mixed with NR for about 12-15 minutes. Some chemicals for rubber compounding were mixed as follows: 5 phr UB 4000; 50 phr Carbon Black; 5 phr ZnO; 1 phr Stearic acid; 3 phr paraffin wax; 2 phr sulfur; 3 phr BHT, and other 0,5 phr resin. The sample formulated in Table 1 were compounded based on the standard procedure ASTM D3182. The rubber mastication and compounding were prepared using an open two-roll mill of 220mm diameter, 560mm length at a speed of 10 r/min and friction ratio of 1:1,15. During the process, the temperature of rolls controlled periodically at less than 50°C to prevent pre-heated and pre-vulcanization of the rubber compound. After a sulfur curative agent was mixed, the sample for testing was pressed and heated using specific molding at 150°C within optimum cure time (t_{90}) about 16-20 minutes. Three replication of each formula were tested.

Table 1. Formulation ternary blends of NR/SBR/NBR.

Sample Formulation	NR (phr)	SBR (phr)	NBR (phr)
$NR/SBR/NBR_{(100/0/0)}$	100	0	0
$NR/SBR/NBR_{(60/40/0)}$	60	40	0
$NR/SBR/NBR_{(60/35/5)}$	60	35	5
$NR/SBR/NBR_{(60/30/10)}$	60	30	10
$NR/SBR/NBR_{(60/25/15)}$	60	25	15
$NR/SBR/NBR_{(60/20/20)}$	60	20	20

2.3. Testing procedures

Physico-mechanical properties were tested one day after heating and pressing at 25-30°C room temperature. Tensile properties such as tensile strength and modulus were measured based on the ISO 37. Universal Testing Machine - Instron 3366 at a crosshead speed of 500mm/min using a dumb-bell shaped cutter sample was used to measure tensile strength and modulus.

The density of ternary rubber blends was conducted based on the ASTM D395 at 25-30°C room temperature. Hardness (International Rubber Hardness Degree/IRHD) was measured based on ASTM D2240 using a Wallace dead load. Hardness was measured by preparing a 6 mm thickness of samples and counted after 5 seconds of loading.

The compression set of the ternary blends was measured at 25% deflection, 70°C for 22 hours based on ASTM D395. The compression set was subjected to examine the extent by which the samples fail to return to its original shape and thickness. The value was calculated between the differences of original thickness and recovered thickness.

Abrasion resistance was tested using Wallace Croydon Akron Abrasion tester following BS 903, Pt.A.9 Method C. The diameter was 15 cm of abrading wheel and 1-inch thickness was placed at an angle 15° to the abrading wheel. A load of 4,5 kg was placed on the disc with 12,7 mm in thickness and 6,35 cm in diameter that has a hole in the center, the result was calculated as volume loss per 1000 revolutions.

To evaluate the chemical resistance of ternary blends of NR/SBR/NR to hydrolysis, degradation, oxidation, and specific hardening or surface attack, in this research, some chemical solution was employed such as 3,5% NaCl; buffer pH 2; buffer pH 4; 10% HCl; 5% H_2SO_4; 10% Detergent; 10% NH_4OH; and 10% NaOH at room temperature. Chemical resistance was measured following the ISO/TR 7620. The test samples with 2 mm in thickness were completely immersed in the chemical solution in a close vessel to avoid evaporation. The samples were immersed at 25-30°C room temperature for 28 days. The volume and hardness of the fresh samples and after immersed samples were measured. The change of hardness was evaluated by subtracting the hardness after immersion with the fresh samples. The percentage change of volume (%V) was calculated based on the different mass of samples in the air after immersion from the initial mass of samples in air and the immersion chemical solution density. Analytical balance 0,0001g was used to weigh the samples.

The thermo-oxidative aging evaluation was tested according to the ASTM D473. The same fresh and aged samples were measured before and after the aging process using Aging Chamber Oven Model VAT, 60, ASLI (China). The vulcanized ternary blends of NR/SBR/NBR were aged using hot air circulation at a test frame of 5 rpm; setting the temperature at 70°C for 24 hours and 72 hours. Hardness and tensile strength were measured before and after aging. The difference of its value was analyzed as the aging resistance of ternary blends of NR/SBR/NBR. The test was conducted in triplicate for each formulation, the average value was reported in this paper.

3. **Results and discussion**

3.1. Physco-mechanical properties

Compared to single blends of NR with 50 phr of carbon black filler, the addition of SBR/NBR improved the hardness and density significantly. Both SBR and NBR had higher hardness than NR, thus the higher loading of NR might increase the hardness of ternary blends as listed in table 2. Along with the increased hardness, the carbon black filler, the ternary blends' density was slightly higher than NR composites. As depicted in table 2, at the same number of carbon black filler, the NBR addition improved hardness and density less significantly. In agreement with previous research, the higher loading of carbon black improved the hardness and density of NR/BR [2,5,19]. It is found that the SBR/NBR addition did not affect the hardness and density significantly [4]. In confirmation of previous research, the hardness value is mainly determined by reinforcing the mix in rubber composites [5,6,10].

The remarkable enhancement of the tensile strength and modulus was caused by the less interaction between NBR, SBR, NR during the mastication process and less fine dispersion of carbon black within the rubber matrixes. During mastication, NR was more dominant than SBR and NR. It is predicted that an excessive 60 phr of NR based on rubber polymer might cause the mastication inhomogeneous. Similar results of NR and BR's blending were less finely masticated [5], thus homogenizer UB4000 was added.

Table 2. The Effect of SBR/NBR loadings on physico-mechanical of NR/SBR/NBR ternary blends.

Samples formulation	Composition NR/SBR/NBR ternary blends					
	100/0/0	60/40/0	60/35/5	60/30/10	60/25/15	60/20/20
Hardness (IRHD)	64	70	71	71	73	74
Density (g/cm^3)	1,137	1,229	1,235	1,235	1,238	1,241
Tensile Strength (MPa)	17,8	15,1	14,1	14,1	14,1	13,9
Modulus at 100% elongation (MPa)	7,4	5,7	5,2	5,2	5,1	5,1
Modulus at 300% elongation (MPa)	9,1	6,3	5,7	5,7	5,5	5,4
Abrasion resistance (DIN.mm^3)	75	71	65	65	64	63
Compression set 25% defl, 28°C, 72 hours (%)	6,8	7,2	7,1	7,2	7,3	7,3

Similarly, the modulus trend at 100% and 300% elongation also significantly decreased due to the high loading of SBR/NBR. Similarly, it is reported that the reactivity of surface interaction caused this decrease in modulus, the weaker SBR/NBR/NR interaction with carbon black, aggregation between NR/SBR/NBR - carbon black, and filler dispersion in the rubber matrix [5,10]. Based on table 2, the formulation of 60/30/10 phr of ternary blends might result in optimum and stable modulus and tensile strength. It implies that 50 phr of carbon black filler had better interaction with the portion of 60/30/20 ternary blends of NR/SBR/NBR.

The affinity of carbon black filler and ternary blends may carbon black and rubber matrixes' interaction by weakening hydrogen bonds and van der walls interaction among them. The improvement of modulus, tensile strength property may be due to the fine dispersion of CB in the ternary blends using homogenizer UB4000 [1]. In addition, rubber filler interaction between CB and NR/SBR/NBR associated with the contact of the surface of filler resulted in a more effective constraint of the motion of rubber chains with improved interfacial bonding between CB and NR/SBR/NBR [2,5]. The loading of SBR/NBR may increase, but at the optimum number of NR, the tensile strength and modulus would remain stable at 60/30/10 phr. It is also reported that an excessive number of NBR and CB led to agglomeration and aggregation of filler - filler or filler-rubber matrix, so the ternary rubber blends phase was at a weak point and decreased the tensile properties [4,10]. However, to ensure this phenomenon and study the morphology of this ternary blends in this research, further observation using SEM might be necessary.

The compression set is the important key parameter in deflection and compressive loads of rubber parts such as rubber pad, damper, idler roller, seal Etc. Compression stress increased with increasing filler content [2,3,19] while the compression set determined the ability of rubber vulcanizates to retain its elastic properties after prolonged compressive loads at certain temperatures [10,20]. The small compression set refers to the elasticity and the ability to maintain original thickness and low damping properties while a large compression set indicates lower stiffness but higher damping properties. The compression set is closely related to the Payne effect, the viscosity of NR/SBR/NBR, and the sulphuric vulcanization system [2,11,13]. It is shown that the compression set of ternary blends increased with the increase of NBR addition until it reached about 7,3%. Compared to NR composites, ternary blends has a high compression set that represents the material blend of NR, SBR, NBR which needs to be improved. In other words, the rigid carbon black fillers in an appropriate amount of SBR/NR, affinity and polarity of SBR/NR with carbon black might restrict the chain mobility of the network of ternary elastomeric rubber blends. Table 2 shows that the lowest compression set was NR composites. It could be explained that the chemical bonds and physical interaction of CB filler rubber were strong enough, so the compression set results to show less permanent deformation [4,19]. The most elastic blends were NR composites, whereas the higher NBR the less elastic NR/SBR/NBR ternary blends are. As similarly reported by previous research, this more permanent deformation of NR/SBR/NBR was due to its relatively higher rigid styrene and acrylonitrile content [2,5].

The abrasion resistance or the volume loss due to mechanical rubbing and erosive action is a major requirement for tire, rubber pad, rail pad, etc. Similar to the decreasing value of tensile strength, the abrasion resistance also experiences a decrease of about 4-12 DIN.mm^3 by the increasing SBR/NBR in ternary blends. The blends have less crosslink density that leads to allowing erosion due to low stiffness [5,14,17]. Even though NBR and SBR have good abrasion resistance as reported by Pashaei, Hosseinzadeh, and Syed [4], table 2 shows different results that the addition of SBR and NBR decreased abrasion resistance. This is because the ternary blends of NBR and SBR were not as homogenous as NBR composites filled carbon black. This phenomenon is also found in research by Susanto and Marlina [8] who blend SBR and NR filled carbon black; Bach, Vu and Vu [11] who blend NR/NBR filled silica. The higher SBR/NBR (20/20 phr) addition influences the abrasion resistance value less significantly. Some papers also reported a similar finding. For instance, acrylonitrile does not directly affect the bonding interaction between filler rubber while NR/BR does due to the less dispersion and affinity of carbon black among the three different ternary polarities of ternary blends (NR, SBR, NBR) [4,5,19]. Consequently, it could not contribute strongly to resist friction and decrease the lost volume.

3.2. *Chemical resistance*

Chemical resistance of ternary blends of NR/SBR/NBR was evaluated to measure the ability to resist the chemical attack with fewer physical appearance changes, surface, volume, and hardness over a period of time. In a chemical environment such as a chemical solution, polymeric materials can react chemically, absorb, dissolve, crack, stress, and plasticize, diffuse solvent into the inner surface [7, 8, 15]. Based on the ISO/TR 7620:2005 (E), the good resistance of ternary blends is showed by the minimum changes of volume and hardness after chemical solution immersion for 28 days at temperature 25-30°C. Table 3 shows the comparative acids, alkalis, and salty resistance of ternary blends NR/SBR/NBR. As listed in table 3, the changes in hardness and volume of ternary blends became smaller because the higher SBR/NBR addition shows that SBR/NBR contributes positively. The complex structure of NR, SBR, and NBR filled carbon black might resist chemical attack stronger than NR filled carbon black composites.

In systematic comparison in table 3, NR filled carbon black composites without SBR/NBR exhibited a minor effect deterioration after 28 days of alkalis and acidic solution. In comparison, the SBR/NBR addition into ternary blends contributed positively to the chemical resistance. Previous research mentioned that the effect of styrene and acrylonitrile enhances acidic resistance [15]. As listed in table 3, the higher NBR added means the lower decrease in hardness value and the less swelling effect. It is clearly shown that each formulation did not affect the increasing volume after immersion. The same results were reported that the adsorption effect on the inner surface of ternary blends was minimized due to the formation of the new surface crust [15]. Compared to oil and organic solvent immersion, the effect on physical properties changing would be more rigorous than a chemical solution since the differences in vapor pressure, chemical properties, and diffusion transport enable to accelerate swelling or shrinkage effect [8,19].

Immersion of the ternary blends in buffer pH2 solution caused a decreased hardness more significantly than the pH4 solution. Whereas, the volume did not significantly increase due to the adsorption effect on the surface. It is highlighted that 10% of HCl immersion caused an extreme increase in hardness at + 15 IRHD on NR composites. The ternary blends with SBR/NBR also exhibit an increase in hardness for only + 6-10 IRHD. This result agrees with the impermeable film on NR composite's surface of the formation of the impermeable film on the surface of NR composite during the chlorination process; hydrochloric acid would cause NR to crust formations of chlorinated rubber membrane with a certain thickness and increasing hardness by Hadi and Kadhim [15]. However, for some time until the membrane dissolves, hydrochloric acid will crack and damage the inner membrane.

Conversely, sulfuric acid destroys the surface of NR composites more strongly than ternary blends. In addition, sulfuric acid immersion contributes to the decreasing in each formulation's hardness more vigorously than other acidic solutions such as buffer pH 2 and pH 4. As reported by Kashani, Ngo, Hemachandra and Hajimohammadi [21], sulfuric acid is very corrosive to tires (natural rubber composites) and may destroy the double back bonds of isoprene.

Table 3. The Effect of SBR/NBR loading on the chemical resistance of NR/SBR/NBR ternary blends.

Samples Formulation	Change In	H_2O	Buffer pH 4	Buffer pH 2	10% HCl	5% H_2SO_4	10% NH_4OH	10% NaOH	10% Detergent	3,5% NaCl
NR/SBR/NBR (100/0/0)	Volume (%)	0	0,13	0,15	0,17	0,31	0,16	0,18	0,02	0,22
	Hardness IRHD	0	-4	-4	+15	-8	-5	-8	0	-2
NR/SBR/NBR (60/40/0)	Volume (%)	0	0,12	0,13	0,15	0,28	0,12	0,12	0	0,1
	Hardness IRHD	0	-2	-2	+10	-5	-1	-2	0	-1
NR/SBR/NBR (60/35/5)	Volume (%)	0	0,08	0,08	0,08	0,19	0,08	0,07	0,02	0,03
	Hardness IRHD	0	-2	-1	+9	-4	-1	-2	0	-1
NR/SBR/NBR (60/30/10)	Volume (%)	0	0,05	0,06	0,08	0,17	0,08	0,06	0,01	0,03
	Hardness IRHD	0	-2	-1	+8	-4	-1	-2	0	-1
NR/SBR/NBR (60/25/15)	Volume (%)	0	0,02	0,03	0,07	0,18	0,01	0,03	0	0,02
	Hardness IRHD	0	-1	-1	+6	-4	-1	-1	0	0
NR/SBR/NBR (60/20/20)	Volume (%)	0	0,02	0,02	0,05	0,17	0,01	0,03	0	0,02
	Hardness IRHD	0	-1	-1	+6	-3	-1	-1	0	0

The resistance of NR composites in alkalis solution was weaker than ternary blends. The high loading of SBR/NBR in NR composites could minimize the hardness and volume changes. Compared to the acidic solution, the changes were relatively smaller than both HCl and H_2SO_4 immersion effects. Similarly, the immersion detergent and NaCl solution did not have any significant effect on physical properties changes. Looking at table 3, the addition of SBR/NBR could positively impact acidic, alkali, and salty resistance. The effect of good compatibility of acrylonitrile, styrene, and isoprene groups filled by carbon black in ternary blends rubber. As explained in the effect of SBR/NBR on physico-mechanical properties, the chemical resistance characteristics might be enhanced by forming more cross-link bonding and complex structure of matrix polymer. This similar clarification has also reported the contribution of styrene groups on chemical resistance [21].

IOP Conf. Series: Materials Science and Engineering **980** (2020) 012002 doi:10.1088/1757-899X/980/1/012002

3.3. Thermal ageing properties

In this work, chemical immersion evaluation and thermal aging properties become key parameters on rubber product evaluation. As designated for rubber products in coal open pit mining usage, acid mine water and sun-light, oxygen, ozone exposure, air oxidation, and heating from the friction and conduction of the machine may reduce the shelf life and shorten the usage period [12,13].

As mentioned before, since the ternary blends were mainly dominated by NR that was susceptible to thermal and air oxidation due to the attack on carbon-carbon double bonds of n-isoprene, the SBR and NBR were added to the blends properly. Similar findings were also reported by Liu, Li, Xu and Zhang [22] that NBR enhances the thermal degradation properties of composites.

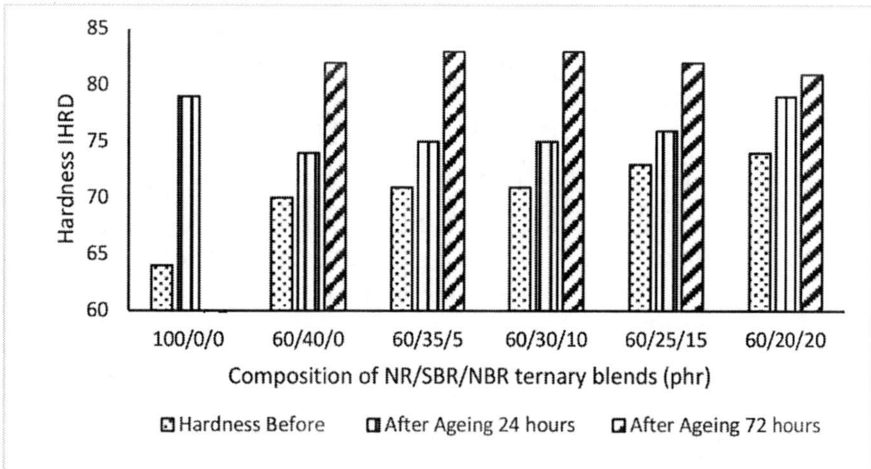

Figure 1. Thermal aging profile: surface hardness changes of NR/SBR/NBR ternary blends.

Figure 1 shows surface hardness changes due to thermal aging at 24 and 72 hours. The graph shows that hardness increased insignificantly as a function of SBR/NBR content as expected. By comparing before and after age 24, 72 hours, it is noticeable that the addition of SBR/NBR contributed positively to the heat resistance of the ternary blends. The NR composites experienced serious damage on the surface, the hardness became remarkably higher by the more extended aging and after 72 hours, the surface became mushy. Previous research mentioned that overheating for a long time could destroy the molecular chain of (n 1,4 cis-isoprene) and influence the movement and mobility of the filler and break the bonding filler- rubber and rubber-rubber [11,19,20]. The main-chain scission polymer backbone or sulfidic cross-linked during the ageing period which led to cleavage of the networking structure and change rubber properties [18]. On the other hand, compared to NR composites without SBR/NBR, the graph shows that the hardness of NR/SBR/NBR ternary blends before and after 24 hours of aging did not change significantly (the increasing hardness is about 3-5 IRHD). It is showed that the higher loading of SBR/NBR, the smaller hardness changes of ternary blends. Higher SBR / NBR can withstand heating longer and prevent surface thermal-oxidative degradation. Similar findings stated that SBR and or NBR might contribute to span the overheating that leads to inhibit overheating and over vulcanization so that the stiffening effect that destroys the surface of ternary blends might be hindered [1,11,20,22].

Tensile strength values decreased by the longer aging time at 70°C approximately 3-7 MPa for each composition. Figure 2 depicts the remarkably decreasing tensile strength after 24 and 72 hours of NR composites without SBR/NBR addition. Meanwhile, in the ternary blends when SBR and SBR/NBR were blended with NR, the changes of tensile strength could be minimized by the more extended periods. Similarly to the hardness changes, the aging may deteriorate the tensile strength of the ternary blends. The decrease in the tensile strength value due to the longer aging period is less significant with the higher the SBR / NBR loading. This is in agreement with research findings by Zainal, Santiagoo, Ayob, Ghani, Mustafa, and Othman [20] on the improvement of thermal properties on the blending of propylene and acrylonitrile butadiene rubber using carbon black bagasse. Besides that deterioration factor as explained in hardness changed, tensile strength dropped due to the over vulcanization. Over a long time, it may lead to serious cross-link density that exceeds a critical point. Consequently, the chain segment becomes less effective, less mobile, and distracts the normal orientation of the network chain and stress concentration, so that the tensile strength was dropped [1,2,18].

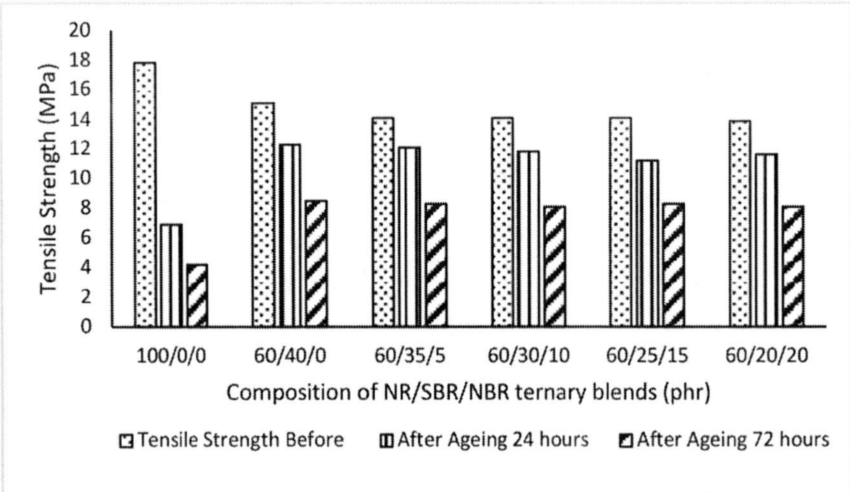

Figure 2. Thermal aging profile: tensile strength changes of NR/SBR/NBR ternary blends.

Another aspect that may exacerbate the thermal aging effect was evaporation. The number of volatile matter compounded in this research such as aromatic oil, paraffin wax, and other chemicals might evaporate and migrate from inner to its surface of the polymer thus it weakened the tensile strength. Theoretically, less finely carbon black dispersion within polymeric ternary blends influences the surface damage so that the optimum amount of filler should be dispersed properly [5,10]. The processing-ability also affects the homogeneous polymer matrix and the tensile and thermal oxidative degradation of the ternary blends.

4. Conclusion

SBR/NR addition into NR polymeric materials reinforced carbon black was successfully blended by using UB4000 as a homogenizer. The increased values of hardness, density, and compression set were more significant than NR composites in the same number of 50 phr carbon black. Furthermore, with the addition of NBR which was higher than the SBR in the ternary NR mixture, the parameter value experienced a slight increase. Conversely, the NBR addition negatively affects tensile strength, modulus 100% and 300% elongation, and abrasion resistance of the ternary blends.

The addition of SBR / NR indicated that the ternary mixture became less elastic and had more permanent deformation. The compression set value increased by the SBR/NBR addition. Agglomeration and less fine carbon black dispersion due to NBR/SBR might reduce the chemical bonding of filler-rubber and rubber-rubber within the ternary blends. The nitrile and styrene groups are more rigid than isoprene, influencing the tensile and physical properties. It is found that NBR/SBR also enhanced the chemical resistance of the ternary blends. In the systematic evaluation by observing the changes of hardness and volume using chemical solutions within 28 days, Changes in hardness and volume can be categorized as not having a significant effect as a substitute for 5% H2SO4 and 10%

HCl solutions. Thermo-oxidative aging at 70°C temperature influenced the increasing hardness and decreasing tensile strength of the ternary blends by the more extended period time. The SBR/NBR loading could minimize the hardness and tensile changes. Compared to NR filled carbon black composites, ternary blends had higher heat resistance. It is suggested to study the processing-ability and morphology of the ternary blends for further research.

Acknowledgment
The authors would like to express their gratitude to the Palembang Institute for Industrial Research and Standardization, Ministry of Industry for financial support, laboratory instruments, and other facilities. The authors also gratefully acknowledge all team members for the hard work during finish the research.

References

[1] Jovanović S,. Samaržija-Jovanović S, Marković G, Jovanović V, Adamović T and. Marinović-Cincović M 2016 *Composites Part B: Engineering* **98** 126-33
[2] Xue X, Yin Q, Jia H, Zhang X, Wen, Ji Q and Xu Z 2017 *App. Surface Sci.* **423** 584-91
[3] Mao N D, Thanh T D, Thuong N T, Grillet A C, Kim N H and Lee J H 2016 *Composites Part B: Engineering* **93** 280-8
[4] Pashaei S, Hosseinzadeh S and Syed A A 2017 *Polymer Composites* **38(4)** 727-35
[5] Abdelsalam A. A, Araby S,. El-Sabbagh S H, Abdelmoneim A and Hassan M A 2019 *J. Thermoplastic Composite Mat* 0892705719844556
[6] Husnan M H, Ismail and Shuib R 2018 *Mat. Sci. and Eng.* **309** 012028
[7] Chauhan N. P, Meghwal K, Gholipourmalekabadi M and. Mozafari M, in *Polyaniline blends, composites, and nanocomposites* 2018 (Elsevier,) pp 149-74
[8] Susanto T and Marlina P 2019 *Jurnal Dinamika Penelitian Industri* **30(2)**
[9] Susanto T presented at the *MATEC Web of Conf.* 2018 (*unpublished*)
[10] Burgaz E, Gencoglu O and Goksuzoglu M 2019 *Res. on Eng. Struc. and Mat.* **5(3)** 233
[11] Bach Q V, Vu C M and Vu H T 2019 *Silicon*, 1-11
[12] Bugarić U, Tanasijević M, Polovina D, Ignjatović D and Jovančić P 2014 *Technical Gazette* **21(5)** 925-32
[13] Yang G 2014 *Sensors & Transducers* **181(10)** 210
[14] Wisojodharmo L, Fidyaningsih R, Fitriani D, Arti D and Susanto H 2017 *MS&E* **223(1)** 012013
[15] Hadi F and Kadhim R, presented at the IOP Conference Series: *Mat. Sci. and Eng*, 2020 (*unpublished*)
[16] Abdul Rashid E S, Muhd Julkapli N B and. Abdul H Y W 2018 *J. Appl. Polymer Sci.* **135(32)** 46594
[17] Bhawal P,. Das T K, Ganguly S, Mondal S and Das N C 2019 *Polymer Bull.* **76(4)** 1877-97
[18] Saha T and Bhowmick A K 2019 *Rubber Chem. and Tech.* **92(2)** 263-85
[19] Liu P, Zhang X, Jia H, Yin Q, Wang J, Yin B and Xu Z 2017 *Composites Part B: Engineering* **130** 257-66
[20] Zainal M, Santiagoo R, Ayob A, Ghani A A, Mustafa W A and Othman N S 2019 *J. Thermoplastic Comp. Mat.* 0892705719832072
[21] Kashani A, Ngo T D, Hemachandra P and Hajimohammadi A 2018 *Constr. and Build. Mat.* **171** 467-73
[22] Liu J, Li X, Xu L and Zhang P 2016 P*olymer testing* **54** 59-66

IOP Conf. Series: Materials Science and Engineering **980** (2020) 012003 doi:10.1088/1757-899X/980/1/012003

The effect of pumice and clay composition in natural rubber-ethylene propylene diene monomer blends towards its curing characteristics and physic-mechanical properties

Rahmaniar[1]*, T Susanto [1], H A Prasetya[1], P Marlina[1], M Purbaya[2], M Chalid[3] and A Hasan[4]

[1] Palembang Institute for Industrial Research and Standardization, Ministry of Industry, Indonesia

[2] Sembawa Research Centre, Indonesian Rubber Research Institute (IRRI), Indonesia

[3] Department of Metallurgical and Materials Engineering, Universitas Indonesia, Indonesia

[4] Department of Chemical Engineering, Politeknik Negeri Sriwijaya, Indonesia

Email: rahmaniar.een0810@gmail.com

Abstract. Since some adverse effects of using carbon black as petroleum-based filler in rubber compounding, mineral fillers may be a feasible alternative filler. This research studied the effect of the composition of pumice and clay as mineral filler in natural rubber (NR) and ethylene propylene diene monomer (EPDM) blending towards its curing characteristics and physical-mechanical properties. The effect of reinforced filled in EPDM-NR (70/30) phr blends was observed by varying 50, 70, and 90 phr of filler. While the filler was designed as the composition of pumice/clay (P/C): 20/80; 50/50; 80:20 (weight/weight). The rubber, fillers, and chemical were compounded based on ASTM D-3182. The results showed that the loading of P/C fillers in EPDM-NR blends increased scorch time and decreased cure rate index. It was found that the optimum number of filler could enhance density, modulus. Remarkably, the higher pumice composition caused the abrasion resistance and compression set improved significantly, whereas the overloading pumice and clay cause the physic-mechanical properties remained stable or evenly drop. FTIR spectra of the binary blends confirmed that the fine fillers dispersion was obtained due to some interaction bonding between silanol, aluminol of fillers with the EPDM-NR.

1. Introduction

Natural rubber is a polymer material with high elasticity that is widely used for tyres, rubber pad, rail pad, gasket, seal, rubber dock fender Etc. The advantages of using natural rubber as the main polymer are excellent tensile properties, abrasion resistance and highly availability in nature [1]. However, some negative effects of using natural rubber are less chemical resistance, low physic mechanic properties, and oil resistance for rubber product applications are urgent to be addressed [1,2]. Generally, in order to improve those properties, specific rubber products were made using a combination of synthetic polymers such as ethylene diene, styrene butadiene, chloroprene, or even combined with other thermoplastic polymer such as poly-vinyl chloride, polyethylene low molecular or polypropylene [1-3]. In addition, to increase physical properties such as hardness, density, and other tensile properties some fillers were blended during rubber compounding. In this research, EPDM, petroleum-based synthetic polymer, was studied since it was blended with NR to produce some rubber products application that needs high abrasion and heat resistance, and specific compression set value. However, in order to improve physical properties such as density and hardness, the compatible reinforcing needs to be added as a filler in rubber compounding. The previous research stated unfilled EPDM, NR or EPDM-NR binary blends are weak and lost its physical and tensile properties [1,4-6]. EPDM blends with NR have also been studied and stated that to obtain better tensile and mechanical properties, the plasticizer and fillers need to be compatible with each other with the rubber matrix [7, 8].

Commercial reinforcing fillers such as carbon black are usually added into the rubber phase matrix to reinforce the polymer's properties of polymer and reduce cost. Some non-reinforcing fillers such as Calcium Carbonates is added in rubber compounding to enlarge the products volume. However, some negative effect have risen due to petroleum based on carbon black such as environmental pollution, high price and low sustainability recently [9]. Thus, some deposited mineral from earth has been studied as either reinforce or non-reinforce filler for rubber compounding. Some previous research using clay from montmorillonite, kaolinite, quartz, etc. to improve physical properties NR blending [9]. It is found that the strengthening effects are determined by its large surface area, porous, size and surface polarity of the fillers [4,6,10]. The type and amount of filler determine the rubber products' physical-mechanical properties, and thermal and aging resistance. Since many deposited minerals have very specific physical and chemical properties, a study on the effect of using some minerals needs toward its physic-mechanical properties is urged. The type, amount and its compatibility of filler with polymers determine the specification and standard of the rubber final products' specification and standard.

In this research, similarly with the zeosil as reported by [3,11], pumice and clay were studied due to the physical and chemical properties of those minerals that may be reinforced filler in rubber compounding. Montmorillonite, bentonite and other clay commonly used as filler in rubber compounding [5,9,12,13]. It is reported that clay has good positive effects on the physical properties of natural rubber composites. Meanwhile, pumice is a new alternative reinforced filler that has not been explored widely. Pumice fine powder is a natural mineral that has high availability from explosive eruptions [14-16]. Pookmanee, Wannawek [16] stated that generally pumice in Indonesia has chemical composition as follows 70% SiO_2; 14% Al_2O_3; 4.39% Fe_2O_3; 4.39 %; 3.77% Na_2O; 4.03% K_2O; 2.6% CaO, 0.71% TiO_2, Etc. This mineral contain also reported by Bahri and Rahmaniar [12]. Naturally, the pumice has a crystalline structure due to alumina, silanol, and hydroxyl groups [15-17]. It has a highly porous glass that has huge surface area [15]. The polar groups on the surface can be mixed easily with the polar polymer [17]. Because of its properties, pumice has mostly been used as an adsorbent, poly aniline mixtures, and other industrial application [17]. In addition, pumice also has excellent chemical resistance and thermal stability [16,17]. The clay added in this research was obtained from Banyuasin. The density was 1.6 g/cm^3 that dominated by kaolinite and a small amount of smectite, quartz, crysto-ballite and quartz in a very small portion [18,19]. Thus, the pumice and clay mixtures may have positive effects on physic-mechanical properties that are not had by carbon black, calcium carbonates, kaolin and other commercial fillers.

Bearing in mind that lack of information and research on pumice usage reinforces filler in NR/EPDM blending, this research would be valuable as reinforcing filler in polymer blends. Therefore, based on some excellent properties of pumice, clay and natural rubber properties, this research studied to prepare a pumice-clay based composite of EPDM-NR. This study emphasized on the formulation of rubber compounding particularly the effect of the number of pumice and clay as fillers towards it density, modulus, abrasion resistance, and compression set. Therefore, some information and knowledge on developing reinforced fillers based on mineral may be published.

IOP Conf. Series: Materials Science and Engineering **980** (2020) 012003 doi:10.1088/1757-899X/980/1/012003

2. Materials and methods

2.1. Materials

The materials used in this study were Ribbed Smoked Sheet (RSS) which is produced in Indonesia. Polymers: Natural Rubber types of RSS-1 (Ribbed Smoke Sheet) was obtained from PT Air Muring Indonesia and Ethylene Propylene Diene Monomer (Bayer M Ltd/ THEP130865W) spec as follows ethylene 67%; ENB 43%; density 1.3 g/cm^3; hardness 65±5 Shore A; compression set 40%. Fillers: Pumice was obtained from the local market, Sumatera and Clay was obtained Banyuasin, Sumatera. Plasticizers / Lubricants / Peptiser: parrafinic oil , brown factice, Stearic Acid, paraffin wax (Oleo-chemicals Industry, PT Sumi Asih). Activators: ZnO (NC 105 Global Chemical Co, Ltd). Other chemicals as follow CBS N-cyclohexyl-2-benzothiazyl sulphenamide; DPG iphenyl guanidine, and sulfur curative agent (Brataco Chemical).

2.2. Pumice and clay preparation

Pumice rock was bought from the local market, Sumatera, Indonesia. The pumice rock was dried at sun light, crushed and ground until obtaining a fine pumice powder. Then, it was sieved at size 200 mesh, dried at the temperature of 70-80°C using air oven and then desiccated to maintain the moisture content before compounded. The density of pumice used in this research was 0.71 g/cm^3.

Clay was obtained from a local pottery market in Banyuasin, Sumatera Regency, Indonesia. Similar to the pumice preparation, the clay was dried, ground to be clay powder, and then sieved at 200 mesh size powder. Both of clay and pumice were maintained the moisture content before compounded with polymers and other chemicals.

2.3. Rubber compounding

The vulcanizate was compounded based on the formulation as follows: 40 phr of EPDM; 60 phr of natural rubber type: Ribbed Smoke Sheet (RSS-1) ; 2 phr of paraffinic oil, 2phr of brown factice, 4 phr of Zinc Oxide; 1.5 of Stearic Acid; 0.5 of Paraffin Wax; 1.5 of CBS; 0.5 phr of DPG; 2.5 phr of Sulphur. The total number of filler was varied 50; 70; and 70 phr for rubber compounding. Meanwhile, the filler was a mixture of pumice and clay that also varied by mixing pumice/clay (weight/weight) as follows 20/80; 50/50; and 80/20 w/w. Polymer and other chemicals were compounded using laboratory mill XK-160 (type: open two roll mill, Shanghai Rubber, China) based on the standard procedure of ASTM D3182. The preparation of vulcanized sheets standard was done using the ASTM D3184-89 procedure. In order to examine the tensile properties, the compounded rubber sheet was vulcanized for t90 at 150°C using heating press hydraulic (YG-220, Shanghai Rubber Machine Worker, China). These procedures were replicated three for each formula in order to obtain highly accurate data.

2.4. Testing procedures

Curing characteristics such as ML, MH, cure time and scorch time were determined using a Monsanto moving to die rheometer (HT8756P) according to ASTM D5289. Then the samples were tested at a vulcanization temperature of 155°C.

For physical-mechanical and tensile properties, the compounded rubber was sheeted and cooled for about 24 hours at room temperature. After that, the sample was heated and pressed using hydraulic press machine YG-220, China, for about 20 minutes at 155°C. The samples were tested based on the standard procedures as follows: modulus (ASTM D412); abrasion resistance (ASTM D5963); Compression set at 25% deflection, 70°C for 22 hours (ASTM D395); and density (ASTM D297). The experiments were conducted three replication to obtain accurate results.

In order to observe the functional group interaction between clay, pumice, natural rubber, and EPDM, the spectra of FTIR was pictured. The FTIR of Perkin Elmer System 2000 was used in a range 4000 cm^{-1} to 500 cm^{-}. The evaluation of the spectra was following the ASTM D3677 procedure.

3. Results and discussion

3.1. Rheological analysis

Table 1. Curing characteristics of EPDM-NR composites filled by pumice and clay.

Pumice/Clay (w/w)	Total Filler (phr)	ML (kg.cm)	MH (kg.cm)	ΔF (kg.cm)	t_{90} (min)	ts_2 (min)	CRI (/min)
20/80	50	0.04	18.21	18.17	20.17	14.38	17.29
	70	0.02	19.40	19.38	20.43	14.78	17.70
	90	0.05	19.52	19.47	20.38	14.92	18.29
50/50	50	0.52	16.44	15.92	14.77	2.67	8.26
	70	0.28	16.64	16.36	16.18	3.60	7.95
	90	0.19	17.45	17.26	16.58	3.95	7.92
80/20	50	0.02	20.90	20.88	7.20	1.60	17.86
	70	0.21	23.29	23.08	12.15	3.08	11.03
	90	0.27	24.15	23.88	12.58	3.90	11.52

The rheology analysis on the effect of filler number and the composition of pumice and clay filler in EPDM and NR blending are shown in table 1. It lists the torque differences that reflect the crosslinking degree of the matrix polymer and also the curing rate index which describes the filler's processing ability of the filler within the polymers. The higher loading of total filler of pumice and clay caused the increase of MH (maximum torque), indicating viscosity due to the high number filler. Similarly, the increased loading of total filler also contributed to the increasing of torque difference (ΔF = MH - ML) that is assumed as cross-link density.

It could be explained that silanol and alumina from both clay and pumice could make interaction among the ethylene and isoprene groups within the rubber matrixes as reported by Ramesan, George [17]. Pumice and clay with a specific area and large porous also could adsorb some chemicals such as sulphur curing agents, plasticizers and chemicals during the rubber compounding [6]. In addition, the high number of pumice and clay also may inhibit the curative agent and other chemicals dispersion within the rubber matrixes as reported Sengupta, Esseghir [10]. The existence of excess sulphur and precursor cross linking also inhibit the curing process. Therefore, it affect the cure time of the rubber blending. In addition both pumice and clay surfaces have a high number of hydroxyl groups and water (as hydrate) that may span the vulcanization of EPDM-NR blending [1,4,6].

As the pumice and clay increased, the CRI relatively decrease. The EPDM and NR have a huge molecular weight that needs some extra energy to melt mixing during the mastication and compounding with the big-sized filler such as pumice and clay [5,13,20]. Hence, the more shear force and heat were needed during melting the mixture due to the high viscosity and the presence of plasticizer and softener in the compounding.

Those higher viscosity of EPDM-NR filled with pumice and clay may shorten the scorch time as listed in table 1. Compared to previous research by [5,13,20], this finding met an agreement that the clay loading on EPDM-NR blending influenced the curing and rheological characteristics. The higher total filler that had a big size of 200 mesh of pumice and clay could also limit the rubber matrixes' molecular movement within the rubber matrixes. Thus, it would affect the physical properties like 300% modulus and density as depicted in figure 1. The pumice and clay molecule restriction could affect the stiffness and rigidity of the rubber products as the density value shown in figure 1.

The value of t_{90} (scorch time) explained the processing-ability of the filler and polymer of the rubber mixing the mastication and compounding process particularly. This process was connected with the amount of heat that could be energized before the rubber cross link-state transformation. The increasing viscosity of filler may shorten the scorch time as reported by some research by Ismail and Mathialagan [6]. The excess number of filler loading within EPDM-NR blending may cause the decreasing of cure time. The competition for cross-linking agents between NR and EPDM and the overload pumice and clay may cause a decrease of cure time. The competition for cross-linking agents between NR-EPDM and the overload pumice and clay casue the decreasing the rate of curing due to the less mobility of the curative agent in highly dense of EPDM, NR, pumice, and clay.

Process-ability and blending of fillers within the EPDM-NR is affected by the presence of pumice and clay [6,10]. Pumice and clay has hydroxyl groups that may enhance the rubber filler networking due to the hydrogen bonding and van der walls effect. The same explanation was also reported by Azizli, Abbasizadeh [21]. It was shown that the stronger bonding interaction between EPDM, NR and pumice and clay fillers that was determined by ΔF (MH-ML) in table 1. In addition, the pumice and clay dispersion in a sufficient composition of EPDM-NR matrixes also contribute to the cross link density as assumed by the ΔF.

3.2. Physical properties

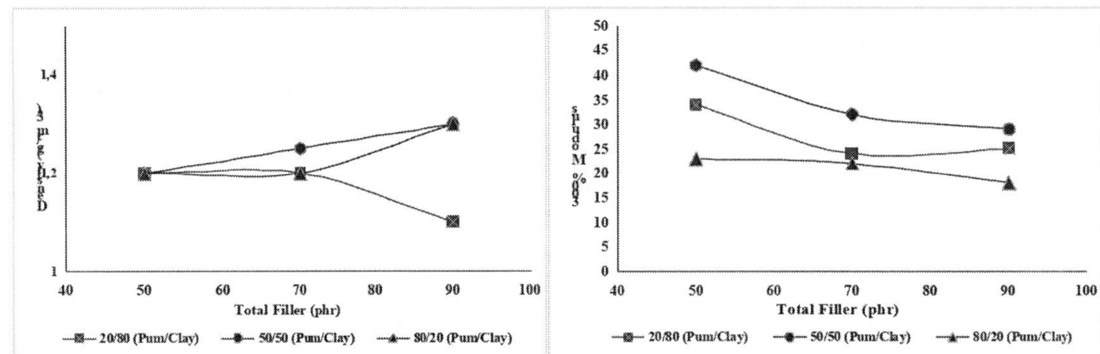

Figure 1. Physical properties (density and 300% modulus) of EPDM-NR blending filled by pumice and clay.

The presence of pumice and clay contributed positively to the increasing of density as shown in figure 1, whereas it decreased the modulus of the polymer blending. The density increased as the loading of pumice and clay, but at an optimum number of filler the density tended to remain stable. Conversely, the modulus of the polymers experienced the decreasing by the increasing filler. The overloading of pumice and clay cause the aggregation of the filler and some chemicals may agglomerate due to the presence of excessive filler [5,11,13]. It might interrupt the filler matrix bonding of oligomer of alumina, silica, isoprene from the filler and polymer matrixes. Thus it reduced the modulus properties as shown in figure 1. As confirmed by the cure rate index in table 1, the modulus might be represented as cross-linking density and its stiffness [6]. Compatibility between rubber and the fillers of mixture between pumice and clay influences the rubber compounding's processing ability. The fine filler dispersion may enable the stronger rubber filler interaction that lead to improve the modulus and other tensile properties [4,6]. However, due to the big size of pumice and clay fillers, at a specific portion of excessive fillers, the pumice and clay may agglomerate and its aggregate could decrease the strength of filler rubber interaction and even rubber-rubber networking bonding [5,13,20]. Thus, physically, the rubber products are more fragile and stiff.

3.3. Mechanical properties

The effect of pumice and clay loading towards its abrasion resistance and compression set is shown in figure 2. Pumice and clay loading had a less significant effect on abrasion resistance and compression set value as reported by some research [22,23]. Due to the scrapping, the volume loss, mechanical rubbing and erosive action is a key mechanical properties for rubber products such as tire, rail pad, rubber pad, etc. Figure 3 illustrates that filler loading of pumice and clay increase the abrasion resistance of the rubber blending. Unlike the modulus trend, better abrasion resistance could be obtained by improving of the filler rubber interfacial adhesion and rubber-rubber networking bonding. The value of abrasion resistance in figure 2 indicated the loss material during the testing, so the lower value indicated the higher mechanical properties. The pumice and clay addition of 50 up to 90 phr might fill the optimum space on the EPDM-NR composites matrix finely, so that its strong bonding interaction could resist the friction and decrease the lost volume. The homogenous particle size of clay and pumice at 200 mesh also contributed to the strength cross-link density and its hardness so that the abrasive treatment could be resisted [4, 6]. As shown in figure 2, due to the differences bulk density of pumice and clay, the abrasion resistance of composites containing high number of pumice had the lowest value of abrasion resistance. This finding is in agreement with some research by Tabsan, Wirasate [22].

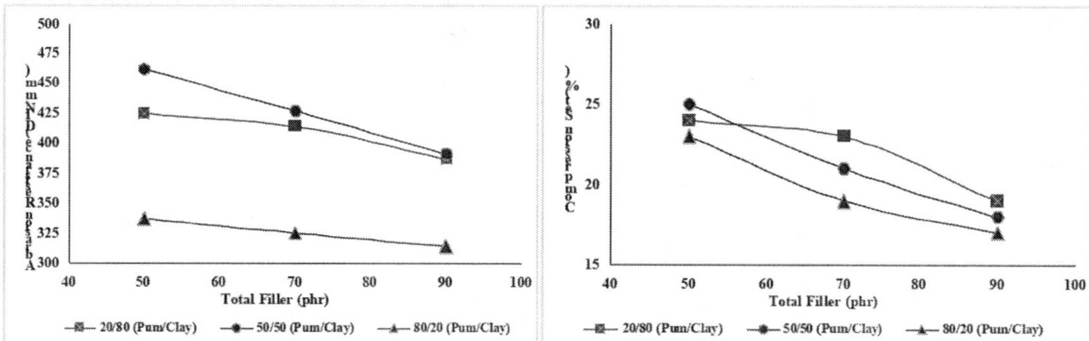

Figure 2. Mechanical properties (abrasion resistance and compression set) of EPDM-NR blending filled by pumice and clay.

Another important mechanical property is the compression set. The compression set's value was measured by evaluate the changes in thickness and elasticity of vulcanization after pressed at a certain deflation and temperature [2,13]. This property contributed to flow, stress relaxation, recover from compression or setting during rubber product usage. The residual deformation or the percentage of deflection after compressed 25% of its original thickness. These properties were very important in rubber padding for heavy construction or transportation. Figure 2 shows the deflection percentage of NR/EPDM filled pumice and clay rubber products that recover to its early shape and how much the rubber failed to return to its previous shape. Generally, the lower percentage indicated a good compression set. It was clearly seen that the compression set value of the rubber less than 30%. It indicated the compression set of this ternary blends complied good the specification standard of rail rubber pad. However, for static usage such as gasket or rubber sealing, the loss of ability to seal may cause leakage, therefore the low compression set is needed for this application. Apart from shrinkage and swelling properties by contacting with the fluids, more prolonged usage and increasing temperature increase the compression set. It was found that the increase of pumice clay total filler caused a positive effect on the improvement of compression set. Similarly to abrasion resistance, the higher number of pumice also gave a lower value of the compression set. The bulk density and chemical properties pumice and clay may enhance the decreasing of mechanical properties of the rubber products due to deterioration in rubber filler interaction bonding and finely filler rubber dispersion [17].

Aforementioned the density and modulus changes by the increasing number of pumice and clay, the compression set also shown a similar trend as shown in figure 2. The formation of more cross-linking bonds, the stronger interaction could resist the changes during compress and set at a certain condition. The previous studies also reported that the number filler and its type of reinforcing influence the mechanical properties [10,13,20,21]. Those results are in consistent with the 300% modulus in figure 1, which the rubber stiffness was reduced, so the rubber became more elastic and the compression sets improved.

3.4. FTIR spectra

FTIR spectra were pictured in figure 3. Compared with EPDM-NR filled carbon black, the spectra of EPDM-NR filled pumice and clay have a higher intensity and more peaks in specific wavenumber. The spectra showed that the higher number of total filler loading might increase the peaks' intensity at specific wavenumber. As depicted in figure 3, adsorption bands derived from mixture clay and pumice were observed as follows: 450-460 cm^{-1} (Si–O–Al), 750-770 cm^{-1} (Si–O–H) , 1010-1300 cm^{-1} (Si–O–Si) and 3400-3450 cm^{-1} (OH). Those adsorption bands also explained as the same functional groups [15-17]. The intense high peaks approximately at 3450, 3000, 1200, 1000 cm^{-1} were the vibration of -OH, C-H, C=C, and Si-OH respectively. It was highlighted that pumice and clay were dispersed in EPDM/NR caused new changes of spectra in 1000-4000 cm^{-1} and some new adsorption bands. This was due to the presence of aluminol, silanol, and the interaction between other functional groups in pumice and clay within the polaric groups of NR and EPDM. It was noticeable that peak intensity at 3450 cm^{-1} became broader and higher due to the presence of OH from pumice/clay and some hydrogen bonding between silanol and aluminol groups from pumice/clay with isoprene and ethylene propylene. Ramesan, George [17] also state the same clarification on filler dispersion with the rubber matrixes. Thus, the dispersion of pumice/clay within binary blends was obtained.

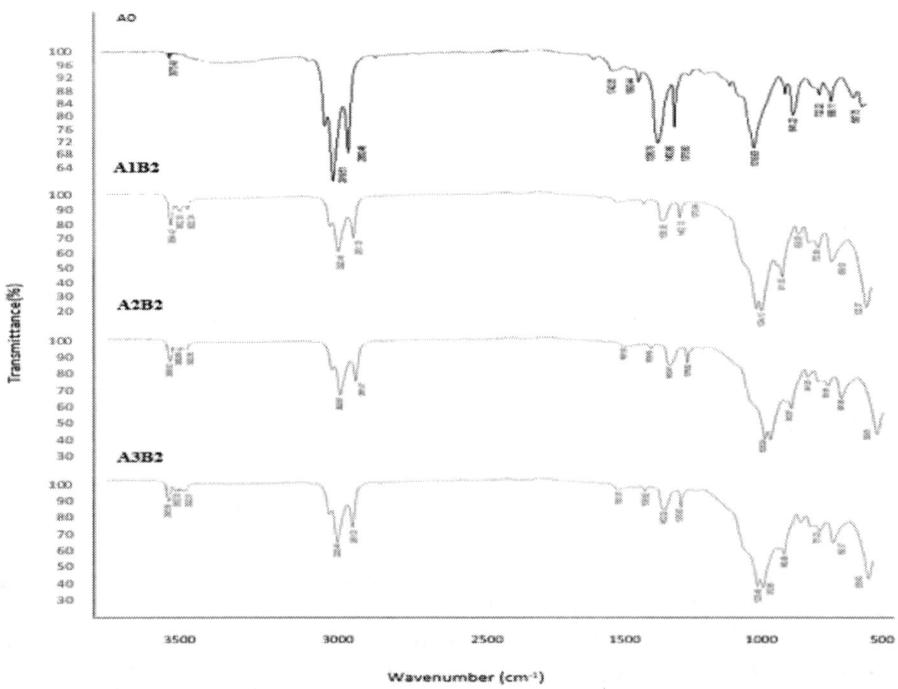

Figure 3. FTIR Spectra of EPDM-NR filled with 70 phr Carbon Black (A0); 50 phr of pumice: clay (50/50 w/w) (A1B2); 70 phr of pumice:clay (50/50 w/w); 90 phr of pumice:clay (50/50 w/w).

4. Conclusions

The binary blending of NR and EPDM filled by pumice and clay was successfully compounded. To sum up, pumice and clay were possible alternative as reinforcing filler in EPDM-NR polymers that contribute positively in density, modulus, abrasion resistance and compression set. Based on the rheological analysis, an optimum number of total filler mixture of pumice and clay could improve the processing ability included its cure rate index and the degree of cross linking formation in the rubber matrixes. The higher loading of total filler pumice and clay might contribute positively and affected the abrasion resistance, compression set and density. However at overloading filler the values might remain stable or evenly drop. Another finding is that the larger pumice composition over clay at NR/EPDM binary blends would improve abrasion resistance and compression set remarkably. The FTIR spectra of EPDM-NR filled by pumice and clay confirmed that pumice and clay might have a strong interaction bonding between silanol, aluminol within EPDM-NR matrixes. However, in order to improve the degree of dispersion of the pumice clays fillers, morphological analysis needs to be evaluated.

Acknowledgment
The authors would like to express their gratitude to the Palembang Institute for Industrial Research and Standardization for financial support, laboratory instruments and other facilities. The authors also gratefully acknowledge all team members for the hard work during finish the research.

References

[1] Rattanasupa B and Keawwattana W 2007 The development of rubber compound based on natural rubber (NR) and ethylene-propylene-diene-monomer (EPDM) rubber for playground rubber mat *Agri. and Natur. Resour.* **41(5)** 239-47
[2] Nabil H, Ismail H and Ratnam C 2014 Simultaneous enhancement of mechanical and dynamic mechanical properties of natural rubber/recycled ethylene-propylene-diene rubber blends by electron beam irradiation *Inter. J. of Poly. Analyz. and Charac.* **19(3)** 272-85
[3] Nabil H and Ismail H 2014 Enhancing the thermal stability of natural rubber/recycled ethylene–propylene–diene rubber blends by means of introducing pre-vulcanised ethylene–propylene–diene rubber and electron beam irradiation *Mater. & Design* **56** 1057-67
[4] Ismail H *et al* 2008 Morphological, thermal and tensile properties of halloysite nanotubes filled ethylene propylene diene monomer (EPDM) nanocomposites *Poly. Test.* **27(7)** 841-50
[5] Alipour A *et al* 2011 Elastomer nanocomposites based on NR/EPDM/organoclay: morphology and properties *Inter. Poly. Process.* **26(1)** 48-55
[6] Ismail H and Mathialagan M 2011 Curing characteristics, morphological, tensile and thermal properties of bentonite-filled ethylene-propylene-diene monomer (EPDM) composites *Poly. Plas. Tech. and Engin.* **50(14)** 1421-28
[7] Rahmaniar and Susanto T 2018 *AIP Conf. Proc.* (AIP Publishing LLC)
[8] Rahmaniar and Susanto T 2019 *IOP Conf. Series: Materials Science and Engineering* (IOP Publishing)
[9] Salam H, Dong Y and Davies I 2015 Development of biobased polymer/clay nanocomposites: A critical review *Fillers and reinforcements for advanced nanocomposites* 101-32
[10] Sengupta S, Esseghir M and Cogen J M 2011 Effect of formulation variables on cure kinetics, mechanical, and electrical properties of filled peroxide cured, ethylene-propylene-diene monomer compounds *J. of App. Poly. Scien.* **120(4)** 2191-200
[11] Alex R and Nah C 2006 Studies on natural rubber/ acrylonitrile butadiene rubber/ organoclay nanocomposites *Plas., Rubb. and Compos.* **35(5)** 219-25
[12] Bahri S and Rahmaniar 2015 Komposit batu apung dan clay sebagai bahan pengisi pada pembuatan kompon lis kaca mobil *J. Dinam. Penel. Indus.* **26(1)** 59-65
[13] Vishvanathperumal S and Gopalakannan S 2019 Effects of the nanoclay and crosslinking systems on the mechanical properties of ethylene-propylene-diene monomer/styrene

butadiene rubber blends nanocomposite *Silicon* **11(1)** 117-35
[14] Njau K N, Minja R J and Katima J H 2003 Pumice soil: A potential wetland substrate for treatment of domestic wastewater *Water Scien. and Techno.* **48(5)** 85-92
[15] Mourhly A *et al* 2015 The synthesis and characterization of low-cost mesoporous silica SiO2 from local pumice rock *Nanomat. and Nanotech* **5** 35
[16] Pookmanee P *et al* 2016 Characterization of diatomite, leonardite and pumice *Materials Science Forum* (Trans Tech Publ.)
[17] Ramesan M T *et al* 2016 Role of pumice particles in the thermal, electrical and mechanical properties of poly (vinyl alcohol)/ poly (vinyl pyrrolidone) composites *J. of Therm. Analys. and Calori.* **126(2)** 511-9
[18] Subandiono R and Badayos R 2005 Morphological characteristics, pedogenic processes and classification of wetland soils in Musi Banyuasin, Palembang, South Sumatra, Indonesia. *Philippine Agricultural Scientist (Philippines) Formerly the Philippine Agriculturist*
[19] Prasetyo B *et al* 2016 Characteristics of rice soils from the tidal flat areas of Musi Banyuasin, South Sumatra *Indo. J. of Agri. Scien.* **2(1)** 10-26
[20] Hejazi I, Sharif F and Garmabi H 2011 Effect of material and processing parameters on mechanical properties of polypropylene/ethylene–propylene–diene–monomer/clay nanocomposites *Mater. & Design* **32(7)** 3803-9
[21] Azizli M J *et al* 2017 Influence of blend composition and organic cloisite 15A content in the structure of isobutylene–isoprene rubber/ethylene propylene diene monomer composites for investigation of morphology and mechanical properties *J. of Compos. Mater.* **51(13)** 1861-1873
[22] Tabsan N, Wirasate S and Suchiva K 2010 Abrasion behavior of layered silicate reinforced natural rubber *Wear* **269(5-6)** 394-404
[23] Arayapranee W 2012 Rubber abrasion resistance *Abrasion Resistance of Materials* 147-66

IOP Conf. Series: Materials Science and Engineering **980** (2020) 012004 doi:10.1088/1757-899X/980/1/012004

Failure analysis of aluminum alloys casting in four-wheels vehicle rims

Surasno[1]* and B Tjahjohartoto[1]
[1]Center for Material and Technical Product (Balai Besar Bahan dan Barang Teknik), Jalan Sangkuriang 14, Bandung 40135 Indonesia

Email: surasno2005@kemenperin.go.id

Abstract. The small industry of casting aluminum alloy produces four-wheeled motorized tire rims that have been used in various vehicles. However, in general, these products do not yet have the quality and safety guarantees required by SNI 1896: 2008. Failure analysis research must be conducted to examine the causes of failure. The research method is to compare the products of small industry with products made by local and imported products by analyzing test results. Observation on visual and penetrant test showed porosity at the center of the rim. Tensile strength was (111-145)MPa much lower than local and imported product (260-285)MPa. The impact value was (500-570) kg (rim broken) and this value was lower than the SNI requirement (936 kg). The content of iron element was (0.401-0.433)%, this content was higher than local and imported rim (0.120-0.135)%. This element will cause a tendency to form micro crack and reduce the strength of mechanical properties. Microstructure observation showed a dendritic structure, porosity, and AlFeSi phase resembles needles. It was concluded that the failure of this rim product is caused by the material and the casting process. Small industries should make efforts to know the casting defects aiming to minimize product failure

1. Introduction

The products of decorative four-wheeled vehicle tire rims have been made by small industries of aluminum casting in Indonesia. However, these products do not have a guarantee of quality and safety for the driver, because it does not meet SNI standards after testing by The Center for Materials and Technical Products (Balai Besar Bahan dan Barang Teknik). Thus, the tire rim is said to be in poor quality or a failed product.

The vehicle's tire rims product when being used and operated on the highway must be able to guarantee the safety of the driver. Tire rims which applied to the vehicles need to have a convincing quality of the product by conducting product research through tests [1]

To discover the product failure on the tire rim, a failure analysis research will be conducted on products that do not meet the SNI 1896: 2008 requirements. The Indonesian National Standard for vehicle tire rim products were on the category of M, N, and O. Rim category M for vehicles with at least four wheels constructed for the carriage of passengers, rim category N for vehicles with at least four wheels constructed for the carriage of goods and rim category O for trailer [2]. While the aluminum casting raw material for rims is regulated by other standards.

According to ASTM B 26 and ASTM B 179, the specifications of chemical composition of aluminum alloy castings, except for aerospace applications, are determined according to the appropriate chemical and spectrometer test methods. The addition of a small amount of modifier and purification element to the melting alloy from the melting furnace is permitted before casting. Pure metals, recycled materials, and master alloys can also be used to make alloys as long as their chemical composition can be analyzed and adjusted to fit the requirements before pouring any casting. Casting products should only be repaired using approved processes such as welding, impregnation, peening, blending, soldering, and so on [1,3].

Aluminum is a mild and light metal which has silver appearance fades because of the thin oxidation that is formed when exposed to air. It also a non-toxic metal which is not pulled out by magnetic fields and has a strength of 49 Mpa to 700 Mpa. Aluminum alloying elements: Copper (Cu) to increase strength and hardness but reduce elongation by 4-6%; Zinc (Zn) increases tensile strength; Manganese (Mg) increases strength in high temperatures; Magnesium (Mg) increases strength and reduces ductility, corrosion resistance, and good weldability; Silicon (Si) increases hardness in heat treatment [3, 4].

Metal casting is the process of making a product by pouring hot molten metal into a box which has a hollowed mold model as demand and then waited until condensing in the mold to cool off. This solid part is known as a casting product. Castings must meet the mechanical properties associated with the ideal microstructure of the chemical composition of materials according to the material standards. [5,6,7].Casting is most often used to make complex shapes that will be difficult or uneconomic to be made by other methods. When the temperature drops below the melting point of the material, the solidification process begins [4, 8].

During the casting process, the rate of growth of defects is still possible. Therefore, it is important for the casting engineer to evaluate the possibility of defects caused by casting, identify the root cause of the casting failure, and take corrective action [10].

The biggest risk of casting failure is that the porosity on the surface in contact with the mold could be caused by the surface moisture of the mold, and the porosity of the core that occurs due to air trapped by the rapid cooling process [11].

Most of the casting defects are related to process parameters. Consequently, it is necessary to monitor the parameters of the casting process so that no defects will occur. This supervision is carried out by someone who must have knowledge of the effects of casting process parameters and their effects on defects. The appropriate identification of casting defects at an early stage is very important to take corrective action [12].

Possible failure castings, including releasing the mold core in a hurry; releasing excessive gas; low pouring temperature; inoculation of unbalanced material; core mold recovery; provision of transfer flow channels; gas supply; gas reduction; slow binder usage; binder reduction; use of coarse sand if necessary; reduction of gas speed and pressure [8].

Throughout metal solidification, some round or oval-shaped cavities could develop on the surface which associated with oxides. These oxides gather into bubbles in the mold cavity which will prevent the liquid metal from filling the space. This type of cavity or blowhole defect becomes a pinhole on the surface and core. Pinholes are very small holes. The core blowhole could only be seen after machining [9].

The phases of the process of making cast rim tires in small industries: heating the furnace to a temperature of 700°C to avoid moisture content; aluminum alloy scrap raw material put into the furnace; measurement of furnace temperature stability; pouring molten aluminum into casting mold; finishing the casting product by machining. Rim casting industry in Gresik is shown in figure 1.

This failure analysis research purposed as an input in solving problems in the small foundry industry to improve sustainable quality. The analysis of the failure of a product or engineering equipment can be carried out with four factors: material usage; product design; equipment operation or use during operation; and environmental conditions [13]. Failure analysis on rim products is carried out on raw materials and the product manufacturing process [14, 15].

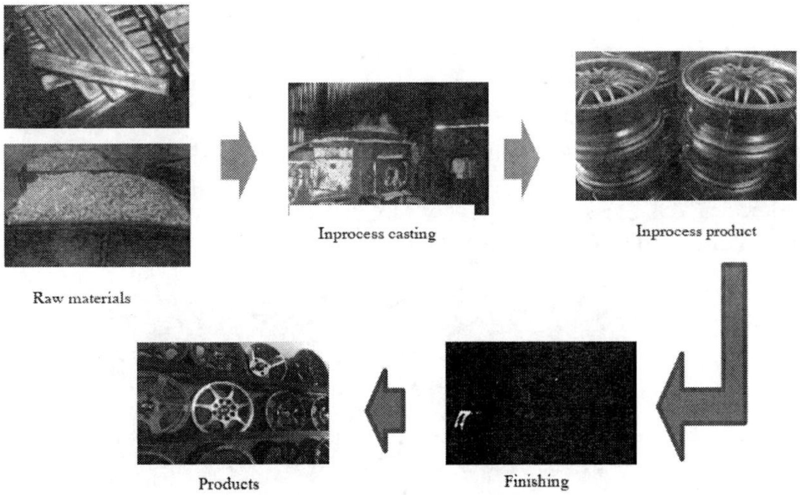

Figure 1. The process of making cast rim tires in small industries

Product testing is done in the form of visual, NDT, chemical composition, hardness, mechanical (impact), macrostructure, and microstructure. The results of the observation data will be analyzed as a statement of the causes of failure so as to improve the sustainability of the quality of the product [15].

2. Materials and methods

2.1 Materials

Five tire rim samples obtained by import (OEM) and local rim industries. The rim which is produced by small industry use raw materials from aluminum-ingots and aluminum scrap is given code EU, EO and YG. The identification of EU, EO, and YG rims are codes given by the rim manufacturer as production identification

2.2 Methods

2.2.1 Casting Analysis. Casting failure analysis was directed at small industrial products. Visual observation method on the surface of tire rim castings by looking for possibilities of surface defects such as shrinkage of casting, porosity, and cracks. Smooth cavities that did not appear visually was checked using the Non Destructive Test (Penetrant Test) method. The penetration test was carried out by applying a penetrant liquid (usually the color is red) on the surface being tested. The liquid penetrant was left at sufficient time to seep into surface opening or defect. The excess penetrant fluid was then removed from the surface. The developer fluid was applied to draw the penetrant fluid to the surface (capillary action) to form a visible indication. The chemical composition of the rim was checked using a spectrometer analysis method. The specimens were taken from the rim and then subjected by an arc spark discharge at the spectrometer equipment. The intensity of emitted radiation caused by the arc spark was then analyzed by this apparatus to show the percentage of the chemical composition.

2.2.2 Tensile and Impact Test. Mechanical testing was done by performing tensile and impact tests and macrostructure observations on the cross section of castings after the impact test. The specimen for the tensile test was taken from the rim (not from extra material during casting). The specimen then subjected to a controlled tension at the tension testing machine 1,000 kg capacity until failure. The impact test was carried out on the rim that had been fitted with tires by dropping a load of 0.6W + 180 kg (W = maximum load for the tire used) on a outer ring flange of the rim from a height of 320 meters at a certain rim angle. Fracture contours were observed with a magnifying lens. Microstructure observation used metallurgy microscope at 50 x magnification to observe porosity, 100 x magnification to observe micro-porosity, and 500 x magnification to observe micro-fissure and matrix structure of the sample.

3. Results and discussion

3.1. Observation of visual and NDT examination (penetrant test)

Through visual observations on the surface at the center of the rim wheels, the defect of cluster porosity appeared. The majority of porosity on the rim surface in the middle was porous castings, which appears that the holes were clustered as shown in figure 2. The porosity defects formed by gas trapped between the surface of the mold and molten metal related to the flow of liquid metal flowing through the mold exert thermomechanical pressure. Relatively large of compressive and shear forces occurred in the surface of mold will form a defect (porosity).

Figure 2. Porosity defect

Irregular porosity was often difficult to diagnose because this defect generally occurred in various positions and was therefore very difficult to be associated with local causes. Porosity can also be trapped beneath the surface of the castings in combination with metal oxides, and can only be seen when machining.

3.2 Observation of mechanical test

Tensile testing of raw materials for imported sample rims resulted in higher tensile strength than local sample rims. However, EU, EO, and YG made from aluminum ingot and aluminum scrap raw materials produced a much lower tensile strength (table 1). Low tensile strength would cause a low impact value, and the hardness value of this rim (66 to 77) Hv10.

Table 1. Mechanical test.

	OEM Local		OEM Import		EU		EO		YG	
Yield Strength, MPa	255	252	255	241	116	110	109	105	78	77
Tensile Strength, MPa	266	260	282	285	144	145	138	126	132	111
Elongation, %	3,7	2,5	11,2	11,5	0,8	0,9	0,9	0,7	2,5	1,6
RA, %	5,2	0,7	16,7	13,7	1,0	1,0	1,0	1,0	3,0	2,0

The impact tests carried out on the rims commonly used for mini bus vehicles with a minimum load stipulation of 936 kg according to SNI standards. Testing of small industrial rim products reached an impact load of 500 to 570 kg of broken rim as in Figure 2, the impact value was lower than that required by SNI standards.

3.3 Observation of Chemical Composition

In the raw material for aluminum-ingot and scrap aluminum alloys, there was an iron content of 0.401-0.433% Fe (table 2), the Fe content was higher than the rim (imported and local). This element would tend to form micro cracks and grew the structure of needles of aluminum, iron, and silicon compounds. This would reduce the strength of mechanical properties (tensile and impact strength). The chemical composition of the raw material for aluminum ingot alloys and aluminum scrap was not in accordance with ASTM B 26 for casting material for rim product.

3.4 Observation of Fractrography

In the macroscopic fractography observation on the cross-sectional surfaces of broken pieces of rim product appeared of coarse grained (brittle), also cavities in the form of porosity defects found (Shown in Figure 3).

Table 2. Chemical composition.

No	Elements	IMPORT	LOCAL	P-EV	P-EO	P-YG
1	Silicon (Si)	7.46	7.12	7.18	7.23	7.33
2	Manganese (Mn)	0.0027	0.0049	0.199	0.216	0.110
3	Chromium (Cr)	0.0083	0.0020	0.0116	0.0127	0.0100
4	Nickel (Ni)	0.0028	0.0043	0.0195	0.0149	0.0173
5	Zinc (Zn)	0.0242	0.0147	0.277	0.241	0.244
6	Copper (Cu)	0.0077	0.0094	1.71	1.54	0.116
7	Lead (Pb)	0.0029	0.0013	0.0175	0.0122	0.0164
8	Tin (Sn)	0.0034	0.0020	0.0079	0.0056	0.0053
9	Iron (Fe)	0.120	0.135	0.432	0.433	0.401
10	Magnesium (Mg)	91.9	0.316	0.175	0.152	0.250
11	Titanium (Ti)	0.127	0.128	0.0433	0.0427	0.0444
12	Cadmium (Cd)	0.0013	0.0014	0.0014	0.0013	0.0013

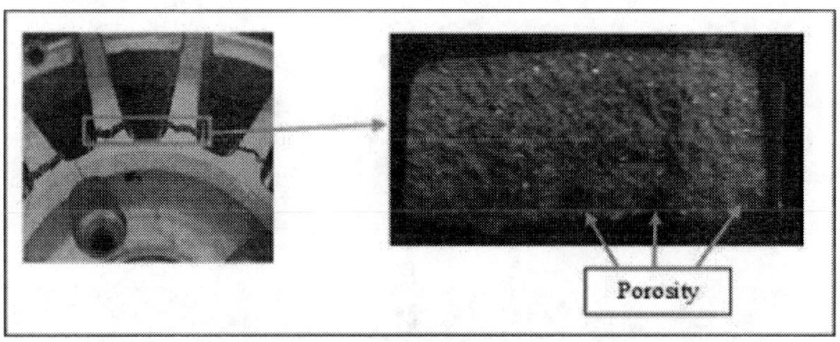

Figure 3. Macroscopic cross section of impact test

3.5 Observation of microstructure

The microstructure of rims from local raw material (aluminum scrap) was in the form of a rough dendritic microstucture containing a sharp needle-shaped phase indication of ferrite. This element was compounded with aluminum and silica to form the FeSiAl compound. Micro cracks (micro fissure) near the porosity cavities was laso found (Shown in figure 4).

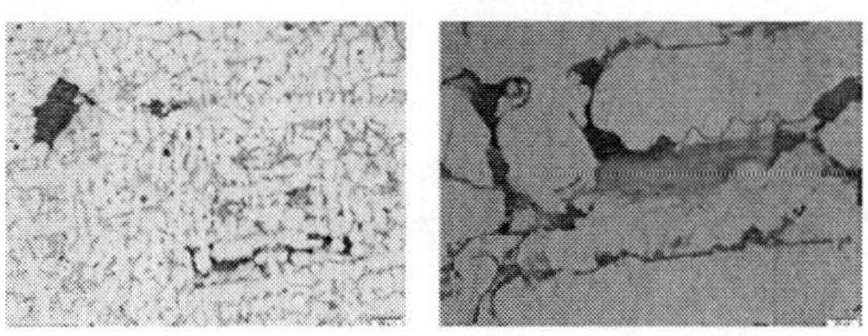

Figure 4. Location of micro fissure and porosity

Through heat treatment, the imported product rim microstructure has a finer dendritic shape than the rim of a local product which has a coarse grain dendritic (figure 5). The microstructure of rim made

IOP Conf. Series: Materials Science and Engineering **980** (2020) 012004 doi:10.1088/1757-899X/980/1/012004

from aluminum-ingot and aluminum-scrap alloys was a coarse-grained dendritic, had a needle-shaped phase structure and porosity (figure 6).

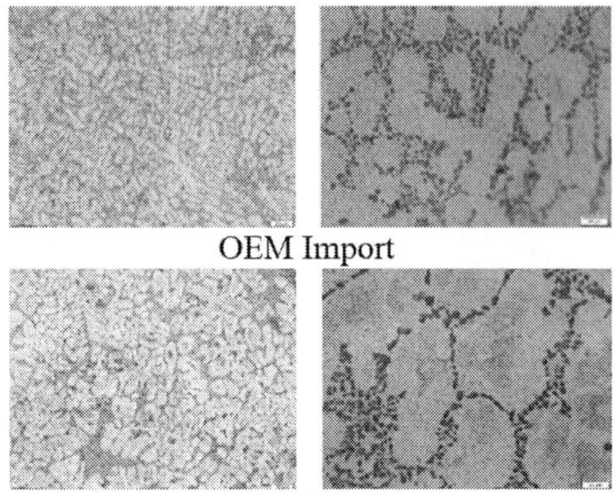

Figure 5. Rim microstructure

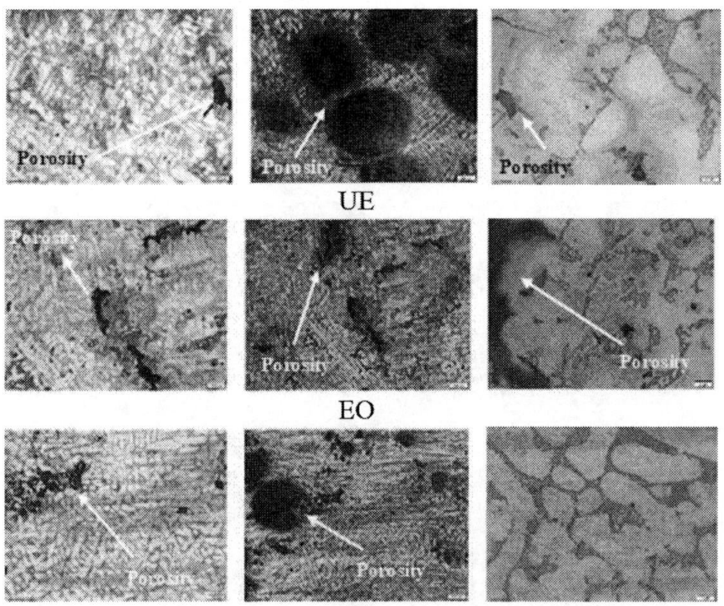

Figure 6. Rim microstructure of aluminum ingot-aluminum scrap

4. Conclusion

Based on observations, the failure analysis of the four-wheeled vehicle tire rim was caused by the porosity on the surface and core materials. Micro fissure in the core of the castings made the value of low mechanical strength. In addition to micro fissures, the structure of ferrite compounds was in the form of AlFeSi compounds which indicated the presence of high Fe content.

To improve the quality of the rim to meet the SNI standard requirements, the local raw material (aluminum ingot-aluminum scrap) must have a low Fe content. To the local aluminum scrap raw material, 99% aluminum (pure) must be added to achieve a balance of chemical alloys to meet the standard of raw material for aluminum casting. To avoid surface porosity, rim molds must meet the criteria for aluminum casting and degassing using nitrogen gas. For the stability of the microstructure, heat treatment tempering is required.

Defects such as porosity and shrinkage can be analyzed by casting computer simulations, as the most efficient and accurate method. The quality and results of casting can be improved efficiently by modifying such techniques as SQC DOE simulation in the shortest time possible and without conducting actual trials at the casting plant.

Efforts to analyze critical defects and possible corrective actions are recommended for casting producers to have knowledge of casting defects with the aim to minimizing the level of product rejection or failure.

References

[1] Ruíz A A B 2015 Standard ASTM B 26 **3(2)** p 54–67
[2] BSN 2008 SNI 1896 Pelek kendaraan bermotor kategori M , N dan O, *Standar Nas. Indonesia*
[3] Adams R C 2001 The materials information company *Technology* **2** p 3470
[4] Anzip A and Suhariyanto 2006 Peningkatan sifat mekanik paduan aluminium A356.2 dengan penambahan manganese (Mn) dan perlakuan panas T6 *J. Tek. Mesin* **8(2)** p 64–68
[5] Warmuzek M 2004 *Aluminum-Silicon Casting Alloys : Atlas of Microfractographs* (Ohio- USA : ASM Int)
[6] Wang W R, Wang W L, Wang S C, Tsai Y C, Lai C. H and. Yeh J W 2012 Effects of Al addition on the microstructure and mechanical property of Al x CoCrFeNi high-entropy alloys *Intermetallics* **28** 34-39
[7] Zolotorevsky V S, Belov N A and Glazoff M V 2007 Structure and microstructure of aluminum alloys in as-cast state *Casting Aluminum Alloys* (Elsevier) p 95 - 182
[8] Mohiuddin M V, Hussainy S F, Krishnaiah A, Laxminarayana P and Sundarrajan S 2014 Experimental study of sand mold process parameters on al-alloy sand castings using DoE *IOSR J. Mech. Civ. Eng.* **11(6)** p 01–06
[9] Zhang B, Cockcroft B S L, Maijer D M, Zhu J D and Phillion A B 2005 Casting defects in low-pressure die-cast aluminum alloy wheels *JOM* J.the Mineral, Metals and Materials Society **57** 36-43 p 1-7
[10] Morgado, T, Dias R, Machado C and Teixeira J P 2017 Casting defects and the mechanical design philosophies *Proc. Int. Conf. on Mechanics and Material Design* (Portugal : INEGI/FEUP) **June** p 1859–1860
[11] Li B and Hu W 2017 Failure analysis of an aluminum alloy material framework component induced by casting defects *IOP Conf. Ser. Mater. Sci. Eng.* **231(1)**
[12] Radford A, Harari J, Northway S and Kroop B 2015 potential causes of failure in an aluminum sand-cast pot from Zambia *Heart Fail.* p 49.
[13] Surasno and Ihlas A 2019 Failure analysis of blade on coconut shell crusher machine *IOP Conf. Ser. Mater. Sci. Eng.* **547(14)** p 012061
[14] Kobayashi T and Toda T 2001 Fracture in complex microstructures *Materials Transactions* **42(1)** p 90 – 99
[15] Brickman B 1997 Pen cap failure analysis and prevention *Am. Soc. Mech. Eng.Congress and Exposition* (Nortbrook : Safety Brief) **25(2)** p 1 - 4

IOP Conf. Series: Materials Science and Engineering **980** (2020) 012005 doi:10.1088/1757-899X/980/1/012005

Current development, potentials, and challenges of biological synthesis of nanoparticle (as a photocatalyst): A review

L Agustina[1,2*], S Suprihatin[1], M Romli[1], and P Suryadarma[1]
[1] Department of Agroindustrial Technology, Bogor Agricultural University, Bogor, Indonesia
[2] Department of Agroindustrial Technology, Lambung Mangkurat University, Banjarbaru, Indonesia

Email: lya.agustina@ulm.ac.id

Abstract. Nanoparticles can be applied potentially in various areas of industrial processes because of their unique mechanical and optical properties, antimicrobial abilities, and catalytic capabilities. This paper aims to analyze the current status of research and development of producing high-value nanoparticles as a photocatalyst using the principle of green chemistry and to identify the potentials and challenges of the new green synthesis in its further developments. The biological method is considered effective and environmentally friendly in producing nanoparticles as a photocatalyst. These efforts can be realized by utilizing natural reducing agents. The bioreductor compounds are available in plants and their waste is in large quantities. It is also reported that some kinds of microorganisms may be used in the biological synthesis of nanoparticles effectively. Some potentials and challenges for it, further research and development were identified, presented, and discussed. Biologically, the synthesis of nanoparticles is considered to be more energy efficient because the process is simpler by utilizing microorganisms, plants and horticultural food waste extracts as a medium for the synthesis of nanoparticles.

1. Introduction

Nanoparticles (NPs) are solid particles that range in size from 1-100 nm. They have different physicochemical properties from other materials and are highly dependent on the size, charge, composition, and coating process that allows them to adhere to these nanoparticles [1]. Over time, the use of nanoparticles has increased so that they can be applied in various fields of science, related to their unique optical properties [2], antimicrobial abilities [3], mechanical properties [4], and catalytic capabilities [5,6]. Nanoparticle synthesis can be carried out either by physical, chemical, or biological methods as well as other hybrid methods.

The physical and chemical methods are very useful for the production of monodispersed nanoparticles. The use of toxic and dangerous chemicals makes this method quite dangerous and can have a considerable impact on the environment. In addition, the use of hazardous chemicals allows the adsorption of the nanoparticles so that they have a dangerous effect when applied to medical purposes. Therefore, it is necessary to develop an environmentally friendly and sustainable method for the synthesis of nanoparticles [7]. Green chemistry demands an approach for the bio-production of nanoparticles through a highly stable and eco-friendly process with no toxic chemical and large scale production. There are various chemical and physical procedures for the synthesis of nanoparticles, either through the wet or dry method approaches. Physical and Chemical processes include vapor deposition, plasma, ultrasonic irradiation, precipitation, microemulsion, reduction methods, sol-gel, and hydrothermal techniques. The increase in production costs will occur because the application of these techniques requires a high amount of energy [8].

Therefore, it is necessary to develop a nanoparticle synthesis that is safe, environmentally friendly, and energy-efficient. To fulfill this, biological synthesis is a promising alternative [9]. The production of nanoparticles based on biological materials such as algae, fungi, viruses, bacteria or plants has been carried out in recent years [10]. This is done because it is known that the metal ion reduction rate is quite high at ambient temperature and pressure conditions, and is capable of releasing some enzymes that will hydrolyze metal ions [7].

2. Biological synthesis

Figure 1 shows the mechanism of nanoparticle formation. The process is initiated by changing the metal ions in solution from a mono- or divalent oxidation state to a zero-valence state [11]. Nucleation will occur due to the reduction of metal ions by biomolecules in the extract of food waste. In the next phase, thermodynamically stable nanoparticles will be formed due to the accumulation of particles on low-energy surfaces. In the next stage, the biomolecules produced by the food waste will also function as natural surfactants for the capping process which will make the nanoparticles more stable and affect the formation of nanoparticles at a later stage.

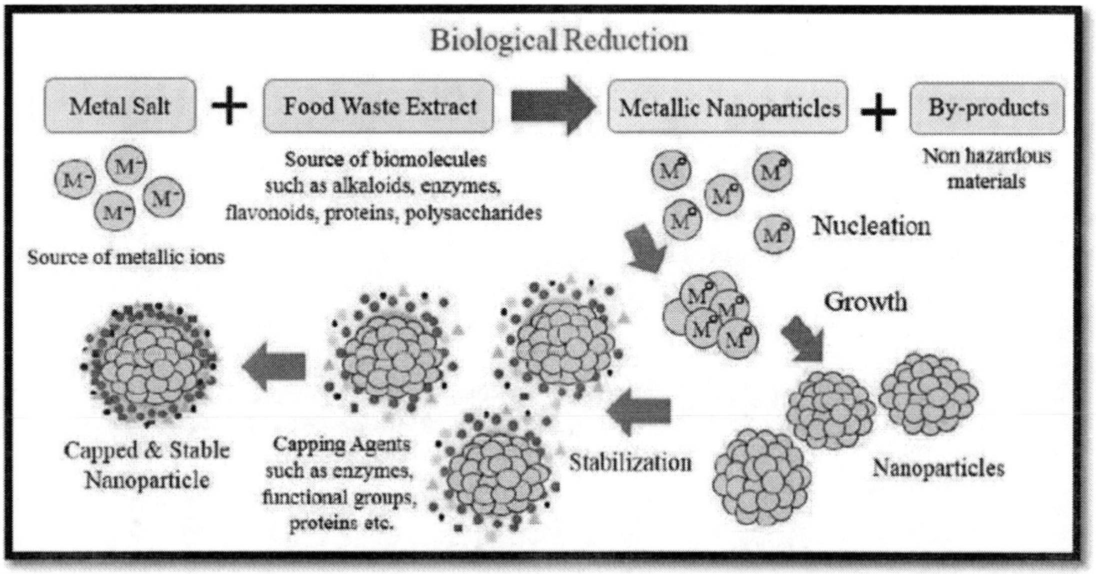

Figure 1. Biological synthesis of metallic nanoparticles using food waste extract [11].

The principle of green chemistry in biological synthesis is basically a process that in its application uses renewable raw materials, optimizes the utilization of waste to produce chemical products, namely nanoparticles. It aims to reduce threats to the safety of the environment and humans. Green chemistry is also known as benign environmental chemistry, clean chemistry, atomic economics, or benign design chemistry. The application of green chemistry is based on twelve principles that can be seen in figure 2.

The cell membrane which is composed of lipids and membranes is the main biological entity [13] of most of the synthetic biochemical conversions that follow the oxidation-reduction mechanism. The dynamic and flexible composition of this membrane means that the effect of the composition is not always constant, thus facilitating the reduction-oxidation reaction. Thus, it is hoped that the synthesis of nanoparticles can use extracellular and intracellular microorganisms in optimal environmental conditions.

IOP Conf. Series: Materials Science and Engineering **980** (2020) 012005 doi:10.1088/1757-899X/980/1/012005

revention

- Prevention takes precedence over remediating waste.

linimize harmful chemicals

- Synthesis of chemicals is sought to use and produce substances with the

esigning safe functional chemicals

- The synthesis process is designed in such a way that the desired results are obtained but with the minimum possible to produce toxic substances.

nergy saving

- Minimizing energy use in the process.

linimize the use of derivatives

- use of reduced additives to reduce the amount of waste generated

Design the degradation

- To simplify the breakdown of chemical products at the end of their functions.
- Monitoring and prevention of substance formation must be carried out directly dangerous at every stage of the synthesis process.
- Minimize the potential for accidents, such as the emergence of emissions of hazardous substances, explosions, and fires.

Figure 2. The application of green chemistry is based on twelve principles [12].

IOP Conf. Series: Materials Science and Engineering **980** (2020) 012005 doi:10.1088/1757-899X/980/1/012005

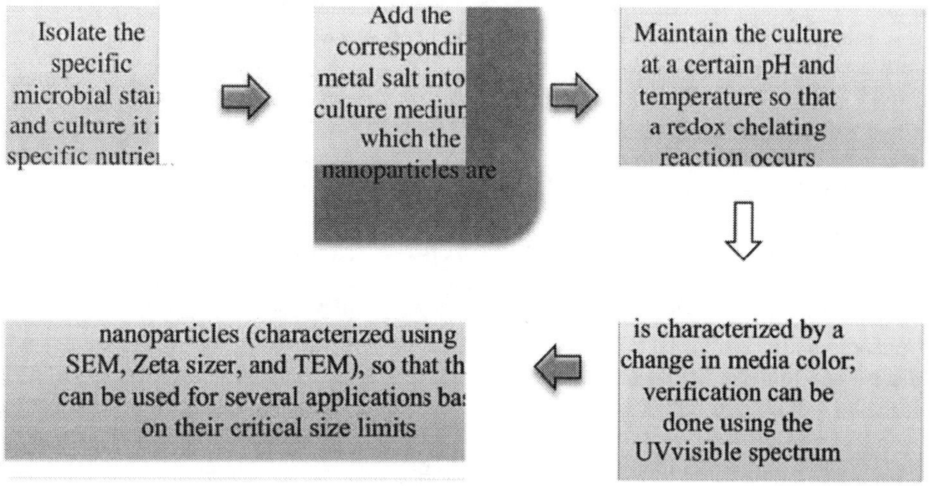

Figure 3. The mechanism of using microorganisms in the synthesis of nanoparticle [13].

Other studies suggest that most of the extracellular synthesis of nanoparticles occurs due to electrostatic attraction between microorganism membranes containing negatively charged phospholipids and a combination of salts which are positively charged metal ions [14]. In the synthesis of nanoparticles, as shown in figure 3, salt is an important ingredient to add. Variations in metal salts and culture parameters need to be investigated further to obtain specific nanoparticles. The color change in the media is an early indicator of the formation of nanoparticles qualitatively

3. Synthesis of nanoparticle (ZnO and TiO)

3.1. Synthesis of ZnO

Potential applications of zinc oxide nanoparticles include high catalytic efficiency, strong adsorption capability, high isoelectric point (9,5), biocompatibility, and fast electron transfer kinetics for biosensing purposes [15]. Biological synthesis is carried out using milk sap from Calotropis procera which will be used as a reducing agent and as a surface stabilizer. The results obtained are that the size of the ZnO nanoparticles is an average of 5-40 nm with a spherical shape [15], and reports for the photoluminescence of the ZnO sample suggest five emitting bands, namely three blue bands at 417 nm 440 and 462 nm, possibly a green band at 520 nm. and one red band at about 675 nm.

Aloe barbadensis Miller leaf extract has been investigated for the synthesis of green ZnO nanoparticles [16]. Experiments were carried out using aloe vera extract which was cooked and not cooked. The synthesis of nanoparticles with different concentrations (50%, 25%, 15%, 10%, 5%) was prepared with distilled water and the volume was made up to 250 ml. Then zinc nitrate was dissolved and stirred at 150°C for 5-6 hours, allowed to cool at room temperature then the supernatant was removed. The resulting product was pale white, it was then centrifuged, and dried at 80°C for 7-8 hours. The results of the characterization of zinc oxide nanoparticles showed different dispersions of poly and particle sizes of 25-45 nm with an average size of 35 nm. The nanoparticles formed are stable, this is due to the interaction of amino groups and carboxylate groups with the zinc surface [16].

The synthesis of ZnO-NPs from *Physalis alkekengi* extract has been investigated [17]. There were two stages, first was to prepare the plant and then synthesize ZnO-NPs. Plant preparation included extraction from *P. alkekengi* using methanol; followed by extracting the Zn produced using sulfuric acid to form zinc sulfate; then reacted with chlorophyllin to form zinc chlorophyllin, and the final step was the synthesis of ZnO-NPs by direct deposition of zinc chlorophyllin in dilute ammonia. The resulting residue was filtered 3 times with filter paper, then the pH was adjusted by adding sodium hydroxide solution (10%) until the pH became 11. The next step was concentrating with an evaporator to 25% of the original volume and extracted 3 times with petroleum ether and a funnel separator. The pH value was again adjusted to 3 by adding uric acid. The product of the reaction was $C_{32}H_{30}ON_4H_2$ $(CO_2H)_2$ [19], which was further dried and reacted again with 150 mL H_2SO_4 (1.00 mol L^{-1}) at 85°C for 360 minutes. Furthermore, filtering and evaporation of up to 20% were carried out. The solution was re-reacted with $C_{32}H_{30}ON_4H_2$ $(CO_2H)_2$ at 60°C for 80 minutes to form $C_{32}H_{30}ON_4Zn$ $(CO_2H)_2$, then heated and added with $NH_3.H_2O$ (1.25 mol L^{-1}) until pH 6. The resulting precipitate was centrifuged, washed with ethanol, and dried. The ZnO nanoparticles produced were porous, rod-shaped which had an average size of 55 nm [17].

Another investigation of synthesis ZnO-NPs from another plant; *Sedum alfredii* also has been conducted [20]. Same with green synthesized of ZnO-NPs from *Physalis alkekengi* extract, there were two phases of the synthesis. The only difference lies in the plants used, namely *Sedum alfredii.* The resulting ZnO-NPs characteristic was pseudo-spherical, with an average particle size of 53.7 nm.

Vimala [21] has conducted other research related to the green synthesis that has been carried out with the use of *Borassus flabellifer* fruit extract for ZnO-NPs. The process stages were generally carried out in the same stages as the previous process. Experiments were carried out at different concentrations, namely 5%, 10%, 15%, 25%, and 50%. The structure of the ZnO-NPs formed was rod-like and porous with an average size of 55 nm.

El-Waseif [22] apart from investigating the synthesis of titanium nanoparticles, also carried out a study on ZnO-NPs using *L. johnsonii.* The results of the analysis show that the absorbance for ZnO-NPs was 492 nm with a spherical shape that had an average size between 4-9 nm.

Table 1. Bio-synthesised of ZnO using plant or microorganism sources.

Nanoparticle	Size & shape	Plant/ microorganism	Year	References
ZnO	Spherical and granular natur, in the range of 5-40 nm.	The latex from *Calotropis procera*	2011	[15]
ZnO	Spherical and stable, 25-45 nm	*Aloe barbadensis Miller* leaf extract	2011	[16]
ZnO	Crystalline poly-dispersed, a mean particle size of 72.5 nm	*Physalis alkekengi* extract	2011	[17]
ZnO	Pseudo-spherical, a mean particle size of 53.7 nm.	*Sedum alfredii*	2011	[20]
ZnO	Porous in nature and rod like structure with an average size of 55 nm.	*Borassus flabellifer* fruit extract	2014	[21]
ZnO	Spherical in shape and polydispersed with maximum particles in size range within 4–9 nm	*Lactobacillus johnsonii*	2018	[22]
ZnS	Spherical, 2-5 nm	*Desulfobacteriaceae*	2000	[23]

3.2. Synthesis of TiO

Titanium oxide is a popular material that is often used as a catalyst for water purification and degradation of organic compounds, such as humic acid in peat swamp water and dyes such as indigo carmine and methylene blue. The use of this material as an ultrafiltration membrane coating used for drinking water treatment as an alternative technology for the conventional separation of organic substances has been investigated [24]. The presence of TiO_2 will increase the ability of the membrane to photocatalytic processes. This is due to the fact that TiO_2 has beneficial properties such as being a semiconductor, chemically and physically stable, has high activity, is resistant to abrasion (scratches) [25], and is relatively inexpensive [26].

It is known that the structure of the titanium oxide crystals greatly affects its catalytic activity (rutile, anatase, or brookite). Likewise, the surface area, size, or porosity of the nanoparticles to be produced. Pure TiO_2 with a metastable crystal anatase structure having a wide band of gab (3.2 eV) has been made as a UV-activated photocatalyst that is predominantly used because of its stability and high oxidation potential and its chemically beneficial properties [27]. Table 2 shows the investigation of the bio-synthesized of TiO2 by some plants and microorganisms.

Table 2. Investigation of TiO_2 biosynthesized using plant or microorganism sources.

Nanoparticle	Size & shape	Plant/ microorganism	Year	References
TiO_2	Spherical 23 ± 2 nm in ranges	*Annona squamosa* peel	2012	[28]
TiO_2	Spherical, 100 to 150 nm	*Nyctanthes arbor-tristis* leaf extracts	2011	[29]
TiO_2	Spherical and quite polydisperse, 36 to 68 nm with calculated average size of 49.5 nm	*Eclipta prostrata* leaf extracts	2012	[30]
TiO_2	Irregular, 25 up to 110 nm	*Catharanthus roseus* leaf extract	2011	[31]
TiO_2	Spherical shape and clusters with an average size of 32.58 nm	*Psidium guajava*	2014	[32]
TiO_2	Irregular shape in size range within 4–9 nm	*Lactobacillus johnsonii*	2018	[33]
TiO_2	150 nm	*Lactobacillus*	2011	[34]
Ti	Spherical, 40-60 nm	*Lactobacillus sp.*	2007	[35]

Research on *Annona squamosa L.* (Annonaceae) to synthesize TiO2-NPs has been carried out [28]. The main ingredient, namely the fresh skin of *A. squamosa*, was processed to become a powder, then 4 g was dissolved in distilled water (40 mL) at room temperature (25°C). The extract obtained was then filtered in stages using nylon mesh (Spectrum) and Millipore hydrophilic filter (0.22 μm). $TiO(OH)_2$ solution was prepared in the Erlenmeyer flask by adding 5 mM $TiO(OH)_2$, 100 mL of distilled water and dissolved for 2 hours. The TiO_2 solution was formulated by adding 20 mL of A. squamosa extract and 80 mL of 5 mM TiO_2 (at room temperature under stirring for 6 hours). The solution was then observed using scanning electron micrographs. Based on the observations, the morphology of TiO_2 nanoparticles was spherical, in the form of aggregates with a size of 40-60 nm. This phytochemical result showed that *A. squamosa* extract was able to reduce titanium ions to metal nanoparticles.

Biological synthesis of TiO_2 NPs nanoparticles using *E. prostrata* leaves by a simple water reduction method has been conducted [30]. An aqueous extract of *E. prostrata* was prepared by boiling 10 g of the ingredients using 100 mL of double distilled water at 60°C for 10 minutes. The solution was then filtered gradually using nylon mesh (Spectrum) and followed by a Millipore hydrophilic filter (0.22 μm) for use in further experiments. Synthesis of TiO_2 nanoparticles was previously carried out by preparing an

Erlenmeyer flask containing 100 mL of $TiO(OH)_2$ (5mM), stirred for 2 hours, then 85 mL 5 mM TiO_2 were taken and mixed with 15 mL of dilute *E. prostrata* extract at room temperature under stirring conditions for 24 hours. The formation of nanoparticles was characterized by a light green color change. The results of characterization using XRD in the rutile [30] phase showed a polydispersion pattern and ranged in size from 36 to 68 nm with an average size of 49.5 nm.

Evaluation of TiO_2 nanoparticles synthesized using *L. johnsonii* in MRS broth at an absorbance of 409 nm was carried out using UV, FTIR, and TEM [22]. Based on the reading of FTIR results, it was found that TiO_2 nanoparticles had a stronger protein ability to bind to metals so that they could prevent particle agglomeration. TEM images of the TiO_2 NPs record an irregular shape with an average size of 4-9 nm.

The synthesis of Titanium nanoparticles was carried out using *Lactobacillus sp* [34] which was started by preparing the filtrate which was diluted 5 times at a pH of 2-4 solution, then added with 10% sugar solution and incubated for 24 hours. After 24 hours, about 20 ml 0.025 (M) of titanium dioxide solution was added and stirred with the help of a magnetic stirrer for 0.5 hours and incubated for 3–4 days (a precipitate would form at the bottom of the Erlenmeyer and a change in pH). Then the filtering process was carried out using Whatman paper and dried under a blow of hot air [34]. The filtered and dried nanoparticles were used for characterization using X-rays and TEM. The synthesized nanoparticles were spherical with a size of 40-60 nm.

4. Potential application and the challenges of metal oxide as nanoparticle

Nanotechnology has a wide potential for application in various fields, including as a catalyst, bio-medicine, electronic components, optical science, engineering, and others. The use of titanium dioxide (TiO_2) as a photocatalyst is considered effective to degrade and inhibit the growth of scattered pollutant parameters either on the surface of the water (waste) or those that spread in the air [36]; therefore TiO_2 is implemented for various purposes, including as a photocatalyst, gas sensor, solar devices and biomaterials [35].

One of the applications of nanoparticles in wastewater treatment was carried out [37] On secondary effluent palm oil mill (POMSE) using a photocatalytic process using ZnO-polyethylene glycol (ZnO-PEG) nanoparticles [37]. The photoreactor for the POMSE treatment process is shown in Figure 4. The photoreactor was equipped with a light source (365 nm UV lamp) with a power of 15 W. The reaction temperature took place at room temperature (25°C) by immersing it in a water bath. For the analysis process of reaction results, 5 ml of sample water was taken at 5-minute intervals and the mixed solution (POMSE and ZnO) was subsequently separated using a Gy Frozen 1736R centrifuge model for 20 minutes.

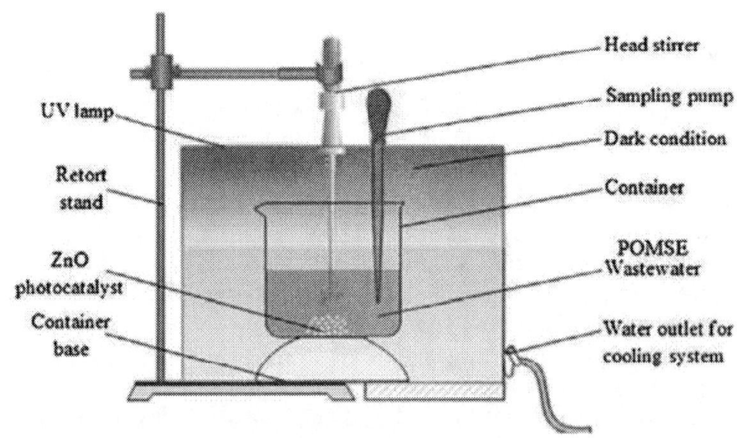

Figure 4. The experimental setup of photocatalysis process [37].

The results obtained were that the optimum processing conditions occurred at pH 6.5 with the addition of a ZnO-PEG photocatalyst of 0.5 g / L in the 50% proportion of POMSE dilution. The color change that occurred was not too large, but the use of PEG polymers in ZnO-PEG photocatalysts was useful in reducing nanoparticle agglomeration. This is due to the adsorption process on the surface of the nanoparticles.

The application of TiO_2 photocatalysts has also been carried out by Chen [38] using NTO photocatalysis as pre-treatment followed by the use of constructed wetlands. The initial stages of the experiment were to first treat household and agricultural wastewater (figure 5, COD was 36.2 ± 7.4) with TiO_2 photocatalysis (coated on α-alumina) before transferring it to a bench-scale wetland system. The hydraulic retention time was 2 days, and it was found that there was a reduction in the levels of halomethane and haloacetic acid so that the value was below the maximum contamination threshold allowed for drinking water.

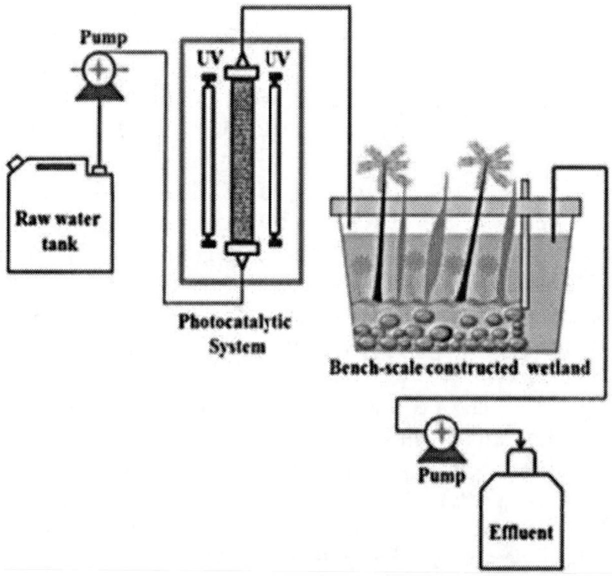

Figure 5. Wetland treatment technologies [38].

According to Hermann [39], factors that influence photocatalysis performances include photoreactor design, light wavelength, light intensity, photocatalyst, initial concentration of reactants, temperature, pH, oxygen content, and the presence of certain ions. Therefore, in order to obtain a better photocatalytic degradation performance, it is recommended to provide a larger surface area for the reaction. A potential photocatalyst with unique properties is TiO_2. It has the ability to mineralize a wide spectrum of pollutants with relatively lower operating costs, and a simple experimental design [40].

TiO_2 NPs can be synthesized biologically using *Lactobacillus sp.* As described above and this topic will become the main focus of the research to be carried out. This research will optimize the photoreactor performance so that it can be applied for the processing of POMSE. The research results will be compared with the results of other studies, such as the results of the study with the use of ZnO NPs by Zainuri [37].

5. Conclusion

Biological synthesis of nanoparticles is considered to be more capable of providing several advantages compared to the physical and chemical synthesis, including being more environmentally friendly with more efficient energy use due to its simpler process. Several studies have been reported and discussed in the description above which describe that microorganisms, plants, and waste extracts from horticultural food can be used as a medium for the synthesis of nanoparticles.

References

[1] Capeness M J, Echavarri-Bravo V and Horsfall L E 2019 Production of Biogenic Nanoparticles for the Reduction of 4-Nitrophenol and Oxidative Laccase-Like Reactions *Front. Microbiol.* **10** 997
[2] Albrecht G, Ubl M, Kaiser S, Giessen H and Hentschel M 2018 Comprehensive study of plasmonic materials in the visible and near infrared: linear, refractory, and nonlinear optical properties *ACS Photon.* **5** 1058
[3] Carvalho P M, Felicio M R, Santos N C, Goncalves S and Domingues M M 2018 Application of light scattering techniques to nanoparticle characterization and development *Front. Chem.* **6** 237
[4] Gao L Z, Fan K L and Yan X Y 2017 Iron oxide nanozyme: a multifunctional enzyme mimetic for biomedical applications *Theranostics* **7** 3207
[5] Siddiqi K S and Husen A 2016 Green synthesis, characterization and uses of palladium/platinum nanoparticles *Nanoscale Res. Lett.* **11** 482
[6] Liu B W and Liu J W 2017 Surface modification of nanozymes *Nano Res.* **10** 1125
[7] Chokriwal A, Madan M S and Abhijeet S 2014 Biological Synthesis of Nanoparticles Using Bacteria and Their Applications *Pharm Tech Res.* **4**
[8] Bandeira M, Marcelo G, Mariana R E, Declan M D and Janaina da S C 2020 Green synthesis of zinc oxide nanoparticles: A review of the synthesis methodology and mechanism of formation *Sustainable Chemistry and Pharmacy* **1** 15 100223
[9] Musarrat J, Dwivedi S, Singh B R, Saquib Q and Khedhairy A A 2011 Microbially synthesized nanoparticles: Scope and applications *Springer-Verlag Berlin Heidelberg, Germany.* DOI 10.1007/978-1-4419-7931-5
[10] Gaidhani S, Singh R, Singh D, Patel U, Shevade K, Yeshvekar R and Chopade B A 2013 Biofilm Disruption Activity of Silver Nanoparticles Synthesized by *Acinetobacter calcoaceticus Mater Lett.* **108** 324
[11] Ghosh P R, Derek F, Shashi B S and Gerrard E J P 2017 Production of High-Value Nanoparticles via Biogenic Processes Using Aquacultural and Horticultural Food Waste *Material.* **10** 852
[12] Anastas P T and Warner J C 1998 *Green Chemistry: Theory and Practice* Eds. (Oxford UK: Oxford University Press)
[13] Malik P, Shankar R, Malik V, Sharma N and Mukherjee T K 2014 Green Chemistry Based Benign Routes for Nanoparticle Synthesis *Nanopart.* 302429
[14] Jain N, Arpit B, Sonali M, Tarafdar J C and Jitendra P 2011 Extracellular biosynthesis and characterization of silver nanoparticles using *Aspergillus flavus* NJP08: A mechanism perspective *Nanoscale* **3** 635
[15] Singh R P, Shukla V K, Yadav R S, Sharma P K, Singh P K and Pandey A C 2011 Biological approach of zinc oxide nanoparticles formation and its characterization *Adv. Mater. Lett.* **2** 313
[16] Sangeetha G, Rajeshwari S and Venckatesh R 2011 Green synthesis of zinc oxide nanoparticles by *aloe barbadensis miller* leaf extract: Structure and optical properties *Mater. Res. Bull.* **46** 2560
[17] Qu J, Yuan X, Wang X and Shao P 2011 Zinc accumulation and synthesis of ZnO nanoparticles using *Physalis alkekengi L. Environ. Pollut.* **159** 1783
[18] Cao X, Ma L Q, Rhue D R and Appel C S 2004 Mechanisms of lead, copper, and zinc etention by phosphate rock *Environ. Pollut.* **131** 435
[19] Masa H, Mojca S and Zeljko K 2007 Isolation of chlorophylls from stinging nettle (*Urtica dioica L.*) *Sep. Sci. Technol.* **57** 37
[20] Qu J, Luo C and Hou J 2011 Synthesis of ZnO nanoparticles from Zn hyper-accumulator (*Sedum alfredii Hance*) plants *Micro Nano Lett.* **6** 174
[21] Vimala K, Sundarraj S, Paulpandi M, Vengatesan S and Kannan S 2014 Green synthesized doxorubicin loaded zinc oxide nanoparticles regulates the Bax and Bcl-2 expression in breast and colon carcinoma *Process Biochem* **49** 160

[22] El-Waseif A A, Hind A A and Dina E E 2018 Biosynthesis and evaluation of TiO_2 and ZnO nanoparticles from *in vitro* stimulation of *Lactobacillus johnsonii JIPBS* **5** 16

[23] Labrenz M, Druschel G K, Thomsen-Ebert T, Gilbert B, Welch S A and Kemner K M 2000 Formation of Sphalerite (ZnS) Deposits in Natural Biofilms of Sulfate-Reducing Bacteria *Science* **290** 1744

[24] Sugita T 2015 *Study of Water Treatment Using Photocatalytic Material* (Jepang: Department of Production Science and Technology Gunma University) chapter 1 pp 15-20

[25] Smith W, Mao S, Lu G, Catlett A, Chen J and Zhao Y 2010 The effect of Ag nanoparticle loading on the photocatalytic activity of TiO_2 nanorod arrays: a article *Elsevier Chemical Physics Letters* **485** 171

[26] Fatimah I, Wijaya K, Narsito and Wang S 2009 Preparation of TiO_2 /Aluminium Pillared Montmorillonite and its Application for Methylene Blue Photodegradation under UV Illumination *World Journal of Chemistry* **4** 21

[27] Miyake M, Takayuki M, Shunsuke N and Yoshikazu K 2015 Water treatment efficacy of various metal oxide semiconductors for photocatalytic ozonation under UV and visible light irradiation *Chemical Engineering Journal* **264** 221

[28] Roopan S M, Bharathi A, Prabhakarn A, Rahuman A A, Velayutham K, Rajakumar G, Padmaja R D, Lekshmi M and Madhumitha G 2012 Efficient phytosynthesis and structural characterization of rutile TiO_2 nanoparticles using *Annona squamosa* peel extract *Spectrochim. Acta A Mol. Biomol. Spectrosc.* **98** 86

[29] Sundrarajan M and Gowri S 2011 Green synthesis of titanium dioxide nanoparticles by nyctanthes arbor-tristis leaves extract *Chalcogenide Lett.* **8** 447

[30] Rajakumar G, Abdul-Rahuman A, Priyamvada B, Gopiesh-Khanna V, Kishore-Kumar D and Sujin P J 2012 *Eclipta prostrata* leaf aqueous extract mediated synthesis of titanium dioxide nanoparticles *Mater. Lett.* **68** 115

[31] Velayutham K, Rahuman A A, Rajakumar G, Santhoshkumar T, Marimuthu S, Jayaseelan C, Bagavan A, Kirthi A V, Kamaraj C and Zahir A A 2011 Evaluation of Catharanthus roseus leaf extract-mediated biosynthesis of titanium dioxide nanoparticles against *Hippobosca maculata* and *Bovicola ovis Parasitol. Res.* **111** 2329

[32] Santhoshkumar T, Rahuman A, Jayaseelan C, Rajakumar G, Marimuthu S, Kirthi A V, Velayutham K, Thomas J, Venkatesan J and Kim S K 2014 Green synthesis of titanium dioxide nanoparticles using *Psidium guajava* extract and its antibacterial and antioxidant properties *Asian Pac. Trop. Med.* **7** 968

[33] Azhar A M, Ladan S A B, Ebrahimi M T and Heydari M 2011 *Lactobacillus*-Mediated Biosynthesis of Titanium Nanoparticles in Mrs-Broth Medium *Brno Czech Republic EU* **9** 21

[34] Prasad K, Jha A K and Kulkarni A R *Lactobacillus* Assisted Synthesis of Titanium Nanoparticles *Nano.* **2** 248

[35] Gong X Q and Selloni A 2005 Reactivity of Anatase TiO_2 Nanoparticles: The Role of The Minority (001) surface *Phys Chem.* **109** 19560

[36] Pan X, Ramirez I M, Mernaugh R and Liu J 2010 Nanocharacterization and Bactericidal Performance of Silver Modified Titania Photocatalyst *Coll Surf B: Biointerfaces* **77** 82

[37] Zainuri N Z, Nur H H H, Dilaelyanam A B S, Amira L D, Nurasyikin M, Norhaniza Y and Abdul W M 2018 Palm oil mill secondary effluent (POMSE) treatment via photocatalytic process in presence of ZnO-PEG nanoparticles *J. of Water Process Engineering* **26** 10

[38] Chen K C, Wang Y H and Lu Y C 2011 Treatment of polluted water for reclamation using photocatalysis and constructed wetlands *Catal. Today* **175** 276

[39] Hermann J M 2005 Heterogeneous photocatalysis: state of the art and present applications *Topics in Catalys* **34** 49

[40] Lazar M A, Varghese S and Nair S S 2012 Photocatalytic Water Treatment by Titanium Dioxide: Recent Updates. *Catalysts* **2** pp 572

Synthesis and characterization of amine-functionalized sugarcane bagasse fiber magnetic nanoparticle biocomposites

R Juwita, C Irawan, R Jelita, and I F Nata*
Department of Chemical Engineering, Faculty of Engineering, University of Lambung Mangkurat, South Kalimantan, Indonesia 70714

Email: ifnata@ulm.ac.id

Abstract. Sugarcane bagasse is one of the by-products in the sugar industry which contains 60% of cellulose. Cellulose can be used as a matrix for biocomposite. The purpose of this research was to produce amine-functionalized sugarcane bagasse fiber magnetic nanoparticle biocomposites (SBB). The SBB was produced from sugarcane bagasse (SB) by solvothermal reaction. The SB was dried and blended into small size (±60 mesh), then lignin was removed with 1% NaOH (w/v) through the delignification. The biocomposites was made by adding delignified SB (SB-D) into a mixture of ethylene glycol, $FeCl_3.6H_2O$, and hexamethylenediamine (HMDA) in solution, and then heated for 6 h at 200 °C. HMDA as an amine source was applied different concentrations (5, 7, and 9 mL). The surface morphology of biocomposites was covered by the magnetic nanoparticles along SB-D which contained amine of about 17.78 mmol/g. The Fe content of SBB was 98.34% which had specific peaks for magnetite at 36°, 43°, and 57° which were measured by X-Ray Diffraction (XRD). The Fourier Transformed Infrared (FT-IR) identified N–H bending vibration on SBB at 1640 cm^{-1}. The iron content and amine group on the surface may affect high adsorption capacity for a wide range of biological pollutants.

1. Introduction

Sugarcane bagasse is a by-product of the extracting liquid sugar cane industry. A factory can produce sugarcane bagasse around 35-40% from the ground sugar cane weight. The sugarcane bagasse needs a further process to gain the added value. Based on a chemical analysis, sugarcane bagasse consists of lignin (22.09%), silicon dioxide (3.01%), cellulose (37.65%), pentose (27.97%), essence (1.81%), and ash (3.28%) [1, 2].

The researchers were interested to study about plant cellulose fibers because they are sustainable, natural, and environmentally friendly. Moreover, there are other advantages of using cellulose fibers such as easy to process, low cost, low energy consumption, lightweight, having excellent specific strength, harmless to the environment, and can be renewed and recycled compared to the conventional synthetic fibers [3]. Cellulose in sugarcane bagasse is coated by lignin which makes a strong structure. For further use as an adsorbent, lignin can disrupt cellulose to bind metal ions. Delignification is a process for the lignin removal. The delignification treatment used is a chemical treatment with NaOH solution. This solution can damage the structure of lignin, crystalline, amorphous parts, and cellulose bloating [4]. Meanwhile, biocomposite is a composite material consisting of natural polymers or bio-fibers (natural fibers) as a reinforcement that can be degraded.

Numerous investigations on the development of sugarcane bagasse have been reported, including the production of microcrystalline cellulose and nanocrystal [5], ethanol [2], biofuel [6] and resin [7].

Especially related to biocomposite materials, many researchers have paid their attention to use sugarcane bagasse as a matrix to form cardanol-formaldehyde composites [8], polyester matrix [9], and polyethylene matrix [10]. The utilization of sugarcane bagasse with magnetic nanoparticles is potential to be developed as biocomposites.

The solvothermal method through a one-step process of surface modification with an amine group has been carried out to synthesize the cellulose-based biocomposites with magnetic nanoparticles [11, 12]. However, there has been no research that develops the modified magnetic nanoparticles based on sugarcane bagasse fibers by a one-step process; thus, further optimization process is necessary to improve the surface functionalization and stability of biocomposites. This research focuses on using sugarcane bagasse fibers to be functionalized by an amine group on magnetic nanoparticle biocomposites. The present work investigates the magnetic nanoparticle biocomposites of sugarcane bagasse which is optimized on the amine content that made through a one-step process of solvothermal method. The surface morphology, iron content, crystalline structure, and functional group of biocomposites are also investigated.

2. *Materials and methods*

2.1. *Materials*

The sugarcane bagasse was obtained from an iced-sugarcane seller in Banjarbaru, South Kalimantan. Besides, iron trichloride hexahydrate ($FeCl_3.6H_2O$), ethylene glycol ($C_6H_6O_2$), sodium acetate anhydrous ($C_2H_3NaO_2$), hexamethylenediamine (HMDA), ethanol (C_2H_5OH), hydrochloric acid (HCl), and sodium hydroxide (NaOH) were purchased from Sigma Aldrich.

2.2. *Delignification of sugarcane bagasse*

Sugarcane bagasse (SB) was washed with distilled water and dried for 3 hours at 80 °C in an oven, mill-grinded to result in SB powder that could pass a 60-mesh sized sieve. The delignification process was carried out by adding the NaOH (1% w/v) solution and 45% w/v of SB powder into the flask and allowed for 2 hours at 80°C until the lignin was removed. The flask and its content were cooled to a room temperature for 2 hours, then the samples were filtered by a filter paper. Distilled water was used to wash the samples until the pH of the filtrate became neutral. At last, the delignified fibers of SB (SB-D) were dried in an oven for 6 hours at 80°C.

2.3. *Synthesis of sugarcane bagasse fiber with magnetic nanoparticle biocomposites*

Solvothermal reaction is one of the methods for SB biocomposites with magnetic nanoparticle synthesis. Firstly, 1.6 g of sodium acetate anhydrous and 0.8 g of iron trichloride hexahydrate were dissolved in the sugarcane bagasse fibers (SBF) (0.5 g) in 24 mL of ethylene glycol by robust stirring at 50°C. The surface of amine-functionalized MNPs were synthesized by adapting the procedure by Wang, et al [13]. When hexamethylenediamine was added, the solution turned into dark orange. Next, the solvothermal was conducted for 6 hours at 200 °C then cooled at a room temperature. HMDA was added by 5, 7, and 9 mL. The SB biocomposites (SBB) were collected from the solution by employing an external magnet. Afterwards, the SBB were rinsed with deionized water followed by ethanol three times for each. The SBB were kept in deionized water for the subsequent use. This synthesis produced 3 types of biocomposites: SBB-5, SBB-7, and SBB-9. SBB-5, SBB-7, SBB-9 got the addition of 5, 7, 9 mL of HMDA, respectively.

2.4. *Characterization*

The investigation of the structural morphology of SB, SBF, SBB-5, SBB-7, SBB-9 was conducted using Field-Emission Scanning Electron Microscopy (FE-SEM, JOEL JSM-6500 LV). XRF measurement was performed on Energy–disperse X-ray Fluorescence while the operation voltage and the current were kept at 20 kV and 77 uA. The surface functional groups on SB, SBF, SBB-5, SBB-7, and SBB-9 was identified by Fourier Transform InfraRed spectrometry (Bio-rad, Digilab FTS-3500). Philips type X'Pert Scan Parameters by using Copper K-alpha (CuKα) radiation performed the X-ray diffraction (XRD) measurement. The operation voltage and the current were kept at 40 kV and 30 mA, respectively. The crystalline index was calculated as equation 1.

$$\mathrm{CrI} = \frac{\mathrm{I_{002}} - \mathrm{I_{am}}}{\mathrm{I_{002}}} \quad (1)$$

where CrI is Crystalline Index (%), I_{002} is intensity of crystal part and I_{am} is intensity of amorph.

2.5. Analysis

The retro-titration method was used to determine the amine content on the samples [14]. 50 mg of samples drop into 0.01 M HCl (25 mL). The mixture was shaken for 2 hours at a room temperature. After the centrifugation, the supernatant (5 mL) was titrated with 0.01 N NaOH. The amine concentration was calculated by equation 2.

$$C_{NH_2} = \left[\frac{(C_{HCl} \times V_{HCl}) - (5C_{NaOH} \times V_{NaOH})}{m_{sample}}\right] \quad (2)$$

where C_{HCl} is HCl concentration (mmol/L), C_{NaOH} is NaOH concentration (mmol/L), the volume of HCl is notated by V_{HCl} (L), the volume of NaOH is notated by V_{NaOH} which was used in the titration of excess non-reacted acid (L), and *m* is the sample mass (g).

3. Results and discussions

The lignin removal from the sugarcane bagasse was confirmed by the morphological structure and color. The original color of sugarcane fibers was cream and changed to darker (closer to grey) after the delignification process (figure 1). Based on the FE-SEM observation, SB looked slightly damaged due to the grinding process (figure 1a), but the lignin was still bound to cellulose and hemicellulose which were arranged in one direction so that the sugarcane fibers looked flat and intact. The contact area and porosity of the material could be increased by grinding. Meanwhile, the delignification process conduced the damage of SB-D from the surface to the inside. Consequently, the lignocellulose structure bonds began to open and the structure became irregular. Rodrigues, et al [7] observed unmodified sugarcane bagasse fibers which had large amount of extractives. After the delignification, the extractives on the surface of fibers were removed. Delignification with NaOH induced the decomposition of hemicellulose, lignin, and silica contained in the SB-D (figure 1b).

The delignification process reduced the size of fiber. The average diameter of bagasse fiber (25 μm) was lower in the raw bagasse. The fiber had smooth surface after the removal of lignin, hemicellulose and pectin [15]. XRF analysis showed a decrease of 31.5% silica in SB-D after treatment. Indeed, SB contained 31.5% Fe which then significantly increased up to 98.7% due to the formation of magnetic nanoparticles. The Fe content may affect the materials as an adsorbent by enhancing the high-adsorption of sorbent capacity for reactivity toward a wide range of biological pollutants [16].

The morphological structure of sugarcane bagasse fiber magnetic nanoparticle biocomposites with different concentrations of HMDA is shown in figure 2. The addition of 5, 7, and 9 mL of HMDA produced biocomposites with the same Fe content around 98.70%. The magnetic nanoparticles were formed on the fiber surface. Hexamethylenediamine played a key role in diminishing the magnetic size during its growth in the solvothermal reaction. The amine contents in biocomposites for the addition of 5, 7, and 9 mL HMDA were 3.21; 17.78; and 3.83 mmol/g, respectively. The different concentrations of amine could be analyzed by the morphological structure of biocomposites. The formation of magnetic nanoparticles for the addition of 5 and 9 of HMDA turned to aggregate and did not distribute on the fiber surface (Figure 2a and 2c). In comparison to 7 mL of HMDA, the magnetic more clearly distributed on the fiber surface (Figure 2b). This is why the amine contents on the surface of magnetic had different amounts. On other hand, the different concentrations of HMDA in this research did not have a big impact on the size of particles. The results obtained were in line with the previous research for magnetic formation by the solvothermal method [13, 17, 18].

XRD analysis was carried out to determine the cellulose crystal structure contained in the samples and the Crystalline Index (CrI) of sugarcane bagasse fibers before and after the delignification.

IOP Conf. Series: Materials Science and Engineering **980** (2020) 012006 doi:10.1088/1757-899X/980/1/012006

Cellulose crystals could be identified at an angle of 2θ between 20°-40° at which there were dominant peaks. Cellulose fiber was composed of several million microfibrils. These microfibrils fiber were divided into two different parts: the amorphous part formed from cellulose chains with crystalline and the flexible masses made of cellulose chains with strong bonds in a rigid linear arrangement. This crystal part was isolated to produce high-quality microcrystalline cellulose.

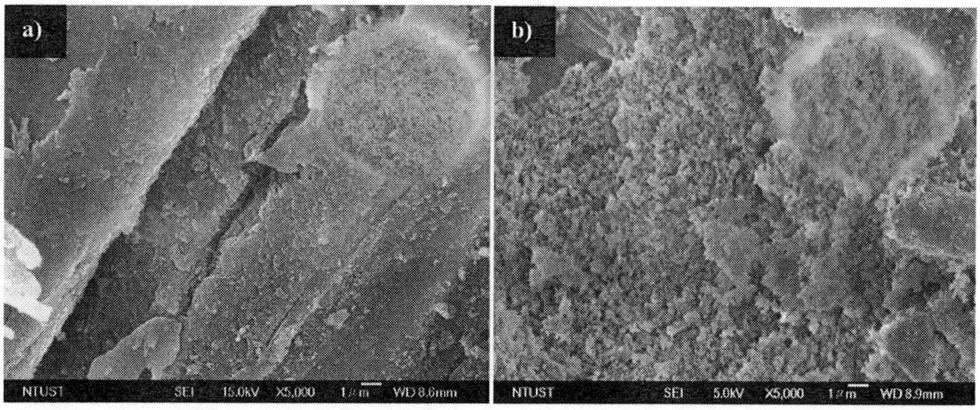

Figure 1. Images for FE-SEM of sugarcane bagasse fibers (a) before (SB) and (b) after delignification (SB-D).

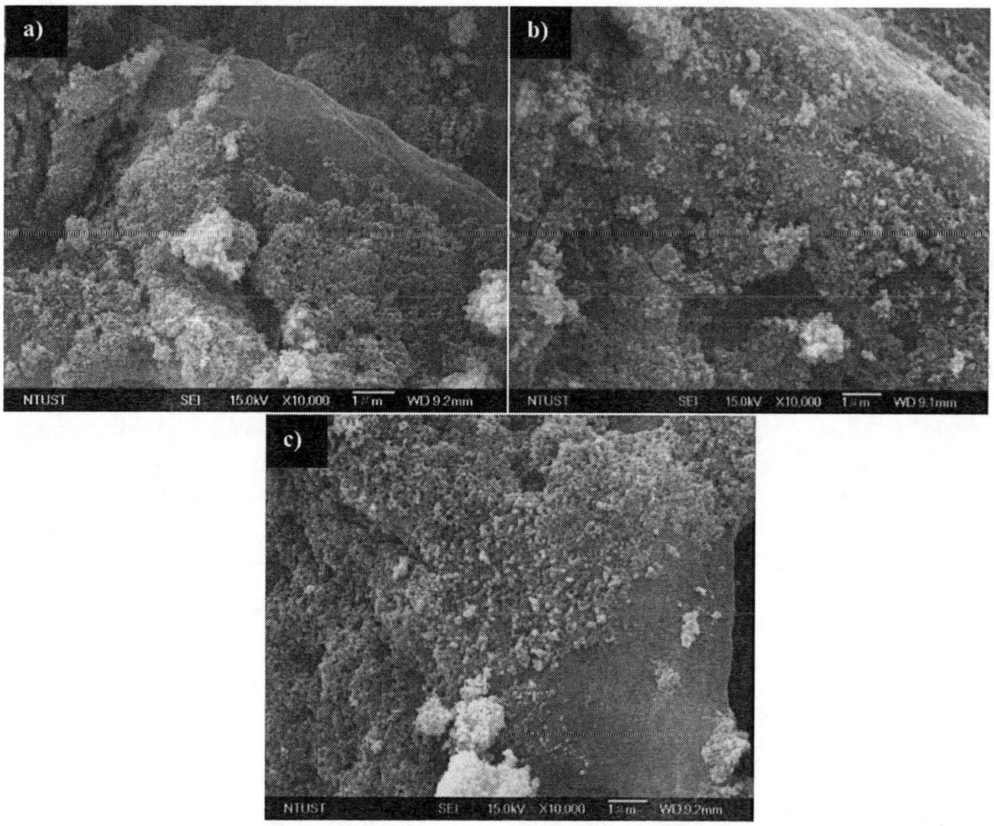

Figure 2. Images for FE-SEM of (a) sugarcane bagasse fiber biocomposite with the addition of 5 mL of HMDA (SBB-5) (b) sugarcane fiber biocomposite with the addition of 7 mL of HMDA (SBB-7) and (c) sugarcane fiber biocomposite with the addition of 9 mL of HMDA (SBB-9).

Table 1. Characterization of SB and SBF.

Sample	Characterization of Peak		CrI (%)
	Amorph (16.78°)	Crystal (21.69°)	
SB	567	782	37,9189
SB-D	683	1035	51,5373

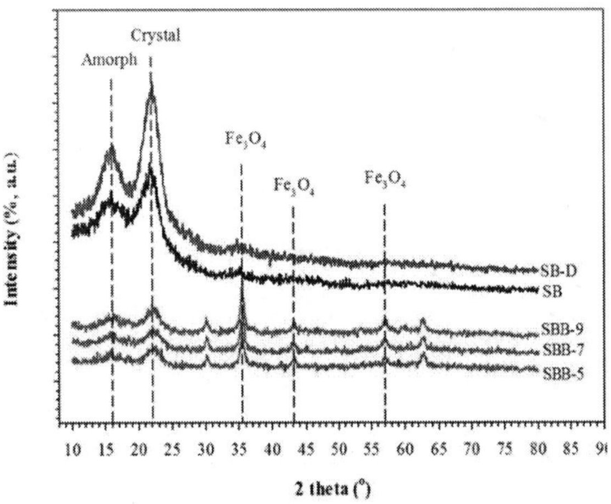

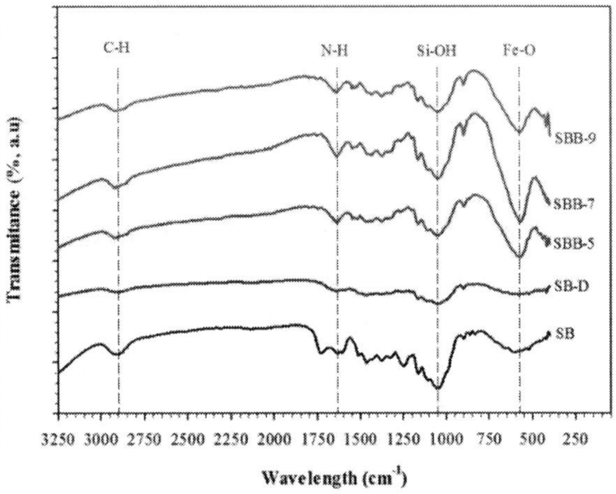

Figure 3. X-Ray Diffraction (XRD) pattern of materials: before (SB) and after delignification (SB-D) and sugarcane bagasse fiber biocomposites of SBB-5, SBB-7, and SBB-9.

Figure 4. FT-IR spectrum of materials: before (SB) and after delignification (SB-D) and sugarcane bagasse fiber biocomposites of SBB-5, SBB-7, and SBB-9.

Cellulose was a parameter that determined the strength of the fibers, hence the product could be influenced by the crystal structure of cellulose. The sugarcane bagasse structure before and after the treatment still had components in amorph form (hemicellulose and lignin) and crystal (cellulose). The characterization of peaks for SB which contained cellulose fibers were identified at 2 tetha (°) = 16.78 in the amorphic form and 21.69 in the crystal form (figure 3). Table 1 shows the increase of SB-D CrI value after the delignification from 37.919% to 51.537% or an increase in crystallinity of 35.91%. This finding was similar to the sugarcane bagasse treatment by alkaline which had crystallinity index about 63.15% (15). It could also be proven from the peak intensity of SB-D which was sharper compared to SB. The treatment of SB with NaOH could change the structure of amorphous cellulose to crystalline cellulose due to the loss of hemicellulose and lignin content after the delignification.

XRD analysis was also used to identify the formation of magnetic nanoparticles in biocomposites, as it clearly appeared peaks at 36°, 43°, and 57°. These specific peaks were identified as magnetite (Fe_3O_4) which was appropriate for the crystalline magnetite standard pattern (JCPDS card 39-0664). figure 3 shows the different crystalline structures in the biocomposites compared to SB and SBF because of the existence of magnetic nanoparticles on SB-D surface. The peak for Fe_3O_4 particles was not found either in SB or SBF. In the case of different concentrations of 5, 7, and 9 mL of HMDA, all biocomposites had the same peak position due to the formation of Fe_3O_4. In addition, SBB-7 had the highest intensity at 36° probably because of the distribution of magnetic nanoparticles on the surface of SBF. This was also confirmed by the FE-SEM images for all biocomposites.

FT-IR spectrum of SB, SB-D, SBB-5, SBB-7, and SBB-9 are shown in figure 4. This analysis was used to detect the functional groups contained in the samples. C-H stretching vibrations as the bonds in SB and SB-D were detected at peak 2900 cm^{-1}. The peak at 1640 cm^{-1} for N-H bending vibration detected the modification of the amine group on the biocomposites. Based on the calculation of amine group, the highest number of amine was on SBB-7 which was also proved by higher intensity peaks on N-H binding compared to the others. The peak at 580 cm^{-1} showed Fe-O stretching band component in Fe_3O_4. This peak was not detected on SB and SB-D because there were no magnetic components. The wavenumber at 1050 cm^{-1} corresponded to Si-OH and O-H band and could be read in all samples, which were also confirmed in the XRF analysis. The peaks at 1525 cm^{-1} and 1260 cm^{-1} in SB showed the aromatic C=C bond of lignin. In SB-D, SBB-5, SBB-7 and SBB-9, the peaks at 1525 cm^{-1} and 1260 cm^{-1} did not exist. This indicated that the lignin content was reduced and the cellulose was separated [15].

4. Conclusions

The synthesis of amine-functionalized sugarcane bagasse fiber magnetic nanoparticle biocomposites was successfully prepared by the solvothermal reaction. The addition of 7 mL of hexamethylenediamine was the best condition to achieve a higher amine content of about 17.78 mmol/g. The amine magnetic nanoparticles consisted of 98.34% Fe and the specific peaks for magnetite were at 36°, 43°, and 57°, measured by X-Ray Diffraction (XRD). The XRD analysis showed the delignification increased the crystalline index (CrI) of BSF up to 51.53%. In addition, the Fourier Transformed Infrared (FT-IR) identified N–H bending vibration on BSB at 1640 cm^{-1}. Due to sustainable, natural and easy process as well as low cost, and low energy consumption of sugarcane bagasse as raw material, it becomes a potential candidate to develop as an adsorbent for high adsorption capacity for a wide range of organic pollutants.

Acknowledgement

The authors are grateful for the financial support for Magister Thesis Research from the Directorate General of Higher Education Ministry of Education and Culture of the Republic of Indonesia with the contract No. 113.1.09/UN8.2/PG/2020.

References

[1] Shankarappa T, Geeta G 2013 Alkali and autohydrolysis pretreatments for effective delignification and recovery of cellulose and hemicellulose in selected agro residues Karnataka Journal of Agricultural Sciences 26 1

[2] Jackson de Moraes Rocha G, Martin C, Soares IB, Souto Maior AM, Baudel HM, Moraes de Abreu CA 2011 Dilute mixed-acid pretreatment of sugarcane bagasse for ethanol production. Biomass and Bioenergy 35 663

[3] Saurabh CK, Mustapha A, Masri MM, Owolabi A, Syakir M, Dungani R, et al 2016 Isolation and characterization of cellulose nanofibers from Gigantochloa scortechinii as a reinforcement material. Journal of Nanomaterials 3

[4] Gunam IBW, Wartini NM, Anggreni A, Suparyana PM 2011 Delignifikasi Ampas Tebu Dengan Larutan Natrium Hidroksida Sebelum Proses Sakaraifikasi Secara Enz imatis Menggunakan Enzim Selulase Kasar Dari Aspergillus Niger Fnu 6018 Jurnal teknologi Indonesia 34 24

[5] Oliveira FBd, Bras J, Pimenta MTB, Curvelo AAdS, Belgacem MN 2016 Production of cellulose nanocrystals from sugarcane bagasse fibers and pith. Industrial Crops and Products 93 48

[6] Ojeda K, Ávila O, Suárez J, Kafarov V 2011 Evaluation of technological alternatives for process integration of sugarcane bagasse for sustainable biofuels production—Part 1. Chemical Engineering Research and Design 89 270

[7] Rodrigues EF, Maia TF, Mulinari DR 2011 Tensile strength of polyester resin reinforced sugarcane bagasse fibers modified by estherification Procedia Engineering 10 2348

[8] Balaji A, Karthikeyan B, Swaminathan J, Sundar Raj C 2017 Mechanical behavior of short bagasse fiber reinforced cardanol-formaldehyde composites Fibers and Polymers 18 1193

IOP Conf. Series: Materials Science and Engineering **980** (2020) 012006 doi:10.1088/1757-899X/980/1/012006

[9] Candido VS, Silva ACRd, Simonassi NT, Luz FSd, Monteiro SN 2017 Toughness of polyester matrix composites reinforced with sugarcane bagasse fibers evaluated by Charpy impact tests Journal of Materials Research and Technology 6 334

[10] de Carvalho Neto AGV, Ganzerli TA, Cardozo AL, Fávaro SL, Pereira AGB, Girotto EM, et al 2014 Development of composites based on recycled polyethylene/sugarcane bagasse fibers Polymer Composites 35 768

[11] Nata IF, Mirwan A, Wicakso DR, Irawan C, Isnaini MD, Fitriani R 2020 Adsorption of Fe^{3+} ion from Aqueous Solution onto Rice Husk Biocomposite Magnetic Nanoparticle IOP Conference Series: Earth and Environmental Science 506 012006

[12] Nata IF, Putra MD, Nurandini D, Irawan C, Fitriani R, Isnaini MD 2018 Rice Husk Fiber Magnetic Nanoparticle Biocomposites: Preparation and Characterization IOP Conference Series: Earth and Environmental Science 175 012005

[13] Wang L, Bao J, Wang L, Zhang F, Li Y 2006 One-pot synthesis and bioapplication of amine-functionalized magnetite nanoparticles and hollow nanospheres Chemistry–A European Journal 12 6341

[14] Nata IF, Sureshkumar M, Lee C-K 2011 One-pot preparation of amine-rich magnetite/bacterial cellulose nanocomposite and its application for arsenate removal RSC Advances 1 625

[15] Mohit H, Selvan V 2019 Thermo-mechanical properties of sodium chloride and alkali-treated sugarcane bagasse fibre Indian Journal of Fibre & Textile Research 44 286

[16] Yam RCM, Mak DMT 2014 A cleaner production of rice husk-blended polypropylene eco-composite by gas-assisted injection moulding Journal of Cleaner Production 67 277

[17] Nata IF, Salim GW, Lee C-K 2010 Facile preparation of magnetic carbonaceous nanoparticles for Pb^{2+} ions removal. Journal of hazardous materials 183 853

[18] Nata IF, Sureshkumar M, Lee C-K 2011 One-pot preparation of amine-rich magnetite/bacterial cellulose nanocomposite and its application for arsenate removal RCS Advances 1 625

IOP Conf. Series: Materials Science and Engineering **980** (2020) 012007 doi:10.1088/1757-899X/980/1/012007

Preparation of magnetic nanoparticle biocomposites using rice husk and sugarcane bagasse fibers as the matrix

Y S Dewi, H Wijayanti, R A Lestari, I F Nata*
Department of Chemical Engineering, Faculty of Engineering, Lambung Mangkurat University, South Kalimantan, Indonesia 70714

Email: ifnata@ulm.ac.id

Abstract. Fiber is one of the important components to construct a matrix structure. Biomass is a potential material as a fiber resource for matrix. Rice husk and sugarcane bagasse consist of 36.6 % and 60% cellulose, respectively. This current research focused on evaluating the production of amine functionalized magnetic nanoparticle biocomposites produced by mixture of rice husk (RH) and sugarcane bagasse (SB) fibers. First, RH and SB fibers were dried and crushed close to 60 mesh for each, followed by a delignification at 80 °C for 2 h in 1% w/v of NaOH. Through by one-step solvothermal process, it was done by adding delignified ratio of RH and SB fibers (1:1; 1:2; 1:3; 1:4) to mixture of ethylene glycol, $FeCl_3.6H_2O$, and 1.6-hexanediamine at 200 °C for 6 h. The magnetic nanoparticles appeared on the fibers surface which contained 97.97% Fe. The magnetite was formed proved by the specific peaks at 36°, 43°, and 57° by X-Ray Diffraction.(XRD). The Fourier Transformed Infrared (FT-IR) identified N–H bending vibration and Fe-O in the biocomposites at 1640 cm^{-1} and 582 cm^{-1}, respectively. The restructure of matrix, iron, and amine groups on the biocomposites surface can affect the adsorption capacity of various waste water treatments.

1. Introduction

One of the biggest natural resources in South Kalimantan is rice. South Kalimantan has a unique type of rice known as "*Beras Banjar*". Cellulose content in the rice husk is 34.34%–43.80% [1]. Besides rice husk, another organic material that contains fiber is sugarcane bagasse. Sugarcane has cellulose content about 35.3%-45.5% [2]. The utilization of rice and sugarcane bagasse will generate waste. A large amount of this waste has impacted the environment. Nowadays, rice husks are only used to compress soft or burn soil. As a result, the combustion activity can cause air pollution. Meanwhile, sugarcane bagasse is usually used for paper making, animal feed [3] or fuel in the inseminator. The fiber content in both materials has the potential to be natural fibers used in biocomposites material production. One of the utilization of biomass waste for various applications is the use of adsorbent as it has good specific strength and is lightweight, environmentally friendly, and easy to get due to its abundant availability in the nature [4].

Through this biomass waste technology approach, it can be isolated to obtain fiber or cellulose. The RH and SB were designed to be a matrix for the adsorbent in ion removal and as hybrid composites. Modification surface of rice husk and sugarcane bagasse fibers was done to increase the properties by chemical treatment technique [5]. Besides, de-polymerizing lignin, hydrolyzing hemicellulose and breaking the covalent bonds between lignocellulosic components were done using aqueous sodium hydroxide (NaOH) solution in the RH and SB fiber treatments [6]. Many researchers have reported that utilization of biomass fibers is in accordance with the function as well as physical and mechanical properties of the material, such as an adsorbent that can absorb dyes [7,8] metal ions Cu (II) and Cd (II) and choromium [9,10] Pb(II) ions [11,12] and Fe(III) ions [13]. The combination of sugarcane bagasse fiber with carbon as an adsorbent as a biocomposite material to absorb color gives good results [14].

The source of fiber material in the composite manufacturing uses the cellulose fibers of RH and SB. The magnetic nanoparticles are attached on the fiber surface of RH and SB in the form of biocomposites by the matrix provided. In addition, the development of magnetic nanoparticle synthesis with a one-step process by modifying functional groups with rice husk fiber as an adsorbent has been investigated [1]. The novelty in this research is the fiber source for the biocomposites is from the combination of rice husk fiber and sugarcane bagasse fiber. There has been no research that develops biocomposites with a variety of these two fiber source compositions through a one-step process. This research is to investigate the best composition of combination of RH and SB fibers for the biocomposites formation. The characterization of biocomposites such as surface morphology, component content, crystalline structure, and functional groups are also investigated.

2. Materials and methods

2.1 Materials

The following are some materials used in this research. The rice husk was obtained in Gambut, Banjar District, and the sugarcane bagasse was obtained from an iced sugarcane seller in Banjarbaru, South Kalimantan. There were also 1.6-hexanediamine ($C_6H_{12}N_2$), ethylene.glycol. ($C_6H_6O_2$), iron (III) chloride hexahydrate ($FeCl_3.6H_2O$), sodium acetate anhydrous ($C_2H_3NaO_2$), ethanol (C_2H_5OH) that were taken from UPT BPPTK LIPI. Last, hydrochloric acid (HCl) and sodium hydroxide (NaOH) were obtained from Sigma Aldrich.

2.2. Delignification

Rice husk (RH) and sugarcane bagasse (SB) were washed to remove impurity particles, dried at 80 °C for 24 h then crushed using a blender, and the RH and SB powder was sieved to pass ± 60 mesh size. Afterward, the dry powder of RH and SB was dissolved in 1% NaOH solution (40% v / v), then stirred at 100 rpm for 2 hours and heated at 80 °C to remove the lignin. The RH and SB fiber powder was later washed with distilled water until the filtrate reached a neutral pH and dried for 3 hours at 80 °C. The results of fibers were in the forms of RH-D and SB-D.

2.3. Preparation of RH and SB fibers with magnetic nanoparticle biocomposites

The one-step solvothermal process was done to synthesize the combined fibers of RH and SB biocomposites that produced the magnetic nanoparticle biocomposites. First, 24 mL of ethylene glycol was added to 1.6 g of sodium acetate anhydrous and 0.8 g of iron (III) chloride were heated at 60 °C and stirred at 200 rpm for 15 minutes. During the time, 7 mL of 1.6 hexanediamine and 0.5 g of RH-D and SB-D were added into the solution with the weight ratios of 1:1; 1:2; 1:3 and 1:4. The mixture of fibers was put into a solvothermal reactor (Teflon Stainless Steel Autoclave) then heated at 200 °C for 6 h. After the reaction, the biocomposites were washed with distilled water and ethanol three times for each to remove any residual chemicals. Finally, it produced four types of biocomposites, namely BM-1:1, BM-1:2, BM-1:3; and BM-1:4.

2.4. Characterization

Analysis of Field-Emission Scanning Electron Microscopy (FE-SEM, JOEL JSM-6500F) was used to observe the morphological structure of RH, BS, RH-D, SB-D, and BM. XRF measured the energy-dispersive X-Ray Fluorescence with a condition operation at 20 kv voltage and 77 UA current. The functional groups of the surface on RH, RH-D, SB, SB-D and BM were identified by Fourier Transform Infra-Red Spectrometry (FT-IR, bio-rad, digilab FTS-3500). Measurement of the X-ray

Diffraction (XRD) by using copper k-alpha (CuKα) radiation was performed using the Rigaku D/max-b XRD. The condition was kept at 400 kv voltage and 100 ma current. Crystalline Index (CrI) was calculated as in equations 1.

$$\mathrm{CrI} = \frac{\mathrm{I_{002}} - \mathrm{I_{am}}}{\mathrm{I_{002}}} \tag{1}$$

where CrI is crystalline index (%), I_{am} is intensity of amorph and I_{002} is intensity of crystal part.

2.5. Analysis
Retro-titration method was conducted to determine the amine contents in the samples [15]. In brief, the analysis started by dropping 50 mg of samples into 0.01 M HCl of 25 mL. The mixture was shaken for 2 hours. Then, 5 mL of supernatant was titrated with 0.01 N NaOH after the centrifugation. The concentration of amine was calculated as in equations 2.

$$C_{NH_2} = \left[\frac{(C_{HCl} \times V_{HCl}) - (5C_{NaOH} \times V_{NaOH})}{m_{sample}}\right] \tag{2}$$

where m is mass of sample (g), C_{NaOH} is concentration of NaOH (mmol/L), V_{NaOH} is volume of NaOH which used in titration of unreacted acid excess (L), C_{HCl} is HCl concentration solution (mmol/L) and V_{HCl} is volume of HCl (L)

3. Results and discussion
Rice husk (RH) and sugarcane bagasse (SB) are biomass cellulose that have a complex structure. In order to collect the cellulose, the materials should be processed through delignification. RH and SB gave differences on the structures of morphology and crystal, and functional groups in the samples. After the delignification process, the color of rice husk fiber changed from the initial color brown to grey. On the other hand, the original color of SB before the treatment with NaOH solution was brown and became brighter after the treatment. The changing of color in the materials is shown in figure 1 and 2 (inset). The micrograph results of rice husk and sugarcane bagasse in 60-mesh size by FE-SEM analysis with 5000x magnification are also shown in figure 1 and figure 2. Based on figure 1a, the surface of rice husk formed a lump which was identified as silica concentrated at a certain point observed by FE-SEM. Meanwhile, the surface of sugarcane bagasse fiber (figure 2a) had the conical protrusions and bright spots at the in-order distance identified as silica which was concentrated in the domical protrusion area, so that the fiber surface looked very undulated and had side-by-side sloping areas [16].

Figure 1. Images of the morphology of rice husk (a) before and (b) after delignification.

Rice husk and sugarcane bagasse fibers were produced through the delignification process by decomposing lignin, hemicellulose and silica contained in the fibers. In this case, the increasing roughness of the surface and the cracking of conical protrusions occurred due to the treatment with NaOH solution. In addition, due to the leaching of lignin, hemicellulose and silica the RH-D and SB-D became shrinking as shown in figure 1b and figure 2b.

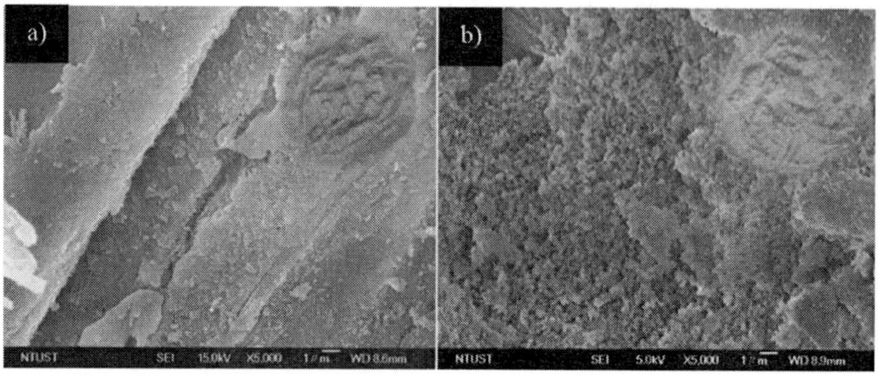

Figure 2. Images of the morphology of sugarcane bagasse (a) before and (b) after delignification.

The morphological structure of magnetic nanoparticle biocomposites with the addition of 1.6-hexanediamine is shown on figure 3. The magnetic nanoparticles were grown on the fiber surface and smaller in size around 30-50 nm. To determine the particle size and to modify the surface spontaneously, hexanediamine played an essential role in the process of magnetic formation. The 1.6-hexanediamnie in the mixture will inhibit the magnetic growth from $FeCl_3$ because the amine groups are kept on the surface that prevents the particle growth [17]. The results of this study regarding the formation of magnetite are in accordance with the findings from [1,18,17].

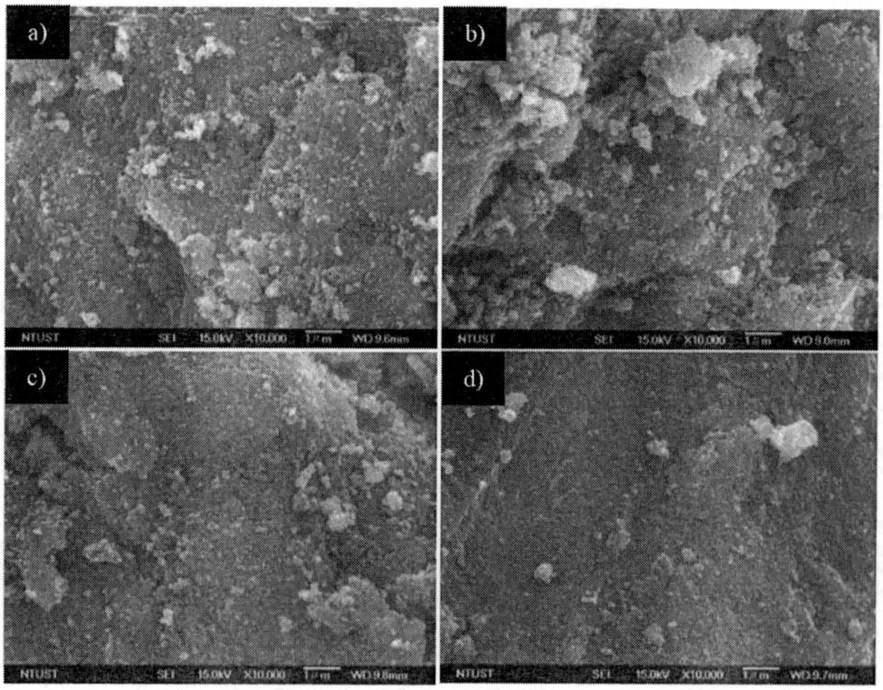

Figure 3. Images of the morphology of rice husk and sugarcane bagasse biocomposites of a) BM-1:1; (b) BM-1:2; (c) BM- 1:3 and (d) BM- 1:4.

A different ratio of RH-D and SB-D in the preparation of biocomposites has shown a number of magnetic that could be produced from these materials. As shown in figure 3, all samples had been distributed well on the surface of fibers. Due to the structure and composition of the fibers, the samples had slightly different amine contents. The BM-1:1, BM-1:2; BM-1:3; and BM-1:4 contained as much as 2.64; 2.30; 2.30; 2.37 mmol/g of amine group, respectively. Further, the iron contents of BM-1:1, BM-1:2; BM-1:3; and BM-1:4 were detected by XRF as much as 97.97%, 97.52%, 96.49% and 95.58%, respectively. As a result, the highest iron and amine contents were found in the biocomposisites with ratio 1:1 or BM-1:1.

Analyses for crystalline index (CrI) and cellulose crystal structure of rice husk (RH) and sugarcane bagasse (SB) were done through XRD. The cellulose crystal displayed dominant peaks shown at an angle of 2 theta between 20°-80° [19]. In order to calculate the intensity of crystal in the samples, it was taken between the angles by XRD data. Meanwhile, the amorph characteristic showed wide diffraction of angle XRD pattern at 2 angle between 0°-20° at which the XRD pattern was seen to diffract widely [20]. Additionally, the lignocelluloses which contained cellulose had peaks in its amorphous form at 16.0° (cellulose I) and at 22.2° (cellulose II) in its crystal form (figure 4). The delignification treatment with NaOH solution of rice husk (RH) and sugarcane bagasse (SB) fibers could increase the intensity or crystal structure of the rice husk (19) and sugarcane bagasse fibers This caused the loss of hemicellulose and lignin content after treatment.

This study also recorded the increased crystalinity index (CrI) values of rice husk and sugarcane bagasse fibers which were 90.32% and 36.8% respectively as displayed in table 1. The crystalline cellulose had been formed due to the increased intensity of the crystal peaks while the peak of amorph showed a reduced polysaccharide structure identified through a broad peak [21]. The crystalline and amorph peaks were used as the indication to measure the organized crystalline cellulose and a less-organized polysaccharide structure.

Table 1. Components of peaks characteristics of RH, RH-D, SB, SB-D.

Samples	Characteristics of Peaks		CrI (%)
	Amorph (16.0 °C)	Crystal (22.2 °C)	
RH	464	665	43.319
RH-D	564	1029	82.447
SB	771	1208	56.680
SB-D	686	1218	77.551

The apparent magnetic on the samples also could be identified by XRD. Based on figure 4, the specific peaks for magnetite was at 36°, 43°, and 57° (JCPDS card 39-0664). This proved that the magnetic was successfully formed in the biocomposites. Based on research of Nata et.al., 2018 (1), the results showed that the formation of magnetite was 30-50 nm size, the iron content in RHB-MH was about 19.85% which confirmed by an EDX analysis and the increased crystalinity index (CrI) of rice husk was about 16.77% confirmed by XRD.

The results of FT-IR spectra of RH, RH-D, SB, SB-D, BM-1:1; BM-1:2; BM-1:3; BM-1:4 can be seen in figure 5. The vibration spectra at 582 cm^{-1} showed that Fe-O in Fe_3O_4 stretched for all the samples of biocomposites. This peak did not appear on the RH and SB fibers. Modification of amine group in the biocomposites for N-H bending vibration was also confirmed by the specific peak at 1640 cm^{-1}. In addition, the wavenumbers of RH and RH-D for C-H stretching vibration are shown at the peak of 2920 cm^{-1}. Last, the wavenumber of Si content in the fibers was detected at 1050 cm^{-1} as Si-OH band.

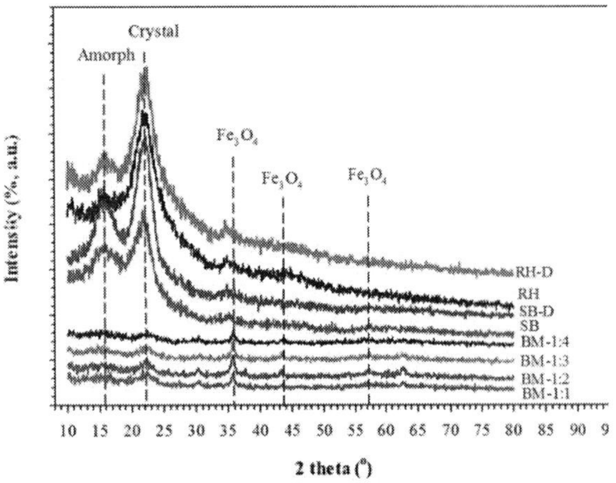

Figure 4. XRD analysis of RH, RH-D, SB, SS-D, BM-1:1; BM-1:2; BM-1:3; BM-1:4.

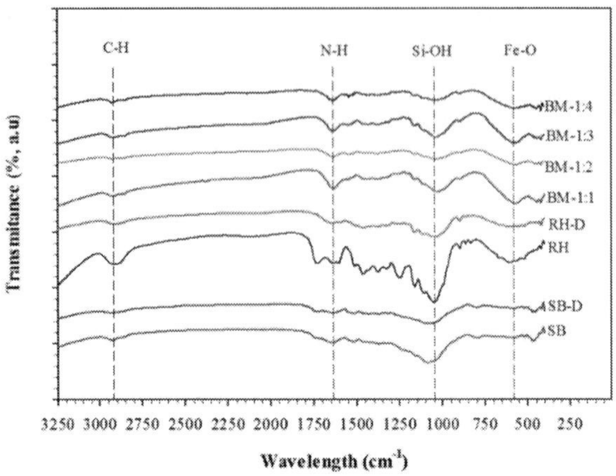

Figure 5. FT-IR spectra of RH, RH-D, SB, SS-D, BM-1:1; BM-1:2; BM-1:3; BM-1:4.

4. Conclusions

Some conclusions are drawn from this study. The combination of rice husk (RH) and sugarcane bagasse (SB) fibers as the matrix for biocomposites was successfully prepared by a one-step hydrothermal process. The magnetic nanoparticles were formed on the surface of rice husk (RH) and sugarcane bagasse (SB) fibers approved by XRD and FE-SEM analyses. The best composition of RH-D and SB-D fibers was achieved in a ratio of 1:1. Moreover, the amine content in the particle was about 2.6 mmol/g which was confirmed by FT-IR. The magnetite was formed in this reaction, and this was proved by the specific peaks at 36°, 43°, and 57° by XRD. The magnetic nanoparticles appeared on the fiber surface that contained 97.97% Fe. All in all, the magnetic nanoparticle biocomposites based on the cellulose is a potential candidate as a matrix for biocomposites.

Acknowledgements

We would like to express our thanks to the DGHE, Ministry of Education and Culture of the Republic of Indonesia for the financial support for magister thesis research with contract No. 113.1.12/UN8.2/PG/2020.

References

[1] Nata IF, Putra MD, Nurandini D, Irawan C, Fitriani R, Isnaini MD 2018 Rice Husk Fiber Magnetic Nanoparticle Biocomposites Preparation and Characterization IOP Conference Series: Earth and Environmental Science **175** 12005

[2] Supranto S, Tawfiequrrahman A, Yunanto DE 2015 Sugarcane Bagasse Conversion to High Refined Cellulose Using Nitric Acid, Sodium Hydroxide and Hydrogen Peroxide as the Delignification Agents. Journal of Engineering Science and Technology **35** 46

[3] Ilindra, Ambuj, Dhake JD 2008 Microcrystalline Cellulose from Bagasse and Rice Straw Indian Journal of Chemical Technology **15** 497

[4] Hameed BH, Ahmad AA 2009 Batch adsorption of methylene blue from aqueous solution by garlic peel, an agricultural waste biomass Journal of hazardous materials **164** 870

[5] Rokbi M, Osmani H, Imad A, Benseddiq N 2011 Effect of Chemical treatment on Flexure Properties of Natural Fiber-reinforced Polyester Composite Procedia Engineering **10** 2092

[6] Karp SG, Woiciechowski AL, Soccol VT, Soccol CR 2013 Pretreatment strategies for delignification of sugarcane bagasse: a review. Brazilian Archives of Biology and Technology **56** 679
[7] Guimarães Gusmão KA, Alves Gurgel LV, Sacramento Melo TM, Gil LF 2012 Application of succinylated sugarcane bagasse as adsorbent to remove methylene blue and gentian violet from aqueous solutions – Kinetic and equilibrium studies Dyes and Pigments **92** 967
[8] Zhang Z, Moghaddam L, O'Hara IM, Doherty WOS 2011 Congo Red adsorption by ball-milled sugarcane bagasse Chemical Engineering Journal **178** 122
[9] Dos Santos VCG, De Souza JVTM, Tarley CRT, Caetano J, Dragunski DC 2011 Copper Ions Adsorption from Aqueous Medium Using the Biosorbent Sugarcane Bagasse In Natura and Chemically Modified Water, Air, & Soil Pollution **216** 351
[10] Mahmood-ul-Hassan M, Suthar V, Rafique E, Ahmad R, Yasin M 2015 Kinetics of cadmium, chromium, and lead sorption onto chemically modified sugarcane bagasse and wheat straw Environmental Monitoring and Assessment **187** 470
[11] Hamza IAA, Martincigh BS, Ngila JC, Nyamori VO 2013 Adsorption studies of aqueous Pb(II) onto a sugarcane bagasse/multi-walled carbon nanotube composite Physics and Chemistry of the Earth, Parts A/B/C **66** 157
[12] Yu J-X, Wang L-Y, Chi R-A, Zhang Y-F, Xu Z-G, Guo J 2013 Competitive adsorption of Pb^{2+} and Cd^{2+} on magnetic modified sugarcane bagasse prepared by two simple steps Applied Surface Science **268** 163
[13] Nata IF, Mirwan A, Wicakso DR, Irawan C, Isnaini MD, Fitriani R 2020 Adsorption of Fe^{3+} ion from Aqueous Solution onto Rice Husk Biocomposite Magnetic Nanoparticle IOP Conference Series: Earth and Environmental Science **506** 012006
[14] Rattanachueskul N, Saning A, Kaowphong S, Chumha N, Chuenchom L 2017 Magnetic carbon composites with a hierarchical structure for adsorption of tetracycline, prepared from sugarcane bagasse via hydrothermal carbonization coupled with simple heat treatment process Bioresource technology **226** 164
[15] Nata IF, Sureshkumar M, Lee C-K 2011 One-pot preparation of amine-rich magnetite/bacterial cellulose nanocomposite and its application for arsenate removal RSC Advances **1** 625
[16] de Carvalho Neto AGV, Ganzerli TA, Cardozo AL, Fávaro SL, Pereira AGB, Girotto EM 2014 Development of composites based on recycled polyethylene/sugarcane bagasse fibers Polymer Composites **35** 768
[17] Irawan C, Nata IF, Lee C-K 2019 Removal of Pb(II) and As(V) using magnetic nanoparticles coated montmorillonite via one-pot solvothermal reaction as adsorbent Journal of Environmental Chemical Engineering **7** 103000
[18] Effendi EZ, Hariadi YC, Salaahuddin MD, Irawan C, Nata IF 2019 Utilization Of Rice Husk Cellulose as a Magnetic Nanoparticle Biocomposite Fiber Source for the Absorpstion of Manganese (Mn) Ions in Peat Water Journal of Science and Applied Chemistry **22** 220
[19] Hashemian S, Hosseini SH, Salehifar H, Salari K 2013 Adsorption of Fe(III) from Aqueous Solution by Linde Type-A Zeolite American Journal of Analytical Chemistry **04** 123
[20] Mohadi R, Hidayati N, Lesbani A 2014 Adsorption Desorption of Chromium (III) Ion on Cellulose From Wood Powder International Journal Of Science and Engineering (IJSE) **7** 77
[21] Ghali L, Msahli S, Zidi M, Sakli F 2011 Effects of Fiber Weight Ratio, Structure and Fiber Modification onto Flexural Properties of Luffa-Polyester Composites Advances in Materials Physics and Chemistry **01** 78

IOP Conf. Series: Materials Science and Engineering **980** (2020) 012008 doi:10.1088/1757-899X/980/1/012008

Molecular dynamics simulation of nanocellulose-oil-water interaction in enhanced oil recovery application

M Ledyastuti* and J Jason

Inorganic and Physical Chemistry Division, Chemistry Department, Institut Teknologi Bandung, Jalan Ganesa no. 10 Bandung 40132

Email: mia@chem.itb.ac.id

Abstract. Enhanced Oil Recovery (EOR) is a way to obtain long hydrocarbon-structured oil by flooding oil reservoirs using water with certain materials. Hydrolyzed polyacrylamides (HPAM) and xanthan gum are commonly used in the EOR process. Both of these materials have several disadvantages, including unstable at high salinity and leave debris in the environment. One alternative to substitute HPAM and xanthan gum is nanocellulose, which is an abundant amount of natural polymer. Molecular dynamics simulations investigate the potential of nanocellulose as an EOR agent. A mixture of decane and naphthalene is used as an oil model, while the SPC/E model is used as a water model. By analyzing the simulation trajectory, the interfacial tension and viscosity values were obtained. The simulation results showed the value of the water-oil interfacial tension increased with the addition of nanocellulose. This was caused by the position of nanocellulose that was always in the aqueous phase during the simulation. The water-oil interfacial tension also increased from 46.94 dyne/cm to 47.96 dyne/cm with the presence of two nanocellulose molecules at 353 K. Water viscosity increased from 3.10×10^{-4} kg/m.s to 3.80×10^{-4} kg/m.s after the addition of one cellulose molecule at 353 K. Nanocellulose can be an EOR agent by increasing the water viscosity but unfortunately the water-oil interfacial tension also increasing. Due to the change of these two opposing properties, nanocellulose cannot optimally improve oil recovery as shown by previous research.

1. Introduction

Enhanced Oil Recovery (EOR) is a way to obtain long hydrocarbon-structured oil by flooding oil reservoirs using water with certain materials. Injected fluids generally can reduce interfacial tension, increase crude oil mobility, or form microemulsions [1]. The ability to reduce interfacial tension is possessed by surfactants such as alcohol propoxy sulfate (APS), $C_{20\text{-}24}$ alpha olefin sulfonate (AOS), and $C_{15\text{-}18}$ internal olefin sulfonate (IOS) [2]. Adding polymers to the water that caused the viscosity increases can do the increase of crude oil mobility. Increased viscosity is expected to have greater thrust to facilitate crude oil mobility to the production well area. Commonly used polymers to increase viscosity are hydrolyzed polyacrilamides (HPAM) and xanthan gum. Unfortunately, these polymers exhibit some disadvantages. HPAM decreases its performance under conditions of high salinity [3]. Meanwhile, xanthan gum can leave debris that will reduce reservoir permeability [4].

Recently, nanocellulose is predicted to be an EOR agent replacing HPAM and xanthan gum. Nanocellulose is cellulose that has a nanometer size in one of its dimensions. It has unique properties including low density, high strength, and stiffness [5-8]. Nanocellulose has emulsifying properties, but it can also be a thickener [5, 9]. Aadland *et al* have reported that adding nanocellulose to the water flooding process can increase oil recovery by 1.2% [1]. This oil recovery is not very effective for an EOR agent [1]. The ineffectiveness can be caused by several factors, including changes in interfacial tension and viscosity that are not significant. The previous research did not discuss in depth how nanocellulose interactions in the water-oil system.

This study aims to look at the interactions that occur when nanocellulose is in the water-oil system. Our study was carried out using a molecular dynamics simulation approach, which allows revealing the position and movement of molecules in detail. Molecular dynamics simulations can be used to complement the explanation of the data obtained from experiments such as how the orientation of molecules on a surface [10-11]. Molecular dynamics simulations can also be used to predict the dissolving energy of a solute in a particular solvent [12-13].

2. Computational methods

The nanocellulose-water-oil interaction was investigated with the GROMACS-2018 software [14]. The force field for all components in the system was using CHARMM36 [15]. In this calculation, three types of water molecules were used, namely SPC, SPC/E, and TIP3P [16]. Preliminary calculations were carried out to determine the water model that was close to the real water molecule. The results of our preliminary calculation were consistent with previous calculations conducted by Vega and de Miguel [17]. They choose that the best model for water is SPC/E based on the surface tension value at 298 K. In this case, we also chose the SPC/E model as the best water model. Our calculation had resulted in 56.69 dyne/cm for the surface tension of the SPC/E model. This study also had proven the number of water molecules did not significantly affect the value of surface tension. In addition, the surface tension values were calculated at 4000, 6000 and 8000 water molecules can be seen in table 1. Based on these results, the next simulation was using 4000 water molecules.

Table 1. Surface tension values using SPC/E model in different number of water molecules at 298 K.

Number of water molecules	Surface tension (dyne/cm)
4000	56.69
6000	56.32
8000	56.27

A mixture of decane and naphtalene was used as an oil model in this study. This mixture was representative of aliphatic and aromatic compounds commonly found in crude oil [18]. The composition of decane and naphtalene used was 1: 1. The oil model was then mixed with water, as shown in figure 1. In the water-oil system that was formed then nanocellulose molecules were added. The selected nanocellulose molecule was glucose dodecamer. Based on previous research, dodecamer glucose was thermodynamically soluble in water [13]. The number of nanocellulose molecules added varies to see how they affected the water-oil system.

The whole system went through the processes of geometry optimization, equilibration, and production. The production stage trajectory was then analyzed to obtain the value of the interfacial tension and viscosity. The geometry optimization stage used the steepest descent algorithm with a maximum force of 100 kJ / mol / nm [19]. Equilibration used NVT and NPT ensembled with a time step of 1.0 fs and a total simulation of 1.0 ns. The Nose-Hoover thermostat was used to keep the temperature constant during the NVT ensemble [20]. Whereas for calculations with the NPT ensemble the Berendsen thermostat and barostat were used [21]. Simulations were carried out at 298 K and 353 K, while the pressure was maintained at 1 bar. The temperature of 353 K illustrated the temperature in the reservoir. The production phase was carried out with an NPT ensemble for 30.0 ns with a Nose-Hoover thermostat and a Parrinello-Rahman barostat [21-22].

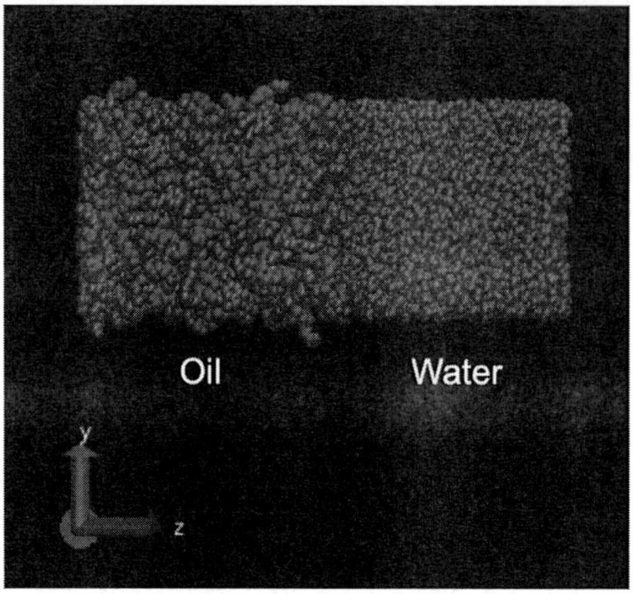

Figure 1. Oil-Water system.

3. Results and discussion

3.1. Interfacial tension

The oil model used in this study was a mixture of 1: 1 decane and naphtalene, which had a density of 891.88 kg/m^3 at 298 K. The value was close to the density of crude oil that ranged around 900 kg/m^3 at room temperature [23]. The analysis of the oil density profile showed that decane and naphtalene were well mixed, as in figure 2. The oil was then mixed with water to form two separated phases. The density profile of this system showed that the water phase and the oil phase were perfectly separate. Based on the trajectory processing of this system simulation, an interfacial tension of 53.51 dyne/cm was obtained at 298 K. If the temperature was raised to 353 K, the interfacial tension decreased to 46.94 dyne/cm. This was caused by the interaction of water-oil, which was getting stronger with rising temperatures. Based on figure 3, naphtalene had a stronger interaction with water. This was consistent with previous research that revealed stronger aromatic compounds in their interactions with water [11].

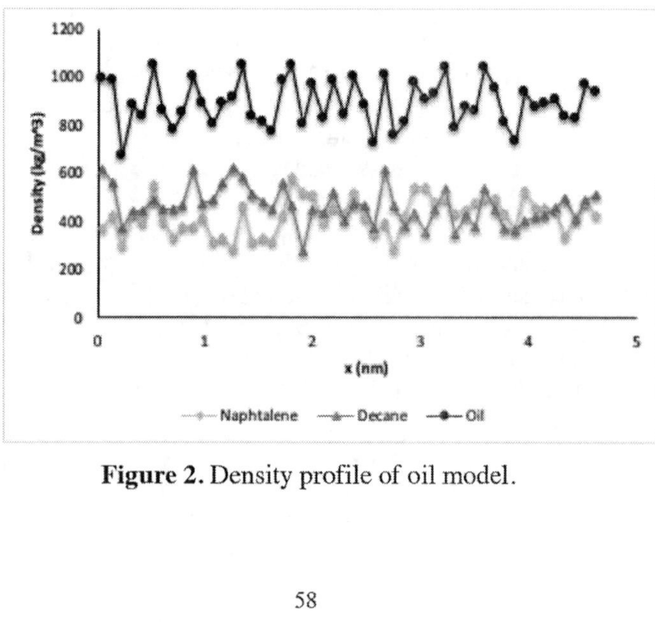

Figure 2. Density profile of oil model.

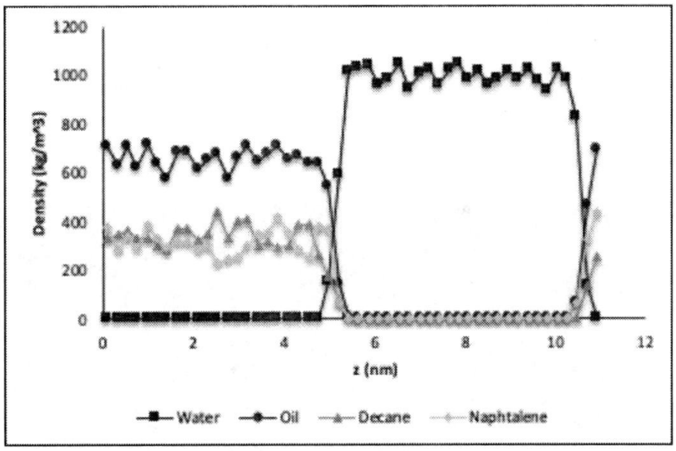

Figure 3. Density profile in water-oil system.

The nanocellulose used in this study had good solubility in water as shown in figure 4. Based on the density profile, nanocellulose was spread evenly along the z-axis. The addition of nanocellulose molecule into the water had increased the surface tension as seen in table 2. The increase of surface tension was probably caused by a stronger interaction between nanocellulose and water. As the data shown in table 2, the surface tension was higher with the addition of nanocelluloses. This proved the presence of nanocellulose strengthened the interaction in the aqueous phase.

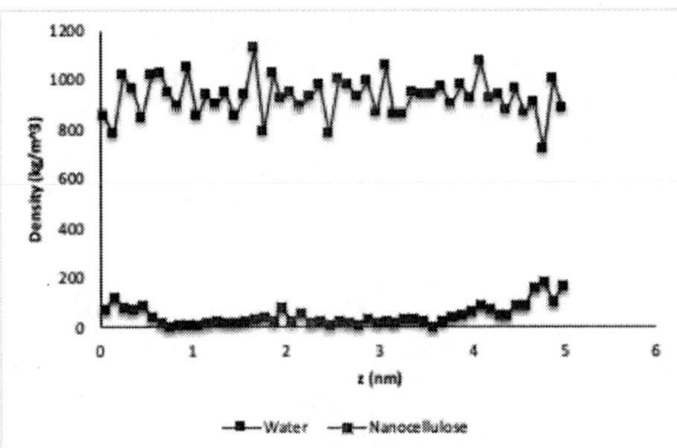

Figure 4. Nanocellulose distribution in the water phase.

Table 2. Surface tension value with nanocellulose variation at 353 K.

Number of Cellulose molecules	Density (kg/m^3)	Surface tension (dyne/cm)
0	962.47	47.81
1	988.35	49.19
2	1012.66	49.93
3	1036.91	50.78

The nanocellulose-water-oil system can be seen in figure 5. The addition of one nanocellulose molecule to the system causes the water-oil interfacial tension to decrease slightly to 46.38 dyne/cm (1.12%) at 353 K. The next addition of two and three molecules increased the interfacial tension to 47.96 dyne/cm (2.17%) and 47.27 dyne/cm (0.70%), respectively. This was likely due to increased interactions in the aqueous phase when the nanocellulose molecule was more than one molecule. Stronger interaction in the aqueous phase caused the weakening water-oil interaction that increased the interfacial tension.

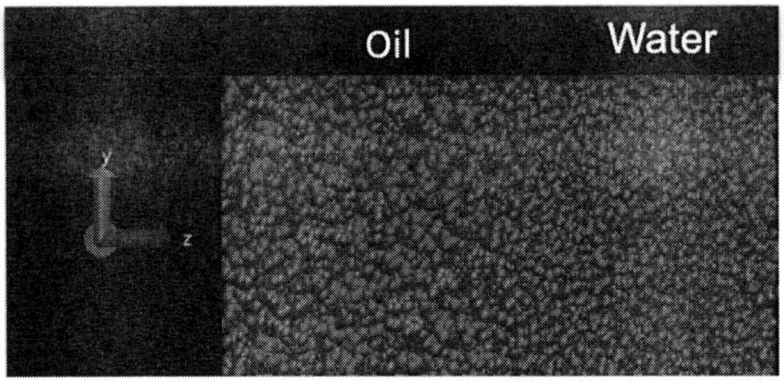

Figure 5. Nanocellulose in the oil-water system.

The addition of nanocellulose, which caused an increase in the water-oil interfacial tension, could also be analyzed based on the position of the nanocellulose during the simulation. Figure 6 showed the position of the nanocellulose throughout the simulation time. In the figure, it appeared that nanocellulose tended to be in the middle of the water phase. This also reinforced the reason for the increased value of the water-oil interface tension. The nanocellulose molecules which tended to be in the water phase caused interactions in the water phase to get stronger and weakened the interaction of the water phase with oil. The hydroxyl groups of nanocelluloses were responsible for the strong interactions with water. This interaction could be reduced by modifying the hydroxyl group to a less polar group.

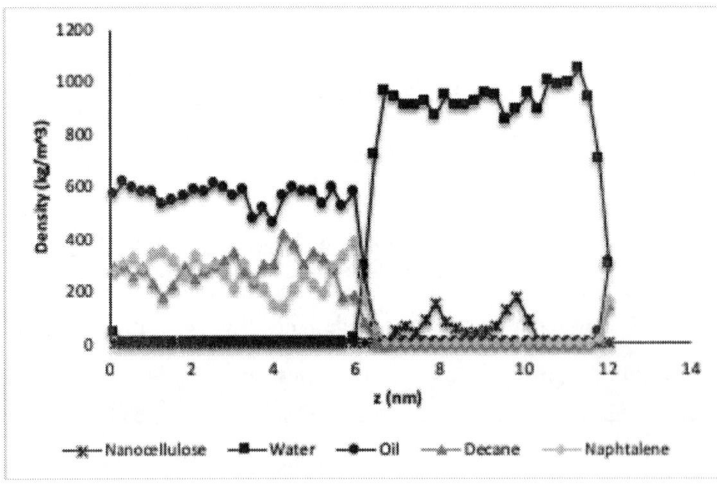

Figure 6. Density profile all components in nanocellulose-water-oil system.

3.2. Viscosity

In addition to the interfacial tension, molecular interaction in the system can also be investigated by viscosity value. Experimental measurements showed the value of water viscosity at 298 K at 1.0 cP, equivalent to 1×10^{-3} kg/m.s [24]. At the same temperature, the results of calculations with the SPC/E water model produced a viscosity of 6.5×10^{-4} kg/m.s. These results were similar to previous calculations [25]. At 353 K, the water viscosity decreased to 3.1×10^{-4} kg/m.s (52.31%). This was because water molecules absorbed additional energy to increase mobility. Adding one nanocellulose molecule to water could increase the viscosity to 3.8×10^{-4} kg/m.s (22.58%). The additions of two and three nanocellulose molecules were way more had the same effect, increasing the viscosity to 4.4×10^{-4} kg/m.s (41.93%). Due to box size limitations, the addition of nanocellulose to the system was only up to three molecules. Higher viscosity had indicated stronger interaction in the presence of nanocellulose molecules in the aqueous phase. Stronger interaction in the aqueous phase could be reflected from the positions of these molecules as shown in figure 7. Distances among nanocellulose molecules were around 2 nm in average during the simulation.

Figure 7. Snapshot of nanocellulose molecules in the aqueous phase.

4. Conclusion

Based on the analysis of our calculation, interaction in aqueous phase was stronger in the presence of nanocellulose molecules. Although the stronger interaction increased and had a good effect on water viscosity, it was not for interfacial tension. By analyzing the molecular dynamics simulation trajectory, the nanocellulose position that always in the aqueous phase makes interactions in the aqueous phase to strengthen while the water-oil interaction weakens. Nanocellulose can be an EOR agent by increasing the water viscosity but unfortunately the water-oil interfacial tension also increasing. Due to the change of these two opposing properties, nanocellulose cannot optimally improve oil recovery as shown by previous research.

References

[1] Aadland R C, Jakobsen T R, Heggset E B, Long-Sanouiller H, Simon S, Paso K G, Syverud K and Torsaeter O 2019 *Nanomaterials* **9** 665

[2] Levitt D, Jackson A, Heinson C, Britton L N, Malik T, Dwarakanath V and Pope G A 2006 Identification and evaluation of high-performance EOR Surfactants, SPE/DOE Symposium on Improved Oil Recovery

[3] Abdel-Azeim S and Kanj M Y 2018 *Energy Fuels* **32** pp 3335-3343

IOP Conf. Series: Materials Science and Engineering **980** (2020) 012008 doi:10.1088/1757-899X/980/1/012008

[4] Sandvik E I and Maerker J M 1977 Application of xanthan gum for enhanced oil recovery *Extracellular Microbial Polysacharides* chapter 19 pp 242-264
[5] Salas C, Nypelo T, Rodriguez-Abreu C, Carrillo C and Rojas O J 2004 *Current Opinion in Colloid & Interface Science* **19** pp 383-396
[6] Edgar C D and Gray D G 2001 *Cellulose* **8** pp 5–12
[7] Siró I and Plackett D 2010 *Cellulose* **17** pp 459–94
[8] Oza K P and Frank S G 1986 *J. Dispers. Sci. Technol.* **7** pp 543–561
[9] Ioelovich M 2016 Fabrication and self-assembly of nanobiomaterials, applications of nanobiomaterials *Nanocellulose-fabrication, structure, properties, and application in the area of care and cure* Volume 1 chapter 9 pp 243-288
[10] Ledyastuti M, Liang Y, Kunieda M and Matsuoka T 2012 *The Journal of Chemical Physics* **137** p 064703
[11] Kunieda M, Nakaoka K, Liang Y, Miranda C R, Ueda A, Takahashi S, Okabe H and Matsuoka T 2010 *J. Am. Chem. Soc.* **132** pp 18281-18286
[12] Shivakumar D, Williams J, Wu Y, Damm W, Shelley J and Sherman W 2010 *J. Chem. Theory Comput.* **6** pp 1509-1519
[13] Santoso A 2019 Pengaruh Derajat Polimerisasi terhadap Energi Bebas Pelarutan Oligomer Glukosa melalui Simulasi Dinamika Molekul *Skripsi* ITB
[14] Lemkul J 2019 *Living J. Comp. Mol. Sci.* **1** p 5068
[15] Lee J, Cheng X, Swails J M, Yeom M S, Eastman P K, Lemkul J A, Wei S, Buckner J, Jeong J C, Qi Y, Jo S, Pande V S, Case D A, Brooks C L, MacKerrel A D, Klauda J B and Im W 2016 *J. Chem. Theory Comput.* **12** pp 405-413
[16] Zielkiewicz J 2005 *J. Chem. Phys.* **123** p 104501
[17] Vega C and De Miguel E 2007 *J. Chem. Phys.* **126** p 154707
[18] Sugiyama S, Liang Y, Murata S, Matsuoka T, Morimoto M, Ohata T, Nakano M and Boek E S 2017 *SPE Journal* SPE-189465-PA
[19] Petrova S S and Solov'ev A D 1997 *Historia Mathematica* **24** pp 361-375
[20] Evans D J and Holian B L 1985 *J. Chem. Phys.* **83** p 4069
[21] Berendsen H J C, Postma J P M, van Gunsteren W F, DiNola A and Haak J R, 1984 *J. Chem. Phys.* **81** pp 3684–3690
[22] Parrinello M and Rahman A 1981 *J. Appl. Phys.* **52** p 7182
[23] Wei Q F, Mather R R, Fotheringham A F, Yang R D 2003 *Marine Pollution Bulletin* **46** pp 780-783
[24] Information on http://www.resources.saylor.org
[25] Markesteijn A, Hartkamp R, Luding S and Werterweel J 2012 *J. Chem. Phys.* **136** p 134104

IOP Conf. Series: Materials Science and Engineering **980** (2020) 012009 doi:10.1088/1757-899X/980/1/012009

Modified physical properties of kaolin by intercalation and exfoliation method

I D G P Prabawa[1]*, R Y Lestari[1], S Hamdi[1] and Sunardi[2,3]

[1]Banjarbaru Institute for Industrial Research and Standardization, Ministry of Industry, Banjarbaru 70714, Indonesia
[2]Department of Chemistry, Mathematics and Natural Science Faculty, Lambung Mangkurat University, Banjarbaru 70714, Indonesia
[3]Wetland-Based Materials Research Center, Lambung Mangkurat University, Banjarbaru 70714, Indonesia

Email: dewa.pprabawa@gmail.com

Abstract. The intercalation method and combination of the intercalation-exfoliation method were applied to modify kaolin. The intercalation process was prepared using urea 60 % as an intercalated agent and the exfoliation process was carried out by ultrasonication. The samples were characterized by X-ray diffraction (XRD), X-Ray Fluorescence (XRF), scanning electron microscope (SEM), and particle size analysis (PSA). Experimental results showed that the physical properties of kaolin significantly improved after modification. This was indicated by the increasing interlayer spacing of modified kaolin, which was confirmed by SEM and XRD analysis. Based on XRF analysis, the loss on ignition value increased after kaolin modification, suggesting that the urea had successfully intercalated into the kaolin interlayer. Kaolin modification with intercalation method produced the most significant physical properties improvement, which increased surface area from 2508.607 to 27297.990 cm^2/g and decreased particle size average from 41.535 to 13.836 μm and had surface area close to illite clay.

1. Introduction

In the last four years (2016-2019), the growth of clay consumption for construction industry materials in Indonesia has increased [1,2,3,4]. In 2019, the import of clay for the construction industry in Indonesia has reached 57,000 tons per year, with a transaction value more than $38 million [4]. Some of the applications clay in the construction industry are for reinforcing filler and fire-resistant material. Illite is one of clay mineral that is widely imported for the construction industry material in Indonesia, such as the fiber cement board industry. The cement board factory with a 36,000 tons/year capacity is estimated to require illite more than 1000 tons/year. The development of the construction industry in Indonesia will increase the import of illite clay for industry materials. Therefore, the development of alternative materials for substitute illite clay in the construction industry is very potential to be studied in Indonesia to reduce dependence on imported materials.

Illite is a clay mineral mica which has a layer structure 2:1. Illite is composed of two silica tetrahedral sheets with one aluminum octahedral sheet, in which the interlayer cation is potassium [5]. Illite has a wide range of physical properties, thus making them applicable to many different construction industry products. The physical properties of illite that are important for the construction industry are high plasticity, fine in particle size, high surface area, good in green and dray strength, good in fire color, and fire strength.

In contrast to illite, kaolin is less reactive when incorporated into the most industrial formulations because it has low surface area, low ion exchange, and particle shape. Illite has a higher surface area than kaolin (i.e., 30 compare 10 m^2/g) and has a smaller particle size than kaolin (i.e., 5.8 compared to 7.3 μm) [6]. This property makes illite very good to be applied as a filler construction material to increase product strength and fire resistance. On the other hand, kaolin is a potential clay mineral in Indonesia with potential reserves of more than 723.56 million tons [7]. Kaolin has some physical properties that are potential for application in construction industry products such as plasticity, green and dray strength, good in fire color and fire strength [5]. Depending upon the application, kaolin must be modified from its natural state by physical or chemical treatments to improve the surface area and reduce particle size.

Kaolin is a clay mineral that has layer structure 1:1 that is combined from a single silica tetrahedral sheet and a single alumina octahedral sheet [5]. Unlike illite, kaolin has no cation interlayer which holds the individual layer in the structure. The layers are held together by hydrogen bonds between hydroxyl groups on the silica in the tetrahedral sheet and the aluminium in the octahedral sheet [8]. because of its structure, kaolin relatively has a low surface area and non-expandable mineral, while only a limited number of polar organic compounds can be directly intercalated [9].

Thus, to improve the physical properties of kaolin, some methods for modification of surface area can be achieved either by the use of acid activation [10], base activation [11], microwave treatment [11], calcination [12], intercalation of organic compounds [13] and exfoliation using ultrasonic wave treatment [14]. Intercalation and exfoliation is a common method used to improve physical properties of organic and inorganic materials. Recently, more studies have used this method for the improvement of kaolinite properties. Zou et al. [8] increase the aspect ratio of kaolin with a combination of the Intercalation method using sodium dodecyl sulfate and ultrasonic exfoliation method for rubber/layered silicate nanocomposites. Zhang et al. [14] using the intercalation-exfoliation method at high temperatures to improve sodium adsorption capacity in kaolin. Experimental results showed that the diameter and pore volume were expanded after modification, which was beneficial to improve adsorption capacity. Guessoum et al. [15] has modified the surface of kaolin by intercalation method using urea then with an ammonium salt and a surface treatment with a silane coupling agent. Then modified and unmodified kaolin were applied as inorganic fillers that are used to improve polypropylene (PP) properties. Experimental results showed that the application of intercalated kaolin as PP filler slightly increased the thermal stability of the composites.

It demonstrated that the intercalation and exfoliation method was able to improve the physical properties of kaolin, especially the surface area. However, there is little literature with study of physical properties modification of kaolin through this method as substitution material for reinforcing filler in the construction industry in Indonesia. Based on the issues in the construction industry described above, the present work aims to study the intercalation method and combination of the intercalation-exfoliation method for improving the physical properties of kaolin as reinforcing filler in the construction industry. The intercalation process was prepared using urea 60 % as an intercalated agent and the exfoliation process was carried out by ultrasonic wave treatment. Afterwards, the properties of modified kaolin are compared with illite clay to determine the potential of modified kaolin as reinforcing filler in the construction industry.

2. Materials and methods

2.1. *Materials*

The kaolin used in this study with a particle size 325 mesh, moisture content 1.6 %, white color, was manufactured by Aneka Kaolin Utama Ltd., Bangka Belitung, Indonesia. Urea (NH_2CONH_2) as an intercalated agent used in this study that is urea fertilizer with nitrogen content 46 % manufactured by Pupuk Kaltim Ltd. Illite clay was provided by Sinar Nusantara Industries Ltd. (one of the particle cement board industries in Indonesia). It was imported from Taiwan, having particle size 325 mesh, moisture content of 2.2 %, and cream color.

2.2. *Kaolinite treatments*

The experimental method for intercalation and exfoliation of kaolin was prepared according to the previous reports [14, 16] with some modified method. In the first experiment, the kaolin modification was prepared by intercalation processes with urea 60 % (KI). 10 g of kaolin was dissolved in 66 mL urea 60 % aqueous solutions, and then the mixture solutions were stirrer by Scilogex magnetic stirrer type MS-PA at a speed rate of 1000 rpm, the temperature of 80° C, for 24 h. After 24 h, the mixture solutions were centrifuged by Ohause centrifuge type FC5707 at a speed rate of 2500 rpm for five minutes and the suspending system was filtrated and washed with aquades to remove residual urea in kaolin. Then, KI was dried in the oven at a temperature rate of 100°C and mashed again up to 325 mesh sizes. In the second experiment, the kaolin modification was prepared by a combination of the intercalation-exfoliation process (KIE). First, Intercalated kaolin was prepared with the same method in the first experiment. Then, 10 g KI was dissolved in 100 ml aquades to form a KI suspension. Furthermore, the KI suspension was exfoliated by ultrasonic treatment with SIBATA ultrasonic cleaner type SU-3THE, at the temperature of 50° C, for 30 minutes. After ultrasonic treatment, the suspension KIE was centrifuged and kaolin filtrated, then dried in the oven at a temperature rate of 100°C and mashed up to 325 mesh sizes.

2.3. *Characterization*

A series of characterizations were carried out to investigate the effect of intercalation and exfoliation modification on the physical properties of kaolin, i.e., X-ray diffraction patterns (XRD), X-ray fluorescence (XRF), scanning electron microscope (SEM), particle size distribution and surface area. The XRD patterns and XRF of the samples were collected by an ARL 9900 series of X-ray spectrometers. The SEM was analyzed with a JEOL JSM-6360LA scanning electron microscope. The particle size distribution and surface area of the samples were measured with laser particle size analyzer LLPA-C10. A series of this characterization was also carried out to determine the physical properties of illite clay.

3. Results and discussion

3.1. *XRD analysis*

Figure 1 shows the XRD patterns of unmodified kaolin, modified kaolin dan illite clay. Pattern (a) in figure 1 displays the 001 reflections of unmodified kaolin at major peak (2θ) 12.32° dan 24.85° with the basal spacing of 7.17 and 3.58 Å, respectively. After kaolin being intercalated with urea (KI), a new reflection appeared at peak (2θ) 8.20° with the basal spacing of 10.76 Å (pattern b), this corresponded to a typical 2θ distance of intercalated kaolin with urea as described before [13]. Intercalation kaolin with urea also decreased the peak of kaolin pattern at (2θ) 12.32°, from 1200 to 800 cps (figure 1, pattern a & b).

The new (2θ) reflections and the expansion of XRD patterns indicated that the intercalation of urea was able to deform hydrogen bonds of the hydroxyl group in the lattice structure of kaolin, and provide a new hydrogen bond structures from Amina group which expanded the basal spacing of intercalated kaolin and make the characteristic of kaolin more swellable [16]. It as described before [15], the basal spacing of the crystal lattice layer of kaolin has been increased by urea intercalation, this formation allowed to provide the expanded basal spacing for cation exchange in the interlayer of kaolin.

Figure 1(c) shows the XRD pattern of modified kaolin with the intercalation and exfoliation process (KIE). After the modification process, a new 001 reflection with significant peak intensity appeared at (2θ) 8.28° with the basal spacing of 10.66 Å. The obtained results suggested that the combination of intercalation and ultrasonic exfoliation process was deformed the lattice structure and expanded the basal spacing of kaolin. Pattern c also shows that the peak intensity of KIE at around (2θ) 12° had decreased significantly compared to the peak intensity of unmodified kaolin at around (2θ) 12° (pattern a).

Based on the expanded results of peak intensity in XRD patterns of kaolin (figure 1) shows the modification of kaolin with a combination of intercalation and exfoliation method (KIE) had a higher peak intensity at (2θ) around 8° than modification with intercalation method (KI). The obtained results suggested that a combination of intercalation and exfoliation methods was more significant to expand the basal spacing of kaolin than only using the intercalation method. The previous study [14] reported that the exfoliation process by ultrasonic treatment supported to break hydrogen bonds and reduced the amount of hydroxyl group in the interlayer of kaolin, which could provide more sites for cation exchange in interlayer kaolin.

Figure 1 shows the XRD pattern of illite clay with the 001 reflections was observed at (2θ) 9.07° with the basal spacing of 9.74 Å. The illite clay had the basal spacing about 10 Å, which the interlayer cation is potassium (K^+). This interlayer cation gives the structure of illite had basal spacing more expanded than kaolin and correlated with higher surface properties of illite clay [5]. Analysis results in figure 1 also show that after the modification process, kaolin had XRD patterns and basal spacings close to illite (figures 1 b, c, and d).

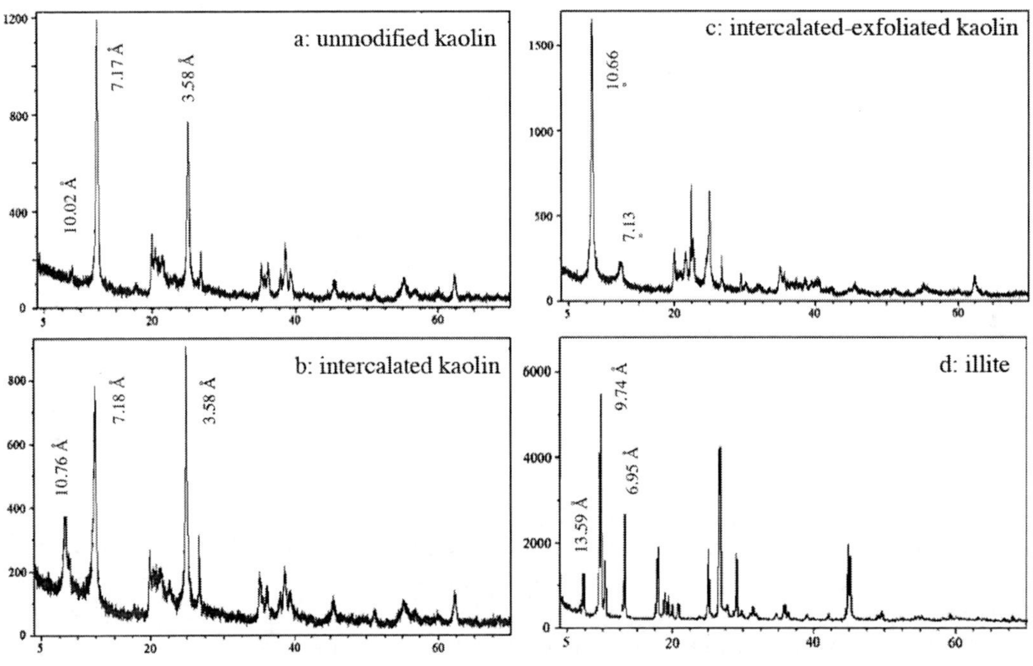

Figure 1. XRD pattern of kaolin, modified kaolin and illite.

3.2. *XRF analysis*

XRF analysis was performed to analyze the compositions of the clay minerals that were present in table 1. The XRF results showed that the compositions of silica (SiO_2) and alumina (Al_2O_3) in kaolin decreased after the modification process. The compositions of silica decreased from 50.02 to 44.95 (KI) and 39.53 % (KIE), while the compositions of alumina decreased from 32.01 to 27.87 (KI) and 22.86 % (KIE), respectively. The intercalation and exfoliation process were contributed to discharge a number of silica and alumina layers in kaolin. As reported before by Aroke and El-Nafaty [17], the intercalation of kaolin with cationic surfactant hexadecyltrimethylammonium bromide (HDTMA-Br) was decreased silica content from 56.29 to 48.42 % and increased Br content up to 6.33%.

On the other hand, the mass of loss on ignition (LOI) of kaolin was gradually increased from 15.29% (K0) to 20.43% (KI) and 35.09 % (KIE) after a modification process. LOI can describe the estimated content of organic matter, inorganic matter and moisture contents in properties of clay [18]. Based on the XRF results, the increase of LOI in modified kaolin indicated the successful intercalation of organic urea in the kaolin.

Table 1. Chemical composition of the unmodified kaolin, modified kaolin and illite.

Chemical compositions (%)	K0	KI	KIE	Illite
SiO_2	50.02	44.95	39.53	54.63
Al_2O_3	32.01	27.87	22.86	28.20
Fe_2O_3	0.75	0.75	0.78	3.64
MgO	0.18	0.15	0.12	0.99
CaO	0.02	0.02	0.02	0.37
P_2O_5	0.03	0.02	0.02	0.03
K_2O	0.95	0.92	0.90	4.64
TIO_2	0.31	0.31	0.31	0.57
LOI	15.29	20.43	35.09	4.90

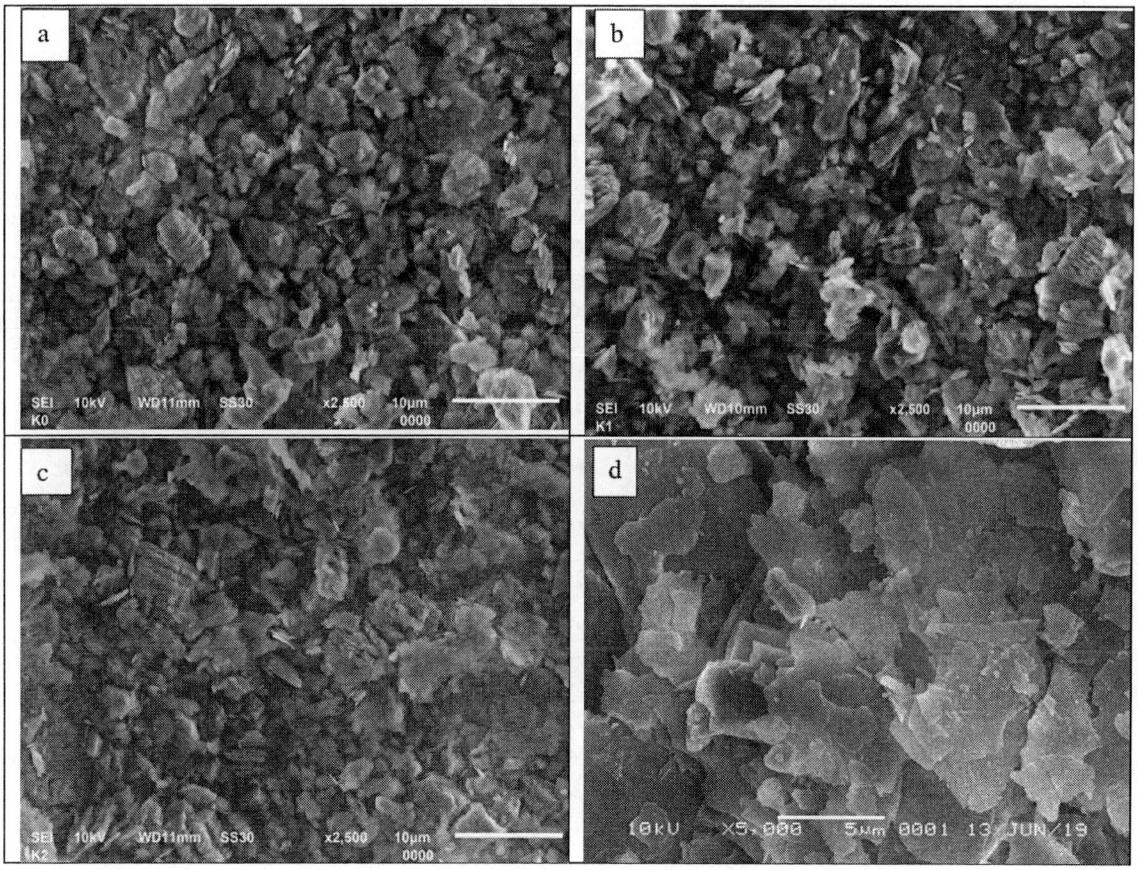

Figure 2. SEM image of (a) unmodified kaolin, (b) intercalated kaolin, (c) intercalated-exfoliated kaolin and (d) illite.

3.3. *SEM analysis*
Figure 2 shows the morphologies of unmodified kaolin, modified kaolin and illite clay. The unmodified kaolin was mainly composed of typical stacks of tetrahedral and pseudo-hexagonal (figure 2a). Compared with unmodified kaolin, the morphology images of intercalated kaolin by urea (figure 2b) showed no significant changes. However, the space among of kaolin particle stacks observed slightly distant after the intercalation process.

Figure 2c displays the morphology of intercalated-exfoliated kaolin, which exfoliation process was treated by ultrasonication. Compared with unmodified kaolin and intercalated kaolin, the morphologi images of intercalated-exfoliated kaolin also showed nearly no changes. However, the type of particle stacks with smaller sizes was more observed than morphology of kaolin particle without ultrasonic treatment.

Previous studies also reported the effect of intercalation and exfoliation process on the SEM morphology of kaolin. Yuan et al. [18] reported that the kaolin particles were mostly transformed to nanorolls after intercalation with cetyltrimethylammonium chloride (CTMACl) in the modified kaolin prepared at 80° C. Zuo et al. [8] reported that a few kaolin particles would transform into nanoscrolls after intercalated with dimethyl sulfoxide (DMSO) and exfoliation treated by ultrasonication.

Figure 2d displays the morphology of illite clay with structure composed of tetrahedral and pseudo-hexagonal particle types, like the kaolin structure. In comparison with unmodified and modified kaolin, the SEM images of illite particle have the sheet forms which the type of layer stacks saw thinner than kaolin. This structure was influenced by the potassium interlayer cation (K^+) in illite that gives the basal spacing structure more expanded than kaolin [5].

3.4. *Particle size analysis*
The results of particle size analysis (PSA) are shown in figure 3 and table 2. After kaolin intercalated with urea and exfoliated with ultrasonication, the particle size distribution peaks shifted to a small size (figures 3 a,b,c) and the mean particle diameter (D_{50}) decreased from 41.535 to 13.836 (KI) and 13.672 μm (KIE) (table 2). These results also showed that the decreased of particle size from KI to KIE has no significant change. These indicated that the exfoliation process in KIE was less significant in changing the particle size of kaolin after the intercalation process.

Compared with unmodified kaolin, the average particle size of modified kaolin decreased nearly three times and their size was closer to the illite clay particle. Moreover, modified kaolin produced the average particle size that was slightly smaller than illite. Figure 3 also showed that the particle size distribution peaks in modified kaolin formed a distribution curve nearly the same as illite.

The smaller of kaolin particle size formed, the higher surface area of kaolin produced. The improved of the surface area of modified kaolin has been confirmed in PSA results, which increased from 2508.607 to 27297.990 (KI) and 24912.340 cm^2/g (KIE). Compared with the surface area of modified kaolin, illite clay has surface area 29664.200 cm^2/g. These results showed that the modification of kaolin via intercalation and combination intercalation-exfoliation process was produced a surface area of kaolin closer to illite clay.

Table 2. The particle size of kaolin, modified kaolin (KI dan KIE) dan illite.

Sampel	D_{10} (μm)	D_{50} (μm)	D_{90} (μm)	$D_{average}$ (μm)
Kaolin	12.183	31.520	81.562	41.535
KI	0.970	5.518	31.375	13.836
KIE	1.061	5.741	31.061	13.672
Illite	0.897	5.430	32.874	14.551

D10, D50, D90 indicate the diameters at 10%, 50% and 90% of the cumulative volume distribution.

IOP Conf. Series: Materials Science and Engineering **980** (2020) 012009 doi:10.1088/1757-899X/980/1/012009

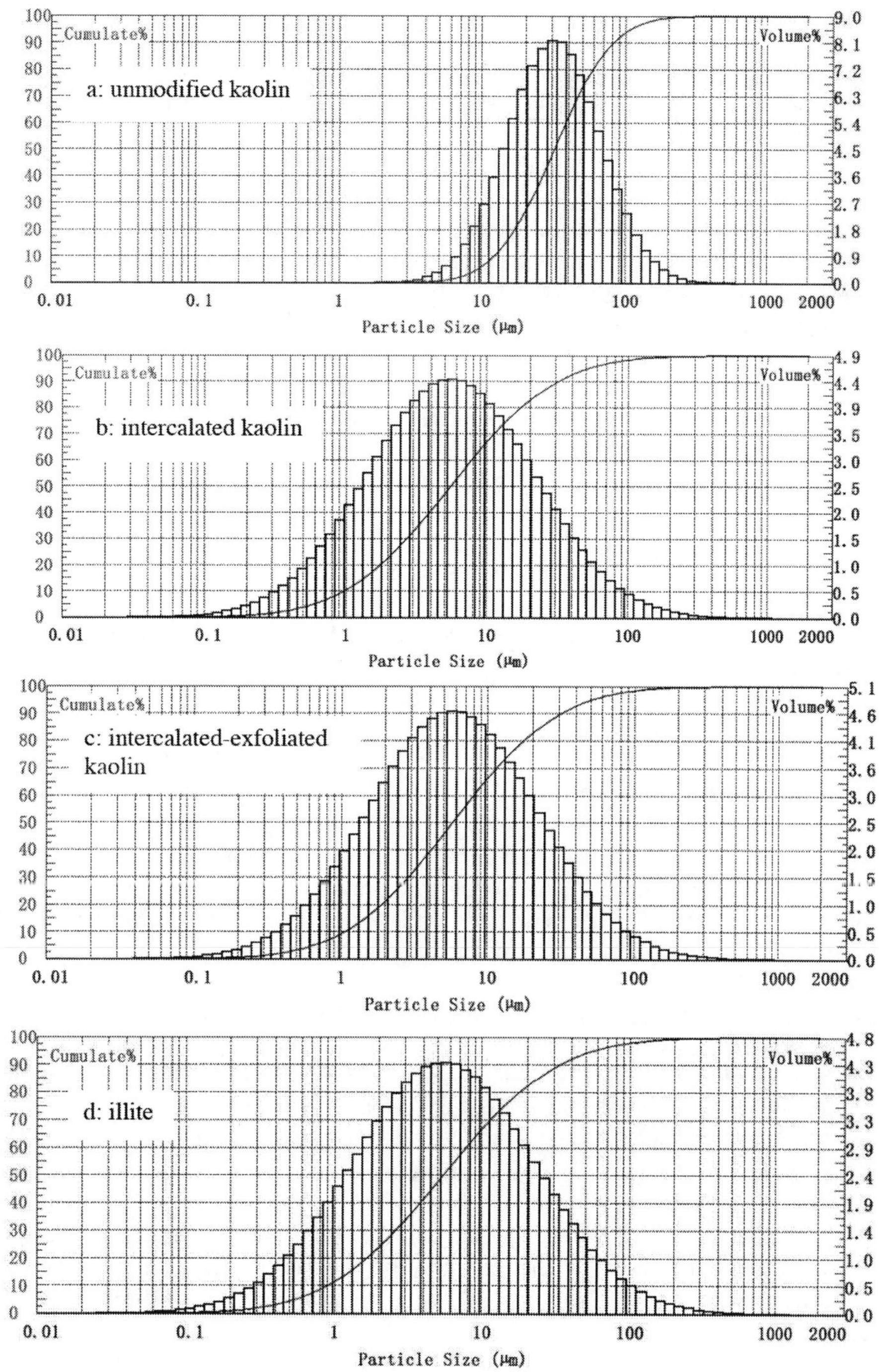

Figure 3. Particle size distribution curve of unmodified kaolin, modified kaolin and illite.

4. Conclusion

Modified physical properties of kaolin for reinforcing filler in the construction industry have been successfully prepared by the intercalation method and combination of the intercalation-exfoliation method. The XRD analysis indicated that the basal spacing of kaolin expanded from 7.17 to 10.76 (KI) and 10.66 Å (KIE) after modification. Based on XRF analyses, it was found that the loss on ignition value increased from 15.29 to 20.43 (KI) and 35.09 % (KIE), which implies successful intercalation of the urea in the interlayer of kaolin. The SEM analysis showed that kaolin has particle stacks that were slightly distantly spaced and few kaolin particles were transformed into smaller sizes after the modification process. The average particle size of kaolin decreased significantly from 41.535 to 13.836 (KI) and 13.672 μm (KIE), and surface area of kaolin increased greatly from 2508.607 to 27297.990 (KI) and 24912.340 cm^2/g (KIE). Based on the XRD pattern, particle size, and surface area analysis showed that the modified kaolin produced physical properties close to illite clay.

Based on the results of this study, the intercalation method with urea 60 % more recommended for prepare kaolin as reinforcing filler than combination intercalation and ultrasonic method. Because the addition of ultrasonic treatment after the intercalation process was not significantly increase the physical properties of kaolin, so the addition of ultrasonic treatment considered less efficient from the aspect of production costs.

Acknowledgments
The authors gratefully acknowledge the financial supports provided by the Banjarbaru Institute for Industrial Research and Standardization, Ministry of Industry, Indonesia. This reaserch also was supported by the Sinar Nusantara Industries Ltd., South Kalimantan, Indonesia.

References

[1] BPS-Statistics Indonesia 2017 Foreign Trade Statistical Buletin Import 2016 (Jakarta: BPS-Statistics Indonesia, Indonesia
[2] BPS-Statistics Indonesia 2018 Foreign Trade Statistical Buletin Import 2017 (Jakarta: BPS-Statistics Indonesia, Indonesia
[3] BPS-Statistics Indonesia 2019 Foreign Trade Statistical Buletin Import 2018 (Jakarta: BPS-Statistics Indonesia, Indonesia
[4] BPS-Statistics Indonesia 2020 Foreign Trade Statistical Buletin Import 2019 (Jakarta: BPS-Statistics Indonesia, Indonesia
[5] Murray H 2009 Applied Clay Mineralogy Occurrences, Processing and Application of Kaolins, Bentonites, Palygorskite Sepiolite, Palygorskite Sepiolite, and Common Clays (Development in Clay Science Volume 2) (Bloomington: Elsevier, Uniteds states) p. 179
[6] Tan X, Zhang G, Yin H, Reed AH and Furukawa Y 2013 Characterization of particle size and settling velocity of cohesive sediments affected by a neutral exopolymer *Int. J. of Sediment Research*. **27(4)** 473–85
[7] Haryadi H 2010 Perkembangan dan prospek bahan galian nonlogam indonesia *J. Teknologi Mineral dan Batubara* **6(1)** 45–63
[8] Zuo X, Wang D, Zhang S, Liu Q and Yang H 2018 Intercalation and Exfoliation of Kaolinite with Sodium Dodecyl Sulfate *J. of Minerals*. **8(112)** 1–11
[9] Sun D, Li Y, Zhang B and Pan X 2010 Preparation and characterization of novel nanocomposites based on polyacrylonitrile/kaolinite *J. of Composites Science and Technology*. **70(6)** 981–8
[10] Yener N, Bic C, Önal M and Sarıkaya Y 2012 Applied Surface Science Simultaneous determination of cation exchange capacity and surface area of acid activated bentonite powders by methylene blue sorption *J. Applied Surface Science*. **258** 2534–9
[11] Bhatnagar A, Hogland W, Marques M, and Sillanpää M 2013 An overview of the modification methods of activated carbon for its water treatment applications *J. Chemestry Engginering*. **219** 499–511
[12] Olu-owolabi BI, Alabi AH, Diagboya PN, Unuabonah EI and Düring R 2017 Adsorptive removal of 2 , 4 , 6-trichlorophenol in aqueous solution using calcined kaolinite-biomass composites *J. Environment Management*. **192** 94–9
[13] Rutkai G, Makó É and Kristóf T 2009 Simulation and experimental study of intercalation of urea in kaolinite *J. Colloid Interface Science*. **334(1)** 65–9

IOP Conf. Series: Materials Science and Engineering **980** (2020) 012009 doi:10.1088/1757-899X/980/1/012009

[14] Zhang X, Liu H, Xing H, Li H, Hu H, Li A, et al 2017 Improved sodium adsorption by modified kaolinite at high temperature using intercalation-exfoliation method *J. Fuel.* **191** 198–203
[15] Guessoum M, Nekkaa S, Fenouillot-Rimlinger F and Haddaoui N 2012 Effects of Kaolin Surface Treatments on the Thermomechanical Properties and on the Degradation of Polypropylene *Int. J. of Polymer Science.* **2012** 1-9
[16] Essawy HA, Youssef AM, El-hakim A and Rabie AM 2009 Exfoliation of Kaolinite Nanolayers in Poly (methylmethacrylate) using Redox Initiator System Involving Intercalating Component *J. Polymer, Plastic, and Technologi Engineering.* **48** 177–84
[17] Aroke U and El-Nafaty U 2014 XRF, XRD and FTIR Properties and Characterization of HDTMA-Br Surface Modified Organo-Kaolinite Clay *Int. J. of Emerging Technology and Advanced Engineering.* **4(4)** 817–25
[18] Yuan P, Tan D, Annabi-bergaya F, Yan W, Liu D, and Liu Z 2013 From platy kaolinite to aluminosilicate nanoroll via one-step delamination of kaolinite : Effect of the temperature of intercalation *J. Applcation of Clay Science.* **83–84** 68–76

IOP Conf. Series: Materials Science and Engineering **980** (2020) 012010 doi:10.1088/1757-899X/980/1/012010

The development of indonesian local clay as a lightweight expanded clay aggregate (LECA) for organic growing medium

Subari[1], Hernawan[1], K Wahyudi[1*], I Rosmayanti[1] and Nurhidayati[2]

[1]Center for Ceramic (Balai Besar Keramik, Bandung)
[2]The Institution of Research and Standardization of Industry Banjarbaru (Balai Riset dan Standardisasi Industri Banjarbaru)

Email : kristantowahyudi76@gmail.com

Abstract. Lightweight Expanded Clay Aggregate (LECA), a clay-based lightweight aggregate expanding after burning at high temperatures, is among the promising material since it can be used as construction materials for buildings and roads, geotechnical fillings, hydroponic growing medium, humidity control, wall damper, heat insulation, water treatment, and others. However, the potential of clay from various places in Indonesia used for LECA as an organic planting medium is still not explored considerably. In this paper, various local clays in Indonesia were studied to obtain a suitable clay as the single raw material for LECA production. Clay raw materials were originated from Purwakarta, Sukabumi, Klaten and Boyolali. The preliminary selection was evaluated based on XRD and XRF test results. The XRD and XRF analysis showed that Boyolali shale and Boyolali bentonite were the most fitted clay as single raw material for LECA production. Both materials were synthesized into LECA through a firing process at 1100°C and 1200°C. Furthermore, the LECA product was characterized by its bloating coefficient and water absorption. The best results were obtained by Boyolali bentonite with a bloating coefficient of 4.63 or can be referred to as medium type and 14.63% for water absorption.

1. Introduction

There are three types of lightweight aggregate. The first type is produced by mechanical methods, grinding the natural raw material to a certain size. The examples material for the first type were pumice and rock tuff. The second type is produced by the combustion method in a rotary kiln, this type comes from clay, shale, or bentonite that is commonly named LECA. And the last is industrial wastes-by products, fly ash or slag [1].

Lightweight Expanded Clay Aggregate (LECA) is a clay-based lightweight aggregate that expands after being burned at high temperatures (1050°C - 1250°C) [2]. The use of LECA is very diverse including construction materials for buildings and roads [3], geotechnical fillings, hydroponic growing media [4], geology, civil and environmental engineering, agricultural and greenhouse issues [5].

LECA has been used as an organic growing media for gardening and agriculture. When compared with other organic growing media such as coconut fiber, charcoal, tile pieces, and so on, LECA has various advantages. LECA can regulate drainage, maintain moisture, protect roots from freezing and store nutrients and help oxygen flow so that plant maintenance will be easier and cheaper. Moreover,

LECA could be applied in urban areas that do not have much green land. Its lightweight properties make LECA as a very good alternative to be used for vertical farming in a building.

LECA structure is formed through 2 (two) simultaneous events during the firing process, the formation of gas, and the formation of a very viscous liquid phase. After a cooling process, the liquid phase changes to glass and form hollow spaces. Gases trapped in the glass phase are the result of organic and inorganic materials decomposition when the material is subjected to high temperatures. The gas formed during the heating process can be in the form of CO, CO_2, SO_2, H_2O, O_2 and N_2. As temperature and gas concentration increase, the pressure of gases in melted glass increases which results in the expansion of material volume known as bloating [3]. Table 1 shows the classification of clays as raw material for LECA based on the bloating coefficient.

Table 1. Classification of clays based on bloating coefficient.

No	Coefficient of Bloating (Kp)		Degree of Bloating
	Vaicklionis [3]	Samer Karem [6]	
1	> 7 – 8	> 4	Good/ self-bloating
2	4 – 5	2.5 - 4	Medium level
3	2 – 2,5	2–2.5	Poor
4	< 2	< 2.5	Not-self bloating

Clay minerals may be grouped into four types, there are kaolinite group, smectite group (montmorillonite group), illite group, and chlorite group [7]. Among all of these groups, montmorillonite is the most swelling clay. Water molecules cause swelling in montmorillonite. The migration of counter-ion, initially bound to montmorillonite surface to the central interlayer plane, leads to swelling in montmorillonite [7]. Montmorillonite is one of mineral contained in clays which could slowly release the crystal of water up to 1200°C provided the gas contributing to the bloating of the clay [8].

LECA chemical composition consists of SiO_2, Al_2O_3, fraction of Fe_2O_3, CaO, MgO and alkalis Na_2O and K_2O [9]. According to Vaicklionis [3] and Rashad [9], self-bloating properties of clay influenced by its ratio of SiO_2 and Al_2O_3 and sum of oxides contains (Fe_2O_3, CaO, MgO, Na_2O and K_2O) or its chemical composition meet requirements as shown in table 2.

Table 2. Chemical composition for self-bloating clay.

No	Composition	Vaicklionis [3] (%)	Rashad [9] (%)
1	SiO_2	50 - 60	53,3 - 70
2	Al_2O_3	15 - 20	15,05 - 27
3	F_2O_3	7 - 8	1 – 14,3
4	CaO	< 8	0,2 – 3,92
5	CaO + MgO	< 5	
6	R_2O	2 - 3	
7	organic	0,5 - 3	

Clay bloatability can be enhanced by an addition of organic materials (heavy fuel oil, wood sawdust, fine turf, coal, waste of rubber) or inorganic material (iron compounds, glass granules, cement, fly ash) [3].

Indonesia has abundant clay mineral natural resources but it is currently used for the production of tiles, tableware, potter, and earthenware ceramics [10]. The utilization of clay as raw material of LECA is an alternative to give value added to Indonesia's local clays. In this study, clay from Purwakarta, Sukabumi, Klaten, and Boyolali were examined. This study aimed to obtain a suitable clay as a single raw material for LECA production.

IOP Conf. Series: Materials Science and Engineering **980** (2020) 012010 doi:10.1088/1757-899X/980/1/012010

2. Materials and methods

2.1. Materials
Pasir Pining clay, Sandaan clay, and Cibodas clay were obtained from the Service Technical Implementation Unit of Ceramic Research and Development, Purwakarta. Melikan clay was obtained from Klaten Pottery Conservation Center laboratory. Clay Sukabumi and bentonite Sukabumi were obtained from Department of Industry and Energy Mineral Resources Sukabumi District. Meanwhile, shale and bentonite Boyolali were obtained from Department of Trade and Industry Boyolali District.

2.2. Method and characterization
The mineral contained in each material was investigated using XRD Bruker D8 Advance, then tested for chemical composition using XRF ADVANT XP Thermo ARL9900. After the chemical composition evaluation, the selected clays were dried under the sun for 5 days, grinded, sieved to 0.14 mm – 18 mm size, added with potable water, formed by manual pelletizer then fired in the rotary kiln, length 3 m, 0,3 rpm at temperature 1100°C and 1200°C.

3. Results and discussion

3.1. Mineralogical characterization of clay raw material
The clay raw materials were originated from Purwakarta (Pasir Pining, Sandaan, Cibodas), Sukabumi (Sukabumi bentonite), Klaten (Melikan) and Boyolali (bentonite and shale). The result of clay raw material characterization using XRD was shown in figure 1 to figure 3. Figure 1 were the results of the XRD characterization of Pasir Pining, Sandaan, Cibodas (Purwakarta) samples. From the three diffractogram pattern, all of them indicate the presence of halloysite ($Al_2Si_2O_5(OH)_4$-$2H_2O$), quartz, muscovite ($KAl_2(AlSi_3O_{10})(F,OH)_2$). Pasir Pining clay presence of feldspar ($KAlSi_3O_8$ – $NaAlSi_3O_8$ – $CaAl_2Si_2O_8$) had melting point below 1200°C but the main mineral was halloysite, usually known as aluminosilicate clay mineral that had softening point above 1300°C and the presence muscovite could increase the percentage of alkali R_2O. Based on mineralogy, those clays had low viscosity and non-carbonaceous [3]. Clays needed any additives to decrease the melting point between 1100°C and 1200°C and source of material which produces gases to support the bloatability. Therefore, those clays were not suitable to be applied as single raw material of LECA.

The XRD characterization for Melikan clay from Klaten was shown in figure 2. The diffractogram pattern according to figure 2 indicated that the main minerals were halloysite, α-quartz, muscovite, calcium and illite. Even though halloysite mineral had a high softening temperature which was above 1300°C, but with the presence of muscovite and calcium; the softening temperature of Melikan clay could be lowered into 1100°C. Besides, in this clay presenced illite mineral as one of non-expanding clay. Therefore, Melikan clay could not be used as a single raw material for LECA products and required other materials such as potassium feldspar, pure kaolin, or other alkalis that had a high melting point or additives to improve the degree of bloating.

Figure 3 illustrated the XRD diffractogram for Gunung Guruh clay from Sukabumi and indicated halloysite, muscovite, montmorillonite, calcite, and small fraction of α-quartz minerals. The presence of calcite shown that clay was carbonaceous clay which could play a role in the bloating process [11]. Montmorillonite as acid clay would provide releasing gases to support the bloatibility. The presence of muscovite provided high alkali R_2O. As a result, by the presence of (1) alkalis as flux and bloating actor; and (2) gases and pore materials, Sukabumi clay had a potential to be developed into LECA product.

The XRD pattern in Boyolali bentonite clay was shown in figure 4. The diffractogram indicated presence of montmorillonite, halloysite, quartz, cristoballite and feldspar minerals. Montmorillonite was contained the acid clay. It released the crystal water slowly up to 1200°C provided the gas contributing to the bloating of the clay [8]. Therefore, Boyolali bentonite clay can be developed into a single raw material of LECA product.

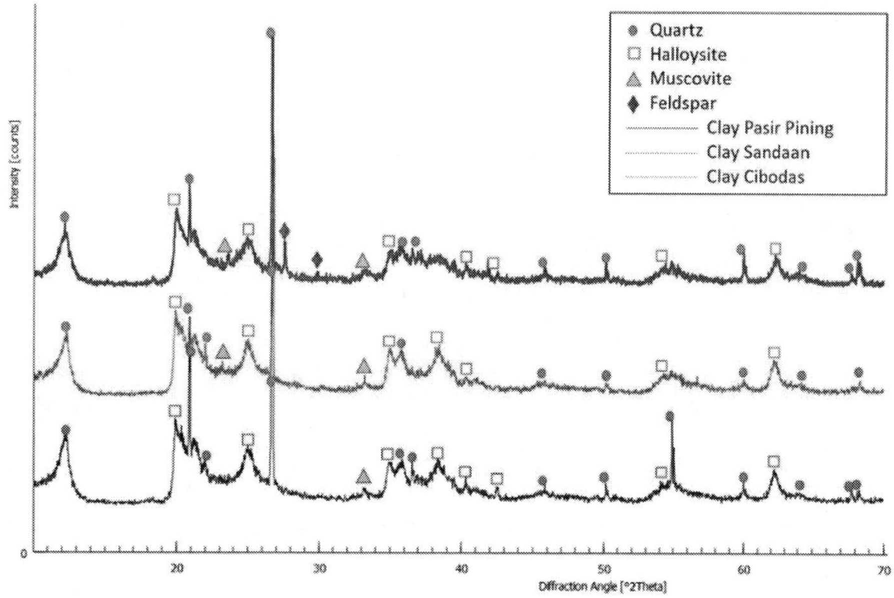

Figure 1. XRD patterns of pasir pining, sandaan and cibodas.

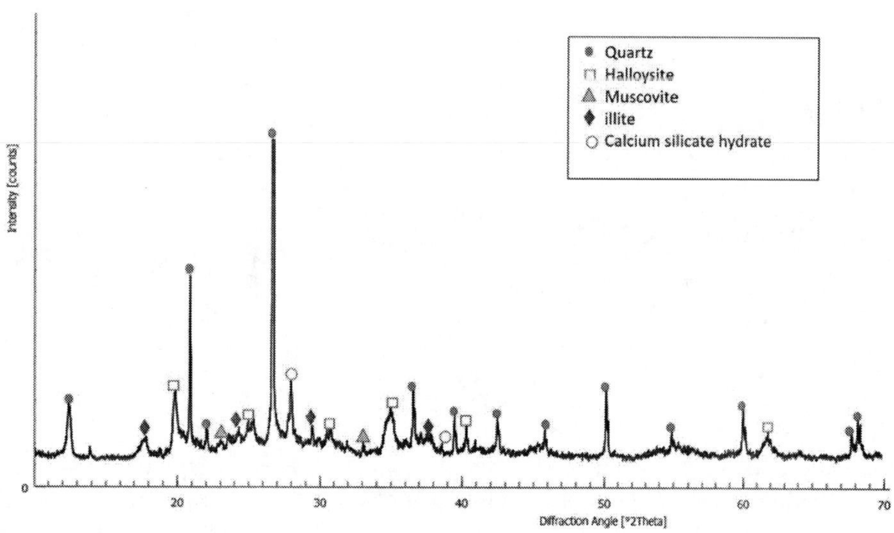

Figure 2. XRD pattern of melikan clay.

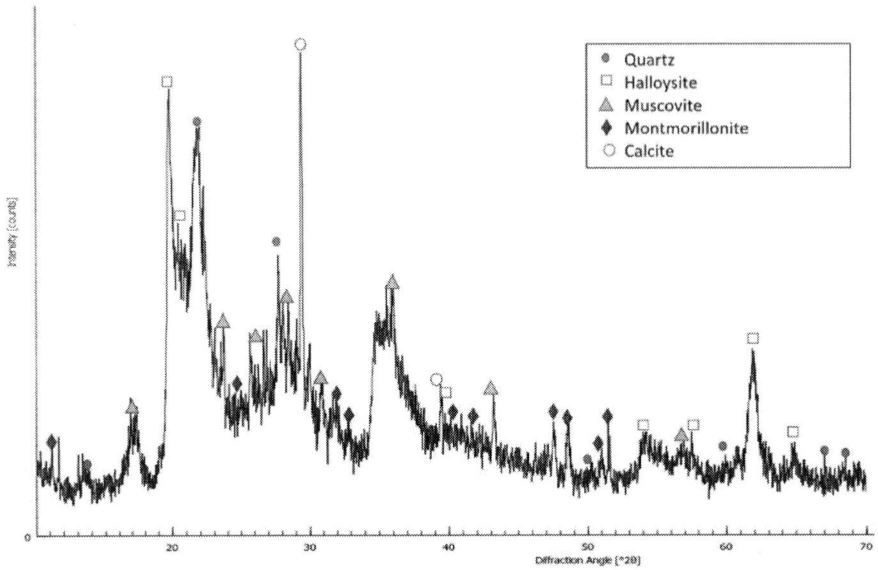

Figure 3. XRD pattern for gunung guruh clay, sukabumi.

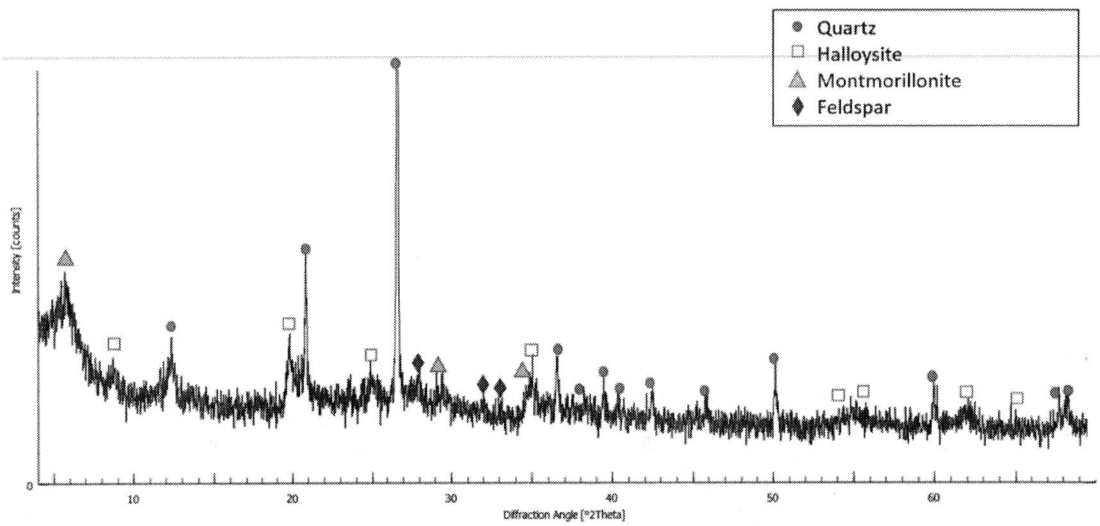

Figure 4. XRD pattern for boyolali bentonite clay.

Based on XRD pattern from all clay sample; Boyolali bentonite, Boyolali shale and Sukabumi clay were selected to be used in preliminary and main research. These raw materials were the most fitted to develop into LECA product since they only required fewer additives or even used as single raw material. Moreover, the XRD diffactogram analysis must be supported by other characterization in order to some materials consider suitable as LECA raw material especially if it is used as single raw material.

3.2. Chemical characterization of clay raw material

The clay ability to expand was closely related to the oxide composition. The X-RF results were presented in table 3. Sukabumi bentonite clay had high SiO_2 that was above 60%, 15% Al_2O_3, 2,8%

Fe_2O_3, 1,74% R_2O (less than 2%), and (CaO + MgO) more than 5% that is 12,49%. (CaO + MgO) more than 5% also appeared in Sukabumi clay that is 13,36 %. Moreover, it could be seen that Sukabumi bentonite clay and Sukabumi clay had a total flux of 16,90% dan 17,37%, respectively. Sukabumi bentonite clay and Sukabumi clay had a lower SiO_2/fluxing ratio of 3.79 and 3.12, respectively. The lower ratio indicated both clays were non-bloating raw material. That meant both clays had lower viscosity that material could not trap a sufficient amount of gas and prevented from expanding during firing [11]. Therefore, clays did not suitable to be applied as raw material of LECA. However, Sukabumi bentonite clay and Sukabumi clay still could be develop as raw material of LECA by using additives to decrease the total flux and to increased SiO2/fluxing ratio.

Theoretically, Boyolali bentonite and shale had met the standard requirements for LECA product. These raw materials had (CaO and MgO) less than 2% and total flux less than 10% as could be seen at table 3. In this study, Boyolali bentonite and shale were qualified to be used as single raw material for LECA production.

Table 3. Chemical composition of bentonite, shale and clay.

No	Oxide	Boyolali Bentonite (%)	Boyolali Shale (%)	Sukabumi Bentonite (%)	Sukabumi Clay (%)
1	SiO_2	51.17	61.20	64.00	54.28
2	Al_2O_3	17.62	17.63	15.18	9.65
3	Fe_2O_3	7.32	8.14	2.67	1.17
4	MgO	1.46	1.90	-	4.42
5	CaO	1.52	1.44	12.49	8.94
6	K_2O	1.89	2.09	1.74	1.10
7	Na_2O	0.91	1.54	-	1.74
8	TiO_2	0.72	0.77	0.32	0.05
9	SO_3	0.35	0.35	0.63	-
10	MnO	0.14	0.07	0.14	-
11	P_2O_5	0.08	0.12	2.83	-
12	LoI	14.37	5.62	-	17.97
13	Total flux	13.10	15.11	16.90	17.37
14	Si/Flux	3.91	4.05	3.79	3.12

3.3. Bloating coefficient analysis

According to Vaickelionis [3] the suitable clay for LECA should have bloating coefficient (Kp) at least 2 and initial melting temperature less than 1300 °C [3]. Bloating coefficient analysis was performed on Boyolali bentonite dan Boyolali Shale. Table 4 shows bloating coefficient calculation results. Both materials were synthesized into LECA through firing process at 1100°C and 1200°C. Boyolali bentonite clay firing at 1200°C gave the best result with Kp > 5 implying medium level of bloating coefficient.

Boyolali bentonite gave a better bloating coefficient result than Boyolali shale with firing process at 1200°C. The firing process held important part that can affect bloating phenomenon. Bloating phenomenon occurred at LECA product as the forming of black core due to the reduction of Fe_2O_3 to FeO at high temperature [8]. The reduction also affects the increase of internal gas pressure as shown in following chemical reaction;

$$6Fe_2O_3 (s) \rightarrow 4Fe_3O_4 (s) + O_2 (g)$$
$$2Fe_3O_4 (s) \rightarrow 6FeO (s) + O_2 (g)$$

Table 4. Bloating coefficient (Kp).

Parameter	LECA			
	Bentonite 1100	Bentonite 1200	Shale 1100	Shale 1200
V_1 (cm^3)	1,77	1,77	1,77	1,77
V_2 (cm^3)	2,15	8,18	1,77	1,77
Kp	1,22	4,63	1	1

Table 5. Water absorption result.

Parameter	LECA			
	Bentonite 1100	Bentonite 1200	Shale 1100	Shale 1200
W (g)	92,71	116,04	108,11	106,22
D (g)	90,09	101,24	99,68	100,68
Water Absorption (%)	2,91	14,62	8,46	7,49

Another reduction process occured in LECA due to organic and carbon oxidation in natural materials as shown in following chemical reaction [12]:

$$3Fe_2O_3\ (s) + C \rightarrow 2Fe_3O_4\ (s) + CO\ (g)$$
$$Fe_3O_4\ (s) + C \rightarrow 3FeO\ (s) + CO\ (g)$$

Those processes produce CO gas that can increase internal pressure and FeO. FeO as flux supposed to decrease the melting point and influence viscosity of material. Kang Hoon Lee, et al, divided the bloating mechanism into three stages: (1) the formation of gas inside material body, (2) the formation of pores by internal pressure and (3) the formation of pores by internal pressure between small and large pores. LECA formation occured with the forming of gas and pore due to internal pressure. Based on chemical composition results, Boyolali bentonite and shale have LoI of 14,37% and 5,62%, respectively. Boyolali bentonite with higher LoI produced more gases than Boyolali shale. Therefore, this significant point caused the bloating coefficient of Boyolali bentonite was higher than Boyolali shale.

3.4. Water absorption analysis

As shown in table 5, the water absorption of LECA which was made of Boyolali bentonite increased as the firing temperature increase. At a temperature of 1100°C, glass phase formation occured on the surface of the bentonite material before bloating. Through bloating,the volume of material expands, porosity got larger and gases release as the temperature rises. Then LECA from bentonite with a temperature of 1200°C had the biggest water absorption, 14.62 %. A different case occured in Boyolali shale, at a temperature of 1100°C the material was still large porous but with increasing temperature the phase glass of the flux will begin to close the pores. Therefore, increasing the temperature would decrease in water absorption properties from 8.46% to 7.49%.

Based on XRD diffractogram pattern and chemical composition analysis, Boyolali bentonite and shale could be used as single raw material for LECA. Both materials were synthesized into LECA through a firing process at 1100°C and 1200°C. Furthermore, the LECA product was characterized by its bloating coefficient and water absorption. The best results waere obtained by Boyolali bentonite with a bloating coefficient of 4.63 implying medium type and water absorption 14.63%.

4. Conclusion

In this study, the lightweight expanded clay aggregate had been produced using Boyolali bentonite as single raw material without any additives for expanding the bloating coefficient. The water absorption increased with increase firing temperature. These result showed that Indonesia's clay which currently used as a raw material for ceramic industry could also be used as a growing medium LECA.

Acknowledgments

This research was supported by Center of Ceramic (Balai Besar Keramik), Ministry of Industry.

References

[1] Chandra S and Berntsson L 2002 *Lightweight Aggregate Concrete* Chalmers University of Technology Göteborg, Sweden by Noyes Publications
[2] Ayati B, Ferrándiz-Mas V, Newport D and Cheeseman C 2017 *Use Of Clay In The Manufacture Of Lightweight Aggregate* Construction and Building Materials **162** 124-31
[3] Giedrius V, Aras K, Danute V 2011 *Production of Expanded Clay Pellets by Using Non-selfbloating Clay* Lakes Sapropel and Glycerol, Material Science (Medziagotyra) **17(3)**
[4] Garcia C M, Palomino T C, et al 2014 *Porosity Of Expanded Clay Manufactured With Addition Of Sludge From Brewing Industry* Int J Energy Environm Eng **5** 341-47
[5] Boudaghpour S, and Hashemi S 2008 *A Study on Light Expended Clay Aggregate (LECA) in a Geotechnical View and its Application on Greenhouse and Greenroof Cultivation* International Journal Of Geology **2(4)**
[6] Khafaji S K T and Al-Majed E A 2016 *Synthesis Of Light Expanded Clay Aggregates From Iraqi Raw Materials* International Journal of Scientific & Engineering Research **7(4)** April
[7] Faheem U 2018 *Montmorillonite: An Introduction to Properties and Utilization*
[8] Wie Y M and Lee K G 2019 *Optimum Bloating-Activation Zone of Artificial Lightweight Aggregate by Dynamic Parameters* Materials Journal MDPI
[9] Rashad A M 2018 *Lightweight Expanded Clay Aggregate As A Building Material – An Overview, Construction And Building Materials*
[10] https://kemenperin.go.id/artikel/20565/Ekspor-Tembus-USD-25-Juta,-IKM-Gerabah-dan-Keramik-Hias-Masih-Prospektif downloaded at December 2019.
[11] Fakhfakh E, Hajjaji W, Medhioub M, Rocha F, López-Galindo A, Setti M, Kooli F, Zargouni F, Jamoussi F 2007 *Effects Of Sand Addition On Production Of Lightweight Aggregates From Tunisian Smectite-Rich Clayey Rocks* Applied Clay Science **35(2007)** 228–37
[12] Lee K H, Lee J H, Wie Y M and Lee K G 2019 *Bloating Mechanism of Lightweight Aggregates due to Ramping Rate* Advances in Materials Science and Engineering, Volume 2019, Article ID 2647391

IOP Conf. Series: Materials Science and Engineering **980** (2020) 012011 doi:10.1088/1757-899X/980/1/012011

The Effect of TiO_2 additives on the antibacterial properties (*Escherichia coli* and *Staphylococcus aureus*) of glaze on ceramic tiles

E Maryani[1]*, N S Nurjanah[2], E P Hadisantoso[2] and R B Wijayanti[1]
[1]Center for Ceramic, Bandung, Indonesia
[2]State Islamic University of Sunan Gunung Djati, Bandung, Indonesia

Email: maryani_eneng@yahoo.co.id

Abstract. Antibacterial glaze coating on ceramic tiles with TiO_2 additive was conducted to meet environmental sanitation needs. Raw materials used were commercial glaze 107, silica sand, and 0-5% weight of TiO_2 powder. The glaze mixture was added with distilled water and ground in an alumina pot mill for 24 hours. The glaze mass was coated on the green body, dried and then burned at a temperature of 1200°C with a holding time of 2 hours. On the glazed tiles, there was a layer of glass with an amorphous structure based on the X-Ray Diffraction results. The addition of TiO_2 caused an increase in opacity but decreased visual white color. Glaze antibacterial properties were tested for *Escherichia coli* and *Staphylococcus aureus* bacteria using the disc diffusion method. The addition of TiO_2 caused resistance to two types of bacteria, both for raw and burned glaze. The TiO_2 inhibition was similar to amoxicillin as a positive control. It had a greater resistance to *S. aureus* than *E. coli* bacteria. The largest inhibitory diameter value of 7.7 mm obtained by 5% TiO_2 glaze powder burnt against *S. aureus* bacteria. Glaze coating on the tile caused a greater inhibition diameter of 24.75 mm for 1% TiO_2 burned glazed.

1. Introduction

Recently, pollutants and bacteria that can harm humans have been growing both in type, quantity and quality. Bacteria evolved to be more resistant and able to develop faster [1]. The spread of bacteria can occur in our environment, so that many industries are competing to launch products that are claimed to have antibacterial properties, including ceramic tiles. Antibacterial ceramic tile is necessary especially for hospitals, schools, restaurants, industrial facilities, public places, and playrooms [2,3].

Antibacterial ceramic tiles can be obtained by applying an antibacterial glaze or coating. Glaze is an impervious layer or coating of a vitreous substance which has been fused to a ceramic body through firing. Antibacterial glaze is obtained by adding active substances such as TiO_2, ZnO, Ag, Cu, its composite, etc [2-11]. In this study, TiO_2 was used as an active ingredient to obtain an antibacterial glaze. Titanium oxide is usually used in the manufacture of glazes to provide color and as an opacifier.

Hazmaliza [4] has examined the antibacterial properties of glazed tiles with titanium dioxide as an antibacterial active ingredient. Antibacterial properties were tested against *Escherichia coli* and succeeded in reducing the growth of the *E. coli* colonies even to death for 10% TiO_2 nanoparticles after 2 hours. However, the resulting tile glaze showed cracks for the addition of 10% and 15% TiO_2. Therefore in this study, a maximum of 5% TiO_2 was added to minimize glaze cracking. In addition, antibacterial tests were carried out on two types of pathogenic bacteria, namely *E. coli* which represents gram-negative bacteria and *Staphylococcus aureus* which represents gram-positive bacteria. *E. coli* bacteria play a role in the process of putrefaction, but if found in excess amounts in humans it causes diarrhea and fever. Bacteria *S. aureus* causes serious skin conditions such as boils, pimples, and itching [12]. The disk diffusion method (Kirby-Baeur) was used to analyze the antibacterial properties of both types of bacteria. This method tests the sensitivity limit of the active substance (titanium dioxide) to bacteria based on the ability of the active substance to diffuse in solid media that has been inoculated by the bacteria [13].

2. Materials and methods

2.1. Glaze suspension production and its coating on tile

Materials used for the production of glaze suspension were technical grade materials, those were: 250g of glaze 107, 250g of silica powder, 400 mL of distilled water and active substance of TiO_2 which varies from 0% - 5%. All materials were mixed and ground for 24 hours using a pot mill which contained 1 kg of alumina ball. The glaze suspension was filtered with a 200 mesh filter and coated on raw tile by dipping method. The tiles that had been coated were dried in the wind and then put in an oven at 105°C for 4 hours. Besides being coated on the tile, the glaze suspension was dried in an oven at 110°C for 4 hours and then crushed with a mortar. Glazed tile and glaze powder were burned in an electric furnace with a combustion route 27-40°C: 1 hour, 40-220°C: 1 hour, 220-400°C: 1 hour, 400-580°C: 1 hour, 580-760°C: 1 hour, 760 - 940°C: 1 hour, 940-1175°C: 1 hour, 1175°C held for 1 hour, 1175-1200°C: 1 hour, 1200°C held for 2 hours, and 1200-40°C: 6 hours.

2.2. Test of antibacterial property of glazed powder and tile

The materials tested for antibacterial properties were glazed tile, dried glaze powder, and glaze powder which had been burned at 1200°C. They were suspended and then tested for antibacterial property aseptically. The disk diffusion method (Kirby-Baeur) was used to analyze the antibacterial properties of both types of bacteria namely *E. coli* ATCC 11229 and *S. aureus* ATCC 6538.

2.2.1. Device sterilization. All instruments used for antibacterial testing were washed at first, dried and sterilized in an autoclave at 121°C for 120 minutes. Sterile petri dishes were then wrapped in a paper.

2.2.2. The making of 2 (two) types of solid media. The first was Mueller Hinton Agar (MHA) for gram-positive bacteria (*S. aureus*). MHA media were made from a mixture of Mueller Hinton Broth (MHB) of 4.2 grams and Bacteriological Agar of 4 grams. The next was Nutrient Agar (NA) media for gram negative bacteria (*E. coli*). NA media were made from a mixture of Nutrient Broth of 1.6 grams and Bacteriological Agar of 4 grams. Each mixture of the two types of media was dissolved with 200 mL aqua bides in an Erlenmeyer and then homogenized and sterilized in a 121°C autoclave for 120 minutes. After that, the sterilized media was poured into a sterile petri dish. The petri dish were shaken in a counterclockwise way at first and then in a clockwise way 5-10 times. The media was left to solidify and stored in the refrigerator.

2.2.3. Liquid media. Mueller Hinton Broth (MHB) was made with a composition of 21 gram / 1L aqua bides for gram-positive bacteria. Nutrient Broth (NB) was made with a composition of 8 gram / 1L aqua bides for gram-negative bacteria. Each liquid media was made in Erlenmeyer and then homogenized and sterilized in an autoclave for 2 hours at 121°C.

2.2.4. Bacterial culture. Inoculated test bacteria were taken out 1 isolate using a sterile ose wire, then suspended into a test tube containing 5 mL of liquid media. Furthermore, the suspension of the test bacteria was incubated in an incubator for 24 hours at 37°C. Then, the turbidity of bacterial suspension was standardized as equal to 0.1 Mc Farland.

2.2.5. Antibacterial test carried out using the disc diffusion method. The suspension of test bacteria was taken out using a cotton swab and applied to the agar medium. Subsequently, the suspension of the positive and negative control test samples was dripped on different disk paper by 10 μL and left for some time. Then, the disc paper was placed regularly on the agar media, then incubated for 18-24 hours at 37°C. After that, it was observed and measured the inhibition zone formed using the calipers. Inhibition was determined by subtracting the diameter of formed inhibition zone by the diameter of disc paper (disc paper diameter = 6 mm).

2.3. Characterization
The instruments carried out to characterize the results of this study are: X-Ray Fluorescence (XRF) to analyze glaze raw material, X-Ray Diffraction (XRD) PANalytical X'Pert HighScore to analyze the types of formed crystal on the surface of glazed tile and glaze powder, Scanning Electron Microscope (SEM) Jeol to analyze the morphology of glazed tile surface.

3. Results and discussion

3.1. Chemical analysis of glaze raw materials
The glaze raw materials used were glaze 107, silica sand, and technical TiO_2. Glaze 107 is a transparent-type (colorless) commercial glaze. The main chemical compounds in glaze 107 are SiO_2 and Al_2O_3 with consentrations of 70.55% and 13.09%, respectively (table 1). The compound of SiO_2 is one of the acid oxides which can be used as a glass layer and hardener. Silica has a melting temperature of 1610-1710°C. The compound of Al_2O_3 (neutral oxide) was used as hardener and thickener for suspension glaze material to form a harder glaze layer. Besides that, there are basic oxide chemicals which can act as fusion agents in glazes [12]. The smelting material is K_2O 4.25%, CaO 3.04%, Na_2O 2.12%, and MgO 0.19%. Glaze 107 also has a yellowish brown color like clay because there is a chemical compound of Fe_2O_3 0.772%.

The active material added to obtain antibacterial property was technical TiO_2 powder with TiO_2 content of 98.18% as shown in table 2. The TiO_2 powder had an anatase crystalline phase. Anatase is a TiO_2 crystal phase which shows the best characteristics as a photocatalyst, antibacterial and optoelectronic [14].

3.2. Whiteness index analysis of glazed tiles
After the glaze tile was burnt, the visual appearance of the surface was observed and the whiteness index was measured using a brightness meter. The brightness test worked by firing UV light/visible light on the surface of the glaze tile. The whiteness index was determined by lightness, i.e. the ability of the test material (tile surface) to reflect light. The whiteness index value (W *) which is getting closer to 0 indicates the whiteness of the test material. The whiteness index value of glazed tile with variations of TiO_2 addition can be seen in table 3. The more TiO_2 powder added indicated the whiteness index value was getting away from zero, so the appearance was increasingly less white.

The brightness index test results were in line with the results of visual tests as shown in figure 1, the more TiO_2 was added, the color of glaze surface was more yellowish. This happened because of the change in the TiO_2 crystal phase from the anatase phase to the rutile phase at a temperature of 700°C. This rutile phase was very easy to form complex compounds with other elements in glaze materials such as SiO_2, Al_2O_3, and Na_2O.

3.3. Mineral analysis of glaze powder and glazed tile
Mineral analysis was carried out for dry glaze powder and glaze powder and glazed tile that were burned at 1200°C, with the addition of 1% TiO_2. Dry glaze powder (not burned) contained crystals that came from the used raw material. Dry glaze powder contained alpha quartz and anatase as shown in figure 2. Quartz alpha came from the silica sand used which was one of the SiO_2 crystals that were stable at room temperature. Anatase was a form of TiO_2 crystal derived from the used TiO_2 powder.

Table 1. The results of chemical analysis of glaze material 107.

Components	Consentrations (% weight)
SiO_2	70,55
Al_2O_3	13,09
CaO	3,40
MgO	0,109
Na_2O	2,12
K_2O	4,25
Fe_2O_3	0,772
TiO_2	0,0810
P_2O_5	0,0205
SO_3	0,0204
MnO	0,0109

Table 2. The results of chemical analysis of TiO_2 powder.

Components	Consentrations (% weight)
TiO_2	98.18
Al_2O_3	0.0193
CaO	0.0143
MgO	0.0832
SiO_2	0.138
K_2O	0.238
Fe_2O_3	0.0205
Na_2O	0.0575
SO_3	0.0282
P_2O_5	0.261

Table 3. Brightness (W*) test results of combustion of glaze tile in temperature of 1200°C with variations of TiO_2 addition.

TiO_2 (%)	W* value
0	-5,92
1	-7,36
2	-11,10
3	-29,80
4	-47,22
5	-54,59

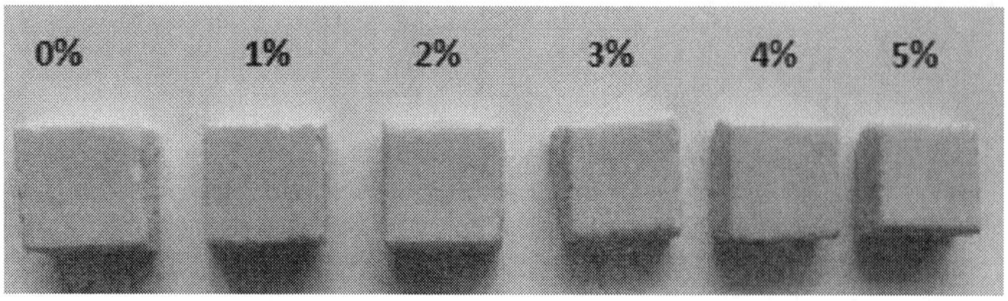

Figure 1. Glaze tiles results from combustion with variations of TiO_2 addition from 0-5%.

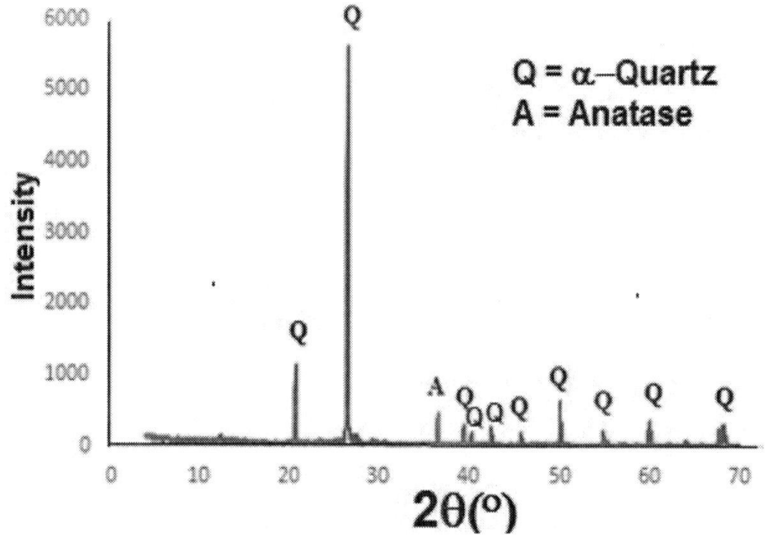

Figure 2. XRD diffractogram of dry glaze powder.

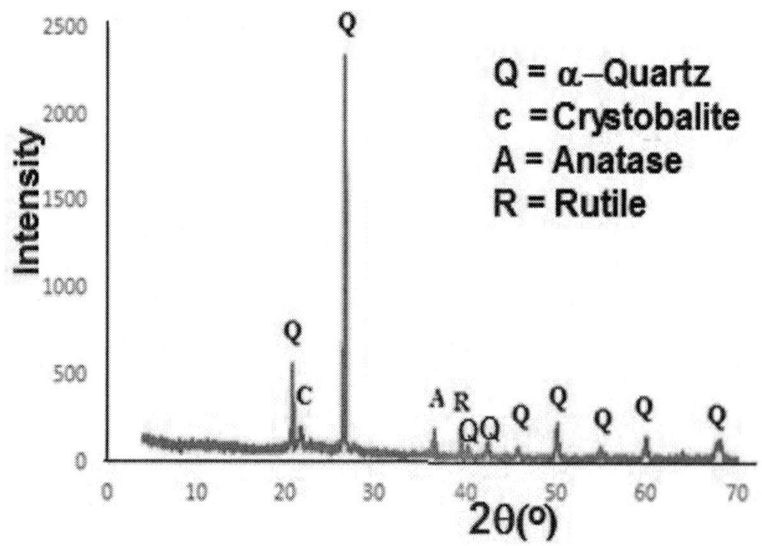

Figure 3. XRD diffractogram of glaze tiles burned at 1200°C.

After being burned at 1200°C, glaze powder contained new crystals, those were cristobalite and rutile, in addition to the crystals found in the raw powder which were alpha quartz and anatase, as shown in figure 3. Crystobalite was formed due to changes in the quartz phase to trydymite (amorphous) and trydymite turnsed into crystobalite. As a result of heating at 800°C, quartz turned into trydymite and when the temperature was raised to 950°C, crystobalite is formed. Rutile was formed due to anatase heating at a temperature of 600-700°C until it transformed into rutile.

The XRD diffractogram pattern for glazed tiles with 1% TiO_2 burned at 1200°C can be seen in figure 4. There were diffraction peaks for anatase, rutile, mullite, and quartz-quartz. Mineral Mullite was formed from the combustion of feldspar and clay in raw tile material. Tile raw material was usually in the form of silica, feldspar, and clay. Mullite was formed from clay ($Al_2O_3.2SiO_2.2H_2O$) which underwent heating to a temperature of 1000°C, was going to release 2 molecules of water to obtain a mullite crystal ($Al_2O_3.2SiO_2$). The XRD diffractogram pattern of glazed tiles (figure 4) was different from the glaze powder (figures 2 and 3), where the peaks that occurred tended to be short with a widening pattern or crystal that occurred were generally amorphous. This was due to the glaze layer applied to the tiles which tended to be thin and fused, resulting in an amorphous crystal structure.

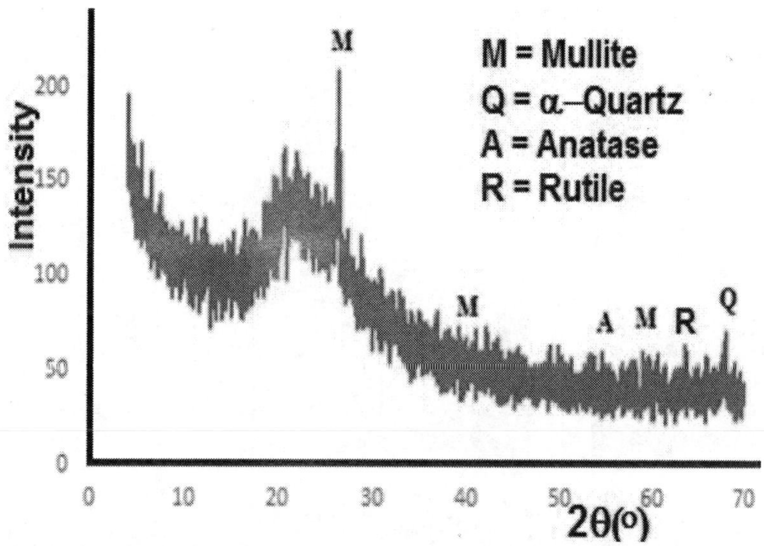

Figure 4. XRD diffractogram of glaze tiles burned at 1,200°C.

3.4. Morphological analysis

SEM analysis aimed to determine the morphology of glaze tiles surface of TiO_2 1%. Morphological observation of glaze tiles using SEM with 5,000x magnification can be seen in figure 5a. It shows Titanium dioxide particles that were shaped like chunks and were disorganized.

Agglomeration or buildup of titanium dioxide particles occurred on the surface of the glaze tile. The factors causing irregularities in particle shape and particle buildup were inhomogeneous glaze suspension and uneven application of glaze. As a result, when the application of antibacterial testing on 4-sided glaze tiles had a different diameter of inhibition. In figure 5, it can be seen on the 0% TiO_2 glaze tiles that there were also titanium dioxide particles. This material was obtained from glaze 107 that had titanium dioxide as much as 0.0810%.

3.5. Test for antibacterial properties

In this test, samples of raw glaze powder, combustion glaze powder and glazed tiles were used. While the tested bacteria were *E. coli* gram-negative bacteria and gram-positive *S. aureus* bacteria. The second reason, these bacteria were used because they can easily grow in general conditions. Aseptic test for antibacterial property was carried out to avoid contamination from other bacteria.

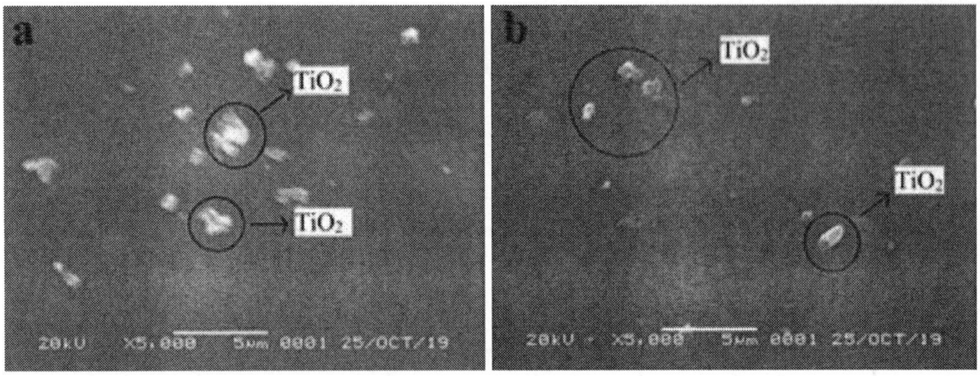

Figure 5. SEM result (a) glazed tiles of TiO_2 1% and (b) glazed tiles of TiO_2 0%.

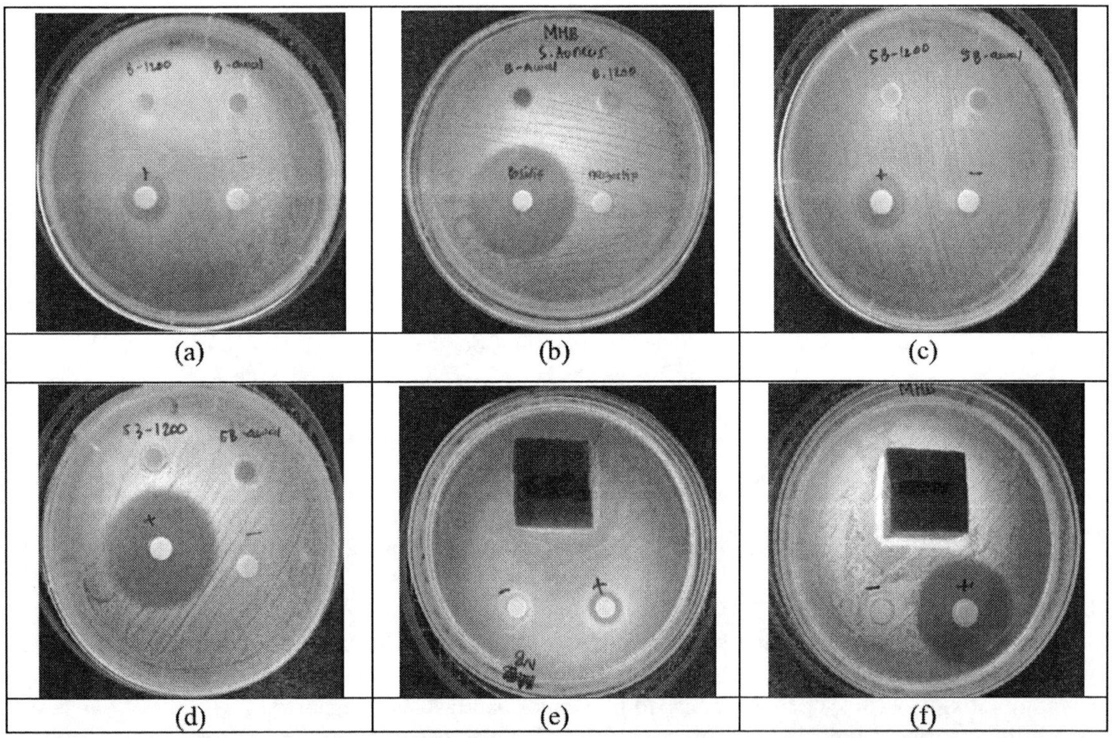

Figure 6. (a) glaze 1% TiO_2 with *E. coli* bacteria; (b) glaze 1% TiO_2 with *S. aureus* bacteria; (c) 5% TiO_2 glaze with *E.coli* bacteria; (d) 5% TiO_2 glaze with *S. aureus* bacteria; (e) 1% TiO_2 glazed tile with *E. coli* bacteria; (f) 1% TiO_2 glazed tile with *S. aureus* bacteria.

In this antibacterial test, it needed a media to breed the desired bacteria. For *E. coli* bacteria, the culture was made in NB (Nutrient Broth) media because NB media is a medium that is commonly used for the growth of heterotropic microorganisms. For *S. aureus* bacteria, the culture was made in the MHB (Mueller Hinton Broth) medium because MHB is resistant to gram-negative and selective to gram-positive bacteria, then MHB is used for the growth of *S. aureus*. MHB contains casein hydrolyzate as a source of proline amino acids that can optimize the growth of *S. aureus* bacteria.

In testing the antibacterial property, two controls were used. Those were positive control in the form of amoxicillin and negative control in the form of sterile aquabides. Positive control served as a comparison of the antibacterial performance of glaze and glazed tiles with commonly used antibiotics. While negative controls were used to determine the effect of solvents on the growth of test bacteria because sterile aquabides were used in making glaze suspensions as solvents. Here is a picture of the results of the antibacterial test, see figure 6.

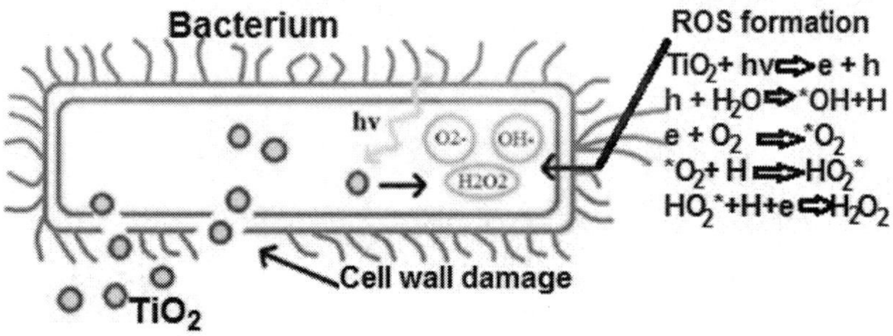

Figure 7. ROS (Reactive Oxygen Species) antibacterial mechanism [15,16].

The results of the antibacterial test were tested qualitatively and quantitatively. The qualitative test was carried out by observing the clear zone which indicated the inhibition of bacterial growth due to the presence of the active substance TiO_2. The quantitative test was measuring the diameter of clear zone or the inhibitory bacteria. Glaze samples and glazed tiles were shown to have antibacterial properties that were characterized by the presence of clear zones until inhibition diameter values were obtained (figure 6). Antibacterial activity of powder and glaze tiles was due to the photocatalytic properties of TiO_2. TiO_2 had a band gap of 3.2 eV and could produce OH^*, O_2^*, and H_2O_2 from the ROS (Reactive Oxygen Species) reaction [4,16]. These radicals will react with organic compounds contained in the cell membrane (peptidoglycan wall) until the bacterial cell undergoes lysis. And Ti^{2+} ions surround the peptidoglycan wall area and bind with organic compounds and lyse the bacterial cell wall. TiO_2 and ZnO compounds have the same antibacterial activity mechanism, namely through the ROS reaction which produces radical compounds (OH^*, O_2^{-*}, and H_2O_2) that can damage bacterial cell walls. This mechanism can be illustrated in figure 7 [15,16].

The diameter of inhibition for *E. coli* and *S. Aureus* bacteria can be seen in table 4. The addition of TiO_2 caused an increase in the inhibitory power for both types of bacteria, both for raw and burnt glazes. The inhibition of the TiO_2 glaze burned decreased for a concentration of 1%. This was due to the phase transformation from part of the anatase TiO_2 to rutile due to the combustion process above 700°C, as confirmed by the XRD diffractogram in figure 4. However, for the 5% TiO_2 concentration, the decrease in inhibition did not occur because the remaining anatase phase was still high so that the decrease in activity due to rutile was not significant. Glaze coating that contained TiO_2 on the surface of ceramic tiles caused an increase in the inhibition of TiO_2 against both types of bacteria because with coating, the contact surface area between TiO_2 and bacteria became larger.

The inhibition of TiO_2 compounds against *S. aureus* bacteria was higher than *E. coli* for all glaze samples as shown in table 4. This was because the cell wall structure of gram-positive bacteria was simpler than -negative one, as shown in figure 7, so that it was easily lysed by radical compounds resulting from the ROS reaction. Gram-positive bacteria only had a thick peptidoglycan layer on their cell walls, whereas negative bacteria had an outer membrane layer consisting lipopolysaccharide (figure 8). Liposaccharide molecules carried a negative charge so that the cell wall of gram-negative bacteria tended to be more negative than -positive one. The ROS reaction of TiO_2 produced several negatively charged species such as *OH and O_2*. These radical compounds would be more difficult to penetrate on the cell membrane of gram negative bacteria, so that these bacteria became more resistant [17].

Table 4. The diameter of inhibition of *S. aureus* and *E. coli* bacteria in glaze samples.

No	Glaze Sample	Diameter of bacterial inhibition	
		E. coli	*S. aureus*
1	Glaze powder with 1% TiO_2	6.1	7
2	Burnt Glaze powder at 1200°C with 1% TiO_2	6	6.5
3	Glaze powder with 5% TiO_2	6.2	7.1
4	Burnt Glaze powder at 1200°C with 5% TiO_2	6.3	7.7
5	Burnt glazed tile with 1% TiO_2	24.1 x 24.15	25.4 x 24.1

Note : Diameter paper disk = 6 mm
Size of tile = 24 mm x 24 mm

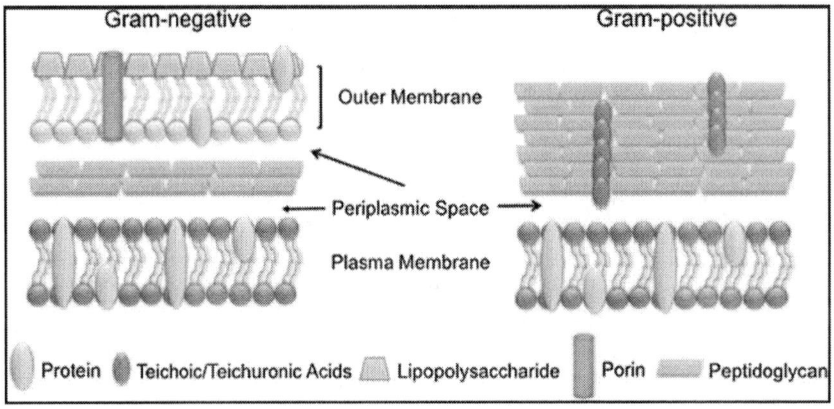

Figure 8. Comparison of bacterial gram-positive and -negative cell wall structure [17].

4. Conclusion

An optimum whiteness value of -5.92 was obtained for tiles without TiO_2. Addition of TiO_2 to the glaze showed antibacterial properties both for *E. coli* and *S. aureus*. Antibacterial properties of TiO_2 glaze against *S. aureus* bacteria were more active than *E. coli*.

Reference

[1] Levy S B 2001 Antibacterial Household Products: Cause for Concern *Emerging Infectious Diseases* **7(3)** 512-5

[2] Noguera J F, Moreno A, Gozalbo A and Orts M J 2010 Development of ceramic glaze compositions with bactericidal and fungicidal properties *Qualicer'10 Castellon (Spain)*.

[3] Miola M, Perero S, Ferraris S, Battiato A, Manfredotti C, Vittone E, Vento D D, Vada S V, Fucale G and Ferraris M 2014 Silver nanocluster-silica composite antibacterial coating for materials to be used in mobile telephones *Applied Surface Science* **313** 107-15

[4] Hasmaliza M, Foo H S and Mohd K 2016 Anatase as antibacterial material in ceramic tiles *Procedia Chemistry* **19** 828-34

[5] Zhang J, Wang A, Zhang C and Wang Y 2015 Inorganic antibacterial agents using in building materials - mechanisms, safety and long term effect : a review *Annales de Chimie Science des Materiaux*.

[6] Chanrawangyot S, Rattanachan S T, Watcharenwong A and Fangsuwannarak T 2017 Antibacterial Activity of ZnO nanoparticles coated on ceramic tiles prepared by sol-gel method *Journal of Metals* **27(2)**

IOP Conf. Series: Materials Science and Engineering **980** (2020) 012011 doi:10.1088/1757-899X/980/1/012011

[7] Sun C, Li Y, Su Q, Wang Y and Liu X 2018 Durable and washable antibacterial copper nanoparticles bridged by surface grafting polymer brushes on cotton and polymeric materials *Hindawi Journal of Nanomaterials.*

[8] Tsou H K and and Hsieh P Y 2017 Anticorrosive, antimicrobial and bioactive Titanium Dioxide coating for surface-modified purpose on biomedical material *Application of Titanium Dioxide Chapter 6 Intech* 103-22

[9] Nikiforov A, Heyse P, Xiaolong D, and Leys C 2017 Antimicrobial biomedical materials: engineering *Encyclopedia of Plasma Technology* 26-35.

[10] Sun S Q, Sun B, Zhang W and Wang D 2008 Preparation and antibacterial activity of Ag-TiO_2 composite film by liquid phase deposition (LPD) method *Bull. Mater. Sci.* **31(1)** 61-6

[11] Niederhäusern de S and Bondi M 2013 Self-cleaning and antibacteric ceramic tile surface *Int. J. of App. Ceramic Tech.: Ceramic Prod. Dev. and Commercialization* **10(6)** 949-56

[12] Ambasr A 2006 *Pengetahuan Keramik* (Yogyakarta: Universitas Gadjah Mada).

[13] Putri S A 2019 *Karakterisasi dan Uji Sifat Antibakteri* (Bandung: Universitas Islam Negeri Sunan Gunung Djati).

[14] Wang P, Wang D, Li H, Xie T, Wang H and Du Z 2007 A facile solution-phase synthesis of high quality water-soluble anatase TiO_2 nanocrystals *J. of Colloid and Interface Sci.* **314** 337-40

[15] Taran M, Rad M and Alavi M 2018 Biosynthesis of TiO_2 and ZnO nanoparticles by halomonas elongate IBRC-M 10214 in different conditions of medium *BioImpacts* **8(2)** 81-9

[16] Yuningrat N W, Oviantari M V and Gunamantha I M 2015 Fotodegradasi senyawa organik dalam lindi dengan menggunakan katalis TiO_2 terimobilisasi pada plat kaca *Jurnal Sains dan Teknologi* **4(2)** 642-54

[17] Slavin Y N, Asnis J, Hafeli Urs O and Bach H 2017 Metal nanoparticles : understanding the mechanisms behind antibacterial activity *Journal of Nanobiotechnology* **15(65)** 1-20

IOP Conf. Series: Materials Science and Engineering **980** (2020) 012012 doi:10.1088/1757-899X/980/1/012012

The effect of solvent in the hydrogenation of lauric acid to lauryl alcohol using Ru-Fe/TiO_2 catalyst

Ibrahim[2], M Riski[2] and Rodiansono[1,2*]

[1]Catalyst & Catalysis Research Group (CATRe), Laboratory of Physical Chemistry
[2]Department of Chemistry, Faculty of Mathematics and Natural Sciences, Lambung Mangkurat University, Banjarbaru Kalimantan Selatan, Indonesia

Email: rodiansono@ulm.ac.id

Abstract. We describe the effect of solvent in the hydrogenation of lauric acid to lauryl alcohol using bimetallic ruthenium-iron supported on titanium oxide (5.0Ru–xFe/TiO_2 (5.0 is loading amount of Ru (wt%) and x = 1.0; 3.0; 5.0 is the loading amount of Fe (wt%). Ru–xFe/TiO_2 catalyst was synthesized via the hydrothermal method at a temperature of 150 °C for 24 hours, then followed by reduction with H_2 at a temperature of 500 °C for 3 h. The catalytic reactions were carried out in a batch reactor system in various organic solvents (*e.g.,* 2-propanol, water, 1,4-dioxane, tetrahydrofuran, and its mixture solvent). Results of XRD analysis of 5.0Ru–xFe/TiO_2 showed a typical diffraction peak at $2\theta = 44.14°$, which can be attributed as metallic Ru(101). The best solvent used was 1,4-dioxane: water (4.5:0.5 ml) at 170 °C, initial H_2 pressure 4.0 MPa and a reaction time of 5 h, providing 39.4% lauric acid conversion and 5.8% yield of lauryl alcohol using 5.0Ru–1.0Fe/TiO_2 catalyst. The yield of lauryl alcohol could be enhanced up to 33% (at 49.2% conversion) after the reaction time was prolonged to 30 h under the same reaction conditions.

1. Introduction

Fatty acids and its common derivatives fatty alcohols, typically in the form of long-chain fatty alcohols such as lauryl, octyl, cetyl, and stearyl alcohols, are important intermediate in the fragrances, pharmaceuticals, detergents, emulsifiers, lubricants, cosmetics, shampoos, and conditioners industries [1]. Fatty acids contain carboxylate group, which the carbonyl group in carboxylate group has weak polarizability and has, consequently, lower reactivity than that of the common carbonyl compounds such as aldehyde and ketone. Therefore, extreme conditions are typically required where control of the selectivity of the process to the desired fatty alcohols is difficult. Bimetallic copper-chromite or zinc-chromite were the first commercial catalysts for the synthesis of fatty alcohols by hydrogenation of fatty acids under harsh temperature (e.g., up to 300 °C) and pressure (*e.g.,* up to 30 MPa of H_2) conditions, which are potentially harmful for the environment [3]. Commercial ruthenium-based catalysts, such as Ru/C, ruthenium dioxide, Ru/γ-Al_2O_3, and bimetallic Ru-Sn//γ-Al_2O_3, has been frequently used for the acid hydrogenations under a similar high temperature and initial H_2 pressure (typically, 150-250 °C and 5-95 MPa) [2,4,19,20]. Therefore, the development of alternative catalysts for more facile and selective hydrogenation of fatty acids to fatty alcohols at lower pressures and temperatures is of great interest and is greatly challenging work.

In the past decade, there has been a substantial increase in the use oxophilic metals (*e.g.,* Mo, W, and Y) or electropositive metals (*e.g.,* Sn, Zn, Co, Ni) as well as the use of reducible oxide supports such as TiO_2 or MoO_3 to enhance the reactivity and selectivity of platinum group metal catalysts in the carboxylic acid hydrogenation to alcohols. For example, yttrium modified titanium oxide supported ruthenium (Ru-Y/TiO_2) catalyst demonstrated high selectivity towards 1-hexadecanol (99%) in the hydrogenation of methyl hexadecanoate at 240 °C, 5 MPa H_2, and after 10 h [9]. Wang et al. have reported the use Ir-MoO_x/SiO_2 catalyst at 100 °C for hydrogenation of acetic acid to ethanol and initial H_2 pressure 60 bar, giving 80% conversion of acetic acid and 63% selectivity to ethanol [21]. Tomishigie and co-worker developed the combination of noble metal catalyst (M_1= Pd, Ru, Ir) with oxophilic oxides M_2O_x (M_2= W, Y, Mo) for effective hydrogenation and hydrogenolysis reactions. The modified M_1–M_2O_x catalysts demonstrated higher activity and selectivity than that of unmodified ones [17]. Moreover, Luo et al. reported the hydrogenation of hexadecanoic acid to alcohol using intermetallic Ru_3Sn_7/SiO_2 catalyst at 240 °C, H_2 40 bar for 4 h, resulting in high conversion (99.0%) and selectivity to alcohol (99.7%) (11). Huang et al. studied the supported Ru-M (M = Cu, Co, Ni) for methyl palmitate hydrogenation in n-heptane at 270 °C, H_2 55 bar after 7 h, giving 1-hexadecanol (83%) [7], while bimetallic Ru-Zn/C catalyst only gave 7.2% conversion even after the temperature reaction increased to 300 °C [6].

In continuation of our standing interest in the liquid phase hydrogenation of biomass-derived carboxylix acids, we found that the presence of second electropositive metals (e.g. Sn) could enhance the activity and selectivity of titanium oxide-supported Ni catalyst for the hydrogenation of typical biomass-derived levulinic acid to g-valerolactone [13,14] and hydrogenation of dodecanoic acid to dodecane-1-ol [16]. In the present paper, we describe our extended studies on the hydrogenation of lauric acid into lauryl alcohol using bimetallic Ru-Fe/TiO_2 catalysts. The addition of Fe to Ru/TiO_2 catalysts improved the catalytic activity and selectivity towards lauryl alcohol. A 49.2% conversion of lauric acid and 33.0% yield of lauryl alcohol were obtained using 5.0Ru-1.0Fe/TiO_2 catalyst at 170°C, 40 bar H_2, and reaction time of 30 h in 1,4-dioxane: water (4.5:0.5 ml). The effect of reaction temperature, initial H_2 pressure, and reaction time on the yield of lauryl alcohol was also discussed systematically.

2. Materials and methods

2.1. Material

$RuCl_3$ (99,0%; Tokyo Chemical Industry), $FeCl_3 \cdot 6H_2O$ (99%; Merck Millipore), 2-propanol (98%, Merck Millipore), TiO_2 anatase, ethanol (96%; Merck Millipore), ethylene glycol (EG) (99.5%; Merck Millipore), $NaBH_4$ (95%; Tokyo Chemical Industry), NaOH (99%; Merck Millipore), H_2 gas (99.99%), N_2 gas (99.99%), lauric acid (97%; Tokyo Chemical Industry).

2.2. Catalyst preparation

A typical procedure for the synthesis of 5.0Ru-1.0Fe/TiO_2 (5.0 and 1.0 was the loading amount of Ru and Fe, respectively) described as follows [15]: $RuCl_3$ (0.371 mmol mmol) was dissolved in deionized water (symbolized as solution A) and $FeCl_3 \cdot 6H_2O$ (0.045 mmol) was dissolved in ethanol/ethylene glycol (2.0/1.0 volume ratio) at room temperature (symbolized as solution B). Solutions A, B, and TiO_2 (as support material; 1.0 g) at room temperature were mixed; the temperature was subsequently raised to 50°C under gentle stirring for 12 h. The pH of the mixture was adjusted to 12 through the dropwise addition of an aqueous solution of NaOH (3.1 M). The mixture was then placed into a sealed-Teflon autoclave for the hydrothermal process at 150 °C K for 24 h. The resulting black precipitate was filtered, washed with distilled water, and then dried under vacuum overnight. Prior to the catalytic reaction, the obtained black powder was treated under hydrogen at 500 °C for 3 h.

2.3. Catalyst characterization

The X-ray diffraction (XRD) analysis was performed on a Miniflex 600 Rigaku instrument with Cu as monochromatic source of CuKα radiation (λ = 0.15444 nm). The XRD was operated at 40 kV and 15 mA with a step width of 0.02°, a scan speed of 4° min^{-1} ($\alpha 1$ = 0.154057 nm, $\alpha 2$ = 0.154433 nm), solar slit 1.25°, and using a Ni Kβ filter.

The BET surface area (S_{BET}) and pore volume (V_p) were measured using N_2 physisorption at -196 °C on a Belsorp Max (BEL Japan). The samples were degassed at 200 °C for 2 h to remove physisorbed gases before the measurement. The amount of nitrogen adsorbed onto the samples was used to calculate the Brunauer–Emmett–Teller (BET) surface area via the BET equation. The pore volume was estimated to be the liquid volume of nitrogen at a relative pressure of approximately 0.995 according to the Barrett–Joyner–Halenda (BJH) approach based on desorption data [10].

The NH_3-TPD was carried out on a Belsorp Max (BEL Japan). The samples were degassed at a high temperature of 100-200 °C for 2 h to remove the analyzed gas before measurement. The temperature was then maintained at 200 °C for 2 h while rinsing with He gas. NH_3 gas (balanced NH_3, 80% and He, 20%) was introduced at 100 °C for 30 min, then evacuated with helium gas to remove the physisorbed also for 30 min. Finally, programmed temperature desorption was carried out at 100-800 °C and the absorbed NH_3 is monitored by TCD.

2.4. Catalytic reaction

Typical procedure for hydrogenation of lauric acid described as follows: Catalysts (0.05 g), lauric acid (2.0 mmol), and 2-propanol/H_2O (5 ml; 4.5: 0.5 volume ratio) as the solvent was placed into a glass reaction tube in an autoclave reactor system of TAIATSU Techno reactor (a Pyrex tube was fitted inside of a sus316 jacket to protect the vessel from corrosion in acidic media. The reactor was flushed with 0.1 MPa H_2 gas for 10 times prior to remove undesired gases. After H_2 was introduced into the reactor with an initial H_2 pressure of 4.0 MPa at room temperature, the temperature of the reactor was increased to 170 °C. After 5 h, the conversion of lauric acid and the yield of lauryl alcohol were determined by gas chromatography (GC) analysis.

2.5. Product analysis

GC analysis of the reactant (lauric acid) and products (lauryl alcohol, 1,1-dodecanediol) was performed on a Perkin Elmer Auto System XL equipped with a flame ionisation detector and Restek Rtx® BAC Plus capillary column. GC analysis was operated at detector and injector temperatures of 250 °C, N_2 as a carrier gas, rates of air and H_2 were 450 ml/min and H_2 45 ml/min, respectively. The used 5.0Ru-1.0Fe/TiO_2 catalyst was easily separated using simple centrifugation or filtration, then finally dried overnight under vacuum at room temperature prior to reusability testing.

3. Results and discussion

3.1. Screening of solvent

In the first experiments, we evaluated the use of various solvents (*e.g.*, 2-propnaol, tetrahydrofuran (THF), 1,4-dioxane and water) as solvent media for the hydrogenation of lauric acid to lauryl alcohol using 5.0Ru-1.0Fe/TiO_2 catalyst and the results are summarised in table 1.

In alcoholic solvent, 2-propanol, 48.7% conversion of lauric acid and 4.3% yield of lauryl alcohol were obtained, whereas a relative large amount of others (*e.g.*, iso-propyl laurate) was observed (entry 1). When 1,4-dioxane was used, the conversion of lauric acid was 41.5% with slightly higher yield of lauryl alcohol (11.4%) (entry 2), meanwhile catalytic reaction in water or in tetrahydrofuran did not occur (entries 3 and 4). This result suggests that the use of polar aprotic solvent (e.g, 1,4-dioxane) can effectively facilitate the hydrogenation. To improve the conversion of lauric acid as well as the yield of lauryl alcohol, further catalytic reactions in various mixture solvent of 1,4-dioxane were also carried out. A significant increase of laurylalcohol yield (15.8%) was obtained when 1,4-dioxane/water (4.5: 0.5 volume ratio) was used as solvent (entry 5). In 1,4-dioxane/THF and 1,4-dioxane/2-PrOH solvent mixtures, the yield of lauryl alcohol was about 10-13% (entries 6 and 7). Therefore, further evaluation was focused to 1,4-dioxane/water for next reaction to obtain high conversion and yield of the product. The amount of water to 1,4-dioxane was varied at range of 1.0, 0.5 ml, 5 drops, and 10 drops and the results are also summarized in table 1, entries 7-10. The highest yield of lauryl alcohol was 15.8% at luric acid conversion of 39.4% in 1,4-dioxane/water (4.5: 0.5 ml). The relatively high yield of lauryl alcohol in 1,4-dioxane/water mixture solvent can be attributed to the presence of an H-bonded water molecule or single chemisorbed water on metallic surfaces that can dramatically reduce the energy span of the reaction pathway, hence enhancing the catalytic activity [12]. Alternatively, the increase of surface concentration in hydrogen atom due to dissociated water and strong interaction between water with substrate by hydrogen bonding lowers the activation energy barrier and leads to high hydrogenation rates [5,18].

Table 1. Results of lauric acid hydrogenation over 5.0Ru-1.0Fe/TiO_2 catalyst in various solvents.

Entry	Solvent	Conversion[a] (%)	Yield[a] (%)			
			Lauryl alcohol	Undecane [b]	Dodecane [b]	Others [c]
1	2-Propanol (2-PrOH)	48.7	9.3	0.1	0.0	17.0
2	1,4-Dioxane	41.5	11.4	0.1	0.0	13.0
3	Water	Trace	Trace	Trace	Trace	Trace
4	Tetrahydrofuran (THF)	2.1	1.1	0.0	0.0	0.0
5	1,4-Dioxane: THF[c]	29.0	10.7	0.1	0.1	0.0
6	1,4-Dioxane: 2-PrOH[c]	41.5	13.9	0.1	0.0	7.2
7[d]	1,4-Dioxane: Water (4.5: 1.0)	38.4	15.5	0.3	0.3	0.0
8[d]	1,4-Dioxane: Water (4.5: 0.5)	39.4	15.8	0.2	0.1	7.0
9[d]	1,4-Dioxane: Water (4.5: 5 drops)	34.0	12.8	0.3	0.2	0.0
10[d]	1,4-Dioxane: Water (4.5: 10 drops)	37.8	12.3	0.0	0.0	0.0

Reaction conditions: lauric acid (0.4 g), catalyst (0.02 g), H_2 (4.0 MPa), solvent (5 ml); 170 °C; reaction time (5 h). [a]Conversion and lauryl alcohol yield are determined by GC using internal standard technique. [b]Yields of undecane, dodecane, and iso-propil laurate were calculated by GC area ratio. [c]Others are included laurate ester (e.g., iso-propyl laurate) or lauric acid dimer (e.g., lauryl laurate).[d]Values in the parenthesis are volume ratio of 1,4-dioxane to water (ml).

3.2. Effect of Fe loading amount

To study the insight into the role of Fe on the catalytic performance of Ru/TiO_2 catalyst in lauric acid hydrogenation, the amount of Fe-loaded was varied at range of 1.0, 3.0, and 5.0 wt% to the amount of Ru and the results are summarized in table 2. A reference catalyst of 5.0Ru/TiO_2 synthesized and reduced with H_2 at 500 °C for 3 h and showed low conversion of lauric acid (9.8%) as well as the low yield of lauryl alcohol (6.0%) (entry 1). The as-prepared 5.0Ru-1.0Fe/TiO_2 (the catalyst without reduction with H_2) was completely inert for the hydrogenation of lauric acid (entry 2), suggesting the transfrom Ru oxide to metallic Ru(0) was important to obtain the active species in 5.0Ru-(x)Fe/TiO_2 catalyst. Although the conversion of lauric acid increased as the Fe loading amount was increased, the yield of lauryl alcohol was unchanged, whereas the yield of others increased gradually. Therefore, the best results for the amount of Fe co-modifier wass 1.0 wt% to Ru as indicated in table 2, entry 3.

Table 2. Results of lauric acid hydrogenation over 5.0Ru-(x)Fe/TiO_2 catalysts with different loading amount of Fe.

Entry	Catalyst	Conversion[a] (%)	Yield[a] (%)			
			Lauryl alcohol	Undecane[b]	Dodecane[b]	Others[c]
1	5.0Ru/TiO_2	9.8	6.0	1.0	0.0	2.0
2[d]	5.0Ru-1.0Fe/TiO_2	9.3	0.0	0.0	0.0	9.0
3	5.0Ru-1.0Fe/TiO_2	39.4	15.8	0.2	0.1	7.0
4	5.0Ru-3.0Fe/TiO_2	58.7	14.4	0.1	0.0	12.1
5	5.0Ru-5.0Fe/TiO_2	53.1	13.0	0.0	0.0	16.5

Reaction conditions: lauric acid (0.4 g), catalyst (0.02 g), H_2 (4.0 MPa), solvent 1,4-dioxane: water (4.5:0.5 ml); 170 °C; reaction time (5 h). [a]Conversion and lauryl alcohol yield are determined by GC using internal standard technique. [b]Yields of undecane, dodecane, and iso-propil laurate were calculated by GC area ratio.[c]Others are included laurate ester (e.g., iso-propyl laurate) or lauric acid dimer (e.g., lauryl laurate). [d] The catalyst without reduction with H_2 (as-prepared).

3.3. Effect of initial H2 pressure

The effect of initial H2 pressure on the conversion and yield in the lauric acid hydrogenation was evaluated and the results are summarised in table 3. A 36.9% conversion of lauric acid and 2.8% yield of lauryl alcohol were obtained at initial H_2 pressure of 2.0 MPa (entry 1). However, a high yield of others (included laurate ester (*e.g.*, iso-propyl laurate) or lauric acid dimer (*e.g.*, lauryl laurate)) was also obtained, suggesting the main reaction was not hydrogenation owing to lower H_2 presence. Interestingly, the yield of lauryl alcohol remarkably increased to 15.8% at 39.4% conversion of lauric acid when the initial H_2 pressure was increased to 4.0 MPa (entry 2). The yield of others also significantly decreased to 7.0%.

Table 3. Results of lauric acid hydrogenation over 5.0Ru-1.0Fe/TiO_2 catalyst at different initial H_2 pressure.

Entry	Initial H_2 pressure (MPa)	Conversion[a] (%)	Yield[a] (%)			
			Lauryl alcohol	Undecane[b]	Dodecane[b]	Others[c]
1	2.0	36.9	2.8	0.1	0.1	19.0
2	4.0	39.4	15.8	0.2	0.1	7.0

Reaction conditions:lauric acid (0.4 g), catalyst (0.02 g), solvent 1,4-dioxane: water (4.5:0.5 ml); 170 °C; reaction time (5 h). [a]Conversion and lauryl alcohol yield are determined by GC using internal standard technique. [b]Yields of undecane, dodecane, and iso-propil laurate were calculated by GC area ratio. [c]Others are included laurate ester (e.g., iso-propyl laurate) or lauric acid dimer (e.g., lauryl laurate).

3.4. Kinetic profiles

The kinetic profiles of lauric acid hydrogenation to lauryl alcohol in the presence of 5.0Ru-1.0Fe/TiO_2 catalyst are summarized in table 4. At an early reaction time of 2 h, the main product was others (included laurate ester (e.g., iso-propyl laurate) or lauric acid dimer (e.g., lauryl laurate)) (6.9%) while the yield of lauryl alcohol was only 3.3% (entry 1). A remarkable increase of lauryl alcohol yield (15.8%) was obtained after a reaction time of 5 h at 39.4% conversion (entry 2). The yield of lauryl alcohol gradually increased as the reaction time was prolonged up to 30 h, whereas the highest yield of lauryl alcohol (33.4%) was obtained at 49.2% conversion of lauric acid (entry 5). These results suggest that the slow activation of lauric acid by catalyst is the main factor for the hydrogenation reaction to corresponding lauryl alcohol as well as the use of appropriate solvent as mentioned above. Alternatively, the applied reaction temperature higher than 170 °C may be enhanced the conversion and yield of targetted product. The catalytic reactions under temperature of 200 °C using the same catalyst were being investigated and will be described in our succeeding paper.

Table 4. Kinetic profiles of lauric acid hydrogenation in the presence of 5.0Ru□1.0Fe/TiO_2 catalyst.

Entry	Reaction time (h)	Conversion[a] (%)	Yield[a] (%)			
			Lauryl alcohol	Undekana[c]	Dodekana[c]	Others[c]
1	2	11.3	3.3	0.0	0.0	6.9
2	5	39.4	15.8	0.2	0.1	7.0
3	15	41.9	27.2	0.4	0.1	5.3
4	24	42.8	30.1	1.1	1.8	5.5
5	30	49.2	33.4	3.7	2.5	4.7

Reaction conditions:lauric acid (0.4 g), catalyst (0.02 g), solvent 1,4-dioxane: water (4.5:0.5 ml); 170 °C; reaction time (2-30 h). [a]Conversion and lauryl alcohol yield are determined by GC using internal standard technique. [b]Yields of undecane, dodecane, and iso-propil laurate were calculated by GC area ratio. [c]Others are included laurate ester (e.g., iso-propyl laurate) or lauric acid dimer (e.g., lauryl laurate).

3.5. Structure-activity relationship

On the basis of spectrometric characterization results, surface acidity, and reaction profiles, we discuss the structure of active Ru and the presence of FeO_x species in Ru-xFeO_x/TiO_2 catalyst after reduction with H_2 at 500 °C for 3 h. The as-prepared 5Ru-1.0FeO_x/TiO_2 was a referenced catalyst without reduction with H_2 and inactive for hydrogenation of lauric acid (table 1, entry 2). The XRD patterns of Ru-xFeO_x/TiO_2 catalysts before and after reduction with H_2 at 500 °C for 3 h (figures 1 (a) and (b)) suggested that the formation of metallic Ru(0) was observed at $2\theta = 44.34°$ corresponding to Ru(101) (21). On the other hand, the metallic or oxide Fe species were unable to detect due to its crystallite sizes lower than that the detection limit of XRD.

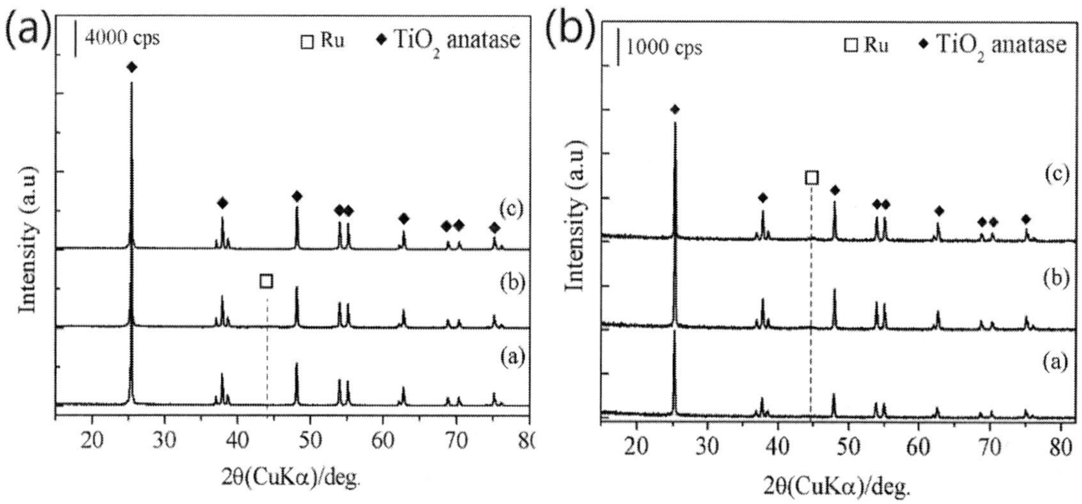

Figure 1.XRD patterns of (a) before and (b) after reduction with H_2 at 500 °C for 3 h with different amount of Fe, (a) 5.0Ru-1.0Fe/TiO_2, (b) 5.0Ru-3.0Fe/TiO_2, and (c) 5.0Ru-5.0Fe/TiO_2.

The interaction between active surface of catalyst and lauric acid substrate and lauryl alcohol product was also monitored by ex-situ ATR-FTIR analysis and the results are shown in figure 2. A specific absorbance of -OH was observed on recovered-5.0Ru-1.0Fe/TiO_2 catalyst (figure 2f), the absorbance of -OH totally disappeared after the recovered-5.0Ru-1.0Fe/TiO_2 catalyst was dried at room temperature under vacuum for 7 h (figure 2g).

Interestingly, a sharp absorbance peak was observed at a wave number of 1728.1 cm^{-1} which can be attributed as the finger print of carboxylix acids suggesting the lauric acid absorbed strongly on catalyst surface and the absorbed carboxylic acid remained even after dried under vacuum at room temperature (figures 2f and 2g). In addition, the results also suggest to remove all absorbed molecule on the catalyst surface that may affect to the quantify both remained reactant (lauric acid) and the formed product (lauryl alcohol) using GC analysis and finally may deviate the number of conversion as well as the yield. Therefore, it is suggested to dilute the reaction mixture by addition of appropriate solvent before GC analysis and it is being investigated in our research group.

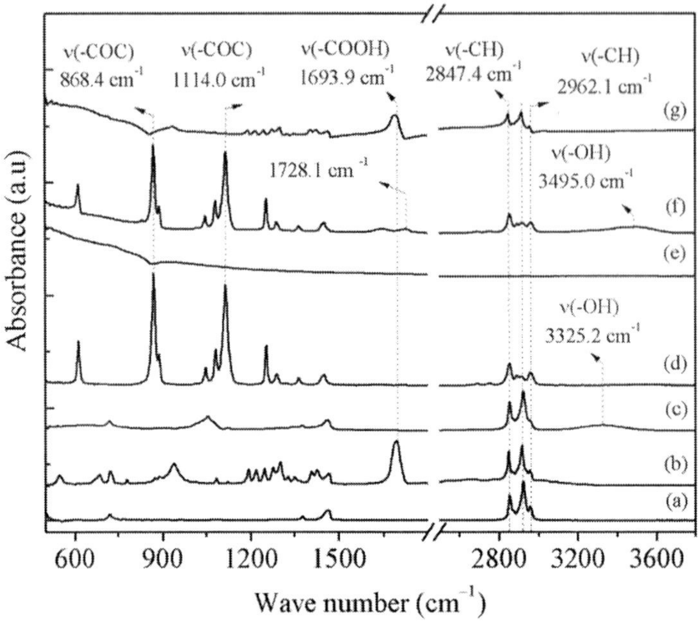

Figure 2. ATR-FTIR spectra of (a) dodecane, (b) lauric acid, (c) lauryl alcohol, (d) 1,4-dioxane, (e) fresh 5.0Ru-1.0Fe/TiO_2 catalyst, (f) recovered-5.0Ru-1.0Fe/TiO_2 catalyst, (g) recovered-5.0Ru-1.0Fe/TiO_2 catalyst, dried at room temperature under vacuum.

4. Conclusions

We described the hydrogenation of lauric acid to lauryl alcohol in the presence of 5.0Ru–(x)Fe/TiO_2 catalysts. The best solvent used was 1,4-dioxane: water (4.5:0.5 ml) at 170 °C, initial H_2 pressure 4.0 MPa and a reaction time of 5 h, providing 39.4% lauric acid conversion and 5.8% yield of lauryl alcohol using 5.0Ru–1.0Fe/TiO_2 catalyst. The yield of lauryl alcohol could be enhanced up to 33% (at 49.2% conversion) after the reaction time was prolonged to 30 h under the same reaction conditions.

Acknowledgments

The authors acknowledge the RISET DASAR Hibah Berbasis Kompetensi (HIKOM) (contract number DIPA-042.06-1.401516/2019), the RISET DASAR (contract number DIPA-042.06-1.401516/2019) from Ministry of Research, Technology, and Higher Education, Indonesian Government, which all financially supported this work. We would also like thank to Professor Shogo Shimazu for kind help in XRD measurements.

References

[1] Biermann U, Bornscheuer U, Meier MAR, Metzger JO and Schäfer HJ 2011 *Oils and fats as renewable raw materials in chemistry Vol 50 Angewandte Chemie - International Edition* p 3854–71
[2] Carnahan JE, Ford TA, Gresham WF, Grigsby WE and Hager GF 1955 Ruthenium-catalyzed Hydrogenation of Acids to Alcohols *J. Am. Chem. Soc.* **77** 3766–8
[3] Connor R, Folkers K and Adkins H 1932 The preparation of copper-chromium oxide catalysts for hydrogenation *Journal of the American Chemical Society* **54** 1138–45
[4] Cui X, Li Y, Topf C, Junge K and Beller M 2015 Direct Ruthenium-Catalyzed Hydrogenation of Carboxylic Acids to Alcohols *Angew. Chemie. - Int. Ed.* **54** 10596–9
[5] Delbecq F, Loffreda D and Sautet P 2010 Heterogeneous catalytic hydrogenation: Is double bond/surface coordination necessary? *J. Phys. Chem. Lett.* **1** 323–6

[6] Gao X, Tong D, Zhong H, Jin B, Jin F and Zhang H 2016 Highly efficient conversion of fatty acids into fatty alcohols with a Zn over Ni catalyst in water *RSC Advances* p 27623–6

[7] Huang C, Zhang H, Zhao Y, Chen S and Liu Z 2012 Diatomite-supported Pd-M (M=Cu, Co, Ni) bimetal nanocatalysts for selective hydrogenation of long-chain aliphatic esters *Journal of Colloid and Interface Science* p 60–5

[8] JCPDS-ICDD 1991 *Powder diffraction files* JCPDS-International center for diffraction data (JCPDS-ICDD)

[9] Lin Q, Zheng H, Zheng G, Li X and Lou B 2014 Effects of Yttrium Doping on the Performance of Ru-Based Catalysts for Hydrogenation of Fatty Acid Methyl Ester *Int. J. Org. Chem.* **04** 219–24

[10] Lowell S, Shields JE, Thomas MA and Thommes M 2004 *Characterization of Porous Solids and Powders: Surface Area, Pore Size and Density - Kluwer Academic* (Netherlands: Dordrecht)

[11] Luo Z, Bing Q, Kong J, Liu JY and Zhao C 2018 Mechanism of supported Ru3Sn7 nanocluster-catalyzed selective hydrogenation of coconut oil to fatty alcohols *Catal. Sci. Technol.* **8** 1322–32

[12] Michel C, Gallezot P. Why Is Ruthenium an Efficient Catalyst for the Aqueous-Phase Hydrogenation of Biosourced Carbonyl Compounds? ACS Catal [Internet]. 2015 Jul 2;5(7):4130–2. Available from: https://pubs.acs.org/doi/10.1021/acscatal.5b00707

[13] Rodiansono R, Astuti MD, Hara T, Ichikuni N and Shimazu S 2016 Efficient hydrogenation of levulinic acid in water using a supported Ni–Sn alloy on aluminium hydroxide catalysts *Catal. Sci. Technol.* **6** 2955–61

[14] Rodiansono R, Astuti MD, Ghofur A and Sembiring KC 2015 Catalytic hydrogenation of levulinic acid in water into γ-valerolactone over bulk structure of inexpensive intermetallic Ni-Sn alloy catalysts *Bull. Chem. React. Eng. Catal.* **10** 192–200

[15] Rodiansono R, Khairi S, Hara T, Ichikuni N and Shimazu S 2012 Highly efficient and selective hydrogenation of unsaturated carbonyl compounds using Ni-Sn alloy catalysts *Catal. Sci. Technol.* **2** 2139–45

[16] Rodiansono R, Pratama MI, Astuti MD, Abdullah A, Nugroho and Susi S 2018 Selective Hydrogenation of Dodecanoic Acid to Dodecane-1-ol Catalyzed by Supported Bimetallic Ni-Sn Alloy *Bull. Chem. React. Eng. Catal.* **13** 311–9

[17] Tamura M, Tamura R, Takeda Y, Nakagawa Y and Tomishige K 2015 Insight into the mechanism of hydrogenation of amino acids to amino alcohols catalyzed by a heterogeneous MoOx-modified Rh catalyst *Chem. - A Eur. J.* **21** 3097–107

[18] Thibault ME, Dimondo D V., Jennings M, Abdelnur PV, Eberlin MN and Schlaf M 2011 Cyclopentadienyl and pentamethylcyclopentadienyl ruthenium complexes as catalysts for the total deoxygenation of 1,2-hexanediol and glycerol *Green Chem.* **13** 357–66

[19] Toba M, Tanaka SI, Niwa SI, Mizukami F, Koppány Z, Guczi L, et al 1999 Synthesis of alcohols and diols by hydrogenation of carboxylic acids and esters over Ru-Sn-Al2O3 catalysts *Applied Catalysis A: General* p 243–50

[20] Trivedi BC, Grote D and Mason TO 1981 Hydrogenation of Carboxylic Acids and Synergistic Catalysts *J. Am. Oil. Chem. Soc.* **58** 17–20

[21] Wang Z, Li G, Liu X, Huang Y, Wang A, Chu W et al 2014 Aqueous phase hydrogenation of acetic acid to ethanol over Ir-MoOx/SiO2 catalyst *Catal. Commun.* **43** 38–41

IOP Conf. Series: Materials Science and Engineering **980** (2020) 012013 doi:10.1088/1757-899X/980/1/012013

Selective hydrogenation of levulinic acid to γ-valerolactone using bimetallic Pd-Fe catalyst supported on titanium oxide

A P Damayanti[1], H P Dewi[1], Ibrahim[1] and Rodiansono[1,2]*
[1]Department of Chemistry, Faculty of Mathematics and Natural Sciences, Lambung Mangkurat University, Banjarbaru Kalimantan Selatan
[2]Catalysis for Energy and Environment (CATRe), Laboratory of Physical Chemistry

Email: rodiansono@ulm.ac.id

Abstract. The synthesis of γ–valerolactone (GVL), a versatile precursor in the manufacture of high-value chemicals such as polymer plasticizer, solvent, jet fuel, and, agrochemicals, has been targeted by many research groups. In this study, we report the catalytic performance of iron-modified palladium supported on titanium oxide (denoted as Pd(5.0)–Fe(5.0)/TiO_2; 5.0 was the loading amount of Pd and Fe, respectively) in the selective hydrogenation of levulinic acid (LA) to γ-valerolactone (GVL) in stainless-steel batch reactor system. Pd(5.0)–Fe(5.0)/TiO_2 catalyst was synthesized by a simple hydrothermal method at 150°C for 24 h, then followed by reduction with H_2 at 500°C for 3 h. The X-ray diffraction (XRD) patterns of Pd(5.0)–Fe(5.0)/TiO_2 before and after reduction showed the formation of metallic Pd at 2θ = 40.34° and 47.2° corresponding to Pd(111) and Pd(200), respectively. The hydrogenation of LA to GVL effectively occurred in the H_2O solvent, whereas in 2-propanol the formation of ester was the main side product. The highest yield of GVL (52.5%) was obtained over Pd(5.0)–Fe(5.0)/TiO_2 catalyst at a temperature of 170 °C, initial H_2 pressure of 4.0 MPa, solvent H_2O 3 ml at a reaction time of 7 h. The yield of GVL slightly increased to 63.3% when the reaction time was prolonged to 15 h.

1. Introduction

Biomass feedstock valorizations are currently being explored for the production of bio-based platform chemicals, fuels, and various commodity products by using bimetallic-based heterogeneous catalysts. The use of bimetallic catalysts is a promising option for biomass feedstock upgrading, as the interaction between metals can modify the surface properties of the catalyst that meet with the complex surface [1–3]. Furfural (FFald) and levulinic acid (LA) have been identified to be the most promising platform molecules that can turn into a variety of value-added derivative molecules including polymer, plasticizer, solvent, jet fuel precursor, and agrochemicals [4]. FFald and LA were produced effectively from C-5 sugars in hemicellulosic biomasses, such as xylan and arabinose or C-6 sugars in form of glucose or fructose via acidic hydrolysis [5].

In the hydrogenation of LA into γ-valerolactone, heterogeneous catalysts of the monometallic bimetallic precious metal transition have been widely used to achieve high yields of GVL. This hydrogenation reaction is important because the resulting products are used in the chemical industry in the production of high-yield chemicals. Additionally, the study is interesting because there is a combination of two metals with just the right reaction condition. Bourne et al. [6] used Ru/SiO_2 catalyst under supercritical-CO_2 and H_2 up to 4.5 MPa at 473 K to give >99% GVL yield, while Wettstein et al. employed bimetallic Ru-Sn for LA hydrogenation to give GVL in *o-iso*-butylphenol at a relatively high reaction temperature of 453 K. They claimed that the formation of intermetallic such as Ru_2Sn_3 and Ru_3Sn_7 slowed the leaching of Sn into the reaction solution [7]. Recently, Luo et al. reported bimetallic Au-Pd/TiO_2 and Ru-Pd/TiO_2 catalysts for hydrogenation of LA into GVL. The formation of bimetallic alloy or dilution of Au or Ru by Pd improved the selectivity and stability of the catalysts [8]. We have reported the catalytic performance of the bulk structure of intermetallic Ni-Sn catalysts in LA hydrogenation and showed high selectivity in aqueous phase hydrogenation of LA into GVL at 433 K, 4.0 MPa for 6 h and produced GVL yield of >99%. The activity and selectivity of used bulk Ni-Sn catalysts slightly decreased after several reaction runs that might be due to the leaching of metals (Ni and Sn) into the reaction solution [9]. In our most recent extended work, Ni-Sn alloy supported on aluminum hydroxide (denoted as prepared Ni-Sn(*x*)/AlOH; *x* = Ni/Sn molar ratio, *7.9* to *1.0*) demonstrated highly active and stable catalysts for aqueous-phase hydrogenation of LA into exclusive GVL product. The used solvent reaction and the high dispersion of Ni-Sn alloy on aluminum hydroxide (bayerite and gibbsite) facilitated reaction rate at a much lower temperature than that of bulk ones. Moreover, the Ni–Sn(*1.4*)/AlOH catalyst was reusable for at least six consecutive runs without any significant loss of activity and selectivity [10].

In the present paper, we report the catalytic performance of iron-modified palladium supported on titanium oxide (denoted as Pd(5.0)–Fe(5.0)/TiO_2; 5.0 was the loading amount of Pd and Fe, respectively) in the selective hydrogenation of levulinic acid to γ-valerolactone in stainless-steel batch reactor system in various solvents. The purpose of this work is to study the effect of Fe loading amount to Pd (%wt), the solvent used, and initial H_2 pressure on the conversion of LA and the yield of GVL. The structure-activity relationship of Pd(5.0)-Fe(5.0)/TiO_2 catalyst in LA hydrogenation to GVL is discussed.

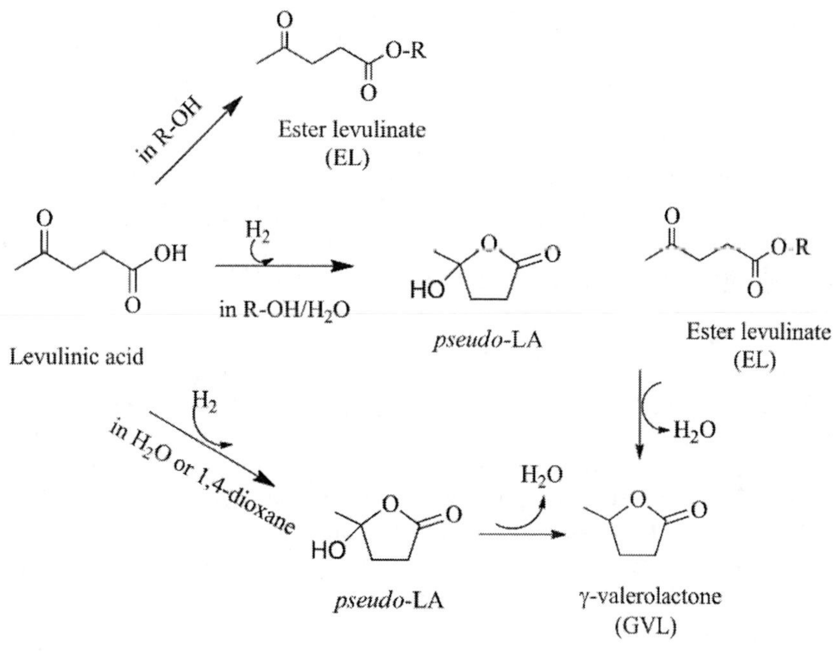

Figure 1. Possible reaction pathways for hydrogenation of levulinic acid in various solvents in the presence of bimetallic Pd(5.0)-Fe(x)/TiO_2 catalysts [10].

2. Materials and methods

2.1. Material

Palladium (II) chloride ($PdCl_2$) (98%; Tokyo Chemical Industry), Iron (III) chloride hexahydrate ($FeCl_3 \cdot 6H_2O$) (99%; Merck Millipore), 2-propanol (98%, Merck Millipore), TiO_2 anatase, ethanol (96%; Merck Millipore), ethylene glycol (EG) (99.5%; Merck Millipore), Sodium tetra hydroborate

($NaBH_4$) (95%; Tokyo Chemical Industry), NaOH (99%; Merck Millipore), H_2 gas (99.99%), N_2 gas (99.99%), lauric acid (97%; Tokyo Chemical Industry).

2.2. *Catalyst preparation*

A typical procedure for the synthesis of Pd(5.0)–Fe(5.0)/TiO_2 (5.0 was the loading amount of Pd and Fe, respectively) described as follows (11): $PdCl_2$ (0.0638 g) was dissolved in deionized water (denoted as solution A) and $FeCl_3 \cdot 6H_2O$ (0.0605 g) was dissolved in ethanol/ethylene glycol (2.0/1.0 volume ratio) (denoted as solution B) at room temperature. Solutions A, B, and TiO_2 (as support material; 1.0 g) were mixed at room temperature; the temperature was subsequently raised to 50°C under gentle stirring for 12 h. The pH of the mixture was adjusted to 12 through the dropwise addition of an aqueous solution of NaOH (3.1 M). The mixture was then placed into a sealed-Teflon autoclave for the hydrothermal process at 150 °C K for 24 h. The resulting black precipitate was filtered, washed with distilled water, and then dried under vacuum overnight. Prior to the catalytic reaction, the obtained black powder was treated under hydrogen at 500 °C for 3 h.

2.3. *Catalyst characterization*

The X-ray diffraction (XRD) analysis was performed on a Miniflex 600 Rigaku instrument with Cu as a monochromatic source of CuKα radiation (λ = 0.15444 nm). The XRD was operated at 40 kV and 15 mA with a step width of 0.02°, a scan speed of 4° min^{-1} ($\alpha1$ = 0.154057 nm, $\alpha2$ = 0.154433 nm), solar slit 1.25°, and using a Ni Kβ filter. The BET surface area (S_{BET}) and pore volume (V_p) were measured using N_2 physisorption at -196 °C on a Belsorp Max (BEL Japan). The samples were degassed at 200 °C for 2 h to remove physisorbed gases prior to the measurement. The amount of nitrogen adsorbed onto the samples was used to calculate the Brunauer–Emmett–Teller (BET) surface area via the BET equation. The pore volume was estimated to be the liquid volume of nitrogen at a relative pressure of approximately 0.995 according to the Barrett–Joyner–Halenda (BJH) approach based on desorption data (12). The NH_3-TPD was carried out on a Belsorp Max (BEL Japan). The samples were degassed at an elevated temperature of 100–200 °C for 2 h to remove physisorbed gases prior to the measurement. The temperature was then kept at 200 °C for 2 h while flushed with He gas. NH_3 gas (balanced NH_3, 80% and He, 20%) was introduced at 100 °C for 30 min, then evacuated by helium gas to remove the physisorbed also for 30 min. Finally, temperature-programmed desorption was carried out at a temperature of 100–800 °C and the desorbed NH_3 was monitored by TCD.

2.4. *Catalytic reaction*

The typical procedure for hydrogenation of lauric acid is described as follows: Catalysts (0.04 g), levulinic acid (0.23 g), and H_2O (5 ml) as the solvent, and 1,6-hexanediol (0.025 g) were placed into a glass reaction tube in an autoclave reactor system of TAIATSU Techno reactor (a Pyrex tube was fitted inside of a sus316 jacket to protect the vessel from corrosion in acidic media. The reactor was flushed with 1.0 MPa H_2 gas 10 times before removing undesired gases. After H_2 was introduced into the reactor with an initial H_2 pressure of 4.0 MPa at room temperature, the temperature of the reactor was increased to 170 °C. After 7 h, the conversion of levulinic acid and the yield of γ-valerolactone were determined by gas chromatography (GC) analysis.

2.5. *Product analysis*

GC analysis of the reactant (levulinic acid) and product (γ-valerolactone) was performed on a Perkin Elmer Auto System XL equipped with a flame ionization detector and Restek Rtx® BAC Plus capillary column. GC analysis was operated at the detector and injector temperatures of 250 °C, N_2 as a carrier gas, rates of air, and H_2 were 450 ml/min and H_2 45 ml/min, respectively. The used 5.0Pd–5.0Fe/TiO_2 catalyst was easily separated using either simple centrifugation or filtration, then finally dried overnight under vacuum at room temperature prior to the reusability testing.

3. Results and discussion

3.1. Screening of solvent

In the first experiments, we evaluate the use of various solvents as solvent media for the hydrogenation of levulinic acid (LA) to γ-valerolactone (GVL) using Pd(5.0)-Fe(5.0)/TiO_2 catalyst as shown in figure 1 and the results are summarised in table 1. Three types of solvents such as 1,4-dioxane, 2-propanol, and water which are corresponding to polar aprotic, polar semi-protic, and polar protic, respectively, were employed. In polar aprotic solvent (1,4-dioxane), 29.3% yield of γ-valerolacton at 29.3% conversion of levulinic acid was obtained (entry 1). In the polar semi-protic solvent of 2-propanol, the products were GVL (27.4%) and isopropyl levulinate (39.8%) at the relatively high conversion of levulinic acid (*c.a.* 67.2%) under the same reaction conditions (entry 2). In H_2O, 52.4% yield of GVL and 10.4% yield of *pseudo*-LA were obtained at 62.7% conversion of LA (entry 3). After reaction temperature was prolonged to 15 h, conversion of LA as well as the yield of GVL slightly increased to 68.6% and 63.3%, respectively, suggesting the extended reaction time caused a further reaction of *pseudo*-LA to the final product of GVL (entry 4). In the H_2O/2-propanol and H_2O/1,4-dioxane mixture solvents, the hydrogenation of levulinic acid occurred with unsatisfactory results (entries 5 and 6) as well as in absence of solvent (entry 7).

Table 1. Results of levulinic acid hydrogenation over Pd(5.0)–Fe(5.0)/TiO_2 catalyst in various solvents.

Entry	Solvent	Conversion[a] (%)	Yield[a] (%)		
			GVL	*pseudo*-LA[b]	Ester[b]
1	1,4-Dioxane	29.3	29.3	0.0	0.0
2	2-Propanol (2-PrOH)	67.2	27.4	0.0	39.8
3	H_2O	62.7	52.4	10.4	0.0
4[c]	H_2O	68.6	63.3	5.3	0.0
5	H_2O: 2-PrOH (3: 2 in ml)	26.7	7.4	0.0	19.3
6	H_2O/1,4-Dioxane (3:2 in ml)	38.3	30.2	8.1	0.0
7	Neat (without solvent)	21.8	11.2	10.6	0.0

Reaction conditions: catalyst 0.04 g; levulinic acid 0.23 g; solvent 3 ml; initial H_2 pressure 3 MPa; temperature 170 °C; reaction time 7 h. [a]Conversion and yield were determined by GC analysis using an internal standard technique. [b]Yields of *pseudo*-LA and ester were determined by GC area. [c]Reaction time was 15 h.

The relatively high yield of γ-valerolactone in water solvent could be attributed to the presence of H-bonded water molecule or single chemisorbed water on metallic surfaces that could dramatically reduce the energy span of the reaction pathway, hence enhancing the catalytic activity [13]. Alternatively, the increase of surface concentration in hydrogen atom due to dissociated water and strong interaction between water with the substrate by hydrogen bonding lowered the activation energy barrier and led to high hydrogenation rates [14,15].

3.2. Effect of Fe loading amount

The influence of Fe intercalation on the catalytic performance of the Pd/TiO_2 catalyst in levulinic acid hydrogenation was evaluated and the Fe amounts were 1.0, 3.0, and 5.0 %wt to the amount of Pd and the results are summarised in table 2.

Table 2. Results of levulinic acid hydrogenation over Pd(5.0)–Fe(x)/TiO_2 catalysts with different loading amounts of Fe.

Entry	Catalyst[a]	Conversion[b] (%)	Yield[b] (%)	
			GVL	*pseudo*-LA[c]
1	Pd(5.0)/TiO_2	25.9	22.7	3.0
2	Pd(5.0)-Fe(1.0)/TiO_2	11.6	11.6	0.0
3	Pd(5.0)-Fe(3.0)/TiO_2	22.7	22.7	0.0
4	Pd(5.0)-Fe(5.0)/TiO_2	62.7	52.4	10.4
5[d]	Pd(5.0)-Fe(5.0)/TiO_2	35.0	35.0	0.0
6[e]	Pd(5.0)-Fe(5.0)/TiO_2	7.1	1.5	5.6

Reaction conditions: catalyst 0.04 g; levulinic acid 0.23 g; solvent 3 ml; initial H_2 pressure 3 MPa; temperature 170 °C; reaction time 7 h. Conversion and yield were determined by GC analysis using an internal standard technique. [d]temperature reaction was 150 °C. [e]The *as-prepared* Pd(5.0)-Fe(5.0)/TiO_2 catalyst (without reduction with H_2 at 500 °C for 3 h).

A reference catalyst of Pd(5.0)/TiO_2 was synthesized and reduced with H_2 at 500 °C for 3 h and showed low conversion of levulinic acid (25.9%), whereas the yields of γ-valerolactone and *pseudo*-LA were 22.7% and 3.0%, respectively (entry 1). The addition of 1wt% Fe to Pd (5.0) / TiO2 reduced the conversion significantly in more than half (entry 2). The conversion of LA increased to 22.7% over Pd(5.0)-Fe(3.0)/TiO_2 catalyst and no *pseudo*-LA was observed under the same reaction conditions (entry 3). A remarkable increase in conversion of LA (62.7%) and yield of GVL (52.4%) as well as the amount of *pseudo*-LA (10.4%) were obtained over Pd(5.0)-Fe(5.0)/TiO_2 catalyst (entry 4). The conversion of LA decreased to 35% when the reaction temperature was reduced to 150 °C without decreasing the selectivity of GVL (entry 5). In addition, the as-prepared Pd(5.0)-Fe(5.0)/TiO_2 catalyst (without reduction with H_2 at 500 °C) was completely inert for the hydrogenation of LA (entry 6).

3.3. Effect of initial H_2 pressure

The effect of initial H_2 pressure on the conversion and yield in the levulinic acid hydrogenation was evaluated using Pd(5.0)-Fe(5.0)/TiO_2 catalyst and the results are summarised in table 3. At initial H_2 pressure of 1.0 MPa, an 18.8% yield of GVL was obtained (entry 1). Both conversions of levulinic acid and yield of GVL increased gradually when the initial H_2 pressure was increased up to 3.0 MPa (entries 2 and 3) then tended to remain unchanged at initial H_2 of 4.0 MPa which decreased in yield of *pseudo*-LA (entry 4).

Table 3. Results of levulinic acid hydrogenation over Pd(5.0)-Fe(5.0)/TiO_2 catalyst at different initial H_2 pressure.

Entry	Initial H_2 (MPa)	Conversion (%)	Yield (%)	
			GVL	*pseudo*-LA
1	1.0	18.8	18.8	0.0
2	2.0	20.8	20.8	0.0
3	3.0	62.7	52.4	10.4
4	4.0	56.5	52.5	3.7

Reaction conditions: catalyst 0.04 g; levulinic acid 0.23 g; solvent 3 ml; initial H_2 pressure 1.0 – 4.0 MPa; temperature 170 °C; reaction time 7 h. Conversion and yield were determined by GC analysis using an internal standard technique.

3.4. Structure-activity relationship

Based on XRD analysis and reaction profiles, we discuss the structure of active Pd and the presence of FeO_x species in bimetallic Pd(5.0)-Fe(x)/TiO_2 catalysts after reduction with H_2 at 500 °C for 3 h. The *as-prepared* Pd(5.0)-Fe(5.0)/TiO_2 (without reduction with H_2) was as referenced catalysts and inactive for hydrogenation of lauric acid (table 1, entry 6). The XRD patterns of Pd(5.0)-Fe(x)/TiO_2 catalysts before and after reduction with H_2 at 500 °C for 3 h (figures 2 (a) and (b), respectively) suggest that the formation of metallic Pd(0) was observed at 2θ = 40.34° and 47.2° corresponding to Pd(111) and Pd(200), respectively [16]. On the other hand, the metallic or oxide Fe species were unable to detect due to its crystallite sizes were lower than that the detection limit of XRD.

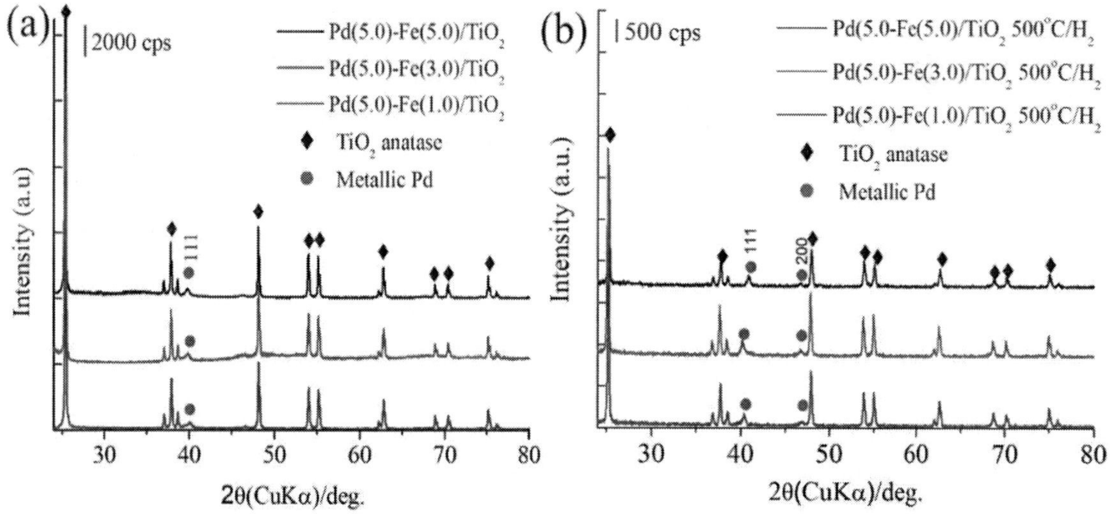

Figure 2. XRD patterns of (a) before and (b) after reduction with H_2 at 500 °C for 3 h with different amounts of Fe.

4. Conclusion

A series iron-modified palladium supported on titanium oxide (Pd(5.0)-Fe(x)/TiO_2 (5.0 and x are loading amount of Pd and Fe, respectively) catalysts were successfully synthesized using a simple hydrothermal method at 150 °C for 24 h followed by reduction with H_2 at 500 °C for 3 h. Results of XRD analysis of Pd(5.0)-Fe(x)/TiO_2 catalysts suggest that the formation of metallic Pd(0) was observed at 2θ = 40.34° and 47.2° corresponding to Pd(111) and Pd(200), respectively. The hydrogenation of LA to GVL effectively occurred in the H_2O solvent, whereas in 2-propanol the formation of ester (iso-propyl levulinate (39.8%) was the main side product at the relatively high conversion of levulinic acid (67.2%). The highest yield of GVL (52.5%) was obtained over Pd(5.0)–Fe(5.0)/TiO_2 catalyst at a temperature of 170 °C, initial H_2 pressure of 4.0 MPa, solvent H_2O 3 ml at a reaction time of 7 h. The yield of GVL slightly increased to 63.3% when the reaction time was prolonged to 15 h.

Acknowledgments

The authors acknowledge the RISET DASAR Hibah Berbasis Kompetensi (HIKOM) (contract number DIPA-042.06-1.401516/2019 and DIPA-042.06-1.401516/2020), the RISET DASAR (contract number DIPA-042.06-1.401516/2019) from the Ministry of Research, Technology, and Higher Education, Indonesian Government, which all financially supported this work. We would also like thank to Professor Shogo Shimazu for the kind help in XRD measurements.

References

[1] Alonso D M, Bond J Q, Dumesic J A 2010 *Green Chem.* **12** 1493–513
[2] Pritchard J, Filonenko G A, Van Putten R, Hensen E J M, Pidko EA 2015 *Chemical Society Reviews* **44** 3808–33
[3] Besson M, Gallezot P, Pinel C 2014 *Chem Rev* **114** 1827–70
[4] Mariscal R, Maireles-Torres P, Ojeda M, Granados I S and M L 2016 *Energy Environ Sci.* **9**

1144–89
[5] Gallezot P 2012 *Chem Soc Rev* **41** 1538–58
[6] Bourne RA, Stevens J G, Ke J, Poliakoff M 2007 *Chem Commun.* **2** 4632–4
[7] Wettstein S G, Bond J Q, Alonso D M, Pham H N, Datye A K, Dumesic JA 2012 *Appl Catal B Environ* 117–118
[8] Luo W, Sankar M, Beale A M, He Q, Kiely C J, Bruijnincx PCA 2015 *Nat Commun.* **6** 1–10
[9] Rodiansono R, Astuti M D, Ghofur A, Sembiring K C 2015 *Bull Chem React Eng Catal.* **10** 192–200
[10] Rodiansono R, Astuti M D, Hara T, Ichikuni N, Shimazu S 2016 *Catal Sci Technol* **6** 2955–61
[11] Rodiansono R, Khairi S, Hara T, Ichikuni N, Shimazu S 2012 *Catal Sci Technol* **2** 2139–45
[12] Lowell S, Shields J E, Thomas M A, Thommes M 2004 *Characterization of Porous Solids and Powders: Surface Area, Pore Size and Density, Kluwer* (Academic Publishers: Dordrecht, The Netherlands)
[13] Michel C, Gallezot P 2015 *ACS Catal* **5** 4130–2
[14] Delbecq F, Loffreda D, Sautet P 2010 *J Phys Chem Lett.* **1** 323–6
[15] Thibault M E, Dimondo D V., Jennings M, Abdelnur PV, Eberlin MN, Schlaf M 2011 *Green Chem.* **13** 357–66
[16] JCPDS-ICDD 1991 *Powder diffraction files*, JCPDS-International center for diffraction data (JCPDS-ICDD)

IOP Conf. Series: Materials Science and Engineering **980** (2020) 012014 doi:10.1088/1757-899X/980/1/012014

Synthesis and characterization of magnetic adsorbent based on Fe_2O_3-fly ash from Pulang Pisau's power plant of Central Kalimantan

D A P Wardani*, L Rosmainar, R M Iqbal and S N Simarmata
Department of Chemistry, Faculty of Mathematics and Natural Science, University of Palangka Raya, Kampus Tunjung Nyahu, Palangka Raya 73112, Kalimantan Tengah, Indonesia

Email: dayupwardani@mipa.upr.ac.id

Abstract. One of the biggest producers of fly ash in Central Kalimantan is Pulang Pisau's power plant. Fly ash, which is produced from burning coal or biomass, is a mixture of oxides with less combustible carbon and other minor inorganic compounds so that it has a dominant-negative surface charge and is a promising adsorbent. Modification of fly ash with Fe_2O_3 can affect the structure and morphology of the adsorbent, and the magnetic properties of the adsorbent can help in the process of separating the adsorbent after treatment. This study aims to study the synthesis and magnetic characteristics of magnetic adsorbent based on Fe_2O_3-fly ash from Pulang Pisau's power plant. The proximate test results obtained showed that the SiO2 compound contained in fly ash was the highest with a percentage level of 32%. The optimum composition of fly ash and Fe_2O_3 ratio is 1: 2. Morphology of magnetic adsorbent based on Fe_2O_3-fly ash was observed using SEM EDX. The results showed Fe_2O_3 particles impregnated to the pores of fly ash particles which had 29,11% of O and 70,22% of Fe. The magnetic properties of magnetic adsorbent based on Fe_2O_3-fly ash from Pulang Pisau's power plant were simply proven using an external magnetic field.

1. Introduction

Fly ash in this research was coal ash waste that was used as a source of energy in the industry. Steam power plant is a producer of fly ash in very large numbers. One of the biggest producers of fly ash in Central Kalimantan is Pulang Pisau's power plant. Coal production in Indonesia in 2015 was 425 million tons and estimated to be in the range of 400 million tons/year until 2019 (Direktorat Jenderal Mineral dan Batubara, 2016). If all the coal is consumed domestically and the average ash content is 8% (ash content generally varies from 3 to 20% by weight), then there are around 32 million tons/year of coal ash produced. Most of the ash is fly ash while the rest is bottom ash (bottom ash). Meanwhile, the global production of fly ash from industry is estimated at 800 million tons/year in 2010 [1, 2].

Fly ash is categorized as hazardous and toxic waste so that it is not allowed to be disposed of without being processed [3]. This is what causes the use of fly ash in the world is still very low at only 20-30%. Fly ash, which results from burning coal or biomass, is a mixture of oxides with less combustible carbon and other minor inorganic compounds so that it has a dominant-negative surface charge and is a promising adsorbent. Many applications in wastewater treatment have been reported, most of them on a fairly large scale. Fly ash is widely used in the adsorption of heavy metals [4] and fly ash has been reported as an adsorbent for dyes from the textile industry waste batik [5].

Iron oxide and hydroxide are interesting subjects because they can be utilized in the fields of geochemistry, mineralogy, metallurgy, and environmental protection [6]. Iron oxide exists in a variety of polymorph variations. Anhydrous iron oxides including hematite (α-Fe_2O_3), maghemite (γ-Fe_2O_3), and β-Fe_2O_3 are less common. Fe_3O_4 (magnetite) and Fe1-xO (wüstite) contain both ferrous and ferrous iron [7]. Fe_2O_3 is a conventional semiconductor material, which has been widely used as an active catalytic, magnetic, nonlinear optical, and gas sensitive material. These properties are caused by the high surface area and porosity of the Fe_2O_3 particles. Fe_2O_3 is commonly found in regional soils, and can also be easily synthesized in a laboratory in a short time. Synthesis of Fe_2O_3-Fly Ash is reported to be able to increase the ability of Pb metal ion adsorption due to increased surface area [8]. Maghemite (γ-Fe_2O_3) is an inexpensive magnetic material, and if the composition of fly ash has magnetic properties, it will be easier to separate the multiphase system by using an external magnetic field. This shows the regeneration of absorbent substances, waste management, and types of separation which are of safe value.

Fly ash contains some silicon and aluminum. In addition, fly ash can also contain calcium, iron, potassium, magnesium, titanium, sulfur, and so on. These elements are reported in their oxide form, for example, silicon in the form of SiO_2 and iron in the form of Fe_2O_3. The oxides can be either amorphous or crystalline.

Fly ash which is classified as N class is a natural calcined material. Class F of fly ash is produced from the burning of anthracite or bituminous coal which is pozzolanic while class C comes from the burning of lignite or sub-bituminous type coal which is pozzolanic and cementitic. Further research shows that the CaO content in class F is below 20%, whereas in class C it is more than 20%. The high calcium and SO_3 content in fly ash can also be caused by the cleaning process that occurs in the chimney's power plant, especially the removal of flue gases from sulfur gases. One of the cleanings can be done by adding calcium which will bind sulfur gases. High calcium content can reduce the compressive strength of geopolymers if the fly ash is used as raw material for geopolymers. According to Damilola et al. [9], this can be overcome by adding NaOH to the water glass, both of which are geopolymers activators, to avoid decreasing the workability of samples containing lots of calcium, thereby increasing the compressive strength of geopolymers [9].

The particle size and shape characteristics of fly ash depend on the source and uniformity of the coal, the degree of destruction before combustion, the combustion environment, the uniformity of combustion, and the type of collection system used. The fly ash from the coal combustion powder has gone through a liquid stage at high temperatures, and subsequently, this is generally round in shape with diameters ranging from below 1 μm to above 150 μm [10]. The size distribution of fly ash particles can be measured in various ways but is usually measured by sieve analysis, a method suitable for particles larger than about 45 μm [11, 12]. The particle size distribution shows how the highest contribution to total mass comes from fly ash particles less than 38 in m size, which is broadly consistent with some of the studies described by Helmuth [13].

Physico-chemical surface characteristics of fly ash, such as site density, acidity constant, specific surface area, and surface electrical characteristics, metal adsorption constant, determination of capacity, strength, and potential of organic compounds bind the surface of fly ash. Surface characteristics such as specific surface area and zero charge point pH (pH ZPC) have been assessed [14–16]. The specific surface area from <1 m^2/g to tens of m^2/g depends on carbon content [14–16]. Research has shown that pH ZPC is between 2.4 and 7.0. Alkaline metric titrations are used to determine site density and solid surface acidity constants [17], [18].On other hand, Magnetic minerals in nature can be classified in the oxide family iron-titanium, iron sulfide, and iron oxide hydroxides. The iron-titanium oxide family is most often found in rocks. This family consists of minerals that meet the triangle diagram (ternary diagram) with edge members consisting of TiO_2, FeO, and Fe_2O_3.

Hematite is the most prominent iron ore. Most high-grade iron ore is subject to simple dry process beneficiation to meet the size requirements. This involves multi-level crushing and screening to obtain chunks of products (−31.5 + 6.3 mm) and plates (about −6.3 mm). Low-grade hematite ore needs to be increased to achieve the required iron content, which involves a more complicated ore beneficiation process. Comminution rates are needed for low-grade hematite ore similar to high-grade ore to provide the same product [19]. Hematite (α-Fe_2O_3), is the most thermodynamically stable iron oxide phase in the family and iron oxide has many advantages, such as natural abundance, biocompatibility, and low cost [20]. Hematite has been reported to be able to bind with Sb (III) closely through the inner-sphere complex. Hematite nanoparticles can remove Sb (III) metal and are easily separated after treatment [21].

Various chemical methods have been developed to synthesize maghemite nanoparticles, including precipitation, thermolysis, and electrochemistry. From these methods, the electrochemical method is simple because this method is easy to control and the equipment used is simple. In the electrochemical method, the reaction that occurs is the oxidation-reduction reaction on the electrode surface, where the reaction is influenced by several factors such as the electrode used (type, surface area, the distance between the two electrodes), electrolyte as a precursor (type and concentration), potential difference, temperature, pH and current density [22].

The electrochemical synthesis of maghemite nanoparticles was successfully carried out using the $FeCl_3$ solution. It was concluded that with increasing current density, the average size of the resulting maghemite particles gets smaller [22]. The synthesis of maghemite by the electro-oxidation method was also carried out using a solution of tetraoctylammonium bromide dissolved in DMF (N, N-dimethylformamide). The resulting maghemite particles have an amorphous size of 3-8 nm deposited in the anode [23]. Based on the above explanation, our research aims to synthesize a potential magnetite adsorbent based on a combination of Fe_2O_3 and fly ash from Pulang Pisau's power plant.

2. Materials and methods

2.1. Equipment and materials

The types of equipment that we used in this work were X-Ray Diffraction (XRD) (Rhigaku Miniflex600), Spectrophotometer Scanning Electron Microscopy (SEM) (Hitachi SU3500), and glassware. The materials that we used in this work were fly ash from Pulang Pisau's power plant of Central Kalimantan, Fe_2O_3, and methanol.

2.2. Methods

The method in this study was divided into three stages, first, the characterization of coal and fly ash from Pulang Pisau's power plant of Central Kalimantan, second, The synthesis of magnetic adsorbent based on Fe_2O_3-fly ash, and third, the structural characterization of magnetic adsorbent based on Fe_2O_3-fly ash.

2.2.1. The characterization of coal and fly ash from Pulang Pisau's power plant of Central Kalimantan. Coal was tested proximately and ultimately to find out its characteristics, then analysis was also carried out on fly ash to determine the composition of the oxide compounds contained by methods that were compliant with ASTM D 3-682 13 standards.

2.2.2. The synthesis of magnetic adsorbent based on Fe_2O_3-fly ash. Fly ash and Fe_2O_3 were mixed into a beaker with a ratio of 1: 1, 1: 2, 1: 3, and 1: 4, then each solid mixture was dispersed into 50 mL of methanol and stirred on a hotplate stirrer for 2-3 hours until homogeneous. Then, the homogeneous mixture was allowed to stand until the filtrate and sedimentary phases were formed, then decanted to separate the filtrate and Fe_2O_3-fly ash. Fe_2O_3-fly ash deposition was heated into the oven at 75 °C for 1 hour to vaporize the methanol which was still trapped on the surface of the Fe_2O_3-fly ash composite.

2.2.3. The structure characterization of magnetic adsorbent based on Fe_2O_3-fly ash. Fe_2O_3-fly ash adsorbent was characterized by Panalytical X-Ray Diffraction (XRD) using a Cu-Kα light source ((λ = 1.5406 Å) which was operated with a current of 30 mA and a voltage of 40 kV. The analysis rate was

1 ° C.min- 1 and step size 0.02°. The morphological of magnetic adsorbent based on Fe_2O_3-fly ash was characterized by Scanning Electron Microscopy (SEM).

3. Results and discussion

3.1. The structure characterization of coal and fly ash from Pulang Pisau's power plant of Central Kalimantan

The ultimate analysis of coal was used to study the properties and compositions in coal from Pulang Pisau's power plant of Central Kalimantan. The result of the analysis is shown in figure 1. According to figure 1, the coal from Pulang Pisau's power plant contained volatile matter which reached 38.53%. It means that the matters contained in the coal were easy to vaporize. The second highest composition was fixed carbon that reached 36.23%. It was because the coal consisted of hydrocarbon chains with many carbon compositions. On the other side, the lowest matter contains in coal was ash which reached 4.01%. The ash that was produced from the steam power plant gave some problems because it was hazardous waste which consisted of two types; fly ash and bottom ash.

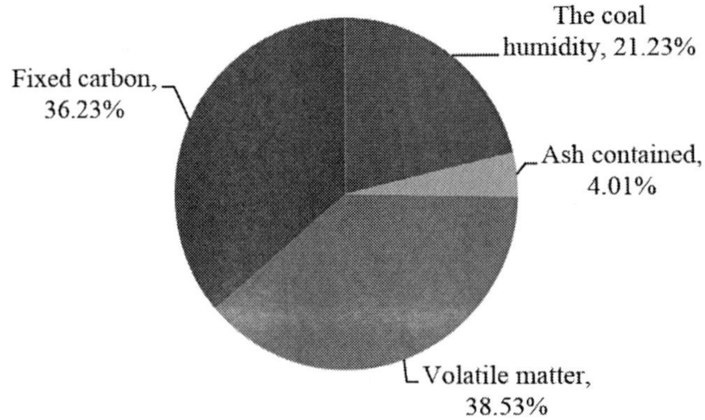

Figure 1. The ultimate analysis result of coal from Pulang Pisau's power plant.

The next analysis for the coal is proximate analysis and the result is shown in figure 2. The composition of C, H, O, and N shows that the coal generally consisted of organic matter formed by hydrocarbon chains.

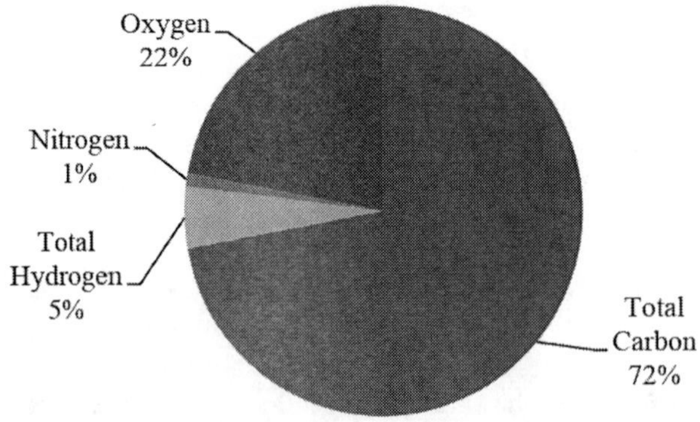

Figure 2. The proximate analysis result of coal from Pulang Pisau's power plant.

The results of electricity production at the power plant produced side products in the form of fly ash. Fly ash was characterized to identify compounds contained in the fly ash sample. Figure 3 shows the oxide compounds contained in the fly ash of Pulang Pisau's power plant.

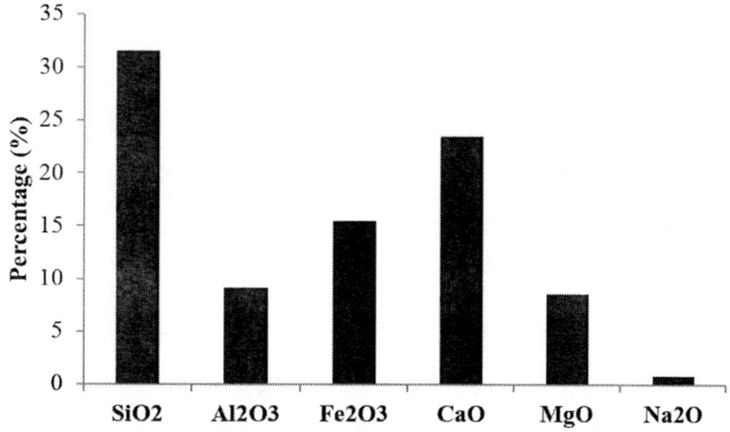

Figure 3. The compounds compositions in fly ash from Pulang Pisau's power plant.

The fly ash from Pulang Pisau's power plant had 32% of SiO_2, 23% of CaO, 15% of Fe_2O_3, and Al_2O_3, MgO, and Na_2O below 10%. According to its classification, the class of the fly ash from Pulang Pisau's power plant was classified as class C because the amount of SiO_2, Al_2O_3, and Fe_2O_3 was more than 50% but less than 70% of the total components. The abundant component of metal oxide in the fly ash had the potential to be used as an adsorbent feedstock that could be improved by its physical and chemical properties by composting fly ash with Fe_2O_3.

The next characteristic studied is the particle size identified in various circumstances, namely AR, ADB, DB, and DAFB. The highest particle size was 98 nm and the most explosive was 18 nm. All particle measurements show that the Pulang Pisau's power plant fly ash was classified as nanomaterial (material that has a size of less than 100 nm). The particle size of the fly ash is shown in table 1.

Table 1. The particle size of fly ash from Pulang Pisau's power plant.

Sample	Particle Size (nm)
AR	98
ADB	96
DB	76
DAFB	18

3.2. The study of synthesis and characterization of magnetic adsorbent based on Fe_2O_3-fly ash from Pulang Pisau's power plant of Central Kalimantan

This section describes several studies related to the synthesis and characterization of magnetic adsorbent based on Fe_2O_3-fly ash including XRD confirmation of fly ash and the influence of Fe_2O_3 and the fly ash composition variation, morphological analysis using SEM, and the magnetic properties of magnetic adsorbent based on Fe_2O_3-fly ash using an external magnetic field.

3.2.1. XRD confirmation of fly ash and the influence of Fe_2O_3-fly ash composition variation. The analysis using X-ray diffraction was conducted to find out the types of minerals contained in fly ash and Fe_2O_3 which is shown by a peak at 2θ. The qualitative analysis result of fly ash and magnetic adsorbent based on Fe_2O_3-fly ash refers to the database of JCPDS No. 39-1364, 46-1045, 02-0431, which is the minerals database of maghemite (γ-Fe_2O_3), quartz (SiO_2), and mullite ($Al_6Si_2O_{13}$). The 2θ database showed in table 2.

Table 2. 2θ database of quartz (SiO_2), mullite ($Al_6Si_2O_{13}$), dan maghemite (γ-Fe_2O_3).

Mineral	2θ Database (°)
Quartz (SiO_2)	20,82; 26,62; 36,52; 39,44; 40,26; 42,41; 45,75; 50,10; 54,83
Mullite ($Al_6Si_2O_{13}$)	16,5; 23,4; 29,5; 31,2; 33,1; 35,4; 39,2; 41,0; 48,3; 52,8; 59,8
Maghemite (γ-Fe_2O_3)	18,3; 23,77; 26,1;30,3; 35,7; 37; 42,2; 53,6; 57,3; 62,8; 72,3; 74,8

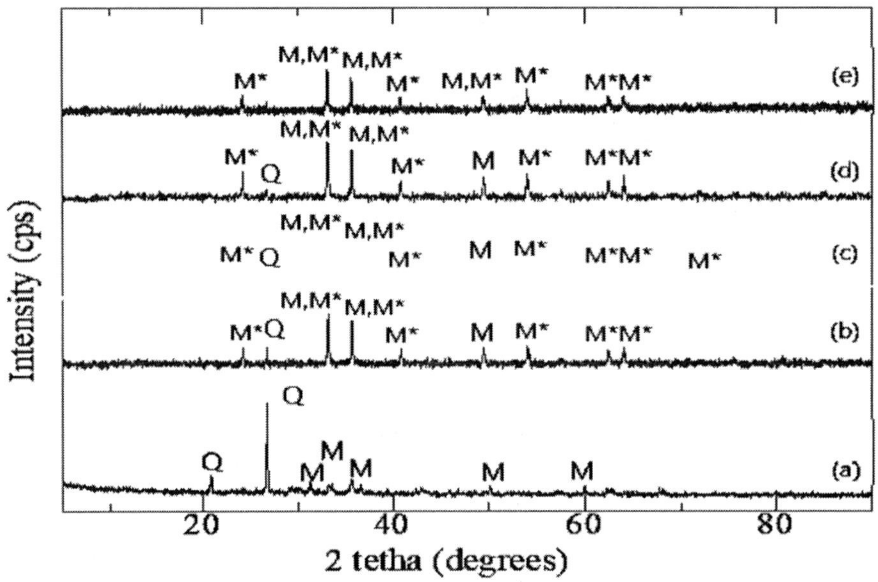

Figure 4. X-ray difractogram of (a) fly ash from Pulang Pisau's power plant of Central Kalimantan; magnetic adsorbent based on Fe_2O_3-fly ash in ratio (b) 1:1; (c) 2:1; (d) 3:1; (e) 4:1. Note: Q= Quartz; M= Mullite; M*=Maghemite.

Figure 4(a) shows the X-ray diffractogram of the fly ash from Pulang Pisau's power plant of Central Kalimantan and figures 4(b), (c), (d), and (e) shows the X-ray diffractogram of magnetic adsorbent based on Fe_2O_3-fly ash composition variation. Figure 4(a) is dominated by the characteristic mineral peak of quartz at 2θ = 20.82°; 26.62° and mullite at 2θ = 31.2°; 33.1°; 35.4°; 39.2°; 59.8°. Figures 4(b), (c), (d), and (e) shows that the characteristic mineral peak of maghemite appeared at 2θ= 26.1°;30.3°; 35.7°; 42.2°; 53.6°; 62.8°; 65.7°;72.3°. According to the result, the optimum of Fe_2O_3-fly ash is showed in figure 4(c) magnetic adsorbent based on Fe_2O_3-fly ash 2:1, because the characteristic mineral peak of fly ash still appeared and the maghemite formed in the pores structure of fly ash.

The addition of Fe_2O_3 in the composition to form maghemite (γ-Fe_2O_3) in the pores structure of fly ash increased the peak. It indicates that Fe_2O_3 in the pores formed perfectly. But, the addition decreased the peak intensity of quartz and mullite. The decreases in peak intensity occurred because of destructive interference increased caused by the Fe_2O_3 particles impregnation in the pores of fly ash. In general, the introduction of material into the interlayer space would increase the phase that eliminated each other which was scattered by the wall and that which was scattered by the interlayer space, thereby reducing the intensity of scattering in the reflection of Bragg.

3.2.2. Morphological analysis using SEM EDX. Analysis using SEM EDX instrumentation was carried out to study the morphological of magnetic adsorbent based on Fe_2O_3-fly ash. Figure 5 shows the morphological of magnetic adsorbent based on Fe_2O_3-fly ash using SEM, in which there were Fe_2O_3 particles formed in the pores of fly ash. Small spherical particles with agglomerated particles in the fly ash pores can be observed in figure 5.

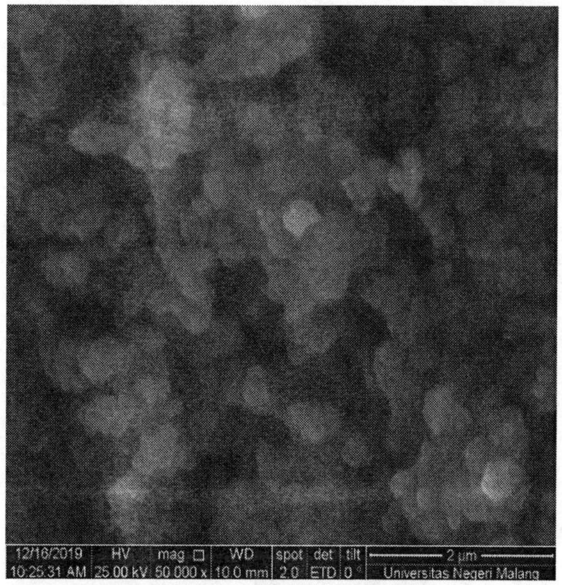

Figure 5. The morphological of magnetic adsorbent based on Fe_2O_3-fly ash.

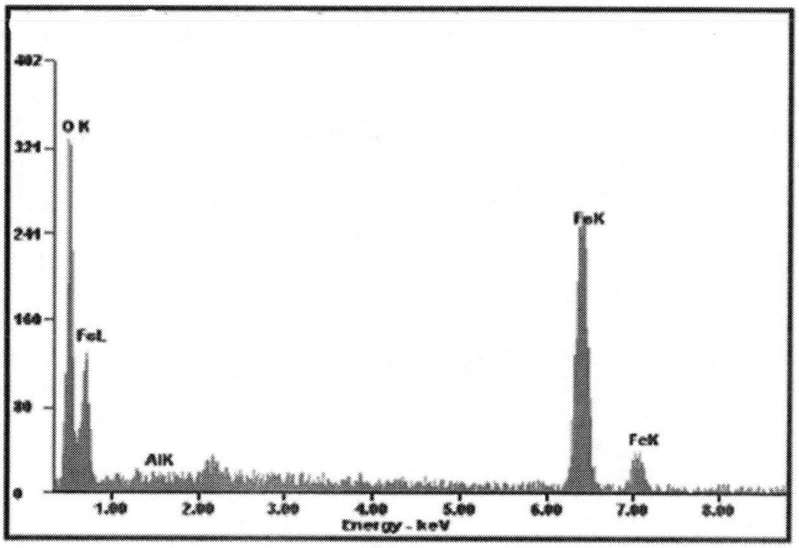

Figure 6. The result of magnetic adsorbent based on Fe_2O_3-fly ash using SEM EDX.

Figure 6 shows that the pores of magnetic adsorbent based on Fe_2O_3-fly ash contained elements of O, Al, and Fe. The percentage of the elements are written in table 3. Magnetic adsorbent based on

Fe_2O_3-fly ash had 29.11% O, 00.67% Al, and 70.22% Fe. It proves that Fe_2O_3 impregnated to the pores of fly ash.

Table 3. The elements contained in magnetic adsorbent based on Fe_2O_3-fly ash.

Element	Wt (%)
OK	29.11
AlK	00.67
FeK	70.22

3.2.3. The magnetic properties of magnetic adsorbent based on Fe_2O_3-fly ash using an external magnetic field. One of the characteristics of maghemite iron oxide (γ-Fe_2O_3) is magnetic properties. If there is a stronger magnetic field effect, the maghemite will be attracted. Figure 7 shows a simple magnetic properties test using an external magnetic field. Fe_2O_3 has ferromagnetic properties, where pure Fe_2O_3 has no superparamagnetic properties because Fe_2O_3 tends to agglomerate so that the particle size becomes larger (> 25 nm). Meanwhile, it is estimated that the analysis results of the magnetic adsorbents based on Fe_2O_3-fly ash would have superparamagnetic properties and had a size of <25 nm which occurred because Fe_2O_3 was distributed into the fly ash pores.

Figure 7. The magnetic properties test using an external magnetic field.

4. Conclusion

Based on the results of the research, it can be concluded that proximate test results show that the total carbon reached 53%. The level of fly ash resulting from coal combustion at Pulang Pisau's power plant of Central Kalimantan was 4.01%. The SiO_2 compounds contained in the fly ash of Pulang Pisau's power plant were the highest with a percentage level of 32%. The addition of the composition of Fe_2O_3 indicated a better formation of Fe_2O_3 in the fly ash pores. However, the addition of this composition decreased the intensity of the characteristic peaks of the constituent minerals of fly ash. The decrease in intensity occurred because of the decreasing crystallinity of the constituent minerals of fly ash. The optimum composition of fly ash and Fe_2O_3 ratio was 1:2. The analysis result using SEM-EDX shows that Fe_2O_3 impregnated to the pores of magnetic adsorbent, which had 29.11% of O and 70.22% of Fe. Magnetic adsorbent based on Fe_2O_3-fly ash from Pulang Pisau's power plant of Central Kalimantan had magnetic properties.

Acknowledgments

The research team would like to thank the Chemistry Education Laboratory of Universitas Palangka Raya. The authors also thank LPPM Universitas Palangka Raya for the research grant under the PDP scheme.

IOP Conf. Series: Materials Science and Engineering **980** (2020) 012014 doi:10.1088/1757-899X/980/1/012014

References

[1] Panias D, Giannopoulou I P and Perraki T 2007 Effect of synthesis parameters on the mechanical properties of fly ash-based geopolymers *Colloids Surfaces A Physicochem. Eng. Asp.* **301** 1–3 246–54
[2] Álvarez E, and Ayuso 2008 Environmental, physical and structural characterisation of geopolymer matrixes synthesised from coal (co-) combustion fly ashes *J. Hazard. Mater.*, **154** 1–3 175–83
[3] Bankowski P, Zou L and Hodges R 2004 Using inorganic polymer to reduce leach rates of metals from brown coal fly ash *Miner. Eng.* **17** 2 159–66
[4] Visa M 2011 Heavy metals removal on dye-modified fly ash substrates in *World of coal ash conference (WOCA), Denver, USA.*
[5] Taufiq A, Hidayat P and Hidayat A 2018 Modified coal fly ash as low cost adsorbent for removal reactive dyes from batik industry in *MATEC Web of Conferences* **154** 1037
[6] Guskos N 2002 Photoacoustic, EPR and electrical conductivity investigations of three synthetic mineral pigments: hematite, goethite and magnetite *Mater. Res. Bull.* **37** 6 1051–61
[7] Navrotsky A, Mazeina Z and Majzlan J 2008 Size-driven structural and thermodynamic complexity in iron oxides *Science (80-.).* **319** 5870 1635–38
[8] Shakhapure J, Vijayanand H, Basavaraja S, Hiremath V and Venkataraman A 2005 Uses of α-Fe_2O_3 and fly ash as solid adsorbents *Bull. Mater. Sci.* **28** 7 713–8
[9] Damilola O M 2013 Syntheses, characterization and binding strength of geopolymers: A review *Int. J. Mater. Sci. Appl.* **2** 185–93
[10] Hewlett P and Liska M 2019 *Lea's chemistry of cement and concrete* (Butterworth-Heinemann).
[11] Wesche K 1991 *Fly ash in concrete: properties and performance* (CRC Press).
[12] Rhodes M J 2008 *Introduction to particle technology* (John Wiley & Sons).
[13] Association P C 1956 *Soil-cement laboratory handbook* (Portland Cement Assoc.)
[14] Lin C J and Chang J E 2001 Effect of fly ash characteristics on the removal of Cu (II) from aqueous solution *Chemosphere* **44** 5 1185–92
[15] Lin R B and Shih S M 2003 Characterization of $Ca(OH)_2$/fly ash sorbents for flue gas desulfurization *Powder Technol.* **131** 2–3 212–22
[16] Smedley P L and Kinniburgh D G 2002 Appl geochem *Appl Geochem* **17** 5 517–68
[17] Ricou P H, Lecuyer I and Le Cloirec P 2001 Experimental design methodology applied to adsorption of metallic ions onto fly ash *Water Res.* **35** 4 965–76
[18] Rao M, Parwate A V and Bhole A G 2002 Removal of Cr^{6+} and Ni^{2+} from aqueous solution using bagasse and fly ash *Waste Manag.* **22** 7 821–30
[19] Jankovic A 2015 *Developments in iron ore comminution and classification technologies.* (Elsevier Ltd.).
[20] Liu X Y, Wang K X and Chen J S 2016 Template-directed metal oxides for electrochemical energy storage *Energy Storage Mater.* **3** 1–17
[21] Wu F, Sun F, Wu S, Yan Y and Xing B 2012 Removal of antimony(III) from aqueous solution by freshwater cyanobacteria Microcystis biomass *Chem. Eng. J.* **183** 172–9
[22] Park H, Ayala P, Deshusses M A, Mulchandani A, Choi H and Myung N 2008 Electrodeposition of maghemite (γ-Fe_2O_3) nanoparticles *Chem. Eng. J.* **139** 1 208–12
[23] Pascal C, Pascal J L, Favier F, Elidrissi M L, Moubtassim and Payen C 1999 Electrochemical synthesis for the control of γ-Fe_2O_3 nanoparticle size. Morphology, microstructure, and magnetic behavior *Chem. Mater.* **11** 1 141–7

IOP Conf. Series: Materials Science and Engineering **980** (2020) 012015 doi:10.1088/1757-899X/980/1/012015

The identification of the components, molecular structures and quality of stamp ink from gambier extract

S Silfia*, S Sofyan, F Failisnur dan G Yeni

Institution of Research and Standardization of Industry - Padang
Jl. Raya LIK No. 23 Ulu Gadut, Padang, Indonesia

Email: silfiabintiarsul@gmail.com

Abstract. Stamp ink is a major product of gambier plant *(Uncaria gambir)* extracts. Its level of tannins is usually high, making it an ideal raw material in the production of dyes and stamp ink. Hence, it is necessary to identify these color forming components in stamp ink. The aim of this study was to identify the components, molecular structures, and the quality of gambier stamp ink complexed with $FeSO_4$. Five samples of the ink were complexed with 30%, 25%, 20%, 15% and 10% $FeSO_4$. The Gas Chromatography - Mass Spectrometry (GC-MS) showed the optimum results with 25% $FeSO_4$ treatment. It contained 11.44% pyrocatechol, 3.74% glycerol, and 2 Propanol, 1 Amina 81.98%, with clearer printings on paper compared with other treatments.

1. Introduction

Gambier extract contains about 20 - 55% of tannin, 7 -33% of catechins and 20 -30% of pyrocatechol [1,2]. These three compounds are natural phenols with flavonoid structures [3–6]. It is a multifunctional plant used as dyes in textile industries [7,8], vegetable tanners [9–12], raw materials in pharmaceutical and cosmetic products [13], and inks [6,14,15].

Ink is a colored liquid, used for personal needs, office administration, organization, and institution. The stamp, usually used with ink, represents the identity of an institution. In addition, some of its components are toxic in nature, which could have a negative impact on one's health [4,6,15,16]. However, gambier extract was used as the raw material in this study due to its high tannin content, environmentally friendly, and readily available, with a production of about 14,220 tons in 2016 [17].

There are various studies on the use of tannins for ink production from other natural ingredients such as lemongrass oil, coffee grounds, resident fruit, flaxseed oil, and soybean seeds [18–21]. However, only gambier's tannin complexed with $FeCl_3$ forms a blackish-blue color, red color with NaOH, and brownish color with $Al_2(SO_4)_3$ [15,22,23]. Additionally, ink with NaOH and $Al_2(SO_4)_3$ produces compounds with OH and CO groups from phenol and alcohol respectively. Furthermore, homogeneous particle size distribution usually comes up with a clear printing on papers [15].

The use of $FeCl_3$ as a complexing agent gives blackish-blue color with high intensity, which is also hazardous to health [16]. Hence, $FeSO_4$ was used as the complexing compound in this study due to its safe nature [24]. $FeSO_4$ was also used as a mordant in the coloring of textile with gambir [25–30].

2. Experimental

2.1 Materials and tools

The research was conducted at the laboratories of the Research Institute and Industrial Standardization in Padang, and the Post Harvest Center in Bogor. The materials used include; raw gambier (*Uncaria gambier* Roxb) from South Pesisir Regency farmers, technical ethanol 70%, glycerin, and $FeSO_4$ produced by Bratachem. The equipments used were; glassware, 100 mesh filters, seals, stamp pads, Gas Chromatography-Mass Spectrometry (GC-MS).

2.2 Research implementation

There were two steps involved in making stamp ink, these were the re-extraction of gambier, and the addition of $FeSO_4$ as complexing compounds. The first step re-extraction of gambier this process was carried out by smoothing gambier, then dissolved in hot water at about 90-100°C with a ratio of 1: 4. This was subjected to heating at 60°C for 5 hours and filtered to remove the impurities. The solution was decanted for 12 hours, its sediment was taken, the precipitate was dried at room temperature, and then sieved with a 100 mesh sieve. Then, the gambier granules were dissolved in technical ethanol 70% at a ratio of 1: 4 and stirred at a speed of 250 rpm at 60°C for 5 hours. Furthermore, it was allowed to stay for 24 hours, then filtered and both the filtrate and sediment were removed according to the recommended method [15].

In the second step, stamp ink formulations were added to $FeSO_4$ as complexing compounds. The formula used in making the stamp ink determines its level of compactness when printed on papers. The study was conducted with various percentages of $FeSO_4$ complexing compounds, such as 30%, 25%, 20%, 15%, and 10%. Also, 15% H_2O, 15% glycerin, and 10% violet crystals was added to each treatment and the solution stirred continuously at 250 rpm for 5 hours at 60°C until it was homogeneous in nature.

2.3 Observation

The stamp ink was subjected to analysis to determine its components and molecular structures using Gas Chromatography-Mass Spectrometry (GC-MS) QP2010S Shimadzu, and Agilent HP 1 column of 30 meters long with helium carrier gas used with a pressure of 13.7 kPa. In addition, the helium flow rate was 20 mL / min, and the column temperature ranged from 70 ° C to 310° C [30].

3. Results and discussion

3.1 Chromatogram as the main component of stamp ink

The constituent components of stamp ink with the use of 30% $FeSO_4$ were expressed and identified in GC-MS, as shown in figure 1a. The chromatograms showed the presence of five compounds which include; Pyrocatechol, 1,3 benzenediol, 4 Methyl Catechol, Glyceraldehyde and N- (Methyl-D2) - Aniline. However, the three dominant components are pyrocatechol (42,50%), 4 methyl catechol (16,17%) and glyceraldehyd (32,13%), with pyrocatechol as the largest (table 1). Identification of each compound in the chromatogram was carried out by matching the MS spectrum of each peak with Wiley's database as shown in figures 1b and 1c. Then, figure 1b shows the peak spectrum of pyrocatechol with a treatment of 30% $FeSO_4$, while figure 1c is the pyrocatechol reference spectrum for Wiley's database.

The chromatogram with 25% $FeSO_4$ treatment showed the presence of four compounds, which were; pyrocatechol (11,44%), glycerol(3,75%) dan 2 Propanol, 1 Amino(81,98%) (table 1) and a chromatogram with an unknown name, as shown in figure 2a. Also, the highest chromatogram was 2 Propanol, 1 Amino, matched with the reference spectrum according to the Wiley's database at 13,517 minutes as shown in figures 2b and 2c.

IOP Conf. Series: Materials Science and Engineering **980** (2020) 012015 doi:10.1088/1757-899X/980/1/012015

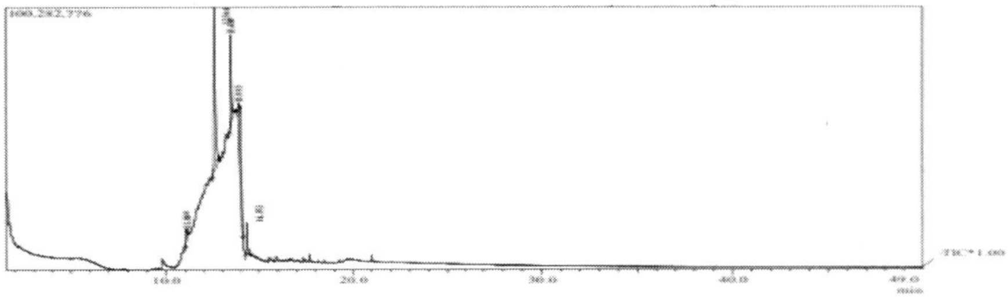

Peak Report TIC

Peak#	R.Time	Area	Conc%	Name
1	11.089	25328660	3.54	N-(METHYL-D2)-ANILINE
2	12.564	304343945	42.50	1,2-Benzenediol (CAS) Pyrocatechol
3	13.480	115776970	16.17	4 METHYL CATECHOL
4	13.933	230106091	32.13	Propanedioic acid (CAS) Malonic acid
5	14.353	40623830	5.67	1,3-Benzenediol, 4-ethyl- (CAS) 4-Ethylresorcinol
		716179496	100.00	

Figure 1a. Stamp ink chromatogram with 30% $FeSO_4$.

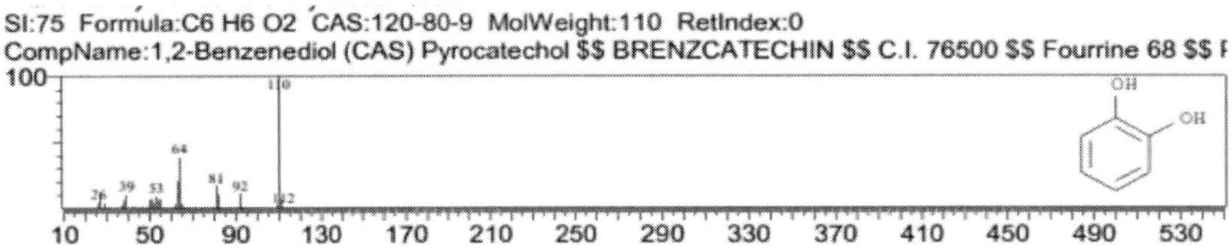

Figure 1b. Pyrocatechol peak spectrum according to Wiley's database of stamp ink spectra with 30% $FeSO_4$.

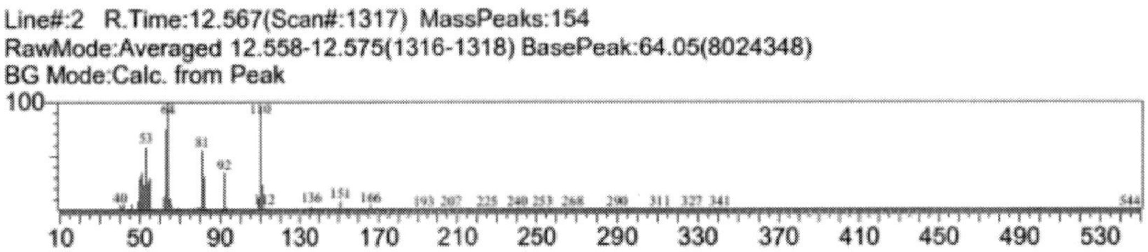

Figure 1c. Pyrocatechol reference spectrum according to Wiley's database of stamp ink spectra with 30% $FeSO_4$.

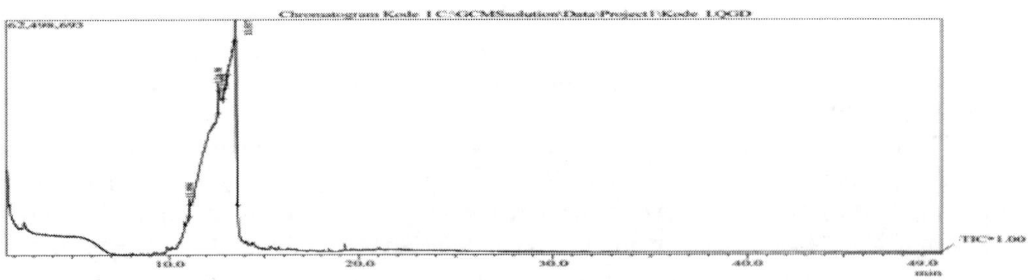

Peak Report TIC

Peak#	R.Time	Area	Conc%	Name
1	11.050	5138235	3.78	Benzenamine, N,N-dimethyl- (CAS) Dimethylaniline
2	12.630	15572600	11.44	1,2-Benzenediol (CAS) Pyrocatechol
3	12.925	3817307	2.80	
4	13.517	111578112	81.98	2-Propanol, 1-amino- (CAS) 1-Amino-2-propanol
		[illegible]	100.00	

Figure 2a. Stamp ink chromatogram with 25% $FeSO_4$.

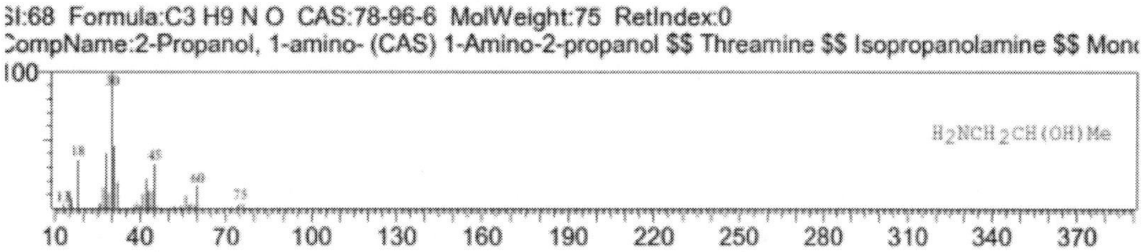

Figure 2b. Peak 2-Propanol,1-amino spectrum according to Wiley's database of stamp ink spectra with 25% $FeSO_4$.

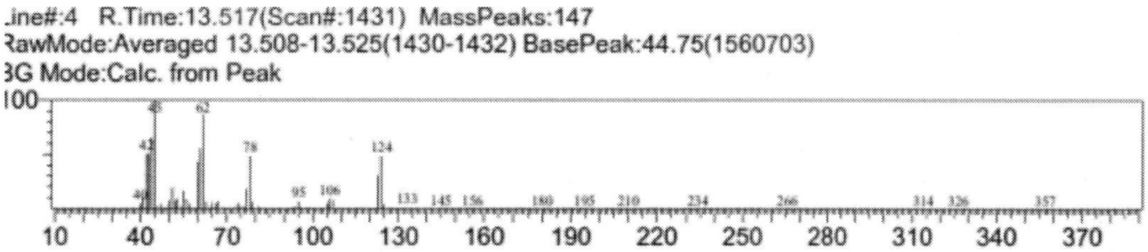

Figure 2c. The 2-Propanol,1-amino reference spectrum according to Wiley's data base of stamp ink spectra with 25% $FeSO_4$.

The peak of the chromatogram with 20% $FeSO_4$ treatment showed the presence of four compounds; pyrocatechol, glycerol, methyl catechol and methyl ester. However, the three dominant ones as shown in figure 3a are pyrocatechol (60,76%), glycerol (31,74%), and methyl catechol (6,86%) (table 1). Also, the peaks from the molecular structure analysis are shown in figures 3b and 3c. In addition, the dominant peak was pyrocatechol, matched with the reference spectrum according to Wiley's data base at a retention time of 12,620 minutes.

There were fifteen compounds in the chromatogram with 15% $FeSO_4$, which included; silanediamine, phenol, glycerol, pyrocatechol, quinoline, N-methylindole, 3-methylquinoline, 3-methylindole, 1,3-dienyl-benzene, palmitic acid, octadecanoate, benzenamine, 4-hydroxycyclohexanone (appeared twice with different retention times), and one unknown chromatogram (figure 4a). However, the three dominant ones were phenol(3,32%), glycerol(80,73%), and pyrocatechol (4,20%) (table 1), with glycerol the highest, as shown in figure 4b. This was matched with the reference spectrum according to Wiley's database at a retention time of 11,243 minutes, as shown in figure 4c.

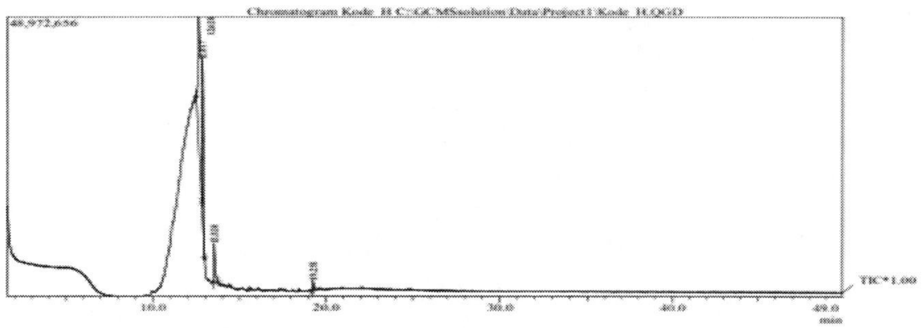

Peak Report TIC

Peak#	R.Time	Area	Conc%	Name
1	12.620	294088729	60.76	1,2-Benzenediol (CAS) Pyrocatechol
2	12.817	153639449	31.74	1,2,3,4-TETRAHYDROXYBUTANE
3	13.520	33190725	6.86	4 METHYL CATECHOL
4	19.251	3113959	0.64	10-Octadecenoic acid, methyl ester (CAS) METHYL OCTADEC-10-ENOATE
		484032862	100.00	

Figure 3a. Stamp ink chromatogram with 20% $FeSO_4$.

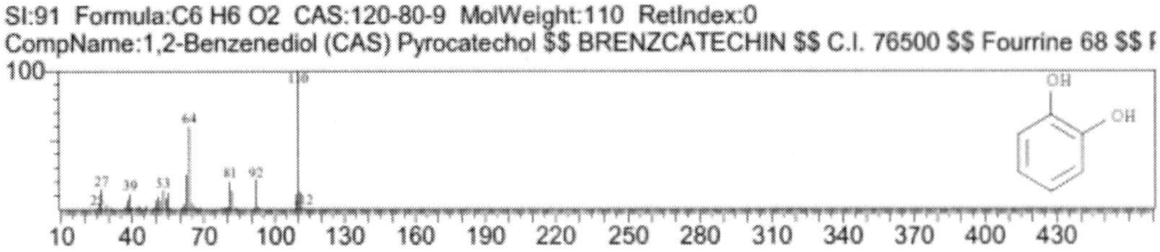

Figure 3b. Pyrocatechol Peak Spectrum according to Wiley's data base of stamp ink spectra with 20% $FeSO_4$.

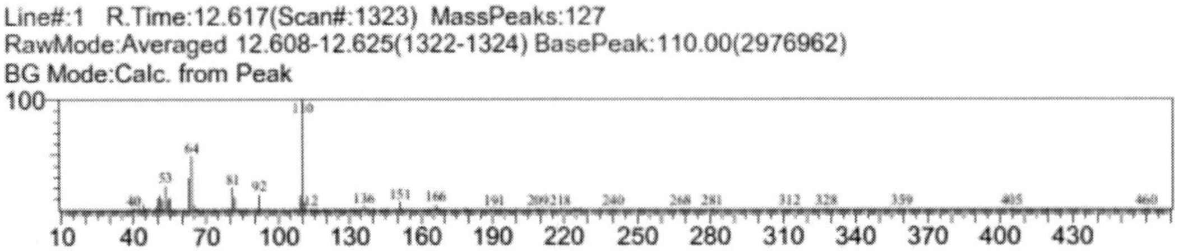

Figure 3c. Pyrocatechol reference spectrum according to Wiley's data base of stamp ink spectra with 20% $FeSO_4$.

The components that appeared in the stamp ink with the $FeSO_4$ complexed were pyrocatechol, 4 methyl catechol, 1,3 benzendiol, phenol (phenolic compounds) and glicerol, gliceraldehyde 2-propanol, 1-amine classified as alcohol compounds [30]. The function of phenolic compounds as reductants/free radical scavengers, in plant physiology played a role in the formation of color/pigmentation, causing tastes such as sweet and bitter, for growth, reproduction and for plant resistance to pathogens and predators [25,27,28]. Glycerol, gliceraldehyde (alcohol which forms a structure with an O-H group and 2 propanol, 1 amino and bound to an amine group) [26,31].

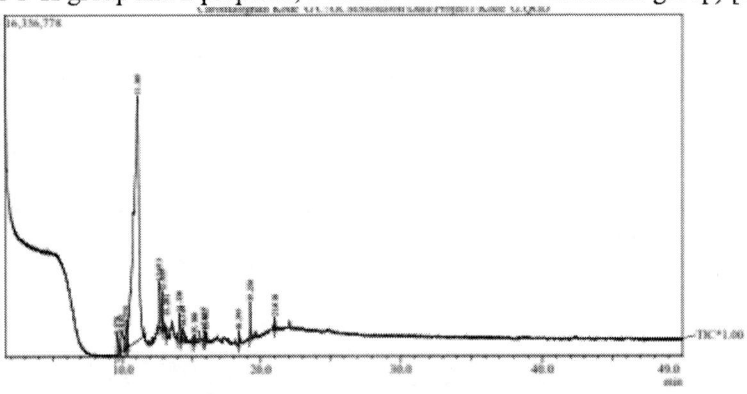

Peak Report TIC

Peak#	R.Time	Area	Conc%	Name
1	9.578	8549370	2.17	Silanediamine, 1,1-dimethyl-N,N'-diphenyl- (CAS) Bis(phenylamino)dimethyl
2	9.892	13115608	3.32	Phenol (CAS) Izal
3	10.275	9881151	2.50	
4	11.105	318726273	80.73	1,2,3-Propanetriol (CAS) Glycerol
5	12.693	16571401	4.20	1,2-Benzenediol (CAS) Pyrocatechol
6	12.935	5267787	1.33	Quinoline (CAS) B 500
7	13.201	1062065	0.27	1H-Indole, 1-methyl- (CAS) N-Methylindole
8	14.130	6822587	1.73	Quinoline, 3-methyl- (CAS) 3-Methylquinoline
9	14.369	4504981	1.14	1H-Indole, 3-methyl- (CAS) 3-Methylindole
10	15.180	1303425	0.33	(5-METHYL-HEPTA-1,3-DIENYL)-BENZENE
11	15.877	1732344	0.44	Cyclohexanone, 4-hydroxy- (CAS) 4-Hydroxycyclohexanone
12	15.983	2323224	0.59	Cyclohexanone, 4-hydroxy- (CAS) 4-Hydroxycyclohexanone
13	18.393	1198135	0.30	Hexadecanoic acid (CAS) Palmitic acid
14	19.250	2870100	0.73	METHYL 9,9-DIDEUTERO-OCTADECANOATE
15	21.018	857526	0.22	Benzenamine, 4,4'-methylenebis(N,N-dimethyl-
		394785977	100.00	

Figure 4a. Stamp ink chromatogram with 15% $FeSO_4$.

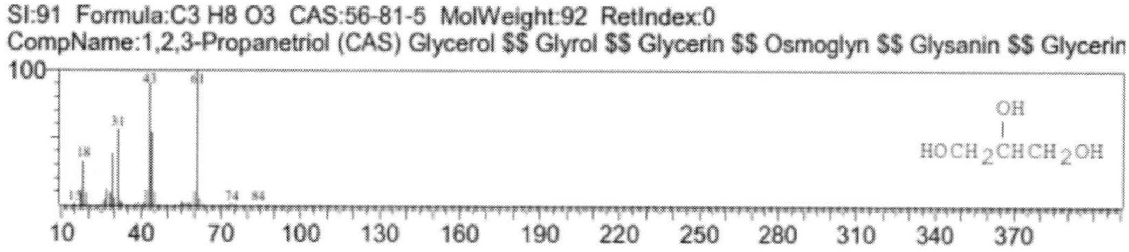

Figure 4b. The glycerol peak spectrum according to Wiley's database of stamp ink spectra with 15% $FeSO_4$.

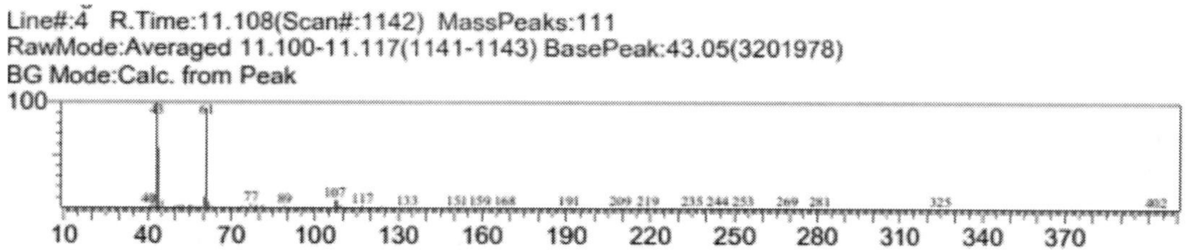

Figure 4c. The glycerol reference spectrum according to Wiley's database of stamp ink spectra with 15% $FeSO_4$

There were five compounds in the chromatograms with 10% $FeSO_4$ treatment, namely; glycerol, pyrocatechol, phenethyl alcohol, palmitic acid and methyl ester, as shown in figure 5a. However, the two dominant ones were pyrocatechol(1,63%) and glycerol(96,87%) (table 1). In terms of its molecular structure, glycerol had three hydroxyl (O-H) groups which easily absorbed water from the air [29]. The dominant peak was glycerol, matched with the reference spectrum according to Wiley's database at a retention time of 11,243 minutes. These are shown in figure 5b and 5c.

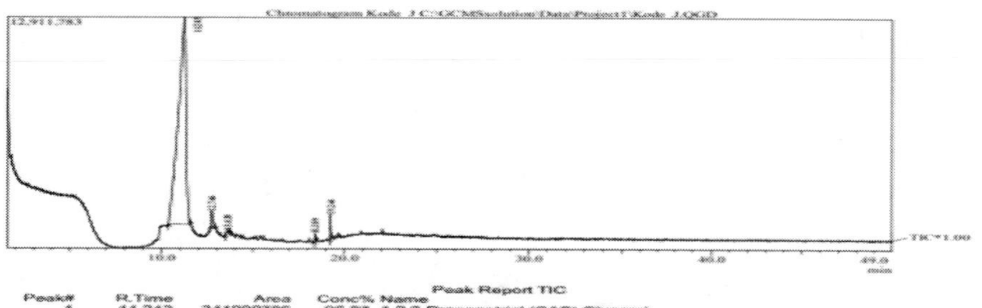

Peak Report TIC

Peak#	R.Time	Area	Conc%	Name
1	11.243	341090786	96.87	1,2,3-Propanetriol (CAS) Glycerol
2	12.744	5747173	1.63	1,2-Benzenediol (CAS) Pyrocatechol
3	13.620	1977943	0.56	Phenethyl alcohol, 2,5-dihydroxy-.alpha.-methyl-, (-)- (CAS) (-)-1-(2,5-DIHYD
4	18.389	827939	0.24	Hexadecanoic acid (CAS) Palmitic acid
5	19.244	2477591	0.70	9-Octadecenoic acid, methyl ester, (E)- (CAS) Methyl elaidate
		352121432	100.00	

Figure 5a. Stamp ink chromatogram with 10% $FeSO_4$

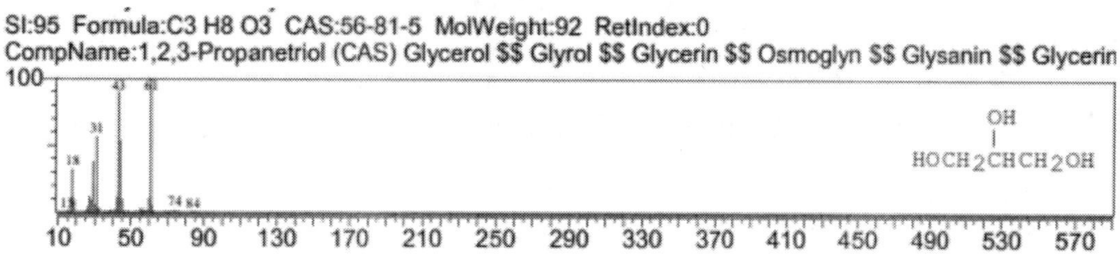

Figure 5b. The glycerol peak spectrum according to Wiley's database of stamp ink spectra with 10% $FeSO_4$

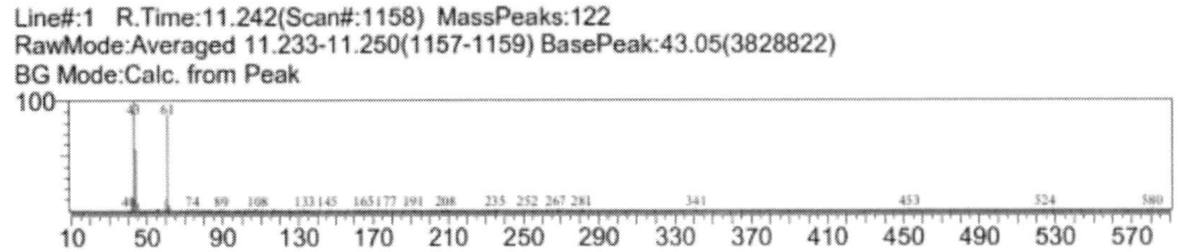

Figure 5c. The glycerol reference spectrum according to Wiley's database of stamp ink spectra with 10% $FeSO_4$

The compound identified at a retention time of about 12 minutes was pyrocatechol, also the highest compound detected in the 30% $FeSO_4$ treatment with a percentage of 42.50 %. This was followed by 60.76% in the 20% $FeSO_4$ treatment, 11.44% in the 25% $FeSO_4$ sample, 4.20% in 15% $FeSO_4$ and 1.63% in the 10% $FeSO_4$ treatment.

Also, the peak which appeared at a retention time of about 11 minutes was glycerol. However, its highest composition of 96.87% was found in the 10% $FeSO_4$ treatment. This was followed by 80.73% in the 15% $FeSO_4$, 31.74% in the 20% $FeSO_4$ and 3.75% in the 25% $FeSO_4$ treatment. In terms of its molecular structure, glycerol had three hydroxyl (O-H) groups which easily absorbed water from the air [29].

The methyl catechol appeared at a retention time of about 13 minutes, which was dominantly found in the 25% $FeSO_4$ treatment at 81.98%, and then followed by 16.17% in the 30% $FeSO_4$ and then 6.86% in the 20% $FeSO_4$. Also, phenol was detected at a retention time of about 9 minutes with 3.23% in the 15% $FeSO_4$ treatment. Usually, phenols are important compounds in plant physiology due to its role in pigmentation, taste, growth, reproduction, as well as plants' ability to resistance to pathogens and predators [25].

Table 1. The analysis results on the molecular structure of stamp ink with the use of $FeSO_4$.

No.	$FeSO_4$ Concentration (%)	RT (min)	Compound Name	Molecular Formula	Concentration (%)
1	30%	12.564	Pyrocatechol	$C_6 H_6 O_2$	42.50
		13.480	4 methyl catechol	$C_7H_8O_2$	16.17
		13.933	Gliceraldehid	$C_3 H_6 O_3$	32.13
2	25%	12.630	Pyrocatechol	$C_6 H_6 O_2$	11.44
		13.517	2 Propanol, 1-Amine	C_3H_9NO	81.98
		12.620	Glycerol	$C_6H_6O_2$	3.75
3.	20%	12.620	Pyrocatechol	$C_6H_6O_2$	60.76
		12.817	Glycerol	$C_4 H_{10} O_4$	31.74
		13.520	4 Methyl catechol	$C_7 H_8 O_2$	6.86
4	15%	9.892	Phenol	$C_6 H_6 O$	3.32
		11.105	Glicerol	$C_4 H_{10} O_4$	80.73
		12.693	Pyrocatechol	$C_6H_6O_2$	4.20
5	10%	11.243	Glycerol	$C_4 H_{10} O_4$	96.87
		12.744	Pyrocatechol	$C_6 H_6 O_2$	1.63

The dominant molecular structure of stamp ink appearing with the $FeSO_4$ complexed is shown in table 2, namely pyrocatechol, phenol, methyl catechol (classified as a flavonoid compound/O-H group). The O-H groups of phenolic compounds were reacted with metal compounds to form complexes that caused color in the ink [16,26,31]. Furthermore, there were 2-Propanol-1amine compounds, gliceraldehyd and glycerol which awee compounds that had straight-chain O-H groups were easy to absorb water [29].

Table 2. Molecular structure of the stamp ink.

Pyrocatechol	4 Methyl Catechol	Phenol
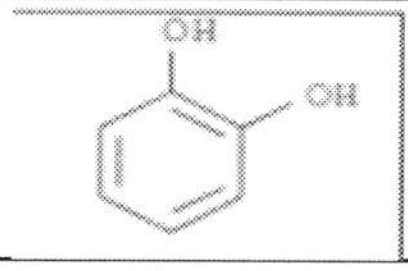	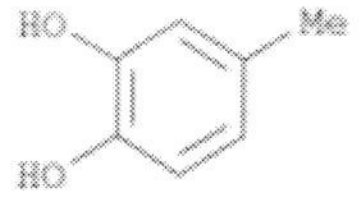	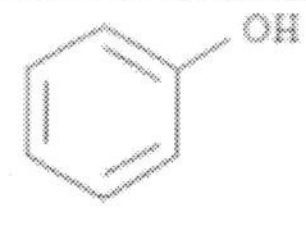
2- Propanol, 1-Amine	Glyceraldehyd	Glycerol
$H_2NCH_2CH(OH)Me$	$HOCH_2CH(OH)CHO$	OH $HOCH_2CHCH_2OH$

The printing results of stamp ink

The printing result of the stamp ink on a paper is in figure 6. The 25% $FeSO_4$ treatment, having the lowest glycerol content of 3.75%, produced the best result, in terms of quality, on the paper. Its absorption capacity was low due to the low glycerol content, there producing clearer results compared with other treatments, as shown in figure 6. Visually, the resulting color matches those on the market today, namely purple. Stamp ink from gambier with $Al_2(SO_4)_3$ complexing produced a different color, namely redish yellow with a slightly faded appearance [14].

Figure 6. The printed results from stamp ink with FeSO4 treatment.

4. Conclusion

Based on the results, the 25% $FeSO_4$ treatment was best for the stamp ink, which contained 11.44% pyrocatechol, 3.74% glycerol and 81.98% 2- propanol,1-amina. The printings on the papers were very clear and had a higher quality due to its low glycerol.

Acknowledgment

The authors are grateful to Mr. Hendri Muchtar, Hendri Yuhelminof, and Awilda Mike Pertiwi for their assistance and contributions towards the successful completion of this research.

References

[1] I. L. Kayaputri, D. M. Sumanti, M. Djali, R. Indiarto, and D. L. Dewi.2014 *Chim. Nat. Acta*, vol. **2**, no. 1, pp. 83–90..
[2] D. Frinanda, Efrizal, and R. Rahayu. 2014 *J. Biol. Univ. Andalas (J. Bio. UA.)*, vol. **3**, no. September, pp. 231–237.
[3] N. Hidayah. 2016 *J. Sain Peternak. Indones.*, vol. **11**, no. 2, pp. 89–98..
[4] F. Rosella. 2015 *J. Ilm. Mhs. Univ. Surabaya*, vol. **4**, no. 1, pp. 1–12.
[5] D. M. Physka. 2018 in skripsi pp 1-.66.
[6] I. T. Anova and H. Muchtar. 2017 *J. Litbang Ind.*, vol. **7**, no. 2, pp. 101–109.
[7] F. Failisnur, S. Sofyan, S. Silfia, S. Salmariza, and A. Ardinal. 2018 *J. Litbang Indsutri*, vol. **8**, no. 2, pp. 77–82..
[8] Failisnur, Sofyan, and W. Hermianti 2017 *J. Litbang Ind.*, vol. **7**, no. 1, pp. 19–28.
[9] Ardinal, S. Salmariza, and A. Kasim. 2014 *J. Litbang Ind.*, vol. **4**, no. 1, pp. 59–66.
[10] Ardinal and Salmariza. 2017 in *Prosiding Seminar Nasional Kulit, karet dan Plastik ke-6*, vol. **6**, pp. 23–36.
[11] I. Juliyarsi, D. Novia, and J. Helson. 2013 *J. Peternak. Indones.*, vol. **15**, no. 1, pp. 35–45.
[12] S. Sutyasmi. 2017 *Maj. Kulit, Karet, dan Plast.*, vol. **33**, no. 1, pp. 11–18.
[13] G. Yeni, K. Syamsu, E. Mardliyati, and H. Muchtar. 2017 *J. Litbang Indsutri*, vol. **7**, no. 1, pp. 1–10.
[14] S. Silfia, S. Sofyan, and F. Failisnur. 2020 *IOP Conf. Ser. Mater. Sci. Eng.*, vol. **845**, no. 1, pp. 1–8, 2020.
[15] S. Silfia, F. Failisnur, and S. Sofyan. 2018 *J. Litbang Ind.*, vol. **8**, no. 1, pp. 31–38.
[16] H. Muchtar, I. T. Anova, and Gustri Yeni. 2015 *J. Litbang Indsutri*, vol. **5**, no. 2, pp. 131–139..
[17] A.Ardinal and S. Sy. 2019 *J. Litbang Indsutri*, vol. **9**, no. 2, pp. 141–149.
[18] C. Aydemir, S. Yenidoğan, A. Karademir, and E. Arman Kandirmaz. 2018 *J. Appl. Biomater. Funct. Mater.*, vol. **16**, no. 3, pp. 137–143.
[19] K. Harismah, D. Vitasari, A. M. Fuadi, and T. Setyawan. 2016 *4th Univesity Res. Coloquium 2* pp. 526–531.
[20] Rahmad, Ehsan, R. Srimulyani, and Heryani. 2018 *J. Aktual.*, vol. **9**, no. 1, pp. 17–31.
[21] A. Rengganis, A. Yulianto, and I. Yulianti. 2017 *J. Mipa*, vol. **40**, no. 2, pp. 92–96..
[22] S. Silfia, H. Muchtar, and F. Failisnur. 2015 *J. Litbang Ind.*, vol. **5**, no. 1, p. 53.
[23] H. Muchtar, I. T. Anova, and Ardinal. 2014 *J. Litbang Indsutri*, vol. **4**, no. 2, pp. 89–96.
[24] R. D. Pangesti, E. Cahyono, and E. Kusumo. 2017 *Indones. J. Chem. Sci.*, vol. **6**, no. 3, pp. 270–278.
[25] A. A. Dahlia and Hasnawati. 2014 *J. Fitofarmaka Indones.*, vol. **1**, no. 1, pp. 24–30.
[26] Kartina and M. A. Mohammad Wahyu Agang. 2019 *J. Biota*, vol. **4**, no. 1, pp. 16–23.
[27] A. iqonil Mabruroh. 2015 in skripsi pp 1-101.
[28] S. Nindya Wulan, F. Miskah Yumna, and W. Anjas. 2018 *IJOBB*, vol. **2**, no. 1, pp. 30–34..
[29] S. H. Ningsih. 2015 in *Skripsi*, pp. 1–116.
[30] Riska Amelia Septiani. 2017 in skripsi pp 1-48.

[31] H. Wahyuni, T. Hanum, and Murhadi. 2017 *J. Teknol. Ind. Has. Pertan.*, vol. **22**, no. 1, pp. 40–51.

Properties of cellulose and modified cellulose-alginate for rifampicin drug delivery

R J D Arianto[1] and Sunardi[1,2]*

[1]Department of Chemistry, Mathematics and Natural Science Faculty, Lambung Mangkurat University, Banjarbaru 70714, Indonesia
[2]Wetland-Based Materials Research Center, Lambung Mangkurat University, Banjarbaru 70714, Indonesia

Email: sunardi@ulm.ac.id

Abstract. Oral drug delivery of rifampicin as tuberculosis healing treatment has several challenging issues such as poor solubility in water and short biological half-life resulting in some unfavorable side effects. Various polymeric materials have the prospective to overcome the obstacles correlated with rifampicin oral drug delivery to deliver controlled release and protect this drug from the severe gastric environment. In this study, alginate (Alg), alginate-cellulose (Alg-Cel), and alginate-$C_{16}TMABr$ modified cellulose (Alg-MCel) beads were prepared and evaluated as a potential agent for drug delivery of rifampicin. Numerous parameters were investigated, such as beads size, gel fraction, swelling ratio, encapsulation efficiency, and release percentage. The results showed that the addition of cellulose and $C_{16}TMABr$ modified cellulose into the alginate improves the encapsulation efficiency and controlled release of rifampicin.

1. Introduction

As the first treatment of Tuberculosis, rifampicin application is convinced to be the most effective way. It is related to anti-bacterial activity on the macrocyclic functional group of rifampicin molecules, which able to binding RNA polymerase enzyme of bacteria inhibiting the initiation of the Bacterial RNA synthesis process [1]. However, rifampicin has a short biological half-time (2-5 hours) and poor solubility in an aqueous environment [2]. Both necessitate a long-term (6-8 months) consumption of rifampicin on a high dose continuously to avoid bacterial resistance to the drug and maintain the drug concentration within a therapeutic range [3]. Furthermore, this consumption design has been resulting in some unfavorable side effects.

In attempts to decreasing the side effects, various controlled delivery systems have been developed. Drug encapsulation method by beads made from natural polymers, such as alginate, as one of the drug delivery systems has been developed numerously because of its advantages on biodegradability, biocompatibility, and non-toxic properties. Alginates is a polysaccharide consisting of β-D-mannuronic acid and α-L-guluronic acid copolymer. Alginate has a nano-porous structure and swelling ability at certain pH conditions, which suitable to use as a drug preparation with a controlled-release system [4]. However, the high number of alginate porosity conduces an attenuation of endurance and mechanical stability that leads to the increasing of the drug diffusion rate [5].

In order to enhance the mechanical properties of alginate beads, several developments have been focused on composite synthesis processes of alginate with other polymers such as chitosan [6], clay

[7], and cellulose [8]. Compared with other composite materials, cellulose is a biopolymer with the highest abundance and availability also requires a relatively low cost on the preparation [9]. Based on the report by Ooi et al., cellulose was able to enhance the mechanical properties of hydrogel up to around four times [10]. The enhancement of alginate mechanical properties by cellulose has been resulting in a more rigid structure of hydrogel and preventing drug release before the requiring time.

However, the hydrophilic nature of cellulose could weaken the interaction between beads and drugs that have a hydrophobic nature. Therefore, the surface modification of cellulose is needed to alter the cellulose into a hydrophobic material. The modification could be done with a chemical approach like a cationic surfactant, such as cetyltrimethylammonium ($C_{16}TMA^+$), adsorption at the surface of cellulose [11]. In this research, the effects of cellulose and $C_{16}TMABr$ modified cellulose addition on the preparation of alginate beads are studied regarding rifampicin encapsulation and release ability of beads.

2. Materials and methods

2.1. Materials
Sodium Alginate was purchased from Kanto Chemical (Japan). Cetyltrimethylammonium bromide ($C_{16}TMABr$), calcium chloride dihydrate, acetic acid glacial, and rifampicin were purchased from Merck (Germany). Cellulose was isolated from nypa (*Nypa fruticans*) frond by alkaline hydrogen peroxide treatment with some modification.

2.2. Cellulose modification with $C_{16}TMABr$
Cellulose was modified by suspending 1 g of cellulose into 250 mL of water (0,4 wt % suspension). The suspension was slowly added into 4.0 mM $C_{16}TMABr$ solution, and the mixture was stirred at 60°C for 3 hours. The process was continued by stirring at room temperature overnight. The suspension obtained was filtered by Whatman No. 42 Filter paper, then the residue was washed with distilled water to remove bromide ion and dried to a constant weight at 65°C for 20 hours.

2.3. Beads preparation
Beads were prepared with three different compositions, which were alginate beads (Alg), alginate-cellulose beads (Alg-Cel), and alginate-modified cellulose bead (Alg-MCel). The staple solution of beads was prepared by adding 0.1 g of cellulose or modified cellulose into 20 mL of 3% sodium alginate solution and stirred at room temperature for 1 hour. Then, 20 mL of 0.5 mg/mL rifampicin solution was added to the mixture under stirring for 10 minutes. The homogenous mixture then added dropwise into a 1.5% solution of calcium chloride ($CaCl_2$) through a syringe device. The spherical beads obtained were cured for 1 hour, then washed and dried at 45°C for 15 hours.

2.4. Characterization
2.4.1. FTIR Spectroscopy. The structural changes of cellulose modification products were analyzed as FTIR spectra using a Bruker Alpha ATR/FTIR. The spectra were recorded in the range of 4000–500 cm^{-1}.

2.4.2. Beads size measurement. The size of dried alginate beads from each formula was measured as the average value of 25 beads' size using a digital caliper.

2.4.3. Gels fraction. The dried beads were first extracted in 5 % acetic acid solution using autoclave for 2 hours at 120°C and 1 bar pressure. After this, the extracted beads were dried at 45°C to a constant weight. The gel fraction (G%) than was determined as the ratio of the dry gel before (W_i) and after (W_F), using the equation 1.

$$G\% = \frac{W_f}{W_i} \times 100\% \quad (1)$$

2.4.4. Swelling behavior. The swelling behavior was studied by suspending 40 mg (W_d) dried alginate-cellulose beads in 4 mL simulated gastric fluid (SGF) pH 1.2 for 2 hours followed by suspension in 4 mL simulated intestinal fluid (SIF) pH 6.8 during 4 hours. The weight of beads was measured at one-hour intervals after removing the liquid from the surface of the beads using filter paper (W_s). This analysis was done in triplicate. The swelling ratio of beads (S%) was calculated using the equation 2.

$$S\% = \frac{W_S - W_d}{W_d} \times 100\% \quad (2)$$

2.4.5. Encapsulation efficiency. The concentration of rifampicin before (W_a) and after (W_b) beads gelation process was determined by measuring the absorbance of rifampicin that dissolved at $CaCl_2$ solution after the gelation process using UV-Vis spectrophotometer (Jenway 7315 Model) at 475 nm (Thomas *et al.*, 2018). The method for rifampicin concentration determination used calibration curve arranged from a series of solutions with a known concentration of rifampicin. The encapsulation efficiency value (EE%) then calculated using the equation 3.

$$EE\% = \frac{W_a - W_b}{W_b} \times 100\% \quad (3)$$

2.4.6. In-Vitro release of rifampicin. Thirty milligrams (30 mg) of rifampicin loaded beads were placed into a dialysis bag with MW cut off of 12,000 Da. Then, the beads were soaked into 9 mL of release medium at 37°C under stirring. The first release medium was SGF for 2 hours, followed by SIF for 4 hours. At one-hour intervals, 3 mL of medium was withdrawn for analysis and replaced by the fresh medium with an equal volume. The amount of rifampicin in the release medium was calculated using spectrophotometry method at 475 nm by UV-Vis spectrophotometer (Jenway 7315 Model). This analysis was done in duplicate.

3. Results and Discussion

3.1. Cellulose and modified cellulose analysis using FTIR spectroscopy
The FTIR Spectra analyzed by FTIR instrument was used to investigate the structural changes of cellulose due to the surface modification with $C_{16}TMABr$. Figure 1 shows the spectra of unmodified cellulose and modified cellulose. A new modest peak at 1455 cm^{-1} was found in modified cellulose, which corresponded to the trimethyl group of the quaternary ammonium [12]. This peak proved the appearance of quaternary ammonium in cellulose as the surfactant is a quaternary ammonium salt. There are also two small peaks at 2897 & 2858 that correspond to asymmetrical and symmetrical CH_2 stretches of the long alkyl chain as an attribute of $C_{16}TMABr$ surfactant [13]. Moreover, the peak intensity at 1052 cm^{-1} had a slight increase as an indication of the long alkyl chain attached to the hydroxyl groups of cellulose [14].

However, the low amount of changes that were shown in spectra indicated the poor condition of $C_{16}TMABr$ adsorption on the cellulose surfaces (figure 2). Based on the report by Zainuddin et al. [16], $C_{16}TMABr$'s groups depicted by FTIR spectra might show a low intensity as the concentration of $C_{16}TMABr$ used was lower or even higher than the concentration needed to construct the material with the surfactant critical micelle concentration on its surfaces. At the higher concentration, the aggregation between surfactant molecules on the cellulose surface could be conducted. This condition was caused by the increase of surface hydrophobicity, which led to the adsorption of excess surfactant through the hydrophobic interaction between surfactants [15]. The aggregated condition of surfactant may lead to the desorption of C_{16}TMABr from the cellulose surfaces during the rinsing step [16].

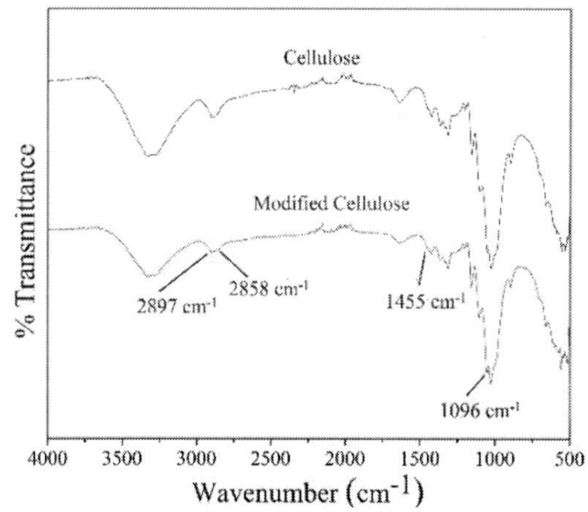

Figure 1. FTIR spectra of unmodified and modified cellulose.

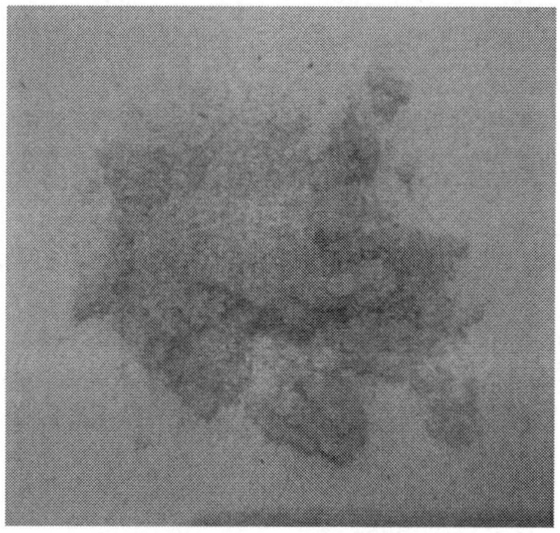

Figure 2. C_{16}TMABr modified cellulose product.

3.2. Morphology and size

All beads obtained were orange as the color of rifampicin as shown by figure 3.Alg-Cel and Alg-MCel beads were darker than Alg beads because of the existence of cellulose which was increasing the crosslinking density. Based on the data in table 1, the diameter of Alg-Cel and Alg-MCel beads was considerably bigger than Alg beads. Overall, the beads from each formula were relatively homogenous in size, as shown by the coefficient of variation between standard deviation and mean size that was less than 0.33 [6].

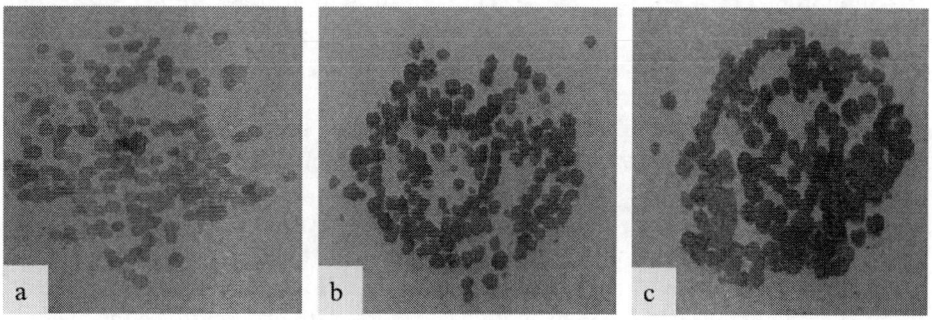

Figure 3. The photographs of Alg beads (a), Alg-Cel beads (b), & Alg-MCel beads (c).

3.3. Gel fraction

The result of the gel fraction analysis of the beads is shown in table 1. The gel fraction of the beads represented the amount of insoluble part of the polymer due to the formation of crosslinking between the polymer chains [17]. The obtained result showed that the presence of cellulose affected the gel fraction of the beads. The cellulose conduced the amount of crosslinking that was formed through hydrogen bonding interaction between alginate and cellulose molecules. These interactions could enhance the chemical resistance and stability of the beads in the acidic environment

However, the gel fraction of Alg-MCel beads was slightly lower than Alg-Cel beads. The adsorption of C_{16}TMABr surfactant by hydroxyl groups as anionic sites on the cellulose surface led to the decline of Alg-MCel beads gel fraction. The presence of surfactant might disrupt the hydrogen

IOP Conf. Series: Materials Science and Engineering **980** (2020) 012016 doi:10.1088/1757-899X/980/1/012016

bonding interaction between cellulose and alginate materials. It led to the reduction of crosslinking formation and lowering the amount of insoluble fraction.

Table 1. Characteristic of beads with three different formulation.

Beads Formulation	Average size	Gel Fraction	Encapsulation efficiency
Alg	0,89 ± 0,05 mm	30 %	47,37 %
Alg/Cel	1,13 ± 0,09 mm	36 %	61,20 %
Alg/Mcel	1.31 ± 0,07 mm	32 %	60,83 %

3.4. Encapsulation efficiency

The presence of cellulose in the beads provided a significant effect in the drug encapsulation, as shown in table 1. The encapsulation value of Alg/Cel and Alg/MCel beads attained over 60 % of total rifampicin that was added, which is higher than Alg beads that just attained 47,374 %. The encapsulation efficiency was calculated as the ratio between the weight of the encapsulated drug and the drug fed. The appearance of free rifampicin in $CaCl_2$ solution was occurred due to the diffusion of the drug while the gelation process of alginate that was forming pores on the surface of the beads [18]. The appearance of cellulose seemed to influence the amount of non-encapsulated drugs as cellulose was able to enhance the structural stability of the beads. Thus the pores formed on the surface of the beads could be decreased.

Meanwhile, the encapsulation efficiency of Alg/MCel Beads was relatively constant compared to Alg/Cel beads. These results might happen because of the presence of surfactant on the cellulose surface that was higher than the surfactant critical micelle concentration [16]. The presence of surfactant on the surfactant critical micelle concentration would increase the hydrophobicity of the cellulose surfaces. While at the higher concentration, above its critical micelle concentration, the surfactant might aggregate and formed positive hydrophilic sites on the cellulose surfaces [15]. The formation of hydrophilic sites led to the decrease of rifampicin binding on the cellulose surface.

3.5. Swelling behavior

The swelling ratios of the beads on the SGF and SIF medium are shown in figure 4. There was a soar for all variation from 135 – 210 % in the second hour of SGF treatment to more than 2000 % in the first hour of SIF treatment. This increase showed the swelling behavior of the alginate beads was depended on the pH condition of medium.

The swelling in SGF was occurred due to the absorption of water into the beads. However, at low pH value, alginate beads were able to maintain the structure because of the change of carboxylate groups (-COO) of alginate that was protonated and generating alginate acid, thus strong hydrogen bonding interactions between the carboxylate group of alginate molecules was formed [19]. The swelling ratio values of Alg, Alg-Cel, Alg-MCel beads in SGF were up to 134,96%, 126,09, and 209,76%, respectively. The swelling ratio of Alg-Cel beads that was the lowest showed that cellulose enhances the network structure of alginate in acidic medium. However, the swelling ratio of Alg-MCel that was the highest might occur due to the presence of surfactant that interfered the hydrogen bonding interactions of cellulose and loosened up the polymer composite structure.

Meanwhile, the significant rise of the swelling ratio of the beads in SIF was resulted by the partial ionization of carboxylate groups of alginate molecules along with the ionic exchange between the Ca^{2+} ion as the crosslinking agent of alginate structure and the Na^{+} ion present in SIF. Both led to the partial degradation of the beads due to sodium alginate salt formation, which is soluble in water and could cause the release of the drug [8]. The swelling ratio values of Alg, Alg-Cel, Alg-MCel beads in SGF were up to 134,96%, 126,09, and 209,76%, respectively. The ratios of Alg-Cel and Alg-MCel in this medium were relatively lower than Alg beads, which showed that the appearance of cellulose might able to enhance the chemical stability of the beads.

IOP Conf. Series: Materials Science and Engineering **980** (2020) 012016 doi:10.1088/1757-899X/980/1/012016

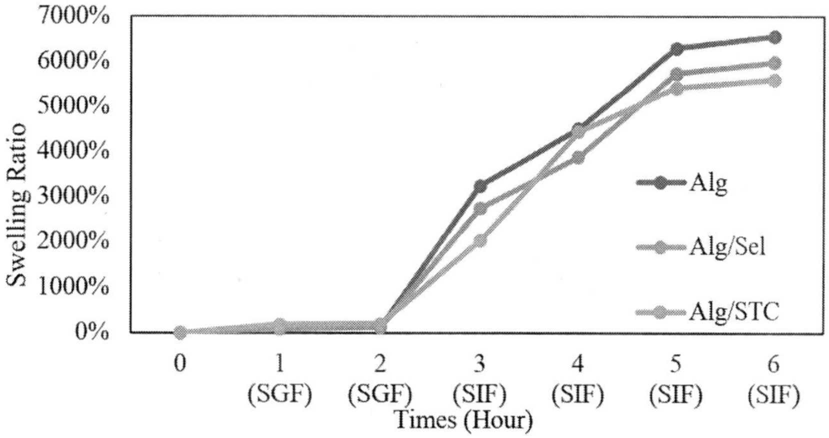

Figure 4. The swelling ratio of the beads in SGF and SIF.

3.6. In-vitro release of rifampicin

The rifampicin percentage value of the beads is shown in figure 5. The number of drugs found in the release medium could depict of drug release behavior in the gastrointestinal tract. The dissolution rate of drug, which is a kinetic property of the drugs, could be controlled by the diffusion process of the drug [20]. That is, the drug delivery system might control the dissolution rate through the control of drug diffusion.

In the SGF medium, the drug release process occurred because of bead swelling, which led to drug diffusion into the SGF solution. The release percentage values of Alg, Alg-Cel, Alg-MCel beads in SGF were up to 3,22%; 1,23%; and 1,57%, respectively. The release percentage value of Alg-Cel in SGF was the lowest among the others, which was appropriate with the swelling ratio of Alg-Cel in SGF that was lowest compared to the other beads. Meanwhile, the release percentage value of Alg/MCel beads was considerably lower than Alg beads, which showed the discrepancy with the swelling ratio value of both beads. This result indicated the effect of cellulose modification by the surfactant, which could maintain the drug binding in the drug system due to the increase of hydrophobicity on cellulose surface [16].

The drug release into solution also occurred in SIF with the release percentage value of Alg, Alg-Cel, Alg-MCel beads in SGF were up to 4,85%; 2,60%; and 3,34%, respectively. The release percentage value of Alg beads was the highest among the others, which was appropriate with the swelling ratio result. Generally, the addition of cellulose into alginate beads decreased the release percentage of rifampicin in both of release medium. These declines had resulted in the presence of cellulose that enhanced the chemical stability of the beads. Thus, the swelling of the beads was decreased, and the diffusion of the drug from the system was controlled. However, the results of Alg/MCel beads which did not show the increase of release percentage could be led by the set on the hydrophobic site formation process was not in optimum condition. Therefore, the beads ability to maintain drugs during the diffusion process decreased. Overall, the lower result in acidic medium and the higher result for all beads in pH-neutral medium indicated the ability of the system to protect drug in the gastric environment as well as the desirable release behavior in the intestinal fluid [21], which was the main criteria for oral drug delivery system.

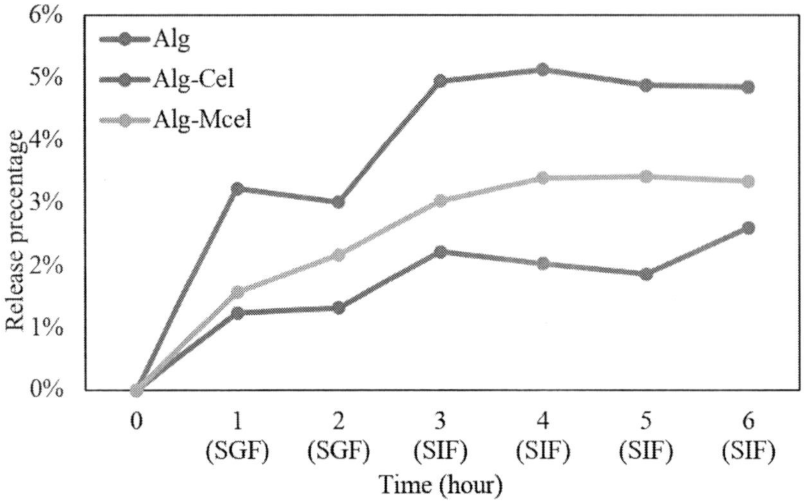

Figure 5. The drug release percentages of the beads in SGF and SIF.

4. Conclusion

In this study, alginate beads with three different formulations of cellulose addition were prepared and characterized. Based on the results, the addition of cellulose and modified cellulose was revealed that it induced the improvement in bead properties such as beads size, gel fraction, swelling behavior, encapsulation efficiency, and drug release behavior. The addition of modified cellulose provided the best results in the size and swelling ratio of the beads. However, the gel fraction, encapsulation efficiency, and release percentage value of Alg/MCel beads showed relatively homogenous results compared to Alg/Cel beads. These results indicated that Alg-Cel beads & Alg-MCel beads had a prospect to provide a controlled release of rifampicin oral delivery system.

Acknowledgement

The authors would like to say thank to Biomaterials Chemistry Laboratory, Faculty of Mathematics and Natural Sciences, Lambung Mangkurat University for the laboratory facilities and Ministry of Research, and Technology Republic of Indonesia for the financial support of Fundamental Research Grant 2019–2020.

References

[1] Singh G, Raghuvanshi H K, Anand A, Pundir R, and Dwivedi H 2010 Targeted Delivery of Rifampicin by Niosomal Drug Delivery System *JPR* **3** 1152-54

[2] Sabitha P, Ratna J V, and Reddy K R 2010 Design and Evaluation of Controlled Release Chitosan-Calcium Alginate Microcapsules of Antitubercular Drugs for Oral Use *Int. J. Chem. Technol. Res.* **2** 88-98

[3] Kasim N A, Whitehouse M, Ramachandran C, Bermejo M, Lennernäs H, Hussain A S, Junginger H E, Stavchansky S A, Midha K K, and Shah V P 2004 Molecular Properties of Who Essential Drugs and Provisional Biopharmaceutical Classification *Molecular pharmaceutics* **1** 85-96

[4] Lin N, Huang J, Chang P R, Feng L, and Yu J 2011 Effect of Polysaccharide Nanocrystals on Structure, Properties, and Drug Release Kinetics of Alginate-Based Microspheres *Colloids and Surfaces B: Biointerfaces* **85** 270-79

[5] Moebus K, Siepmann J, and Bodmeier R 2009 Alginate–Poloxamer Microparticles for Controlled Drug Delivery to Mucosal Tissue. *European Journal of Pharmaceutics and Biopharmaceutics* **72** 42-53

IOP Conf. Series: Materials Science and Engineering **980** (2020) 012016 doi:10.1088/1757-899X/980/1/012016

[6] Tahtat D, Mahlous M, Benamer S, Khodja A N, Oussedik-Oumehdi H, and Laraba-Djebari F 2013 Oral Delivery of Insulin from Alginate/Chitosan Crosslinked by Glutaraldehyde *International journal of biological macromolecules* **58** 160-68
[7] Uyar G, Kaygusuz H, and Erim F B 2016 Methylene Blue Removal by Alginate–Clay Quasi-Cryogel Beads *Reactive and Functional Polymers* **106** 1-7
[8] Thomas D, Latha M, and Thomas K K 2018 Synthesis and in Vitro Evaluation of Alginate-Cellulose Nanocrystal Hybrid Nanoparticles for the Controlled Oral Delivery of Rifampicin *Journal of Drug Delivery Science and Technology* **46** 392-99
[9] Yu X, Tong S, Ge M, Wu L, Zuo J, Cao C, and Song W 2013 Adsorption of Heavy Metal Ions from Aqueous Solution by Carboxylated Cellulose Nanocrystals *Journal of Environmental Sciences* **25** 933-43
[10] Ooi S Y, Ahmad I, and Amin M C I M 2016 Cellulose Nanocrystals Extracted from Rice Husks as a Reinforcing Material in Gelatin Hydrogels for Use in Controlled Drug Delivery Systems *Industrial Crops and Products* **93** 227-34
[11] Qing W, Wang Y, Wang Y, Zhao D, Liu X, and Zhu J 2016 The Modified Nanocrystalline Cellulose for Hydrophobic Drug Delivery *Applied Surface Science* **366** 404-09
[12] Abitbol T, Marway H, and Cranston E D 2014 Surface Modification of Cellulose Nanocrystals with Cetyltrimethylammonium Bromide *Nordic Pulp & Paper Research Journal* **29** 46-57
[13] Salajková M, Berglund L A, and Zhou Q 2012 Hydrophobic Cellulose Nanocrystals Modified with Quaternary Ammonium Salts *Journal of Materials Chemistry* **22** 19798-805
[14] Song Y, Zhang L, Gan W, Zhou J, and Zhang L 2011 Self-Assembled Micelles Based on Hydrophobically Modified Quaternized Cellulose for Drug Delivery *Colloids and Surfaces B: Biointerfaces* **83** 313-20
[15] Hu Z, Ballinger S, Pelton R, and Cranston E D 2015 Surfactant-Enhanced Cellulose Nanocrystal Pickering Emulsions *Journal of colloid and interface science* **439** 139-48
[16] Zainuddin N, Ahmad I, Kargarzadeh H, and Ramli S 2017 Hydrophobic Kenaf Nanocrystalline Cellulose for the Binding of Curcumin *Carbohydrate polymers* **163** 261-69
[17] Mozalewska W, Czechowska-Biskup R, Olejnik A K, Wach R A, Ulański P, and Rosiak J M 2017 Chitosan-Containing Hydrogel Wound Dressings Prepared by Radiation Technique *Radiation Physics and Chemistry* **134** 1-7
[18] Sankalia M G, Mashru R C, Sankalia J M, and Sutariya V B 2005 Papain Entrapment in Alginate Beads for Stability Improvement and Site-Specific Delivery: Physicochemical Characterization and Factorial Optimization Using Neural Network Modeling *Aaps pharmscitech* **6** E209-E22
[19] Deepathomas D, Latha M, and Kurienthomas K 2018 Zinc Alginate Beads for the Controlled Release of Rifampicin *Orient. J. Chem* **34** 428-33
[20] Siegel R A, and Rathbone M J 2012 Fundamentals and Applications of Controlled Release Drug Delivery (Vol. 3) ed J Siepmann, *et al* (New York: Springer) p 30
[21] Thomas D, Latha M, and Thomas K K 2017 Synthesis and in Vitro Evaluation of Iron Cross Linked Alginate Nano Particle for Controlled Drug Release *Lung Pulm Respir Res J* **1** 111-17.

IOP Conf. Series: Materials Science and Engineering **980** (2020) 012017 doi:10.1088/1757-899X/980/1/012017

Sodium alginate-chitosan nanocomposite as a novel carrier agent for cinnamaldehyde: characterisation and release studies

S T S Wong*, A Kamari and J Jumadi

Department of Chemistry, Faculty of Science and Mathematics, Universiti Pendidikan Sultan Idris, 35900 Tanjong Malim, Perak, Malaysia.

Email: susanawongsiewtin@hotmail.com

Abstract. The ultimate aim of this study was to synthesize sodium alginate-chitosan (SA-Chi) nanocomposites for the first time in larvicide formulation. The physicochemical properties of the nanocomposites were characterized by Fourier transform infrared (FTIR) spectroscopy, X-ray diffraction (XRD) spectroscopy, thermogravimetric analysis (TGA) and differential scanning calorimetry (DSC). The feasibility of SA-Chi nanocomposites to encapsulate and load cinnamaldehyde, a common essential oil used against mosquitoes larvae, was evaluated in terms of encapsulation efficiency (EE) and loading capacity (LC). There were three mass ratios of SA-Chi nanocomposites to cinnamaldehyde, namely 50:1,100:1 and 125:1. Research findings suggested that the mass ratio of nanocomposites to drugs, significantly affected the functional groups availability, crystallinity property and thermal stability of nanocomposites. The EE and LC of cinnamaldehyde-loaded nanocomposites at 125:1 were determined as 91.33% and 75.25%, respectively. The release profile data of the nanocomposite were well fitted to Korsmeyer-Peppas kinetic model with correlation coefficient (R^2) of 0.9231 for the mass ratio of cinnamaldehyde-loaded nanocomposites at 125:1. The cinnamaldehyde-loaded nanocomposites at 125:1 had prolonged the drug release for 24 hours. Overall, results from this study highlight the potential of SA-Chi nanocomposites as novel nanocarriers to deliver cinnamaldehyde in mosquito larvicide formulations.

1. Introduction

When technology development induces a new innovation of materials, a drug carrier system is one of the most attentive products. Drug carrier systems are able to encapsulate drugs in order to enhance drug delivery process to targeted sites [1]. However, the amount of drug carrier systems has to be increased if the system is unable to encapsulate a fixed quantity of drugs. Therefore, the nanotechnology becomes a brand new innovative direction of the drug carrier system. The recent common materials for drug carrier systems are mostly from natural polymers, lipid, and polysaccharides, or synthetic polymers. A lot of studies have proved the unusual behaviour of atoms and molecules when the products are invented in nanoscale [2]. Large surface area to volume ratio, tuneable surface, better strength, and improved chemical reactivity are the benefits of using nanoscale in drug carrier systems [3-5]. Hence, the utilisation of nanosized drug carrier system is highly concerned by researchers and scientists in order to achieve effective drug delivery in the shortest time with the lowest possible drug dosage.

Hydrogel nanocomposites is the most effective drug carrier system in the water environment because of its good interaction with the environment and this could intensify the effectiveness of the drug delivery process [6]. In this study, sodium alginate (SA) and chitosan (Chi) were chosen to be the drug carrier system for cinnamaldehyde in the water environment. SA is the sodium salt of alginic acid and it is mainly from the cell wall of brown algae which are from class *Phaeophyceae.* In a textile printing industry, SA has been used as a stabilizer or an emulsifier because it contains chains of mannuronic acid and guluronic acid [7]. Other than that, SA has been used widely in wound dressing and adsorbent materials because of its non-toxic and non-inflammatory [8]. The liquid-gel behaviour of SA allows more reactive sites for drugs to be delivered in the water environment. Chi is a linear polysaccharide which has deacetylated D-glucosamine and *N*-acetyl-D-glucosamine [9]. Chi can be obtained from hard shells of shellfish, and it has been an ideal material in drug delivery due to its properties of biodegradability, biocompatibility, and high membrane permeability [10]. This is because Chi has primary amino groups that enable controlled drug release, mucoadhesion, and in situ gelation [11]. Chi is also being applied widely in industry food, agriculture, textile, and paper.

There are various plant extracts have been studied few years back and essential oils (EO) is the topic the most studied by the researchers. EOs are the concentrated, volatile, aromatic compound of a plant and they have been extracted through steam distillation. There are a lot of studies related to applications of EO in various fields have been reported in recent years. According to Seenivasan et al., cinnamaldehyde has been applied against larvicidal activity of *Culex* mosquitoes [12]. However, most of the EOs has low water solubility. In this study, EO from bark of cinnamon tree has been chosen to be encapsulated in the carrier system. Cinnamon tree is from *Lauraceae* family and it is originated in Sri Lanka, India, and Myanmar then being cultivated in South America and the West Indies [13]. Cinnamaldehyde is the EO distilled from the bark fragments and it has been used in food, liquor, perfumes and drugs [14]. Not only those, cinammaldehyde has been chosen as a natural larvicide formulation because it is effective against bacteria, microbes and fungi [15]. A drug carrier system is required for cinnamaldehyde to perform the larvicidal effects in water environment due to its low water solubility (1420 mg/L at 25 °C). Knockdown effects of cinnamaldehyde could not reach mosquito larvae because they are living beneath water surface. Therefore, a drug carrier system is required to deliver cinnamaldehyde in water environment in order to reduce the number of larvae.

Drug delivery system is designed to carry drug in possible amount and deliver drug in the shortest period. In this study, attention is highly paid towards biodegradable polymer materials. Formation of SA-Chi is by ionic linkage between carboxyl groups of sodium alginate and amino groups of Chi after sodium alginate is induced gelation with calcium ions [16]. In this stage, formation of gel aggregates occurs and this interaction has been favoured effective drug encapsulation and drug delivery [17,18]. Thus, cinnamaldehyde can be encapsulated successfully and can be released in higher effective rate with SA-Chi nanocomposites compared with SA or Chi alone. The novelty of this study is the application of SA-Chi nanocomposites in larvicide formulations. In this system, the period of mosquito control could be extended and with high effectiveness.

2. Materials and methods

2.1 Materials

Chi (MW 600,000-800,000, degree of deacetylation ≥75%) and Tween 20 were purchased from Acros Organic. Food grade SA powders which are comprised of equal amount of mannuronic acid and guluronic acid units and calcium chloride were purchased online. Glacial acetic acid was purchased from ChemAR. Cinnamaldehyde (min. 99%) was purchased from ShangHai Kaiyi Chemical Co.

2.2 Synthesis

This method was proposed by Loquercio *et al.* [19] with some modifications. 4 µL of cinnamaldehyde was added in 20 mL of 20 ppm acetonenitrile. 2.0 mL of the mixture solution was added into a beaker that contained 250 mL of 0.06 % (w/v) sodium alginate solution and 1.5 mL of Tween 20. Tween 20 was added as an emulsifier in order to disperse hydrophobic particles. The solution was stirred for 90 min and sonicated for 15 min (5 mins × 3 cycles). 15 mL of 0.2 % (w/v) calcium chloride was added to the mixture solution, stirred, and sonicated as described earlier. 0.5 g of Chi was dissolved in 200 mL of 1 % (v/v) acetic acid and then it was added to the mixture. The mixture solution was stirred and

IOP Conf. Series: Materials Science and Engineering **980** (2020) 012017 doi:10.1088/1757-899X/980/1/012017

sonicated to increase the uniformity and dispersity of particles. The nanoparticle solution was centrifuged at 4000 rpm for 15 min. A low speed filter paper was used to remove any excess polymer. Retentate was frozen and lyophilised. The nanocomposite obtained was kept in an air-tight container. The volume of cinnamaldehyde was calculated accordingly based on the ratio of nanocomposites:drugs, namely 50:1, 100:1 and 125:1.

2.3 Characterisation

FTIR spectra of precursors and the nanocomposites were recorded on a Shimadzu IRTracer-100 FTIR spectrophotometer in the wavenumber ranged between 4000 to 400 cm^{-1} with over 32 cumulative scans to identify the functional groups and chemical bonds between molecules. XRD diffractograms of the samples were obtained using a Rigaku Miniflex II diffractometer equipped with a rotating anode and a CuKα source ($\lambda = 0.1504$ nm) for analyse the crystallinity of the nanocomposites. To evaluate the thermal properties of the nanocomposites, TGA and DSC analysis were performed under argon atmosphere at a heating rate of 10 °C/min and temperature range of 25-1000 °C using a TGA/DSC 1 Mettler Toledo analyser.

2.4 Encapsulation efficiency and loading capacity

5 mg of nanocomposite was dissolved in 5 mL of 95 % (w/v) acetonitrile and this was conducted in triplicates. The solution was left in the dark environment for 48 h for allowing cinnamaldehyde to be encapsulated into the nanocomposites. The solution was vortexed for 10 s and being filtered. The concentration of encapsulated cinnamaldehyde was measured using an Agilent Cary 60 Ultraviolet-Visible spectrophotometer at 280 nm. The encapsulation efficiencies and loading capacities were determined according to the following equations 1 and 2 [20,21].

$$\text{Encapsulation efficiency} = \frac{\text{Amount of encapsulated cinnamaldehyde}}{\text{Total amount of cinnamaldehyde}} \times 100\ \% \qquad (1)$$

$$\text{Loading capacity} = \frac{\text{Amount of encapsulated cinnamaldehyde}}{\text{Mass of nanocomposite}} \times 100\ \% \qquad (2)$$

2.5 In vitro release profile of the nanocomposite and fitted release kinetic models

The in vitro release of cinnamaldehyde from nanocomposite was conducted by separating cinnamaldehyde-loaded SA-Chi nanocomposite from the aqueous nanoparticulate suspension medium through ultra-centrifugation. Cinnamaldehyde-loaded nanocomposite (20 mg) was then placed in a dialysis membrane bag with a molecular cut-off of 5 kDa, tied and placed into 200 mL of deionised water at pH 7.0. The entire system was kept at 25 ± 0.5 °C with continuous magnetic stirring (100 rpm). At certain time intervals, 20 mL of the release medium was taken and 20 mL of deionised water was added into the system to maintain the sink conditions. The amount of cinnamaldehyde released from the nanocomposite was measured by using an Agilent Cary 60 Ultraviolet-Visible spectrophotometer.

The release mechanisms of cinnamaldehyde from the nanocomposites were studied using zero order, first order, Higuchi and Korsmeyer-Peppas kinetic models. The correlation coefficient (R^2) and release exponent (n) of Korsmeyer-Peppas model were also determined by regression analysis, as described by Ritger and Peppas [22].

3. Results and discussions

3.1 Characterisation

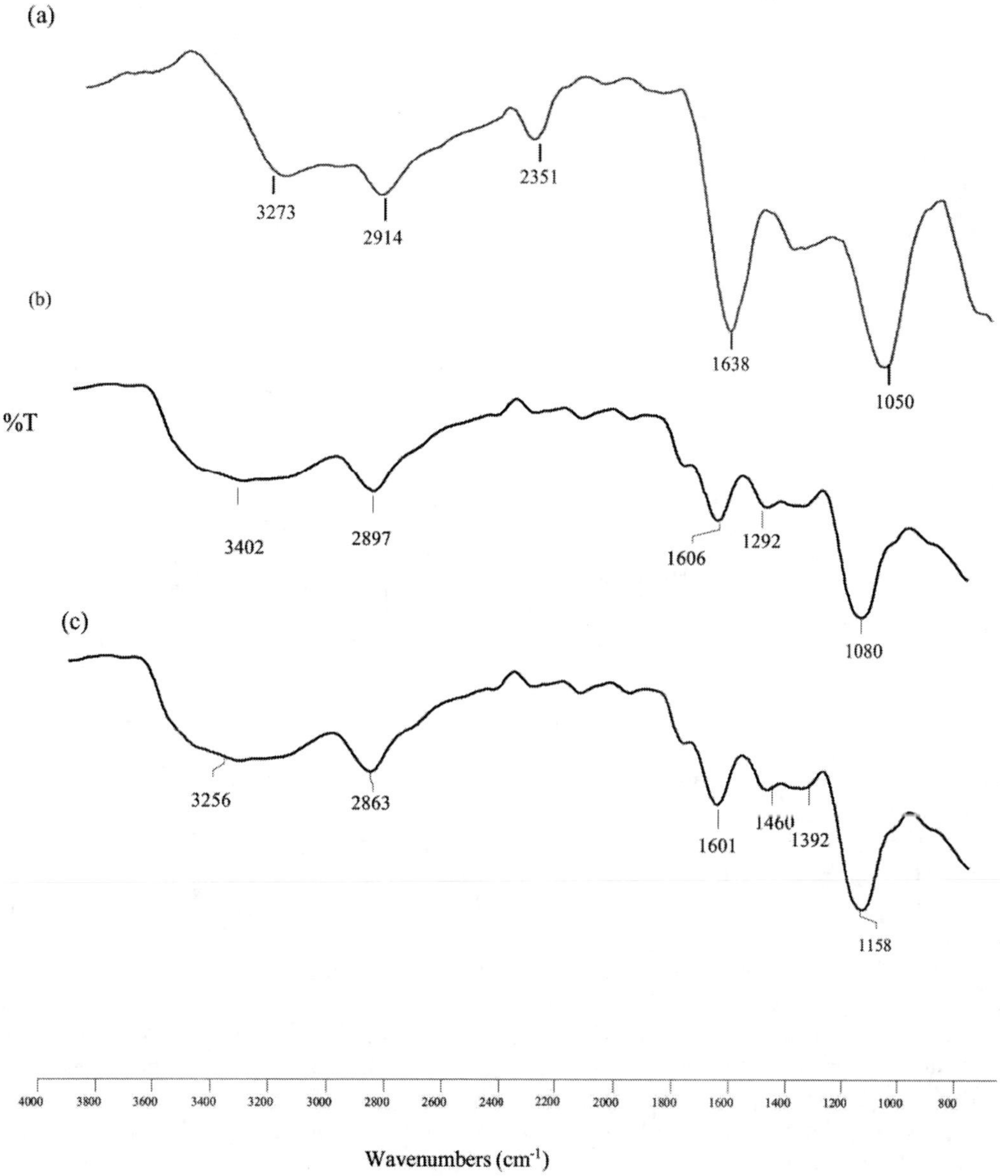

Figure 1. FTIR spectra of (a) cinnamaldehyde, (b) SA-Chi nanocomposite, and (c) SA-Chi nanocomposite loaded with cinnamaldehyde.

The spectra of cinnamaldehyde, SA-Chi nanocomposite, and SA-Chi loaded with cinnamaldehyde have been shown in figure 1. From figure 1(a), the absorption bands of cinnamaldehyde had been exhibited at 3273, 2914, 2351,1638 and 1050 cm^{-1} which corresponded to O-H stretching, C-H stretching, O=C=O stretching, C-H bending, and CO-O-CO stretching. Meanwhile, the broad and intense peak at 3325 cm^{-1} indicated that hydrogen bonds between –OH and $-NH_2$ in Chi and –C=O and –OH of sodium alginate [23-25]. The intensity of C-H bending at 1390 cm-1 decreased and shifted to 1404 cm-1 due to strong interactions in the nanocomposite. There was a marginal reduction in the absorption intensity of –CO stretch band at 1290 and 1300 cm^{-1}. Additionally, the band represents C=O of sodium alginate at 1041 cm^{-1} became more prominent and shifted after interaction with $NH3^+$ of Chi. It is clear that the absorption bands at 3325, 2902, and 1604 cm^{-1} which represent the stretches of O–H, C–H, and C–O of nanocomposite (figure 1a) were shifted to 3323, 2899, and 1595 cm^{-1} (figure 1c), respectively following encapsulation of cinnamaldehyde in the nanocomposite. From figure 2(a), it is obvious that SA-Chi nanocomposite has a crystalline characteristic. There was no peak of any other phases was observed in the diffractogram indicating high purity of the nanocomposite. The broad peak of X-ray diffraction patterns indicated the significantly small size of the resulting crystallites [26]. The intensity of the reflection peak at 23.46° 2θ increased by two-folds following the encapsulation of cinnamaldehyde into nanocomposite (figure 2(b)). This increment could be related to additional reflection of X-ray caused by cinnamaldehyde molecules. A similar observation was reported by Lim and Ahmad whereby no peak being detected following encapsulation of imidacloprid into SA-Chi nanocomposite [27].

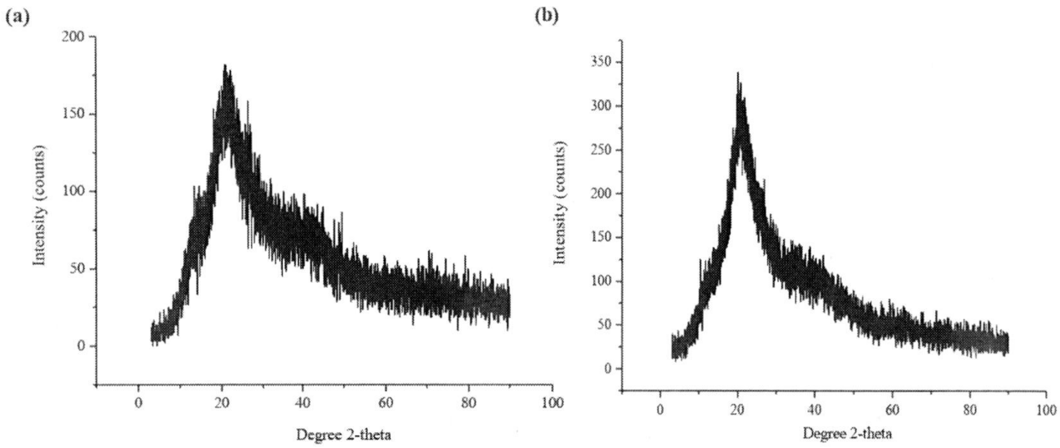

Figure 2. XRD diffractograms of SA-Chi nanocomposite (a) before and (b) after the encapsulation of cinnamaldehyde.

The TGA thermograms of nanocomposite and cinnamaldehyde-loaded nanocomposite are presented in figure 3(a). SA-Chi nanocomposite shows three thermal degradation steps. In the first step (108-275 °C), 26 % of weight loss occurred due to the elimination of loosely bonded water. The next degradation took place at 275 °C which could be due to nanocomposite decomposition and scission of the polymer chain. With increasing the temperature, the weight of nanocomposite decreased slowly in the last step. However, cinnamaldehyde-loaded nanocomposite exhibited four thermal degradation steps and this showed that the thermal stability of nanocomposite increased after encapsulation of cinnamaldehyde [27]. The first thermal degradation required a higher temperature for the release of absorbed water in the nanocomposite or vaporization of volatile components. Considering water molecules were bonded to carboxyl, amine, and hydroxyl groups of Chi, SA, and cinnamaldehyde, the degradation step at 345 °C involved 17 % of weight loss and this could be related to incomplete removal of water in the previous step. For the last step of degradation, the decomposition of polyelectrolyte composition might take place.

From figure 3(b), the nanocomposite had three endothermic peaks (60, 747, 934 °C) and two exothermic peaks (256, 367 °C), while two endothermic peaks (60, 747 °C) and four exothermic peaks (256, 367, 502, 876 °C) were seen for cinnamaldehyde-loaded nanocomposite. At 60 °C, a broad endothermic peak that could be related to water elimination was observed in nanocomposite and cinnamaldehyde-loaded nanocomposite. A higher enthalpy is required to overcome decomposition of amine unit by observing an increase in exothermic peaks at 256 and 367 °C. The second endothermic peaks represented at 747 °C could be ascribed to the dissociation of the nanocomposites. Heat energy is absorbed by the nanocomposite at 934 °C that may be due to bonds cleavage and this corroborates with figure 3(b) which is about 10 % of nanocomposite residue remained. In the context of cinnamaldehyde-loaded nanocomposite, an additional exothermic reaction occurred at 502 °C that could be associated with strong hydrogen bond formation after crosslinking. Furthermore, cinnamaldehyde-loaded nanocomposite altered the enthalpy of nanocomposite by releasing heat at 876 °C instead of absorbing heat.

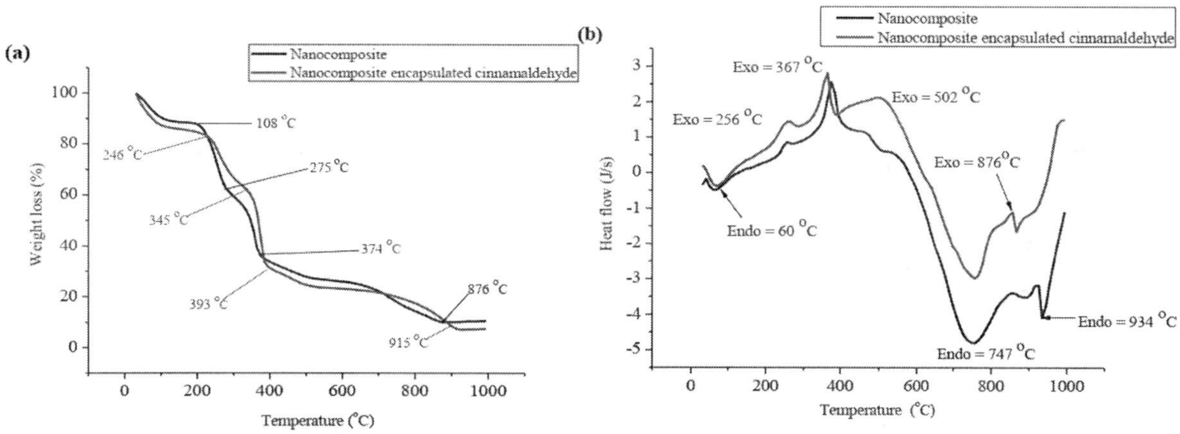

Figure 3. (a) TGA thermograms and (b) DSC thermograms for SA-Chi nanocomposites before and after encapsulation of cinnamaldehyde.

3.2 Encapsulation efficiency and loading capacity

Table 1 reports the encapsulation efficiencies and loading capacities of the nanocomposites loaded cinnamaldehyde at different mass ratios. The encapsulation efficiency of the nanocomposites loaded cinnamaldehyde at 50:1, 100:1 and 125:1 was 37.07, 76.58 and 92.76 %, respectively. When the mass ratio of nanocomposite/cinnamaldehyde had been increased, the encapsulation efficiency would be also increased. By increasing the amount of SA-Chi, there are more spaces were available to be fulfilled by cinnamaldehyde molecules. The cinnamaldehyde encapsulation efficiencies obtained from this study were higher than 27.42 % for chitosan/tripolyphosphate nanocomposite at an initial cinnamaldehyde concentration of 0.4 ppm as reported by Joghataei et al. [28].

As shown in table 1, the loading capacity of the nanocomposite increased as the mass ratio was increased. The highest loading capacity was obtained for 125:1 ratio. Following cross-linking reaction between SA and Chi, the egg-box structure of polymer chains will be formed and agglomeration of the nanocomposite occurred due to adhesive properties of Chi[29]. Hence, more cinnamaldehyde will be encapsulated in the nanocomposite.

Table 1. Encapsulation efficiencies and loading capacities of cinnamaldehyde loaded SA-Chi nanocomposites in different mass ratios.

Ratio of cinnamaldehyde loaded SA-Chi nanocomposite	Encapsulation efficiency (%)	Loading capacity (%)
50:1	30.02	12.20
100:1	71.22	13.90
125:1	91.18	74.80

IOP Conf. Series: Materials Science and Engineering **980** (2020) 012017 doi:10.1088/1757-899X/980/1/012017

3.3 Release profiles of cinnamaldehyde from the nanocomposite

As depicted in figure 4, cinnamaldehyde exhibited three release stages while cinnamaldehyde loaded nanocomposites had four release stages. The total hours for cinnamaldehyde release were 48. The nanocomposites were able to prolong the release of cinnamaldehyde to 52, 60 and 70 hours for 50:1, 100:1, and 125:1 respectively.

The cinnamaldehyde showed a slow release rate for the first 21 h. Nearly 10 % of cinnamaldehyde was released in the next hour. A slow delayed release of cinnamaldehyde was observed until 100 % release has been reached. The results taken at specified intervals showed that cinnamaldehyde-loaded nanocomposites had burst releases of about 44 % for the first 20 h. The release of the cinnamaldehyde from the nanocomposites obtained a steady state from this hour and after about 24 h, about 58 % of cinnamaldehyde was released. A slow release trend was observed for 46 h duration (36th to 70th h). The decrease of cinnamaldehyde diffusion from the nanocomposites could be ascribed to the dense structure of polyelectrolyte complex. Lim and Ahmad also reported SA-Chi nanocomposite could release imidacloprid over 120 h [27]. Therefore, the release profile of the prepared SA-Chi nanocomposites showed that cinnamaldehyde could have a sustained release from synthesized nanocomposites.

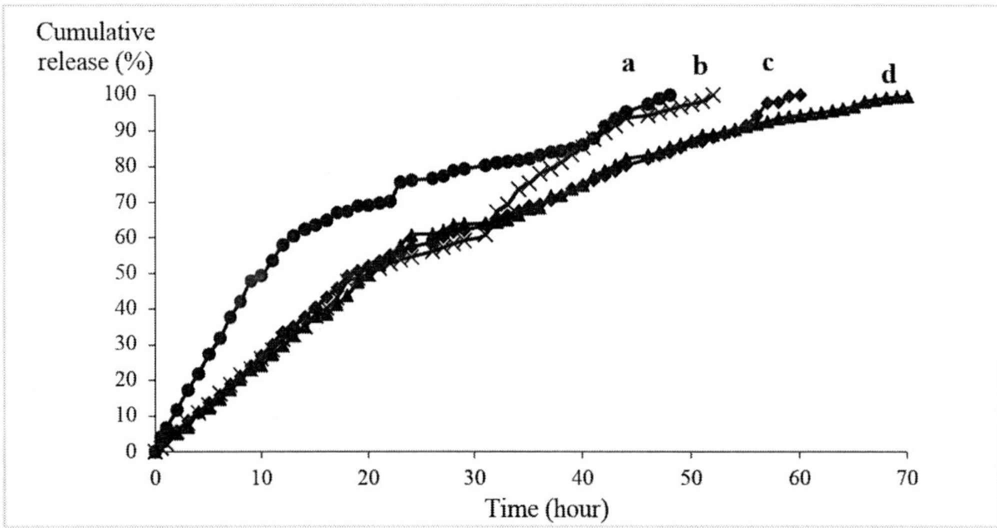

Figure 4. In vitro release of (**a**) cinnamaldehyde and cinnamaldehyde-loaded SA-Chi nanocomposite with different mass ratios of SA-Chi nanocomposites to cinnamaldehyde (**b** 50:1, **c** 100:1 and **d** 125:1) in deionized water at 25 °C.

3.4 Best fitted release kinetic models

From table 2, it was found that Korsmeyer-Peppas model provided a higher correlation coefficient ($R^2 = 0.9814$) than other kinetic models. Therefore, the drug release profile of the nanocomposites followed non-Fickian mechanism/anomalous diffusion (n is between 0.45 and 0.89) which indicated diffusion through the nanocomposite leads drug release process.

Table 2. Correlation coefficient (R^2) of the fitted kinetic models for the release of cinnamaldehyde.

SA-Chi/cinnamaldehyde	Kinetic models				
	Zero order	First order	Higuchi	Korsemeyer-Peppas	
	R^2	R^2	R^2	R^2	n
50:1	0.9633	0.7344	0.9247	0.9841	0.8252
100:1	0.9412	0.8563	0.9456	0.9795	0.8378
125:1	0.9158	0.9517	0.9153	0.9837	0.7250

4. Conclusion

In this study, the potential of SA-Chi nanocomposites as drug carriers for larvicide formulation was investigated by carrying out characterization studies. There were major changes to the properties of functional groups, thermal stability, and crystallinity of nanocomposites following interaction with cinnamaldehyde. The SA-Chi nanocomposites encapsulated cinnamaldehyde successfully at a different mass ratio of the nanocomposite to cinnamaldehyde. The kinetics profile of cinnamaldehyde was best described by Korsmeyer-Peppas model. These key properties enable SA-Chi nanocomposite as an ideal nanocarrier to carry hydrophobic larvicides. This research creates impacts on the environment because it is biodegradable and low-cost. It is worth to study on other water-insoluble larvicides.

Acknowledgments

We thank Ministry of Education Malaysia and Universiti Pendidikan Sultan Idris for providing financial support (FRGS 2019-0003-102-02) and Rising Star UPSI 2019-0121-103-01) to this project. We thank the laboratory assistants from the faculty for their technical support.

References

[1] Bruschi M L 2015 *Strategies to modify the drug release from pharmaceutical systems* (United Kingdom: Woodhead Publishing) chapter 2 15-28
[2] Simonazzi A, Cid G A, Villegas M, Romero Al, Palma S D, Bermudes J M. *Nanotechnology applications in drug controlled release* (William Andrew Publishing) chapter 3 81-116
[3] Ding C and Li Z 2017 *Mater Sci Eng C Mater Biol Appl.* **76** 1440-53
[4] ud Din F, Aman W, Ullah I, Qureshi O S, Mustapha O, Shafique S and Zeb A 2017 *Int. J Nanomedicine* **12** 7291
[5] Mukherjee B, Satapathy B S, Bhattacharya S, Chakraborty R and Mishra V P 2017 *Pharmokinetic and Pharmacodynamic Modulations of Therapeutically Active Constituents From Orally Administered Nanocarriers Along with a Glimpse of Their Advantages and Limitations Nano-and Microscale Drug Delivery Systems* ed A M Grumezescu (Asterdam: United Kingdom/United State Elsevier) chapter 19 357-375
[6] Patra J K, Das G, Fraceto L F, Campos E V, del Pillar Rodriguez-Torres M, Acosta-Torres L S, Diaz-Torres L A, Grillo R, Swamy M K, Sharma S and Habtemariam S 2018 **16** 71
[7] Rao S S, Rekha P D, Anil S, Lowe B and Venkatesan J 2019 *Drug Delivery and Biomedical Applications* (London: Academic Press) chapter 21 495-512
[8] Kim H J, Lee H C, Oh J S, Shin B A, Oh C S, Park R D, Yang K S and Cho C S 1999 *J. of Biomater. Sci.* **10** 543-56
[9] Hamman J H 2010 *Mar. Drugs* **8** 1305-22
[10] Kashyap P L, Xiang X and Heiden P 2015 *Int. J. of Biol. Marcomol.* **77** 36-51
[11] Bernkop-Schnurch A and Dunnhaupt 2012 *Eur J Pharm Biopharm* **81** 463-69
[12] Seenivasan P, Tennyson S, Jayakumar M and Prabu A 2019 *Res J Life Sci Bioinform Pharm Chem* Sci 5(3) 601-08
[13] Cardoso-Ugarte GA, Lopez-Malo A and Sosa-Morales M E 2016 *Food Preservation, Flavor and Safety* (London: Academic Press) chapter 38 339-47
[14] da Silva M L, Bernardo M A, Singh J and de Mesquita M F 2019 *The Roles of Functional Food Security in Global Health* (London: Academic Press) chapter 33 565-76
[15] Thomas J and Kuruvilla K M 2012 Cinnamon *Handbook of herbs and spices* ed K V Peter (United States of America: Woodhead Publishing Limited) pp 182-96
[16] Gombotz W R and Wee S 1998 *Adv. Drug Delivery Rev.* **31** 267-85
[17] George M and Abraham T E 2006 *J. Control. Release* **114** 1-14
[18] Hoare T R and Kohane D S 2008 *Polymer* **49** 1993-2007
[19] Loquercio A, Castell-Perez E, Gomes C and Moreira R G 2015 *J. Food Sci.* **80** N2305-15
[20] Tu J, Boyle A L, Friedrich H, Bomans P H, Bussmann J, Sommerdijk N A 2016 *ACS Appl. Mater. Interfaces* 32211-19
[21] Guo J, Giusti M M and Kaletunc G 2018. *Int Food Res J* **107** 414-22

[22] Ritger P L, Peppas N A 1987 *J Control Release* **5** 37-42
[23] Kulig D, Zimoch-Korzycka A, Jarmoluk A, Marycz, K 2016 Polymers **8** 167
[24] Ji M, Sun X, Guo X, Zhu W, Wu J, Chen L, Wang J, Chen M, Cheng C and Zhang Q 2019 Food Hydrocoll **90** 515-22
[25] Nalini T, Basha S K, Sadiq A M M, Kumari V S and Kaviyarasu K 2019 *J Drug Deliv Sci Tec* **52** 65-72
[26] Trivedi M K, Branton A, Trivedi D, Nayak G 2015 *Pharm Anal Acta* **6**
[27] Lim G P, Ahmad M S 2017 *J Ind Eng Chem*
[28] Joghataei M, Hosseini SF, Arab-Tehrany E 2019 *J Food Process Preserv* **e13972**
[29] González-Rodrıguez ML, Holgado MA, Sanchez-Lafuente C et al 2002 *Int J Pharm* **232** 225-34

IOP Conf. Series: Materials Science and Engineering **980** (2020) 012018 doi:10.1088/1757-899X/980/1/012018

Co (II) desorption from silica gel and mercapto-silica hybrid

D R Mujiyanti[1,2]*, U Irawati[1], and N M Akhir[1]
[1]Department of Chemistry, Faculty of Mathematics and Natural Sciences, Lambung Mangkurat University, Jalan A. Yani km 36 Banjarbaru Indonesia
[2]Laboratory of Inorganic Chemistry, Laboratory of Faculty of Mathematics and Natural Sciences, Lambung Mangkurat University, Banjarbaru Indonesia

Email: drmujiyanti@ulm.ac.id

Abstract. This paper aims to study Co (II) desorption from silica gel (SG) and mercapto-silica hybrids (MSH) by many desorbing agents. SG was prepared from sodium silicate from rice husk ash, whereas MSH was SG modified by organic compounds 3- (trimethoxysilyl) -1-propanthiol. The characterization of adsorbents covered by an X-ray diffractometer and an FTIR spectrophotometer. Optimization of contact time for adsorption of Co (II) on HMS and the procedure of desorption agents (H_2O, HCl, $Na_2EDTA.2H_2O$) at different contact times and concentrations via batch methods. The results showed that the optimum contact time for Co (II) adsorption on MSH was 120 minutes. The most appropriate desorbing agents for SG and MSH were HCl and $Na_2EDTA.2H_2O$. The optimum contact times for Co (II) desorption on SG and MSH were 90 and 120 minutes, respectively. The optimum concentration of HCl for desorbing Co (II) on SG was 0.025 M. While, the optimum concentration of $Na_2EDTA.2H_2O$ for Co (II) desorption on MSH was 0.1 M.

1. Introduction

Abundant rice husk from the rice mills had been used as a fuel for combustion of red brick, cooking, media hydroponic plants, or burned away. Another application, waste rice husk ash (ASP) has a great potential as the source of silica-based adsorbent because numerous studies showed that the ASP contains sufficiently high points of silica, equivalent to 98% (w/w) [1]. Mujiyanti, et al. [2] reported that the silica content in the rice husk ash from the peat, Banjar Regency, South Kalimantan extended 95.6% (w/w).

Expenditure of the adsorbent adsorbs heavy metals, became a smaller amount of consideration. The adsorbent can be reused through recovery. Natalia [3] has investigated for desorption studies hybrid silica on metal sulfonate with Ca (II) and Cd (II) ions via technique column. The results showed that the result of desorption was Cd (II) ion, 98% with 0.1 M HCl and 100% with 0.05 and 0.1 M $Na_2EDTA.2H_2O$. On the other side, Ca (II) ion desorbed on HCl 0.5 M, amounting to 78.13% and for $Na_2EDTA.2H_2O$, Ca (II) ion desorbed amounting to 88.66%.

Furthermore to adsorbent recovery, desorption can be carried out for the recovery of adsorbed metals. Through desorption studies, we can also find out the adsorption mechanism between the metal and the adsorbent. Mujiyanti (2007) conducted a study of the multi-metal adsorption-desorption of Ag (I), Pb (II), Cr (III), Cu (II), and Ni (II) through a column system in mercapto-silica hybrids (MSH) using three types of eluents, HCl, H_2O, and $Na_2EDTA.2H_2O$. The desorption with H_2O % was variety 0.8-31.0% for monometallic solutions and 0.3-24.4% for multi-metal solutions. Meanwhile, the contribution of the adsorption mechanism for the formation of complex bonds with $Na_2EDTA.2H_2O$ in

IOP Conf. Series: Materials Science and Engineering **980** (2020) 012018 doi:10.1088/1757-899X/980/1/012018

monometallic solutions alternated from 0.3-29.4% and multi-metal solutions reached from 0.2-14.3%. The desorption process indicated that the adsorption mechanism tends to the formation of hydrogen bonds. This is maintained by the formation of an octahedral aqua complex from the five metal ions in water.

Mujiyanti, et.al [4], stated that Co (II) adsorption on silica gel (SG) optimum at pH 5 and contact time of 75 minutes, while the MSH takes place optimally at pH 3, but the optimum time on MSH could not be determined. The contact time of adsorption was only until 90 minutes in a 50 mL solution of Co (II) 100 mg/L at pH 3 and the amount of adsorbed metal showed a tendency to rise. Therefore, one of the initial steps in this paper was to study the effect of the optimum contact time for Co (II) adsorption by MSH. Furthermore, this paper will observe some aspects of Co (II) desorption from SG and MSH.

2. Materials and methods

2.1. Materials

The material used was rice husk ash from peat area, South Kalimantan as a source of silica, NaOH (pa Merck), HCl 37% (pa Merck), 3- (trimethoxysilyl) -1-prophantiol (pa Merck), the stock solution of Co (II) 10000 ppm (pa Merck), $Na_2EDTA{\bullet}2H_2O$ (pa Merck).

2.2 Determining optimum contact time of Co(II) adsorption

The procedure of determining optimum contact time of Co (II) adsorption on the MSH referred to research that had been conducted by Mujiyanti,et.al [4]. A total of 100 mg of MSH interacted with 50 mL solution of Co (II) 100 mg / L at pH 3. The solution was stirred with a variation of the contact time 30, 60, 90, 120, and 150 minutes, and strained with filter paper. The filtrate was analyzed by Atomic Absorption Spectrophotometry (AAS).

2.3 The adsorption of Co (II) on silica gel and mercapto-silica hybrid

The adsorption process referred to research by Mujiyanti,et.al [4] in optimum conditions, pH = 5 and contact time 75 minutes for SG. For MSH accomplished at pH 3 and contact time obtained in determining the optimum contact time adsorption Co (II) at MSH. A total of 1 g of MSH contacted with 50 mL solution of metal ions Co (II) 100 mg/L at the optimum condition. The mixture was filtered then the filtrate was analyzed by AAS. The same treatment was also examined for SG.

2.4 The desorption of Co(II) Ion by silica gel and mercapto-silica hybrid

2.4.1 Determining desorption agent of Co(II) desorption. The desorption agent used in this study were H_2O, HCl 0.1 M and 0.1 M $Na_2EDTA{\bullet}$ $2H_2O$. Total of 25 mL desorption agent contained in erlenmeyer and MSH residues that have adsorbed metal ions Co (II), then solution shaken by contact time for 120 minutes. The mixture strained then the filtrate was analyzed by AAS. The same treatment was also examined for SG.

2.4.2 Determining optimum contact time of Co (II) desorption. Desorption was done as in the determination of desorption agent in variation of 30, 60, 90, 120, and 150 minutes.

2.4.3 Determining optimum concentration of desorption agent for Co (II) desorption. Desorption was done as in the determination of desorption agent and contact time, but consuming the suitable agent with a concentration of 0.025; 0.05; 1; 1.5; and 2 M for optimum contact time.

3. Results and discussion

3.1 Determining optimum contact time of Co(II) adsorption

IOP Conf. Series: Materials Science and Engineering **980** (2020) 012018 doi:10.1088/1757-899X/980/1/012018

Interaction time between adsorbent and the metal ion was one of the factors that play a role in determining the optimum conditions for interaction of metal ions and the equilibrium state of the adsorbent [5]. Figure 1, showed that the number of ions Co (II) adsorbed MSH on contact time 30, 60, 90, 120, and 150 minutes, respectively

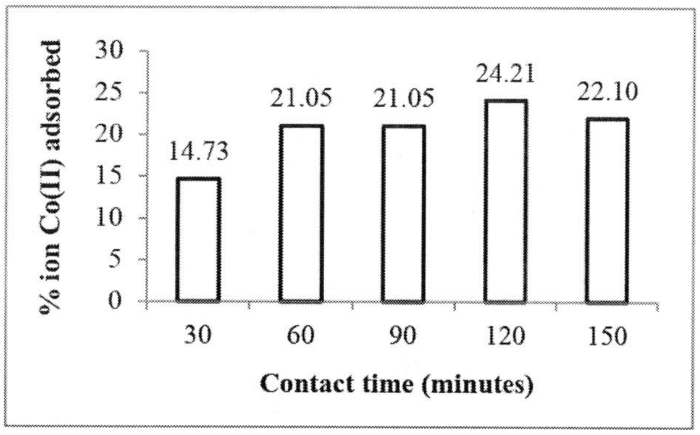

Figure 1. The chart of % ion Co(II) adsorbed with the various contact time.

The quantity of Co (II) ions adsorbed was enlarged with an accumulative interval of stirring (30-120 minutes). The extended contact time of MSH and metal ion approved diffusion and annealing process taken more immaculate adsorbate molecules and the amount of Co (II) ions that absorbed increasingly. Adsorption was assessed by the maximum point when the contact time of 120 minutes, in which the percentage of adsorption MSH compared to Co (II) ion was 24.21%. Adsorption equilibrium was characterized by the lack of the concentration of metal ion alterations both on the surface of the adsorbent and metal ion in solution [6], in which the rate of adsorption was equal to the rate of desorption [7]. After passing through 120 minutes of the percentage of Co (II) ions adsorbed tended constantly, even declining. This trend suggested that the active sites on the surface of the adsorbent had been saturated by metal adsorbates and MSH no longer able to adsorb Co (II) ions. Metal ions that had been adsorbed could be desorbed from the surface of the adsorbent. This could occur if the interaction between Co (II) metal ions with the active site on MSH was fragile.

3.2 The desorption of Co(II) ion by silica gel and mercapto-silica hybrid

3.2.1 Determining desorption agent of Co(II) desorption. Desorption is the opposite of the adsorption process. Desorption is the process of releasing the adsorbate from the adsorbent into a solution [8]. The desorption process can provide information about the adsorption mechanism that plays a role in the adsorption of Co (II) by SG and MSH adsorbents. Co (II) metal ion, which had been adsorbed by the adsorbent, desorbed by the eluents; H_2O, 0.1 M HCl and 0.1 M $Na_2EDTA\cdot 2H_2O$ by batch method. The relationship between percent Co^{2+} which wass desorbed with desorption agent types is showed in figure 2.

Distilled water (H_2O) as a solvent metal ion was attached to the surface of the adsorbent. Desorption result with distilled water did not have a significant effect to release ions Co (II), just 0.76 and 1.47% for SG and MSH. Co(II) metal ion desorbed on surface SG and MSH by HCl solution was successively as much as 103.46 and 54.53%. HCl was a strong acid group that releases H^+ ions to replace metal ions adsorbed by ion exchange process or through hydrogen bonding interactions. Desorption percentage by HCl on SG exceed 100% due to the addition of HCl was expected to dissolve some of the SG, so that Co(II) less reacted with SG. Desorption with $Na_2EDTA\cdot 2H_2O$ solution indicated that as many as 93.84 and 67.40% of ions Co(II) previously adsorbed by SG and MSH desorbed. The results showed that the adsorption of Co (II) in the adsorbent occurred such as through the formation of a complex between Co(II) with active sites, mercapto (-SH) in the adsorbent.

EDTA is a strong ligand, so when the ability of the ligand adsorbent active group is weaker than the EDTA then the metal ions on the adsorbent will be adsorbed by the formation of a stable complex [3].

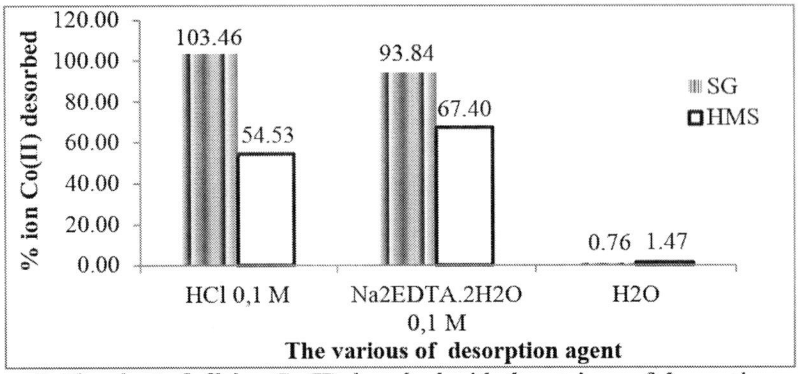

Figure 2. The chart of % ion Co(II) desorbed with the various of desorption agent.

From figure 2, showed that the main interaction on SG and MSH was chemisorption. The adsorption mechanism of Co (II) cations on SG adsorbent was the most dominant estimated by ion exchange and /or hydrogen bonding, because the percentage of desorption with HCl equated to other desorbing agents maximum. However, the adsorption mechanism on MSH adsorbent, assumed mostly through the complex formation, an interaction was stronger than ion exchange interactions and the formation of hydrogen bonds.

3.2.2 Determining optimum contact time of Co (II) desorption. Determination of the optimum contact time of desorption aimed to obtain information on how long it takes to achieve optimum Co (II) desorption from the adsorbent. The variation of contact time was from 60 to 180 minutes. This research was using the desorbent agents 25 ml of 0.1 M HCl and 0.1 M $Na_2EDTA{\bullet}2H_2O$ as. The amount of Co (II) depleted for each variation of contact time is shown in figure 3.

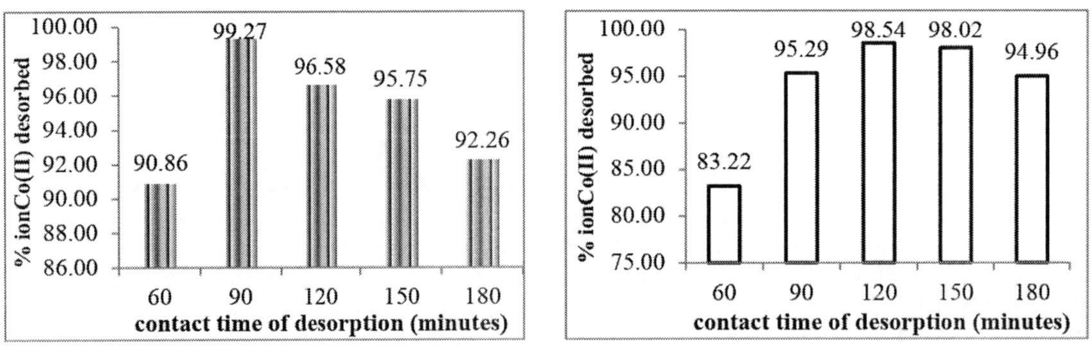

(A) HCl (B) $Na_2EDTA{\cdot}2H_2O$

Figure 3. The relationship of Co(II) desorpted in various contact time of desorption.

From figure 3 it was shown that the extended contact time of metal ions using 0.1 M HCl and 0.1 M $Na_2EDTA{\bullet}2H_2O$ in SG and MSH, the number of Co (II) ions were separated greatly. However, at a certain moment the addition of time would decrease the percent desorption Co (II) ion which was separated from the adsorbent. Figure 3(A) indicated that the optimum contact time of desorption achieved after SG interacted with HCl for 90 minutes. This was due to the H^+ ion from HCl was strong enough to replace metal ions attached to the adsorbent SG [9]. While the pattern of desorption time relationship with the percentage of Co (II) ions was separated on MSH by $Na_2EDTA{\cdot}2H_2O$ showed in figure 3(B). Desorption of MSH continued to rise until it reached a maximum desorption at 120 minutes, which amounted to 98.54%. For the interaction time of 150 and 180 minutes, separate Co (II) ions decreased. According to Khasanah [10] a similar tendency was possible due to a decrease in the surface of the adsorbent that had been saturated by Co-EDTA complex was formed.

3.2.3 *Determining the optimum concentration of desorption agent for Co (II) desorption.* This research also studied the effect of the concentration of desorption agent compared to Co (II) desorption (variation of the concentration of the agent 0.025 M to 0.2 M). The relationship between the concentration of HCl with Co (II) ions which desorbed on SG can be seen in figure 4(A). by desorption at 97.82 %.

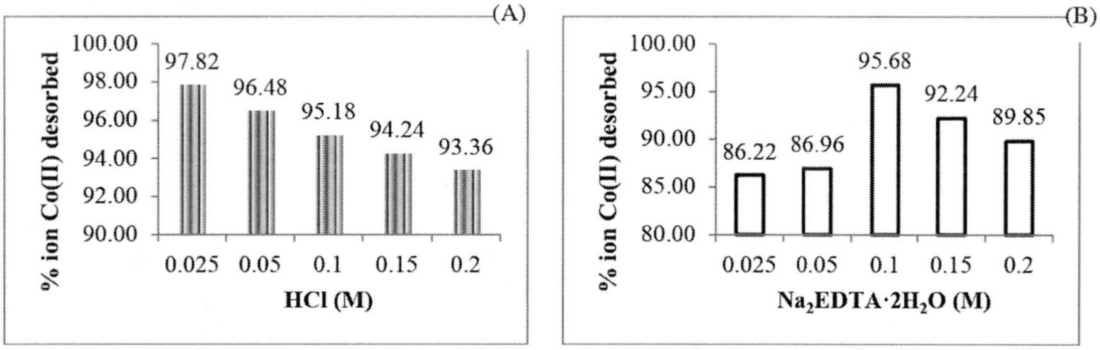

Figure 4. The relationship % ion Co(II) desorpted on various concentration of desorption agent.

According to Natalia [3], one factor was the effectiveness of competition desorption H^+ ions with adsorbed metal, the concentration of the solvent greatly gave affect the effectiveness of desorption. At the appropriate concentration, the ability of H^+ ion to replace the presence of metal ions on the adsorbent being highest. Figure 4(A) showed the increasing of concentration of HCl had lower percentage desorption of Co (II) ions. It was possible that the solvent concentration was too high affecting Co (II) which were separated partially settled down as $CoCl_2$. From this research, it was found that the maximum percentage of Co (II) ions was desorbed at a concentration of 0.025 M for HCl as a desorption agent for the Co (II) attached to SG. Figure 4(B) shows that the accumulated concentration of Na2EDTA • 2H2O caused the desorbed Co (II) ion to increase at a certain concentration. At a concentration of 0.1 M, the optimum amount of Co (II) ions was absorbed at 95.68%. The increase in the desorbed Co (II) ion is proportional to the increase in the concentration of Na2EDTA • 2H2O. But after the concentration had exceeded the optimum, a decrease Co (II) ion was desorbed. This phenomenon was possible due to the partial of deprotonitation $Na_2EDTA{\cdot}2H_2O$ therefore the acid groups ionized carboxyl was less active and caused the formation of the coordination bonding between EDTA and metal ions was not intensive.

4. Conclusion

Based on the results of the research, adsorption of Co (II) ion on MSH had reached a maximum point at a contact time of 120 minutes as much as 24.21%. The maximum desorption process of Co (II) ions from SG with HCl and in MSH with $Na_2EDTA{\cdot}2H_2O$ as a desorbing agent. The contact time used in desorption of Co (II) ions on SG by 0.1 M HCl reached a maximum in 90 minutes, as much as 99.27% and MSH by 0.1 M $Na_2EDTA{\cdot}2H_2O$ was 120 minutes, as much as 98.54%. The concentration of 0.025 M HCl as desorbent ion Co (II) in SG which maximally absorbed Co(II) was 97.82% and the concentration of $Na_2EDTA{\cdot}2H_2O$ in the desorption of ion Co(II) from MSH was 95.68%.

References

[1] Nuryono 2004 Pengaruh konsentrasi NaOH pada peleburan abu sekam padi cara basah. *Prosiding Seminar Nasional Hasil Penelitian MIPA 2004* Semarang
[2] Mujiyanti D R, Astuti MD, & Umaningrum D 2010 *Pembuatan Silika Amorf pada Limbah Sekam Padi Gambut di Kabupaten Banjar Kalimantan Selatan* Laporan DIPA (Banjarbaru : FMIPA, Universitas Lambung Mangkurat)
[3] Natalia V 2009 Study of desorption of Ca(II) and Cd(II) in sulfonate silica hybrid adsorbent using column technic Skripsi (Yogyakarta: FMIPA Universitas Gadjah Mada)
[4] Mujiyanti D R, Komari N and Sari N I 2013 Kajian Adsorpsi hibrida merkapto-silika dari abu sekam padi terhadap ion Co(II) *Jurnal Kimia Valensi* **3(2)** 71
[5] Buhani, Narsito, Nuryono and Kunarti E S 2009 Hibrida amino-silika dan merkapto-silika sebagai adsorben untuk adsorpsi ion Cd(II) dalam larutan *Indo. J. Chem.* **9(2)** 170
[6] Lesbani A 2011 Studi interaksi vanadium dan nikel dengan pasir kuarsa *Jurnal Penelitian Sains* **14(4)** 43-6
[7] Riapanitra A, Setyaningtyas T and Riyani K 2006 Penentuan waktu kontak dan pH optimum penyerapan metilen biru menggunakan abu sekam padi *J. Molekul* **1(1)** 44
[8] Triani L 2006 *Desorpsi Ion Logam Tembaga (II) dari Biomassa Chlorella Sp yang Terimobilisasi dalam Silika Gel*. Skripsi. (Semarang: FMIPA, Universitas Negeri Semarang)
[9] Sembiring Z, Suharso, Regina, Marta F, & Murniyarti 2008 Studi Proses Adsorpsi-desorpsi Ion Logam Pb(II), Cu(II) dan Cd(II) terhadap Pengaruh Waktu dan Konsentrasi pada Biomassa *Nannochloropsis Sp*. yang Terenkapsulasi Aqua-gel silika dengan Metode Kontinyu. *Prosiding Seminar Nasional Sains dan Teknologi* (Bandar Lampung) p591-607
[10] Khasanah D R 2006 Studi Penggunaan Na_2EDTA dalam Desorpsi Ion Logam Kobalt(II) pada Kitin Terdeasetilasi dari Limbah Cangkang Kepiting Hijau (Scylla serrata) dalam Medium Air Skripsi (Semarang: FMIPA, Universitas Negeri Semarang)

IOP Conf. Series: Materials Science and Engineering **980** (2020) 012019 doi:10.1088/1757-899X/980/1/012019

The exploration of banana bunch as a new vegetable tanning agent

T Maryati[1], A Pertiwiningrum[2], Z Bachrudin[2], and R Yuliatmo[1]*
[1]Department of Leather Processing Technology, Politeknik ATK Yogyakarta, Yogyakarta, Indonesia
[2]Faculty of Animal Science, Universitas Gadjah Mada, Yogyakarta, Indonesia

Email: ragilyuliatmo@atk.ac.id

Abstract. Recently, the green tannery is an issue that has been developed because it is known as a contributor of hazardous waste. The substitution of mineral tanning agents, such as chrome, with vegetable tanning agent is the best choice. Banana bunch has the potential to be an environmentally friendly tanning agent. This study aims to explore the banana bunch as a vegetable tanning agent for rabbit skin. The banana bunch was extracted using ethanol. The tannin quality of banana extract was determined by both qualitative and quantitative methods. The quality of leather tanned by the banana bunch was determined by chemical and physical properties. The qualitative results of the tannin content showed that the banana bunch had the tannin compounds, while quantitatively the tannin content was 4.1%. Banana bunch has an inhibitory potential, indicating that the banana bunch can be an anti-bacterial. Besides, the chemical analysis showed that the tanning by banana bunch had a water content of 14.90±0.01%, a water-soluble content of 3.55±0.01%, the pH value of 3.42±0.03, and the degree of tannage of 52.47±2.29%. Also, the physical quality of rabbit skin showed the tensile strength of 302.00 kg/cm^2 and the elongation of 174.12±6.44%. Based on these results, it shows that the extract of the banana bunch has the potential as a vegetable tanning agent.

1. Introduction

The leather industry is included in the sustainable industry. Wu [1] stated that this industry is one of the most effective ways to recycle raw hides for a by-product of animal husbandry, such as skin and hide. The tannery and leather industry is one of the development industries in Indonesia, along with the Indonesian growing fashion industry. Based on BPS [2], the growth of the leather, leather goods, and footwear industry has recently increased by 18.78%. However, this industry is not without problems. The tannery is a conventional industry which contributes hazardous waste and harms the environment [3]. It is due to the tanning agent derived from dangerous materials, such as chromium. Environmentally friendly tanning agents have begun to develop to overcome these problems. One of the best ways is to use a vegetable tanning agent. The requirement is to have tannin as a crosslinking agent in the hydroxyl group on the skin. Generally, vegetable tanning material commonly used in the tanning industry is mimosa, which is derived from acacia wood. Unfortunately, these chemicals are still imported and expensive. Also, acacia wood is an endangered plant [4,5].

Recently, the supply of vegetable tanning resources is imported from abroad and does not meet the local needs, so that the development of the leather industry is limited. Hence, finding more tannin resources to meet future industrial production needs is an essential goal [6]. The banana bunch contains tannin that can be used as a tanning agent. Banana bunch is known as a market waste that has not been handled properly. The banana bunch is easy to obtain, inexpensive, sustainable, and environmentally friendly so that it can reduce the use of hazardous chemicals in the tanning process for leather products [7]. This study aims to investigate banana bunches as a vegetable tanning agent. Therefore, the use of the banana bunch as a vegetable tanning agent is interesting to explore.

2. Materials and methods

2.1. Extraction of tannins

Three banana species (Kepok, Raja, and Ambon) were extracted to get the tannins according to China [8]. The collected banana bunches were dried under the shed for seven days and cut into small chips. The chipped pieces were ground in a milling machine. About 10 g of milled barks were soaked in 200 mL of ethanol 70% in a glass beaker covered with an aluminum foil. The mixture was placed on a water bath maintained at 70°C for 60 minutes. The sample mixture was stirred using an overhead stirrer connected to the beaker's opening through a small hole made on the aluminum foil. The extraction process continued for 4 hours, and then the filtrates were collected.

2.2. Qualitative analysis of tannins

A qualitative analysis of tannin in the banana bunch extract was done according to Ezeonu [9] by using ferric chloride, gelatine protein, ammonia, and chloric acid as the indicators. 1 mL of banana bunch extract was taken and put into each of three different test tubes and 10 mL of distilled water was added to each tube, followed by the addition of 2 drops of ferric chloride. The appearance of green coloration which disappeared on standing showed the presence of tannins.

The principle of the gelatine test for tannin quality was carried out by taking 1 mL of banana bunch extract and putting it into each of the 3 test tubes, then 1% gelatine solution was added and diluted with distilled water. The sedimentation formation indicated the presence of tannins. Besides, the principle of the tannin test with ammonia was carried out by taking 1 mL of banana bunch extract and putting it into each of the 3 test tubes. The ammonia was added then exposed to the air. The appearance of green coloration which disappeared on standing showed the presence of tannins. Next, the principle of the chloric acid test for tannin quality was carried out by taking 1 mL of banana bunch extract and putting it into each of the 3 test tubes, then 3 drops of chloric acid were added to each tube.

2.3. Total tannins (quantitative analysis)

The total tannin was determined quantitatively using Guo [6]. The extract of banana bunch was diluted with distilled water 20 times. 1 mL of the sample was added to 0.5 mL of Follin Denis reagent and 1 mL of 20% sodium carbonate then homogenized. The results were measured with a spectrophotometer with a wavelength of 720 nm. The total tannin content was expressed as the number of mg of tannic acid equivalent per gram of extract.

2.4. Antibacterial activity assay

The antibacterial assay method was administered according to Udkhiyati [10]. Besides, Mueller Hinton Agar (MHA) was prepared for the *Staphylococcus aureus* growth medium in a concentration of 1.5 x 108 CFU/mL. Fifty microliters of banana bunch extract and 50 µL of chloramphenicol (as the positive control) were inoculated in a petri dish. The petri dish was incubated in an incubator for 24 hours at 37°C. Next, antibiotic chloramphenicol as a positive control was dissolved in dimethyl sulfoxide (DMSO) with a concentration of 100 ppm. The measured zone of inhibition is the radius in mm, which is the clear area around the well.

2.5. Tanning process

The dried banana bunch tannin extracts were used in the tanning of rabbit skins, and the tanning process is displayed in table 1. The steps in the tanning process: soaking to restoring the water content in the skin; liming-unhairing to remove the hair and soluble protein; deliming to remove the lime; bating to remove the globular protein; pickling to prepare the skin to enter the tanning process by decreasing the pH; tanning to make the skin more stable; neutralizing to remove some of the remaining free acids that came from acidification or when aging; fatliquoring to lubricate hydrated leather fibers during the tanning process; and the last was fixation.

Table 1. Technological parameter in tanning process of rabbit skins.

Process	Materials	Dosage (%)	Temperature (°C)	pH	Time/min
Soaking	Water	300	25		
	Alkyl sulfate	2			30
	Sodium hydroxide	1		12	90
	Bactericide	0.03			5
Liming-unhairing	Water	150			
	Sodium sulfide	0.3			30
	Calcium hydroxide	1	25		60
Deliming	Water	150	25		
	Ammonium sulfate	1		7	30
Drain wash					
Bating and degreasing	Water	100			
	Pancreatic enzyme	0.3			30
	Degreasing agent	1			30
Pickling	Water	100			
	Salt	8			10
	Formic acid	1.5		3.8-4.0	120
Tanning	Treatment 1: Banana bunch extract	25	70		120
	Treatment 2: Mimosa (Commercial vegetable tanning agent)	25	70		120
Neutralizing	Water	100			
	Alkyl sulfate	1			30
	Sodium bicarbonate	0.3	25	6	30
Retanning	Water	50	50		5
	Bleaching syntan	4			30
Fatliquoring	Water	100	60		
	Synthetic oil	10			90
Fixation	Oxalit acid	1		3.8-4	40
	Anti-fungi	0.03			15

2.6. Chemical characteristics of tanned skin

The chemical characteristics of tanned skin consist of water content, water-solubility, and degree of tannage [11]. The water content was determined by weighing 5 grams of skin, then the skin was dried in a drying cabinet at 100°C ± 2°C. The resulting moisture content was expressed as a percent of the skin. The water solubility was determined by putting 50 ml of the extract in a glass cup and evaporated with a water bath until dry. Next, the sample was heated in a drying cabinet at 100°C ± 2°C until its weight remained. The result was a water-soluble substance and was expressed as a percent of the skin. The degree of tannage was the ratio between the content of tannin bound and the content of raw leather substances in %.

2.7. Physical characteristics of tanned skin
The physical analysis consisted of the tensile strength and the elongation. The tensile strength test was conducted by pulling the skin to reach the desired force or until the skin sample broke, expressed in N/mm^2. Elongation is the increase in skin length when the skin is stretched to break divided by its original length, expressed in % [12].

3. Results and discussion

3.1. Characteristics of banana bunch
The qualitative analysis results of tannin are shown in table 2. This qualitative analysis is the initial stage to identify the tannin characters from natural materials, such as the banana bunch. The tannin identification was done by adding ferric chloride that resulted in a change in color from brown to dark green and murky. It was estimated that the phenol group in tannins could bind with ferric chloride to form the green color. Ezeonu [9] stated that the formation of dark green or dark blue in the extract after adding to ferric chloride causes tannin to form complex compounds with Fe^{3+} ions. The tests were done by adding gelatin solution that resulted in the sediment. Further, Thomas [13] and Zhang [14] pointed out that tannins can precipitate proteins, such as gelatin. The identification of banana bunch tannin extracts including condensed tannins or catechol with the addition of concentrated hydrochloric acid and the heating process formed a reddish-brown color. The catechol tanning agent is usually reddish-brown and will enlarge its molecules when heated in an acidic solution [15,16].

Table 2. Characteristics of banana bunch.

Banana species	Qualitative Indicators for tannin				Tannin content (%)
	Ferric chloride	Gelatine	Ammonia	Chloric acid	
Kepok	Dark Green	Sediment	Green	Reddish brown	4.10
Ambon	Green	Sediment	Green	Reddish brown	1.34
Raja	Green	Sediment	Green	Reddish-slight brown	0.44

Table 3. Physicochemical properties of tanned leather.

Properties	Leather tanned by banana bunch	Leather tanned by mimosa
Water content (%)	14.90±0.01	12.09±0.01
Water-soluble content (%)	3.55±0.01	7.74±0.04
pH value	3.42±0.03	4.41±0.02
Degree of tannage (%)	52.47±4.27	57.00±1.27
Tensile strength (kg/cm^2)	302.00±2.29	239.55±6.09
Elongation (%)	174.12±6.44	173.68±4.88

Meanwhile, the quantitative analysis of tannin aimed to confirm and quantify the tannin content in natural ingredients. This analysis was carried out through the spectrophotometric method using the Folin denis reagent. The principle of the Folin denis method is the formation of blue complexes whose absorption can be measured in the visible area. According to Ezeonu [9], the formation reaction that occurs is the reduction of oxidation in which tannin as a reducing agent and Folin denis as an oxidizer. The oxidation reaction will form a blue color that can be read at maximum waves. In this study, there were three species of banana tested (Kepok, Raja, and Ambon). The total values of tannin content in each Kepok, Raja, and Ambon banana bunch consecutively were 4.10, 1.34, and 0.44%. The tannin content of the banana bunch is almost the same as *Salix folium* extract (4.0%) but smaller than that of

Acacia seyal bark extract (28.9%) [17]. Zhang [14] reported that the phenolic component in bananas varied depending on the banana variety. According to Griyanitasari [18], the tannin content of a tanning agent is considered good if the value is higher than 0.4% so that the tannin content in the banana bunch is considered suitable for use as a new vegetable tanning agent.

As a tanning agent, the banana bunch can also function as an anti-bacterial. This supports the aim of tanning the skin, which is protecting the skin from physical, chemical, and biological disorders, including leather damage by bacteria. The ability of banana bunch as an anti-bacterial can be seen from the presence of clear zones on the medium containing *Staphylococcus aureus* (figure 1). *Staphylococcus aureus* is the majority of the microbial species isolated from hides/skins [19]. Based on the observation results shown in figure 1, there is a clear zone with a diameter of 11.52 mm. It shows that the banana bunch was successfully inhibited bacterial growth. It is almost the same as Udkhiyati's [10] results who used chitosan as an antibacterial agent (10-15 mm). This can be caused by the presence of secondary metabolites such as flavonoids, tannins, saponins, and alkaloids which act as anti-bacterial and influence the inhibition of bacterial growth.

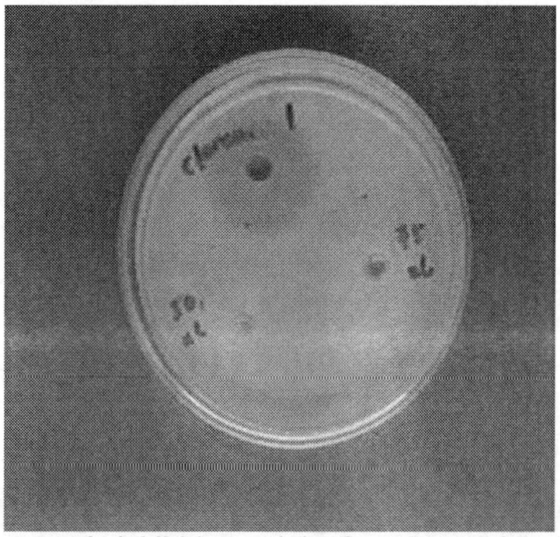

Figure 1. Bacteria inhibition activity from banana bunch extract.

3.2. Phsycochemical properties of rabbit skin tanned by banana bunch

The chemical properties of the rabbit skin tanned by the banana bunch consisted of moisture content, water-soluble content, pH value, and degree of tannage. The moisture content of rabbit skin was 12.09%. Based on the Indonesian National Standards or SNI, this rabbit skin tanned by banana bunch was classified as a leather standard because the requirement of moisture content in leather is a maximum of 18% [11]. High water content (>18%) causes the quality of vegetable tanned skin to be easily overgrown by mold and inhibits the penetration of chemicals at a later stage. Nurbalia [19] explained that the water content of the skin is influenced by the nature of penetration of tanning agent material into the skin, wherein the tannin content begins to diffuse into the skin and attaches itself to the fibers. At the same time, it will release a certain amount of water.

Water-soluble content is a tanning material that is bound or not bound to collagen but is still soluble by water expressed in percent (%). The water-soluble substances in the treatment with banana bunch were 8.74%. This value approaches the skin standard in Indonesia based on SNI [11]. This obtained value is higher than skin tanned by *Acacia xanthophloea* extract (4.4%). Low water solubility indicates that leather tanned with tannins from the studied plants has good water resistance [8]. Covington [3] also stated that low solubility in water is shown by the level of water-soluble content. Besides, the pH value on the skin was 3.42. This obtained value approaches the skin standard based on SNI, which requires the skin to have a pH value ranging from 3.5 to 7 [11]. The pH of the skin tanned with vegetable material is acidic with a low pH, which can prevent the growth of bacteria [20]. In the meantime, the degree of tannage of leather tanned with banana bunch was 52.47%. The result from this sample was almost similar to that of tanned leather with *Acacia xanthophloea*, 52.6% [8]. The value of the degree of tanning will affect the physical properties of the skin [21].

Table 3 shows the obtained elongation value by the tanned leather with the banana bunch of 174.12%. The high elongation is influenced by the process carried out that results in the stretch quickly, increase in length, and not easily torn, damaged, or changing shape [21]. The tensile strength of tanned leather with this banana bunch is 302 kg/cm^2. This value meets the Indonesian leather standard based on SNI, which is a minimum of 75 kg/cm^2[11].

4. Conclusion

Based on the qualitative and quantitative tannin tests, the extract derived from Kepok banana had the best characteristic than other species (Raja and Ambon). Banana bunch extract could inhibit bacteria, such as *Staphylococcus aureus*. The rabbit skin tanned by banana bunch extract had almost the same chemical and physical properties as the skin tanned by the commercial tanning agent (mimosa) and is classified as the Indonesian leather standard (SNI). The banana bunch extract is considered suitable for use as a new vegetable tanning agent.

Acknowledgments

This work was jointly sponsored by the UP2M (Unit for Research and Community Services) Research Grant from Politeknik ATK Yogyakarta, Ministry of Industry, Republic of Indonesia.

References

[1] Wu X, Qiang X, Liu D, Yu L and Wang X 2020 An eco-friendly tanning process to wet-white leather based on amino acids *J. Clean. Prod.* 122399

[2] BPS 2019 Berita Resmi Statistik Pertumbuhan Ekonomi Indonesia Triwulan III-2019 *Ber. Resmi Stat.*

[3] Covington A D 2009 *Tanning Chemistry: The Science of Leather* (Nothampton, UK: RSC Publishig)

[4] Yates C J and Broadhurst L M 2002 Assessing limitations on population growth in two critically endangered Acacia taxa *Biol. Conserv.* **108** 13–26

[5] Monks L and Coates D 2002 The translocation of two critically endangered Acacia species *Conserv. Sci. West. Aust.* **4** 54–61

[6] Guo L, Qiang T, Ma Y, Wang K and Du K 2020 Optimisation of tannin extraction from Coriaria nepalensis bark as a renewable resource for use in tanning *Ind. Crops Prod.* **149** 112360

[7] Lertchunhakiat K, Keela M, Yodmingkhwan P, Sirirotjanaput W and Rungroj A 2016 Comparisons of Physical Characteristics of Crossbred Boer Goat Fur Skin Tanned by Coffee Pomace and Gros Michel Banana Bunch *Agric. Agric. Sci. Procedia* **11** 143–7

[8] China C R, Hilonga A, Nyandoro S S, Schroepfer M, Kanth S V., Meyer M and Njau K N 2020 Suitability of selected vegetable tannins traditionally used in leather making in Tanzania *J. Clean. Prod.* **251** 119687

[9] Ezeonu C S and Ejikeme C M 2016 Qualitative and Quantitative Determination of Phytochemical Contents of Indigenous Nigerian Softwoods *New J. Sci.* **2016** 1–9

[10] Udkhiyati M and Silvianti F 2019 The utilization of chitosan as natural antibacterial for vegetable tanned leather *Mater. Sci. Forum* **948 MSF** 212–6

[11] BSN 2011 Standard Nasional Indonesia SNI 4593:2011: Kulit Jaket Domba/kambing *Badan Stand. Nas.* 31

[12] Maryati T, Pertiwiningrum A and Bachruddin Z 2019 The Effects of Tanning with Kepok Banana (Musa Paradisiaca L) Bunch on the Physical Quality of Rabbit Skin *Int. J. Recent Technol. Eng.* **8** 12564–7

[13] Thomas A W and Frieden A 1973 The Gelatin-Tannin Reaction *J. Biophys. Biochem. Cytol.* **219**

IOP Conf. Series: Materials Science and Engineering **980** (2020) 012019 doi:10.1088/1757-899X/980/1/012019

214–9
[14] Zhang X, Do M D, Casey P, Sulistio A, Qiao G G, Lundin L, Lillford P and Kosaraju S 2010 Chemical modification of gelatin by a natural phenolic cross-linker, tannic acid *J. Agric. Food Chem.* **58** 6809–15
[15] Ping L, Pizzi A, Guo Z D and Brosse N 2011 Condensed tannins extraction from grape pomace: Characterization and utilization as wood adhesives for wood particleboard *Ind. Crops Prod.*
[16] Hoong Y B, Pizzi A, Md. Tahir P and Pasch H 2010 Characterization of Acacia mangium polyflavonoid tannins by MALDI-TOF mass spectrometry and CP-MAS 13C NMR *Eur. Polym. J.*
[17] Hussein S A 2017 Utilization of Tannins Extract of Acacia seyal Bark (Taleh) in Tannage of Leather *Int. Conf. Technol. Innov. Soc.* **08**
[18] Griyanitasari G, Pahlawan I F and Lestari U R 2020 Application of Tegeran (Maclura cochinchinensis) bark as natural tanning agent for the quality of tilapia fish skin *Maj. Kulit, Karet, dan Plast.* **35** 55
[19] Nurbalia E 2016 Characteristics Identification Gambier (Uncaria Gambier Roxb) As Leather Tanning Agent *Int. Conf. Technol. Innov. Soc.* 62–7
[20] Yuliatmo R, Fitriyanto N A, Bachruddin Z and Erwanto Y 2017 Increasing of angiotensin converting enzyme inhibitory derived from Indonesian native chicken leg protein using Bacillus cereus protease enzyme **24** 1799–804
[21] Wibowo R L M S A, Anggriyani E and Yuliatmo R 2018 The influence of sodium chloride replacement with potassium chloride as a curing agent on the quality of tanned pufferfish (Arothon reticularis) skin *Leather Footwear J.* **18** 101–8

IOP Conf. Series: Materials Science and Engineering **980** (2020) 012020 doi:10.1088/1757-899X/980/1/012020

Application of cocoa pod husk (*Theobroma cocoa* Spp) for natural dyes powder on silk batik cloth

Masiswo*, A Haerudin, Isnaini, D W Lestari, G B Mandegani, Y Satria, T K Arta, and V Atika
Center for Handycraft and Batik, Ministry of Industry, Yogyakarta, Indonesia

Email: masiswo@gmail.com

Abstract. The chocolate industry (*Theobroma cocoa*) in Indonesia reaches a total production of approximately 500 thousand tons. It has the potential to be used as natural dyes for the batik industry. This study aims to utilize cocoa pod husk powder on silk batik. The method was carried out by varying the concentration of maltodextrin as filler powder by 5%, 10%, and 15%, the final mordant variations (alum, calcium oxide, and ferrous sulfate) on silk cloth, and the natural dye solution of cocoa pod husk. The natural dye solution for cocoa pod skin was tested for phytochemistry to see the tannin and flavonoid content, the results of dyeing application on silk batik cloth were tested for the quality of its color resistance to washing, color strength, and color differences. The colorfastness test result showed a good category with an average value of 4-5 and the result of the color strength test showed that the best maltodextrin was 5% with a natural dye concentration of 15g / L. The results of alum mordant, calcium oxide, and ferrous sulfate (tunjung) produced colors that led to light and dark with a predominance of brown, red, and blackish brown.

1. Introduction

Natural-colored batik is increasingly popular in the world since the recognition of batik as an intangible cultural heritage of Indonesia. Apart from continuing cultural promotion, there is also an awareness of lifestyle trends in the use of environmentally friendly products. Yet on the other side, natural dyes used for batik are still limited, therefore an exploration of new sources is necessary. Cocoa pod waste is one of the sources that can be made into powder. Natural dyes powder is more practical in use and easy for shipping. Indonesia is still one of the largest cocoa producers in the world. In 2015, 593.331 tons were produced from an area of 1.709.284 hectares. Cocoa -producing regions are spread from East Java and Central Java, South Sulawesi, Southeast Sulawesi, Central Sulawesi, and Sumatra. And Sulawesi Island is the largest contributor, with the number of 62.95%, followed by Sumatera Island with 22.88% and Java Island with 5.42% [1]. Cocoa production tends to produce waste, one of which is cocoa pod waste. These wastes accumulate to rot in the area of cocoa plantations which can cause unpleasant odors to the local environment [2], although some are used as absorbents, animal feed, and others. The estimated availability of cocoa pods is around 516.259 tons. Cocoa pods contain theobromine alkaloids (3.7-dimethylxanthine) with 0.17-0.22% and tannins with 0.84% [3]. Generally, the structure of cocoa consists of four parts, the cocoa pod husk, placenta, pulp, and seeds. About 75% of one whole cocoa fruit is in the form of rinds, 23% cocoa beans, and 2% placenta [4]. The current utilization of cocoa pod is for animal feed [5], organic fertilizer, and adsorbents [6].

The main source of plant pigment is contained in many parts of plants, such as leaves, stems, bark, flowers, fruit, root bark, fruit peels, and other parts, with different amounts and types of color-bearing compounds. The use of natural dyes now has a great demand for the batik craft industry because it is more environmentally friendly. Some types of pigments that we find around us are in the form of chlorophyll which is found in many green leaves, carotenoids, tannins, and anthocyanins. The final mordant or fixation process in the natural coloring process usually uses metal salts such as alum $KAl(SO_4)_2.12H_2O$, calcium oxide $Ca(OH)_2$, and ferrous sulfate/*tunjung* ($FeSO_4$). Besides strengthening bonds, metal salts also function to change the color of natural dyes according to the type of metal salt that binds them [7]. Textile that has been dyed with natural dyes needs to be strengthened to stay attached to the textile and not fade.

Cocoa pod husk can be used as natural dyes for batik products and non-textile natural fibers and has a fairly good fastness (value 4) [3]. Research in the process of taking dye is limited to the use of extraction technology, the disadvantages of extraction technology include the lack of stability of the dye solution during storage, which is marked by a change in the color of the solution due to the influence of bacteria and molds. Another disadvantage is that the extracted dye solution is impractical to use, as well as for dye manufacturers who are constrained in terms of shipping out of the region. Based on those shortcomings, a natural coloring powder from cocoa pods was made.

This study aims to utilize cocoa pod husk powder on silk batik for the batik industry. Natural dyes powder is more practical in storage and less moldy than in solution.

2. Materials and methods

2.1. Materials

Silk fabric, cocoa pod husk solution, batik wax, alum, calcium oxide, ferrous sulfate, and maltodextrin.

2.2. Methods

Cocoa pod husk solution available with 3 types of pH (acidic, neutral, and basic) was extracted at 100°C in 3 hours. The making of powder was processed by mixing a solution of cocoa pod husk in the amount of 5 liters each with maltodextrin (variation 5%, 10%, and 15% w/v) while stirred until homogeneous then filtered. The processing process was carried out at an inlet temperature of 240°C and an outlet temperature of 80°C, and the dye powder formula was 15g/L of water. Application of coloring in silk batik cloth utilized variations of the final mordant (alum, calcium oxide, and ferrous sulfate). The results of coloring applications on batik were tested for the quality of colorfastness against washing, color strength, and color differences (CIE L* a* b*).

3. Results and discussion

The cocoa pods husk used for the experiment were taken from farmers in the Gunung Kidul area. Then, the experiments were carried out in the Central Laboratory of Crafts and Batik process, Yogyakarta while the powderization was applied at the Center for Agro Industry in Bogor. The results of the experiments are as follows.

3.1. The cocoa pods husk solution and powder solution

Data on the tannin and flavonoid contents of the extraction solution from cocoa pods husk made in powder are shown in table 1.

The result of the extraction solution of cocoa pods husk was dried using a spray dryer to powder by adding maltodextrin. Spray drying is a product processing technology by changing the form of liquid into dry particles through a heat-drying spray media. The aim is to reduce the moisture content of the material so that the material becomes more durable, reducing the volume of material to facilitate and save transportation, pack strength, and storage costs. Spray-drying drying technology is also used to convert reactive substances into more stable materials so that the product lasts longer. Spray dryer drying process with the addition of certain solvents, the liquid spray is brought into contact with the heating gas to evaporate the solvent where the resulting product is in the form of a particle dryer in the form of extract powder [8]. The results of the calculation of the yield of cocoa pods husk are shown in table 2.

Table 1. The test results for tannin and flavonoid levels of natural dyes cocoa pods husk extract.

No	Solution sample	Tanin (μg/Ml)	Flavonoid (μg/Ml)	Total Tanin and Flavonoid (μg/Ml)
1	Acidic Cocoa (AC)	1260.00	3474.98	4734.98
2	Neutral Cocoa (AN)	880.25	1292.65	2172.9
3	Basic Cocoa (BC)	1038.00	3163.85	4201.85

Table 2. Yield of cocoa powder.

Variation maltodextrin	Acidic Cocoa (AC)		Neutral Cocoa (AN)		Basic Cocoa (BC)	
	Gram powder	% powder (b/vv)	Gram powder	% powder (b/vv)	Gram powder	% powder (b/vv)
5%	118.7	2.37	96.7	1.93	77.6	1.55
10%	189.6	3.80	182.0	3.64	143.9	2.88
15 %	328.9	6.58	266.9	5.34	204.9	4.10

Table 2 shows that the highest yield of powder was 328.9 with the addition of 15% maltodextrin. The effect of variations in the concentration of maltodextrin as a filler in the manufacture of cocoa pods husk solution was quite influential on the yield of the dye powder produced. The higher the percentage of maltodextrin used, the more amount of dye powder obtained. Conversely, the less the percentage of maltodextrin used, the less cocoa dyes produced. The use of maltodextrin as a filler used for the manufacture of natural dyes powder with spray dryer technology was quite influential on the physical form of the resulting powder.

3.2. Application of cocoa pods husk powder for coloring on silk batik cloth

Cocoa pods husk powder was applied for coloring on silk batik cloth by doing three variations of mordant namely alum, calcium oxide, and ferrous sulfate. The mordant served to bind the cocoa dyes in the fiber so it did not fade easily and direct the color. The cocoa pods husk powder used for coloring silk batik cloth was 15g/L.

3.2.1. Color difference test (CIE L.a.b.). The color difference test L, a, b was carried out from the results of the application of cocoa skin coloring on silk cloth. The color difference test method was based on the brightness (lightness) of color saturation (chroma) and color hue (hue) with L*, a*, and b* notation. The L* notation had values from 0 to 100. The higher L* value means the color approaches white and the lower L* value indicates the color is black/dark.

The notation a* represents the chromatic color of the red-green mixture, with a value of a+ (positive) from 0 to100 indicating the direction of red, and a- (negative) value from 0 to100 indicating the direction of green. The b* notation represents the chromatic color mixture of blue-yellow, with a value of b+ (positive) from 0 to 100 indicating the direction of the yellow color and a value of b- (negative) from 0 to 100 for the direction of blue [9].

The results of coloring with cocoa pods husk powder on silk batik cloth can be seen in table 3. While the results of different color test L*, a*, b* application of natural dye coloring of cocoa pods husk powder on silk batik cloth that produced the level of value notation are shown in table 4.

Table 3, shows that the use of alum mordant, calcium oxide and ferrous sulfate produced different color directions. Ferrous sulfate mordant directed the intensity of colors that tended to be dark, alum directed the intensity of colors that tended to be bright and calcium oxide directed the intensity of colors between light and dark. The darkest color was found in the results of the powder with a mixture of maltodextrin 5% and neutral pH because the active dyes were larger and were absorbed more in fabric. Rusdi [10] states that the addition of a filler to the solution affects the particle size and color produced in a textile fabric. Fewer fillers make the color intensity darker, conversely more fillers make the color intensity whiter.

The L* notation indicates the direction of dark brown or away from white, leading to dark/black. The L* notation of the mordant variation (alum, calcium oxide, ferrous sulfate) for all treatments that had the lowest value of 52.01 was found in the neutral 5% pH concentration of maltodextrin with ferrous mordant. While the L* notation with alum mordant for all types of color treatment lowest value 63.31 was found in the mixing of 5% maltodextrin concentration, the basic pH was the same as the L* notation with calcium oxide mordant but with the lowest value of 61.42. The a* notation all produced an a+ value indicating that the color of cocoa pods husk contained a reddish-brown color. The highest a+ value of 19.23, was found at the base pH of 5% maltodextrin concentration with calcium oxide mordant. Note b* of all test samples produced positive values which showed that the color of the cocoa pods husk powder contained a yellowish-brown color, while the highest positive b notation value of 17.08 was obtained from the treatment of acid pH, 5% maltodextrin concentration, and with Alum mordant.

3.2.2. The colorfastness test to washing. The colorfastness test for washing was applied from the results of the application of cocoa pods husk powder coloring on silk batik cloth. The test method was based on colorfastness to washing [11]. The results of the colorfastness test for washing from the cocoa pods husk powder application test sample on silk cloth are shown in table 5.

The mordants treatment of the test sample produced an average value of 4-5 indicating a good category. This was due to the color fastness of the sample with fewer mordants so that there appeared to be no degradation of color deterioration in the test sample. The color absorbed in the textile fiber was the optimal color because the mordant bonded with the dye in the silk batik cloth fibers well and were not easily separated in washing. Ardhianti [12] states that for dyes used for dyeing to have good colorfastness, a mordant process is needed to lock the dyes that enter the fibers to increase the colorfastness. This mordant process will condition the coloring agent which has been absorbed in the material at a certain time and undergoes a complex reaction between the material with the coloring agent and the material used for mordant.

3.2.3. Color strength test. The color strength test method on silk batik cloth was measured at the maximum wavelength, at the wavelength with the smallest reflectance value (%R) indicating the maximum wavelength and if converted to the K/S value would produce the largest K/S value. The higher K/S value means the absorption of dyes by a larger material or the color is darker and vice versa, the lower K/S value means less absorption of dyes so that the color is brighter [13]. The results of the color strength test from the application of cocoa pods husk powder coloring are shown in table 6.

IOP Conf. Series: Materials Science and Engineering **980** (2020) 012020 doi:10.1088/1757-899X/980/1/012020

Table 3. Results of coloring of cocoa pods husk powder on silk batik cloth.

No	Sample	Mordant		
		Alum	Calcium oxide	Ferrous Sulphate
1	15gpl/NC5			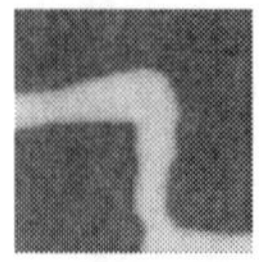
2	15gpl/AC5			
3	15gpl/BC5			
4	15gpl/NC10			
5	15gpl/AC10			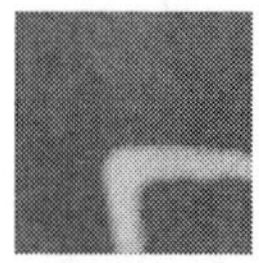
6	15gpl/BC10			
7	15gpl/NC15			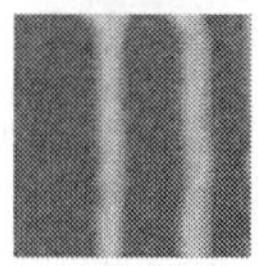
8	15gpl/AC15			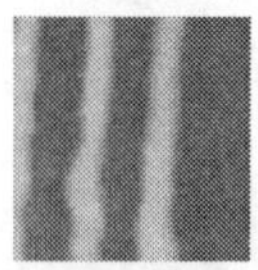
9	15gpl/BC15	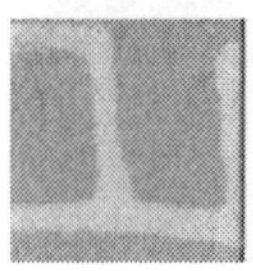		

Table 4. Test results for different colors of the application of cocoa pods husk powder on silk batik cloth.

No	Sample	Alum			Calcium oxide			Ferrous Sulphate		
		L*	a*	b*	L*	a*	b*	L*	a*	b*
1	15gpl/BC5	63.31	17.74	8.74	61.42	19.23	9.87	52.76	8.13	3.86
2	15gpl/BC10	70.45	13.98	7.13	68.61	16.53	8.83	60.50	9.67	5.39
3	15gpl/BC15	73.61	12.03	6.63	71.41	14.53	9.29	64.62	9.59	6.46
4	15gpl/AC5	68.84	14.02	17.08	72.05	11.78	15.34	57.88	7.05	6.18
5	15gpl/AC10	74.57	11.01	12.68	72.35	12.76	14.09	62.61	7.81	7.60
6	15gpl/AC15	76.43	9.11	11.95	74.45	10.92	14.42	65.44	7.29	7.97
7	15gpl/NC5	69.13	12.99	12.61	63.50	16.01	16.99	52.01	7.09	3.49
8	15gpl/NC10	69.64	12.63	11.61	65.22	14.31	16.60	52.18	6.81	3.00
9	15gpl/NC15	73.01	10.54	11.05	72.11	12.32	13.47	60.19	7.16	5.69

Table 5. Test results of the color fastness of silk batik fabric against washing.

No	Sample	Silk batik		
		Alum	Calcium oxide	Ferrous Sulphate
1	15gpl/NC5	4	4	4 -5
2	15gpl/NC10	4	4	4
3	15gpl/NC15	4	4 – 5	4 – 5
4	15gpl/AC5	4 – 5	4 – 5	4
5	15gpl/AC10	4	4	4 -5
6	15gpl/AC15	4 – 5	4	4
7	15gpl/BC5	4	4	4 -5
8	15gpl/BC10	4 -5	4 – 5	4 -5
9	15gpl/BC15	4 -5	4	4

Table 6. Results of the color strength test of the application of cocoa pods husk powder on silk batik cloth.

No	Sample	Silk batik (K/S)		
		Alum	Calcium oxide	Ferrous Sulphate
1	15gpl/NC5	0.35	0.55	1.05
2	15gpl/AC5	0.22	0.37	0.45
3	15gpl/BC5	0.36	0.39	0.80
4	15gpl/NC10	0.42	0.51	0.86
5	15gpl/AC10	0.12	0.18	0.42
6	15gpl/BC10	0.18	0.18	0.48
7	15gpl/NC15	0.20	0.22	0.32
8	15gpl/AC15	0.11	0.15	0.34
9	15gpl/BC15	0.17	0.21	0.20

The average K/S results were obtained for the use of maltodextrin and acidic, basic, and neutral pH to alum, calcium oxide, and ferrous sulfate mordants to illustrate the level of color strength. The concentration of maltodextrin on the value of color strength from the application of cocoa pods husk powder coloring on silk cloth batik which got the highest value of 1.05, powder with 5% maltodextrin concentration, neutral pH with tide mordant is shown in table 6. Kembaren [14] said that the lower use of maltodextrin filler concentration will result in better color strength, while the higher usage of maltodextrin filler concentration, the lower the color strength level will be.

Variations in the treatment of the mordant substances of alum, calcium oxide, and ferrous sulfate on the dyeing results of silk batik cloth have been shown to increase the value of different color strength. The variation of mordant material also influenced the color direction with different strength intensities. The highest strength value was the coloration of the sample which was fixed with a color intensity which led to dark/black. Because at the time of immersion with ferrous sulfate mordant, a reaction occurred between tannins and flavonoids from dyes with Fe^{2+} metal from ferrous sulfate mordant materials which produced Ferro tannate complex salts.

Table 6 shows the average K/S levels of pH (acidic, basic, neutral), the use of maltodextrin variations, and mordant variations. The results of the mean calculations are shown in table 7.

Table 7. Average K/S values of silk batik fabric.

No	Variation	Average value
1	Acidic pH	0.26
2	Basic pH	0.33
3	Neutral pH	0.50
4	5 % maltodextrin	0.50
5	10 % maltodextrin	0.37
6	15 % maltodextrin	0.21
7	Alum mordant	0.24
8	Calcium oxide mordant	0.31
9	Ferrous sulfate mordant	0.54

The dark color of the application of the natural dye of cocoa pods husk powder in silk, which was seen that the addition of 5% maltodextrin filler, gave a higher color strength value of 0.50 K/S values. The greater amount of maltodextrin produced a lower color strength value. This was because the amount of filler added affects the concentration of the dyes produced, the more amount of maltodextrin used reduced the concentration of the dye found in the color powder so that less color was absorbed in fabrics. In the manufacture of powder products, maltodextrin fillers were added to increase the volume and weight of the powder produced and the drying speed [15].

The treatment of the use of ferrous sulfate mordant produced the greatest value of 0.54 K/S on silk batik cloth. This was because the ferrous metal donor (Fe^{2+}) performed a complex metal reaction between tannins from natural dyes with complex salts (Ferro tannates) and formed covalent bonds so that there was a good bond between the dye with fibers and produced a high degree of color strength [16]. This shows that natural cocoa pods husk powder had a good affinity for protein fiber.

4. Conclusion

Cocoa pod husk powder can be used as a dye on silk batik and is ready to use for the batik industry. The use of maltodextrin 5% filler for cocoa pods husk powder was the best. This means that the fewer fillers will further enhance the color strength. Cocoa pods husk powder which gave the best color pigment was a neutral pH condition. The use of ferrous sulfate mordant led to dark/black color intensity, while alum mordant led to bright/yellow color intensity.

Acknowledgments
Our gratitude goes to the head of the Yogyakarta Center for Handicraft and Batik, cocoa farmers in Gunung Kidul, and team researcher of making cocoa pods husk powder for natural dyes.

References
[1] Direktorat Jenderal Perkebunan 2018 *Statistik Perkebunan Indonesia Komoditas Kakao 2017-2019* Jakarta
[2] S N Wulan 2001 Kemungkinan Pemanfaatan Limbah Kulit Buah Kakao (*Theobroma cacao* L) Sebagai Sumber Zat Pewarna (β-karoten) *J. Teknol. Pertan.*
[3] Pujilestari T, Farida, E Pristiwati, A Haerudin, and V Atika 2016 Pemanfaatan Zat Warna Alam Dari Limbah Perkebunan Kelapa Sawit Dan Kakao Sebagai Bahan Pewarna Kain Batik *Dinamika Kerajinan dan Batik* **33(1)**
[4] Zain M 2009 Substitusi Rumput Lapangan dengan Kulit Buah Coklat Amoniasi dalam Ransum Domba Lokal *Journal of Animal Science and Technology*
[5] Kamelia M and F Fathurohman 2017 Pemanfaatan Kulit Buah Kakao Fermentasi Sebagai Alternatif Bahan Pakan Nabati Serta Pengaruhnya Terhadap Pertumbuhan Ternak Entok (*Cairina muschata*) *Jurnal Tradisi Biologi* **08(1)**
[6] Purnamawati H and Utami B 2014 Pemanfaatan Limbah Kulit Buah Kakao (*Theobroma cocoa L*) Sebagai Adsorben Zat Warna Rhodamin B *Prosiding Seminar Nasional Fisika dan Pendidikan Fisika*
[7] Samanta A K and Agarwal P 2009 Application of natural dyes on textiles *Journal of Fibre and Textile Research.*
[8] Parastiwi E, Rahmad A and Hendri S E S 2015 Bubuk ekstrak sari buah dan sayur: teknologi produksi menggunakan metode spray drying *Prosiding SENTIA Politeknik Negeri Malang (Malang).*
[9] Hunter Associates Laboratory I 2008 CIE L* a* b* Color Scale **8**
[10] Rusdi M 2017 Karakteristik Ukuran Partikel dan Indeks Polidispersitas Formulasi Nanoemulsi Pewarna Alam Ekstrak Kayu Secang (*Caesalpinia Sappan Linn*) *Jurnal Pertanian Terpadu*
[11] SNI ISO 105-C06 2010 *Textile-Test method of color fastness-Part C06: Colorfastness to domestic and commercial washing* (Jakarta: Badan Standardisasi Nasional)
[12] Ardhianti N I D and Sari D C 2011 Pengaruh Penggunaan Fiksator Kapur, Tawas dan Tunjung Pada Proses Pewarnaan Kain Batik Dengan Zat Warna Alam Larutan Soga
[13] Kanaya D 2005 *Bahan Ajar Kimia Zat Warna* (Bandung: Sekolah Tinggi Teknologi Tekstil)
[14] Kembaren R B et al 2014 Ekstraksi Dan Karakterisasi Serbuk Nano Pigmen Dari Daun Tanaman Jati (*Tectona Grandis Linn. F*) *Jurnal Kimia dan Kemasan*
[15] Gonnissen Y, Remon J P and Vervaet C 2008 Effect of maltodextrin and superdisintegrant in directly compressible powder mixtures *European Journal of Pharmaceutics and Biopharmaceutics* **68** 277-282
[16] Haerudin A, and Satria 2018 Ekstraksi Limbah Serutan Kayu Matoa (*Pometia pinnata*) sebagai Zat Warna Alam pada Pewarnaan Kain Batik Serat Protein

IOP Conf. Series: Materials Science and Engineering **980** (2020) 012021 doi:10.1088/1757-899X/980/1/012021

Biduri (*Calotropis gigantea*) leaves extract as natural dyes and ultraviolet protector applied on silk fabric with an exhaust dyeing method

J Nugraha*, A Sukmawati, A S Mulyawan, D Sugiyana
Centre for Textiles (Balai Besar Tekstil) , Jalan Ahmad Yani No. 390 Bandung, Indonesia

Email: jaka@kemenperin.go.id

Abstract. Biduri (*Calotropis gigantean*) leaves extract can be natural dyes and ultraviolet protector agents on the silk fabric. In this research, silk fabric with ultraviolet protection performance was developed using biduri leaves extract in a simple dyeing process. This study aimed to obtain a colored silk fabric from biduri leaves extract with good UV protection properties. The methodology consisted of the extraction process using an aqueous method, dyeing process with exhaust dyeing method, and post mordanting with several mordant. Evaluation of dyeing results included color strength (K/S), color brightness (L*), color direction (a*, b*), ultraviolet protection factor (UPF), and durability after washing. The result showed brightness was inversely proportional to the color strength, and the direction of the resulting color were greenish and yellowish. Dyeing with dyes from the biduri leaves extract with $FeSO_4$ mordant were effectively reflected by the improvement of the UPF up to 31 from an initial value of 5.4. UPF value after 5 times washing down to 29.2 and at 26.8 washing 10 times. Biduri (*Calotropis gigantea*) leaves showed a high potential to be utilized as natural dyes and UV protector agents simultaneously on the silk dyeing process.

1. Introduction

Indonesia has huge biodiversity with a high potential for utilization, one of which is biduri. However, exploration and utilization of this plant have not been optimized. Biduri *(Calotropis gigantea)* is one of the shrub plants/reeds plants that could grow widely on the coast, the mountains, and even on hard and calcareous soil [1]. In the textile sector, especially in functional textile, the utilization of biduri plants still limited to the biduri fibres such as material for filler and insulative filler [2], oil-absorbing [3], and composite [4]. Another part that may potentially be used in textiles is leaves. Alkaloids, glycoside, saponin, fats and oil, tannin and phenolic, flavonoids are bioactive which contained in biduri leaves [5]. Flavonoids, tannin, and phenolic are commonly known as a natural pigment for dyeing textile material. Its purposes not only for dyes but also for enhancement of other functional properties such as antimicrobial and UV protector [6,7].

Ultraviolet light contained in solar radiation with wavelength 200-400 nm, is considered to be one of the causes of disease in humans, such as skin cancer, wrinkles and DNA cell damage [7,8]. Protection against ultraviolet radiation can be achieved by using various types of sunscreens and physical barriers, such as clothing. Clothing materials can be chemically treated to make them able to absorb or reflect ultraviolet radiation. One of the fabrics that can be used for ultraviolet protection is silk fabric, however its ability to protect ultraviolet radiation requires more treatment to enhance the protection performance through the immobilization of such UV protector agents [7].

In previous research, one of genus of biduri, milkweed (*Calotropis procera*) leaves extract has been used as natural dyes to dye cotton and served as a potential source of natural dyes [8]. Moreover, Gajanan (2018) has reported biduri leaves extract has an ability to absorb in the entire UV range and considered as a proper wide spectrum sunscreen and also in anti-aging cream preparation. In this research, we aim to produce a colored silk fabric with good UV protection properties from biduri leaves extract using a simple and common dyeing method in the textile wet process. Evaluation to the silk fabric was carried out to its color value, UV protection performance, and durability after washing.

2. Materials and methods

2.1. The materials and apparatus

100% silk fabric that was ready to dye produced by Centre for Textiles, Bandung, Indonesia, with EPI 86, PPI 80, weight 67.11 g/m^2. Fresh biduri leaves were harvested from Bandung, West Java during January-February 2020. $Al_2(SO_4)_3$, $FeSO_4$, and $CuSO_4$ (Merck) were used as mordant. Teepol was used as a wetting agent. Aqua demineralization (Brataco) was also used for boiling, washing and dilution of whole processes.

The apparatus used in this research included: laboratory glassware (Iwaki Pirex), string hotplate (Heidolph), oven (Heraeus), analytical balance (Chyo JL-180), launder-o-meter (SDL Atlas), padder and dryer (Werner Mathis AG), spectrophotometer (X-Rite Color-i7), spectrophotometer UPF (JASCO V-750). All works on this research were conducted in textile chemistry research laboratory and textile chemistry testing laboratory Centre for Textiles.

2.2. Extraction biduri leaves

Biduri leaves were collected and washed correctly with water. The leaves material was shade-dried for five days, roasted at 105°C for 1 hour, then uniformly ground using the mechanical grinder and passed through the sieve number 60. Biduri leaves powder were boiled by aqua demineralization in 100°C with liquor ratio 1 : 10 for 60 minutes and then filtered. The filtrate was then used in the dyeing process.

2.3. Dyeing procedure

The dyeing process using biduri leaves extract included dyeing, washing, rinsing, and mordanting. The dyeing process was conducted using a laboratory exhaust dyeing machine launder-o-meter at 40°C for 30 minutes with a liquor ratio at 1 : 30. In order to reduce the surface tension of the fabric, 2 mL/L wetting agent was added into dye bath. After 30 minutes process, the fabric washed and rinsed prior to the mordanting process. Furthermore, the fabric immersed into a solution containing 30 g/L mordant respectively and padded with 80% WPU for 3 dips and 3 nips, then drying on 100°C for 2 minutes. Finally, the fabric samples were washed thoroughly with cold water, squeezed, and dried.

2.4. Color measurement

As preparation for colour measurement, fabric samples were scanned with 350 dpi for recording the sample photographs. Color measurement procedures followed SNI-ISO 105-J03 2016 (Textile – Test for colourfastness – part J03; Calculation of colour difference). Color values were evaluated by means K/S and CIELab color difference value on spectrophotometer. K/S value was automatically assessed by kubelka-munk equation 1.

$$K/S = \frac{(1-R)^2}{2R} \tag{1}$$

where K is the light absorption coefficient, S is the scattering coefficient and R is the reflectance of the dyed samples at the wavelength of the maximum absorption.

2.5. UPF measurement and washing durability

Dyed silk fabric samples were prepared for UV protection factor (UPF) measurement, which was carried out by following a standard procedure AS/NZS 4399: 2017 (Sun protective clothing – Evaluation and classification*)* on wavelength interval 200 – 00 nm. UPF factor each sample determined by total transmittance spectra and automatically assessed by following equation 2.

$$UPF = \frac{\sum_{290}^{400} E_\lambda S_\lambda \Delta_\lambda}{\sum_{290}^{400} E_\lambda S_\lambda T_\lambda \Delta_\lambda} \quad (2)$$

Where E_λ is CIE reference erythema dose spectrum, S_λ is the radiation intensity distribution of sunlight, T_λ is the difuse transmittrance spectrum (%), and Δ_λ is the bandwith. UPF classification based on AS/NZS standard 4399: 2017 is shown in table 1.

Table 1. UPF classification based on AS/NZS standard 4399: 2017 [1,2].

UPF rating	UV protection	UV blocked
15-24	Good	93.3 – 95.9%
25-39	Very good	96 – 97.4%
40-50, 50+	Excellent	> 97.5%

In order to find out the effect of multiple washing to UPF value, washing durability test for sample fabric was conducted following SNI-ISO 105-C06:2010 (Textile – Test for color fastness – part C06; Color fastness to domestic and commercial laundering).

3. Results and discussion

3.1. Effect of mordant on color strength and color value

Table 2 shows the L*, a*, b*, and K/S value of silk fabric dyed with biduri leaves extract dye solution at different mordant. L*, a*, and b* refer to the three-axis of the CIE-Lab color system. The L* value indicated the perceiving of brightness with interval scale value from 0 (black) until 100 (white). The higher the L* reading, the lighter the color. a* and b* value indicating color direction, +a* for red, -a* for green, +b* for yellow, and -b* for blue. The color strength value increased with the addition of dyes and mordant, while brightness decreased. Mostly, color direction of resulted silk fabric was greenish and yellowish, addition $Al_2(SO_4)_3$ and $CuSO_4$ mordant generated color direction to be brighter and darker, respectively. $FeSO_4$ mordant generated darkest color as the effect of conversion from part of ferro became ferri after its reaction with oxygen in the air. Ferro and ferri forms on silk fabric produced overlapping spectrum and changed the maximum wavelength, thus affected the color of silk fabric became darker [7].

Mordant was able to increase the complexity among color molecules and silk fabric, therefore the color intensity would also increase. As the result, mordanting usually generates darker color of fabric compared than that of fabric without mordant [12]. The darkest color was found in silk fabric which using $FeSO_4$ mordant, whilst the lightest was achieved in silk fabric which using $Al_2(SO_4)_3$ mordant. Silk fabric with $Al_2(SO_4)_3$ mordant was darker than that of undyed fabric, but lighter compared to fabric without mordant. This phenomenon can be explained as follows, iron and copper ions tend to form coordination complex between dye and amino acid or carboxylic groups, while as aluminum ion tends to form bonding with dyes rather than fibre, therefore reduce interaction among dyes and silk fabric [7].

Table 2. Color strength and color values of dyed silk with biduri leaves extract.

Mordanting agent	Mordant concentration (g/L)	Color strength (K/S)	Color value L*	a*	b*	Color obtained
-	Undyed	0.74	88.89	0.06	12.87	
Without	-	1.85	82.95	-1.21	23.9	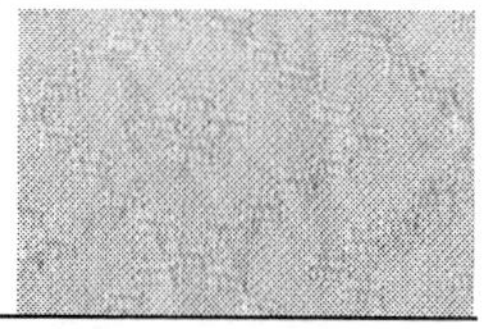
$Al_2(SO_4)_3$	30	0.83	88.35	-1,43	14.63	
$FeSO_4$	30	6.35	69.75	5.06	25.17	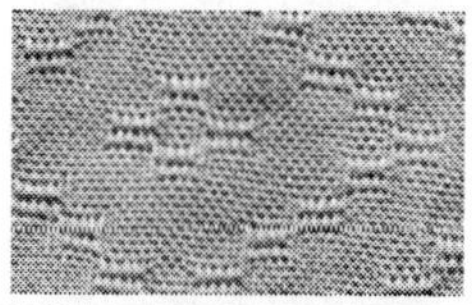
$CuSO_4$	30	1.51	83.12	-7.62	1.51	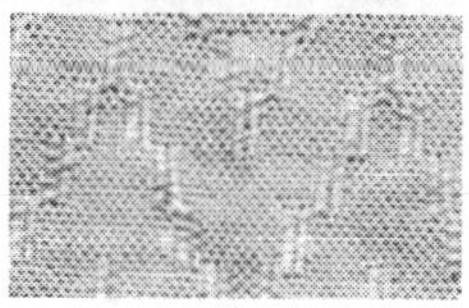

3.2. UPF value

The UV protection factor value are presented in table 3. The UPF value on silk fabric after dyed result with and without mordant increased compared than that of standard fabric (undyed). Silk fabric naturally had a low UPF value and increased significantly after dyeing with biduri leaf extract. This indicated that silk fibre had little resistance to exposure to ultraviolet rays and biduri leaf extract can increase its UPF value significantly. The addition of mordant could relatively increase the UPF value of silk fabric colored with biduri leaf extract. Mordant can react with dye molecules and silk fabric, so that the dye fixed on the fabric. The result showed that the mordant used in this study could reduce the transmission of ultraviolet light through silk. However, based on table 3, the highest UPF enhancement found in the sample fabric which using $FeSO_4$ as mordant and classified as "very good" UPF.

3.3. UPF after washing durability

After the washed treatment, sample fabric run into decrease UV protection performance as shown in table 4. The highest decline occurred in the sample fabric without mordant down to 37.3% from initial UPF value after 5 times washed. This decline indicates the use of mordant is required in dyeing with natural dyes, thus dyes may bind to the fibre perfectly.

Table 3. UPF test result of silk dyed with biduri leaves extract.

Mordanting agent	Mordant concentration (g/l)	UPF	Class	Enhancement (%)
-	undyed	5.4	No class	-
Without	-	9.2	No class	70.37%
$Al_2(SO_4)_3$	30	x6.4	No class	18.52%
$FeSO_4$	30	31.175	Very good	477.31%
$CuSO_4$	30	12.05	No class	123.15%

Table 4. UPF test result of silk dyed with biduri leaves extract after washed.

Mordanting agent	Mordant conc. (g/L)	UPF	UPF 5x washes	Decline (%)	Class.	UPF 10x washes	Decline (%)	Class.
without	-	9.2	6.7	37.3	No class	-	-	-
$Al_2(SO_4)_3$	30	6.4	6.1	4.7	No class	5.8	9.4	No class
$FeSO_4$	30	31.2	29.2	6.3	Very good	26.8	14.0	Very good
$CuSO_4$	30	12.1	11.2	7.1	No class	8.0	33.6	No class

4. Conclusion

Biduri (*Calotropis gigantea*) leaves can be utilized as natural dyes and UV protector agents simultaneously on the silk dyeing process. The color strength and color values may change according to mordant used, which $FeSO_4$ mordant showed the most significant influence to color strength of the resulted fabric. Mostly, the color direction produced greenish and yellowish in whole resulted silk fabrics. Moreover, the use of $FeSO_4$ mordant significantly increasing UPF value up to 31.2 and keeping UPF value on "very good" classification after 10 times washing treatment.

References

[1] M D Sukardan, D Natawijaya, P Prettyanti, Cahyadi and E Novarini 2017 Karakterisasi serat dari tanaman biduri (*Calotropis gigantea*) dan identifikasi kemungkinan pemanfaatannya sebagai serat tekstil *Arena Tekstil* **81(2)** 51-62

[2] A W Sana, N Noerati, D Sugiyana and M D Sukardan 2020 Aplikasi serat alam biduri (*Calotropis gigantea*) sebagai bahan pengisi insulatif pada jaket musim dingin *Arena Tekstil* **35(1)** 1-12

[3] Zheng Y, Zhu Y, Wang A and Hu H 2016 Potential of *Calotropis gigantea* fiber as an absorbent for removal of oil from water *Ind. Crops Prod.* **83** 387-90

[4] Sayanjali J M, Hasani H, Zadhoush A and Safi S 2018 Effect of alkali treatment on mechanical properties of the green composites reinforced with milkweed fibers *J. Text. Inst.* **109** 24-31

[5] Singh S, Mishra R M and Shrivastava M P 2014 Preliminary phytochemical screening of *Calotropis gigantea* leaf **4** 2-4

[6] Koh E and Hong K H 2019 Preparation and properties of wool fabrics dyed with spent coffee ground extract *Text. Res. J.* **89** 13-9

[7] Mongkholrattanasit R, Kryštůfek J, Wiener J and Viková M 2011 Ultraviolet protection properties of silk fabric dyed with eucalyptus leaf extract *J. Text. Inst.* **102** 272-79

[8] Hussaan M, *et al.* 2017 Microwave-assisted enhancement of milkweed (*Calotropis Procera* L.) leaves as an eco-friendly source of natural colorants for textile *Environ. Sci. Pollut. Res.* **24** 5089-94

[9] Saewan N and Jimtaisong A 2013 Photoprotection of natural flavonoids *J. Appl. Pharm. Sci.* **3**

IOP Conf. Series: Materials Science and Engineering **980** (2020) 012021 doi:10.1088/1757-899X/980/1/012021

129-41
[10] Wungkana I and Suryanto E 2013 Aktivitas antioksidan dan tabir surya fraksi fenolik dari limbah tongkol jagung (*Zea Mays* L.) *Pharmacon J. Ilm. Farm. – Unsrat* **2** 149-55
[11] J Nugraha and E Y Rakhmatiara 2020 Pemanfaatan daun rami sebagai bahan zat warna alam dan fungsionalisasinya pada pencelupan kain kapas dan sutera *Arena Tekstil* **35(1)** 29-38
[12] Chattopadhyay S N, Pan N C, Roy A K, Saxena S and Khan A 2013 Development of natural dyed jute fabric with improved colour yield and UV protection characteristics *J. Text. Inst.* **104(8)** 808-18
[13] Bhattacharya S D and Shah A K 2000 Metal ion effect on dyeing of wool fabric with catechu *J. Soc. Dye. Colour*

IOP Conf. Series: Materials Science and Engineering **980** (2020) 012022 doi:10.1088/1757-899X/980/1/012022

Improving aluminum strength with chemical modification based on titanium and boron elements

E Rahman*, R Kumar, V Monandes and R Yadi
Product Development and Industrial Standardization Center of Pekanbaru

Email: eddifa-rahman@kemenperin.go.id

Abstract. In general, Al5TiB is added to hypoeutectic Al-Si alloys with Si levels <12%. The addition of Al5Tib grain refiner to hypoeutectic Al-Si alloy has been shown to be effective in increasing the mechanical properties of the alloy. While the addition of Al5TiB grain refiner to the Al-11% Si alloy master has never been done. This study aims to determine the effect of adding Al5TiB grain refiner to the microstructure and mechanical properties of Al-Si eutectic alloys (Si 11%). The mechanical property to be known after the addition of Al5TiB grain refiner is hardness. In this research Al-11% Si material was used. Al5TiB grain refiner added to Al-11% Si material to determine changes in microstructure and the value of violence that occurs. The results showed that the addition of Al5TiB would produce a large number of fine grains with a uniform distribution and shortening the distance between dendrite arms. The results of the hardness test showed that the hardness tended to increase with increasing levels of Al5TiB grain refiner added. The decreasing in the value of hardness in the addition of Al5TiB grain refiner with 0,15% is due to the addition of Al5TiB grain refiner has exceeded its maximum limit.

1. Introduction

Aluminum is one of the elements that has many benefits in human life. Light weight with higher strength value makes aluminum an alternative to the future of wood. In line with this, Indonesia is a very rich area of raw materials that are made of aluminum bauxite based on data from the central bureau of statistics (BPS) in 2014 – 2017. In 2017, the potential for bauxite in Indonesia reached 2.7 metric tons [1] Bauxite is one of the most potentially mining goods in Indonesia.

In 2017, bauxite production reached 1.2 billion tons and decreased when compared to production in the year 2014. Despite the decline in production but still one of the most promising potential in the future. When compared with the potential to export aluminum raw materials to several countries in the world, aluminum raw materials also have a very promising potential. In 2018, Indonesia was able to export aluminum raw materials around 8.6 million tons [2].

In recent years, China is intensively importing aluminum as raw materials which will be used as the main raw material in vehicle manufacturing. Although in the current global economic crisis, it is predicted that there will be a decline in automotive sales, but the development of the automotive industry still making enough progress rapidly [3]. Based on data compiled from trademap.org, it can be seen that the export of aluminum as raw materials from Indonesia, taken as a whole by China. This is in line with data from the CAAM National Bureau of Statistics, Company Reports, Public Sources, CM Group which states that China's total state vehicle production in 2017 reached 31.4 million units which increased by 4.7% from the previous year. Because of the massive increase in vehicle production in

China, the need for aluminum automatically will also increase. In 2017 China's aluminum consumption for vehicles reached 3.9 Metric Tons (MT) and increased by 0.3 MT from before. This shows that there is a tendency for aluminum levels in vehicles to increase due to the many advantages of aluminum. This is arguably an opportunity for the Indonesian state to have abundant natural resources for the aluminum raw material itself.

The increasing use of aluminum in a vehicle can result in a reduction in weight in the vehicle. This will automatically cause the use of fuel in the car will be more efficient. Seeing this fact, it is very clear that the level of aluminum usage will continue to increase every year.

With the many needs for aluminum in the transportation sector, especially for vehicle engine components, it is absolutely necessary to have aluminum with good quality and good characteristics in accordance with needs. However, the use of aluminum alone is not enough to get certain desired characteristics. Therefore, aluminum needs to be combined with silicon because this alloy has good characteristics as vehicle engine components such as pistons and engine blocks.

Although a lot of aluminum used for making automotive products, however aluminum also has disadvantages. So it has limitations in the application like hardness, wear-resistance and strength. This is very important to consider in product manufacturing automotive especially in products that are experiencing loading and experiencing friction with other components. Some aluminum-silicon applications in automotive components with a silicon content of 7% [3]. To complement the properties of silicon aluminum alloys with better quality where the mechanical properties according to needs, there are several parameters that can be used. One way is to add other elements in the silicon aluminum alloy. In this process, aluminum-silicon is combined with certain elements in order to improve the mechanical properties of the product.

The process of adding other elements is one of the methods used to obtain good mechanical properties of a liquid metal (aluminum silicon). This process is done by adding grain refiner elements to aluminum alloys. One of the refining elements is a mixture of aluminum, titanium and boron (Al5TiB). Al5TiB alloy acts as a grain refiner in the alloy. Al5TiB grains refiner on silicon aluminum alloys are used to control the grain structure. Grain refiners produce nucleating sites for the formation of primary dendrites and produce large numbers of fine and uniform grains distribution. In Al-Si (aluminum silicon) alloys containing Fe, finer grains will produce smaller intermetallic and more uniform distribution. With the occurrence of grain reduction and uniform distribution, the mechanical properties of the product will be much better than before [4].

2. Materials and methods

2.1 Data and sources of data

The data presented in this study were obtained by conducting research on Aluminum Silicon alloys added with other elements namely Titanium and Boron. The main material in the form of aluminum-silicon eutectic master alloy is Al-11% Si and the temperature used wass ± 700°C. Each sample was added Al5TiB content of 0.05%; 0.1%; 0.15%. By doing this research it is expected to know the effect of adding Al5TiB to the morphology of the alloy and its effect on the hardness of the Al-Si alloy. The hardness test specimens and specimens for microstructure testing using HRC (*Mobile Hardness Tester*) machine were made as seen in figure 1. The microstructure was obtained by using SEM (Scanning Electron Microscope) tool. The test results are presented in tabular form and then continued with data analysis related to the test results. In addition, the effect on the microstructure of the alloy was also investigated because the shape of the microstructure is very influential on the strength value of a material or alloy material.

2.2 Data analysis

Data analysis was performed by comparing the results obtained in theory with the results obtained from testing. Then drawn conclusions from the results of testing that had been done. The conclusion obtained was the effect of giving other elements to an alloy both in the form of altered violence and changes in the morphology of the microstructure.

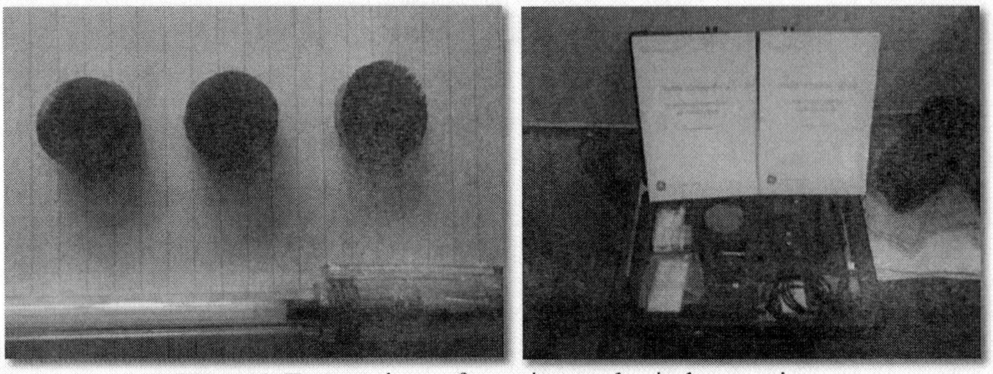

Figure 1. Test specimens for testing mechanical properties.

3. Results and discussions

Hardness testing was carried out on the test specimen with the aim to see the effect of chemical modification on the strength increase of aluminum material. The strength aspect that was seen was the value of the hardness of the test specimen. Before testing the effect of the added elements, a hardness test for each test sample that had not yet been added was executed first. It aimed to get the value of violence due to the influence of the addition of these other elements.

Hardness testing values for each sample can be seen in table 1. For each sample, hardness testing was performed as many as 8 points for each specimen that was carried out randomly, then the average hardness value was taken. Hardness testing was performed for all four variables (0% Al5TiB, 0.05% Al5TiB, 0.1% Al5TiB, and 0.15% Al5TiB), all four variables were tested using Hardness Mobile Tester. From the hardness test data obtained, then plotted in the form of graphs of Al5TiB levels versus hardness values (HRC). The graph obtained is seen in figure 2, while microstructure of hardness values for each Al5TiB addition variable describe in figure 3.

Table 1. Testing values for sample hardness after adding Al5TiB.

Al_5TiB levels	Hardness rating for each test point (HRC)								Hardness average
	I	II	III	IV	V	VI	VII	VIII	
0%	24.4	26.1	29.8	20	29.5	21.3	27.8	24.3	25.4
0.05%	25.4	28.5	28.2	24.3	22.4	28.4	27.3	24.4	26.1
0.10%	25	25.6	22.1	29.1	28.4	27.9	28	26.8	26.6
0.15%	22.3	21.8	21.3	23.1	23.5	22.9	25	25.7	23.2

There was clearly an increase in the value of hardness after the addition of other chemical elements into the alloy. The chemical elements of Titanium and Boron are believed to be able to improve the morphology of the microstructure of aluminum and silicon alloys. The results obtained from the test are in accordance with the theory that the Al5TiB alloy acts as a grain [4].

Grain refiners produce nucleating sites for the formation of primary dendrites and produce large numbers of fine and uniform grains distribution. In Al-Si alloys containing Fe, finer grains will produce smaller intermetallic and more uniform distribution. With the occurrence of grain reduction and uniform distribution, the mechanical properties of the product will be much better than before [4].

The test results were plotted in graphical form to show the improvement of mechanical properties that had occurred in the study. Figures 2 and 3 show that as the aluminum and boron elements increased in silicon aluminum alloys, the grain structure would become smaller and increase the hardness value. But there was a maximum level that limits the number of alloys which, if achieved, would actually reduce the hardness of the alloy.

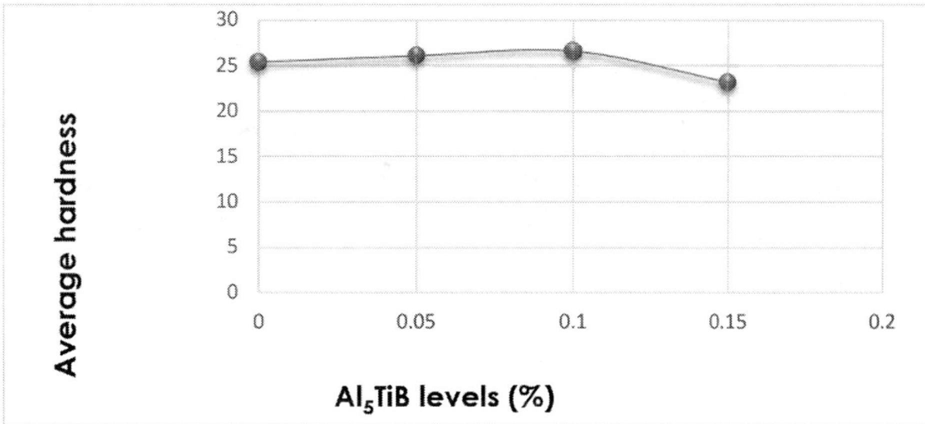

Figure 2. Hardness values for each Al5TiB addition variable.

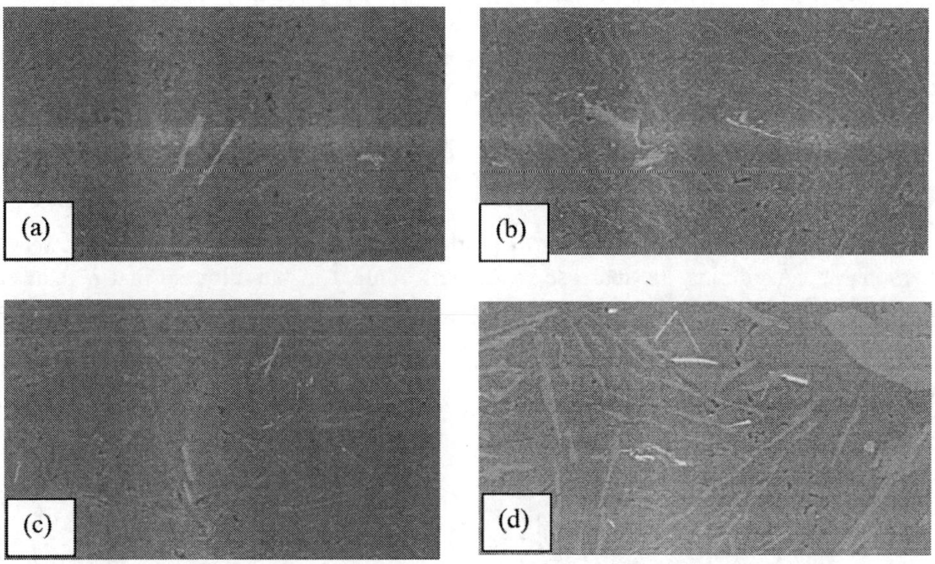

Figure 3. Microstructure of Hardness values for each Al5TiB addition variable
(a) microstructure Al_5TiB level 0, (b) microstructure Al_5TiB level 0.05,
(c) microstructure Al_5TiB level 0.1 and (d) microstructure Al_5TiB level 0.15.

The hardness value would tend to increase with increasing levels of Al5TiB in the alloy. These data are in accordance with the research conducted by Suhariyanto which stated that the results of the hardness test showed that Ti was quite influential on the mechanical properties of the alloy, namely the increase in the value of the hardness. Titanium combined with boron or carbon is an Al-Si alloy element that serves to refine grain (grain refiner) in cast aluminum, because Ti C or TiB can be the core of aluminum alloys. This is what affects the changes in mechanical properties and microstructure [5].

With the finer grain, the propagation of the blockage will be more difficult, so it has greater resistance, because it requires more energy to damage the fine grain, this is indicated by the increase in tensile strength and hardness. Besides that, grain refiner can also reduce hot tearing and porosity, and can tighten grain pressure.

In the SEM test results, the difference was seen between silicon crystals without the addition of Al5TiB with the addition of Al5TiB of 0.05%. This can be seen in figure 4.

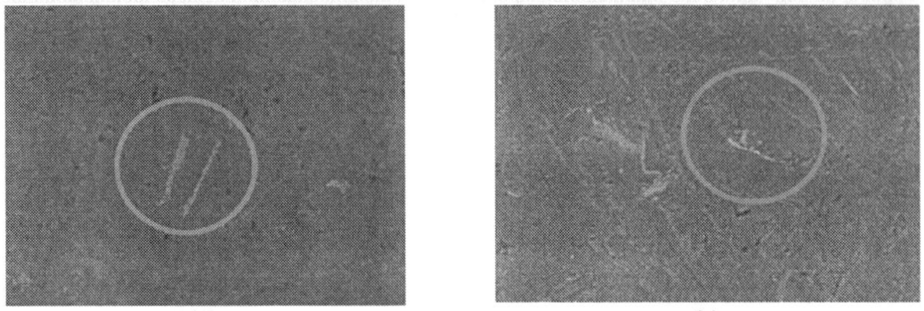

(a) (b)

Figure 4. Al-11% Si microstructure changes a) Without the addition of Al5TiB b) With the addition of 0.05% Al5TiB.

From the picture above it can be seen that there was a change in the microstructure of Al-11% Si alloy. The addition of 0.05% Al5TiB content reacted in the alloy to produce a finer and more evenly distributed grain structure. The crystalline silicon, which was originally shaped as long and very coarse needles, had become finer and shortened. This was the reaction given by Al5TiB as an item refiner.

The fine graining element reacted by wrapping silicon crystals to prevent grain growth from becoming rough. This can be interpreted that the fine graining element can act as an element that can control the amount of grain structure produced [4]. Al5TiB refiner grains on aluminum-silicon alloys were used to control the grain structure. Grain refiners produced nucleating sites for the formation of primary dendrites and produce large numbers of fine and uniform distribution grains. This has led to an increase in the value of violence.

On the addition of 0.1% Al5TiB levels there was also an increase in the value of violence. This was very reasonable considering the increase in Al5TiB content in the alloy so that it reduced the grain structure and had an effect on the price of violence which also rose. This can be seen in figure 5.

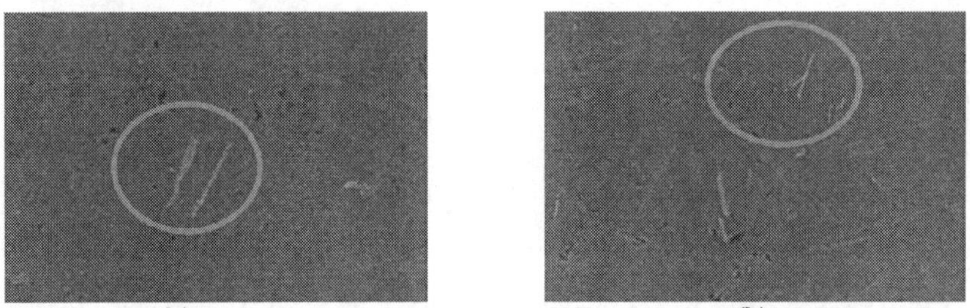

(a) (b)

Figure 5. Al-11% Si microstructure change a) Without the addition of Al5TiB b) With the addition of 0.10% Al5TiB

From the picture above it can be seen that there was a change in the microstructure of the Al-11% Si alloy. Modified silicon crystals were more and more evenly distributed. The grain structure at the addition of 0.10% Al5TiB became smoother when compared to the addition of 0.05% Al5TiB. This has resulted in an increase in the value of violence from before. This is in line with research conducted by Imanuel Ginting in his journal entitled "Strengthening with the fineness of grains on Al-9.4% Si alloys".

In that study, it was said that along with the addition of grain refiners in the alloy, it would cause the price of violence would increase and the grain structure tends to be smoother [6].

When referring to the value of the DAS (Dendrite Arm Spacing) or the distance between the dendritic arms, the value of this DAS depends on the speed of the freezing process of the casting process. DAS value has a great effect on the mechanical properties of a metal alloy.

Increased arm spacing indicates rough microstructure so that its mechanical properties are not good. While the small DAS value indicates a delicate micro structure with good mechanical properties. In the research conducted by Budi Lesmana, it can be seen that the addition of grain refiner can produce a faster freezing rate. If the freezing rate is fast, the resulting DAS value will be smaller. Therefore, the value of violence is higher than before. So it is very clear that the addition of grain refiner in the alloy is very influential on the value of the DAS which can directly improve the mechanical properties of an alloy [7].

In the Al-11% Si alloy added by Al5TiB as much as 0.15%, it can be seen that the results are contrary to previous data. In this condition, there was a decrease in the value of violence from an average of 26.6 HRC to 23.2 HRC. From the microstructure image obtained, it can be seen that the silicon particles were charged after adding Al5TiB of 0.15%. This had become the main cause of the decline in the value of violence found. For more details, can be seen in figure 6.

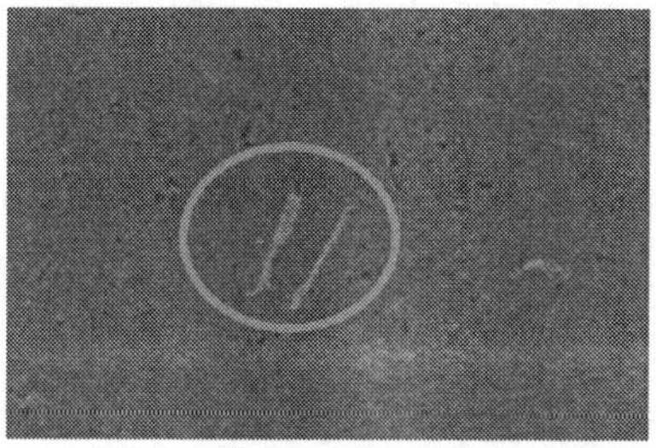
(a)

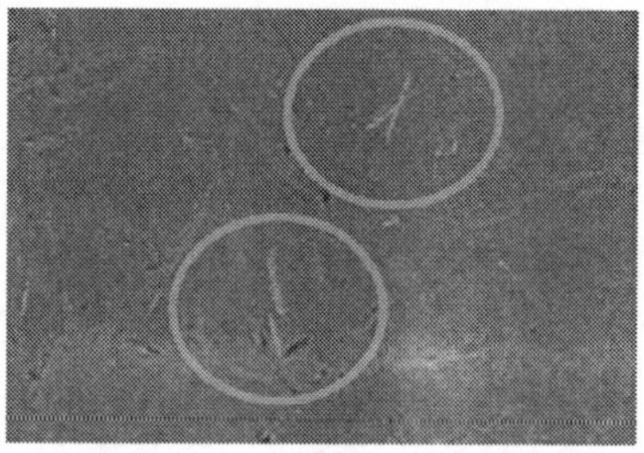
(b)

Figure 6. Al-11% Si microstructure changes a) Without the addition of Al5TiB b) With the addition of 0.15% Al5TiB.

From figure 6 the change in silicon crystals in Al-11% Si alloy can be seen. The structure of silicon which was originally shaped as short needles turned into a structure of longer needles at the addition of 0.15% Al5TiB. Besides that, the grain size tended to be bigger than before. Grain structure that did not spread evenly also had a very large impact on the decrease in the value of the hardness found in the test. This was because the added Al5TiB content had exceeded the ideal limit so that the addition process did not improve the mechanical properties but worsen it.

This is in accordance with research conducted by Ahmad Supriyadi [8], that giving Ti-B more than the right percentage would reduce its mechanical strength because Ti-B inoculants produce impurity. The research also shows that if the addition of TiB is greater than 0.12%, it will reduce the value of violence.

In addition, the decrease in the value of violence can also be explained by research conducted by Dobrzanski [9], who explained that the titanium element has a maximum concentration of 0.14%. If it has passed these numbers, the tendency for the process of imperfect modification will be even greater. This has caused the value of violence to be lower.

4. Conclusion

The addition of Al5TiB grain refiner grading 0.05% caused a significant change in the Al-11% Si alloy microstructure. The effect on the alloy was that the silicon crystal which was originally coarse became finer with the addition of the Al5TiB grain refiner. This change in microstructure caused an increase in hardness of 2.756% of Al-11% Si alloy without the addition of Al5TiB grain refiner. The addition of

Al5TiB grain refiner grading 0.10% caused significant changes in the microstructure of Al-11% Si alloys. The effect on the alloy was that the crystalline structure of silicon became smoother when compared to the addition of an Al5TiB grain refiner of 0.05%. In addition, the grains were more evenly distributed than before. This had led to an increase in the value of hardness of 4.724% of Al-11% Si alloy without the addition of Al5TiB grain refiner. The addition of Al5TiB grain refiner grading 0.15% caused changes that were contrary to changes in the micro-alloy structure of Al-11% Si. The effect on alloys was that silicon crystals which were originally refined now even experienced uneven distribution and distribution. This happened because the addition of Al5TiB grain refiner had exceeded the maximum limit. Therefore, the hardness value decreased very dramatically by 8.661% of Al-11% Si alloy without the addition of Al5TiB grain refiner.

Acknowledgments
The author would like to thank colleagues in the Product Development and Technology Transfer Section (PP & AT) Pekanbaru Industrial Standardization and Product Development Center for all their assistance so as to complete the writing of this scientific paper. Hopefully this paper can be used properly.

References
[1] Badan Pusat Statistik 2017 *Produksi Barang Tambang Mineral 2015-2018* (Jakarta: BPS)
[2] Trade Map 2020 List of supplying markets for a product imported by Indonesia. https://www.trademap.org/Country_SelProductCountry_TS.aspx. **October**
[3] Nanda I P and Angga Afrinaldi 2012 The Effect of Adding Modifier Strontium (Sr) on Hardness and Microstructure of Master Alloy Al-7%Si *Proc. SNTTM XI & Thermofluid IV.* (Yogyakarta: UGM) **XI** p1508-11
[4] Nanda I P 2010 Pengaruh Penambahan Fe dan Sr Terhadap Pembentukan Fasa Intermetalik dan Nilai Fluiditas Metode Vakum Pada Paduan Al-7% Si dan Al-11% Si *Disertasi* (Depok : FT-UI)
[5] Suhariyanto 2003 Perbaikan Sifat Mekanik Paduan Aluminium (A356) Dengan Menambahkan TiC *Jurnal Teknik Mesin* 20–24
[6] Ginting I Penguatan dengan Penghalusan Butir Pada Paduan Hypoeutektik Al-Si 9,4% Si *Buletin IPT No. 5* Vol. **III(5)** 13-18
[7] Lesmana B 2008 Studi Pengaruh Fading Penambahan Grain Refiner 0.019 Wt% Ti Terhadap Karakteristik Paduan AC4B Hasil Low Pressure Die Casting *Skripsi* (Depok: FT UI)
[8] Supriyadi A 2011 Pengaruh Penambahan Grain Refiner Ti-B Terhadap bahan ADC12 Pada Pengecoran HPDC Untuk Peningkatan Kualitas Sepatu Rem Sepeda Motor Produk IKM *Orbith* **7(3)** 393-400
[9] Dobrzański L A, Krupiński M, Zarychta P and R Maniara 2007 Analysis of Influence of Chemical Composition of Al-Si-Cu Casting Alloy on Formation of Casting Defects *J Achievement in Materials and Manufact. Eng.* 53–56.

IOP Conf. Series: Materials Science and Engineering **980** (2020) 012023 doi:10.1088/1757-899X/980/1/012023

Redistillation of wood vinegar from peat swamp species

R S Wahyuningtyas

Banjarbaru Environment and Forestry Research Development Institute, Banjarbaru, South Kalimantan

Email : renisetyowahyuningtyas@gmail.com

Abstract. This study aimed to determine the characteristics of wood vinegar (WV) from 3 native peat swamp forest species: gerunggang (*Cratoxylon arborescens*), mertibu (*Dactylocladus stenotachys*), and meranti bunga (*Shorea teysmanniana*) after redistillation. The redistillation was carried out in a laboratory by heating 550 mL of wood vinegar in the distillation flask at a temperature of 100℃ for 7 hours. The parameters observed were the yield, the density, the acetic acid content, the pH, and the percentage of hydroxyl scavenging. The results showed that WV of *D. stenotachys* had the highest values of yield (%), pH, and, percentage of hydroxyl scavenging among other WV types. However, its acetic acid content was the lowest when compared with that of *S. teysmanniana* WV and *C. arborescens* WV. Besides, both *S. Teysmanniana* WV and *C. arborescens* WV were considered to have better potential as an antioxidant, an antimicrobial, and a food preservative. The pH levels of all purified WV in this study were classified as very acid indicated by the high quality of WV. Furthermore, *D. stenotachys* WV and *C. arborescens* WV fulfilled the requirement of Japanese WV standard. The redistillation of WV from the three species of peat swamp forest successfully produced WV with a clearer, transparent, and pale-yellow color.

1. Introduction

The problems that often appear after the fire in peat swamp forest (PSF) are numerous wood piled up from dead and fallen trees on the forest floor. Accumulation of this wood waste actually can become a potential fuel and, worse, potentially trigger the occurrence of bigger fires in the future [2]. To minimize the risk of the fires, one method that can be attempted is by reducing the volume of dead wood waste to be utilized as the raw material for charcoal and wood vinegar [8].

Wood vinegar (WV), also called pyroligneous acid, is a brown flavorful liquid produced by the distillation of wood in the absence of air condition [22]. In the pyrolysis process, thermal degradation of biomass initiated by evaporation of water vapor followed by hemicellulose decomposition, subsequently cellulose decomposition (at 180℃ – 350℃), and terminated by lignin decompositin (at 160℃ – 625℃). The decomposition of hemicellulose and cellulose will produce carboxylic acids and carbonyl compounds, whereas the decomposition produces phenolic compounds [20]. The higher the pyrolysis temperature, the volume of WV produced becomes greater [15].

The pyrolysis process produces a liquid which is called by pyrolysis oil, bio-oil, bio-crude, bio-fuel, oil, tar oil, or wood vinegar [8]. Generally, the color of WV is brownish yellow because it still contains tar and other volatile compounds. Furthermore, Fachraniah [6] explain that the yellow and brownish colors in WV are caused by carbonyl compounds. WV has many uses, such as for the food industry and health as well as an insecticide, pesticide, and a crop fertilizer. To remove tar and reduce the strong smoke odor in WV, a purifying process is absolutely necessary. The purpose of WV purification is to remove tar and to produce WV with prominent functional properties, such as having a clearer color, high acetic acid, lower pH, and less smoke odor [6].

Several experiments to produce WV from three native species of PSF, namely mertibu *(Dactylocladus stenotachys)*, meranti bunga (*Shorea teysmanniana)* and gerunggang (*Cratoxylon arborescens*) have been successfully done. However, the WV produced from the pyrolysis process is crude WV with a red brownish color and has a strong smoke odor [2]. Redisstilation is a technique for refining WV by separating the solution based on different boiling points. In a redistillation process, some components can evaporate faster than others [17]. The redistillation of WV is carried out by removing unwanted and dangerous compounds, such as poliaromatic hydrocarbon (PAH) and tar. By controlling the boiling temperature, it is expected that a clear liquid smoke which is free of tar and benzopyrene can be obtained. This WV purification process usually uses a temperature lower than the pyrolysis temperature, which is around 100 ℃ depending on the type of wood used for WV material [4].

Because the information about WV from PSF, both crude WV and purified WV, is very limited, hence this current study attempts to find the physical quality of WV from the redistillation of three types of peat swamps forest species: mertibu *(Dactylocladus stenotachys*), meranti bunga (*Shorea teysmanniana)*, and Gerunggang (*Cratoxylon arborescens*). The results of this study are expected to give beneficial information to support the development of WV from wood waste to become a more useful product.

2. Materials and methods

2.1. Production of wood vinegar

Three types of WV were produced at Tumbang Nusa research station, Central Kalimantan Province. The equipment for the WV pyrolysis consisted of a drum with a capacity of 200 L, an iron plate, and a galvanis pipe. The raw materials for WV were some dead woods from 3 species of mertibu (*Dactylocladus stenotachys* Oliv. Family *Melastomataceae)*, gerunggang (*Cratoxylon arborescens* Bl. Family *Guttiferae*), and meranti bunga (*Shorea teysmanniana* Dyer Ex. Brandis Family *Dipterocarpaceae*).

The production of WV started by selecting some wood sticks (9 – 20 cm diameter size) and collecting them near the kiln. Then the wood sticks were cut 20-35 cm long and chopped into 4-6 parts. These wood pieces were arranged in a stove with a standing position until the drum kiln was full then closed tightly. The combustion was done with a big flame at a temperature of 400 ℃ for approximately 8 hours. The smoke from this combustion was then captured and channeled through a galvanized pipe which was cooled from a water channel. The smoke from this pyrolysis was collected to make WV [2].

2.2. Purification of wood vinegar (WV)

The purification of WV was carried out in the main laboratory of the Faculty of Mathematics and Natural Sciences, Lambung Mangkurat University. The equipment consisted of a distillation device, a scale, a measuring cup, and a pycnometer, while the required material was WV from *C. arborescens*, *D. stenotachys* and *S. teysmanniana* resulting from the pyrolysis process.

The process of purifying WV began with putting 550 mL of WV into a distillation flask. Then it was heated with a heating mantle at a temperature of 100 ℃ for 7 hours. The steam was formed then flowed into the cooling pipe and collected as the WV distillate.

Afterward, the resulting distillate was analyzed for measuring several parameters, such as the yield, the density, the pH, the acetic acid content, and the percentage of hydroxyl scavenging.

2.3. Analysis procedure

2.3.1. Yield. The yield was obtained by measuring the purified WV (distillate) with a measuring cup. Then the distillate volume was compared with the initial volume of WV before the purification. It was calculated by the following formula:

$$Yield\ (\%) = \frac{Purified\ wood\ vinegar\ volume}{Wood\ vinegar\ volume} x\ 100\% \quad (1)$$

2.3.2. Density. The specific gravity was measured by putting the purified WV into a pycnometer until exceeding the mark. Then it was closed to avoid the pressure from the air bubbles and weight. The density of liquid smoke was calculated by the following formula :

$$Density = \frac{(Pycno\ weight{+}sample)-(Empty\ pycno\ weight)}{(Pycno\ weight{+}distilled\ water)-(Empty\ pycno\ weight)} \quad (2)$$

2.3.3. pH. To check the pH levels, a pH meter was employed to obtain duplicate results.

2.3.4. Acetic acid. The concentration of acetic acid was measured by the titration method. 5 mL of WV was put into a measuring tube, and the water was added until the volume was 100 mL. Next, 10 mL of this liquid was taken and put into an erlenmeyer tube. 2 drops of phenolphthalein solution was added, and the solution was titrated with NaOH 0.1 M. This NaOH solution was used to redden the solution then recorded.

2.3.5. Percentage of hydroxyl scavenging. Hydroxyl scavenging is the percentage which states the ability of a compound to scavenge hydroxyl radicals. The hydroxyl radical scavenging activity test refers to the deoxyribose method [7] as follows. *Determination of operating time.* The solution consisted of 600 μL of deoxyribose solution 2.5 mM, 300 μL of $FeCl_3$ 1 mM, 300 μL of EDTA 1mM, 300 μL of H_2O_2 20 mM, 300 μL of vitamin C 1 mM, and 4500 of buffer fosfat pH 7.4. The solution was incubated for 30 minutes at 37 °C. Next, 1 mL of TCA 5% and 1 mL of TBA 1% were added at a temperature of 80 °C for 30 minutes. After cooling, the absorption of the solution was read at a wavelenght of 532 nm. *Sample solution.* 600 μL of deoxyribose solution 2.5 mM, 300 μL of $FeCl_3$ 1 mM, 300 μL of EDTA 1mM, 300 μL of H_2O_2 20 mM, 300 μL of vitamin C 1 mM, and 4500 buffer fosfat pH 7.4 were put into a closed reaction tube. The solution was incubated for 30 minutes at 37 °C. Then 1 mL of TCA 5% and 1 mL of TBA 1% were added at a temperature of 80 °C for 30 minutes. After cooling, the absorption of the solution was read at a wavelenght of 400-600 nm. *Control solution.* The solution consisted of 600 μL of deoxyribose solution 2.5 mM and the distillated WV from *D. stenotachys, C. arborescens, S. teysmanniana* as much as 200 μL, 400 μL, 600 μL, 800 μL, and 1000 μL, then 300 μL of $FeCl_3$ 1 mM, 300 μL of EDTA 1mM, 300 μL of H_2O_2 20 mM, 300 μL of vitamin C 1 mM, and the buffer fosfat at pH 7.4 were added to each tube until the final volume reached 6 mL. The mixture was incubated for 30 minutes at 37 °C. Next, 1 mL of TBA 1% and 1 mL of TCA 5% were added at a temperature of 80 °C for 30 minutes. After cooling, the absorption of the solution was read with a maximum wavelength from the optimization results.
The result analysis was calculated using following formula in equation 1.

$$\%\ scavenging = \frac{Absorbance\ of\ control\ solution - Absorbance\ of\ sample\ solution}{Absorbance\ of\ control\ solution} \text{x } 100\% \quad (3)$$

The hydroxyl scavenging analysis was carried out at the Faculty of Medicine, Lambung Mangkurat University.

IOP Conf. Series: Materials Science and Engineering **980** (2020) 012023 doi:10.1088/1757-899X/980/1/012023

3. Results and discussion

3.1. Yield (%)

The highest volume of distillate was produced from *D. stenotachys* WV (395 mL) followed by *C. arborescens* (332 mL), whereas the lowest was from *S. teysmanniana* (325 mL) (table 1).

Table 1. The yields of purified wood vinegar from 3 native species of PSF.

Type of WV	Volume (mL)		Yield of distillate	Tar	Yield of WV
	WV	Distillate	(%)	(mL)	(%)
S. teysmanniana	550	325	59.1	240	5.87
C. arborescens	550	332	60.4	80	4.98
D. stenotachys	550	395	71.8	320	6.12

When compared with the yield of WV from the pyrolysis, the distillate yields from the three native species of PSF were much higher. This happened because the purification process was carried out in a closed distilation device and using a low temperature (100 ℃). Therefore, only a little gas wasted out from the distillation process. On the other hand, the temperature used was higher (400 ℃) in the pyrolysis process. Besides, the resulting products were not only WV but also charcoal, ash, and uncondensed gases (CO, CO_2 and CH_4) [2,15]. Therefore, the WV yield produced from the pyrolysis is generally lower than puryfied WV.

The WV purification aims to separate tar from WV based on their different boiling points [17]. In a purification process of wood vinegar, the distillate will evaporate more quickly and cool down becoming clearer WV. Meanwhile, substances with a higher boiling point will settle and produce a dark brown liquid called tar. In this research, the highest tar yield was resulted in the purified VW of *D. stenotachys* (320 mL), followed by *S. teysmanniana* (240 mL), and *C. arborescens* (40 mL) (table 1).

Table 2. Physical properties of woods from three native species of PSF.

Species	Content (%)			Specific gravity	Strenght grade
	Cellulose	Lignin	Pentosan		
S.teysmanniana	-	-	-	0.83 (heavy)	III-II
C. arborescens	53.1 (high)	22.2 (med)	18.4	0.47 (light)	III-IV
D. stenotachys	56.2 (high)	24.1 (high)	14.4	0.53 (light)	III

Sources : [3,13,19]

The amount of tar in WV of *D. stenotachys* and *S. teysmanniana* is expected to have a relation with higher wood density of both types of WV than that of *C. arborescens*. Martawijaya [13] stated that wood density is the ratio of weight and volume of wood in dry air condition with the moisture aproximately 15%. Woods which have greater specific gravity are generally also stronger and harder [3]. The thermal decomposition biomass process in hardwood will also process more slowly, thus the steam generated from the combustion will be abudant and condensed in WV. However, Komarayati [9] provide a differet assumption. They argue that a higher density material will cause the rate of burning becomes longer so that the smoke produced is less. However, several studies have shown that woods with high lignin content generally produce more smoke so that more WV will be produced [5, 9,21].

IOP Conf. Series: Materials Science and Engineering **980** (2020) 012023 doi:10.1088/1757-899X/980/1/012023

3.2. Density

The density was measured to determine the mass of volume from each purified WV. The analysis showed that there was no significant decrease in the density of WV before and after purification (figure 1). Similarly, the results of [17]'s study showed that the density of galam (*Melaleuca leucadendron*) WV did not have much difference before and after the purification process.

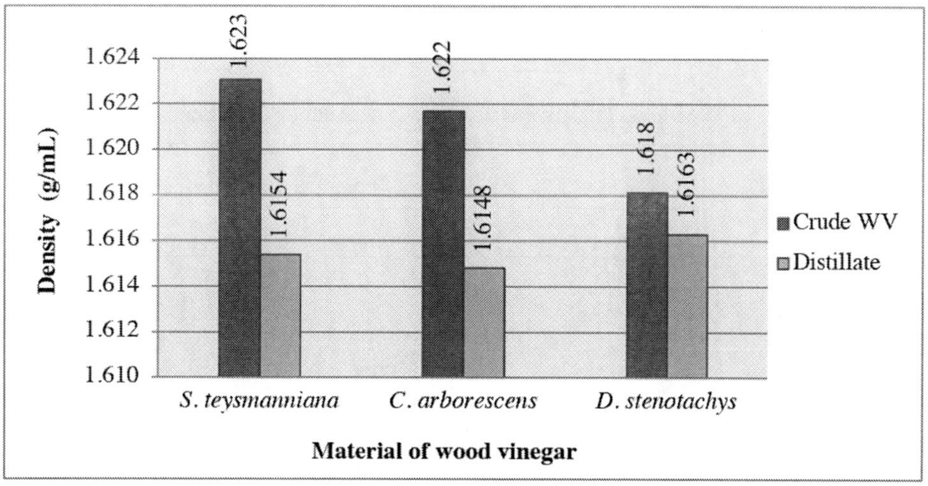

Figure 1. Density of wood vinegar before and after redistillation process.

The average density of the WV distillate from 3 species of PSF was 1.6 g/mL, indicating that this density was the highest among the density of galam WV (1.000-1.002 g/mL) [17]; the density of sawdust WV (1.004-1.018 g/mL) [14], the density of coconut shell WV (1.026 g/mL) [6]; the density of nyamplung (*Calophylum inophylum*) shell WV (1.009 g/mL) [21], the density of *Shorea* spp. WV (0.9818 g/mL) [16], and the density of ironwood WV (0.967 g/mL) [12].

The density of these three PSF species has fulfilled the density requirements of the Japanese WV standard i.e > 1.001 g/mL [21].

3.3. Acetic acid content

The acetic acid contents in the purified WV from the three PSF species were different. The highest acetic acid level was found in WV of *S. teysmanniana* (1.389 mg/L) followed by the acetic acid level in *C. Arborescens* WV (1.157 mg/L), while the lowest level was obtained in *D. stenotachys* WV (0.570 mg/L). The differences in acetic acid contents in the three types of WV indicated that the types of materials and the chemical characteristics of the ingredients affected the yield of acetic acid. Although *D. stenotachys* wood had a high lignin content, but the acetic acid from the resulting WV was actually the lowest among the other types (table 3).

According to Wibowo [21], acetic acid is formed due to the pyrolysis of wood components, such as cellulose, hemicellulose, and lignin. The acid compound is the most dominant compound in WV [1]. The results of Komarayati's [9] study also revealed that there was a difference in the acetic acid contents in bamboo species. In specific, the acetic acid contents in WV of betung bamboo, black bamboo, and spotted bamboo were 31.37%, 85.39%, and 53.37%, respectively.

The acetic acid from WV has function to accelerate the plant growth and to prevent plant diseases [23], as an antibacterial, and to form the flavor of smoke product [6]. The inhibition process of acetic acid against the bacterial growth occurs by disrupting the synthesis of membrane stability enzymes through an acidification of the cytoplasm of bacterial cells [20]. The result of [10]'s research has showm that the high acetic acid and phenol contents in WV can inhibit the development of bacteria *Eschericia coli* in vitro. In other words, WV has a potential to be developed as an antioxidant and antimicrobial agent for medicine or food preservation.

In this study, the acetic acid content of *S. teysmanniana* WV was the highest of the other two types, indicating that *S. teysmanniana* WV has a better potential as an antioxidant, antimicrobial, and food preservative agent. To determine the effectiveness of the antimicrobial of *S. teysmanniana* WV, it certainly needs to be supported by further research.

3.4. pH
The degree of acidity (pH) is one parameter that can be used to determine the quality of WV. The purified WV in this study was clasified as very acid. The pH of *S. teysmanniana* WV was 1, while pH in each *D. stenoctachys* and *C. arborescens* WV was 2 (table 3).

Table 3. The properties of wood vinegar distillate.

Parameters	Wood vinegar distillate			Quality standard of WV distillate from Japan [1]
	D. stenoctachys	*S. teysmanniana*	*C. arborescens*	
- Density (g/mL)	1.6163	1.6154	1.6148	> 1.001
- pH	2 ± 0	1 ± 0	2 ± 0	1.3 – 3.7
- Color	Colorless – pale yellow	Colorless – pale yellow	Colorless – pale yellow	Colorless – pale yellow, brownish pale-red
- Transparency	Transparant	Transparant	Transparant	Transparant
- Floating matters	No floating matters	No floating matters	No floating matters	No floating matters
- Acetic acid (mg/L)	0.57 ± 0.043	1.389 ± 0.00	1.157 ± 0.2	1 – 18%
- *hydroxyl scavenging* (%)	12.815 ± 0.550	12.663 ± 0.368	7.786 ± 0.178	

According to Achmadi [1], lower pH value and higher acetic acid level indicate a better quality of WV. A low pH value indicates a high quality of wood vinegar because harmful microbes or bacteria cannot live and reproduce properly at a low pH [5].

The pH value of redistillated WV which was clasified as very acid indicated that the WV from the three native peat swamp trees had very high quality. Indeed, pH shows a level of process of breaking down the chemical components of wood into organic acids [17]. Very low pH (very acidic) occurs due to the complete decomposition or the decomposition of chemical components within an airless pyrolysis tube. The pH WV from *D. stenoctachys* and *C. arborescens* in this research has fullfiled the Japanese standard requirement i.e: 1.3-3.7 [23].

3.5. Hydroxyl scavenging
Hydroxyl scavenging indicates the ability of a compound to scavenge hydroxyl radicals. The greater the value of % scavenging from purified WV, the greater are the hydroxyl radical scavenging activities.

In this research (table 3), the highest hydroxyl scavenging value obtained was in *D. Stenotachys* WV (12.815 % ± 0.550). The second was obtained in *S. teysmanniana* WV (12.663 % ± 0.368). Meanwhile, the lowest hydroxyl scavenging value was shown in *C. arborescens* WV (7.786 % ± 0.178).

The compounds in WV which suppose to have the scavenging activity are the polyphenolic compounds. Generally, the compounds have many hydroxyl clusters that function as a free radical scavenger. According to Li [10], phenolic compounds or polyphenols are an important group of compounds in plants. Polyphenols have at least 8000 different structures including simple phenol, phenolic acids, cumarins, isocumarins, naphtoquinones, xanthones, stilbenes, phlovonoids, and lignins.

In wood vinegar, phenolic compounds are produced from the process of breaking ether bonds and lignin carbon bonds. Polyphenols have a strong ability to clean free radicals, hence it can be used as a reducing agent and an antioxidant [21].

The lowest scavenging activity was found in *C. arborescens* when compared with other species in this study, and it is actually difficult to explain. This might occur because the flavonoid content in *C. arborescens* stems is quite low only 3.8 ± 0.4 g/mL of quercetin equivalent (QE), which is much smaller than the flavonoid contents in *Mangifera casturi* fruits (30.0 ± 1.2 g/mL QE), *Setnochlaena palustris* fern leaves (14.5 ± 0.7 g/mL QE), and *Eurycoma longifolia* root (6.1 ± 0.8 g/mL QE) [17].

Although there has been no data regarding the flavonoid content in *D. Stenotachys* and *S. teysmanniana,* it is supposed that the flavonoid content in both species are greater. Therefore, the hydroxyl scavenging acticitivies in *D. Stenotachys* and *S. teysmanniana* were higher.

4. Conclusion

The redistillation of the wood vinegar from three peat swamp forest species (*D. stenotachys, S. teysmanniana,* and *C. arborescens*) produced various characteristics of the wood vinegar. The results of this study indicated that WV of *D. stenotachys* showed the highest yield value (%), pH, and hydroxyl scavenging (an ability to ward off free radicals). *D. stenotachys* WV showed the highest yield (71.8%), meaning that it succeeded to remove more tar than the other two types. However, the acetic acid content of *D. stenotachys* WV was much lower than that of *S. teysmanniana* and *C. arborescens*. Therefore, WV from *S. teysmanniana* and *C. arborescens* tended to be more apropriate to utilize as an antioxidant, an antimicrobial, and a food preservative. Besides, the three types of purified WV in this study were classified as vey acid, indicating that the quality of purified WV was very high. Of the three types of purified WV, Both purified WV of *D. stenotachys* and *C. arborescens* have fulfilled the requirement of Japanese WV standard. All in All, the purification of the WV from the three peat swamp species have successfully improved the quality of WV as their reddish brown color became clearer and transparent with pale yellow color.

References

[1] Achmadi S S, Kusumaningrum H D, and Anggara I 2015 Redistilat asap cair cangkang Kelapa Sawit sebagai bahan pengawet bakso sapi *J. Teknol. Dan Industri Pangan*, ***26*(1)** 1–8 https://doi.org/10.6066/jtip.2015.26.1.1

[2] Ariani R, Supriadi R and Purwanto B 2016 Uji coba pemanfaatan limbah kayu pasca terbakar untuk arang dan cuka kayu di KHDTK Tumbang Nusa. *Galam*, ***2*(2)** 19–24 Retrieved from https://studylibid.com/doc/692818/bpk-banjarbaru---jurnal-galam-vol-2-no-2-2016.indd

[3] Arsad E 2013 Prospek kayu kualitas rendah dan kurang dikenal sebagai substitusi kayu komersial *Jurnal Riset Industri Hasil Hutan*, ***5*(1)** 45–53 Retrieved from http://ejournal.kemenperin.go.id/jrihh/article/view/1212/1031

[4] Darmadji P 2002 Optimasi pemurnian asap cair dengan metoda redistilasi *Jurnal Teknologi Dan Industri Pangan*, ***13*(3)** 267–71 Retrieved from http://103.10.105.65/index.php/jtip/article/view/5312

[5] Dewi J, Gani A, and Nazar M 2018 Analisis kualitas asap cair tempurung kelapa dan ampas tebu sebagai bahan pengawet alami pada tahu *Jurnal IPA Dan Pembalajaran IPA*, ***02*(02)** 106–12 Retrieved from http://e-repository.unsyiah.ac.id/JIPI/article/view/12743

[6] Fachraniah, Fona Z, and Rahmi Z 2009 Peningkatan kualitas asap cair dengan distilasi *Jurnal Rekasi (Journal of Science and Technology)*, ***7*(14)** 1–11 Retrieved from http://jurnal.pnl.ac.id/wp-content/plugins/Flutter/files_flutter/1396325107AsapCair.pdf

[7] Halliwell B and Gutteridge J M 1999 *Free radicals in Biology and medicine*. (Third Edition, Ed.). (New York - USA: Oxford Universuty Press)

IOP Conf. Series: Materials Science and Engineering **980** (2020) 012023 doi:10.1088/1757-899X/980/1/012023

[8] Kartal S N, Terzi E, Kose C, Hofmeyr J, and Imamura Y 2011 Efficacy of tar oil recovered during slow pyrolysis of macadamia nut shells *International Biodeterioration & Biodegradation*, ***65*(2)** 369–73 https://doi.org/10.1016/j.ibiod.2010.08.011

[9] Komarayati S and Wibowo S 2015 Karakteristik asap cair dari tiga jenis bambu. *Jurnal Penelitian Hasil Hutan*, ***33*(2)** 167–74 Retrieved from http://ejournal.forda-mof.org/ejournal-litbang/index.php/JPHH/article/view/824

[10] Li Z, Zhang L, Chen G, Wu L, Liu B, Li Y, … Wang Z 2018 A new method for comprehensive utilization of wood vinegar by distillation and liquid–liquid extraction *Process Biochemistry*, *75*(**April**) 194–201 https://doi.org/10.1016/j.procbio.2018.08.012

[11] Lipinski B 2011 Hydroxyl radical and its scavengers in health and disease. Review Article. *Oxidative Medicine and Cellular Longevity*, *2011*. https://doi.org/10.1155/2011/809696

[12] Mahdie M F, Violet and Helmi M 2020 Rendement and characteristics of wood vinegar produced from Ironwood delinquent waste through clay kiln charcoaling furnace *Journal of Wetlands Environtmental Management*, ***8*(2)** 140–8 Retrieved from http://ijwem.ulm.ac.id/index.php/ijwem/article/view/231/158

[13] Martawijaya A, Kartasujana I, Kadir K and Prawira S A 2005 *Atlas Kayu Indonesia Jilid I* (Edisi Keti). (Bogor-Indonesia: CV. Miranti) Retrieved from https://www.academia.edu/31409093/Atlas_Kayu_Indonesia_Jilid_I

[14] Nurhayati T, Pasaribu R A and Mulyadi D 2006 Produksi dan pemanfaatan arang dan cuka kayu dari serbuk gergaji kayu campuran *Jurnal Penelitian Hasil Hutan*, ***24*(5)** 395–411 Retrieved from http://ejournal.forda-mof.org/ejournal-litbang/index.php/JPHH/article/view/4080/3580

[15] Ratnawati and Hartanto S 2010 Pengaruh suhu pirolisis cangkang sawit terhadap kuantitas dan kualitas asap cair *Jurnal Sains Materi Indonesia*, ***12*(1)** 7–11 Retrieved from http://jurnal.batan.go.id/index.php/jsmi/article/view/4551/3966

[16] Satriadi T 2012 Rendemen dan kualitas cuka kayu dari 3 jenis kulit meranti (Shorea spp.) *Enviro Scienteae*, ***8*(2)** 102–7 Retrieved from https://ppjp.ulm.ac.id/journal/index.php/es/article/view/2075

[17] Setiawati E 2014 Pengaruh redestilasi cuka kayu Galam (*Melaleuca leucadendron* Linn) terhadap pengawetan ikan *Jurnal Riset Industri Hasil Hutan*, ***6*(2)** 13–22 https://doi.org/http://dx.doi.org/10.24111/jrihh.v6i2.1233

[18] Suhartono E, Viani E, Rahmadhan M A, Gultom I S, Rakhman M F, and Indrawardhana D 2012 Total flavonoid and Antioxidant Activity of Some Selected Medicinal Plants in South Kalimantan of Indonesian *APCBEE Procedia*, ***4*** 235–9 https://doi.org/10.1016/j.apcbee.2012.11.039

[19] Supriadi A and Abdurachman 2018 Sifat pemesinan lima jenis kayu asal Riau *Jurnal Penelitian Hasil Hutan*, ***36*(2)** 85–100 Retrieved from http://ejournal.forda-mof.org/ejournal-litbang./index.php/JPHH/article/view/3875/4282

[20] Suryani R, Rizal W A, Pratiwi D and Prasetyo D J 2020 Karakteristik dan aktivitas antibakteri asap cair dari biomassa kayu putih (*Melaleuca leucadendra*) dan kayu jati (*Tectona grandis*) *Jurnal Teknologi Pertanian*, ***21*(2)** 106–17 Retrieved from https://jtp.ub.ac.id/index.php/jtp/article/view/855/1037

[21] Wibowo S 2012 Karakteristik asap cair tempurung Nyamplung *Jurnal Penelitian Hasil Hutan*, ***30*(3)** 218–27 Retrieved from https://media.neliti.com/media/publications/125884-ID-none.pdf

IOP Conf. Series: Materials Science and Engineering **980** (2020) 012023 doi:10.1088/1757-899X/980/1/012023

[22] Yang J F, Yang C H, Liang M T, Gao Z J, Wu Y W and Chuang L Y 2016 Antibacterial activity of wood vinegar from Litchi chinensis *Molecules*, ***21*(1150)** 1–10 https://doi.org/10.3390/molecules21091150

[23] Yatagai M 2000 *Utilization of charcoal and wood vinegar in Japan* Graduate School of Agricultural and Life Sciences The University of Tokyo

Redistillation and characterization of liquid smoke from ulin wood (*Eusideroxylon zwageri Teijsm.* & Binn.) and its ability as a chitosan solvent

A B Junaidi*, A Nursyifa and Abdullah
Department of Chemistry, Faculty of Mathematics and Natural Science, Lambung Mangkurat University Jl. A. Yani KM. 35,8 Banjarbaru, Kalimantan Selatan

Email : abjunaidi@ulm.ac.id

Abstract. The purpose of this research was to study the solubility of chitosan in the liquid smoke as a preliminary study for the application of liquid-chitosan combination as a bioimmunizer and growth booster in plants and as a preservative in fish processing. This research focused on redistillation and characterization of the liquid smoke from ulin wood (*Eusideroxylon zwageri Teijsm.* & Binn.) and its ability as a chitosan solvent. The purpose was to obtain the scientific data on the yield redistillation, characterization of the liquid smoke (pH, total acid, total phenol, and composition), and its ability as a chitosan solvent. This liquid smoke was the side product of ulin wood charcoal in South Kalimantan. The results showed that the yield of liquid smoke from redistillation was 76.14% and the characteristics were containing 2.21 pH, 3.38% g/mL of total acid, 3.05 g/L of total phenol, and other chemical compositions such as carboxylic acid (73.08%), alcohol (14.73%), carbonyl (9.43%), phenolic (1.56%), and ether (1.21%). This research also uncovered that the solubility of chitosan in the liquid smoke increased along with the higher value of the liquid smoke concentration, and the liquid smoke concentration of 5% was able to dissolve more than 50 g/L of the chitosan.

1. Introduction

Liquid smoke and chitosan have been widely recognized and applied in various fields of technology. Liquid smoke is widely applied in the fields of food science (preservatives, flavors, and color), health, and agriculture [1], whereas chitosan is widely applied in the fields of pharmaceuticals, agriculture, biochemistry, preservation, health, and as adsorbent. Purba et al. [2] have reviewed the utilization of liquid smoke and chitosan as a preservative in tofu and meatballs.

Liquid smoke is the result of incomplete combustion (pyrolysis) from organic matter/biomass (wood and agricultural waste) that produces smoke (steam) and is condensed to form a liquid. This liquid smoke contains three major groups of compounds, namely carboxylic acids, carbonyl, and phenol [3]. Meanwhile, chitosan is a natural cationic biopolymer abundant in nature with the structure of β-(1.4)-2-amino-2-deoxy-D-glucose. Chitosan can be obtained from the results of chitin deacetylation that is contained in shrimp, insect, fungi, Mollusca, and arthropod shells through various isolation processes of deproteination, demineralization, depigmentation, and deacetylation [4].

One of the most frequent problems in chitosan applications is its solubility. Chitosan is not water-soluble and some of the solvents (alcohol, acetone) are slightly soluble in strong acids (HCl, HNO_3, H_3PO_4), and easily soluble in weakly dilute acids (acetic acid and formic acid). Chitosan is soluble in organic acid through the protonation of free amino groups ($-NH_2$) to cationic amino [1].

Liquid smoke is potential as a solvent for chitosan because one of the main components contained in it is the carboxylic acid group. Based on the acid content possessed by liquid smoke and the solubility property of chitosan, it can be predicted that chitosan can be dissolved in liquid smoke easily. Thus, in some applications using a combination of chitosan and liquid smoke can be done at once without the addition of a solvent. In this research, the liquid smoke is the side product of ulin wood charcoal (pyrolysis) [5].

Wood pyrolysis generally produces the liquid that still contains many carcinogenic substances (PAH and tar), thus it is important to do a redistillation [6]. Redistillation is a process of separating the components in a mixture based on their boiling points [7]. This is why the redistillation process is necessary to remove the tar compound contained in the liquid smoke. The characteristics of the resulting liquid smoke depend on the pH value, total acid, total phenol, and composition so that it is essential to characterize the liquid smoke. Based on the above description, this research aims to study the redistillation and characterization of liquid smoke from ulin wood and its ability as a chitosan solvent.

2. Materials and methods

2.1. Experimental principle

The redistillation of liquid smoke resulted in residue and distillate. The residue was a blackish-brown viscous liquid (figure. 1.a) which was tar and distillate (figure. 1.b.) containing a group of acidic, carbonyl, and phenol compounds. The distillate was characterized by physical and chemical properties included pH, total acid, total phenol, and composition. The total acid was analyzed by a titration method, while the total phenol was analyzed using UV-Vis Spectrophotometry, and the composition was analyzed using GC-MS. The presence of acidic compounds in liquid smoke made the chitosan soluble. The solubility of chitosan in liquid smoke was analyzed by the gravimetric method.

2.2. Reagents and instruments

Here are the list of reagents and instruments used in this study. Liquid smoke ulin wood: wood charcoal production in Takisung, Tanah Laut, South Borneo-Indonesia [5]; Chitosan (medium molecule weight; 200-800 cPs; deacetylation degree 80-85%): Sigma-Aldrich; NaOH pellets: Merck; Na_2CO_3: Merck; indicator Phenolphthalein (PP): Merck; ethanol p.a: Merck; gallic acid p.a; Merck; reagent Follin-Ciocalteu 10%: Merck; Thermometer: Beurer; water bath: Daihan Labtech; centrifuge: Hettich; analytical balance: Shimadzu; vortex shaker: Thermo scientific; buret: Duran; oven: Memmert; a set of distillation tools: Duran; Gas Chromatography- Spectroscopy Mass (GC-MS): Shimadzu QP2010S; UV-Vis Spectrophotometer: Hitachi, Japan.

2.3. Redistillation of liquid smoke

As much as 500 mL of liquid smoke was redistilled at a temperature of 25-140°C, and the resulting distillate was put in an Erlenmeyer and the yield was calculated with the equation 1.

$$\text{Yield (\%)} = \frac{\text{Volume of condensate (mL)}}{\text{Volume of sample (mL)}} \times 100\ \% \quad (1)$$

2.4. Characterization of the liquid smoke

2.4.1. Degree of Acidity (pH) and Total Acid. Measurement of the degree of acidity (pH) of liquid smoke was done using a pH meter. 5 ml of distilled liquid smoke was put into a 50 ml measuring flask, then the aquadest was added to the boundary marker. A total of 10 mL of solution was taken, 2-3 drops of PP indicator were added, and titrated with 0.1 N NaOH. The total acid was determined by the equation 2.

$$\%\text{Total Acid} = \frac{\text{Volume of NaOH} \times \text{ N NaOH} \times \text{MW acetic acid}}{\text{Volume of liquid smoke} \times 1000} \times 100\% \quad (2)$$

2.4.2. Total phenol content. 0.1 mL of liquid smoke was added with ethanol p.a in a 10 mL flask. As much 0.5 mL of solution was taken then added with 0.75 mL of 10% Follin-Ciocalteu reagent and 2 mL of Na2CO3 2%, shaken in a vortex shaker and allowed to stand for 18 minutes. The absorbance was determined with spectrophotometer UV-Vis at 700-800 nm wavelength. The standard curve making was done using the gallic acid.

2.4.3. GC-MS analysis. 1 μL of sample was injected into GC-MS in an oven at a temperature of 50-300°C with a temperature rise rate of 5°C/min, 13 kPa of a pressurized helium carrier gas, a total rate of 40 mL/min, and a split ratio of 66.8. The chromatogram was then compared to the data bank.

2.5. Determination of the solubility of chitosan in liquid smoke

As much as 0.5 grams of chitosan was dissolved in 10 mL of liquid smoke, heated in a water bath at 60°C for 30 minutes. After that, it was centrifuged at a rate of 1500 rpm for 20 minutes, then filtered and dried in an oven at 100°C for 24 hours until the weight was constant. The solubility was calculated based on the following equations 3, 4, and 5.

$$W_{\text{dissolved chitosan}} = W_{\text{sample}} - W_{\text{insoluble chitosan}} \quad (3)$$

$$\%\ \text{Solubility chitosan (b/b)} = \frac{\text{weight of dissolved chitosan (g)}}{\text{weight of sample chitosan (g)}} \times 100\,\% \quad (4)$$

$$\text{Solubility (g/L)} = \frac{\text{amount of soluble chitosan (g)}}{\text{volume of solvent (L)}} \quad (5)$$

3. Results and discussion

3.1. Yield and physical characteristics of the redistilled liquid smoke

Redistillation of the liquid smoke was carried out to separate the heavy fractions (PAH and tar) contained in the liquid fumes from the light fractions (phenol, carbonyl, and acid groups) based on their boiling points. Based on table 1, it can be seen that the obtained value for the redistillation yield of liquid smoke is 76.14%. This shows that the residue as the result from redistillation using wood smoke in this study is quite high (23.86%). The high residuals obtained indicated that the tar content in the ulin wood smoke was quite a lot. The liquid smoke before and after redistillation can be seen in figure 1.

(a) **(b)**

Figure 1. Liquid smoke before **(a)** and after **(b)** redistillation

Figure 1 shows that the liquid smoke color has changed from (a) to (b) after redistillation. The change from dark brown (before redistillation) to clear yellow (after redistillation) might because the compounds of the light fraction of liquid smoke (phenolic, acid, and carbonyl) have separated from the heavy fraction compounds (PAH and tar) based on the boiling points. The boiling points of the PAH were above 200°C (e.g. naphthalene 218°C, anthracene 340°C, pyrene 404°C dan benzopyrene 495°C). Besides, the liquid smoke from redistillation in this study had a distinctive smell (scent). This indicated that there was a group of phenolic compounds, carbonyls, and acids that had an important role in producing the smoke scent. Muhammad et al. [8] pointed out that the presence of acid compounds mixed with carbonyl and phenol will cause a strong scent.

3.2. Characteristics of the redistillated liquid smoke

Table 1. Yield and physical characteristics of the redistilled liquid smoke.

Redistillation temperature (°C)	Characteristics of liquid smoke					
	Yield (%)	pH	Total acid (%g/mL)	Total phenol (g/L)	Color	Scent
90-140	76,14	2,21	3,38	3,05	yellow	typical liquid smoke

3.2.1. Degree of acidity (pH) and total acid. Based on the data in table 1, the characteristics of liquid smoke as the result from redistillation was having pH of 2.21 and total acid as much as 3.38% g/mL. In comparison, the results of research by Achmadi et al. [3] showed higher values (pH 3.2; total acid 9.2%) which indicated that in this present study the degree of acidity was high (low pH), but the total acid was low. In other words, the acid compound in the liquid smoke had a large Ka (acid equilibrium constant) value in this study.

The pH value and total acid from the redistillation results showed the decomposition of the chemical compounds which were present in the raw material when the pyrolysis produced the groups of organic acid compounds. The dominant organic acid compound present in liquid smoke is usually a carboxylic acid group. Carboxylic acids are formed from components in the liquid smoke raw materials such as cellulose and hemicellulose [9].

3.2.2. Total phenol content. Based on table 1, it can be seen that total phenol content obtained was 3.05 g/L. When compared with the results of other studies using different raw materials such as coconut shell and coconut fiber [10], bamboo [11], various wood [12], the liquid smoke had a different total phenol content. This indicated the influence of lignin content on raw materials. Indeed, the more amount of lignin decomposed into groups of phenol compounds causes higher levels of phenol in the liquid smoke. The amount of decomposed lignin depends on the amount of lignin in the feedstock and the pyrolysis temperature used [13]. Another suspected influential factor was the pyrolysis temperature. The higher the pyrolysis temperature, the more lignin broke down so that the phenol content in the liquid smoke was higher, too.

3.2.3. Composition analysis of liquid smoke using GC-MS. GC chromatogram and the components of the liquid compounds from the redistillation result based on MS data can be seen in figure 2 and table 2.

Based on the data in table 2, 12 compounds were identified in the liquid smoke of ulin wood. The identification results of compound components in the ulin wood smoke could be divided into some groups, namely acidic (acetic acid, propanoic acid, and citric acid), phenolic (o-Guaiakol), alcohols (methyl alcohol and ethyl alcohol), ketones (acetone and 3.5-Cycloheptadienone), ethers (ethylene glycol diglycidyl ether), and esters (butyrolactone, γ-valerolactone, and ethyl heptanoate) groups. The main component of the liquid wood smoke composition was acetic acid which reached 71.57%. Acetic acid is a compound that is very suitable for dissolving chitosan [4]. Liquid smoke with high acetic acid content has a great potential to be a solvent for chitosan. Similarly, the results of Rinaldi et al. [14] showed that acetic acid became the dominant component with a percentage of more than 50%. However, a different finding was found by Janairo and Amalin [15] where the main content of liquid

IOP Conf. Series: Materials Science and Engineering **980** (2020) 012024 doi:10.1088/1757-899X/980/1/012024

smoke was phenolic. Furthermore, the data in table 2 also show that in the redistilled wood smoke there is no compound of heavy fractions (PAH and tar) found, such as pyrene, perylene, benzo(a) anthracene, and benzo(a)pyrene.

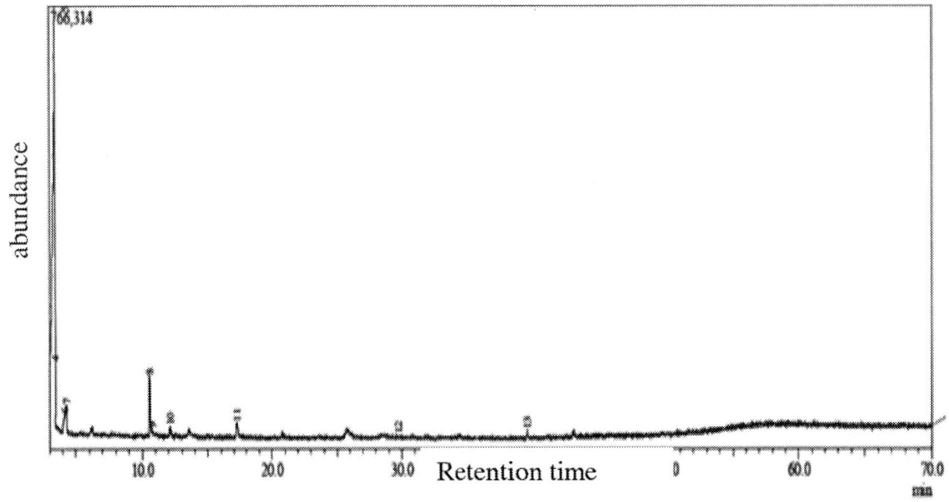

Figure 2. GC Chromatogram of ulin wood liquid smoke from the redistillation result.

Table 2. The components of ulin wood liquid smoke resulting from from the redistillation result.

Peak	Retention time (minute)	% Area	Molecular formula	Presumption of compound
1	2.320	12.19	CH_4O	Methyl alcohol
2	2.416	2.54	C_2H_6O	Ethyl alcohol
3	2.510	5.80	C_3H_6O	Acetone
4	2.962	4.77	$C_2H_4O_2$	Acetic acid
5	3.446	66.80	$C_2H_4O_2$	Acetic acid
6	4.125	0.85	$C_6H_{10}O_4$	Ethylene glycol diglicidyl ether
7	4.294	1.08	$C_3H_6O_2$	Propanoic acid
8	10.582	3.14	$C_4H_6O_2$	Butyrolactone
9	10.775	0.43	$C_6H_8O_7$	Citric Acid
10	12.132	0.49	$C_5H_8O_2$	γ-Valerolactone
11	17.225	1.21	$C_7H_8O_2$	*o*-Guaiacol
12	29.710	0.35	C_7H_8O	3,5-Cycloheptadienone
13	39.417	0.36	$C_9H_{18}O_2$	Ethyl heptanoate

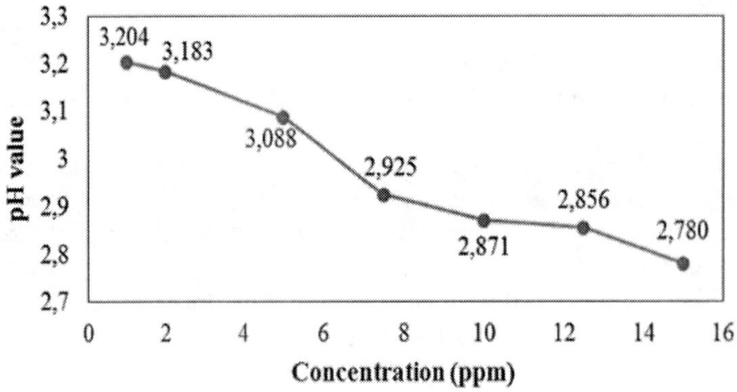

Figure 3. The graphic of correlation between pH value and liquid smoke concentration.

3.3. The solubility of chitosan in the liquid smoke

Figure 3 shows that the higher the liquid smoke concentration, the lower was the pH value of liquid smoke. However, the decrease in pH of the solution was not significant when compared with the increased concentration of liquid smoke in the solution. This indicated that the acids contained in the liquid smoke were weak. Weak acids only slightly release the $[H^+]$ at equilibrium because they have a small Ka value (Ka value of acetic acid is 1.8×10^{-5}). Therefore, in this study the measured pH did not change too much even though the acid concentration changed significantly.

Table 3. The degree of chitosan solubility in the liquid smoke concentration.

Liquid smoke concentration (v/v)	Analysis Parameter		
	pH of dilution	Chitosan solubility (g/L)	pH of solution
1.0%	3.204	3.7	5.141
2.0%	3.183	4.2	4.541
5.0%	3.088	>50	*
7.5%	2.925	>50	*
10.0%	2.871	>50	*
12.5%	2.856	>50	*
15.0%	2.780	>50	*

The data in table 3 shows that the liquid smoke resulting from redistillation with a concentration of 5.0% is able to dissolve more than 50 g/L of chitosan, while the smoke concentrations of 1% and 2% can only able to dissolve the maximum of 3.7 g/L and 4.2 g/L chitosan. The lowest solubility of chitosan in the liquid smoke was at the concentrations of 1.0% and 2.0%. This probably occurred due to the effect of the addition of chitosan which caused the pH of liquid smoke concentration of 1% and 2% to increase from 3.204 and 3.183 to 5.141 and 4.541. Besides, at the concentrations of 5.0%-15.0%, the addition of chitosan did not significantly affect the pH value of the solution. Meanwhile, based on the chitosan solubility data above, a smoke concentration of more than 5.0% can dissolve chitosan significantly. In conclusion, this liquid smoke has the potential to be a good solvent for chitosan.

4. Conclusion

In conclusion, the results of this study revealed that the redistillation yield of liquid smoke was 76.14 %. Furthermore, this study defined some characteristics of the liquid smoke, namely containing 2.21 pH, 3.38% g/mL of the total acid, and 3.05 g/L of the total phenol. The GC-MS analysis results also showed that the twelve compound components could be divided into several composition groups, namely acids (73.08%), alcohols (14.73%), phenolic (1.21%), and carbonyls (5.19%). Besides, the concentration of liquid smoke has a correlation with the degree of solubility of chitosan. The liquid smoke concentration of 5% (v/v) was able to dissolve more than 50 g/L of chitosan.

Acknowledgments

The researchers are thankful to the Ministry of Research, Technology, and Higher Education and Lambung Mangkurat University. This study was funded by the Ministry of Research, Technology, and Higher Education through PDUPT grant 2018-2019 with a contract number 122.5/UN8.2/PP/2019 and by Lambung Mangkurat University through the Lecturers' Obligation for Conducting Research or *Dosen Wajib Meneliti* Program 2020 with a decree number 701/UN8/PP/2020.

IOP Conf. Series: Materials Science and Engineering **980** (2020) 012024 doi:10.1088/1757-899X/980/1/012024

References

[1] Qavami N, Badi H N, Labbafi M R, Mehregen M, Tavakoli M and Mehrafarin A 2017 *Trakia J. Sci.* **1** 83-91
[2] Purba R, Suseno S H, Izaki A F, and Muttaqin S 2014 *Int. J. Appl. Sci. Tech.* **4** 212-7
[3] Achmadi S S, Mubarik N R, Nursyamsi R and Septiaji P 2013 *J. Appl. Sci.* **13** 401-8
[4] Junaidi A B, Rusdiarso B, and Kartini I 2009 *Indones. J. Chem.* **9(3)** 369-72
[5] Junaidi A B, Abdullah, and Santoso U T 2019 *Konversi* **8** 39-43
[6] Sasongko P, Mushollaeni W and Herman 2014 *Buana Sains* **14** 193-7
[7] Saloko S, Darmadji P, Setiaji B and Pranoto Y 2014 *J. Food Bioscience*. **8** 71-9
[8] Muhammad D R A, Darmadji P and Pranoto Y 2011 *J. Tek. Hasil Pertanian* **IV** 104-12
[9] Rusdi, M. 2017 *J. Pertanian Terpadu* **4** 143-57
[10] Budaraga I K, Arnim, Marlida Y and Bulanin U 2016 *Int. J. Adv. Sci. Eng. Inf. Tech.* **6** 306-15
[11] Triastuti W E, Budhi P A, Agustiani E, Hidayat R A, Retnoningsih R and Nisa' A A 2019 *IPTEK J. Proc. Series* **3** 114-7
[12] Komarayati S, Gusmailina and Efiyanti L 2018 *J. Penelitian Hasil Hutan* **36** 219-38
[13] Sari T I, Amalia A K and Rahmawati 2009. *Jurnal Teknik Kimia*. **16** 44-7.
[14] Rinaldi, Alimuddin A and Panggabean A S 2015 *Molekul*. **10** 112-20
[15] Janairo J I B and Amalin D M 2018 *International Food Research Journal* **25** 213-6

IOP Conf. Series: Materials Science and Engineering **980** (2020) 012025 doi:10.1088/1757-899X/980/1/012025

The use of FTIR spectroscopy in combination with chemometrics for the authentication of milk fat from palm oil

A Windarsih[1*], Irnawati [2,3], and A Rohman[2,4]
[1]Research Division for Natural Product Technology (BPTBA), Indonesian Institute of Sciences (LIPI), Yogyakarta, 55861, Indonesia
[2]Department of Pharmaceutical Chemistry, Faculty of Pharmacy, Universitas Gadjah Mada, Yogyakarta, 55281, Indonesia
[3]Faculty of Pharmacy, Universitas Halu Oleo, Kendari, 93232, Indonesia
[4]Institute of Halal Industry nd System (IHIS), Universitas Gadjah Mada, Yogyakarta, 55281, Indonesia

Email: anjarwindarsih2@gmail.com

Abstract. This research was aimed to develop Fourier transform infrared spectroscopy (FTIR) combined with chemometrics of linear discriminant analysis (LDA), partial least square (PLS), and principal component regression (PCR) for authentication of milk fat from palm oil adulterant. FTIR spectroscopy and LDA have been successfully used to detect the presence of palm oil in MF. All the adulterated samples were clearly separated with authentic MF shown by the Cooman's plot. Chemometrics of PLS at the wavenumber of 3033-692 cm^{-1} using first derivative spectra was successfully applied for the quantification of palm oil in MF. The suitability of the model was presented by its high R^2 value both for calibration and validation models, accounting for 1 and 0.9998 respectively and its lower RMSEC (root mean square error of calibration) and RMSEP (root mean square error of prediction) value, accounting for 0.154 and 0.743 respectively. Quantification of palm oil was also successfully performed using chemometrics of PCR. The model showed high R^2 in both calibration (0.9998) and validation (0.9997) values with lower RMSEC (0.671) and RMSEP (0.905) values. It can be concluded that a combination of FTIR spectroscopy with chemometrics could be used for the authentication of milk fat adulteration.

1. Introduction

Milk fat is one of the high-quality fats which have many applications in food products. It has various compositions of nutrition which are important for human health. It contains essential fatty acids and rich in fat-soluble vitamins including vitamins A, D, E, and K [1]. Moreover, it also provides proteins, minerals, lactose, and triglycerides [2]. Milk fat is considered as the most valuable fat among other fats and it has a higher price compared to other fats and oils. Unethical sellers may take advantage by mixing it with other fats or oils having a lower price for economic reasons. This fraudulent practice is a serious problem, therefore, an analytical method capable of ensuring quality toward the authenticity of milk fat is highly required [3].

Several methods have been developed for fats analysis including milk fat either conventional method or advanced method using certain instruments. Analysis of fatty acid composition in milk fat has been performed using gas chromatography using a detector of flame ionization detector (FID) and mass spectrometer detector (MS) [4,5]. High-performance liquid chromatography has also been explored for milk fat analysis by determining the triglycerides composition [6]. These methods offer reliable and reproducible analytical methods for milk fat authentication. However, there are some disadvantages of using them for instance, they require a long time, need several steps of sample preparation, and require many solvents. Therefore, a rapid and reliable detection method such as FTIR spectroscopy is important to be investigated.

FTIR spectroscopy is a fingerprint technique that offers several advantages such as rapid analysis [7], it does not require sample preparation meaning that samples can be directly measured using attenuated total reflectance (ATR) technique by directly placing the samples on ATR crystal, it requires less solvent (considered as green analytical chemistry), and it is easy to be coupled with advanced statistical techniques such as chemometrics of multivariate analysis [8,9]. Chemometrics has been widely used for food and pharmaceutical product authentication. Chemometrics is divided into two categories, namely chemometrics of pattern recognition such as discriminant analysis (DA) and principal component analysis (PCA) and chemometrics of multivariate calibration or regression such as multiple linear regression (MLR), partial least square (PLS), and principal component regression (PCR) [10].

FTIR spectroscopy has wide applications in food analysis including milk, butter, cheese, fat, and oil [11,12]. For instance, FTIR spectroscopy using attenuated total reflectance (ATR) technique has been used for the detection and quantification of urea in milk. Combination with chemometrics of SIMCA (soft independent modeling class analogy) perfectly classified between milk samples and urea [13]. Quality analysis of raw milk has been carried out using FTIR spectroscopy by determining some parameters such as fat protein, lactose, and total solids. Based on the results, it showed that FTIR is a fast and reliable analytical technique for authentication of raw milk [14]. The objective of this research was to develop FTIR spectroscopy coupled with chemometrics of multivariate analysis for the authentication of milk fat from palm oil.

2. Materials and methods

2.1. *Materials*
Three commercial pure milk fat samples were obtained from different markets, namely in Bantul, Sleman, and Yogyakarta. The pure palm oil used was purchased from a market in Yogyakarta. The milk fat samples were high-quality food-grade while the palm oil used was commercially produced by industry. There were 42 adulterated samples and they were prepared manually in binary mixtures of milk fat and palm oil. Acetone and n-hexane were purchased from Mercks (Germany).

2.2. *Samples preparation*
A pure sample of milk fat and palm oil was prepared in a glass vial. Adulterated samples of milk fat with palm oil were prepared by making binary mixtures of samples with various adulterant concentrations in the range of 0-100% w/w with a total weight of 2 grams for each sample.

2.3. *FTIR measurement*
Each sample was measured using FTIR spectroscopy (Thermo Nicholet iS10) using ATR (attenuated total reflectance) technique. The acquisition of spectra was performed in the wavenumber region of 4000-650 cm^{-1} using a resolution of 8 cm^{-1} and a scan number of 32. Background spectra were carried out prior to each sample measurement.

2.4. *Chemometrics analysis*
Chemometrics pattern recognition was carried out using discriminant analysis (DA). The variables used for creating the DA model were absorbance in the region of 1500-900 cm^{-1}. DA result was evaluated using the Cooman plot. Chemometrics multivariate calibration for quantitative analysis was carried out using partial least square (PLS) and principal component regression (PCR) techniques.

Samples were divided into calibration and validation models. Wavenumber optimization was performed to obtain the best calibration and validation model in the range of 4000-650 cm^{-1}. The spectra were also derivatized using Savitzky-Golay either the first or second derivative. The models were chosen by evaluating the coefficient of determination (R^2) and root mean square error of calibration (RMSEC) and root mean square error of prediction (RMSEP).

3. Results and discussion

3.1. FTIR spectra analysis

FTIR spectroscopy is a fingerprint technique which means that no samples are having identical FITR spectra. Figure 1 shows the FTIR spectra of pure milk fat and pure PO. General investigation through the spectra showed that the spectra were very similar because the major component in fat and oil is fatty acids [7]. Further investigation revealed some differences in several peaks especially in the fingerprint region (1500-900 cm^{-1}). The functional groups of milk fat and palm oil observed using FTIR spectroscopy are shown in table 1. Even though deep investigation could differentiate between milk fat and palm oil spectra, however, in adulterated samples of milk fat with PO, it is very difficult to differentiate between authentic milk fat spectra and adulterated milk fat spectra. Therefore, an advanced statistical method such as chemometrics was performed to detect and quantify the adulterant (PO) in milk fat.

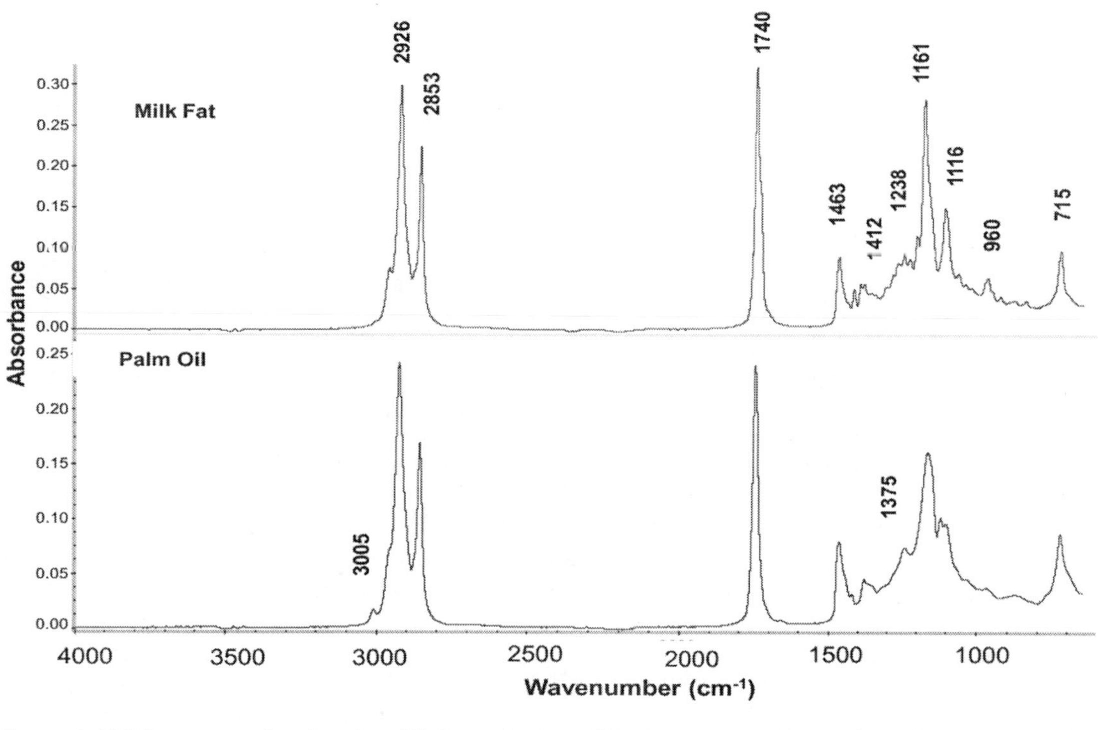

Figure 1. FTIR spectra of authentic milk fat and palm oil in the wavenumber region of 4000-650 cm^{-1}.

3.2. *Chemometrics analysis*

Chemometrics analysis was successfully used for managing the huge data extracted from FTIR spectra. The absorbance values were used for variables in chemometrics models because they are directly associated with concentration. Chemometrics of discriminant analysis perfectly classified between pure milk fat sample and adulterated milk fat with palm oil. All the adulterated samples were correctly separated from a pure sample of milk fat as can be seen in the Cooman plot (figure 2).

Discriminant analysis is one of the supervised pattern recognitions in the chemometrics technique which is widely used in sample differentiation. The x-axis showed the Mahalanobis distance which demonstrated the distance among samples. Samples containing a higher level of adulterants showed higher Mahalanobis distance from pure milk fat samples indicating a greater difference between them.

Table 1. Functional groups identification of FTIR spectra observed in milk fat and palm oil [15].

Wavenumber region (cm^{-1})	Functional groups and vibration modes
3005	=CH, stretching vibration
2926 and 2853	$-CH_3$, $-CH_2$, and -CH asymmetric, and symmetric stretching vibrations
1740	carbonyl (C=O), stretching vibration
1463	methylene ($-CH_2$), bending vibration
1412, 1375, 1235, 1161	bending vibration (wagging, twisting, scissoring) of $-CH_3$, $-CH_2$, and CH
1116	C-O, stretching vibrations
960	–HC=CH–(*trans*), bending out of the plane
715	–HC=CH–(*cis*), bending out of the plane

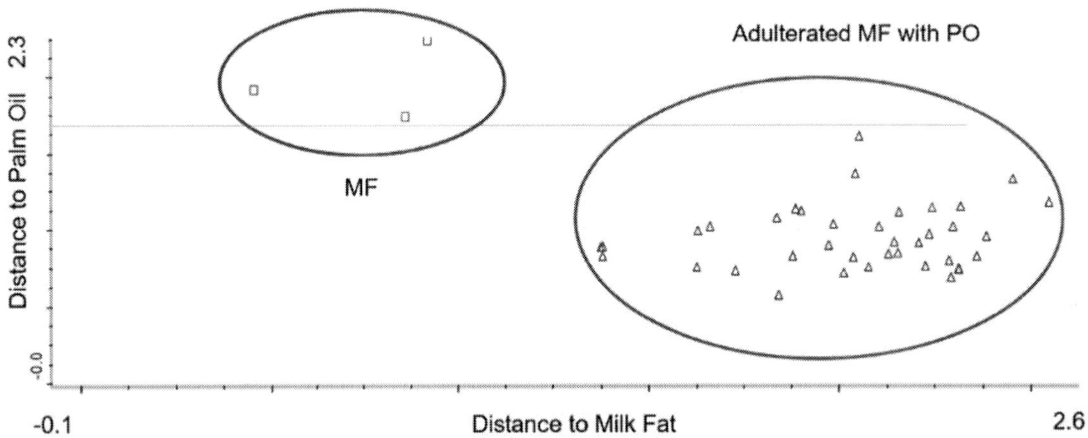

Figure 2. Discriminant analysis of milk fat adulterated with palm oil.

Chemometrics of multivariate calibration namely partial least square and principal component regression were investigated for palm oil quantification in milk fat. The values of coefficient of determination (R^2) both in calibration and validation models were used for model evaluation because they are related to the model accuracy. Moreover, the root mean square error of calibration (RMSEC) and root mean square error of prediction (RMSEP) were used to evaluate the model error. A model with a high value of R^2 and low value of RMSEC and RMSEP was chosen for analysis because it showed good accuracy, precision, and predictability [10,16]. PLS model was successfully applied using the first derivative FTIR spectra at the wavenumber region of 3033-692 cm^{-1} (figure 3). Spectra derivatization could provide better resolution in overlapping spectra so that it generated a good dataset for creating a model. The obtained R^2 and RMSEC values of the PLS calibration model were 1 and 0.154, respectively. The model is then validated using an external validation technique and the R^2 and RMSEP of the validation model were evaluated. Validation is important to be carried out to evaluate the validity of the PLS calibration model and to avoid model overpredicting. The result of the validation model demonstrated a high R^2 value (0.9998) and a low RMSEP value (0.743).

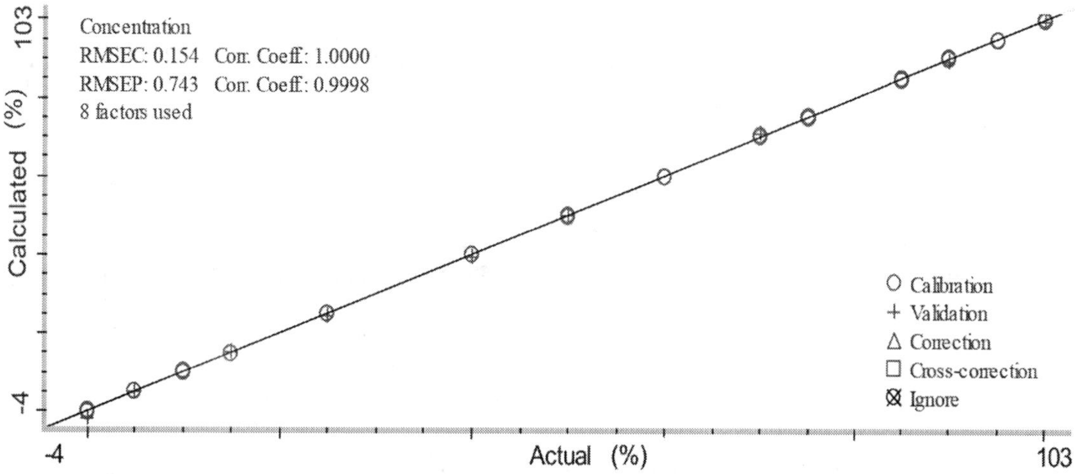

Figure 3. PLS calibration model for milk fat authentication from palm oil.

Another chemometrics model for quantification, namely PCR also demonstrated good accuracy and precision model for milk fat authentication. PCR was successfully used for palm oil quantification in binary mixtures with milk fat. PCR model was created using the first derivative spectra at the wavenumber combination of 3033-2770 cm^{-1} and 1806-692 cm^{-1}. The model had a high value of R^2 either in calibration (0.9998) or validation (0.9997). The obtained RMSEC and RMSEP values were 0.671 and 0.905, respectively. The difference between PLS and PCR models is in their algorithm for creating a model. PLS searched for latent variables to create a calibration model while PCR created principal components (PCs) then directly used them to build a regression model. From the above results, it is suggested that FTIR spectroscopy and chemometrics analysis offer potential advantages for milk fat authentication.

4. Conclusion

FTIR spectroscopy is a fingerprint technique that can be used for the authentication of milk fat. Combination with chemometrics of multivariate analysis such as linear discriminant analysis, partial least square, and principal component regression offered a rapid and reliable analytical method for adulterant detection and quantification in milk fat. It can be concluded that the combination of FTIR spectroscopy and chemometrics is a promising method for future authentication of fats and oils.

Acknowledgments

The authors thank the Ministry of Research/National Agency of Research and Innovation for financial support through the PDUPT scheme with contract number 2729/UN1.DITLIT/DIT-LIT/PT/2020, Indonesian Institute of Sciences and Integrated Research and Testing Laboratory for facilitating instrument.

References

[1] Sasaki R, Umezawa M, Tsukahara S, Ishiguro T, Sato S and Watanabe Y 2015 Assignment of milk fat fatty acid propyl esters by GC-FID analysis with the aid of Ag-ion solid-phase extraction *J. Oleo Sci.* **64** 1251–58

[2] Depeters E J, German J B, Taylor S J, Essex S T and Perez-Monti H 2001 *Fatty* acid and triglyceride composition of milk fat from lactating holstein cows in response to supplemental canola oil *J. Dairy Sci.* **84** 929-36

[3] Komponen P, Lemak B, Daripada B, Haiwan L, Nina Naquiah A N, Marikkar J M N, Mirghani M E S, Nurrulhidayah A F and Yanty N A M 2017 Differentiation of fractionated

components of lard from other animal fats using different analytical techniques *Sains Malaysiana* **46** 209–16

[4] Simionato J I, Garcia J C, Dos Santos G T, Oliveira C C, Visentainer J V and De Souza N E 2010 Validation of the determination of fatty acids in milk by gas chromatography *J. Braz. Chem. Soc.* **21** 520–4

[5] Amores G and Virto M 2019 Total and free fatty acids analysis in milk and dairy fat *Separations* **6** 1-10

[6] Robinson N P and MacGibbon A K H 1998 The composition of new zealand milk fat triacylglycerols by reversed-phase high-performance liquid chromatography *J. Am. Oil Chem. Soc.* **75** 993–9

[7] Rohman A, Windarsih A, Riyanto S, Sudjadi, Shuhel Ahmad S A, Rosman A S and Yusoff F M 2016 Fourier Transform Infrared Spectroscopy Combined with Multivariate Calibrations for the Authentication of Avocado Oil *Int. J. Food Prop.* **19** 680-7

[8] Lerma-García M J, Ramis-Ramos G, Herrero-Martínez J M and Simó-Alfonso E F 2010 Authentication of extra virgin olive oils by Fourier-transform infrared spectroscopy *Food Chem.* **118** 78–83

[9] Guntarti A, Ahda M, Kusbandari A and Prihandoko S 2019 Analysis of lard in sausage using Fourier transform infrared spectrophotometer combined with chemometrics *J. Pharm. Bioallied Sci.* **11** S594–600

[10] Worley B and Powers R 2013 Multivariate analysis in metabolomics *Curr. Metabolomics* **1** 92–107

[11] Saputra I, Jaswir I and Akmeliawati R 2018 Identification of pig adulterant in mixture of fat samples and selected foods based on FTIR-PCA wavelength biomarker profile *Int. J. Adv. Sci. Eng. Inf. Technol.* **8** 2341–48

[12] Du L, Lu W, Gao B, Wang J, Yu L, Editor G and Cetó X 2019 Authenticating raw from reconstituted milk using fourier transform infrared spectroscopy and chemometrics *J. Food Qual.* **2019** 1–6

[13] Jha S N, Jaiswal P, Borah A, Gautam A K and Srivastava N 2015 Detection and quantification of urea in milk using attenuated total reflectance-fourier transform infrared spectroscopy *Food Bioprocess Technol.* **8** 926–33

[14] Kučević D S, Glamočić D M, Trivunović S J, Jajić I M, Perišić B and Krstović S Z 2017 Validation and application of FTIR spectroscopy in raw milk analysis *Acta Period. Technol.* **48** 167–75

[15] Windarsih A, Irnawati and Rohman A 2020 Application of FTIR - ATR spectroscopy and chemometrics for the detection and quantification of lard oil in bovine milk fat *Food Res.* **4** 1732–38

[16] Zettl D, Bandoniene D, Meisel T, Wegscheider W and Rantitsch G 2017 Chemometric techniques to protect the traditional Austrian pumpkin seed oil *Eur. J. Lipid Sci. Technol.* **1600468** 1–11

IOP Conf. Series: Materials Science and Engineering **980** (2020) 012026 doi:10.1088/1757-899X/980/1/012026

Antibacterial activity assay of essential oils from limau kuit peel against *Staphylococcus aureus*

A Irwan[1]*, N Humaida[1] and H S Nur[2]
[1]Department of Chemistry, Faculty of Mathematics and Natural Sciences, University of Lambung Mangkurat, Indonesia
[2]Department of Biology, Faculty of Mathematics and Natural Sciences, University of Lambung Mangkurat, Indonesia

Email : airwan@ulm.ac.id

Abstract. This research aims to determine the antibacterial activity of essential oils from Limau Kuit (a local variety of limes of South Kalimantan) peel against *Staphylococcus aureus* ATCC25923, the minimum inhibitory concentration, and to identify compounds of essential oils contained. The essential oil has been isolated by hydro steam distillation and then was analyzed by using GC-MS to identify major compounds of the essential oils which were supposed as antibacterial agents. The result of this research showed the oils had antibacterial activity against *Staphylococcus aureus* ATCC25923 at concentrations 100%, 75%, 50%, and 25% with minimum inhibitory concentration 25% wet and dry sample. There was no significant difference in MIC between the two samples. At low concentrations, essential oils from wet samples had better inhibition power than dry samples up to a concentration of 75%. But at a concentration of 100%, their inhibitory power tended to decrease. Based on GC-MS data were demonstrated of 15 compounds in the wet sample and 16 compounds in the dry sample. Five major compounds of essential oil from Limau Kuit peel were limonene, γ-terpinene, β-pinene, α-pinene, and sabinene. Meanwhile, it was known that compounds limonene, α-pinene, γ-terpinene, β-pinene, α-terpineol, sabinene, and terpinen-4-ol had antibacterial activity.

1. Introduction

South Kalimantan possess local oranges are known as Limau Kuit. This orange has several peculiarities that can distinguish it from other types in terms of fruit aroma, fruit size, and what stands out is the surface morphology of wrinkled rind that is often associated with kaffir lime. The Banjar tribe community in South Kalimantan uses this orange as a chili sauce and food flavoring ingredient [1]. Part of the lime is the fruit peel usually only becomes waste, even though the orange peel consists of essential oils containing component compounds that can be utilized [2].

Orange peel contains essential oils consisting of various groups of compounds such as terpenes, sesquiterpenes, aldehydes, esters, and sterols. Orange peels have different compound contents depending on variety caused that the aroma is different [3]. Essential oils or known as essential oils, etherical oils, or volatile substances are volatile compounds that do not dissolve in water [4]. Essential oils and extracts from various kinds of plants have been widely used in the health field [5].

Existing components in essential oils can be used to reduce the development and spread of bacteria [6]. The antibacterial activity of essential oils is supposed to be in the presence of phenol compounds, terpenes, and aldocetone [7]. One of the bacteria due to some infections in humans and resistant to several types of drugs is *Staphylococcus aureus*. This bacterium is a common pathogen in humans frequently found as common biota on the skin [8,9].

A recent study on Limau Kuit is the use of ethanol extract from the peel against *S. aureus* [10] with concentrations of 100%, 75%, 50%, and 25%, resulting in inhibition zone diameter of 14 mm, 10.7 mm, 9.6 mm, and 0 mm, respectively. Based on the results of the study, it is urgent to determine the antibacterial activity of essential oils of the lime peel against *S. aureus* by taking into account the diameter of the inhibitory zone and the value of the minimum inhibitory concentration (MIC). Limau Kuit used were taken from the Astambul area, Kabupaten Banjar. The essential oils extracted by the steam-water distillation method then was analyzed by using GC-MS to identify the existing compounds.

2. Materials and methods

The Limau Kuit used in this research was taken from Astambul, Kabupaten Banjar, South Kalimantan, *S. aureus* ATCC 25923 bacterial culture was obtained from the The Institution of Research and Standardization of Industry Banjarbaru. The products were prepared by hydro-steam distillation with two sample preparations, wet/fresh and dry peel samples. Chemicals in the analytical grade were sodium sulfate anhydrous (Merck), H_2SO_4 (Merck), $BaCl_2$ (Merck), DMSO (Merck), and other materials were paper discs, chloramphenicol, NaCl (Oxoid), BHI Agar, BHI Broth, MH Agar (Oxoid), and distilled water.

2.1. Samples preparation

There were 2 variations of preliminary preparation, namely fresh fruit peel and dried fruit peel. There were 2 variations of preliminary preparation, namely fresh fruit peel and dried fruit peel. The fresh fruit rind was prepared by separating from the flesh of the fruit and then cut to the side size of ± 1 cm. Part of them then dried until withered for the dried sample.

2.2. Purification of extracted oils

The purification was conducted to remove the remaining water in oils. Anhydrous sodium sulfate was added as much as 1-3 g/L into oils in beaker glass, stirring it gently until the Na_2SO_4 anhydrous settles or agglomerates later, separated using filter paper. Purification results were stored again for antibacterial activity tests and composition analysis.

2.3. Antibacterial activity assay

Samples of oils from wet and dried peel were diluted to certain concentrations prior antibacterial assays. Dilution of essential oil was using DMSO (v/v) to make 75%, 50 %, 25% and 10% of concentration. Each concentration was then labeled as test tube II, III, IV, and V respectively. Test tube I was 100% of essential oils.

The BHI Agar media was prepared by weighing 0.94 g BHI Agar into an Erlenmeyer flask, adding 20 mL of distilled water, heated to boiling, and dissolved. The media was then sterilized using an autoclave at 121°C for fifteen minutes then poured into the petri dish aseptically. Media BHI Broth was made by weighing as much as 1.85 g of BHI Broth put into an Erlenmeyer flask adding 50 mL of dissolved distilled water. A total of 10 mL of media was put into each test tube. The media was then sterilized using an autoclave of 121°C for 15 minutes. Media MH Agar was made by weighing 3.8 g MH. To be put into the Erlenmeyer flask, 100 ml of distilled water added, heated to boiling, and dissolved. The media were sterilized using an autoclave at 121ºC for 15 minutes. Then it was poured into the Petri dish aseptically.

2.4. Preparation of S. aureus ATCC 25923 bacterial suspension

Bacterial culture obtained from the Research Institute and Industrial Standardization was taken as much as 1 ose and then inoculated into the BHI Agar medium and incubated in an incubator for 24 hours at ± 30°C. After incubating for 24 hours the rejuvenating bacteria were taken 1-2 ose then inoculated into BHI Broth media and incubated in an incubator for 24 hours at ± 30°C. The suspension of the bacteria was equated to the standard McFarland I Solution (3x108 CFU.mL-1).

2.5. Antibacterial activity assay and determination of MIC values
The method used for the antibacterial activity test was the Antimicrobial Paper Discs Method using MH Agar media. *S. aureus* ATCC 25923 bacteria as much as 0.2 mL were spread into Petri dishes containing MH Agar media with sterile cotton swabs. Paper discs containing essential oils of lime peel concentration 100%, 75%, 50%, 25%, 10%, negative control, and positive control were placed on the test media incubated for 24 hours at ± 30°C, then the inhibition zone was measured using Calipers. The negative control used was DMSO, while chloramphenicol was used as a positive control. The formation of inhibitory zones around disc paper showed antibacterial activities.

Determination of the MIC (Minimum Inhibition Concentration) value aims to determine the lowest concentration of essential oils which can show antibacterial activity which is characterized by the formation of inhibitory zones. The method used to determine the MIC value was the same as the determination of the antibacterial activity of the essential oils of lime skin, namely the Paper Disc Plate method. The MIC value could be seen from the inhibition zone formed at the lowest concentration, while quantitatively determined using the Regression line by carotemplating the inhibitory diameter of essential oils in mm (Y axis) to the log concentration (X axis) the intersection of the equation obtained by linear regression $Y = aX + b$. The value of Y was the dependent variable while X was the independent variable. MIC value was obtained from the calculated X value with $Y = 0$.

2.6. GCMS analysis
The components of the essential oils were analyzed by gas chromatography-mass spectrometry (GC-MS), using a Shimadzu QP2010S GC-MS Series. Samples were injected on a capillary column (RTX 5 MS). The carrier gas was Helium, at the flow rate of 0.55 ml/min. The inlet injection temperature was 300°C with the inlet split ratio of 139. The GC oven temperature was kept at 50°C for 3 min and programmed to 250°C at a rate of 3°C/min, held for 5 min and finally at 10°C/min and programmed to 300°C, held for 3 min. The GC/MS interface (Auxs) temperature was set at 300 °C. The Mass Spectrometer condition was as follows: EL Source temperature 250°C and Ele energy 70 eV. The MS library was Wiley 229 and NIST62.

3. Results and discussion
Essential oils of Limau Kuit peel resulting from steam-water distillation were separated between the water layer and the oil layer using a pipette. Essential oils were in the upper layer while the water in the lower layer. The essential oil was purified with anhydrous sodium sulfate to remove the remaining water in order to get the essential oil which was free of water then the yield was calculated. The essential oil yield showed the amount of oil produced from the distillation process in a certain amount expressed in percent (%). Wet samples produced yields of 0.5047% (w/w) and dried samples produced yields of 0.2739% (w/w).

The content of the essential oil compound Limau Kuit peel was analyzed using GC-MS Shimadzu QP2010S series. The results of GC-MS analysis of the components of wet and dried Limau Kuit peel essential oil obtained as chromatogram and mass spectra.

Based on the table 1, the five compounds which had relatively higher levels than other compounds making up the essential oils of Limau Kuit in wet samples are limonene (63.42%), γ-terpinene (18.50%), β-pinene (8.40%), α-pinene (1.51%), and sabinene (1.35%) whereas for dry samples (table 2) are limonene (62.59%), γ-terpinene (16.94%), β-pinene (9.09%), α-pinena (1.75%), and sabinene (1.30%).

The antibacterial activity test results of essential oils of Limau Kuit peel after incubated 1x24 hours at ± 30°C showed that essential oils were able to inhibit the growth of *S. aureus* ATCC 25923 bacteria, seen from the inhibition zone produced with variations in the concentration of 100%, 75%, 50%, and

IOP Conf. Series: Materials Science and Engineering **980** (2020) 012026 doi:10.1088/1757-899X/980/1/012026

25% while the 10% concentration there was no inhibition zone. The diameter of inhibitory zone of essential oils of lime skin could be seen from the presence of clear zone around the paper disc that showed no bacterial growth (figure 1). The measurement results of inhibition zone formed compared with the standards of the Clinical and Laboratory Standards Institute (CLSI) are shown in (table 3).

Table 1. The composition of Limau Kuit essential oils in wet samples.

Peak	Retention Time (menit)	Relative Concentration (%)	Formula	Compound	SI *(Similarity index)* (%)
1	10,037	0,42	$C_{10}H_{16}$	*α*-thujene	93
2	10,321	1,51	$C_{10}H_{16}$	*α*-pinene	94
3	11,841	1,35	$C_{10}H_{16}$	sabinene	94
4	12,057	8,40	$C_{10}H_{16}$	*β*-pinene	96
5	12,463	1,27	$C_{10}H_{16}$	Mircene	96
6	13,002	0,25	$C_8H_{16}O$	octanal	90
7	13,507	0,39	$C_{10}H_{16}$	α-terpinene	94
8	13,833	1,23	$C_{10}H_{14}$	cimene	93
9	14,160	63,42	$C_{10}H_{16}$	limonene	95
10	15,118	18,50	$C_{10}H_{16}$	*γ*-terpinene	96
11	15,997	1,00	$C_{10}H_{16}$	*α*-terpinolene	95
12	19,258	0,57	$C_{10}H_{18}O$	terpinen-4-ol	95
13	19,783	0,72	$C_{10}H_{18}O$	*α*-terpineol	95
14	20,001	0,33	$C_{12}H_{24}O$	dodecanal	89
15	28,136	0,63	$C_{15}H_{24}$	germacrene	90
Total		100			

Table 2. The composition of Limau Kuit essential oils in dried samples.

Peak	Retention Time (menit)	Relative Concentration (%)	Formula	Compound	SI *(Similarity index)* (%)
1	10.040	0.44	$C_{10}H_{16}$	*α*-thujene	92
2	10.324	1.75	$C_{10}H_{16}$	*α*-pinene	93
3	11.843	1.30	$C_{10}H_{16}$	sabinene	94
4	12..063	9.09	$C_{10}H_{16}$	*β*-pinene	96
5	12.463	1.27	$C_{10}H_{16}$	mircene	95
6	13.004	0.49	$C_8H_{16}O$	octanal	91
7	13.508	0.34	$C_{10}H_{16}$	α-terpinene	94
8	13.833	1.15	$C_{10}H_{14}$	cimene	94
9	14.163	62.59	$C_{10}H_{16}$	limonene	96
10	15.118	16.94	$C_{10}H_{16}$	*γ*-terpinene	96
11	15.467	0.30	$C_{10}H_{18}O_2$	linalool oxide	91
12	15.997	0.78	$C_{10}H_{16}$	*α*-terpinolene	95
13	19.280	1.11	$C_{12}H_{24}O$	terpinen-4-ol	94
14	19.729	1.23	$C_{10}H_{18}O$	*α*-terpineol	96
15	20.001	0.61	$C_{10}H_{20}O$	decanal	91
16	28.140	0.61	$C_{15}H_{24}$	germacrene	90
Total		100			

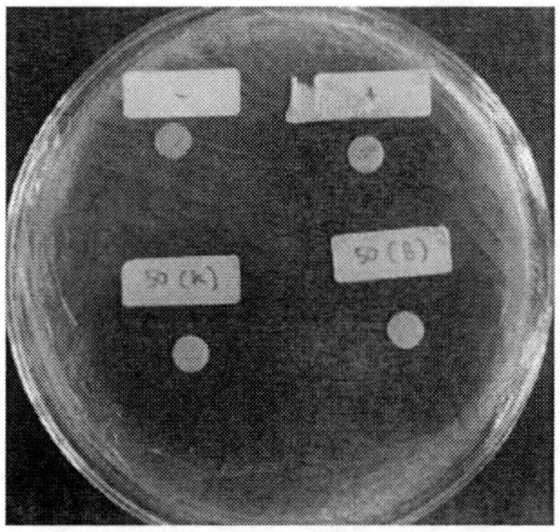

Figure 1. Diameter of inhibition zones of Limau Kuit essential oils against *S. aureus* ATCC 25923.

Table 3. The results of the measurement of inhibition zones of antibacterial activity of essential oils of Limau Kuit against *S. aureus* ATCC 29523.

Concentration	Samples	Inhibition Zones (mm) Sensitivity		
		Oils	Positive Control (Chloramphenicol 30μg)	Negative Control (DMSO)
100%	Wet	19.5 (I)	30.5 (S)	-
	Dry	26.8 (S)		
75%	Wet	38.3 (S)	28 (S)	-
	Dry	39 (S)		
50%	Wet	27.9 (S)	29.8 (S)	-
	Dry	26.3 (S)		
25%	Wet	32.9 (S)	31.7 (S)	-
	Dry	10.8 (R)		
10%	Wet	-	31.8 (S)	-
	Dry	-		

R = resistent, S = Sensitive, I = Intermediate

Based on the table 3, the lowest concentration of essential oils of lime peel in inhibiting the growth of *S. aureus* ATCC 25923 was at a concentration of 25%. In this study, the inhibition zone diameter did not increase in proportion to the increase in concentration of essential oils. This could occur due to several factors including the volume of absorption of essential oils on paper discs, the thickness of agar media, and the diffusion ability of essential oils on agar media [5,11]. The mechanism of essential oil as an antibacterial based on microstructural observations was by increasing permeability and disrupting the integrity of cell membranes. The hydrophobic nature of essential oils allows essential oils to accumulate in cell membranes, disrupt structure, and increase membrane permeability. Damage from intracellular components and a decrease in the workings of enzymes and loss of cell contents in large numbers caused cell death [7,12,13,14,15,16,17,18]. Essential oil compounds that had antibacterial activity were limonene, α-pinene, β-pinene, γ-terpinene, α-terpineol, sabinene, and terpinene-4-ol.

4. Conclusion

Limau Kuit peel essential oils have antibacterial properties against *S. aureus* ATCC 25923 bacteria. The MIC value of citron peel essential oil was at a concentration of 25% of wet and dried samples with inhibition zone diameters of 32.9 mm and 10.8 mm. Dried and wet sample of Limau Kuit peel essential oils had 5 dominant components with relatively greater concentration values, namely limonene, γ-terpinene, β-pinena, α-pinena, and sabinene. Essential oil compounds that had antibacterial activity were limonene, α-pinene, β-pinene, γ-terpinene, α-terpineol, sabinene, and terpinene-4-ol.

References

[1] Kurniawati Y 2017 Pengaruh perlakuan pada daun limau kuit terhadap karakteristik minyak atsiri dengan metode ekstraksi soxhlet menggunakan pelarut n-heksana *Skripsi* (Banjarbaru: Universitas Lambung Mangkurat)

[2] Sugiantoro, Jayuska A and Alimuddin A H 2016 Biotransformasi limonena dari minyak atsiri kulit jeruk pontianak menggunakan jamur *Rhizopus oligosporus* dalam media air kelapa *JKK.* **5(3)** 40-4

[3] Hidayati 2012 Distilasi minyak atsiri dari kulit jeruk pontianak dan pemanfaatanya dalam pembuatan sabun aromaterapi *Biopropal Industri.* **3(2)** 39-49

[4] Kurniawan A, Kurniawan C, Indraswati N and Mudjijati 2008 Ekstraksi minyak kulit jeruk dengan metode distilasi, pengepresan dan leaching *Widya Teknik.* **7(1)** 15-24

[5] Srisukh V, Tribuddharat C, Nukoolkarn V, Bunyapraphatsara N, Chokephaibulkit K, Phoomniyom S, Chuanphung S and Srifuengfung S 2012 Antibacterial activity of essential oils from *Citrus hystrix* (makrut lime) against respiratory tract pathogens *Science Asia.* **38** 212-7

[6] Stefanakis M K, Touloupakis E, Anastasopoulos E, Ghanotakis D, Katerinopoulos H E and Makridis P 2013 Antibacterial activity of essential oils from plants of the genus Origanum *Food Control.* **34** 539-46

[7] Fei Lv, Liang H, Yuan Q, Li C 2011 In vitro antimicrobial effects and mechanism of action of selected plant essential oil combinations against four food-related microorganisms *Food Research International.* **44** 3057-64

[8] Pradani N R 2012 *Uji aktivitas antibakteri air perasan jeruk nipis (Citrus aurantifolia,* Swingle) *terhadap pertunbuhan bakteri Staphylococcus aureus secara in vitro Skripsi.* (Jember: Universitas Jember)

[9] Chouhan S, Sharma K and Guleria S 2017 Antimicrobial activity of some essential oils-present status and future perspectives *Medicines.* **4(58)** 1-21

[10] Ariyani H, Nazemi M, Hamidah and Kurniati M 2018 Uji efektivitas antibakteri ekstrak kulit limau kuit (*Citrus hystrix* D.C) terhadap beberapa bakteri *Journal of Current Pharmaceutical Sciences.* **2(1)** 136-41

[11] Dewi F K 2010 *Aktivitas antibakteri ekstrak etanol buah mengkudu (Morinda citrifolia* Linnaeus*) terhadap bakteri pembusuk daging segar Skripsi.* (Surakarta: Universitas Sebelas Maret)

[12] Cox S D, Mann C M, Marrkham J L, Gustafon J E, Warmington J R and Wyllie S G 2001 Determining the antimicrobial actions of tea tree oil *Molecules.* **6** 87-91

[13] Burt S 2004 Essential oils: their antibacterial properties and potential applications in foods - a review *International Journal of Food Microbiology.* **94** 223-53

[14] Oyedemi S O, Okoh A O, Mabinya L V, Pirochenva G and Afolayan A J 2009 The proposed mechanism of bactericidal action of eugenol, α-terpineol, γ-terpinene against *Listeria monocytogenes, Streptococcus pyogenes, Proteus vulgaris* and *Escherichia coli African Journal of Biotechnology.* **8(7)** 1280-86

[15] Huang D F, Xu J G, Liu J X, Zhang H and Hu Q P 2014 Chemical constituents, antibacterial activity and mechanism of action of the essential oil from *Cinnamomum cassia* bark against four food-related bacteria *Microbiology.* **83(4)** 357-65

[16] Raut J S and Karuppayil S M 2014 A status review on the medicinal properties of essential oils *Industrial Crops and Products.* **62** 250-64

IOP Conf. Series: Materials Science and Engineering **980** (2020) 012026 doi:10.1088/1757-899X/980/1/012026

[17] Li Z H, Cai M, Liu Y S, Sun P L and Luo S L 2019 Antibacterial activity and mechanisms of essential oil from *Citrus medica* L. *var. sarcodactylis Molecules.* **24(1577)** 2-10
[18] Guimaraes, A. C., L. M. Meireles, M. F. Lemos, M. C. C. Guimaraes, D. C. Endringer, M. Fronza, & R. Scherer. Antibacterial activity of terpenes and terpenoids present in essential oil *Molecules*. **24(2471)** 1-12

IOP Conf. Series: Materials Science and Engineering **980** (2020) 012027 doi:10.1088/1757-899X/980/1/012027

The ulitization of durian wood (*Durio zibethinus*) and corn cob (*Zea mays*) biochar on corn yields in acid sulphate soil

E Setiawati[1]*, W A Yusuf[2]

[1] Institute of Research and Standardization of Industry, Ministry of Industry, Jl. Panglima Batur Barat No.2, Banjarbaru South Kalimantan, Indonesia

[2] Indonesian Swampland Agricultural Research Institute, Jl. Kebun Karet Banjarbaru, South Kalimantan, Indonesia

Email: evy.kemenperin@gmail.com

Abstract. Biochar addition can be used to overcome the problem of acid sulphate soil. This study was aimed to analyze the effect of biochar dose on plant height and corn yield in acid sulphate soils. Durian wood waste and corn cob biochar used in this study was pyrolyzed for 2 hours at 550 °C temperature and grinded until around 0.4-1.0 mm. The experiment was conducted in a screen house at the Indonesian Swampland Agriculture Research Institute. The experimental design used was a Completely Randomized Design consisting of six doses of biochar treatment of with 3 replications. The total of 18 experimental pots was obtained with the following details: B1 0 t/ha, B2 4 t/ha, B3 8 t/ha, B4 12 t/ha, B5 16 t/ha and B6 20 t ha. Biochar treatment significantly affect plant height age of 2nd, 4th, 6th, 8th, 10th and 12th weeks. The longer the age, the higher the corn plant height. The corn cob length, corn cob diameter, dry weight of seed, and the dry water content significantly different in all treatments. The biochar dose significantly affected the corn yields. The best treatment of biochar in acid sulphate soil was obtained in 12 t ha^{-1}.

1. Introduction

Biochar is a subtle-grained carbonization product that has a high organic carbon content and low degradation susceptibility, obtained through biomass and biodegradable waste pyrolysis [1]. Biochar can be used for energy-related intention related to agriculture and environmental conservancy. Various kinds of biochar utilization continue to develop, especially in fields such as farming, industry, and operation linked with natural surroundings [2]. A modern approach to biochar production and utilization promising long-range agricultural advantages [3]. Biochar can be returned to the soil to restock nutrients and minerals. Hence it can increase the soil physicochemical features such as density, porosity, and pH. Biochar has also been found to absorb heavy metal and release macro- and micro-nutrients slowly [4] which can contribute to increasing fertilizer efficiency.

Biochar is produced through pyrolysis, it is heated without enough oxygen for complete combustion, removes various liquids and gases from the feedstock, and leaves a residue as biochar. Slow pyrolysis is the most effective pyrolisis because it results in a high yield of approximately 35.0% by weight of dry biomass [5]. There are many kinds of pyrolysis feedstocks, such as municipal waste, agricultural biomass, and forestry residue. Nowadays, the forestry residues can be used as a sustainable and reliable pyrolysis feedstock. The forestry residue has the most interesting characteristics as pyrolysis biomass since it has a sustainable supply system for large scale base, and raw materials suitability because of its features (low humidity, metal, and ash content) [4]. Biochar derived from wood has a large potency to increase C storage in tropical soils because of its high C concentration, aromatic character, and low H/C ratio. The high of aromatic C content is due to the high amount of lignin and cellulose in wood biomass [6].

As soil amendment, biochar has been developed as a sustainable practice that simultaneously can mitigate climate change and improve the marginal soil quality [7]. There are lots of marginal land in Indonesia, such as tidal land, acidic and peat soils. The use of marginal land for crop cultivation has several obstacles that must be faced. Acid sulphate soils have been seen as one of the difficult soil to manage for crop production because of their low fertility, such as low soil pH and available phosphate and high Fe content [8]. The corn cultivation is very potential to be developed in acid sulphate soil. Soil acidity in acid sulphate soils is a main obstacle in the corn cultivation because it needs optimum soil pH at 5.6-7.5 for best growth. The high acidity in the soil will cause a deficiency of macronutrients such as potassium, magnesium, calcium, and phosphorus. The corns begin to suffer at pH 5.6, and can not survive at pH 4.0 [9]. Liming improves soil for the microbe habitat. They accelerate the decay of plant residues and release micronutrients, N, and P in the soil. Liming the acid sulphate soil can reduce the solubility of manganese and aluminum at low pH. Biochar, as liming material, can improve plant growth and crop yield. The fertilizer application can increase the corn yield as 4.4 t ha^{-1} [10]. Fertilizers that are combined with biochar can numerically increase corn yields. Based on Brantley [11], corn yields at fertile silt loam can increase at a dose of 5 and 10 Mg ha^{-1} of woodchip biochar in combination with N fertilizer. The addition of cow manure biochar at 15 t ha^{-1} and 20 t ha^{-1} in combination with NPK fertilizer significantly increased the corn yield of 150% and 98%, respectively in sandy soil [12]. This study was aimed to analyze the effect of biochar dose on plant height and corn yield in acid sulphate soils.

2. Materials and methods

2.1. Biochar production and properties

The method of biochar production was adopted from Setiawati [13]. Durian wood waste and corn cob as feedstock were collected from Banjarbaru, South Kalimantan, Indonesia. The feedstock was air-dried to remove moisture. Dry wood waste was then pyrolyzed at 550 °C with a holding time of 2 hours. The hot biochar was then quickly immersed by water. The biochar produced was then air-dried and grinded until around 0.4-1.0 mm to yield Durian Wood Biochar (DWB) and Corn Cob Biochar (CCB). Proximate and ultimate analyses of biochar included moisture content (MC, ASTM D.3173), pH (1:5 for biochar: water), volatile matter (VM, ASTM D.3175), fixed carbon (FC, ASTM D.3172), ash content (ASTM D.3174). Carbon (C), hydrogen (H), nitrogen (N), and sulfur (S) were analyzed using ASTM D.5373, while oxygen (O) was computed according ASTM D.3176 by the sum of percentage of C, H, N, S, ash, subtracted from 100. The exchange cation capacity (CEC) was analyzed by using 1 M NH_4OAc extraction at pH 7.0. The biochar morphologies were analyzed by scanning electron microscope (SEM) method at 1000x magnification in the scales of 50 μm with an acceleration voltage of 15 kV. Fourier Transform Infrared (FT-IR) Spectrophotometer to analyze functional groups.

2.2. Soil used and experimental design

The acid sulphate soil was collected from Barambai, Barito Kuala District, South Kalimantan. The 30-50 cm depth layer soil had the following characteristics: pH (H_2O) 3.47; soil organic carbon (SOC) 4.50%; total N 0.246%; P_2O_5 105.15 mg $100g^{-1;}$ K_2O 45.05 mg $100g^{-1}$; 64.23 mg $100g^{-1}$; exchangeable cations of Ca, Mg, K, Na, respectively, 0.21, 0.50, 0.82, and 0.52 me $100g^{-1}$ and CEC 20.34 me $100g^{-1}$. The soils were incorporated with 1 t ha^{-1} dolomite and 20 t ha^{-1} cow manure in all treatments one week before transplanting. Each biochar (DWP and CCB) was added to these soil based on each treatment rate.

The treatments of biochar added to soil were (1) Soil control (no biochar/B1); (2) Soil + 4 t ha^{-1} of biochar (B2); (3) Soil + 8 t ha^{-1} of biochar (B3); (4) Soil + 12 t ha^{-1} of biochar (B4); (5) Soil+16 t ha^{-1}

of biochar (B5); (6) Soil + 20 t ha^{-1} of biochar (B6). Each of the soil was put in a pot at the screen house. The hybrid corn seed, Bima 20 Uri, was then spread to the soils at the pot. The soil pH was analysed at the 4th, 8th, and 12th weeks. The growth of plant height was observed at the 2nd, 4th, 6th, 8th, 10th, and 12th weeks. The corn yield parameters included diameter parameters of cobs, cob length, dry seed shell weight, and dry water content.

2.3. Experimental design
The experimental design used in this study was a Completely Randomized Design (CRD) consisting of 6 (six) doses of biochar treatment of DWB and CCB with 3 replications. The analysis of data used SAS 9.1 software for windows by using the analysis of variance at the level of confidence at 5 %. Difference between treatments for each parameter was analyzed by the Duncan's Multiple Range (DMRT) Test with a 5% confidence level [14] .

3. Results and discussion

3.1. Biochar characteristics
Table 1 showed that the moisture content of corn cob durian (6.18%) was lower than the durian wood biochar (7.66%). The value of water content in biochar could be influenced by the drying process of biochar just after being immersed in water. Biochar for agricultural purposes should have a low water content because it could accelerate the absorption of nutrients and/or water [15]. The value of water content affected the ability of biochar to hold water. Biochar had the high ability to retain water to maintain soil moisture and create environmental carrying capacity [16].

Some research results concerning the pH value of biochar stated that biochar used as a soil conditioner should have an alkaline pH or above 7 [17,18]. These alkaline pH values indicated that both durian wood and corn cob biochars could also be used to reduce soil acidity. The corn cob biochar had a higher pH than that of durian wood (9.35 vs. 8.93, respectively), as well as a much higher CEC (25 vs. 24 me $100g^{-1}$, respectively). The high pH was due to the nutrient content of corn cob was higher than durian wood. Furthermore, high ash content of corn cob biochar also played a vital role in increasing the pH of biochar. High CEC in biochar indicated the ability to retain nutrients and water so that the water availability of plants was sufficient, especially for the process of plant physiology which necessary for plant height growth [19].

In biomass, volatile matters were methane, hydrocarbons, hydrogen, carbon monoxide, and non-combustible gases such as carbon dioxide and nitrogen. The content of VM durian wood biochar and corn cob biochar was (15.21 and 17.13)%. The VM in biochar was composed of compounds other than water, ash, and carbon. It was consists of elements of hydrogen, hydrocarbons, CO_2, CH_4, methane, and carbon monoxide [15]. Therefore, the high or low value of VM in biochar was based on the chemical composition of extractive substances from its feedstock. Fixed-C showed that carbon had been converted to a more stable carbon which cannot be easily degraded. In line with the previous studies [13,15,20], the biochar FC of durian wood and corn cob was 71.9% and 70.37%.

The higher C concentration with a rise in pyrolysis temperature was due to a higher polymerization rate, resulted in a more condensed carbon structure in biochar. A similar result was reported for biochar produced by Domingues [6], biochar C contents in durian wood and corn cob were 79.88% and 81.13%. Yang [21] and Khodadad [22] found that biochar derived from wood at higher temperatures was more stable because it contains large amounts of aromatic organic matter compared to biochar made from agricultural residues when pyrolyzed at high temperature. An increase in the aromatic character of biochars was associated with dehydration reactions and removal of O and H functional groups, as well as the formation of aromatic structures [23]. The durian wood biochar had a higher H content than durian wood (11.09% vs. 10.23%, respectively), as well as a much higher O content (3.41% vs. 1.61%, respectively).

IOP Conf. Series: Materials Science and Engineering **980** (2020) 012027 doi:10.1088/1757-899X/980/1/012027

Table 1. Biochar characteristics.

Testing parameter	Durian wood	Corn cob
Moisture content (%)	7.66	6.18
pH	8.93	9.35
Ash content (%)	5.23	6.32
Volatile matter (%)	15.21	17.13
Fixed carbon (%)	71.9	70.37
C (%)	79.88	81.13
H (%)	11.09	10.23
O (%)	3.41	1.61
N (%)	0.36	0.65
S (%)	0.03	0.06
CEC (me $100g^{-1}$)	24.0	25.0

As seen in figure 1, there was the similarity of functional groups between durian wood biochar and corn cob biochar, which consisted of conjugated C-H bend (aromatic) at 862.88 cm^{-1} - 758.74 cm^{-1}, while the other peaks showed the presence of cluster C=O stretch (in ketone) 1686.43 cm^{-1}, C=C stretch (aromatic) at 1570.71 cm^{-1} - 1481.99 cm^{-1}, -CH3 bend (in alkane) at 1481.99 cm^{-1}, and C-O stretch (in ether, esther, carboxylic acid, alcohol) at 1385.56 cm^{-1} - 111.68 cm^{-1}. The higher number of C–H bend in durian wood biochar than corn cob biochar indicated that the durian wood biochar was more aromatic. The aromatic carbonyl (C-O) and carboxyl groups (C=O) were attributed to retained oxygen-containing organic groups in biochar produced. The broad peak of durian wood biochar at 3472.39 cm^{-1} came from the feedstock, which was commonly found in lignin of phenolics (O-H stretch) [24].

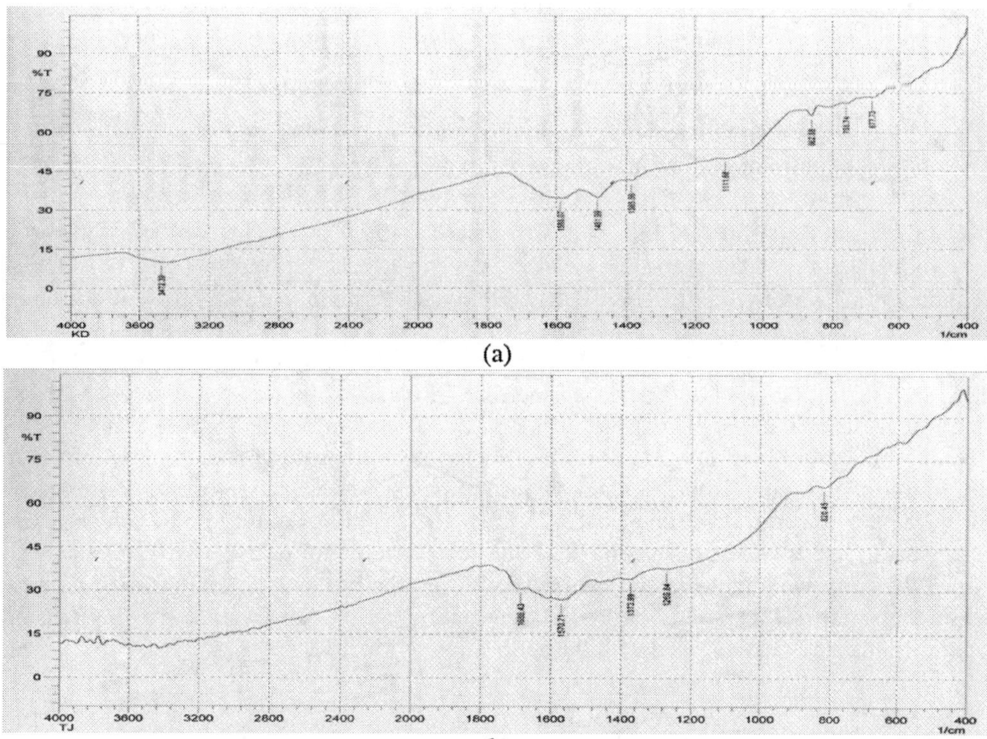

(a)

(b)

Figure 1. FTIR spectrum of (a) durian wood biochar; (b) corn cob biochar.

IOP Conf. Series: Materials Science and Engineering **980** (2020) 012027 doi:10.1088/1757-899X/980/1/012027

3.2. Soil pH and growth plant height

Based on the results of a variety of analysis, it showed that the various doses of durian wood biochar and corn cob biochar on corn plants significantly affected the height of corn plants aged 2nd, 4th, 6th, 8th, 10th and 12th weeks of observation. The longer the observation, the higher the corn plant height. It could be stated that biochar addition can significantly influence plant growth [25]. The obtained results of plant growth indicated that the ameliorative effect of biochar was largely related with pH increase (table 2). The soil pH increased at biochar dose starting from 4 t ha^{-1} (B2). In accordance with Manickam [7], the application of 4 t ha^{-1} biochar combined with bio-fertilizer yielded a high increase in soil pH and crop yields.

Table 2. Soil pH of various biochar doses.

Treatments	pH					
	4th week planting		8th week planting		12th week planting	
	DWB	CCB	DWB	CCB	DWB	CCB
B1	3.85^{b}	3.93^{b}	4.85^{c}	5.05^{b}	5.21^{c}	5.41^{b}
B2	5.00^{a}	4.97^{ab}	5.76^{b}	5.48^{ab}	5.54^{bc}	5.76^{ab}
B3	5.34^{a}	5.45^{a}	5.81^{ab}	5.78^{a}	5.61^{b}	5.58^{b}
B4	5.24^{a}	5.29^{a}	5.88^{ab}	5.61^{ab}	5.68^{b}	5.85^{ab}
B5	5.24^{a}	5.62^{a}	5.84^{ab}	6.03^{a}	5.86^{ab}	5.98^{ab}
B6	5.39^{a}	5.60^{a}	6.02^{a}	6.12^{a}	6.10^{a}	6.51^{a}

Remarks: Numbers followed by the different letter within each column were significantly different based on DMRT $\alpha = 5\%$

Table 3. Plant height of corn plant.

Treatments	Plant Height (cm)											
	2nd week planting		4th week planting		6th week planting		8th week planting		10th week planting		12th week planting	
	DWB	CCB	DWB	CCB	DWB	CCB	DWB	CCB	DWB	CCB	DWB	CCB
B1	35.8^{c}	35.8^{c}	67.0^{d}	67.7^{d}	98.7^{e}	98.7^{d}	121.8^{c}	123.7^{cd}	144.2^{a}	145.7^{a}	115.0^{e}	114.0^{c}
B2	34.0^{c}	35.3^{c}	58.3^{e}	59.3^{e}	101.0^{d}	102.9^{c}	119.2^{d}	124.9^{cd}	137.3^{c}	138.1^{b}	130.7^{d}	133.9^{b}
B3	39.3^{b}	40.2^{b}	76.4^{c}	76.0^{c}	109.0^{c}	110.2^{b}	120.7^{cd}	121.9^{d}	132.3^{d}	135.8^{b}	140.0^{b}	143.9^{a}
B4	39.7^{b}	40.6^{b}	78.0^{b}	78.9^{b}	112.7^{b}	112.7^{b}	126.3^{b}	128.3^{ab}	140.0^{b}	142.3^{a}	144.9^{a}	149.6^{a}
B5	43.7^{a}	44.3^{a}	79.0^{b}	79.3^{b}	112.0^{b}	112.6^{b}	126.0^{b}	126.5^{bc}	140.0^{b}	143.4^{a}	130.3^{d}	133.6^{b}
B6	45.2^{a}	46.4^{a}	83.3^{a}	81.6^{a}	124.0^{a}	124.7^{a}	132.7^{a}	130.9^{a}	141.2^{b}	144.4^{a}	135.0^{c}	143.9^{a}

Remarks: Numbers followed by the different letter within each column were significantly different based on DMRT $\alpha = 5\%$

Based on table 3, the results showed an increase in plant height at 2nd, 4th, 6th, 8th, 10th and 12th weeks of observation in all biochar doses, except in B2. The plant height decreased in 4 t ha^{-1} at all of age planting periods in both DWB and CCB. Biochar dose of 4 t ha^{-1} had the least effect to plant growth. In line with Alvum-Toll [26], the low dose of biochar was not significantly different from the zero dose in plant height. During the first 1-2 weeks after planting, seeds provided food reserves for plant growth. In line with the development of the root system and decrease in food reserves in the seeds then the roots were immediately obtained following their functions. The growing point was 2-3 cm below the surface of the soil in the 3-4 week planting periods. In the first growing season at 3-4 week periods, plant height increased with the application of biochar.

3.3. Corn yield

Based on the study results, it showed that the corn cob length, corn cob diameter, dry weight of seed, and dry water content significantly different in all treatments. The biochar dose significantly affected the corn yields. The cornob length was (11.20-14.14) cm for DWB treatment and (10.98-114.67) cm for CCB treatment. The diameter of corn cobs was (3.14-3.55) cm for DWB and (3.23-3.67) cm for CCB. The dry weight of seed was (26.77-50, 68) gr for DWB and (27.17-51.23) gr for CCB. The dry water content was (13.17-15.50) % for DWB and (13.24-14.56) % for CCB.

The biochar could increase crop yield. The presence of cow manure as compost could improve productivity. The compost showed the existence of microbiological characteristics that the potential to be used as a soil conditioner. The combination of compost and biochar seemed to increase added value. It was proved in this study, which B2-B6 had a better yield performance than B1. In line with Trupiano [27] which stated that a combination of biochar and organic fertilizers (compost) which provides important nutrients for plants could increase soil fertility and crop yield. Compared to B1, the increasing of corn cob length was 12.28% - 26.5% for DWB and 25.3%-33.6% for CCB (figure 2a), the corn cob diameter was 1.8%-13.1% for DWB and 8.4%-13.6% for CCB (Figure 2b), the dry weight of seed was 13.1-89.3% for DWB and 27.2-88.6% for CCB (Figure 2c), while the dry water content of corn was decreased by 5.6%-15.1% for DWB and 4.0%-9.1% for CCB.

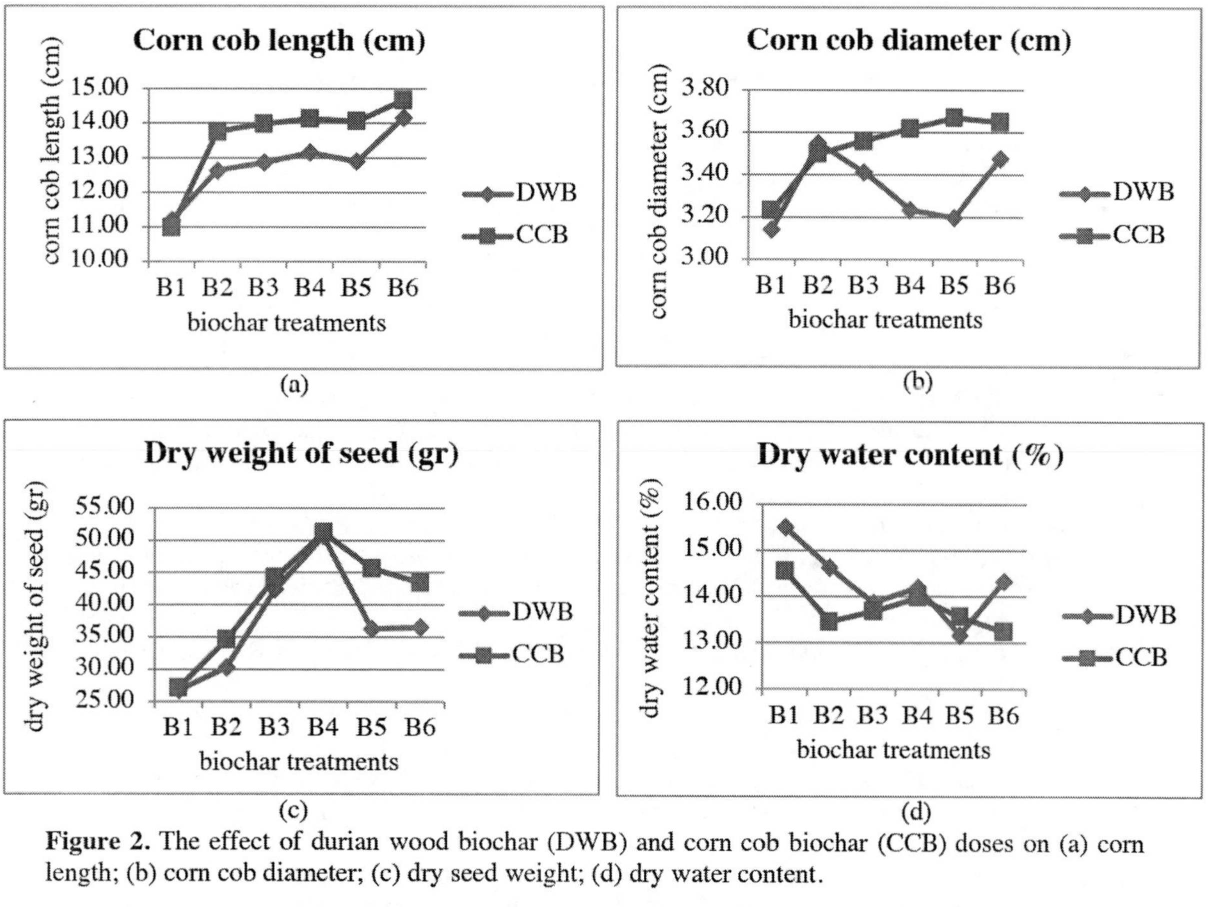

Figure 2. The effect of durian wood biochar (DWB) and corn cob biochar (CCB) doses on (a) corn length; (b) corn cob diameter; (c) dry seed weight; (d) dry water content.

Figure 2e showed that the highest dry weight of seed in acid sulphate soil was obtained from 12 t ha^{-1} (B4) biochar dose, which increased by 89.3% for durian wood biochar and 88.6% for corn cob biochar, compared with the treatment without biochar (B1). Ngongo [28] state that the optimum dose of bamboo biochar on dry weight of seed was 10.44 t ha^{-1}. Bamboo biochar on 10 t ha^{-1} could increase the growth of corn crops in dryland [29].

4. Conclusion

Biochar treatment significantly affected plant height age of 2nd, 4th, 6th, 8th, 10th and 12th weeks. The longer the age, the higher the corn plant height. The corn cob length, corn cob diameter, dry weight of seed, and the dry water content significantly different in all treatments. The biochar dose significantly affected the corn yields. The cornob length was (11.20-14.14) cm for DWB treatment and (10.98-114.67) cm for CCB treatment. The diameter of corn cobs was (3.14-3.55) cm for DWB and (3.23-3.67) cm for CCB. The dry weight of seed was (26.77-50, 68) gr for DWB and (27.17-51.23) gr for CCB. The dry water content was (13.17-15.50) % for DWB and (13.24-14.56) % for CCB. The best treatment of biochar in acid sulphate soil was obtained in 12 t ha^{-1}.

Acknowledgments

The authors would like to thank all the parties who are giving the facilities/laboratories and support in this research. The author also thank Indonesia Endowment Fund for Education for funding the research.

References

[1] International Biochar Initiative 2015 Standardized product definition and product testing guidelines for biochar that is used in soil Version number 2.1

[2] Saletnik B, Zagula G, Bajcar M, Tarapatskyy M, Bobula G and Puchalski C 2019 Biochar as a multifunctional component of the environment-A review *Appl. Sci.* **9** 1139

[3] Timmons D, Lema-Driscoll A and Uddin G 2017 *The economics of biochar carbon sequestration in Massachusetts* (Boston: University of Massachusetts) p 62

[4] Verma M, Godbout S, Brar S K, Solomatnikova O, Lemay S P and Larouche J P 2012 Biofuels production from biomass by thermochemical conversion technologies *Int. J. Chem. Eng.*, **2012** 542426

[5] Tomczyk A, Sokołowska Z and Boguta P 2020 Biochar physicochemical properties: pyrolysis temperature and feedstock kind effects *Rev. Environ. Sci. Biotechnol.* **19(1)** 191–215

[6] Domingues R R, Ttrugilho P F, Silva C A, de Melo I C N A, Melo L C A, Magriotis, Z M, Sánchez-Monedero M A 2017 Properties of biochar derived from wood and high-nutrient biomasses with the aim of agronomic and environmental benefits *PLoS One* **12(5)** 1–19

[7] Manickam T, Cornelissen G, Bachmann R T, Ibrahim I Z, Mulder J and Hale S E 2015 Biochar application in Malaysian sandy and acid sulfate soils: Soil amelioration effects and improved crop production over two cropping seasons *Sustainability* **7** 16756–70

[8] Shamshuddin J, Azura A E, Shazana M A R S, Fauziah C I, Panhwar Q A and Naher U A 2014 Properties and management of acid sulfate soils in Southeast Asia for sustainable cultivation of rice, oil palm, and cocoa *Advances in Agronomy* (Cambridge: Academic Press) chapter 3 pp 91–142

[9] Brewbaker J L 2003 *Corn production in the tropics: The Hawaii experience* (Manoa: University of Hawaii) p 76

[10] Raihan H S 2000 Pemupukan NPK dan ameliorasi lahan pasang surut sulfat masam berdasarkan nilai uji tanah untuk tanaman jagung *J. Ilmu Pertan.* **9(1)** 20–8

[11] Brantley K E, Savin M C, Brye K R and Longer D E 2015 Pine woodchip biochar Impact on soil nutrient concentrations and corn yield in a silt loam in the Mid-Southern U.S. *Agriculture* **5** 30–47

[12] Uzoma K C, Inoue M, Andry H, Fujimaki H, Zahoor A and Nishihara E 2011 Effect of cow manure biochar on maize productivity under sandy soil condition *Soil Use Manag.* **27(2)** 205–12

[13] Setiawati E, Prijono S, Mardiana D, Prayogo C and Soemarno 2019 Impact of pyrolysis temperature and water quenching on hydrophilicity of biochar derived from durian wood waste *Biosci. Res.* **16(2)** 2047–62

[14] Harter H L 1960 Critical values for Duncan's New Multiple Range Test *Biometrics* **16(4)** 671–85

[15] Iskandar T and Rofiatin U 2017 Karakteristik biochar berdasarkan jenis biomassa dan parameter proses pyrolisis *J. Tek. Kim.* **12(1)** 28–34
[16] Shafie S T, Mohd.Salleh M, Hang L L, Rahman M, Azlina W and Ghani W A W A k 2012 Effect of pyrolysis temperature on the biochar nutrient and water retention capacity," *J. Purity, Util. React. Environ.* **1(6)** 323–37
[17] Lehmann J 2007 Bio-energy in the black *Front Ecol Env.* **5(7)** 381–7
[18] Blok C, Regelink I, Hofland-Zijistra J, Streminska M, Eveleens B and Bolhuiss P 2016, *Perspectives for the use of biochar in horticulture* (Netherland: Wageningen UR Greenhouse Horticulture) p 40
[19] Ndruru J I, Nelvia and Adiwirman 2018 Application of biochar and liquid Smoke to the growth of upland rice (Oryza sativa. L) on ultisol medium *J. Agroteknologi* **9(1)** 9–16
[20] Chowdhury Z Z, Karim Z M, Ashraf M A and Khalid K 2016 Influence of carbonization temperature on physicochemical properties of biochar derived from slow pyrolysis of durian wood (*Durio zibethinus*) sawdust *BioResources* **11(2)** 3356–72
[21] Yang H, Yan R, Chen H, Lee D H and Zheng C 2007 Characteristics of hemicellulose, cellulose and lignin pyrolysis *Fuel* **86(12–13)** 1781–88, 2007
[22] Khodadad C L M, Zimmerman A R, Green S J, Uthandi S, and Foster J S 2010 Taxa-specific changes in soil microbial community composition induced by pyrogenic carbon amendments," *Soil Biol. Biochem* **43(2)** 385–92
[23] Lehmann J and Joseph S 2012 *Biochar for environmental management: Science and technology*, 2nd ed (London: Earthscan) p 944
[24] Poletto M and Zattera A J 2013 Materials produced from plant biomass. Part III: Degradation kinetics and hydrogen bonding in lignin *Mater. Res.* **16(5)** 1065–70
[25] Seresmic S I, Zivanov M S, Milosev D S, Vasin J R, Ciric V I, Vasiljevic M B and Vujic N J 2015 Effects of biochar application on morphological traits in maize and soybean *Matica Srp. J. Nat. Sci.* 129 17–25
[26] Alvum-Toll K, Karlsson T and Strom H 2011 *Biochar as soil amendment-A comparison between plant materials for biochar production from three regions in Kenya* (Sweden: Swedish University of Agricultural Sciences) p 68
[27] Trupiano D, Cocozza C, Baronti S, Amendola C, Vaccari F P, Lustrato G, Lonardo S D, Fantasma F, Tognetti R, Scippa G S 2017 The effects of biochar and its combination with compost on lettuce (*Lactuca sativa* L.) growth, soil properties, and soil microbial activity and abundance *Int. J. Agron.* D 3158207
[28] Ngongo P M 2018 Utilization of bamboo biochar andphonska fertilizer on cultivation of maize (*Zea mays* L. *SEAS (Sustainable Environ. Agric. Sci.* **2(1)** 67–71
[29] Situmeang Y P, Adnyana I M, Subadiyasa I N N and Merit I N 2015 Effect of dose biochar bamboo, compost, and phonska on growth of maize (*Zea mays* L.) in dryland *Int. J. Adv. Sci. Eng. Inf. Technol.* **5(6)** 433–39

IOP Conf. Series: Materials Science and Engineering **980** (2020) 012028 doi:10.1088/1757-899X/980/1/012028

The utilization of activated carbon from cassava stems on the glucose and cholesterol adsorption

S Mutiaradini[1], L Efiyanti[2*], G Pari[2], and B M Soebrata[1]
[1] IPB University, Jl. Tanjung Dermaga Bogor 16680, Indonesia
[2] Forest Product Research and Development Center, Jl. Gunung Batu No. 5 Bogor 16118, Indonesia

Email: lisnaefiyanti@gmail.com

Abstract. The utilization of cassava stems from waste as an activated carbon raw material for glucose and cholesterol adsorption has been investigated. This study aims to know about the application of activated carbon from cassava stems waste on glucose and cholesterol adsorption with the variations of adsorbent weight and reaction contact time. The process of carbon acquiring was done by the pyrolysis method at a temperature of 400°C for 4 hours. After that, the process was followed by the chemical activation process using KOH as an activator. The carbon obtained was washed in KOH solution with a ratio of 1:2 w/w for ±24 hours, then filtered, washed again, and heated with an activation reactor at a temperature of 800°C. The application of activated carbon was carried out on glucose and cholesterol adsorption to produce glucose and cholesterol adsorption of 3.33 mg/g and 6.84 mg/g respectively. The optimum contact time for glucose and cholesterol adsorption was 45 minutes with the use of 0.1 g adsorbent.

1. Introduction

The cassava plants grow widely in Indonesia with a production range of about 1,9 million tons [1]. In general, the cassava plants only used in tuber part as an additional staple food, while the cassava stems become waste that not used optimally. This problem encourages the utilization of cassava stems to have added value as a raw material of activated carbon. It is expected that the waste can be an alternative precursor with low cost, high abundant, and renewable sources for activated carbon production.

Activated carbon is a product from pyrolysis and activation reaction with several methods, like the physical, chemical, or a combination of both. This process produces the activated carbon that has specific properties, such as high surface area, good surface reactivity, high physical-chemical stability, high pore size, and superior adsorption capacity [2]. The activated carbon with good properties is considered to be a promising material for industrial applications, such as for the adsorption process. Several findings of activated carbon application as adsorbent are for metal adsorption [3], methylene blue dan acid blue 29 [4], and synthetic organic contaminants [5,6].

On its development, the activated carbon is not only for metal or dyes adsorbent, but it can be used for health, including for poisoning removal pharmaceutical waste removal, such as antibiotics, beta-blocker, psychiatric drugs [7], and drugs delivery [8]. This research is aimed to study the potentials of activated carbon from cassava stems waste on glucose and cholesterol adsorption. This is due to the two compounds have an important role in the human body, while it is necessary to avoid body system disorder. As we know that the excess glucose in our body causing diabetes mellitus disease with high rate of mortality [9]. While the cholesterol compound has a function in the digestion process, but if it is too high, it will cause a blockage in the blood vessels causing various disease, like hypertension, arteriosclerosis, and coronary heart disease, otherwise if it is too low, it causes anemia [10].

In the removal process of glucose and cholesterol, several materials have been used such as polysaccharide [11], chitosan [12,13,14], and alumina [15] but activated carbon utilization for this purpose is still limited. The production and application of activated carbon for glucose and cholesterol adsorption necessary to study the potential of cassava stems activated carbon as an adsorbent. This adsorption process was carried out by using various contact time and adsorbent weight to measure the optimum adsorption capacity.

2. Materials and methods

2.1 Materials and equipment

The materials used in this research were cassava stems waste, D-glucose (Merck, Darmstadt, Germany), KOH, $FeCl_3.6H_2O$, ethanol 95% (v/v), cholesterol powder, phenol 5% (b/v), CH_3COOH, H_2SO_4(p), $Na_2S_2O_3$ 0.1N, I_2 0.1 N. The equipments used were Unitronic OR P shaker water bath (J.P. Selecta, Barcelona, Spain), rotor, furnace, analytical balance, porcelain curs, oven, 60 and 100 sieve mesh, glassware, and UV-Vis Spectrophotometer (Shimadzu-1700, Kyoto, Japan).

2.2 Method

2.21 Synthesis and characterization of activated carbon. The synthesis of activated carbon from cassava stems was carried out according to Prapagdee [3] method. Cassava stems were cleaned and chopped into the size of ±5cm, then dried with air dry process. The pyrolysis process was done at a temperature of 400°C for 4 hours, then the equipment was cooled. The carbon was collected and calculated. The carbon as a pyrolysis product was activated using a KOH activator with a ratio of 1:2 (w/w) (carbon soaked into KOH solution for 24 hours). The carbon was then dried for 24 hours using dry air process. The carbon produced was heated in an activation reactor at a temperature of 800°C and the activated carbon was obtained. The quality of activated carbon was analyzed based on Indonesian National Standard (SNI 06-3730 1995), including the moisture, ash content, volatile matter content, fixed carbon, and iod adsorption.

2.2.2 Glucose adsorption analysis. The glucose adsorption process using activated carbon was started with glucose content measurement by quantitative method (phenol sulphate method) based on [16]. The glucose standard solutions were made with a concentration of 5, 25, 50, 75, dan 100 μg/mL. The activity test of glucose adsorption using activated carbon was conducted with a contact time and adsorbent weight variations. The contact time variations were 30, 45 and 60 minutes while the adsorbent weights were 0.1; 0.3 and 0,5 g. The activated carbon was placed into a 10mL glucose solution with a concentration of 100 μg/mL, then shaken using shaker equipment follows the time contact and adsorbent weight variations. After the completed process, the solution filtered and then the glucose content was measured. The glucose measurement was done by 1 mL phenolic 5% and 5 mL H_2SO_4(p) added into blanc, standard, and sample solutions. The solutions were shaken and cooled for 10 minutes at room temperature. Then, the solutions were heated in the water bath for 15 minutes. The absorbance measurement analyzed by using UV-Vis spectrophotometer was done at 485nm wavelength.

The equation 1 of adsorption capacity [2]

$$Q = \frac{V\,(Co-Ca)}{m} \tag{1}$$

Q is adsorption capacity (mg/g), V is solution volume (L), Co is glucose/cholesterol initial concentration (μg/mL), Ca is glucose/cholesterol final concentration (μg/mL) and m is adsorbent weight (g).

2.2.3 Cholesterol adsorption analysis. Cholesterol adsorption process by activated carbon was done by using FeSac reagent. The process was initiated with FeSac reagent synthesis using 2.5 g of $FeCl_3.6H_2O$ diluted in 25 mL glacial acetic acid. Then 1 mL of iron solution was placed into the H_2SO_4 solution. Meanwhile, a cholesterol solution was prepared by dissolving 25 mg of cholesterol in 250 mL of glacial acetic acid and dilute into concentrations (5, 25, 50, 75, and 100 μg/mL). The cholesterol adsorption process was conducted similarly with glucose adsorption, but 2 mL of FeSac added into blanc, standard and filtered solution then those were shaken for 20 minutes and the absorbance was measured using spectrophotometer UV-Vis at 560 nm wavelength.

3. Results and discussion

3.1 Characterization of activated carbon-based on Indonesian National Standard
The synthesis of activated carbon in this research was started with the pyrolysis method at a temperature of 400°C. In this process, several stages of biomass raw material decomposition depended on pyrolysis temperature. In the initial stage at low temperature, the cellulose content tended to first decompose, then followed by hemicellulose and lignin at higher temperatures resulted in carbon as a product. The carbon product used as a precursor for the activating process with KOH activator and heated at a temperature of 800°C resulted in activated carbon with more excellent properties. The activated carbon from biomass generally prepared by chemical methods [17]. Table 1 shows the activated carbon quality based on proximate analysis using Indonesian Standard Method (SNI 06-3730 (1995). From the table, we can see that the quality of activated carbon from cassava stems met SNI (moisture <15%; ash content <10%; Volatile matter content <25%, fixed carbon >65% and iod adsorption >750 mg/g).

Table 1. The characterization of activated carbon.

	Moisture (%)	Ash (%)	Volatile Matter (%)	Fixed Carbon (%)	Iod Adsorption (mg/g)
Carbon	2.49	4.38	30.17	65.45	173.136
Activated carbon	14.2	7.75	24.85	67.4	979.119
SNI	<15	< 10	< 25	> 65	> 750

In general, the pyrolysis from biomass that involved decomposition reaction would produce carbon and liquid smoke as by-product. Carbon as a pyrolysis product could be activated with a chemical or physical activator. The activation process would produce activated carbon with better characteristics. From figure 1, it can be seen that the proximate analysis of activated carbon was different than those of without the activation process. The activation process with KOH as an activator compound could reduce the impurity and ash content, also open carbon pores [18]. This resulted in activated carbon with higher iod adsorption. With the utilization of a temperature 800°C, the potassium boiling point could be reached and the diffusion of potassium compound into the surface internal structure would break the chemical bond on biomass and cause the volatile matter apart from the surface. KOH activator utilization play an important role in the forming and developing of pore structure on the surface of activated carbon, it will be accessible and adsorbate molecules easier to diffuse [19]. The carbon activation process using KOH forms micropores more than other activators, so this is an advantage for adsorption [20], like in methylene blue adsorption [17]. KOH and heating process used at a temperature of 800°C, and it's increasing the surface area and pore of activated carbon so it is potential to become CO_2 adsorbent [21]. In other research, the activated carbon used for rhodamine B adsorption with an adsorption capacity of 123.46 mg/g was reached in 12 minutes contact time and can be regenerated until 7 times [22].

3.2 Glucose and cholesterol adsorption

Glucose and cholesterol have hydrocarbon groups so that most of the activated carbon surface can function as effective adsorption sites for glucose (table 2). The difference in the presence/absence of OH groups will affect the saturation capacity, the more OH groups presence, the higher the saturation. Molecule charges contribute to the electrochemical resistance on pore channels. The adsorbate polarizability can affect toward adsorption process. The aromatic compounds, in general, has a higher of PKa and usually do not have charges so the polarity is more dominant than charges. The polarity influenced by molecule size, the higher molecule size, the higher the polarity of the compound. The adsorption capacity of activated carbon to glucose is lower than cholesterol, it is caused by OH groups, size, and mass molecule difference from both of the compounds. The compound with higher molecule size tends to has a higher affinity and more adsorbed in adsorbent surface than the pore, while the compound with lower molecule more adsorbed into pores [4].

Table 2. Characteristics of compounds studied.

Compounds	Chemical formula	Structure	Mass (g/mol)	Λ_{maz} (nm)
Glucose	$C_6H_{12}O_6$	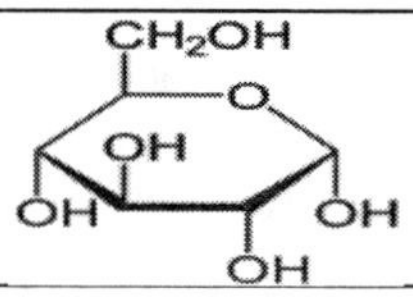	180,156 g/mol	485
Cholesterol	$C_{27}H_{46}O$	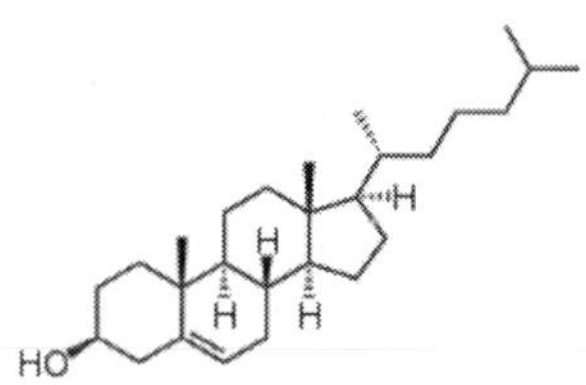	386.65 g/mol	560

The increasing of oxygen in the external surface of activated carbon causes a higher negative charge and it will increase water molecule bond and form a barrier on the hydrophobic molecule adsorption process [5]. The negative surface and acidity tend to associate with low adsorption capacity. The carbon surface with negative charge tends to more beneficial for adsorption of adsorbate with positive charges. In this study, the maximum glucose and cholesterol adsorption capacities were 3.3 mg/g and 6.84 mg/g respectively. In other research, the activated carbon utilization with KOH activator showed the adsorption capacity of methylene blue of 642-683 mg/g [23] and H_2S of 98 mg/g [19]. The cholesterol removal from dairy products reached 97% (2.2 mg/g to 0.1 mg/g) using combination methods with supercritical and alumina as an adsorbent but the weight of alumina used is 50 g [15] . The activated carbon with KOH and EDTA combination also can adsorb the phenolic compound with an adsorption capacity of 194 mg/g [24] and used for butyric acid adsorption with 98% adsorption [25].

The contact time variations showed different adsorption results. From figures 2 and 3, we could see that the increasing contact time used can increase the adsorption capacity but in a certain time the adsorption will decreased, this is due to the fast adsorption on adsorbent in initial stages until equilibrium reached. After the optimum time passed through, the adsorption would be decreased due to the active sites was occupied with adsorbate molecules and the desorption process reached. The presence of empty sites on adsorbent in initial reaction causing the adsorption process to increase, but later then there was a rejection force between adsorbate and adsorbent molecules. In general, when the adsorption doses increased, the adsorption process will be increased. This can be explained by an increasing number of activated carbon sites, so the adsorbate will be easier to adsorbed, but if it is too much, the saturation will be reached so the adsorption will be not optimal [22].

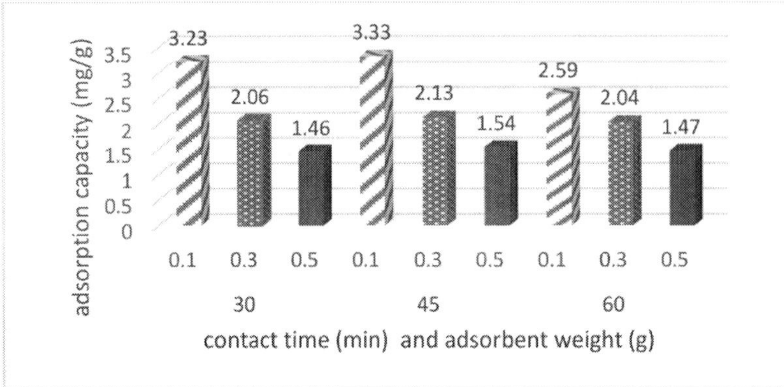

Figure 1. The glucose adsorption capacity of activated carbon.

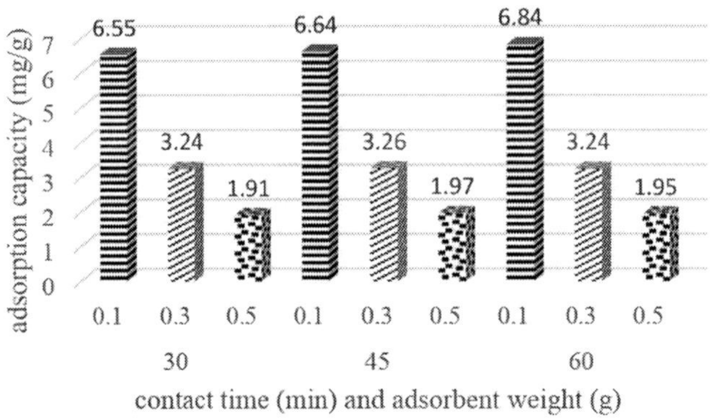

Figure 2. The cholesterol adsorption capacity of activated carbon.

4. Conclusion

The activated carbon from cassava stems waste is potential to be utilized on glucose and cholesterol adsorption process. The adsorption capacity of activated carbon on glucose and cholesterol adsorption were 3.33 mg/g and 6.84 mg/g respectively. The optimum contact time for glucose and cholesterol adsorption was 45 minutes with the addition of 0.1 g adsorbent.

References

[1] Susanti A A and Waryanto B *Statistik Pertanian 2017 (Agricultural Statistics)* (Kementerian Pertanian Republik Indonesia)

[2] Marrakchi F, Hameed B H and Bouaziz M 2020 Mesoporous and high-surface-area activated carbon from defatted olive cake by-products of olive mills for the adsorption kinetics and isotherm of methylene blue and acid blue 29 *Biochem. Pharm.* **8(5)** 104199

[3] Prapagdee S, Piyatiratitivorakul S and Petsom A 2014 Activation of cassava stem biochar by physico-chemical method for stimulating cadmium removal efficiency from aqueous solution *Environ. Asia* **7(2)** 60-9

[4] Bazan-Wozniak A and Pietrzak R 2020 Adsorption of organic and inorganic pollutants on

activated bio-carbons prepared by chemical activation of residues of supercritical extraction of raw plants *Chem. Eng. J.* **393** 124785

[5] Partlan E, Ren Y, Apul O G, Ladner D A and Karanfil T 2020 Adsorption kinetics of synthetic organic contaminants onto superfine powdered activated carbon *Chemosphere* **253** 126628

[6] Huan Li Y, Chang F M, Huang B, Song Y P, Zhao H Y and Wang K J 2020 Activated carbon preparation from pyrolysis char of sewage sludge and its adsorption performance for organic compounds in sewage *Fuel* **266** 117053

[7] Sbardella L, Comas J, Fenu A, Rodriguez-Roda I and Weemaes M 2018 Advanced biological activated carbon filter for removing pharmaceutically active compounds from treated wastewater *Sci. Total Environ.* **636** 519-29

[8] Bielicka A *et al* 2013 Carbon materials as new nanovehicles in hot-melt drug deposition 2013 *J. Phys. Condens. Matter* **25(35)**

[9] Phetsang S, Jakmunee J, Mungkornasawakul P, Laocharoensuk R and Ounnunkad K 2019 Sensitive amperometric biosensors for detection of glucose and cholesterol using a platinum/reduced graphene oxide/poly (3-aminobenzoic acid) film-modified screen-printed carbon electrode *Bioelectrochem.* **127** 125-35

[10] Huang Y *et al* 2017 Ultrasensitive cholesterol biosensor based on enzymatic silver deposition on gold nanoparticles modified screen-printed carbon electrode *Mater. Sci. Eng. C Mater Biol. Appl.* **77** 1-8

[11] Saputro P and Estiasih T 2015 Pengaruh polisakarida larut air (PLA) dan serat pangan umbi-umbian terhadap glukosa darah: kajian pustaka *J. Pangan dan Agro.* **3(2)** 756-62

[12] Hargono H, Abdullah A, and Sumantri I 2008 Pembuatan kitosan dari limbah cangkang udang serta aplikasinya dalam mereduksi kolesterol lemak kambing *Reaktor* **12** 53-7

[13] Kurniasih M, Setyaningtyas, Kartika D, Badriyah E H and Riyani K 2017 Adsorpsi Kolesterol lemak sapi dengan n-metil kitosan *J. Rekay. Kimia Lingk.* **12(2)** 103-11

[14] Kusuma S H 2016 Kemampuan kitin dari cangkang kepiting bakau (*scylla* spp) dalam menurunkan kadar kolesterol jeroan sapi. *J. Ilmiah Mahas. Kegur. dan Ilmu Pendi. Unsyiah* **1(1)**

[15] Mohamed R S, Saldaña M D A, Socantaype F H and Kieckbusch T G 2000 Reduction in the cholesterol content of butter oil using supercritical ethane extraction and adsorption on alumina *J Superc. Fluids* **16(3)** 223-5

[16] Wiyantoko M, Rusitasari B and Putri R 2017 Identifikasi glukosa hasil hidrolisis serat daun nanas menggunakan metode fenol-asam sulfat secara spektrofotometri uv-visibel 2017 *Pros. Sem. Nas. Kimia. FMIPA UNESA* 124-31

[17] Yagmur E, Gokce Y, Tekin S, Semerci N I and Aktas Z 2019 Characteristics and comparison of activated carbons prepared from oleaster (*Elaeagnus angustifolia* L.) fruit using KOH and ZnCl2 *Fuel* **267** 117232

[18] Apriani R, Faryuni I D, Wahyuni D and Kunci K 2013 Pengaruh konsentrasi aktivator kalium hidroksida (KOH) terhadap kualitas karbon aktif kulit durian sebagai adsorben logam fe pada air gambut*Prisma. Fisika* **1(2)** 82-86

[19] Wang H, Han J, Bo Z, Qin L, Wang Y and Yu F 2019 Non-thermal plasma enhanced dry reforming of CH4 with CO2 over activated carbon supported Ni catalysts *Mol. Catal.* **475** 110486

[20] Mistar E M, Alfatah T and Supardan M D 2020 Synthesis and characterization of activated carbon from Bambusa vulgaris striata using two-step KOH activation *J. Mater. Res. Technol* **9(3)** 6278-6286

[21] Yu J H, Joo S Y, Sim T I, Hong S C, Kim O S and Kang J M 2019 Post-KOH activation of nitrogen-containing porous carbon with ordering mesostructure synthesized through a self-assembly *Chem. Phys. Lett* **739** 137028

[22] Xiao W *et al* 2020 Preparation and evaluation of an effective activated carbon from white sugar for the adsorption of rhodamine B dye *J. Clean. Prod.* **253** 119989

IOP Conf. Series: Materials Science and Engineering **980** (2020) 012028 doi:10.1088/1757-899X/980/1/012028

[23] Efimov M N, Vasilev A A, Muratov D G, Baranchikov A E and Karpacheva G P 2019 IR radiation assisted preparation of KOH-activated polymer-derived carbon for methylene blue adsorption *J. Environ. Chem. En* **7(6)** 103514

[24] Lv S, Li C, Mi J and Meng H 2020 A functional activated carbon for efficient adsorption of phenol derived from pyrolysis of rice husk, KOH-activation and EDTA-4Na-modification *Appl. Surf. Sci* **510** 145425

[25] Talebi A, Razali Y S, Ismail N, Rafatullah M and Tajarudin H A 2020 Selective adsorption and recovery of volatile fatty acids from fermented landfill leachate by activated carbon process *Sci. Total Environ.* **707** 134533

IOP Conf. Series: Materials Science and Engineering **980** (2020) 012029 doi:10.1088/1757-899X/980/1/012029

Synthesis and property of porous material for sustainable resources-based biosensor: A review

N A Saputra[1], G Pari[1*], S Darmawan[1], D Hendra[1] and M Harsini[2]

[1] Forest Product Research and Development Center, Gunung Batu No. 5 Bogor, Indonesia
[2] Airlangga University, Airlangga No. 4-6 Surabaya, Indonesia

Email: gustanp@yahoo.com

Abstract. This paper describes the biosensor synthesis from teak lignocellulosic material. Nanocarbon was produced in drum-kiln at a temperature of 400-500^{0}C for 7-8 hours. The activation procedure at a temperature of 800^{0}C for 1 hour using H_2O and KOH activator and characterized by proximate analysis. Molecularly imprinted polymer (MIP) and non-imprinted polymer (NIP) techniques were applied for biosensor synthesis, while potential measurements and detection limits were used to measure product performance. This study shows that the optimal formula was a mixture of 15% MIP, 45% carbon, and 40% paraffin which produced a Nernst factor of 49.7 mV/decade and detection limit of 1.02 x 10^{-6} M. Related references were reviewed in this paper.

1. Introduction

Remarkable potential of lignocellulose material has been observed for the biosensor production purpose. In previous studies, bamboo and teak materials have been studied using Zn, Ni, and Cu as an intercalation material [1]. 15% KOH activator (w/w) was used in the process of activated charcoal into activated carbon [2,3] while charcoal production used simple kiln-drum technology at a temperature of 400-500^{0}C for 6-8 hours. The characterization of both materials shows that the teak profile was superior to bamboo [1]. Properties of crystallinity degree, dielectric profile, and conductivity were the important parameters of the final product target. Activated carbon conductivity is believed to have relatively stable properties in the case of giving and receiving electrons - thus it can be used as a biosensor material. Another advantage of lignocellulose-based activated carbon is its high abundance and known as renewable materials.

One of the techniques used for biosensor production is Molecularly Imprinted Polymer (MIP) by considering the materials, apparatus, and operating units needed. Previous procedures have been carried out by Díaz-García & Fernández-González [4] including template design (target template, functional monomer); scaffold (organic polymers, inorganic polymers, polymerization techniques, cross-links), and morphological design. The nature of the scaffold greatly affects the integrity of the circuit, where biosensor accessibility is determined by the site. The macro-micro-nano pores contained in activated carbon are deciding factors in the process of moving the target ions to be captured by the template. However, it is well understood that carbonaceous materials have a wide variety of pore structures based on raw materials, production techniques, and their activation. This fact is the background of researchers to conduct research related to the pore structure of carbonaceous materials.

Enzymatic carbon electrode paste has been proven to increase electrochemical reversibility through oxidative activation on the surface of the base media and optimize biosensor response to molecular targets [5]. To be more advanced, the development of the detection of the cholinesterase enzyme which can reduce the performance of acetylcholine continues to be carried out. The nanoporous matrix produces a low linear detection range of 10^{-12} to 10^{-6} [6]. However, variations in nanoscale structure materials are still being debated, especially in the form of base materials including graphite powder, carbon nano-tubes or carbon fiber. Mobilization of electrons towards the electrodes and good biocatalytic activity are the basis of biosensor efficiency. The aim of this knowledge is to approach the sensitivity, stability, and reversibility of the electrochemical reaction in the biosensor. Vamvakaki, Tsagaraki & Chaniotakis [7] prove that carbon fiber is superior to carbon nanotubes and carbon fiber. The surface area of carbon fiber provides high sensitivity as a glucose biosensor. Indeed, a summary of the sol-gel method in the preparation of nanomaterials has been previously established, including great attention to biosensors [8]. Application of biosensors based on sustainable green materials that is increasingly being developed include: hydrogen peroxide[9], glucose determination [10–14], pathogen online detection [15], herbicide [16], melamine [17,18], ester-flavour food [19], biosensor performance and energy storage [20], hemoglobin glycation [21], acetylcholine [22], urine glucose [23] or dopamine detection [24]. This paper presents the results of previous research and a reference analysis was carried out to complement the research results.

2. The previous experiments

This paper presents previous study results on the potential of lignocellulosic stuff as a biosensor material [2]. In fact, this study is a sequence of observations on the characterization of the carbon nanostructures of lignocellulosic material [1]. References were added to this paper by including the keywords activated carbon and biosensor. At least 35 references were added to this paper and a comparison of results was put forward as a scientific consideration.

This paper discusses carbonized *Tectona grandis* using kiln-drums at 400-500^{0}C for 7-8 hours. The carbonization product was hydrolyzed using a stainless steel reactor and the product was immersed in 15% *w/w* of potassium hydroxide (KOH) for 24 hours. Activation at 800^{0}C using activator-H_2O was carried out for 60 minutes. The activation product was doped with Ni with a weight ratio of 1:5, then purified at 900^{0}C for 6 minutes. This final product was then called AC-Ni.

Molecularly Imprinted Polymer (MIP) technique was used taking into account its simplicity and the equipment used. 0.8 mmol of monomer methacrylic acid was mixed into a 100 mL benzene solution, the crosslinker used was ethylene glycol dimethacrylate of 2.4 mmol and 0.2 mmol of melamine. A 1% benzoyl peroxide initiator was added later. The chemical mixture was refluxed at 80 ^{0}C for 17 hours and the precipitate was dried in the laboratory. Filtrate washing was carried out by acetic acid, ethanol, and water at 70^{0}C. *Non-Molecularly Imprinted Polymer* (NIP) polymer was carried out with the same procedure, without the addition of melamine.

The biosensor electrode was then formed into the micropipette tube, by mixing paraffin so that it became a carbon paste phase. 40 g of MIP/NIP was dispersed to 0.2 mL DOP and mixed into a 60 mg PVC solution in THF. The mixture was homogeneous. The micropipette tube which was filled with carbon paste was then dipped in MIP solution and dried. As a link between the electrodes with a potentiometer, a silver wire was used. Several treatments were applied to determine the electrode response (table 1).

Table 1. The composition of the electrode component.

Electrode	MIP (%)	AC (%)	Parafin (%)
E-1	0	60	40
E-2	0[a]	60	40
E-3	5	55	40
E-4	10	50	40
E-5	15	45	40
E-6	20	40	40

[a] Content of 15% NIP.

3. The quality of activated carbon

Proximate analysis is often used to assess thermal performance in the carbonization process, activated carbon product (AC). Yield (Y,%), moisture content (MC,%), volatile matter (VM,%), ash content (AsC,%), and fixed carbon (FC,%) become parameters in studying the characteristics of carbonization products. Some studies even add Iod (I_2) index parameters as reference results. Extensive thermal decomposition causes the material to release VM from the original material, although some of it remains in the solid phase. The higher temperature treatment, the less Y-AC obtained [25], even the power used affects Y, producing an effect similar to temperature [26]. Temperature alteration causes the decomposition of material to run and change composition in the material. Y-AC decreases when applied temperature treatment is above 700^0C [25]. Carbon purification above 700^0C provides an opportunity for carbon material to be discharged, in the form of CO, CO_2, H_2, CH_4 or reaction products C and H_2O. Increasing temperature treatment demonstrates kinetics reaction theory where an increase in temperature will trigger a faster reaction speed and cause precursor degradation. Even in chemical reagents, the degraded VM attains 96% [3]. Contrary to chemical activation, there is a chance of Y-AC addition from activator atom being attached to the AC surface, such as potassium atoms [3]. Simultaneously, both temperature and activation time affect the AC production process. What is debatable is the nominal activated precursor, the extent to which the number affects the quality of the AC. This point relates to the ability of activator (steam/chemical) to reach precursor. The nature of the original material has an indispensable influence, although it is not a significant factor compared to the treatment applied to convert the original material to AC. For example, sulfur content of origin material will determine sulfur compounds released during carbonization, which affects acid rain.

In the form of AC, the product has a high resistivity to chemical reaction [25], due to a stable double bond. Another property, in the form of AC, mineral content remains even after the carbonization and activation process. Generally, AC has a higher mineral value than the precursor. C content in AC influences the mechanical properties which are in line with mineral property. Materials with high AsC have the potential to inhibit the porous formation and reduce adsorption ability. The target of AC for waste handling applications is low AsC and MC content. AsC slightly prevents additional catalytic processes while MC anticipates the pores which are not locked [25]. However, there is a condition where MC-AC has a relative increase in chemical activation. Some water vapor molecules may be trapped in the pores and biochar hygroscopic properties when the activation process was carried out [3]. Precursor structure plays a role toward MC, like bamboo which has more cavity structures.

The indisputable fact is increasing FC-AC. Both physical and chemical activators, plus heat treatment, release functional groups of oxygen and hydrogen into volatile compounds [3]. Internal heat reaction causes only atoms to remain in solid phase while the VM and other substances are degraded in stages. Another parameter is I_2 index, which shows the ability of adsorption that increases dramatically compared to the original material. I_2 is a synthetic dye that is very concentrated and difficult to absorb. Thus, the ability of porous carbon is to be tested against I_2, furthermore, an increase in the I_2 index is a hallmark of good adsorption ability. The intensity of macro and micro

Table 2. List of activated carbon precursors and their quality.

Researcher	Yield (%)	MC (%)	VM (%)	AsC (%)	FC (%)	Iod (mg.g^{-1})	Material
This paper	28.9-46.6	1.1-13.7	5.6-16.02	2.62-8.9	81.4-85.4	264.4-865.2	*T.grandis*
Lam, Su, et al. [26]	81-87	2.4-2.7	5.0-5.9	3.0-3.6	88.1-89.3	-	Palm kernel shell
Gonsalvesh et al. [27]	29.3-38.6	3.0-4.3	4.1-15.6	28.0-51.1	43.8-67.5	-	Tobacco
[28]	21.6	7.17	5.13	5.85	89.02	814	Date stone
[29]	12-17	3.0 ± 0.4	16.0 ± 2.1	4 ± 0.3	77 ± 2.8	-	Waste palm shell
[25]	68.4-97.8	2	22	4	73	9.-47.0	Waste palm shell
[3]	24.9-80.1	2.55-7.72	2.01-14.71	4.79-16.45	74.47-90.64	222.3-833.6	Teak, bamboo
[30]	46.51-97.36	2.53-5.71	4.39-18.04	36.55-50.56	37.99-52.25	363.5-668.6	sawdust

pore formation by heat is responsible for improving the adsorption properties of the product. A clear sign that internal pore formation occurs is when cellulose, hemicellulose, and lignin are overhauled to become volatile and mild gases by treatment. Finally, it can be notified that the quality of AC is influenced by precursor, carbonization, and activation treatment and target of AC production (table 2).

4. The structure and surface morphology of activated carbon

The surface area of AC of this paper was 280.26 m^2. g^{-1}, 917.11 m^2. g^{-1} and 692.22 m^2. g^{-1}, while the degree of crystallinity was 25.98%, 36.10%, and 38.88%, respectively for charcoal, AC-steam, and AC-KOH [2]. Surface morphology is part of AC that reacts with reactants, thus good morphology indicates the optimum performance of electrons surrounding the AC surface while the degree of crystalline provides notification of arrangement of carbon at regular distances and leaves amorphous behavior. This surface structure and morphology formation involve dehydration and elimination reactions during the carbonization and activation process. Again, the profile of the original material and treatment have different effects on the structure and morphology of targeted AC.

During the activation process, whether physical or chemical, the activation time is important in the preparation of macro-meso-micropores (figure 1). In AC form, pores are more uniform with smaller pores found on the inside and or around larger pores. Under certain conditions, steam activation results in a superior product compared to steam activation that contains N_2 [31]. An increase in activation time pushes the AC to be more porous than before. But at a certain time limit, it triggers cracks in the internal pore which causes pore deformation and decreases porous properties [26]. This phenomenon is difficult to predict, but knowledge of precursor profiles becomes mandatory. Porous structure causes AC to provide more sites for adsorption performance. Adsorption efficiency of AC is determined by porosity, surface character, pore size [26], and pore distribution. The interconnection of the aromatic ring layer is restored during the activation process to form a macro-micro-nano pore. Therefore, the structure and morphology of AC are more onward compared to the original material. Carbonization and activation methods will determine the course of the structure of carbon material to be more advanced. Among them is the microwave technique that is superior to the carbon stockpile technique or activation in a modified reactor. The thermal efficiency of microwaves can minimize cracks that release carbonized product content [29]. The hierarchical structure of carbon material follows a sequence of precursors, carbonized products (biochar), and activated carbon. The last product has a high profile compared to the previous structure. For AC application, pore size is the target achieved in the synthesis procedure. The gas molecular distance was 0.4-0.9 nm while the molecular size of liquid was > 27 nm [32]. Thus 3 types of pore distribution from AC: macropore (> 50 nm), mesopore (2-50 nm), and micropore (<2 nm) become unique properties for various applications. Precursors that have dominant cellulose content produce high micropore distribution, while lignin material forms macropores [31]. Mesopores are known to be formed through a volatile overhaul of the process of carbonization and pyrolysis in stages. Exposure to steam forms micro and mesopores in carbon fiber which was a product of the interaction between water molecules and carbon atoms [33]. Isotherm deformation at relatively low pressures was the formation of micro-pores typical for adsorbents [31]. A phenomenon of structure and morphology differences in the same treatment was affected by the chemical composition of the precursor. Even in one individual plant, there are a variety of different organic chemical contents, for example, roots, stems, branches, twigs, or leaves.

The reaction by N_2 ran slower than the reaction by water vapor. N_2 blast was superior in structural but less porous than H_2O activation [31]. This statement is marked by yield (%) of AC-steam products against AC-N_2 product, the solid phase was overhauled to become volatile which was heated up by heat. This condition continues to change with increasing temperature and activation time. But this approach needs to get further clarification which is more comprehensive. On the other hand, organic chemical composition affected resistance in the process of activation by N_2. The tar content of the original material had a stable reaction compared to material that had dominant furfural [31]. The ultimate analysis shows the reactions that occurred in the furnace. An increasing portion of O_2 in AC indicates carbon was oxidized during the activation process, and this was important in pores formation of AC. Referring to the above statements, precursors with high furfural might have more oxygen functional groups. There were different conditions when using KOH as an activation agent. Element K diffused into the deepest carbon layer and reconstructed the internal pores [34], whether by forming new pores or cracking pores that had been formed before. Thus, AC product by KOH activation was superior to steam and steam-N_2.

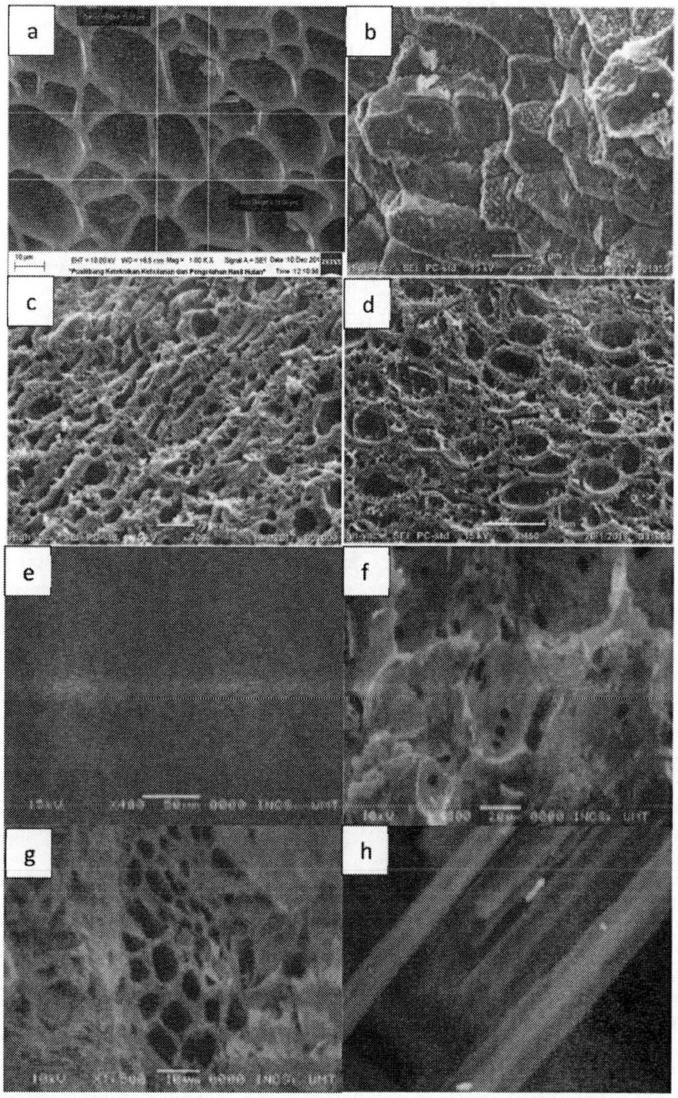

Figure 1. Surface morphology of different activated carbons (AC).
(a) Pari, Santoso, Hendra, et al. [2]
(b) (c) (d) Lee et al. [25]
(b) Waste palm shell (WPS)
(c) Carbonized WPS
(d) AC-WPS
(e) (f) (g) Lam et al. [35]
(h) Nieto-Delgado, et al [36]

In all heat treatment techniques, every single step of activation changed significantly in morphology compared to the precursor. Thermal deformation caused the inner structure to shrink and rearranged the chemical bond. Macropore structural deformation was formed due to differences in thermal decomposition in a biopolymer that formed a cell wall [36] and the S2 layer played a key role related to its most dominant layer in the cell wall. The target application determined if the activation technique used was drawn backwards. Simply, physical activation was superior to chemical activation, information of functional groups, although it was not a rule. Oxidation of carbon surfaces during steam activation was responsible for this point.

The degree of crystallinity demonstrates the order of the aromatic layer. Carbon reconstruction was characterized by increasing crystalline properties and leaving irregular (amorphous) properties. The transformation of the degree of crystallinity is shown in **Figure 2**. Changes in the structure of precursors, carbonization products, AC, and AC-Ni show improved character. The intense peak changes on theta-43 become a characteristic of carbonous material. Ni metal atoms were successfully intercalated into carbon atoms, crystal structures of 100 and 102 [2]. Thermal treatment played an important role in a better and more conductive arrangement of aromatic layers (symmetrical and crystalline).

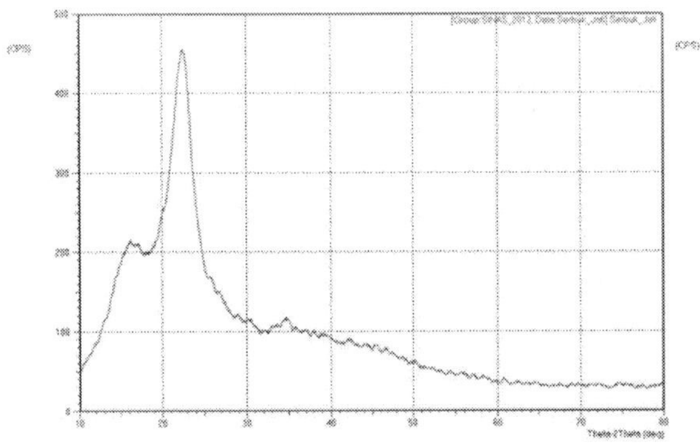

(a)

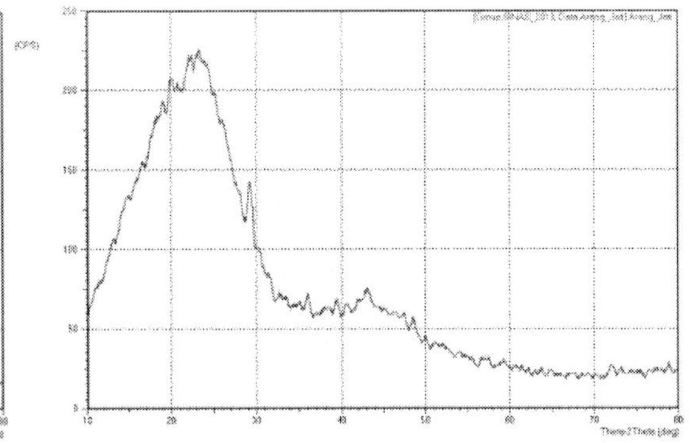

(b)

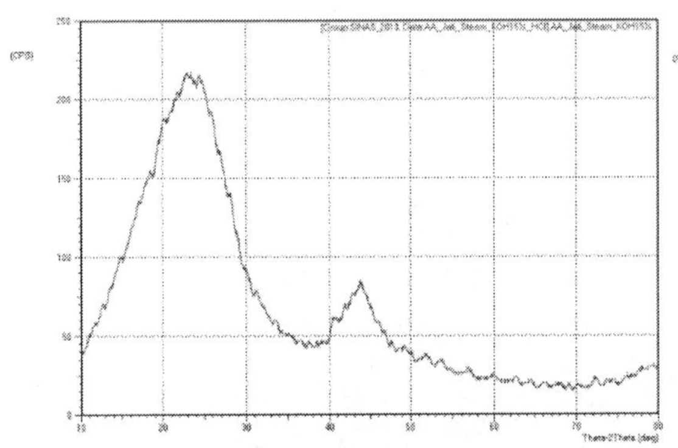

(c)

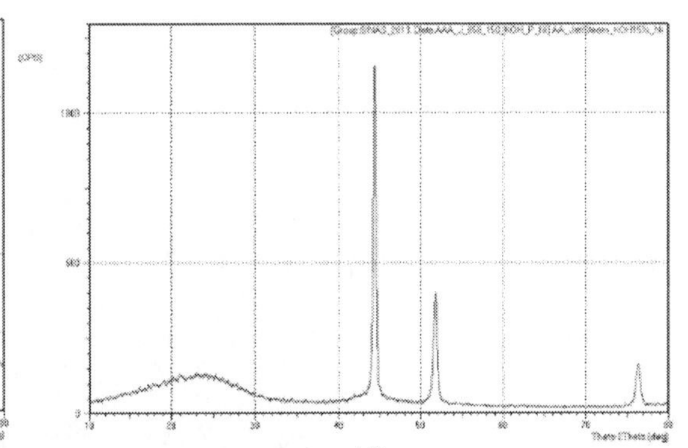

(d)

Figure 2. *Tectona grandis* diffractogram: (a) precursor (b) carbonized (c) AC-HOH&KOH (d) AC-Ni [2].

5. The electrical properties of activated carbon electrodes

The third generation is deemed necessary to get supporting material for the development of integrated circuits in the future. Nano-sized activated carbon is a solution that is offered with the availability of resources for production and low-cost matter. The superiority of this porous material allows electron transfer between electrodes to run effectively between redox reactions.

Resistance and dielectric behavior represent AC electrical properties, while supercapacitor electrode performance by potential measurement and detection limits calculation. KOH activator in this paper had significantly reduced the value of AC resistance, it was noted of 0.0 – 0.78 Ω [2]. Potassium from KOH, which was an alkali metal, had been shown to increase the conductivity behavior of AC. Intercalation of Ni and activator-KOH increased the ability to conduct electricity of

AC, even silver which had the most conductive conductor properties. The surface area, total volume, and relatively small temperature changes become inherent properties of the AC to improve the conductor level. Electron transfer was performed at range 18.0 – 2,733.3 S/m [2]. This paper demonstrated good carbon paste and MIP at slope 54 with 49.7 mV/decade Nernst factor and detection limit of 1.02 x 10^{-6} M [2] for melamine solution 10^{-8} – 10^{-2} M at pH 4.

Rovina et al. (2015) conducted a study with a wider detection limit of 9.6 x 10^{-16} while developing melamine detection in the food industry. Green resource-based melamine biosensor was used as an instrument due to its simplicity, rapidity, stability, selectivity, sensitivity, wide detection distance, and affordable method. The AC-based biosensor was reported to increase stability and reduce detection limits 1000-fold, 10^{-12} – 10^{-6} [6].

6. Conclusion

Sustainable resources *Tectona grandis* have tremendous potential as biosensor material. The structural conversion of *Tectona grandis* to activated carbon is a mandate for the application of biosensor. A variety of activated carbon production techniques is an opportunity, where the stable nature of the chemical reaction, the porous structural hierarchy, and decreased detection limits are inherent products. This paper has presented KOH as an activator and Ni metal as a doping agent with a ratio of 1: 5 in activated carbon production, as a follow-up of previous research. *Molecularly Imprinted Polymer* (MIP) technique was used due to the ease and simplicity of the equipment. The ideal composition was achieved with carbon, MIP, and paraffin portions of 45%: 15%: 40%, respectively, for melamine solution of 10^{-8} – 10^{-2} M at pH 4.

Acknowledgement:

Thanks to all author who's contributed equally as main contributor to the paper writing.

References

[1] Pari G, Santoso A, Hendra D, Buchari B, Maddu A, Rachmat M, Harsini M, Heryanto T and Darmawan S 2013 Characterization of Carbon Nano Structure from Lignosellulose Material *J. Forest Products Research* **31** 75–91

[2] Pari G, Santoso A, Hendra D, Bukhari, Maddu A, Rachmat M, Harsını M, Safitrı B A, Heriyanto T and Darmawan S 2016 Possibile Application of Ligno-cellulose Nano Carbon as Biosensor *J. Forest Products Research* **34** 111–26

[3] Pari G, Santoso A, Buchari D H and Maddu A 2016 The Potency of Nano Carbon Structure Made from Bamboo and Teak Wood Lignocellulose *J. Forest Products Research* **34** 309–22

[4] Díaz-García M E and Fernández-González A 2004 Molecularly Imprinted Polymers *Encyclopedia of Analytical Science: Second Edition* pp 172–82

[5] Motta N and Guadalupe A R 1994 Activated Carbon Paste Electrodes for Biosensors *Anal. Chem.* **66** 566–71

[6] Sotiropoulou S and Chaniotakis N A 2005 Lowering the detection limit of the acetylcholinesterase biosensor using a nanoporous carbon matrix *Anal. Chim. Acta* **530** 199–204

[7] Vamvakaki V, Tsagaraki K and Chaniotakis N 2006 Carbon Nanofiber-Based Glucose Biosensor *Anal. Chem.* **78** 5538–42

[8] Díaz-García M E and Laíño R B 2005 Molecular imprinting in sol-gel materials: Recent developments and applications *Microchim. Acta* **149** 19–36

[9] Song J, Xu J M, Zhao P S, Lu L De and Bao J C 2011 A hydrogen peroxide biosensor based on direct electron transfer from hemoglobin to an electrode modified with Nafion and activated nanocarbon *Microchim. Acta* **172** 117–23

[10] Şavk A, Aydın H, Cellat K and Şen F 2020 A novel high performance non-enzymatic electrochemical glucose biosensor based on activated carbon-supported Pt-Ni nanocomposite *J. Mol. Liq.* **300** 112355

[11] Koskun Y, Şavk A, Şen B and Şen F 2018 Highly sensitive glucose sensor based on monodisperse palladium nickel/activated carbon nanocomposites *Anal. Chim. Acta* **1010** 37–

43
[12] Khalifa A M, Abdulateef S A, Kabaa E A, Ahmed N M and Sabah F A 2020 Study of acidosis, neutral and alkalosis media effects on the behaviour of activated carbon threads decorated by zinc oxide using extended gate FET for glucose sensor application *Mater. Sci. Semicond. Process.* **108** 104911
[13] Im J S, Kim J G, Bae T S, Yu H R and Lee Y S 2011 Surface modification of electrospun spherical activated carbon for a high-performance biosensor electrode *Sensors Actuators, B Chem.* **158** 151–8
[14] Kim J H, Lee D, Bae T S and Lee Y S 2015 The electrochemical enzymatic glucose biosensor based on mesoporous carbon fibers activated by potassium carbonate *J. Ind. Eng. Chem.* **25** 192–8
[15] Upadhyayula V K K, Deng S, Smith G B and Mitchell M C 2009 Adsorption of Bacillus subtilis on single-walled carbon nanotube aggregates, activated carbon and NanoCeram^{TM} *Water Res.* **43** 148–56
[16] Tucci M, Grattieri M, Schievano A, Cristiani P and Minteer S D 2019 Microbial amperometric biosensor for online herbicide detection: Photocurrent inhibition of Anabaena variabilis *Electrochim. Acta* **302** 102–8
[17] Rovina K, Siddiquee S and Wong N K 2015 Development of melamine sensor based on ionic liquid/nanoparticles/chitosan with modified gold electrode for determination of melamine in milk product *Sens. Bio-Sensing Res.* **4** 16–22
[18] Alam M M, Asiri A M, Uddin M T, Inamuddin, Islam M A, Awual M R and Rahman M M 2019 One-step wet-chemical synthesis of ternary ZnO/CuO/Co3O4 nanoparticles for sensitive and selective melamine sensor development *New J. Chem.* **43** 4849–58
[19] Radulescu M C, Bucur M P, Bucur B and Radu G L 2019 Ester flavorants detection in foods with a bienzymatic biosensor based on a stable Prussian blue-copper electrodeposited on carbon paper electrode *Talanta* **199** 541–6
[20] Madhu R, Veeramani V, Chen S M, Veerakumar P, Liu S Bin and Miyamoto N 2016 Functional porous carbon-ZnO nanocomposites for high-performance biosensors and energy storage applications *Phys. Chem. Chem. Phys.* **18** 16466–75
[21] Wang X, Su J, Zeng D, Liu G, Liu L, Xu Y, Wang C, Liu X, Wang L and Mi X 2019 Gold nano-flowers (Au NFs) modified screen-printed carbon electrode electrochemical biosensor for label-free and quantitative detection of glycated hemoglobin *Talanta* **201** 119–25
[22] Wang L, Chen X, Liu C and Yang W 2016 Non-enzymatic acetylcholine electrochemical biosensor based on flower-like NiAl layered double hydroxides decorated with carbon dots *Sensors Actuators, B Chem.* **233** 199–205
[23] Hossain M F and Park J Y 2016 Plain to point network reduced graphene oxide-activated carbon composites decorated with platinum nanoparticles for urine glucose detection *Sci. Rep.* **6** 1–10
[24] Rahman S F, Min K, Park S H, Park J H, Yoo J C and Park D H 2016 Selective determination of dopamine with an amperometric biosensor using electrochemically pretreated and activated carbon/tyrosinase/Nafion®-modified glassy carbon electrode *Biotechnol. Bioprocess Eng.* **21** 627–33
[25] Lee C L, Yek P N Y, Lam S S, Chuah J H, Liew R K, Park Y-K and Osman M S 2019 Microwave steam activation, an innovative pyrolysis approach to convert waste palm shell into highly microporous activated carbon *J. Environ. Manage.* **236** 245–53
[26] Lam S S, Su M H, Nam W L, Thoo D S, Ng C M, Liew R K, Yuh Yek P N, Ma N L and Nguyen Vo D V 2019 Microwave Pyrolysis with Steam Activation in Producing Activated Carbon for Removal of Herbicides in Agricultural Surface Water *Ind. Eng. Chem. Res.* **58** 695–703
[27] Gonsalvesh L, Yperman J, Carleer R, Mench M, Herzig R and Vangronsveld J 2016 Valorisation of heavy metals enriched tobacco biomass through slow pyrolysis and steam activation *J. Chem. Technol. Biotechnol.* **91** 1585–95
[28] Akila E, Pugalendhi S, Subramanian P, Duraisamy M R and Surendrakumar A 2019

IOP Conf. Series: Materials Science and Engineering **980** (2020) 012029 doi:10.1088/1757-899X/980/1/012029

Production and characterization of activated carbon from date stones by single step steam activation. *Environ. Ecol.* **37** 194–7
[29] Lam S S, Yuh Yek P N, Ok Y S, Chong C C, Keey Liew R, Tsang D C W, Park Y-K, Liu Z, Wong C S and Peng W 2019 Engineering pyrolysis biochar via single-step microwave steam activation for hazardous landfill leachate treatment *J. Hazard. Mater.* **390** 121649
[30] Pari G, Tohir D, Mahpudin and Ferry J 2006 Activated Charcoal from Wood Sawdust as Adsorbent Material for Frying Oil Refinery *J. Forest Products Research* **24** 309–22
[31] Savova D, Apak E, Ekinci E, Yardim F, Petrov N, Budinova T, Razvigorova M and Minkova V 2001 Biomass conversion to carbon adsorbents and gas *Biomass and Bioenergy* **21** 133–42
[32] Danish M and Ahmad T 2018 A review on utilization of wood biomass as a sustainable precursor for activated carbon production and application *Renew. Sustain. Energy Rev.* **87** 1–21
[33] Manickam S S, Waisi B I, Nijmeijer A, McCutcheon J R and Benes N E 2019 Activated Carbon Nanofiber Nonwovens: Improving Strength and Surface Area by Tuning Fabrication Procedure *Ind. Eng. Chem. Res.*
[34] Rostamian R, Heidarpour M, Mousavi S F and Afyuni M 2015 Characterization and sodium sorption capacity of biochar and activated carbon prepared from rice husk *J. Agric. Sci. Technol.* **17** 1057–69
[35] Lam S S, Liew R K, Wong Y M, Yek P N Y, Ma N L, Lee C L and Chase H A 2017 Microwave-assisted pyrolysis with chemical activation, an innovative method to convert orange peel into activated carbon with improved properties as dye adsorbent *J. Clean. Prod.* **162** 1376–87
[36] Nieto-Delgado C, Partida-Gutierrez D and Rangel-Mendez J R 2019 Preparation of activated carbon cloths from renewable natural fabrics and their performance during the adsorption of model organic and inorganic pollutants in water *J. Clean. Prod.* **213** 650–8

IOP Conf. Series: Materials Science and Engineering **980** (2020) 012030 doi:10.1088/1757-899X/980/1/012030

The physicochemical characteristics of cassava starch modified by ultrasonication

E M Satmalawati, Y Pranoto, D W Marseno and Y Marsono
Food Science, Faculty of Agriculture Technology Gadjah Mada University, Indonesia

Email: mulatendah@gmail.com, pranoto@ugm.ac.id, djagal@ugm.ac.id, yustimar49@yahoo.co.id

Abstract. In this study, cassava starch was modified using physical modification by ultrasonication. Ultrasonication, known as one of the advanced oxidation methods, is environmentally friendly. Modified starches using ultrasonication are still relatively lacking. This research aims to evaluate the stage of oxidation through carbonyl and carboxyl contents, amylose content, and physicochemical characteristics (swelling power, solubility, paste viscosity, brightness) of ultrasonicated starch. Ultrasonication was carried out in combination amplitude (60% and 80%) and process duration (10, 20, 30, and 40 minutes) consisting of eight combinations. The results showed that ultrasonication did not significantly affect the degree of oxidation. Ultrasonication has more effect on starch depolymerization, mainly on amylose content, solubility, and paste viscosity. Considering the issue of environmentally friendly modification that continues to be campaigned for, ultrasonication can be the method of choice for obtaining starch with the characteristics as needed.

1. Introduction

The availability of carbohydrate sources as the main menu as well as the needs of the food industry is in line with population growth and increasing food processing technology, especially the application of starch in the manufacture of food products to reformulate them and enhance their properties. Starch production has significantly grown recently, increased from 60 million tons in 2007 to 90 million tons in 2016. Cassava is one of the four main sources of starch in the world, second after corn, in addition to other sources of starch (wheat and potatoes) [1]. Cassava has the potential to be developed because of its high carbohydrate content (73.7-84.9%) and the abundance of cassava in Indonesia.

Starch is the most abundant component in several grains (70-80%), roots, and tuber [2]. Starch is included among biopolymer groups. The two main components of starch are amylose (linear chain with few branching points which has a role in the amorphous region of the granules with glucose units in α-(1–4) bonds) and amylopectin (branch component which is responsible in the starch crystalline lamella with glucose units in α-(1–4) bonds in the linear sections and by α-(1–6)bonds in the branching points [3].

The issue of safe food and the environment as a result of the modification process is still being pursued. Chemical modification can leave residue on the product and has the potential to cause environmental pollution problems while many physical modifications are made to avoid the negative effects of using unsafe chemical compounds on the final product. Also, the use of chemical reagents in the process should be decreased for consumers and the environment's safety, and lower production costs [4].

Ultrasonication is a starch modification method that causes physical depolymerization. This process reduces the use of chemicals, minimizes waste, and energy consumption. The effects of ultrasonic radiation generate degradation of starch granules with perceptible cracks and pores on the surface but the shape of starch granules does not change [5]. The ultrasound process does not require the addition of oxidants or catalysts and does not produce waste [6]. The Ultrasonication process uses wavelengths of 20-100 kHz which is beyond the range human can hear. The advantages of ultrasonication are higher selectivity and quality, less chemical usage, and processing period. Ultrasonication is also one of the environmentally friendly technologies that are relatively new, fast, and usable in food [7].

Based on the opportunity to increase the use of cassava starch which has certain physicochemical characteristics that can meet the increasing industrial demand by considering food and environmental safety claims, the objective of this study is to modify cassava starch using ultrasound technology. Some physicochemical characteristics of modified starches are evaluated and explained to better understand the level based on different methods of starch modification.

2. Material and methods

2.1. Materials

Starch was obtained from cassava (var. adira-4) with a harvest age of 10 to 11 months obtained in Yogyakarta, Indonesia. Maximum starch yield extraction(30%–35%) was done by adjusting the ratio of cassava pulp to water to 1: 4 (w/w) through the stages that consisted of stripping, grating, dissolving, extracting, filtering, settling, and drying (in a cabinet drier at 50°C for 24 h to yield a 9%–10% moisture level). The starch was then grounded in a mortar, sieved (60 mesh), placed in plastic packaging with silica gel wrapped in aluminum foil, put into a container, and kept at room temperature until further use.

2.2. Sonication process

The ultrasound process uses an ultrasonication probe (ø7 mm). Starch suspension of 5% (w/v) was kept in a glass cell and homogenized with continuous stirring for 5 minutes before ultrasonication at a constant temperature of 20 ±1°C. The ultrasonication process was carried out using variations of amplitude (60% and 80%) at four different time process (10; 20; 30; and 40 min). After the ultrasonication process, the starch suspension was then precipitated and dried with a dryer at a temperature of 50^0 C for 24 hours.

2.3. Carbonyl and carboxyl contents

The carbonyl content was examined according to the method by Smith (1967)[8] with slight modifications and the carboxyl content was examined using the method by Chattopadhyay *et al.* (1997) [9].

2.4. Amylose content

Amylose content was obtained using the method described by Juliano *et al.* (1981) [10].

2.5. Swelling power and solubility

The swelling power was determined using the procedure described by Adebowale *et al.* (2002) [11]: sample (1.0 g and had been weighed) was placed into a test tube, which had been cleaned, dried, and weighed first. The sample and tube weight were calculated together as W1. Distilled water (50 ml) was added to the sample until a slurry formed. The mixture was then homogenized and heated at 95°C for 30 min in a boiling bath. The mixture was rested until the temperature reached 30±2°C and then centrifuged (500 rpm, 15 min) until the gel and aliquots were separated. Aliquots (5 ml) were dried at

110°C until a constant weight was achieved. The residue obtained from dried supernatant could be expressed as the solubilized starch in water. Solubility was calculated as g per100 g of starch on a dry weight basis. The gel and test tube from the above experiment (after centrifugation) with its retained water were quantitatively weighed and taken together as W2. The swelling power was calculated with the equation:

$$\text{Swelling power } (\%) = W2 – W1/\text{weight of starch (dry weight basis)} \quad (1)$$

2.6. Paste clarity

Paste clarity was determined using the procedure described by Craig *et al.* (1989) [12]. Put into a test tube was 0.2 g of starch in 20 mL of distilled water. For 30 minutes the tube was heated in a tub of boiling water, stirred individually every 5 minutes, then cooled at room temperature and stored for a few days to be controlled. Percent transmittance (% T) was determined at a wavelength of 625 nm against water blank and determined using GENESYS 10S UV-VIS Spectrophotometer (USA).

2.7. Pasting properties

Rapid Visco Analyser (RVA 4500 Perten Instrument-Australia) was used to determine pasting properties of the starch with a Standard Analysis 1 profile and the result was expressed in cP (centiPoise) with pasting temperature, peak viscosity, breakdown, final viscosity and setback as parameters. The sample was weighed directly in the aluminum RVA canister, and 25 mL of distilled water was then added to the canister. The temperature of the sample was held at 50°C for 2 min, increased to 95°C in 3.5 min, and then held at 95°C for2.5 min. The temperature of the sample was decreased to 50°C in 4 min and then held at 50°C for 1 min. The rotating speed was held at 960 rpm for 10 s, and it was held at 160 rpm during the remaining process. All parameters were then recorded.

3. Result and discussion

3.1. Carbonyl and carboxyl contents

Carbonyl and carboxyl content of native and ultrasonication starches are presented in table 1 and table 2. The initial group resulting from the starch oxidation process was carbonyl. Carbonyl was a main functional group formed during the oxidation process (a minor amount of carboxyl groups was also formed). Thus, the degree of oxidation was determined from the formation of the carbonyl and carboxyl groups. The sonication process would produce acoustic cavities in the aqueous mixture causing microbubbles. The microbubbles spreading released high energy which was converted to pressure and temperature. This process generated the degradation of the polymer as well as the acceleration of catalytic reactions [13]. As a result of this treatment, the collapsing microbubbles caught the nearby polymer chain in a high gradient shear field generating the formation of long-chain radicals and breakage of macromolecular C-C bonds[7].

Table 1. Carbonyl level of cassava starch after ultrasound process at various amplitudes and time.

	Time (min)			
Amplitudo	10'	20'	30'	40'
60%	0.05 ± 0.01aA	0.06 ± 0.04aA	0.06 ± 0.02aA	0.28 ± 0.03bA
80%	0.23 ± 0.03bA	0.22 ± 0.03bA	0.20 ± 0.02bA	0.11 ± 0.00aA

Each value is presented as mean ± standard deviation.
Values with different letters are significantly different ($p < 0.05$).
a,b: indicates time; A,B: indicates amplitudo

The use of ultrasonication in the liquid system causing cavities formation is a phenomenon of bubbles formation, growth, and finally breaking. When ultrasonic waves propagated, the bubbles oscillated and broke apart, causing heat, mechanical effects (pressure, turbulence, and shear stress) [14], and chemical effects (the formation of free radicals) [15]. Ultrasonication generated water hydrolysis in oscillating bubbles and free radicals (H^+ and OH^-) that could be captured in several chemical reactions in the formed liquids. However, the ability of ultrasonication in oxidizing starch was low, this is indicated by the absence of carboxyl groups (table 2) as a reference for further oxidation of starch. Overall, ultrasonication did not significantly affect the carboxyl group (similiar with native).

Table 2. Carboxyl level of cassava starch after ultrasound process at various amplitudes and time

	Time (min)			
Amplitudo	10'	20'	30'	40'
60%	0.007± 0.000aA	0.007 ± 0.001aA	0.005 ± 0.001aA	0.006 ± 0.001aA
80%	0.007 ± 0.004aA	0.006 ± 0.002aA	0.007 ± 0.00aA	0.005 ± 0.001aA

Each value is presented as mean ± standard deviation.
Values with different letters are significantly different ($p < 0.05$).
a,b: indicates time; A,B: indicates amplitudo

3.2. Amylose content

Depolymerization occurred during the starch oxidation process through the hydrolysis of the glycosidic linkage. This is indicated by the increased level of amylose which was the result of breaking the branches of amylopectin into a straight structure and read as an appearance of amylose. The ultrasound process had a significant effect on starch amylose levels. Depolymerization of amylopectin and amylose molecules due to the hydrolysis of starch resulted in the formation of smaller size chains [16]. The results of amylose content analysis are shown in table 3.

Table 3. Amylose content (%) of cassava starch after ultrasound process at various amplitudes and time.

	Time (min)			
Amplitude	10'	20'	30'	40'
60%	37.84 ± 0.00aA	40.20 ± 0.50bA	42.50 ± 0.62cA	44.13 ± 0.28dA
80%	39.75 ± 0.31aA	41.68 ± 0.18bA	43.85 ± 0.21cA	41.23 ± 0.25bA

Each value is presented as mean ± standard deviation.
Values with different letters are significantly different ($p < 0.05$).
a,b: indicates time; A,B: indicates amplitudo

The results showed that the time of the ultrasonication process had a significant effect on the amylose content. The ultrasound process would produce cavities in the solution and cause microbubbles[13]. The spreading of microbubbles then released high energy which was converted to high temperatures and pressures, this process caused polymer degradation. Cavitation is the breakdown of micro-sized bubbles (microbubbles) and spreads when sound waves pass through the solution. The energy generated from the cavitation process can break covalent bonds in polymeric material [17,18].

3.3. Swelling and solubility

Swelling power and solubility indicate the amount of starch chains interaction in crystalline and amorphous regions. The magnitude of the interaction is affected by the ratio of amylose-amylopectin, phosphor content, amylose-amylopectin characteristics such as molecular weight, branching or length, and branch conformation[19].

Swelling power analysis results for all treatments were greater than native (35.39%). The swelling power and solubility of all samples can be seen in table 4 and table 5. The ultrasonication process did not significantly affect swelling power. This can be correlated with the carboxyl group which was not produced during the process. Carboxyl groups tend to be negatively charged which would provide an electronegative repulsive force, which affects the size of the swelling power.

The solubility of ultrasonicated starches was greater than native, as well as swelling power. Solubility represents the amount of starch molecule that solubilizes at a certain temperature [20]. Solubility increased with the increase of amplitude and the time process. The increase in solubility can be related to the depolymerization and structural weakening of starch granule due to the oxidation process[11]. Ultrasonication starch has an amorphous tendency.

Table 4. Swelling power (%) of cassava starch after ultrasound process at various amplitudes and time.

	Time (min)			
Amplitude	10'	20'	30'	40'
60%	37.65 ± 0.96aA	38.36 ± 2.26aA	38.36 ± 1.02aA	37.63 ± 0.07aA
80%	41.88 ± 2.19aA	41.07 ± 1.53aA	43.33 ± 2.65aA	42.22 ± 0.25aA

Each value is presented as mean ± standard deviation.
Values with different letters are significantly different ($p < 0.05$).
a,b: indicates time; A,B: indicates amplitudo

Table 5. Solubility (%) of cassava starch after ultrasound process at various amplitudes and time.

	Time (min)			
Amplitude	10'	20'	30'	40'
60%	0.55.± 0.016aA	0.56 ± 0.04aA	0.53 ± 0.00aA	0.55 ± 0.00aA
80%	0.61 ± 0.04aB	0.75 ± 0.01aB	0.68 ± 0.04 aB	0.80 ± 0.07B

Each value is presented as mean ± standard deviation.
Values with different letters are significantly different ($p < 0.05$).
a,b: indicates time; A,B: indicates amplitudo

3.4. Paste clarity

Ultrasonication starch produced higher clarity than native starch (56.16%) and was more stable during storage. The results of the starch paste clarity analysis are shown in figure 1. Decreased clarity was caused by the formation of gels from the joining of hydroxyl groups between starch chains through hydrogen bonds, this mechanism is called retrogradation.

The clarity of ultrasound starch paste was relatively low, with a maximum number of 66% which was the result of a combination of 60% amplitude treatment at 40 minutes processing time. The clarity of starch paste is influenced by the carboxyl groups formed during the oxidation process. The negatively charged carboxyl group provides a repulsive force to prevent retrogradation [20, 21,22,23]. In this study, the ultrasound process could not produce a carboxyl group so that retrogradation could not be prevented, this is indicated by a low clarity value and unstable paste during storage.

3.5. Paste viscosity

One of the parameters in the starch modification is the decrease of viscosity. Ultrasonication causes depolymerized starch through hydrolysis of glycosidic bonds in starch granules, which tends to produce lower starch viscosity as a result of starch molecular weight loss [20,24,25]. The results showed the viscosity of the paste decreased. The lowest viscosity was produced from a combination of 80% amplitude treatment at 10 minutes processing time (6159 cP). Paste viscosity of native and ultrasonicated starch can be seen in figure 2a and 2b.

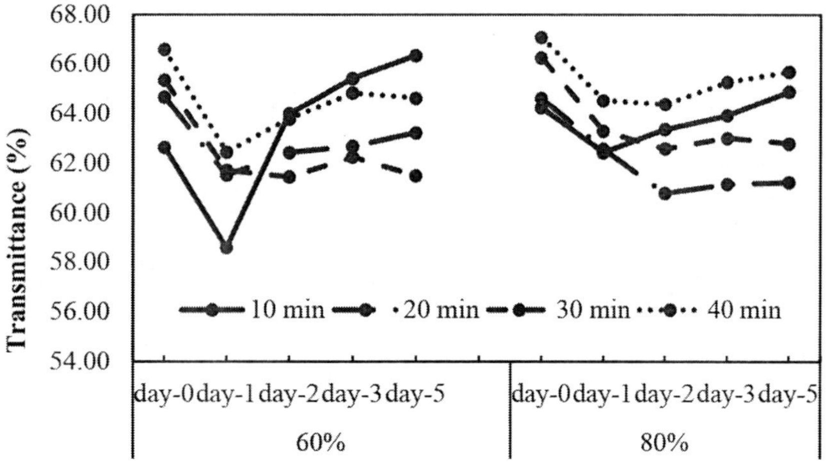

Figure 1. Paste clarity of ultrasound starches.

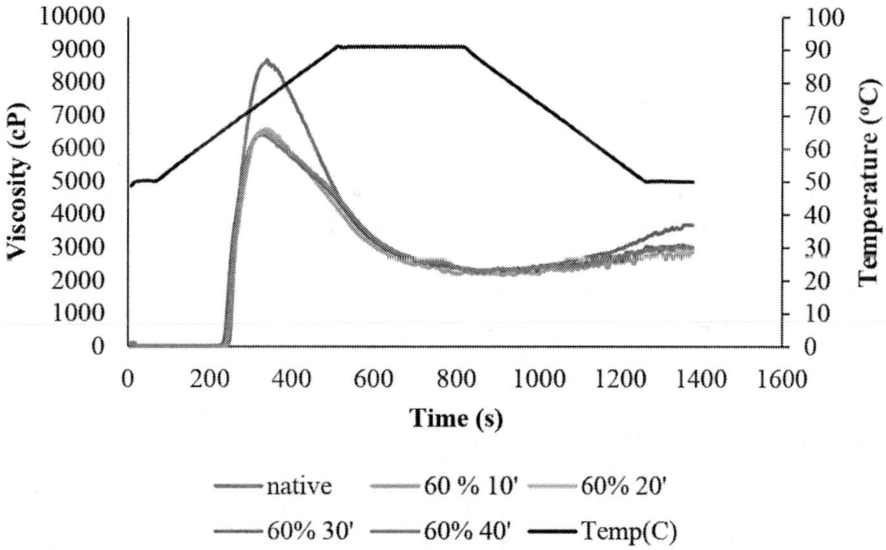

Figure 2a. Amilograph curve of native and ultrasound starches at 60% amplitudo.

The amplitude and processing time of the ultrasonication process affected the decrease in viscosity. However, in this study, ultrasonication tended to result in a decrease in the paste viscosity which can be related to the presence of starch depolymerization mainly in the chain of amylopectin branches into a straight structure with lower molecular weight and amorphous so that the ability of water absorption increases which results in a decrease in paste viscosity.

The application of ultrasonication to the liquid system generates acoustic cavities in the aqueous mixture and causes bubbles formation, growth, and breaking. Cavitation is the breakdown of micro-sized bubbles (microbubbles) and spreads when sound waves pass through a solution. The energy produced from the cavitation process ruptures the covalent bonds in polymeric substance [18,19]. It was also found that the viscosity and molecular weight of starch produced decreased after ultrasonication [26].

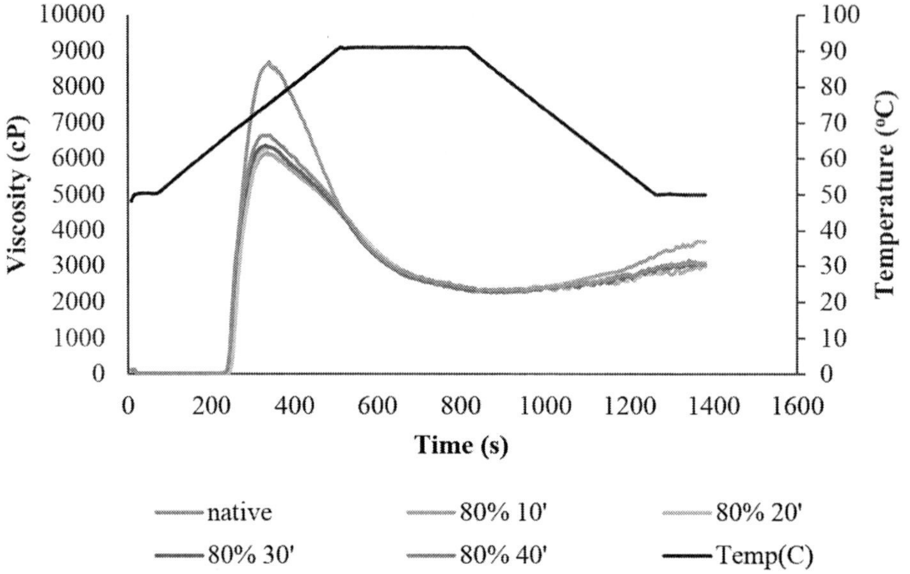

Figure 2b. Amilograph curve of native and ultrasound starches at 80% amplitudo.

4. Conclusion
Ultrasonication did not significantly affect the degree of oxidation. The low degree of oxidation affected the swelling power and the clarity of starch paste because it was associated with carboxyl groups that were not formed during the ultrasonication process. Ultrasonication had more influence on the degradation of starch polymers which is indicated by significant amylose content, low viscosity, and increased solubility with the production of amorphous starch.

Acknowledgments
Thank you for funding support to Directorate of Research and Comunity Service, Ministry of Research Technology and Higher Education

References
[1] Vilpoux O F and Brito V H 2019 *Starch Extracted from Corms, Roots, Rhizomes, and Tubers for Food Application* (Elsevier Inc.)

[2] Zeoula L M, Martins A D S, Nunes I and Regina C 1999 Solubility and degradability of starch of different feeds *Solub. e Degradab. Rumin. do Amido de Diferent. Aliment. Rumin.* 898–905

[3] Perez S and Bertoft E 2010 The molecular structures of starch components and their contribution to the architecture of starch granules: A comprehensive review *Starch* **62** 389–420

[4] Schmiele M, Sampaio U M, Gomes P T G and Clerici M T P S 2019 *Physical Modifications of Starch*

[5] Sujka M 2017 Ultrasonic modification of starch – impact on granules porosity *Ultrason.-Sonochem.* **37** 424–29

[6] Guo W, Shi Y, Wang H, Yang H and Zhang G 2010 Intensification of sonochemical degradation of antibiotics levofloxacin using carbon tetrachloride *Ultrason.-Sonochem.* **17 (4)** 680–84

[7] Monroy Y, Rivero S and García M A 2018 Microstructural and techno-functional properties of

cassava starch modified by ultrasound *Ultrason.-Sonochem.* **42** 795–804
[8] Smith R J 1967 Characterization and analysis of starches In *Starch Chemistry and Technology 2* (USA Academic Press.)
[9] Chattopadhyay K and Singhal R S 1997 Optimisation of conditions of synthesis of oxidised starch from corn and amaranth for use in film-forming applications. *Carbohydr. Polym.* **34 (4)** 203–12
[10] Juliano B B and Perez C M 1981 International cooperative testing on the amylose content of milled rice *Starch* **33(5)** 157–62
[11] Adebowale K O, Afolabi T A and Lawal O S 2002 Isolation, chemical modification, and physicochemical characterisation of Bambarra groundnut (*Voandzeia subterranean*) starch and flour *Food Chem.* **78** 305–11
[12] Craig R C S P C, Maningat S A S, Seib C C and Hoseney P A 1989 Starch paste clarity *Cereal Chem.* **66** 173–82
[13] Majid I, Nayik G A and Nanda V 2015 Ultrasonication and food technology : A review *Cogent. Food. Agric.* **20(1)** 1–11
[14] Yusaf T and Al-juboori R A 2014 Alternative methods of microorganism disruption for agricultural applications *Appl. Energy* **114** 909–23
[15] Lateef A, Oloke J K and Prapulla S G 2007 The effect of ultrasonication on the release of fructosyltransferase from *Aureobasidium pullulans* CFR 77 *Enzym. Microb. Technol.* **40** 1067–70
[16] Castanha N, da Matta-junior M D and Augusto P E D 2017 Potato starch modification using the ozone technology *Food Hydrocoll.* **66** 343–56
[17] Farzi M, Saari M M, Emam-djomeh Z and Mohammadifar M A 2011 Original article Effect of ultrasonic treatment on the rheological properties and particle size of gum tragacanth dispersions from different species *Int. J. Food Sci. Technol.* **46** 849–54
[18] Zhang L, Ye X, Ding T, Sun X, Xu Y and Liu D 2013 Ultrasound effects on the degradation kinetics, structure and rheological properties of apple pectin *Ultrason.-Sonochem.* **20(1)** 222–31
[19] Chan H, Bhat R, Leh C and Karim A A 2012 Effect of deproteinization on degree of oxidation of ozonated starch *Food Hydrocoll.* **26** 339–43
[20] Wang L and Wang Y J 2003 Phsycochemical properties of common and waxy corn starches oxidazed by different level of sodium hypoclorite *Carbohydr. Polym.* **52** 207–17
[21] Sandhu H P S, Manthey F A and Simsek S 2012 Ozone gas affects physical and chemical properties of wheat (*Triticum aestivum* L .) starch *Carbohydr. Polym.* **87(2)** 1261–68
[22] Tharanathan 2005 Starch valuae addition by modification *Food Sci.* **45** 371–84
[23] Sandhu S, Kaur K S, Singh M and Lim N 2007 Comparizon of native and oxydized normal and waxy corn starches: Phsycochemical, thermal, morphological and pasting properties *LWT.* **41** 1000–10
[24] Takizawa I, Silva F F, Konkel G O and Demiete F E 2004 Characterization of tropical starch modified with potassium permanganat and lactic acid *Brazil. Arch. Biol. Technol.* **47** 921–31
[25] Demiete C, Wosiacki I M, Cereda G and Mestres M P 2005 Viscographic characteristics of chemically modified cassava starches assesed by RVA *Sci. Agr. Sci. Eng. Ponta Grossa* **11** 7–17
[26] Iida Y, Tuziuti T, Yasui K, Towata A and Kozuka T 2008 Control of viscosity in starch and polysaccharide solutions with ultrasound after gelatinization *Innov. Food Sci. Emerg. Technol.* **9** 140–6

IOP Conf. Series: Materials Science and Engineering **980** (2020) 012031 doi:10.1088/1757-899X/980/1/012031

Morphological, thermal, and pasting properties of cocoyam (*Xanthosoma sagittofolium*) starch from three locations in Mollucas Islands

C G C Lopulalan[1,3], Y Pranoto[2], Y Marsono[2] and D W Marseno[2]
[1] Faculty of Agriculture, Pattimura University, Jl. Mr. Putuhena – Poka, Ambon, Maluku, Indonesia
[2] Faculty of Agricultural Technology, Gadjah Mada University, Jl. Flora No. 1, Bulaksumur, Yogyakarta, 55281 Indonesia
[3] Doctoral Student, Food Science Study Program, Faculty of Agricultural Technology, Gadjah Mada University, Jl. Flora No. 1, Bulaksumur, Yogyakarta, 55281 Indonesia

Email: djagal@ugm.ac.id

Abstract. This study investigated the morphological, thermal, and pasting properties of cocoyam starch from three locations in Mollucas islands. The islands are Buru, Porto, and Saumlaki. The results showed that starch granules were round and polygonal in shape, and the size varied significantly in range of 11,42 – 12,99 µm, while the range of crystallinity of the three cocoyam starches studied was 11,81% -16,61% and crystal type was A, this was a type for tubers. The thermal properties of cocoyam starch measured by DCS showed the results of onset (To), peak (Tp), endset (Te) and enthalpy (ΔH) in the range of To: 76-77°C, Tp: 78-80°C, Te: 82-86°C and Δ H (J / g): -130.0 - (-) 175.76 respectively. The FTIR analysis showed the functional group profiles of native starches studied had the highest absorption spectra of 3286.70-3325.86 cm^{-1} (O-H group vibrations). The peak, trough, breakdown, setback, and final viscosities were in the range of each PV: 4601 - 5155 RVU, TV: 2623 - 2874 RVU, BV: 1978 - 2287 RVU, FV: 4026 - 4176 RVU, SV: 1240 - 1403 RVU, Pt (Min) 7.87 - 8.40, PT (° C): 79.90 - 81.70.

1. Introduction

Keladi (*Xanthosoma sagittofolium*) is one of the tubers plant in Mollucas islands. Mollucas people only consume it in a simple processed form such as boiling or frying the tubers. This is conducted to complement the carbohydrate needs. It is known that it meets the daily energy needs of around 2100 kcal which is contributed by 10-15% protein, 35-45% fat, and 40-50% carbohydrate for all age groups, except for pregnant and lactating women [1]. Three sampling locations in this study were spread over 3 districts in Mollucas Islands. The distance between one district and another is limited by the sea. This location was the reason for studying characteristics of cocoyam starch from different locations on one island. Research and utilization of natural starch cocoyam still have not been done optimally, an important role of starch which is much needed in the world of food processing industries including starch can be used as raw materials and additives such as thickening agents, gelling agents, film-forming (filming agent), and

IOP Conf. Series: Materials Science and Engineering **980** (2020) 012031 doi:10.1088/1757-899X/980/1/012031

stabilizing agent. Cocoyam contains carbohydrates which are quite high, with its main component being starch which is more than 30%. The main components of starch are amylose (30%) and amylopectin (70-80%) also contain minor components such as lipids, protein, phosphorus, and minerals. Physical and chemical characteristics of cocoyams are influenced by variety, place of grow, and soil fertility that will affect the use in food or non-food [2,3,4,5]. This research aimed to study the physical, thermal, and pasting characteristics of cocoyam native starches from three different locations on Mollucas islands.

2. Materials and methods

2.1. Material

Three cocoyam starches were taken from 3 locations in Mollucas islands, i.e Buru Island, Saumlaki Island and Saparua/Porto Island. Furthermore, the codes samples in this study were based on the location of the sample: Buru Starch, Saumlaki Starch, and Porto starch. The cocoyam tubers were peeled and washed with clean water. Then they were thinly sliced with a size of about 2 mm and washed with clean water repeatedly up to 8 times. Then the thin slices were extracted using a blender, squeezed using a filter, and precipitated for 8 hours drained and dried sediment using a dryer cabinet with a temperature of 50°C for 6 hours. Futhermore, the dried starch was reduced in size using a blender and 80 mesh sieve to get the cocoyam starch.

2.2. Method

Analysis of morphological characteristics of starch: granule shape with Scanning Electron Microscope, particle size with HORIBA Laser Scattering Particle Size Distribution Analyzer LA-960; crystallinity properties of starch granules with X-Ray Diffraction (XRD); FTIR spectroscopy; thermal characterictic with Differential scanning calorumetry (DSC); pasting properties with Rapid Visco Analyzer (RVA).

3. Results and discussion

3.1. Morphological properties

(a) Buru starch (b) Porto starch (c) Saumlaki starch

Figure 1. Scanning electron micrographs (SEM) of three native cocoyam starch.

Morphological characteristics of cocoyam starch from three locations in Mollucas island include the shape and size of starch particles. The shape of the cocoyam starch produced by scanning electronic microscopy on the three starches studied was rounded as well as polyhedral and truncated granules similar. This is in line with what was investigated by [2] on several cocoyam starches from Malawi [4]. The particle size of the three starches is in the range of 2.87 μm - 12.99 μm and the largest size (75%) is 11.4 μm - 12.9 μm. The particle size and shape of starch is affected by variety, growing environment, soil quality [6]. Based on the particle size produced in the study are in small and medium sizes, this size is larger than the size of cocoyam and is the same as the size of the cassava and potato particles produced in the study [2,7].

3.2. X-Ray diffraction pattern

The crystalline structure of starch granules which forms a semi-crystalline system consisting of crystalline and amorphous regions studied using X-ray reactions [8]. According to Zhou [9], the starch crystal component is the sharp peak and the amorphous component as the dispersive peak is shown by the X-ray diffraction pattern. The range of starch crystallinity based on literacy is 15% to 45%, this value is based on the source and moisture content of starch and the measurement methods used [10]. The results of this study showed the degree of crystallinity of the three cocoyam starches was 11.81% Saumlaki starch, 13.69% Buru starch, and 16.61% Porto starch. This result was still very low when compared with the crystallinity value of other tuber starches, such as arrowroot starch 20.01% [11]. The difference in the level of crystallinity could be attributed to the amylose content in this starch, with previous reports showing that the percentage of low crystallinity in starch with high amylose content. The results from value of 2Θ showed that cocoyam starch was included in type A in the type of crystallinity with a value of 2Θ each cocyam starch is 15.18°, 17.13°, 18.03°, 22.86° 2θ (strong peak) [12]. These results are in line with Alvani [11] about Garut tuber and some of the other tubers from Srichuwong [13] research.

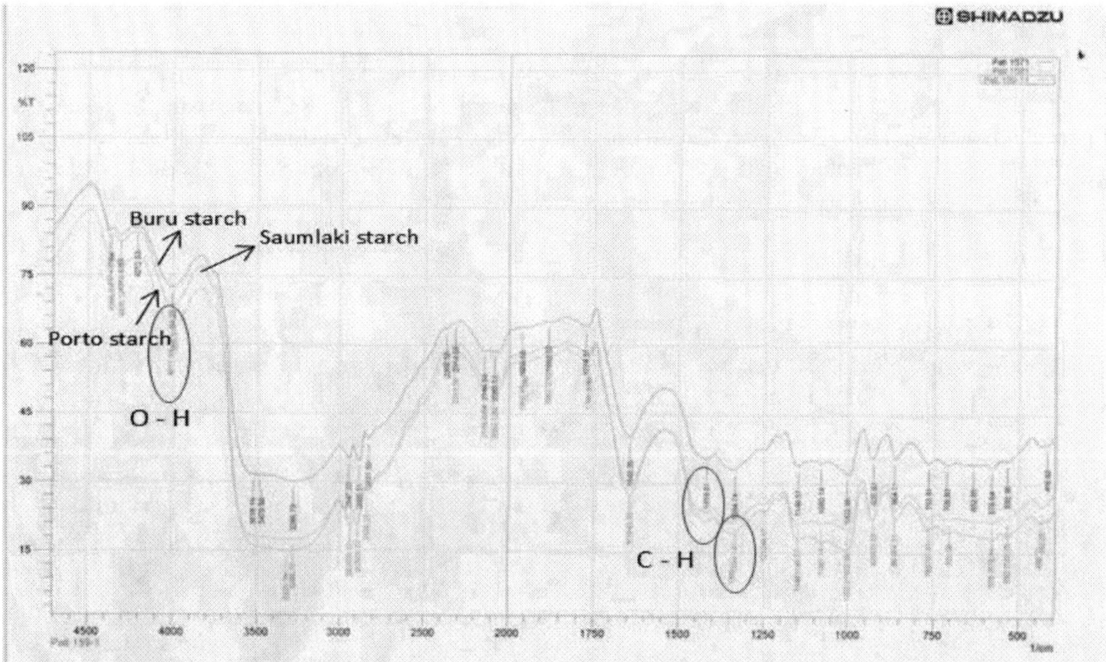

Figure 2. FT-IR spectra of three native cocoyam starch

3.3. FTIR of cocoyam starches

Spectrometer analysis was used to determine the chemical structure of natural starch and its changes after modification with various methods of starch modification. The results of FTIR analysis on the three starches showed the O-H group peaks at peaks of 3000 - 3600 cm^{-1} as stated by Xu [14]. It can be seen in figure 2 that the peak of the O-H group in the three cocoyam starches ranged from 3286.70 cm^{-1} - 3518.16 cm^{-1} while the C-H group was at the peak of 1419.61 cm-1 (Buru starch) and 1334.74 cm-1 (Saumlaki starch and Porto starch). As stated by Skoog *et al.*, 1998 in Fiqtinovri [15] that the C-H bond was found at the peak of 1491.61 and 1365.60 cm-1.

3.4. Thermal properties

Table 1 presents that gelanitinization temperature values of the three cocoyam starchs studied were in the range of 76°C - 77°C with enthalpy values of 13.0-017.5 J/g. Gelanitinization temperature values obtained were in the range of values studied by [2] in Malawi cocoyam (61,8 – 83,7°C), [5] in *Xanthosoma sagittifolium, Colocasia esculenta, (*78,0 77,2 °C). The gelatinization temperature of cocoyam starch results of the study was also higher compared to the gelatinization temperature of cassava starch and corn starch but the transition enthalpy starch values studied were higher than 2 references of cocoyam starches. The thermal characteristics of taro starch are higher than that of cassava and corn starch, showing that in the interior of the cocoyam starch granules there is a strong bonding strength [2].

Table 1. Thermal properties of native cocoyam starch from three locations in Mollucas Island: onset (T_0), peak (Tp), and conclucion (Tc) and transition energi (ΔH).

Thermal Properties	Native cocoyam starch		
	Buru starch	Porto starch	Saumlaki starch
To (°C)	76	77	76
Tp (°C)	78	80	78
Tc (°C)	82	86	86
Δ H (J/g)	14,3	17,5	13,0

To = Onset Temperature; Tp = Peak Temperature; Tc = Conclusion Temperature; ΔH = Enthalpy of Gelatinization.

3.5. Pasting properties

Table 2. Pasting properties of native cocoyam starch from three locations in Mollucas Island.

Pasting Properties	Buru starch	Porto starch	Saumlaki starch
Peak viscosity (RVU)	5155,00	4601,00	5126,00
Trough viscosity (RVU)	2868,00	2623,00	2874,00
Breakdown viscosity (RVU)	2287,00	1978,00	2252,00
Final viscosity (RVU)	4176,00	4026,00	4114,00
Setback viscosity (RVU)	1308,00	1403,00	1240,00
Peak time (min)	7,87	8,40	8,13
Pasting temperature (°C)	79,90	81,70	80,20

Paste viscosity values of cocoyam native starch analyzed using RVA are presented in table 2. Peak viscosity was measured when the starch paste reached maximum viscosity during the heating phase, viscosity at 95°C or hot paste viscosity. Breakdown viscosity showed stability of viscosity towards heating, final viscosity after being maintained at 50°C. The viscosity of the setback indicated the tendency of starch to undergo retrogradation. Gelatinization temperature (°C) is the temperature at which the viscosity value begins to read which indicated the starch had begun gelatinizing. In general, native starch had a gelatinization profile with a maximum peak high enough with a sharp decreased in viscosity (breakdown viscosity) during the heating process. This is in line with the peak viscosity and viscosity breakdown values by the three cocoyam native starches (table 2). This showed that the cocoyam pasta was less stable in the heating process. Starch viscosity will increase again in the cooling process through the process of reassembling amylose, and amylopectin molecules by hydrogen bonds [11]. The three cocoyam starches studied had a peak of high viscosity with a sharp decrease, gelatinization temperature of 79.90°C (Buru starch), 81.70°C (Porto starch), and 80.20°C (Saumlaki starch), the initial temperature of starch gelatinization obtained was still in the temperature range reported by Pérez [16], which is 67.75°C – 81.40°C. The setback viscosity of three cocoyam starch was relatively high to facilitate retrograde of cocoyam starch. This condition indicated that the type A starch gelatinization profile was characterized by a high peak viscosity value and a fairly sharp breakdown viscosity. Gelatinization profiles obtained on the three cocyam starches could be included in a group with arrowroot starch, tapioca starch, potatoes, sweet potatoes, sago, waxi corn and waxy barley Collado *et al.*, 2001; Wattanachant *et al.*, 2002; Singh *et al.*, 2005 *in* [11].

4. Conclusion

The granular starch granules form was rounded as well as polyhedral and truncated granules. OH group in the three cocoyam starches ranged from 3286.70 cm-1 - 3518.16 cm-1 while the CH group was at the peak of 1419.61 cm-1 (Buru starch), and 1334.74 cm -1 (Saumlaki starch and Porto starch). X-ray diffraction pattern showed crystalline of cocoyam starch classified as type A which was characterized by peaks at 2 tetras 15.18°, 17.13°, 18.03°, 22.86°. Type A gelatinization profile with high peak viscosity and sharp breakdown viscosity.

References

[1] Institute of Medicine (U.S.) and Subcommittee on Technical Specifications for a High-Energy Emergency Relief Ration 2002 *High-Energy, Nutrient-Dense Emergency Relief Food Product* (Washington (DC): National Academies Press (US))

[2] Mweta D E, Labuschagne M T, Bonnet S, Swarts J and Saka J D K 2010 Isolation and physicochemical characterisation of starch from cocoyam (*Colocasia esculenta*) grown in Malawi *J. Sci. Food Agric.* **90** 1886–96

[3] Lauzon R D, Shiraishi K, Yamazaki M, Sawayama S, Sugiyama N and Kawabata A 1995 Physicochemical properties of cocoyam starch *Food Hydrocoll.* **9(2)** pp 77–81

[4] Mepba H D, Eboh L, Eko C B and Ukpabi U J 2009 Composition and pasting properties of starch from two cocoyam cultivars *J. Food Qual.* **32(4)** pp. 522–37

[5] Pérez E, Schultz F S and de Delahaye E P 2005 Characterization of some properties of starches isolated from *Xanthosoma sagittifolium* (tannia) and *Colocassia esculenta* (taro) *Carbohydr. Polym.* **60(2)** pp. 139–45

[6] Cornejo-Ramírez Y I, Martínez-Cruz O, Del Toro-Sánchez C L, Wong-Corral F J, Borboa-Flores J and Cinco-Moroyoqui F J 2018 The structural characteristics of starches and their functional properties *CyTA - J. Food* **16(1)** pp 1003–17

[7] Martínez P, Peña F, Bello-Pérez L A, Núñez-Santiago C, Yee-Madeira H and Velezmoro C 2019 Physicochemical, functional and morphological characterization of starches isolated from three native potatoes of the Andean region *Food Chem.* **X(2)** p 100030

[8] Xu Y 2013 Resistant starch content, molecular structure and physicochemical properties of starches in Virginia-grown corn, potato and mungbean *J. Cereals Oilseed* **4(1)** pp 10–8

[9] Zhou F, Liu Q, Zhang H, Chen Q and Kong B 2016 Potato starch oxidation induced by sodium hypochlorite and its effect on functional properties and digestibility *Int. J. Biol. Macromol.* **84** pp 410–17

[10] Alvani K, Qi K, Tester R F and Snape C E 2011 Physico-chemical properties of potato starches *Food Chem.* **125(3)** pp 958–65

[11] Faridah D N, Fardiaz D, Andarwulan N and Sunarti T C 2014 Physicochemical characterisation of arrowroot starch (*Maranta arundinaceae*) *AGRITECH.* **34(1)** p 8

[12] Lee H J 2007 The isolation and characterisation of starches from legume grains and their application in food formulations (School of Applied Sciences Science Engineering and Technology Portfolio RMIT University) p 217

[13] Srichuwong S, Sunarti T, Mishima T, Isono N and Hisamatsu M 2005 Starches from different botanical sources I: Contribution of amylopectin fine structure to thermal properties and enzyme digestibility *Carbohydr. Polym.* **60(4)** pp 529–38

[14] Xu Y, Miladinov V and Hanna M A 2004 Synthesis and characterization of starch acetates with high substitution *Cereal Chem. J.* **81(6)** pp 735–40

[15] Fiqtinovri S M, Benita A M, Marseno D W and Pranoto Y 2020 Physical, amilograph and morphological properties of Lai Seed (*Durio Kutejensis*) starch assylation using acetate anhydrat *agriTECH.* **40(1)** p 74

[16] Pérez E and Lares M 2005 Chemical composition, mineral profile, and functional properties of Canna (*Canna edulis*) and Arrowroot (*Maranta* spp.) Starches *Plant Foods Hum. Nutr.* **60(3)** pp 113–16

IOP Conf. Series: Materials Science and Engineering **980** (2020) 012032 doi:10.1088/1757-899X/980/1/012032

Packaging and storage of cocoa beans fermented with *Lactobacillus plantarum* HL-15 in Agricultural Technology Park Nglanggeran, Yogyakarta

T Marwati[1*], T F Djaafar[1*], S D Indrasari[1*], S Widodo[1*], N Cahyaningrum[1*], A Fajariyah[1*], Sulasmi[1*], D E Susanto[2*], R Yanti[2] and E S Rahayu[2*]

[1] Yogyakarta Assessment Institute for Agricultural Technology, Indonesia
[2] Department of Food and Agricultural Product Technology, Faculty of Agricultural Technology, Universitas Gadjah Mada, Yogyakarta, Indonesia

Email: watipasca2@gmail.com
*Main contributors

Abstract. In Gunungkidul, there is Agricultural Techno Park (ATP) Nglanggeran which processes fermented cocoa beans from local farmers for making chocolate products. This research aimed to obtain a suitable packaging method for fermented cocoa beans during storage in ATP Nglanggeran. This research was carried out using a completely randomized factorial design. The first factor was the addition of starter (B), which consisted of B1 (with *Lactobacillus plantarum* HL 15) and B2 (without *Lactobacillus plantarum* HL 15). The second factor was the packaging methods (K), which consisted of K1 (vacuumed PP plastic 0.8 mm in thickness), K2 (non-vacuumed PP plastic 0.8 mm in thickness), K3 (plastic container), K4 (vacuumed PP plastic 0.8 mm in thickness stored within the plastic container), and K5 (nylon sack). The third factor was the storage time (P), which were P0 (0 month), P1 (1 month), and P2 (2 months). The result showed that the samples with the addition of the starter had a lower number of fungal colonies than those without the addition. Besides, the packaging using vacuumed PP plastic 0.8 mm in thickness stored within a plastic container gave the best result on minimizing the total fungal growth contamination and the peroxide value.

1. Introduction

Indonesia is the third-largest cocoa producer in the world after Ghana and Ivory Coast. The top five importing countries for Indonesian cocoa are Malaysia, America, China, India, and the Netherlands. In addition to the increasing open export opportunities, the domestic cocoa bean market is still quite large, including for the cocoa processing industry in Java [1]. The production in Yogyakarta in 2018 was 1,362 tons, and it was predicted to have an increase of 10% in 2019 and 2020 up to 1,498 tons, and 1,648 tons, respectively [2]. One of the cocoa productions centers in the Special Region of Yogyakarta is in the

Patuk District of Gunungkidul Regency. In Nglanggeran Village, Patuk Subdistrict, Gunungkidul, there is an Agricultural Techno Park (ATP) that processes fermented cocoa beans from local farmers for making chocolate products. ATP Nglanggeran needs to be supported by some studies on fermentation technology, packaging, and storage of cocoa beans to be able to produce quality and safely processed chocolate.

Several critical stages of the contamination of mycotoxin-producing fungi in handling cocoa beans are fermentation, drying, and storage [3, 4]. During the fermentation process, several types of microbial growth occur, such as lactic acid bacteria (BAL), acetic acid bacteria (BAA), and yeast. These microorganisms have their respective roles in the fermentation process of cocoa beans [5]. In the fermentation process, the fermentation box also plays an important role. The use of fermentation box (single box with a hand crank design) requires more efficient material and space, needs to have better aeration and optimal temperature (>40 °C), and needs to be able to produce the lowest fungi level (1.33%) in the five-day fermentation [6]. If the fermentation, drying, and storage conditions are not controlled, fungus contamination will occur then cause damage and even produce toxins.

Aspergillus sp., *Fusarium* sp., *Trichoderma* sp., and *Penicillium* sp. are able to produce mycotoxins and cause uncontrolled fermentation processes [7]. Mycotoxin-producing fungi can be inhibited by the addition of *Lactobacillus plantarum* HL-15 culture, which is isolated from cocoa beans during fermentation [8]. To extend the shelf life of the culture and for an easier application, the starter culture is produced in a dry form. The method of drying the *L. plantarum* HL-15 culture can be performed with a spray dryer with a rice flour matrix [9], a rack-type dryer with a rice flour matrix [10], or an oven dryer with a tapioca matrix [11]. The use of dry starter culture of *L. plantarum* HL-15 in the fermentation process can provide increased value added to dried cocoa beans [12].

Lactic acid bacteria that have been proven to have the power to inhibit fungi from organic acid production are from the species *Lactobacillus fermentum* and *Lactobacillus plantarum* [13]. The fermentation process with the addition of lactic acid bacteria shows an increase in the fermentation index which indicates that the fermentation process can be accelerated [14]. Improvement of the fermentation process of dried cocoa beans can be done by adding cane drops, yeast, and lactic acid bacteria [15]. The aroma of unfermented cocoa beans can be enhanced by the fermentation process with the addition of *S. cerevisiae, L. plantarum,* and *A. aceti* cultures [16]. The application of *Lactobacillus fermentum* in the fermentation process can produce good quality cocoa which is free of mycotoxins [17]. Controlling mycotoxin-producing fungal contaminants to improve the quality of cocoa beans can use lactic acid bacteria [18,19]. Besides, *L. plantarum* HL-15 can inhibit fungal growth and synthesis of oxytocin A in cocoa beans during fermentation and drying [20]. The addition of *L. plantarum* HL 15 as a culture starter in the old and new fermentation boxes has a smaller level of fungus contamination compared to the treatment without *L. plantarum* HL 15 as the culture starter [21].

In the cocoa beans, *Eurotium* sp was still detected to be carried by the cocoa powder being produced. The presence of *Eurotium* in cocoa beans will be followed by the growth of *Aspergillus* and *Penicillium* [21]. In the storage of cocoa beans, there were also found some fungi (*Aspergillus, Paecilomyces, Talaromyces, Penicillium, Pseudophitomyces, Cimplicillium,* and *Aspergillus carbonarius)* [3]. Besides, there were found fungus *A. flavus, A. niger, A. fumigatus, Penicillium* sp, *Fusarium* sp, *Trichoderma* sp., *Rhizopus* sp., *Mucor* sp., and *Verticillium* sp. in the dried cocoa beans at the farmer level. Meanwhile, at the trader level, *A. flavus*, *A. niger*, *Penicillium* sp., *Fusarium* sp., *Trichoderma* sp., and *Mucor* sp. were found, and at the exporter level, there were found fungi *A. flavus*, *A. niger, Penicillium* sp, *Trichoderma viridae,* and *Geotrichum* sp. The total population of mycotoxigenic fungi at the levels of farmers, traders and exporters were respectively 1.4 x 109 cfu/ml, 6.5 x 107 cfu/ml, and 6.0 x 105 cfu/ml [22]. The *Penicillium* species which was found mostly in the cocoa belongs to *P. citrinum* which is likely to have a capability in the production of OTA [23]. OTA was detected in the beans at all stages of post-harvesting operations, such as pod-opening, fermentation, drying, and storage [24]. Black *Aspergillus* was the dominant species of *Aspergillus* in cocoa beans, and *A. carbonarius* was the OTA producer among the black *Aspergillus* [25].

Packaging plays an important role in maintaining the water content of fermented cocoa beans to keep the seeds from being attacked by the fungi during storage. Proper packaging and storage of cocoa beans are as important as proper fermentation and drying processes. According to Owusu [26], cocoa beans must be packaged in clean bags that are strong enough and properly sealed. The bags used must be made of non-toxic food-grade material. After the drying and sorting process, cocoa beans must be put in an appropriate bag and stored. Furthermore, packaged cocoa beans must be placed in a weather-resistant storage warehouse, free from pests, away from smoke, and other odors. Packaged cocoa beans also need to be placed on the surface of the soil and away from the wall.

The quality of dried cocoa beans in the international trade is assessed by the level of the total percentage of fungi, unfermented beans, purple seeds, seeds that are attacked by insects, and germinated seeds. In the recent cocoa trade, from the scientific point of view, much emphasis is put on free fatty acid content (FFA) which is influenced by many factors, such as humidity, oxygen, and insect infestation [27]. The facts show that the high water content of cocoa beans causes fungal infections, which are related to FFA values in cocoa beans. The increase in FFA during storage can be related to the activity of the lipase enzyme which is naturally present in cocoa beans [28]. The enzyme becomes active due to changes in the moisture content of the seeds and high-temperature storage environment [29]. Therefore, storage management plays an important role to maintain the quality of fermented cocoa beans.

This study aims to determine the effect of adding *Lactobacillus plantarum* HL-15 culture to the fermentation process and packaging method on the quality of cocoa beans during storage in ATP Nglanggeran. This starter of *Lactobacillus plantarum* HL-15 was added at the beginning of the fermentation process to minimize the total fungal growth in the cocoa beans. The packaging methods were also studied to select which one had the best effect on maintaining the quality of cocoa beans.

2. Materials and methods

2.1. Materials

The fermentation and packaging of cocoa beans were carried out at UPH Ngudi Raharjo II Gunungkidul Yogyakarta, while the storage was carried out at ATP Nglanggeran. The cocoa beans were obtained from Gunungkidul. The starter culture of *Lactobacillus plantarum* HL-15 was obtained from FNCC Gadjah Mada University, Yogyakarta [6]. The packaging materials were PP 0.8 mm plastic, nylon sacks, raffia, and plastic containers. Chemicals for conducting the chemical and microbiological analyses included solid KI (MERCK), KIO3 1% (MERCK), HCl 2N (MERCK), sodium thiosulfate (Na2S2O3) MERCK brand, 1% starch, chloroform (CH3Cl), methanol, Na2SO4 anhydrous (MERCK), glacial acetic acid, DG-18 (OXOID) plating media, chloramphenicol, glycerol, sodium chloride (NaCl) MERCK brand, and 0.05% Bayclin brand disinfectant (purchased in a local supermarket in Yogyakarta).

2.2. Methods

Fresh cocoa fruits, after harvesting, were carefully broken down using a wooden beater, or by hitting one fruit with another fruit. Then the cocoa beans were removed from the fruit and separated from the pith attached to the seeds. Afterward, the sorting of cocoa beans was done by separating the fully ripe cocoa beans from young seeds and deformed seeds due to pests, fungi, or sprouting.

The study of this fermentation process was carried out using a field scale with a 40 kg capacity of fermentation box. The fermentation process of cocoa, using starter culture, is a modification of the fermentation method performed by [14,15,16,17,18,19,20]. The fermentation was carried out for five days. On the third day, the cocoa seeds were turned upside down so that they were evenly fermented. On day 5, the cocoa beans were removed from the fermentation box, soaked, washed, and ready to be dried. Next, the drying was done by a combination of manual drying and rack type dryer at the temperature between 55-66°C to reach 6-7% water content. After that, the dried cocoa beans were packaged using the packaging materials and then stored at room temperature in ATP Nglanggeran.

This research was carried out using a completely randomized factorial design. The first factor was the addition of the starter (B), which were B1 (with *Lactobacillus plantarum* HL-15) and B2 (without *Lactobacillus plantarum* HL-15). The second factor was the packaging methods (K), which were K1 (vacuumed PP plastic 0.8 mm in thickness), K2 (non-vacuumed PP plastic 0.8 mm in thickness), K3 (plastic container), K4 (vacuumed PP plastic 0.8 mm in thickness stored within a plastic container), and K5 (nylon sack). The third factor was the storage duration (P), which were P0 (0 month), P1 (1 month), and P2 (2 months) as seen in table 1.

Table 1. The experimental design of packaging and storage of cocoa beans.

Packaging methods (K)	Addition of the starter (B)	
	With *L. plantarum* HL-15 (B1)	Without *L. plantarum* HL-15 (B2)
Vacuumed PP plastic 0.8 mm in thickness (K1)	v	v
Non-vacuumed PP plastic 0.8 mm in thickness (K2)	v	v
Plastic container (K3)	v	v
Vacuumed PP plastic 0.8 mm in thickness stored within a plastic container (K4)	v	v
Nylon sack (K5)	v	v

2.3. Analyses of dried cocoa beans
A number of analyses conducted on the dried cocoa beans before and during the storage: (1) the quality assessment based on SNI 2003: 2828; (2) moisture content; (3) seed water activity; (4) peroxide value; and (5) total mushroom contamination. The data were analyzed using SPSS software version 22.0. One-way analysis of variance (ANOVA) with a significance level of 5% was used.

3. Results and discussion

3.1. Temperature and relative humidity of storage room
The storage of fermented cocoa beans was carried out at ATP Nglanggeran, Gunungkidul. The optimal room temperature for storing cocoa beans is below 30°C with a maximum relative humidity (RH) of 75% [30]. The temperature and RH of the room during the storage were quite fluctuating, with the lowest temperature of 26.6 °C and the highest temperature of 28.8 °C, and with the lowest relative humidity of 60% and the highest of 69%. The range of temperatures and RH values did not exceed the upper limit of temperature and relative humidity for optimal cocoa bean storage [30]. This indicated that the temperature and relative humidity of the room in ATP Nglanggeran for two months of storage of cocoa beans was always in the optimal range.

3.2. The quality of dried cocoa beans based on the Indonesian National Standards (SNI)

The analysis results of the quality of cocoa beans which include moldy seeds, levels of unfermented beans (slaty), levels of germinated seeds, levels of insect-containing seeds, broken seeds, and flat beans can be seen in table 2. In the first month of storage, the cocoa beans without or with the addition of starter *L. plantarum* HL-15, in all packaging methods, did not show fungal growth. This is in line with research by Yanti [31] which showed that the addition of local microbes in the form of lactic acid bacteria did not exhibit fungal growth in cocoa beans. In the cocoa bean sample, after two months of storage, it was found that the growth of this external fungus was characterized by grayish-white spots and grayish-white fine fibers on the surface of the dried cocoa beans as the result of the fermentation using culture or without the addition of culture.

In all samples of the cocoa beans, the number of seeds germinated and contaminated by insects show a 0% level. This means that no sample had been germinated or contaminated by insects. Seed germination is undesirable in dried cocoa bean products because germination shows that the seeds are still alive, the fermentation is not going well, and the flavor-forming precursor compounds cannot appear. The seeds that are contaminated by insects are also undesirable because they will reduce the quality of the cocoa beans, especially physically. Some examples of insects that commonly attack cocoa beans include fig moth (*Cadra cautella*), gray moth (*Cadra elutella*), *Araecerus fasciculatus, Lasioderma serricorne*, red flour beetle (*Tribolium castaneum*), and khapra beetle (*Trogoderma granarium*). Damage caused by the insects varies from the cavity on the surface of the seeds, dirt in the form of larvae, cocoon shells, to silk spun threads [32].

From all the parameters being analyzed, the values obtained were compared with The Indonesian National Standards for cocoa beans. From this comparison, it can be seen that the quality of the samples of cocoa beans was homogeneous. The samples of 0 month, the first month, and the second month were all included in the I-B class. Meanwhile, the maximum percentages for the moldy seeds, the slaty seeds, the insect-containing seeds, the impurities, and the germinating seeds were 2%, 3%, 1%, was 1.5%, and 2%, respectively.

3.3. The moisture content of cocoa beans

The moisture content of the cocoa beans during storage is shown in table 3. From table 3, it can be seen that statistically, the factor of addition of *L. plantarum* HL-15 starter and the packaging methods did not significantly influence the water content. After two months of the storage, there was a significant increase in water content, but it still met the export quality standards. The standard moisture content of export quality cocoa beans is a maximum of 7.5%. If it is higher than this value, cocoa beans will not be safe to be stored for a long time, but if the moisture content is too low, cocoa beans tend to be fragile [33]. Besides, if the moisture content is too low (<5%), the cocoa beans will become brittle and break easily during transportation [34].

3 4. The cocoa beans water activity

The moisture content of cocoa beans during storage is shown in table 4. Aw value is closely related to the potential for growing microorganisms including fungi that can damage dried cocoa beans during storage. Statistically, the factors that has a significant effect on Aw sample are packaging and the duration of storage. Because the p values of both are less than 0.05, while the starter factor had a p-value of 0.707, it can be concluded that it does not have a significant effect. However, the interaction between the starter and packaging factors significantly influences the Aw values of the samples. In conclusion, the addition of a starter will have a significant effect on Aw samples if followed by packaging variations.

IOP Conf. Series: Materials Science and Engineering **980** (2020) 012032 doi:10.1088/1757-899X/980/1/012032

Table 2. The quality of dried cocoa beans based on the Indonesian National Standard including moldy seeds, levels of unfermented (slaty) beans, levels of germinated seeds, levels of insect containing seeds, broken seeds, and flat beans.

Cocoa beans			Analysis results								
Month	Fermentation	Packaging	Number of seeds / 100 grams	Moldy seeds (%)	Unfermen-ted (slaty) beans (%)	Germinated seeds (%)	Insect containing seeds (%)	Broken seeds (%)	Flat beans (%)	Quality	Size group
0	With *L. plantarum* HL-15		102 ± 1.41	0	0	0	0	0.5	0	I-B	B
	Without *L. plantarum* HL-15		106 ± 0	0	0	0	0	0.47	0	I-B	B
1	With *L. plantarum* HL-15	Vacuumed PP plastic 0.8 mm in thickness	105 ± 2.83	0	0	0	0	0.47	0	I-B	B
		Non-vacuumed PP plastic 0.8 mm in thickness	101 ± 5.66	0	0	0	0	0	0	I-B	B
		Plastic container	111 ± 2.83	0	0	0	0	0	0	I-B	C
		Vacuumed PP plastic 0.8 mm in thickness stored within a plastic container	109 ± 14.14	0	0	0	0	0	0	I-B	B
		Nylon sack	109 ± 3.54	0	0	0	0	0	0	I-B	B
	Without	Vacuumed PP plastic 0.8 mm in thickness	110 ± 0	0	0.46	0	0	0.46	0	I-B	B

IOP Conf. Series: Materials Science and Engineering **980** (2020) 012032 doi:10.1088/1757-899X/980/1/012032

	L. plantarum HL-15	Non-vacuumed PP plastic 0.8 mm in thickness	121 ± 5.66	0	0	0	0	0	0	I-B	S
		Plastic container	108 ± 1.41	0	0	0	0	0	0	I-B	B
		Vacuumed PP plastic 0.8 mm in thickness stored within a plastic container	118 ± 2.12	0	0	0	0	0.42	0	I-B	C
		Nylon sack	111 ± 1.41	0	0	0	0	0	0	I-B	C
2	With *L. plantarum* HL-15	Vacuumed PP plastic 0.8 mm in thickness	110 ± 1.41	0,9	0,46	0	0	0	0.92	I-B	B
		Non-vacuumed PP plastic 0.8 mm in thickness	108 ± 2.83	0.46	0	0	0	0.91	0.46	I-B	B
		Plastic container	118 ± 3.54	1.72	0.42	0	0	0.44	1.74	I-B	C
		Vacuumed PP plastic 0.8 mm in thickness stored within a plastic container	102 ± 6.36	0	0.47	0	0	0	0.95	I-B	B
		Nylon sack	112 ± 1.41	0.89	0.44	0	0	0.44	0.44	I-B	C
	Without	Vacuumed PP plastic 0.8 mm in thickness	109 ± 2.12	0.92	0	0	0	0.47	1.37	I-B	B

L. plantarum HL-15	Non-vacuumed PP plastic 0.8 mm in thickness	121 ± 2.83	1.67	0.41	0	0	0.42	0.42	I-B	S
	Plastic container	112 ± 0.71	0	0	0	0	0	0	I-B	C
	Vacuumed PP plastic 0.8 mm in thickness stored within a plastic container	112 ± 1.41	0,9	0	0	0	0	0.45	I-B	C
	Nylon sack	103 ± 4.24	0	0,5	0	0	1	0	I-B	B

Table 3. The moisture content of cocoa beans during storage.

Cocoa beans		Moisture content (%) during storage		
		0 Month	1 Month	2 Months
With *L. plantarum* HL-15	Vacuumed PP plastic 0.8 mm in thickness	6.24^{aA}	6.18^{aA}	7.05^{aB}
	Non-vacuumed PP plastic 0.8 mm in thickness	6.24^{aA}	6.18^{aA}	6.73^{aB}
	Plastic container	6.24^{aA}	6.56^{aA}	7.39^{aB}
	Vacuumed PP plastic 0.8 mm in thickness stored within a plastic container	6.24^{aA}	6.02^{aA}	6.39^{aB}
	Nylon sack	6.24^{aA}	6.30^{aA}	7.20^{aB}
Without *L. plantarum* HL-15	Vacuumed PP plastic 0.8mm in thickness	7.02^{aA}	6.36^{aA}	6.91^{aB}
	Non-vacuumed PP plastic 0.8 mm in thickness	7.03^{aA}	7.02^{aA}	7.79^{aB}
	Plastic container	7.03^{aA}	6.28^{aA}	7.23^{aB}
	Vacuumed PP plastic 0.8 mm in thickness stored within a plastic container	7.03^{aA}	7.55^{aA}	7.07^{aB}
	Nylon sack	7.03^{aA}	6.28^{aA}	7.06^{aB}

Remark: Figures followed by different lowercase letters (a, b, c) in the same column and figures followed by different capital letters (A, B) in different columns show significant difference.

The highest average Aw values are found in the samples with K5 packaging variations, followed by K1, K3, K4, and K2, respectively. The K5 package is a nylon sack packaging as used by cocoa farmers in the Gunungkidul area to store their dried cocoa beans. In terms of air permeability, this type of packaging is easier to penetrate than other types of packaging because nylon sacks have gaps on wider surfaces. Thus, the dried cocoa beans inside will be more susceptible to contact with the surrounding air. The statistical analysis shows that the packaging treatment using 0.8 mm un-vacuum PP (K2) plastic is actually able to maintain the Aw value better than the 0.8 mm PP plastic package that is vacated (K1). This can be seen from the average Aw value of each treatment each month, where for K2, the mean value is lower than K1, as well as the change in Aw that occurs, which also has a smaller minimum value for K2 when compared to K1. Through Duncan's post hoc test, it is found that the variation of K2 packaging is significantly different from K1.

3.5. The peroxide value in cocoa beans

The peroxide value in the cocoa beans during storage is shown in table 5. From table 5, it can be seen that the type of packaging and the storage duration significantly affect the peroxide value while the addition of a starter has no effect on it. The longer the seeds are stored, the larger would be the average peroxide value, which indicates an increase in fat damage.

Table 4. The cocoa bean water activity during storage.

	Cocoa beans	Water activity during storage		
		0 Month	1 Month	2 Months
With *L. plantarum* HL-15	Vacuumed PP plastic 0.8 mm in thickness	0.73^{bcA}	0.74^{bcAB}	0.78^{bcB}
	Non-vacuumed PP plastic 0.8 mm in thickness	0.71^{aA}	0.72^{aAB}	0.77^{aB}
	Plastic container	0.75^{bcA}	0.75^{bcAB}	0.76^{bcB}
	Vacuumed PP plastic 0.8 mm in thickness stored within a plastic container	0.71^{abA}	0.72^{abAB}	0.74^{abB}
	Nylon sack	0.78^{cA}	0.79^{cAB}	0.76^{cB}
Without *L. plantarum* HL-15	Vacuumed PP plastic 0.8 mm in thickness	0.74^{bcA}	0.75^{bcAB}	0.77^{bcB}
	Non-vacuumed PP plastic 0.8 mm in thickness	0.72^{aA}	0.72^{aAB}	0.75^{aB}
	Plastic container	0.74^{bcA}	0.75^{bcA}	0.76^{bcB}
	Vacuumed PP plastic 0.8 mm in thickness stored within a plastic container	0.76^{abA}	0.76^{abAB}	0.74^{abB}
	Nylon sack	0.75^{cA}	0.75^{cAB}	0.71^{cB}

Remark: Figures followed by different lowercase letters (a, b, c) in the same column and figures followed by different capital letters (A, B) in different columns show significant difference.

The lowest peroxide value is found in the cocoa beans with K4 packaging, while the highest is in the cocoa beans with K5. Indeed, K4 packaging has a barrier to environmental conditions. It is oilier than other variations of packaging so that it has the ability to inhibit the greatest oxidation of fat as well. For example, it can inhibit direct contact with light that can cause photooxidation and inhibit direct contact with air that can trigger autoxidation. This is in contrast to the variation of K5 packaging, where nylon sack packaging is the least airtight packaging so that air can easily enter and make contact with the seeds.

The peroxide limit in dried cocoa beans is 5 meqiv O2/Kg sample [35]. In fact, the peroxide values obtained from the cocoa seeds for all the treatments are below the threshold.

3.6. The total fungal contamination of cocoa beans

The total fungal contamination of cocoa bean during storage is shown in table 6. Table 6 records that the type of packaging has not shown an effect on the total fungal colonies, both on fermented cocoa beans by adding *L. plantarum* HL-15 or without adding *L. plantarum* HL-15 for the two-month storage. Meanwhile, the storage time has a significant effect on the growth of fungal colonies, both on fermented cocoa beans with the addition of *L. plantarum* starter and without the addition of plantarum HL-15 starter. It is obvious that the number of fungal colonies increases significantly after the two-month storage.

High air humidity can trigger mold growth [30]. Hence, besides using vacuum packaging, other treatments should be taken during storage to maintain the quality of cocoa beans. In visual observation based on SNI, there was no fungus, but based on the results of microbial observations in the laboratory analysis with pour plate method, there were still internal mushroom populations. Therefore, a good handling is necessary during the process of storing dried cocoa beans so that the growth of internal mold can be prevented and does not damage the dried cocoa beans and reduce its quality.

Table 5. The peroxide value in cocoa beans during storage.

Cocoa beans		The peroxide value (5 meqiv O_2/Kg sample) during storage		
		0 Month	1 Month	2 Months
With *L. plantarum* HL-15	Vacuumed PP plastic 0.8 mm in thickness	0^{abA}	0.46^{abB}	0.59^{abC}
	Non-vacuumed PP plastic 0.8 mm in thickness	0^{bA}	0.55^{bB}	0.65^{bC}
	Plastic container	0^{abA}	0.47^{abB}	0.58^{bC}
	Vacuumed PP plastic 0.8 mm in thickness stored within a plastic container	0^{aA}	0.38^{aB}	0.49^{aC}
	Nylon sack	0^{cA}	0.80^{cB}	1.07^{cC}
Without *L. plantarum* HL-15	Vacuumed PP plastic 0.8 mm in thickness	0^{abA}	0.48^{abB}	0.52^{abC}
	Non-vacuumed PP plastic 0.8 mm in thickness	0^{bA}	0.66^{bB}	0.61^{bC}
	Plastic container	0^{abA}	0.47^{abB}	0.60^{abC}
	Vacuumed PP plastic 0.8 mm in thickness stored within a plastic container	0^{aA}	0.39^{aB}	0.41^{aC}
	Nylon sack	0^{cA}	0.63^{cB}	0.69^{cC}

Remark: Figures followed by different lowercase letters (a, b, c) in the same column and figures followed by different capital letters (A, B) in different columns show significant difference.

Table 6. The total fungal contamination of cocoa beans during storage.

Cocoa beans		Total fungal contamination (log CFU/g) during storage		
		0 Month	1 Month	2 Months
With *L. plantarum* HL-15	Vacuumed PP plastic 0.8 mm in thickness	2.0^{aA}	2.16^{aA}	3.14^{aB}
	Non-vacuumed PP plastic 0.8 mm in thickness	2.18^{aA}	2.38^{aA}	3.51^{aB}
	Plastic container	2.0^{aA}	2.32^{aA}	3.27^{aB}
	Vacuumed PP plastic 0.8 mm in thickness stored within a plastic container	ND	ND	3.25^{aB}
	Nylon sack	2.22^{aA}	2.54^{aA}	3.26^{aB}
Without *L. plantarum* HL-15	Vacuumed PP plastic 0.8mm in thickness	2.53^{aA}	3.16^{aA}	3.41^{aB}
	Non-vacuumed PP plastic 0.8 mm in thickness	2.0^{aA}	2.52^{aA}	3.21^{aB}
	Plastic container	2.0^{aA}	2.05^{aA}	3.35^{aB}
	Vacuumed PP plastic 0.8 mm in thickness stored within a plastic container	2.0^{aA}	2.18^{aA}	3.0^{aB}
	Nylon sack	2.26^{aA}	2.34^{aA}	3.42^{aB}

Remark: Figures followed by different lowercase letters (a, b, c) in the same column and figures followed by different capital letters (A, B) in different columns show significant difference.

The primary role of vacuum packaging is to remove oxygen to help prevent food spoilage and to minimize the absorption of water vapor by the packaged product. This is related to the amount of free water (aW) in food products that can be used by microorganisms such as fungi and bacteria to grow [36].

4. Conclusion

This research concluded that the addition of *Lactobacillus plantarum* HL-15 starter during the fermentation of cocoa beans did not have a statistically significant effect on the growth of fungi that contaminated the dried cocoa beans during the two months of storage. Packaging with 0.8 mm PP plastic that was vacuumed and put into a plastic container had a significant effect on the value of peroxide but did not have a significant effect on the total fungus contamination during the 2 months of storage.

Acknowledgments
The authors express their gratitude to the Indonesian Agency for Agricultural Research and Development that has supported funding for this work through a research project entitled "The assessment of Post-Harvest Technology of Cocoa for Increasing Added Value Supports the Development of ATP (Agricultural Technology Park) of Nglanggeran".

References
[1] Statistics Indonesia 2017 *Indonesian Cocao Statictic* (Jakarta: Badan Pusat Statistik)
[2] Statistics Indonesia 2018 *Indonesian Cocao Statictic* (Jakarta: Badan Pusat Statistik)
[3] Akinfala T O, Houbraken J, Sulfok M, Adedeji A R, Odebode A C, Krska R and Ezekiel C N 2019 Moulds and their secondary metabolites associated with the fermentation and storage of two cocoa bean hybrids in Nigeria *Int. J. Food Microbiol.* **316** 108490
[4] Fossi B, Ngah B, Nchanji G, Wanji S and Ndjouenkeu R 2016 Lactobacilli cultures against ochratoxin a producing moulds isolated from cocoa in the south west region of Cameroon *Br. Microbiol. Res. J.* **15(1)** 1–16
[5] Ho V T T, Zhao J, Srzednicki G and Fleet G 2013 *Int. Symp. on Agri-Foods for Health and Wealth* pp 83–91
[6] Hatmi R U, Kobarsih M and Cahyaningrum N 2015 *Int. Symp. on Food and Agro-biodiversity* (ISFA 2014) 3 pp 371-82
[7] Marwati T 2017 *Proc. of the Int. Conf. on Food Science and Nutrition* (Kinabalu: Faculty of Food Science and Nutrition Universiti Malaysia Sabah) p 5
[8] Marwati T, Khusna R N B, Djaafar T F and Rahayu E S 2017 *Proc. of the 15th Asean Conf. on Food Science and Technology* (Ho Chi Minh City: Science and Technology Publishing House) 2 pp 295-302
[9] Marwati T, Lamadoken I P P, Djaafar T F, Utami T and Rahayu E S 2018 Production of dried *Lactobacillus plantarum* HL-15 culture for inhibition growth of mycotoxin producing fungi *JAST B* **8** 396-403
[10] Marwati T, Puspitarum M C, Djaafar T F, Utami T and Rahayu E S 2019 *Pros. Sem. Nas. Hasil Penelitian Pangan dan Hasil Pertanian 2018* (Departemen Teknologi Pangan dan Hasil Pertanian Fakultas Teknologi Pertanian Universitas Gadjah Mada) pp 304-16
[11] Marwati T, Djaafar T F, Setiawan E E, Utami T and Rahayu E S 2020 *Proc. of 10th Asian Conf. of Lactic Acid Bacteria* vol. 2 (Digital Press Life Sciences UGM Digital Press)
[12] Purwaningsih, Marwati T and Dfaafar T F 2019 *Nilai Tambah Biji Kakao Fermentasi dengan Perlakuan Penambahan Starter Kering* (Research Fair Unisri - Strategi Hilirisaasi Hasil Penelitian untuk Meningkatkan Daya Saing Bangsa dalam Era Revolusi Industri 4.0) 3(1) pp 676-82
[13] Ruggirello M, Nucera D, Cannoni M, Peraino A, Rosso F, Fontana M, Cocolin L and Dolci P 2019 Antifungal activity of yeasts and lactic acid bacteria isolated from cocoa bean fermentations *Food Res. Int.* **115** 519-25
[14] Widianto D, Pramita A D and Wedhastri S 2013 Perbaikan proses fermentasi biji kakao kering dengan penambahan tetes tebu, khamir dan bakteri asam laktat *J. Teknosains* **3(1)** 1-80
[15] Kresnowati M T A P, Suryani L and Affifah M 2013 Improvement of cocoa beans fermentation by Lab. starter addition *J. Med. Biol. Eng.* **2(4)** 274-78
[16] Hernani and Haliza W 2017 Optimasi komposisi nutrien pembentukan komponen citarasa pada fermentasi biji kakao asalan *J. Pascapanen* **10(2)** 74-82

[17] Misnawi, Jauhari A and Budi I A 2016 *Aplikasi Agensia Hayati Lactobacillus fermentum untuk Menghasilkan Kakao Berkualitas Baik dan Bebas Mikotoksin* (Laporan Kegiatan KKP3N Pusat Penelitian Kopi dan Kakao Jember)

[18] Rahayu E R, Marwati T, Djaafar T F, Hatmi R U and Budiwati 2016 *Teknologi Pengendalian Kontaminan Mikotoksin Menggunakan Bakteri Asam Laktat untuk Meningkatkan Mutu Biji Kakao* (Laporan kegiatan KKP3N Universitas Gadjah Mada Yogyakarta)

[19] Marwati T *et al* 2017 *Pengkajian Teknologi Pascapanen Kakao untuk Peningkatan Nilai Tambah* (Laporan Akhir Tahun BPTP Yogyakarta)

[20] Triyadi R 2017 *Penghambatan Pertumbuhan Jamur dan Sintesis Okratoksin A menggunakan Lactobacillus plantarum Hl-15 selama Fermentasi Biji Kakao* (*Theobroma cacao* Linn.) (Thesis Master Fakultas Teknologi Pertanian Universitas Gadjah Mada Yogyakarta)

[21] Djaafar T F, Monika D C, Marwati T, Triwitono P and Rahayu E S 2020 *Proc. of 10th Asian Conference of Lactic Acid Bacteria* (UGM Digital Press 2)

[22] Asrul 2009 Populasi jamur mikotoksigenik dan kandungan aflatoksin pada beberapa contoh biji kakao (*Theobroma cacao* L) asal Sulawesi Tengah *Agroland* **16** 258–67

[23] Moulina M N, Setyabudi S, Salleh B and Rahayu E S 2014 Penicillium species from cocoa, coffee beans, and dried cassava in Yogyakarta Indonesia and their ochratoxin production *Indonesian Food and Nutri. Prog.* **13(1)** 1-10

[24] Dano S D, Manda P, Dembele A, Abla A M K, Bibaud J H, Gouet J Z and Sika C B Z 2013 Influence of fermentation and drying materials on the contamination of cocoa beans by ochratoxin A. *Toxins* **5** 2310-23

[25] Nugroho A D, Setyabudi S, Salleh B and Rahayu E S 2013 Ochratoxigenic black *Aspergilli* isolated from dried agricultural products in Yogyakarta *Indonesian J. Food Sci. Eng.* **3** 472-80

[26] Owusu S K 2013 *Identification of the Critical Points for Polycyclic Aromatic Hydrocarbon Contamination Along The Cocoa Processing and Storage Chain In Ghana* (Thesis Submitted to the Department of Materials Engineering Kwame Nkrumah University of Science and Technology in Partial Fulfillment of the Requirements for the award of Master of Science Degree in Environmental Resources Management)

[27] Jonfia-Essien W A and Navarro S 2012 Effect of storage management on free fatty acid content in dry cocoa beans *J. Life Sci.* **6(4)** 398-404

[28] Pontillon J 1998 Le beurre de cacao et les matières grasses en chocolaterie *Cacao, Chocolate, Production, Utilisation, Caractéristiques Techniques et Documentation* (Paris: Lavoisier) pp 257-69

[29] Minifie B W 1989 *Chocolate, Cocoa and Confectionery* 3rd ed (Science and Technology) p 885

[30] Dumadi S 2011 The moisture content increase of dried cocoa beans during storage at room temperature *JITE* **1(12)** 45–54

[31] Yanti N A, Jamili and Susilowati P E 2014 *Pros. Semi. Nas. Biologi 2014* (Biologi - Penelitian, Pengembangan dan Pembelajarannya) pp 450-58

[32] Siregar A Z 2007 *Kakao yang Nikmat Sulit Dirawat* (Medan: Universitas Sumatera Utara Press.)

[33] Ndukwu M 2009 Effect of drying and drying air velocity on the drying rate and drying constant of cocoa bean.agricultural engineering international *The CIGR E. J. Manuscript* **XI**

[34] Afoakwa E, Budu A S, Brown H M, Takrama J F and Akomanyl 2014 Changes in biochemical and physico-chemical qualities during drying of pulp pre-conditioned and fermented cocoa (*Theobroma cacao*) beans *J Nutrition Health Food Sci.* **2(3)** 1–6

IOP Conf. Series: Materials Science and Engineering **980** (2020) 012032 doi:10.1088/1757-899X/980/1/012032

[35] Herlina R, Indra T M and Esti R S 2016 *Kandungan Komponen Asam Lemak Biji Kakao (Theobroma cacao* L.*) Hasil Fermentasi dan Non Fermentasi* (Bandung: Posiding Farmasi)
[36] Eskin M and Robinson D S 2010 *Food Shelf Life Stability, Chemical, Biochemical and Microbiological Changes* (USA-Taylor and Francis)

IOP Conf. Series: Materials Science and Engineering **980** (2020) 012033 doi:10.1088/1757-899X/980/1/012033

Nixtamalization application to shelf life of corn flour

N Musita

Bandar Lampung Industrial Research and Standardization Center, Jl. By Pass Soekarno-Hatta Km.1 Rajabasa. Bandar Lampung. Indonesia

Email: nantimusita@gmail.com

Abstract. Corn is very potential to be developed as a source of flour and starch for the raw material of the food industry and others, but the functional properties of corn flour are less favored in its application to processed products because the lack of texture and the durability of flour is lack due to the fat content of corn. One way to improve it is by the process of nixtamalization. The purpose of this study was to evaluate the effect of nixtamalization on aflatoxin content and shelf life of corn flour. The treatments used were long immersion in water or without nixtamalization (W1), and soaking time in a solution of $Ca(OH)_2$ consisting of 8 hours (W2), 16 hours (W3), 24 hours (W4), and 32 hours (W5), washed. and ground to form a smooth dough (mass). The experiments were arranged with a single treatment, with variation in the duration of immersion in lime solution (control, 8, 16, 24, and 32 hours). This treatment was repeated 3 times. The results showed a decrease in level of type B1 aflatoxin with increasing immersion time, namely 85 μg/kg (0 hours), 69 μg/kg (8 hours), 62 μg/kg (16 hours), 55 μg/kg (24 hours), and 52 μg/kg (32 hours). The shelf life of corn flour through observation of critical moisture content is 426 days (0 hours), 430 days (8 hours), 489 days (16 hours), 408 days (24) hours, and 462 days (32 hours).

1. Introduction

Corn is one of the strategic commodities and has economic value as well as having opportunities to be developed because of its position as the main source of carbohydrates and protein which substitutes rice [1]. Indonesia's corn production in 2018 is more than 28.61 million tons and the growth of Indonesia's corn production is predicted to be a surplus until 2021, which is 32.65 million tons. While corn consumption in 2018 is predicted to reach 20.35 million tons. The amount consists of animal feed consumption of 14.27 million tons and household consumption and the needs of the food industry weighing 6.08 million tons [2]. Therefore, increasing production must also increase by using corn by the community.

Corn has long been used as a staple food in several regions in Indonesia, and processed and consumed as snacks. Corn can also be stored or processed into corn flour. However, corn flour when used as a raw material in food processing cannot produce good rheological quality, especially in the texture. In addition, so far the consumption of corn as a staple food is still identified with a low standard of living. Corn is reported to contain carotenoids both function as pro-vitamin A and non-pro-vitamin A, both of which are functional because they have an important beneficial role in maintaining health [3].

In addition to the importance of conducting studies on the potential of processed corn products as a functional food, another important thing that must be considered is the food safety problem of the processed corn products. Corn is one of cereals products which is easily infected by mycotoxins, especially those producing aflatoxin [4]. Publications on techniques to improve the quality of corn flour, aflatoxin content and techniques to reduce the aflatoxin content of corn-based food products are still rarely obtained.

In order to increase the potential of corn commodity, related to the planned diversification and national food security program based on local commodities, it is necessary to study more deeply about the processing of products based on corn starch, such as corn noodles, high-calorie fibrous biscuits, and tortilla chips.

Corn is very potential to be developed as a source of starch for the basic ingredients of the noodle industry and high-fiber biscuits, but the process of extracting corn starch requires a lot of water availability, requires heat energy for drying, therefore utilization as corn flour can be used as an alternative for corn utilization with a simpler process to be developed commercially at the household or medium industry level. However, the functional properties of natural corn meal are less favored in its application to processed products because the resulting texture is not good, as well as the durability of flour which is not durable due to the fat content of corn.

To produce a good texture requires starch that has low viscosity. One way to reduce viscosity is by the process of nixtamalization. According to Mendez-Montealvo [5], nixtamalization process, namely boiling corn grains in lime water followed by soaking for several hours, washed and milled to form a dough (mass). Corn processing through the process of nixtamalization reportedly very beneficial because it can help provide a safe and quality food. These advantages include increasing the product crispness, increasing the availability of niacin and protein digestibility and reducing the content of pathogenic bacteria [6,7,8]. It was further reported that with or without spontaneous fermentation, nixtamalization could be used to improve the functional properties of corn which could ultimately improve the rheological properties of corn flour.

Some researchers report that corn is one of cereals products which is easily infected by mycotoxins [9,10]. The dominant mycotoxin in maize is aflatoxin [4]. Aflatoxin has long been proven to be toxic and carcinogenic to humans [11]. The aflatoxin content in cereal-based foods can be reduced through cooking under alkaline pH conditions [12] and the fermentation process [13,14]. This study was aimed at evaluating the effect of nicstamalization on shelf life and aflatoxin content of corn flour.

2. Material and methods

2.1. Materials and equipments

The main raw material used was dried corn grain DK hybrid variety, yod solution, $Ca(OH)_2$, dimethyl sulfoxide (DMSO), ion free water, aquades, methylene chloride, diethyl ether, methanol, NaOH (aw 0.06), $MgCl_2$ (aw 0.32), K_2CO_3 (aw 0.44), KI (aw 0.69), NaCl (aw 0.75), KCl (aw 0.84), $BaCl_2$ (aw 0.90), $K_2Cr_2O_7$ (aw 0.98), selenium, H_2SO_4, Zn, and 67% NaOH solution, H_3BO_3, phenolptalin blue and red 1%.

Equipment used were pans, stoves, blander, scalecs, grinding machines, sieves, ovens, slide glass, hemacytometers, test tubes, vortices, volumetric tubes, and other supporting tools.

2.2. Research method

This research was conducted by identifying the shelf life and aflatoxin content of corn flour made by nixtamalization. The treatments used were long immersion in water or without nixtamalization (W1), and soaking time in a solution of $Ca(OH)_2$ consisting of 8 hours (W2), 16 hours (W3), 24 hours (W4), and 32 hours (W5).). The study was conducted with 3 replications. The samples were observed for shelf life and aflatoxin levels. Then the data were analyzed descriptively and the data were presented in tables and graphs.

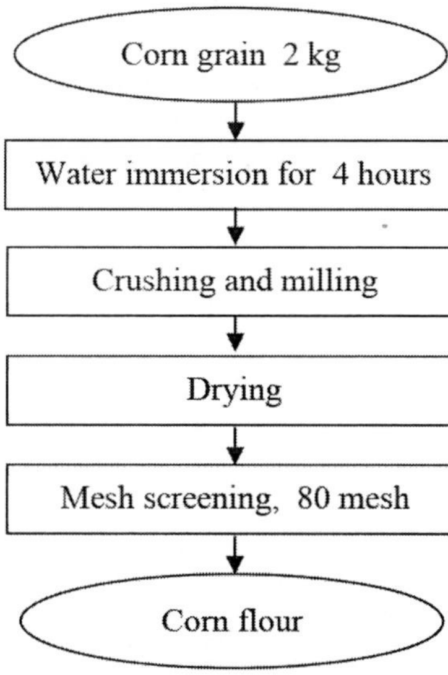

Figure 1. Production of corn flour without soaking in $Ca(OH)_2$ solution [15].

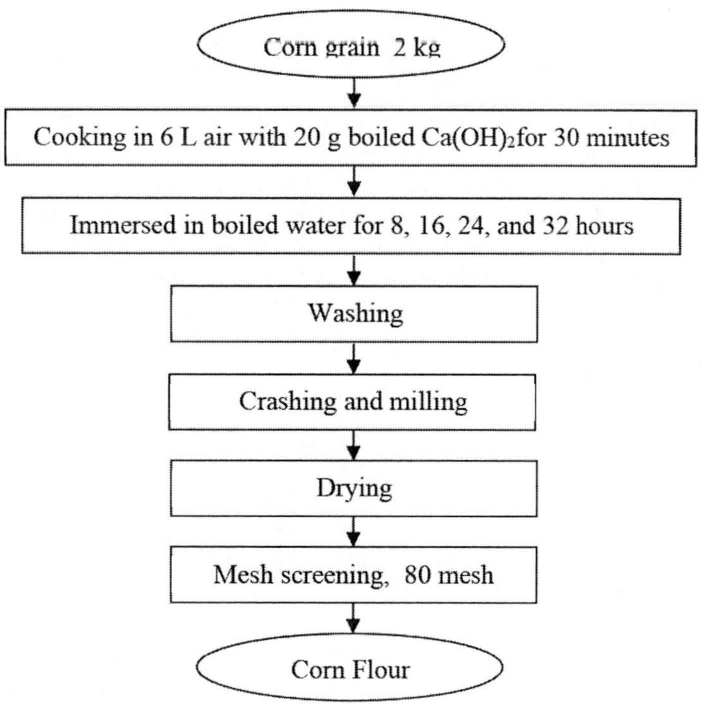

Figure 2. Production of corn flour with nixtamalization process[16].

IOP Conf. Series: Materials Science and Engineering **980** (2020) 012033 doi:10.1088/1757-899X/980/1/012033

2.2.1. Nixtamalization process. The process of nixtamalization was carried out according to Rooney and Serma-Saldivar (2003) with slight modifications to the number of samples and the steps involved. A total of 2 kg of dry shelled corn was put into a pot containing 6 liters of water and 20 g of $Ca(OH)_2$ to boil for 30 minutes. Then immersed in boiling water for 8 hours (W2), 16 hours (W3), 24 hours (W4), and 32 hours (W5). Next, the corn was washed with water to remove the remaining $Ca(OH)_2$ contained in the corn kernels. Meanwhile, for control, flour was processed by means of corn in water for 4 hours, then washed and drained.. Washed corn was then ground until crushed with a grinding machine. The corn which had been instilled and ground was then dried to be made into flour. Flour drying could be done at a temperature of 55°C to a moisture content of about 5%. Then the flour was mashed and sieved in an 80 mesh filter. The flow chart of the process of making corn flour without nixtamalization and by means of nixtamalization can be seen in figure 1 and figure 2.

2.3. *Observation*

2.3.1. Shelf life analysis [17]. The shelf life of corn flour was determined by the Accelerated Shelf Life Testing (ASLT) method with a critical water content model approach. The main principle of this model is to determine the equilibrium moisture content of cornmeal stored at various RH.

$$\theta gain = \frac{\ln\left(\frac{me - mi}{me - mc}\right)}{\frac{k}{x}\frac{A}{Ws}\frac{Po}{b}} \quad (1)$$

Information :

$\theta gain$	= Estimated time of shelf life (days)
me	= equilibrium moisture content (% bk)
mi	= initial moisture content (% bk)
mc	= critical moisture content (% bk)
Ws	= Dry weight of material (g)
A	= The surface area of the package (m^2)
k/x	= Bottled water vapor permeability (g/m^2.hari.mmHg)
Po	= saturated water vapor pressure (mmHg)

a. Initial moisture measurement. The initial moisture content is the percentage of the original moisture content in a material. The initial moisture content of the corn starch product was determined at the beginning of storage. Dry clean empty cup in an oven at approx. 105°C. Cooled in a desiccator for about 15 minutes and weighed. A total of 2 grams of sample in a cup was put in an oven at 105°C for 6 hours until it reached a constant weight. The plate containing the sample was cooled in a desiccator and then weighed. The initial moisture content could be calculated using the formula:

$$Kadar\ air = \frac{A - B}{C} x\ 100\% \quad (2)$$

Information:
A = Plate and sample weight before drying (g)
B = Weight of the plates and samples after drying (g)
C = Weight of the sample before drying (g)

b. Measurement of critical water content. Critical water content is the amount of critical water content possessed by a product in its critical condition. The critical condition itself is defined as a condition where the product is at the limit of consumer acceptance, in other words it has begun to be rejected

[17]. The critical water content was obtained by storing the sample at room temperature with RH conditions of 76% using saturated NaCl solution. Periodically (every 24 hours) an organoleptic acceptance test was carried out on the appearance of the product. Every day the average acceptance score was calculated, until it reached a value of 2 (dislike), it was determined that the product was in a critical condition. The measurement of critical water content was carried out using the gravimetric method as was done in the initial water content measurement.

c. Determination of the isothermic sorption curve. Isothermic sorption curve is a curve that illustrates the relationship between water content in foodstuffs and water activity (aw) or relative humidity (RH) of storage in the storage room. Isothermic sorption curve obtained by storing corn samples into a humidity chamber which has a certain aw. Furthermore, an amount of water was added and stirred until saturated. The humidity chamber was closed and left for 24 hours at a temperature of 30°C. The packaged product (as much as 5 grams) was hung in the humidity chamber containing a saturated salt solution. Samples were weighed periodically (every 24) hours until a constant weight obtained (meaning that the equilibrium water content has been reached). Samples that had reached constant weight were measured for moisture using the oven method and expressed on a dry basis as in the initial water content measurement. Isothermic sorption curves were created by plotting water content and balance water activity. Water activity (aw) was calculated by dividing the RH value of each humidity chamber by 100.

d. Determination of supporting parameters. The supporting parameter is a variable that is considered to influence the process of estimating shelf life. Parameters must be known because they are very influential in the process of estimating shelf life. The supporting parameters include:
- Packaging permeability value $(\frac{k}{x})$, obtained from literature references
- The value of saturated vapor pressure (Po) at 30^{0}C is obtained from the Labuza table
- The value of b (slope of the curve) is obtained from the gradient of the selected isothermic sorption equation model
- The value of the cross-sectional area (A) is obtained by multiplying the dimensions of the package
- The total solids value (Ws) is obtained by correcting for the total weight of the sample minus the initial moisture content.

2.3.2. Analysis of aflatoxin content. Analysis of aflatoxin levels B1, B2, G1, G2 was carried out on all treatments of nixtamalized corn flour. using the TLC Desitometer (Thin Layer Chromatopgraphy Desitometer) method. Densitometric TLC is a method analysis which is still being used up to this time. Aflatoxin analysis was performed using the TLC plate stationary phase silica gel 60 F_{254} size 20 x 10 cm with chloroform-ethyl acetate as mobile phase (7: 3). Detection and quantitation were carried out using a TLC scanner densitometry, fluorescence detector, at maximum excitation wavelength 354 nm and 400 nm of emissions.

This method has a detection limit (LOD) for aflatoxins B1 is 9.62 pg and for aflatoxin G1 amounting to 10.9 pg. Meanwhile, limits quantitation (LOQ) for aflatoxin B1 and G1 respectively amounted to 32.08 pg and 36.41 pg [18].

3. Results and discussion

3.1. Shelf life analysis
The results of the observations on the variables needed in estimating the shelf life of corn starch that were stabilized with different soaking times are presented in table 1. In table 1, it can be seen that there was a change in the value of the estimating factors for the shelf life of corn flour due to the difference in immersion time in the $Ca(OH)_2$ solution, namely an increase in the initial water content, critical water content and equilibrium water content. During storage in various RH conditions there would be interactions between the product and its environment. Water vapor would move from the

environment to the product or vice versa until equilibrium was reached. This water vapor transfer occurred as a result of differences in RH environment and product, where water vapor would move from high RH to low RH.

Table 1. Value of the factors in the treatment.

Faktor-faktor	Nilai				
	W1	W2	W3	W4	W5
Initial moisture content (%)	6,9986	7,0134	7,0602	8,1712	8,3217
Critical moisture content (%)	12,1415	11,6292	11,5738	11,4917	11,9250
Balanced moisture content (%)	12,9598	12,2288	11,9646	11,9047	12,2518
Isothermic curve slope	14,67	13,64	13,23	12,84	12,91
Packaging permeability (H_2O/m^2.hari.mmHg)	0,5	0,5	0,5	0,5	0,5
Packing area (m^2)	0,012	0,012	0,012	0,012	0,012
Dry weight (g)	2,7900	2,7895	2,7881	2,7548	2,7503
Saturated vapor pressure 30°C (mmHg)	31,824	31,824	31,824	31,824	31,824
Estimated shelf life (days)	426	430	489	408	462

The achievement of equilibrium conditions between the sample and the environment was indicated by a constant sample weight. According to [19], the water content of equilibrium in a food material will be achieved, which is indicated by the constant weight of the material. The weight of the material is said to be constant if the difference in weight between three consecutive weights is not more than 2 mg/g for RH conditions ≤ 90% and not more than 10 mg/g for RH conditions> 90%.

In this study, the samples stored at RH 7 had a tendency to decrease the weight of the material. While the samples stored at RH 7, 32, 69.76, 84, 90, and 98 tended to increase the weight of the material. This was because at RH 7 the sample underwent a desorption process while on other RH the sample underwent absorption. The desorption process is the process of releasing material moisture into the environment and absorption is the process of absorbing water vapor by materials from the environment. The equilibrium moisture content obtained from each experiment is plotted with the aw or RH value of the environment, thus forming a curve called an isothermic sorption curve.

Table 1 also shows that the higher the slope value of the isotherm curve, the dry weight of the sample, the difference between the equilibrium moisture content value and the initial moisture content, the longer the shelf life of a product. Meanwhile, the higher the difference between the value of equilibrium water content and critical water content causes a decrease in shelf life. Because of the same packaging permeability, package area, and saturated vapor pressure, the most influential factors are the interaction between the initial moisture content, the equilibrium water content, the critical water content, the sample weight, and the isothermic curve slope. According to [20], the gradient or slope of the curve (b) is determined from the straight line formed on the curve of the isothermic sorption equation model that is formed.

Other supporting parameters that must be known are packaging permeability $(\frac{k}{x})$, packaging surface area (A), and pure vapor pressure on storage. Each type of packaging has a different water vapor permeability. Packaging permeability is the speed of water vapor transmission through a unit area of material with a certain thickness due to differences in water vapor pressure between the product and the environment at a certain temperature and humidity [21].

In table 1, it can be seen that the shelf life of maize flour with treatment W1 to W5 was 426 days, 430 days, 489 days, 408 days, and 462 days, respectively. The shelf life of the coating flour based on corn flour was 209 days, calculated through the critical water content approach [22]. When compared with the research of [22], the shelf life of corn flour with a critical moisture content approach, this study had a slightly longer shelf life. This might be due to differences in initial moisture content, packaging permeability, and packaging area.

Based on these data, it can be seen that although the nixtamalization process was able to reduce the fat content of corn flour [23], the fat content did not affect the estimated shelf life of corn flour from each treatment using the critical water content approach method. This is because in the process of determining the shelf life of corn flour through the critical water content approach, the determining factor is the ability of the product to absorb water from the environment. According to [17], models of critical moisture content are usually used for food products that are relatively easily damaged due to absorption of moisture from the environment. Arrhenius method may be more suitable to see the shelf life of corn flour which is influenced by the fat content contained in corn. Arrhenius model is generally used to estimate the shelf life of food products that are sensitive to changes in temperature, including food products that are prone to rancidity (due to fat oxidation), changes in color by browning reaction, or damage to vitamin C. Examples of products that can be determined its shelf life with the Arrhenius model is flour, nuts, and other products that contain high fat or contain reducing sugars and proteins that enable fat oxidation or browning reactions [26].

3.2. Aflatoxin content

The results of the analysis of aflatoxin content of corn flour (table 2) showed a decrease in aflatoxin content with increasing immersion time. In corn flour W1 (control) showed the highest content of aflatoxin, then decreased in the soaking time W2 (8 hours), W3 (16 hours), W4 (24 hours), and W5 (32 hours). This shows that the process of nixtamalization can reduce aflatoxin content in corn flour.

Table 2 also shows that the aflatoxin content of corn flour from the five treatments were quite high, exceeding the minimum content required by SNI, which was a maximum of 50 ppb for aflatoxin B1. Content of aflatoxin B2 (< 1), G1 (< 4), and G2 (< 1) were also positive, although relatively low. This situation was caused because the corn samples used were no longer good and has been stored for a long time, so that the process of nixtamalization could not eliminate the aflatoxin. According to [25], aflatoxin has a stable nature of physical, chemical, heat-stable treatment, the melting point is quite high, namely above 237°C.

Table 2. The content of aflatoxin in nixtamalization corn flour at different immersion times.

Aflatoxin	B1 (μg/kg)	B2 (μg/kg)	G1 (μg/kg)	G2 (μg/kg)
W1	85	< 1	< 4	< 1
W2	69	< 1	< 4	< 1
W3	62	< 1	< 4	< 1
W4	55	< 1	< 4	< 1
W5	52	< 1	< 4	< 1

Corn quality requirements in SNI 4483: 2013, states that the maximum aflatoxin content for corn as animal feed is Quality I 100 ppb and Quality II is 150 ppb. If referring to the BPOM RI Regulation on the maximum limit of aflatoxin in food, the level exceeds the maximum level required, which is 20 ppb for aflatoxin B1 and total aflatoxin 35 ppb in seeds, flour and corn starch, as well as corn flakes. WHO, FAO and UNICEF set a limit on the aflatoxin content of carbohydrate sources of food no more than 30 ppb, while the European Commission set a maximum limit of 4 ppb total aflatoxin for cereal products [26].

Aflatoxins are secondary metabolites of *Aspergillus flavus* and *Aspergillus parasiticus* which are found in corn, peanuts, soybeans, rice and other agricultural commodities starting from planting, harvesting, storing in storage or processing. The climatic conditions in Indonesia strongly support the growth of *Aspergillus sp* and the formation of aflatoxins. There are at least 13 types, the most important are B1, B2, G1, G2, M1, and M2. The type of aflatoxin is based on its intensity with UV light, namely B1 and B2 (which gives blue color).

IOP Conf. Series: Materials Science and Engineering **980** (2020) 012033 doi:10.1088/1757-899X/980/1/012033

Aflatoxin G1 and G2 are only produced by A. Parasiticus, which gives a green color. The four aflatoxins are found together in varying proportions in corn kernels. Aflatoxin B1 is produced by both species, the most toxic compounds, potentially stimulating cancer, especially liver cancer. The mildest toxin attack is mild abrasions due to tissue death (necrosis). Exposure to high levels can cause cirrhosis, carcinoma of the liver, and digestive disorders, absorption of food ingredients, and nutrient metabolism. This toxin in the liver will be reacted to epoxide which is very reactive to the compounds in the cell. Carcinogenic effects occur because the base N guanine on DNA will be bound and disrupt the work of genes.

Aflatoxin M1 and M2 are the metabolic products of aflatoxin B1 and B2 detected in cow's milk, whose feed contains aflatoxin B1 and B2, whose danger levels are lower than their initial form. The negative impact of aflatoxin on body health is quite significant, various reports mention that aflatoxin is carcinogenic and teratogenic, and consumption of food contaminated with low concentrations of aflatoxin that can continuously damage the liver and reduce the immune system in the body.

4. Conclusions

The results showed a decrease in level of type B1 aflatoxin with increasing immersion time, namely 85 μg/kg (0 hours), 69 μg/kg (8 hours), 62 μg/kg (16 hours), 55 μg/kg (24 hours), and 52 μg/kg (32 hours). The shelf life of corn flour through observation of critical moisture content was 426 days (0 hours), 430 days (8 hours), 489 days (16 hours), 408 days (24) hours, and 462 days (32 hours).

References

[1] Anonim 2010 *Pencapaian Produksi Jagung di Lampung* (Retrieved from www.translampung.com)

[2] Kementerian Pertanian 2018 *Produksi Jagung Diprediksi Surplus Hingga 2021* (Pusat Data dan Informasi Kementerian Pertanian Republik Indonesia)

[3] Menkir A, Liu W, White W S, Mziya-Dixon B and Rocheford T 2008 Carotenids diversity in tropical-adapted yellow maize inbred lines *Food Chem.* **109** 521-29

[4] Brera C, Catano C, De Santis B, Debegnach F, De Giacomo M, Pannunzi E and Miraglia M 2006 Effect of industrial processing on the distribution of aflatoxins and zearalenone in corn-milling fractions *J. Agri. Food Chem.* **54(14)** 5014-9

[5] Mendez-Montealvo G, Shanchez-Rivera M, Paredes-Lopez O and Bello-Perez L A 2006 Thermal and rheological properties of nixtamalized maize starch *Int. J. of Bio. Macromol.* **40** 59-63

[6] Sefa-Dedeh S, Cornelius B, Amoa-Awua W, Sakyi-Dawson E and Afoakwa E O 2004 The microflora of nixtamalized corn *Int. J. of Food Microbio.* **96** 97-102

[7] Bharati P and Vaidehi M P 1989 Enrichment of sorghum in grains *Food and Nutri.Bul.* **11(2)** 53-5

[8] Vivas N E, Waniska R D and Rooney L W 1987 Thin porridges (atole) prepared from maize and shorgum *Cereal Chem.* **64(6)** 384-9

[9] Fernandez-Ibanez V, Soladado A, Martinez-Fernandez A and Roza-Delgado B 2009 Application of near infrared spectroscopy for rapid detection of aflatoxin B1 *Food Chem.* **113** 629–34

[10] Fandohan P, Hell K, Marasas W F O and Wingfield 2008 Infection of maize by Fusarium spesies and contamination with fumonisin in Africa *Afri. J. of Biotech.* **2(12)** 570-9

[11] IARC 1993 International agency for research on cancer. somenaturally occurring substances: Food items and constituents, heterocyclic aromatic amines and mycotoxins. *Risk. Chem. Hum.* **56** 359-444 (IARC Monogr. Eval. Carcinog)

[12] Torres P, Guzmn-Ortiz M and Ramrez-Wong B 2001 Revising the role of pH and thermal treatments in aflatoxin content reduction during the tortilla and deep frying processes *J. Agri. Food Chem.* **49(6)** 2825-9

[13] Shetty P H, Hald B and Jepersen L 2007 Surface binding of aflatoxin B_1 by *Saccharomyces*

strains with potential decontaminating abilities in indigenous fermented foods. *Int. J. of Food Microbiol.* **113** 41-6
[14] Shetty P H and Jespersen L 2006 Saccharomyces cerevisiae and lactic acid bacteria as potential mycotoxin decontaminating agents *Trends in Food Sci. and Techno.* **17** 48-55
[15] Suarni 2009 Prospek pemanfaatan tepung jagung untuk kue kering (*cookies*) *J. Litbang Pertanian* **28(2)** 63-71
[16] Rooney L W and Serma-Saldivar S O 2003 Food use of whole corn and dry milled fraction In White P J and Johnson L A *Corn: Chemistry nd Technology* Ed d Secon (Minnesota: American Association of Cereal Chemists Inc.) 495-536
[17] Labuza T P 1982 *Shelf Life Dating of Food. Food and Nutritions* (Press Inc. Connecticut) In Sugiyono, Setiawan E, Syamsir E and Sumekar H 2011 Pengembangan produk mi kering dari tepung ubi jalar (*Ipomoea batatas*) dan penentuan umur simpannya dengan metode isoterm sorpsi *J. Tekno. dan Ind. Pangan* **22(2)** 164-9
[18] Yohanes A 2005 *Optimasi dan Validasi Metode Kromatografi Lapis Tipis Densitometri untuk Penetapan Kadar Aflatoksin pada Makanan yang Mengandung Kacang Tanah* (Skripsi Universitas Indonesia)
[19] Lievonen S M and Ross Y H 2002 Nonenzymatic browning in amorphous food models: Effects of glass transition and water. *J. of Food Sci.* **67(6)** 2100-6
[20] Rahayu W P and Arpah 2003 *Penuntun Teknis: Penetapan Kadaluarsa Produk Industri Kecil Pangan* (Departemen Teknologi Bogor)
[21] Robertson G L 1993 *Food Packaging Principles and Practices* (New York: Marcell Dekker Inc.)
[22] Sugiyono, Setiawan E, Syamsir E and Sumekar H 2011 Pengembangan produk mi kering dari tepung ubi jalar (*Ipomoea batatas*) dan penentuan umur simpannya dengan metode isoterm sorpsi *J. Tekno. dan Ind. Pangan* **22(2)** 164-169
[23] Musita N 2018 Kajian kadar aflatoksin dan proksimat tepung jagung nikstamalisasi pada berbagai lama perendaman *Pros. Sem. Nas. I Hasil Litbangyasa Industri* (Palembang)
[24] Kusnandar F 2006 *Disain Percobaan dalam Penetapan Umur Simpan Produk Pangan dengan Metode ASLT (Model Arrhenius dan Kadar Air Kritis)* (Bogor: Modul Pelatihan Pendugaan dan Pengendalian Umur Simpan Bahan dan Produk Pangan 7-8 Agustus 2006)
[25] Villaeli A 2010 *Ragam Aflatoxin Sebagai Salah Satu Cemaran Alamiah Bahan Pangan* (Departement of Chemistry University Negeri Yogyakarta)
[26] Visconti A 1998 *New European Union Regulation For Aflatoxin ini Food Stuffs Micotoxicology* (Newsletter) 4(2) 1

IOP Conf. Series: Materials Science and Engineering **980** (2020) 012034 doi:10.1088/1757-899X/980/1/012034

The assessment of good manufacturing practices (GMP) implementation and critical control point (CCP) determination on the cocoa powder processing in Agricultural Techno Park Nglanggeran, Yogyakarta

T F Djaafar[1]*, T Utami[2]*, T Marwati[1]*, P C Pramesi[2], R Wikandari[2] and E S Rahayu[2]*

[1] Department of Postharvest, Yogyakarta Assessment Institute for Agricultural Technology, Indonesia

[2] Department of Food Science, Faculty of Agricultural Technology, UGM Yogyakarta, Indonesia

Email: titiekfd1212@gmail.com

Abstract. Cocoa powder is a cocoa-based food product. Indonesian cocoa powder has a big potential in the ASEAN (*Association of Southeast Asian Nations*) market. Agricultural Agro Techno Park is a company that produces cocoa powder in Gunungkidul, Yogyakarta. This company realizes the importance of improving their product safety. Therefore, the research aimed to determine the safety quality of cocoa powder produced by Agricultural Agro Techno Park by assessing its Good Manufacturing Practice (GMP) implementation and determining the critical control points. The results of assessment of GMP implementation showed that it was categorized in Level IV (the lowest level). Thus, it is strongly recommended to do some corrective actions that identified some incompatibilities before planning the Hazard Analysis and Critical Control Point (HACCP). The flow chart assessment showed that there were two stages that might affect the cocoa powder quality. The stages were receiving raw material and the storage at a low temperature. From the determination of the critical control points which was conducted to improve the cocoa powder quality, the same two stages were also found as the critical control points that might affect the final product quality.

1. Introduction

Cocoa is one of the plantation commodities which has a big role in the Indonesian economy. Cocoa powder is one of the food products made of cocoa beans. Production of sweet cocoa powder in Indonesia in 2017 reached 26,011,959 kg while unsweetened cocoa powder reached 11,039,647 kg [1]. Indonesian cocoa powder has a high potential, especially in ASEAN (*Association of Southeast Asian Nations*) countries.

As one of the ASEAN countries, Indonesia needs to improve national and even international competitiveness for the quality of chocolate production. The Ministry of Industry has issued a lot of new policies including the obligation to apply the Indonesian National Standard for 134 industrial products, one of which is cocoa powder [2]. The fulfilment of Indonesian National Standard requires the implementation of a quality assurance system.

One of the quality assurance that can be applied is the implementation of Good Manufacturing Practices (GMP) in cocoa powder processing. The Good Food Production Method is an important factor in fulfilling the food quality and safety standards [3]. This guideline is implemented to meet the specified requirements to produce quality food products that consumers demand. Implementing a good food production method can be an essential requirement in implementing a Hazard Analysis and Critical Control Point (HACCP) for the product safety guarantee system. In this guarantee system, the hazard analysis and the determination of critical control points or control of a stage that can affect the product safety.

Gunungkidul Regency is one of the cocoa-producing district in Yogyakarta. Cocoa beans produced by farmers are a mixture of unfermented, under fermented (not fully fermented) and fermented (perfect fermented) cocoa beans so that the quality varies [4,5]. In Nglanggeran Village, Patuk Sub-district, Gunungkidul, the Agricultural Techno Park (ATP) processeed the fermented cocoa beans from local farmers into cocoa powder and candies. The factory's production capacity of 5 kg fermented cocoa beans was relatively small, but this factory realized the importance of a food quality assurance and a food safety system. The industry has implemented the good processing food production method and plans to apply HACCP as a food safety assurance system. However, they did not yet have two critical components which should be included in the requirements of good processing food production method namely, flow chart and process control. Based on this case, four research objectives were formulated in this current study, namely (1) to study the implementation of GMP, (2) to study each processing stage based on the flow chart as a first step in determining the standard procedure, (3) to review the intermediate products and the final product through analysis of the water content and water activity (AW) as an effort to control the process, and (4) to determine the critical control point (CCP) as the initial stage of making the HACCP plan for the cocoa powder production.

2. Material and methods

2.1. *Material*
The material used for this research was cocoa powder obtained from ATP Nglanggeran, Gunungkidul.

2.2. *Methods*
The assessment was conducted in 2017-2018, starting from the field observation, interview, and sampling at the production stages of cocoa powder including steaming, roasting, deshelling, compressing, and final cooling. The samples from each production stages were analyzed for their water content and water activity (Aw).

2.2.1. The assessment of general description and building layout. Observation and interview methods were used to review the general description of ATP [6]. The observation was carried out by directly observing the processing of cocoa powder at the research location and the factory environment. The interview was conducted to the employees to get a complete explanation related to the raw material, process, and administrative activities. The general descriptions included location, building, factory employees and cocoa powder processing. Meanwhile, the building layout included processing area layout, processing lines, and processing area based on risk.

2.2.2. The assessment of GMP. The GMP assessment was conducted based on the Regulation of the General Director of Agro Industry Number 30/IA/Per/12/2011 concerning The Technical Guidelines for Evaluating Good Processed Food Production Method Application.

2.2.3. The assessment of the processing and the flow chart. The flow chart was made based on the data from the observation and the interview. After that, the assessment to the processing was conducted in order to improve the process control.

2.2.4. Water content and water activity analyses for intermediate and final products. The sampling was carried out three times by taking several 500 g samples. The water content analysis was conducted based on [7] while the water activity analysis was done based on the method presented by [8].

2.2.5. Critical control point determination. Critical control point was determined on the basis of the CODEX Alimentarius, namely determination of product description, making and verification of flow chart, hazard analysis, and critical control point determination [9, 10]. The results of this water activity analysis were to determine the presence of potential hazards of microorganisms [11].

3. Results and discussion

3.1. General description and building layout assessment

Agricultural Techno Park (ATP) is a chocolate processing factory, located in Nglanggeran Village, Patuk District, Gunungkidul Regency. This factory is located in the tourism area of Embung and Purba Mount, Nglanggeran, so that it becomes a destination for tourists from various countries. It produces cocoa powder and Etawa goat milk powder, 3 in 1 cocoa powder, chocolate candies, chocolate *dodol*, and Etawa goat milk ice cream.

During the study, there were 13 employees who worked in the factory consisting of 10 process employees, two janitors and night guards, and an administrative staff. The process employees were divided into three groups: five people for processing chocolate candies and cocoa powder, two persons for processing of goat milk powder, and two persons for processing chocolate *dodol*. One employee helped the processing of all products and as a shopkeeper.

The processing of cocoa powder at ATP was not done every day. The processing was carried out either when the product was sold out or if there were orders from consumers. The employees worked from 8:00 a.m. to 04.00 p.m. for five working days. The lunch break was from 12.00 a.m. to 01.00. p.m. During the break time, all the equipment was turned off for the sake of the maintenance.

The ATP building consisted of two floors. The processing rooms and outlets were on the first floor, while the meeting room was on the second floor (figure 1). There were three rooms on the first floor, and the processing room doors were restricted to employee access. The first processing room was for the raw material storage, steaming, roasting and deshelling. The second processing room was for pasting, compression, crushing, grinding and sifting. In this room, there was a cooler for storing cake and cocoa powder. The third room was for product-packaging activities. The separation of the processing rooms was carried out based on a risk analysis where the first room as a low risk area while the second and the third rooms as a high-risk area. There was an additional process after deshelling, which was separating nibs from cocoa skin manually. On the second floor, there were two meeting rooms and facilities for visitors, such as a prayer room (*mushola*) and two toilets.

The cocoa powder processing line had been effective, putting the machines in order according to the processing stages. However, it would be better if the refrigerator was placed close to the crushing machine to shorten the material transfer. The refrigerator was also used for tempering chocolate candies so that it required a large space because the chocolate molds cannot be stacked. Therefore, is necessary to add shelves to the refrigerator so that the chocolate molds, cocoa cake and cocoa powder can be stored in the refrigerator.

3.2. The assessment of GMP

The assessment results showed that the Good Manufacturing Practices (GMP) implementation in the cocoa powder processing at ATP was in the level IV category with one major incompatibility, five serious incompatibilities and three critical incompatibilities. The detailed assessment results of the GMP implementation are explained as follows.

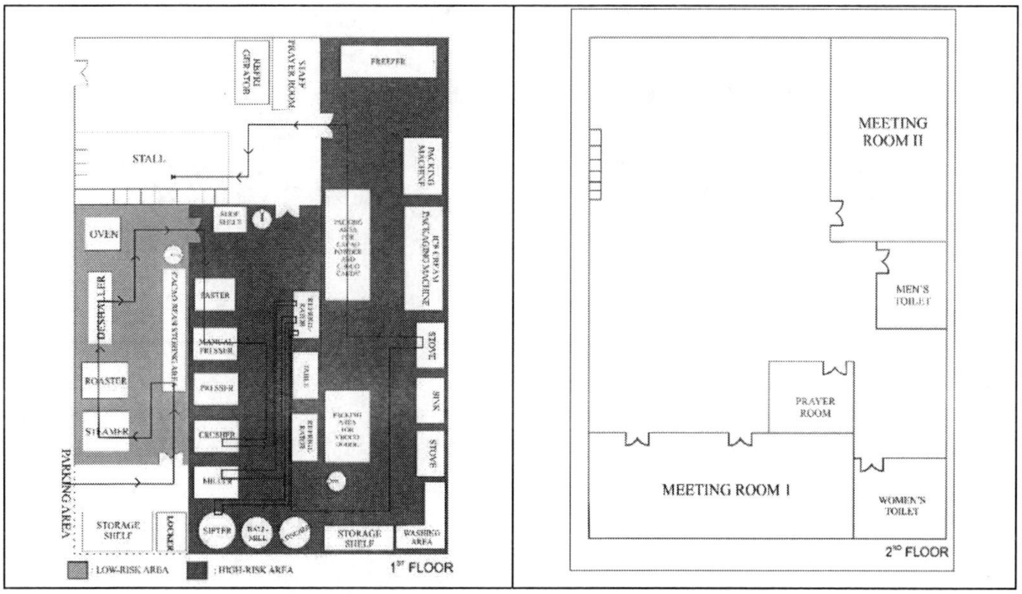

Figure 1. The ATP building layout in Nglanggeran, Gunungkidul, Yogyakarta.

3.2.1. Aspect 1: Location and factory environment. The location and environment around ATP production had fulfilled the requirements in the reference regulation. It was well-maintained and clean due to the regular cleaning (figure 2). The factory building was higher than the surroundings that made the wind did not bring much dust into the building, so that it made the building easier to clean up.

Figure 2. The ATP building in Nglanggeran, Gunungkidul, Yogyakarta.

3.2.2. Aspect II: Building and facilities. The ATP building was made of 60x60 cm ceramic tiles that had bright colors, gypsum plafond, and brick walls painted with bright colors, too. The building components had met the requirements for good food processing, easy to clean and to maintain.

Regarding the building facilities, here is the description. The building had five windows in the first room that could be opened (facing outside of the building) and one was closed (facing the second room). Besides, there were two doors in the first room (figure 3). One door faced outside of the building which was always opened because the additional processing stage after deshelling was done in that room. Another door connected to the second room but was always closed. Meanwhile, the two and the three processing rooms had the same number of windows, vents and doors (figure 3). There were two windows between the rooms that could not be opened, eight ventilations facing outside of the building and could be opened, and two doors to connect the processing rooms with the outlet which were always closed. The second and the third rooms were equipped with air conditioners.

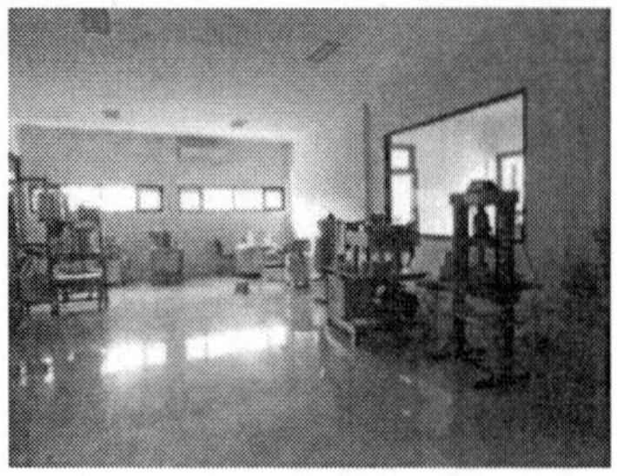

Figure 3. The first (left) and the second (right) processing rooms at ATP Nglanggeran, Gunungkidul, Yogyakarta.

However, from this aspect there were two serious incompatibilities with SPP-IRT regulation. The first incompatibility found in the factory building structure was the meeting point between the wall and the floor that formed a dead angle. This caused a lot of dirt and puddle in that space that was difficult to clean. A corrective action that can be done is by using a hospital plint (figure 4), a patch that can be put on a dead angle to form a basin so the space is easier to clean.

The second incompatibility was the dirty windows. The corrective action suggested is by cleaning the windows regularly. It can be done using a bristle brush covered with a patchwork so that the sharp part of the brush does not damage the surface of the wall.

(Anonim. -)

Figure 4. Hospital plint.

3.2.3. Aspect III: Equipment. The ATP equipment had fulfilled the requirements. All the equipment was clean and made of stainless steel. Besides, there was also a maintenance for the equipment, and the cleaning devices were available. All equipment was cleaned by the employees every day using hot water, detergent and alcohol. This cleaning activity was done before the working time ended.

3.2.4. Aspect IV: Water supply or clean water supply facilities. Agricultural Technology Park had an adequate supply of clean water coming from wells and stored in some water tanks. The use of this facility had been good and followed the reference regulation. The factory used clean water to clean the equipment and for toilet facilities. Meanwhile, the employees and the visitors used mineral water as the drinking water.

3.2.5. Aspect V: Hygiene and sanitation facilities. In this aspect, a serious incompatibility was found based on SPP-IRT regulation. There was an unavailability of several sanitation facilities, such as hand dryers and soaps for washing hands. Although hand-washing sinks were available in each processing room, soaps and hand dryers were not provided. In this case, one corrective action suggested is providing soaps and hand dryers. The dryers should be wiped dryers as they are easy to absorb water and to wash so that the regular cleaning can be done more easily.

3.2.6. Aspect VI: Health and hygiene of employees. The incompatibility found in this aspect was the absence of a person in charge of employee hygiene. Even though the employees had taken good care of their personal hygiene, such as wearing special work clothes when processing, washing their hands when starting to work in the food processing and after handling the raw material, there should be one person in charge of employee hygiene. This person's tasks are related to supervising personal hygiene, reminding employees when they do not apply personal hygiene and documenting a report if there is any employee is absent due to illness. These responsibilities can be given to an employee based on his or her work performance and working experiences in the production. It is recommended that there should be one person from each processing group to be responsible for the employee hygiene.

3.2.7. Aspect VII: The Maintenance and hygienic sanitation program. The incompatibility found in this aspect was quite critical. There was a pet—a cat, that freely came inside the building. The cat was in an open area that was used to separate hulls and nibs. Thus, the recommended corrective action is that the door that leads to the production area should always be closed.

3.2.8. Aspect VIII: Storage. For this aspect, the factory had implemented it well and met the good food processing requirements. The packaging materials, the food additives, and the final products were stored separately in clean and closed containers. The equipment was stored closely. The containers used for food storage, packaging and final products were not placed against the wall and did not touch the floor directly, so that contamination from dust on the floor or wall did not occur.

3.2.9. Aspect IX: Process control. One incompatibility with SPP-IRT regulation was also found in this aspect. The factory did not make any process flow chart yet. This flow chart functions as the main guideline and operating standard in processing cocoa powder. Therefore, this flow chart is basically important in the industry. The recommended corrective action is to make a flow chart. The designing of a recommended flow chart is explained further in the flow chart assessment section.

3.2.10. Aspect X: Product label and description. In this aspect a critical incompatibility with SPP-IRT regulation was the product label did not list the ingredients and the production code. Production code is crucial because it can improve traceability in a claim from consumers and form a systematic record. Regarding this case, the corrective action that can be done is adding information about the list of ingredients and the production code.

3.2.11. Aspect XI: Supervision by the person in charge. In the cocoa powder processing, there was no regular internal control carried out, including monitoring and correction. So far a supervision had been conducted through documentation in notebooks about the raw material and the products. Internal control such as whether the process specifications had reached had not been implemented. Consequently, corrective actions could not be established if product specifications did not meet. The recommended corrective action for the factory is to fill in the forms that contain the required data to facilitate the recording and supervision by the person in charge. By utilizing such forms, the person in charge can record the results of the process supervision in an organized manner.

3.2.12. Aspect XII: Product withdrawal. One incompatibility with SPP-IRT regulation found in this aspect was that the product withdrawal procedure had not been established. The product sales of ATP was limited to sales at the outlet only, so this had not become a serious problem. Although there were only a few cases found regarding the product withdrawal from the market, it is still better if the withdrawal plan preparation is carried out as a corrective measure. Moreover, the product withdrawal procedure can be applied in case the company will expands its market.

3.2.13. Aspect XIII: Recording and documentation. The ATP had conducted good documentation dealing with the production and had good records (recorded in a notebook). The notebook had been stored for one year or similar to two times of the food production. However, the requirements for selecting the raw material were not yet documented. Last, it is recommended to record the results of processing in a structured form.

3.2.14. Aspect XIV: Employee training. Employees who worked at ATP had attended a food safety training in ATP. The topic of training was The Planning of Hazard Analysis and Critical Control Point in Chocolate Powder Processing. By attending the training, the employees were expected to be able to understand the importance of food quality and safety and be able to implement the system in the processing of cocoa powder.

Based on the assessment of each aspect as explained above, it is necessary to improve the aspects that did not follow the regulation and to maintain those that met the regulation to improve the implementation of GMP. The main corrective action that must be taken is to make a flow chart as the main guideline for the processing as an effort to improve the supervision process. The aspects to be in the flow chart should not only include the specification of the process, but also the monitoring method and the corrective action when there is any incompatibility that serve as an internal control. In addition, by making a complete flow chart, the GMP assessment will be better as the incompatibility was reduced in two aspects, so that a significant change in the results was seen.

3.3. The assessment of flow chart

The twelve stages of cocoa powder processing were carried out at the ATP Nglanggeran. They were receiving raw material, steaming, roasting, deshelling, pasting, pressing, breaking, grinding, sifting, cooking, cooling and packaging. This cocoa powder processing generally took 2-5 days depending on the cooling time of the powder. The flow chart of the cocoa powder processing is presented in figure 5.

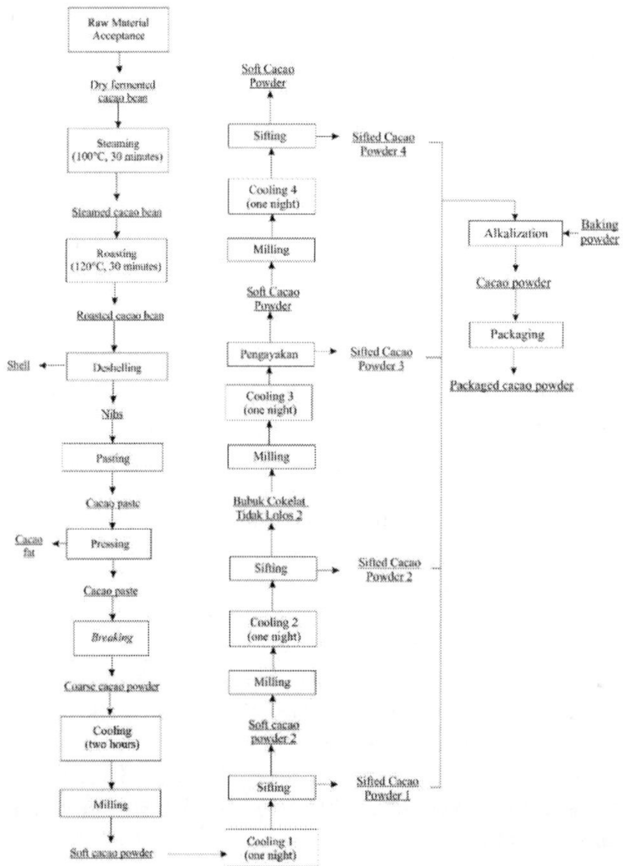

Figure 5. The flow chart of cocoa powder processing at ATP Nglanggeran, Gunungkidul.

3.3.1. Receiving raw material. The cocoa powder processing began with receiving the fermented cocoa beans produced by the local cocoa farmers. The purchasing of cocoa beans was done when the supply had run out. The cocoa beans were placed in an open plastic container in the first processing room, and the maximum length of storage was three weeks. The requirements for the raw material had been conducted, such as the cocoa beans had been fermented for five days, had been washed and dried to a maximum 7.5% water content.

The study at this stage focused on the method of storing cocoa beans after being received. The cocoa beans were stored in an open container in processing room 1, resulting in moisture and Aw of cocoa beans due to the absorption of water vapor from the environment. Although the received raw materials had been dried to 7.5% water content based on the regulation, when the storage was not good, an increase in water content could occur that allowed microorganism contamination.

The next investigation was the specifications of cocoa beans that were accepted by the industry. According to [12], in the ATP's cocoa powder, there was still found *Eurotium* sp. molds which commonly grew during storage. [5] also suggested that drying fermented cocoa beans used for the raw materials for making cocoa powder in Gunungkidul can be done through two methods: using machine drying or manual drying (without machine). When using machines, the drying time to reach water content of 7.5% is faster than that of the manual drying, so molds cannot grow. Therefore, it is recommended for ATP to add another specification for the received fermented cocoa beans, namely the beans should be dried with a drying machine. The method of storing raw materials also needs to be improved. The use of open containers could increase the potential for physical hazards (dust and hull) and biological hazards (insects) because in processing room 1 there was a door facing outside of the building which was always opened. To avoid the increase in water content and the emergence of these hazards, the seed storage can be done with a plastic vacuum then put it in a plastic box and closed.

3.3.2. Steaming. Steaming is an optional step in the cocoa powder processing. This step was done to clean the surface of the cocoa bean shell and to deactivate enzymes. The steaming machine capacity in ATP was 10 kg. The steaming process was carried out at a temperature of 80-100 °C for 30 minutes.

The steaming is an important stage in cocoa powder processing in the industry. In US patent 6025002 published on February 15, 2000, it is explained that steaming cocoa beans at 100-140 °C for 10 to 120 minutes is an economical process in the cocoa beans processing that maximizes chocolate flavor and suppresses off-flavor [13]. Steaming serves to clean the surface of cocoa beans and can improve the quality of the cocoa powder. Thus, the industry has chosen the right decision to carry out the processing stage of steaming and producing high quality cocoa powder.

3.3.3. Roasting. The roasting of cocoa beans was done at 120°C for 30 minutes with the capacity of roasting machine of 5 kg. The purpose of this stage was to develop the distinctive taste and aroma of chocolate, reduce water content, kill microbes, reduce acidity, and make nibs crispier.

This stage is a fundamental process for developing chocolate flavor during fermentation and drying [14]. [15] suggested that the roasting temperature of 120 °C for 30 minutes can produce roasted cocoa beans with color characteristics (Browning Index and L * value) that can be accepted by consumers. [16] also suggested that roasting at 120 °C for 50 minutes produces a higher chocolate flavor and lower acidity. However, the cocoa beans from each of those studies had several differences such as the types of cocoa fruit, fermentation treatment, and the drying method. Therefore, further research wa needed to obtain specification of cocoa bean roasting that was used in ATP.

3.3.4. Deshelling. The machine used for separating the cocoa beans shell was a desheller machine (capacity of 40-50 kg) which was capable of carrying out two processes, namely separation of the cocoa bean shell and breaking. When the roasted cocoa beans entered the machine, the seeds were crushed and then the shell would be exhaled by air so that the mild shell was blown into the drainage area while the heavy nibs fell into the nibs container. To produce cleaner nibs, manual removal of the cocoa beans shell was carried out.

3.3.5. Pasting. At this stage, the nibs were ground to produce cocoa paste (cocoa liquor). There was no definite process specification because the process took place depending on the nibs resulted from deshelling. The machine was equipped with a temperature regulator to produce the heat which was capable of melting cocoa butter and resulting in a semi-liquid cocoa paste.

3.3.6. Pressing. At this stage, the cocoa paste was pressed to separate the cocoa butter and the cocoa cake (cocoa mass). The resulting cocoa fat was used in the chocolate candy processing, while the cocoa cake was used as the raw material of cocoa powder. There were two machines used for the pressing stage, namely: manual and automatic machines. Similar to the pasting process, there was no exact process specification because it took place depending on the number of intermediate products produced from the previous process.

3.3.7. Breaking. The breaking stage was done using a breaker machine to crush the cocoa cake into coarse cocoa powder. Before being put into the machine, the cocoa cake was crushed manually to ease the machine's work.

3.3.8. Cooling 1. At ATP, there were two refrigerators and the temperatures were at 9-12 °C. The first refrigerator was used to store the finished products while the second one was for storing the intermediate products. The cooling stage was conducted for two hours. The cooling time was dependent on the processing schedule for the other products. This was due to the limited number of employees who processed and be responsible for processing the other products, so that sometimes the cooling stage was done overnight.

The assessment was carried out to check the purpose of processing and the length of processing time. The purpose of cooling was to prevent cake formation during the grinding stage. This was in accordance with the literature because cocoa beans relatively change to be smooth because of the effect of fat content. If the refining temperature is below 34 ° C, the glyceride fraction in cocoa butter becomes unstable and causes the powder to coagulate again to form lumps and cause an optimum sifting stage [17]. During the breaking process, there can be an increase in the temperature of the coarse cocoa powder due to friction among the powders. This is why a cooling step is needed to reduce the temperature. However, there had not been a control to the cooling process to prevent excessive processing time and to stop the cooling stage when the coarse cocoa powder reached the temperature below 34 °C.

Cocoa powder is hygroscopic which is able to absorb water from the surrounding air. If this process is intended to reduce the temperature until the clots do not form, then the time needed should be shorter that there is no increase in water content. Therefore, further research is required to determine the appropriate cooling time.

3.3.9. Grinding. In this grinding stage, the coarse cocoa powder was refined so that it was able to shape the sifting process. After grinding, the cocoa powder was put in the refrigerator to prevent clotting. The grinding process was conducted four times until the powder did not pass the sifter.

3.3.10. Cooling 2. This second cooling stage was done between the grinding and the sifting stages. Cocoa powder that did not pass the sifter was put in the refrigerator for one night. The result of the study in this part was the same with the result in cooling 1 stage, namely the control to the processing time had not been done efficiently at this cooling stage.

3.3.11. Sifting. At this stage, the cocoa powder was sifted to produce 200 mesh cocoa powder based on the Indonesian National Standard (SNI 3747:2013). After the cooling stage, the sifting was repeated especially to the cocoa powder that did not pass the sifter in order to get the better yield of cocoa powder. The process was carried out four times. Further, the cocoa powder that could not be refined was used as the raw material for chocolate *dodol* or chocolate dumplings by home industries around ATP.

3.3.12. Alkalization. In this part, the cocoa powder was added to baking soda as an alkalizing agent and then heated. The alkalization process aimed to reduce the acidity level and to increase the color darkness. The addition of some amount of alkalizing agent needs to be considered. According to [18], baking soda in the manufacture of cocoa powder can increase the pH and make the powder color darker but there is a decrease in flavone compounds and antioxidant activity. Stahl said that the final pH of chocolate powder $\geq$ 7.25 indicates the loss of flavone compounds. Therefore, it is recommended to readjust the dosage of baking soda in the processing to prevent the decrease in flavone compounds and the antioxidant activity of the cocoa powder.

3.3.13. Packaging. The final stage of cocoa powder processing was the packaging. The cocoa powder was packaged manually with the capacities of 100 grams and 200 grams. The manual packaging took longer time so this stage was considered less efficient. In addition, human error could also occur. Last, the packaging used was aluminum foil and was labelled.

3.4. Water content and water activity (Aw) analyses

The results of water content and water activity analyses during cocoa powder processing are presented in figure 6 and figure 7. In figure 6, it can be seen that the water content of cocoa beans increased during steaming and decreased after roasting, deshelling and pressing compared to the water content of fermented dried beans (7.5%). The roasting process at high temperatures caused the cocoa bean water to evaporate that made the water content decreased. However, an increase in the water content occurred again in the cocoa powder product. This was due to the absorption of water vapor from the environment during the treatment process which took a long cooling time as previously explained.

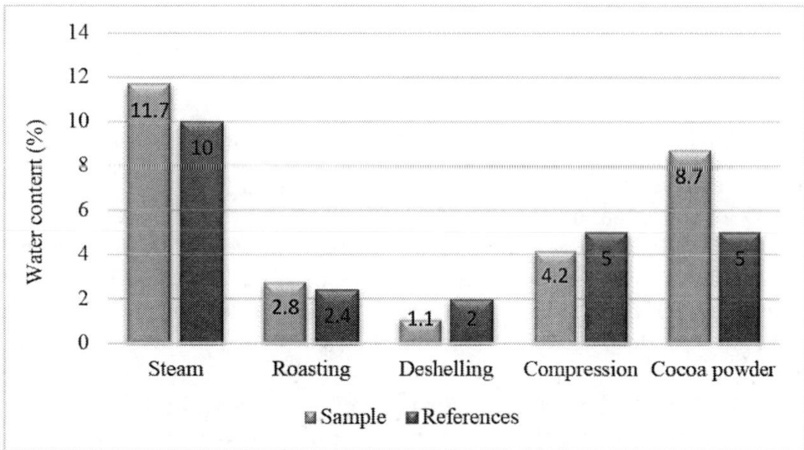

Figure 6. Water content during processing of cocoa powder.

Based on the results of this study, three stages showed different water contents which were contrast with some previous literature, namely steaming, roasting, and alkalization. In terms of steaming stage, the water content obtained was 11.69%. Meanwhile, according to [19], the steaming treatment temperature of 95 ° C for 15 minutes increased the water content of cocoa beans up to 10%. The difference in these water content results was due to the use of different process specifications. In ATP, the process applied higher temperature and the longer time than that of the previous study. It made the exposure to water vapor in cocoa beans was also higher.

In terms of roasting stage, ATP did the roasting of cocoa beans at 120 °C for 30 minutes and then produced the roasted cocoa beans with a water content of 2.78%. Meanwhile, the previous literature applied a roasting temperature of 130 ° C for 30 minutes and obtained the water content of 2.36% [20]. The use of high temperature will reduce the water content of products. Therefore, a strict supervision is necessary at this stage. Changes in processing specification cause a decrease in product quality and safety. If the temperature and/or the time is used excessively, charred beans result in a decrease in volatile components. Conversely, lower temperature and shorter time will increase the potential hazard of microorganisms because this stage is a process that involves heat to reduce water content and A_w.

In this study, the water content of cocoa powder produced was 8.69%, while the water content of cocoa powder according to SNI 3747: 2013 should be 5% [21]. This discrepancy occurred due to the repeated grinding and sifting stages, and the cocoa powder was stored in the refrigerator for two hours to overnight. Storage in humid condition (refrigerator) can increase the water content of chocolate powder which has hygroscopic characteristic.

In figure 7, it can be seen that the levels of Aw in the roasting, deshelling and compression stages decreased. This was because of the high heat exposure to cocoa beans which could evaporate water and finally could reduce A_w. In the deshelling stage, Aw slightly increased due to the nib separation from the cocoa shell. The nibs had higher water content compared to the roasted cocoa beans. Decreased Aw levels occurred again during the compression stage because of the separation of cocoa cake from cocoa butter.

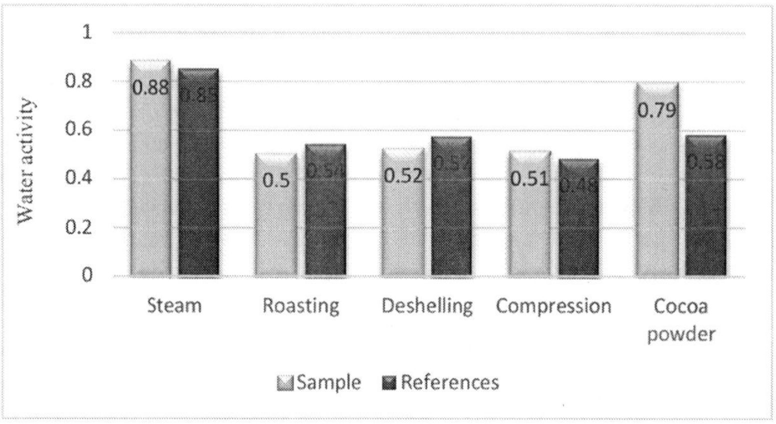

Figure 7. Water activity of the product during processing.

Compared with the results of the previous literature, this current study found three products had slightly higher Aw content, namely steam seeds, cocoa cake and cocoa powder. According to [22], the A_w level limit for steamed seeds was from 0.4 to 0.85. The Aw levels of roasted cocoa beans and nibs in this study were also lower than the literature. The A_w level of roasted cocoa beans was 0.50, while the research conducted by [23] recorded that the Aw level in the roasted cocoa beans was around 0.54 and the nibs was around 0.52. In figure 7, it can also be seen that the A_w level of cocoa powder was higher (0.79) than the results of [23]'s study who stated that the A_w level of cocoa powder was 0.58. This difference occurred due to the cocoa powder processing at ATP Nglanggeran where the powder was stored at prolonged cold temperature resulting in absorption of water vapor from the surrounding air.

3.5. Potential hazards and hazard analysis

The potential hazards and the hazard analysis are presented in table 1. The hazards that potentially occurred in the cocoa powder processing in ATP Nglanggeran were divided into three on the basis of their causes: physical, chemical and biological hazards. The potential physical hazards in each processing were found because the transfer of products between one to the next stage was still done manually. Moreover, in the first room processing, there were door facing faced outside of the building and windows that were often opened. Such situation could increase some potential physical hazards to enter during the processing such as small twigs, leaves, and insects. These hazards can be prevented by closing the door that faced outside of the building. In addition, the microbiological hazard identification was carried out based on the analysis of A_w levels and water content of the cocoa powder (figure 6 and figure 7). The contamination was found from six types of microorganism in the cocoa powder

IOP Conf. Series: Materials Science and Engineering **980** (2020) 012034 doi:10.1088/1757-899X/980/1/012034

processing, namely *Aspergillus carbonarius, Aspergillus parasiticus, Aspergillus flavus, Aspergillus ochraeous, Aspergillus niger,* and *Pencillium citrinum.*

Table 1. Hazard analysis on the cocoa powder processing at ATP Nglanggeran.

Processing Stages	Types of Hazards	Hazard Justification	Hazard Evaluation (Hazard Signification)			Preventive Action
			P	Se	HS	
Receiving Raw material	B: *Aspergillus sp* dan *Penicillium*	- From raw material	H	H	S	- Select raw material - Close raw material containers
	C: Ochratoxin A & citrinin		H	H	S	
	P: foreign object	- Contamination from the processing environment	L	M	TS	- Store raw material in different rooms
Steaming	B: *Aspergillus sp & Penicillium*	- From raw material	H	H	S	- Follow the steaming procedure of cocoa beans
	C : Ochratoxin A dan citrinin		H	H	S	
Roasting	B: *Aspergillus sp & Penicillium* C : Ochratoxin A & citrinin	- From raw material	H	H	S	- Follow the roasting procedure to fulfill stage specification
Deshelling	P: foreign object & epidermis	- From factory employees - From processing environment	L	M	NS	- Further separation is done in closed room
Pasting/grinding	P: foreign object	- From factory employees	L	L	NS	- Maintain employees' personal hygiene and working area sanitation
Compression	P: foreign object	- From factory employees	L	L	NS	- Maintain employees' personal hygiene and working area sanitation
Crushing cake	P: foreign object	- From factory employees	L	L	NS	- Maintain employees' personal hygiene and working area sanitation
Cooling	P: foreign object	- From factory employees	L	L	NS	- Maintain employees' personal hygiene and working area sanitation
Grinding	P: foreign object	- From factory employees	L	L	NS	- Maintain employees' personal hygiene and working area sanitation
Sifting	P: foreign object	- From factory employees	L	L	NS	- Maintain employees' personal hygiene and working area sanitation
Packaging	P: foreign object	- From factory employees	L	L	NS	- Maintain employees' personal hygiene and working area sanitation

B: Biological Hazard; **P**: Physical Hazard; **C**: Chemical Hazard; **P**: Possibility; **Se**: Severity; **HS**: Hazard Signification; **L**: Low; **M**: Medium; **H**: High; **S**: Significant; **NS**: Not Significant

The chemical hazards were identified from the presence of toxin molds that grew on the product. In the microbiological hazard analysis, five types of molds producing Ochratoxin A were found; *Aspergillus carbonarius, Aspergillus parasiticus, Aspergillus flavus, Aspergillus ochraeous* and *Aspergillus niger*, and one type of mold producing citrinin mycotoxin, namely *Pencillium citrinum*.

However, in some previous literature, the presence of these microbes had not shown the potential for toxin hazards when viewed from the low water activity of the product (A_w of cocoa powder was 0.79). This type of mold will produce toxins at A_w more than 0.8 [24,25,26].

Each processing stage had the potential hazards. At the stage of receiving raw material, the physical hazard came from some foreign objects. This potential physical hazard was caused by the use of open containers for storing the raw material (figure 8) and the processing rooms which had doors opened and faced outside of the building. However, this hazard had a low severity so that it was included in the not significant hazard category. Additionally, the stages of receiving raw material, evaporation and roasting had potential biological hazard. It came from the mycotoxin-producing molds and the chemical hazards from mycotoxin contamination. This happened because the dry fermented cocoa beans and the steamed cocoa beans had A_w levels of 0.85 and 0.88. In these Aw levels, the five types of molds identified could grow and produce mycotoxins. This hazard had a high severity because it produced mycotoxins, such as ochratoxin A and citrinin. According to [27], ochratoxin A can an increase in fat vacuoles (more fat in the liver) and mitochondrial swelling in the liver cells. It also has bad effect on kidney, that is necrosis in the tubules. [27] further claimed that the citrinin mycotoxins produced by *Penicillium citrinum* can affect the kidneys and cause tubular damage similar to glomerulonephrosis kidney disease. Thus, the level of potential hazard from mold contamination was high due to the high A_w levels.

Figure 8. The raw materials (the fermented cocoa bean storage at the ATP Nglanggeran.)

3.6. Critical control point determination

The determination was done by describing the product. The steps were forming the HACCP team, describing the product, compiling and verifying the flow chart, conducting the hazard analysis and critical control point determination. Detailed descriptions of the cocoa powder product are presented in table 2 and figure 9.

Table 2. Description of the cocoa powder produced by ATP Nglanggeran.

Product description	Cocoa powder "original taste"
Composition	Dry fermented cocoa beans and an alkalizing agent
Primer package	Aluminium foil
Secondary package	----
Storage method	----
Storage condition	Stored and closed, dry, dark and relatively in cold condition
Distribution method	Sale on line and off line (outlet)
Expired time	Six months after production
Consumer target	All ages
How to consume	Brewed or mixed as coloring and flavor enhancers in bread products

Figure 9. Cocoa powder produced by ATP.

Table 3. Critical control point determination of the cocoa powder produced by ATP Nglanggeran, Gunungkidul, Yogyakarta.

Processing Stages	Types of Hazard	Hazard Justification	P1	P2	P3	P4	CCP	Reason for Decision
Receiving Raw material	Biological : *Aspergillus* sp and *Penicillium*	From raw material	√	√			√	This step was done for raw material grading with low biological and chemical hazards
	Chemical : Ochratoxin A and Citrinin		√	√			√	
Steaming	Biological: *Aspergillus sp* and *Penicillium*	From surrounding air and raw material condition that supported the growth of fungi and production of mycotoxins	√		√	√		At the roasting stage, hazard could be eliminated or reduced to an acceptable level
	Chemical : Ochratoxin A and citrinin		√		√	√		
Roasting	Biological: *Aspergillus sp* and *Penicillium*	From surrounding air and raw material condition that support the growth of fungi and production of mycotoxins	√	√			√	Roasting was a stage to reduce the potential hazard to an acceptable level
	Chemical : Ochratoxin A and citrinin		√	√			√	

P1: potential hazard 1; **P2:** potential hazard 2; **P3:** potential hazard 3; **P4:** potential hazard 4; **CCP**: critical control point.

Based on the hazard analysis, there were three stages which were considered having significant hazards and needed to be reviewed as the critical control points. The critical control point determination for the cocoa powder processing is presented in table 3. The stage of receiving raw materials was determined as a critical control point. According to [12], the cocoa powder produced by ATP contained Eurotium sp. molds even after the processing. This indicated that the processing was not able to reduce or eliminate molds that had contaminated raw materials. The discovery of the molds strengthened the results of hazard identification and analysis. Therefore, selecting raw materials which are free from mold contamination is needed. Besides, Eurotium fungi could cause biodeterioration, and generally the growth of the fungi are followed by the growth of Aspergillus and Penicillium which were found from the hazard identification results in this study. This might cause economic losses. Another concern was the possibility of the mycotoxin hazard presence [28].

The determination of a critical control point at the roasting stage was carried out based on the critical control point tree in the second decision, that the roasting stage was specifically designed to eliminate or reduce the hazards that might occur to an acceptable level. Thus, the roasting stage was done to reduce the biological and chemical hazards identified by decreasing A_w levels where the molds (hazardous microorganism) could not grow. Therefore, it is highly necessary to add a microorganism boundary specification to the requirements for receiving dry fermented cocoa beans as the raw material by the ATP Nglanggeran.

4. Conclusion

Based on the findings of this study, the GMP implementation in ATP Nglanggeran was classified as level IV in accordance with the regulation of home industry food product certification (SPP-IRT). Therefore, several corrective actions are required before planning HACCP so that the product safety can be guaranteed. As stated in the flow chart, two processing stages that affected the cocoa powder quality and safety (receiving raw material and cooling 2) also need some improvements. At the stage of receiving raw material, an additional specification is required, such as a method for drying raw material. Additionally, the cooling 2 stage requires a new processing time setting that can increase the water content and A_w of cocoa powder. These two stages are critical control points for the cocoa powder processing. Further research needs to study on the cooling process time that can reduce the temperature of the cocoa powder but does not increase the water content and A_w.

Acknowledgments

The Authors would like to deliver their thanks to the Indonesian Agency for Agricultural Research and Development that has supported the funding for this work through a research collaboration project entitled "The Control Technology of Mycotoxin Contaminants Using Lactic Acid Bacteria to Improve Cocoa Beans Quality".

References

[1] Badan Pusat Statistik 2017 *Statistik Indonesia* 2017 https://www.bps.go.id/publication/2017/07/26/b598fa587f5112432533a656/statistik-indo Download 6 Januari 2020

[2] Herjanto E 2011 Mandatory application of SNI in industrial sector: effectivity and aspects in its implementation *J. Ris. Industri* **2** 121-30

[3] Balai Besar Pengawasan Obat dan Makanan 2012 Peraturan Kepala Badan Pengawas Obat Dan Makanan Republik Indonesia Nomor Hk.03.1.23.04.12.2206 Tahun 2012 Tentang Cara Produksi Pangan Yang Baik Untuk Industri Rumah Tangga

[4] Marwati T *et al* 2018 Fermentasi biji kakao pada akhir panen: Studi kasus di Gunung Kidul Yogyakarta Pros. Seminar Nasional Hasil Penelitian Pangan dan Hasil Pertanian 2018 "Inovasi Teknologi Pengolahan Pangan dan Hasil Pertanian untuk Meningkatkan Daya Saing Global" Yogyakarta, 30-31 Agustus 2018

[5] Djaafar TF, Elghina L, Widodo S, Marwati T, Utami T, and Rahayu E S 2019 Study of good handling practices and critical control point determination of dried fermented cocoa bean in Gunung Kidul Regency, Yogyakarta 2nd International Conference on Agriculture Postharvest Handling and Processing. IOP Conf. Series: Earth and Environmental Science 309(2019) 012015 doi:10.1088/1755-1315/309/1/012015

[6] Creswell J W 2012 *Educational Research: Planning, Conducting, and Evaluating Quantitative And Qualitative Research* (4th ed.) (Boston: MA Pearson)

[7] Association of Official Analytical Chemists (AOAC) 2006 *Official Methods of Analysis, the Association of Official Analytical Chemists* 14th ed. (Arlington: AOAC, Inc.)

[8] Susanto A 2009 Pros. Seminar Nasional Teknologi Peternak dan Veteriner Halaman 826-836.

[9] Mortimore, Sarah and Wallace C 2001 *HACCP* (Oxford: Blackwell Science Ltd.)

[10] Surak, John G and Wilson S 2014 *The Certified HACCP Auditor Handbook, Third Edition* (Wisconsin: American Society for Quality, Quality Press)

[11] ICMSF (International Commission on Microbiological Specifications for Foods) 1998 *Microorganisms in foods 4: Application of the Hazard Analysis Critical Control Point (HACCP) System to Ensure Microbiological Safety and Quaity* (London: Chapman and Hall)

[12] Djaafar T F, Monika D C, Marwati T, Triwitono P and Rahayu ES 2020 Proc. of 10th Asian Conference of Lactic Acid Bacteria (UGM Digital Press **2**

[13] Holsche, W J, Konopka U, Vitzthum O G, Bolenz S, Grosso M C, and Koch K D 2000 *Method of Removing Off-Notes from Cocoa Product* (US Patent 6025002)

[14] Toker O S, Palabiyik I, Pirouzian H R, Aktar T and Konar N 2020 Chocolate aroma : Factor, importance and analysis *Trends in Food Science & Technology* **99** 580-92

[15] Afoakwa E O, Budu A S, Brown H M, Takrama J F and Ansah E O 2014 Effect of roasting conditions on the browning index and appearance properties of pulp pre-conditioned and fermented cocoa (*Theobroma Cocoa*) beans *J. Nutr. Health Food Sci.* **2(1)** 1-5

[16] Rocha I S, de Santana L R R, Soares S E and Bispo E S 2017 Effect of the roasting temperature and time of cocoa beans on the sensory characteristics and acceptability of chocolate *J. Food Sci. Technol.* **37(4)** 522-30

[17] Mulato S, Widyotomo S and Handaka 2004 Desain teknologi pengolahan pasta, lemak, dan bubuk cokelat untuk kelompok tani Warta Penelitian dan Pengembangan Pertanian Badan Litbang Pertanian Departemen Pertanian. http://pustaka.bogor.net. (Diunduh tanggal 19 Juni 2020)

[18] Stahl L, Miller K B, Apgar J, Sweigart D S, Stuart D A, McHale N, Ou B, Kondo M and Hurst W J 2009 Preservation of cocoa antioxidant activity, total polyphenols, flavan-3-ols, and procyanidin content in food prepared with cocoa powder *J. Food Sci.* **74(6)** C456-C61

[19] Kopp G M 2014 Process for Producing High Flavor Cocoa European Patent Application (EP 2241190A1)

[20] Azizah S 2005 Uji kinerja mesin sangrai tipe silinder horizontal berputar untuk penyangraian biji kakao "Under Grade" (Skripsi Universitas Jember)

[21] SNI 3747 2013 *Kakao Bubuk* (Badan Standar Nasional Indonesia)

[22] Copetti M V, Lamanaka B T, Frisvad J C, Pereira J L and Taniwaki M H 2011 Mycobiota of cocoa: from farm to chocolate *Food Mycrobio.* **28(8)** 1499-504

[23] Nascimento M S, Reolon E M, Santos A R B, Moreira V E, da Silva I F and da Silva N 2011 *Enterobacteriae* in processed cocoa products *Dev. Do Inst. Adolfo Lutz J* . **70(1)** 81-5

[24] Ardhana M M and Fleet G H 2003 The microbial ecology of cocoa bean fermentations in Indonesia *Inter. J. of Food Micro.* **86** 87-99

[25] Taylor S L 2005 *Advances in Food and Nutrition Research* (Houston: Gulf Publishing Company)

[26] Passamani F R F, Hernandes T, Lopez N A, Bastos S C, Santiago W D, Cardoso M G and Batista L R 2014 Effect of temperature, water activity, and pH on growth and production of ochratoxin A by *Aspergillus niger* and *Aspergillus carbonarius* from Brazilian Grapes *J. of Food Protect* **77** 1947-52

[27] Rahayu E S, Sarjono S and Samson R A 2014 *Jamur Benang* (*Mold*) *Pada Bahan Pangan* (Yogyakarta: Kanisius) p 284

[28] Arvanitoyannis I S 2012 *Modified Atmosphere and Active Packaging Technologies* (Boca Raton: CRC Press)

IOP Conf. Series: Materials Science and Engineering **980** (2020) 012035 doi:10.1088/1757-899X/980/1/012035

Quality attributes of probiotic-enriched chocolate: A preliminary study

A D Saputro[1*], F I P Hati[1], W A Yuda[2], R Yanti[1], T Marwati[3], T F Djaafar[3], T Utami[4] and E S Rahayu[4]

[1] Department of Agricultural and Biosystems Engineering, Faculty of Agricultural Technology, Universitas Gadjah Mada, Yogyakarta, Indonesia

[2] Cokelat nDalem, Yogyakarta, Indonesia

[3] Yogyakarta Assessment Institute for Agricultural Technology, Indonesia

[4] Department of Food and Agricultural Product Technology, Faculty of Agricultural Technology, Universitas Gadjah Mada, Yogyakarta, Indonesia

Email: arifin_saputro@ugm.ac.id

Abstract. Research on the development of foods containing probiotics is growing, such as research in the incorporation of probiotics in chocolate. Nevertheless, the quality attributes of the probiotic-enriched chocolate have not been fully explored. This present study is a preliminary study to the impact of ingredients proportion and processing method on the quality attributes of probiotic milk chocolate, more specifically on moisture content, appearance, hardness, rheological behavior, and microstructural characteristics. In this study, four different formulations and processing methods were involved. Skim milk powder was used as a freeze-dried probiotic replacer as it has high similarity in terms of physical characteristics. The result showed that chocolate containing freeze-dried probiotic powder replacer (Choc B) tended to have the comparable hardness and L*values to the chocolate reference (Choc A). However, hardness and L* values of milk chocolates enriched with freeze-dried probiotic powder replacer formulated with milk butter (Choc C and Choc D) were lower than that of chocolate reference (Choc A). Concerning the moisture level and flow parameters, Choc C exhibited higher moisture content and flow parameters values than other chocolates. Microstructural visualization showed that all chocolates exhibited a low level of agglomeration degree with particle sizes in the range of 19-32 μm.

1. Introduction

Chocolate is one of the most popular confectionery products in the world, and regardless of the level of chocolate consumption per capita chocolate is highly demanded by consumers at various ages both in developed and developing countries [1,2]. The popularity of chocolate is high due to its characteristics in terms of smoothness, melting properties, flow behavior, taste, and aroma. Apart from this, chocolate, especially dark chocolate, also contains many antioxidants and polyphenols beneficial for human health [3,4] and becomes one of the reasons people eating chocolates.

Nowadays, public awareness of healthy products is increasing. This condition encourages chocolate producers to continuously innovate and develop their products. Some research innovations have been performing, for example, making chocolate with reduced-fat/sugar [5], replacing sucrose with alternative ones considered as healthy sugar [6,7], and another attention-gaining way is adding probiotic

IOP Conf. Series: Materials Science and Engineering **980** (2020) 012035 doi:10.1088/1757-899X/980/1/012035

to chocolate [8-11]. The incorporation of probiotics to chocolate is vastly challenging. This is because probiotics have to survive during food processing [12]. Besides, adding probiotics might change the characteristics of chocolate. This does not only influence consumer preference but also the processing method and handling properties applied [13,14]. Therefore, investigating the impact of probiotics addition on the chocolate characteristics is the main aim of this study.

This study is a preliminary study of the addition of freeze-dried probiotics into chocolate. Thus, in this study, instead of using freeze-dried probiotics, a material replacer with high similarity is used. In this preliminary study, skim milk powder is added as a freeze-dried probiotic replacer considering its similarity in terms of physical characteristics. The formulation and processing method are the two main factors used as variables in this study.

2. Materials and Methods

The ingredients used were cacao mass made of beans obtained from Gunung Kidul, Yogyakarta, while sucrose, cacao butter, full cream milk powder, skim milk powder, and milk butter were purchased in local Supermarkets in Yogyakarta.

2.1 Chocolate processing

Chocolate samples were produced following this formulation (table 1). Chocolates were processed using a stone melange suitable for the small scale chocolate producer [15]. The method used can be seen in table 2.

Table 1. Chocolate formulation.

Ingredient	Proportion (gr)			
	Choc A	Choc B	Choc C	Choc D
Cacao mass	250	250	250	250
Sucrose	250	250	250	250
Cocoa butter	250	250	250	250
Full cream milk powder	250	200	150	150
Skim milk powder	-	50	50	50
Milk butter	-	-	50	50

Table 2. Chocolate processing method.

Processing stages	Choc A	Choc B	Choc C	Choc D
Mixing process	All ingredients	All ingredients, except skim milk powder, were mixed	All ingredients, except skim milk powder and milk butter, were mixed	All ingredients, except skim milk powder and milk butter, were mixed
Grinding process	17 hours	17 hours	17 hours	17 hours
Sequence of tempering temperatures	50°C - 27°C - 29°C	50°C - 27°C - 29°C	50°C - 27°C - 29°C	50°C - 27°C - 29°C
2nd mixing process	-	Skim milk was added (before the end of the tempering process).	Milk butter and skim milk were added (before the end of the tempering process).	Milk butter and skim milk were mixed. The mixture was ground for 1 hour and added before the end of the tempering process.
Molding process	5-gr mould	5-gr mould	5-gr mould	5-gr mould

2.2 Analytical methods

2.2.1 Moisture content. Moisture content was determined by the thermogravimetric method. Approximately 3 grams of chocolate were weighed. Afterward, the sample was heated at 105°C for 24 hours. Subsequently, it was cooled in the desiccator for 1 hour and weighed again as the final mass of the sample. Water evaporated in this process was stated as moisture content (%). Measurements were done in triplicate.

2.2.2 Particle size. Particle size observation was carried out using an OptiLab Microscope. About 0.5 grams of molten chocolate was diluted in 10 ml of cooking oil and heated in the oven at 55°C for 1 hour. A drop of diluted chocolate was then put on the glass slide. Subsequently, a coverslip was placed on the sample before the observation. Measurements were done in triplicate.

2.2.3 Color. The color of the chocolates was measured using a colorimeter (Minolta CR-400) which was calibrated using a white reference standard. The color parameters were expressed in the L*a*b* color space system. L* represents lightness (luminance ranging from 0 (black) to 100 (white)), a* indicates the level of green to red, and b* represents blue to yellow. Measurements were done in triplicate.

2.2.4 Hardness. The hardness of chocolates was determined at room temperature of 20°C. Brookfield Texture Analyser with probe TA39 with a diameter of 2 mm was used. The probe speed was 0.5 mm/second with a pressing target of 3 mm. Measurements were done in triplicate.

2.2.5 Rheological behavior. Casson yield value, Casson viscosity, and thixotropy were determined using Viscometer (Brook Field type of RV 9). The obtained data were plotted to the Casson model. Casson yield value and Casson plastic viscosity were determined using equation 1, while thixotropy was determined by calculating the difference between ramp up and ramp down shear stress at 5 s^{-1}.

$$\sqrt{\tau} = \sqrt{\tau CA} + \sqrt{\mu CA} \,.\, \sqrt{\gamma} \tag{1}$$

T is shear stress, τCA is Casson yield value (Pa), μCA is Casson viscosity (Pa.s) and γ is the shear rate.

2.3 Data analysis
The data of this study were analyzed using SPSS software version 22.0. The researchers used one-way analysis of variance (ANOVA) with a significance level of 5% to test differences in moisture content, color, and hardness. For testing the homogeneity of variances, the Levene Test was performed. Meanwhile, Tukey test was used to determine the differences among samples and Principal Component Analysis (PCA) was used to visualize relationships among the samples.

3. Results and discussion

3.1 The relationship between chocolate characteristics
Principal Component Analysis (PCA) plots in figure 1 (a) and (b) show a general overview of the relationship among chocolates. PCA explained 88% of the variance in the first to factors, with PC 1: 67% and PC 2: 21%. The chocolates were clustered into three groups. In the same group, Choc A and Choc B had a propensity to have similar characteristics, characterized by their relatively high values of hardness and lightness (L*) exhibiting relatively high positive values in PC 2. In the second cluster, Choc C was characterized by its high value of Casson Yield Value, Casson Viscosity, Thixotropy, and moisture content exhibiting high values in PC 2. The third cluster (Choc D) was characterized by its low value of hardness exhibiting high negative values in PC 1 and PC 2.

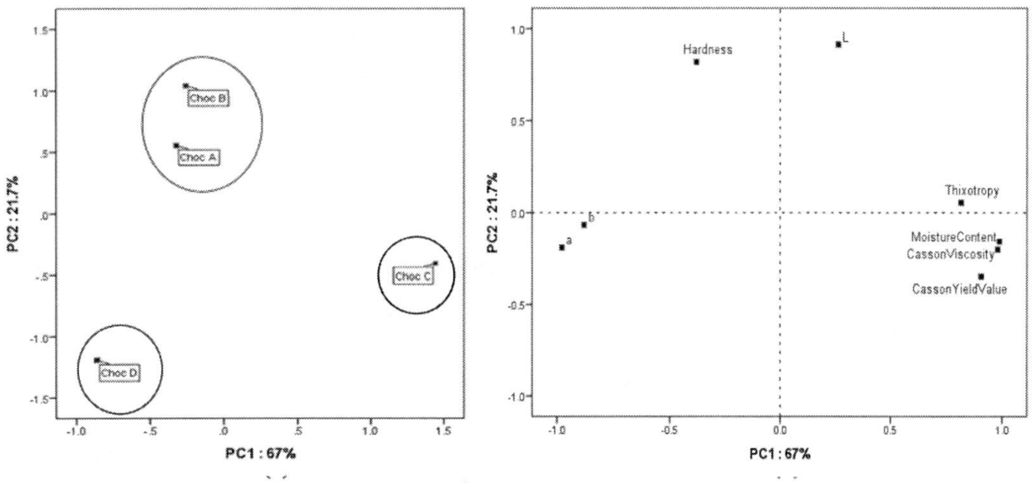

Figure 1. PCA score plots of chocolates (a) and PCA loading plots of quality characteristics of chocolates (b).

3.2. Physicochemical characteristics

3.2.1 Moisture content. The moisture content of chocolates affects the level of chocolate grittiness, hardness, flow properties, and influences the color of chocolates, to some extent [1,13,16,17]. Chocolate containing high moisture content tends to have higher hardness and viscosity value. Due to agglomerates formed, high moisture content chocolate also exhibits a higher level of grittiness [2,18]. Besides, due to the occurrence of sugar bloom which is commonly observed in high-moisture chocolate, moisture content also determines the appearance of chocolate [1].

The results in table 3 shows that the chocolate samples tended to have a relatively higher moisture content (3.06%-3.88%) than dark chocolates sweetened with sucrose (0.7%) [18] and dark chocolate sweetened with palm sap sugar (1%-1.5%) [16]. The relatively high level of moisture observed in the chocolate could be due to the high moisture content of the ingredients used and the processing method applied. The latter determines the amount of moisture that can be evaporated during chocolate processing [19]. Aside from this, the method used for moisture determination might also affect the values.

Table 3. Moisture content, color, and hardness of probiotic-enriched chocolate produced using alternative processing.

Chocolate Samples	Moisture Content (%)	Color			Hardness (N/mm^2)
		L*	a	b	
Choc A	3.12 ± 0.10^{a}	38.65 ± 0.56^{a}	18.06 ± 0.53^{a}	18.70 ± 0.94^{a}	5.93 ± 0.37^{b}
Choc B	3.12 ± 0.26^{a}	39.53 ± 1.21^{a}	18.08 ± 0.64^{a}	19.17 ± 0.86^{a}	4.94 ± 1.89^{ab}
Choc C	3.88 ± 0.08^{b}	38.66 ± 0.76^{a}	17.57 ± 0.13^{a}	18.33 ± 0.46^{a}	2.86 ± 1.23^{a}
Choc D	3.06 ± 0.08^{a}	37.75 ± 1.07^{a}	18.41 ± 0.52^{a}	19.31 ± 0.81^{a}	3.06 ± 0.31^{ab}

Different superscripts in the same column indicate significant differences ($p < 0.05$) among samples

3.2.2 Particle size. The particle size of chocolate affects the grittiness of the chocolate. Besides, as discussed in several works, particle size influences the color, hardness, and viscosity of chocolate [1,16]. In this study, the particle size of chocolate was measured and visualized using a normal light microscope.

In general, it could be concluded that the range of chocolate particle sizes among samples was comparable (figure 2). It can be seen in figure 2 that Choc A, Choc B, Choc C, and Choc D had a particle size range of 25-27 µm, 20-27 µm, 19-29 µm, and 22-32 µm, respectively.

Furthermore, it can be seen in figure 2 that all chocolates exhibited agglomeration. This phenomenon occurred due to the presence of relatively high moisture which sticks the particles to each other. In the previous study by Saputro [18] reported that due to high moisture content, palm sap sugar-sweetened chocolate exhibited particle agglomeration.

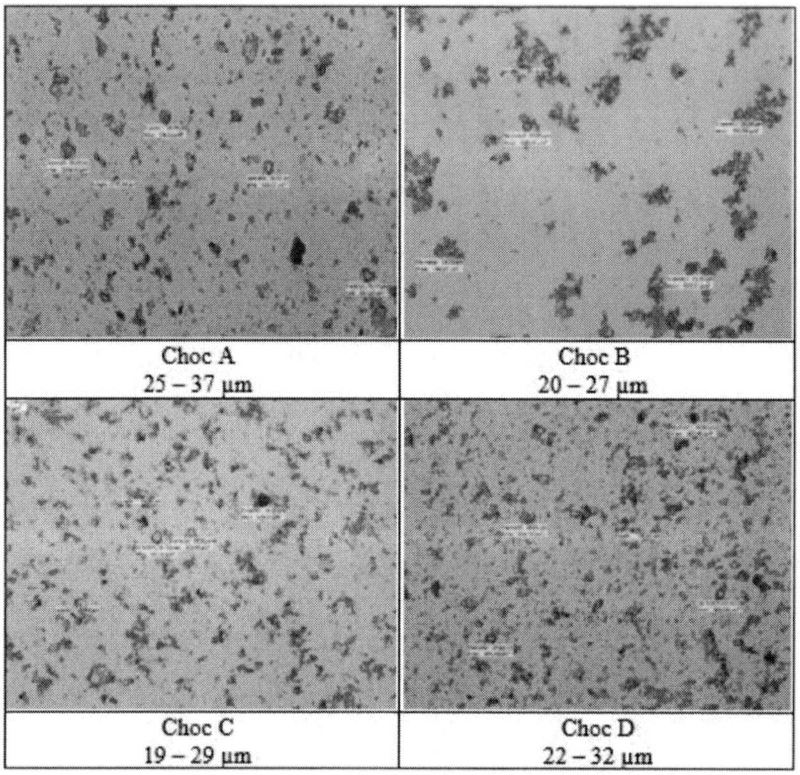

Figure 2. The particle size of probiotic-enriched chocolates processed using alternative processing

3.2.3 Color. Color, in combination with the glossiness, determines the appearance of chocolate that influences consumer preferences. Color is not only influenced by the particle size of the chocolate [20], but also by the surface roughness of the chocolate [21]. Afoakwa [20] reported that chocolate with fine particles has more particle interactions than that with coarser particles, resulting in a lighter color. It can be seen in table 3 that the color of the chocolates was not significantly different ($p<0.05$). Since the particle size of the chocolates were in a similar range figure 2, slight difference values in the color parameters could be associated with the surface roughness of the chocolates. In his study, Briones [21] found that chocolate with a smooth surface tended to have a brighter color.

3.2.4 Hardness. Similar to the color of chocolate, with the same level and type of fat, the hardness of chocolate is determined by particle size. It was reported by Afoakwa [20] that smaller particle size resulted in higher particle-particle interactions, creating harder chocolate. Besides, hardness, to some extent, is also influenced by moisture content. The higher the moisture content, the higher the hardness was observed in palm sap sugar-sweetened chocolate [18]. It can be seen in table 3 that the hardness of Choc A, Choc B, and Choc D was not significantly different ($p<0.05$). On the other hand, Choc C, even though the value was quite low (2.86 ± 1.23^a N/mm^2), was comparable to Choc B and Choc C ($p<0.05$).

This phenomenon showed that in this study, moisture content only had a little influence on the hardness of chocolate. With regard to the influence of particle size, since all the chocolates exhibited a comparable range of particle sizes (figure 2), it seemed that particle size did not have a significant effect.

3.2.5 Rheological behavior. The rheological behaviour of chocolate determines the mouthfeel of chocolate during consumption [5]. To easily quantify the value of rheological behavior parameters, the Casson model is commonly used. It can be seen in figure 3 that all chocolates, except Choc C, exhibited a comparable Casson yield value and Casson viscosity, while the thixotropy value of the chocolates varied among samples. Moreover, it can be seen in figure 3 that the Casson yield value and Casson viscosity of Choc C were higher than those of other samples. This phenomenon could be due to the presence of higher moisture content in Choc C (table 3). Beckett [2] stated that for every 0.3% extra moisture left in chocolate, an extra of 1% fat should be added to reach similar flow properties to the one without 0.3% extra moisture. Due to the presence of relatively high moisture in all chocolates, homogeneous chocolates could not be created, resulting in chocolates with high thixotropy values.

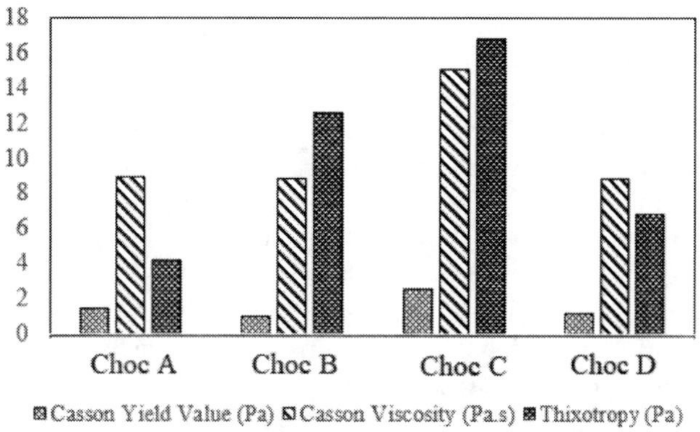

Different superscripts per parameter indicate significant differences ($p < 0.05$) among samples

Figure 3. Comparison of flow properties of probiotic-enriched milk chocolates produced using alternative processing.

4. Conclusion

The incorporation of freeze-dried probiotic powder replacer and the applied processing method influenced the characteristics of milk chocolate to some extent. From the study, it was found that milk chocolate made without milk butter exhibited the closest characteristics to the chocolate reference. Moreover, the addition of a freeze-dried probiotic powder replacer at the end of the tempering process was the best option. To have a better understanding of the impact of the processing method, further study related to viability assessment of probiotics is required.

Acknowledgements

This research was funded by the Government of Indonesia through Indonesia Endowment Fund for Education (Lembaga Pengelola Dana Pendidikan, LPDP). Title of the project was the prospect of probiotic chocolate industrialization as an innovative functional food product, as an effort to use and commercialize the potential of local natural resources.

References

[1] Afoakwa E O 2016 *Chocolate Science and Technology* 2nd Edition (John Wiley & Sons)
[2] Beckett S T 2009 *Industrial Chocolate Manufacture and Use* 4th Edition (Blackwell Publishing Oxford UK)
[3] Afoakwa E O 2010 *Chocolate Science and Technology* (Wiley Blackwell Oxford UK)

[4] Muhammad D R A, Saputro A D, Rottiers H, Van de Walle D and Dewettinck K 2018 Physicochemical properties and antioxidant activities of chocolates enriched with engineered cinnamon nanoparticles *European Food Res. and Tech.* **244(7)** 1185-202

[5] Do T-A L, Hargreaves J M, Wolf B, Hort J and Mitchell J R 2007 Impact of particle size distribution on rheological and textural properties of chocolate models with reduced fat content *Food Engineer. and Physic. Proper.* **72** E541 - E52

[6] Shah A B, Jones G P and Vasiljevic T 2010 Sucrose-free chocolate sweetened with *Stevia rebaudiana* extract and containing different bulking agents – effects on physicochemical and sensory properties *Int. J. of Food Sci. and Tech.* **45** 1426–35

[7] Aidoo R P, Appah E, Van Dewalle D, Afoakwa E O and Dewettinck K 2017 Functionality of inulin and polydextrose as sucrose replacers in sugar-free dark chocolate manufacture–effect of fat content and bulk mixture concentration on rheological, mechanical and melting properties *Int. J. of Food Sci. & Tech.* **52** 282-90

[8] Konar N 2013 Influence of conching temperature and some bulk sweeteners on physical and rheological properties of prebiotic milk chocolate containing inulin *European Food Res. and Tech.* **236** 135-43

[9] Mirković M, Seratlić S, Kilcawley K, Mannion D, Mirković N and Radulović Z 2018 The sensory quality and volatile profile of dark chocolate enriched with encapsulated probiotic *Lactobacillus plantarum* bacteria *Sensors* **18** 2570

[10] Possemiers S, Marzorati M, Verstraete W and Van de Wiele T 2010 Bacteria and chocolate: a successful combination for probiotic delivery *Int. J. of Food microbio.* **141** 97-103

[11] Konar N, Toker O S, Oba S and Sagdic O 2016 Improving functionality of chocolate: A review on probiotic, prebiotic, and/or synbiotic characteristics *Trends in Food Sci. & Tech.* **49** 35-44

[12] Silva M P, Tulini F L, Marinho J F, Mazzocato M C, De Martinis E C, Luccas V and Favaro-Trindade C S 2017 Semisweet chocolate as a vehicle for the probiotics Lactobacillus acidophilus LA3 and Bifidobacterium animalis subsp. lactis BLC1: Evaluation of chocolate stability and probiotic survival under in vitro simulated gastrointestinal conditions *LWT* **75** 640-7

[13] Saputro A D, Van de Walle D, Caiquo B A, Hinneh M, Kluczykoff M and Dewettinck K 2019 Rheological behaviour and microstructural properties of dark chocolate produced by combination of a ball mill and a liquefier device as small scale chocolate production system *LWT* **100** 10-19

[14] Aidoo R P, Afoakwa E O and Dewettinck K 2014 Optimization of inulin and polydextrose mixtures as sucrose replacers during sugar-free chocolate manufacture – Rheological, microstructure and physical quality characteristics *J. of Food Engineer.* **126** 35-42

[15] Hinneh M, Van de Walle D, Haeck J, Abotsi E E, De Winne A, Saputro A D, Messens K, Van Durme J, Afoakwa E O and De Cooman L 2019 Applicability of the melanger for chocolate refining and Stephan mixer for conching as small-scale alternative chocolate production techniques *J. of Food Engineer.* **253** 59-71

[16] Saputro A D, Van de Walle D, Aidoo R P, Mensah M A, Delbaere C, De Clercq N, Van Durme J and Dewettinck K 2017 Quality attributes of dark chocolates formulated with palm sap-based sugar as nutritious and natural alternative sweetener *European Food Res. and Tech.* **243** 177-91

[17] Saputro A D, Van de Walle D, Hinneh M, Van Durme J and Dewettinck K 2018 Aroma profile and appearance of dark chocolate formulated with palm sugar–sucrose blends *European Food Res. and Tech.* **244** 1281-92

[18] Saputro A D, Van de Walle D, Kadivar S, Sintang M D B, Van de Meeren P and Dewettinck K 2017 Investigating the rheological, microstructural and textural properties of chocolates sweetened with palm sap-based sugar by partial Replacement *European Food Res. and Tech.* **243(10)** 1729-38

IOP Conf. Series: Materials Science and Engineering **980** (2020) 012035 doi:10.1088/1757-899X/980/1/012035

[19] Saputro A D, Van de Walle D, Kadivar S, Mensah M A, Van Durme J and Dewettinck K 2017 Feasibility of a small-scale production system approach for palm sugar sweetened dark chocolate *European Food Res. and Tech.* **243** 955-67
[20] Afoakwa E O, Paterson A, Fowler M and Vieira J 2008 Particle size distribution and compositional effects on textural properties and appearance of dark chocolates *J. of Food Engineer.* **87** 181–90
[21] Briones V, Aguilera J M and Brown C 2006 Effect of surface topography on color and gloss of chocolate samples *J. of Food Engineer.* **77** 776–83

IOP Conf. Series: Materials Science and Engineering **980** (2020) 012036 doi:10.1088/1757-899X/980/1/012036

The effect of tiwai onion extract drink on the malondialdehyde levels in mice (*Mus musculus* L.)

S H Saputra[1]*, S Ismail[2], and P E Yustini[1]
[1]Research and Standardization Industry Center, Samarinda
[2]The Faculty of Medicine, Mulawarman University

Email: surotohs.65@gmail.com

Abstract. Tiwai onion contains secondary metabolic compounds of phenol, alkaloid, terpenoid, and has antioxidant and antimicrobial bioactivity. The antioxidant of tiwai onion can be used for supplement drinks. The purpose of this study was to obtain the best treatment of tiwai onion extract drinks on malondialdehyde levels in mice (*Mus musculus* L.). This study used a completely randomized design with one factor consisting of 8 levels of treatments: 1) new oil (mb) as the control, used cooking oil (mj) as the control, sugarless tiwai onion extract drink (ebt) with the doses of 3) 2% ebt, 4) 4% ebt, and 5) 8% ebt, and tiwai onion extract with sugar (meb) with the doses of 6) 2% meb, 7) 4% meb, and 8) 8% meb. The data obtained were analyzed using the analysis of variance and Duncan's Multiple Range Test. The malondialdehyde level in mice was a test parameter. The results showed that the best treatment was obtained at a 2% dose for each of sugarless tiwai onion extract drink and tiwai onion extract drink with sugar with an average of malondialdehyde levels in the mice of 26.380% and 20.215%, respectively.

1. Introduction

Nowadays, most humans tend to consume fast food but seldom to have some exercises and to consume foods containing a lot of polyunsaturated fatty acids (atherogenic). This can cause hypercholesterolemia indicating by the increase in low density of the lipoprotein levels or Low-Density Lipoprotein (LDL), triglycerides, and malondialdehyde. The final product of the lipid oxidation process is malondialdehyde, a membrane precursor, which plays a role in a process for the degradation of polyunsaturated fatty acids (arachidonic acid).

Malondialdehyde (MDA) was a lipid peroxidation biomark that contributes to the oxidative damage to DNA, namely oxidative stress. High levels of free radicals can cause lipid damage [1]. There are two types of free radicals: hydroxyl and hydroperoxide. Hydroperoxide is the main product of lipid peroxidation including malondialdehyde (MDA), propanol, hexanal and 4-hydroxynonenal (4HNE). MDA is the most mutagenic lipid peroxidation product and included as a lipid peroxidation product [2]. DNA changes induced by MDA can cause cancer and other generative diseases. In other words, MDA in the body can cause cancer, diabetes, liver, cardiovascular, Parkinson's and Alzheimer's disease [3].

The use of cooking oil repeatedly can produce free radicals absorbed in the fried food. If eaten, the radicals will enter the body. Free radicals can damage lipids to form lipid peroxidation and to produce malondialdehyde (MDA). Nevertheless, antioxidants can reduce the malondialdehyde level in the blood serum. Antioxidants in the form of supplements of vitamin C, vitamin E and multimineral, bioflavonoids, and tomatoes can significantly reduce the plasma MDA levels [4,5]. Several types of secondary metabolic compounds have been proven to have antioxidant properties, such as lycopene, polyphenols, flavonoids, chlorophyll, and some vitamins C and E. Tiwai onion (*Eleutherine americana* Merr.) is a plant that has some secondary metabolic properties i.e. alkaloids, polyphenol, and flavonoids that are useful as the antioxidant. Based on these facts, this current study is carried out to investigate the effect of tiwai onion extract drinks on the malondialdehyde levels in mice (*Mus musculus* L.)

2. Materials and methods

2.1. Materials

The materials used in this study included mice (*Mus musculus* L.), some tubers of tiwai onion (*Eleutherine americana* Merr.), white sugar, water, filter cloth, cooking oil, fried oil (used), filter paper, styrofoam, alumunium sheets, alcohol, anesthetics, Trichloroacetic Acid (TCA), Thiobarbituric Acid (TBA) Hydroclorid Acid, Tetraetoksipropan (TEP).

2.2. Equipment

The required equipment consisted of a blender, gas stove, digital scale, and other cooking utensils, syringes, oral syringes, Sensi-branded gloves, vacutainer, tourniquet, cooler box, centrifuge, micropipettes, microtubes, hot plate magnetic stirrers, and spectrophotometers.

2.3. Methods

2.3.1. The making of tiwai onion extract. The making of tiwai onion extract was carried out in the laboratory of Samarinda Research and Industry Standardization Center and made by the following process. First, 500 grams of clean tiwai onion was put in a blender then blended with 1 liter of water until the form was mush. Then the tiwai onion mush was heated at a temperature of 50^0C for 10 minutes. After that, the mush was filtered using a filter cloth so that the extract of tiwai onion was obtained. Next, the 400 grams of onion extract filtrate was placed on the stove and cooked with a small flame. The 600 grams of white sugar was added while stirring the extract until completely dissolved and the heat reached a temperature of 50^0C for 10 minutes. The sugarless tiwai onion extract filtrate and the tiwai onion extract filtrate with sugar were the ingredients of the research.

2.3.2. Experimental design. The experimental design in this study applied a completely randomized design with one factor consisting of 8 eight levels of treatments. Level 1 used new cooking oil (labelled mb) while used cooking oil (labelled mj) was for Level 2. The next levels of treatments used sugarless tiwai onion extract drinks (labelled ebt) with the concentrations of 2% (Level 3), 4% (Level 4), and 8% (Level 5). The other levels of treatments used tiwai onion extract with sugar (labelled meb) with the concentrations of 2% (Level 6); 4% (Level 7), and 8% (Level 8).

Each treatment level was conducted as follows. First, in each treatment 3 groups or cages were made so that there were a total of 24 groups or cages of mice for all the treatments. Each group or cage consisted of 6 mice so that there were a total of 144 mice. Second, all the treatment ingredients were administered to the mice as the experimental animals orally (in vivo) by giving them 0.5 ml/20 g body weight (bw) of mice. Third, the treatment ingredients were given by the following doses: 1) mb = new cooking oil of 0.5 ml/20 g bw; 2) mj = used cooking of 0.5 ml/20 g bw; 3) mj + ebt = 0.5 of used cooking oil and 0.5 ml of ebt 2%/20 g bw (c_1); 4) mj + ebt = 0.5 ml of used cooking oil and 0.5 ml of ebt 4%/20 g bw (c_2); 5) j + ebt = 0.5 ml of used cooking oil and 0.5 ml of ebt 8%/20 g bw (c_3); 6) mj + meb = 0.5 ml of used cooking oil and 0.5 ml of meb 2%/20 g bw (d_1); mj + meb = 0.5 ml of used cooking oil and 0.5 ml of meb 4%/20 g bw (d_2); and 8) j + meb = 0.5 ml of used cooking oil and 0.5 ml of meb 8%/20 g bw (d_3).

2.3.3. In vivo test. The mice were acclimated for one week after being grouped in each group/cage. They were given a BR-branded pellet feed and clean water for drinking. They were placed at the Pharmacology Laboratory of the Faculty of Medicine, Mulawarman University. The treatments were given every day in accordance with the predetermined doses for three weeks. The observation was also carried out during the 3 weeks. The data taken from the observation were analyzed using one way ANOVA to see the comparison of the malondialdehyde levels in the blood serum of mice from each group of cooking oil (mb), used cooking oil (mj), sugarless tiwai onion extract (c-c2-c3), and tiwai onion extract with sugar (d_1, d_2, d_3).

2.3.4. Data analysis. The observation data were analyzed using one way ANOVA to determine the comparison of the malondialdehyde levels in the mice blood serum from each group of cooking oil (mb), used cooking oil (mj), onion extract without sugar (2% ebt, 4% ebt, 8% ebt), and tiwai onion extract with sugar (2% meb, 4% meb, and 8% meb).

3. Results and Discussion

The results of this study was to determine the malondialdehyde (MDA) levels in the mice (*Mus musculus* L.) blood serum. Malondialdehyde (MDA) is the compound resulted from lipid peroxidation which is generally used as the indicator of oxidative stress. The analysis results of the MDA levels in the mice blood serum from the treatments using tiwai onion extract drinks are presented in table 1.

Tabel 1. The analysis of MDA levels in the blood serum of mice from the treatments using tiwai onion extract drinks.

Treatment	Week/average value		
	I	II	III
cooking oil(mb)	10.727	20.847	25.767
used cooking oil(mj)	40.021	54.072	60.767
Tiwai onion extract (ebt)	-	-	-
mj-ebt 2%	29.066	23.453	26.380
mj-ebt 4%	25.606	27.362	35.276
mj-ebt 8%	27.606	27.362	35.273
Tiwai onion extract drink	-	-	-
mj-meb 2%	19.377	18.893	20.215
mj-meb 4%	14.533	17.264	25.767
mj-meb 8%	16.609	26.384	34.356

Table 1 shows the malondialdehyde (MDA) levels in the blood serum of the mice that got new and used cooking oil increased until the third week. The malondialdehyde levels in the blood serum of the mice that got used cooking oil were higher with a range of 40.021 to 60.767 compared to that of the mice given new oils with a range of 10.727 to 25.767. This indicates that the lipid peroxidation or MDA continued to increase in the mice blood serum after being given the new oil (mb) and the used cooking oil (mj). The high peroxidation values also indicate that oil has been oxidized that was marked with rancid taste and smell [6,7,8]. Fat or oil at high temperatures (200-250^{0}C) will become toxic in the body and causes various diseases. Peroxide levels increase due to oxidized cooking oil after being heated at 200-250°C [9,10].

The malondialdehyde (MDA) levels in the blood serum of the mice after getting sugarless tiwai onion extract drink (mj + ebt) and tiwai onion extract drink with sugar (mj + meb) in Week I of observation decreased with a range values of 25.606 to 29.066 and 14.533 to 19.377, respectively. These values were much lower compared to the MDA levels in the blood serum of the mice that got used cooking oil (mj) with a value of 40.021. Similarly, MDA levels in the mice blood serum decreased in Week II in the range of values of 23.453 to 27.362 for mj + ebt and 17.264 to 26.384 for mj + meb when compared to that of used cooking oil (mj) with a value of 54.072. Also, there was a decrease in the MDA levels of the blood serum in Week III with a range of 26.380 to 35.276 from the mice with mj + ebt and 20.215 to 34.356 from the mice with mj + meb, compared to that of the mice with used cooking oil (mj).

Lipid peroxidation can damage the cells in the membrane which is caused by the increased ROS production. Henderomartono [11] showed that an increase in ROS in the cell membrane could increase the formation of MDA. The product from lipid oxidation caused by free radicals in the body is malondialdehyde (MDA) [12]. MDA is an indicator to determine the oxidative stress in the body and as a biomarker of oxidative stress. ROS production that exceeds the cell's antioxidant capacity can cause an increase in the oxidative stress along with the occurrence of lipid peroxidation in cell membranes. Therefore, the increased MDA due to lipid peroxidation in oxidative stress is in line with the lipid peroxidation in cell membranes. Thus, the increased MDA is the result of lipid peroxidation [13,14]. These results prove that the malondialdehyde levels in the mice blood serum decrease by giving tiwai onion extract. It shows the active role of tiwai onion extract as an antioxidant in preventing lipid peroxidation in cell membranes.

Some research has been done to investigate about tiwai onion and its ability to reduce lipid peroxidation. For instance, H.S. and Sampepana [15] reported that water and methanol extracts of tiwai onion contain secondary and primary metabolic aldehydes-ketones, phenols, glycosides, tannins, flavonoids and carboxylic acids, carbohydrates, and proteins. Tiwai onion also contains flavonoids, alkaloids, tannins, seteroids, and triterpenoids [16], and water and methanol extracts of tiwai onion contain antioxidant activity [17]. Besides, Saputra [18] believes that tiwai onion extract as an antioxidant can inhibit the speed of the occurrence of lipid peroxidation in cooking oil from coconut until the shelf life of 3 months which still meets the requirements of SNI: 3741: 1995. The quality standard [19] showed that the cooking oil is maximum 2 mg/kg. Further, Saputra [20] confirms that water and ethanol extracts have antioxidant activity, free radical scavenging capacity, good reducing power, and reactive oxygen species. Percent of antioxidant activity of tiwai onion syrup, when stored for 2 days, is 64.47% [21]. Another study reveals that ethanol extract of tiwai onion has antioxidant activity as much as IC_{50} 25,339 µg/ml [22]. Indeed, tiwai onion has several benefits as an herbal remedy for breast cancer, hypertension, diabetes, cholesterol, stroke, and anti-acne [23]. Winarsi [24] also found that the treatment of vitamin E decreases MDA level in the blood serum. Primary antioxidants, mostly phenolic compounds, are a group of compounds that stop the formation of free radicals in lipid oxidation including tocopherol, buthylated hydroxytoluene (BHT), buthylated hydroxyanisole (BHA), and tertiary-buthyl hydroquinone (TBHQ) [25]. Some natural compounds that have antioxidant effects are phenols, polyphenols and flavonoids, consisting of flavonols, isoflavones, flavones, catechins, and flavonoids as well as polypfunctional organic acids. Phenolic compounds are bioactive components that are widely found in plants [26]. Tiwai onion (*E.americana* Merr.) syrup has phenolic compounds, flavonoids, and antioxidant activity [21]. It is estimated that the decrease in MDA levels in blood serum is due to the primary antioxidant effect which acts as a one-atom free radical scavenger and, together with the SOD enzyme, is able to minimize and prevent lipid peroxidation reactions so that the formation of malondialdehyde products can be suppressed. The provision of vitamins can reduce MDA levels in the blood serum of the white mice with diabetes mellitus. If MDA in the body is formed as a result of the oxidative stress due to an imbalance between the formation of ROS and antioxidants, free radical level will be higher [27].

High MDA level can cause high level of damage to cell membranes and lipoproteins. According to a study by Makaryani *et al.* [28], consuming green grass jelly can reduce the MDA levels in the students of

IPB who consumed fried food. Last, it has been revealed that garlic extract can reduce MDA levels in the objects after getting exposure to cigarette smoke [29].

4. Conclusion

The results of this study revealed that both of tiwai onion extract drinks with and without sugar could reduce the malondialdehyde (MDA) levels in the mice blood serum. The best dosage for reducing MDA levels up to the third week was the 2% for each of sugarless tiwai onion extract drink and tiwai onion extract drink with sugar.

Aknowledgements

The authors would like to deliver many thanks to the staff of the Biopharmaceutical Laboratory of the Medical Faculty of Mulawarman University and the staff of the Research and Standardization Industry Center of Samarinda, who assisted the research process. The authors did understand each other's positions and did not have any conflicts of interests.

References

[1] Wahyuni S. 2012 *Malondialdehyde Monograph Precursor to Oxidative Stress* (Bali: Udayana University Press)
[2] Esterbauer H, Eckl P and Ortner A 1990 Possible mutagens derived from lipids and lipid precursors *Mutation Research* **238(3)** 223–33
[3] Ayala A, Munos M F and Aguelles 2014 Lipid peroxidation: production, metabolism, and signaling mechanisms of malondialdehyde and 4-hydroxy-2-nonenal *Hindawi Publishing Corporation Oxidative medicine and cellular longevity* **360438** 1-31
[4] Ramatina, Amalia L, Ekayanti I 2014 Effect of antioxidant supplements on malondialdehyde plasma level among college students of IPB *Journal Nutrition and Food* **9(1)** 35-42.
[5] Wahyuni, Asj'ari S R, Sadewa A 2008 Study of the ability of tomato juice to inhibit increased levels of plaque malondialdehyde after high impact aerobic exercise *Journal Health* **1(2)** 123-32.
[6] Ketaren S 1986 *Introduction to Food Oil and Fat Technology* (Jakarta:UI Press)
[7] Siti N W, Tri Dewanti W, Kuntanti 2001 A study of the level of damage and food safety of used cooking oil (a study of the different types of cooking oil and fried foodstuffs) *Research Report* (Malang: Faculty of Agricultural Technology Brawijaya university)
[8] Nawar W W 1985 Lipids Food Chemistry 2nded, ed O R Fennema (New York, Basel: Marcel Dekker Inc.)
[9] Nurhayati H, Supriningrum R and Caesariana N 2015 Determination of levels of free fatty acids and peroxide numbers in cooking oil used by fried food traders on the street A.W. Sjahranie samarinda. *Journal Scientific Manuntung* **1(1)** 25-30
[10] Suroso A S 2013 Studies of consumable cooking oil in terms of peroxide numbers, acid numbers and water content *Journal Indonesia Pharmacy* **3(2)** 77-88
[11] Henderomartono 2000 The world health report 2000 World Health Organization 1-10
[12] Jeyabalan A and Caritis S N 2006 Antioxidant the prevention of preeclamsia unresolved issues.*The England Journal of Medicine* **354(17)** 1841-3
[13] Maslachah L, Sugihartuti R and Kurniasanti R 2008 Inhibition of production of reactive oxygen species superoxide radicals (O2-) by antioxidant vitamin E (-tocopherol) in white rats (*Rattus norvegicus*) receiving electric shock stressors *Veterinary Media* **24(1)** 21-6
[14] Soviana E, Rachmawati B and Nyoman W 2014 The effect of β-carotene supplement on blood glucose levels and malondialdehyde levels in Streptozotocin-induced Spague Dawley rats. *Indonesian Nutrition Journal* **2(20)** 41-6

[15] Saputra H S and Sampepana E 2007 Chemical content analysis and utilization of tiwai onion for industrial raw materials *Journal Industrial Technology Research* **1(1)** 25-33
[16] Sukrasno, Umemi Sadaruddin, 2006. Micropagance traditional medicine research and the development of tiwai onion/sabrang onion (*Eleutherine americana* L.) as an herbal medicine. ITB. Bandung.
[17] Saputra S H 2007 Bioactive analysis and utilization of tiwai onions (*Eluetherine americana* Merr.) For food additives *Journal of Industry Technology Research* **1(2)** 24-30.
[18] Saputra S H 2012 Tiwai onions (*Eleutherine americana* Merr.) As a preservative, antioxidant and food coloring *Journal of Industrial Technology Research* **6(12)** 102-10
[19] National Standardization Agency 1995 SNI. 375-1995 Cooking oil quality standards
[20] Saputra S H 2010 Tiwai onion extract as an antioxidant in coconut oil *Journal Industrial Technology Research* **4(8)** 14-9
[21] Saputra S H, Sampepana E and Susanty A 2018 The effect of bottle, temperature and circular storage packaging of flash drink tiwai extracts *(Eleutherine americana* Merr) on secondary metabolic and microba *Journal Industrial Technology Research* **12 (2)** 156-65
[22] Kuntorini E M and Astuti M D 2010 Determination of the antioxidant activity of Dayak bulb (*eleutherine Americana* Merr.) ethanol extract *Science and Applied Chemisty* **4(1)** 15-22
[23] Galingging RY 2009 Dayak onion (*Eleutherine palmifolia* L. Merr) as a multifungsional medicinal plant. Research and development news *Agricultural research and development agency* **15** 10-6
[24] Winarsi H 2005 Effect of soybean sprout protein extract supplementation on levels of IL-1 beta of type 2 diabetics *Journal Technology and Food Industry* **21(1)** 6-10
[25] Kochhar S P and Rossel, J B 1990 Detection, estimation and evaluation of antioxidant in food sytems *Food Antioxidant* 19-64
[26] Pratt D E and B J F Hudson 1990 Natural antioksidant not exploited commercialy Food Antioksidant ed B J F Hudson (New York:Elsevier Applied Science) pp 171-92
[27] Vera B, Al Azhar D, Karmi T F, Rindy G and Sabri M 2018 The effect of vitamin E to malondialdehyde (MDA) serum level in diabetes mellitus induced white rat (*Rattus norvegicus*) *JIMVET* **2(1)** 70-6
[28] Makaryani I, Amalia L, Ramadhani N R, Pertiwi KI, Aprillia D D 2014 The effect of antioxidant feeding on plasma malondialdehyde levels of fried food students. *Journal Indonesia Nutrion Clinic* **10(04)**169-79
[29] Duwairoh A M, Wirjatmadi B, Adriani M 2018. The Effect of Solo Garlic Extract in Decreasing Malondialdehyde (MDA) Levels due to E-Cigarette Exposure *Journal Medical Scientific Wijaya Kusuma* **7(2)** 149-57

IOP Conf. Series: Materials Science and Engineering **980** (2020) 012037 doi:10.1088/1757-899X/980/1/012037

The potentials and prospects of tidal swamp local brown rice as a nutrition source of iron and zinc and the parent of rice breeding

I Khairullah[1*] and M Sarwani[2]
[1] Indonesia Swampland Agricultural Research Institute (ISARI), Jl. Kebun Karet, Loktabat Utara, Tlp/Fax 05114772534 Banjarbaru 70712 South Kalimantan
[2] Indonesia Center for Agricultural Land Resources Research and Development (ICALRD), Jl. Tentara Pelajar No. 12 Bogor 16114

Email: izhar.balittra@gmail.com

Abstract. Brown-rice originating from local tidal swamp rice is identified as having high iron (Fe) and zinc (Zn) content. This paper aims to elaborate the potentials and prospects of Fe and Zn content of brown rice in tidal swamp local rice, especially from South Kalimantan. This research was carried out at tidal swampland of Belandean Experimental Station, Barito Kuala District South Kalimantan. A total of 71 local varieties of tidal swamp rice originating from the tidal swampland were planted in a 2 m x 5 m plot. The planting system was followed by the South Kalimantan farmers' traditional system with three times transplanting. The harvested rice was threshed and then milled which was still aleuron. The analysis of Fe and Zn of the brown rice was carried out in the laboratory of The Center for Soil and Agroclimate Research (at present ICALRD) Bogor. The results of the analysis of Fe and Zn levels of rice showed that levels of Fe and Zn of brown rice varied greatly. Fe levels ranged from 11-83 ppm, where Fe levels of local varieties were higher than high yielding varieties. Likewise, levels of Zn with a wide enough interval, which ranged from 20-108 ppm Zn. Among 71 brown rice local varieties, Siam Panangah showed high levels of Fe and Zn, which were 54 ppm Fe and 108 ppm Zn. The brown rice can be used as a source of nutrition Fe and Zn and as the parents in rice breeding with high Fe and Zn content.

1. Introduction

Nutrient deficiency of iron and zinc in Indonesia is still high and is one of the problems that occur in children. The percentage of iron (Fe) nutrient deficiency was 50% and zinc (Zn) deficiency was 17% [1]. Iron (Fe) and zinc (Zn) nutrition is a micronutrient that functions in the development of the brain, especially in functioning of neurotransmitter that plays an important role in increasing intelligence especially for children [2]. One source of nutrition for Fe and Zn was the consumption of rice from local varieties of tidal swamp rice. Brown rice from several local varieties of tidal swamp rice showed that Fe and Zn levels were high.

Local varieties of tidal swamp rice are still dominating paddy fields in tidal swamplands, especially in South Kalimantan. Tidal wetlands in South Kalimantan are around 191,740 ha of the total land area, about 146,612 ha had been cultivated, while the remainder was not planted or cultivated. The utilization of

146,612 ha of land is for paddy fields which are more dominated by local varieties [3]. In 2001, around 92% of the land was grown by rice, with local rice varieties dominating, which was around 96%, the rest were planted by high yielding varieties twice a year. The average productivity of rice (local and high yielding varieties) was 3.08 in 2000 t/ha [4]. The large percentage of local rice varieties planted in the tidal swamplands area might be due to high adaptability, as well as ease of cultivation in the field. According to Wiggin [5], local varieties had several advantages from the interests of farmers, which were easily obtained in almost all places, only require very minimum maintenance, and tall plant height so that farmers did not need to bend in harvesting by using *ani-ani* tool, a kind of harvested tool using hand.

Tidal wetland is one part of swampland that had specific hydrological and soil properties. Based on hydrologic nature or type of overflow, it is divided into types A, B, C, and D, while based on the type of soil, it is categorized into potential soils, acid sulfate, saline, and peat land. Type A land is the land that is always waterlogged and daily tidal drainage; Type B land is only affected by large tides, but drained daily; type C land has never been fulfilled even though tides is large, permanent drainage, tides affect it indirectly, ground water near surface soils <50 cm; type D land has never been flooded with the tide, and groundwater > 50 cm from ground surface, and drainage was limited [6].

Potential land is the land that includes potential acid sulfate, pyrite content of 2% at a depth <50 cm from the soil surface. The acidic sulfate soil has a pyrite layer > 2% at a depth <50 cm, which is divided into potential acid sulfate soil if the pyrite layer has not been oxidized, and actual acid sulfate soil if pyrite has been oxidized to a soil pH <3.5. Peatland was a land that is formed by organic matter with organic carbon content of 12-18% if the matter is saturated with water, and 10% if it has never been saturated [7]. Thus, the diversity of tidal swamplands could be seen, and it potentially allows for a diversity of food crops, especially local rice that could grow adaptively on the land.

Its hydrological nature was influenced by tidal diurnal and flooded over 3 months, while the characteristics of soil was the presence of pyrite minerals as a form of acid sulfate soils and the peat as the basic ingredients of peat soil. Under natural conditions, tidal swamplands were one of the ecosystems that had a high level of biological diversity including a variety of plants, fish, and livestock that were typical of swamplands [8].

The development of agriculture in tidal swamplands faces some problems including water management, soil fertility, and soil acidity [9]. Besides, the high level of Fe, Al, Mn, P deficiency, low base cations, and high salinity also encounter [10,11]. Land problems that are often encountered in tidal swamplands were the presence of pyrite layers (FeS_2). It was under conditions of inundation the soil pH increase that caused an increase in Ferro concentration to reach thousands of mg L^{-1} in the soil solution [12,13].

Tidal swampland has a diversity of plants that has superior properties, such as tidal swamp local rice that is able to adapt to inundation conditions, low pH, and high levels of soil Fe. This local rice is also tolerant of Fe toxicity [14,15]. Local varieties of tidal swamp rice are widely grown in tidal swamp rice fields, including in South Kalimantan. Various local varieties on the tidal swampland of South Kalimantan are found under various local names. Naming could be based on plant-specific characteristics such as grain size, grain color, number of grains in panicles, and so on. Khairullah [16] compiled it with several names groups such as Siam, Bayar, Pandak, and Lemo.

Adaptive rice varieties in tidal swamplands, especially for acid sulfate soil, has the potential to contain high levels of Fe and Zn because the growth environment is rich in Fe and Zn soil. Conversely, varieties that are not adaptive would languish their growth and might even die, partly because the plant is Fe toxicity. In general, local varieties of tidal swamp rice has been adaptive to grow environmental conditions so that they could contain high Fe and Zn in their brown rice [17].

This paper aims to elaborate the potentials and prospects of Fe and Zn levels of brown rice in tidal swamp local rice, especially those from South Kalimantan tidal swamplands.

1.1. Potential of local rice varieties

Local varieties of tidal swamp rice still dominate rice fields on tidal swampland. For example, in South Kalimantan, tidal swampland is around 172,117 ha, and 84.3% or 145,168 ha have been developed. About 92.5% of the land is planted with rice, where once a year local rice varieties are dominating about 96.24%, the rest are planted with high yielding varieties (HYV's) twice a year. Average rice productivity (local and HYV's) in 2000 was 3.08 t/ha [4]. The large percentage of local rice varieties planted might be due to good adaptability, in addition to the ease of cultivation at the farmer field. According to Wiggin [5], local varieties has several advantages seen from the farmers' interests which are easily obtained in almost all places, only require minimum maintenance, and height plant so that farmers do not need to bend in harvesting with *ani-ani* tool.

Local rice varieties on tidal swampland, for example in South Kalimantan, are inseparable from early history of land clearing by the people. The utilization of tidal swampland for agriculture, especially lowland rice by local farmers began spontaneously since hundreds of years ago. Reclamation and opening of tidal swampland began first in the area around Banjarmasin conducted by Banjarese farmers from Hulu Sungai Selatan since 1920 along with the opening of the road. Until 1965, about 65,000 ha of tidal swampland in South and Central Kalimantan had been reclaimed into paddy fields [18].

The opened paddy field was originally called *Bayar* rice, because the rice planted was *Bayar* variety. Furthermore, segregation of local *Bayar* varieties is to bring up the names of new local varieties, namely *Bayar Putih, Bayar Kuning, and Bayar Melintang.* The naming of Bayar Putih and Bayar Kuning is related to color of light yellow (white) and yellow straw. The description of Bayar varieties included 9-10 months, nurseries 4-5 months, including photoperiod sensitive varieties, sensitive to water systems, easy to fall down or lodged, and can not stand drought, and plant height 160-170 cm, yield 65-70%, productivity 2.5-3.0 t/ha.

The segregation of local Siam varieties initially gave names of Siam Halus (small), Siam Manangah (medium), and Siam Ganal (bold). This naming is associated with the grain form of the variety. Siam Halus is the most popular because it is considered more delicious, and its price is more expensive. From Siam Halus then it is known as Siam Karangdukuh, which is a Siam variety that shows its superiority in the Karangdukuh area village. The popularity of Siam Karangdukuh lasted long enough and spread widely in the farmer paddy field. The next, come to a new variety called Siam Unus, which is thought to be a segregation of Siam Karangdukuh.

Unus Kuning, Siam Unus variety then becomes very popular in South Kalimantan replacing Siam Karangdukuh, because of its small, slender in grain form, translucency in the hulled rice, the very white color of boiled rice with good eating quality, as well as its expensive price. The segregation of Siam Unus variety gives rise to Siam Unus Halus, Siam, and Siam Unus Putih varieties. All of which refer to the nature of the shape and color of the grain. Siam Unus Halus variety is furthermore popular among farmers and consumers particularly in South Kalimantan [19].

2. Materials and methods

The research was conducted in acid sulfate tidal swampland in Belandean Experimental Station, ISARI in WS 2001 to DS 2002. A total of 71 local varieties and five high yielding varieties were planted in a 2 x 5 m^2 plot. The local varieties came from the tidal swampland of South Kalimantan. The planting was carried out based on the local farmers' planting, namely with the seedling system up to three times. The seedlings (Banjar language - *teradak*) was carried out in October; first transplanting was called *meampak*) in November; transplanting II (*melacak*) in January; and transplanting III (planting) in April. Planting this way was done because the local rice varieties were classified as photo period varieties. The land management and preparation were done by using a *tajak* tool. Weeds were cut down, allowed to mine on the water then spun and spread back to the land when it had undergone a perfect decomposed. The fertilization only used urea at a rate of 100 kg/ha given one week after planting. The planting was maintained by weeding and preventing pest attacks.

The rice harvested for each variety then threshed and dried, and a manual milled to get brown rice. The brown rice sample was taken as much as 100 g to be analyzed Fe and Zn contents. Analysis of Fe and Zn contents of brown rice was carried out at the Indonesian Center for Soils and Agricimate Research laboratory (at present the Indonesian Center for Agricultural Land Resources Research and Development – ICALRD, Bogor. The data analysis used frequency distribution method to determine groups based on a range of Fe and Zn contents.

3. Results and discussions

The frequency distribution Fe and Zn content of 71 local brown rice varieties showed in Figure 1 dan Figure 2. The frequency distribution of brown rice Fe content (%) consisted of seven groups, namely group 1 (11-21 ppm), group 2 (11-21 ppm), group 3 (33-43 ppm), group 4 (44-54 ppm), group 5 (55-65 ppm), group 6 (66-76 ppm), and group 7 (77-83 ppm). Group 1, 2, 3, 4, 5, 6, and 7 consisted of 41%, 38%, 13%, 4%, 0%, 3%, and 1% of 71 local varieties, respectively (figure 1). Meanwhile the frequency distribution of brown rice Zn content (%) consisted of seven groups too, namely group 1 (20-32 ppm), group 2 (33-45 ppm), group 3 (46-58 ppm), group 4 (59-71 ppm), 5 (72-84 ppm), group 6 (85-97 ppm), and group 7 (98-110 ppm). Group 1, 2, 3, 4, 5, 6, and 7 consisted of 51%, 40%, 4%, 1%, 3%, 6%, and 4% of 71 local varieties, respectively (figure 2).

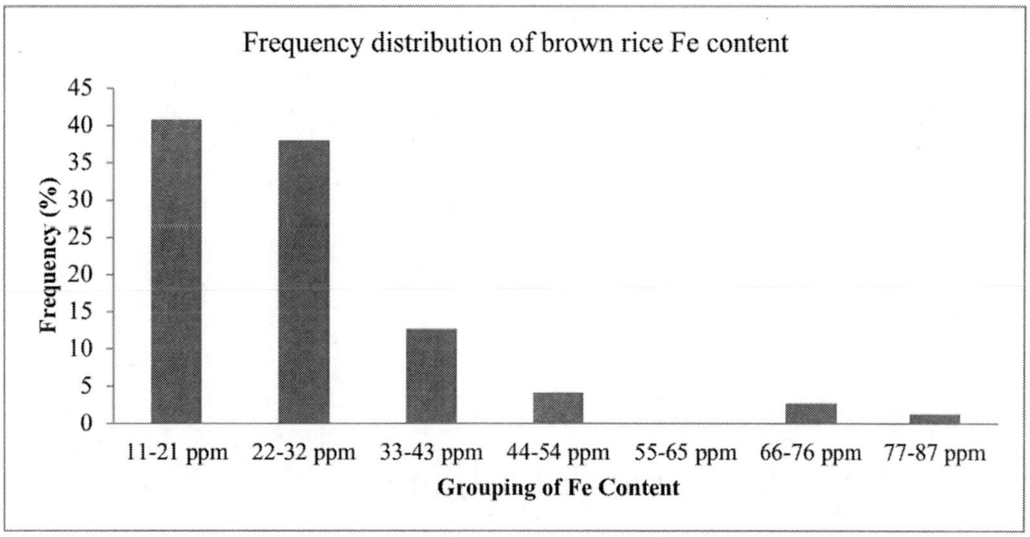

Figure 1. Distribution of Fe content of brown rice of local tidal swamp rice varieties.

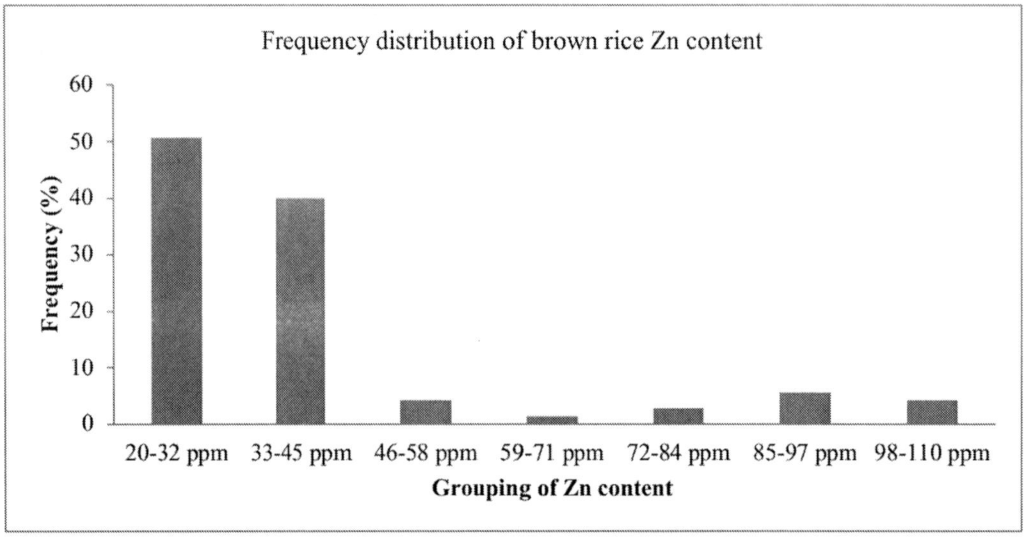

Figure 2. Distribution of Zn content of brown rice of local tidal swamp rice varieties.

The results of Fe and Zn contents analysis on 71 local varieties of rice showed that Fe and Zn content varied greatly (figures 3 and 4). Fe content of 71 brown rice of local varieties of tidal swamp rice ranged from 11-83 ppm, where the lowest levels were indicated by Kutut variety (11 ppm Fe), and the highest was shown by Siam Pandak variety (83 ppm Fe). The highest frequency was in range 11 - 21 ppm Fe followed by 22 - 32 ppm Fe with a frequency 40.8% and 38.0% respectively. This means that around 75% of local varieties tested for Fe content of brown rice were in the range 11 - 32 ppm Fe.

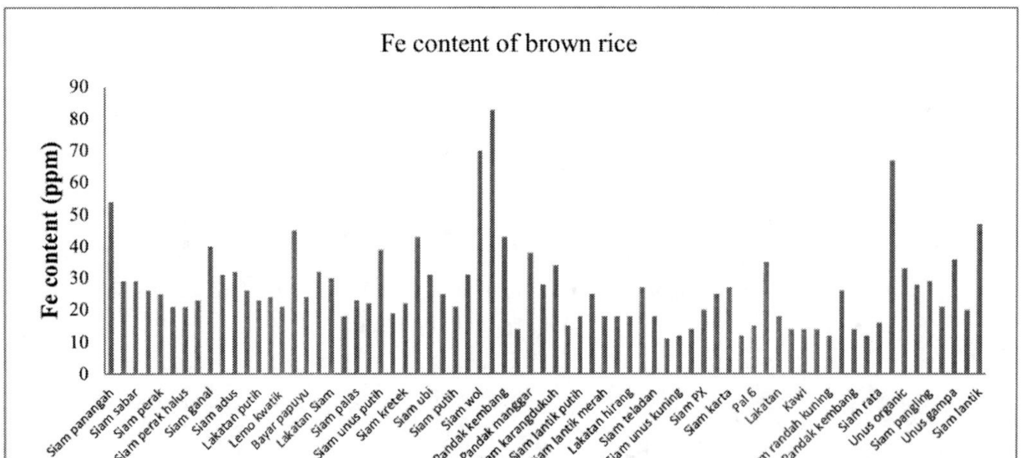

Figure 3. Fe content (ppm) of brown rice local tidal swamp varieties.

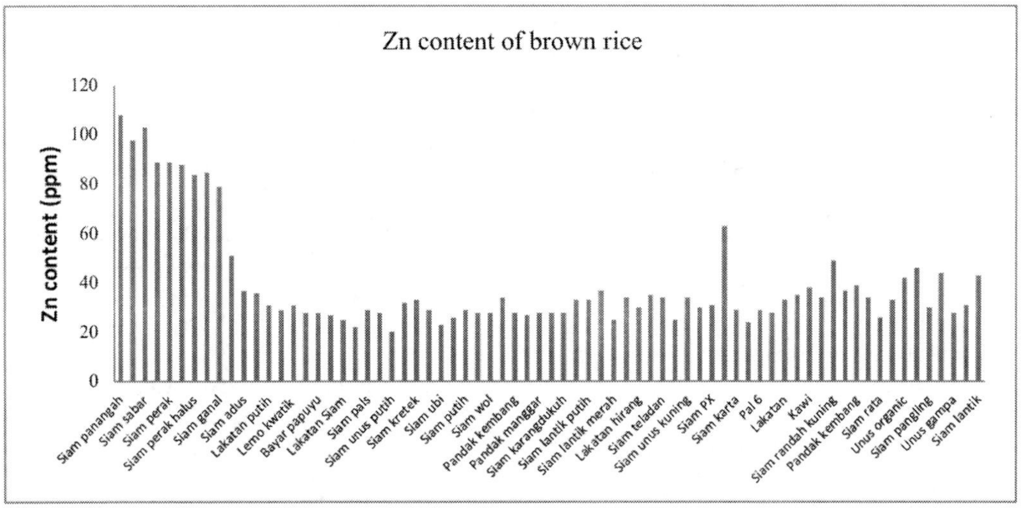

Figure 4. Zn content (ppm) of brown rice tidal swamp varieties.

The Fe and Zn content variation for HYV's (Margasari, Martapura, Mendawak, Kapuas, Banyuasin, IR66) or promising lines (GH 47, GH 137, GH 173) were not as wide as local varieties. The HYV's (Margasari 43 ppm dan Banyuasin 44 ppm Fe) showed Fe higher than Martapura, Mendawak, Kapuas, and IR66 varieties. The promising lines that had higher Fe content showed by GH 47 (70 ppm Fe) than GH 137 (32 ppm Fe) and GH 173 (32 ppm Fe). The lowest Fe content showed by Kapuas variety (20 ppm Fe). Meanwhile, the highest Zn content was Martapura variety 65 ppm Zn and the lowest Zn content was Mendawak variety 32 ppm Zn. For promising lines, the highest Zn content was GH 137 (63 ppm Zn) and the lowest was GH 173 (28 ppm Zn) (figure 5).

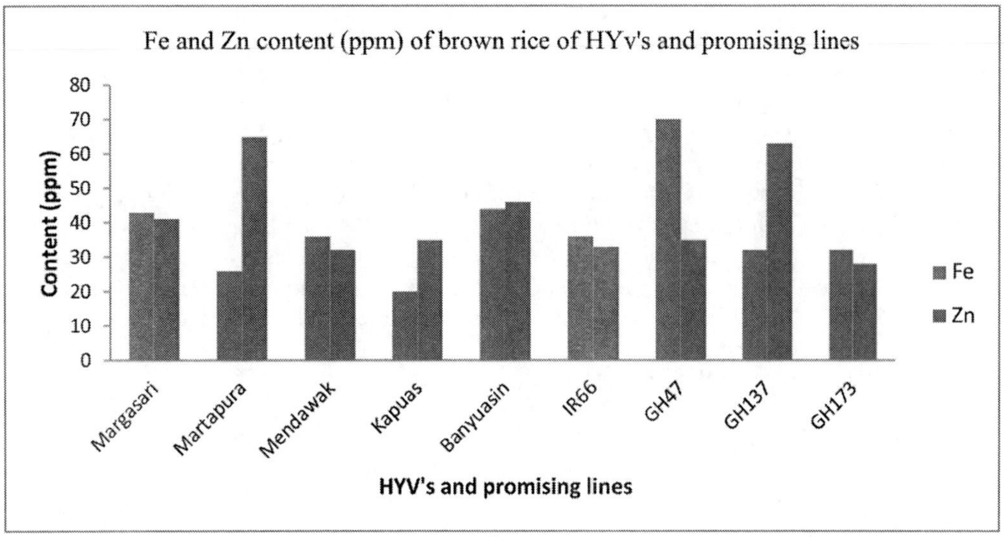

Figure 5. Fe and Zn (ppm) content of brown rice HYV's and promising lines planted same with local varieties.

The results of Yustisia [20] showed that levels of Fe and Zn of brown rice were varied from 5 HYV's (Ciherang, Widas, IR64, Cisokan, and Cimelati) planted in inseptisol soil (10.84 - 19.80 ppm Fe and 19.64 - 24.55 ppm Zn). The highest Fe content was indicated by Widas variety and the lowest was Cisokan. The highest Zn level was shown by Ciherang variety and the lowest was by Cimelati. The results of other studies of Fe and Zn levels of red rice and white rice using the co-AANI method showed that Fe levels were higher than Zn levels. Fe content of white rice was 46.53 ± 1.51 µg/g and red rice 61.07 ± 3.43 µg/g Fe. While Zn contents of white rice were 44.75 ± 0.53 µg/g and brown rice 19.27 ± 1.79 µg/g Zn [21]. While Indrasari [22] showed that the mean content of Fe and Zn in HYV's were 11.7 ppm Fe and 23.9 ppm Zn. Of the 71 brown rice local varieties of tidal swamp rice from South Kalimantan that was analyzed by its content of Fe and Zn, only Siam Panangah showed high levels of Fe and Zn, i.e. 54 ppm Fe and 108 ppm Zn.

3.1. Prospects of local varieties rice development

From the results it can be seen that there are several local varieties of tidal swamp rice from South Kalimantan which have higher Fe and Zn levels than the broken rice of other varieties. It is suspected that the high level of Fe content in rice had to do with the soil where the rice was planted. The location of this test was in acid sulphate soil containing high levels of Fe. The results of Dianawati [23] research showed that iron content in rice correlates with iron content in soil where rice is planted and has no correlation with iron content in water used to irrigate the rice.

If the brown rice was milled to milled rice it will reduce the content of Fe and Zn about 67% Fe and 13.2% Zn, respectively [24]. When referring to nutritional adequacy rate for Fe of 13 mg and Zn of 15 mg/capita/day [25], some local varieties that containing high Fe and Zn in groups 44-54 ppm, 66-76 ppm, and 77-87 ppm Fe in Figure 1 and group 59-71 ppm, 72-84 ppm, 85-97 ppm, and 98-110 ppm Zn in Figure 2 fulfill the mineral or nutrient adequacy of Fe and Zn.

The information on Fe and Zn content from local varieties of brown rice is very useful for practical purposes: (1) as a consumption material, especially for the mineral or nutrient intake needs of Fe and Zn; in addition (2) it is also useful for rice plant breeders for improving high yielding varieties with high Fe and Zn content in rice, which will help reduce iron and zinc intake from other food sources.

4. Conclusions

Local varieties of tidal swamp rice have great prospects for development because some of them have high Fe and Zn content in brown rice. The results of the analysis of Fe and Zn content of brown rice of 71 local tidal swamp rice varieties of South Kalimantan showed that content of Fe and Zn of brown rice varied greatly. The Fe content ranged from 11-83 ppm, where Fe content of local varieties were higher than high yielding varieties. Zn content was also with a ranged wide enough, which ranged 20 - 108 ppm Zn. Siam Panangah variety showed high content of Fe and Zn, which had 54 ppm Fe and 108 ppm Zn, respectively. The high Fe and Zn content can be used as parents in crossing to get high yielding varieties of rice with high Fe and Zn content. Moreover, it can also be used as direct consumption material to meet the intake of Fe and Zn minerals. The varieties tested were relatively tolerant to be planted in acid sulfate tidal swampland with high soil Fe content.

References

[1] Hidayati L, Hadi H and Kumara A 2010 Kekurangan energi dan zat gizi merupakan faktor risiko kejadian stunted pada anak usia 1-3 tahun yang tinggal di wilayah kumuh perkotaan Surakarta *J. Kesehatan* **4** 89-104

[2] Almatrsier S 2010 *Prinsip Dasar Ilmu Gizi* (Jakarta: PT Gramedia Pustaka Utama)

[3] Dinas Pertanian TPH Kalsel 2006 *Laporan Tahunan 2006* (Dinas Pertanian Tanaman Pangan dan Hortikultura Provinsi Kalimantan Selatan Banjarbaru)

[4] Zauhari R M 2001 *Pengembangan Lahan Basah di Dalam Otonomi Daerah* (Makalah pada Lustrum ke-8 Fakultas Pertanian Universitas Lambung Mangkurat Banjarbaru)

[5] Wiggin G 1976 *Buginese Agriculture in the Tidal Swamps of South Sumatera* (Lembaga Pusat Penelitian Pertanian Bogor)

[6] Kselik R A L 1990 *Workshop on Acid Sulphate Soils in the Humid Tropics* (Bogor: AARD-LAWOO)

[7] Widjaja-Adhi I P G 1986 Pengelolaan lahan rawa pasang surut dan lebak *J. Litbang Pertanian* **5 (1)**

[8] Silvius M J, Steerman A P J M, Berczy E T, Djuhara E and Taufik A W 1987 *The Indonesian Wetland Inventory, A Preliminary Compilation of Existing Information on Wetlands of Indonesia* (Bogor: PHPA AWB/Interwader Edwin)

[9] Maas A, Afandie R and Suryanto 1992 *Potensi dan Kendala Reklamasi Lahan Pasang Surut* (Makalah pada Pertemuan Nasional Pengembangan Pertanian Lahan Pasang Surut Cisarua 3-4 Maret 1992)

[10] Ikehashi H and Ponnamperuma F N 1978 *Varieties Tolerance of Rice for Adverse Soils* (Los Banos: Soil and Rice International Rice Research Institute)

[11] Ponnamperuma F N 1977 *Physico-Chemical Properties of Submerged Soil in Relation to Fertility* (IRRI Research Paper Series No. 5 Int. Rice Res. Ins. Los Banos The Philippines)

[12] Widjaja-Adhi I P G, Suriadikarta D A, Sutriadi M T, Subiksa I G M and Suastika I W 2000 *Pengelolaan Pemanfaatan dan Pengembangan Lahan Rawa* 127-64 In Adimihardja A, Amien L I, Agus F and Jaenuddin D *Sumberdaya Lahan Indonesia dan Pengelolaannya* (Pusat Penelitian Tanah dan Agroklimat Bogor)

[13] Jumberi A and Alihamsyah T 2005 *Pengembangan Lahan Rawa Berbasis Inovasi Teknologi* In Ar-Riza *et al* 2005 *Proc. Inovasi Teknologi Pengelolaan Sumberdaya Lahan Rawa dan Pengendalian Pencemaran Lingkungan* (Banjarbaru: Puslitbang Tanah dan Agroklimat) pp 11-42

[14] Khairullah I, Humairie R, Imberan M, Subowo S and Sulaiman S 2003 *Varietas Lokal Padi Pasang Surut Kalimantan Selatan: Karakterisasi dan Pemanfaatan* In Kasno A *et al Proc. Perhimpunan Ilmu Pemuliaan Indonesia (PERIPI)* (Balai penelitian Kacang-Kacangan dan Umbi-Umbian Malang)

[15] Khairullah I dan M Saleh 2014 *Sumberdaya lokal tanaman pangan lahan rawa Biodiversiti Rawa: Eksplorasi Penelitian dan Pelestariannya* Penyunting : Mukhlis *et al* Badan Penelitian dan Pengembangan Pertanian IAARD Press Jakarta 366 hal

[16] Khairullah I, Imberan M and Sulaiman S 1997 *Keragaan Agromorfologi Lima Varietas Padi Lokal yang Populer di Lahan Pasang Surut Kalimantan Selatan* In Maamun M Y, Arriza I, Simatupang R S, Noor M, Simanungkalit D and Rayitno B M *Proc. Pembangunan Pertanian Berkelanjutan Menyongsong Era Globalisasi* (Banjarmasin 13-14 Maret 1997)

[17] Khairullah I, Mawardi and Sarwani M 2006 *Karakteristik dan Pengelolaan Lahan Rawa: 7 Sumberdaya Hayati Pertanian Lahan Rawa* (Balai Besar Penelitian dan Pengembangan Sumberdaya Lahan Pertanian) pp 203-28

[18] Idak H 1982 *Perkembangan dan Sejarah Persawahan di Kalimantan Selatan* (Banjarmasin: Pemda Tk. I Kalimantan Selatan) p 40

[19] Khairullah I 2007 *Pros. Pertanian Lahan Rawa "Revitalisasi Kawasan PLG dan Lahan Rawa Lainnya untuk Membangun Lumbung Pangan Nasional"* (Kuala Kapuas: Badan Penelitian dan Pengembangan Pertanian) pp 339-48 ISBN 978-979-8253-63-8

[20] Yustisia, Tohari, Shiddieq D and Subowo G 2012 Pengkayaan besi (Fe) dan seng (Zn) dalam beras dan karakter penentu varietas padi sawah efisien pada tanah vertisol dan inseptisol *Agrotrop.* **2(1)** 67-75

[21] Mulyaningsih T R 2009 Kandungan unsur Fe dan Zn dalam bahan pangan produk pertanian, peternakan, dan perikanan dengan metode ko-AANI *J. Sains dan Tekno. Nuklir Indonesia* **10** 71-80

[22] Indrasari S D, Hanarida I and Daradjat A A 2002 *Indonesian Final Report Year I Breeding for Iron Dense Rice: A Low Cost Sustainable Approach to Reduce Anemia* (Asia International Food Policy Research Institute (IFPRI) & Indonesian Center Food Crop Research and Development (ICFCR))

[23] Dianawati N and Sugiarso R D 2015 Penentuan kadar besi selama fase pematangan padi menggunakan spektrofotometer UV-Vis *J. Sains dan Seni* **4** 2337

[24] Indrasari S D, Wibowo P and Daradjat A A 2008 *Proc. Sem. Nas. BB Padi* pp 1457-72

[25] Muhilal, Sadono A, Krisdinarmurtirin, Husaini, Sugih R and Khumaidi M 1989 *Angka Kecukupan Gizi Rata-rata yang Dianjurkan (AKG)* (Widyakarya Pangan dan Gizi Lembaga Ilmu Pengetahuan Indonesia)

IOP Conf. Series: Materials Science and Engineering **980** (2020) 012038 doi:10.1088/1757-899X/980/1/012038

Cadmium binding to antioxidant enzymes: in silico study

N Komari[1]* and E Suhartono[2]
[1] Department of Chemistry, Faculty of Mathematics and Natural Science, Lambung Mangkurat University, Banjarmasin 70712, South Kalimantan, Indonesia
[2]Department of Medical Chemistry/Biochemistry, Faculty of Medicine, Lambung Mangkurat University, Banjarmasin 70712, South Kalimantan, Indonesia

Email: nkomari@ulm.ac.id

Abstract. Cadmium (Cd) is a heavy metal that can be used for industry. Cd is toxic and can damage the kidneys, neurotoxic, glucose metabolism disorders and others. The reaction of Cd with protein causes oxidative stress, which is an imbalance of oxidants and antioxidant enzymes. Antioxidant enzymes include Superoxide dismutase (SOD), Catalase (CAT), and Glutathione reductase (GSR). There are not many studies that explain the interactions between Cd and antioxidant enzymes. For this reason, this research was carried out using in silico. The structure of the enzymes was obtained from the RCSB Protein Data Bank (http://www.rcsb.org) with the following code SOD (PDB ID 1E9Q), CAT (PDB ID 3RGP), and GSR (PDB ID 1GRT). Cd interactions with these enzymes were used by MIB: Metal Ion-Binding site prediction and docking server (http://bioinfo.cmu.edu.tw/MIB/). The interactions between Cd and amino acids of targeted protein were visualized on UCSF Chimera 1.14. The results of the study identified amino acid residues involved in the mechanism of binding of Cd with antioxidant enzymes. SOD binding sites with Cd were GLU-121 and SER-142; CAT binding sites were amino acid residues of HIS-230, ASP-228, HIS-181, HIS-109 and ASP-178; GSR binding sites were HIS-503 and TYR-502.

1. Introduction

Cadmium (Cd) is a heavy metal located in group 12 period 5 in the periodic system. This metal is widely used in various industries such as the battery industry, metal coating, the paint industry and others [1,2]. Cd is a pollutant that can disrupt ecosystem stability and diversity. The disruption of the ecosystem is determined by factors in the amount of pollutants, toxicity, and bioaccumulation [3].

Cd enters the body through water, food, and air. Furthermore, Cd is metabolized through a biotransformation mechanism involving cytochrome P-450 in the mitochondria. Cd binds covalently to various ligands, for example -OH, -COO-, -C=O, -SH, -S-S-, $-NH_2$ and -NH, which are contained in the residues of the amino acids making up proteins and enzymes. Cd bonding with these ligands will inhibit the activity of antioxidant enzymes, so oxidative balance disorders occur [4,5].

Previous studies have suggested that administration of Cd to teeth can cause a decrease in minerals Ca, P, Mg, and Zn [6]. Cd can also affect glucose metabolism *in vitro* [7,8] and induce neurotoxic [9] by involving oxidative stress. Oxidative stress is an imbalance between oxidants and enzymatic antioxidants, which includes Superoxide dismutase (SOD), Catalase (CAT), and Glutathione reductase (GSR). Superoxide dismutase works by catalyzing superoxide ($\cdot O_2$) to peroxide molecule, whereas CAT decomposes peroxide acid substrate into water and oxygen [10,11].

Tribowo [12] states that Cd can reduce the activity of SOD and CAT ovary enzymes. However, the research has not explained in detail the pathomechanism. Therefore, it is necessary to explain in detail about pathomechanism and interaction site of antioxidant enzymes that are SOD, CAT and GSR.

2. Materials and methods

This study tested 3 antioxidant enzymes, namely Superoxide dismutase (SOD), Catalase (CAT), and Glutathione reductase (GSR). The structure of the enzymes was obtained from the RCSB Protein Data Bank (http://www.rcsb.org) with the following code SOD (PDB ID 1E9Q), CAT (PDB ID 3RGP) and GSR (PDB ID 1GRT). Cd interactions with these enzymes were used by MIB: Metal Ion-Binding site prediction and docking server (http://bioinfo.cmu.edu.tw/MIB/). The interactions between Cd and amino acid of targeted protein were visualized on UCSF Chimera 1.14 [13].

3. Results and discussion

The SOD enzyme has the PDB code 1E9Q. This enzyme was released on December 3, 2000 with a resolution of 1.75 Å. SOD enzyme from the *Boss taurus* has sequence length of 151 with the name of the SOD1 gene with E.C: 1.15.1.1 [14]. The CAT used in this study was Catalase with PDB code 3RGP. This enzyme was released on May 11, 2011 with a resolution of 1.88 Å. An enzyme was a catalase from *Boss taurus*, with sequence length of 499 and has 4 chains, A, B, C and D whose gene name is CAT with E.C: 1.11.1.6 [15]. The GSR enzyme has PDB code 1GRT. This enzyme was released on June 16, 1997 with a resolution of 2.30 Å. GSR originates from *Homo sapiens* with sequence length of 478 with chain A, with E.C: 1.6.4.2 [16]. Each enzyme is shown in figure 1. Interactions of antioxidant enzymes with cadmium occurred in several amino acid residues. The results of the interaction can be seen as in table 1.

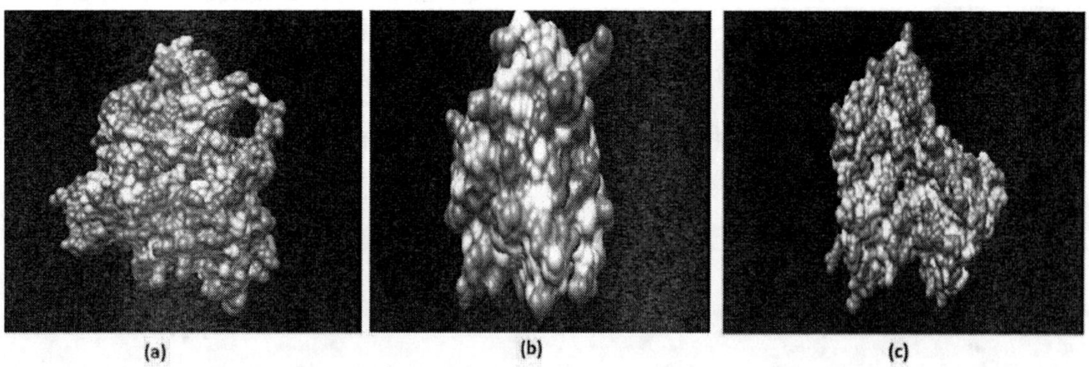

Figure 1. Antioxidant enzymes: (a) Catalase, (b) Superoxide dismutase and (c) Glutathione reductase.

The SOD enzyme showed that 2 amino acid residues interacted with cadmium. GLU-121 and SER-142 residues formed metal bonded with Cd with bond lengths of 2.326 Å and 2.421 Å (table 1). These metal bonds were very strong and caused Cd to be strongly bound by enzymes. These 2 amino acid residues formed the binding site of the SOD enzyme.

Table 1. Amino acid interaction with Cd.

Antioxidant Enzyme	Amino acid residues	Interactions	Distances
SOD	GLU-121	Metal interaction	2.326 Å
	SER-142	Metal interaction	2.421 Å
CAT	HIS-230	Metal interaction	2.510 Å
	ASP-228	Metal interaction	2.327 Å
	HIS-181	Metal interaction	2.312 Å
	HIS-109	Hidrophobic interaction	--
	ASP-178	Hidrophobic interaction	--
GSR	HIS-503	Metal interaction	2.264 Å
	TYR-502	Hidrophobic interaction	--

The binding site of Cd to SOD and amino acid residues which were formed in bond formation is presented in figure 2. Cd was bound to the surface of enzymes which were generally dominated by polar amino acid residues.

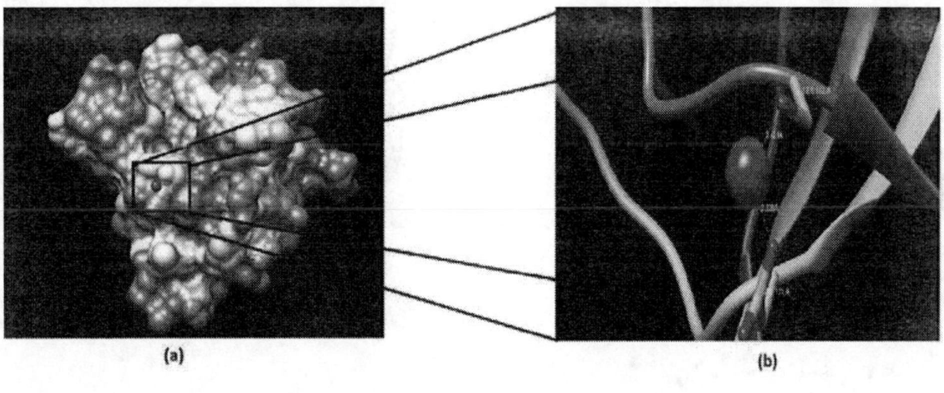

Figure 2. Interactions of Cd and the SOD enzyme: (a) Cd was bound to the surface; (b) amino acid residues bound to Cd.

Table 1 shows that 5 amino acid residues in the CAT interacted with Cd. HIS-230, ASP-228, and HIS-181 residues formed metal bonds with Cd of 2.510 Å; 2.327 Å; and 2.312 Å respectively. These metal bonds were very strong and caused Cd to be strongly bound by enzymes. HIS-109 and ASP-178 residues formed ionic bonds with Cd. This bond supported the strength of existing metal bonds, so that the interaction of Cd and enzymes became very stable. These 5 amino acid residues formed the binding site of CAT. The place where Cd was bound to CAT and the amino acid residues that were involved in forming bonds is shown in figure 3. Cd was bound to the surface of enzymes which were generally dominated by polar amino acid residues.

IOP Conf. Series: Materials Science and Engineering **980** (2020) 012038 doi:10.1088/1757-899X/980/1/012038

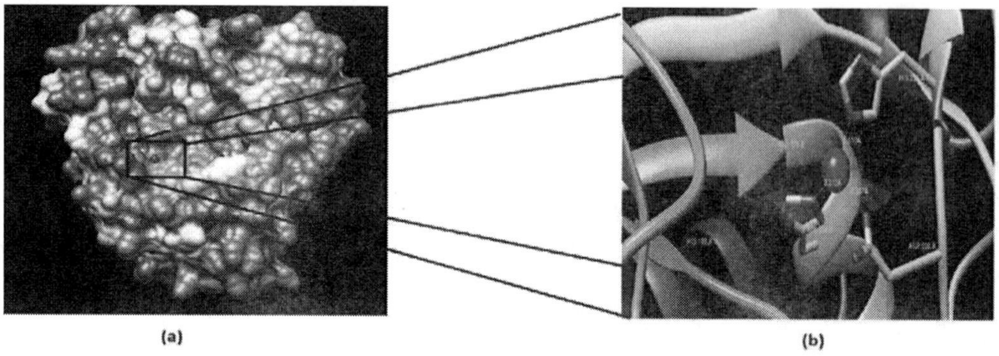

Figure 3. Interactions of Cd and catalase enzymes: (a) Cd bound to the surface; (b) amino acid residues bound to Cd.

Table 1 also shows the 2 amino acid residues of the GSR enzyme that bound to Cd. HIS-503 residue formed a metal bond with Cd with a bond length of 2.264 Å and TYR-502 residue formed an ionic bond with Cd. These two residues became the binding site of the GSR enzyme to act with Cd. Figure 4 shows that Cd was bound to the surface of an enzyme that was dominated by polar amino acid residues.

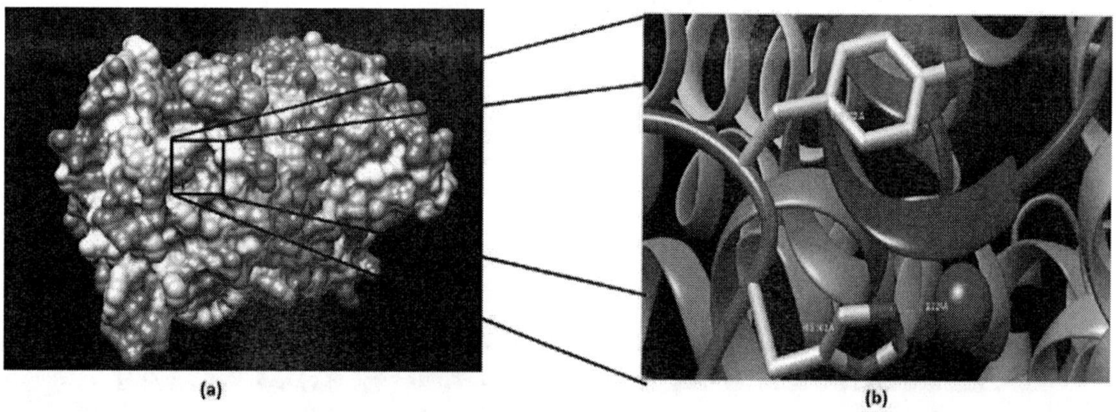

Figure 4. Interactions of Cd and GSR enzyme: (a) Cd bound to the surface; (b) amino acid residues that bind to Cd.

In the enzyme with PDB ID 4EVD, Yokohama [17] illustrates that Cd generally interacts with HIS, GLU and ASP residues. While, Leinala [18] in PDB ID 1NP8 shows that amino acid residues that bind to Cd include: ASP, GLU, THR, HIS, ALA and LYS. This is generally a polar amino acid.

4. Conclusion

The results of the study identified amino acid residues involved in the mechanism of binding of cadmium with antioxidant enzymes. SOD binding sites with Cd were GLU-121 and SER-142; CAT binding sites were amino acid residues of HIS-230, ASP-228, HIS-181, HIS-109 and ASP-178; GSR binding sites were HIS-503 and TYR-502. Then, the in silico study can predict the oxidative mechanism of Cd against antioxidant enzymes. Therefore, it can explain the occurrence of oxidative stress due to the interaction of Cd with SOD, CAT, and GSR.

References

[1] Liu J, Qu W and Kadiiska M B 2009 *Toxicol. Appl. Pharmacol.* **238(3)** 209-14
[2] Sharma H, Rawal N and Mathew B 2015 *Int. J. of Nanotech. and Nanosci.* **3** 1-9
[3] Andjelkovic M *et al* 2019 *Int. J. Environ. Res. Public Health* **16(274)** 1-21
[4] Marisa D, Aflanie I, Muthmainah N and Suhartono E 2019 *J. of Physics: Conf. Series* **1374** 012034
[5] Suhartono E, Thalib I, Aflanie I, Noor Z and Idroes R 2018 *IOP Conf. Series: Materials Science and Engineering* 350(1)
[6] Ratnasari D, Permana M, Widodo, Yanuar I, Dewi N and Suhartono E 2016 *Int. J. of Pharm. and Clinic. Research* **8(8)** 1240-44
[7] Suhartono E, Bahriansyah M and Triawanti 2016 *Int. J. of Pharm. and Clinic. Research* **8(4)** 248-53
[8] Suhartono E, Iskandar and Santosa P 2015 *Int. J. of Pharmaco. and Pharmace. Sci.* **7**(11) 17-20
[9] Suhartono E, Iskandar, Hamidah S and Arifin Y 2016 *Int. J. of Toxic. and Pharm. Res.* **8(1)** 1-6
[10] Budianto W, Khotimah H and E Suhartono 2019 *Indonesian Biomedic. J.* **11(2)** 159-66
[11] Faisal M, Maulida S, Mairudi S and Suhartono E 2019 *AIP Conference Proceedings* 2108
[12] Tribowo J A, Arizal M H, Nashrullah M, Aditama M R and Utama D G 2014 *Int. J. of Chem. Engin. and App.* **5(3)** 254-58
[13] Pettersen E F, Goddard T D, Huang C C, Couch G S, Greenblatt D M, Meng E C and Ferrin T E 2005 *J. Comput. Chem.* **25(13)** 1605-12
[14] Hough M A, Strange R W and Hasnain S S 2000 *J. Mol. Biol.* **304** 231
[15] Purwar N, McGarry J M, Kostera J, Pacheco A A and Schmidt M 2011 *Biochem.* **50** 4491-503
[16] Stoll V S, Simpson S J, Krauth-Siegel R L, Walsh C T and Pai E F 1997 *Biochem.* **36** 6437-47
[17] Yokoyama H, Tsuruta O, Akao N and Fujii S 2012 *Biochem. Biophys. Res. Commun.* **422** 745-50
[18] Leinala E K, Arthur J S, Grochulski P, Davies P L, Elce J S and Jia Z 2003 *Proteins* **53** 649-55

IOP Conf. Series: Materials Science and Engineering **980** (2020) 012039 doi:10.1088/1757-899X/980/1/012039

Investigation on the antibacterial activity of the methanol extract of purun tikus root (*Eleocharis dulcis*)

K Rosyidah[1,2*], L A P Sari[1] and T Rohman[1]
[1] Department of Chemsitry, Faculty of Mathematics and Natural Science, Lambung Mangkurat University, Banjarmasin, South Kalimantan, Indonesia
[2] Laboratory of Organic & Biochemistry, Faculty of Mathematics and Natural Science, Lambung Mangkurat University, Banjarmasin, South Kalimantan, Indonesia

*Email: krosyidah@ulm.ac.id

Abstract. One of the vegetations growing in tidal swampland and lebak is purun tikus (*Eleocharis dulcis)* which belongs to the *Cyperaceae* family. This research studied the antibacterial activity of methanol extract from purun tikus root on *Staphylococcus aureus* ATCC 25923 and *Escherichia coli* ATCC 25922 bacteria. The chemical compounds of purun tikus root were extracted using the maceration method with methanol solvent, followed by thickening the filtrate using a rotary evaporator. The results of phytochemical screening showed that purun tikus root contains phenolic, tannin, flavonoid, and terpenoid compounds. The result showed that this extract has the activity to inhibit the growth of *S. aureus* ATCC 25923 and *E. coli* ATCC 25922 bacteria at a concentration of 2500 μg/ml and 3000 μg/ml, respectively. The Minimum Inhibitory Concentration (MIC) value of the methanol extract from purun tikus *(E. dulcis)* root for both *S. aureus* ATCC 25923 and *E. coli* ATCC 25922 bacteria was 6,25 μg/ml. Based on the linear regression calculation, *S. aureus* ATCC 25923 and *E. coli* ATCC 25922 inhibited at a concentration of 0.96 μg/ml and 1μg/ml, respectively. The GC-MS spectrum showed that methanol extract of purun tikus contained compounds of the furfural group, namely furfural, 5-methyl furfural, and furfuryl alcohol.

1. Introduction

The southern Kalimantan region consists mostly of swamps. The dominant growth in the tidal and tidal swamps is the purun tikus (*Eleocharis dulcis*). Society in general uses purun tikus leaves as handicrafts in the form of bags and mats [1]. Other benefits of these plants as animal feed, especially for swamp buffaloes such as in Pandak Daun Village, South Kalimantan [2]. Research by Asikin and Thamrin [2] shows that purun tikus can act as a natural controller for white rice stem borer pests. In addition to being a trap plant, purun tikus also acts as a place to live parasitoids and predators. Baehaki et al. [3] reported the antibacterial test of ethanol, ethyl acetate, and n-hexane extracts of purun tikus leaves against the bacteria *Pseudomonas aeruginosa, Bacillus subtilis*, and *Vibrio cholera*. All extracts showed antibacterial activity against all test bacteria. Of the three bacterial species, the highest antibacterial activity is against *B. subtilis* bacteria when compared with *P. aeruginosa* and *V. cholera* bacteria. Zhan et al. [4] also reported the presence of antibacterial activity from methanolic extracts of purun tikus tuber skin against three pathogenic bacteria *Staphylococcus aureus, Escherichia coli,* and *Listeria monocytogenes*. The compounds in the tubers of these aquatic plants also have antitumor and antioxidant activity [5].

Based on the description above, further research is carried out on purun tikus especially the root as an antibacterial. The use of these bacteria in research is because *S. aureus* bacteria represent gram-positive while *E. coli* bacteria represent gram-negative. Both of these bacteria are pathogens that cause disease in animals and humans.

This article will discuss the methanol extract of purun tikus root as an antibacterial and analyze its chemical content using phytochemical screening methods and GC-MS analysis.

2. Materials and methods

2.1 Materials

The materials used the root of purun tikus, originating from Purun village South Kalimantan, *S.aureus* ATCC 25923 and *E.coli* ATCC 25922, aqua dest, FeCl3 1% (Merck), Mg powder (Merck), concentrated HCl (Merck), 2N HCl (Merck), Mayer reagents, Wagner reagents, Dragendroff reagents, glacial acetic acid (Merck), methanol (pa), NA (nutrient agar), NB (nutrient broth), disc paper, 70% alcohol, sterile cotton, cotton swab, chloramphenicol, DMSO.

2.2 Extraction

The extraction was followed from the research by Rosyidah [18]. The root of purun tikus was washed with running water then cut into pieces and dried. Purun tikus root which had been cleaned, dried at room temperature without direct sunlight exposure. The stages were intended so that the active compound in the sample was not damaged. Also, to reduce water content, stop enzymatic reactions, and prevent mold growth so that it can be stored for a long time [6]. The dried sample was then mashed using a blender to form a powder, aiming to expand the surface of the sample. The smaller the size, the greater the surface area so that the extraction process would be more effective. Purun tikus root powder obtained as much as 109.02 g.

Desiccated deserts were macerated in methanol solvent for 1x24 hours. Purun tikus root powder macerated with 1.5 L methanol solvent for 1 x 24 hours at room temperature, placed in a closed container, and protected from sunlight. Methanol is the most widely used solvent for the extraction process of natural product compounds because it can dissolve all classes of secondary metabolites. This maceration process produces brown Maserati. The filtrate was filtered and the methanol vaporized. The residue was macerated again with a new methanol solvent until clear. The results of Maserati were filtered. The filtrate was evaporated using a pressure rotary vaporizer at 40°C. Methanol extract obtained as much as 5.5g brown with a yield of 5.04%.

2.3 Phytochemical screening

This screening test included metabolite compounds of phenol, flavonoids, tannins, alkaloids, terpenoids and steroids [18].

2.4 Antibacterial test

Zhan [4] shared the antibacterial test. A total of 1-2 ose test bacteria were inoculated into Nutrient Agar slants, then incubated in an incubator at 37°C for 18-24 hours. Antibacterial testing was carried out by the agar diffusion method, using agar nutrient (NA) media. NA media on Petri dishes scratched with test bacterial. A blank disc paper with a diameter of 6 mm was dropped with 50 μl of the sample with concentrations of 3000, 2500, 2000, 1500, and 1000 μg/ml, then placed on NA media that had been inoculated with test bacteria. Incubation was carried out at 37°C for 1 x 24 hours. Observations were made on the formation of inhibition zones around the disc paper. The same procedure was carried out on positive control (chloramphenicol) and negative control (DMSO). The method used for MIC analysis was the agar diffusion method. The extract was then determined by the MIC value using variations in the concentration of 100, 500, 250, 125, and 6.25 μg/ml.

3. Result and discussion

The results of the phytochemical test showed that the extract of the purun tikus root contained several classes of secondary metabolite compounds which can be seen in table 1. Secondary metabolite compounds contained in methanol extracts of purun tikus root are phenols, tannins, flavonoids, and terpenoids. The same results were also shown by others who reported that methanol extracts of purun tikus leaves contain secondary metabolites such as phenolic, tannin, flavonoid, and terpenoids [7].

Table 1. Result of phytochemical screening.

Phytochemical test	observed	result
phenol	green color	+
flavonoid	red tomato's color	+
tanin	dark blue color	+
terpenoid	red color	+
steroid	blue or purple color	-
alkaloid	White precipitate (Mayer reagent)	-
	Brown sediment (Wagner reagent)	-
	Orange sediment (Dragendroff reagent)	-

The antibacterial activity test in this study used the disc diffusion method (paper disc). The inhibitory activity against bacterial growth was characterized by clear zones formed around the test extract shown in figure 1. The extract used for the activity test was first dissolved with DMSO which also negatively control. A negative control was a solvent that did not affect the activity test [8]. While the positive control used was chloramphenicol. Chloramphenicol was a broad-spectrum bacteriostatic antibiotic that was active against gram-positive and gram-negative organisms [9].

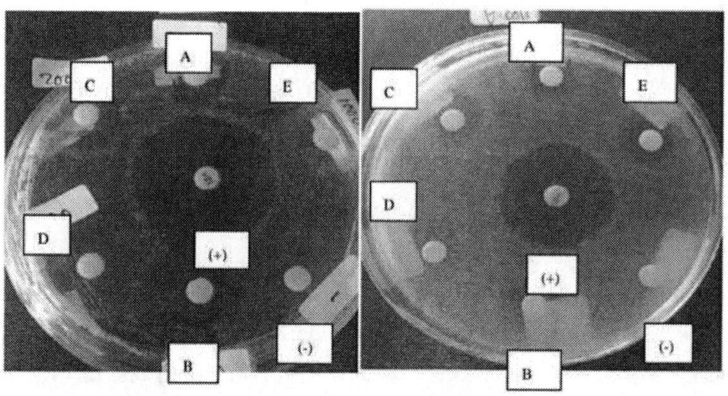

Figure 1. Test of the antibacterial activity of methanol extract of purun tikus root against *S. aureus* ATCC 25923 and *E. coli* ATCC 25922.
A: 1000 μg / ml
B: 500 μg / ml
C: 250 μg / ml
D: 125 μg / ml
E: 6.25 μg / ml
(+):chloramphenicol
(-): DMSO

The measurement of inhibition zone diameter showed that the methanol extract of root purun tikus (*E. dulcis*) against *S. aureus* and *E. coli* bacteria showed antibacterial activity. The inhibition zone diameter around 5 mm shows that the inhibitory activity is categorized as weak [10].

The methanol extract of purun tikus root can inhibit the growth of *S. aureus* (left) and *E. coli* (right) bacteria, it may be caused by the content of secondary metabolites such as flavonoids, tannins, phenolics, and terpenoids as evidenced by the results of phytochemical tests. Baehaki et al. [3]

reported secondary metabolite compounds such as flavonoids, tannins, and terpenoids able to inhibit the growth of *E. coli* bacteria.

In table 2, it can be seen that *S.aureus* was the most sensitive bacterium for methanol extract of purun tikus root. The results of research conducted by other reported that the bacterium *S. aureus* has a larger inhibitory diameter than that of the *E.coli* bacteria [4]. Inhibition zones formed are also influenced by factors such as organism sensitivity, culture media, incubation conditions, microorganism concentration, and media composition [11].

Table 2. Antibacterial activity of methanol extract of purun tikus root.

Extract Concentration (µg/ml)	Inhibition zone (mm)	
	S. aureus ATCC 25923	*E. coli* ATCC 25922
3000 µg/ml	12.9	10.2
2500 µgm/l	13.3	9.8
2000 µg/ml	11.2	7.1
1500 µg/ml	10.0	6.7
1000 µg/ml	7.2	6.3
500 µg/ml	6.9	4.8
250 µg/ml	5.2	4.6
125 µg/ml	3.8	3.5
6,25 µg/l	2.8	0.7
Chloramphenicol (30 µg)	22.7	23.6
Kontrol negatif	-	-

The content of secondary metabolites such as flavonoids, phenols, tannins, and terpenoids are known to have the ability to be antibacterial. The phenol group can damage the cell membrane, activate enzymes, and denaturation proteins so that the cell wall is damaged due to decreased permeability [12]. Flavonoid compounds are antibacterial compounds that can denaturation bacterial cell proteins and damage cell membranes [13] function again so that protein damage or denaturation occurs. The denaturation causes protein coagulation and disrupts the metabolism and physiological function of bacteria [14].

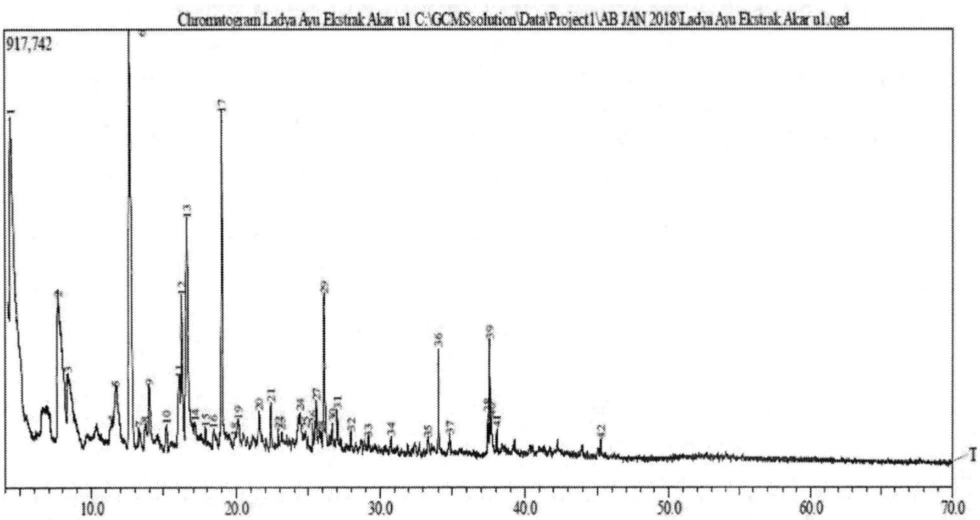

Figure 2. Chromatogram gas chromatography of methanolic extract of purun tikus root.

Tannins have a mechanism of coagulating and denaturing binding proteins with proteins that form H^+ ions and cause pH to become acidic so that proteins are denatured. Acidic conditions inactivate enzymes in bacteria and cause impaired metabolism and cell damage and even death [15]. Whereas terpenoids have antibacterial activity associated with damage to cell membranes by lipophilic compounds [14].

Inhibitory zone ability is also influenced by differences in the structure of gram-positive and gram-negative bacterial cell walls. Bacterial cell walls consist of layers of peptidoglycan that differ between gram-positive bacteria and gram-negative bacteria. Gram-positive bacterial cell walls contain 90% and a thin layer of theatric acid while in gram-negative bacteria the cell walls only contain 5-10% peptidoglycan, the rest consists of protein, lipopolysaccharides, and lipoproteins [16].

Based on GC-MS analysis (figure 2), methanol extract of purun tikus root contained three main compounds. They were furfural (15.02%) as peak 1, 5-methyl furfural (11,82%) as peak 2 and furfuryl alcohol (17.64%) as peak 6. The furfuryl alcohol compound was a reduced furfural. The 5-methyl furfural compound was reported to be present in the methanol extract of *Cyperus rotundus* tubers from the same family as purun tikus [17].

4. Conclusion

The methanol extract of purun tikus root is active antibacterial against gram-positive and gram-negative bacteria. *Staphylococcus aureus* bacteria which represented gram-positive and *E. coli* bacteria as gram-negative. The MIC value of *S. aureus* and *E. coli* bacteria was 6.25 μg/ml. Phytochemical screening results indicated the presence of phenol compounds, flavonoids, tannins, and steroids. But GC-MS analysis showed that the methanol extract contained the main components of furfuryl alcohol, furfural, and 5-methyl furfural compounds.

Acknowledgments

Thank you to the DIPA of Lambung Mangkurat University 2017 Fiscal Year Number: SP-DIPA-042.01.2.400957/2018 dated December 5, 2017, for funding this research.

References

[1] Sunardi and Istikowati W T 2012 Analisis kandungan kimia dan sifat serat tanaman purun tikus (*Eleocharis dulcis*) *Bioscientiae* **9(2)** 15–25
[2] Asikin, S and Thamrin M 2012 Manfaat purun tikus (*Eleocharis dulcis*) pada ekosistem sawah rawa. *J. Lit. Pertanian* **31(1)** 35-42
[3] Baehaki A, Herpandi and Putra A A 2018 Antibacterial activity of extract from swamp plant eleocharis dulcis *Orien. J. of Chem.* **34(1)** 573- 75
[4] Zhan G, Pan L Q, Mao S B, Zhang W, Wei Y and Tu K 2014 Study on antibacterial properties and major bioactive constituents of chinese water chestnut (*Eleocharis dulcis*) peels extracts/ fractions *Eur. Food. Res. Tech.* **238(5)** 789-96
[5] Zhan G, Pan L Q, Tu K and Jiao S 2016 Antitumor, antioxidant and nitrite scavenging effects of chinese water chestnut (Eleocharis dulcis) peel flavonoids *J. of Food Sci.* **81(10)** 2578-86
[6] Lenny S and Zuhra C F 2006 Isolasi dan uji bioaktivitas kandungan kimia utama puding merah (*Graptophyllum pictum* L. Griff) dengan metode uji *brine shrimp J. Komun. Penel. MIPA* **17(5)** 56-9
[7] Rosyidah K, Rohman T and Fitriani R 2018 Aktivitas antioksidan ekstrak metanol daun purun tikus (*Eleocharis dulcis*) *J. Kimia dan Pend. Kimia* **3(3)** 135-40
[8] Natheer S E, Sekar C, Amutharaj P, Rahman M S A and Keroz K 2012 Evaluation of antibacterial activity of *Morinda cirifolia, Vitex trifolia and Chromolea odorata Afri. J. of Pharm. and Pharm.* **6(11)** 783-8
[9] Brooks G F, Morse S A, Butel J S, Carroll K C and Mietzner T A 2013 *Mikrobiologi Kedokteran* 25th edition (Jakarta: EGC)

IOP Conf. Series: Materials Science and Engineering **980** (2020) 012039 doi:10.1088/1757-899X/980/1/012039

[10] Davis W W and Stout T R 1971 Disc plate method of microbiological antibiotic essay *J. of Microbio.* **22(4)** 659-65
[11] Srivastava P, Logesh A R, Upreti D K, Dhole T N and Srivastava A 2013 In vitro evaluation of some indian lichens against human pathogenic bacteria *Mycosphere* **4(4)** 734-43
[12] Purwatiningsih I K, Suranindiyah Y Y and Widodo 2014 Aktivitas senyawa fenol dalam buah mengkudu (*Morinda citrifolia*) sebagai antibakteri alami untuk penghambatan bakteri penyebab mastitis. *J. Bulet. Petern.* **38(1)** 59-64
[13] Ngajow, Mercy, Jemmy A and Vanda S K 2013 Pengaruh antibakteri ekstrak kulit batang matoa (*Pometia pinnata*) terhadap bakteri *Staphylococcus aureus* secara in vitro *J. MIPA UNSRAT Online* **2(2)** 128-32
[14] Heni, Arreneuz S and Zaharah T A 2015 Efektivitas antibakteri ekstrak kulit batang belimbing hutan (*Baccaurea angulata* merr.) terhadap *Staphylococus aureus* dan *Escherichia coli J. Kimia Khatulis.* **4(1)** 84-90
[15] Mayanti T, Julaeha E and Putri Y 2011 Isolasi dan karakterisasi senyawa antibakteri dari fraksi etil asetat kulit batang *lansium domesticum* corr. CV Kokossan (Bandung: Universitas Padjajaran Fakultas MIPA) pp 10-11
[16] Brooks G F, Carroll K C, Morse S A, Mietzner T A and Butel J S 2013 *Medical Microbiology* 26th edition (E-book: The McGraw-Hill)
[17] Kusuma A V C, Chozin M A and Guntoro D 2017 Senyawa fenol dari tajuk dan umbi teki (*Cyperus rotundus* L.) pada berbagai umur pertumbuhan serta pengaruhnya terhadap perkecambahan gulma berdaun lebar *J. Agro. Indo.* **45(1)** 100-7
[18] Rosyidah K, Puspa L A and Rohman T 2018 Aktivitas antioksidan ekstrak metanol daun purun tikus (*Eleocharis dulcis*) *J. Kimia dan Pend. Kimia* **3(3)** 135-40

IOP Conf. Series: Materials Science and Engineering **980** (2020) 012040 doi:10.1088/1757-899X/980/1/012040

The antioxidant activity of white kapul (*Baccaurea macrocarpa*) fruit rinds

M D Astuti[1,2*], W F Ana[1], K Rosyidah[1], Rodiansono[1]
[1]Department of Chemistry, Faculty of Mathematics and Natural Sciences, Lambung Mangkurat University, Jalan A. Yani km 36 Banjarbaru Indonesia
[2]Laboratory of Organic and Biochemistry, Faculty of Mathematics and Natural Sciences, Lambung Mangkurat University

Email: mdastuti@ulm.ac.id

Abstract. The antioxidant activity of white kapul (Baccaurea macrocarpa) fruit rinds was investigated in this research. Baccaurea macrocarpa fruit rinds were extracted with n-hexane, ethyl acetate, and methanol, consecutively. All extracts were determined for their antioxidant activity based on the DPPH method. The yields from hexane, ethyl acetate, and methanol extract were 0.14%, 0.64%, and 0.94%, respectively. The highest antioxidant activity was observed in methanol extract (IC50 22.968 ppm), followed by the activity from ethyl acetate extract (IC50 29.741 ppm), and hexane extract (IC50 141.931 ppm). As a comparison, the IC50 of vitamin C was 5.019 ppm.

1. Introduction

The genus *Baccaurea* (Euphorbiaceae) is a tropical fruit tree that spreads from India, Borneo, Sumatra, Peninsular Malaysia, Thailand, Philippines, to the Pacific island [1]. *Baccaurea* genus consists of 43 species, such as *B. motleyana* (rambai), *B. angulata* (belimbing darah), *B. macrocarpa* (kapul or tampoi), *B. brevipes* (rambai tikus), *B. velutine*, *B. reticulata*, *B. lanceolata* (limpasu), *B. polyneura* (jentik), and *B. ramiflora*.

There is a lot of research on *Baccaurea* genus. Nutrient composition of *B. angulata* on whole fruit, skin, and berry i.e moisture, ash, protein, total fat, carbohydrate, and crude fiber have reported [2]. Skin or fruit rind's of *B. angulata* contained the highest total phenol (16.58 mgGAE/g), total flavonoid (31.05 mgQE/g) dan total antocyanin (0.72 mgc-3-g/100g) than the whole fruit and berry of *B. angulata* [2]. Fruit seed of *B. macrocarpa* contained fiber 2.2%, fat 1.1%, carbohydrate 34.6%, protein 1.5%, moisture 61.9% and vitamin C 1.5% [3]. The pericarp contained the highest of total phenol (60.04±0.53 mgGAE/g), total flavonoid (44.68± 0.67 mgQE/g), total antocyanin (1.23±0.20 mgc-3-gE/100g), and total carotene (0.81±0.14 mgCE/g) than the flesh and seed of *B. macrocarpa* [4].

The whole fruit, skin, and berry of *B. angulata* have antioxidant activity by DPPH method, TEAC/ABTS assay and FRAP assay [2]. The juice of pulp, whole fruit, and skin of *B. angulata* inhibits lipid peroxidation and induces the increase in antioxidant enzyme activities [5]. Methanol extract of flesh, pericarp, and seed of *B. macrocarpa* and *B. lanceolata* showed antioxidant activity [4].

B. macrocarpa or known as white kapul produces edible fruits (pulp or berry) and their fruit rinds are not used. Therefore, the research aimed to evaluate the antioxidant activities of *B. macrocarpa* fruit rind's extract. Fruit rinds were extracted with n-hexane, ethyl acetate, and methanol, consecutively. Antioxidant activity was evaluated by the DPPH method.

2. Materials and methods

2.1 Materials

Fruits of *B. macrocarpa* were collected from Banjar, South Kalimantan, methanol p.a (Merck), 2,2-Diphenyl-1-Picryl Hydrazil (Aldrich), methanol (technical), ethyl acetate (technical), n-hexane (technical). All technical solvents were distilled prior to use. Ultra Violet-Visible Spectrophotometer (GENESYS 10S), rotary vacuum evaporator, and glasses apparatus were used.

2.2 Preparation of fruit rind's

Fruits of *B. macrocarpa* were washed with water and the fruit rinds were separated from the edible parts. The fruit rinds were cut into small pieces and dried up at room temperature. The air-dried fruit rinds were powdered with the milling machine.

2.3 Extraction of fruit rind's

Approximately, 500 g of the air-dried powdered fruit rinds were extracted with n-hexane (ratio solvent 1:5) for 24h and filtered. Remaceration was repeated 2 times. The residue was extracted with ethyl acetate for 24h and filtered. Remaceration was repeated 2 times. The residue was extracted with methanol and filtered. Remaceration was repeated 2 times. The extracts were concentrated with a rotary vacuum evaporator to get 0.68 g of hexane extract, 3.21 g of ethyl acetate extract, and 4.71 g of methanol extract. The percentage yield was determined as the following equation 1.

$$Yield\ (\%) = \frac{mass\ of\ crude\ extract\ (g)}{mass\ of\ fine\ powder\ (g)} x100 \qquad (1)$$

2.4 Antioxidant activity

The antioxidant activity based on the free radicals of DPPH was determined according to the method described by Molyneux [6] with slight modification. Different concentrations of hexane, ethyl acetate, and methanol extract in the methanol solvent (2 mL) were mixed with 2 mL DPPH 0,1mM. After incubation for 30 minutes in the dark at room temperature, the absorbance was recorded at a maximum wavelength (516 nm). Ascorbic acid was used as a standard. The controls contained all reagents except the extract or standard (ascorbic acid). The decreased absorbance DPPH indicates an increase of DPPH radical scavenging activity. The scavenging of free radical DPPH was estimated based on the following equation 2.

$$Inhibition\ of\ free\ radical\ (\%) = \frac{Ao - As}{Ao} x100\% \qquad (2)$$

Ao was the absorbance of control reaction and As was the absorbance of solutions with the sample extract or standard. The value was presented in triplicates analysis. The IC_{50} was the inhibition concentration that could scavenge 50% DPPH. The IC_{50} value for each extract was graphically determined by plotting the inhibition of free radical DPPH percentage (y-axis) and extract concentration (x-axis).

3. Results and discussions

3.1 Extract yields

Fruit rinds of *B. macrocarpa* were extracted with solvents with different polarities, consecutively. Extraction was done at room temperature. The extraction of fruit rinds of *B. macrocarpa* initiated with n-hexane, followed by ethyl acetate, and terminated by methanol. The gradual extraction would extract compounds with different polarities.

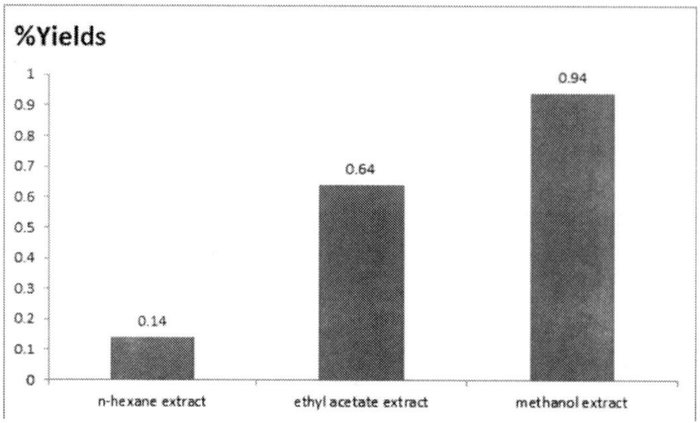

Figure 1. Yields of various extracts.

Table 1. Antioxidant activity of extract *B. macrocarpa* and reference standard ascorbic acid.

Extract/Standard	Concentration (ppm)	Inhibition of free radical (%)	IC50 (ppm)
n-Hexane	40	4.716	141.931
	60	12.301	
	80	20.650	
	100	28.298	
	120	39.771	
	140	51.052	
Ethyl acetate	15	18.986	29.471
	20	31.169	
	25	45.207	
	30	52.938	
	35	57.514	
Methanol	5	11.975	22.968
	10	23.949	
	15	35.142	
	20	45.406	
	25	54.888	
	30	62.219	
Ascorbic acid	1	7.279	5.019
	2	15.342	
	3	27.436	
	4	38.634	
	5	50.504	

Figure 1 shows that methanol extract has the highest yield of the other extracts. This means that there were more polar chemical compounds than non-polar compounds. Total yields from various extracts were 1.72%. The obtainable total yields from *B. macrocarpa* rinds were less than the yield from methanol extract of *B. angulata rinds*, i.e 31.0% [7]. This is due to different extraction methods and species of *B. angulata* fruit rinds were extracted in an oven shaker set at 250 rpm and 37°C [7] while *B. macrocarpa* in this research was extracted at room temperature.

3.2 Antioxidant activity

Antioxidant activity was measured at a wavelength of 516 nm, which was its maximum absorption of DPPH. Free radical inhibition of various extracts and reference standard (ascorbic acid) are shown in table 1. All extracts had an antioxidant activity with IC_{50} vary from 22.968 to 141.931 ppm, and IC_{50} ascorbic acid was 5.019 ppm. Methanol extract had the highest antioxidant activity, followed by ethyl acetate extract, and n-hexane extract. The antioxidant activity of all extracts was lower than ascorbic acid as a standard. The extract was a mixture of various compounds while the ascorbic acid was a pure chemical compound.

Methanol and ethyl acetate extract might contain polar and semipolar phytochemical compounds, and n-hexane extract contained non-polar phytochemical compounds. Some polar phytochemical compounds that might be in methanol extract and responsible for antioxidant activity were phenolic glycoside and phenols because it is suspected that *B. macrocarpa* also had phenolic glycoside compounds as well as in *B. ramiflora*. Some phenolic glycosides have been isolated from *B. ramiflora* were 4'-O-(6-O-vanyloil-β-D-glucopiranosyltachioside D, 6'-O-vanyloilpicraquasiosidetachioside, 6'-O-vanyloilicarisida B_5, 6'-O-vanilloylisotachioside and 6'-O-vanilloyltachioside [8,9]. Some phenols compounds in *B. ramiflora* were epichatecin, aviculin [9].

4. Conclusion

The highest antioxidant activity was observed in methanol extract, followed by the activity from ethyl acetate extract and hexane extract.

Acknowledgments

This work was financially supported by DIPA Fund Year 2019 Faculty of Mathematics and Natural Sciences, Lambung Mangkurat University.

References

[1] Haegens and Peter R M A 2000 Taxonomy, Phylogeny, and Biogeography of Baccaurea, Distichirhops, and Nothobaccaurea (Euphorbiaceae) *Supplement* **12(1)** 1-218

[2] Ibrahim D, Hazali N, Jauhari N K, Omar M N, Yahya M N A, Ahmed I A, Mikail M A 2013 Physicochemical and antioxidant characteristics of *Baccaurea angulata* fruit juice extract *African Journal of Biotechnology* **12** 5333-38

[3] Tirtana E, Idiawati N, Warsidah and Jayuska A 2013 Analisa Proksimat, Uji Fitokimia dan Aktivitas Antioksidan pada Buah Tampoi (*Baccaurea macrocarpa*) *Jurnal Kimia Khatulistiwa* **2(1)** 42-5

[4] Bakar M F A, Ahmad N E, Karim F A, Saib S 2014 Phytochemicals and Antioxidative Properties of Borneo Indigenous Liposu (*Baccaurea lanceolata*) and Tampoi (*Baccaurea macrocarpa*) Fruits *Antioxidants* **3** 516-25

[5] Mikail M A, Ahmed I A, Ibrahim M, Hazali N, Abdul Rasad M S B, Abdul Ghani R, Hashim R, Abdul Wahab R, Jauhari Arief S, Isa M L, Draman S, Yahya M N A 2015 *Baccaurea angulata* fruit inhibits lipid peroxidation and induces the increase in antioxidant enzyme activities *European journal of nutrition* **55(4)** 1435-44

[6] Molyneux P 2004 The use of the stable free radical diphenylpicrylhydrazyl (DPPH) for estimating antioxidant activity Songklanakarin. *J. Sci. Technol.* **26(2)** 211-9

[7] Ahmed I A, Mikail M A, Ibrahim M, Hazali N, Abdul Rasad M S B, Abdul Ghani R, Abdul

IOP Conf. Series: Materials Science and Engineering **980** (2020) 012040 doi:10.1088/1757-899X/980/1/012040

Wahab R, Jauhari Arief S, Adros Yahya M N 2015 Antioxidant activity and phenolic profile of various morphological parts of underutilised *Baccaurea angulata* fruit *Food chemistry* **172C** 778-87
[8] Lin Y F, Yi Z and Zhao Y 2003 *Chinese Dai Medicine Colourful Illustrations*. Yunnan National Publishing House, Kunming
[9] Yang X W, Wang J S, Ma Y L, Xiao H T, Zuo Y Q, Lin H, He H P, Li L, Hao X J 2007 Bioactive Phenols from the Leaves of *Baccaurea ramiflora Planta Med* **73** 1415-7

IOP Conf. Series: Materials Science and Engineering **980** (2020) 012041 doi:10.1088/1757-899X/980/1/012041

Structural characterization and antioxidant activity of liquid sugar from Alabio potato using enzymatic hydrolysis processes

VS Angkasawati, I Kamaliyah, MD Putra, IF Nata, and C Irawan*
Department of Chemical Engineering, Faculty of Engineering, Lambung Mangkurat University, Jl. A. Yani Km. 35.5, South Kalimantan, Indonesia 70714

Email: cirawan@ulm.ac.id

Abstract This study aims to produce functional antioxidant liquid sugar from Alabio potato (*Dioscorea alata L.*). This study focuses on the effect of the percentage of Alabio potato starch in the liquefaction process, the concentration of β-glucosidase enzymes, and the determination of the optimum concentration of cinnamon extract in its activity as an antioxidant liquid sugar. The amount of glucose/reduced sugar was analyzed by the DNS method. The optimum percentage of glucose was obtained from the liquefaction results with 10% starch and saccharification with a concentration of β-glucosidase enzyme of 1.2 mL/kg of starch with a liquid sugar concentration of 59.16 mg/mL. FTIR analysis showed that antioxidant liquid sugar had hydroxyl functional groups of C-O, C-H and O-H at wavelengths of 1560.11 cm-1, 1720.22 cm-1, and 3520 cm-1. Antioxidant liquid sugar activity showed the optimum performance at cinnamon extract concentration was 5 grams. DPPH free radicals of 56.72% were consumed for 5 minutes and reached 57.60% for 10 minutes. Liquid glucose produced could be a substitute for glucose sources with antioxidant content from the cinnamon extract.

1. Introduction
The production of sweet potato in South Kalimantan, especially in Hulu Sungai Selatan regency has an average of 19.4 ton/ha/years during 2014–2018. Most of the sweet potato is cultivated in the swampy land. One of the sweet potato types is Alabio yam *(Dioscorea alata L).* The Alabio yam could be a substitute for food because it has a fairly good nutritional composition. Purple Alabio sweet potato has a starch content of 11,07% and white sweet potato of 11,30%. The total sugar found in red Alabio yam is 4.48% and the white Alabio yam is 2,80%. Each year the production of Alabio sweet potatoes has increased in the planting area so that the number of sweet potatoes produced has also increased. In 2015 the Alabio yam production reached 17 tons.ha^{-1} [1].

The making of liquid glucose was carried out in two main stages, which are liquidation and saccharification with the help of the enzyme α-amylase and β-glucosidase. Enzymatic glucose making is an appropriate technology that can be developed directly in the community because it does not require many chemicals and comes from natural ingredients. Purple Alabio sweet potato contains anthocyanin which can be used as a natural coloring agent in food [2].

Starch is present in a variety of staple foods such as potatoes, corn, rice, wheat, and pasta [3]. Making liquid glucose is expected to be able to overcome the high domestic sugar imports. The food and beverage industry is currently growing rapidly, where a lot of food industries are starting to innovate their products. The food and beverage industry is starting to switch to using liquid sugar because it has several advantages, namely, it dissolves easily, is more practical, and has a more attractive appearance compared to crystal sugar. Liquid sugar made from the Alabio yam is expected to replace the production

IOP Conf. Series: Materials Science and Engineering **980** (2020) 012041 doi:10.1088/1757-899X/980/1/012041

of crystalline sugar and the availability of raw material sources for making liquid sugar is quite abundant and easily cultivated. This research is expected to provide benefits to the community to produce food diversity and increase the economic value of the Alabio yam by utilizing the sweet potato meat to become liquid glucose so that it can raise food as a local characteristic.

Starch is a polysaccharide found in grains and tubers and fruits. Generally, starch is used as a modified food ingredient [4], tubers, for example, become sweet after being boiled due to the starch contained in them breaks down into simple sugars such as glucose. If the starch is heated, the molecule will split into smaller molecules such as sugar called dextrin. Then dextrin decomposes again into maltose and then into glucose [5].

2. Materials and methods

2.1 Materials

The materials used in this research were Alabio sweet potato, α-amylase enzyme, β-glucosidase enzyme, distilled water, calcium carbonate ($CaCO_3$), sodium carbonate (Na_2CO_3), sodium hydroxide (NaOH), dinitro salicylic reagent (DNS), potassium sodium tartrate tetrahydrate ($H_2C_4H_4O_6$), 2,2-diphenyl-l-picrylhydrazil (DPPH), activated carbon, methanol (CH_3OH), standard glucose ($C_6H_{12}O_6$). All the chemical used is pure grade analysis.

2.2 Preparation of raw materials

The Alabio sweet potato was peeled then cleaned and soaked with water then mashed using a blender then filtered using a porous cloth. The Filtrate was added to the container and allowed to stand for 1-2 days to get starch deposits. The immersion water was released. The starch was dried using an oven at 80°C for 3 hours until it got a constant weight, then mashed and sieved with a 60 mesh sieve.

2.3 Liquidation process

The Alabio sweet potato starch was weighed and put in a three-neck flask. Added to the starch was a solution of Ca^{2+} (80 ppm) of 50 mL with starch percentages of 1%, 5%, 10%, 15% and 20% (w / v). The solution was heated to a temperature of 85°C and then the enzyme α-amylase 330 units / g was added. After the addition of the enzyme and 1 hour of the liquidation process, filtering was carried out with 100 circles of filter paper to get the filtrate free of unreacted starch residue. The filtrate obtained was then analyzed based on the content of dextrin levels formed using the DNS method.

2.4 Saccharification process

Dextrin with a pH of 3 - 3.5 was added to the enzyme β-glucosidase 293 units/kg of starch. Then the solution that had been added with the β-glucosidase enzyme was in the shaker for 60 hours at a rate of 150 rpm on temperature 60°C. Glucose content was analyzed again using DNS.

2.5 Neutralization, blanching, and evaporation processes

Liquid glucose from saccharification which was acidic was neutralized by adding 0.1 M Na_2CO_3 base to pH 6-7 then the bleaching process was carried out by mixing liquid glucose with activated carbon as much as 2% of the weight of liquid glucose. The bleaching temperature was set to 80°C for 2 hours. The last step was evaporation to get the viscosity of liquid glucose with the desired thickness.

2.6 Preparation of cinnamon extract and combined with glucose

Cinnamon was washed and oven-dried at 80°C for 2 hours. It was then mashed using a blender, sifted to get a size of 60 mesh. Cinnamon extract was made from 3, 5, and 10 grams of cinnamon powder dissolved in 100 mL water, heated at 85°C for 15 minutes. Then the cinnamon extract was mixed with liquid glucose (% v / v) in a ratio of 1: 1.

2.7 Analysis

Analysis of reduced glucose/sugar content was done using the Dinitrosalisylic (DNS) method. The antioxidant capacity used the method of 2,2-diphenyl-1-picrylhydrazil (DPPH). Scanning Electron Microscope (SEM, JEOL JSM 6500 LV) was carried out to see the morphological structure of Alabio sweet potato starch. Fourier Transform Infra-Red (FTIR) was performed to determine the functional groups of the sample [6].

3. Result and discussion

3.1 Preparation of Alabio sweet potato starch

The Alabio sweet potato starch was obtained through the process of immersion and refinement, which could be seen in figure 1. The results were obtained in the form of flour with a pink color. The yield of flour obtained from the Alabio sweet potato meat was 40%. The components contained in the Alabio sweet potato are shown in table 1, the results show that the Alabio sweet potato starch had potentials as a substrate for hydrolysis. The carbohydrate content of the Alabio sweet potato was 90.77% as a source of dextrin production.

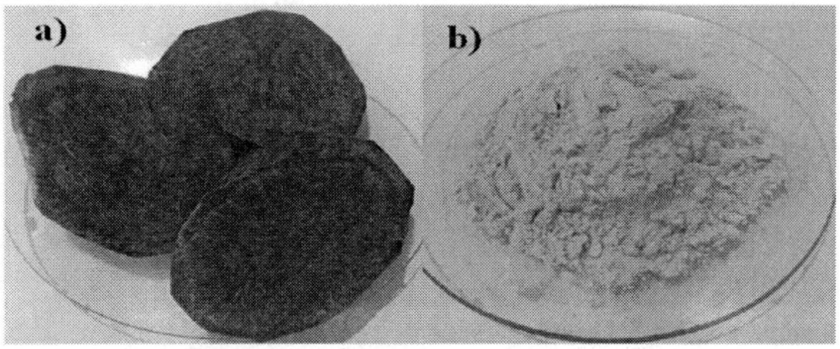

Figure 1. a) The Alabio sweet potato and b) Starch from the Alabio sweet potato

Table 1. Components of the Alabio sweet potato starch.

Components	Content (%)
Water	6,54
Ash	0,28
Protein	2,09
Fat	0,32
Carbohydrate	90,77

The results of observations with SEM of the Alabio sweet potato starch can be seen in figure 2. Observation of starch granules in figure a) using a magnification of 500x and in figure b) using a magnification of 1,500x shows that the starch granules of the Alabio sweet potato had a round shape with a size that was not homogeneous, about 10 - 40 µm. For other types of starch granules, such as cassava starch, it has a size of 4-35 µm [7].

3.2. Production of dextrin from Alabio flour meats by process liquidation

Liquidation is the process of hydrolysis of starch into dextrin by adding the enzyme α-amylase. The hydrolysis process melts the high viscosity starch gel. The solution used during the hydrolysis process to dissolve the starch of Alabio Sweet Potato was Ca^{2+} 80 ppm. Ca^{2+} solution was used as a cofactor to optimize the work of the α-amylase enzyme which converted the Alabio sweet potato meat starch into dextrin. The addition of enzymes to the hydrolysis reaction is beneficial for the process efficiency of the liquidation process [8]. The liquidation process is said to be good if the viscosity of the resulting dextrin is low. Based on the results of the study, the best dextrin was obtained at 10% starch concentration. This is based on a visualization analysis of the product produced, the comparison between starch and Ca^{2+} solution that produce the most dextrin solution with low viscosity. A comparison of product results and volumes resulting from the liquidation process can be seen in table 2 and figure 3.

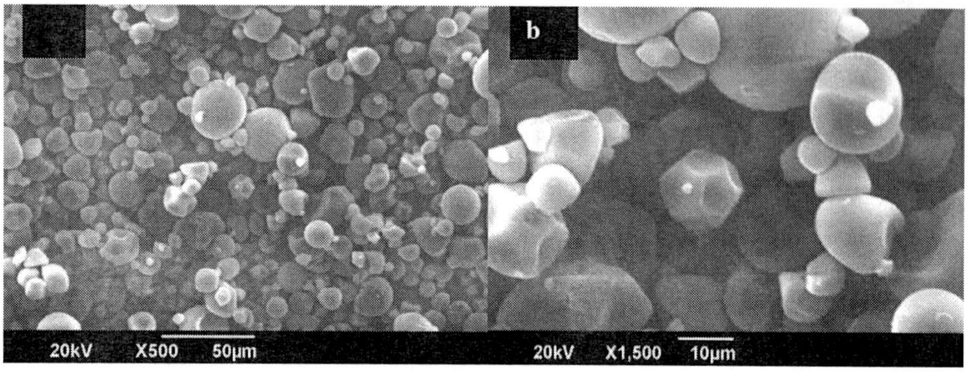

Figure 2. SEM images of a) 500x magnification of the Alabio sweet potato starch and b) 1,500xmagnification of sweet potato starch.

Table 2. Production of the liquidation process in the Alabio sweet potato starch.

The concentration of starch (%)	Result analyst	
	Production (mL)	Viscosity
1	49	Low
5	47	Low
10	46	Low
15	36	High
20	36	High

Based on the literature [6], it is calculated that the percentages of dextrins 5 and 10% have low viscosities, while the percentages of dextrins 15 and 20% have high viscosities. The dextrin content obtained from the liquidation process was analyzed by the DNS method. To get the reduced sugar concentration in the solution, a standard solution of glucose was previously made as a basic curve to determine the concentration of the sample by extracting the absorbance values obtained from the obtained line equations.

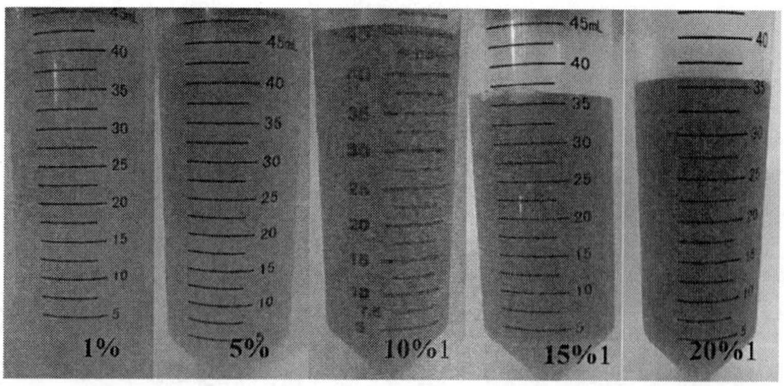

Figure 3. The liquidation process products with starch concentration of 1%, 5%, 10%, 15%, and 20% (w / v) temperature at 85 °C for one hour.

The higher concentration of glucose in the solution, the higher the absorbance value which is marked by the increasingly orange-brown discoloration of the final DNS analysis. After calculating and measuring with different starch concentrations, the content of dextrin at each Alabio Sweet Pot starch concentration can be seen in figure 4.

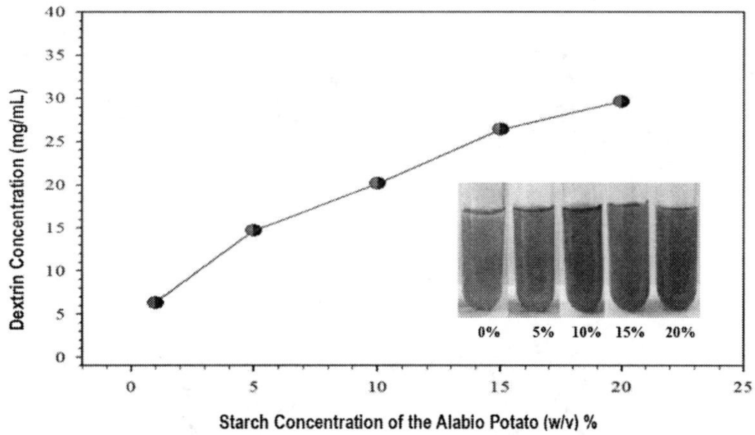

Figure 4. Relationship of dextrin concentration (mg / mL) to starch concentration (%) temperature at 85 °C for one hour.

The higher concentration of starch, the higher the total reducing sugar value was obtained. At starch concentrations of 10% and 15%, there was an increase in concentration from 20.11 mg / mL to 26.38 mg / mL, but the volume of the solution obtained decreased from 46 mL to 36 mL. It occurred because the starch concentration was high enough to make the enzyme unable to work optimally, so gelatinization began to increase. Based on the observations of the results of the dextrin obtained, a concentration of 10% starch was the optimum condition in the liquidation process, because it produced the largest product with low viscosity.

3.3 Conversion of dextrin to liquid glucose by the process of saccharification and addition of antioxidant sources from cinnamon extract

The saccharification process was the process of dextrins being converted into glucose after the best results from the 10% starch liquidation process. A comparison of glucose yields from products produced from the saccharification process with variations in the concentration of the β-glucosidase enzyme can be seen in figure 5. From the process of liquidation to saccharification, the reduced sugar concentration increased approximately 2 times the concentration of the liquidation product which indicated that the enzyme β-glucosidase had worked well. The enzyme concentration of β-glucosidase 1.2 mL / kg of starch was the optimum result of the saccharification process because it had the highest glucose concentration of 59.16 mg / mL. According to Ruiz [9] the saccharification process for 60 hours, there was an increase in high glucose levels of 46.9 mg / mL with a source of cassava starch. The resulting product meets the minimum requirements for reduced sugar content from liquid glucose based on the Indonesian National Standard (SNI), which is 30 mg / mL shown by the red line in figure 5.

For the liquid glucose products to be neutral, the results of liquid glucose from the saccharification process were neutralized by adding sodium carbonate (Na_2CO_3). To produce a clear product, the bleaching process was carried out by adding activated carbon. Activated carbon has a strong adhesion or absorption ability. After going through the stages of neutralization, bleaching, and evaporation, liquid glucose became more concentrated and aromatic and changed color. This happened because the impurities in liquid glucose were bound by activated carbon and some of the water content contained in liquid glucose was lost by evaporating when heating at 80 ^{0}C for 2 hours. Liquid glucose products before and after evaporation are shown in figure 6.

IOP Conf. Series: Materials Science and Engineering **980** (2020) 012041 doi:10.1088/1757-899X/980/1/012041

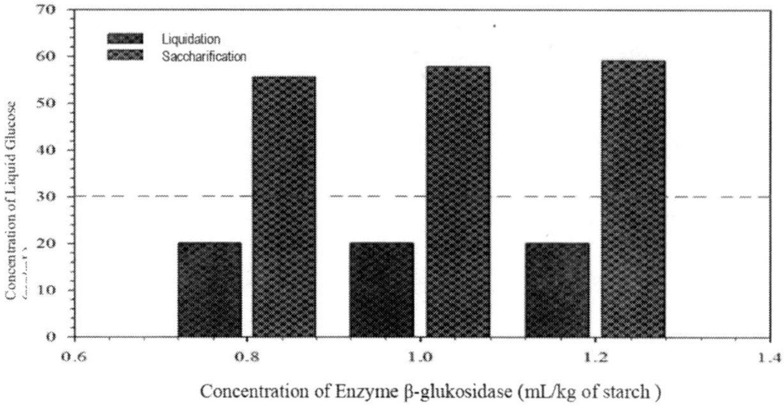

Figure 5. Relationship of dextrin concentration (mg / mL) to enzyme β-glukosidase concentration (mL/kg of starch).

Figure 6. Liquid glucose before a) and b) After neutralization process blanching and evaporation

To produce liquid sugar with functional value, a cinnamon extract was added as a source of antioxidants with cinnamon extract concentration of 3, 5, and 10 grams. These additions gave liquid glucose antioxidant properties and created a fragrant aroma in liquid glucose. The capacity of antioxidants was measured quantitatively by measuring the capture of DPPH radicals by antioxidant compounds in cinnamon so that the value of free radical activity would be known. When the sample reacted with DPPH free radicals, there would be a transfer process of hydrogen atoms from antioxidant compounds in cinnamon to make DPPH stable and form reduced DPPH. The reaction was characterized by a change in color from purple to yellow [10]. Hydrogen/electron donated by antioxidants could prevent oxidation / neutralize oxidized compounds by binding to free radicals and reactive molecules [11]. The results of the analysis of liquid glucose with cinnamon extract on antioxidant activity can be seen in figure 7.

The optimum antioxidant activity of liquid glucose resulted in 5-gram cinnamon variation because 56.72% of free radicals from DPPH had been consumed for 5 minutes and reached 57.60% for 10 minutes. The red line in figure 7 shows the antioxidant activity without adding liquid glucose which was 60.3%. In 5-gram cinnamon variation was observed a high activity that almost reached the dotted line. 5-gram cinnamon variation can be said to be optimum where cinnamon extract works optimally in liquid glucose. Reduction in absorbance indicated scavenging activity (antioxidant activity) of the sample. The decrease in absorbance occurred due to the addition of electrons from antioxidant compounds to unpaired electrons in the nitrogen group in the structure of DPPH compounds. The intensity of the purple color would decrease when the DPPH radical bound with hydrogen. The stronger the antioxidant activity in the sample, the higher decrease in purple intensity and the percentage decrease in absorbance [12]. Fourier Transform Infra-Red (FT-IR) analysis was performed to find out the functional groups of compounds with those shown by the frequency absorbed by the compound. The amount of frequency that passed through a compound was measured as a percent of transmittance [13]. FT-IR analysis on commercial liquid glucose, saccharification results, cinnamon extract, and various concentrations of the cinnamon extract can be seen in figure 8.

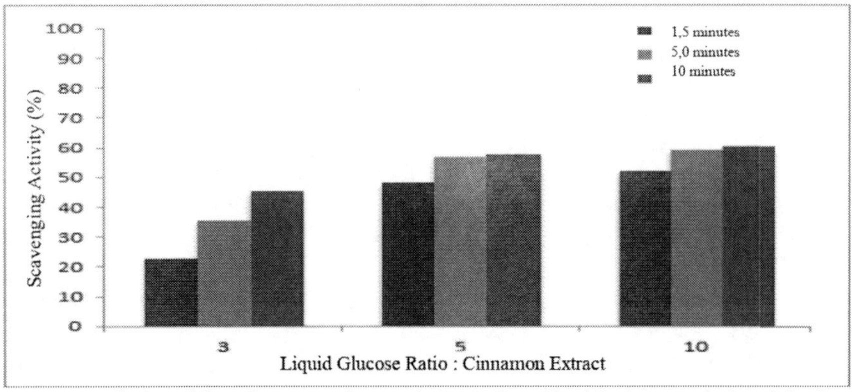

Figure 7. Relationship of scavenging activity (%) to liquid glucose ratio with cinnamon extract.

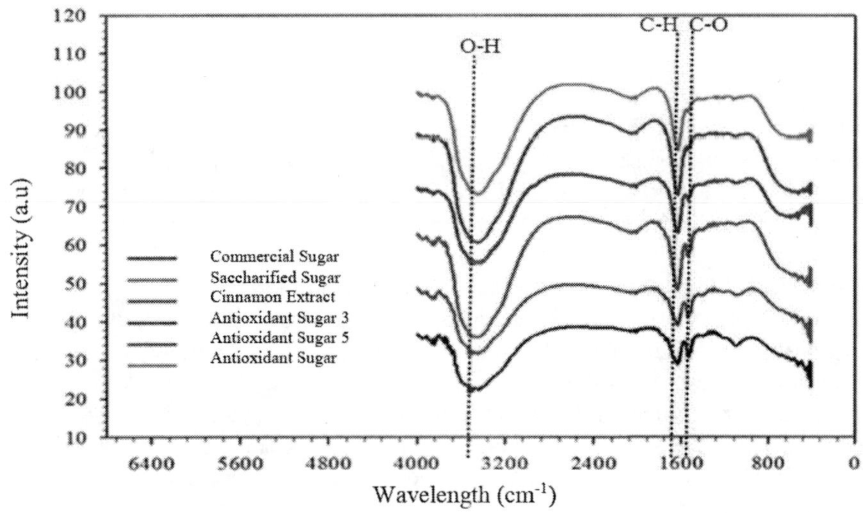

Figure 8. FT-IR for several types of samples.

Liquid glucose is a monosaccharide containing the hydroxy group -CHO aldose group [14]. The glucose function group is $C_6H_{12}O_6$. FT-IR analysis for commercial glucose and saccharification results did not show any difference, qualitatively the glucose produced was appropriate because it showed the same functional groups at each wave peak, ie CO, CH, and OH groups could be read at a wavelength of 1560.11 cm-1, 1720.22 cm-1, and 3520 cm-1. The difference between several types of samples from the FT-IR analysis occurred because the intensity of each sample was different in the O-H group. The intensity of the largest O-H group was found in the cinnamon extract samples. The cinnamon extract showed the intensity of O-H functional groups that were sharper because they contained O-H groups which were owned by flavonoid compounds which were a source of antioxidants in the cinnamon extract [15].

4. Conclusion

The Alabio purple potato can be used as liquid glucose which is shown from the high carbohydrate content that is 90.77%. In the liquidation process, the optimum results were obtained at a starch concentration of 10% (w / v) with a TRS of 20.11 mg/mL. In the saccharification process, the optimum results were obtained at the enzyme concentration of β-glucosidase 1.2 mL/kg of starch with a TRS of 59.16 mg/mL. Based on cinnamon extract, the best results were obtained in the variation of 5 grams of cinnamon with free radical consumption of 56.72% within 5 minutes and reached 57.60% after 10 minutes.

Acknowledgments

This work was financially supported by a grant of Directorate of Research and Community Service, Directorate General of Higher Education, The Ministry of Education and Culture of the Republic of Indonesia, the fund for Thesis Magister Research Grant 2020 with contract No. of 113.1.07/UN8.2/PG/2020.

References

[1] Istiqomah, Hafizah N and Puadi K 2018 Pengaruh Penambahan bokashi eceng gondok terhadap masa vegetatif Ubi Alabio (*Dioscorea Alata* L) *Jurnal Ziraa'ah* **14(2)**

[2] Robi'a and Sutrisno A 2015 Karakteristik sirup glukosa dari tepung ubi ungu (kajian suhu likuifikasi dan konsentrasi α-amilase) *Jurnal Pangan dan Agroindustri* **3**

[3] Kumar V B, Pulidindi I N and Gedanken A 2015 Renewable energy *Selective Conversion of Starch to Glucose Using Carbon Based Solid Acid*

[4] Dudu O E, Li L, Oyedeji A B, Oyeyinka S A and Ma Y 2019 Structural and functional characteristics of optimised dry-heat-moisture treated cassava flour and starch *International Journal of Biological Macromolecules* **133**

[5] Lacerda L D, Leite D C and Silveira N P d 2019 Relationships between enzymatic hydrolysis conditions and properties of rice starches *Journal of Cereal Science* **8**

[6] Yulimasita D D, Fitria A A and Nata I F 2016 *Production of Antioxidant Liquid Glucose From Nagara Sweet Potatoes With Enzymatic Process* (Banjarbaru: Lambung Mangkurat University)

[7] Li C, Zhou D, Fan T, Wang M, Zhu M, Ding J, Zhu X, Guo W and Shi Y-C 2020 Structure and physicochemical properties of two waxy wheat starches. *Journal Pre-proofs*

[8] Cardona M J, Tozzi E J, Karuna N, Jeoh T, Powell R L and M J McCarthy 2015 A process for energy-efficient high-solids fed-batch enzymatic liquefaction of cellulosic biomass. *Bioresource Technology* **198** 9

[9] Ruiz M I, Sanchez C I, Torres R G and Monila D R 2011 Enzymatic hydrolysis of cassava starch for production of bioethanol with a colombian wild yeast strain *Article* **22** 12

[10] Molyneux P 2014 The use of the stable free radical diphenylpicrylhydrazyl (dpph) for estimating antioxidant activity *Songklanakarin J Sci Technol* **26(2)** 211-9

[11] Bendra B, Fetouani S, Laffray X, Vanclooster M, Sbaa Mand Aleya L 2012 Effects of irrigation on soil physico-chemistry: a case study of the triffa plain (Morocco) *Irrigation and Drainage* **61(4)** 507-19

[12] Ghasemzadeh A, Jaafar H Z E and Rahmat A 2016 Variation of the phytochemical constituents and sntioxidant activities of Zingiber officinale var. rubrum theilade associated with different drying methods and polyphenol oxidase activity *Molecules* (Departemen of Crop Science Faculty of Agriculture Universiti Putra Malaysia)

[13] Sabrina Q 2011 *Kajian Sifat Optis Pada Glukosa Darah* Fakultas Sains dan Teknologi (Jakarta: Universitas Islam Negeri Syarif)

IOP Conf. Series: Materials Science and Engineering **980** (2020) 012041 doi:10.1088/1757-899X/980/1/012041

[14] Anggraeni P, Addrojah Z and Anggoro D D 2013 Hidrolisis selulosa eceng gondok (*eichhornia crassipe*) menjadi glukosa dengan katalis arang aktif tersulfonasi. *Jurnal Teknologi Kimia dan Industri* **2(3)** 63-9
[15] Adawiah A, Sukandar D and Muawanah A 2015 Aktivitas antioksidan dan kandungan komponen bioaktif sari buah namnam. *Jurnal Kimia VALENSI* 127-33

IOP Conf. Series: Materials Science and Engineering **980** (2020) 012042 doi:10.1088/1757-899X/980/1/012042

Antioxidant activity of *Porphyridium cruentum* water extracts for cosmetic cream

S Agustina*, N N Aidha, E Oktarina and I Setiawati

Badan Penelitian dan Pengembangan Industri, Kementerian Perindustrian

Email: tinaratujaya@yahoo.com

ABSTRACT. *Porphyridium cruentum* is one of the microalgae that has antioxidant capacity. These antioxidants are useful for the cosmetics and pharmaceutical industries. The purpose of this study was to obtain antioxidants from *P. cruentum* which will be used in cosmetic products as a cream. *P. cruentum* was extracted by maceration method for one hour with water as solvent (5 g/L) and encapsulated using maltodextrin with a spray dryer. The variable concentration of maltodextrin was 50 g/L, 100 g/L, and 150 g/L. Characterizations observed for encapsulated microalgae extracts were the antioxidants capacity and proximate analysis; hence for cosmetic cream were the antioxidants capacity and microbiological contaminant. The best result was *P. cruentum* extract encapsulated with maltodextrin 50 g/L which gave antioxidant capacity as AEAC 7.07 mg vit. C/100 g sample and IC_{50} 484.09 mg/mL. The cosmetic cream with addition encapsulated *P. cruentum* extract gave antioxidant capacity as AEAC on 8.12 mg Vit. C/g sample and IC_{50} 29.08 mg/mL. The microbial and heavy metals contaminant for cream were below the threshold according to BPOM regulation. Based on these results, the *P. cruentum* has the potential to be used in the cosmetics industry.

1. Introduction

Porphyridium cruentum is one of the unicellular red alga which belongs to Rhodophyta, which lives naturally on neutral water and seawater [1]. *Porphyridium cruentum* lives solitary and colony. As colony *P. cruentum* capsulated with mucilage, a sulfated polysaccharide soluble in water [2–4]. The dominant pigment in *P. cruentum* is phycoerythrin and phycocyanin, which contain phycobilin. *P. cruentum* have chlorophyll a and d, and lack of chlorophyll b. All of these pigments contributed to photosynthesis [4,5]. Phycobilisome (PBS) located on the outer surface of thylakoid; therefore, several mechanisms performed to extracting the PBS. The membrane cell of *P. cruentum* constructed by two layers, the first layer is pectic, and the second layer is cellulosic microfibrils [2,5].

Biomasses compositions of *P. cruentum* are 32.1% (w/w) carbohydrate; 34.1% protein; Fatty acid consist of 1.6% for hexadecenoic acid; 0.4% of linoleic acid; 1.3 % of arachidonic acid; and 1.3% for eicosapentaenoic acid. Minerals on 100 g dry biomass are Ca (4960 mg), K (1190 mg), Na (1130 mg), Mg (629 mg), and Zn (373 mg). *P. cruentum* also contains tocopherol, flavonoid, alkaloid, phenol hydroquinone, and carotenoid [2]. With those compositions, *P. cruentum* has pharmaceutical properties like an antibiotic, antihyperglycemic, and antioxidant [3,6,7]. *P. cruentum* also has been already used as a cosmetic compound for conditioning agents (Asta technology), moisturizing agent, self-tanning agent, anti-acne agent, and anti-redness (Green-tech); and anti-wrinkle agent or anti-aging, film former and moisturizing agent (Givaudan active beauty) [8].

Antioxidant properties have a beneficial role in the cosmetic area by combating the damage caused by a free radical agent (Radical Oxygen Scavenger and Radical Nitrogen Scavenger). The accumulation of uncontrolled free radicals can lead to severe diseases like skin cancer [9]. Free radicals also can lead the skin to wrinkle; therefore, the antioxidant activity correlates with anti-aging. Antioxidant properties of *P. cruentum* have several mechanisms to neutralize free radical, like hydrogen transfer (proton donating ability). Algae pigments have potential as an electron-directing agent in trans-membrane migration of electrons against oxidative stress by metal reduction [10]. Phenolic compounds have redox properties, which engage in absorbing and neutralizing free radicals, quenching singlet, and triplet oxygen or decomposing peroxides [9]. Antioxidant activity of protein contributed by functional groups or the residue of amino acids [11]. Polysaccharide contributed to antioxidant activity by combining several factors like the oligomeric, sulfate, and glycoprotein components of the polymer [7]. Those bioactive compounds were unstable and sensitive to the process condition. One of the technologies that can be used to protect the bioactive compound from the uncontrolled condition was encapsulation.

The encapsulation process can be done by several methods; one of the methods that generally used is spray drying. Encapsulation will convert the bioactive compound from liquid form to solid form (better mixing and prevent lumping), coating the bioactive compound by carrier material to prevent evaporation and to degradation of the bioactive compound by undesirable circumstances while on process production [12]. Encapsulation by spray drying was influenced by several factors like inlet and outlet temperature, aspirator rate, nozzle rate, pump rate, and materials concentration used for encapsulation [13].

Encapsulation with spray drying can maintain the antioxidant capacity of *P. cruentum* [14]. The encapsulated material has to have high emulsifier ability, able to maintain the flavor of a compound, accelerate the drying rate of a compound, and able to release the encapsulated compound. Maltodextrin was used as encapsulated materials because it is soluble in water, not toxic, and able to protect the bioactive compound from degradation [15]. This research aims to obtain antioxidants from *P. cruentum*, which will be applied to cosmetic products as cream.

2. Materials and methods

2.1. Materials

Porphyridium cruentum procured from Balai Besar Perikanan Budidaya Air Payau (BBPBAP) Jepara. The material used were maltodextrin (Bratachem), 2,2-diphenyl-1-picrylhydrazyl (Sigma), butylated hydroxytoluene (Sigma), ethanol (Merck), acetic acid (J.T. Baker), potassium sulphate (Merck), $CuSO_4$ (Merck), H_3BO_3 (Merck), sodium hydroxide (Merck), sulfuric acid (Merck), chloric acid (Merck), hexane (J.T. Baker), isopropyl alcohol (Merck), methylene chloride (Merck), diethyl ether (Merck), acetonitrile (Merck), HNO_3 (Merck), aquadest, and aquabidest (Ikapharmindo Putramas).

2.2. Instrumentation

The instruments used were homogenizer (Miyamoto Riken), spray dryer (Eyela), and centrifugation (Kokusan), Kjeldahl (Kjeltec FOSS), oven (Memmert), Atomic Absorbance Spectrophotometry (AAS) (Shimadzu), Soxhlet evaporator (Buchi), spectrophotometer UV-VIS (Shimadzu UV-1280), Inductively Coupled Plasma- Optical Emission Spectrometry (ICP-OES) and Scanning Electron Microscope (SEM) (Jeol JSM-6510LA).

2.3. Procedure

2.3.1. Extraction of Porphyridium cruentum. P. cruentum that has been separated by maceration media with water (1% w / v) 190 rpm, for 1 hour at room temperature. Separation with debris was done by centrifugation at 6000 rpm for 15 minutes.

2.3.2. Encapsulation of Porphyridium cruentum Extract. P. cruentum extracts were mixed with maltodextrin, as a matrix for spray-dried process, for 60 minutes. The ratio of variation on *P. cruentum* extract (mL): maltodextrin (g) were 20:1 (F50), 10:1 (F100) and 6.7:1 (F150). Spray dry running with condition inlet 120°C - 130°C, outlet 41°C - 44°C, blower 1.01 m^3/min – 1.04 m^3/min and atomizing 200 kPa - 230 kPa.

2.3.3. Determination of antioxidant capacity. Antioxidant capacity were determined by Ascorbic acid Equivalent Antioxidant Capacity (AEAC) method using spectrophotometric on 517 nm, by measured the capacity of free radical content which reacted with 2,2-diphenyl-1-picrylhydrazyl. As blank and standard were aquadest and butylated hydroxytoluene, respectively [16].

2.3.4. Analysis of moisture content and ash content. Analysis of moisture content and ash content by gravimetric was regarding SNI 01-2891-1992 point 5.1 and 6.1, respectively [17].

2.3.5. Analysis of lipid, protein and carbohydrate content. Lipid content was determined by Soxhlet-hydrolysis method. The analysis was conducted by warping the sample by filter paper then Soxhlet with hexane. Determination of protein contents by biuret method (Kjeltec) referred to SNI 01-2891-1992 point 7.1. Carbohydrate content was determined by calculation. Carbohydrate = 100 – (%water content + %ash content + %protein + %lipid) [17].

2.3.6. Analysis of morphology. Rheology of encapsulated *Porphyridium cruentum* was determined using Scanning Electron Microscopic (SEM), by coating the sample with gold (Au) and observed it at 500x and 4000x times magnifications [18].

2.3.7. Physical evaluation cream. Analysis of pH was using a pH meter that already calibrated using a buffer solution of pH 4 (four) and pH 7 (seven). The organoleptic analysis was examined based on the sensing process, to observe the physical stability of the cream on transformation in texture, color, and odor during storage time. The observations were done for 8 weeks, on room temperature [19].

2.3.8. Microbial and heavy metal contamination analysis. Analysis of Total Plate Count was determined by pour plate which referred to ISO 21149:2006 [20]. Analysis of mold and yeast were determined by pour plate that referred to ISO 16212:2008 [21]. Detection of microbial contamination by *Pseudomonas aeruginosa*, *Candida albicans*, and *Staphylococcus aureus* were according to ISO 22717:2015 [22], ISO 18416:2015 [23], and ISO 22718:2015 [24], respectively. Lead and cadmium were detected with ICP-OES; mercury and arsenic were detected with AAS.

3. Results and discussions

3.1. Ash and moisture content

The moisture content ranged from 6.24/100 g - 7.59 g/100 g, with the lowest result was F50. The ash content ranged from 14.77/100 g - 32.70 g/100 g, with the lowest result was F150 (figure 1). The highest for moisture content and ash content were F100. Fuentes [2] observed that the ash contents of *Porphyridium cruentum* were ranging from 16.8 g to 23.6 g. The moisture content of *Porphyridium cruentum* was low, with a mean of 3.24 g, a value in the range of general recommendations which was less than 10% [2]. Maltodextrin encapsulation will hold the bioactive compound of *Porphyridium cruentum* and prevent it from loss by high temperature. The water could cause the variable of moisture content to escape through the surface of a particle through diffusion, avoiding the crust of particle [25], based on its encapsulation form. The encapsulation shape of F100 was round and has less dents than F50 and F150, but the encapsulated shape and agglomerate formed was irregular, which tended to break, causing higher moisture and ash content.

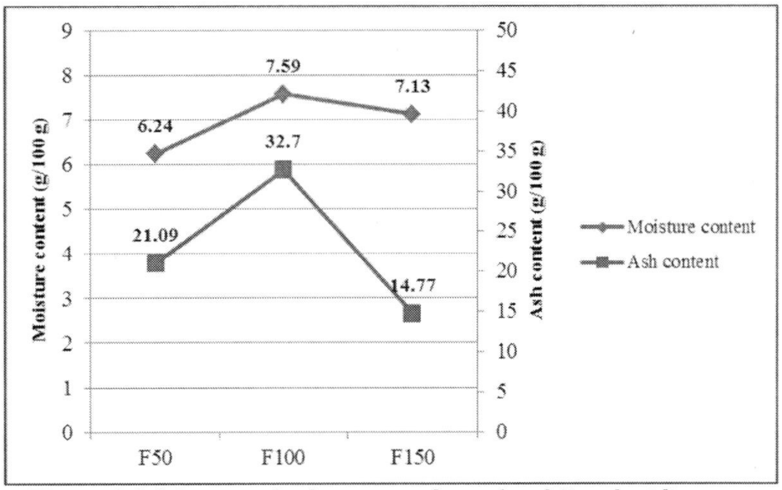

Figure 1. Moisture and ash content of powdered *Porphyridium cruentum*.

3.2. Total fat, protein and carbohydrate

The *Porphyridium cruentum* was polar extracted by water, with fatty acid that consisted of a non-polar compound like hexadecenoic acid, linoleic acid, arachidonic acid, and eicosapentaenoic acid [2] might lead the low result or undetected of total fat. The protein content of *Porphyridium cruentum* was around 0.07 g/100 g - 0.34 g/100 g (figure 2). Research by Safi [28] concluded that *Porphyridium cruentum* didn't need complicated extraction for the protein due to its fragility of pseudo-cell wall. Alternatively, several methods for cell disruption and extraction of *Porphyridium cruentum* has been developed using high-pressure cell disrupter [26] and pressurized liquid extraction (PLE) [27]. *P. cruentum* protein contribute 35% of biomass, the protein consists of 38.7% essential amino acid, and 61.0% non-essential amino acids, with optimum pH for extraction were 12-14 [28,29]. The low concentration of protein in this research might be due to several factors like the cell lysis before extraction because of the fragility of its cell wall and also can be due to the growing condition (pH medium, medium composition, and harvested time). The recommended time to harvested *Porphyridium cruentum* is 12-16 days, after that the *Porphyridium cruentum* tend to rupture and dissolve in the medium. Due to its diameter which ranging between 6 and 10 µm, the high-speed centrifugation was the optional step to separate biomass from the medium [30–32]. From SEM analysis, we concluded that rupture on F150 more than F50 causes the high protein concentration on F150. The carbohydrate of *Porphyridium cruentum* were around 30% [2], the addition of maltodextrin $(C_6H_{12}O_5)nH_2O$ that is starch hydroxylate (polymer of saccharide) will add the carbohydrate content.

3.3. Antioxidant capacity

The antioxidant capacity, which represents by the inhibition value, describes the ability of a compound as antioxidant to scavenge DPPH as free radical. Purple chromogen of DPPH reduced by antioxidant or reducing compound to pale yellow hydrazine. The reaction mechanism is based on an electron transfer reaction, while the hydrogen atom transfer reaction is a marginal reaction pathway, on the type strong hydrogen-bond-accepting solvent [33]. The inhibition value ranged from 0.87 % – 4.80 % (figure 3). The highest antioxidant content, near the raw material, was on F50 which give 4.80 % on sample concentration 5%. Research by Hasanah [14], non-encapsulated and encapsulated *P. cruentum* on 6.25 ppm gives inhibition of 3.05% and 2.85%, respectively. The capsulation was done with the addition of 75 g/L maltodextrin and observed with DPPH assay [14]. The addition of maltodextrin can maintain the antioxidant capacity of the bioactive compound on the encapsulated form. Encapsulation can prevent degradation because of light, oxygen, and slow down the evaporation rate. Encapsulation can be used in the spray drying process for the heat-labile compound [12,15].

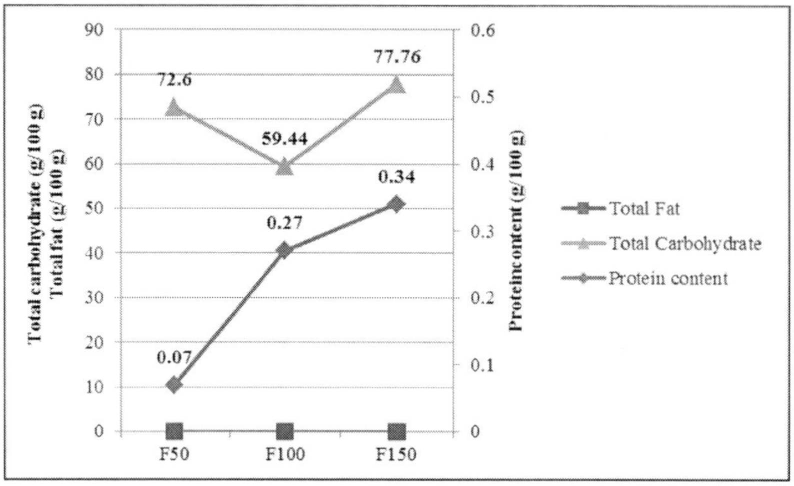

Figure 2. Total fat, carbohydrate and protein of powdered *Porphyridium cruentum.*

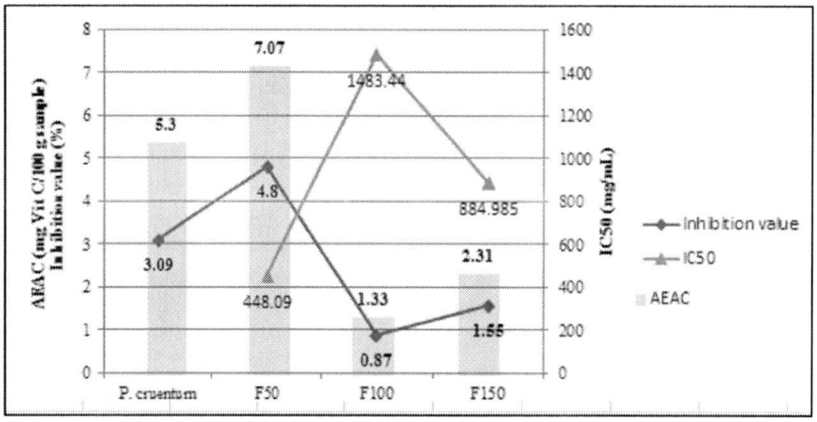

Figure 3. Antioxidant activity of powdered *Porphyridium cruentum.*

From our previous research on the encapsulation of *Spirulina* on maltodextrin as cosmetic raw materials, the protein showed a significant result with the antioxidant activity (as inhibition capacity) [34]. Because of the high protein content in *Spirulina* around 60-70% (including PBS), this contribute to its antioxidant capacity. Phycocyanin and phycoerytrhin contains linear tetrapyrrols, with double bonds and reducing ability, that contributing to the total antioxidant activity [11,28]. But that trend wasn't showed by *Porphyridium cruentum.* Thos can be due to *Porphyridium cruentum* which has multiple compounds that contribute to its antioxidant activity besides PBS, like polysaccharide. Polysaccharide on *Porphyridium cruentum* consists of xylose, glucose, galactose, arabinose, rhamnose, and mannose. The cell-wall complex also contains sulfate, glucuronic acid, and a non-covalently glycoprotein [35]. The polysaccharide of dry matter *Porphyridium cruentum* (2 mg/mL) give inhibition rate until 20%, detected with TBA assay [7].

The susceptibility of maltodextrin was unable to form an emulsion, unable to form crosslinks with bioactive compounds, and influenced by the viscosity. Therefore, with a high concentration of maltodextrin will lead lower antioxidant activity.

3.4. Analysis morphology

The morphological characteristic of encapsulated *Porphyridium cruentum* extract shown in figure 4. F50 and F100 had the particle with smoother and spherical surface rather than F150, which had dent surface. F50 had isotropic globular form and separate dispersedly, rather than F100 which gave anisotropic globular form and agglomerate. F150 also had anisotropic wavy form and tended to rupture. The increase of maltodextrin concentration gave wavy particle and cracked surface. The increase in surface roughness might increase the distance between surface, thus reducing the adhesive force and eventually decrease the wall deposition [25].

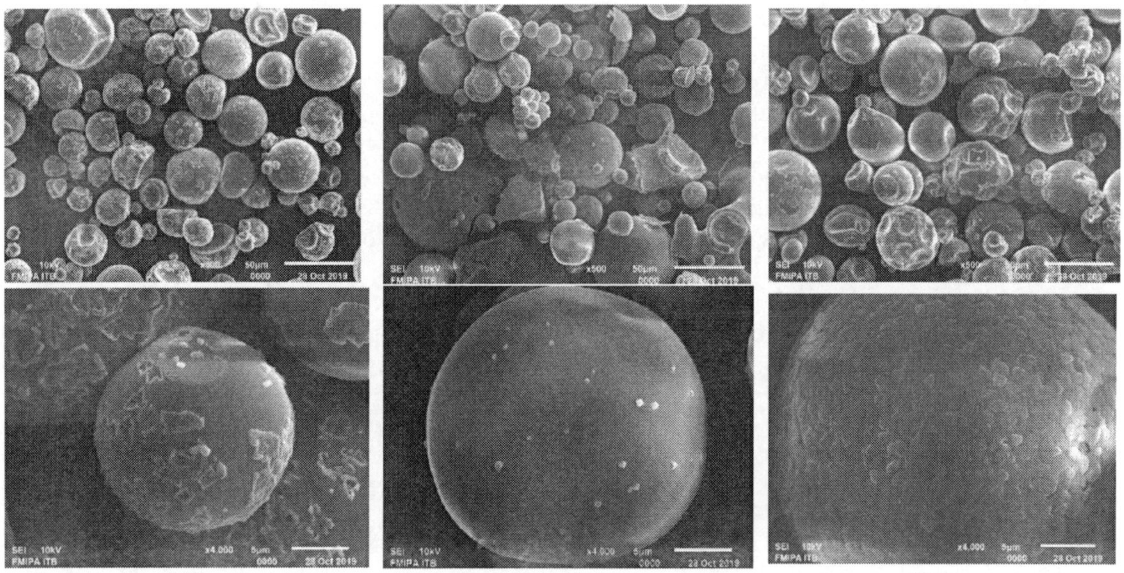

F50 F100 F150

Figure 4. SEM from encapsulated *Porphyridium cruentum* extract.

Table 1. The analysis result of cosmetic cream using encapsulated *Porphyridium cruentum* extract.

Parameter	Unit	Result	BPOM (2011)
IC50	mg/mL	29.08	
AEAC Sample concentration 10%	mg Vit. C/100 g sample	8.12	
Inhibition value Sample concentration 10%	%	10.22	
Total Plate Count	CFU/g	5.9×10^2	$< 10^3$
Yeast and Mould	CFU/g	$< 1.0 \times 10^1$	$< 10^3$
P. aeruginosa	Per g	Negative	Negative
C. albicans	Per g	Negative	Negative
S. aureus	Per g	Negative	Negative
Mercury (Hg)	mg/kg	n.d.	< 1
Lead (Pb)	mg/kg	n.d. (LOD 0.08)	< 20
Arsenic (As)	mg/kg	n.d.	< 5
Cadmium (Cd)	mg/kg	n.d. (LOD 0.02)	

n.d. = not detected

IOP Conf. Series: Materials Science and Engineering **980** (2020) 012042 doi:10.1088/1757-899X/980/1/012042

3.5. Encapsulated Porphyridium cruentum extract as cream for cosmetic

P. cruentum extract was applied as a cosmetic cream compound as natural antioxidant sources [36,37]. One of the antioxidants compound in *Porphyridium cruentum* is Phycoerythrin. According to Gianeti [38], the application of antioxidant substances in cosmetic formulations able to protect the skin against oxidative damage by UV radiation and prevent signs of aging [38]. Cosmetic creams with *Porphyridium cruentum* extract were analyzed for the antioxidant capacity and microbiological contamination according to BPOM. The analysis results for cosmetic cream with encapsulated *Porphyridium cruentum* extract were shown on table 1. The increasing of antioxidant activity on cream was supported by others compound and also the formation of emulsion. Emulsion can maintain antioxidant activity and prolonged the antioxidant stability [39].

Observation on cosmetic creams was performed on pH and organoleptic. The pH value of cosmetic creams was still in the normal pH range. According to the Indonesian National Standard (SNI) 16-4399-1996 [40], the pH of the cream is in the range of skin pH, which is around 4.5 - 8.0. The cream pH can affect the stability and contamination of the cream, and also influence the skin as a topical cream. If the pH of cream not in between the range of skin pH, it will trigger skin irritation. According to the Regulation of the Republic of Indonesia Drug and Food Supervisory Agency (BPOM) concerning Amendments to BPOM Regulation Number HK.03.1.23.07.11.6662 in 2011, the limit of microbial contamination as Total Plate Count no more than 10^3 colonies/g; Yeast and Mold no more than 10^3 colonies/g [41]. The heavy metals content was undetected on the cream. Based on the analysis above, it can be concluded that encapsulated *Porphyridium cruentum* can be applied as natural antioxidant on cream as cosmetic.

Evaluation of cosmetic creams also observed organoleptically (table 2). Visually, the cream color with the addition of *Porphyridium cruentum* extract which contain phycoerytrhin as natural antioxidant was white (figure 5). We assume the low concentration of phycoerythrin may affect the color of the cream, further analysis needed to strengthen the hypothesis. The color of cosmetic creams was stable until eight weeks, and shown no changes. The smell of the cream wasn't given by rotten odor until the 8th week. This shows that the oil phase in the cream wasn't undergoing oxidation. Discoloration and odor caused by oxygen from the air, can oxidize the fat or oil phase; besides that, light can also induce oxidation reactions [19]. The benefit for using natural antioxidant which also have antibacterial activity is the inhibition on microbial growth on the cream thus can be related by the reduction of concentration of preservative for the cream. *Porphyridium cruentum* extract which contains phenolic compound exhibited antioxidant and antimicrobial capacity, due its functional group. On pH range of the cream, it could maintain the stability of color, odor, texture and homogeneity until eight weeks.

Table 2. Organoleptic observations on *Porphyridium cruentum* cream.

Weeks	Condition			
	Color	Odor	Texture	Homogeneity
1	White	No odor	Smooth	Homogeneous
2	White	No odor	Smooth	Homogeneous
3	White	No odor	Smooth	Homogeneous
4	White	No odor	Smooth	Homogeneous
5	White	No odor	Smooth	Homogeneous
6	White	No odor	Smooth	Homogeneous
7	White	No odor	Smooth	Homogeneous
8	White	No odor	Smooth	Homogeneous

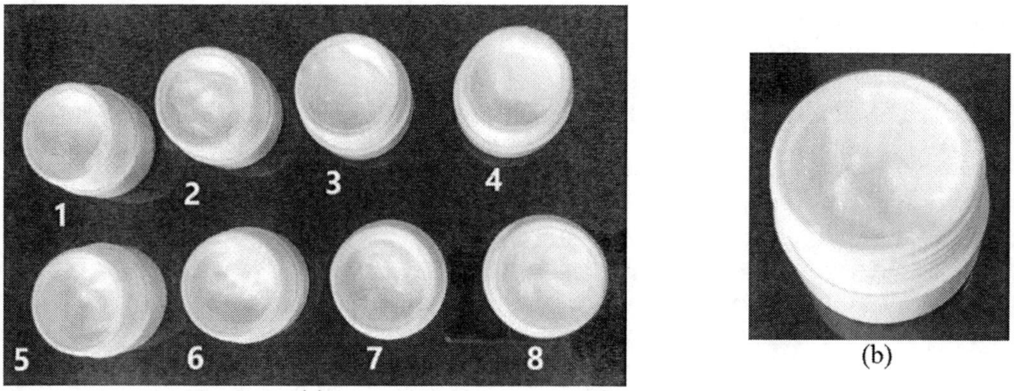

(a) (b)

Figure 5. *Porphyridium cruentum* cream on week 1 until 8 (a) and close view from *Porphyridium cruentum* cream on week 8.

4. Conclusion

The bioactive compound from *Porphyridium cruentum* could be extracted by maceration and encapsulated with maltodextrin by drying method with a spray dryer. The identification results on *Porphyridium cruentum* extracts indicated that it had capacity as antioxidant; the pinkish color on *Porphyridium cruentum* extract powder came from phycoerythrin. The best condition process obtained when *Porphyridium cruentum* extract encapsulated with maltodextrin 50 mg/L, with an antioxidant capacity as AEAC on 7.07 mg Vit. C/100 g sample and IC_{50} 484.09 mg/mL.

Applications in cosmetic creams showed antioxidant capacity as AEAC on 8.12 mg Vit. C/100 g sample and IC_{50} 29.08 mg/mL. Organoleptic analysis of cosmetic creams showed stable results for color, odor, texture, and homogeneity during eight weeks of storage at room temperature. Microbial and heavy metals contamination analysis were below the threshold according to BPOM regulations.

Acknowledgements

This research was supported by Industrial Research and Development Agency (BPPI) and Balai Besar Perikanan Budidaya Air Payau (BBPBAP) Jepara for the cooperation.

References

[1] Jones R F, Speer H L and Kury W 1963 Studies on the Growth of the Red Alga Porphyridium cruentum *Physiol. Plant.*
[2] Rebolloso Fuentes M M, Acién Fernández G G, Sánchez Pérez J A and Guil Guerrero J L 2000 Biomass nutrient profiles of the microalga Porphyridium cruentum *Food Chem.* **70** 345–53
[3] Borowitzka M A and Borowitzka L J 1988 *Microalgal Biotechnology* (Cambridge University Press)
[4] Gantt E 1969 Properties and Ultrastructure of Phycoerythrin From Porphyridium cruentum *Plant Physiol.* **44** 1629–38
[5] Sharma O D 1986 *Textbook of Algae* (Mc. Grow Hill Publishing Company Limitted)
[6] Setyaningsih I, Salamah E, and Rahman D A 2013 Komposisi kimia dan aktivitas antihiperglikemik biomassa dan polisakarida ekstraseluler dari mikroalga *Porphyridium cruentum J. Pengolah. Has. Perikan. Indones.* **16(1)** 79-85
[7] Tannin-Spitz T, Bergman M, Van-Moppes D, Grossman S and Arad S 2005 Antioxidant activity of the polysaccharide of the red microalga Porphyridium sp. *J. Appl. Phycol.* **17** 215–22
[8] SpecialChem 2020 Porphyridium cruentum extract
[9] Punampalam R, Khoo K S and Sit N W 2018 Evaluation of antioxidant properties of phycobiliproteins and phenolic compounds extracted from Bangia atropurpurea *Malaysian J.*

IOP Conf. Series: Materials Science and Engineering **980** (2020) 012042 doi:10.1088/1757-899X/980/1/012042

Fundam. Appl. Sci. **14** 289–97
[10] McDougal D H and Gamlin P D 2010 The influence of intrinsically photosensitive retinal ganglion cells on the spectral sensitivity and response dynamics of the human pupillary light reflex *Vision Res.* **50** 72–87
[11] Patel S N, Sonani R R, Jakharia K, Bhastana B, Patel H M, Chaubey M G, Singh N K and Madamwar D 2018 Antioxidant activity and associated structural attributes of Halomicronema phycoerythrin *Int. J. Biol. Macromol.* **111** 359–69
[12] Ré M I 1998 Microencapsulation by spray drying *Dry. Technol.* **16** 1195–236
[13] Hariyadi P 2017 Pengering semprot : *Food Rev. Indones.* **XII** 50–3
[14] Hasanah 2011 *Mikroenkapsulasi biomasa Porphyridium cruentum* (Bogor: IPB)
[15] Kenyon M M 1995 Modified Starch, Maltodextrin, and Corn Syrup Solids as Wall Materials for Food Encapsulation *ACS Symposium Series* vol 590, ed S J Risch and G A Reineccius (American Chemical Society)
[16] Van De Velde F, Tarola A, Güemes D and Pirovani M 2013 Bioactive compounds and antioxidant capacity of camarosa and selva strawberries (Fragaria x ananassa Duch.) *Foods* **2** 120–31
[17] Badan Standardisasi Nasional 1992 SNI 01-2891-1992 Cara Uji Makanan dan Minuman
[18] Kurniasih R A, Purnamayati L, Amalia U and Dewi E N 2018 Formulation and characterization of phycocyanin microcapsules within maltodextrin- alginate *Agritech* **38** 23–9
[19] Dewi R, Anwar E and Yunita K S 2014 Uji Stabilitas Fisik Formula Krim Yang Mengandung Ekstrak Kacang Kedelai (Glycine Max) *J. Pharm. Sci. Res.* **1** 194–208
[20] International Organization for Standardization 2006 ISO 21149:2006 Cosmetics — Microbiology — Enumeration and detection of aerobic mesophilic bacteria
[21] International Organization for Standardization 2008 ISO 16212:2008 Cosmetics — Microbiology — Enumeration of yeast and mould
[22] International Organization for Standardization 2015 ISO 22717:2015 Cosmetics — Microbiology — Detection of Pseudomonas aeruginosa
[23] International Organization for Standardization 2015 ISO 18416:2015 Cosmetics — Microbiology — Detection of Candida albicans
[24] International Organization for Standardization 2015 ISO 22718:2015 Cosmetics — Microbiology — Detection of Staphylococcus aureus
[25] Zhang L, Zeng X, Fu N, Tang X, Sun Y and Lin L 2018 Maltodextrin: A consummate carrier for spray-drying of xylooligosaccharides *Food Res. Int.* **106** 383–93
[26] Jubeau S, Marchal L, Pruvost J, Jaouen P, Legrand J and Fleurence J 2013 High pressure disruption: A two-step treatment for selective extraction of intracellular components from the microalga Porphyridium cruentum *J. Appl. Phycol.* **25** 983–9
[27] Gallego R, Martínez M, Cifuentes A, Ibáñez E and Herrero M 2019 Development of a Green Downstream Process for the Valorization of Porphyridium cruentum Biomass *Molecules* **24**
[28] Safi C, Charton M, Pignolet O, Pontalier P and Vaca-garcia C 2013 Evaluation of the protein quality of Porphyridium cruentum *J. Appl. Phycol.* **25** 497–501
[29] Matos Â P 2019 *Microalgae as a Potential Source of Proteins*
[30] Agustini N W S and . K 2017 Potency of Endo-Exopolysaccharide from Porphyridium cruentum (S.F.Gray) Nägeli as Antioxidant (DPPH) and Biological Toxicity (BSLT) *KnE Life Sci.* **3** 147
[31] Usov A I 2011 *Polysaccharides of the red algae* vol 65 (Elsevier Inc.)
[32] Csogör Z, Kiessling B, Perner I, Fleck P and Posten C 2001 Growth and Product Formation of Porphyridium purpureum *J. Appl. Phycol.* **13** 317–24
[33] Foti M C, Daquino C and Geraci C 2004 Electron-transfer reaction of cinnamic acids and their methyl esters with the DPPH(*) radical in alcoholic solutions *J Org Chem.* **69** 2309–14
[34] Agustina S, Aidha N N and Oktarina E 2019 Effect of maltodextrin concentration on the characteristic of phycocyanin powder as a functional food *AIP Conf. Proc.* **2175**

[35] Geresh S and Arad S 1991 The extracellular polysaccharides of the red microalgae: Chemistry and rheology *Bioresour. Technol.* **38** 195–201
[36] Nur M M A 2014 Potensi Mikroalga sebagai Sumber Pangan Fungsional di Indonesia (overview) Potency of Microalgae as Source of Functional Food in Indonesia (overview) *Eksergi* **XI** 1–6
[37] Sedjati S, Yudiati E, Metode M and Pembahasan H 2012 Profile of Polar and Non-Polar Pigment from Marine Microalgae Spirulina sp. and Their Potential as Natural Coloring *Ilmu Kelaut.* **17** 5–8
[38] Gianeti M D and Maia Campos P M B G 2014 Efficacy evaluation of a multifunctional cosmetic formulation: The benefits of a combination of active antioxidant substances *Molecules* **19** 18268–82
[39] Shin G H, Kim J T and Park H J 2015 Recent developments in nanoformulations of lipophilic functional foods *Trends Food Sci. Technol.* **46** 1–14
[40] Badan Standardisasi Nasional 1996 SNI 16-4399-1996 Sediaan tabir surya
[41] Badan Pengawas Obat dan Makanan Republik Indonesia 2011 PerKa BPOM Hk.03.1.23.07.11.6662 2011 Persyaratan cemaran mikroba dan logam berat dalam kosmetika

IOP Conf. Series: Materials Science and Engineering **980** (2020) 012043 doi:10.1088/1757-899X/980/1/012043

Physicochemical properties of fermented rice bran in optimal lactic acid bacteria growth

K Nisa[1*], S Handayani[1], V T Rosyida[1], S Nurhayati[1], W Apriyana[1], A W Indrianingsih[1], C Darsih[1], D A Ekamuri[2], D A S Ningrum[2] and Siswanti[2]
[1]Research Division for Natural Products Technology, Indonesian Institute of Sciences, Yogyakarta, Indonesia
[2]Faculty of Agriculture, Universitas Sebelas Maret, Surakarta, Indonesia

Email: khoi001@lipi.go.id

Abstract. This work aimed to characterize the physicochemical properties of fermented rice bran in optimal incubation time with two lactic acid bacteria. Fermented rice bran was prepared by inoculation of rice bran using *Lactococcus lactic* and *Lactobacillus plantarum* (6.0-7.5 x 10^4 CFU/mL) at 37°C for 96 h in a solid state condition. Physicochemical composition of fermented rice bran was analyzed by proximate, color test, reducing sugar, solubility, phenolic content and morphology image. Fermentation of rice bran with lactic acid bacteria had increased its protein content and phenolic content (25-70%). In contrast, fat content, carbohydrate, and reducing sugar of rice bran were decreasing during fermentation. SEM analysis indicated changes in rice bran surface structure after enzymatic process compared to unfermented rice bran.

1. Introduction

More than half of people in the world are having dietary demands on rice. As a staple food, rice is growing abundantly in China, India, Indonesia, Japan, Thailand, Philippines, Bangladesh, Vietnam, and Burma. Rice bran, outer layer of rice, is a by-product of rice milling industry which commonly used as animal feed. Since it is a good source of protein, mineral, fatty acids, and dietary fibre, public started to give more attention and generated a wide range of research activities [1]. Nowadays, dietary fiber sources become highly and popular demand to the development of functional foods. Moreover, rice bran is the subject of different researches interesting due to its abundance bioactive compounds that reveals great potential to be used in biotechnological processes. Several active compounds such as oryzanols, tocopherols, tocotrienols, phytosterols, and nucleotides have been identified in rice bran which is widely used to enhance the functionalities of some food stuffs and improve their capacity against chronic diseases [2-3]. The fermentative process implies employing microorganisms to get resultant transformations of their metabolic activity as well as increasing the availability of nutrients in raw materials. Biochemical changes associated with microbial metabolism and enzyme actions during fermentation are the main factors affecting the nutritional and textural properties of the fermented product. Many rice bran fermentation processes have been studied during the years. Several active metabolites are produced during fermentation of rice bran like γ -oryzanol, inositol, ferulic acid, vanillic acid, protocatechuic acid, and phytic acid [4-5].

Several investigations of bioactivity effect on fermented rice bran for promoting health have been performed previously using bacterial or fungal agents This work was performed to study the influence of lactic acid bacteria (LAB) in physicochemical properties of modified rice bran by a solid-state fermentation treatment.

2. Materials and Methods

2.1.Materials

The materials were purchased from a rice processing company located in Yogyakarta, Indonesia. The by-product of rice milling i.e rice bran was re-sorted by sifting into powder form using a standard sieve (60 mesh). Starter bacteria which used in this work were *Lactobacillus lactic* FNCC 0080 and *Lactobacillus plantarum* FNCC 0027.

2.2.Lactic acid fermentation of rice bran

Rice bran fermentation was performed with *L. lactic* and *L. plantarum* in static condition. Before used, rice bran (30g) was sterilized (121°C; 15 min) by adding of 50 mL distilled water in flasks. The sterile rice bran powder was cooled. Each LAB was poured into the flask in 6% ratio of total volume. The mixture was then incubated for 96 h at 37°C. After 96 h (4 days), the fermented rice bran was harvested and dried for 60°C.

2.3.Physicochemical characterization of fermented rice bran

The dried fermented rice bran was characterized for their physicochemical properties. The water content, ash content, protein, and lipid were examined based on AOAC (1984) method [6]. The crude protein was expressed as total nitrogen and multiplied in a factor 6.25 by Kjeldahl method [7]. The sample was dried at 105°C until obtained a constant weight to determine its water content. Rice bran sample (1 g) was incinerated in a muffle furnace (550°C; 5h) to measure the ash content. Crude lipid was analyzed by extracting samples (5 g) using petroleum ether in a Soxhlet apparatus. The pH value was measured in pH meter (Eutech PC700, NY, US). Reducing sugar was determined by Nelson Somogyi method [8]. The physical characteristics of fermented products were also identified by Scanning Electron Microscopy (Hitachi SU3500).

2.4.Total phenolic contents determination

Dried samples (10 g) were extracted with methanol by ultra-sonication at room temperature. Total phenolic content (TPC) of fermented rice bran extracts was determined by Folin-Ciocalteu reaction method. A stock solution (0.5 mL), distilled water (7.5 mL), and Folin-Ciocalteu reagent (0.5 mL) were mixed and continued by incubation. After 8 min, 20% Na_2CO_3 solution (1.5 mL) was added and re-incubated for 2h. The absorbance of the mixture was read using a UV-Vis spectrophotometer (Hitachi HALO RB-10) at 765 nm. TPC value was expressed as gallic acid equivalents of dry weight (mg GAE/g dw) [9].

3. Results and discussion

3.1.Physicochemical characterization

Lactic acid fermentation needs nutrients and medium for LAB to grow which can be provided by rice bran. Therefore, the effectiveness of fermentation depends on the starter used. Many bioactive compounds could release through bio-synthesis process which structural breakdown rice bran cell walls stimulated by fermentation [10]. The physicochemical properties of fermented rice bran are presented in table 1.

The moisture content of unfermented rice bran was lower than that of fermented rice bran. It might be caused fermentation process involved LAB media contained amount of water although it was solid state condition. In contrast, the ash content decreased due to the structure breakdown of lignocelluloses caused by enzymatic process. Amount of fat tended to be reduced after incubation, but the protein content was not significantly changed. Microbial fermentation can produce some proteolytic enzymes, which can stimulate the degradation of peptides and amino acids, hence support the other microorganisms to grow [11].

Table 1. Physicochemical properties of unfermented rice bran (URB), 96 h fermented with *L. plantarum* (RBLP) and *L. lactic* (RBLL).

Physicochemical properties	URB	RBLP	RBLL
Moisture content	5.27% ± 0.07	8.60% ± 0.07	8.72% ± 0.02
Ash content	11.50% ± 0.07	8.46% ± 0.03	10.31% ± 0.02
Protein	11.78% ± 0.06	11.46% ± 0.03	11.89% ± 0.28
Lipid	15.79% ± 0.08	14.88% ± 0.08	14.83% ± 0.12
Carbohydrate	55.67% ± 0.10	56.64% ± 0.11	54.25% ± 0.16
Reducing Sugar	1.16% ± 0.01	0.74% ± 0.001	0.63% ± 0.004
pH	6.41	4.55	4.57
Solubility	0.151%	0.130%	0.136%
Total Phenolic Content	43.68 mg GAE/g	77.98 mg GAE/g	54.77 mg GAE/g

During 96 h incubation, the pH of rice bran significantly decreased from 6.4 to 4.5. The amount of organic acids will be increased by microbial activity of LAB in fermented rice bran which each LAB has different capabilities in producing organic acids. Homolactic fermentation emerged to be the process which generates lactic acid as the main product yet detects of very low ethanol levels in the broth [12]. Consequently, pH of products was decreased since all the available carbohydrate was consumed and organic acids would be produced [13]. Glucose is the main sugar released during thermal/enzymatic cellulose hydrolysis. A significant decrease in glucose levels after 96 h incubation was presented by fermented rice bran using *L. plantarum* and *L. lactic* compared with unfermented rice bran. During fermentation, LAB hydrolyzed the substrate carbohydrates into glucose, which is used as its nutrients [14].

The physicochemical properties of fermented rice bran were also studied by evaluation of phenolic contents. The phenolic content of rice bran was enhanced during fermentation. Fermented rice bran with *L. plantarum* gave the higher phenolic content than fermented rice bran with *L. lactic*. Activity of enzymes produced by microorganism such as β-glucosidase, α-amylase and laccase, together with other enzymes affected in the improvement of phenolic contents in plants. Furthermore, cleavage of the bonds between phenolic compounds with other substances during the fermentation process releasing monomers of phenolic compounds [15].

Based on the light performance, food products can be categorized as opaque, transparent/translucent. Rice bran powder might be categorized as opaque foods since it is seen wholly by reflected light. Color pattern and stability is important indicator of the quality of food product.

Table 2. Color test of unfermented rice bran (URB), 96 h fermented with *L. plantarum* (RBLP) and *L. lactic* (RBLL).

Parameter	Value		
	URB	RBLP	RBLL
L*	67.5 ± 0.6	62.8 ± 0.2	62.9 ± 0.2
a*	3.6 ± 0.3	5.2 ± 0.1	5.4 ± 0.2
b*	20.6 ± 0.4	23.5 ± 0.2	23.6 ± 0.4

Table 2 reveals changes in the lightness (L*), yellowness (b*) and redness (a*) values of rice bran after and before fermentation. The lightness value of rice bran decreased after 96 h incubation with LAB. In contrast, the redness and yellowness level significantly increased. The release of some compounds by enzymatic process might effect to the color of rice bran product. Changes in a* and b* values after fermentation was correlated with changes in the rice bran pigments such as carotenoids and polyphenols (anthocyanin) [16]. The surface morphology images of fermented rice bran were carried out using SEM on voltage acceleration 5 kV and 5.4 to 5.5 mm working distance as shown in figure 1.

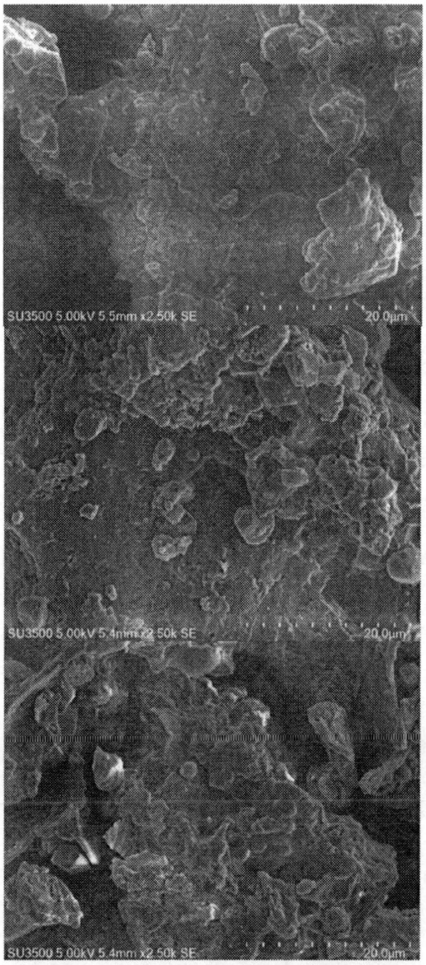

Figure 1. The morphology images of unfermented rice bran (top), fermented rice bran with *L. plantarum* (middle) and fermented rice bran with *L. lactic* (bottom).

SEM images revealed changes on surface structure of rice bran after enzymatic process compared by unfermented rice bran. The surface morphology of unfermented rice bran exhibited as a compact and well-organized surface, whereas the fermentation product appeared more hollow and cracked morphology. Cracks on rice bran surface structure might be affected by enzymes activity in degradation of rice bran cell wall such as cellulase and xylanase. Xylan is major component of hemicelluloses in rice bran which found between lignin and lower cellulose fibers compounds. A number of monomers like xylose, xylooligosaccharide, and arabinose is generated by the hydrolysis of xylan involving the activity of xylanase. Several studies reported that hydrolysis enzymes such as cellulase, xylanase, and those both combinations on rice stalk caused cracks on the surface structure of rice bran [17-18].

IOP Conf. Series: Materials Science and Engineering **980** (2020) 012043 doi:10.1088/1757-899X/980/1/012043

4. Conclusion

This study revealed the altering of the physicochemical properties on the fermentation of rice bran using LAB for 96 h. Fermented rice bran had significantly increased protein content, solubility, and phenolic content. While, fat content, carbohydrate, and reducing sugar of rice bran were decreasing during fermentation. It suggested that the utilization of cellulosic and hemi-cellulosic enzymes produced by LAB during fermentation could improve the nutrition value of rice bran.

Acknowledgements

This work was funded by INSINAS research project no. 06/INS-1/PPK/E4/2018, Ministry of Research, Technology and Higher Education of the Republic of Indonesia.

References

[1] Mccaskill D and Zhang F 1999 *Food Technol.* **53** 50
[2] Ardiansyah H, Shirakawa H, Koseki T, Ohinata K, Hashizume K and Komai M 2006 *J. Agric. Food Chem.* **54** 1914
[3] Iqbal S, Bhanger M I and Anwar F 2005 *Food Chem.* **93** 265
[4] Soi-Ampornkul R, Junnu S, Kanyok S, Liammongkolkul S, Katanyoo W and Umpornsirirat S 2012 *Food and Nutri. Sci.* **3** 135
[5] Oduguwa O O, Edema O O and Ayeni A O 2008 *Bioresour. Technol.* **99** 1816
[6] AOAC 1984 *Official Methods of Analysis (14th ed.)* (Washington DC: Association of Official Analytical Chemists) p 249
[7] Kjeldahl J 1883 *Ency. of Food Sci.* 439
[8] Hatanaka C and Kobara Y 1980 *Agric. Biol. Chem.* **44** 2943
[9] Nisa K, Nurhayati S, Apriyana W and Indrianingsih A W 2017 *IOP Conf. Ser.: Earth Environ. Sci.* 101 012002
[10] Katina K, Liukkonen K H, Kaukovirta-Norja A, Adlercreutz H, Heinonen S M and Lampi A M 2007 *J. Cereal Sci.* **46** 348
[11] Schieberle P 1996 *Adv. Food Sci.* **18** 237
[12] Wee Y J, Kim J N and Ryu H W 2006 *Food Technol. Biotechnol.* **44** 163
[13] Liang S, McDonald A G and Coats E R 2014 *Waste Manag.* **34** 2022
[14] Gilbert H J and Hazlewoo G P 1993 *J. Gen. Microbiol.* **139** 187
[15] Zheng Z and Shetty K 2000 *J. Agr. Food Chem.* **48** 895
[16] Muna I, Jamilah B, Hasanah M G, Ahmed M and Kharidah M 2015 *Food Sci. Technol.* **52** 5534
[17] Subramaniyan S and Prema P 2002 *Critical Rev. Biotechnol.* **22** 33
[18] Medeiros R G, da Silva Jr. F G, Báo S N, Hanada R and Filho E X F 2007 *Braz. Arch. Biol. Technol.* **50** 231

IOP Conf. Series: Materials Science and Engineering **980** (2020) 012044 doi:10.1088/1757-899X/980/1/012044

The effect of heat treatment of java plum seed extract on its polyphenolics content and antioxidant activities

R Rohadi*, **I Iswoyo and D Larasati**
Department of Processing Agricultural Product, Faculty of Agricultural Technology, Semarang University, Jl. Artery Soekarno-Hatta, Semarang 50196, Indonesia

Email: rohadijarod_ftp@usm.ac.id

Abstract. Butylated hydroxy anisole (BHA) is widely added in lipid and food lipid as a preservative and it is effective for inhibiting lipid peroxidation. However, BHA raised doubt would be the health impact for consumers. The methanolic extract of Java Plum seed (MEJS) contains a diverse group of phenolic compounds, it has potential as a natural antioxidant. The objective of the research was to determine the effect of heating treatment on total phenolics content and antioxidant activity of MEJS. Heating treatment applied on MEJS at 110, 120, 130, 140 and 150°C/10 minutes and heating time (10, 15, 20, 25 and 30 minutes/110°C). The results showed that heating of the extract at the temperature rose to 130°C should decrease the total phenolic content (TPC), total flavonoid content (TFC), and total tannin content (TTC) slightly, coincided with a decrease of the reducing power, but it could increase the free radical scavenging activity. Heating treatment at 110°C for about first ten minutes had a positive impact on increasing of its phenolics content and antioxidant activities. However, heating further at a higher temperature (>130°C) gave a bad influence on the free radicals scavenging activity. The extract heated at the 130°C/10 minutes gave the best EC_{50} value as 140 ppm.

1. Introduction

An artificial antioxidant such as butylated hydroxy anisole (BHA) and butylated hydroxytoluene (BHT) is widely used in lipid and food lipid as a preservative and they are effective for inhibiting the lipid oxidation [1,2]. However, artificial antioxidants have negative impacts on health. Therefore, that is required to replace it with an alternative antioxidant that is safe for human health. The Java Plum seeds (JS) extract contained a diverse group of phenolic compounds as such as (+)- catechin, gallic acid, ellagic acid, quercetin, (+)-epicatechin, rutin, kaempferol, and galloysl glucose group [3,4], meanwhile the methanolic extract of Java Plum seed (MEJS) containing rutin and (+)- catechin in abundant [5]. Many Java plum phenolic compounds mentioned above have been reported to possess potent antioxidant activity [4,6,7,8,9,10,11].

The antioxidant effect of plant phenolic compounds has been studied in relation to the retardation of lipid oxidation in the last decades [2,12]. Previous research showed that the MEJS was strong to act as reducing agents, radical scavenging, and hydrogen donors [4,5]. As a food additive, an antioxidant must be stable against food processing in particular heat treatment such as heating [13,14]. The heat treatment is commonly applied to food processing for various purposes including inactivation of enzyme, and improving the color and texture. Its effectiveness is strongly influenced by the intensity of heating, methods, temperature, and long periods [14].

The heat treatment at the high temperature (> 180°C) could decrease the total phenolic content (TPC), flavonoid content (TFC), and free radical scavenging both 1,1-diphenyl 2-picrylhydracyl (DPPH) and 2,2'-azino-bis (ABTS) radicals at the lower temperature 100-120 °C [15]. Heat treatment (autoclave) on grape seed extract (100 °C/0-60 min.) caused hydrolysis of gallocatechin (70%), polyphenolic (61%), epicatechin (65%), procyanidin B1 (75%), and procyanidin B2 (73%). However, autoclave heat treatment could increase the gallic acid content (71%), gallocatechin (100 %), and epicatechin-gallate (ECG) content to 129% in grape pomace. Autoclave heat treatment affects the antioxidant activity of both grape seed extract and grape pomace [16].

An antioxidant was sensitive toward heat treatment. Heating at the high temperature could decrease the antioxidant activities, even destructive to the structure of the chemical compounds [15,16,17]. Previous research showed that the antioxidant activity of white tea extract is still stable on heating up to 110°C, but its antioxidant activity declined by heating further [18]. Before the extract is applied to the food process, its antioxidant stability on heat treatment must be characterized. This study aimed to determine the effect of heat treatment on the total polyphenolics and antioxidant activity of MEJS.

2. Materials and methods

2.1. Materials

Java Plum (*Syzygium cumini* Linn.) seeds powder (60 mesh, 10% water content). Chemicals: methanol > 99.5% (Merck), gallic acid and tannic acid hydrate (Sigma-Aldrich, Belgium), (+)-catechin (Sigma Chemical Co. St. Louis USA), hydrochloric acid (HCl), ferro chloride ($FeCl_2$), ferry chloride ($FeCl_3$), ammonium thiocyanate, potassium ferricyanide (III), trichloroacetic acid (TCA), phospotungstic-acid, 1,1-*diphenyl 2-picrylhydracylradical* (DPPH) (Sigma-Aldrich Chemical Co.), Whatman filter paper (Whatman International, Ltd. England), Folin-Ciocalteu reagent and buffer phosphate. All reagents and chemicals were an analytical grade. The equipment used was analytic scales of Shimadzu AUW 120 (Shimadzu, Kyoto Japan), a rotary vacuum evaporator (IKA-RV10 Basic), freeze dryer (Virtis SP Scientific Sentry 2.0), oven Memmert UN 110 (Memmert, Germany), vortex (Velp Scientifica Europe), water-bath shaker (Julabo SW 22), UV-Visible spectrophotometer (UV1601 Shimadzu, Japan).

2.2. Sample preparation

The Java Plum fruits (*Genthong* var.) harvested on October 2018 (on the season), at the optimum maturity stage from a local farmer, at Semarang regency Central Jawa, Indonesia. After picking up the Java Plum fruits were stored at the cold storage (-18°C) until the use on May 2019. The Java Plum seeds were generated from Java Plum fruits that had been separated from the pulp. The seeds were cut into pieces with a sharp knife then dried using a cabinet dryer (55±5°C). That dried seeds were taken kernel and separated from husk and seed coats. The kernel was milled with a cutting mill then it was sieved so it was obtained Java Plum seeds powder (60 mesh, <10% water content. Java Plum seeds powder (JSP) was packed with airtight packaging and stored in the dark and dry room until it was used for the next step.

2.3. Proximate assay

Briefly, the JSP were analyzed proximately which including the analysis of water content assay referred to gravimetric methods of analysis AOAC [19], protein content assay by Kjeldahl extraction methods of analysis AOAC [19], lipid content by Soxhlet extraction methods of analysis AOAC [19], the crude fiber content was estimated as standard procedure in methods of analysis AOAC [19], ash content was analyzed on dry matter basis in methods analysis of AOAC [19], and carbohydrate total was analyzed by the difference of analysis AOAC [19].

2.4. JSP extraction
JSP extraction was done using modified methods as described by Rohadi *et al.* [10]. Briefly, approximately 30 g of JSP was extracted using aqueous methanol 50% (v/v), at ratio materials: solvent (1:10), macerated on water–bath shaker (40 ±1°C/6 hrs. 100 rpm), thereafter the mixture filtered using Whatman filter paper. Thereafter the filtrate obtained was collected in the Erlenmeyer 1000 milliliters volume. The residue was extracted twice by using the same methods. The extracts were collected then concentrated using a rotary vacuum evaporator (IKA-RV 10 Basic Germany) to obtain the methanol free extract. The methanol free extract was freeze-dried (Virtis SP Scientific Sentry 2.0) for solvent residue removal and then the methanolic extracts of Java Plum seeds (MEJS) were obtained. Thereafter it kept in the refrigerator for further analysis. This extraction procedure was repeated three times.

2.5. Heat treatment of MEJS
Approximately 0.5 g of each sample was heated with an oven furnace (Memmert, Germany) at the temperature of 110, 120, 130, 140, and 150°C as long as for 10 minutes. Thereafter it was put out of the oven then it put into a desiccator until reached equilibrium temperature before using the next step. In the other part, with the same methods, as many 0.5 g of each sample was heated on an oven furnace (Memmert, Germany) at temperature 110°C in varied heating time (10, 15, 20, 25, and 30 minutes).

2.6. Total phenolic assay
The total phenolic compounds were determined using the Folin–Ciocalteu modified methods [20]. Briefly, 0.5 mg MEJS was diluted in 1 ml of methanol, and 0.5 ml of Folin -Ciocalteu reagent (1:1) was added and the mixture was homogenized. After 8 minutes, 4.5 ml of 2% of sodium carbonate (Na_2CO_3) was added and the mixture was mixed. Thereafter the mixture was stored in the dark room at the room temperature. As a result of the reaction, color was developed and absorbance at λ –765 nm was measured after 60 minutes (UV-1601 Shimadzu, Japan) against the blank. The same procedure was repeated using a standard solution of gallic acid. The result was expressed as a gram of gallic acid equivalent (GAE) per 100 grams of MEJS.

2.7. The total flavonoid assay
The total flavonoid compounds were determined using a method referred Ebrahimzadeh *et al.* [20]. Briefly amount 0.5 mg MEJS was diluted in 1.5 ml methanol, 0.1 ml aluminum chloride 10%, 0.1 ml 1 M potassium acetate, and 2.8 ml distilled water and then homogenized. After that, mixture incubated at the room temperature for 30 min. The absorbance was measured at 415 nm by spectrophotometer (UV-1601 Shimadzu, Japan). The total flavonoid content was calculated as (+)-catechin equivalent (CE) from a calibration curve as described previously.

2.8. The total tannin assay
The total tannins compound was determined using a method referred Palici *et al.* [21]. Briefly amount 2 mL of MEJS (0.01 – 0.001 %) added 1 mL of phospotungstic-acid and 17 mL 50 % sodium bicarbonate (Na_2CO_3). Thereafter the mixture incubates as long 2 min. and then measured its absorbance. The absorbance was measured at λ= 750 nm by spectrophotometer (UV-1601 Shimadzu, Japan). The total flavonoid content was calculated as a tannic acid equivalent (TAE) from a calibration curve as describes previously.

2.9. DPPH radical scavenging activity assay
The DPPH radical scavenging activity was assayed using methods referred Vasi and Austin [7]. Briefly amount 0.5 ml MEJS solution on various concentrations (25, 50, 100, 200, and 400 ppm) was added 0.5 ml of DPPH solution (100 µM) and then incubated at room temperature (37±2°C) for 15 min. the absorbance was recorded at λ = 517 nm. The reaction was then subjected to absorbance measuring at λ = 517 nm. Radical scavenging activity (RSA) was expressed as the percentage of inhibition of ˙DPPH free radical and calculated as equation 1.

$$\text{Inhibition RSA-DPPH (\%)} = 1 - [\frac{Abs.sample}{Abs.control}] \text{ x } 100\ \% \qquad (1)$$

2.10. Reducing power assay
The reducing power activity was assayed using methods referred Vasi and Austin [7]. Briefly amount 2.5 ml MEJS solution in various concentration (25, 50, 100, 200 and 400 ppm) was mixed with 2.5 ml buffer phosphate (0.2 M/pH= 6.6) and 2.5 ml 1 % potassium ferricyanide solution, incubated at 50°C (in a water bath) for 20 min. and then received to terminate the reaction, 1 ml of TCA (10%) was added to each reaction. After that the solution then centrifuges (3.000 rpm/10 min). The upper layer of the mixture (2.5 ml) was mixed with 2.5 ml distilled water and 0.5 ml iron (III) chloride solution (0.1%) and subjected to absorbance measuring at λ = 700 nm. A higher absorbance indicates a higher total reducing power.

2.11. Statistical analysis
All the experiments in this study were done in triplicate and the results were expressed as mean ± SD. The data were analyzed through one-way of variance (ANOVA) using the statistical package for social sciences (SPSS) for Windows version (22.0) was used to analyzed data (SPSS Inc., Chicago, IL). Statistical significance was declared at (p <0.05).

3. Results and discussions
As the first of characteristic parameter of JSP revealed chemicals content were water content 9.1±0.1%, lipid 0.25 ± 0.04%, protein 3.89 ± 0.02%, crude fiber 2.14 ±0.07%, ash 1.07 ± 0.01% and carbohydrate (by difference) was 85.73 ± 0.1%. Previous research showed that Java plum seed was a solid waste (20.65%) of fresh fruits. And then JSP was rich in dietary fiber, sucrose (0.17%), fructose (2.78%), glucose (2.24%), and source of minerals as such as potassium 8,813 ppm, magnesium 2,162, ferrum 137, sodium 115 ppm, and some trace minerals as like calcium, phosphor and cuprum [9]. Yield extraction of JSP was 12.84 ± 0.8%, less than 13.89 ± 0.5% [9]. Its differences suggest to be due to the difference of freshness of sample and particles size. The smaller size of JSP particles (increasing the mesh size) obtained a higher yield [9]. Vasi and Austin, reported that yield extraction of Java plum seed with 50% ethanol was 12.96% [7].

3.1. Phenolic compounds
The phenolics compound of MEJS before heat treatment was expressed in three ways, these were; total phenolic, flavonoid and total tannins. The total phenolics was 34.73 ± 0.3 (g-GAE/100 g) and referred from a calibration curve y_1= 7.145x – 0.034 R^2=0.986. The total flavonoid was 7.12±0.17 (g-CE/100 g) and referred from a calibration curve y_2 = 0.0012x – 0.0038, R^2 = 0.998 and the total tannins was 29.81 ±0.15 (g-TAE/100 g) and referred from a calibration curve y_3 = 0.84x + 0.031, R^2 = 0.996. Yield extraction of JSP was 12.84 ± 0.8%, less than 13.89 ± 0.5 % [9]. Its differences were caused due to the difference of temperature maceration, size particles and freshness of the sample. In addition, the use of (+)- catechin was as a standard on the total flavonoid assay of the sample exactly. Therefore (+)- catechin content on the MEJS was abundant than quercetin [5]. Many publications reported using (+)- catechin as a

standard on the total flavonoid assay [22,23]. Vayuparp and Laksanalamal [1] reported that the grape seeds extraction using 50% ethanol, in maceration (50°C/6 hrs.1:10) obtained a yield as 14.86±0.03% and total phenolic as 32.86±0.04% (g-GAE/100 g dry extract).

3.2. Phenolic compounds and antioxidant activity assay

Table 1 showed the effect of the difference of heat treatment on total phenolics content (TPC), total flavonoids content (TFC), total tannins content (TTC), and in vitro antioxidant activity immediately (0 days) after the heat treatment.

Table 1. The effect of heat treatment on phenolic compounds and in vitro antioxidant activity of MEJS.

Temperature (°C)/10 min.	Phenolic compounds			Antioxidant Activity*	
	TPC (g-GAE/100g)	TFC (g-CE/100 g)	TTC (g-TAE/100 g)	RSA-DPPH (%)	FRAP (OD)
30/C	31.76 ± 0,17^{b}	7.12 ± 0.20^{a}	29.81 ± 0.15^{a}	81.16 ± 0.1^{d}	2.26 ± 0.002^{a}
110	33.93 ± 0.10^{a}	6.95 ± 0.17^{b}	26.26 ± 0.20^{b}	83.15 ± 0.1^{c}	2.13 ± 0.001^{b}
120	31.15 ± 0.20^{b}	6.41 ± 0.10^{c}	21.59 ± 0.10^{c}	84.73 ± 0.1^{b}	2.09 ± 0.005^{c}
130	27.34 ± 0.10^{c}	5.47 ± 0.10^{d}	18.07 ± 0.10^{d}	84.94 ± 0.1^{a}	1.75 ± 0.002^{d}
140	19.54 ± 0.20^{d}	5.17 ± 0.22^{e}	12.71 ± 0.20^{e}	79.41 ± 0.1^{e}	1.34 ± 0.003^{e}
150	17.84 ± 0.30^{e}	3.84 ± 0.20^{f}	10.07 ± 0.14^{f}	76.74 ± 0.1^{f}	1.24 ± 0.004^{f}

Mean in the same column with difference alphabetical letters are significantly different ($p<0.01$)
*Antioxidant activity was measured at 400 ppm extract. c = control

The effect of heat treatment (oven) on phenolic compounds and in vitro antioxidant activity of the samples revealed interesting phenomena. In general, the increase of heating temperature (110-150°C) caused an extensive decrease in the TPC, TFC, and TTC of the samples as 48.7%, 46.06 %, and 66,2% respectively compared to the control (table 1). Similarly, the data mentioned above, heating time (10 minutes the second and so on) of MEJS caused a decrease in the TPC, TFC, TTC, and free radical scavenging activity (table 2). However, heating for about ten minutes first had a positive impact on increasing in its phenolic content and antioxidant activities (table 2).

Meanwhile, oven treatment means raising the temperature to 130°C/10 minutes could increase the free radical scavenging by 4.7 % (significantly). It might suggest that the heat treatment of MEJS (30-130°C) liberated phenolics compounds and thus increased the amounts of activity at these range temperatures, but raising the temperature to 150°C/10 minutes let the free radical scavenging declined. The decreasing of TPC, TFC, and TTC samples accordance with the decrease of the total reduction power. As the changes in the phenolic compound were not related to the changes in antioxidant activity. This was in line with what was stated by Chamorro *et al.* [16] that modifying the phenolic profile content due to different heat treatment methods was not related to changes in antioxidant activity. Similarly, with the statement mentioned above, Pop *et al.* [24] reported that heat treatment (70-100°C/15 min.) on the grape seed extraction was not significantly affected the radical scavenging activity on its extracts.

Anthocyanin degradation increased in temperature and time of pasteurization of Jamun (*Eugenia jambolana*) fruit juice [25]. Heating treatment (60-125°C/8-72 hr.) decreased procyanidin concentration significantly ($p<0.05$) in both blueberry and grape pomace [26]. It was suggested that a decrease in the TPC, TFC, and the TTC assay (Table 2) could be attributed to the degradation of proanthocyanins present in MEJS powder which in having higher molecular weight into the release of simple phenolic compounds like quercetin, rutin, and (+)- catechin. That's why the radical scavenging activity increasing while the total phenolic compounds assay declined.

Table 2. The effect of heating time on phenolics changes and antioxidant activity of MEJS.

Time-heating (min./110°C)	Phenolic compounds			Antioxidant Activity*	
	TPC (g-GAE/100g)	TFC (g-CE/100 g)	TTC (g-TAE/100 g)	RSA-DPPH (%)	FRAP (OD)
C	31.76 ± 0.17^{b}	7.12 ± 0,20^{c}	29.81 ± 0.15^{c}	89.04±0.16^{b}	2.66± 0.01^{f}
10	33.27 ± 0.14^{a}	8.40 ± 0.20^{a}	32.65 ± 0.10^{a}	89.11±0.10^{ab}	2.68 ± 0.01^{e}
15	31.67 ± 0.17^{b}	7.70 ± 0,10^{b}	31.51 ± 0.06^{b}	89.32±0.10^{a}	2.83 ± 0.01^{d}
20	31.05 ± 0.10^{c}	7.20 ± 0,10^{c}	29.64 ± 0.17^{c}	88.39±0.16^{c}	2.86 ± 0.01^{c}
25	30.60 ± 0.10^{d}	6.71 ± 0,10^{d}	28.15 ± 0.20^{d}	87.24±0.01^{d}	2.87 ± 0.01^{b}
30	29.82 ± 0.16^{e}	6.20 ± 0,20^{e}	25.43 ± 0.10^{e}	87.03±0.16^{d}	2.88 ± 0.01^{a}

Mean in the same column with difference alphabetical letters are significantly different (p<0.01)
*Antioxidant activity was measured at 400 ppm extract. c = control

3.3. DPPH radical scavenging activity assay

DPPH free radical scavenging activity expressed as (% inhibition) was used for the assay in vitro antioxidant of MEJS samples (both heated and control). The effect of heating treatment against in vitro antioxidant activity of the samples served in figure 1. In general, the results showed that the increase of heating temperatures (110-130°C/10 min.) affected the increase of free radical scavenging activity from the samples significantly (p<0.01). While the increasing of the sample's concentration (25-400 ppm) in each heating treatment gave a positive correlation (r=0.89-0.99) with the increase of antioxidant activity. However, the further increase in heating temperature (140-150°C) led to a decrease in the free radical scavenging activity significantly (p<0.01). It suggested that heating temperature had a considerable influence on the changes of phenolic compounds. Heating treatment at the lower temperature (< 130°C/10 min) induced breakdown the proanthocyanins are converted to the oligomeric and monomeric aglycone. On the other hand, the heating at the higher temperature (> 130°C/10 min.) induced the degradation of the monomeric aglycone, therefore the effect led to the decrease of the antioxidant activity [16]. On behalf of increasing the antioxidant activity on the samples heated at high temperature, Miranda *et al.* [27] told that it might be due to the generation and accumulation of Maillard derived melanoidins.

According to Chamorro [16], the type of the samples (an extract or not extract) influenced the type of the ongoing reaction. Pasteurization, sterilization, or thermosonication are less able to maintain anthocyanin retention than sonication process in Jamun fruit juice [25]. In addition, the anthocyanin degradation will increase together with the increase of temperature and time of pasteurization. However, we could not establish a relationship between the changes of phenolic compound and the antioxidant activity [11,15,25,28,29,30].

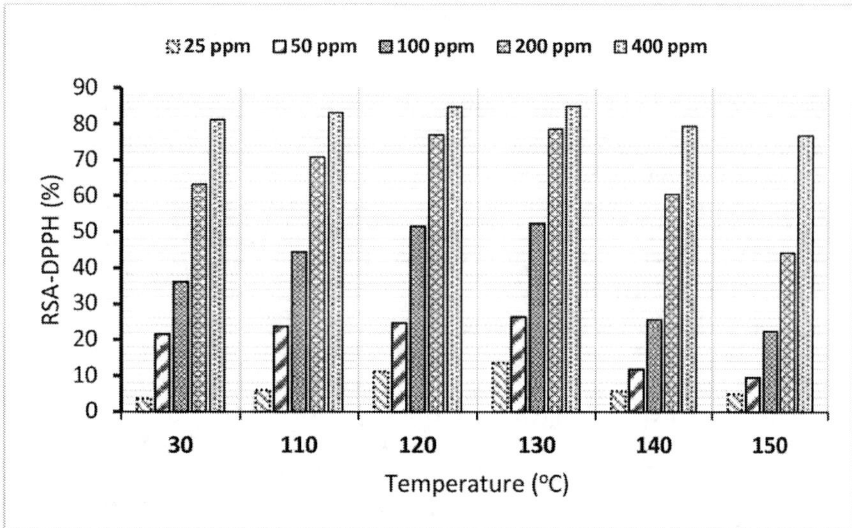

Figure 1. Free radicals scavenging activity of MEJS (25-400 ppm) in an air oven at various temperatures heating (110-150°C/10 min.). There is a significant effect of temperature heating against antioxidant activity ($p<0.05$). Each value is expressed as mean ± SD, n=3.

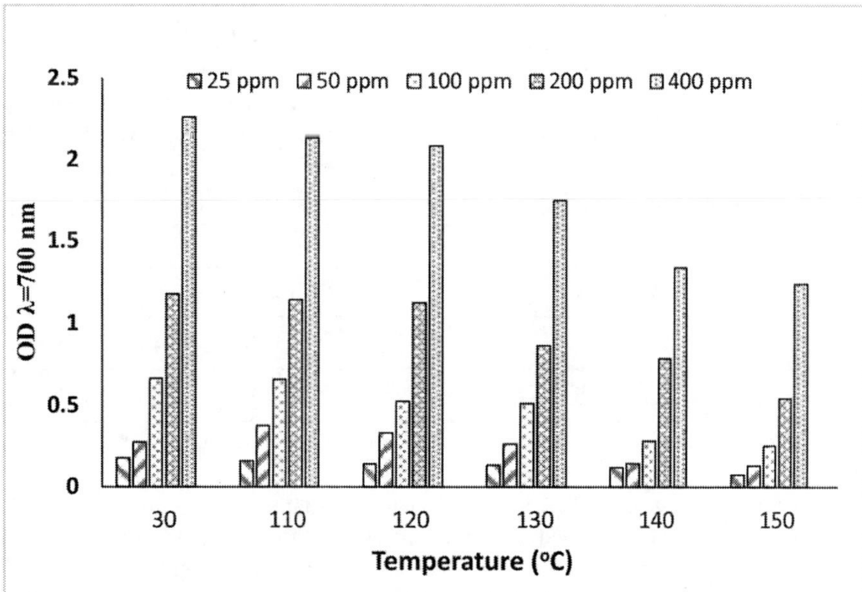

Figure 2. The reducing power activity of MEJS (25-400 ppm) in an air oven at various temperatures heating (110-150°C/10 min.). There is a significant effect of temperature heating against antioxidant activity ($p<0.05$). Each value is expressed as mean ± SD, n=3.

3.4. Reducing power assay
The total reduction ion Ferric (Fe^{+3}) expressed as (optical density-OD) was used for the assay in vitro antioxidant of MEJS samples (both heated and control). Effect of heating treatment against in vitro antioxidant activity of the samples served in figure 2. In general, the results showed that the increase of heating temperatures (110-150°C/10 min.) affect the decrease of the reducing power activity from the samples significantly (p<0.01). Dry heating at a higher temperature (> 110°C) gave a negative correlation (r=0.8-0.96) with the reducing power. This finding is in line with previous research done by Reda [17]. While increasing of the sample's concentration (25-400 ppm) in each heating treatment gave a positive correlation (r=0.98-0.99) with the increase of radical scavenging activity (figure 1), it was in the contrary with what had mentioned above, Nikousaleh & Prakash [31] reported that the heat treatment on clove exhibited significantly higher reducing power activity. The decrease of the optical density (OD) value of the samples means were less able to the reduction of ferric ion (Fe^{3+}).

3.5. Half maximal effective concentration (EC_{50} value) in antioxidant properties
The antioxidant properties were normally expressed as EC_{50} value (mg-extract per Liter) for comparison. Efficacy in antioxidant properties inversely correlated with EC_{50} value. EC50 value of the heated of MEJS were 200 ppm, 178.2; 156.1; 149; 22, and 250.1 ppm with the heating temperature 30, 110, 120, 130, 140 and 150°C respectively. On behalf to antioxidant efficacy, heated the samples with 110, 120, and 130°C/10 minutes were more effective as evidence by the lower EC_{50} value compared to the control. Nevertheless, heating further at a higher temperature (>130°C) gave a bad influence on the antioxidant activity.

The enhanced antioxidant efficacy after heated might be attributed to the degradation of proanthocyanins present on MEJS into release of a free simple bioactive compounds like rutin, quercetin and (+)- catechin as aglycone. The effect of the heating applied on foods perhaps could decrease some of the identified phenolic compounds, but in the other hand could increase others depend on the methods, temperature, and time. But, in general, heating treatment demonstrated a slight decrease in some of the identified phenolic compounds [28]. In this study, there was a negative correlation between the TPC, TFC, and TTC with the free radical antioxidant activity. An antioxidant efficacy (EC50) of the samples are presented in table 3.

Table 3. Antioxidant efficacy (EC_{50}) values of MEJS at the different heating temperatures derived by RSA-DPPH methods (ppm).

Sample	(°C)					
	30	110	120	130	140	150
MEJS	200	178.2	156.1	149	221	250.1

4. Conclusion
The heating treatment that was applied to MEJS at the temperature risen to 130°C should decrease the TPC, TFC, and TTC slightly coincided with the decrease of the reducing power. The contrary heating treatment risen to 130°C should increase the free radical scavenging activity of the extract. However, heating further at a higher temperature (> 130 °C) gave a bad influence on the free radicals scavenging. Antioxidant activity of Java Plum (*S. cumini* Linn.) seed extract was unstable by heating at high temperature more than 130°C/10 minutes. The extract heated at the 130°C/10 minutes gave the best EC_{50} value as 140 ppm. Concerning of the effect thermal treatment on the changes of the phenolics compound,

finally the present study revealed that there was no established relationship between the changes of phenolic compound and the antioxidant activity.

Acknowledgments
The authors are grateful to the Director of Research and Devotion the Community, The Directorate General Strengthening Research and Development, Ministry of Research, Technology and Higher Education Republic of Indonesia for financial support of the work. The authors are grateful to the Chief of The Research Institute and Devotion the Community, Semarang University, who has facilitated the signing of the contract research by No. 020/USM.H7.LPPM/L/2019.

References

[1] Vayupharap B and Laksanalamal V 2012 Recovery of antioxidant from grape seeds and its application in fried food *J. Food Process Technol.* **3(4)** 1-6

[2] Shahidi F and Ambigaipalan P 2015 Phenolics and polyphenolics ini foods, beverages and spices: Antioxidant activity and health effects-A review *J. Funct. Foods.* **18** 820-97

[3] Swami S B, Thakor N S J, Patil M M and Haldankar P M 2012 Jamun (*Syzygium cumini* L.): A review of Its food and medicinal uses *Food. Nutr. Sci.* **3** 1100-17

[4] Rydlewski A A, de Morais D R, Rotta E M, Clause T and Visentainer J V 2013 *Determination of Bioactive Compounds, Antioxidant Activity and Physical and Chemical Composition Of Different Parts of Four Brazilian Fruits* (Colombo: State University of Maringa Postgraduate Program)

[5] Rohadi, Santoso U, Raharjo S and Falah I I 2017 Determination of antioxidant activity and phenolic compounds of methanolic extract of Java Plum (*Syzygium cumini* Linn. (Skeel)) seed. *IFNP* **13(1)** 9-20

[6] Benherlal P S and Arumughan C 2007 Chemical composition and in vitro antioxidant studies on *Syzygium cumini* fruit *J.Sci* **87(14)** 2560-69

[7] Vasi S and Austin A 2009 Antioxidants potential of Eugenia jambolana Lam. seeds *J. Biol. Sci.* **9(8)** 894-98

[8] Peixoto M P G and Freitas L A P 2012 Spray-dried extracts from *Syzygium cumini* seed: physicochemical and biological evaluation *Rev. Bras. Farm.* **23(1)** 145-52

[9] Rohadi, Raharjo S, Falah I I and Santoso U 2016 Antioxidant activity of Duwet (*Syzygium cumini* Linn.) extract on lipid peroxidation in vitro *J. Agritech.* **36(1)** 30-7

[10] Rohadi, Raharjo S, Falah I I and Santoso U 2017 Methanolic extract of Java Plum (*Syzygium cumini* Linn) seed as natural antioxidant on lipid oxidation of oil-in water emulsions *IFRJ* **24(4)** 1636-43

[11] Rohadi, Iswoyo and Larasati D 2019 *Conf. Universitas Muhammadiyah Semarang* e-ISSN:2654-3168 p-ISSN: 2654-3257

[12] Mielnik M B 2006 Grape seed extract as antioxidant in cooked, cold stored Turkey Meat. LWT *J. Food Sci. Technol* **39(3)** 191–98

[13] Micali M and Fiorino M 2016 Thermal processing in food industries and chemical transformation In Micali M (*ed.*) 2016 *The Chemistry of Thermal Food Processing Procedures*, *Chemistry of Foods*

[14] Zhou L Y, Li W, Pan W J, Sajid H, Wang Y, Guo W Q, Cai Z N, Yang W W, Wang D D and Chen Y 2017 Effects of thermal processing on nutritional characteristics and non-volatile flavor components from Tricholoma lobayense *Emir. J. of Food and Agri.* **29(4)** 285-92

[15] Hihat S, Remini H and Madani K 2017 Effect of oven and microwave drying on phenolic compounds and antioxidant capacity of coriander leaves *Intern. Food Res. Journ.* **24(2)** 503-9

[16] Chamorro S, Gon˜i I, Viveros A, Hervert-Hernandez D and Brenes A 2012 Changes in polyphenolic content and antioxidant activity after thermal treatments of grape seed extract and grape pomace *Eur. Food Res. Technol.* **234(1)** 147–55
[17] Reda S Y 2011 Evaluation of antioxidants stability by thermal analysis and its protective effect in heated edible vegetable oil *Ciênc. Tecnol. Aliment. Campinas.* **31(2)** 475-80
[18] Rohadi and Wahjuningsih S B 2019 The Effect of thermal treatment on tea (*C. sinensis* Linn.) extract, type of white tea on the stability of its antioxidant activity *J. Indus. Has. Perkeb.* **14(1)** 41-49
[19] Latimer G W Jr. 2005 *AOAC Official Methods of Analysis* 18thed (Maryland: AOAC International)
[20] Ebrahimzadeh M A I, Khalili M and Dehpour A A 2015 Antioxidant activity of ethyl acetate and methanolic extract of two marine algae *Nannochloropsis oculata* and *Gracilaria gracilis* an in vitro assay *Brazil. Journ. of Pharm. Scien.* **54(1)** ISSN 2175-9790.
[21] Palici I, Tita B, Ursica L and Tita D 2005 Method for quantitative determination of polyphenolic compounds and tannins from vegetal products *Acta Univ. Cibiniensis Ser. F. Chem.* **8** 21-32
[22] Abdeltaif S A, Sir Elkhatim K A and Hassan A B 2018 Estimation of phenolic and flavonoid compounds and antioxidant activity of spent coffee and black tea (processing) waste for potential recovery and reuse in Sudan *Recy.* **3(27)** 1-9
[23] Fattahi S, Zabihi E, Abedian Z, Pourbagher R, Motevalizadeh A A, Mostafazadeh A and Akhavan-Niaki H 2014 Total phenolic and flavonoid content of aqueous extract of *Stinging Nettle* and in vitro antiproliferative effect on Hela and BT-474 cell lines *IJMCM* **3(2)** 102-7
[24] Pop C, Rotar-Ancuta M, Salanta L, Socaci S, Ranga, Floricuta and Socaciu C 2015 Thermal stability study of the grape seeds extracts. *Bull. UASVM Food Scien. and Techno.* **71(1)**
[25] Shaheer C A, Hafeeda P, Kumar R, Kathiravan R, Kumar D and Nadanasabapathi S 2014 Effect of thermal and thermosonication on anthocyanin stability in jambu (*Eugenia jambolana*) fruit juice. *IFRJ* **21(6)** 2189-94
[26] Khanal, Ramesh C, Howard L R and Prior R L 2010 Effect of heating on the stability of grape and blueberry pomace procyanidins and total anthocyanins *Food Res. Intern.* **43** 1464-69
[27] Miranda M, Maureira H, Rodriguez K and Vega-Galvez A 2009 Influence of temperature on the drying kinetics, physicocemical properties and antioxidant capacity of Aloe Vera (*Aloe Barbadensis* Miller) gel *J. Food. Eng.* **91** 297-304
[28] Pereira E, Barros L, Antonio A L, Cabo-Verde S, Santos-Buelga C, Ferreira I C and Rodriques P 2017 Is gamma radiation suitable to preserve phenolic compounds and to decontaminate mycotoxins in aromatic plants? A Case-study with *Aloysia citrodora* Palau *Molecules* **22(347)** 1-13
[29] Naveed S, Rizwan E and Sajid A 2018 Effect of gamma irradiation on phytochemical content and antimicrobial activities of selected herbs *Chem. Xpress* **11(1)** 1-15
[30] Sari B L, Rahayu D P, Rohdiana D, Nurlita S and Sahara P S 2018 Influence of gamma ray irradiation on total flavonoid and tannin content of white tea (*C. sinensis* L.) and loranthus of tea (*Scurulla atropurpurea* BL. Dans) *J. Farm. Indo.* **8(1)** 1-9
[31] Nikousaleh A and Prakash J 2016 Antioxidant components and properties of dry heat treated clove in different extraction colvents *J. Food Sci. Tech.* **53(4)** 1993-2000

Empty fruit bunches, potential fiber source for Indonesian pulp and paper industry

L Indriati[1]*, N Elyani[1], and S F Dina[2]
[1]Center for Pulp and Paper, Jl. Raya Dayeuhkolot no. 132, Bandung, Indonesia
[2]Center for Industrial Research and Standardization of Medan, Jl. Sisingamangaraja no. 24, Medan, Indonesia

Email: liesinag@gmail.com

Abstract. Empty fruit bunches (EFB) is one of solid waste from crude palm oil (CPO) mill. It's about 20-22% from fresh fruit bunches (FFB) or equal to the CPO yield. Indonesian CPO production in 2019 was 51.8 million tons; increased by 9% compared to 2018. So, about 51.8 million tons of EFB was available in 2019. EFB is a potential lignocellulosic material for pulp and paper. The utilization of EFB fibers for papermaking has been studied extensively. The use of EFB for printing and writing papers tends to cause pitch problems and poor visual appearance. The most potential use of EFB fibers is for packaging papers. EFB fiber is classified as short up to moderate fibers, so its usage must still be mixed with other fibers to improve the sheet strength. Many experiments have been done regarding the mixture of EFB fibers with other fibers such as OCC (Old Corrugated Containers) and UBC (Used Beverage Cartons) recycled fibers. Other types of paper from EFB are still being developed, including for nursery bags and active packaging papers.

1. Introduction

Indonesian pulp and paper industry experienced a period of very rapid growth in the era of 80s. At present, Indonesia has no less than 80 integrated and non-integrated pulp and paper industries spread across Java, Sumatera, and Kalimantan islands. Indonesian pulp and paper production ranks 10th in the world for pulp and 6th for paper. Whereas in Asia it is ranked 3rd for pulp and 4th for paper [1].

Overall, Indonesian paper production in the last few years increased, especially for industrial paper types. Based on the data from Indonesian Ministry of Industry, in 2019, the industry of paper, paper goods, printing, and reproduction of recording media grew by 8.86%. This growth must be supported by the sustainable availability of its raw materials, especially fiber which is the largest proportion in papers (70-95%) [2].

The fiber source for paper can be generated from wood, non-wood, or secondary fibers from recycled papers. For the printing paper industry, the need for bleached short-fiber wood pulp can be fulfilled by the local pulp mills which generally using Acacia and Eucalyptus as their raw material [3]. Whereas long fiber wood pulp both bleached and unbleached is obtained from imports. Almost all packaging paper industry uses secondary fibers from recycled paper. Even if there is a slight use of unbleached long fiber wood pulp, this pulp is fully obtained from imports [4].

In recent years, the need for recycled paper increased in line with the increasing packaging paper production [5]. In addition to facing unstable paper quality problems due to the variety of recycled paper quality used, the packaging paper industry is also faced with the problem of supply of recycled paper that is constrained by the inadequately of local recycled paper availability, so that must be obtained from imports. The increasing stringent regulations on an imported recycled paper have made the availability of imported used paper even more limited [6]. Therefore, the sustainable alternative sources of fiber for pulp and paper is needed.

Indonesia has a lot of non-wood fiber sources for pulp and paper; some of which have been used commercially such as straw, bagasse, and bamboo. However, due to its decreasing availability, there are no more pulps that are produced from non-wood raw materials. Some non-timber plants also have potential such as abaca, flax, kenaf, roselle, and many more [7], but until now they have not been produced commercially due to the insecure sustainability of its' supply.

Oil palm empty fruit bunches are one of the solid wastes produced from CPO (crude palm oil) mills which have not been utilized commercially. Currently, its use is only limited as fuel at the CPO mill itself or as compost in oil palm plantations. Even though it has been known that EFB is a potential lignocellulosic material for pulp and paper [8], it is still not utilized commercially. In the future, oil palm empty fruit bunches are expected to be a potential alternative of non-wood raw material [9-13]. Especially for pulp and paper manufacturing, there have been many studies conducted, even in Malaysia, there have been industries that process oil palm empty fruit bunches commercially as raw material for pulp and paper [10,14].

Indonesia, as the largest palm oil-producing countries in the world, has the opportunity to utilize this oil palm empty fruit bunches as raw material for pulp and paper. This paper outlines how big the potential of EFB is in Indonesia and some types of paper that can be produced from EFB fibers [11,15].

2. Availability of EFB in Indonesia

Oil palm is one of the commodities of plantation products that has an important role in economic activities in Indonesia because of its ability to produce vegetable oil which consists of crude palm oil (palm oil obtained from the extraction or pressing of oil palm fresh fruit bunch) and palm kernel oil (palm kernel oil). As the largest palm oil-producing country in the world, the plantation area and production of oil palm in Indonesia continue to increase (figure 1). The oil palm plantation in Indonesia is spread in 25 provinces. However, only five provinces are the largest palm oil producers, respectively starting from the highest production capacity namely Riau, Central Kalimantan, North Sumatra, West Kalimantan and South Sumatra provinces [16].

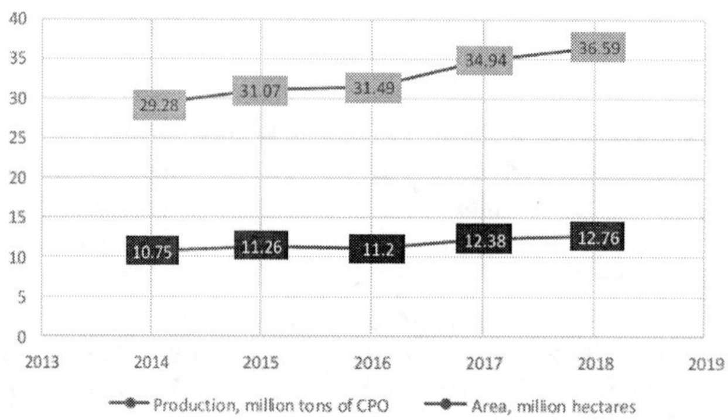

Figure 1. Growth of area and production of oil palm plantations Indonesia, 2014-2018 [16].

From figure 1 above, it can be seen that over the past five years, CPO production has increased by almost 25%. From the processing of oil palm fresh fruit bunches (FFB), several side products that are potential to provide added value are generated, including empty fruit bunches (EFB). In figure 2, it can be seen that the mass balance in the processing of oil palm fresh fruit bunch. As a side product, the amount of EFB generated is almost the same as the amount of crude palm oil produced [12]. Each processing of 1 ton of fresh fruit bunches will produce 21–23% of oil palm empty fruit bunches or 210-

230 kg [17,18]. From figure 1, it can be predicted that the number of EFB obtained will be around 8.0498 million tons during 2018 or equivalent to 26,830 tons/day.

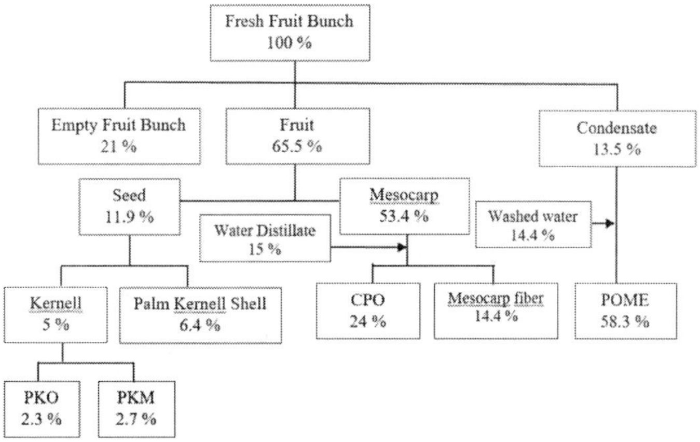

Figure 2. Mass balance in palm oil fresh fruit bunch processing [17].

3. Characteristics of EFB fibers

The characteristics of fiber affect the properties of paper/paperboard produced and the ease of fiber to be processed before it is formed into paper or paperboard. The morphological and chemical characteristics of EFB fibers, and how it affects the pulp and paper produced are discussed in the following description.

Figure 3. Three types of EFB fibers [19].

There are 3 forms of EFB fibers based on the number of ripe fruits in fresh fruit bunches (FFB), i.e. type 1, type 2, and type 3 as shown in figure 3. In type 1, the percentage of ripe fruit is only 40%, while type 2 and type 3 have 75% and 90% of ripe fruit respectively. These forms affect the integrity and compactness of the EFB and whether or not the EFB is easy to be deciphered. The more ripe fruit in the EFB, the easier it is to be deciphered [20].

Bulk density of EFB is 177.98 kg/m^3 [19], lower than that of *Acacia mangium*, i.e. 461 kg/m^3 [21], 390 kg/m^3 [22], or *Acacia auriculiformis*, i.e. 410 kg/m^3 [22]. Fiber and non-fiber content of EFB are 72.67% and 27.33% for the base section of EFB, and 62.47% and 37.53% for the tip of EFB [19, 23-24].

3.1. Fibers morphology

The morphology EFB fiber is shown in table 1. It can be seen that the average of EFB fiber length is in the range of 0.76 mm up to 1.81 mm. Based on IAWA (International Association of Wood Anatomy) classification, EFB fiber is classified as very short (0.5-0.7 mm) up to moderate (0.9-1.6 mm) fiber [11].

The average fiber length of EFB is 0.87 mm and the largest proportion, i.e. about 91%, is in the range of 0-1 mm [25].

The EFB fiber length is also affected by the pulping process. The average fiber length of EFB semi-chemical pulp is 0.767 mm, while the EFB thermo-mechanical pulp is 1.023 mm at the temperature of 166°C with hot air-dried, and 0.737 mm at 140°C with warm air-dried. Meanwhile, the EFB chemical pulp with Kappa number of 9.6 and yield of 31.1% has a low average fiber length of 0.556 mm [26].

Table 1. Morphology of EFB fiber.

Parameter	[10]	[27]	Base section [19, 28]	Tip [19, 28]	[29]
Fiber Length, L (mm)					
- Range		0.27 – 1.6	0.63 – 1.81	0.46 – 1.27	
- Average	0.99	0.77	1.20	0.76	0.99
Diameter (μm)					
- Fiber diameter (D)	19.1	12.66	16.89	14.34	
- Lumen (l)		6.52	8.04	6.99	
Fiber wall thickness (W)	3.38[a]	3.07	3.49	3.68	3.38[a]
Runkle ratio (2W/l)		0.94	0.87	1.05	
Felting power (L/D)		60.82	79.95	53.00	
Flexibility ratio (l/D)		0.52	0.54	0.49	
Muhlstep ratio [100% x(d^2 - l^2)/ d^2]		73.48			
Fiber coarseness, s mg/m	0.107				1.37[b]
Rigidity index $(T/D)^3$ x 10^{-4}	55.42				55.43

Note: [a]Cell wall thickness (T); [b]Fiber coarseness mg/m

Table 2. Morphology of EFB compared to some hardwoods.

Parameter	EFB [27]	*Acacia mangium* [22]	*Acacia auriculiformis* [22]	Wheat straw [30]
Fiber Length, L (mm)				
- Range	0.27 – 1.6			
- Average	0.77	0.86	0.84	1.18
Diameter (μm)				
- Fiber diameter (D)	12.66	17.07	14.47	13.60
- Lumen (l)	6.52	13.30	11.32	5.68
Fiber wall thickness (W)	3.07	1.89	2.15	3.96
Runkle ratio (2W/l)	0.94	0.28	0.38	1.59
Felting power (L/D)	60.82	50.54	57.04	86.76
Flexibility ratio (l/D)	0.52	0.78	0.78	0.42
Muhlstep ratio [100% x(d^2 - l^2)/ d^2]	73.48	39.42	38.46	
Rigidity index $(T/D)^3$ x 10^{-4}	0.24	0.11	0.15	

The fiber length is the main morphological characteristics of a fiber since it affects the sheet strength. The longer the fiber, the higher the strength. The Runkle ratio of EFB fiber is 0.94, lower than 1. This indicates that EFB fiber has a high lumen diameter with a thin wall. It means that the EFB fiber can be easily flattened during the refining process. Muhlstep ratio and felting power of EFB fibers, namely 73.48 and 53.0-79.95 respectively shows that the EFB fiber is easy to flatten and form inter-fiber bonding [27]. In general, the morphology of the base section of EFB fiber is better than its tip section [19].

Although Acacia wood species is classified as short fibers, the average fiber length is slightly higher than EFB fibers (table 2). The lumen diameter is much greater than that of the EFB fiber. With the lower fiber wall thickness, the Runkle ratio of Acacia wood species is much lower than that of EFB fiber. This shows that the fiber of Acacia wood species is easier to be refined compared to EFB fiber. The felting power of EFB fiber is almost the same as Acacia wood species, but the Muhlstep ratio is almost twice. The Muhlstep ratio affects the ability of fiber to be flattened in the refining process. The lower the Muhlstep ratio, the easier the fiber to flatten so that the refining process becomes easier [27].

Compared to wheat straw fiber, EFB fiber length is lower, with a Runkle ratio below 1. While, the wheat straw fiber has a Runkle ratio of greater than 1, which indicates that the fiber will be difficult to flatten during the refining process. Wheat straw fiber has a higher felting power than EFB fiber [30]. The rigidity coefficient (W/D) of wheat straw fiber is 0.29, higher than EFB fiber (0.24). According to fiber requirements for pulp raw materials issued by the Directorate General of Forestry in 1976, fibers with a rigidity coefficient greater than 0.2 are included in the lowest grade (IV) [22].

3.2. Chemical composition

EFB has almost the same cellulose content as those of hardwoods, while lower in lignin content, and relatively higher in pentosans and hemicellulose content [31]. Table 3 shows the same tendency, with the extractive levels also almost equivalent. Compared to other non-wood fibers, EFB is relatively equivalent to bamboo except for the ash content, and quite equivalent to wheat straw but better than rice straw.

Table 3. Chemical composition of EFB compared to other fibers.

Parameter	EFB [23, 29, 32- 35]	Wheat straw [30, 36]	Hardwoods [36-38]	Bamboo [36, 39]	Rice straw [40]
α-cellulose. %	37.26-63	39.8-43.2	42.0-54.9	33.8-51.6	28-36
Hemicellulose, %	14.6-37	28.95-34.2	27.1	27.2	23-28
Pentosans, %	29.4		15.4-17.1		
Lignin, %	17-31.7	19.8-21.12	22.7-35.2	23.2-32.5	12-14
Extractives, %	1.3-4.2		1.54-4.47	6.00-9.23	
Ash, %	1.2-6.7		0.5-2.9		

Based on the classification of Indonesian hardwood chemical components established by the Ministry of Agriculture in 1976 [37], the classification of EFB chemical components can be seen in table 4.

From table 4, it can be seen that the cellulose content of EFB is a very wide range, ranging from high to low class. However, the general value is around 40% which shows that the EFB cellulose content is categorized in the moderate class. In terms of extractives content, EFB has relatively high extractives content. In the pulping process, it will cause an increase in the consumption of cooking chemicals and affects the quality of the paper produced.

Table 4. Classification of EFB chemical components.

	High	Moderate	Low
Cellulose, %	>45	40-45	<40
Lignin, %	>33	18-33	<18
Pentosans, %	>24	21-24	<21
Extractives, %	>4	2-4	<2
Ash, %	>6	0.2-6	<0.2

▒: Class of EFB chemical components

The unique characteristic of non-wood fibers requires special attention to the pulping process, including by conducting a pre-treatment before the pulping process. This pre-treatment can be in the form of physical/mechanical action, hydrothermal, chemical, and/or biological treatment. Each pre-treatment has a specific impact on the non-wood chemical contents such as cellulose, hemicellulose and lignin as well as other chemical components. However, the purpose of this pre-treatment is basically to carry out further non-wood processing processes that are easier and more efficient as well as improves the quality of the final products [31]. Pre-treatment of EFB by immersing in sodium hydroxide solution in a certain concentration have been reported effective to increase to pulp yield of EFB refiner mechanical pulp (RMP) [41].

Figure 4. EFB thermo-mechanical pulp.

4. Paper products from EFB fiber

One of the potentials uses of EFB fiber is as pulp raw material for paper and paperboard. Before being made into paper, EFB fibers must first go through the pulping process. Experiments about EFB pulping have been carried out chemically, semi-chemically, and mechanically. Figure 4 shows the EFB pulp produced by thermo-mechanical pulping process. The EFB pulp bleaching process has also been investigated in both bio-bleaching and TCF (Totally Chlorine Free) processes [19]. The description below is limited to the explanation of some research results that have been carried out in relation to the use of EFB fibers in papermaking processes.

4.1. Writing and printing papers

Writing and printing paper is a type of paper used for writing and printing purposes. In addition, to fulfill the specified writing and printing properties, this type of paper must also have good aesthetic properties that are expressed in its optical properties. This paper is generally white, so it should be made from bleached pulp or deinked pulp from recycled fibers. The experiments of writing-printing papermaking from EFB bleached pulp mixed with deinked pulp have been done on a laboratory scale. In general, the quality of the paper produced was quite good, but the use of EFB pulp reduced the brightness of the sheet [42].

EFB fiber contains about 4.5% residual oil that found on the surface and tip of the fiber [43]. Actually, EFB does not contain oil, but in the process of boiling and releasing the fruit from its' bunches in the CPO mill, the oil enters into the EFB by osmosis through the fiber wall [19]. The presence of oil content in EFB has the potential to pitch problems that cause stains and even a pinhole on the paper produced. Therefore, the oil content of EFB must be removed as much as possible. The pre-treatment of EFB fibers before the pulping process using lipase enzymes can be done to reduce this pitch problem [28,44]. Lypase enzyme can be utilized as well for pitch control in the papermaking process [45]. Besides, mechanical pitch removal using a multistage cleaning system can also be applied both in pulp mill and paper mill [46].

The experiments to improve the cellulose characteristics from EFB have been done extensively. A study regarding the recovery of high-purity lignin and cellulose from EFB using ethanol as a solvent has also been carried out [47]. The oxidation of EFB fiber using sodium hydroxide followed by the treatment with sodium hypochlorite increased the cellulose and hemicellulose content of pulp, while its' lignin content was decreased [48].

4.2. Packaging papers

Based on its fiber characteristics, the use of EFB fiber for paper packaging is considered the most appropriate. A fiber with high wall thickness tends to be more rigid so that their usage is more ideal for corrugating medium paper or folded paper toweling [49]. Some research results in relation to this topic have been reported.

Industrial growth in Indonesia should be supported by the provision of industrial product packaging, including those made of paper and paperboard. The type of papers commonly used as raw material for industrial packaging are liner and corrugating medium paper. Liner paper is the inner and outer lining of a corrugated board which is a raw material for corrugated carton box while corrugating medium paper is a corrugated layer between two or more liner papers.

At present, the main raw material for liner and corrugating medium papers is recycled fibers from OCC (Old Corrugated Container). The main problem in the use of waste paper as a raw material for paper is the difficulty to obtain a stable and uniform quality due to the extremely varied quality of waste papers. This includes how many times the fiber from the waste papers has been recycled. EFB fiber is a very promising alternative as a substitute for recycled fiber [11,17,50].

The experiments on the manufacturing of liner and corrugating medium papers have been reported both at the laboratory and factory scale, as well as the manufacturing of corrugated board and corrugated carton boxes from those papers [19,20,28,51,52]. An experiment of liner and corrugating medium papermaking using EFB mechanical pulp has been investigated as well [23].

Generally, in those experiments, the EFB fibers were mixed with recycled fibers in various compositions, namely 50-60% EFB pulp for liner paper and 70-80% EFB pulp for corrugating medium paper, blended with OCC recycled fiber [20]; 6-50% EFB pulp blended with OCC recycled fiber for corrugating medium papermaking [52]; 25-75% EFB pulp blended with OCC, OCP (old copier paper) and ONP (old newsprint paper) recycled fibers [50]. The corrugating medium paper made from a mixture of EFB and OCC pulp is claimed to fulfill the specification of Indonesian National Standard (SNI) for corrugating medium-class B [20].

The manufacture of paper bags from EFB pulp has also been investigated. This type of papers is used as packaging for industrial products such as cement, or food products such as tea and sugar, or other materials in the form of powder [53]. In the experiment of cement sack papermaking, EFB pulp was added to substitute some of the unbleached long fiber pulp from *Pinus radiata*. The results showed that the use of 20% EFB pulp did not significantly affect the strength of the sheet.

The papermaking experiment from EFB pulp mixed with microbial cellulose (MC) from nata de cassava on various compositions have been also carried out [54]. The addition of MC from nata de cassava increased the strength properties of the paper sheet.

Mixing EFB fiber with UBC (Used Beverage cartons) recycled fibers can also be done since the UBC recycled fiber is categorized as high-grade fiber. UBC is a term for aseptic carton consisted of several laminated layers which have been used as a beverage packaging. It is a post-consumer waste which is consisted of paperboard (75%), and the rest of about 25% are polyethylene and aluminium foil. Since there is no contact between the beverage with the paperboard, so no wet strength chemicals are added. Therefore, the UBC repulping process in a paper mill can be done easily [55]. The use of EFB mixed with UBC recycled pulp for liner and corrugating medium papermaking has been tried and paper with good quality were resulted.

4.3. Other potential papers from EFB

In addition to its use for writing and printing as well as industrial papers, the following are some of the experiments that have been carried out and/or in progress related to the development of special papers from EFB.

4.3.1. Active packaging paper. Active packaging is one type of smart packaging commonly used as food product packaging with the aim of increasing durability and extending product shelf life while maintaining its quality [56]. Active packaging technology includes the ability to release or absorb the components from or into food products or the surrounding environment for the purpose of increasing the durability and extending the shelf life of food products [57].

The use of paper for active packaging has not been much developed yet, but its utilization as a passive packaging has been applied extensively, both for food and non-food products packaging. The research on the development of ethylene absorbent active paper has been carried out by superimposing specific chemical additives on the paper surface [58, 59]. Ethylene is a plant hormone in the form of gas which its level will increase with the increasing of plant maturity. In fresh fruits and vegetables, this hormone will trigger the process of ripening and aging so as to shorten the shelf life of fruits and vegetables. Therefore, the presence of ethylene in fruits and vegetables or in its surrounding environment should be controlled [60].

Corrugated carton boxes are widely used as fresh fruit and vegetable packaging. The control of fruit ripening or aging of vegetables is done by providing respiration holes on the walls of the box and storage in the controlled room condition. It has been previously discussed that EFB has the potential to be used as a source of fiber in the manufacture of liners and corrugating medium papers which are the raw materials for corrugated carton box manufacturing. The development of ethylene absorbent active paper made from EFB fiber, besides increasing the EFB added value as a by-product of CPO mill, it is also a challenge as well as an opportunity to support increased exports of Indonesian fruits and vegetables.

The use of EFB for manufacturing anti-microbial active paper has been reported. The experiments were carried out by mixing EFB pulp with citronella leaf pulp. Although the properties of the paper produced have not optimal yet, especially in terms of its strength, the paper showed a quite good anti-microbial properties and potential to be developed for food packaging paper [41].

The food packaging from EFB paper that is coated with chitosan has been developed. The chitosan coated onto paper surface increased the physical properties of paper, i.e. tensile strength, tensile modulus and elongation, and imparts antimicrobial properties, both Gram-positive and Gram-negative bacteria [61].

4.3.2. Nursery bag paper. The use of plastic packaging has become one of the contributors to environmental pollution that are quite troubling because it takes between 100-500 years to be decomposed perfectly. Indonesia is declared as the second-largest consumer country of plastic waste after China. Likewise, the use of polybags for nurseries on plantation and agriculture, or households needs, continue to increase and contribute to this pollution. Therefore, various efforts to control these surges have been made, including using packaging made from bio-plastic which is easier to decompose and/or paper.

Paper products have a high tendency to absorb liquid due to its lignocellulosic fiber content. Although these properties are useful for certain types of paper such as tissue, some paper products are required to be resistant to liquid penetration. Packaging paper is one type of paper that requires a certain level of resistance to water penetration, therefore the hydrophilic nature of the fiber must be controlled by the addition of certain chemicals that can resist water penetration into the paper [62].

Base paper for sacks (paper bag) must have sufficient strength properties to withstand the load in it. The strength of the paper is influenced by the type of fiber used and the inter fiber bond and the addition of reinforcing additives. In addition, paper bags must also be able to maintain its strength when it is wet, has a certain porosity/permeability, and especially for nursery needs, paper bags must be biodegradable and safe for the environment.

Utilization of unbleached EFB pulp mixed with other virgin pulp or recycled fibers in papermaking has been investigated. The results showed how potential the EFB as an alternative source of short fiber for paper [52,53,63]. Thus, EFB is a potential fiber source for nursery bag papermaking that more environmentally friendly instead of the polybag. Below are some pictures of EFB papers and paperboards (figure 5) and corrugated boxes produced from a mixture of EFB pulp and OCC recycled fiber (figure 6).

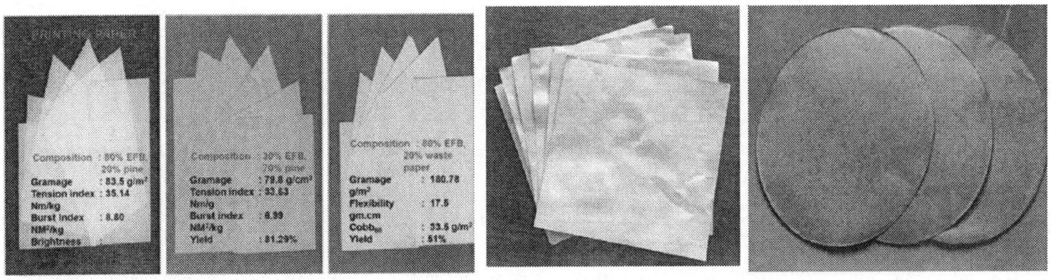

(a) (b) (c)

Figure 5. Papers made from EFB (a) EFB mixed with other fibers [28]; (b) 100% EFB paperboards; (c) 100% EFB papers.

Figure 6. Corrugated box made from EFB papers [28].

Experiments on the utilization of EFB fiber for dissolving-pulp has been carried out using soda-anthraquinone (soda-AQ) cooking process after being prehydrolized at 150°C for 90-180 minutes. The optimum condition is achieved at prehydrolisis stage for 180 min followed by soda-AQ with alkali active of 20%. An EFB dissolving-pulp with kappa number of lower than 10 and high viscosity has resulted. However, the brightness of pulp still needs to be improved by using the proper bleaching process [64].

5. Conclusion

Up till now, the utilization of EFB for high value-added products has not yet been done commercially despite the potential availability of EFB in Indonesia from year to year continues to increase in line with the expansion of oil palm plantation and increasing CPO production. The utilization as a raw material for pulp and paper has been extensively studied both for writing and printing papers as well as for packaging papers, even some special types of paper from EFB have also started to be developed. With its high considerable potential of availability and promising research facts, EFB can be used commercially as an alternative raw material for the pulp and paper industry in Indonesia.

Acknowledgments

Gratitude is delivered to the Director of Center for Pulp and Paper, Ministry of Industry, who has given his supports up to this paper completed and submitted to be presented and published. Appreciation is also conveyed to the co-authors and other parties for their contributions to the completion of this paper.

References

[1] https://www.idnfinancials.com/archive/news/22291/Indonesias-pulp-and-paper-are-in-the-top-10-in-the-world Accessed at October 1st, 2020
[2] file:///C:/Users/Lies%20I/Downloads/prev-Analisis%20Perkembangan%20Industri%20Edisi%201%20-%202020-covered%20(1).pdf Accessed at October 1st, 2020

[3] http://apki.net/?p=1638 Accessed at October 1st, 2020
[4] https://neraca.co.id/article/18741/industri-kertas-dipaksa-impor-serat-panjang-dari-eropa-gara-gara-perbedaan-iklim Accessed at October 1st, 2020
[5] https://ekonomi.bisnis.com/read/20190805/257/1132666/tak-terpengaruh-perang-dagang-permintaan-kertas-masih-tumbuh Accessed at October 1st, 2020
[6] https://swa.co.id/swa/trends/pelaku-industri-pulp-dan-kertas-butuh-kepastian-aturan-impor-kertas-bekas Accessed at October 1st, 2020
[7] https://republika.co.id/berita/p4yawm453/tanaman-serat-untuk-pulp-dan-kertas-berkualitas Accessed at October 1st, 2020
[8] https://www.qureta.com/post/bahan-baku-alternatif-pembuatan-kertas Accessed at Ocotber 1st, 2020
[9] Ali M M, Muhadi N 'Atirah, Hashim N, Abdullah A F and Mahadi M R 2020 Pulp and paper production from oil palm empty fruit bunches: A current direction in Malaysia *Journal of Agricultural and Food Engineering* **2** 1–9
[10] Daud W R W and Law K N 2011 Oil palm fibers as papermaking material: Potentials and challenges *BioResources* **6(1)** 901–917
[11] Indriati L and Elyani N 2018 Empty fruit bunches as packaging papers raw material *Konversi* **7(2)** 45–54
[12] Irwansyah F S, Yanto J, Azis W D I and Kelvin 2018 optimization of empty fruit bunches (EFB) as a solution to deforestation as a result of the exploitation from the paper industry in Riau of conferences *IOP Conference Series: Materials Science and Engineering* **434(1)**
[13] Tutik R, Siti P and Wiyarsi M A 2012 Karakter ekstrak zat warna kayu secang (*Caesalpinia Sappan* L) sebagai indikator titrasi asam basa *Prosiding Seminar Nasional Penelitian, Pendidikan Dan Penerapan MIPA* 1–9
[14] Wan Daud W R, Shafie R and Nasrullah R 2017 Oil palm lignocellulosics: a potential papermaking material for Malaysia *Proceedings of International Workshop on Non Wood Pulping and Papermaking Technology* 33–40
[15] Rame 2018 Oil Palm Empty Fruit Bunches (OPEFB): Existing Utilization and Current Trends Bio Refinery in Indonesia *E3S Web of Conferences* **31** 1–5
[16] Anonymous 2019 Statistik kelapa sawit Indonesia 2018 In *Badan Pusat Statistik*.
[17] Hambali E and Rivai M 2017 The Potential of Palm Oil Waste Biomass in Indonesia in 2020 and 2030 *IOP Conference Series: Earth and Environmental Science* **65(1)**
[18] Rahmadi R, Awaluddin A and Itnawita 2014 Pemanfaatan limbah padat tandan kosong kelapa sawit dan tanaman pakis-pakisan ubtuk produksi kompos nebggubakan aktivator EM-4 *Jurnal Online Mahasiswa Fakultas Matematika dan Ilmu Pengetahuan Alam* **1(2)** 245–53.
[19] Erwinsyah, Afriani A and Kardiansyah T 2015 Potency and opportunity of oil palm empty fruit bunch as a raw material for pulp and paper *Jurnal Selulosa* **5(02)** 79–88
[20] Erwinsyah T, Kardiansyah A, Afriani S, Sugesty T, Hidayat R M 2014 Production of the old corrugated container from the empty fruit bunched at industrial scale *Prosiding Seminar Teknologi Pulp Dan Kertas 2014* 33–42.
[21] Le Dinh K, Harwood C E, Duc Kien N, Baltunis B S, Hai N D and Thinh H H 2012 Growth and wood basic density of acacia hybrid clones at three locations in Vietnam *New Forests* **43(1)** 13–29
[22] Sunarti S, Praptoyo H and Nirsatmanto A 2016 Characteristics of fiber from the wood of Acacia auriculiformis x Acacia mangium hybrid with regard to pulp *Jurnal Pemuliaan Tanaman Hutan* **10(2)** 135–143
[23] Erwinsyah, Sugesty S and Hidayat T 2012 Kraft liner and corrugating medium from mechanical pulp empty fruit bunches of palm oil *Jurnal Selulosa* **2(01)** 8–13
[24] Risdianto H, Kardiansyah T, Sugiharto A 2016 Empty fruit bunches for pulp and paper production: The current state in Indonesia *Palpu Chongi Gisul/Journal of Korea Technical Association of the Pulp and Paper Industry* **48(6)** 25–31

[25] Gonzalo A, Sanchez J L, Escudero E, Marín F and Fuertes R 2007 Pulp and paper production from EFB using a semichemical process *Engineering, Pulping and Environmental Conference 2007, TAPPI 4*(August 2016) 2295–321
[26] Harsono L, Mulyantara A T, Rizaluddin A., Nakagawa-izumi H, Ohi K N 2015 Properties of fibers prepared from oil palm empty fruit punch for use as corrugating medium and fiberboard *Japan Tappi Journal* **69(12)** 1349–59
[27] Anggraini D and Roliadi H 2011 manufacture of pulp from empty oil-palm fruit bunches for paperboard at small-scale endeavor *Pusat Litbang Keteknikan Kehutanan dan Pengolahan Hasil Hutan* **53** 160
[28] Erwinsyah & Afriani A 2014 Empty fruit bunch: Its characteristics and potency as raw material for pulping *Prosiding Seminar Teknologi Pulp dan Kertas 2014* 23–32.
[29] Rahmasita M E, Farid M & Ardhyananta H 2017 Analisa morfologi serat tandan kosong kelapa sawit sebagai bahan penguat komposit absorpsi suara. *Jurnal Teknik ITS 6*(2)
[30] Singh S, Dutt D and Tyagi C H 2011 Complete characterization of wheat straw (Triticum aestivum PBW-343 L. Emend. Fiori & PAOL.) - A renewable source of fibers for pulp and paper making. *BioResources* **6(1)** 154–77
[31] Mohieldin S D 2015 Pretreatment approaches in non-wood plants for pulp and paper production : a review *Journal of Forest Products & Industries* **3(2)** 84–8.
[32] Ansell M P and Mwaikambo L Y 2009 The structure of cotton and other plant fibers. In *Handbook of Textile Fiber Structure* (Vol. 2, Issue December 2009 Woodhead Publishing Limited
[33] Daud W R W, Wahid K A, & Law K N 2013 Cold soda pulping of oil palm empty fruit bunch (OPEFB) *BioResources* **8(4)** 6151–60
[34] Dewanti D P2018 Cellulose potential of empty fruit bunches waste as the raw material of bioplastics. *Jurnal Teknologi Lingkungan* **19(1)** 81
[35] Sudiyani Y, Styarini D, Triwahyuni E, Sudiyarmanto Sembiring K C, Aristiawan Y, Abimanyu H and Han M H 2013 Utilization of biomass waste empty fruit bunch fiber of palm oil for bioethanol production using pilot - Scale unit *Energy Procedia* **32** 31–8
[36] Chen W, Yu H, Liu Y, Hai Y, Zhang M, & Chen P 2011 Isolation and characterization of cellulose nanofibers from four plant cellulose fibers using a chemical-ultrasonic process *Cellulose* **18(2)** 433–42
[37] Sokanandi A, Pari G, Setiawan D and Saepuloh S 2014 Komponen kimia sepuluh jenis kayu kurang dikenal : kemungkinan penggunaan sebagai bahan baku pembuatan bioetanol *Jurnal Penelitian Hasil Hutan* **32(3)** 209–20
[38] Wibisono H S, Jasni and Arsyad W O M 2018 chemical composition and natural durability of eight wood species tested under the shade *Jurnal Penelitian Hasil Hutan* **36(1)** 59–65
[39] Wahab R, Mustafa M T, Mohamed A, Samsi H W and Mat Rasat M S 2016 Extractives, Holocellulose, α-Cellulose, Lignin, and Ash Contents in 3 Year-Old Bamboo Culms Gigantochloa Brang, G. Levis, G. Scortechinii and G. Wrayi *Research Journal of Pharmaceutical, Biological and Chemical Sciences* **4(2)** 1346–54.
[40] Nurwahdah, Naini A A, Nadia A, Lestari R Y and Sunardi 2015 Pretreatment lignoselulosa dari jerami padi dengan deep eutectic solvent untuk meningkatkan produksi bioetanol generasi dua *Jurnal Riset Industri Hasil Hutan* **10(1)** 43–54
[41] Nur Aqeela M Y, Ainun Z M A and Jawaid M 2018 Effect of pretreatment concentration on pulp blending between oil palm empty fruit bunch and citronella leaf fibers in terms of pulp and paper properties *IOP Conference Series: Materials Science and Engineering* **368(1)**
[42] Rismijana J and Cucu 2010 Oil palm empty fruit bunches pulp as writing-printing paper raw material. *Prosiding Seminar Teknologi Pulp Dan Kertas* 96–101.
[43] Ibrahim Z, Aziz A A, Ramli R, Jusoff K, Ahmad M and Jamaludin M A 2015 Effect of treatment on the oil content and surface morphology of oil palm (*Elaeis guineensis*) empty fruit bunches (EFB) fibers *Wood Research* **60(1)** 157–66
[44] Sudarmin A L, Titin F S, Hermanto G K and Kathomdani P D 2010 Pitch control using pressure

reactor on woods raw materials *Prosiding Seminar Teknologi Pulp dan Kertas* 74–9
[45] Elyani N, Rismijana J, Wirawan S K and Cucu 2010 Handling of sticky and pitch in the waste paper enzimatically *Prosiding Seminar Teknologi Pulp Dan Kertas 2010* 91–5.
[46] Indriati L 2010 The multistage cleaning system removal for hardwood pulp pitch *Berita Selulosa* **45(1)** 9–15.
[47] Mondylaksita K, Ferreira J A, Millati R, Budhijanto W, Niklasson C and Taherzadeh M J 2020 Recovery of high purity lignin and digestible cellulose from oil palm empty fruit bunch using low acid-catalyzed organosolv pretreatment *Agronomy* **10(5)**
[48] Isroi I and Cifriadi A 2018 The oxidation of cellulose from oil palm empty fruit bunch by using *Jurnal Selulosa* **8(2)** 51–60
[49] Rainey T J and Covey G 2016 Pulp and paper production from sugarcane bagasse *Sugarcane-Based Biofuels and Bioproducts* **October 2017** 259–80
[50] Rushdan, Ibrahim 2003 Structural, mechanical and optical properties of recycled paper blended with oil palm empty fruit bunch pulp *J. Oil Palm Research* **15(2)** 29–36.
[51] Rafidah D, Ainun Z M A, Hazwani H A, Rushdan I, Luqman C A, Sharmiza A, Paridah M T and Jalaluddin H 2017 Characterisation of pulp and paper manufactured from oil palm empty fruit bunches and kenaf fibers *Pertanika Journal of Tropical Agricultural Science* **40(3)** 449–57
[52] Rushdan I, Latifah J, Hoi W K and Mohd Nor, M Y 2007 Commercial-scale production of soda pulp and medium paper from oil palm empty fruit bunches *Journal of Tropical Forest Science* **19(3)** 121–6
[53] Dina S F, Hidayat T, Indriati L, Asid D S, Supriyatna D and Bunyamin A 2005 The optimation of sack kraft paper quality from oil palm empty *Berita Selulosa* **39(2)** 37–44
[54] Syamsu K, Haditjaroko L, Pradikta G I, and Roliadi H 2014 Campuran pulp tandan kosong kelapa sawit dan selulosa mikrobial nata de cassava dalam pembuatan kertas (the mixture of empty oil palm fruit bunch pulp and microbial cellulose from Nata de Cassava for paper manufacture) *Jurnal Ilmu Pertanian Indonesia* **19** 14–21
[55] Abreu M. 1999 Recycling of Tetra Pak Aseptic Cartons *Tetra Pak Canada Inc.* **6**
[56] Biji K B, Ravishankar C N, Mohan, C O and Srinivasa Gopal T K 2015 Smart packaging systems for food applications: a review *Journal of Food Science and Technology* **52(10)** 6125–35
[57] Schaefer D and Cheung W M 2018 Smart Packaging: Opportunities and Challenges *Procedia CIRP* **72** 1022–27
[58] Jaimun R and Sangsuwan J 2019 Efficacy of chitosan-coated paper incorporated with vanillin and ethylene adsorbents on the control of anthracnose and the quality of Nam Dok Mai mango fruit *Packaging Technology and Science* **32(8)** 383–94
[59] Jaimun R, Sangsuwan J, Intipunya P and Chantrasri P 2018 Active wrapping paper against mango anthracnose fungi and its releasing profiles *Packaging Technology and Science* **31(6)** 421–31
[60] Dafri M, Ratianingsih R and Hajar H 2018 Penanganan produksi buah pisang pasca panen melalui model pengendalian gas etilen *Jurnal Ilmiah Matematika dan Terapan* **15(2)** 173–87
[61] Prasetiyo K W, Zulfiana D, Anita S H, Fatriasari W, Suryanegara L, Masruchin N and Gutari S 2020 Effects of chitosan coating on the physical, mechanical and antimicrobial properties of food packaging paper *Jurnal Sains Materi Indonesia* **21(2)** 62–8
[62] Rahmaninia M, Javid E and Varshoei A 2016 Process variables and the performance of cationic rosin as an internal sizing agent in recycled corrugated container pulp *BioResources* **11(2)** 5333–42
[63] Khairul Hafizuddin A B, Rohaizu R and Wan Rosli W D 201. The use of oil palm empty fruit bunches (OPEFB) fibers as partial replacement for imported recycled fibers. *Key Engineering Materials* ***737 KEM*** 335–40.
[64] Harsono H, Putra A S, Maryana R, Rizaluddin A T, H'ng Y Y, Nakagawa-izumi A and Ohi H 2016 Preparation of dissolving pulp from oil palm empty fruit bunch by prehydrolysis soda-anthraquinone cooking method *Journal of Wood Science* **62(1)** 65–73

The sustainability of the iron industry based on local wisdom in the Barito watershed

Hartatik [1*], H O Soffian[2], Sunarningsih[1], N N Susanto[1], and R B Sulistiyo[1]
[1]Balai Arkeologi Kalimantan Selatan (The Archaeological Research Office South Kalimantan), Jl. Gotong Royong RT 03/06 Banjarbaru, Indonesia
[2]Pusat Penelitian Arkeologi Nasional (The National Research Centre of Archaeology), Jl. Raya Condet Pejaten No. 10, Jakarta, Indonesia

Email: tatitatibalar@gmail.com

Abstract. Iron has been important in human civilization since the abandonment of stone tools. The traditional technology of iron ore smelting in Indonesia has been known for hundreds of years, but research on the local traditional iron industry is still very minimal. This study aims to explain the iron industry in the southeast Kalimantan before modern times until the present using ethnohistory and experimental archaeological approaches. The research conducted in 2017-2019 at the upstream of Barito watershed, North Barito Regency, Central Kalimantan, has found 19 sites with the findings were furnaces, air pipes, iron ore fragments, charcoal, and slag. The traditional iron-making began with the process of mining, followed by iron ore smelting and forging. The archaeological experiments by smelting furnace produced the raw iron by a percentage of 40-60% from ore. The experiments produced the ingots with Fe content of about 80% (XRF analysis) and showed good ore quality. However, since the early 20th century the iron-making industry has been abandoned because of the arrivals of cheaper iron from China. Recently, the remaining iron industry consists of only the blacksmiths who produce agricultural equipment from scrap iron, some only gild to sharpen machetes and 'mandau'.

1. Introduction

After the stone age period, iron was the most dominant type of metal as a material for human equipment. Iron ore is abundant on the surface of the earth to the earth's core and celestial bodies that fall to the earth (meteors). Iron has been processed and used by humans for thousands of years ago, but the massive use of iron has occurred since the discovery of blast furnace technology in the United States from the early 17th century, but in this 21st century, China is leading the iron and steel production [1]. Iron is obtained from the stones containing the iron element (Fe), then it is melted and forged. Iron ore smelting expertise had been owned by people who live near iron ore sources, such as the Angkor iron and Cambodia which were mined by the Kui people on the border with Laos. Burmese and Siamese iron is mined from the mountains near the border of Burma and Siam, then processed in blacksmith villages [2]. In Malaysia, the smelting furnaces and tuyere fragments were found at the Lembah Bujang and Kampung Gading sites dating from 535 BCE-1500 CE [3,4]. The furnaces with associations of laterite ore deposits were also found at the Ban Kao Din Tai site, (Bangkok), Thailand with the site dating from the 4th-15th century [5]. The iron smelting technique in Xinjiang, China is much older at 1000 BCE [6], while in Khasi Hill, India at 353 BCE – 128 CE [7]. In Sumatra, iron ore mines were located in Gunung Besi, West Sumatra, which then moved to Salimpaung in the late 18th century, Beliton and Bangka [2]. The iron ore from Lake Matano, Luwu (South Sulawesi) is the best because of the nickel content that makes solid and corrosion-resistant iron [8,9]. Based on the OXIS (Origins of Complex Society in South

Sulawesi) research, it is known that the oldest iron foundries in Matano Lake were from the years of 1000-1250 CE [10].

People in Kalimantan are familiar with iron weapons such as mandau, blowguns, machetes, *jamiya*, and the spearhead. The narrative story states that the ancestors have the expertise to choose rocks that contain iron. The oral story of the greatness of iron in Kalimantan is well known, but until early 2017, the sites for making iron and iron ore sources have never been researched and scientifically informed. Archaeological research on traces of ironwork began with information that people in the Montalat watershed are still salvaging ironstones. The research has found some evidence of iron ore smelting sites such as iron slag mounds, iron ore fragments, air pipes made of clay (tuyere), charcoal, and dust furnaces [11,12].

There are four technique stages of iron metals making, mining of iron ore, breaking iron ore into small pieces, smelting ore to separate iron minerals from their impurities, and the final stage is forming the desired object [13]. The result of geological and mineral research found that the reserves of laterite iron ore in the southeastern part of Kalimantan is the highest compared to other places in Indonesia [14]. Laterite types are formed due to weathering, decomposition, and accumulation of residues in environments with a humid tropical climate that is very commonly found in Africa and Southeast Asia in the form of eluvial and alluvial deposits. The dominant mineral types in laterite are limonite and hematite [8,15].

Indonesia has an iron-making history, ore, and fuel sources, but until now we still import almost all raw iron as a material for the national steel industry. Blacksmith craftsmen in Indonesia still use traditional techniques that last for generations. On the other hand, nature in the southeast of Kalimantan provides a source of iron ore that can be processed not only for the local iron industry but also for the national iron industry. Therefore, this study aims to explain the traditional iron industry that once existed in southeastern Kalimantan, especially in the Barito watershed in the past and its sustainability until the present.

2. Materials and Methods

This research is qualitative and quantitative descriptive with deductive inductive reasoning. Primary data were obtained from archeological research in 2017-2019 using survey, excavation, and experimental methods. Tools used in surveys and excavations were data recording equipment (cameras and voice recorders), notebooks, GPS, compass, scraper to extract findings in the soil, buckets to dispose of soil, meter as a measuring instrument, scales, threads, and pegs to layout dig box, water-pass to measure the height of the ground surface, and Munsell soil color charts to determine the type and color of the soils. Medical kits were also prepared, especially basic medicines and first aid such as bandages, red drugs, alcohol, and disinfectants.

The Archaeological experiment and mineral analysis used the square-shaped furnace from the Buren Benangin as a basis for preparing the equipment. Iron ore smelting experiments required a place in open land, an electric blower with a capacity of 2 inches and 3000 RPM (rotation per minutes) as a substitute for double piston gray, air pipes from the iron pipe with a size of 2 inches, large pliers to lift iron, wooden square to solidify the walls of the furnace, water and water barrel for emulsifying the soil and safety if fire spread, thermometer with the ability to measure up to 1763 °C, metal detectors, charcoal from *halaban* or *ulin* wood, limestone as flux, iron ore in the form of laterite rocks, clay for making furnaces, and magnets to find out the iron content (figure 1).

2.1. Archaeological survey and excavation

A survey was conducted to gain a physical description of data in the field by directly observing objects and interviewing informants who knew about the object. The excavation was a way to obtain archeological data that was buried in the ground through systematic soil excavation [16]. The excavation was based on surface findings and alleged archeological data placed in the ground obtained from the survey. The excavation consisted of the stages of preparation, excavation, recording data, and findings analysis.

IOP Conf. Series: Materials Science and Engineering **980** (2020) 012046 doi:10.1088/1757-899X/980/1/012046

Figure 1. Materials and tool instruments for the archaeological experiment. (Source: Balai Arkeologi Kalsel)

The preparation phase included licensing with agencies, customs, and landowners, checking equipment, laying out the digging box according to the direction of the wind, determining the highest point (datum point) for the reference depth of findings, and determining the type of excavation (system grid, lot, spit, or layer). The excavation process consisted of extracting soil per layer or certain depths (spit) depending on the condition of the finding (for these Buren sites with spit depth per 10 cm and the soil layer), recording the findings, and ending with returning the soil into the dug box. The handling of basic level findings in the form of cleaning of attached soil, basic classification, and labeling was done in the field. The taxonomy and typology phases were carried out in the base camp or the artifact workspace at the office, while mineral and radiocarbon analyzes were carried out in the laboratories. The mineral analysis using X-RF (X-ray fluorescence) technique was conducted at the Mineral and Materials Laboratory of the Faculty of Mathematics and Natural Sciences (MIPA), University of Negeri Malang, while for dating analysis radiocarbon (AMS) was used at Waikato University, New Zealand.

2.2. Archaeological experiment and mineral analysis

The stages in archeological experiments include:

1. Preparing tools and materials
2. Constructing smelting furnaces and installation of air pipes
3. Drying ore and breaking into small pieces
5. Inserting charcoal into the furnace, ore fragments, and flux through the top of the furnace
6. Lightening the furnace, turn on the electric blower to supply air into the furnace through the air pipe (PVC)
7. Smelting iron ore until it reaches the melting point of iron (1530 °C)
8. Taking the raw material through the bottom of the furnace

The experiment began with two questions about the iron content in laterite ore and smelting furnaces which had square shape interior. Several variables were the objects of this experiment, independent variables and dependent variables. The independent variable was the actor or research subject who determined the shape of the melting furnace: round or square, the amount and type of iron ore, fuel (charcoal), air intake, and the time needed to melt. The dependent variable in the form of physical data was the amount and quality of raw material (ingot) and iron slag produced.

The experimental furnace was made based on data from Buren Benangin, whose top was a round cone and the inside was square. The furnaces were built by yellow and black clay taken from the outcrop of the Montalat River at RT 01, Pelari Village, Kecamatan Gunung Timang District, Kabupaten Barito Utara. One of the aims of this experiment was to prove the difference in the melting point between the two different shapes of furnaces. Therefore, four furnaces have been made in two different types, square

IOP Conf. Series: Materials Science and Engineering **980** (2020) 012046 doi:10.1088/1757-899X/980/1/012046

and spherical inside shape. The question would be answered by the first stage of smelting, from furnace A (square) and B (spherical).

The two hypotheses formulated to be tested in the iron smelting experiments were:

1. Laterite ore from the Buren sites has a low iron (Fe) content so to obtain raw iron (ingot) requires a lot of ore.
2. The interior furnace with a square shape requires more time to reach the melting point than the furnace which has a spherical shape because the air circulation in the square is not smooth.

The first hypothesis about red laterite (hematite) was tested in furnaces A and C, while brown laterite in furnaces B and D. Experimental materials in the form of laterite were taken from Buren Benangin and Semayap River, while red laterite (hematite) was taken from Buren Benangin and Temelalo. The four-furnace experiment required about 250 kg of *halaban* wood charcoal.

There were two methods used to measure temperature quantitatively, thermocouple measuring instrument and steel-stirr-up. Thermocouple measuring devices functioned as temperature sensors. The K thermocouple sensitivity type starts at 0°C-1767°C, so temperatures above 1767°C cannot be measured. Steel stirrups were inserted into the smelting furnace to help the limitations of thermocouple devices. The iron would melt at a temperature of 1538°C and boil at 2800°C so that the comparison between the two tools would be obtained by cross-checking the obtained temperature.

3. Results and discussion

3.1. The results of archaeological survey and excavation

Archaeological research of ancient metals was started based on the ironstones (the Dayak Taboyan community term for iron ore) information in Montalat River. The Montalat is a branch of the upper Barito River located in Kecamatan Gunung Timang and Montalat, Kabupaten Barito Utara, Central Kalimantan. The upstream is situated on the Muller Mountains (Kabupaten Murung Raya, Central Kalimantan) and empties into the Java Sea.

The Dayak Taboyan community called the location of the former iron ore smelting as '*Buren*', which was marked by a mound of iron slag. Archeological research in 2017 s.d. 2019 found the Buren sites along the Teweh River and Montalat River, both of them were tributaries of the Barito. The survey found 19 sites that were identified by the remaining findings of iron ore smelting activities such as iron slag, the raw material (ingot), iron ore fragments, and burning clay. The fragments of burning clay indicated forms of smelting furnaces and air pipes (tuyere) beside it in almost intact forms.

Figure 2. The Buren Benangin with findings of furnaces, ore, tuyere, and slag. (Source: Balai Arkeologi Kalsel)

The excavation of test pits was carried out at four sites, Buren Benangin, Buren Mejahing, Buren Temelalo, and Buren Jaga. The excavations at those sites found six smelting furnaces, thousands of iron slags, the fragments of air pipes from clay (tuyeres), and iron ores. The other furnaces were found during the survey. Through the excavations in Buren Benangin (figure 2), Buren Temelalu, and Buren Jaga, two furnaces were found in each site. The smelting furnaces were made of clay with round shape and narrow top. The outside part of the seven furnaces had a similar roughly round shape but the inside was square with compaction walls. The furnaces were destroyed on the top and it is estimated that the furnace's height was intact at around 1 meter.

According to the ethnohistorical data of Dayak Taboyan, smelting has been abandoned for the past four generations or around 170 years ago. This is in line with the results of radiocarbon analysis (^{14}C) from charcoal samples that show the youngest period of 1820 (Buren Jaga) and the oldest from 1360 (Buren Benangin). The community still treasures legacy weapons such as saber, machete, *jamiya* (small knife) which were made from Montalat also known as Montalat iron weapons [12].

The sustainability of the iron industry in Montalat was supported by natural resources of laterite ore found along the Montalat River, Teweh River, and its tributaries. The data of archaeological survey and excavation depicted that the type of iron ore found in the Buren sites had the same type of rock in the surounding river and hill, such as the Benangin River near Buren Benangin, and the hill around Buren Japus. The ore type was laterite as the result of ultra-base rocks weathering.

Local peoples informed that the source of ore from Montalat River was precisely in Maninyau cascade. Whenever the water discharge was high, the stones could not be taken even though by diving, because of the deep and heavy current. Maninyau cliff looked like riverbanks with a type of brownish-red laterite rock which contained a lot of iron, during the dry season research in June 2019 [12]. In general, there are three types of iron ore in Indonesia, namely primary iron with an average of iron content (Fe) of 47.14%, iron sand 47.08%, and laterite with a Fe content of 30.26% [18]. Laterite with iron (Fe) content of 40-60% was found in the Kalimantan region [14], higher than the average Fe content in laterites which is 30.26%. The good quality ore for smelting contains hematite (Fe2O3), magnetic (Fe2O4), siderite (FeCO3), and limonite/goethite (FeO) minerals [19].

3.2. Archaeological experiment and mineral analysis

The two hypotheses formulated to be tested in the iron smelting experiments are:

1. Laterite ore from the Buren sites has a low iron (Fe) content so to obtain raw iron (ingot) requires a lot of ore.
2. The interior furnace with a square shape requires more time to reach the melting point than the furnace which has a spherical shape because the air circulation in the square is not smooth.

Several variables were the objects of this experiment, independent variables and dependent variables. The independent variable was the actor or research subject who determined the shape of the melting furnace: round or square, the amount and type of iron ore, fuel (charcoal), air intake, and the time needed to melt. The dependent variable in the form of physical data was the amount and quality of raw material (ingot) and iron slag produced.

The stages in archeological experiments include:

1. Preparing tools and materials
2. Constructing smelting furnaces and installation of air pipes
3. Drying ore and breaking into small pieces
5. Inserting charcoal into the furnace, ore fragments, and flux through the top of the furnace
6. Lightening the furnace, turn on the electric blower to supply air into the furnace through the air pipe (PVC)
7. Smelting iron ore until it reaches the melting point of iron (1530° C)
8. Taking the raw material through the bottom of the furnace

The experimental furnace was made based on data from Buren Benangin, whose top was a round cone and the inside was square. The furnaces were built by yellow and black clay taken from the outcrop of the Montalat River at RT 01, Pelari Village, Kecamatan Gunung Timang District, Kabupaten Barito Utara. One of the aims of this experiment was to prove the difference in the melting point between the two different shapes of furnaces. Therefore, four furnaces have been made in two different types, square and spherical inside shape. The question would be answered by the first stage of smelting, from furnace A (square) and B (spherical).

The first hypothesis about red laterite (hematite) was tested in furnaces A and C, while brown laterite in furnaces B and D. Experimental materials in the form of laterite were taken from Buren Benangin and Semayap River, while red laterite (hematite) was taken from Buren Benangin and Temelalo. The four-furnace experiment required about 250 kg of *halaban* wood charcoal.

The ore was roasted before being put into the smelting furnace to reduce water content. This process was carried out in the open place by burning charcoal over the wet ore until the iron was almost dry, then the ore was broken into small pieces.

Figures 3 and 4. The iron ore smelting experimen process
(Source: Balai Arkeologi Kalsel)

After the charcoal, iron ore and flux were put into the furnace then it was ignited (figure 3), and temperature measurements were taken to control the heat (figure 4). There were two methods used to measure temperature quantitatively, thermocouple measuring instrument and steel stirrup. Thermocouple measuring devices functioned as temperature sensors. The K thermocouple sensitivity type starts at 0°C-1767°C, so temperatures above 1767°C cannot be measured. Steel stirrups were inserted into the smelting furnace to help the limitations of thermocouple devices. The iron would melt at a temperature of 1538°C and boil at 2800°C so that the comparison between the two tools would be obtained by cross-checking the obtained temperature.

The comparison of melting temperature between furnaces A, B, C, and D before using the blower was below 1000 ° C with the highest temperature amid an average temperature of 800 ° C. The comparison after using a blower, however, the temperature was above 1000°C with the highest temperature above 1767°C where the thermocouple could not measure it. An air pump blower was useful to increase the temperature in the melting furnace. The iron ore smelting process took between 4,5 and 9,5 hours, with the results shown in table 1.

Experimental results in table 1 show that the red laterite ore from Buren Temelalo (furnace A) and Buren Benangin (furnace C) produced the crude iron percentage of 60.4% and 50.9%. The percentage above 50% of iron ore is high. The blackish-brown laterite from Semayap River (furnace B) and Buren Benangin (furnace) produced a fairly low percentage of raw iron at 46.9% and 21.7%. Based on the smelting results, it is concluded that ancient iron smelters had the expertise to choose the type of red laterite with high hematite elements similar to those found in the Buren Benangin and Temelalo. Hematite is one of the good constituent minerals of laterite iron ore, besides magnetic, siderite, and limonite [19]. It was found from the XRF analysis that the laterite iron ore from Buren Benangin contained a very high iron (Fe) element with 96.69%, as well as laterites taken directly from Maninyau which had 90.55% iron content (table 2) based on the percentage relative to another mineral or real content of iron. Thus, the hypothesis that states that the laterite type of iron ore from the Montalat watershed has a low iron content so they need to gain large amounts of iron ore apparently cannot be accepted.

Table 1. Comparison of ingredients and result of smelting experiment.

Furnace/ Basic Form	Ore		Char-coal (kg)	Flux (kg)	Time (hour)	Smelting Result (Ingot %)	Annotation
	Source	Weight (Kg)					
A Square	Laterite[a)] Buren Temelalo	10 kg	58.4	-	4.5	6.04 kg (60.4 %)	iron was separated from slag but not perfectly
B Rounded	Laterite[b)] Semayap River	10 kg	48.8	-	4.5	4,69 kg (46.9 %)	iron was separated from slag but not perfectly
C Square combined A & B	Laterite[a)] Buren Benangin	5,3 kg)	82.1	22,5	9.5	2.7 kg (50,9)	Iron was perfectly separated from the slag
D Square	Laterite[b)] Buren Benangin	7.8 kg)	61.4	22,5	9.5	1.7 kg (21.7%)	iron was perfectly separated from the slag

[a)] red laterite with heavier weights than brown laterite (hematite)
[b)] brown laterite with small iron content

Tabel 2. Results of XRF analysis of mineral content based on the real content.

No.	Sample	Content (%l)							
		Fe	Al	Si	K	Ca	Mn	P	Others Mineral
1.	Ingot Furnace A (from Buren Temalalu)	79,21	5	8.4	2.3	1.28	1.0	0.34	1.34
2.	Ingot Furnace B (from Semayap River)	84.24	4	7.3	1.7	0.74	0.3	0.39	1.33
3.	Ingot Buren Benangin	86.79	3	6.9	0.38	0.57	0.66	0.26	1.44
4.	Red Laterite Buren Benangin	96.69	-	1.2	-	0.19	1.64	0.2	0.08
5.	Brown Laterit Maninyau (Montalat River)	90,55	1	4.3	0.35	0.43	1.66	0.26	1.45

The quality of laterite was also seen from the Fe content in raw iron (ingots) produced from the smelting furnace. The XRF analysis of ingot samples from smelting furnace A shows that red laterite from Buren Temelalo contained iron (Fe) as much as 79.21% rel, while aluminum, potassium, calcium, manganese, each between 1-8% rel, whereas phosphorus, titanium, chromium, bromine, rubidium, and europium were respectively below 1% rel. Ingot from furnace B (laterite material from the Semayap River) contained 84.24% rel iron (Fe), greater than ingot from furnace A. There were different types of minerals found in furnace A and furnace B, titanium (Ti) existed in ingots furnace A but not in furnace ingot B. Likewise the lanthanum mineral (La) was in furnace ingot B but not in furnace ingot A. Even though iron ore and ingot materials were from the same material but the iron minerals were different.

This difference in mineral types was most likely influenced by the quality of the sample and the smelting process where the iron mineral was separated from other minerals that were bound to each other.

The results of the iron ore smelting experiment also prove the second hypothesis that the shape of the inside of the furnace has no effect on time and result from smelting, but only affects the capacity (table 1). The square shape contained more iron ore and charcoal material than the round shape. Thus, the hypothesis that a furnace with a square inside takes more time to reach the smelting point than a furnace with a round inside is incorrect.

3.3. The sustainability of the iron industry on the Barito watershed

Industrial archeology aims to explain the development of industrial processes to arouse the enthusiasm of educated generations to continue the research and utilization [20]. Documentation and preservation of archeological sites need to be carried out as evidence of human civilization. The important value of information content in industrial sites can be utilized in the global era, for example as educational material and evoke awareness of the social and cultural problems.

The sustainable iron industry in the Barito's upstream is not directly proportional to the availability of material sources. The ethnohistorical studies from documents mentioned that the iron ore smelting industry in upstream of Barito had stopped at the end of the 19th century due to the entry of cheaper iron from China [11]. Now, the remaining iron industry is in the form of blacksmiths who make plantation and agricultural equipment using scrap metals such as cars, boat plates, and railroads. Recently, every one village in the upper Barito, only has one or two blacksmiths, who have a type of work for sharpening blunted machetes.

The wrought iron industry is still ongoing in the downstream part of the Barito, in Nagara (Hulu Sungai Selatan, South Kalimantan). There are still hundreds of blacksmiths in the area, especially in Tumbukan Banyu and Sungai Pinang villages. They no longer use raw iron from smelting, but scrap metal from car pears and plate iron [21]. In the past, Nagara blacksmith obtained raw iron from iron ore smelting craftsmen from Dusun Hulu or Barito upstream, but since it had stopped, Nagara's blacksmith craftsmen have used scrap metal as their raw material. The raw materials are supplied by iron traders from Banjarmasin.

In the South Kalimantan precisely Sebuku Island, Kotabaru, there was an iron smelter, namely PT SILO. During the year of 2004 until 2014, the company had exported 9-10 tons of laterite iron ore to China annually. After limiting the export of raw minerals (without processing) in early 2014, the company tried to build a smelter that produced laterite concentrate with Ferrum (Fe) content of 53% [22]. However, shortly afterward, in 2017 the company stopped operating due to the licensing issues with the Gubernur Provinsi Kalimantan Selatan (South Kalimantan Provincial Government) [23]. There is no more iron ore processing industry in Kalimantan, so raw iron has been imported for the national steel materials industry and it continues until now [24. This is an irony because the source of iron ore is available but almost 100% raw iron in the national steel industry for PT Krakatau Steel is imported from China. It is proven from geological and mineral surveys that the source of iron ore in South Kalimantan is around 500 million tons [25].

The traditional technique of ironworking in Indonesia known today is the iron forging carried out by blacksmiths. Iron forging techniques are needed to condense iron particles so that a solid, strong iron surface is obtained. There are several blacksmith technology studies to get more effective performance and results, for example using a closed furnace with a blower [26]. There are also hitting machine technology and carbonization technology to simplify the production process and produce quality products with a strong surface and high wear resistance [27]. However, the results of these studies have not been applied by the blacksmith.

In the iron industry, iron forging techniques are still used as a method of compacting iron particles to make them more homogeneous, solid, and robust. In fact, the surviving iron industry in Kalimantan consists only of blacksmiths who utilize scrap metal into new tools. The decreasing number of blacksmiths has been caused by the lower public orders due to a large number of cheaper manufactured iron tools on the market. Blacksmith has a high risk in its technical work, so it is necessary to think of practical technologies that will simplify the production process. These practical technologies include hitting machines instead of hammers and sledgehammer, or robotic blacksmiths to make large-sized metals that cannot be done by a human. Blacksmith robots that began to be applied in modern countries is an improvised blacksmith's art with new digital capabilities. The basic principle is the automation of the process of forming parts, but with a blacksmith's basic approach, to make parts larger, more efficient, and reproducible [28]. The modern blacksmith's model is a hope that someday Indonesia can realize it. Hopefully, the government will encourage to revive the iron-making industry and modern blacksmith-based smithing technology to meet the national iron needs.

4. Conclusion

The experiment of simple furnaces found that the shape of the smelting furnace had no effect on the time to reach of smelting point and smelting result, but it gave the effect of the amount or volume of material that was inserted into the furnace. Square shaped furnaces could contain more iron ore and charcoal than round furnaces. The results of experiments in furnaces A and C with hematite laterites (red laterites) content showed the high percentage of the production of raw iron (between 50 to 60%), while brown laterites (furnaces B and D) produced quite low raw iron (ingots) which was between 21 to 46%. In the Barito watershed, there were two types of laterite, red laterite which had more weights with a high content of iron elements, and brown laterite which contained little iron.

The iron ore smelting industry in the Barito's upstream has stopped, but the iron forging industry in the old village with its center on the banks of the Nagara River continues until the present. In Indonesia, especially in Kalimantan, there is no more raw iron produced from iron ore smelting, so blacksmiths use the scrap metal. Blacksmiths who are using scrap metal have an impact on cultural sustainability as well as the utilizing of reused items that reduce metal waste on the earth. The existence of reused items is a continuation of the traditional iron industry chain. Learning from the history of our ancestors and the existence of an abundance of iron ore, reviving the iron ore smelting industry and modern forging is an important necessity.

Acknowledgments

Thanks to the technical team research such as Gauri V.D, Sundoko, Adurrasyid, and Ismail; The Gunung Timang District officials and Gunung Timang people that helped during the field research, as well as the head of the South Kalimantan Archeological Office (Balai Arkeologi Kalimantan Selatan) and the National Archaeological Research Center (Pusat Peneliltian Arkelogi Nasional) that provided funding for this research.

Reference

[1] Kurunov I F 2010 Blast-furnace smelting in China, Japan, North America, Western Europe, and Russia *Metallurgist* **54** (1–2) 114–26

[2] Reid A 2011 *Asia Tenggara dalam Kurun Niaga 1450-1680 Jilid 1(Southeast Asia in the Age of Commerce 1450-1680* vol 1 trans. Mochtar Pabotinggi (Jakarta: Yayasan Pustaka Obor Indonesia) p 123

[3] Muztaza N M, Saidin M M, Azwin I N and Rosli S 2012 3D GPR mapping for excavation plan in Jeniang Kedah Malaysia *International Journal of Environmental Science and Development* **3** (6) 574–78

[4] Mokhtar N A M and Saidin M 2018 Budaya material industri besi di Kompleks Sungai Batu, Lembah Bujang Kedah *Jurnal Arkeologi Malaysia* **31(2)** 29–39

[5] Pryce T O and Natapintu S 2010 Smelting iron from Laterite : Technical possibility or ethnographic aberration? *Asian Perspectives* **48(2)** 249–64

[6] Mei J, Wang P, Chen K., Wang L, Wang Y and Liu Y 2015 Archaeometallurgical studies in China: Some recent developments and challenging issues *Journal of Archaeological Science* **56** 221–32

[7] Prokop P and Suliga I 2013 Two thousand years of iron smelting in the Khasi Hills Meghalaya

North East India *Research Communication* **104(6)** 761–8
[8] Do M 2013 Iron-nickel alloy smelting production in Luwu, South Sulawesi during th e pre-islamic period *Dissertation* (London: UCL Institute of Archaeology) p 30-8
[9) Adhityatama S, Triwurjani R, Suryataman S, Hadi P, Utomo B B 2017 Underwater archaeological study on prehistoric material culture in Matano Lake South Sulawesi Indonesia *Journal of Southeast Asian Archaeology* **37** (December) 37-49
[10] Bulbeck F D and Prasetyo B 2000 Two millennia of socio-cultural development in Luwu, South Sulawesi, Indonesia *World Archaeology* **32(1)** 121–37
[11] Hartatik and Sofian H O 2018 Jejak pengerjan logam kuna di hulu DAS Barito Kalimantan Tengah: Kajian Arkeometalurgi *Purbawidya* **7** (21) 119–36
[12] Hartatik, Sofian H O, Sunarningsih, Dhaneswara G V and Sulistyo R B 2019 *Teknik Pengerjaan Alat Logam Kuno Dan Pemanfaatan Situsnya Di DAS Montalat Kabupaten Barito Utara Kalimantan Tengah: Studi Eksperimental Dan Arkeologi Publik* (Banjarbaru: Balai Arkelogi Kalimantan Selatan)
[13] Killick D 2014 From Ores to Metals *Archaeometallurgy in Global Perspective: Methods and Syntheses* ed Roberts et al pp. 11–45
[14] Haryadi H and Saleh R 2012 Analisis keekonomian bijih besi Indonesia *Jurnal Teknologi dan Batubara* **8(1)** 1–16
[15] Santoso B and Subagio S 2018 Pemodelan nikel laterit berdasarkan data resistivitas di daerah Kabaena Selatan Kabupaten Bombana, Provinsi Sulawesi Tenggara. *Jurnal Geologi dan Sumberdaya Mineral* **19(3)** 148–61
[16] Renfrew C and & Bahn P 2012 *Archaeology, Theories, Methods, and Practice* (London: Thames & Hudson) pp 348-351
[17] Busuttil C 2013 Experimental archaeology *Malta Archaeological Review* **2008-2009(9)** 60–6
[18] Bakri S and Sanwani E 2019 Studi transformasi goetit menjadi hematit secara mekanokimia untuk benefisiasi bijih besi laterit *Jurnal Teknologi Mineral dan Batubara* **15(3)** 179-88
[19] Schumann W 1992 *Rock Minerals and Gemstones* (London: Harper Collins Publisher) pp 103-07
[20] Martin P E 2009 Industrial Archaeology *International Handbook of Historical Archaeology* ed Majewski *et al.* (London: Springer Science & Business Media) pp 285–98
[21] Hendrawan A 2015 Pengaruh proses sepuh terhadap kekerasan mata kapak hasil pandai besi di Kabupaten Hulu Sungai Selatan Kalimantan Selatan *Jurnal Poros Teknik* **7(1)** 47–53
[22] Ummah A R 2014 Kuota ekspor bijih besi pt silo disetujui. *Investor Daily Indonesia* (April 1) retrieved from https://investor.id/energy/kuota-ekspor-bijih-besi-pt-silo-disetuju
[23] Hanafi I 2018 Produksi smelter PT SILO senilai Rp 2,2 triliun terhenti sejak November. *Antara News* (February 1) retrieved from https://www.antaranews.com/berita/682392/produksi-smelter-pt-silo-senilai-rp22-triliun-terhenti
[24] Irfan M A 2019 Penguatan industri besi baja menuju ketahanan nasional. *Investor Daily Indonesia* ed Gora Kunjana November 2 retrieved from https://investor.id/opinion/penguatan-industri-besibaja-menuju-ketahanan-nasional.
[25] Pardiarto B 2009 Tinjauan rencana pembangunan industri besi baja di Kalimantan Selatan. *Buletin Sumberdaya Geologi* **4(2)** 1–13
[26] Sodikin I, Waluyo J and Pratiwi Y 2016 Rancang bangun tungku pemanas untuk pande besi yang ramah lingkungan guna meningkatkan kapasitas produksi alat pertanian. *Simposium Nasional 15 RAPI*, 458–63. Retrieved from https://publikasiilmiah.ums.ac.id/handle/11617/8078.
[27] Khairil, Irwansyah, Edhy, H S and Rizal S 2012 Fundamental study on the carbonization characteristics of low rank coal under low temperature and its application on traditional blacksmith *Advanced Materials Research* **576** 615–8
[28] Daehn G S 2019 Robotic blacksmithing: a technology that could revive US manufacturing *The Conservation* (Newsletter December 17) retrieved from https://theconversation.com/robotic-blacksmithing-a-technology-that- could-revive-us-manufacturing

IOP Conf. Series: Materials Science and Engineering **980** (2020) 012047 doi:10.1088/1757-899X/980/1/012047

Online monitoring for processes and condition of a machine using smart management card

I A V Damanik*, R D Rumbara and N Juhana
Balai Besar Logam dan Mesin, Jl. Sangkuriang No.12 Bandung

Email: damanik@kemenperin.go.id

Abstract. This work aims to propose a solution on online monitoring for machining in the manufacturing process through web service, so that all data usage and machine conditions are obtained and stored on the server as Big-Data. The methodology in this work is RFID-based sensors on machines with Card Management System (CMS). Machining data of each machines that are monitored through a self-made application, connected to the internet networking and can be accessed online from everywhere by using laptop or mobile devices. Those machine activities as data record, stored on the server as big data that further can be used as Big-Data analytic. The result will be a system model which is able to monitor all machines in real-time and online by using internet connectivity, and also to find out the current status of the machine whether available or not for a new process. In addition, we can also know the process of the work being done by the machine with the use of RFID Card. This real-time data are also used to prevent sudden engine damage by doing regular maintenance according to work-hour of machines and also to maintain the sustainability of the process to keep it on schedule.

1. Introduction

Information Technology (IT) is developing rapidly in the 20th century especially in Industry 4.0 which provides a lot of benefits and conveniences to industrial sector and provides short and long-term solutions in terms of increasing productivity. As new technologies are introduced that also backed up Digital Transformation (DT) in terms of using digital records and slowly reducing the use of paper, the machinery sector get benefits from its optimization as well. Modern machinery development today is very complex and relatively expensive, concerning the operation and the maintenance [1]. Operators and engineers must ensure that the machine is monitored properly so that damage does not occur to the engine. Current information and problem solving actions on production machines are needed quickly, to avoid both machine failures and decreased productivity [2]. The real fact on the process floor is delayed information about the machine problem. Hence, both engineers and supervisors will be unable to know the problem in advance; unless by checking the machine directly to the shop floor; and also could not fulfill the schedule. The supervisor would have difficulties in making the daily report of the machine. These problems require a monitoring system that capable of providing information in real-time and online. With internet connection, the information needed can be obtained quickly anywhere. Monitoring aims to provide information about the current status of an object with a certain time period. Generally, monitoring is carried out for specific purposes, for instance observation, examination, control and correction of all processes activities, as a basis for determining actions to be taken next. Also, documentation of machine operational data, such as what work is done, how long it needs to be done, and also to find out whether a process has been completed or is still in progress. These data are stored in a database on the server. The database not only provides information about the condition of the machine online and in real time but can also provide trend data for taking preventive and corrective maintenance steps on a regular basis [3-7].

Another common and ubiquitous problem is that previous collecting data are still not available to be used as reference to a further analysis in determining the trends of a process and operating a machine. By 2020, data will have been generated, imitated, and consumed by users over 40 Zettabytes (or 40 trillion gigabytes) [8].The unavailable collecting data and separated data at many points inflict an incorrect decision due to the lack of reference data. Industry 4.0 presence through the application of technologies such as Internet of Things (IoT), Cloud Computing, Cyber-Security, Big Data analytics provided a lot of progress in terms of production step-up capabilities and also lowering operational costs. One of the IoT item is Smart Management Card (SMC) so-called RFID system which has an important role in terms of online and real-time monitoring in Industry [9]. Radio-Frequency identification (RFID) systems known as powerfull method for many applications such as line production monitoring, the supply chain, public transportation, health industry, etc. [10–12].

This work aims to propose a solution on online monitoring for the machining process in the manufacturing process display through the web service. This solution result in all usage data and machine conditions are able to be obtained and are stored on the server as Big-Data. A number of data obtained by using an RFID system will be used as information through Big-Data, thus the management level can get real-time information to carry out further both analysis and results which become the basis to make up a decision on the machine such as maintenance and especially increasing machine productivity.

2. Materials and methods

RFID system is faster and does not make contact between the reader unit and tagged object compared to other types of identification methods recently [13]. All materials used in building a monitoring RFID system are obtained by assembling and using existing electronic components; making applications and systems by utilizing machines in the workshop from conventional to CNC and placing devices (hardware) consisting of electronic components and sensors in the machine and data communication is built with wireless method. Overall, the Smart Management System development is the use of RFID Card as the most important part of the type of card that can be both written and read by the system built, and the card is used to provide the code for each work, and also replace work cards that are usually used by operators.

2.1. Front- end

The front-end scheme is a communication step that aims to register an RFID card in the database. The process that occurs in this section is how to attach identification code to the RFID card after it is registered on the writer device so that it can send a signal to the PC, as figure 1. Subsequently the PC will send the status of card to the server (database) and the card is completely registered. When this card is attached to one of the readers on a machine, the reader will send a signal to the server about the existence of the card. The card identity contains the job information such as job number, Company name and date of collection.

2.2. Back-end

In contrast with the Front-end, the back-end system is when the RFID card is used in the workshop. As can be seen in figure 2, RFID cards that have been registered (front-end) will be used as marking to execute the machining process, so that all activities in each machines or job process will be transferred online; how long the duration of work, and still ongoing or completed.

The method is that a card must be swiped twice (first as ON sign and second as OFF sign). When this card is only once attached to the reader and is not re-attached again for the second time which indicating of a machining process is complete, then no other card will be recognized in the monitoring system although it is pasted on the machine.

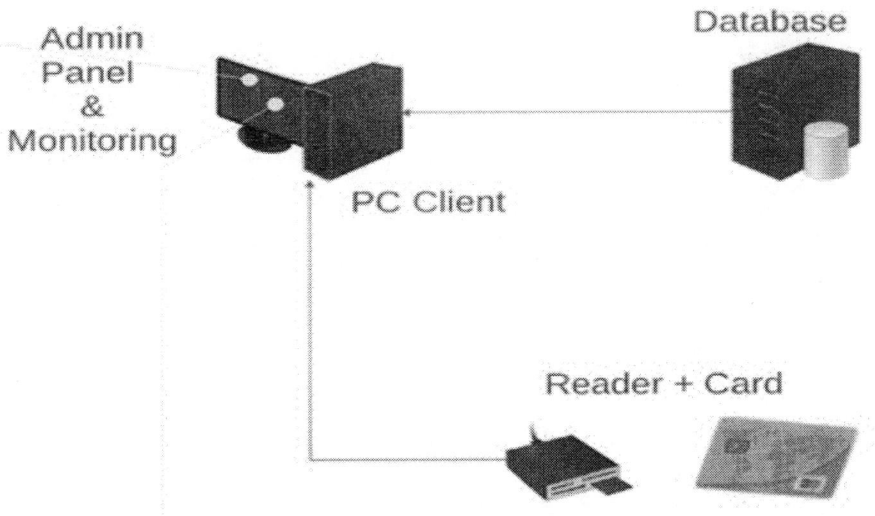

Figure 1. Smart management system- front end.

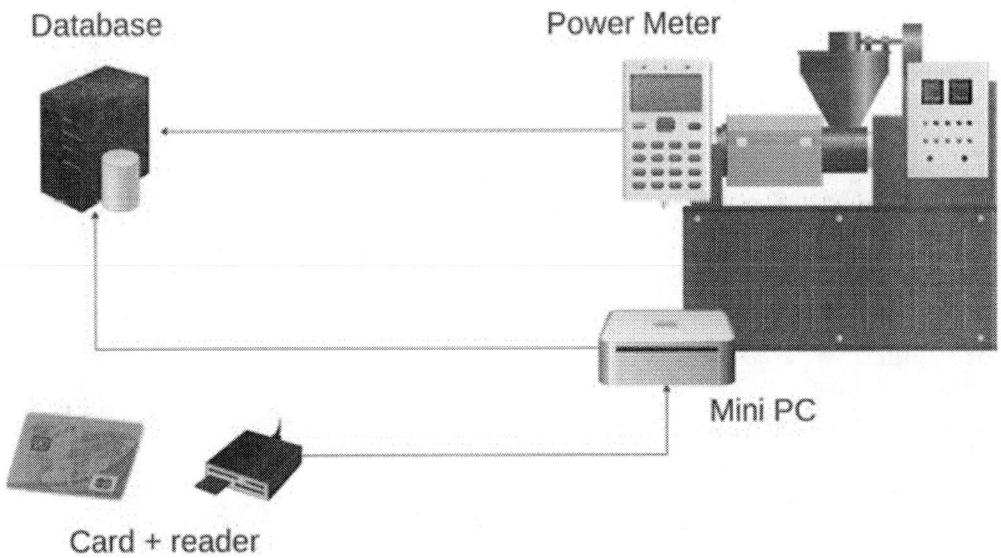

Figure 2. Smart Card Management System – Back end.

2.3. Smart card specification

Smartcard Reader is a device that functions to read information stored on a smartcard as shown in figure 3. In its implementation, a smartcard reader will operate if detect the valid card according to the workflow setting that has been determined. The HF RFID standard including ISO/IEC 18000 standard is a reference for RFID tag specification [14]. Operators or users will be able to monitor all activity on the machine by online for all the time.

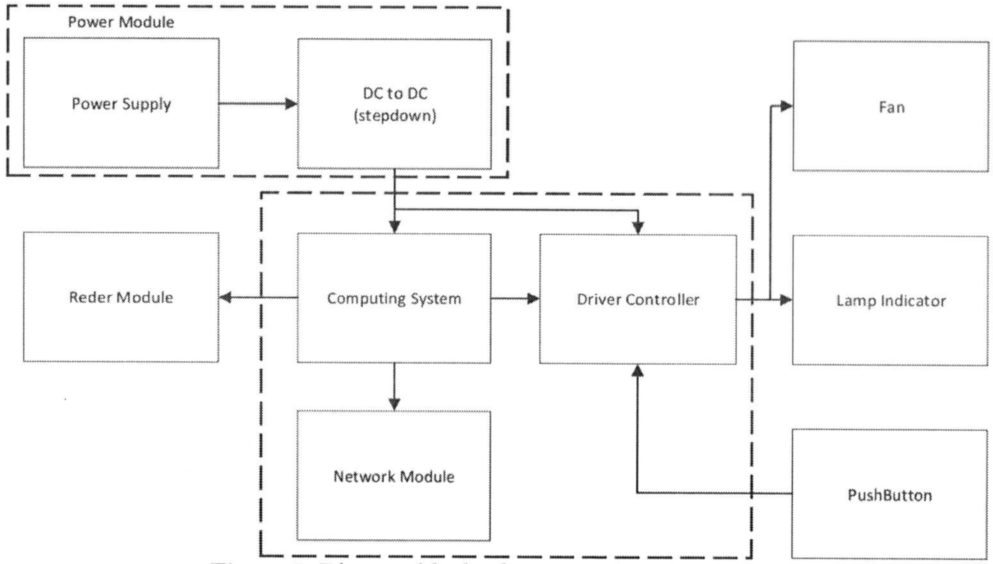

Figure 3. Diagram block of Smartcard reader system

- Power module:
  This device is a module that functions to convert AC voltage to DC and provide power to the reader, this module consists of a power supply and DC to DC Step down to convert 220VAC to 12VDC and 5VDC.
- Reader module:
  Reader module is an antenna device smart card reader (card reader) that functions to connect the card with a computing system.
- Main control module:
  This module functions as a storage of logic data, memory, and work settings for reader devices (computing). This module consists of a computing unit, network module, and driver controller.
- Network module:
  This module is a communication interface module between the reader and external devices and applications. This module consists of 3 types of interfaces, among others, USB, Wifi, and Ethernet.
- Driver controller:
  This module functions as an I/O module that controls the push button system, indicator and fan.
- Lamp indicator:
  The lamp indicator functions as a work indicator of the smartcard reader.
- Fan:
  Fan is a device that functions to maintain the temperature of the smartcard reader.
- Push button:
  This module functions as input from the system to turn on and turn off the reader

2.4. Database and server
All Activities data for number of machines that is monitored by using a Card Management System (RFID) will be stored in a database on the server. The database built is able to accommodate data with a duration of delivery to the server per 5 seconds for approximately 3 years period. Server's setup depends a lot on how much and size each data will be stored.

Table 1. Hardware and other part specifications.

Product	Specification
PC Server	HPE ProLiant DL20 Gen9; Intel Xeon E3-1220v6 Quad Core; 16GB
HDD - Add on Server	Harddisk HPE 1.2TB SAS 10K
RAM - Add on Server	HPE DL20 8GB (1x8GB) Dual Rank x8 DDR4
Routerboard CCR1036-12G-4S	1U rackmount, 12x Gigabit Ethernet, 4xSFP cages, LCD 36 cores x 1.2GHz CPU, 4GB RAM
Module Transceiver Fiber Optic SFP	Konverter fiber (optic-RJ45) untuk Mikrobits SFP Transceiver (RJ45)
WD My Cloud	My Cloud EX2 Ultra Device; 4 TB; Ethernet cable;
Kabel LAN	Kabel UTP Belden CAT 5
PRW	Data on

2.5. *Server specifications*

Server specifications in table 1 is selected based on the need for storing the RFID Smart Card system activities. Determining the specifications is very important to accommodate all data that will be accepted as a database that ultimately forms a Big Data Methods.

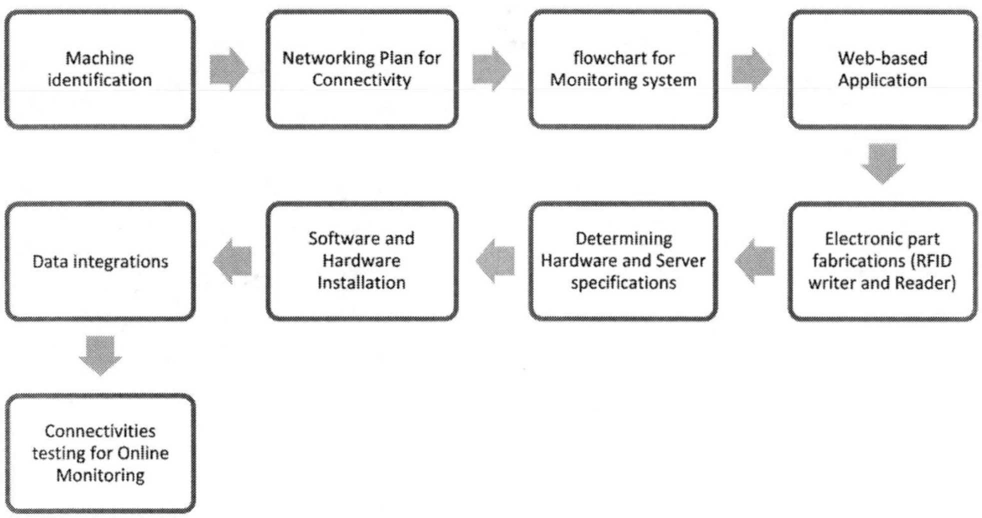

Figure 4. Work flow for online monitoring based Smart Management Card – RFID.

Online monitoring system begins with a machine identification activity to determine which machine will be used in the system as shown in figure 4. These activities intend to set out the topology of the system and make a networking plan for connectivity. Monitoring system is built based on web browser application by using local internet and collect all data in server. Electronic part fabrications (RFID writer and Reader) is adjusted to meet the needs of data captured to be delivered to the network. Then,

IOP Conf. Series: Materials Science and Engineering **980** (2020) 012047 doi:10.1088/1757-899X/980/1/012047

determining Hardware and Server specifications stage will be easily done by matching as needed. Since hardware and software are complete, both will be installed to the system until it is able to integrate the data. Connectivity testing for Online Monitoring are the last but not the least on this work.

3. Results and discussions

The means of the built-in online monitoring system is obtaining an integrated system from the use of a Card Management System (CMS) on a real-time machine condition that can be accessed online through an internet-based application.

3.1. RFID card management system

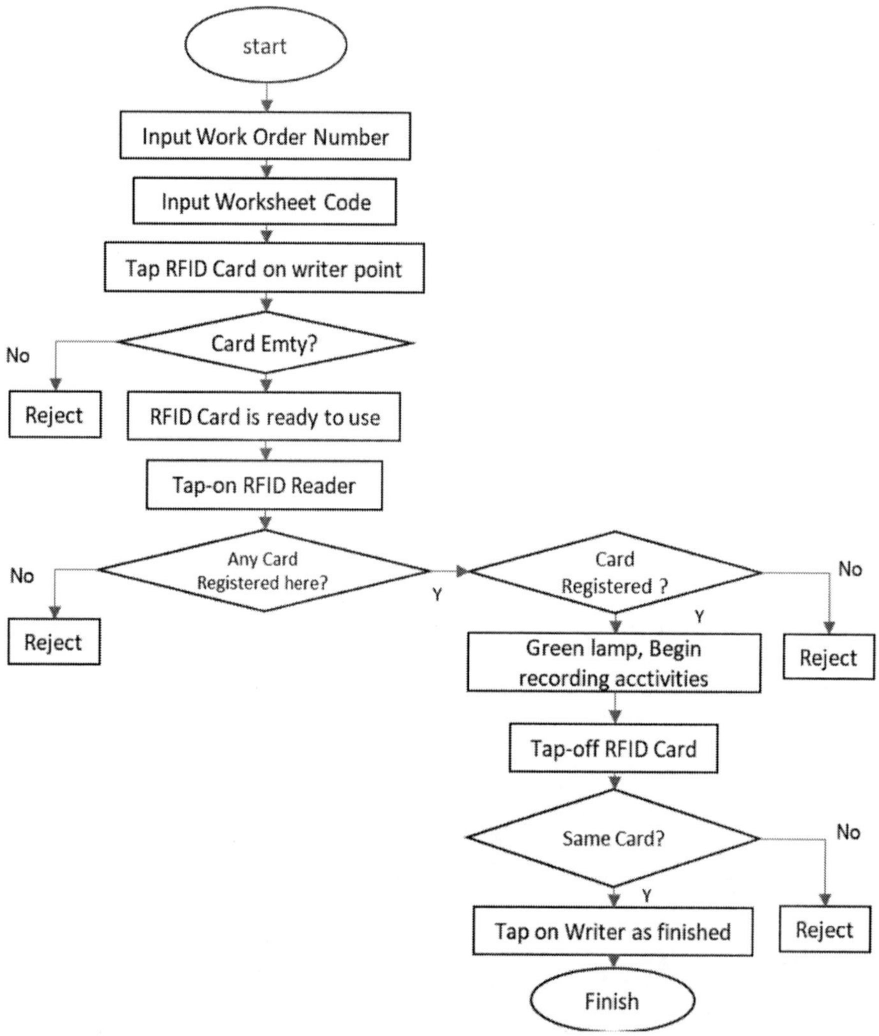

Figure 5. RFID card management system flowchart.

Flowchart in figure 5 shows the stages how to use RFID Smart Management card in an online monitoring system for machine process. All stages cover the process begin with how to register the work order that obtained from marketing section at the beginning, get the number and register it to the system. Users must ensure that the cards used are in good condition and still empty so it is possible to write some information into the RFID Card.

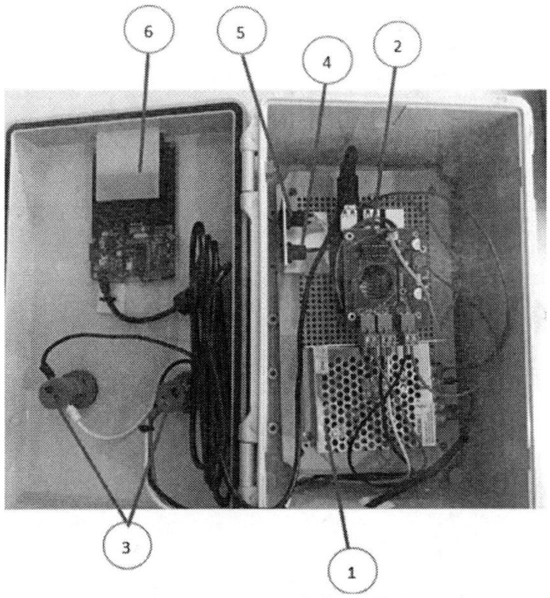

Figure 6. Components of the RFID reader.

Caption in figure 6:

1. Power Module
2. Main control module
3. Indicator lamp
4. Power on button
5. Power off button
6. Smart card Reader

3.2. Dashboard for online monitoring using RFID card system

Both writer and reader send all information as database to a server. All information can be seen on the dashboard in real-time and online which can be accessed using electronic devices (laptop, PC, mobile devices) connected to the internet network. An online built system based on nodejs programming language on the web-server allows the user to monitor machine activities directly, and to find out work is being accomplished by the machine as well. RFID Card is designed to be more easily used. Firstly, user do the registration of Work Order "Surat Perintah Kerja" to database server which is usually in the form of a certain number made from the marketing department. In figure 7, the dashboard shows the registration is complete with SPK number, description, and the name of the consumer company. To ensure the SPK has been registered for the customer, just click on the list menu that shows that the SPK registration status. The application also provides information about the date of registering the work paper and the status of the work completed or still in progress.

The second step is determining how many worksheets are used to complete one SPK that has been previously registered as it can be seen in figure 8. The number of worksheets varies according to the needs and level of difficulties of a machine process. The number of worksheet is closely related to the number of RFID cards that will be used because it functions as a substitute for work cards in the form of paper that is commonly used. RFID Card serves as a marker on which machines the job will be carried out and at the same time to replace work cards used by the operator to record machining activities (starting and ending in hours).

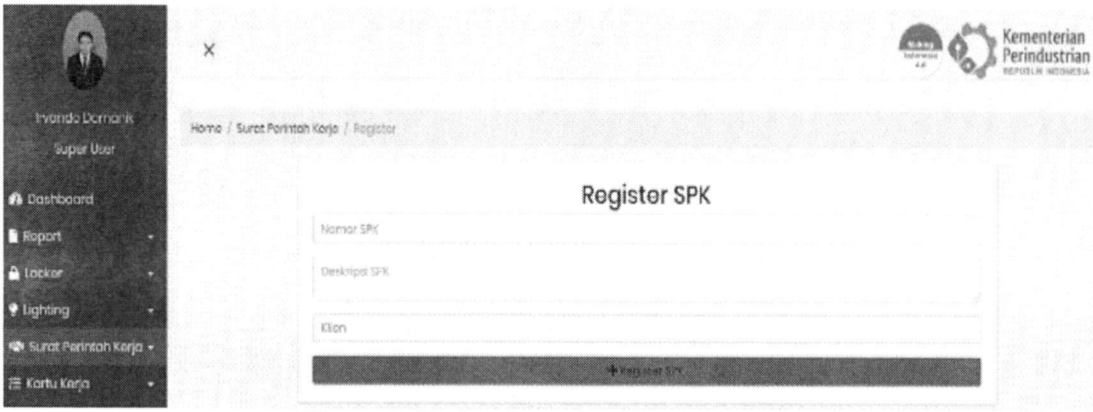

Figure 7. Work order registration dashboard.

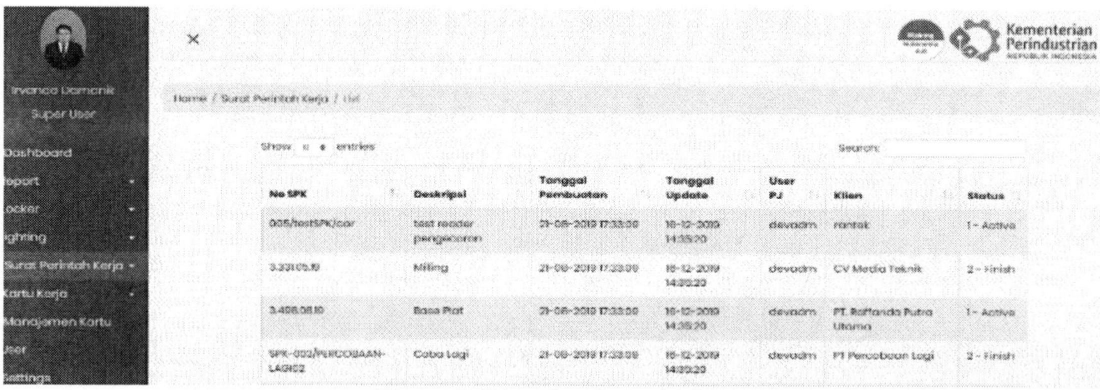

Figure 8. List of registered jobs.

Next step, RFID card is written by swiping it to the writer devices so the identification data is entered into the card. RFID cards containing data are ready to be used by operators for the registration function on each machine.

When an object is going to be processed on machine A, then just look at the application whether machine A is being used for another job or in idle status. After deciding which machine is used for this process, the operator attaches the RFID card to the reader on the machine, and the machine begin to record all activities. RFID Cards used are that have been registered ones. So, if you use a card that has not been registered, the reader will not recognize the card and the data will not be sent to the server. Other indicators to know wether the RFID card has registered or not is three various colours of lamp indication. RFID reader will light on red if detect unregistered card while green indicates valid card.

The main dashboard as shown in figure 9 provides information about all machines status. The classification based on whether the machine is ready for use, being used, or in a damaged condition. Status information is presented in colour display that describes the description of each machine:

- Red = ready to use
- Green = in use
- Black = cannot be used (broken, dead or in maintenance)

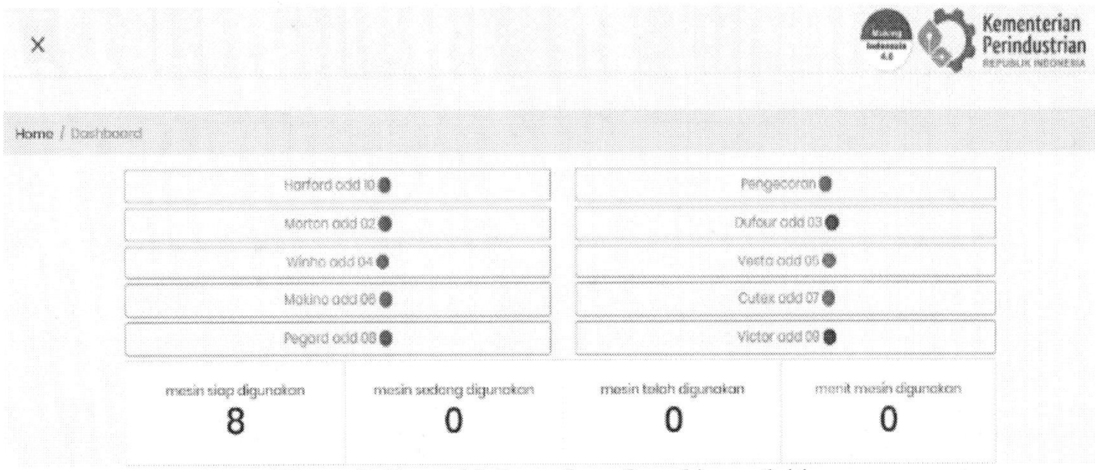

Figure 9. General information of machine activities.

The colour indicators provides machine information so that the process engineering can prepare the machine schedule according to the readiness of each tools and equipments. So, there are three main parts in this work :

- Part 1: Card Management System (CMS) Application and Smart Card reading module
- Part 2: Monitoring system Application, Smart Card Data Transmission Module and Database and Data Integration Module
- Part 3: Web Services and Data Integration Modules

These three main parts are combined to form an online monitoring of processes and conditions of a machine using a smart management card on an integrated web-based online.

3.3. Data security.

Most of the manufacturing industry is currently facing their own digital transformation by using automations as solutions to increase productivities at workshop. Online monitoring is one of obvious example for digital transformation implementation that has a very high risk in terms of cyber-attack. This puts the manufacturing industry at serious risk of cyber-attacks which have the potential to derail their operations due to the amount of data or company time lost.

This online monitoring applications equipped by local-server that built with such a suitable language programming. Using local-server can prevent your data from being stolen even though the risk of data loss may still occur. Moreover, this application provide a large data storage spaces so the user be able to back-up their data every time depend on how long the programmer set the system send data to the storage. The more data sent, the greater the storage allocation needed.

4. Conclusion

Machining process activities can be monitored in real-time and online by using internet connectivity which is displayed on a dashboard, including the status of the machine; ready for use, in use or damaged. The RFID Card Management system is able to replace paper as a working paper communication media as well as being a job tracing media on each machine in the workshop. The RFID system also provides detailed information on the length of working time on each machine which is displayed in dashboard that can be monitored online. All RFID system reading data results are stored in the server as work data that will be used as Big-data.

Acknowledgments

In completing this paper, the author would like to thank the authority of the Metal Industry Development Center (MIDC/BBLM) for providing funding and other laboratory facilities in the implementation of

IOP Conf. Series: Materials Science and Engineering **980** (2020) 012047 doi:10.1088/1757-899X/980/1/012047

this project. We also would like to thank the head of MIDC for allowing us to do research there, and specifically for the teammates: Anugrah Erick, Sony, Ali Firmanysah, Nana Juhana, Agus S, and Sandy Sitorus who always struggle together. The author will never be able to solve this without all of your guidance.

References

[1] Roy R, Stark R, Tracht K, Takata S, and Mori M 2016 *"Continuous maintenance and the future – Foundations and technological challenges,"* CIRP Ann. - Manuf. Technol., **65(2)** 667–88
[2] Guo R and Jin Y 2019 *"Phase Identification and Online Monitoring for the Uneven Batch Processes,"* IEEE Access, vol. 7, pp. 81351-63
[3] Yew H T, Ng M F, Ping S Z, Chung S K, Chekima A, and Dargham J A 2020 *"IoT Based Real-Time Remote Patient Monitoring System,"* Proc. - 2020 16th IEEE Int. Colloq. Signal Process. its Appl. CSPA 2020, no. Cspa, 176–9
[4] Wang L, Yan Y, Hu Y and Qian X 2013 Intelligent condition monitoring of rotating machinery through electrostatic sensing and signal analysis *IEEE Int. Conf. Smart Instrumentation, Meas. Appl. ICSIMA* 2013 pp 26–7
[5] Juliasari N, Hartanto E D and Mulyati S 2016 *Monitoring Suhu dan Kelembaban pada Mesin Pembentukan Embrio Telur Ayam Berbasis Mikrokontroler Arduino UNO J. TICOM* **4(3)** 109–13
[6] Zhang Y and Ruan J 2018 *Large-scale machinery monitoring system based on the visual reality* Proc. 2018 IEEE 3rd Adv. Inf. Technol. Electron. Autom. Control Conf. IAEAC 2018, pp 863–67
[7] Zhang P, Li J, Yang T and Huang S *2012 Flexible and smart online monitoring and fault diagnosis system for rotating machinery* Proc.2012 Int. Conf. Comput. Distrib. Control Intell. Environ. Monit pp 343–47
[8] Sivarajah U, Kamal M M, Irani Z and Weerakkody V 2017 Critical analysis of Big Data challenges and analytical methods J. Bus. Res. **70** 263–86
[9] Nayyar A and Kumar A 2019 *A Roadmap to Industry 4.0: Smart Production, Sharp Business and Sustainable Development* Mourad Amer, International Experts for Research Enrichment and Knowledge Exchange (IEREK), Cairo, Egypt p 89
[10] Han S, Lim H S and Lee J M 2007 An efficient localization scheme for a differential-driving mobile robot based on RFID system IEEE Trans. Ind. Electron **54(6)** 3362–69
[11] Glidden R, Bockorick C and Cooper S 2004 *Design of ultra-low-cost UHF RFID tags for supply chain applications* IEEE Commun. Mag **42(8)** 140–51
[12] Meillere S, Barthelemy H and Martin M 2006 13.56 MHz CMOS transceiver for RFID applications Analog Integr. Circuits Signal Process **49(3)** 249–56
[13] Hwang Y S and Lin H C 2009 A new CMOS analog front end for RFID tags IEEE Trans. Ind. Electron **56(7)** 2299–2307
[14] ISO/IEC 18000-3 2001 Information Technology-Radio Frequency Identification for Item Management—Part 3: *Physical Layer, Anti Collision System and Protocol Values at 13.56 MHz,*

IOP Conf. Series: Materials Science and Engineering **980** (2020) 012048 doi:10.1088/1757-899X/980/1/012048

Improving the stability of catechin from gambier in β-cyclodextrin and nanoemulsion-based inclusion complexes

G Yeni[1]*, K Syamsu[2], O Suparno[2], E Mardliyati[3], Silfia[1], E Syafri[4], N Nazir[5], A Fudholi[6]

[1] Institute for Research and Standardization of Industry, West Sumatra 25164, Indonesia
[2] Department of Agroindustrial Technology, Agricultural Engineering and Technology, IPB University, West Java 16680, Indonesia
[3] Agency for the Assessment and Application of Technology (BPPT), Jl. MH. Thamrin 8, Jakarta 10340, Indonesia
[4] Department of Agroindustrial Technology, Agricultural Polytechnic, Payakumbuh, West Sumatra 26271, Indonesia
[5] Andalas University, Agricultural Technology, West Sumatra 25163, Indonesia
[6] Solar Energy Research Institute, Universiti Kebangsaan Malaysia, 43600 Bangi, Selangor, Malaysia

Email: gustry87@gmail.com

Abstract. The application of catechin is limited due to low stability, however nanoencapsulation technology can improve catechin stability. This study aimed to determine the effect of the types and concentrations surfactants in the catechin encapsulation process on β-CD. Concentrations of surfactants used were Poloxamer 188 (NKP, 0.5%-1.5%), Tween 80 (NKT, 2,5% and 3%) without surfactants was used to control (NKB). The catechin nanoemulsion complex formed was inclusion them into β-CD (ratio 1:1), and dried by spray drying. The stability test showed a sample of NKP 1% was more stable with the lower turbidity and viscosity values, namely 175 NTU and 0.93 cP, NKT 2.5% (118 NTU and 0.94 cP), NKB (461 NTU and 4.0 cP). The size of the sample particle decreased according to an increase in the surfactant concentration, where the NKP 1% (37 nm) produced smaller particles, the appearance of clear and yellowish suspension. NKP 1% had the highest EE value, followed by that NKT 2.5%, and NKB, i.e. 91.9%, 89.5%, 77.4%, respectively. Sample NKP 1% had a morphology shape with compact structures and the highest crystallinity degree (92.4%). This research showed the use of surfactants could improve the stability of catechins compared to that without surfactants.

1. Introduction

Uncaria gambier Roxb. (gambier) is one of the Indonesian community plantation commodities that has high economic value and perspective for multi-use commercial development in the future. Catechin is one of the main components of the gambier plant. Its amount in gambier extract depends on the processing. Gambier contains catechin phenolic that has a role as an antioxidant [1,2].

Phenolic compounds have antioxidant properties by scavenging radical oxygen species [3–5]. The instability of catechin is shown by the change in its color from yellowish-white to brown when it is left in the open air. The instability of its active compounds can decrease bioactivity during the storage so that the application process is still difficult to be applied [2,6,7].

Nanoencapsulation technology can improve bioavailability. It also can control and improve the release properties of the active compound [8–11]. Previous researchers have encapsulated active compounds of catechins using a variety of different methods and coatings. But no one has researched the catechin encapsulate from gambier catechin raw materials and uses a size reduction method through the process of emulsification using surfactants Poloxamir 188 and Tween 80 before being coated on β-cyclodextrin (β-CD).

The technology of the encapsulation matrix method of inclusion with β-CD has become a popular method in increasing the stability and solubility of poorly soluble active compounds.This method has a unique characteristic β-CD, as host, has a cavity, in which the outer surface is hydrophilic and the inner surface is lipophilic (hydrophobic). This characteristic makes it possible to select active compounds (guest) to be inserted [6,12–14].

The formation of the complex inclusion to produce stable products requires high efficiency of active compound adsorption into β-CD. Adsorption of active compounds into the β-CD can be done if the guest molecules are significantly smaller than the cavity of β-CD. An active compound with nano-size can get inclusion into β-CD properly because it possesses a large surface so that it can be absorbed into the β-CD perfectly and can improve the stability of the active compounds [6,7,15,16].

To produce nanoparticle of catechin from gambier, a processing technology that does not damage the particles is needed. The process of emulsification of nanoparticle catechin using surfactants can increase the stability of emulsion products. Surfactants have good control during nano-capsulate formation. The surfactants have a function to reduce surface tension and interfacial tension. The emulsion process to produce nano-sized particles is affected by current power stirring [15,17–20].

Increased catechin stability was needed in storage and application. Therefore, it is needed to find the appropriate type of surfactant to produce catechin nanoencapsulation products from gambier which was stable in storage. This study aimed to determine the effect of the types and concentrations of Poloxamer 188 and Tween 80 surfactants in the catechin encapsulation process on β-CD and analysed the potential nanoemulsion-based inclusion complexes formed that can increase its stability.

2. Experimental

2.1. Materials

Catechin from gambier was made used the method of [2] β-cyclodextrin, and Tween 80 (Sigma-Aldrich), Polaksamer 188 (PT. BASH Indonesia), distilled water, technical ethanol, deionized water were purchased from Betrachem-Bogor, chemicals for analysis by Merck of Darmstadt, Germany.

2.2. Methods

2.2.1. Determination of the encapsulation efficiency of nanoencapsulation of catechin- β-cyclodextrin. Nanoencapsulated catechin (25 mg) was dissolved in 50 mL buffer phosphate pH 7.2. The mixture was shaken for 24 hours and filtered. The filtrate's absorbance was read using a UV-VIS spectrophotometer, at the maximum wavelength was (λ) 279 nm. Encapsulation efficiency (EE) was calculated by the equation 1.

$$EE = \frac{\text{x mgL}^{-1}\text{ x 1 L x 1000 mL}^{-1}\text{x vol.extraction x a mg x b mg}^{-1}}{\text{mass of first catechin (mg)}} \text{ x } 100\% \quad (1)$$

x is concentration of formula, a is total mass of nanoparticles that obtained, and b is mass of nanoparticles used for the determination of efficiency.

2.2.2. Production of encapsulation of catechin in β-cyclodextrin without surfactants. Nanoencapsulate of catechin without surfactants was made using a modified method by [21]. Catechin was dissolved in ethanol 70% (1:10), stirred with a magnetic stirrer at 50 rpm for 5 min and then sonicated for 5 minutes. β-cyclodextrin was diluted with distilled water whose temperature is 50±5°C (4:1), stirred with a magnetic stirrer at 700 rpm for 5 minutes. Catechin solution was inclusion on the β-CD (molar ratio of 1:1) with a magnetic stirrer for 24 hours. The complex inclusion solution of catechin-β-CD evaporated its solvent with spray dryer at the condition of inlet temperature of 175±5°C and the outlet temperature of 90±5°C, a solution flow rate 10 $mL.min^{-1}$.

2.2.3. Production of encapsulation of catechins in β-cyclodextrin through an emulsion process. The making of catechin nanoencapsulation with surfactants used a modified method by [15], [22]. β-cyclodextrin (40 mg) was dissolved in 10 mL of deionized water at a temperature 50±5°C (water phase). Nanoemulsion catechin used surfactants Poloxamer (NKP), Tween surfactant (NKT) and without surfactant (NKB) were dropped on β-CD (3 $mL.min^{-1}$) while stirred with a magnetic stirrer (700 rpm). After combining the nanoemulsion into β-CD, it is continued with magnetic stirring at 400 rpm for 18±2.0 hours. Inclusion complex of catechin-β-CD which was formed from the evaporated solvent with a spray dryer at an inlet temperature of 175±5°C, the outlet temperature of 80±5°C and flow rate solution 10 $mL.min^{-1}$.

3. Results and discussion

3.1. Preparation of nanoencapsulation catechin conventionally (without surfactants)

The results showed that particle preparation of catechin from gambier used in this process produced catechin, with an average diameter particle size was big enough (607 nm), with low homogeneity polydispersity index (IP) 2.96 (table 1). The polydispersity index indicated the size of the mass distribution of molecules in the sample. IP value was less than 0.3 indicating that the sample has a size distribution which is more uniform or homogeneous [20,24,25].

The process of micelles formation by emulsions surfactants and particle size that was produced was strongly influenced by the mixing process or ultrasonication. Formation of catechin nanoemulsion was conducted by using a surfactant Poloxamer 188 ($HO(C_2H_4O)_{38}(C_3H_6O)_{29}(C_2H_4O)_{38}H$) and Tween 80 ($C_{64}H_{124}O_{26}$). The results of emulsion treatment with the smallest particle size catechin will be included on β-CD.

The process of making nanoencapsulation with active compounds by inclusion is influenced by the composition of the materials and methods used. The composition of a material affects the equilibrium complex formation of active compounds. The composition of material between catechin and β-CD equilibrium formation of inclusion complex occur in comparison molarity of 1:1, wherein catechin used was 300 mg (molecular weight of catechin = 290.26 g. mol^{-1}), and β-CD was 1144 mg (molecular weight β-CD=1134.99 $g.mol^{-1}$). In this condition, the mole fraction of catechin bounded with β-CD was 0.85 and the value of fixed constant ($Ka.M^{-1}$) was 21.800 M^{-1} [13,15,21].

Table 1. The results of research formation nanoencapsulation of catechin conventionally (without surfactants).

Process	Turbidity (NTU	Viscosity (cP)	Mean of particle size diameter (nm)	Diameter (nm)			IP
				(10%)	(50%)	(90%)	
Preparation of catechin solution							
Stirring 700 rpm, 30 minutes, sonication 10 minutes	461	7.36	607	1 699	4 679	8 915	2.96
Formation of the catechin-β-CD inclusion complex							
Stirring 50 rpm for 24 hours	461	5.17	286	339	977	2 985	0.92

The reduction of particle size of catechin (tabel 1) can be done using ultrasonication technique and surfactants. Ultrasonic phase will reduce the particle size and the big energy of ultrasonic will be able to break down particles into smaller sizes and uniform. The weakness of using the ultrasonic technique is that long processing times, can produce heat and cause a decreasing activity of catechin. At the phase of the ultrasonication process, is where surfactants had a function to make effective collisions between molecules so that the large particles broke down into smaller (nano) [13,22].

Surfactants in reducing particle size depended on the type and concentration of it that used. Surfactants have surface activity (surface-active agent) which can reduce the surface tension between the air-liquid and liquid-liquid. The role of surfactants in lowering the interfacial tension with the surrounding particles took place during the merging process of surfactant (liquid phase) and the active compound (organic phase) through dropping. To produce nanoemulsion surfactants were dispersed by adjusting the number of droplets of the organic phase to the liquid phase [17,22,26,27].

The particle size affected the absorption efficiency of the quest into β-CD. To produce the perfect complexation with β-CD the size of the active compound molecules had to be smaller than the cavity of β-CD. The cavity of β-CD had a diameter of about 0.6-0.65 nm and volume about ±26.2 nm. Guest molecules which were larger than the β-CD cavity would result in the low and weak form of the complexs [7,13]. Large particle size also caused catechin molecules encapsulated in β-CD to become low (tabel 1). To improve the encapsulation high efficiency on the formation of catechin nanoencapsulation-β-CD a decreasing particle size of catechin was conducted.

3.2. Preparation of nanoencapsulation catechin conventionally (without surfactants).

3.2.1. Production of catechin nanoemulsion with surfactant. Production of catechin nanoemulsion done by phase inversion method was with the addition of the organic phase (catechin in 70% ethanol solution) into the liquid phase (surfactant and deionized water). The results of the catechin nanoemulsion analysis are shown in table 2.

The test results of emulsions (table 2) showed that the smallest particle size was obtained from the use of Poloxamer 1% (NKP) and Tween 2.5% (NKT) concentration, with an average diameter of the particle, sizes 84 nm (IP = 0.29) and 22 nm (IP = 0.25). A decreasing particle size predicted the value of turbidity, viscosity and particle size diameter (nm) which were produced. The particle size was directly proportional to the turbidity value, the smaller the particle size the lower the turbidity value of the suspension and the lower the viscosity the smaller the particle size.

Table 2. Value of turbidity, viscosity, particle size diameter (nm) and polydispersity index (IP) emulsion catechin.

Type of surfactants	Turbidity (NTU)	Viscosity (cP)	Mean of particle size diameter (nm)	Diameter (nm)			IP
				(10%)	(50%)	(90%)	
Poloxamer 188 (%)							
0.5	461	4.0	185	135	204	295	0.47
1	175	0.93	84	17	36	93	0.29
1.5	185	2.67	166	98	170	295	0.11
Tween 80 (%)							
1	461	6.4	Solution was turbid and had sediment				
1.5	461	5.7	Solution was turbid and had sediment				
2	406	3.0	245	246	676	1 863	0.75
2.5	118	0.94	22	7	11	21	0.25
3	199	1.17	151	41	170	490	0.37

Reduction size of the catechin particles at the ultrasonication stage due to the effects of cavitation produced vibration energy of sound waves spreading through the liquid medium in a cycle of high and low pressure (frequency of 20-40 kHz), simultaneous collision among the particles. At low pressure, bubbles were formed. They absorbed energy and then broke out at high pressure and released high of mechanical and thermal energy. The outbreak of the bubble caused the particle size became smaller and decreased the value of turbidity size [5,20,25,28,29].

The use of Poloxamer or Tween at lower concentration produced larger particle size with a high value of turbidity and viscosity (tabel 2). This condition might be due to the equilibrium between the surfactant and catechin that has not happened yet. High concentration surfactant produced a high value of turbidity and viscosity but the emulsion that was formed occurred separately between 2 phases in storage. It was possibly caused by a complex form already exceed the concentration of critical micelle concentration (CMC).

Emulsions that had a large particle size occurred when the surfactant concentration was lower than CMC and the surfactant used was not enough sufficiently contribute to reducing the interfacial energy between the active compounds and solvents. Conversely, the use of high surfactant concentrations caused clustered particles to form aggregates and larger sizes. Increasing particle size indicated that the emulsion was not balanced due to a tendency to agglomerate [10,15,26].

3.2.2. The formation of inclusion nanoemulsion complex of catechin-β-cyclodextrin (β-CD). The result of the inclusion complex formation of catechin-β-CD using surfactants Poloxamer (NKP) (table 3) produced the clear suspension, yellowish color with the smallest average of particle size (37 nm). Meanwhile, the surfactant Tween (NKT) produced a color slightly darker and an average diameter of the catechin particles which was bigger (177 nm) than that of the first nanoemulsion (Tween concentration 2.5%) (22 nm). This showed that catechin with a complex Tween surfactant had an unstable emulsion. The nanoparticle has limitations, its size and large surface area can lead to aggregation of the particles and physical handling is difficult in liquid form [10,15].

The result of the formation of inclusion complexes of catechin-β-CD samples NKB (without surfactant) produced complex turbid solution and sedimentation. This could be due to the large size of catechin particles (254 m) and their less homogeneous so it caused catechin inclusion could not take place perfectly in β-CD. The formation of an inclusion complex of the active compound with β-CD could form sediment due to a lack of sufficient power of combining between the head groups of active compounds with NKB.

Table 3. Particle size distribution of inclusion nanoencapsulate complex catechin-β-CD based on treatment variations.

Sample	Mean of diameter (nm)	Diameter (nm)			IP
		(10%)	(50%)	(90%)	
NKB	254	123	324	676	0.39
NKP	37	7	12	28	0.28
NKT	177	65	214	408	0.27

Complex inclusion of active compounds with β-CD occurred because the availability of cavity ring-shaped donat with the outside part was hydrophilic and with the inside part was hydrophobic. A hydrophobic cavity was an ideal place for a molecule that was less soluble in water by protecting the hydrophobic part. Catechin was a compound that was less soluble in cold water. β-CD had a role in shaping complex host-guests, where most of the surfactant put the hydrophobic edge truss into the cavity of β-CD. At the time of formation of inclusion, a complex active compound with β-CD occurred because of the driving force by H-bonds and ionic repulsion between the head and tail. The magnitude of each impulse greatly affected the stability of the surfactant-β-CD solution that occurs [2,7,15,26].

3.3. Test of Formation Catechin Inclusion Complex with β-Cyclodextrin (β-CD)

3.3.1. Morphology and structure of nanoencapsulation particle of catechin. The results of the nanoencapsulation morphology test using SEM showed that catechin had an irregular form of crystal particles with large sizes (figure 1). NKB samples had irregular morphology with a wrinkled surface and empty capsules. This might happen during the release of active compounds while the solvent evaporated.

NKP samples showed a shape of surface which is spherical, dense and homogeneous. Meanwhile, the NKT samples had a rounded, solid, and a bit rough surface. It also showed a round structure that had a cavity or which is empty. The NKT cavity structure was allegedly occurred due to the influence of heat during the process of solvent evaporation.

Figure 1. Morphology of nanoencapsulation catechin NKB, NKP and NKT result SEM photograph (1500X).

Complex morphology on the sample NKB could be attributed to the crimped rate of solvent evaporation during the drying process to generate heat spray dryer. This caused active compounds that were not resistant to heat disappeared or separated from the capsules. The samples with crimped morphology were possibly caused by the rapid evaporation of the points of fluid during the drying process using a spray dryer. The releasing of active compounds could be caused by large size particles which are trapped outside of the β-CD. The heat in the evaporation from the solvent could cause the re-releasing of active compounds of β-CD because the heat in the evaporation from solvent could occur re-releasing of the active compounds of β-CD [7,13].

3.3.2. Characteristics of stability of nanoencapsulate catechin. Evaluation of complex formation using XRD showed the diffraction pattern of the material of catechin powder (K sample) showed an amorphous structure with a low intensity of less crystal compared to that of NKB, NKP, and NKT samples. This characteristic was organic compounds and amorphous that showed the structure of unstable compounds. The lower the appeared top the intensity indicated the lower of its stability.

The diffraction pattern of NKB, NKP and NKT samples showed the process of complex formation which is seen from the form top (figure 2). The diffraction pattern of the NKB sample had the highest top intensity compared to that of NKT and NKB. However, the diffraction pattern of NKT and NKB was not significantly different in the crystal size. NKT sample gave a new climax at 2Θ 37.8, and NKP sample at 2Θ 37.9°. The new climax maybe came from each surfactant that was used.

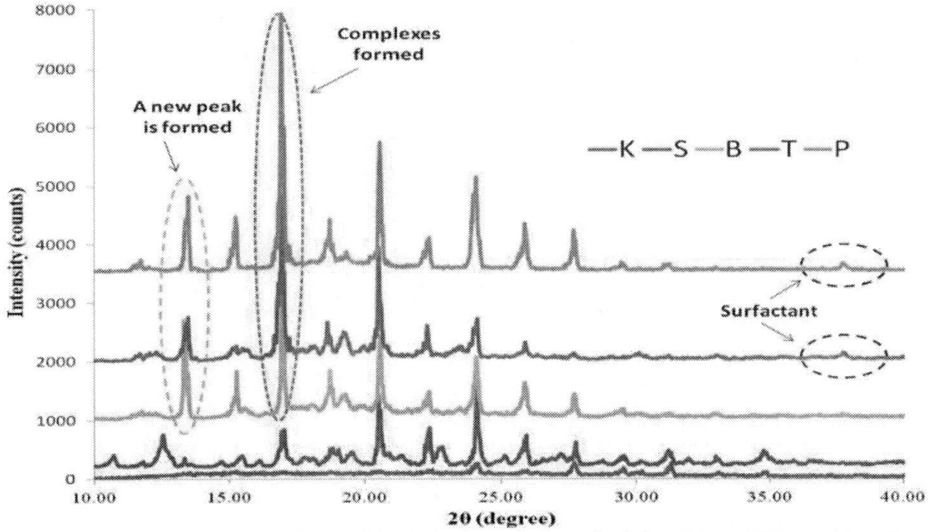

Figure 2. XRD patterns of catechin (K), β-cyclodextrin (S), NKB, NKP, and NKT.

Complex formation using XRD was done by observing the changing of the top particle of the guest molecule (active compounds). The appeared top in a complex can identify the characteristics of the sample and transition phase from a material as amorphous or crystal. The indication of complexation was a top with a narrow and sharp top. The higher the top the more complex the crystal. Low crystal was showed by imperfect complex and amorphous.

A new peak was found at XRD pattern, where the new climax maybe came from each surfactant that was used. It means there was an overlapping among the diffraction patterns of physical compounds, from active compounds and β-CD, with diffraction patterns that had lower climax intensity [11,19,27].

The results showed that the complex inclusion formation of catechin β-CD with surfactant could produce particle structure became more crystal. Nanoencapsulate catechin which was treated with NKP (table 4) produced crystal with a crystallinity degree in the highest degree and it indicated high stability as well.

Table 4. Crystallinity degree of raw materials and nanoencapsulation catechin.

Treatment	Number of total area (2Θ, 0° to 60°)	The area total (2Θ, 0° to 60°)	The degree of crystallinity (%)
Catechin (K)	41259	27342	66.3
NKB	81852	63161	77.2
NKP	93720	86957	92.4
NKT	79555	66609	83.7

3.3.3. Eficiency of catechin nanoencapsulate encapsulation. The results of EE analysis of catechin nanoencapsulate based on treatment variation showed that EE value from NKP sample was higher than that of the NKT and NKB, with treatment average values 91.9%, 89.5%, and 77.4% (figure 3). NKB sample had the lowest EE value, which may be due to the large particle size of catechin so that absorption of catechin at β-CD became lower too.

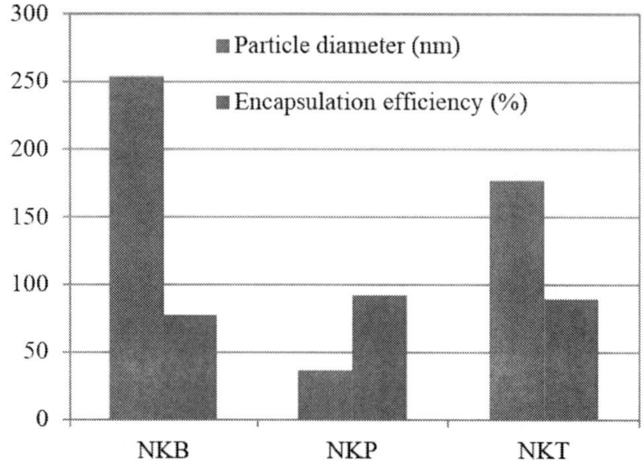

Figure 3. Effect of particle size to encapsulation efficiency of catechin at β-cyclodextrin.

The particle size at catechin nanoencapsulate formation had opposite relation with EE. Samples with the highest EE value had a smaller particle size and can be more encapsulated at β-CD (NKP). In contrast, NKB samples with large diameter made a total catechin particle that could be included at β-CD became lower. The small number of catechin particle that got inclusion was showed by a separation between two phases in complex solution after the homogenization process. Sedimentation phase was predicted to contain incomplex catechin.

Analysis of EE of catechin nanoencpsulation was conducted to see the total amount of absobable catechin or those that were found inside the capsule. EE value of catechin nanoencapsulate directly affected the trapped catechin in β-CD got protection from environment influence (temperature, light). Particle size affected the EE of active compounds at β-CD. The smaller the particle size, the higher the EE value produced. This is because the small size of catechin particles can be included perfectly to β-CD. Particle size affected complex formation with β-CD. If guest molecules that were included were single molecules then the formed complex can change the physical characteristics and chemical molecules with higher solubility and stability.

The EE value can be influenced by the particle size of the active compound on β-CD. The smaller the particle size, the higher the EE value produced. This is because the small size of catechin particles can be included perfectly to β-CD. Particle size affected complex formation with β-CD. If guest molecules that were included were single molecules then the formed complex can change the physical characteristics and chemical molecules with higher solubility and stability [7,9,13].

Value of EE can be affected by the heat produced during the drying process and can result in a broken capsule. It was predicted that catechin were released from the capsule as it was seen from the morphology of NKT sample after SEM test. A temperature produced by the drying process can cause the loss or release of active compounds. This can be caused by the trapped particles outside of β-CD. The effect of the drying process is the release of active compounds which cause EE values lower [30,31].

4. Conclusion

Polaksamer 188 and Tween 80 surfactants functioned to form emulsions and reduce the particle size of catechin. Polaksamer 188 surfactants at a concentration of 1% produced the smallest and uniform particle size. Complex catechin nanoencapsulate inclusion of nanoemulsion with Poloxamer (NKP) at β-CD produced a stable complex solution with a particle size of 37 nm (IP 0.28). Catechin had the highest nanoencapsulate efficiency value (91.9%) comparing using Tween (NKT) (89.5%) and without using surfactant (NKB) (77.4%). Complex inclusion formation using Poloxamer surfactant also produced nanoencapsulate, with the highest crystallinity degree (92.4%), and indicated the highest stability.

Acknowledgments
The authors are grateful to Mr. Hendri Muchtar, Tita Aviana dan Riv'atul Aliyah for their assistance and contributions towards the successful completion of this research.

References

[1] Rauf A, Rahmawaty and A Z Siregar 2015 *Procedia Chem.* **14** 3–10
[2] Yeni G, Syamsu K, Suparno O, Mardliyati E and H Muchtar. 2014, *Int. J. Appl. Eng. Res.* **9(24)** 24565–78.
[3] Esfanjani AF, Assadpour E and S M Jafari 2018 *Trends Food Sci. Technol.* **76** 56–66
[4] Munin A and F Edwards-Lévy 2011 A *review* **3(4)**
[5] Ullah S and C G Hyun 2020 *Indones. J. Chem* **20(3)** 716-21
[6] Rezaei A, Fathi M and S M Jafari 2019 Elsevier B.V **88**
[7] Ho S, Thoo Y Y, Young D J and L F Siow 2017 *LWT - Food Sci. Technol.* **85** 232–239
[8] Fitriana W D, Ersam T, Shimizu K and S Fatmawati 2018 *Orient. Pharm. Exp. Med.* **18(4)** 299–
307.
[9] Dasgupta N, Ranjan S, Mundra S, Ramalingam C and A Kumar 2016 *Int. J. Food Prop.* **19(3)** 700–8
[10] Ko W C, Chang C K, Wang H J, Wang S J and C W Hsieh 2015 *Food Chem* **172** 497–503
[11] Lestari A D N, Siswanta D, Martien R and Mudasir M 2020 *Indones. J. Chem.* **20(4)** 929
[12] Ezhilarasi P N, Karthik P, Chhanwal N, and C Anandharamakrishnan 2013 *Food Bioprocess Technol.* **6(3)** 628–647.
[13] Gharibzahedi S M T and S M Jafari 2017 Elsevier Inc.
[14] Ho S, Thoo Y Y, Young D J, and L F Siow 2013 *Lwt-Food Sci. Technlogy* **86** 555–565
[15] Zhou C, Cheng X, Yan Y, Wang J and J Huang 2013 *Langmuir* **29** 13175–13182
[16] Zu Y, Li N, Zhao X, Li Y, Ge Y, Wang W, Wang K and Liu Y 2014 *Int. J. Pharm.* **471(1–2)** 366–376.
[17] Yang Y and McClements D J 2013 *Food Hydrocoll.* **30(2)** 712–720
[18] Tojo C, de Dios M and F Barroso 2011 *Materials* (Basel) 55–72
[19] Takahashi A I, José F, Veiga B and H G Ferraz 2012 *Int. J. Pharm. Sci. Rev. Res.*, **12(1)** 8–15
[20] Abbas S, Hayat K, Karangwa E, Bashari M and X Zhang 2013 *Food Eng. Rev.* **5(3)** 139–157
[21] Krishnaswamy K, Orsat V and K Thangavel 2012 *J. Food Eng.* **111(2)** 255–264
[22] Chowdary K P R, Rao K S P and D Madhuri 2011 *Int. J. Chem. Sci.* **9(2)** 677–686
[23] Krishnaswamy K, Orsat V and Thangavel K 2012 *J. Food Eng* **111(2)** 255–264
[24] Danhier F, Magotteaux N,Ucakar B, Lecouturier N, Brewster M and V Préat 2009 *Eur. J. Pharm. Biopharm.* **73(2)** 230–238
[25] Paximada P, Echegoyen Y, Koutinas A A, Mandala I G, and J M Lagaron 2017 *Food Hydrocoll.* **64** 123–132
[26] Surassmo S, Min S G, Bejrapha P and M J Choi 2010 *Food Res. Int.* **43(1)** 8–17
[27] Yang L J, Chen W, Ma S X, Gao Y T, Huang R, Yan S J and J Lin J 2011 *Carbohydr Polym.* **85(3)** 629–637
[28] McClements D J and S M Jafari 2018 *Adv. Colloid Interface Sci.* **251** 55–79
[29] Taurozzi J S, Hackley V A, and M R Wiesner 2011 *Nanotoxicology* **5(4)** 711–729
[30] Katouzian I and S M Jafari 2016 *Trends Food Sci. Technol.* **53** 34–48
[31] Tshweu L, Katata L, Kalombo L and H Swai 2013 *J. Nanoparticle Res.* **15(2040)** 1–10

IOP Conf. Series: Materials Science and Engineering **980** (2020) 012049 doi:10.1088/1757-899X/980/1/012049

Optimization of linear transducer calibration system using laser interferometer based on the Abbe principle

A Rahman*, E Pratiwi, N Alfiyati, O Novyanto, and O Hedrony
National Standardization Agency of Indonesia (BSN), Kompleks Puspiptek Building 420, South Tangerang, Banten, 15314, Indonesia

Email: ardi.rahman@bsn.go.id

Abstract. The validity of the measurement results can be guaranteed by performing calibration using a measurement standard traced to the international system of units (SI). The linear transducer is a measurement standard that is used to calibrate calibration tester, dial gauge tester, and gauge block in the calibration and testing laboratories. The critical position of the linear transducer in the traceability chain leads to the need for an optimal system to calibrate it with the smallest error and highest accuracy. As an optimization step, this research was carried out to improve linear transducer calibration performance using a laser interferometer combined with a universal length machine (ULM). The laser interferometer, ULM, and linear transducer were aligned with the Abbe and cosine principles. The results show a significant reduction of the linear transducer measurement error in the forward and backward directions by 548% and 119%, respectively. The measurement uncertainty could also be decreased by 8% to 0.18 μm.

1. Introduction

The measurement process requires four components: measuring objects, measuring instruments, measurement agents, and conditioned environment. Imperfections of the four components can cause errors so that no measurement process can provide absolute accuracy [1]. In the manufacturing, food, beverage, chemical, and cosmetics industrial processes, dimensional quantities are among the four most widely used quantities. Therefore, the equipment for dimensional quantities becomes one of the types of equipment that need to be given priority in its maintenance. Calibration plays an essential role in controlling the quality of measuring instruments used in daily life. Calibration provides knowledge about measuring instrument errors so that the reading can be corrected, along with the measurement uncertainty. As mentioned, calibrated measuring instruments can guarantee the validity of quality control, such as in the manufacturing process, the production process of agricultural and forestry tools, and measuring noise and environmental pollution [2–4].

One of the industrial and laboratory calibration measuring equipment that must be calibrated, especially in length metrology, is a linear transducer, better known as a linear variable displacement transducer (LVDT). Measuring equipment and standards such as calibration tester, inclination table, electronic level, dial gauge tester, roughness, autocollimator, and gauge block are calibrated using LVDT [5–7]. Linear transducers are even used for routine tests of asphalt mixtures, quality control of bridges with vibration measurements, fuel control at power plant facilities, and an online adaptive calibration system [8–10].

A laser interferometer is widely applied as a precision measuring instrument. Based on the Michelson interferometer, the laser is used as a measurement to reveal the phenomenal gravitational wave [11]. Besides, the laser interferometer is also used for measuring shifts with high accuracy [12,13]. This article describes a laser interferometer as a measurement standard to calibrate linear transducer / LVDT. In the process, cosine error was reduced with proper optical alignment. Also, the Abbe principle for reducing sinus errors was applied.

2. Materials and methods

Inside the linear transducer, there is a periodic grating scale made of optics. Figure 1(a) shows the multilevel scale shape contained in the linear transducer. The LVDT position information was obtained by calculating the number of steps passed relative to their initial position. It worked based on the principle of interferential scanning, which would determine the position information. Figure 1(b) shows the principle of interfering scanning principle where it utilized diffraction and interference from light that was stratified to produce a signal to determine the position/distance measured—a light source (see figure 1(b) for scanning a silent reticle [14,15]. Reflections from multilevel scales on measuring standards would cause diffraction and light interference, converted by photocells into electronic signals.

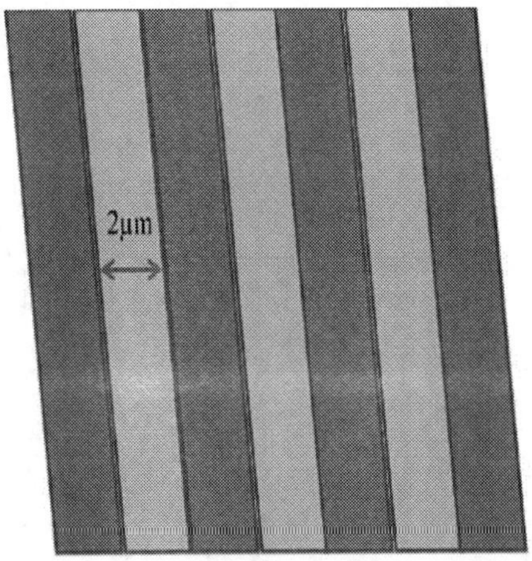

(a)

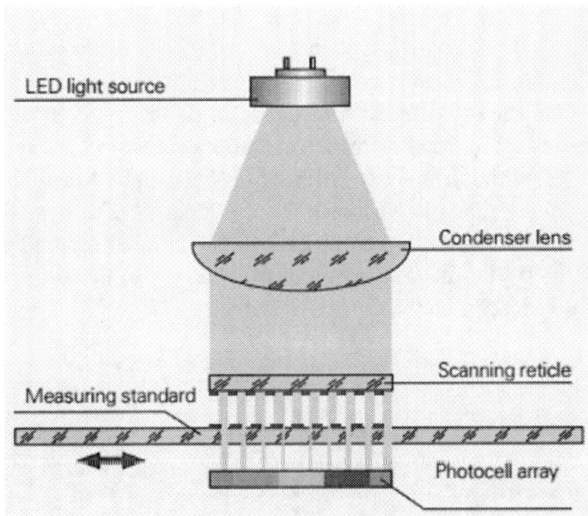

(b)

Figure 1. (a) Grating scale inside the Linear transducer; (b). Sensing principle of LVDT [15].

Laser interferometer, which was to calibrate the linear transducer, worked with two types of polarized wavelengths and had a very close frequency difference between 1.5 to 2.0 MHz [16]. A beam splitter separated the two rays, and the first beam was directed towards the target while the second beam was deflected into a retroreflector. The second beam was reflected by the retroreflector and was forwarded to the counter located inside the laser housing. The target reflected the first ray to the counter, so the difference in the length of the two rays' path can be recorded (see figure 2). In the linear transducer calibration process, cosine error could occur due to the incompleteness of measuring lines and dimension lines. This error would cause the measurement data to be larger than the actual dimensions [17].

In measurements using a laser interferometer, cosine error and Abbe error can occur. The cosine error occurs when the moving axis and laser beam axis are not parallel in practical. The cosine error is caused by the inconsistency of the measuring line and the dimension line so that it forms an angle θ, as shown in figure 2. This error causes the measurement data to be larger than the actual dimension. Reducing cosine error is done by perfecting optical alignment, where the beam splitter and target must be placed in a straight line so that the incoming beam from the laser will be reverse parallel to the counter. The value of the deviation from the dimension line getting smaller is the indicator of the decreasing of cosine error. The other error, related to angle is the sine error or Abbe error [18]. It occurs when the axis of measurement (target/retroreflector) is offset from the scale axis. The angular motion

IOP Conf. Series: Materials Science and Engineering **980** (2020) 012049 doi:10.1088/1757-899X/980/1/012049

error, e.g., pitch and yaw, causes the measurement error due to displacement. The sine error causes the measured displacement to appear longer or shorter than the actual displacement, depending on angular motion. Usually, the angular motion error of the mechanical stage can be within 20 to 30 arcsec. The Abbe error does not depend on the displacement value. Making the axis of measurement in the same axis as the scale will neglect the Abbe error.

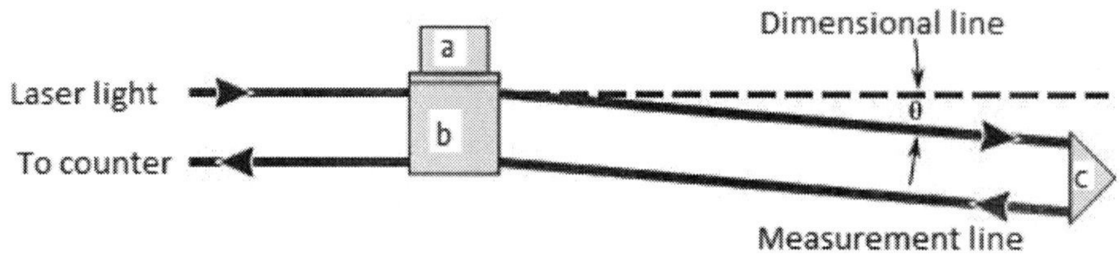

Figure 2. Cosine error in the alignment of laser interferometer optical components [17], where: (a) retroreflector; (b) beam splitter; (c) target.

The research equipment that was used in the study is shown in figure 3. Heidenhain has made a 25 mm length of a linear transducer with the type of MT 2501. It was calibrated at the SNSU-BSN Laboratory in South Tangerang, Indonesia, using a laser interferometer made by Agilent with a type of 5519B, which had a resolution of 0.01 μm. The linear transducer was associated with a universal length measuring machine system. The universal length measuring machine shaft of motion was placed on a laser interferometer retroreflector. The transducer and shaft of motion's linear motion would cause a change in the distance of the retroreflector to the stationary beam splitter. This change in the distance would be seen on the computer screen.

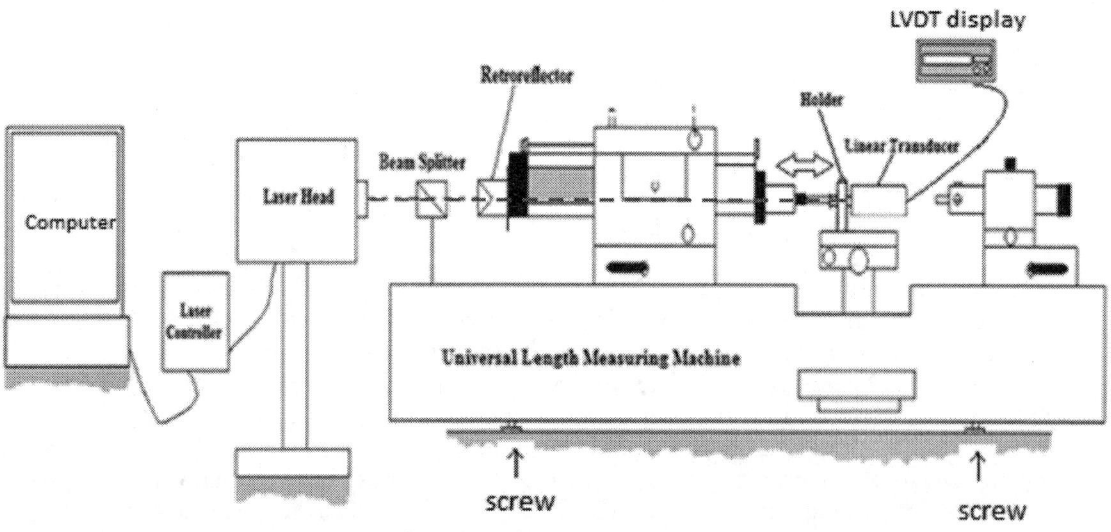

Figure 3. The configuration of linear transducer calibration using the laser interferometer.

A level indicator made by Digi-fitting type DWL 280 with a resolution of 0.01° helped set the slope of the universal table measuring machine. The tilted universal table measuring machine, the beam splitter's pedestal, caused the laser beam to not align with the linear transducer movement. In this research, the slope of the universal measuring machine was adjusted by turning the two screws located at the bottom of the universal length measuring machine (see figure 3) so that the appointment of level indicators at both ends of the measuring machine showed the same value. As a result, the beam splitter position would be parallel to the target (retroreflector). It was indicated by the beam returning to the counter on the laser head was correctly received / not deviated, or it can be said that angle θ approaches zero (see figure 2). Finally, the cosine error would be reduced significantly.

Several publications mentioned some works in designing the system based on the Abbe principle [19–22]. The same principle was applied to the research scheme, as shown in figure 3, the dashed red line shows that the axis of measurement (laser head-beam splitter-retroreflector) and the scale (linear transducer) were on the same axis. It can be said that the configuration in the research scheme had followed the Abbe principle. In the sequel, Abbe error, or other terms, sine error can be ignored.

The laser interferometer used in this study was calibrated by NMIA, Australia so that the laser displacement readings had been traced to the international system of units (SI). The calibration process was carried out by comparing the linear transducer and laser interferometer readings to produce a correction value. The universal length machine's rod moved to press the linear transducer so that the linear transducer could move forward and backward. On the same hand, the retroreflector also moved forward or backward following the universal length machine's rod movement. In this research, the percentage of cosine error decreases were calculated by comparing the forward and backward corrections of the experimental results with the previous linear transducer calibration data. It was done to see how optimal the effort had been made. The previous linear transducer calibration was done by measurement of the linear transducer without using a laser interferometer. On the contrary, the laser interferometer acted as a calibration standard in the experiments.

3. Results and discussion

The correction of forward and backward directions for previous calibration and the experiment results are shown in figure 4 and figure 5, respectively. In general, forward and backward corrections for previous calibration data had a sharper slope compared to the results of experiments that had been carried out. The vast range of correction data in the previous calibration was indicated by the vast range of correction data, namely 1.66 μm in the forward direction and 0.76 μm in the backward direction. It was very contradictory compared to the range of correction data in the experiment, namely 0.62 μm in the forward direction and 0.07 μm in the backward direction. Some references [17,23,24] explain that this phenomenon shows that there had been a considerable cosine error in the previous calibration data. The Experiments that had been done show that the laser interferometer's optimal optical alignment could significantly reduce cosine errors. Note that the zero reading of the linear transducer was the starting point of the calibration process so that the forward correction would be zero.

Another interesting thing was that both forward and backward directions at the linear transducer readings above 17 mm indicated a nearly horizontal curve. Correction values tend to be constant. If this constancy was around zero correction value, probably this indicated a better thing. However, based on figure 4 and figure 5, this constancy was at a correction value above 1.2 μm.

Table 1 shows the error calculation decreased at each measurement point. The percentage of error gives the idea of how significant the laser interferometer's optical alignment of the experiments is. The maximum correction value of the previous calibration data was 1.34 μm and 1.82 μm, respectively, in the forward and backward directions, while in this experiment was 0.57 μm and 0.73 μm, respectively, in the forward and backward directions. The maximum values in the forward and backward directions for both methods (previous and experimental data) were located at the nearly same point around the linear transducer reading of more significant than 20 mm.

There was a more than ten times decrease in cosine error in the forward direction as the result of experiments. It was achieved at a 2 mm linear transducer reading. Figure 4 at the 2 mm point also reinforced this fact, where the previous calibration data correction reached the highest positive correction point. On the other hand, the 2 mm point correction in this experiment was close to zero. In contrast, a decrease in cosine error in the backward direction of 169% occurred in the linear transducer reading of 23 mm. Figure 5 shows that at the 23 mm linear calibration reading, the calibration correction value of the previous data was in the lowest valley while in the experimental curve was uphill.

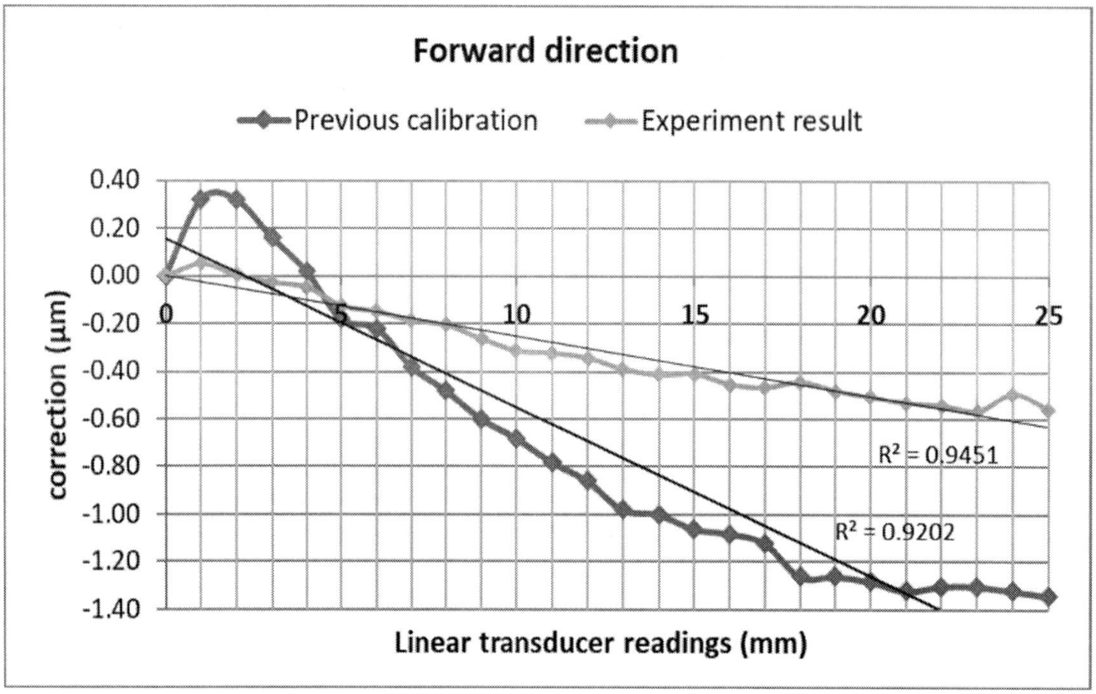

Figure 4. The comparison of previous calibration data and experiment results in the forward direction.

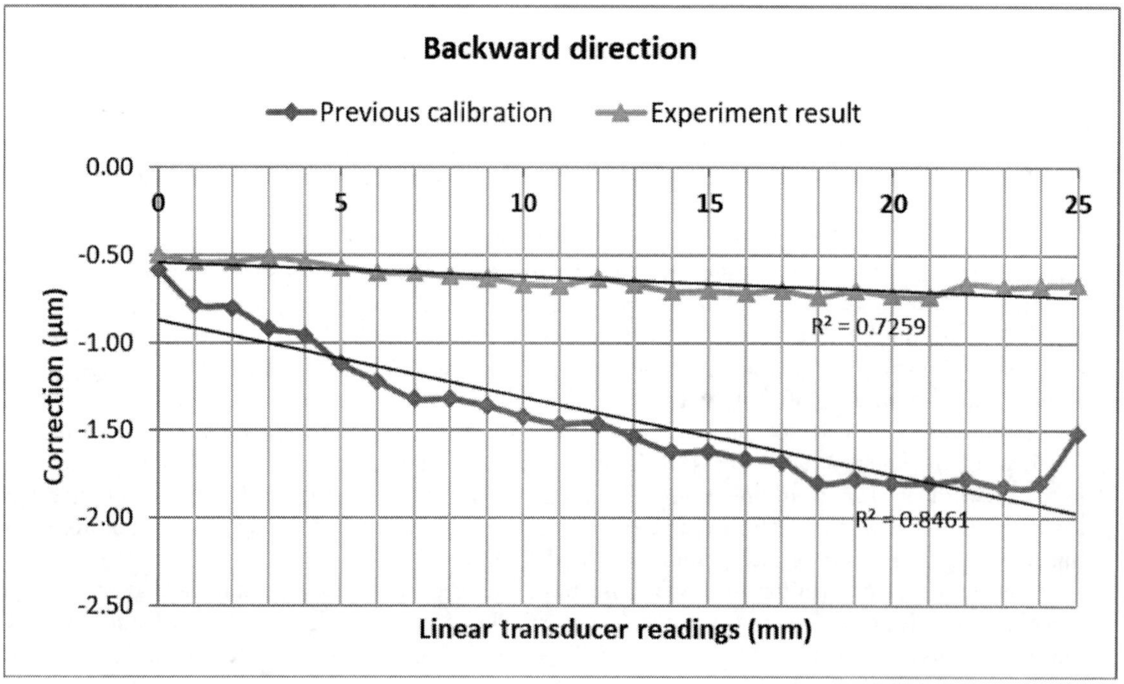

Figure 5. The comparison of previous calibration data and experiment results in the backward direction.

Table 1. The decreased of error at each measurement point.

Linear transducer readings (mm)	Experiment results		Previous calibration		Cosine error decrease (%)	
	Forward correction (μm)	Backward correction (μm)	Forward correction (μm)	Backward correction (μm)	Forward	Backward
0	0.00	-0.49	0.00	-0.58	0	17
1	0.05	-0.54	0.32	-0.78	495	46
2	0.00	-0.53	0.32	-0.80	10064	50
3	-0.03	-0.51	0.16	-0.92	710	81
4	-0.05	-0.53	0.02	-0.96	143	80
5	-0.12	-0.56	-0.18	-1.12	46	98
6	-0.15	-0.59	-0.22	-1.22	48	105
7	-0.19	-0.60	-0.38	-1.32	104	121
8	-0.21	-0.62	-0.48	-1.32	134	114
9	-0.26	-0.63	-0.60	-1.36	127	116
10	-0.31	-0.66	-0.68	-1.42	117	115
11	-0.32	-0.67	-0.78	-1.46	143	119
12	-0.34	-0.63	-0.86	-1.46	151	132
13	-0.39	-0.66	-0.98	-1.54	152	133
14	-0.41	-0.70	-1.00	-1.62	142	131
15	-0.41	-0.70	-1.06	-1.62	159	131
16	-0.45	-0.71	-1.08	-1.66	138	134
17	-0.47	-0.70	-1.12	-1.68	141	141
18	-0.44	-0.73	-1.26	-1.80	184	146
19	-0.48	-0.70	-1.26	-1.78	163	155
20	-0.51	-0.73	-1.28	-1.80	153	147
21	-0.53	-0.73	-1.32	-1.80	149	147
22	-0.54	-0.66	-1.30	-1.78	140	169
23	-0.57	-0.68	-1.30	-1.82	130	169
24	-0.49	-0.67	-1.32	-1.80	168	168
25	-0.56	-0.67	-1.34	-1.52	140	128
		Average (%)			548	119

Table 1 also shows the average percentage decrease in cosine error at forward and backward directions were 548% and 119%. As mentioned before, the maximum correction value after this research was smaller than 0.80 μm, a significant decrease compared to previous calibration data, the maximum correction value reaching 1.80 μm.

Table 2 shows eight sources of uncertainty. The measurement uncertainty analysis was based on [25]. The readability of the laser interferometer came from the resolution of the laser interferometer itself. The calibration certificate of laser interferometer from NMIA Australia contributed to two sources of uncertainty: scale and air compensation errors and the laser compensation wavelength. Abbe error still appeared in the budget due to the imperfect axis alignment that could still occur. Linear transducer contributed to two sources: readability (from the resolution of LVDT) and repeatability (gained by repetitive measurement on the same point of reading). Temperature difference came from the variation of temperature in the laser interferometer system and linear transducer. The ultimate source, geometric and periodic error, came from the linear transducer sensing system (see figure 1). At some points, it drifts the reading so that it will cause measurement bias.

Table 2. Linear transducer calibration by using a laser interferometer uncertainty budget.

No.	Uncertainty sources	Distribution type	Standard Uncertainty (μm)
1	Readability of laser interferometer	Rectangular	0.0058
2	Scale error and air compensation error (from laser calibration certificate)	Normal	0.0402
3	The wavelength of laser compensation	Normal	0.0000
4	Abbe error	Rectangular	0.0087
5	Readability of linear transducer	Rectangular	0.0058
6	Repeatability of linear transducer	Type-A	0.0095
7	Temperature difference	Rectangular	0.0035
8	Geometric and periodic errors	Rectangular	0.0808
	Combined uncertainty (μm)		0.09
	Expanded uncertainty (μm), k = 2		0.18

The most significant source of uncertainty was obtained from geometric and periodic errors. The second significant source was the scale and air compensation errors from the estimated influence of temperature, humidity, and air pressure when measuring. Finally, the expanded uncertainty of the calibration system of 0.18 μm was obtained with a confidential interval of 95%. Measurement uncertainty of the previous calibration was 0.19 μm. It can be said that the uncertainty was decreased up to 8%.

4. Conclusion

A calibration system of the linear transducer using a traceable laser interferometer has been realized concerning the Abbe principle. It showed a significant reduction of the linear transducer measurement error in the forward and backward directions by 548% and 119%, respectively. The measurement uncertainty could also be decreased by 8% to 0.18 μm. This article is expected to be useful for metrologists and industries directly related to the application of dimensional measurement. As a suggestion, it is necessary to develop a new system that can vertically calibrate linear transducers to eliminate the geometric effects due to gravity.

Acknowledgments

The national standardization agency of Indonesia (BSN) supported this research. The authors thank the SNSU Length BSN staff for the fruitful discussion and suggestion for this research.

References

[1] BIPM 1936 *Bureau international des Poids et Mesures* vol 138

[2] Sunoj S, Igathinathane C, Saliendra N, Hendrickson J and Archer D 2018 Color calibration of digital images for agriculture and other applications *ISPRS J. Photogramm. Remote Sens.* **146** 221–34
[3] Patiño-Herrera R, Catarino-Centeno R, González-Alatorre G, Gama Goicochea A and Pérez E 2017 Enhancement of the hydrophobicity of recycled polystyrene films using a spin coating unit *J. Appl. Polym. Sci.* **134** 45365
[4] Trisna B A, Rahman A, Nugraha A R, Darmayanti N T E and Pusaka J 2018 Current situation and prospect of nanometrology and its standardization in Indonesia *Mapan - J. Metrol. Soc. India* **33** 469–80
[5] Peingbangyang S and Wongsaroj J 2017 characteristics of linear variable differential transformer (LVDT) probe for gauge blocks calibration *J. Phys. Conf. Ser.* **901**
[6] Nasreen S, Rokunuzzaman M and Biswas S 2017 assessment of surface roughness using lvdt: a convenient and inexpensive way of measuring surface irregularities *Int. Conf. Mech. Ind. Mater. Eng. 2017* **2017** 6
[7] Novyanto O and Alfiyati N 2016 An approach method to calibrate the autocollimator with small angle measurement range *Mapan - J. Metrol. Soc. India* **31** 9–15
[8] Hamam T, Levenberg E and Zelnik-Manor L 2016 Development of an optical displacement transducer for routine testing of asphalt concrete *J. Mater. Civ. Eng.* **28** 04016066
[9] Skifton R, Schoensee L, Bartz A, Jaques B, Davis K, Calderoni P and Crawford A 2017 In-pile fuel rod deformation measurements using miniaturized LVDT technology
[10] Santhosh K V. and Roy B K 2017 Online implementation of an adaptive calibration technique for displacement measurement using LVDT *Appl. Soft Comput. J.* **53** 19–26
[11] Abbott B P *et al.* 2017 A gravitational-wave standard siren measurement of the Hubble constant *Nature* **551** 85–98
[12] Azaryan N, Budagov J, Gayde J C, Di Girolamo B, Glagolev V, Lyablin M, Mergelkuhl D and Shirkov G 2017 The innovative method of high accuracy interferometric calibration of the Precision Laser Inclinometer *Phys. Part. Nucl. Lett.* **14** 112–22
[13] Byman V, Jaakkola T, Palosuo I and Lassila A 2018 High accuracy step gauge interferometer *Meas. Sci. Technol.* **29**
[14] Halbgewachs C, Kentischer T J, Sändig K, Baumgartner J, Bell A, Fischer A, Funk S, Heidecke F, Scheiffelen T, Schmidt W, Spanner E, Speckbacher P and von der Lühe O F 2016 Qualification of HEIDENHAIN linear encoders for picometer resolution metrology in VTF Etalons *Ground-based Airborne Instrum. Astron. VI* **9908** 99084H
[15] Heidenhain J 2016 *Length Gauges*
[16] Packard H 1988 *5528 A Laser Measurement System: User's Guide* (Santa Clara, CA: HP Co.)
[17] Rahman A 2012 Cosine error reduction on the universal measuring machine calibration using displacement laser interferometer *Widyariset* **15** 657–64
[18] Novyanto O and Alfiyati N 2020 Measurement optimization of high accuracy check master based on interferometry method in National Standardization Agency of Indonesia: A preliminary study for measuring range of 10 mm to 400 mm *J. Phys. Conf. Ser.* **1568**
[19] Wang Z, Wang D, Dong H and Yu S 2019 An invariant method updating Abbe principle for accuracy test and error calibration of rotary pairs in machine tools *Int. J. Mach. Tools Manuf.* **141** 46–58
[20] Liu H, Xiang H, Chen J and Yang R 2018 Measurement and compensation of machine tool geometry error based on Abbe principle *Int. J. Adv. Manuf. Technol.* **98** 2769–74
[21] Ren G, Qu X and Chen X 2020 Performance evaluation and compensation method of trigger probes in measurement based on the Abbé principle *Sensors (Switzerland)* **20** 1–13
[22] Li Y T and Fan K C 2018 A novel method of angular positioning error analysis of rotary stages based on the Abbe principle *Proc. Inst. Mech. Eng. Part B J. Eng. Manuf.* **232** 1885–92
[23] Haitjema H, Schellekens P H J and Wetzels S F C L 2000 Calibration of displacement sensors up to 300 μm with nanometre accuracy and direct traceability to a primary standard of length

IOP Conf. Series: Materials Science and Engineering **980** (2020) 012049 doi:10.1088/1757-899X/980/1/012049

Metrologia **37** 25–33
[24] Dai G, Koenders L, Pohlenz F, Dziomba T and Danzebrink H U 2005 Accurate and traceable calibration of one-dimensional gratings *Meas. Sci. Technol.* **16** 1241–9
[25] Anon 2008 *JCGM 100:2008 GUM 1995 with minor corrections Evaluation: Evaluation of measurement data – Guide to the expression of uncertainty in measurement*

Development of a rapid-test method for the determination of calcium, zinc, phosphorus, and sulfur in automotive engine oil by WD-XRF (wavelength dispersive x-ray fluorescent)

D Cahyadi[1,2]*, E Susilowati[1], M Arsyansyah[1], and S Febriany[1]

[1]Center for Material and Technical Product, Ministry of Industry, Bandung
[2]The Faculty of Mechanical and Aerospace Engineering, Bandung Institute of Technology

Email: denb4t@gmail.com

Abstract. This study describes a rapid method for the determination of calcium, zinc, phosphorus, and sulfur in lubricating oil products using WD-XRF (Wavelength Dispersive X-ray Fluorescent). Currently, there are several normative references appointed by the mandatory SNI (Indonesian standard) used for elemental analysis in automotive engine oils, i.e., ICP-AES (Inductively Coupled Plasma - Atomic Emission Spectroscopy), AAS (Atomic Absorption Spectrophotometry) for calcium, zinc, or magnesium; ED-XRF (Energy Dispersive X-ray Fluorescent) for sulfur, lamp method for sulfur, and quinoline phosphomolybdate method for phosphorus. The wet chemical analysis principle used in those reference methods (AAS, ICP-AES, and conventional method) needs a sample preparation step before the main instrument can use to measure an oil sample. It is a time-consuming process and commonly consists of a few steps such as dilution, weighing, heating, and destruction. The statistical characteristics (limit of detection, the limit of quantification, precision, accuracy, and linearity) obtained in this experiment were relatively acceptable. Elemental analysis by WD-XRF needs no sample preparation process, fast, less chemical reagents, multi-elements, simultaneously, and simple. The WD-XRF method potentially could be used as an alternative method for the determination of certain elements (Ca, Zn, P, S) or more as specified in the mandatory SNI for automotive engine oil products.

1. Introduction

Lubrication is one of the critical factors in maintaining engine performance in optimum conditions. The lubricating oil can protect an engine from degradation factors such as wear, heat, corrosion, oxidation, and various contaminant [1]. The quality of oil used in engine parts will affect fuel consumption and exhaust gas. Optimum lubrication can reduce gas emission produced by a combustion engine [2]. Good lubricating oil could preserve engine performance, cost efficiency, less air pollution, and increase safety.

The number of motor vehicles that continues to grow every year will boost the need for lubricants in Indonesia. According to BPS (Statistics Indonesia) data, in 2018 the total number of four types of motorized vehicles (passenger cars, bus cars, freight cars, and motorbikes) reached 146,858,759 vehicles [3]. The need for lubricants for such large numbers of motor vehicles needs to be balanced by reliable lubricant quality.

On September 5, 2018, the Indonesian Ministry of Industry published "Permenperin No. 25 (2018), tentang Pemberlakuan Standar Nasional Indonesia Secara Wajib" which set the mandatory SNI

(Standar Nasional Indonesia) regulation for seven types of automotive engine oil products. Products included in this regulation are: (1) four-stroke gasoline engine oil (SNI 7069.1:2012), (2) four-stroke gasoline engine oil for motorcycle (SNI 7069.2:2012), (3) two-stroke gasoline engine oil with air cooler (SNI 7069.3:2016), (4) two-stroke gasoline engine oil with water cooler (SNI 7069.4:2017), (5) Diesel engine oil high rotation (SNI 7069.5:2012), (6) manual transmission gear oil (SNI 7069.6:2017), and (7) automatic transmission oil (SNI 7069.7:2017). This regulation has been applied effectively since September 10, 2019[3].

The specific element's content is included in the mandatory SNI specifications. The elements that should be determined in lubricating oil products are calcium, magnesium, zinc, phosphorus, and sulfur. According to the SNI, each element could be analyzed by a few different methods. Table 1 shows the reference methods used in the SNI. Metals content in lubricating oil provided valuable information about its composition, sedimentation of metal parts, or degradation of additive in an engine [4]. Elements contained in automotive lubricating oils were found as impurities or additives. Calcium and zinc in automotive lubricating oil were used as a detergent additive ingredient and anti-corrosion agent [5]. Phosphorus in engine oil was seen as anti-wear/antioxidant additive that, during operation, could be volatilized [6]. Sulfur in engine oil was found as mineral base oil impurities [7]. Heavy metals (Al, B, Cr, Cu, Fe, Pb, Mn, Mo, Ni, Na, Si, Ag, Sn, Ti, V, Cd, and As) in used oil can provide valuable information about wear metals and contaminants [8].

Table 1. Testing method

SNI Specification	ASTM testing method				
	Calcium	Zinc	Phosphorus	Sulfur	Magnesium
7069.1:2012	D4628/ D5185	NS*	D4047/ D4951	D2622/ D4951	D4628/ D5185
7069.2:2012	D4628/ D5185	D4628/ D5185	D4047/ D4951	NS	D4628/ D5185
7069.3:2016	D4628/ D5185	NS	NS	NS	NS
7069.4:2017	D4628/ D5185	NS	NS	NS	NS
7069.5:2012	D4628/ D5185	D4628/ D5185	D4047/ D4951	D2622/ D4951	D4628/ D5185
7069.6:2017	NS	NS	D4047	D1266	NS
7069.7:2017	D4628/D5185	D4628/D5185	D4047	D4294/ D6481	NS

**NS: Not Specified*

The Determination methods of elements content in lubricating oil according to ASTM standards are as follows: (a) Calcium, magnesium, and zinc; tested by AAS (Atomic Absorption Spectrophotometry) [9] or ICP-AES (Inductively Coupled Plasma – Atomic Emission Spectroscopy) [10]; (b) Phosphorus tested by quinoline phosphomolybdate method (titration) [11] or ICP-AES [12]; and (c) Sulfur tested by lamp-method [13], WD-XRF [14], ICP-AES [12] or ED-XRF [15,16]. The wet chemical analysis technique was used in AAS, ICP-AES, and titration.

Elemental analysis in lubricating oil product using AAS method has a good detection limit [17]. Still, it needs a time-consuming sample preparation process, and cannot be determined multi-elements simultaneously. For ICP-AES, although it has a very low detection limit and could measure multi-elements at once [8], it needs some expensive consumable supporting materials and precise sample preparation process. Titration method for phosphorus determination by quinoline phosphomolybdate is a low-cost method (no need an expensive instrument like AAS, ICP-AES, nor XRF), but it needs a complicated preparation process and more chemicals.

Among the other available analytical techniques, XRF spectrometry potentially can be used as a single method for determing all of the specific elements required by the mandatory SNI specification simultaneously. WD-XRF (Wavelength Diffraction X-Ray Fluorescence) method has developed for the analysis of Al, Ba, Ca, Cr, Cu, Fe, Mg, Mn, Ni, P, Pb, S, Si, Ti, V, and Zn in various petroleum liquid products with no significant loss of accuracy[18]. The WD-XRF method offer advantages by its minimum preparation, fast, and has an excellent detection limit compared to other elemental analysis techniques.

This study focuses on determining the concentration of selected elements (Ca, Zn, P, and S) in automotive engine oils by using the WD-XRF method for described a simple and fast testing method. Statistical parameters for method validation is also presented [19].

2. Materials and methods

2.1. Materials

ASI (Analytical Services Incorporation) LOE17 was used as Oil Reference Material. Commercial, automotive engine oil that available in markets in Bandung region was used as a lubricating oil sample. Plastic X-ray cell of 25 mm diameter and a plastic film chemically resistance (Mylar®, with 6 μm thickness) was used as a base holder material.

2.2. Instrument

ASI (Analytical Services Incorporation) LOE17 was used as Oil Reference Material. Commercial, automotive engine oil that available in markets in Bandung region was used as a lubricating oil sample. Plastic X-ray cell of 25 mm diameter and a plastic film chemically resistance (Mylar®, with 6 μm thickness) was used as a base holder material.

2.3. Procedures

About 50 mL standard or sample was added to a cell cup then placed in an autosampler holder. The measurement of metals content was performed by WD-XRF Analyzer and controlled by PC. The calibration standards were prepared from CRMs (Certified Reference Materials) diluted by white oil; at least five standards solution were measured to obtain the calibrations curves. The measurement of samples was carried out after the calibration curves were made. Every sample was measured at least three times (triple) to have a representative result. The Limit of Detection (LD) and the Limit of Quantification (LQ) value was determined by measuring a new solution (white oil) or spiked sample for five times. Measured CRM determined the Accuracy and Precision for five times, then compared the value obtained from the measurement with the actual number stated in the certificate. These experiments were evaluated using CRM as a control to ensure the accuracy of the analysis.

3. Result and discussion

3.1. Time of analysis

A simple procedure was applied in this experiment. Lubricating oil sample from the original sealed bottle poured into a cell cup up to volume marker (about 50 ml), then it placed in autosampler holder. The measurement proceeds directly after the cell cup placed in the autosampler holder. The whole process for conducting elemental analysis for a single sample took about 3-5 minutes or less than 5 minutes.

The common elemental analysis method consisted of three necessary steps: sample preparation, measurement, and calculation. The modern analytical instrument commonly had equipped by software that could be used for calculating the instrument data to generate the measurement result, so the count took place during the measurement process. The main issues in the analysis duration were the sample preparation process. The wet chemical sample preparation commonly needed a few procedures, such as weighting, dilutions, and digestion. This process could become an uncertainty contributor and affected the accuracy of the method. The time needed during a sample preparation process (i.e. weighthing, dissolved, making standard solution, and dilution) was varied, depended on the sample characteristics (i.e. solubility, and suspended solid) and instruments (i.e gas type, light source, and compatibility) used. A fast analytical method or rapid-test would have a lot of advantages if applied in-field inspection and quality control processes in the factory. Less chemical means more contribution to green industrial production. Time efficiency would be to reduce the cost of analysis and human resources.

3.2. Limit of detection and limit of quantification

Limit of Detection (LD) or Limit of Quantification (LQ) was commonly defined as the lowest concentration of an analyte that can be detected. Several approaches can be used to estimate LD and LQ. i.e., visual assessment, signal to noise ratio, calculated from the standard deviation of the blank, calculated from the calibration line at low concentration [20]. LD was meant to know the limit test for impurities, and the LQ was to determine the quantitative analysis for the impurities quantification of the main component [21].

The LD and LQ were calculated from the measurement of a spiked sample. The sample was measured seven times under reproducibility conditions. It was calculated by concentration corresponding to the blank or spiked sample plus three standard deviations, as shown in equation 1, and LQ is corresponding to ten standard deviations, as shown in equation 2 [20].

$$L_D = x + 3\sigma \quad (1)$$

$$L_Q = x + 10\sigma \quad (2)$$

where x is the mean concentration of the blank or spiked sample, and σ is the value of the standard deviation of the measurements. Although by this method it was difficult to prove that a low concentration of analyte signal was different from the blank signal, it had the benefit as the simplest and fast method that could be used in LD and LQ determination. Table 2 shows that LD (%mass) for Ca = 0.00955, P = 0.00800, S = 0.06907, and Zn = 0.03075; and LQ (%mass) for Ca = 0.01693, P = 0.01277, S = 0.07508, and Zn = 0.03442.

Table 2. L_D and L_Q measurement.

	Element (%mass)			
Replicates	Ca	P	Zn	S
1	0.00589	0.00549	0.06590	0.02900
2	0.00863	0.00588	0.06710	0.02890
3	0.00548	0.00582	0.06810	0.02940
4	0.00621	0.00522	0.06660	0.02940
5	0.00633	0.00591	0.06610	0.03000
6	0.00567	0.00737	0.06610	0.02830
7	0.00651	0.00599	0.06560	0.02920
Sum (Σ) =	0.04472	0.04168	0.46550	0.20420
Mean (x) =	0.00639	0.00595	0.06650	0.02917
Standard Deviation (σ) =	0.00105	0.00068	0.00086	0.00053
Limit of Detection (L_D) =	0.00955	0.00800	0.06907	0.03075
limit of Quantification (L_Q) =	0.01693	0.01277	0.07508	0.03442

3.3. Precision and accuracy

The precision of the method was calculated by analyzing a series of samples from a homogenous solution. The Collaborative International Pesticides Analytical Council Limited (CIPAC) guidelines stated that the repeatability precision would be acceptable if several replicated samples were not less than five replicates and the value of %RSD calculated from measurement less than CV-Horwitz multiple by 0.67[22]. Data from table 2 shows that the %RSD of analysis for Ca, P, S, and Zn were below the 0,67 x CV-Horwitz value. %RSD was calculated by equation 3, and CV-Horwitz obtained using equation 4.

$$\%RSD = \frac{\sigma}{x} \times 100\% \quad (3)$$

$$CV\text{-}Horwitz = 2^{(1-0.5 \log C)} \quad (4)$$

Where, x is measurement results, σ is standard deviation of measurement, and C is measurement results in percent (%).

The measurement result generated the accuracy of the testing method to the actual value stated in CRMs. The accuracy was evaluated by using z-scores (z), a statistical parameter that indicates how many standard deviations of the measurement results compared with the certified/true value of an element in CRMs, z-scores was calculated by the equation 5.

$$z = \frac{x-X}{\sigma} \quad (5)$$

where, X is the certified value of CRMs (%)

The interpretation of the z-score is satisfactory if the score not more than 2 (|score|≤2), unsatisfactory if more than 3 (|score|≥3), and questionable if the score between 2 and 3 (2<|score|<3)[23]. According to table 3, the accuracy of the measurement of sulfur, phosphorus, and zinc was satisfied, but for calcium, it needed to be improved.

Table 3. CRM measurement for precision and accuracy.

Replicates	Element Content (% mass)			
	Ca	P	S	Zn
1	0.0529	0.0253	0.1041	0.0216
2	0.0491	0.0228	0.0968	0.0199
3	0.0472	0.0212	0.0958	0.0200
4	0.0463	0.0212	0.0949	0.0196
5	0.0456	0.0215	0.0945	0.0195
6	0.0448	0.0208	0.0946	0.0194
7	0.0460	0.0204	0.0942	0.0193
8	0.0457	0.0210	0.0931	0.0193
9	0.0455	0.0206	0.0931	0.0195
10	0.0448	0.0202	0.0930	0.0192
Sum (Σ) =	0.4679	0.2150	0.9541	0.1973
Average (x) =	0.0468	0.0215	0.0954	0.0197
Standard Deviation (σ)=	0.0025	0.0015	0.0033	0.0007
%RSD =	0.0533	0.0706	0.0345	0.0358
CV-Horwitz =	3.1710	3.5647	2.8485	3.6111
0,67 x CV Horwitz =	2.1245	2.3883	1.9085	2.4194
Reference value (X) =	0.0400	0.0200	0.0900	0.0200
z-score =	2.7248	0.9876	1.6455	0.3826

IOP Conf. Series: Materials Science and Engineering **980** (2020) 012050 doi:10.1088/1757-899X/980/1/012050

3.4. Calibration curve and linearity

The linearity of response to the analyte concentration should be calculated at least three measurement data. The curved showed no significant deviation from linearity if the correlation coefficient (R2) value was above 0.99 [22]. Interferences and sample matrix need to consider in constructing a working range because obstructions can cause non-linear responses, and sample matrix could change the ability of the method to extract/recover the analyte from the sample. The working range and linearity for this method were assessed by visual inspection of plotted data in the calibration curve supported by statistical linear regression. The calibration curve needed to be established individually for each element and matrix to get a quantitative analysis. Working range proposed for a sulfur determination as shown in figure 1 was between 0.002 to 5 mg/kg with linearity coefficient (R^2) value 0.99713, for calcium shown in figure 2 was between 0.0025 to 0.5 mg/kg with linearity coefficient (R^2) value 0.99964, for zinc shown in figure 3 was between 0.05 to 0.25 mg/kg with linearity coefficient (R^2) value 0.99643, and for phosphorus shown in figure 4 was between 0.002 to 5 mg/kg with linearity coefficient (R^2) value 0.99713.

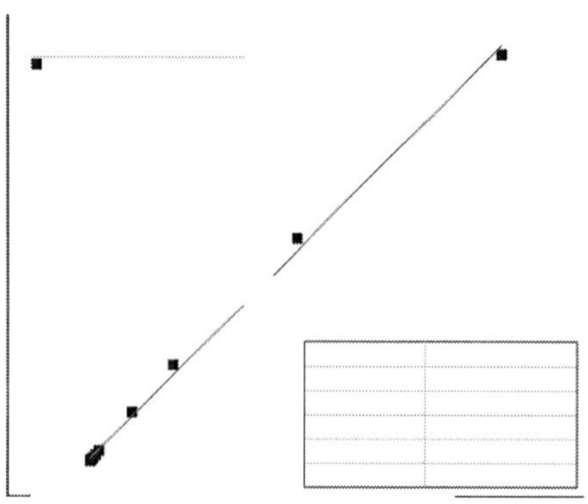

Figure 1. Calibration curve of s

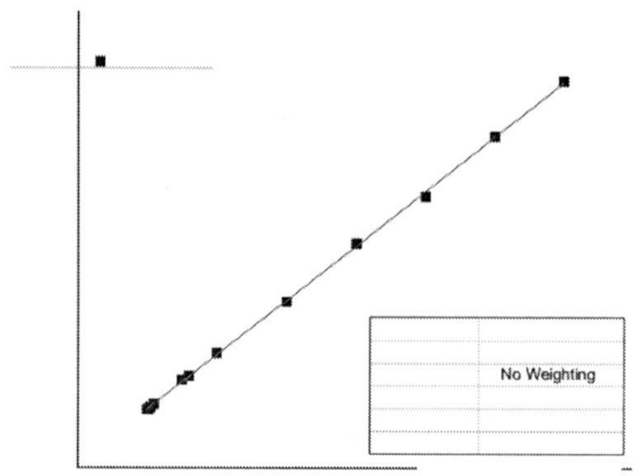

Figure 2. Calibration curve of calcium

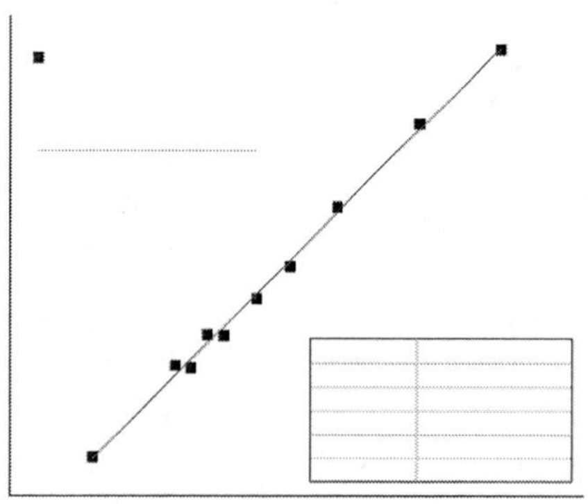

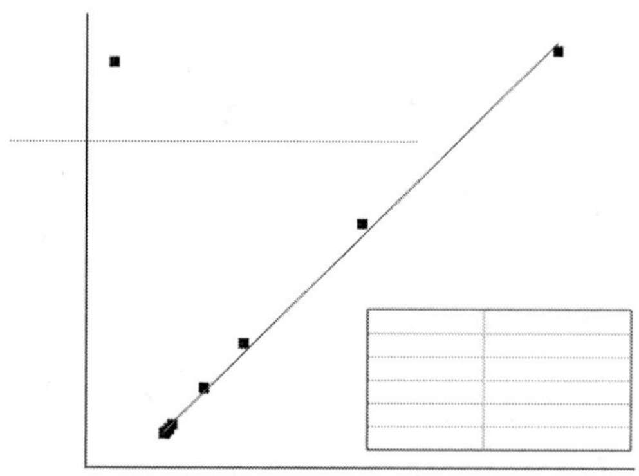

3.5. Uncertainty of measurement

Sources of uncertainty in XRF spectrometry according to IAEA (International Atomic Energy Agency) consists of six components, i.e., calibration of the spectrometer, instability of spectrometer (detector and electronics), sample preparation (heterogeneity of the material and non-uniformity of sample thickness), spectral data processing with the fitting program, quantification (determination of the absorption correction factor), and uncertainty in the determination of total mass per unit area[24]. The uncertainty contributors in XRF measurement were influenced by the calibration curve, variation of replicates, sample preparation, and influence from operators. Uncertainty contributors from calibration measurements and laboratory analysts were negligible [25].

The estimation of uncertainty is simple in principle. The following approached could use to the quantification of uncertainty associated with a measurement result in different circumstances, i.e., using data from in-house and collaborative method validation studies, Quality Control, Proficiency Testing, relevant prior studies, and the use of formal uncertainty propagation principles [26]. Uncertainty estimation for elemental analysis of lubricating oil by WD-XRF base on the testing process indicated that the best available contributor was from the individual repeatability of the measurement.

3.6. Qualitative and quantitative measurement

WD-XRF is a powerful tool in elemental analysis problems. The method is proposed for qualitative and quantitative analysis of solid material (thin or bulk, alloys or rocks, etc.), with advantages in sample preparation (without treatment, fusion, etc.) and measurement (single or multi-element) [27]. Qualitative and quantitative analysis can be used for the determination of major, minor, and trace elements in various kinds of samples. The qualitative analysis (scanning mode) can be done satisfactorily for automotive engine oil samples, for quantitative analysis depends on the availability of oil standards (CRMs, SRM). Table 4 shows the analysis results for various types of automotive engine oils. The twelve samples sequentially analyzed by WD-XRF and finished in 50 minutes.

Table 4. Analysis of various engine oil samples.

Nr.	Sample ID	Elements (% mass)			
		Ca	P	S	Zn
1	N2	0.2917	0.0954	0.2228	0.1109
2	LOE17.15	0.1037	0.1018	0.1980	0.1904
3	Oil F1	0.2346	0.0901	0.3596	0.1019
4	Oil F2	0.1875	0.0717	0.2688	0.0809
5	Oil F3	0.2532	0.0989	0.2825	0.1126
6	Oil F4	0.2307	0.0708	0.2106	0.0778
7	Oil F5	0.2810	0.0856	0.2604	0.0956
8	Oil F6	0.2807	0.0848	0.2552	0.0944
9	Oil F7	0.2559	0.0789	0.2303	0.0869
10	Oil F8	0.3732	0.0973	0.3009	0.1108
11	Oil F9	0.4225	0.0751	0.2258	0.0854
12	Oil F10	0.3635	0.0762	0.2172	0.0852

4. Conclusion

WD-XRF method allows the simultaneous determination of certain elements (calcium, zinc, phosphorus, and sulfur) in lubricating oils quantitatively. Complicated sample preparation was not needed, and it was simple in terms of its analysis procedures (no need to burn the sample, heated, weighted, destruction, etc.), along with less chemical reagents used. WD-XRF offered a fast (rapid-test), low-cost (all elements by single method), and a more environmentally friendly process in elemental analysis of automotive engine oils. The performance characteristics (detection limit, quantification limit, precision, linearity, and uncertainty) obtained in this experiment shown that this method can confidently be applied for the determination of certain elements (Ca, Mg, P, S) in the automotive engine oils. More elements also have the potential to be analyzed by this method.

Acknowledgment
The authors would like to show their gratitude to Ilham, Elva Istifani, and Indra Hadiwijaya for their support in preparing and analyzing lubricating oils used in this study. We thank B4T (Center for Material and Technical Product) for providing sample and testing laboratory. Special thanks delivered to LPDP for funding.

References
[1] A Wolak, Zając G and Gołębiowski W 2019 Determination of the content of metals in used lubricating oils using AAS *Pet. Sci. Technol.* **37(1)** 93–102
[2] Santos N D S A, Roso V R, and Faria M T C 2020 Review of engine journal bearing tribology in start-stop applications *Eng. Fail. Anal.* **108** 104344
[3] Kemenperin 2018 *Permenperin No.25 Tahun 2018*
[4] Stan C, Andreescu C and Toma M 2018 Some aspects of the regeneration of used motor oil *Procedia Manuf.* **22** 709–13
[5] Stout S A, Litman E and Blue D 2018 Metal concentrations in used engine oils: Relevance to site assessments of soils *Environ. Forensics* **19(3)** 191–205
[6] Selby T W, Bosch R J and Fee D C 2005 Phosphorus additive chemistry and its effects on the phosphorus volatility of engine oils *J. ASTM Int.* **2(9)** 153–68
[7] Brown S F 2015 Base oil groups: Manufacture, properties, and performance *Tribol. Lubr. Technol.* **71(4)** 32–5
[8] Zali M A, Ahmad W K W, Retnam A and Catrina N 2015 Concentration of Heavy Metals in Virgin, Used, Recovered and Waste Oil: A Spectroscopic Study *Procedia Environ. Sci.* **30** 201–04
[9] ASTM D4628-16 2016 Standard Test Method for Analysis of Barium, Calcium, Magnesium, and Zinc in Unused Lubricating Oils by Atomic Absorption Spectrometry *ASTM Int.* **05.02**
[10] ASTM D5185-18 2018 Standard Test Method for Multielement Determination of Used and Unused Lubricating Oils and Base Oils by Inductively Coupled Plasma Atomic Emission Spectrometry (ICP-AES) *ASTM Int.* **05.02** 800-800–6
[11] ASTM D4047-00 2000 Standard Test Method for Phosphorus in Lubricating Oils and Additives by Quinoline **(05)** 1–4
[12] ASTM D4951-02 2008 Standard Test Method for Determination of Additive Elements in Lubricating Oils by Inductively Coupled Plasma Atomic Emission Spectrometry *Man. Hydrocarb. Anal. 6th Ed.* **05** 766-766–5
[13] ASTM D1266-98(2003) 2008 Standard Test Method for Sulfur in Petroleum Products (High-Temperature Method) *ASTM Int.* **98** 272-272–5
[14] ASTM D2622-03 2008 Standard Test Method for Sulfur in Petroleum Products by X-Ray Spectrometry *ASTM Int.* **05** 402-402–4
[15] ASTM D6481-99(2004) 2010 Standard Test Method for Determination of Phosphorus , Sulfur , Calcium , and Zinc in Lubrication Oils by Energy Dispersive X-ray Fluorescence *ASTM Int.* **99** 1–5
[16] ASTM D4294-16e1 2016 Standard Test Method for Sulfur in Petroleum and Petroleum Products by Energy Dispersive X-ray Fluorescence Spectrometry *ASTM Int.* **05.02**
[17] Fernández-Feal M 2018 Study of Metal Concentration in Lubricating Oil with Predictive Purposes Curr. *J. Appl. Sci. Technol.* **27(6)** 1–12

IOP Conf. Series: Materials Science and Engineering **980** (2020) 012050 doi:10.1088/1757-899X/980/1/012050

[18] Gazulla M F, Orduña M, Vicente S and Rodrigo M 2013 Development of a WD-XRF analysis method of minor and trace elements in liquid petroleum products *Fuel* **108(March)** 247–53

[19] BSN 2018 *SNI ISO/IEC 17025:2017, Persyaratan umum kompetensi laboratorium pengujian dan laboratorium kalibrasi*

[20] Shrivastava A and Gupta V 2011 Methods for the determination of limit of detection and limit of quantitation of the analytical methods *Chronicles Young Sci.* **2(1)** 21

[21] Magnusson B and U Örnemark (eds.), 2014 *Eurachem Guide: The Fitness for Purpose of Analytical Methods – A Laboratory Guide to Method Validation and Related Topics*

[22] CIPAC 2003 *Guidelines on method validation to be performed in support of analytical methods for agrochemical formulations* **3807**

[23] Rao, T N 2018 Validation of Analytical Methods *Validation of Analytical Methods, Calibration and Validation of Analytical Methods - A Sampling of Current Approaches* IntechOpen

[24] Markowicz A and Haselberger N 2004 Quantifying uncertainty in nuclear analytical measurements, XRF Analysis of Intermediate Thickness Samples *IAEA-TECDOC-1401* **July** 37–8

[25] Fiamegos Y and de la Calle Guntiñas M B 2018 Validation strategy for an ed-xrf method to determine trace elements in a wide range of organic and inorganic matrices based on fulfilment of performance criteria *Spectrochim. Acta - Part B At. Spectrosc.* **150(July)** 59–66

[26] Eurachem 2012 *Quantifying Uncertainty in Analytical Measurements Third Ed.* EURACHEM

[27] Sitko R and B Zawisza 2012 *Quantification in X-Ray Fluorescence Spectrometry X-Ray Spectroscopy* ed D S K Sharma (Ed. InTech)

IOP Conf. Series: Materials Science and Engineering **980** (2020) 012051 doi:10.1088/1757-899X/980/1/012051

Phenol degradation by fenton reaction in air injection using plasma electrolysis method

J Z Wahono[1], R D Yusharyahya[1], Harianingsih [1,2] and N Saksono[1]
[1]Departement of Chemical Engineering, Universitas Indonesia, Kampus Baru UI. Depok 16242, Indonesia
[2]Departement of Chemical Engineering, Universitas Negeri Semarang, Kampus Sekaran, Semarang 50229, Indonesia

Email: nelson@che.ui.ac.id

Abstract. Phenolic compounds are pollutants which have high toxicity and could be harmful to the environment even at low concentrations. The degradation of phenolic has been conducted by various methods including photocatalytic and ozonation. However, these methods require high costs and large energy. Therefore, plasma electrolysis by the Fenton reaction may be an alternative method for phenol degradation. The purpose of this study is to determine the optimum conditions for phenol degradation by adding Fe^{2+} ions. The result shows that the most effective phenol degradation was obtained by using an optimum power of 700 watts with an air injection flow rate of 0.2 L/min and addition Fe2+ 20 ppm ions for 30 min experiments. Based on these conditions, the concentration of phenol in waste was reduced to 0.425 mg/L and the highest percentage of phenol degradation (99.58%) was observed. The final phenol concentrations obtained in this study met the quality standards for clean water at a concentration of 1 mg/L as well as the quality standards for drinking water treatment at a concentration of 0.5 mg/L.

1. Introduction

Phenol, also called carbolic acid or benzonal, is a white crystalline compound with a distinctive odor. This compound is usually used as an antiseptic or better known as trichlorophenol (TCP). Previous studies have shown that the effects of phenol exposure cause a decrease in fetal weight, growth retardation, and abnormal development in offspring [13,14]. Thus, it is very important to treat wastewater containing phenolic compounds prior to the final disposal. The final phenol concentrations of wastewater should meet the quality standards for clean water based on the Regulation of the Minister of the Environment (0,1 mg/L) and quality standards for the drinking water treatment process based on LH PERMEN No. 5 of 2014 (0,5 mg/L) [14]. In this study, phenol degradation was investigated by adding Fe^{2+} to produce the Fenton reaction. The Fenton reaction involves a transformation reaction between Fe^{2+} and Fe^{3+}. In addition, the presence of Fe^{2+} and Fe^{3+} ions can encourage H_2O_2 that is formed due to the incorporation of fellow •OH to decompose back into •OH [9]. Some methods have been carried out for phenol degradation in organic and inorganic waste, such as oxidation [19], photocatalyst [17], ozonation [16], and adsorption [8]. However, these methods are often considered inefficient in breaking down liquid waste because the equipment is quite expensive, results in low waste degradation, and requires quite a long time [6]. Further, plasma electrolysis

IOP Conf. Series: Materials Science and Engineering **980** (2020) 012051 doi:10.1088/1757-899X/980/1/012051

technology can be an alternative method because of its high ability for phenol degradation [7]. Plasma electrolysis or Contact Glow Discharge Electrolysis (CGDE) is a sophisticated technology from conventional electrolysis in which the electrochemical process is applied a high enough voltage to both the electrodes and the electrolyte surface that surrounds it to produce plasma in solution [5].

2. Materials and methods

2.1. Material
Materials used for phenol degradation included artificial waste with a phenol concentration of 100 mg/L. The electrolyte used was 0.02 M Na_2SO_4, for the hydroxyl radical production test, 10 M H_2SO_4 and 0.01N $KMnO_4$ was used and for the analysis of the percentage of phenol degradation, the phenol reagent was used.

2.2. Instrumentation
Phenol degradation in artificial waste was observed using the Contact Glow Discharge Electrolysis (CGDE) plasma electrolytic reactor. The reactor is made of a 2000 mL transparent pyrex glass beaker and is covered with 10 mL acrylic glass, equipped with a cooling jacket and operating temperature range at 50-60°C. The CGDE reactor is a batch reactor consisting of two electrodes; cathode and anode. The cathode used in this study was stainless steel with a length of 17.5 cm and a diameter of 6 mm, while the anode used was tungsten (tungsten) with a length of 15 cm and a diameter of 0.5 mm and an anode depth of 15 mm which can be seen in figure 1.

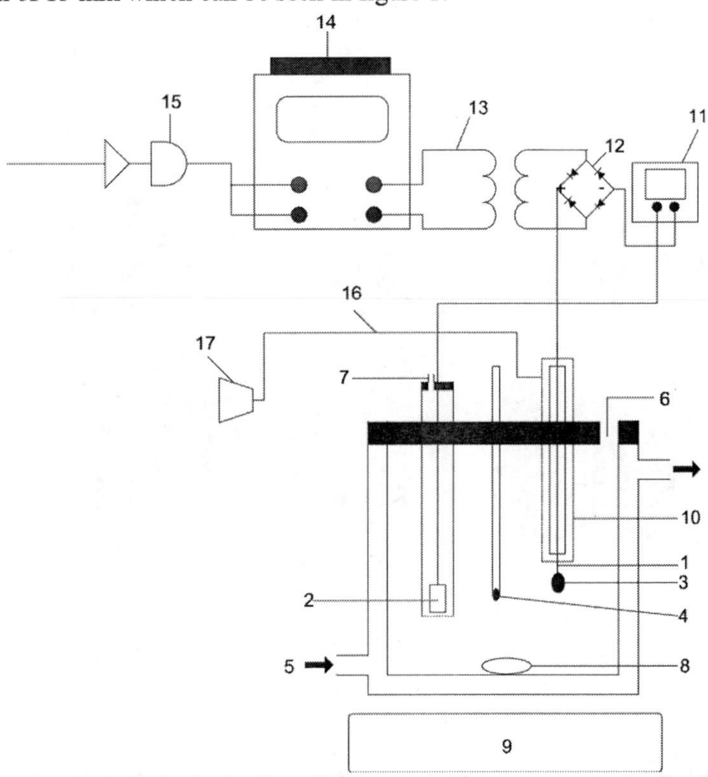

Note: 1. Anode, 2. Cathode, 3. Glow Discharge, 4. Thermometer, 5. Cooling Water, 6. Gas Products and Sampling Ports, 7. Gas Products for Cathodes, 8. Stirers, 9. Baffles , 10. Quartz Glass & Air Injection, 11. Multimeters, 12. Bridge Diodes, 13. Transformers, 14. Slide Regulators, 15. MCB.

Figure 1. CGDE reactor.

2.3. Procedure

Phenol degradation testing was carried out in 3 stages. The first stage was testing the effect of air injection flow rates (0, 0.2, 0.3, 0.4, 0.6, 0.8 L/min), on various powers (500, 600, 700 watts) as seen from the percentage of degradation and energy. The second stage was looking at influence power seen from the percentage of degradation and energy. The third step was testing the effect of adding Fe^{2+} ions with variations in the air injection rate. Measurement of the concentration of the degraded phenol waste was done by measuring the absorbance before and after the plasma electrolysis process. To test the measurement of phenol concentrations, a UV-Vis spectrophotometer at a wavelength of 500 nm according to SNI 06-6989.21-2004 was used. The data read on the UV-Vis spectrophotometer was the absorbance. Therefore, a standard or calibration curve from the calibration solution was needed.

3. Results and discussion

3.1. The effect of air injection flowrate

The effect of air injection flow rate (0.0, 0.2, 0.3, and 0.4 L/min) on the phenol degradation was carried out to test CGDE performances. This test used 100 ppm of phenol waste, 0.02 M Na_2SO_4 as an electrolyte solution, and a temperature range of 50-60°C was maintained by using cooling water circulation. During the process, 1.5 L of the solution was stirred at 2 mot motions speed for 90 min of the experiment. The sample was taken out every 0, 15, 30, 45, 60, 75, and 90 min. The amount of energy used in degrading phenol is a factor that determines the effectiveness of the process. So that the power used in this study was equated for each variation of the air injection rate so that the results of the percentage of degradation could be compared in a balanced manner. The effect of air injection flow rate on phenol degradation in various power (500, 600, and 700 watt) can be seen in figure 2.

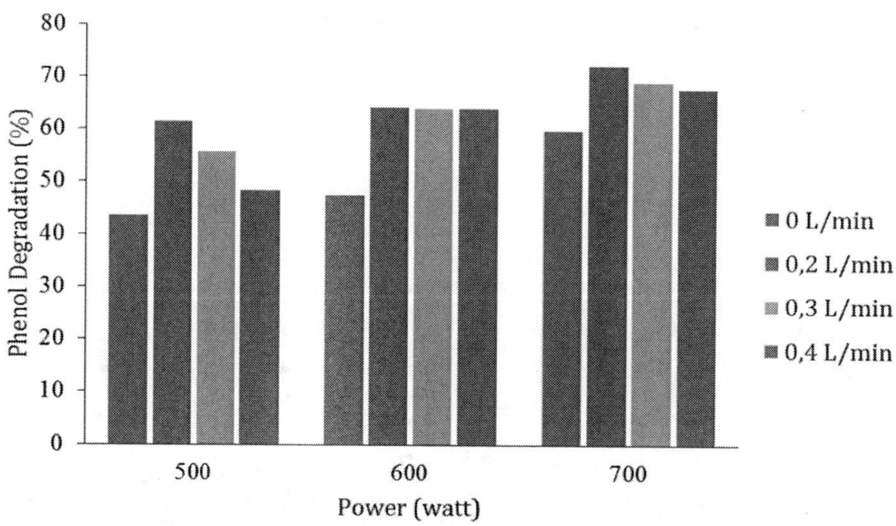

Figure2. Percentage of phenol degradation with the variation of air injection flow rate at each power (Phenol 100 ppm; Na_2SO_4 0.02 M; temperature 55°C; anode depth 1.5 cm; process time 15 minutes).

As can be seen in figure 2, the percentages of phenol degradation using 0, 0.2, 0.3, and 0.4 L/min of air injection at a power of 500 watts were 43.50; 61.50; 55.75; and 48.27%, respectively. The result shows that the percentage of phenol degradation increased with the addition of air injection, especially in the flow rate variation of 0.2 L/min. This trend also occurred while the power of 600 and 700 watts applied (figure 2). This was due to the higher air injection flow rate that caused instability in the plasma or passed its optimal point, thus decreased the phenol degradation rate. However, the addition of air injection at 0.2 L/min could increase the degradation rate in comparison with and without air injection. This shows that air injection was very important to degrade phenol compounds because the addition of bubbles increased the average length of the free electron path and electron velocity so that excitation and ionization of water molecules occurred in bubbles and solution to produce more radicals [2].

The existence of plasma is strongly influenced by the stability of the air envelope produced at the electrodes. The more stable the gas casing, the better the plasma (plasma becomes brighter and more effective). To increase the stability of the gas envelope, air was injected into the plasma area thereby reducing the evaporation process and reducing energy consumption. This was due to the formation of a gas casing with the Joule Heating process which required high energy consumption. This air bubble and gas envelope could also reduce the breakdown voltage needed for the initiation process in the formation of plasma. This was caused by the dielectric strength of the gas which was smaller when compared to the liquid.[1]. The addition of air injection could make plasma fade but the rate of waste degradation continued to increase. This mechanism is in good agreement with the previous work, as depicted in figure 3 [10].

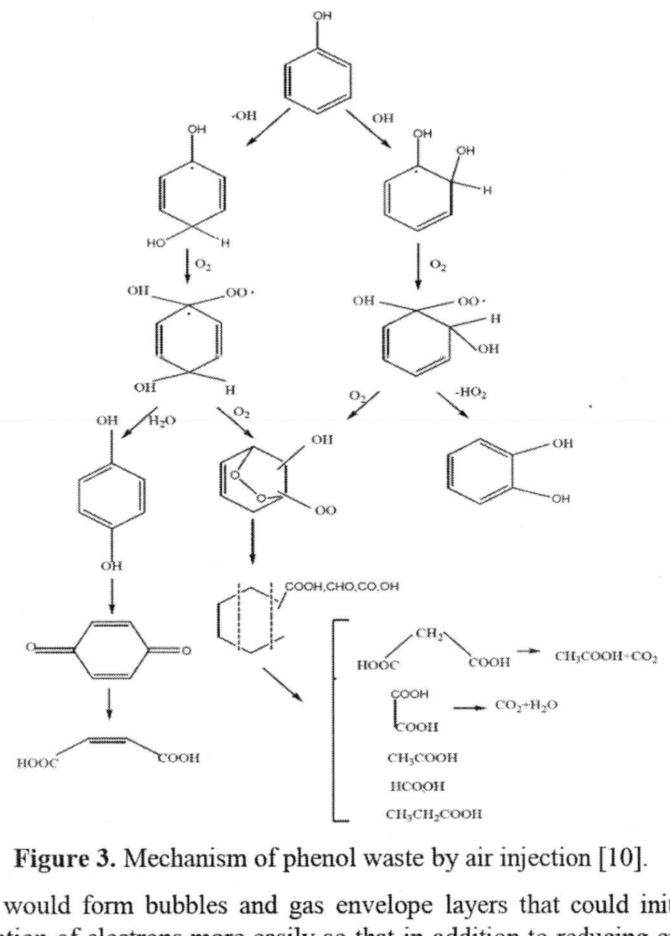

Figure 3. Mechanism of phenol waste by air injection [10].

The injected air would form bubbles and gas envelope layers that could initiate the process of ionization and excitation of electrons more easily so that in addition to reducing energy consumption, it could also affect the formation of active species other than •OH [15]. Based on these results, it can be seen that the most effective degradation process was the use of an air injection of 0.2 L/min. This shows that the small air injection rate was able to stabilize the plasma and supply oxygen and nitrogen gas in the air added to the plasma to produce reactive species in processes other than •OH. The following equation represents the possible mechanism reaction that might occur when air injection was applied [23].

$$e + O_2 \rightarrow O\bullet + O\bullet + e \quad (1)$$

$$O\bullet + H_2O \rightarrow \bullet OH + \bullet OH \quad (2)$$

$$e + H_2O \rightarrow H\bullet + \bullet OH + e \quad (3)$$

$$\bullet OH + O_2 \rightarrow HO_2\bullet + O\bullet \quad (4)$$

$$2N_2 + 2H_2O + 5O_2 \rightarrow 4HNO_3^- \quad (5)$$

Based on equation 1 to equation 5, it can be seen that the use of air injection produced other reactive species other than •OH, namely $HO_2\bullet$, •O, and HNO_3^- which could help the phenol degradation process. Phenol degradation test on the variation of power over time was carried out to determine the performance of plasma electrolysis. The variations of power used were 500, 600, and 700 watts with a concentration of 0.02 M Na_2SO_4 electrolyte solution and with a concentration of phenol waste of 100 ppm. CGDE performance testing with the effect of power on the percentage of degradation was carried out at a variation of 0.2 L/min. The air injection rate of 0.2 L/min was obtained based on variations in the air injection rate which had the highest degradation percentage compared to other air injection rates.

3.2. Power variation of phenol degradation

Figure 4 shows that the use of 600 watts power could increase 8.65% of phenol degradation in the first 15 min. However, the phenol degradation increased significantly (16.12%) in the first 15 min while applying 700 watts power. Previous research reported that the plasma produced at 500 watts and 600 watts of power was smaller than the plasma produced at 700 watts of power.

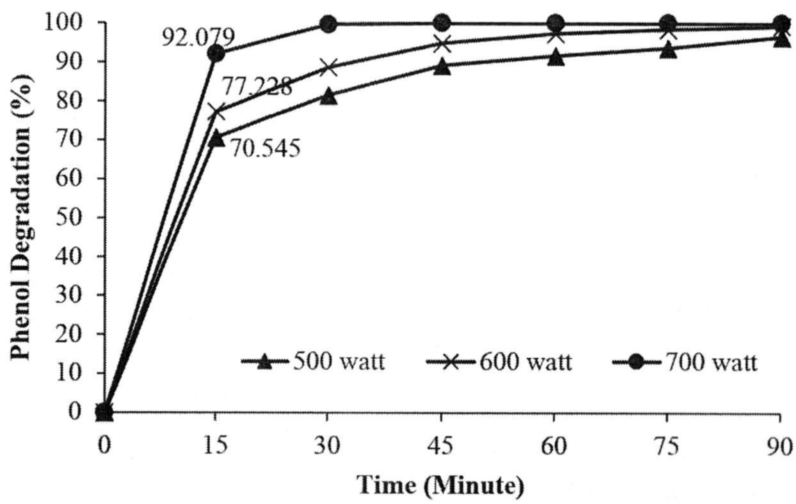

Figure 4. Percentage of power variation degradation (Phenol 100 ppm; Na_2SO_4 0.02 M; temperature 55°C; anode depth 1.5 cm; air injection flow rate 0.2 L/min, Fe^{2+} 20 ppm ions).

Based on figure 4, with a plasma produced at 700 watts or more, more OH will be generated. This was because the increase in power was affected by the increase in voltage. The increase in voltage applied to the system could increase the production of H_2O_2 in solution. At a large potential difference, the energy given to the ions to move carried electrons toward the larger anode, and electrons capable of jumping through the envelope of gas became larger. The collision of electrons with the gas envelope was what caused the formation of •OH and H_2O^+ gas ions. This H_2O^+ gas ion was used to decompose water molecules (H_2O) into •OH which played a role in degrading waste. This mechanism can be seen in equations 6 and 7 [4].

$$H_2O_{gas} + e^- \rightarrow H_2O^+_{gas} \quad (6)$$

$$H_2O^+_{gas} + H_2O \rightarrow \bullet OH + H_3O^+ \quad (7)$$

Thus the greater the voltage the more intensive plasma produced and the more formation of H_2O_2. In addition, figure 4 shows that in a longer electrolysis process, the degradation rate of phenols tended to be sloping in the last 60 to 90 min. It indicated that the interaction between •OH and phenol waste occurred quickly at the beginning of the experiment, but the concentration of phenol waste began to run out in the last 60 and 90 min so that •OH tended to degrade intermediate compounds (organic) produced from the previous process.

3.3. The effect of Fe2+ ions addition phenol degradation

The absence and presence of Fe^{2+} ions in the wastewater were also investigated and carried out with a phenol concentration of 100 ppm, with optimum power of 700 watts, a maintained operating temperature of around 50-60°C, anode depth of 1.5 cm, Na_2SO_4 0.02 M, and the volume of the waste solution of 1.5 L. In this study, variations in air injection flow rate (0 L/min; 0.2 L/min; 0.3 L/min; and 0.4 L/min) was used to obtain the optimum flow rate with the addition of Fe^{2+} ions. The study was carried out for 90 min of the experiment and the samples were taken out every 15 min.

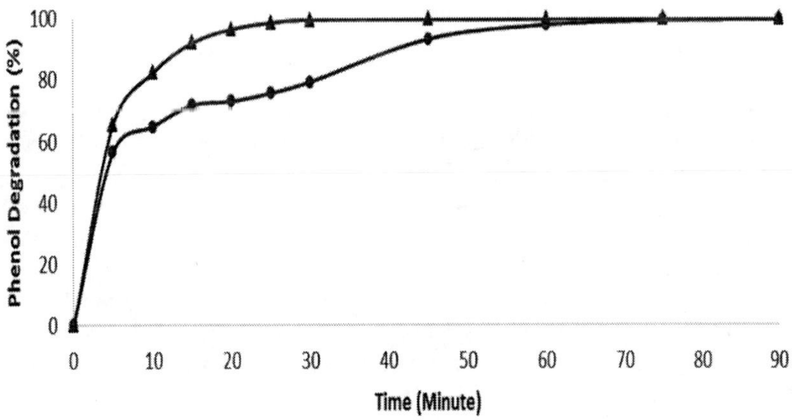

Figure 5. Comparison of phenol degradation without addition of Fe^{2+} Ion (●) and with addition of Fe^{2+} Ion (▲) 20 mg/L (air injection flow rate 0.2 L/min and power 700 watts).

The Fenton reaction involves a transformation reaction between Fe^{2+} ions and Fe^{3+} ions. It can be seen in equations 8 to equation 12, the presence of Fe^{2+} and Fe^{3+} ions could encourage H_2O_2 formed by combining fellow •OH to decompose into •OH and could catalyze H_2O_2 to produce more •OH [9].

$$Fe^{2+} + H_2O_2 \rightarrow Fe^{3+} + \bullet OH + OH^- \qquad k=76\ M^{-1}s^{-1} \quad (8)$$

$$\bullet OH + Fe^{2+} \rightarrow Fe^{3+} + OH^- \qquad k=4.3 \times 10^8\ M^{-1}s^{-1} \quad (9)$$

$$Fe^{3+} + H_2O_2 \rightarrow Fe^{2+} + HO_2\bullet + H^+ \qquad k=0.01\ M^{-1}s^{-1} \quad (10)$$

$$HO_2\bullet + H_2O_2 \rightarrow \bullet OH + H_2O + O_2 \qquad k=3\ M^{-1}s^{-1} \quad (11)$$

$Fe^{3+} + {\bullet}H \rightarrow Fe^{2+} + H^+$ $k=2 \times 10^6\ M^{-1}s^{-1}$ (12)

This study used the addition of Fe^{2+} ions because the rate constant of Fe^{2+} ions with H_2O_2 (76 $M^{-1}s^{-1}$) was almost four times greater than Fe^{3+} ions with H_2O_2 (0.01 $M^{-1}s^{-1}$) and the oxidative ability of $HO_2{\bullet}$ (3 $M^{-1}s^{-1}$) was lower than •OH (4.3 x 108 $M^{-1}s^{-1}$). The addition of Fe^{2+} ions to the system could have a catalytic effect on the degradation of organic compounds and increased the efficiency of its degradation [20,21]. It can be seen that more Fe^{2+} ions were formed in the CGDE process and the newly formed Fe^{2+} ions reacted with H_2O_2 in solution to produce more •OH. Water molecules in the liquid phase were decomposed by positive ions entering from the plasma into •OH and •H atoms by the CGDE method. OH and H could react with phenols, which led to a decrease in the concentration of phenol compounds (degradation). The mechanism can be seen in equation 13 and equation 14.

$C_6H_5OH + {\bullet}OH \rightarrow {\bullet}C_6H_5(OH)_2$ $k=1,2 \times 10^{10}\ M^{-1}s^{-1}$ (13)

$C_6H_5OH + {\bullet}H \rightarrow {\bullet}C_6H_6OH$ $k=4,2 \times 10^{9}\ M^{-1}s^{-1}$ (14)

From equations 13 and 14, it can be seen that •OH reacted more quickly than •H with phenol, so that •OH played an important role in degrading phenol compounds. However, the effectiveness of •OH in degrading organic waste in the CGDE reactor needed to be increased because the production of • OH was quite high but the interaction between • OH and organic waste was still not effective enough so that at the end of the reaction, • OH in the form of H_2O_2 in the reactor was still much left. This caused the CGDE process to become ineffective since it required a relatively long time to obtain better degradation results, especially for complex organic waste. Surely this condition would cause high energy consumption of the process. To minimize energy use, the effectiveness of the waste degradation process by the CGDE method can be increased by the addition of Fe^{2+} ions through the Fenton reaction [22].

The rate of degradation of phenols by the addition of Fe^{2+} ions was very high compared to without the addition of Fe^{2+} ions in the first 30 and 15 min where a significant slope occurred. It indicated that the role of Fe^{2+}, for •OH recombination to increase degradation rate appeared at the beginning of the CGDE process. Phenol degradation was a rapid reaction which only required 30 min by the addition of 20 mg/L of Fe^{2+} ions.

Figure 5 shows that phenol degradation of 92% occurred in the first 15 minutes of the process. The remainder degraded slowly, taking about half an hour to achieve a greater oxidation efficiency of 99%. There are several reasons for this phenol oxidation rate behavior, but generally, this is thought to be related to the dynamics of hydroxyl radical generation. Thus, it was proposed that initially, the Fe^{2+} ion reacted rapidly with H_2O_2, producing a large amount of hydroxyl radicals which caused the aromatic ring to break quickly [3]. But after that, as the reaction continued, the Fe^{2+} ion was continuously regenerated and reacted again, so that the rate of regeneration kept the production of radicals in the solution proportional to the actual production of H_2O_2 [24]. In the amount of products tested, the reduction in the phenol oxidation rate might be due to the decreased iron activity, because Fe^{3+} ions were formed and reacted with other intermediate compounds [12]. This is consistent with [20] research results which state that •OH produced from the GDP process attacks phenol to form para- and ortho dihydroxycyclohexadienyl radicals, and then dihydroxycyclohexadienyl radicals convert Fe^{3+} ions to Fe^{2+} ions (figure 6).

The newly formed Fe^{2+} ion reacted with H_2O_2 to produce more •OH to increase phenol degradation. In addition, the presence of Fe^{3+} ions accelerated the degradation process to react directly with phenol intermediate products. Para and ortho dihydroxycyclohexadienyl radicals result from the release of the dihydrocyclohexadienyl radical hydrogen atom, reacted again with Fe^{3+} ions to form benzoquinone. Benzoquinone attacks OH and Fe^{3+} ions to form carboxylic acids to form CO_2 and H_2O [9,20,21].

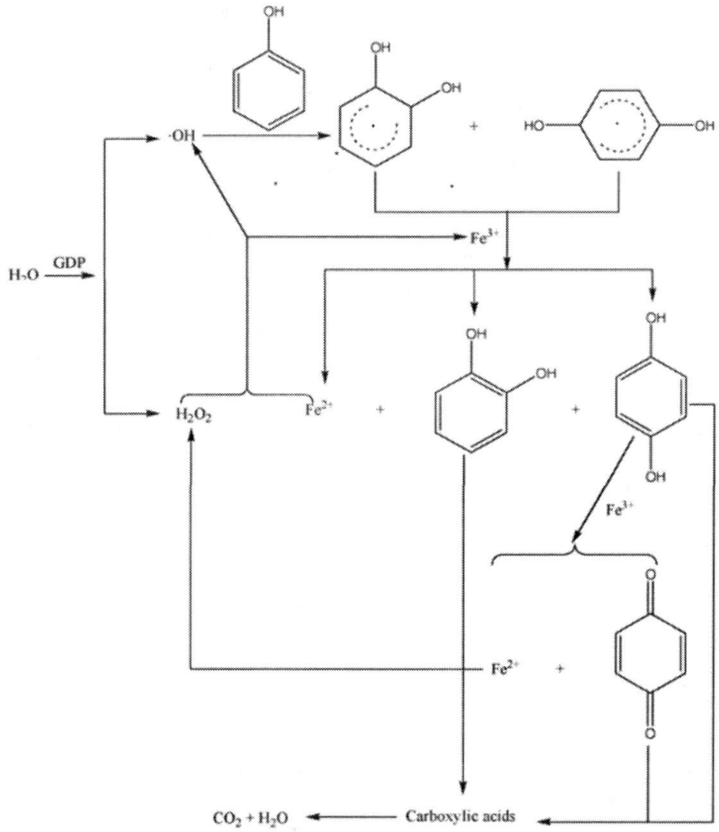

Figure 6. Mechanism of phenol degradation with Fe^{2+} ion.

4. Conclusion

Phenol degradation in artificial waste contained 100 ppm of phenol and the use of air injection with a flow rate of 0.2 L/min could increase the percentage of phenol degradation by 17.05%. The optimum condition for phenol degradation was obtained by using 700 watts of power with an air injection flow rate of 0.2 L/min. The addition of 20 ppm of Fe^{2+} ion increased the phenol degradation up to 99.58% in the first 30 min by using 700 watts power and 0.2 L/min of air injection. Further, the highest phenol degradation (99.89%) was obtained for a total of 90 min of the experiment.

Acknowledgments

This research was partially funded by "Publikasi Terindeks Internasional (PUTI) Prosiding Tahun Anggaran 2020 Contract No. NKB-1181/UN2.RST/HKP.05.00/2020" from Directorate Research and Community Service Universitas Indonesia. The authors declare no competing interests or any conflicts of financial interests.

References

[1] Ahmed M W, Suresh R, Yang J-K, Choi S and Lee H 2016 Effect of water conductivity on the generation of OH• radicals in high frequency underwater capillary discharge *Int. J. Renew. Energy Environ. Eng.* **4** 28-34

[2] Chen Y-S, Zhang X-S, Dai Y-C and Yuan W-K 2004 Pulsed high-voltage discharge plasma for degradation of phenol in aqueous solution *Separat. and Purif. Techno.* **34(1-3)** 5-12

IOP Conf. Series: Materials Science and Engineering **980** (2020) 012051 doi:10.1088/1757-899X/980/1/012051

[3] Gallard H and De Laat J 2000 Kinetic modelling of Fe (III)/H2O2 oxidation reactions in dilute aqueous solution using atrazine as a model organic compound *Water Research* **34(12)** 3107-16
[4] Gao J 2006 A novel technique for wastewater treatment by contact glow-discharge electrolysis *Pak. J. Biol. Sci.* **9(2)** 323-29
[5] Gupta S K S 2015 Contact glow discharge electrolysis: its origin, plasma diagnostics and non-faradaic chemical effects *Plasma Sources Scien. and Techno.* **24(6)**
[6] Gupta S K S 2017 Contact glow discharge electrolysis: a novel tool for manifold applications *Plasma Chem. and Plasma Process.* **37(4)** 897-945
[7] Gupta S K S and Singh R 2016 Cathodic contact glow discharge electrolysis: its origin and non-faradaic chemical effects *Plasma Sources Scien. and Techno.* **26(1)**
[8] Hamadi N K, Chen X D, Farid M M and Lu M G 2001 Adsorption kinetics for the removal of chromium (VI) from aqueous solution by adsorbents derived from used tyres and sawdust *Chem. Engin. J.* **84(2)** 95-105
[9] Jiang B, Zheng J, Qiu S, Wu M, Zhang Q, Yan Z and Xue Q 2014 Review on electrical discharge plasma technology for wastewater remediation *Chem. Engin. J.* **236** 348-68
[10] Jufang Z, Jierong C and Xiaoyong L 2009 Remove of phenolic compounds in water by low-temperature plasma: a review of current research *J. of Water Resource and Protect.*
[11] Lu Q, Yu J and Gao J 2006 Degradation of 2, 4-dichlorophenol by using glow discharge electrolysis *J. of Hazard. Materi.* **136(3)** 526-31
[12] Mijangos F, Varona F and Villota N 2006 Changes in solution color during phenol oxidation by Fenton reagent *Environ. Scien. & Techno.* **40(17)** 5538-43
[13] Permata Y, Cahyani R, Karamah E and Bismo S 2019 Characteristic of phenol and 2, 4-dichlorophenol synthetic wastewater degradation in a DBD (dielectric barrier discharge) reactor *J. of Physics: Conf. Series*
[14] Roosmini D and Rahmawati S 2012 Domestic wastewater services and facilities in Indonesia: policy and regulation role *Water Environment Partnership in Asia*
[15] Ruma H M, Hosseini S, Sakugawa T and Akiyama H 2015 Treatment of wastewater by underwater discharge in gas bubbling water *Inter. J. of Renew. Energy and Environ. Engine.* **3(3)** 189-94
[16] Saksono N, Seratri R T, Muthia R and Bismo S *Phenol Degradation in Waste Water with Contact Glow Discharge Electrolysis (CGDE) System by Using Sodium Sulfate Solution.*
[17] Slamet S, Arbianti R and Marliana E 2007 Pengolahan Limbah Cr (VI) dan Fenol Dengan Fotokatalis Serbuk TiO2 Dan CuO/TiO2 *Reaktor* **11(2)** 78-85
[18] Urszula G, Izabela G, Danuta W and Sylwia Ł 2009 Isolation and characterization of a novel strain of Stenotrophomonas maltophilia possessing various dioxygenases for monocyclic hydrocarbon degradation *Brazil. J. of Microbio.* **40(2)** 285-91
[19] Wahyudi D, Ghaisani S and Bismo S 2019 Degradation of phenol and 2, 4-dichlorophenol wastewater by ozonation in multi-injection bubble column reactor *J. of Physics: Conf. Series*
[20] Wang L and Jiang X 2009 Unusual catalytic effects of iron salts on phenol degradation by glow discharge plasma in aqueous solution *J. of Hazard. Materi.* **161(2-3)** 926-32
[21] Wang X, Zhou M and Jin X 2012 Application of glow discharge plasma for wastewater treatment *Electro. Acta* **83** 501-12
[22] Yang H and Tezuka M 2011 Plasma-induced complete destruction of tetrachlorophenols in an aqueous solution *J. of Physics D: Appl. Physics* **44(15)**
[23] Yasuoka K and Sato K 2009 Development of repetitive pulsed plasmas in gas bubbles for water treatment *Int. J. Plasma Environ. Sci. Technol.* **3(1)** 22-7
[24] Zhao N, Ding L, Bei H, Zheng S and Han B 2018 The role of dissolved oxygen in Fenton system *IOP Conf. Series: Earth and Environmental Science*

IOP Conf. Series: Materials Science and Engineering **980** (2020) 012052 doi:10.1088/1757-899X/980/1/012052

The effect of sodium bicarbonate ratio for the mechanical properties of underarm pads rubber for crutches

Nasruddin*, A T Bondan and S Agustini
Palembang Institute for Industrial Research and Standardization, Ministry of Industry, Indonesia

Email: nas.bppi@gmail.com

Abstract. Research on the effect of the ratio of sodium bicarbonate on the mechanical properties of underarm crutch pads from natural rubber SIR-20 has been carried out. The ratio of sodium bicarbonate used was 1.00 phr; 1.50 phr; 2.00 phr; and 2.50 phr. The rubber compound for underarm crutch pads was made by the compounding process using an open rubber mixing mill and vulcanization at a temperature of 120 °C for 11 minutes. The best formula for curing characteristic rheometer test results at a temperature of 150 °C was URC-04 with S_{max}= 4.95 kg/cm, S_{min} = 0.44 kg/cm, ΔS = 4.51 kg/cm, t_{90} = 6.57 min and ts_2 = 4.13 min. The mechanical test result showed the hardness test before aging and after aging of 31 Shore A, and 32 shore A, tear strength of 8.50 kN/m, elongation at the break before aging and after aging of 300%. The results of the identification of functional groups by FTIR for infrared absorption in each sample generally showed the same pattern with one another. SEM analysis result showed the distribution of materials and the formation of pores by sodium bicarbonate was not evenly distributed with an average diameter of 5 microns.

1. Introduction

Underarm rubber is a soft rubber formed by air bubbles that are trapped between its molecules. Underarm rubber is placed at the top of the crutch as a base to feel comfortable when touching the underarm rubber when used. Crutches are aids for patients who have temporary or permanent health problems especially for walking [1-2]. Materials that are widely used to make underarm rubber for crutches generally come from polyurethane rubber, this is due to lightweight polyurethane rubber with a specific gravity of 36 kg/m^3. Polyurethane resin compositions using polyols have an average molecular weight below 400 [3]. Polyurethane is a synthetic rubber from urethane compounds (ONHCOO), widely used for various products because it has high thermal stability, is resistant to weather, ozone, and radiation [3-4]. Polyurethane made from polyisocyanates, polyhydroxy compounds, and a small number of macromolecular compounds -NHCOO is widely used as an organic polymer material for various purposes such as medical equipment, textile industry, building construction, and so on [5].

The development of polyurethanes for various purposes such as for underarm rubber has been carried out by many industries and researchers. Latinwo [6] developed polyurethane foam with fillers $CaCO_3$ and $CaMg(CO_3)_2$. According to Chen [7] using micro and nano-size $CaCO_3$ fillers can increase foam cell density. Agarry [8], and Usman [9] developed polyurethane foam with $CaCO_3$ fillers. Marhoon [10] used a filler from rice husk ash as a polyurethane filler. Underarm rubber made of polyurethane, based on observations and interviews with users can easily be torn. Tearing can be caused by the structure of the wall between the pores formed by carbon dioxide gas trapped in polyurethane molecules that do not have a strong bond.

Based on the description above, this study used solid natural rubber of the SIR-20 type as an ingredient in underarm rubber. Natural rubber is a hydrocarbon polymer that is naturally elastic in the form of a white liquid such as milk [11]. Natural rubber with the general formula of isoprene (C_5H_8) and with the chemical structure of cis-1,4-polyisoprene, which is not easy to tear, has widely been used for various applications [12-14]. The development of natural rubber into underarm rubber in this study was carried out by adding blowing agent materials, fillers, softeners, and additives.

Blowing agent materials that are widely used to form air cavities into foam are categorized as organic and inorganic materials. Organic blowing agents are for example dichloromethane, freon, pentane [15], while inorganic blowing agents contain elements such as carbon dioxide, nitrogen, and air [16]. The material used to form air cavities in rubber vulcanisate is sodium bicarbonate which is commonly used by industry and researchers. The air cavity formed in the rubber is a polymer foam that has good elasticity, which will return to its original state when under pressure from other objects.

Polymer foam has advantages including, low density, good heat insulation, can be used as a damper, and it has corrosion resistance [17]. According to Hussein [18], the reason for using sodium bicarbonate as a foam material is because the price is cheap and does not pollute the environment. Mohammad [19], used sodium bicarbonate as a blowing agent with a ratio of 4 phr that produced a density of 0.9 with water absorption by cells 1.7 x 10^{-3} g/min, and maximum compression of 105 kPa. Ariff *et al.* [20] used sodium bicarbonate as a foam development agent at vulcanization temperatures of 140, 150, and 160 °C. Najib [21], developed rubber foam using sodium bicarbonate in a ratio (4, 8, 10, and 12) phrase, and the results showed that the increase in sodium bicarbonate could increase the cell volume. Cell volume increased due to the presence of air that came out to form the cell nucleus permanently from the nucleation [22-23].

The development of natural rubber into underarm rubber used zinc oxide activators and stearic acid as co-activators. The activator was used to accelerate the performance of the accelerator during the vulcanization process. Wardana [24] used ZnO and stearic acid as activators and stearic acid as co-activators with results that could increase the efficiency and rate of the air vapor transmission. Duan [25] studied the effects of ZnO nanoparticles and stearic acid in the manufacture of foam rubber with the result of increased tear strength and tensile strength.

Fillers were used in making foam to strengthen the structure of cell wall networks and reduce production costs. Tangboriboon *et al.* [26] used 60 phr $CaCO_3$ fillers with $NaHCO_3$ as a blowing agent that produced a bulk density of 240.5 kg/m^3, a relative foam density of 0.2614, cross-link density of 74.279 mol/m^3, the compression set of 32.60%, and thermal conductivity of 0.066 W/mK. This study used a multi filler consisting of kaolin, $CaCO_3$, and zeosil. Kaolin and $CaCO_3$ are fillers to reduce production costs, and zeosil to improve the mechanical properties of underarm rubber. According to Rosszainily [27], calcium carbonate as a filler can improve the thermal properties of rubber vulcanisate. Zeosil is amorphous precipitated silica that is more easily dispersed in rubber compounds. Zeosil is widely used as a tire filler to reduce hysteresis of compounds and heat buildup while maintaining rigidity. To improve the homogeneity of fillers and additives in the rubber matrix structure, minarex oil was added as a softener [28]. This research studied the effect of sodium bicarbonate ratio by using multi fillers (kaolin, $CaCO_3$, and zeosil) with softener minarex oil on the mechanical properties of the underarm rubber vulcanisate. To determine the maturity of the underarm rubber compound, a curing characteristic rheometer test was performed. The mechanical properties of rubber vulcanisate tested were hardness before aging and after aging, tear strength, elongation at the break before and after aging.

Functional group analysis was done using Fourier transform infrared (FTIR). The uniformity of material distribution and formation of pores by sodium bicarbonate were observed using a scanning electron microscope (SEM). This research is expected to provide information on the development of natural rubber as a raw material for underarm rubber material along with sodium bicarbonate as a blowing agent, and multi-fillers consisting of kaolin, $CaCO_3$, and zeosil.

2. Materials and methods

2.1. Materials

The materials used in this research included SIR-20 natural rubber. Titanium dioxide (TiO_2) MBR-9672 purity 94% min, oil absorption 19 g/100 g max. Pigment & dyestuff [22094-93-5] pigment yellow 81, density 94% min, heat resistance 180 °C, light fastness 7, water resistance 5, oil resistance 5, and alkali resistance 5. Yipin inorganic pigment powder S-130 iron oxide red for color paving brick max moisture 1%, color difference 1.0, predominant particle 0.1-0.3, main purity 99%, water (oil) absorption 15-25%, density 5.0 g/m^3. Zinc oxide (ZnO) purity 99.70%, water-soluble ≤ 0,10%. Stearic acid purity 99% min. Kaolin 325 mesh from Tanjungpandan Belitung Indonesia. Calcium carbonate ($CaCO_3$) purity 98%, pH 9.5%, whiteness 97% min. Zeosil 175P. Pacific rubber processing oil (paraffinic) E-245 density 0.88 g/mL @ 29.5 °C, pour point, -6 °C, aromatics 10%, naphthenic 30%, paraffin 60%, density 0.88 g/ml @ 29.5 °C, pour point, -6 °C, aniline point, 96 °C, viscosity @ 40 °C 30-32 CST. Sodium bicarbonate ($NaHCO_3$) purity 99% min, density 2.16 g/mL at 25 °C (lit.). Vulcanization accelerator MBTS DM/2,2-dithiobis (benzothiazole)120-78-5 purity 98%, ash 0,3%. TMTM (TS) accelerator with the chemical formula $C_6H_{12}N_2S_3$ yellow powder the specific density was 1.37-1.40. Rubber chemicals PVI/rubber retarder anti scorching agent PVI. Rubber vulcanizer insoluble sulfur/OT20 use in tyre purity 99%, ash max 0.15%, fineness (100 mesh) ≥ 99.8%, oil content oil content 20±1%.

2.2. Methods

2.2.1. Mastication process and manufacture of rubber compound . The development of natural rubber into underarm rubber in this study consisted of 4 (four) formulas. Each formula was differentiated based on the ratio of sodium bicarbonate used. Natural rubber for all treatments was weighed according to the ratio as shown in table 1.

Table 1. The ratio of ingredients to underarm rubber.

Material	Ratio of Ingredients			
	URC-01*	URC-02*	URC-03*	URC-04*
NR	100.00	100.00	100.00	100.00
Titanium	5.00	5.00	5.00	5.00
Pigment Yellow	6.00	6.00	6.00	6.00
Pigment Red	0.25	0.25	0.25	0.25
ZnO	5.00	5.00	5.00	5.00
Stearic Acid	2.00	2.00	2.00	2.00
Kaolin	45.00	45.00	45.00	45.00
$CaCO_3$	10.00	10.00	10.00	10.00
Zeosil	12.50	12.50	12.50	12.50
Paraffinic Oil	22.00	22.00	22.00	22.00
$NaHCO_3$	1.00	1.50	2.00	2.50
MBTS	1.00	1.00	1.00	1.00
TMTM	0.50	0.50	0.50	0.50
PVI	0.01	0.01	0.01	0.01
Sulfur	2.00	2.00	2.00	2.00

Annotation: *phr = parts per hundred of rubber
URC = underarm pads rubber for crutches

After each rubber was weighed according to the material ratio in table 1, the mastication process was carried out through milling using an open mill type RSK-300. Next, after mastication (± 15 minutes) titanium oxide was added to make white rubber so easy to stain. After the rubber turned white, color pigments were added (red and yellow composites) until the whole rubber turned yellow. The next process was the addition of ZnO and stearic acid to improve the performance of the

IOP Conf. Series: Materials Science and Engineering **980** (2020) 012052 doi:10.1088/1757-899X/980/1/012052

accelerator in the vulcanization process. Then kaolin, $CaCO_3$, zeosil were added also at the same time as paraffinic oil as a softener and sodium bicarbonate until the mixture became homogeneous. While continuing milling, MBTS, and TMTM as accelerators were added at the same time as PVI/rubber retarder was added. The final process of compounding was adding sulfur as a vulcanizing agent. Each formula was replicated three times to get the best data. Table 1 is the material ratio of each treatment.

2.2.2. Compound vulcanization process. The compounds of each formula were vulcanized and molded at 120 °C for 11 minutes

2.2.3. Testing of rubber compound and rubber vulcanizates. Rubber compound testing was carried out to determine the maturation characteristics of rubber compounds into vulcanizates. The rheometer was used according to ASTM 5289. The test of mechanical properties included hardness (ASTM D.2240-15), elongation et break (ASTM D.412-16), tear strength (ASTM D.624-00 (ra2012)), and ozone resistance at 25 pphm, 20% (D.1149 -16). Identification of functional groups was done using the Agilent array 630 FTIR spectrometer, wavenumber range 7,000-350 cm^{-1} /5,100–600 cm^{-1}, interferometer 60° air bearing interferometer, data integrity such as 21 CFR Part 11 and EU Annex 11. The Fourier transform infrared (FTIR) spectroscopy was used to study functional groups. The identification of the surface of the objects was conducted using scanning electron microscopy (SEM) JSM-6510 LA with the magnification 5 times up to 300.000 times (on 128 mm x 96 mm image size).

3. Results and discussion

3.1. The curing characteristics of rubber compounds

Characteristics of rubber compound maturation for underarm crutch rubber from 4 (four) samples are shown in table 2.

Table 2. Test results for the maturation of rubber compounds for underarm rubber.

Curing Characteristic Rheometer 150 °C	Test Result			
	URC-01	URC -02	URC -03	URC -04
S* Maximum, kg-cm	4.93	5.12	5.00	4.95
S* Minimum, kg-cm	0.66	0.36	0.54	0.44
Δ_S (Kg/cm)	4.27	4.76	4.46	4.51
Opt cure time (t_{90}) min; sec	8.00	10.26	8.52	6.57
Scorch time (ts_2), min; sec	4.16	6.9	5.24	4.13

Note: = Test Method PPK, Using MDR 2000 (Not accredited)

Characteristics of compound maturation for underarm crutch rubber are generally explained by the results of testing maximum torque (Smax), minimum torque (S_{min}), delta torque (Δ_S) (S_{max}-S_{min}), optimal maturation time (t_{90}) and scorch time (t_{s2}). The maximum torque test of rubber compounds for underarm crutch rubber in this study was to study changes in the viscosity value of each difference in the addition of sodium bicarbonate. The change of viscosity was related to the formation of air cavities due to the performance of sodium bicarbonate which released CO_2 gas into the rubber compound during vulcanization.

The CO_2 gas released in the rubber compound from the decomposition of sodium bicarbonate during vulcanization affected the formation of cross bonds and the molecular structure of the underarm rubber vulcanisate. According to Najib [21], cross-linking and foam density are relatively decreased with the energy absorbed by the foam, and at the same time, the formation of foam cells and the increase in the number of cells per unit area also occurs. As shown in Table 2, the highest maximum torque value (5.12 kg-cm) was obtained at the sodium bicarbonate ratio of 1.5 phr (table 1), while the highest minimum torque value (0.66 kg-cm) was observed at URC-01 with sodium bicarbonate ratio 1 phr.

Viewed from an increase in the sodium bicarbonate ratio, an increase in the maximum torque value and non-uniform minimum torque fluctuated with an increase in the sodium bicarbonate ratio. This might be due to the performance of sodium bicarbonate releasing CO_2 gas into the rubber compound matrix when vulcanization occurred. This event resulted in the disruption of the formation of a cross-linking degree during vulcanization. The highest value (4.76 kg/cm) degree of crosslinking formed during vulcanization occurred at a sodium bicarbonate ratio of 2.0 phr (URC-02).

For an optimal maturation time (10.26 minutes) at a sodium bicarbonate ratio of 1.5 phr (URC-02), the maturation before the vulcanization process lasted for 6.6 minutes with a sodium bicarbonate ratio of 1.5 phr (URC-02) with a value of 6.9 minutes. According to Poh [30], a ratio of sodium carbonate exceeding 1.5 phr can increase the speed of vulcanization, but there is a reduction in the scorch time and cure index. The time of maturation and the formation of cross-link during the vulcanization process are influenced by the ratio of sodium bicarbonate. Besides, the speed of maturation of rubber compounds and the number of pores formed by CO_2 affect the strength of the cross-linking formed.

3.2. Mechanical properties of vulcanisate

The test results of the mechanical properties of underarm crutch rubber vulcanisate are shown in figures 1, 2, and 3.

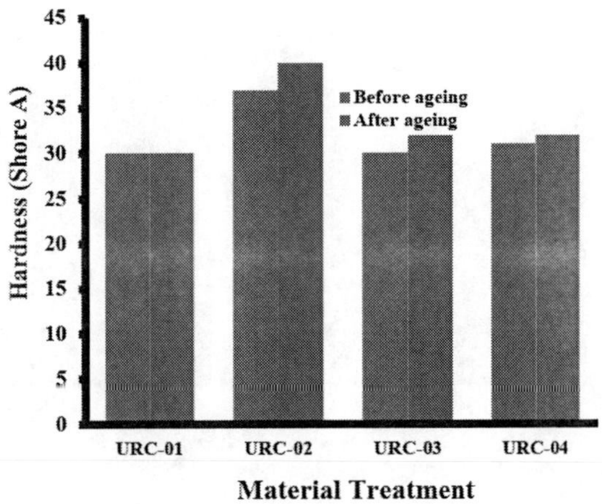

Figure 1. Effect of material ratio to hardness

Figure 1 shows the results of the underarm rubber hardness from 4 formulas. The hardness values for the URC-01 and URC-03 before aging had the same value (30 Shore A), for the URC-04 formula, the value increased to 31 Shore A. The highest hardness value before aging and after aging was obtained at URC-02. The difference in the value of hardness illustrates the difference in the level of material density in the rubber vulcanizates matrix per mm^3.

The level of space density in the rubber vulcanisate matrix was determined by the material ratio (table 1), the distribution of materials, and the vulcanization process. In addition, the effect of sodium bicarbonate which formed air cavities by CO_2 gas affected the density and structure of molecules that made up vulcanisate rubber. Sodium bicarbonate was a blowing agent that produced CO_2 and H_2O when it received heat [31]. The reduced level of crosslinking density in the molecular structure of the rubber compound resulted in a better and more homogeneous formation of foam [32]. The amount of CO_2 gas trapped between vulcanisate rubber molecules was inversely proportional to the hardness value.

The elongations at break test results for the URC-01 formula before and after aging were 220%, and 210.94% respectively, it was the lowest compared to other formulas (figure 2). The differences in the value of the elongation at break were due to the number of air cavities that were formed reduced the density of spaces and bonds between molecules. According to Roy [33], sodium bicarbonate during heating decomposes into sodium carbonate, water, and CO_2. CO_2 gas trapped in the polymer matrix produces foam.

The amount of CO_2 gas formed reduces the strength of the double bond, which in turn decreases the value of the elongation at break. This is because the empty space formed by CO_2 gas has no molecular bonds. The elongation at break value is also influenced by additives such as ZnO. Where the ratio of additives added will affect cross-linking. The ZnO ratio at 5 phr showed a significant reduction in tensile strength, elongation at break, modulus, hardness, and compression [34]. Meanwhile, if the ZnO ratio is more than 7 phr, there will be a decrease in tensile strength, elongation at break, modulus, and hardness [35]. In this study ZnO ratio for all formulas was 5 phr, but the ratio of ZnO to elongation at break was also influenced by the ratio of sodium bicarbonate and other additives (table 1).

The elongation at break value in the URC-02 formula before aging was 360%, while after aging was 348.89%. It was the highest compared to other formulas. As seen from the sodium bicarbonate ratio of 1.5 phr and the filler ratio (table 1), it was possible that in this ratio (URC-02) the amount of CO_2 formed was proportional to the amount of filler available. Whereas in the URC-04 formula with a sodium bicarbonate ratio of 2.5 phr, the elongation value before aging and after aging was the same, which was 300%.

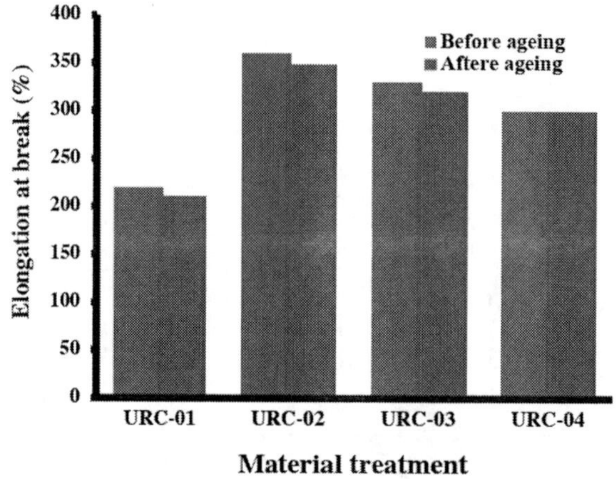

Figure 2. Effect of material ratio to elongation at break.

In the formula URC-04, there was a balance of the amount of CO_2 and the availability of fillers so that the value of elongation at the break before and after aging was also the same. In figure 2 there is a decrease in the value elongation at break from URC-02 to URC-04. An increase in the ratio of sodium bicarbonate decreases the elongation value. Elongation at break is a stretch that occurs when the object is pulled in the opposite direction at the critical point experiencing a break [34]. This is because the bonds formed between material molecules do not have the ability to resist the forces acting on them so that when a strain occurs it breaks.

Tear strength as well as elongation at break is one of the measuring parameter to see the resistance of objects when experiencing bonding movements between molecules until the critical point is torn. The tear strength is the force required to tear a sample, the stronger the bonds between molecules, the higher the tear strength [36]. The tear strength test result for the URC-02 formula was 10.5 kN/m, the highest compared to the other formulas. Meanwhile, for the URC-01 tear strength value was the lowest compared to other formulas (figure 3). Tear strength is influenced by the strength of the cross-linking, where the energy required is directly proportional to the strength of the materials that form the cross-link [37].

The difference in tear strength is also influenced by the amount of air that is present between the material molecules. The amount of air trapped is influenced by the ratio of the constituent material (table 1). The amount of CO_2 as a blowing material in underarm rubber in this study was influenced by the ratio of sodium bicarbonate and other materials that made it up. The amount of CO_2 formed in the underarm rubber vulcanisate in the same space when there is a difference will affect bonds between molecules. In addition, the amount of CO_2 formed in this study affected the density of the cross-linking formed by sulfur.

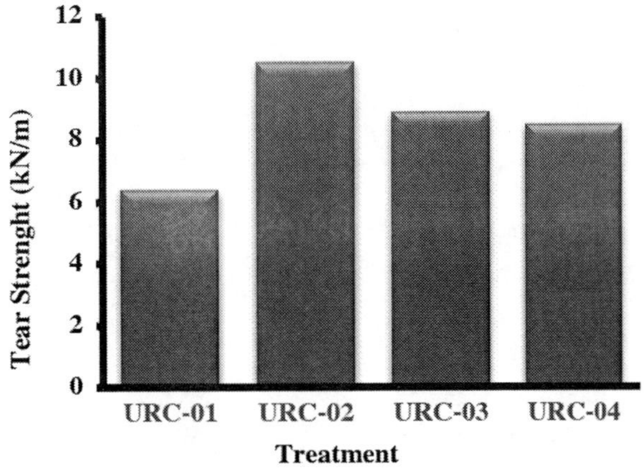

Figure 3. Effect of material ratio to the tear strength

The number of air bubbles trapped reduced the density of the rubber vulcanisate. Low cross-linking density affected tear strength. With the weak cross-linking that formed underarm rubber vulcanisate, the tear strength also became weak. The cross-linking formed was influenced by the ratio of the activator and accelerator. According to Mariano et al. [38], mechanical properties are affected by the amount of accelerator added. Tear strength is the amount of force used to stretch the bond between the material molecules to the critical point until the tear.

3.3. Fourier Transform Infrared

Observation of samples was conducted using Fourier transform infrared (FTIR) and scanning electron microscope (SEM) to determine the wavenumber, chemical composition, and surface morphology of the sample [39]. The results of the characterization of vulcanisate rubber by FTIR using sodium bicarbonate showed a number of bonds formed. The bonds formed were C - H atomic bonds with different peaks in each treatment (table 1) (figures 4, 5, 6, and 7).

The graphs in Fig 4, 5, 6, and 7, show the same pattern between one graph and another. The main functional group are represented by wave number i.e. 3600 cm^{-1} (Uptake of the OH group); 3000 cm^{-1} (absorption of aliphatic C-H); 1525 cm^{-1} (uptake of the -COO group); 1000 - 1100 cm^{-1} (absorption of Si-O interactions); <750 cm^{-1}, fingerprint area according to the functional group identification table. However, based on data from the graph above it can be seen that there was a spectral difference between the four formulas. At wavelengths between 1000 cm^{-1} to 1100 cm^{-1} there is a Si-O-Si bonding group in which at each spectral result, in each different formula, a line formed a valley.

The silica detected in the graph above was silica added as a filler to strengthen the cell wall structure between the spaces formed by CO_2 from the decomposition of sodium bicarbonate by heat. Silica sources detected on the graph came from kaolin, $CaCO_3$, and zeosil (table 1).

3.4. Morphological properties of vulcanizate underarm rubber

Testing of rubber vulcanisate was done using scanning electron microscopy (SEM) to determine the distribution of air cavities that occur by sodium bicarbonate in the rubber compound matrix, and the bonds among substances used [29]. Scanning with SEM was carried out at 5000x magnification as shown in figures 8, 9, 10, and 11.

IOP Conf. Series: Materials Science and Engineering **980** (2020) 012052 doi:10.1088/1757-899X/980/1/012052

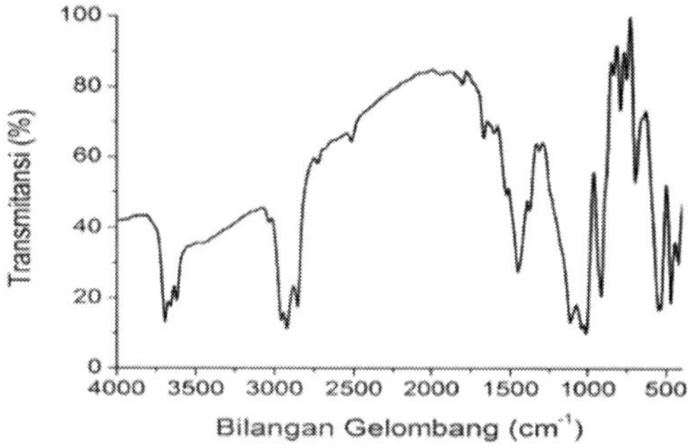

Figure 4. URC-01 axillary rubber vulcanisate wave number.

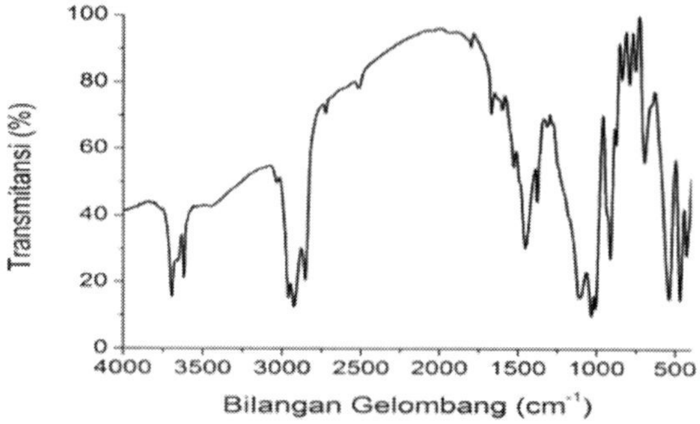

Figure 5. URC-02 axillary rubber vulcanisate wave number.

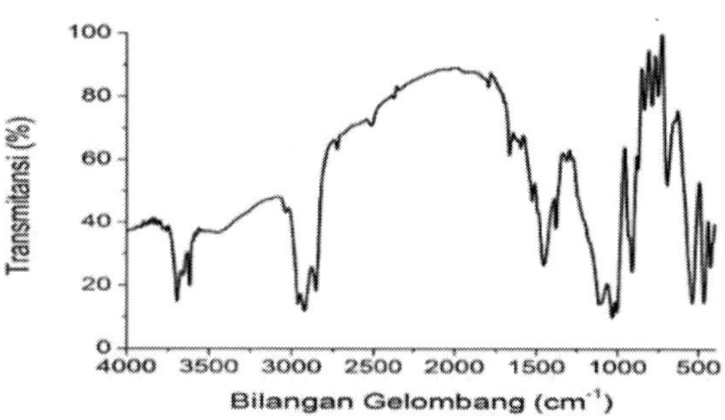

Figure 6. URC-03 axillary rubber vulcanisate wave number.

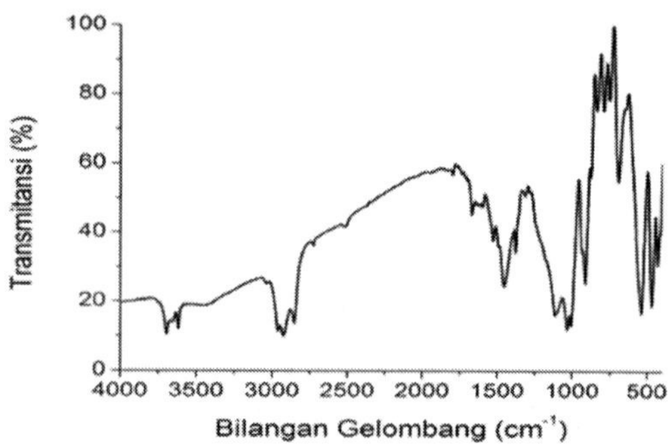

Figure 7. URC-04 axillary rubber vulcanisate wave number.

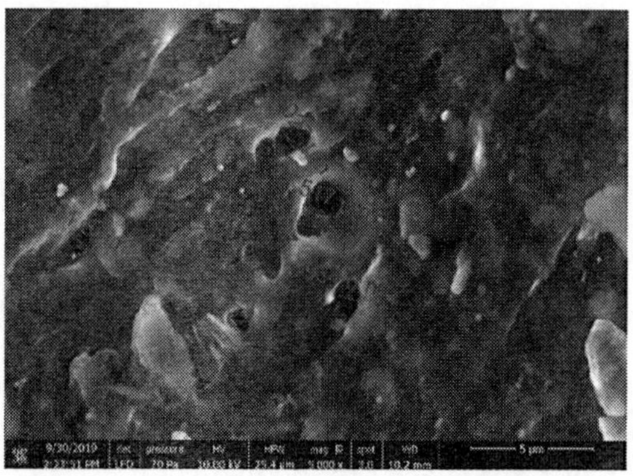

Figure 8. Enlargement of URC 01 underarm rubber morphology with SEM 5000x.

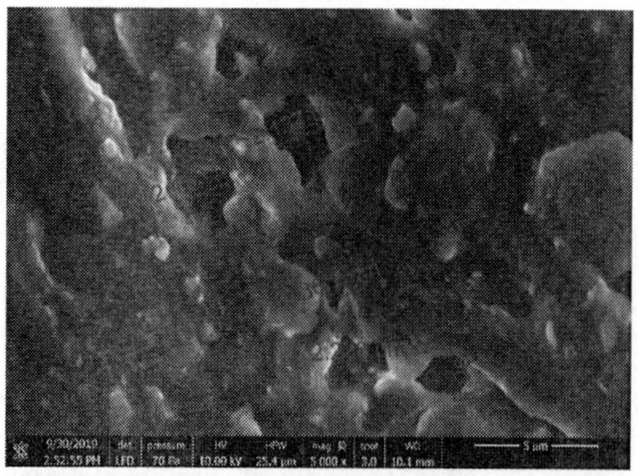

Figure 9. Enlargement of URC 02 underarm rubber morphology with SEM 5000x.

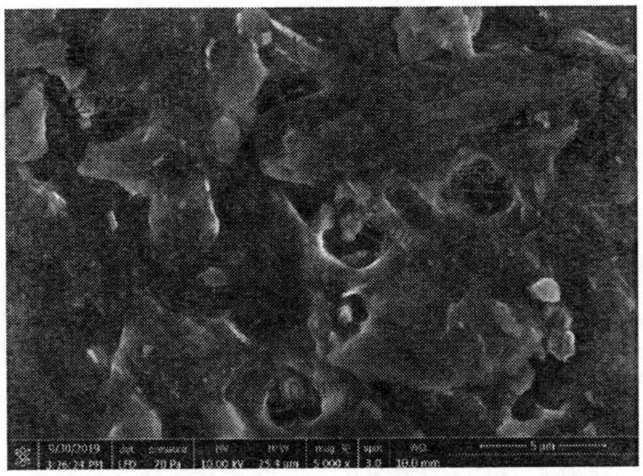

Figure 10. Enlargement of URC 03 underarm rubber morphology with SEM 5000x

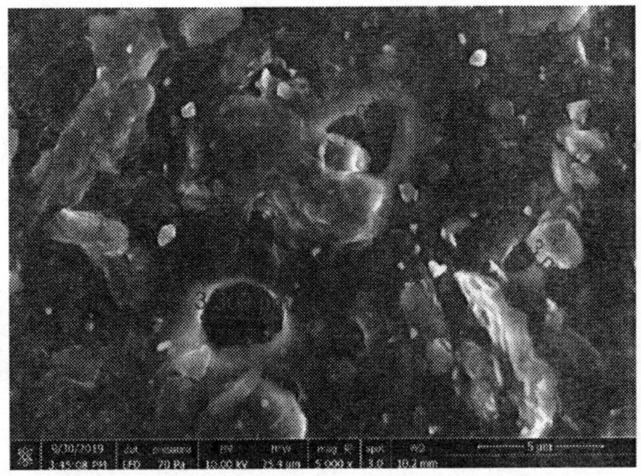

Figure 11. Enlargement of URC 04 underarm rubber morphology with SEM 5000x.

SEM analysis showed that the morphological differences were influenced by the ratio of sodium bicarbonate and additives used. The results of morphological testing using SEM with a magnification of 5000x showed that the air cavity of each sample was not uniform in size. The air cavity occurred due to the interaction between the rubber compound phase and the blowing agent [40]. In addition, the formation of non-uniform air cavities from each sample was influenced by the ratio of the addition of the blowing agent [41].

The results of morphological scanning with an electron microscope on underarm rubber vulcanisate samples using SEM in figures 8, 9, 10, and 11 with 5000 times magnification were intended to see the air cavity formed by CO_2 in rubber vulcanisate resulting from the decomposition of sodium bicarbonate by heat. From the pictures above it can be seen that there were differences in the air cavity formed between the vulcanized rubber molecules and additives. The air cavity formed by CO_2 in the picture above was a hole with a non-uniform diameter. The difference in the distribution of the blowing agent in each part during the formation of foam by CO_2 gas resulted in the formation of different air cavities [42]. This was due to the distribution of CO_2 in the underarm rubber vulcanisate that was not homogeneous, also there was a possibility that fillers and additives added were not homogeneously distributed.

Thus, the effect of the distribution of material influenced cross-linking by sulfur by producing non-uniform double bond strength. As a result, air cavities were formed with varying diameters. In figure 8 air cavity was formed between 1.021 μm – 934.6 nm, in figure 9 air cavity was formed between 1.290 nm - 836.7 nm, in figure 10 air cavity was formed between 1.021 μm - 772.1 nm, and in figure 11 air cavity was formed between 1.802 μm - 983.3 nm. In general, the diversity of air cavities formed by CO_2 on average was 5 μm.

4. Conclusion

The time of maturation and the formation of cross-link during the vulcanization process were influenced by the ratio of sodium bicarbonate. In addition, the speed of maturation of rubber compounds and the number of pores formed by CO_2 affected the strength of the cross-linking formed. The amount of CO_2 gas trapped in the rubber compound matrix was influenced by the ratio of sodium bicarbonate and the density of filler added. The amount of CO_2 formed also affected the mechanical properties of vulcanisate rubber. Besides, based on the results of tests using FTIR there were differences in functional groups, and based on the results of scanning with SEM there were differences in the diameter of the air cavity and the distribution of fillers from each treatment.

Acknowledgments
The authors thanked the Head of the Palembang Industrial Research and Standardization and the rubber laboratory technical team. We also thank the Director of PT. Shima Prima Utama Palembang together with rubber production unit technicians who have provided opportunities and cooperation for the implementation of this research at PT. Shima Prima Utama.

References

[1] Rasouli F, David H, Tim H, Ismet H and Kyle B R 2017 *Int. Conf. on Rehabilitation Robotics (ICORR) QEII Centre* (London: UK) pp 424–29
[2] Frizziero L, Giampiero D, Alfredo L, Giulia A, Grazia C M and Emma V 2019 *Appl. Sci.* **9** 5032
[3] Gogoi, R, Dayal S M, Utpal K N, Rakesh K K and Sarwar m A 2010 *J. Polym. Mater.* **279(1)** 1
[4] Chiu H T, Chiu S H and Wu J H 2003 *Journal of Applied Polymer Science* **89** 959
[5] Yanping Y and Qingxuan Z 2017 *Materials science: Materials Review* **1(1)** 1–8
[6] Latinwo, G K, David S A, Alfred A S and Semiu A K 2010 *Nature and Science* **8(6)** 23
[7] Chen L, Blizard K, Straff R and Wang X 2002 *Journal of Cellular Plastics* **38** 139
[8] Agarry S A, Ganiyu K L, Tinuade J A, and Semiu A K 2015 *American Journal of Polymer Science*. **5(1)** 1
[9] Usman M A, Adeosun S O, and Osifeso G O 2012 *Journal of Minerals & Materials Characterization & Engineering* **11(3)** 311–20
[10] Marhoon I I 2017 *IJSTR* **6**(8) 174
[11] Vijetha P, Prasanna Y K, Kumaraswamy K, Alka K, Pragati S and Satyasree N 2014 *RJPBCS* **5(5)** 851
[12] Rippel M M and Fernando G 2009 *J. Braz. Chem. Soc.* **20(6)** 1024
[13] Puskas J E and Chiang K 2014 *Researchgate*.
[14] Venkatachalam P, Natesan G, Palanivel S and Arjunan T 2013 *African Journal of Biotechnology* **12(12)** 1297
[15] Song W, Barber K and Lee K Y 2017 *Polymer* **118** 97
[16] Xie P, Wu G, Cao Z, Han Z, Zhang Y, An Y and Yang W 2018 *Polymers* **10(5)**
[17] Jin F L, Miao Z, Mira P, and Soo J P 2019 *Polymers* **11** 953
[18] Hussein, M S, Leng TP, Rahmat A R, Zainuddin F and Keat Y C 2018 *Solid State Phenomena* **280** 270
[19] Mohammad, N M, Mazliah Z N, Sharafina M N A, Asyraf M F B, Abdollah A M H and Effendy 2015 *Proc. of Malaysian Int. Tribology Conf.* 86.
[20] Ariff Z M, Zakaria Z , Tay L H and Lee S Y 2008 *Journal of Applied Polymer Science* **107** 2531
[21] Najib, N N, Ariff Z M, Manan N A, Bakar A A and Sipaut C S 2009 *Journal of Physical Scienc.* **20(1)** 13
[22] Pérez, T S, Solórzano E, Hilger A, Manke I and Rodríguez P M A 2018 *Eur. Polym. J.* **109** 169
[23] Kuska R, Milovanovic S., Frerich S, and Ivanovic J 2019 *J. Supercrit. Fluids.* **144** 71
[24] Wardana, A A, Suyatma N E, Muchtadi T R and Yuliani S 2018 *IFRJ* **25(5)** 1837
[25] Duan, Z, Jianzhong M, Chaohua X and Fuquan D 2014 *Journal of Cellular Plastics* **50(3)** 263–77
[26] Tangboriboon N, Sunisa R, Karat P and Anuvat S 2014 *Polimeri* **35(1-2)** 10
[27] Rosszainily, I R A, Salim M A, Mansor MR, Akop M.Z, Putra A, Musthafah M T, Hassan M Z, Rahman R M N and Sudin M N 2016 *Journal of Mechanical Engineering and Sciences* **10(2)** 2043
[28] Nasruddin and Tri S 2018 *Journal of Physics: Conf. Series* **1095** 012027
[29] Hossen F, Ali A, Kudrat E Z, Masuqul H, Rausan Z and Choudhury M Z 2020 *JMSRR* **5(3)** 1
[30] Poh B T and Tan E K 2001 *Journal of Applied Polymer Science* **82** 1352–1355
[31] Bakirtzis D, Delichatsiosa M A, Liodakis S and Ahmed W 2009 *Thermochimica Acta* **486** 11

IOP Conf. Series: Materials Science and Engineering **980** (2020) 012052 doi:10.1088/1757-899X/980/1/012052

[32] Fauzi M S, Du N L, Hakimah O and Supri A G 2015 *Sains Malaysiana* **44(6)** 869
[33] Roy, J K, Ruhul A K, Zakir H K M, Mubarak AK, Sabyasachy M and Sarwaruddin C A M 2012 *International Journal of Polymeric Materials* **61(8)** 571
[34] Chukwu, M N, Ekhator I and Ekebafe L O 2019 *NIJOTECH* **28(3)** 675
[35] Heideman, G, Jacques W M, Noordermeer and Rabin N D 2005 *Rubber Chemistry and Technology* **78(2)** 245–257
[36] Zanchet, A, Aline L, Bandeira D and Fabiula D B S 2017 *Recycling* **2** 1
[37] Gibala, D, Thomas D and Hamed G R 1999 *Rubber Chem Technol.* **72(2)** 357
[38] Mariano, R M, Helson M C, Marcelo R L O, Mayura M M R and Visconte L L Y 2008 *Journal of Applied Polymer Science* **110(4)** 1938
[39] Lopattananon, N, Kuljanee P, Kannika S and Bryan E 2006 *J. Appl. Polym. Sci.* **102(2)** 1974
[40] Ginting, E M, Bukit N, Muliani and Frida E 2017 *IOP Conf. Series: Materials Science and Engineering* **223** 012003
[41] Maged S, Sobhya D E, El-Nasharb and Maziad N A 2003 *Egypt. J. Sol.* **26(1)** 1
[42] Lan M S F D N U, Hakimah O and Supri A G 2015 *Sains Malaysiana* **44(6)** 869

IOP Conf. Series: Materials Science and Engineering **980** (2020) 012053 doi:10.1088/1757-899X/980/1/012053

Characterization of 17-4 PH stainless steel metal injection molding feedstock using mixing torque data

S Virdhian[1*], M Doloksaribu[1], S Supriadi[2], N M Balfas[2], B Suharno[3] and A D Shieddieque[4]

[1] Balai Besar Logam dan Mesin (Metal Industries Development Center), Ministry of Industry, Jl. Sangkuriang 12, Dago, Bandung, Jawa Barat, 40135, Indonesia

[2] Department of Mechanical Engineering, Faculty of Engineering University of Indonesia, Kampus UI Depok, Jawa Barat, 16424, Indonesia

[3] Department of Metallurgy and Materials Engineering, Faculty of Engineering University of Indonesia, Kampus UI Depok, Jawa Barat, 16424, Indonesia

[4] Department of Mechanical Engineering, STT Wastukancana, Jl. Alternative Bukit Indah, Purwakarta, Jawa Barat, 41151, Indonesia

Email: shinta-v@kemenperin.go.id

Abstract. Currently, there are many components produced by a metal injection molding process in automotive, consumer goods, medical, and electronics. Metal Injection Molding process (MIM process) consists of four stages, mixing, injection molding, debinding, and sintering. Feedstock plays critical roles in the MIM process since the feedstock's low quality cannot be corrected later. Feedstocks, which are a mixture of powder and binder, are mixed at an elevated temperature. A feedstock should be homogeneous and has a pseudo-plastic behavior. In the MIM process, the shear rate during injection molding is usually 10 to 10 000 s^{-1}. Within the shear rates range, a maximum viscosity for injection molding was 100 Pas at molding temperatures. In this paper, the rheological characteristic of feedstocks was analyzed using the torque rheometer. The objective of this research was to find the value of viscosity and compare to the Material Safety Data Sheet (MSDS) of the commercial feedstock by using torque mixing data. All the three feedstocks had pseudo-plastic behavior and below 100 Pas within shear rates range. Form the validation of injection molding experiment, feedstock B with solid loading 60 %, and binder system consists of 35 % PP, 64% PW, and 1 % SA showed a good flowability and moldability.

1. Introduction

Metal injection molding (MIM) is a manufacturing technology that uses the shaping advantage of the injection molding process to be applied to metal. The MIM products have characteristics of geometrical complexity combined with high properties because of their near theoretical density and homogenous microstructures [1-2]. MIM process is suitable for most common engineering metals such as carbon steel, stainless steel, tungsten, nickel-based alloys, titanium alloys.

The MIM process consists of four stages, mixing, injection molding, debinding, and sintering. Initially, powder and thermoplastics binder are mixed, kneaded, and pelletized to form the feedstock. Like in the plastic injection molding process, this feedstock is injected into a mold using an injection molding machine. The injection part is called the green part. The binder inside the green part is removed by heating, chemical extraction, or catalytic reaction. The binder removal process is called as debinding process. The final process is the sintering process to produce the last densified parts.

Binders play an essential role in the final quality of MIM products. There are some binder requirements for the MIM process. The binder must wet the powder particles, melt at low temperature for the injection molding process, easily be removed without leaving residues, be decomposed at a temperature that low enough to reduce chemical reaction with metal powder, provide sufficient green strength, provide lubricity, and be environmentally friendly [2]. The binder system consists of more than one component to prevent defects during the binder removal time. The binder system includes low molecular weight, high molecular weight polymer, and surfactant. The multi-components binder allows the binder removed gradually during the debinding process without promoting the defects [3-5]. The surfactant has the function to improve the fluidity of powder and the viscosity of the feedstock. Feedstock plays critical roles in the MIM process since the quality of the feedstock affects the next process.

The MIM feedstock has a pseudo-plastic behavior that the viscosity decreases with increasing shear rate. In the MIM process, the shear rate during injection molding is usually in the range of 10 to 10 000 s^{-1}. Within the shear rates range, a maximum viscosity for injection molding is 100 Pas at molding temperatures [2]. Torque rheometers (TR) are widely applied in polymers and the design of compound. The processing conditions are examined through measured torque and temperature. If the process conditions are the same, different materials flow behavior can be compared relatively [6]. The research objectives are to examine the MSDS rheology data with experimental torque data of commercial feedstock and characterize the torque and rheology properties of the feedstock with the variation of feedstock compositions.

2. Materials and methods

2.1 Materials

In this study, the 17-4 PH stainless steel (17-4 PH SS) powders were mixed with a multi-component binder consisted of paraffin wax (PW), high-density polyethylene (HDPE), polypropylene (PP), and stearic acid (SA). Stearic acid was added to act as a surfactant to the feedstock to improve the powder's wetting. The torque rheology characteristic was compared to the commercial 17-4 PH SS feedstock. Table 1 shows the powder properties of the used powder and industrial feedstock using the Material Safety Data Sheet (MSDS). Table 2 shows the chemical composition of the metal powders using the MSDS data. The morphology of the powder can be seen in figure 1. The morphology of 17-4 PH SS powders was irregular shapes with particle size D_{50} around 7 μm.

Table 1. Powders used in this study.

	Commercial feedstock powder	Used powder
Powder Particle Size	D_{10} 2.8 μm D_{50} 6.9 μm D_{90} 18.6 μm	D_{10} 2.89 μm D_{50} 6.9 μm D_{90} 17.91 μm
Material	17-4PH	17-4PH
Supplier		Epson ATMIX

Table 2. Chemical composition of 17-4 PH SS powder used in this study.

Elements	C	Cr	Co	Cu	Fe	Mn	Si	Nb	Ni	O	P
% Weight	0.04	16.4	0.03	3.96	74.26	0.16	0.39	0.29	4.04	0.35	0.02

IOP Conf. Series: Materials Science and Engineering **980** (2020) 012053 doi:10.1088/1757-899X/980/1/012053

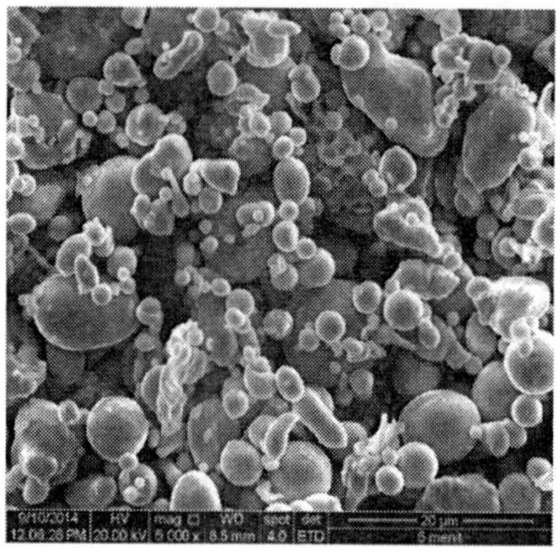

Figure 1. The SEM images of 17-4 PH stainless steel powder.

2.2 Torque rheometers

Torque caused by the resistance of a material to the applied shear is measured by torque rheometer. It is usually used for quality control and compared the rheological characteristics of the polimer and food industries. In this study, the mixing process was carried out using Rheomix from Polylab to prepare the MIM feedstock at different solid loading and binder composition. The mixing temperature was set at 190°C for 1 hour. The feedstocks were tested at different rotation speeds at specific temperatures to obtain the mixing torque data. The compositions of the mixed feedstocks are shown in table 3. Figure 2 shows the Rheomix Polylab that used in this study and the specifications.

Table 3. Feedstock compositions.

Feedstock	Solid loading	Binder system			
		PW (wt %)	PP (wt %)	HDPE (wt %)	Stearic Acid (wt %)
Feedstock A	60	64	0	35	1
Feedstock B	60	64	35	0	1
Feedstock C	55	64	35	0	1

Item	**Rheomix 3000**
Chamber Volume	**625 cm^3**
With Rotor	**310 – 541 cm^3**
Gear ration	**3:2 (optional 2:3)**
Max. Speed	**250 min^{-1}**
Max. Torque	**300 Nm**
Max. Temperature	**400°C**
Temperature Control	**3 zones electric heating and air cooling**

Figure 2. Rheomix polylab.

2.3 Shear rate and feedstock viscosity

In this paper, the commercial feedstock's viscosity and three compositions feedstock viscosity are characterized using torque mixing data, according to Bousmina et.al. [7]. To find the shear rate - viscosity using the simple analogy of Couette for each cylinder and then calculating the overall torque given by the two cylinders in the mixed liquid. The effect of the aperture that exists between the two sets of cylinders is negligible. Then the shear rate and feedstock viscosity equations as following Bousmina et.al. [7]:

$$\dot{\gamma} = 16\,\pi\,N \frac{\beta^2}{(1+\beta)^2(\beta^2-1)} \approx 2\pi N/\ln(\beta) \tag{1}$$

where $\dot{\gamma}$ is shear Rate (s-1), N is rotor rotation (rpm), where β is the ratio of the mixing chamber (R_c)diameter with rotor diameter (R_b).

$$\eta = \frac{T}{N}\frac{(\beta^2-1)}{8\pi^2 L\, R_C^2(1+g^2)} \tag{2}$$

where η is viscosity (Pa.s), T is mixing torque (N.m), L is rotor length (m), g is gear ratio N_2/N_1.

3. Results and discussion

3.1 Rheology characterization of commercial feedstock

The mixing torque is an indicator of the viscosity of the mixture. Figure 3 shows the mixing torque as a function of the mixing time at various rotation speeds at a specific temperature. Figure 3a shows the mixing torque at 170°C and figure 3b at 190°C. In the beginning, when the powder was added to the mixture, the mixing torque increased to a high value because it had not reached a steady state yet. When the mixture became homogenous, the torque reached equilibrium value. Figures 3a and 3b show the effect of the temperature into mixing torque. As the temperature increased, the mixing torque decreased.

Figure 4 shows commercial feedstock's viscosity derived from torque data using the Bousmina model [7]. It can be seen from figure 4 that the viscosity of the feedstock declined with increasing the shear rate. The results showed that the feedstock mixture had pseudo-plastic properties following the characteristics of the MIM feedstock [8]. The viscosity of the commercial feedstock decreased with increasing temperature. Figure 4 also compared the commercial feedstock's viscosity from the MSDS data and from mixing torque data. The results showed the same trendline between the MSDS data and calculated viscosity. However, there was a small gap between the MSDS data's viscosity and calculated from mixing torque data. This result agreed with other researchers that the Bousmina model was best suited for pure polymers [9].

3.2 Rheology characterization of the mixed feedstock

The effects of solid loading and binder composition on the mixing torque and viscosity were examined. Figure 5 shows the mixing torque data versus rotation speed for different compositions. Figure 6 shows the viscosity calculated from mixing torque data for three different feedstocks. The results showed a similar trend for three feedstock compositions, and feedstock A and B had the same solid loading but different binder composition. Feedstock A used the HDPE as a backbone polymer, and Feedstock B uses PP as a backbone polymer. Feedstock A exhibits a higher viscosity than Feedstock B because of the HDPE had a higher viscosity than PP [10-11]. Feedstock C had the lowest viscosity among others because its low solid loading meant lower metal powder content in the mixture. Moreover, all the three feedstocks had pseudo-plastic behavior below 100 Pas within shear rate ranging from 10 to 10 000 s^{-1}. Form the validation of injection molding experiment, feedstock B with solid loading 60 %, and binder system consistsed of 35 % PP, 64% PW, and 1 % SA showed a good flowability and moldability.

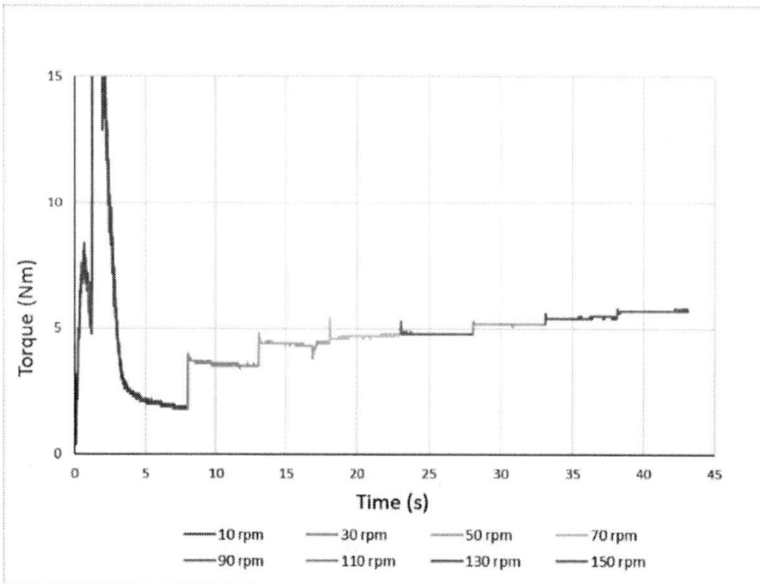

(a)

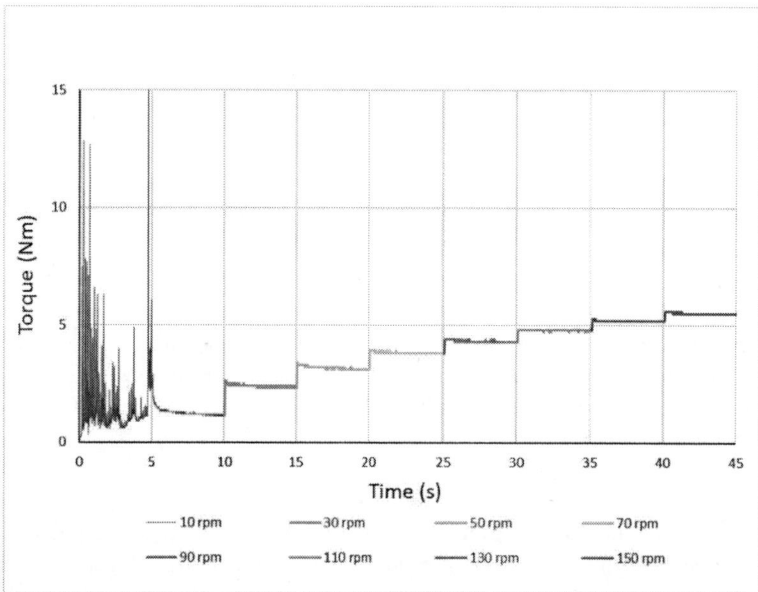

(b)

Figure 3. Torque measurement results of commercial feedstock with the various rotation speed at a temperature a) 170°C and b) 190°C.

IOP Conf. Series: Materials Science and Engineering **980** (2020) 012053 doi:10.1088/1757-899X/980/1/012053

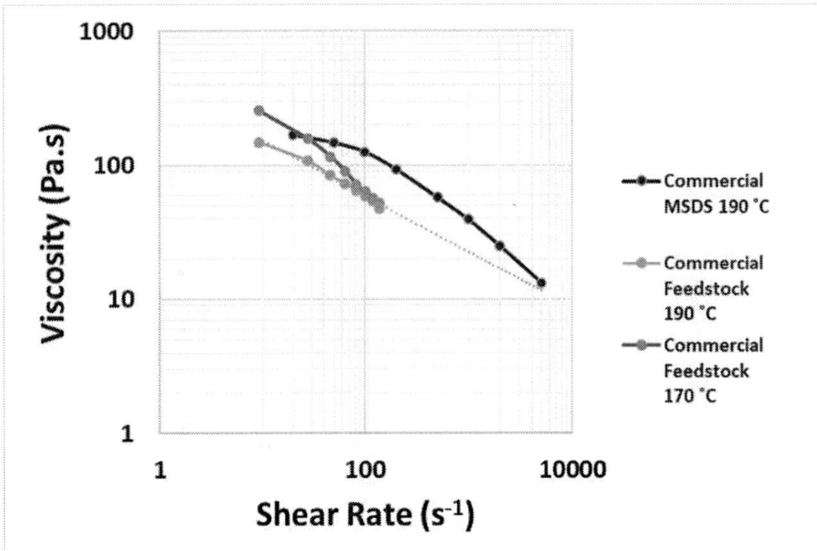

Figure 4. The calculated viscosity from torque data using the Bousmina Model.

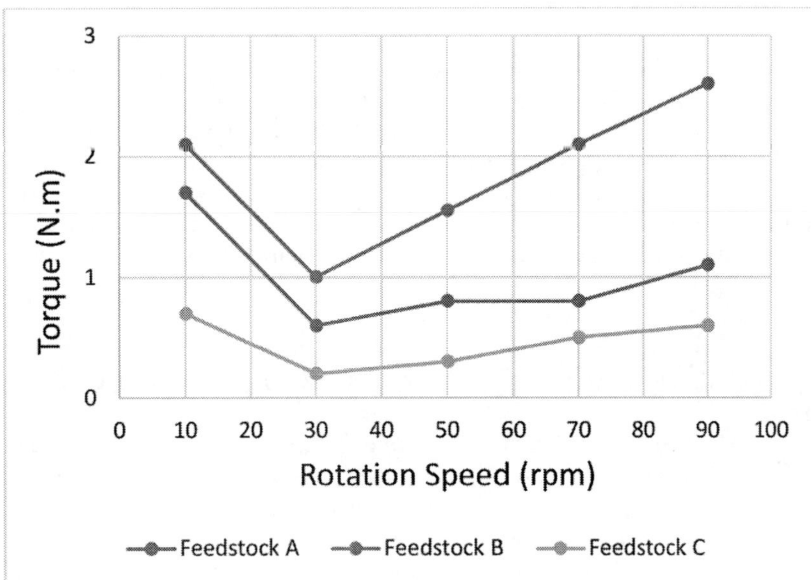

Figure 5. The mixing torque for three different feedstock composition.

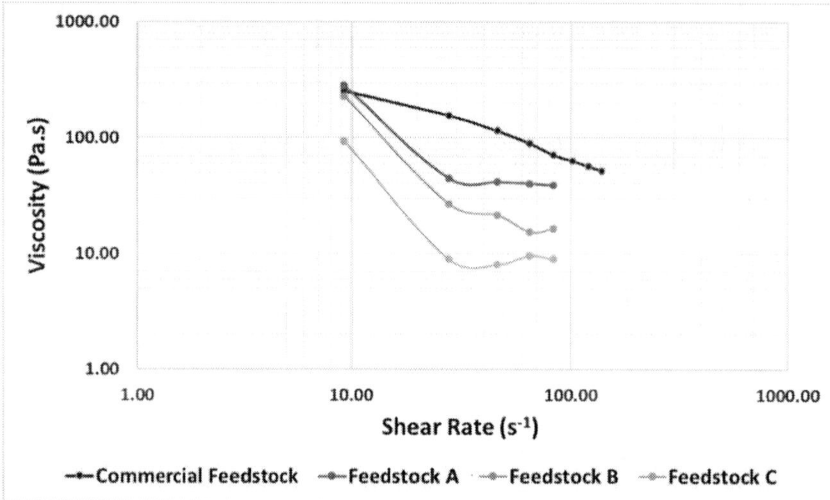

Figure 6. The calculated viscosity from torque data using the Bousmina Model for three different feedstock compositions.

4. Conclusion

The rheological characteristic of feedstocks was analyzed using the torque rheometer. The mixing torque data was used to find the value of shear rate and viscosity. Using the Bousmina Model, the torque data from the torque rheometer could be converted into viscosity value. The viscosity calculated from the mixing data showed a similar trendline with the MSDS data's viscosity value. However, the viscosity value was lower than the MSDS Data. The results in agreement with other researchers that the Bousmina Model was suitable for a pure polymer. The feedstocks (A, B, C) showed pseudo-plastic behavior and had viscosity below the commercial feedstock and fulfilled the requirement for the feedstock viscosity for metal injection molding. From the validation of injection molding experiment, feedstock B with solid loading 60 %, and binder system consisted of 35 % PP, 64% PW, and 1 % SA showed a good flowability and moldability.

Acknowledgments

The author would like to thank Yoshida Shunsuke from Pacific Showa for the powder materials used in this study.

References

[1] Deaney D 2012 *Handbook of Metal Injection Molding* (Cambridge: Woodhead Publishing Series)

[2] German R M 2011 *Metal Injection Molding, A Comprehensive MIM Design Guide Metal Powder Industries Federation* (Princeton New Jersey USA)

[3] German R M and Bose A 1997 *Injection Molding of Metal and Ceramics Metal Powder Industries Federation*

[4] Virdhian S, Osada T, Kang H G, Tsumori F and Miura H 2012 Evaluation and analysis of distortion of complex shaped Ti-6Al-4V compacts by metal injection molding process *Key Engin. Materials* **520** 187-94

[5] Miura H, Shinozaki R, Virdhian S and Osada T 2010 Evaluation and analysis of distortion during MIM process of titanium alloy materials *J. Jpn. Soc. Powder Powder Metall.* **57** 352-6

[6] Gang J, Zhao X L and Wang M M 2011 Evaluation of rheological measurement of polymer blends melts in torque rheometer *Adv. Materials Res.* **314-316** 1424-9

[7] Bousmina M, Ait-Kadi A and Faisant J B 1999 Determination of shear rate and viscosity from batch mixer data *J. of Rheology* **43(2)** 415–33
[8] Schramm G 2004 *A Practical Approach to Rheology and Rheometry* 2nd edition (Gebrueder HAAKE Gmbh)
[9] Santi C R, Jr E H, Correa C A and Vlachopoulos J 2009 Torque Viscometry of Molten Polymers and Composites *Appl. Rheol*ogy **19 (1)** 13148-1-13148-7
[10] Ibrahim M H I, Muhamad N and Sulong A B 2009 Rheological investigation of water atomized stainless steel powder for micro metal injection molding *Int. J. of Mechanical and Materials Eng. (IJMME)* **4(1)** 1-8
[11] Matula G, Tomiczek B, Krol M, Szatkowska A and Sotomayor M E 2018 Application of thermal analysis in the selection of polymer components used as a binder for metal injection moulding of Co–Cr–Mo alloy powder *J. of Thermal Analysis and Calor.* **134** 391–9

The initial study on the synthesis and characterization of $NaZr_2(PO_4)_3$ from West-Borneo $ZrSiO_4$

R Septawendar[1]*, F Arifiadi[1], D Taufik[1,2], K Wahyudi[1], and Suhanda[1]

[1]Researchers at Center for Ceramics, Ministry of Industry of Indonesia
[2]Department of Engineering Physics, Faculty of Industrial Technology, Institut Teknologi Bandung, Gd. TP Rachmat (Labtek VI) Jl. Ganesha 10, Bandung 40132, Indonesia

Email: rifkiseptawendar@yahoo.com

Abstract. Due to its excellent electrical properties, $NaZr_2(PO_4)_3$-type sodium superionic conductor is a preferred material for the application in energy storages such as sodium ion batteries. Here, the initial study on the synthesis of $NaZr_2(PO_4)_3$ from Indonesian West-Borneo Zircon is reported. The synthesis was conducted using our facile technique of a sodium carbonate sintering method. Zircon and the carbonate were mixed at a stoichiometry ratio approach and sintered at 1000°C. The sintered product was leached with water to dissolve the silica compound and continued by filtration and washing to obtain the zirconia compound, followed with phosphoric acid blending until pH 5 to get a precipitate. It was calcined at 800°C. The mineralogy, chemical composition, and particle sizes of the calcined product were evaluated. The synthesized sodium superionic conductor consists of $NaZr_2(PO_4)_3$ with a trigonal structure as the main phase and SiO_2 as the minor phase. The SiO_2 content is accounting 6% of the total. Particle size analysis measured 90% of the total particle size distribution of the calcined NZP is less than 18.283 μm in sizes, while the average particle size is about 5.633 μm. The present study unveils a facile technique in the synthesis of $NaZr_2(PO_4)_3$ from Indonesian West-Borneo $ZrSiO_4$.

1. Introduction

Sodium zirconium phosphate known as NZP is an interesting inorganic material due to its unique crystalline structure and properties, such as low thermal expansion, fast and high ionic conductivity, great chemical stability, and high radiation resistance [1-5]. Thus, NZP based material finds application in superionic conductors, the immobilization of radioactive wastes, sensor, and catalysis [3,4,6-8]. In addition, NZP is also a parent compound of sodium superionic conductor (NASICON) comprising no Si [9-11], a ceramic ion conductor that has a general formula of $Na_{1+x}Zr_2P_{3-x}Si_xO_{12}$ [12-14].

Many methods have been introduced to synthesize NZP materials, such as a conventional solid-state method, solution routes, molten salts, sol-gel processing [3], a microwave technology [1,2,6], hydrothermal synthesis [7], a chemical solution deposition technique [10] and so on. Nevertheless, most of those synthesis methods are commonly used high chemical grades of salts and organometallic compounds for NZP precursor materials and specific instruments, which are less economic and may show some drawbacks when the technologies are applied in a large-scale.

In this paper, the initial study on the synthesis and characterization of NZP from Indonesian West-Borneo zircon ($ZrSiO_4$) is reported using a low-cost, facile technique of a sodium carbonate sintering method. Our previous works demonstrated the method accomplishment in the preparation of nanomaterial ZrO_2 from $ZrSiO_4$ [15-18]. A low-cost, facile technique of sodium carbonate sintering method basically is the alkaline disintegration process of $ZrSiO_4$ compound in a solid form through sintering process at high temperatures. Meanwhile, $ZrSiO_4$ is a valuable and potential material as a ZrO_2 source with its content achieving ~ 65wt.% [15-16]. Thus, the present work focuses on the synthesis of $NaZr_2(PO_4)_3$ from local $ZrSiO_4$ as the ZrO_2 precursor and other technical grade chemicals via a sodium carbonate sintering method. The initial characteristics including crystalline phases, chemical composition, and particle sizes of the synthesized NZP are presented in this paper.

2. Experimental Methods

2.1. Materials

The main chemical as the ZrO_2 source used in this initial study was the micronized Indonesian West Borneo zircon ($ZrSiO_4$) obtained from PT. Monokem Surya Indonesia. The chemical composition of zircon consisted of 64.51% of ZrO_2, 31.31% of SiO_2, and 0.95% of HfO_2 as major components. Meanwhile, other chemicals used were technical grade sodium carbonates (Na_2CO_3), phosphoric acid, carboxyl methyl cellulose (CMC) obtained from a local-chemical market. All chemicals were used without further purification.

2.2. Synthesis of NZP from $ZrSiO_4$

Sodium zirconium phosphate (NZP) was synthesized from $ZrSiO_4$ via a low-cost, facile technique of sodium carbonate sintering method. Zircon was combined with sodium carbonate in solid mass at stoichiometry ratio approach assisted with an organic binder of CMC. Solid mixtures were then sintered at 1000°C in a furnace. The sintered product was then leached by water to dissolve the silica compound, leaving a zirconium compound in residues. The residues were then mixed with phosphoric acid until the pH 5, resulting in white precipitates. The products were filter-pressed, washed several times, dried in an oven, and then calcined at 800°C. The final product was characterized including crystalline phases, chemical composition, and particle sizes.

2.3. Characterization

The mineralogy of the calcined product was investigated by powder X-ray diffraction (XRD) instrument (XRD, D/teX Ultra 250 detector, Rigaku) at 40 Kv and 30 mA with Cu/Kα radiation source. The diffraction patterns were scanned from 2.0000 to 65.0000 (2θ) with an angular step of 0.0100. Furthermore, a Thermo ARL 9900 X-ray fluorescence system was used to analyze chemical compositions of the synthesized NZP. Then, the particle sizes distribution of the synthesized NZP was measured applying a Malvern mastersizer.

3. Results and Discussion

In the synthesis of of $NaZr_2(PO_4)_3$ from Indonesian West-Borneo Zircon through a sodium carbonate sintering method, $ZrSiO_4$ was initially mixed with the alkaline of Na_2CO_3 in a solid mass. This product was then sintered at 1000°C in a furnace, resulting in a mixture of Na_2ZrO_3 and Na_2SiO_3 compounds (as given in the equation 1) [15,17].

$$ZrSiO_4 + 2Na_2CO_3 \xrightarrow{\uparrow at\ T\ 1000℃} Na_2SiO_3 + Na_2ZrO_3 + 2CO_2 \quad (1)$$

However, the separation of Na_2ZrO_3 and Na_2SiO_3 compounds was conducted by water-leaching because of a water-soluble compound of Na_2SiO_3. This process led to the formation of the aqueous solution of Na_2SiO_3 and left residual of Na_2ZrO_3, as shown in the equation 2 [15,17].

$$Na_2SiO_3 + Na_2ZrO_3 \xrightarrow{H_2O} 2\mathrm{Na}^+SiO_2^{2-}{}_{(aq)} + Na_2ZrO_{3_S} \quad (2)$$

The products in the equation 2 were then separated through a filtration technique. The residue of Na_2ZrO_3 was further reacted with phosphoric acid at approximately pH 5 to give white fine precipitates such as Balm-like solids. The precipitates were washed with aquadest for several times to remove the existing silica compound that attached on the precipitate surfaces and then dried in an oven. The final product was the calcined at 800°C. The calcined product was characterized including mineralogy, chemical composition, and particle sizes.

3.1. The result of XRD investigation of the calcined product

The XRD investigation result on the as-synthesized $ZrSiO_4$-NZP based after calcination at 800°C was demonstrated in figure 1. As according to PDF2 No.00-033-1312, sodium zirconium phosphate ($NaZr_2(PO_4)_3$) with a trigonal crystal system was detected by the appearance of its main peaks at the diffraction angles, 2θ, of 19.38°, 20.07°, 23.27°, and 31.04°, corresponding to the (104), (110), (211), (116) crystal planes of the $NaZr_2(PO_4)_3$ structure. Meanwhile, the minor crystalline phase of SiO_2 in the form of cristobalite polymorph was found at a diffraction angle, 2θ, of 21.86°, corresponding to the (101) crystal plane of the tetragonal cristobalite structure. The SiO_2 finding was based on PDF2 No.00-076-0939.

Based on the XRD result in figure 1, it could be concluded that $NaZr_2(PO_4)_3$ NASICON-type was successfully synthesized from local $ZrSiO_4$ through a sodium carbonate sintering method. The residue of Na_2ZrO_3 in the equation 2 was completely reacted with phosphoric acid (H_3PO_4) which resulted a new compound of sodium zirconium phosphate ($NaZr_2(PO_4)_3$) after calcination process at 800°C, as indicated by equation 3.

$$2Na_2ZrO_{3(s)} + 6H_3PO_{4(l)}, pH\ 5 \rightarrow NaZr_2(PO_4)_3 + 3NaH_2PO_4 + 6H_2O \quad (3)$$

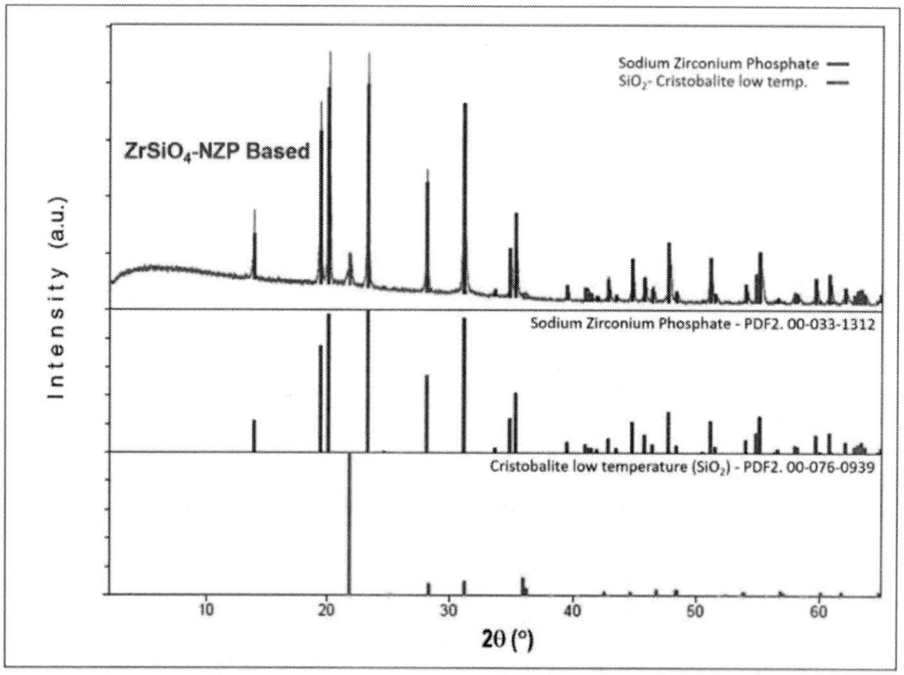

Figure 1. The XRD identification results on the as-synthesized NZP after calcination at 800°C.

However, the SiO_2 existence in NZP was probably caused during removing a soluble silica compound from a sintered product in the preparation of NZP from $ZrSiO_4$. Since the sintered product was leached by water to dissolve the silica compound, there should not have been the material in the surface of the zirconium compound especially after water-washing treatment. Nevertheless, when the silica compound exhibited very fine particles, it would infiltrate into a solid residue of zirconia compound, so the silica compound removing was slightly difficult to be controlled from the zirconia compound. Therefore, the infiltrated silica compound inside the zirconia compound would have transformed to cristobalite in the next calcination process at a higher temperature. Nevertheless, the NZP synthesized from the local raw materials in this work was obtained at a relative low temperature of 800°C.

In addition, our result seems to be different from the finding of Luo *et al.* [1]. They reported that pure NZP phase had been successfully prepared at 1100 °C. They used sodium carbonate (Na_2CO_3), ammonium dihydrogen phosphate ($NH_4H_2PO_4$), and zirconyl nitrate hydrate ($ZrO(NO_3)_2{\cdot}xH_2O$) as NZP precursors and a microwave-assisted solid-state reaction technique. The issue of the different crystallization temperature of the NZP phase commonly should be governed by the particular synthesis conditions, such as the precursor types, pH and media, temperature, etc.

3.2. Chemical compositions of NZP

The chemical compositions of the synthesized NZP from $ZrSiO_4$ were analyzed using a Thermo ARL 9900 X-ray fluorescence system whose result is presented in table 1. The chemical analysis in table 1 shows the SiO_2 presence in NZP sample after calcination at 800°C. It is accounted 6% of the total. This result corroborates the XRD investigation result in the calcined NZP.

Table 1. The chemical analysis results on the as-synthesized NZP after calcination at 800°C.

Compound	m/m (%)	Std.Err.	**Element**	m/m (%)	Std.Err.
P_2O_5	39.93	0.24	**Px**	17.43	0.11
ZrO_2	34.14	0.24	**Zr**	25.27	0.18
Na_2O	13.21	0.17	**Na**	9.80	0.13
SiO_2	6.07	0.12	**Si**	2.84	0.06
SO_3	1.37	0.06	**Sx**	0.547	0.023
HfO_2	0.420	0.021	**Hf**	0.356	0.018
Al_2O_3	0.372	0.019	**Al**	0.198	0.010
MgO	0.144	0.007	**Mg**	0.0867	0.0043
MoO_3	0.135	0.010	**Mo**	0.0898	0.0065
Y_2O_3	0.0890	0.0044	**Y**	0.0701	0.0035
Fe_2O_3	0.0676	0.0034	**Fe**	0.0473	0.0024
CaO	0.0667	0.0033	**Ca**	0.0477	0.0024
K_2O	0.0261	0.0014	**K**	0.0217	0.0011
TiO_2	0.0258	0.0014	**Ti**	0.0155	0.0008
Am_2O_3	0.0247	0.0041	**Am**	0.0225	0.0037
La_2O_3	0.0158	0.0022	**La**	0.0135	0.0018
Yb_2O_3	0.0140	0.0019	**Yb**	0.0123	0.0017
Sc_2O_3	0.0046	0.0010	**Sc**	0.0030	0.0006

As previously mentioned, after having leached a sintered product wisely by water solvent, a soluble silica compound of the sintered product was dissolved in the solution and some of its particles attached in a residue of zirconia compound surface. After purification of the residue by water-washing and filter-press treatment were done, there should be no silica in zirconia compound. However, some of the soluble silica compound particles still persisted inside the zirconia compound residue. This phenomenon was proven by silica yield after calcination process on the as-synthesized NZP at 800°C (table 1). Thus, a soluble silica compound is proposed consisting of fine particles, which may infiltrate into a zirconia compound.

3.3. Particle size measurement

Figure 2 demonstrates a particular particle size distribution of the NZP calcined at 800°C, exhibiting a size range of 0.12 –50 µm with an average particle size (D50) of 5.633 µm. In addition, according to the presented volume density curve in figure 2, the calcined NZP consists of particles sizes below 1.481 µm with the population of 10%. Meanwhile, 90% of the total particle size distribution of the calcined NZP is less than 18.283 µm in sizes. According to figure 2, the particle sizes of NZP calcined at 800°C shows a quite wide distribution range, indicating that the NZP particles exhibit dissimilarity.

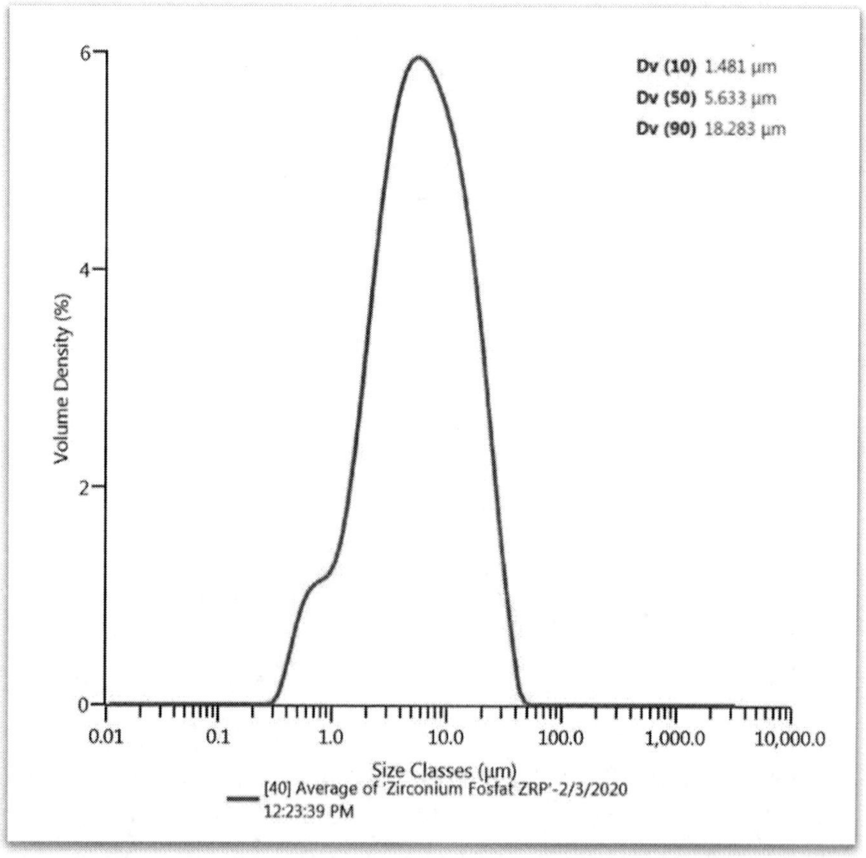

Figure 2. Particle size distribution measured by a Malvern mastersizer of the as-synthesized NZP calcined at 800°C.

4. Conclusion

Sodium zirconium phosphate NASICON-type, $NaZr_2(PO_4)_3$, was successfully synthesized from Indonesian West Borneo Zircon via a facile technique of a sodium carbonate sintering method. The initial study shows that the synthesized NASICON consisted of $NaZr_2(PO_4)_3$ (NZP) with a trigonal structure as the main phase and SiO_2 as the minor phase. Besides, the SiO_2 content is accounting 6% of the total. Meanwhile, particle size analysis measured 90% of the total particle size distribution of the final product of NZP is less than 18.283 μm in sizes, while the average particle size is about 5.633 μm.

References

[1] Luo P, Wang J, Wang J, Li M, Zhong L, Yang K and Wen J 2018 Low-temperature rapid synthesis of sodium zirconium phosphate by microwave-assisted solid-state method *Journal of the Australian Ceramic Society*

[2] Wang J, Luo P, Wang J, Zhan L, Wei Y, Zhu Y, Yang S and Zhang K 2019 Microwave-sintering preparation and densification behavior of sodium zirconium phosphate ceramics with ZnO additive *Ceramics International*

[3] Agrawal D K and Adair J H 1990 Low-temperature sol-gel synthesis of $NaZr_2P_3O_{12}$ *J. Am, Ceram. Soc.* **73** 2153

[4] Shrivastava O P, Kumar N and Chourasia R 2007 Synthesis, crystallographic characterization and ionic conductivity of iron substituted sodium zirconium phosphate $Na_{1.2}Zr_{1.8}Fe_{0.2}(PO_4)_3$ *J. Mater. Sci.* **42** 2551

[5] Sukhanov M V, Pet'kov V I, Firsov D V, Kurazhkovskaya V S and Borovikova E Y 2011 Synthesis, structure, and thermal expansion of sodium zirconium arsenate phosphates *Russian Journal of Inorganic Chemistry* **56** 1351

[6] Vaidhyanathan B, Agrawal D K and Roy R 2004 Microwave-assisted synthesis and sintering of NZP compounds *J. Am. Ceram. Soc.* **87** 834

[7] Yan-Qing W, Zhao-Jia L and Ming-Lin G 2015 Hydrothermal synthesized microflower sodium zirconium phosphate and its transformation to zirconium hydrogen phosphate *Materials Letters* **159** 197

[8] Kimpa M I, Mayzan M Z H, Yabagi J A, Nmaya M M, Isah K U and Agam M A 2018 *IOP Conf. Ser.: Earth Environ. Sci.* **140** 012156

[9] Nagai M, Fujitsu S and Kanazawa T 1980 Ionic Conductivity in the System $NaZr_2(PO_4)_3$-$Na_3Y_2(PO_4)_3$. *Journal of The American Ceramic Society -Discussions and Notes* 63 476

[10] Ihlefeld J F, Gurniak E, Jones B H, Wheeler D R, Rodriguez M A and McDaniel A H 2016 Scaling effects in sodium zirconium silicate phosphate ($Na_{1+x}Zr_2Si_xP_{3-x}O12$) ion-conducting thin films *J. Am. Ceram. Soc.* 1–8

[11] Small L J, Wheeler J S, Ihlefeld J F, Clem P G and Spoerke E D 2018 Enhanced alkaline stability in a hafnium-substituted NaSICON ion conductor *Mater. Chem. A* **6** 9691

[12] Kim J, Jo S H, Bhavaraju S, Eccleston A and Kang S O 2015 Low temperature performance of sodium–nickel chloride batteries with NaSICON solid electrolyte *Journal of Electroanalytical Chemistry* **759** Part 2 201

[13] Pratiwi M, Purwaningsih H, Widyastuti, Fajarin R and Setyawan H 2017 *IOP Conf. Series: Materials Science and Engineering* **202** 012083

[14] Hou M, Liang F, Chen K, Dai Y and Xue D 2020 *Nanotechnology* **31** 132003

[15] Septawendar R, Sutardi S, Karsono K and Sofiyaningsih N 2016 A low-cost, facile method on production of nano zirconia and silica from local zircon in a large scale using a sodium carbonate sintering technology *Journal of The Australian Ceramic Society* **52** 92

[16] Taufik D, Abdullah M, Hernawan, Sutardi S and Septawendar R 2016 Mineral phase, microstructure, and Infra-Red characteristics of calcia-stabilized zirconia nanocrystallines synthesized from local zircon and slaked lime *Journal of Ceramic Processing Research* **17** 1063

[17] Septawendar R, Nuruddin A, Sutardi S, Maryani E, Asri L A T W and Purwasasmita B S 2018 Low-temperature metastable tetragonal zirconia nanoparticles (NpMTZ) synthesized from local zircon by a modified sodium carbonate sintering method *Journal of the Australian Ceramic Society* **54** 643

[18] Septawendar R, Nuruddin A, Sutardi S, Asri L A T W, Abdullah M R, Adhika D R, Maryani E, and Purwasasmita B S 2019 Nano cellulose-assisted synthesis of zirconia nanorods from local zircon-based Zr(OH)4 precursors *Journal of the Ceramic Society of Japan* **127** 864-868

IOP Conf. Series: Materials Science and Engineering **980** (2020) 012055 doi:10.1088/1757-899X/980/1/012055

The effect of chelating agents on the formation of manganese oxide (MnO) in the synthesis of sodium manganese oxide ($Na_2Mn_3O_7$)

H A Marlina[1,8]*, K Sebayang[1,6]*, S Gea[2,7]*, Z Noer[1,5], R Septawendar[3], and B Sunendar[4]

[1]Department of Physics, Faculty of Mathematics and Natural Sciences, Universitas Sumatera Utara, Jl. Bioteknologi No. 1, Medan 20155, Sumatera Utara, Indonesia
[2]Department of Chemistry, Faculty of Mathematics and Natural Sciences, Universitas Sumatera Utara, Jl. Bioteknologi No. 1, Medan 20155, Sumatera Utara, Indonesia
[3]Center for Ceramics, Ministry of Industry of Indonesia, Jl. Akhmad Yani 392, Bandung 40272, Jawa Barat, Indonesia
[4]Advanced Materials Processing Laboratory, Department of Engineering Physics, Institut Teknologi Bandung, Bandung 40132, Jawa Barat, Indonesia
[5]Integrated Research Laboratory, Universitas Sumatera Utara, Jl. Tridarma No. 8, Medan 20115, Sumatera Utara, Indonesia

Emails: keristasebayang@usu.ac.id, s.gea@usu.ac.id, ayuhilda89@gmail.com

Abstract. This current study successfully synthesized $Na_2Mn_3O_7$ in two steps. Firstly, the synthesis of MnO from NaCl and $MnCl_2.4H_2O$ precursors. NaCl was mixed with citric acid as a chelating agent, while $MnCl_2.4H_2O$ was mixed with two types of chelating agents (citric acid and 1% chitosan). The solutions of [NaCl-a chelating agent] and [$MnCl_2.4H_2O$-a chelating agent] were stirred for ±2 hours. The solutions were hydrothermally heated at 150°C for 6 hours and then calcined at 800°C for 1 hour in a nitrogen condition. The MnO presence was then analyzed using an XRD method. As a comparison, another sample was water leached. Then, it was examined by an XRD method. Secondly, the synthesis of $Na_2Mn_3O_7$ from MnO and Na_2CO_3 powder was carried out through a solid method. The calcination was carried out at 800°C for ±3 hours in an oxidizing atmosphere. The XRD results showed the presence of $Na_2Mn_3O_7$ in the final products of both samples. Based on these experimental results, the sample [NaCl-citric acid; $MnCl_2.4H_2O$-1%chitosan] showed higher purity than the sample [NaCl-citric acid; $MnCl_2.4H_2O$-citric acid], so that it had a higher intensity of the $Na_2Mn_3O_7$ phase than the sample [NaCl-citric acid; $MnCl_2.4H_2O$-citric acid]. Besides, the Cl^- ions removal after calcination greatly affected the intensity of the MnO and $Na_2Mn_3O_7$ formed.

1. Introduction

In order to achieve current and future energy needs, humans need electrical energy. Energy resources such as wind, sun, coal, nuclear, and ocean waves do not meet the competent, permanent, consistent, and inexpensive energy storage requirements [1]. As an alternative, electrochemical energy storage systems answer these challenges by providing a constant and economical distribution of energy [2-4].

Due to its high energy density and long life cycle, lithium ion batteries (LIB) is considered the most popular electrochemical energy storage system. However, the lack of availability of their raw materials limits the application of LIB [5]. For this reason, another ion batteries with high abundance and low cost raw materials are required, such as sodium ion batteries (SIB) which attract great attention as alternative energy storage systems [6–8].

Several cathode materials with high energy densities to improve SIB performance have been investigated [7–13]. However, the most interesting thing to study is manganese oxide (MnO) because it is non-toxic and low cost, and it strongly influences high energy density. Therefore, it is ideal for SIB cathode application and the development of SIB [8].

Various Na_xMnO_2 structures are found with changes in the value of x. Generally, a compound with the x value lower than 0.44 has a full tunnel structure. A compound with the mixture of tunnels and layered structures is formed when x value lower than 0.66. Besides that, a compound with the x value lower than or equal to 1 has a fully layered [14]. These structures affect the cathode performance. The S-type tunnel structure of $Na_{0.44}MnO_2$ or $Na_4Mn_9O_{18}$ has a three-dimensional structure which is advantageous for Na-ion transfer. However, theoretically the capacity is only 96 mAh g^{-1} [15] while the capacity of the $Na_{0.7}MnO_2$ structure is 171 mAh g^{-1} [16]. Recently, the large capacity with a value of 245 mAh g^{-1} that supports a high energy density was obtained from the $Na_2Mn_3O_7$ with triclinic structure [17,18].

Herein, we synthesized $Na_2Mn_3O_7$ through a solid method from MnO and Na_2CO_3. MnO was prepared through the hydrothermal process using citric acid and 1% chitosan as the chelating agents. The chelating agent acts as a dispersant [19]. The purpose of this study was to determine the effect of the chelating agents and to determine the effect of the presence of NaCl in the synthesis of MnO and $Na_2Mn_3O_7$.

2. Materials and methods

Synthesis of $Na_2Mn_3O_7$ in this study was conducted in two steps. The initial step was the formation of MnO from $MnCl_2.4H_2O$. Then, the reaction between the MnO and Na_2CO_3 produced $Na_2Mn_3O_7$. The 10 mmol NaCl (Merck Pro Analysis ≥ 99.5%) solution was stirred with 10 mmol of citric acid as a chelating agent under stirring for ± 15 minutes. Meanwhile, 10 mmol of $MnCl_2.4H_2O$ (Merck Pro 99% Analysis) was stirred with different chelating agents (10 mmol of citric acid and 1% chitosan). The solutions of [NaCl-chelating agent] and [$MnCl_2$-chelating agent] were stirred under vigorous stirring for ± 2 hours. Code A was labeled for sample [NaCl-citric acid; $MnCl_2$-citric acid], while code B for [NaCl-citric acid; $MnCl_2$-1% chitosan]. Each sample was put into an autoclave, and the hydrothermal process was carried out at 150°C for ± 6 hours [5], [6]. Then the calcination process was conducted in the nitrogen condition at 800°C for ± 1 hour. MnO phase was analyzed using an XRD instrument. As a comparison, the Cl^- ions in sample B were removed using aquadest [5,20,21]. Furthermore, in the preparation of $Na_2Mn_3O_7$, each MnO powder produced from Samples A and B was mixed with Na_2CO_3 powder (Merck Pro Analysis 99.5%) through a solid method. The calcination was also carried out at 800 °C for ± 3 hours in an oxidizing atmosphere. Characterization using XRD was carried out to determine the phases formed in Samples A and B.

3. Results and discussion

3.1. X-Ray identification on MnO phases

Figure 1 demonstrates the result of XRD analysis of any phases formed in Sample A and Sample B. The mineral phases in Sample A consists of NaCl according to PDF2 number 77-2064, MnO according PDF2 number 89-4835, and $Na_2Mn_3O_7$ according to PDF2 number 78-0193. Based on PDF2 number 77-2064, NaCl appears to be a predominant phase in sample A as shown by the highest intensity (around 1000 counts per second (cts)) from the main peak of NaCl at a diffraction angle 2θ of 31.68°, corresponding to the main crystal plane (200) of the cubic structure of sodium chloride. The high intensity of NaCl is indicated to occur because the Cl^- ions were not removed after calcination in the synthesis of MnO samples [5,20,21].

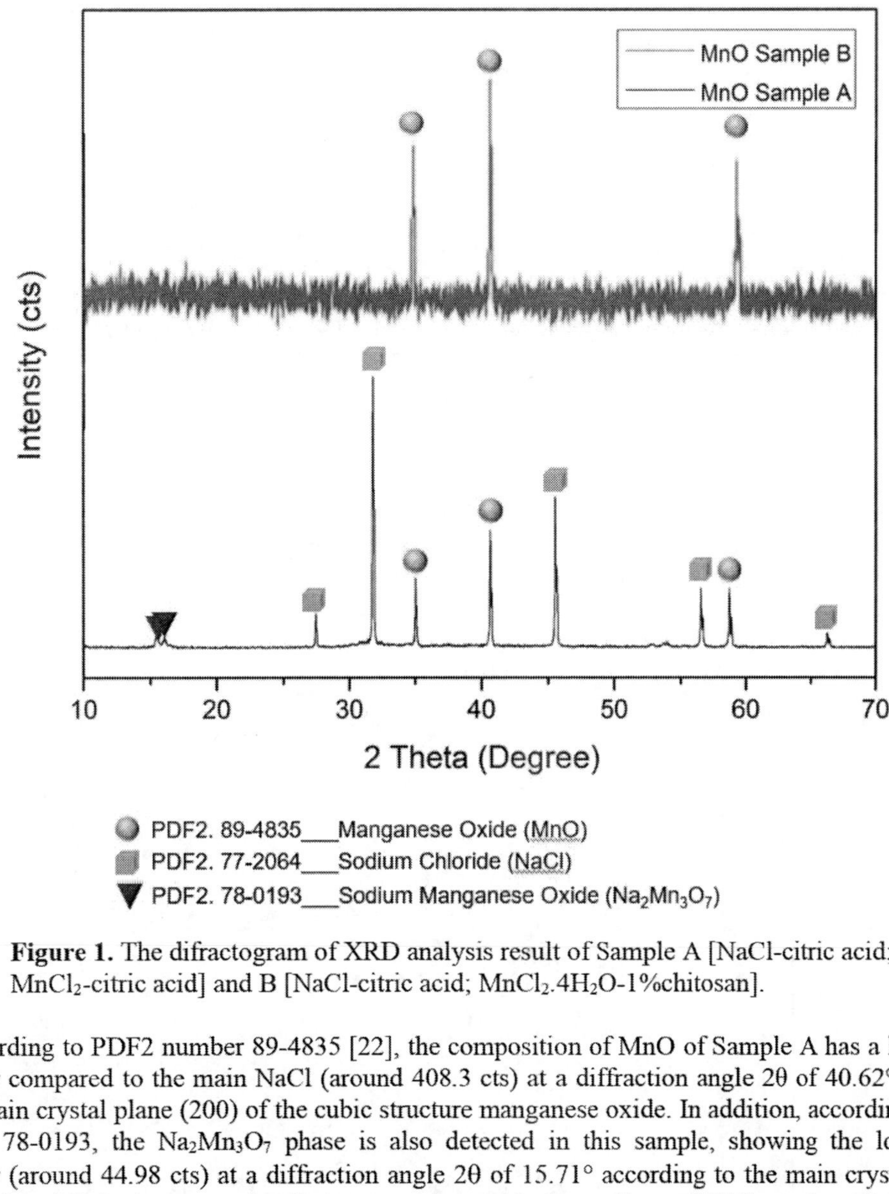

Figure 1. The difractogram of XRD analysis result of Sample A [NaCl-citric acid; $MnCl_2$-citric acid] and B [NaCl-citric acid; $MnCl_2.4H_2O$-1%chitosan].

According to PDF2 number 89-4835 [22], the composition of MnO of Sample A has a lower main intensity compared to the main NaCl (around 408.3 cts) at a diffraction angle 2θ of 40.62° according to the main crystal plane (200) of the cubic structure manganese oxide. In addition, according to PDF2 number 78-0193, the $Na_2Mn_3O_7$ phase is also detected in this sample, showing the lowest peak intensity (around 44.98 cts) at a diffraction angle 2θ of 15.71° according to the main crystal plane (-110) of the triclinic structure of sodium manganese oxide (according to PDF2 number 78-0193) [17], [23]. This proves that the $Na_2Mn_3O_7$ phase began to form in the synthesis of MnO sample A.

The result of XRD analysis of any phases formed in Sample B is displayed in figure 1. According to PDF2 number 89-4835 [22], the MnO compounds appear to be the only phase that has a primary intensity of about 1000 cts at a diffraction angle 2θ of 40.62° according to the main crystal plane (200) of the cubic structure of manganese oxide. In contrast to the synthesis of MnO Sample A, the $Na_2Mn_3O_7$ phase is not detected during the synthesis of MnO Sample B. Likewise, NaCl is lost due to the Cl^- ions removal after calcination [5,20,21].

Based on figure 1, it was indicated that the use of citric acid as a chelating agent in the synthesis of MnO Sample A produced intensity of diffraction peak of MnO which was lower than the use of 1% chitosan as a chelating agent in the synthesis of MnO Sample B. This was assumed to occur because NaCl decreased the MnO intensity in Sample A. The existence of a phase with high concentrations in a sample will result in high peak intensity and causes minor phases that had low peak intensity or even did not appear. If the major phase is removed, the minor phase will appear as a single phase with a higher intensity than before [24]. In addition, the $MnCl_2$ precursor in Sample A during the synthesis process was strongly bonded ionically with citric acid. Citric acid is a strong metal complexing agent, and the coordination compound produced is quite stable. However, the decomposition process of Mn-citrate compounds did not require extra energy or higher temperatures to break the bonds that occurred between Mn^{2+} ions and citrate. At temperatures above 175 °C, Mn-citrate decomposed into Mn-oxaloacetate [25]. Then, it decomposed gradually during the calcination process. The Mn-citrate decomposition process will occur at a relatively lower temperature than Mn-chitosan [26].

Some of MnO compounds formed at relatively lower temperatures would then react with a number of NaO produced from the decomposition of Na-citric acid precursors at a certain temperature, resulting in the $Na_2Mn_3O_7$ compounds at a diffraction angle 2θ of 15.71° according to the main (-110) of the triclinic structure of sodium manganese oxide (according to PDF2 number 78-0193) [17,23]. However, the above phenomenon did not occur in Sample B. Although the Mn precursors in Sample B covalently bounded to each functional group of the chitosan polymer chain, that was the hydrogen bonds between OH from Mn with NH_2 and OH functional groups from chitosan, and covalent coordination between O atoms of chitosan ethoxy groups with Mn metal center [27,28], the decomposition process of Mn-chitosan compounds required a relatively longer time and a relatively higher temperature [29] because of high molecular weights of chitosan [30]. Thus, the MnO crystallization process from the Mn-chitosan decomposition process was relatively slower [31]. Therefore, the MnO phase produced was likely to react with NaO (from the citric acid precursor) at a relatively higher temperature. This phenomenon was supported by the results of the XRD analysis in Sample B, that the MnO phase had a higher peak intensity than the peak intensity of MnO in Sample A. Besides, the NaMnO phase was not detected in Sample B.

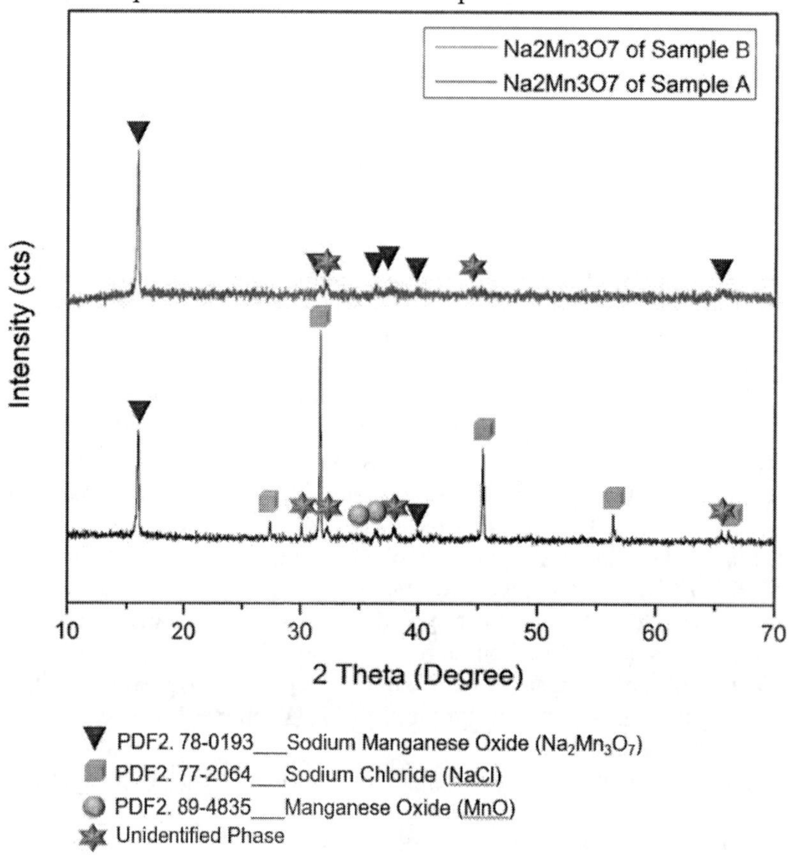

Figure 2. The XRD identification on the $Na_2Mn_3O_7$ phase of Sample A and Sample B.

3.2. X-Ray identification on $Na_2Mn_3O_7$ phases
The result of XRD identification on the $Na_2Mn_3O_7$ phase of Sample A and Sample B is shown in figure 2. In Sample A, according to PDF2 number 78-0193, the intensity of the main $Na_2Mn_3O_7$ increases to around 553.90 cts at a diffraction angle 2θ of 15.71° according to the main crystal plane (-110) of the triclinic structure of sodium manganese oxide [17,23]. However, according to PDF2 number 77-2064, NaCl is still detected [5,20,21] at a diffraction angle 2θ of 31.68° according to the main crystal plane (200) of the cubic structure of sodium chloride with increasing intensity to about 1000 cts. Meanwhile, according to PDF2 number 89-4835 [22], the intensity of the major MnO decreases to around 124.57 cts at a diffraction angle 2θ of 2θ 35.98° according to the main crystal plane (200) of the cubic structure of manganese oxide. In addition, there are unidentified phases at a diffraction angle 2θ of 30.07°; 32.31°; 37.60°; and 65.72°.

In comparison to the synthesis of $Na_2Mn_3O_7$ Sample A, according to PDF2 number 78-0193 the intensity of $Na_2Mn_3O_7$ produced in the synthesis of $Na_2Mn_3O_7$ Sample B appears to be higher, which is around 996.54 cts at a diffraction angle 2θ of 15.71° according to the main crystal plane (100) of the triclinic structure of sodium manganese oxide [17,23]. The $Na_2Mn_3O_7$ peaks produced are also more numerous than Sample A. However, similar to $Na_2Mn_3O_7$ Sample A, in $Na_2Mn_3O_7$ Sample B there are still unidentified phases found at a diffraction angles 2θ of around 31.90° and 44.74°.

4. Conclusions
The $Na_2Mn_3O_7$ phase was obtained through the mixing of MnO and Na_2CO_3 precursors. Meanwhile, the MnO precursor was obtained through the mixing of [NaCl-chelating agent] and [$MnCl_2.4H_2O$-chelating agent]. Citric acid was used as a chelating agent in NaCl, while the chelating agent in $MnCl_2.4H_2O$ consisted of two types, namely citric acid and 1% chitosan. In the mixing of [NaCl-citric acid; $MnCl_2.4H_2O$-citric acid], it produced an MnO phase with a lower peak intensity than that of the mixing of [NaCl-citric acid; $MnCl_2.4H_2O$-1%chitosan]. This further influenced the peak intensity of $Na_2Mn_3O_7$ formed in both samples. The sample [NaCl-citric acid; $MnCl_2.4H_2O$-1%chitosan] showed a higher intensity of the $Na_2Mn_3O_7$ phase than that of sample [NaCl-citric acid; $MnCl_2.4H_2O$-citric acid], which was 996.54 cts at a diffraction angle 2θ of 15.71°. In addition, the Cl^- ions removal after calcination in MnO synthesis not only greatly affected the intensity of MnO, but also affected the intensity of $Na_2Mn_3O_7$ formed and affected its impurities.

Acknowledgments
This work was supported by the 2020 Doctoral Dissertation Research Grant from the Deputy Funds Research Sector Strengthening Research and Development, Ministry of Research and Technology/National Research and Innovation Agency (KEMENRISTEK/BRIN) in Indonesia for the 2020 Budget. The authors also wish to thank the Advanced Materials Processing Laboratory, Department of Engineering Physics, Institut Teknologi Bandung, Indonesia.

References
[1] Xiang X, Zhang K, and Chen J 2015 Recent Advances and Prospects of Cathode Materials for Sodium-Ion Batteries *Adv. Mater.* **27** 5343–64
[2] Jiang Y, Kang L, and Liu Y 2019 A unified model to optimize configuration of battery energy storage systems with multiple types of batteries *Energy* **176** 552–60
[3] Khan K *et al.* 2019 Going green with batteries and supercapacitor: Two dimensional materials and their nanocomposites based energy storage applications *Progress in Solid State Chemistry.*
[4] E. Banguero, A. Correcher, A. Perez-Navarro, E. García, and A. Aristizabal, (2020) Diagnosis of a battery energy storage system based on principal component analysis *Renew. Energy J.* **146** 2438–49
[5] Noer Z, Sembiring T, Sebayang K, Nasruddin M N, Septawendar R, and Sunendar B

2020 The effect of the calcination atmosphere in the formation of mineral sodium titanate," in *AIP Conference Proceedings* **2221** 110027 -1 - 110027-5

[6] Ta A T *et al.* 2020 Hydrothermal synthesis of $Na_4Mn_9O_{18}$ nanowires for sodium ion batteries *Ceram. Int.* **45** 17023-28

[7] Zhang Y *et al.* 2020 $AlPO_4$-coated P2-type hexagonal $Na_{0.7}MnO_{2.05}$ as high stability cathode for sodium ion battery *Chem. Eng. J.* **382** 122697

[8] Zheng X *et al.* 2018 New Insights into Understanding the Exceptional Electrochemical Performance of P2-Type Manganese-based Layered Oxide Cathode for Sodium Ion Batteries *Energy Storage Mater*

[9] Lyu Y *et al.* 2019 Recent advances in high energy-density cathode materials for sodium-ion batteries *Sustain. Mater. Technol.* **21** 1–21

[10] Yanase I and Takano T 2019 CO_2 capture and release of $Na_{0.7}MnO_{2.05}$ under water vapor at 25–150 °C *Inorg. Chem. Commun.* **104** 212-8

[11] Kim H *et al.*, 2015 High energy organic cathode for sodium rechargeable batteries *Chem. Mater.*

[12] Jian Z, Hu Y, Ji X, and Chen W 2017 NASICON-Structured Materials for Energy Storage *Adv. Mater.*

[13] Jiang Y *et al.* 2018 High-capacity and selective ammonium removal from water using sodium cobalt hexacyanoferrate *RSC Adv.* **8(60)** 34573–81

[14] Zhang, Wang W, Wang W, Wang S and Li B 2019 Comprehensive Review of P2 Type $Na_{2/3}Ni_{1/3}Mn_{2/3}O_2$, A Potential Cathode for Practical Application of Na-Ion Batteries *ACS Appl. Mater. Interfaces* **11** 22051–66

[15] He X *et al.* 2016 Durable high-rate capability $Na_{0.44}MnO_2$ cathode material for sodium-ion batteries *Nano Energy* **27** 602–10

[16] Khan M A, Han D, Lee G, Kim Y I and Kang Y M 2019 P2/O3 phase-integrated $Na_{0.7}MnO_2$ cathode materials for sodium-ion rechargeable batteries *J. Alloys Compd.* **771** 987–93

[17] Sada K and Barpanda P 2019 Layered Sodium Manganese Oxide $Na_2Mn_3O_7$ as an Insertion Host for Aqueous Zinc-ion Batteries Krishnakanth *MRS Adv*

[18] Zheng P *et al.* 2019 A high-performance primary nanosheet heterojunction cathode composed of $Na_{0.44}MnO_2$ tunnels and layered $Na_2Mn_3O_7$ for Na-ion batteries *ChemSusChem*

[19] Purwasasmita B S, Larasati L D, Septawendar R, Nugraha A B, Aufan M R, and H. Sosiati H 2013 Synthesis and characterization of zirconia crystal using base hot water treatment (BHWT) method *J. Aust. Ceram. Soc.* **49(2)** 89–94

[20] Budiarti H A, Puspitasari R N, Hatta A M, Sekartedjo, and Risanti D D 2017 Synthesis and Characterization of $TiO_2@SiO_2$ and $SiO_2@TiO_2$ Core-Shell Structure Using Lapindo Mud Extract via Sol-Gel Method *Procedia Eng.* **170** 65–71

[21] Heraldy E, Nugrahaningtyas K D, Sanjaya F B, Darojat A A, Handayani D S, and Hidayat Y 2016 Effect of reaction time and (Ca+Mg)/Al molar ratios on crystallinity of Ca -Mg-Al layered double Hydroxide *IOP Conf. Ser. Mater. Sci. Eng.* **107(1)** 2016

[22] Zheng M *et al.* 2013 A simple additive-free approach for the synthesis of uniform manganese monoxide nanorods with large specific surface area *Nanoscale Res. Lett.* **8(1)** 1–7

[23] Adamczyk E and Pralong V 2017 $Na_2Mn_3O_7$: A Suitable Electrode Material for Na-Ion Batteries?," *Chem. Mater.* **29(11)** 4645–48

[24] René Guinebretière, (2007) *Diffraction, X-ray Materials, Polycrystalline.* London: ISTE Ltd.

[25] Tadkalkar A P, Deshmukh B D, and Bichile G K 2012 Study of excess acoustic parameters of citric acid in ionic solutions *Arch. Phys. Res.* **3(4)** 373–83

[26] Srivastava A, Gunjikar V G, and Sinha A P B 1987 Thermoanalytical studies of zinc citrate, bismuth citrate and calcium citrate *Thermochim. Acta* **117** 201–17

IOP Conf. Series: Materials Science and Engineering **980** (2020) 012055 doi:10.1088/1757-899X/980/1/012055

[27] Wang X, Du Y, Fan L, Liu H and Hu Y 2005 Chitosan-metal complexes as antimicrobial agent: Synthesis, characterization and Structure-activity study *Polym. Bull.* **55(1–2)** 105 13

[28] Webster A, Halling M D and Grant D M 2007 Metal complexation of chitosan and its glutaraldehyde cross-linked derivative *Carbohydr. Res.* **342(9)** 1189–201

[29] Thirumavalavan M, Huang K L and Lee J F 2013 Preparation and Morphology Studies of Nano Zinc Oxide Obtained Using Native and Modified Chitosans," *Materials (Basel)* **6(9)** 4198–212

[30] Bakshia P S, Selvakumar D, Kadirvelu K and Kumar N S 2019 Chitosan as an environment friendly biomaterial – a review on recent modifications and applications *International Journal of Biological Macromol* **150** 1072-83

[31] Ou C *et al.* 2016 Effect of transition metal ions on the thermal degradation of chitosan," *Cogent Chem.* **2(1)**

IOP Conf. Series: Materials Science and Engineering **980** (2020) 012056 doi:10.1088/1757-899X/980/1/012056

The comparison of cathodic and anodic plasma electrolysis performance in the synthesis of biodiesel

N Saksono[1]*, J J C Pranata[1], Y Muharam[1], and Harianingsih [1,2]
[1]Departement of Chemical Engineering, Universitas Indonesia, Kampus Baru UI. Depok 16242, Indonesia
[2]Departement of Chemical Engineering, Universitas Negeri Semarang, Kampus Sekaran, Semarang 50229, Indonesia

Email: nelson@che.ui.ac.id

Abstract. Synthesis of biodiesel using plasma electrolysis is very promising. This study aims to understand the differences between cathodic and anodic plasma performance in synthesizing biodiesel from palm oil. The raw materials used were palm oil, methanol, and KOH catalyst. In the present study, the effects of the type of plasma and the depth of an electrode for biodiesel synthesis were evaluated. The results show that cathodic plasma gave the highest yield of 98.76% and the lowest specific energy consumption of 720 J/ml. Meanwhile, under the same conditions, the anodic plasma gave the highest yield of 96.09% and the lowest specific energy consumption of 910 J/ml. These results indicate that cathodic plasma is better than anodic plasma in synthesizing biodiesel because it gives higher yield and lowers specific energy consumption.

1. Introduction

The synthesis of biodiesel using the plasma electrolysis method is a breakthrough because it can accelerate the rate of synthesis and lower specific energy consumption compared to the conventional methods. Plasma produces a large amount of methoxyl radical ($CH_3O\bullet$) that will react to triglycerides to form biodiesel [1]. Plasma electrolysis process begins with gas formation at the anode and cathode [2]. In such a large scale system, materials of anode and cathode are very important as fundamental characteristics [3]. In our previous research, biodiesel has been synthesized using the anodic plasma electrolysis method [4].

This process could produce reactive species in the anode [5]. Plasma can be formed in either cathodic or anodic, depends on the type of electrode [6]. In general, plasma is more stable and easily formed in the cathode than the anode because the cathode emits more secondary electrons[7]. This study aims to investigate the effect of cathodic and anodic plasma performance on biodiesel synthesis from palm oil. Initially, methanol is mixed with potassium hydroxide to produce species, as shown in equation 1 [8]. The high-energy electrons produced by the plasma will break the water molecules into radicals and transform methoxyl ions into methoxyl radicals in the particular electrode [1]. However, a conventional electrolysis reaction will take place in other electrodes in which plasma does not exist [9]. Using cathodic plasma, the reactions in Equations 2 and 3 occur in the cathode while the reaction in equation 4 occurs in the anode. Further, using anodic plasma, reactions in equations 2 and 3 are in anode and reaction in equation 5 is in cathode.

$$CH_3OH_{(aq)} + KOH_{(aq)} \rightarrow CH_3O^-{}_{(aq)} + K^+{}_{(aq)} + H_2O_{(l)} \quad (1)$$
$$H_2O_{(l)} + e^* \rightarrow H\cdot + OH\cdot + e^- \quad (2)$$
$$CH_3O^-{}_{(aq)} + e^* \rightarrow CH_3O\cdot + e^- \quad (3)$$
$$2H_2O_{(l)} \rightarrow 4H^+{}_{(aq)} + O_{2(g)} + 4e^- \quad (4)$$
$$2H_2O_{(l)} + 2e^- \rightarrow 2OH^- + H_{2\,(g)} \quad (5)$$

The methoxyl radicals produced by those reactions will attack triglycerides from palm oil to produce methyl esters [10]. In general, the mechanism consists of three stages as shown in figure 1 [11]. The first stage is a nucleophilic alkoxide attack producing tetrahedral intermediate. The second stage is the formation of alkylester and anion diglyceride. The third stage is the regeneration of the active species that will react with the second molecule of the other alcohol followed by the recovery of the base catalyst.

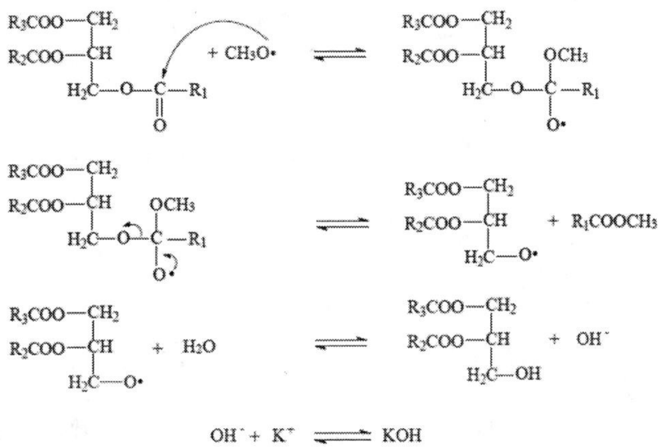

Figure 1. The possible mechanism of biodiesel synthesis using plasma electrolysis.

2. Materials and methods

2.1. Materials

The palm oil used in this study was a commercial cooking oil product in Indonesia. Methanol (analytical grade, Merck) and Potassium hydroxide (Merck) were used as transesterification agent and catalyst, respectively. Water was applied for the cooling process.

2.2. Instrumentation

The main equipment in this research was a 2-liter Plasma Electrolysis batch reactor similar to the one used in the previous research [12]. The reactor scheme can be seen in figure 2. The difference is the type of charge given to the tungsten as the electrode where the plasma is formed. When the tungsten is given a negative charge, it will act as a cathode. Contrarily, when a positive charge is given, it will act as an anode. Stirring used a magnetic stirrer with the addition of baffles inside the reactor. Baffles were used to make stirring more effective. The reactor was equipped with a coolant stream to maintain the temperature inside the reactor at 40°C to reduce the possibility of methanol evaporation from the solution.

Tungsten electrode and stainless steel electrode were given negative and positive charges respectively when the reactor would be used as a cathodic plasma reactor. Meanwhile, the tungsten electrode and stainless steel electrode were given positive and negative charges respectively when the reactor would be used as an anodic plasma reactor.

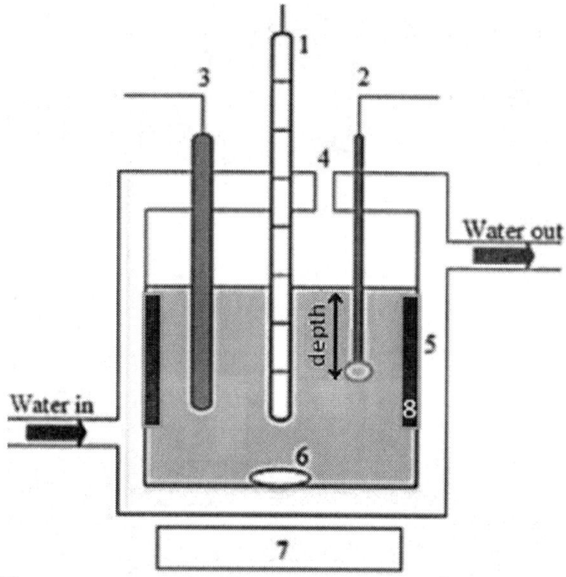

Figure 2. Plasma reactor, 1: Thermometer, 2: Electrode (tungsten), 3: Electrode (stainless steel), 4: Sample hole, 5: Cooling jacket, 6: Magnetic stirrer, 7: Stirrer, 8: Baffle. [12]

2.3. Methods
Potassium hydroxide and palm oil with a ratio of 1:100 were dissolved with methanol. The mixture was then mixed and stirred with oil in a ratio of 1:24 in both reactors until the homogeneous phase was obtained. The electric current applied at the desired voltage so that the plasma was formed and the reaction occurred for an hour.

3. Results and discussion
3.1. Current-Voltage (I-V) characteristics
Some text, figure 3 shows voltage-current profiles from a mixture of palm oil and methanol with a 1:24 molar ratio plus potassium hydroxide as much as 1 wt% of palm oil in experiments with cathodic and anodic plasma.

Zone A is an area of Faraday electrolysis where voltage is proportional to current while Zone C is an area where a gas shield is formed on the electrode surface [12,13]. This gas shield reduces the contact area between the electrode and electrolyte solution so that the current falls. In this zone, spark as a phenomenon of excited electrons also begins to appear [13]. Point B1 and B2 are the points at which the gas shield starts to form, also known as breakdown voltage (V_B). V_B of anodic plasma is lower than V_B of cathodic plasma ($V_{B1} < V_{B2}$) which means that a gas shield is more easily formed in anode than cathode[14]. On the other hand, V_D (Mid-point Voltage) of anodic plasma is higher than V_D of cathodic plasma ($V_{D1} > V_{D2}$) which means that plasma was easier to achieve stability in the cathode than in the anode. It was due to the most energy in the anodic plasma was used to evaporate the solution around the anode. However in cathodic plasma, the most energy was used by electrons to support the kinetic energy to excite and form the plasma faster to obtain stability [5,16]. The value of Townsend secondary electron emission coefficient (γ) for the cathode and anode differs very significantly. The value of γ was about 0.01-0.1 for the cathode metal and 0.0001-0.001 for the electrolyte solution around the anode. The negative electrode (cathode) was surrounded by many positive ions. The positive ions made not only the primary electron but also the secondary electrons were emitted [3,5]. Therefore, secondary electron emission occurred more at cathode than at anode. The more excited the electrons are, the easier the plasma will achieve stability.

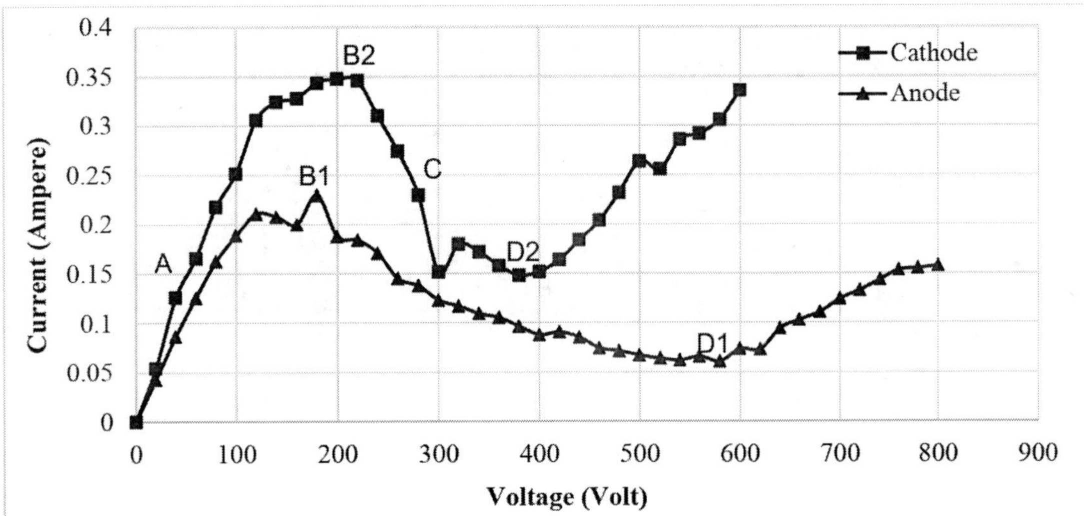

Figure 3. Voltage-current profile on the solution of palm oil-methanol-KOH using cathodic and anodic plasma

3.2. The effect of electrode depth on biodiesel synthesis

The experiment was conducted by varying the cathode and anode depth of 0.5 cm, 1.5 cm, and 3.5cm. The result shows that the deeper electrode would increase hydrostatic pressure, hence gas could reach the surface easily and the higher evaporation energy was required to maintain the size of the gas shield as shown in table 1 [14].

Table 1. Specific energy consumption based on cathode or anode depth.

Plasma	Depth (cm)	Yield (%)	Biodiesel Volume (ml)	Energy (Joule)	Specific Energy Consumption (J/ml)
	0.5	88.54	622	465833	749
Cathodic	1.5	92.81	652	477310	732
	3.5	98.76	694	499339	720
	0.5	92.95	653	528264	809
Anodic	1.5	94.15	661	597168	903
	3.5	96.09	675	613872	910

The increase in energy is marked by the increase of current and brighter plasma. A deeper electrode includes brighter plasma as shown in figure 4. It could enhance the interaction between plasma and other reactants to produce more radicals which initiate the reaction continually so that the conversion increased[15]. Further, this was demonstrated by the increase of biodiesel volume and yield as shown in table 1.

(a) (b) (c)

Figure 4. Cathodic plasma at depth of (a) 0.5 cm; (b) 1.5 cm; and (c) 3.5 cm

The specific energy consumption of biodiesel synthesis could be measured by the amount of electric energy consumed divided by the volume of biodiesel produced. The specific energy consumption of anodic and cathodic plasma was 720-749 J/ml and 809-910 J/ml, respectively as shown in table 1. This energy formed by the plasma method was lower than the conventional method (1168 J/ml) which is in line with previous research [13,16].

Table 1 exhibits the volume of biodiesel increased by 11.6% as the cathode depth increased (from 0.5 to 3.5 cm) while its energy consumption increases by 7.2%. It caused a decrease in specific energy consumption. However, in anodic plasma, the volume of biodiesel increased by only 3.4% and its energy consumption increased by 16.2% as the anode depth increased. The increase in energy consumption in anodic plasma was higher than cathodic plasma [17]. Its plasma was easier to achieve stability at cathode than anode as depicted in the characteristics of voltage currents in figure 3 [18].

(a) (b)

Figure 5. (a) Cathodic plasma and (b) anodic plasma; at depth of 0.5 cm, 600 volt

There was a significant increase in biodiesel volume (694 ml) while using cathodic plasma for 3.5 cm depth and it was relatively higher than biodiesel volume formed by anodic plasma (675 ml). It was probably due to the larger and brighter plasma formed by cathodic plasma than anodic plasma in figure 5. Furthermore, the lighter plasma indicates that the number of reactive species (i.e. methoxyl radical) for biodiesel production increased [19].

3.3. Biodiesel product characterization

The samples were analyzed using gas chromatography. The peaks shown in table 2 represent the compound names and compositions in the samples in general, there was no significant difference in

IOP Conf. Series: Materials Science and Engineering **980** (2020) 012056 doi:10.1088/1757-899X/980/1/012056

product composition which was synthesized by cathodic and anodic plasma in the varying depth of electrodes. The products were dominated by Hexadecanoic acid, methyl ester (methyl palmitate), (Z)-octanes-9-enoic acid, and methyl ester (methyl oleate). However, other compounds such as water and free fatty acid were also detected. These products depended on the natural compounds of feedstock which was dominated by palmitic acid and oleic acid. It also shows that the reaction mechanisms of biodiesel synthesis by using cathodic plasmas are relatively similar. Both cathodic and anodic plasmas were able to convert methanol to methoxyl radicals which initiated biodiesel production.

Table 2. Methyl ester composition in biodiesel products.

No	Compound Name	Composition (%-wt)					
		Cathodic	Plasma	Depth	Anodic	Plasma	Depth
		0.5 cm	1.5 cm	3.5 cm	0.5 cm	1.5 cm	3.5 cm
1	Tetradecanoic acid, methyl ester	0.70	0.73	0.79	0.73	0.73	0.76
2	Hexadecanoic acid, methyl ester	27.44	28.91	30.53	28.04	28.53	29.60
3	Heptadecanoic acid, methyl ester	14.99	15.21	16.69	15.38	15.90	16.13
4	Octadecanoic acid, methyl ester	0.00	0.00	3.64	3.25	3.30	3.39
5	(Z)-octadec-9-enoic acid, methyl ester	36.86	38.79	37.31	34.87	35.54	36.79
6	(9Z,12Z)-octadeca-9,12-dienoic acid, methyl ester	8.19	8.78	9.48	9.17	9.34	9.42
7	Eicosanoic acid, methyl ester	0.29	0.31	0.32	0.00	0.00	0.00
8	Tetracosanoic acid, methyl ester	0.07	0.08	0.00	1.49	0.81	0.00
	Total	88.54	92.81	98.76	92.95	94.15	96.09

4. Conclusion

This research has successfully investigated biodiesel production using cathodic and anodic plasma formation. Anodic plasma was more easily formed than cathodic plasma, but plasma achieved stability faster in cathodic plasma. The highest biodiesel yields of cathodic and anodic plasma were 98.76% and 96.09%, respectively. The specific energy consumptions of cathodic and anodic plasma were 720 J/ml and 910 J/ml respectively. The low specific energy consumption of cathodic plasma was heavily caused by the lower energy consumption, which was 499.339 kJ while the anodic plasma reached 613,872 kJ. Gas chromatography test results show that there is no significant difference between product composition in cathodic and anodic plasma.

Acknowledgments

The research was partially funded by Hibah Dasar Kemenristekdikti 2020 Contract No. NKB-277/UN2.RST/HKP.05.00/2020. The authors declare no competing interests or any conflicts of financial interests.

References

[1] Yan Z C, Li C, and Lin W H 2009 Hydrogen generation by glow discharge plasma electrolysis of methanol solutions *International Journal of Hydrogen Energy,* **34** 48-55

[2] Yang H, An B, Wang S, Li L, Jin W, and Li L 2013 Destruction of 4-phenolsulfonic acid in water by anodic contact glow discharge electrolysis *Journal of Environmental Sciences,* **25** 1063-70

[3] Ni'mah Y L, Cheng J H, Cheng M Y, Su W N, and Hwang B J 2018 Improvement of Cycling Performance of Na 2/3 Co 2/3 Mn 1/3 O 2 Cathode by PEDOT/PSS Surface Coating for Na Ion Batteries *Indonesian Journal of Chemistry,* **18** 127-36

IOP Conf. Series: Materials Science and Engineering **980** (2020) 012056 doi:10.1088/1757-899X/980/1/012056

[4] Saksono N, Anditashafardiani R, Junior A B, and Muharam Y 2019 Effect of anode depth in synthesis of biodiesel using the anodic plasma electrolysis method *International Journal of Technology,* **10** 491-501

[5] Jiang B, Zheng J, Qiu S, Wu M, Zhang Q, Yan Z, and Xue Q 2014 Review on electrical discharge plasma technology for wastewater remediation *Chemical Engineering Journald.* **236** 348-68

[6] Bruggeman P and Leys C 2009 Non-thermal plasmas in and in contact with liquids *Journal of Physics D: Applied Physicsd.* **42** 053001

[7] Wenzel S, Hara T, Janek J, and Adelhelm P 2011 Room-temperature sodium-ion batteries: improving the rate capability of carbon anode materials by templating strategies *Energy & Environmental Science* **4** 3342-45

[8] Lotero E, Liu Y, Lopez D E, Suwannakarn K, Bruce D A, and Goodwin J G 2005 Synthesis of biodiesel via acid catalysis *Industrial & engineering chemistry research* **44** 5353-63

[9] Krčma F, Kozáková Z, Mazánková V, Horák J, Dostál L, Obradović B and Belmonte T 2018 Characterization of novel pin-hole based plasma source for generation of discharge in liquids supplied by DC non-pulsing voltage. *Plasma Sources Science and Technologyd.* **27** 065001

[10] Lee D W, Park Y M and Lee K Y 2009 Heterogeneous base catalysts for transesterification in biodiesel synthesis *Catalysis surveys from Asiad.* **13** 63-77

[11] Gupta S K S 2015 Contact glow discharge electrolysis: its origin, plasma diagnostics and non-faradaic chemical effects *Plasma Sources Science and Technologyd.* **24** 063001

[12] Jin X, Wang X, Yue J, Cai Y, and. Zhang H 2010 The effect of electrolyte constituents on contact glow discharge electrolysis *Electrochimica Actad.* **56** 925-28

[13] Gupta S K S and Singh R 2016 Cathodic contact glow discharge electrolysis: its origin and non-faradaic chemical effects *Plasma Sources Science and Technology.* **26** 015005

[14] Saksono N, Nugraha I and Febiyanti I A 2016 Hydroxyl radical production on contact glow discharge electrolysis for degradation of linear alkylbenzene sulfonate *Environmental Progress & Sustainable Energy.* **35** 962-68

[15] Yan Z, Li C and Lin W 2006 Experimental study of plasma under-liquid electrolysis in hydrogen generation *The Chinese Journal of Process Engineering.* **6**

[16] Shurtleff W and Aoyagi A 2017 *History of Biodiesel-with Emphasis on Soy Biodiesel (1900-2017): Extensively Annotated Bibliography and Sourcebook*: Soyinfo Center.

[17] Jin X, Zhang H, Wang X and Zhou M 2012 An improved multi-anode contact glow discharge electrolysis reactor for dye discoloration *Electrochimica Acta.* **59** 474-78

[18] Yang X, Jiang Z, Ding X, Hao G, Liang Y and Lin J 2018 Influence of Solvent and Electrical Voltage on Cathode Plasma Electrolytic Deposition of Al2O3 Antioxidation Coatings on Ti-45Al-8.5 Nb Alloys *Metals.* **8** 308

[19] Jin X, Wang X, Zhang H and Ren H 2013 Study on the onset of DC diaphragm glow discharge *Electrochimica Acta.* **87** 336-40

IOP Conf. Series: Materials Science and Engineering **980** (2020) 012057 doi:10.1088/1757-899X/980/1/012057

Optimisation and characterisation studies of biodiesel production from black soldier fly larvae fed by soya residue

A Kamari[1]*, S Ishak[1], M I A M Hussin[2], S T S Wong[1], J Jumadi[1] and N M Yahaya[3]

[1] Department of Chemistry, Faculty of Science and Mathematics, Universiti Pendidikan
Sultan Idris, 35900, Tanjong Malim, Perak, Malaysia.

[2] Eko Dinamik Resources, 78, Kampung Baru, 35600, Sungkai, Perak, Malaysia.

[3] MARA Junior Science College Merbok, PT 1408, Jalan Taman Bujang, 08400, Kuala Muda, Kedah, Malaysia.

Email: azlan.kamari@fsmt.upsi.edu.my

Abstract. In this study, a two-step transesterification process was applied to convert crude BSF larval oil into biodiesel. The effects of methanol:oil (molar ratio), catalyst amount (weight %), reaction time (minutes) and temperature (°C) on biodiesel production were investigated. Central Composite Design (CCD) of Response Surface Methodology (RSM) was used to optimise experimental data obtained from these four variables. The prediction of biodiesel yield was made by employing a quadratic model, whereby the R^2 values were greater than 0.99. Based on optimisation studies, a combination of an application of petroleum ether and reaction time of 6 hours was the best to extract crude larval oil with 47% of lipid yield. Meanwhile, methanol:oil molar ratio of 12:1, catalyst (NaOH) concentration of 1.0 weight %, a reaction time of 32 minutes and temperature of 60°C were required to obtain biodiesel with 96% of production yield from BSF crude larval oil. Both BSF crude larval oil and biodiesel were characterised using several analytical instruments such as Fourier Transform Infrared (FTIR) spectrometer and Gas Chromatography – Flame Ionisation Detector (GC-FID). Overall, findings from this study highlight the potential to produce eco-friendly biodiesel from BSF larvae reared on soya residue.

1. Introduction

Fossil fuel is the dominant world energy supply. However, due to the disadvantages of the non-renewable energy resources, the interest in investigating the alternative energy resources known as renewable energy keeps on increasing. Moreover, Soliman et al. [1] stated that renewable energy provides energy security as compared to non-renewable energy as well to promote global energy growth. The renewable energy including solar, hydro, wind, tidal and biomass provides sustainable and environmentally friendly characteristics than non-renewable energy [2]. Biodiesel is one of the potential renewable energies that has been widely reported by researchers and received high consideration due to several advantages such as engine combustion efficiency without engine modification, low sulphur content, high cetane number and biodegradable [3-5]. Moreover, application of biodiesel is also able to reduce the greenhouse hazardous gas emission mainly from fuel combustion such as carbon dioxide, methane, nitrous oxide, ozone and chlorofluorocarbons [6,7].

Biodiesel is a type of diesel fuel derived from the renewable lipid of plants or animals and consists of long-chain fatty acid esters [8]. Vegetable oil is the most famous feedstock category for biodiesel production worldwide. However, the use of vegetable oil as a biodiesel feedstock is restricted and minimal in many countries due to the high total biodiesel production cost [9]. This situation occurred due to the increment of raw material cost for about 60-75% of the total biodiesel production and also the requirement of large arable plant for extensive biodiesel production [10,11]. Additionally, the usage of vegetable oils particularly edible oils may cause a food crisis in the long term [9]. Another issue is the seasonal harvest period for almost oilseeds which make vegetable oils are inefficient for the biodiesel production process [12]. Previous studies by Zheng et al. [11], Nguyen et al. [12], Wong et al. [13] and Ishak et al. [14] have shown that insect larvae can be used as alternative feedstocks for biodiesel production. Although they have reported the success to produce biodiesel from crude larval oil, the influence and optimisation of several crucial experimental parameters on esterification and transesterification processes have been seldom studied statistically. Therefore, this work was devised with an ultimate and novel aim to apply CCD of RSM for optimisation of biodiesel production from BSF larvae reared on soya residue for the first time.

2. Materials and methods

2.1. Cultivation of BSF larvae

A colony of adult BSF (*Hermetia illucens*) was placed in a rearing cage with a size of 150 cm x 75 cm. Soya residue and water were provided in the rearing cage to promote BSF for mating and oviposition process. Soya residue was collected from a food and beverage factory in Rawang, Selangor, Malaysia. The temperature and humidity were kept in the range of 25-32°C and 60-75%. After 20 days, BSL larvae were harvested, rinsed, and inactivated at 100°C for 5 minutes followed by drying in an oven at 60°C for 24 hours.

2.2. Extraction of crude BSF larval oil

About 25 g of dried BSF larvae were placed in three-round bottom flasks containing 250 mL of petroleum ether, ethanol and acetone, separately. The extraction was carried out in a Soxhlet system and set from 1 to 6 hours. Following extraction, the distillation process was performed to remove excess solvent and to obtain crude BSF larval oil. The crude oil was then heated in an oven at 105°C until a constant weight was achieved. It was also treated with 1% (v/v) of concentrated phosphoric (85%, v/v) and 2-4% of softened water at 30°C. Finally, the crude BSF larval oil was centrifuged and high purity of BSF larval oil was collected after the refinement process. The percentage of crude BSF larval oil produced was determined.

2.3. Production of BSF larval biodiesel

In this study, crude BSF larval oil was converted to biodiesel through a two-step transesterification process, which involved acid-catalysed esterification and alkaline-catalysed transesterification [15,16]. The acid-catalysed esterification step was important to reduce the FFA content in the crude oil, meanwhile, alkaline-catalysed transesterification converted the esterified oil to biodiesel. For acid-catalysed esterification procedure, about 15 g of BSF larval oil was placed into a 250 mL of a conical flask, heated to 45°C and added with the mixture solution of methanol and sulphuric. The resulted solution was then stirred for 60 minutes followed by centrifuge at 400 rpm for 15 minutes. The upper layer (esterified oil) was then transferred to a conical flask and dried in an oven at 105°C for 15 minutes. The alkaline-catalysed transesterification was conducted by placing about 15 g of esterified oil into a conical flask, heated to 45°C, added with the mixture solution of methanol and sodium hydroxide and stirred for 15 minutes. The solution was then separated using a centrifuge at 400 rpm for 10 minutes. The upper layer (biodiesel) was collected and the percentage yield was determined.

2.4. *Statistical analyses*

A five-level four-factor CCD was employed and generated thirty experimental runs. Statistical analyses aimed to determine the optimum biodiesel yield (Y) as well as to investigate the effect of experimental variables, namely methanol:oil molar ratio (A), catalyst amount (B), time (C) and temperature (D) on biodiesel production. The experimental data were analysed using RSM and their fitness to the quadratic polynomial equation was examined. The statistical analyses such as analysis of variance (ANOVA) and regression coefficient (R^2) were accomplished based on the response obtained.

2.5. *Characterisation studies*

In this study, several analyses were carried out to characterise the physical and chemical properties of crude BSF larval oil and BSF biodiesel. The analyses were performed using a Thermo Nicolet 6700 FTIR Spectrometer, a Jeol JNM-ECX-500 NMR Spectrometer and a Shimadzu GC-17A GC equipped with a Zebron ZB-FAME column and an FID.

3. Results and discussion

3.1. *Monitoring of BSF larvae growth*

Observation of BSF larval growth development was conducted for 20 days of the cultivation process. Table 1 presents the physical properties of BSF larvae reared on soya residue. Four main physical properties were determined including colour, individual weight, length and width. The BSF larval weight increased up to day 15 of the cultivation process. However, the reduction of BSF larval weight about 14.7% was observed at day 20 as compared to day 15 of cultivation. A similar finding was reported by Wong et al. [14], of which the weight decreased from 5.24 g to 3.70 g for fifth to six instars of BSF larvae.

Table 1. Physical properties of BSF larvae fed by soya residue.

Properties	Cultivation (days)			
	5	10	15	20
	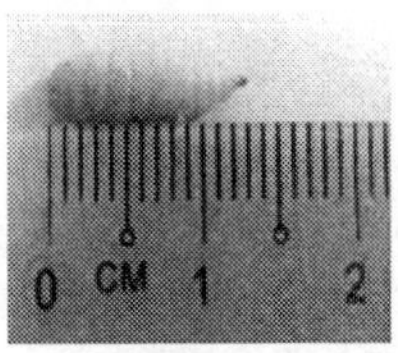	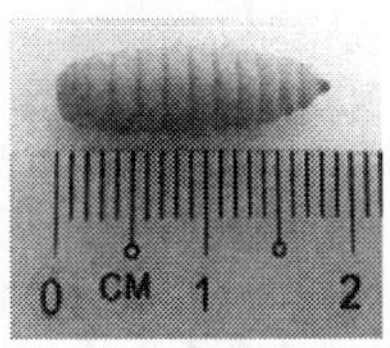	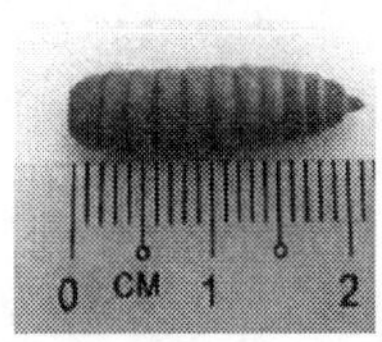	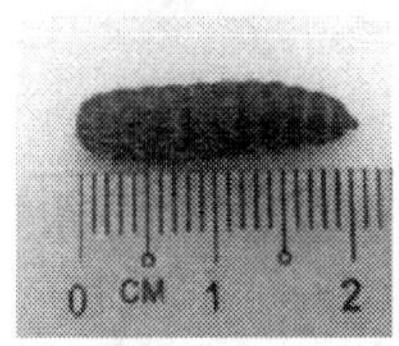
Colour	Creamy	Beige	Dark beige	Black
Individual weight (g)	0.0988 ± 0.15	0.1414 ± 0.17	0.2081 ± 0.26	0.1775 ± 0.15
Length (mm)	14.00 ± 0.58	17.67 ± 0.33	21.33 ± 0.42	19.67 ± 0.28
Width (mm)	3.67 ± 0.19	4.78 ± 0.33	5.87 ± 0.26	4.56 ± 0.21

Values represent mean of three replicates ± standard error.

This scenario was mainly due to the cessation or commonly known as terminal growth period that correlated to the metamorphosis of insect larval life cycle [17]. In this study, changes of BSF larval colour could be observed from creamy to black during the cultivation of BSF larvae within 20 days. Turning of insect larval colour was significantly in accordance with the metamorphosis process [18]. According to Janssen et al. [19], the variations of BSF larval colour was due to iron-polyphenol complexes content in the BSF larvae itself.

3.2. Extraction of crude BSF larval oil

Solvent extraction is a widely used method for the extraction of crude oil. In this study, the effectiveness of lipid extraction was evaluated using three solvents with different polarity properties, namely petroleum ether, acetone, and ethanol. Figure 1 shows the percentage of lipid yield extracted from BSF larvae using the aforementioned solvents. It is apparent that ethanol has resulted in the highest yield of BSF lipid of up to 47%, followed by petroleum ether (44%) and acetone (43%). As reported by López et al. [20], a non-polar solvent such as petroleum ether results in slightly lower yield than ethanol due to its ability to extract triglycerides only. Meanwhile, acetone also has potential in extracting lipids but from this experiment, the application of acetone produced the lowest crude oil yield as compared to other solvents. This outcome was consequently due to polarity properties of acetone itself that was more non-polar than ethanol and less non-polar than petroleum ether. Indeed, polar and non-polar lipids dissolve in acetone but not as high as ethanol and petroleum ether.

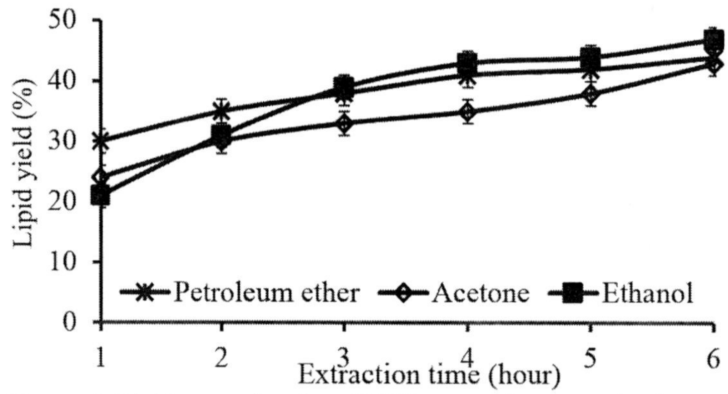

Figure 1. Lipid extraction from BSF larvae using several solvents.

Reaction time has a significant influence on the extraction of crude oil. From figure 1, it is clear that when extraction time was increased, the lipid yield (%) was also increased but with a marginal increment. For example, the percentage yield of lipid extracted by ethanol after 4, 5 and 6 hours of extraction was determined as 42, 44 and 47%, respectively. As discussed by Feng et al. [21], the marginal increment trend might be due to washing and diffusion phases involved in the extraction process. In the context of crude larval oil extraction, washing phase refers to the cleaning of lipid on the insect body surface while the diffusion phase refers to mass transfer of solid-liquid from lipid to extractant solvent. Although ethanol extracted the highest lipid yield, petroleum ether was more favourable to be used as an extractant solvent due to its ability to extract triglycerides only. Therefore, the optimum extraction process was set with a combination of petroleum ether and 6 hours of extraction time.

3.3. Optimisation of transesterification

The design matrix for 30 experiments and experimental yields using CCD are shown in table 2. From Table 2, it is clear that the biodiesel yield was significantly influenced by the combination of each variable. The experimental error and reproducibility of obtained data were indicated by the center point (five replicates).

Table 3 presents an analysis of variance (ANOVA) for the CDD model, which is important for verification of the significance and fitness of the quadratic model. The F-value and Prob>F values for model were determined as 13.20 and <0.0001, respectively, implying that the model was significant at 95% confidence level. ANOVA results identified that A, B, C, D, AB, AC, AD, BC, BD, CD, A^2, B^2, C^2, and D^2 were significant model terms. The lack of fit F-value of 4.71 and p-value of 0.0506 indicates that the model was satisfactorily fitted to experimental data.

The R^2 value of 0.9890 represents the reasonable agreement between the predicted yield and the actual yield. The relationship for predicted yield against actual yield is shown in figure 2 (a), while normal probability against studentised residuals and studentised residuals against run number are shown in figures 2 (b,c). Adequate precision (AP) is defined as a measurement of signal to noise ratio, whereby the range of predicted values at the design points are compared with the average prediction error. The AP value obtained from this study was determined as 34.764 (table 3) which indicates a desirable ratio. An AP value which greater than 4 suggests adequate model discrimination [22].

3.3.1. Effect of methanol:oil molar ratio

The 3D plots of the interaction effect between A: methanol:oil, B: catalyst amount, C: time and D: temperature on biodiesel yield are presented by figures 3 (a,b,c). The plots show that high biodiesel yield (>90%) can be obtained with methanol:oil molar ratio between 8:1 and 10:1. In general, a further increment in methanol:oil molar ratio from 11:1 to 12:1 might reduce biodiesel production [23].

3.3.2. Effect of catalyst amount

The application of catalyst in biodiesel production will accelerate the breaking of triglycerides (TAG) bond during the transesterification process [24]. Figures 3 (a,d,e) present the interaction of catalyst amount between A: methanol:oil, C: time and D: temperature on biodiesel production. Based on 3D plots, high biodiesel production (>90%) can be obtained when the amount of catalyst (weight %) in the range from 0.5 to 1.0. However, Rajan et al. [25] have reported that high catalyst amount may lead to soap formation and therefore difficult to separate glycerol from biodiesel.

Table 2. Design matrix of experiments and experimental yields using CCD.

Run	Methanol:oil (molar ratio)	Catalyst amount (%)	Time (min)	Temperature (°C)	Biodiesel yield (%)	Run	Methanol:oil (molar ratio)	Catalyst amount (%)	Time (min)	Temperature (°C)	Biodiesel yield (%)
1	10:1	1.5	40	45	55	16	10:1	1.5	20	45	55
2	14:1	1.5	40	45	59	17	12:1	1.0	30	75	67
3	10:1	0.5	20	65	65	18	12:1	1.0	30	55	83
4	10:1	0.5	40	65	73	19	10:1	0.5	40	45	44
5	12:1	1.0	30	55	89	20	12:1	1.0	30	35	45
6	10:1	0.5	20	45	35	21	12:1	1.0	30	55	90
7	8:1	1.0	30	55	45	22	14:1	0.5	20	65	68
8	12:1	1.0	30	55	90	23	14:1	1.5	20	65	50
9	12:1	1.0	50	55	58	24	14:1	0.5	40	45	51
10	12:1	1.0	30	55	93	25	12:1	1.0	15	55	64
11	12:1	2.0	30	55	65	26	14:1	0.5	20	45	46
12	10:1	1.5	20	65	69	27	16:1	1.0	30	55	57
13	10:1	1.5	40	65	78	28	14:1	1.5	40	65	50
14	14:1	1.5	20	45	61	29	12:1	0.2	30	55	54
15	14:1	0.5	40	65	78	30	12:1	1.0	30	55	90

3.3.3. Effect of reaction time

Figures 3 (b,d,f) depict the results for interaction study between reaction time with A: methanol:oil, B: catalyst amount and D: temperature on biodiesel yield. It is apparent that high biodiesel production (>90%) could be obtained when the reaction was conducted between 40 to 60 minutes. Extending the reaction time to 80 minutes with increasing catalyst amount up to 1.5 weight % would cause a significant reduction in biodiesel yield. As discussed by Kafuku and Mbarawa [26], acceleration of ester hydrolysis and saponification are likely to occur at a longer reaction time. In contrast, a different trend was obtained for interaction between reaction time with methanol:oil and temperature. The high biodiesel yield (>90%) may be achieved at reaction time between 60 to 80 minutes. This might be due to insufficient reaction time to reach equilibrium (figures 3 (b,f)).

IOP Conf. Series: Materials Science and Engineering **980** (2020) 012057 doi:10.1088/1757-899X/980/1/012057

Table 3. Analysis of variance (ANOVA) for CDD model.

Source	Sum of squares	Degree of freedom	Mean of square	F-value	Prob>F	
Model	7036.32	14	502.59	13.20	< 0.0001	Significant
A-Methanol:oil	7.04	1	7.04	0.18	0.6733	
B-Catalyst amount	126.88	1	126.88	3.33	0.0879	
C-Time	110.96	1	110.96	2.91	0.1084	
D-Temperature	1190.04	1	1190.04	31.26	< 0.0001	
AB	248.06	1	248.06	6.52	0.0221	
AC	10.56	1	10.56	0.28	0.6061	
AD	280.56	1	280.56	7.37	0.0160	
BC	39.06	1	39.06	1.03	0.3271	
BD	517.56	1	517.56	13.60	0.0022	
CD	14.06	1	14.06	0.37	0.5524	
A^2	2067.81	1	2067.81	54.32	< 0.0001	
B^2	1134.23	1	1134.23	29.79	< 0.0001	
C^2	1168.78	1	1168.78	30.70	< 0.0001	
D^2	1510.11	1	1510.11	39.67	< 0.0001	
Residual	571.21	15	38.07			
Lack of fit	516.21	10	51.62	4.71	0.0506	Insignificant
Pure error	54.83	5	10.97			
Cor total	7607.37		29			

R^2: 0.9885 Adjusted R^2: 0.9890 Adequate precision : 34.764

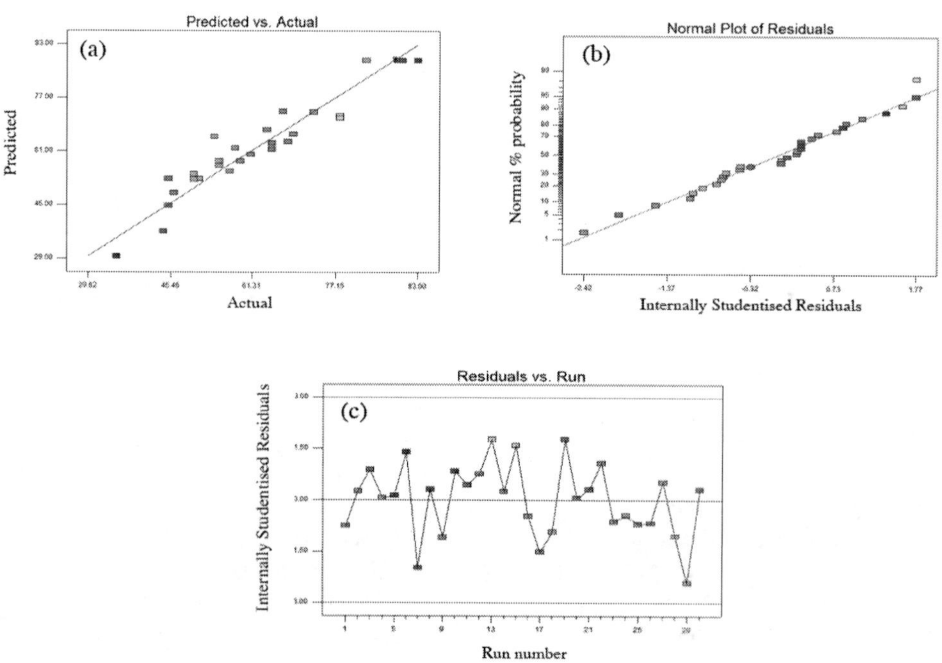

Figure 2. Residual plots (a) predicted against actual plot, (b) normal probability against studentised residuals and (c) studentised residuals against the run number.

IOP Conf. Series: Materials Science and Engineering **980** (2020) 012057 doi:10.1088/1757-899X/980/1/012057

3.3.4. Effect of reaction temperature

Figures 3 (c,e,f) show the interaction between reaction temperature and A: methanol:oil, B: catalyst amount, and C: time on biodiesel yield. From 3D plots, it was clear that high biodiesel yield (>90%) could be obtained between the reaction temperature of 55 and 65°C. A higher reaction temperature of 75°C reduced biodiesel production, which was due to vaporisation of methanol. Furthermore, there was a possibility of side reaction of alkyl ester hydrolysis to occur and to produce acids at the higher temperature [27].

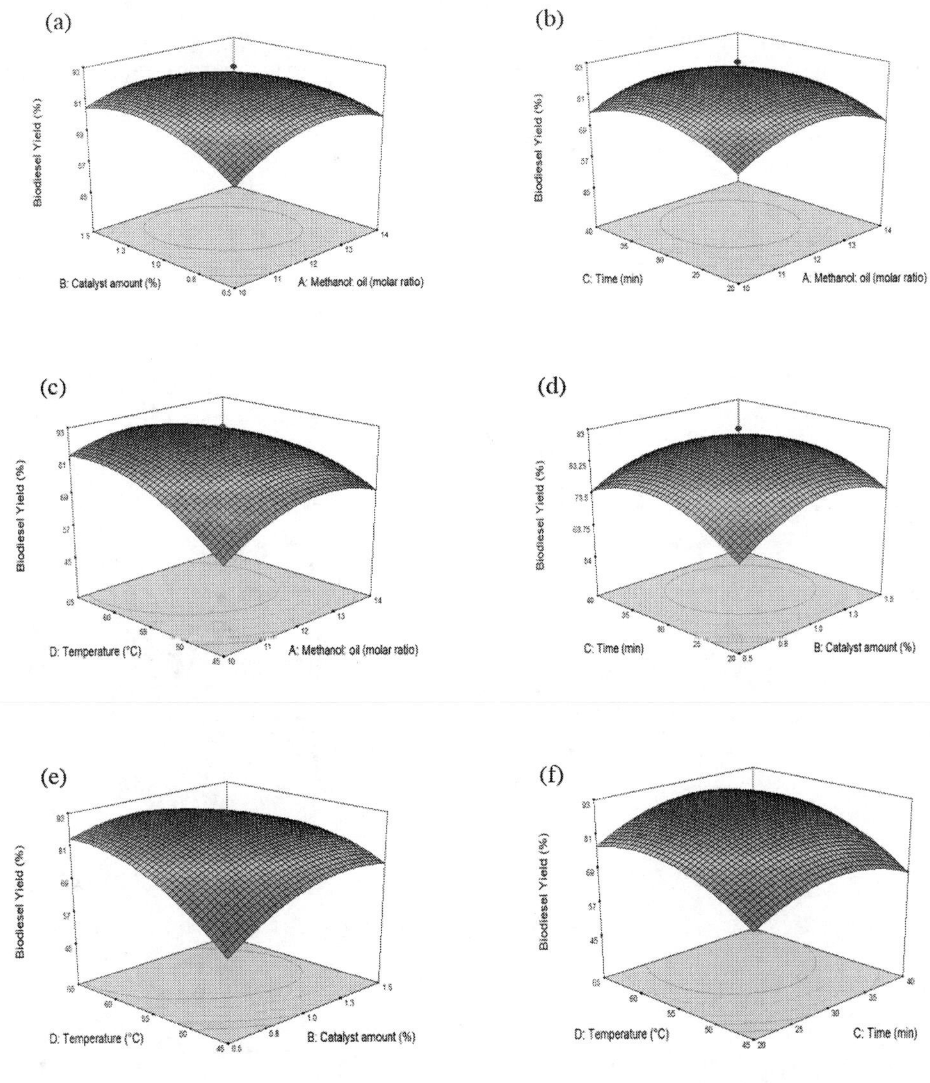

Figure 3. 3D plots represent the interaction between each variable on biodiesel yield: (a) catalyst amount and methanol:oil, (b) time and methanol:oil, (c) temperature and methanol:oil, (d) time and catalyst amount, (e) temperature and catalyst amount, and (f) temperature and time.

3.4. Optimisation of biodiesel production

The maximum biodiesel yield was determined by optimisation study. In this study, biodiesel yield was set to a maximum value, while other variables were set in a range between high and low levels. The optimum biodiesel yield of 95.7% was achieved by the transesterification of BSF larval oil with methanol:oil molar ratio of 12:1, catalyst amount of 1.0 weight %, a reaction time of 32 minutes and reaction temperature of 60°C. RSM with CCD was able to predict the optimisation of biodiesel production and transesterification process with a small percentage error of 1.01%.

3.5. FTIR analysis

FTIR analysis is important to confirm the functional groups that are present in the BSF larval lipid and biodiesel. Figure 4 shows an overlay of FTIR spectra of BSF larval lipid and biodiesel cultivated in soya residue. The absorption bands observed at 2920 and 2851 cm^{-1} can be related to stretches of C-H (alkanes) and sharp bands at 1740 to 1741 cm^{-1} can be assigned to C=O stretches (ketones, carboxylic acid and esters) [28]. The absorption bands observed at 1165, 1149, 1116 and 1111 cm^{-1} can be ascribed to C-O stretches (ester). The peaks at 721 and 720 cm^{-1} can be related to C-H bending vibration (long-chain fatty acids) [29]. In the context of figure 4 (b), the $-CH_2$ and $-CH_3$ bending vibrations of triglycerides were identified at 1456, 1435 and 1374 cm^{-1} [30]. Overall, results obtained from FTIR analysis confirmed the formation of fatty acid methyl esters (FAMEs) in BSF larval biodiesel derived from larval lipid cultivated on soya residue.

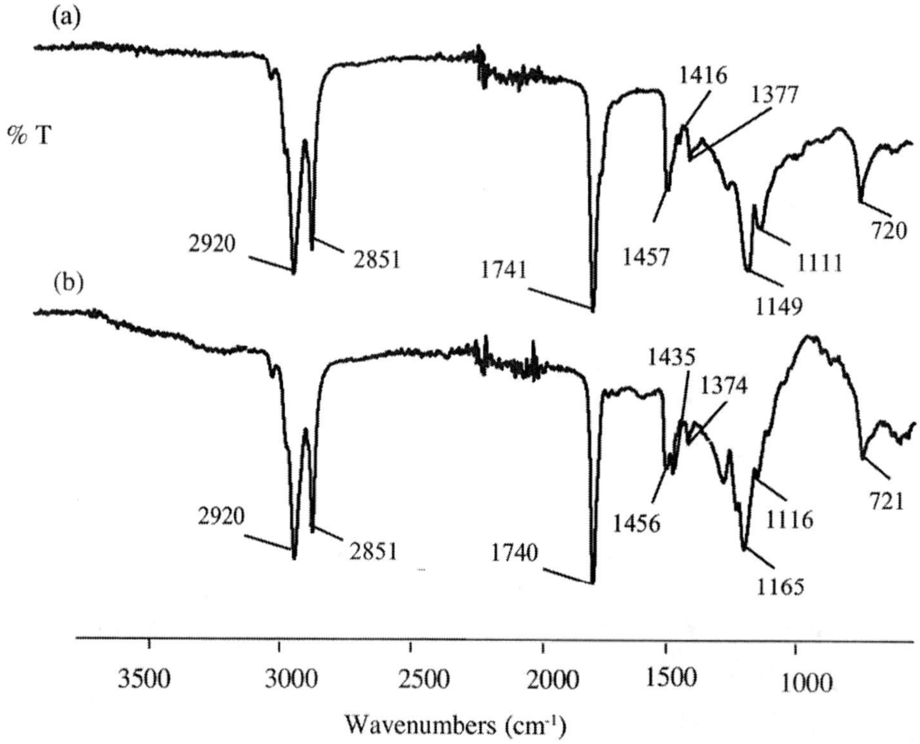

Figure 4. FTIR spectra of crude lipid (a) and biodiesel (b) produced from BSF larvae fed by soya residue.

3.6. GC-FID analysis

In this study, the fatty acid composition of BSF larval biodiesel was analysed using GC-FID. In general, GC-FID was mainly applied for determining the degree of saturation and unsaturation of biodiesel. Figure 5 displays the GC-FID chromatogram of the larval biodiesel produced from BSF larvae fed by soya residue. As can be seen from figure 5, eight FAMEs were detected in the BSF larval biodiesel including lauric acid (C12), myristic acid (C14), palmitic (C16), palmitoleic acid (C16:1), stearic (C18), oleic acid (C18:1), linoleic acid (C18:2) and gondoic acid (C20:1). It is evident from GC-FID analysis that BSF larval biodiesel comprised of saturated (lauric acid, myristic acid, palmitic acid, capric acid and stearic acid) and unsaturated (oleic acid, palmitoleic acid, linoleic acid and gondoic acid) FAMEs.

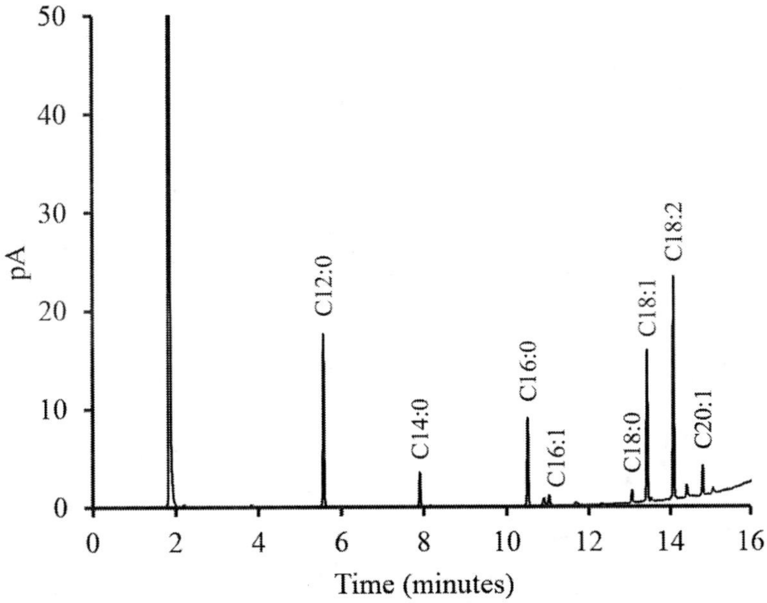

Figure 5. GC-FID chromatogram of biodiesel produced from BSF larvae fed by soya residue.

4. Conclusion

RSM with CCD based on several experimental parameters was successfully employed for the optimisation of biodiesel production from BSF larvae fed by soya residue. At optimum conditions of 12:1 (methanol:oil molar ratio), 1.0 weight % catalyst amount within 32 minutes of reaction time and 60°C of reaction temperature, 95.7% of biodiesel was produced. The physicochemical properties of crude BSF larval oil and biodiesel were characterised using FTIR spectrometer and GC-FID. The FTIR analysis revealed the appearance of absorption bands represent the stretches of C-H, C=O and C-O which confirmed the formation of fatty acid methyl esters in BSF larval biodiesel. Meanwhile, GC-FID proved that BSF larval biodiesel consists of saturated and unsaturated of fatty acid methyl esters.

Acknowledgements

MyBrain15 PhD scholarship provided by the Ministry of Higher Education Malaysia for S Ishak is gratefully acknowledged. The authors also thank Mr Ibrahim Saidin, Mr Mohd Zurin Mahmood, Mr Muhammad Hazim Baharudin, Mrs Johana Jumiran and Mr Mohd Hashimi Ma'ani for their technical assistance.

References

[1] Soliman J L, Lopez N S A and Biona J B M M 2018 *Energy Procedia* **158** 3723-28
[2] Nongbe M C, Ekou T, Ekou L, Benjamin Y K, Le Grognec E and Felpin F-X 2017 *Renew. Energy* **106** 135-41
[3] Knothe G and Razon L F 2017 *Prog. Energy Combust. Sci.* **58** 36-59
[4] Bhuiya M M K, Rasul M G, Khan M M K, Ashwath N and Azad A K 2016 *Renew. Sust. Energy Rev.* **55** 1109-28
[5] Ghazali W N M W, Mamat R, Masjuki H H and Najafi G 2015 *Renew. Sust. Energy Rev.* **51** 585-602
[6] Omidvarborna H, Kumar A and Kim D S 2016 *Environ. Prog. Sust. Energy* **35** 1537-44

IOP Conf. Series: Materials Science and Engineering **980** (2020) 012057 doi:10.1088/1757-899X/980/1/012057

[7] Kirubakaran M and Selvan V A M 2018a *Renew. Sust. Energy Rev.* **82** 390-401
[8] Omidvarborna H, Kumar A and Kim D S 2016 *J. Environ. Chem. Eng.* **4** 4769-75
[9] Abas N, Kalair A and Khan N 2015 *Futures* **69** 31-49
[10] Wang C, Qian L, Wang W, Wang T, Deng Z, Yang F, Xiong J and Feng W 2017 *Renew. Energy* **111C** 749-56
[11] Zheng L, Li Q, Zhang J and Yu Z 2012 *Renew. Energy* **41** 75–9
[12] Nguyen T T X, Tomberlin J K and Vanlaerhoven S 2015 *Environ. Entomol.* **44** 406-10
[13] Wong C H, Rosli S S, Uemura Y, Ho Y C, Leejeerajumnean A, Kiatkittipong W, Cheng C K, Lam M K and Lim J W 2019 *Energies* **12(8)** 1570
[14] Ishak S, Kamari A, Yusof S N M and Halim A L A 2018 *J. Phys.: Conf. Ser.* **1097** 012052
[15] Park J Y, Lee K, Choi S A, Jeong M J, Kim B, Lee J S and Oh Y K 2015 *Renew. Energy* **79** 3-8
[16] Yang S, Li Q, Gao Y, Zheng L and Liu Z 2014 *Renew. Energy* **66** 222–7
[17] Helm B R, Rinehart J P, Yocum G D, Greenlee K J and Bowsher J H 2017 *Proc. Natl. Acad. Sci.* **114(41)** 10924-29
[18] Cammack J A and Tomberlin J K 2017 *Insects* **8** 56
[19] Janssen R H, Canelli G, Sanders M G, Bakx E J, Lakemond C M M, Fogliano V and Vincken J-P 2019 *Sci. Rep.* **9** 2967
[20] López E N, Medina A R and Moreno P A G 2016 *Bioresour. Technol.* **216** 904-13
[21] Feng W, Qian L, Wang W, Wang T, Deng Z, Yang F, Xiong J and Wang C 2018 *Renew. Energy* **119** 12-18
[22] Patel P D, Lakdawala A and Patel R N 2016 *Biofuels* **7** 87-95
[23] Verma P, Dwivedi G and Sharma M 2017 *Fuel* **188** 586-94
[24] Issariyakul T and Dalai A K 2014 *Renew. Sust. Energy Rev.* **31** 446-71
[25] Rajan R R N, Rajesh R and Elayidom M R 2016 *ICEETS* 187-93
[26] Kafuku G and Mbarawa M 2010 *Fuel* **89** 2556-60
[27] Hasni K, Ilham Z, Dharma S and Warman M 2017 *Energy Convers. Manage.* **149** 392-400
[28] Ahmed W, Nazar M F, Ali S D, Rana U A and Khan S U 2015 *J. Energy Chem.* **24(3)** 331-6
[29] Tariq M, Ali S, Ahmad F, Ahmad M, Zafar M, Khalid N and Khan M A 2011 *Fuel Process. Technol.* **92** 336-41
[30] García-Martín J F, Alés-Álvarez F J, López-Barrera M del C, Martín-Domínguez I and Álvarez-Mateos P 2019 *Fuel* **240** 10–15

IOP Conf. Series: Materials Science and Engineering **980** (2020) 012058 doi:10.1088/1757-899X/980/1/012058

Conversion of rubber seed oil into biodiesel with potassium oxide alumina supported by (K_2O/Al_2O_3) catalyst

J A Karo Karo[1*], H Husin[2,3*], F Nasution[2], F T Yani[2], S Maliki[3], D D Prayuda[3], and F Hasfita[4]

[1]Baristan Industri Medan, Badan Penelitian dan Pengembangan Industri Kementerian Perindustrian, Medan, Indonesia

[2]Doctoral Program, School of Engineering, Universitas Syiah Kuala, Banda Aceh, Indonesia, 23111

[3]Reaction Engineering and Catalysis Laboratory, Department of Chemical Engineering, Universitas Syiah Kuala, Banda Aceh, Indonesia 23111

[4]Department of Chemical Engineering, Malikussaleh University, Lhoekseumawe, 24300, Aceh Utara, Indonesia

*Email: justaman.karo@gmail.com; husni_husin@che.unsyiah.ac.id

Abstract. In this paper, we report the synthesis of K_2O/Al_2O_3 for its application as a catalyst in the transesterification reaction for biodiesel production from rubber seed oil. So far, biodiesel production has been done through transesterification of vegetable oils through homogeneous catalysts, such as NaOH and KOH. Heterogeneous catalysts are more advantageous because they are easy to separate from the products, have less corrosion, and can be reused. This study aims to investigate the effectiveness of potassium oxide alumina-supported (K_2O/Al_2O_3) solid catalysts to convert rubber seed oil into biodiesel. The K_2O/Al_2O_3 catalyst was synthesized by impregnating KOH on the support of Al_2O_3 through the inception wetness method. The catalysts were characterized by scanning electron microscopy with energy-dispersive X-ray spectroscopy (SEM/EDX) to detect surface structure, composition, and active phase of the catalyst. EDX analysis indicated that the catalyst consisted of 36.41% potassium, 32.89%, and 30.7% oxygen. The reaction was conducted in a two-neck glass reactor at a temperature of 65 °C, methanol to oil ratios of 10:1, during 1.5 h. The yield of biodiesel was achieved up to 96.9% at the ratio KOH: Al_2O_3 of 7.5: 2.5.

1. Introduction

Nowadays, the inadequacy of fossil fuels along with global warming due to their massive utilization leads to the consideration of alternative fuels to substitute fossil fuels [1-6]. One of the alternative fuels that are concerned to be developed and used as a substitute for fossil fuel is biodiesel. Biodiesel is one of the renewable green fuels and promising energy which has properties similar to commercial diesel. It was non-toxic, biodegradable, and produces lower emissions than petroleum diesel [7-12].

Biodiesel is generally produced via transesterification of vegetable oils and methanol using base-catalyst which is taken at moderate temperatures [13-17]. The products, glycerol and alkyl esters, are separated by settling them after the neutralization of the catalyst. Furthermore, the crude glycerol and biodiesel are purified. However, commercial biodiesel processing requires quite high production costs compared to petroleum diesel. There are two factors that affect the cost of biodiesel production (the price of raw materials and production cost). The reduction of production costs has been tried by simplifying the process steps and minimizing waste flow. One option that has been used is transesterification with supercritical methanol (without using a catalyst) [18]. This supercritical process is very fast, it takes less than 5 minutes and does not involve a catalyst to reduce product purification costs. However, this reaction requires a very high reaction temperature, around: 350-400oC and high pressure (100-250 bar) which also increases the capital cost which is quite high. [19]. The use of heterogeneous catalysts can be one of the solutions to reduce production costs because solid catalysts are easily separated from reaction products and require relatively low operating temperatures around 60-80°C. The use of solid catalysts can be more competitive with commercial diesel. Several alkaline metals such as MgO, $Ca(OH)_2$, and $Ba(OH)_2$ have been tried, all of which dissolve easily in alcohol reactants so the process is homogeneous [20-24].

Various attempts were made to obtain a better solid base-catalyst. The literature shows that there are several heterogeneous catalysts that have been reported to transform the vegetable oil into biodiesel, $CaCO_3$, Mg-Al-Olt-Bu hydrotalcite, zeolite, which are operated at relatively high pressures. Several research groups have been developed a heterogeneous catalyst, such as K_2O/Zeolite, CaO from eggshells, CaO from seashells and reporting that catalysts have a promising performance. Previously, the MgO catalyst supported by mesoporous silica succeeded in increasing the conversion of vegetable oil into biodiesel to 96% (reaction time 5 hours at 220°C) [25-26].

In this study, our research group successfully used the solid catalyst from the natural and synthesis material, coconut coir, palm ash (bunches), CaO, and K_2O from water hyrachin, which show the biodiesel yield from vegetable oils between 87,97% to 98% [27-30]. These results were obtained from several processes that are efficient and economical in the process of making methyl esters. This research explored solid catalyst potassium alumina supplying catalysts for transesterification of rubber seed oil to biodiesel. So far, the development of KOH catalyst Al_2O_3 supported is still limited in the literature. This work is concentrated on the development of alumina-supported potassium oxide (K_2O/Al_2O_3) catalyst for the production of biodiesel from rubber seed oil. The catalyst was prepared through a simple method of inception wetness and analyzed by SEM and EDX instrument. The effect of the KOH to Al_2O_3 ratio has been evaluated against biodiesel yield and also characterized by SNI standard.

2. Material and Methods

2.1 Materials

KOH and Al_2O_3 (as a precursor), methanol (Merck, Germany) and rubber seed oil used in the transesterification reaction into biodiesel were purchased from a market and used as received.

2.2 Catalyst preparation

KOH/Al_2O_3 catalyst at various mole ratios (1:9, 2.5:7.5, 5.0:5.0; 7.5:2.5, 9:1) was prepared by impregnation method. The process started by adding and mixing the KOH, Al_2O_3 and distilled water into the beaker glass. It was carried out using a magnetic stirrer for 6 hours. After that, the process was continued by removing the water content on the mixtures. It was dried in the oven at 120°C for 12 hours. The powder was further calcined on the furnace at a temperature of 600°C for 5 h to convert the KOH into K_2O. Afterward, the catalyst was stored and kept in a desiccator.

2.3 Transesterification, separation, and washing of products

The biodiesel production via transesterification reaction using rubber seed oil and methanol with K_2O/Al_2O_3 catalyst was carried out in the batch stirred reactor. First, the weighed catalyst (4% oil weight), methanol (1: 6 oil), and rubber seed oil were put into the reactor and operated at constant temperature (65oC) and stirred for 1.5 hours. After that, the reaction was stopped and would be continued by separating the catalyst and products with a funnel (deposited it for 48 hours). During the deposition process, two layers from the mixture would be formed. The upper layer was the methyl

IOP Conf. Series: Materials Science and Engineering **980** (2020) 012058 doi:10.1088/1757-899X/980/1/012058

ester (biodiesel), while the downside layer was glycerol. After that, the methyl ester layer was filled into the separating funnel and then continued by washing it with warm water (50°C). During that process, some esters would form an emulsion with water. It took 24 hours to get a good separation of the product (biodiesel). Afterward, a clear yellow ester layer would be formed in the upper layer. Biodiesel yield was calculated using equation 1.

$$\%\,\text{Yield} = \frac{\text{Mass of biodiesel produced}}{\text{Mass of Oil}} \times 100\% \tag{1}$$

2.4 Product analysis

2.4.1. Catalyst characterization. Catalyst characterization - Scanning Electron Microscope (SEM) – Energy Dispersive X-ray Spectroscopy (EDX) was used to analyze and evaluate the morphology structure and chemical elements of the catalyst.

2.4.2. Biodiesel characterization. Biodiesel characterization consists of yield, density, viscosity, acid number, and moisture number were used to evaluate the effect of catalysts on the physical characteristics of the product (biodiesel). Density and viscosity were respectively measured using pycnometer and viscometer cannon Fenske while the acid number was using KOH base titration.

3. Results and Discussion

3.1 Characterization of K_2O/Al_2O_3 catalyst

3.1.1 SEM. The SEM patterns of K_2O/Al_2O_3 are displayed in figure 1. The selected catalyst with a magnification of 3000 times was observed.

On the SEM analysis, it can be seen that the K_2O was distributed on the Al_2O_3 perfectly. The alumina group was formed after calcining the catalyst at 600°C for 3 hours. From figure 1, it could be confirmed that the structure of catalyst (pores) was shaped like a marine coral group. It could provide better contact between catalyst and fluid in the transesterification process.

Figure 1. SEM image of K_2O/Al_2O_3 catalyst at mole ratio 2.5:7.5.

3.1.2 EDX. The transesterification reaction of RSO for biodiesel production was a surface reaction at catalytically active sites on KOH. The active sites (KOH) on the support surface (Al_2O_3) played important roles that can be evaluated by EDX. The results are presented in table 1.

From table 1, it can be seen that the catalyst consisted of 36.42% aluminum, 32.88% potassium, and 30.70% oxygen. Based on the result, the percentage of Al, K, and O_2 elements as well as their theoretical percentage of KOH/Al_2O_3 catalyst (impregnation process) were provided for each catalyst.

3.2 Biodiesel production using K_2O/Al_2O_3

3.2.1 Catalytic activity. The catalytic activity was affected by the mole ratio of catalyst (K_2O/Al_2O_3) and assessed based on the acquisition of fatty acid methyl ester (FAME) and saponification. The effect of various mole ratios on the catalyst performance on biodiesel acquisition is shown in figure 2.

From figure 2, it can be seen that the yield of biodiesel increased and started to decrease when the ratio of K_2O/Al_2O_3 was above 2.5:7.5. It was seen that the yield of biodiesel increased with the content of KOH. In this situation, the K_2O content was distributed well on the support surface to form K_2O/Al_2O_3, so it provided a better intensity to contact in the transesterification reaction process [31-32]. It could be caused by the synergy between K_2O and Al_2O_3 ions. A certain amount of Al_2O_3 supported could enhance the catalytic activity of K_2O.

Table 1. EDX Analysis of elements.

Element	Weight %	Atomic %
Oxygen	30.70	46.69
Aluminum	36.42	32.85
Potassium	32.88	20.46

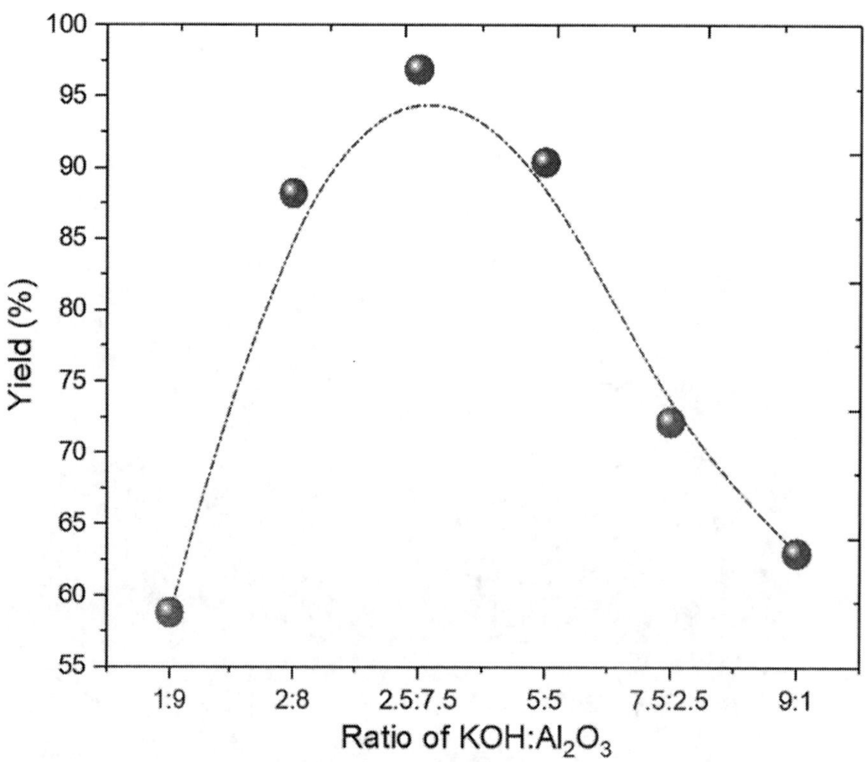

Figure 2. Effect of various ratio of catalyst K_2O/Al_2O_3 on biodiesel yield.

Table 2. Catalyst analysis of K_2O/Al_2O_3 at various condition (impregnation).

Ratio (KOH/Al_2O_3)		Density	Viscosity	Acid Number	Moisture Content
		(gr/mL)	(mm^2/sec)	(mg KOH/gr)	%
	SNI stand. >>	*0.85 - 0.89*	*2.3 - 6.0*	*Maks = 0.8*	*Maks = 0.05*
	Yield (%)				
2.5:7.5	**96.9**	**0.87**	**5.85**	**0.79**	**0.05**
5:5	90.4	0.87	6.59	0.82	0.06
7.5:2.5	70.2	0.89	7.00	0.90	0.06
9:1	63	0.90	7.13	0.96	0.08

3.2.2 Biodiesel properties. Biodiesel product was also analyzed to identify the properties to SNI standard. Density, viscosity, acid number, and moisture content are the most important properties of biodiesel to measure the quality of the product because of their effects on the work of pressurized combustion. The values of density, viscosity, acid number and moisture content of all reactions are presented in table 2.

The combination of K_2O and Al_2O_3 as catalyst significantly affected the parameters. Based on the experiments, K_2O/Al_2O_3 catalyst with a mole ratio at 2.5:7.5 was the best combination to produce biodiesel (from rubber seed oil) that meets the SNI standard. The expulsion of other ratio combinations could be caused by the influence of tri-, di- and mono- glyceride (doesn't react with methanol perfectly) and also by the contamination of glycerine [33].

4. Conclusion

The K_2O/Al_2O_3 catalyst was studied and applied to the transesterification of rubber seed oil into biodiesel. The best results of catalyst were obtained when the K_2O/Al_2O_3 was used at 2.5:7.5 (ratio). The analysis of the catalyst through SEM-EDX characterization found that the catalyst consisted of 36.41% alumina, 32.88% potassium, and 30.70% oxygen, with a catalyst in the form of a pore, like a sea coral structure.

Acknowledgment

The author thanks Professors Grant Research from the Universitas Syiah Kuala, Banda Aceh, Indonesia.

References

[1] Mardian R, Husin H, Pontas K, Zaki M, Asnawi T M and Ahmadi 2018 Hydrogen production from water-glucose solution over NiO/La-$NaTaO_3$ photocatalyst *IOP Conference Series: Materials Science and Engineering* **334** 012013

[2] Husin H, Adisalamun, Sy Y, Asnawi T M and Hasfita F 2017 Pt nanoparticle on $La_{0.02}Na_{0.98}TaO_3$ catalyst for hydrogen evolution from glycerol aqueous solution *AIP Conference Proceedings* **1788** 030073

[3] Husin H, Asnawi T M, Alam P N and Hasfita F 2018 Enhanced photocatalytic hydrogen production from waterethanol solution by Ruthenium doped La-$NaTaO_3$ *IOP Conf. Series: Materials Science and Engineering* **345**

[4] Husin H, Mahidin M, Yunardi Y and Hafita F 2015 Visible Light driven photocatalytic hydrogen evolution by lanthanum and carbon-co-doped $NaTaO_3$ photocatalyst *Key Engineering Materials* **659** 231-6

[5] Husin H, Pontas K, Sy Y, Syawaliah S and Saisa S 2014 Synthesis of nanocrystalline of lanthanum doped $NaTaO_3$ and photocatalytic activity for hydrogen production *Journal of Engineering and Technological Sciences* **46** 318-27.
[6] Husin H, Pontas K, Zaki M, Darmadi D and Masna A 2019 Enhance photocatalytic of hydrogen production from water-glycerol solution over RuO_2-loaded *Journal of Physics: Conference Series* **1402** 055008
[7] Tran N N, Tišma M, Budžaki S, McMurchie E J, Gonzalez O M M, Hessel V and Ngothai Y 2018 Scale-up and economic analysis of biodiesel production from recycled grease trap waste *Applied Energy* **229** 142-50
[8] Zaki M, Husin H, Alam P N, Darmadi, Rosnelly C M and Nurhazanah 2019 Transesterifikasi minyak biji buta-buta menjadi biodiesel pada katalis heterogen kalsium oksida (CaO) *Jurnal Rekayasa Kimia & Lingkungan* **14** 36-43
[9] Sy Y, Nurhazanah N, Maulana A, Mahidin M and Husin H 2019 Study of pH influences on the performance of Na-loaded $NbOPO_4$ solid acid catalyst for biofuel production *Journal of Physics: Conference Series* **1402** 055006
[10] Yanti D, Husin H, Yani F T, Maulana A, Adisalamun A and Husaini H 2020 Transesterification of Pangium Edule Reinw oil to biodiesel using durian rind ash as heterogeneous catalysts *IOP Conference Series: Materials Science and Engineering* **845** 012031
[11] Masjuki H H, Ruhul A M, Mustafi N N, Kalam M A, Arbab M I and Rizwanul Fattah I M 2016 Study of production optimization and effect of hydroxyl gas on a CI engine performance and emission fueled with biodiesel blends *International Journal of Hydrogen Energy* **41** 14519-28
[12] Mazaheri H, Ong H C, Masjuki H H, Amini Z, Harrison M D, Wang C-T, Kusumo F and Alwi A 2018 Rice bran oil based biodiesel production using calcium oxide catalyst derived from Chicoreus brunneus shell *Energy* **144** 10-9
[13] Nurhazanah N, Sy Y, Husin H, Rosnelly C M and Maulana A 2020 $K_2O/NbOPO_4$ solid acid catalyst for biodiesel production *IOP Conference Series: Materials Science and Engineering* **796** 012048
[14] Husin H, Asnawi T M, Firdaus A, Husaini H, Ibrahim I and Hasfita F 2018 Solid catalyst nanoparticles derived from oil-palm empty fruit bunches (OP-EFB) as a renewable catalyst for Biodiesel Production *IOP Conference Series: Materials Science and Engineering* **358** 012008
[15] Husin H, Mahidin and Marwan 2011 Studi penggunaan katalis abu sabut kelapa, abu tandan sawit dan K_2CO_3 untuk konversi minyak jarak menjadi biodiesel *Reaktor* **13** 254-61
[16] Zuhra Z, Husin H, Hasfita F and Rinaldi W 2015 Preparasi katalis abu kulit kerang untuk transesterifikasi minyak nyamplung menjadi biodiesel *Agritech* **35** 69-77
[17] Husin H, Pontas K, Yunardi, Adisalamun, Alam P N and Hasfita F 2017 Photocatalytic hydrogen production over Ni/La-$NaTaO_3$ nanoparticles from NaCl-water Solution in the presence of glucose as electron donor *ASEAN Journal of Chemical Engineering* **17** 27-36
[18] Mortensen P M, Grunwaldt J D, Jensen P A, Knudsen K G and Jensen A D 2011 A review of catalytic upgrading of bio-oil to engine fuels *Applied Catalysis A: General* **407** 1-19
[19] Bournay L, Casanave D, Delfort B, Hillion G and Chodorge J A 2005 New heterogeneous process for biodiesel production: A way to improve the quality and the value of the crude glycerin produced by biodiesel plants *Catalysis Today* **106** 190-2
[20] Abdullah, Rahmawati Sianipar R N, Ariyani D and Nata I F 2017 Conversion of palm oil sludge to biodiesel using alum and KOH as catalysts *Sustainable Environment Research* **27** 291-5
[21] Boonyuen S, Smith S M, Malaithong M, Prokaew A, Cherdhirunkorn B and Luengnaruemitchai A 2018 Biodiesel production by a renewable catalyst from calcined Turbo jourdani (Gastropoda: Turbinidae) shells *Journal of Cleaner Production* **177** 925-9
[22] Boro J, Thakur A J and Deka D 2011 Solid oxide derived from waste shells of Turbonilla striatula as a renewable catalyst for biodiesel production *Fuel Processing Technology* **92** 2061-7

IOP Conf. Series: Materials Science and Engineering **980** (2020) 012058 doi:10.1088/1757-899X/980/1/012058

[23] Bet-Moushoul E, Farhadi K, Mansourpanah Y, Nikbakht A, Molaei R and Forough M 2016 Application of CaO-based/Au Nanoparticles as Heterogeneous Nanocatalysts in Biodiesel Production. *Fuel* **164** 119-27

[24] Buasri A, Worawanitchaphong P, Trongyong S and Loryuenyong V 2014 Utilization of scallop waste shell for biodiesel production from palm oil – optimization using taguchi method *APCBEE Procedia* **8** 216-21

[25] Fitriana N, Husin H, Yanti D, Pontas K, Alam P N, Ridho M and Iskandar 2018 Synthesis of K_2O/Zeolite catalysts by KOH impregnation for biodiesel production from waste frying oil *IOP Conference Series: Materials Science and Engineering* **334** 012011

[26] Yani F T, Ulhaqi R, Pratiwi W P, Pontas K and Husin H 2019 Utilization of water hyacinth-based biomass as a potential heterogeneous catalyst for biodiesel production *Journal of Physics: Conference Series* **1402** 055005

[27] Husin H, Hasfita F and Rinaldi W 2013 Waste shells of cockle (Clinocardium nuttalli) as solid catalysts for transesterification of calophyllum inophyllum L. oil to biodiesel production. In: *Proceedings of The Annual International Conference, Syiah Kuala University-Life Sciences & Engineering Chapter,* pp 14-8

[28] Husin H, Zuhra, Hasfita F and Rinaldi W 2013 Waste shells of cockle (Clinocardium nuttalli) as solid catalysts for transesterification of calophyllum inophyllum L. oil to biodiesel production. In: *The Annual International Conference Syiah Kuala University,* ed Z A Muchlishien (AAC Hall, Syiah Kuala University, Darussalam Banda Aceh, Indonesia: Syiah Kuala University) pp 14-8

[29] Syamsuddin Y and Husin H 2010 Pembuatan katalis padat ZrO_2/Al_2O_3 untuk produksi biodiesel dari minyak jarak *Jurnal Rekayasa Kimia & Lingkungan* **7**

[30] Hani M R, Mahidin M, Husin H, Hamdani H and Khairil K 2018 Oil palm biomass utilization as an energy source and its possibility use for polygeneration scenarios in Langsa City, Aceh Province, Indonesia *IOP Conference Series: Materials Science and Engineering* **334** 012003

[31] Islam A, Taufiq-Yap Y H, Chan E-S, Moniruzzaman M, Islam S and Nabi M N 2014 Advances in solid-catalytic and non-catalytic technologies for biodiesel production *Energy Conversion and Management* **88** 1200-18

[32] Islam A, Taufiq-Yap Y H, Ravindra P, Teo S H, Sivasangar S and Chan E-S 2015 Biodiesel synthesis over millimetric γ-Al_2O_3/KI catalyst *Energy* **89** 965-73

[33] Husin H, Abubakar A, Ramadhani S, Sijabat C F B and Hasfita F 2018 Coconut husk ash as heterogenous catalyst for biodiesel production from cerbera manghas seed oil *MATEC Web Conf.* **197** 09008

IOP Conf. Series: Materials Science and Engineering **980** (2020) 012059 doi:10.1088/1757-899X/980/1/012059

Performance and engine exhaust emissions in a mixture of pertamax with PET plastic oil

K Winangun[1*], W T Putra[1], G A Buntoro[2], A Nirmala[1], and I Puspitasari[3]
[1]Department of Mechanical Engineering, Universitas Muhamadiyah Ponorogo
[2]Department of Informatics Engineering, Universitas Muhamadiyah Ponorogo
[3]Department of Mechanical Engineering, Politeknik Negeri Madiun

Email: kuntang@umpo.ac.id

Abstract. Public knowledge about plastic waste management is still a major problem. This study aims to determine the performance and engine emissions of a mixture of pet (p) and Pertamax (b) plastic oil, the purpose of which is to increase the RON of plastic oil. There were 3 fuels tested, pure plastic oil (p100), 90% Pertamax + 10% plastic oil (bp10), 80% Pertamax + 20% plastic oil (bp20). The highest torque test result was 11.38 nm at p100, the highest power was at bp20 with a value of 8.24 hp. exhaust gas emissions of CO = 4.33% at p100, and the best O_2 at 6.43% at rpm of 8500 with a mixture of bp20 fuel.

1. Introduction

The myriad of human mobility needs now has caused transport companies to compete to produce land, sea, and air transport. A form of transportation that Indonesians often use is a motorcycle. The number of motorcycles in Indonesia increases every year [1]. Increasing transport means increasing air pollution [2]. Therefore, it can be said that air pollution is directly proportional to the number of vehicles used by the people of Indonesia. Air pollution from vehicles is caused by incomplete combustion [3–5]. Mixing fuel with unbalanced air can cause incomplete combustion. Incomplete combustion can also be caused by the type of fuel with the octane number used that does not match the vehicle's compression ratio [6]. In Indonesia, the types of motorcycle fuel oil can be divided into three, namely Premium with an octane number of 88, Pertalite with an octane number of 90, and Pertamax fuel with an octane number of 92. Pertamina has determined this as a fuel that can be used in general [7].

Pertamax is a fuel oil that uses the addition of additives which was issued in 1999, the successor to Premix98. With the element MTBE (Methyl Tetra Butyl Ether) Pertamax is one of the highly recommended fuels used in motor vehicles. The Pertamax prices sold in the market are higher than the Premium. In addition, Pertamax has a higher octane value than the premium [7,8].

Plastic is a type of macromolecule formed by the polymerization process. One type of plastic that can be recycled is PETE or PET (polyrthyleneterephthalate). PET is commonly used for clear plastic bottles, such as mineral water bottles, juice bottles, and almost all types of other beverage bottles. In addition to PET, there are types of HDPE (high-density polyethylene) plastic. HDPE is a type of plastic that is strong and rigid, derived from petroleum, which often forms when blown. HDPE can generally be found in milk bottles, medicine bottles, shampoo bottles, juice bottles, liquid soap bottles, and baby soap bottles. Plastic waste of various types has a negative impact on the environment [9]. One of the uses of plastic waste is to make it into fuel [10–13].

The purpose of this study was to determine the performance and exhaust emissions of engines using PET type plastic oil with a mixture of Pertamax. Mixing plastic oil with Pertamax aims to increase the octane value of plastic oil so that it can be used for motorized vehicles. The fuels used included pure plastic oil (P100), 90% Pertamax mixture with 10% plastic oil (BP10), and 80% Pertamax mixture with 20% plastic oil (BP20). Another objective of this research is to use PET plastic waste as fuel.

2. Materials and methods

2.1. Material
Fuel engines have been used in human activities, as driving power in water pumps, generators, lawnmowers, or as a means of transportation to support the mobility of people and goods [7,8,14,15]. The combustion engine itself is divided into 2 (two) groups, namely gasoline and diesel engines. the difference lies in the fuel/air mixture ignition system, wherein the gasoline engine the fuel/air mixture is burned by sparks from the spark plug or often called the Spark Ignition Engine (SIE). According to [16], factors affecting engine performance include compression rates, air pressure, engine size, and temperature with the combustion process and the quality of the fuel. The result of engine performance is power, torque, fuel consumption.

The result of the combustion of a mixture of fuel and air released by the engine contains many gas components, most of the gas is pollution that is less friendly to the environment. Air pollution that endangers the environment occurs from incomplete combustion processes. Although the exhaust gases of motor vehicles mainly consist of harmless compounds like nitrogen, carbon dioxide, and water vents, They also contain a large number of other compounds that can harm exhaust gases and endanger the health and the environment. The pollutants that are mainly present in the exhaust gases of motor vehicles are carbon monoxide (CO), various subsequent hydrocarbon compounds, various oxides of nitrogen (NO_x) and sulfur (SO_x), and dust particles, including lead (PB) [17,18].

2.1.1 Plastic oil fuel
Alternative fuel is an effort to reduce the type of fuel that is commonly used. This effort is certainly an alternative solution for how to reduce or use other alternative fuels. This does not mean that the use of the fuel is 100% pure. Usage is generally a comparison or mixture with the fuels used daily, if 100% pure alternative fuels are used, it is generally necessary to study further machine construction suitable for the use of these alternative fuel characteristics.

2.1.2 Plastic PETE (POLYETHYLENE)
PETE plastic is one type of plastic that is generally used for clear bottles, such as cooking oil bottles and the like which, when used repeatedly, can remove carcinogens, especially in hot conditions. PETE plastic is recommended to be used once with used bottles and berets. Figure 1 below is a code for the type of plastic commonly used by the community.

Figure 1. Plastic code number.

2.2. Method
This research was conducted experimentally by comparing pure plastic oil fuel (P100) with 10% plastic oil fuel mixture and 90% Pertamax (BP10) and 20% plastic oil fuel mixture and 80% Pertamax (BP20). The properties of plastic oil are shown in table 1. This mixture of fuel oil and Pertamax was tested on a motorized vehicle with the specifications in table 2. Next, the 110 cc engine with plastic oil fuel and a mixture of plastic oil and Pertamax was tested for performance using dyno test with the specifications presented in table 3.

Table 1. Properties of plastic oil and plastic oil mixture with Pertamax.

Parameters	Limit	Results	
		Plastic Oil	20% Plastic Oil 80% Pertamax
Density at 15°C	715-770	750,7	742,6
Colour	Visual	Yellow	Green
Destilasi :			
-Initial Boiling Point (IBP)	ASTM D-86	65°C	45°C
-10%Vol.evap.		92°C	57°C
-50%Vol.evap.		-	92°C
-90%Vol.evap.		-	166°C
-F.B.P (End Point)		-	209°C
-Residu		-	1.3
Sulfur Content	ASTM D-4294	-	0,01
Octane Value	-	-	88,7

Table 2. Engine specifications.

Parameters	Description
Engine type	4-step SOHC engine with 2-valve, with air cooling.
Diameter and steps	50 × 55 mm
Step volume	108 cc (karburator)
Campression ratio	9,2 : 1
Maximum power	8,22 PS /8.000 rpm (carburator)
Maximum torque	0,85 kgf.m/ 5.500 rpm (carburator)
Gas tank capacity	3,5 liter (carburetor)
Spark plug type	NGK CPR8EA-9 atau Denso U24EPR9
Ignition system	DC-CDI
Carburetor type	Keihin AVK22 / throttle body 22 mm

Table 3. Dyno test specifications.

Parameters	Descriptions
Measurement	Speed, Rpm, acceleration, torque, gears, and power
Data transfer	RS232 - USB
Maximum torque	50 Nm
Maximum RPM	20.000 rpm
Maximum power	50 Hp
Maximum speeds	350 km/hour
RPM measurement system	Induction
Torque measurement system	Load cell
Break type	Mechanical disk
Break control	Pneumatic 4 bar

Then the 110 cc engine with plastic oil fuel and a mixture of plastic oil and Pertamax exhaust gas emissions were also tested using dyno test with specifications such as table 4.

IOP Conf. Series: Materials Science and Engineering **980** (2020) 012059 doi:10.1088/1757-899X/980/1/012059

Table 4. Gas analyzer specifications.

Parameters	Measurement Range
CO	0.00 -10.00% vol
CO_2	0.00 – 18.00% vol
HC	0 – 9999 ppm vol
NO	0.00 – 22.00% vol
O_2	0 – 5000 ppm vol
Lambda (γ)	0.500 – 9.99
Rpm Counter	100 - 15000 rpm

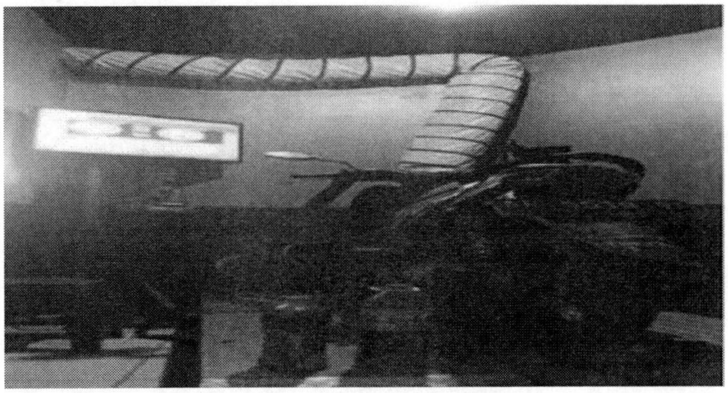

Figure 2. Testing performance.

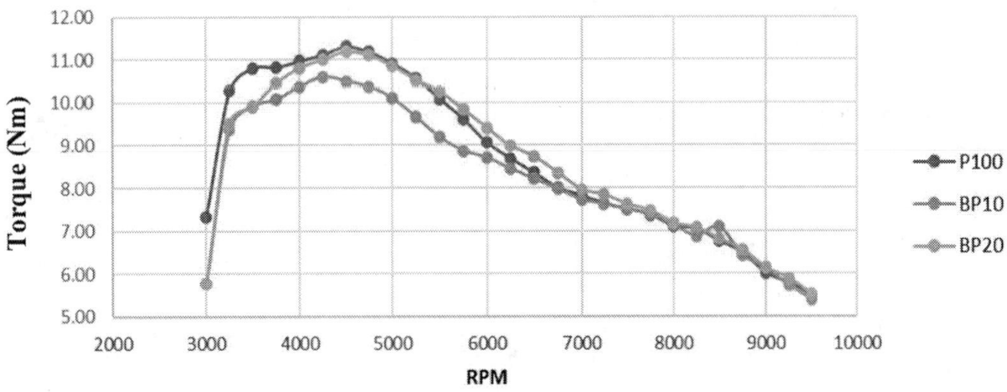

Figure 3. Torque against RPM on P100, BP10, and BP20 fuels.

3. Results and discussions

Data were collected on torque and power using variations in rpm, namely 3000-9500 rpm (figure 2). From several variations in rpm, torque and power would be figured out. Using a certain type of fuel, this test was conducted 5 times to obtain maximum and specific data.

3.1. Torque testing results

Based on figure 3, the torque generated from P100 plastic fuel oil at 3000-9500 rpm. Torque increases slowly until it reaches a peak of 4,500 rpm with an average torque of 11.30 Nm. From the BP10 fuel mix test results, the highest torque is at 4,250 rpm with an average torque of 10.60. When compared with the results of testing for pure plastic fuel oil without mixture, the mixed fuel BP10 experienced an increase in torque by 6.6%. From the results of the BP20 mixed fuel test, the highest torque is at 4,500 rpm with an average torque of 11.18 Nm. When compared to pure plastic fuel oil without mixture, the mixed fuel BP20 has a torque increase of 1.07%. Based on the test results, it can be seen that the highest torque is achieved with P100 plastic fuel oil. This is because plastic fuel oil has a higher density compared to Pertamax and plastic fuel oil.

3.2. Exhaust emission test results

From several fuels that were tested using a gas analyzer, the ratio of exhaust gas emissions (CO, HC, CO_2, O_2) was found.

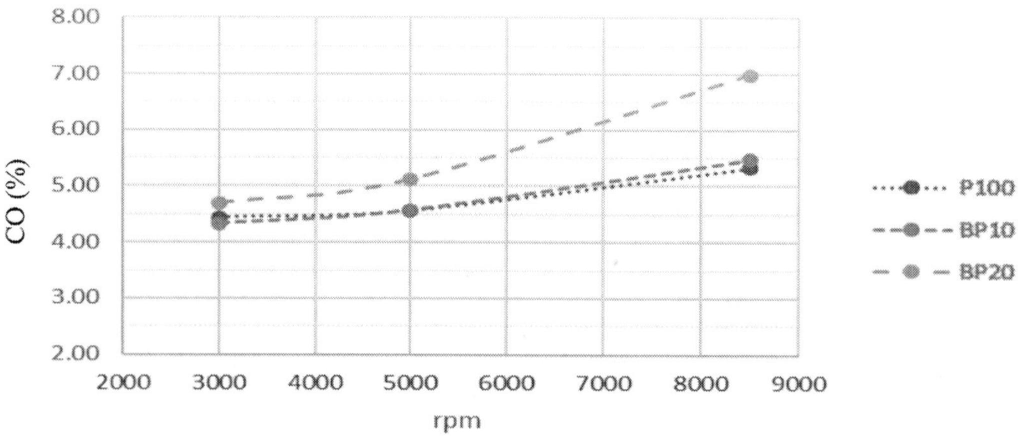

Figure 4. Comparison of exhaust emissions (CO) on fuel mixtures P100, BP10, BP20.

Figure 4 shows that the highest CO value is in the BP20 mixture of 6.98%, while the lowest performance is in the BP10 mixture with a value of 4.33% at 3,000 rpm rotation. The BP10 has a higher yield (5.5% at 8500 rpm) than the P100 because it is blended with alternative fuels. The smallest CO in the cheapest gasoline affects the mixture and the air.

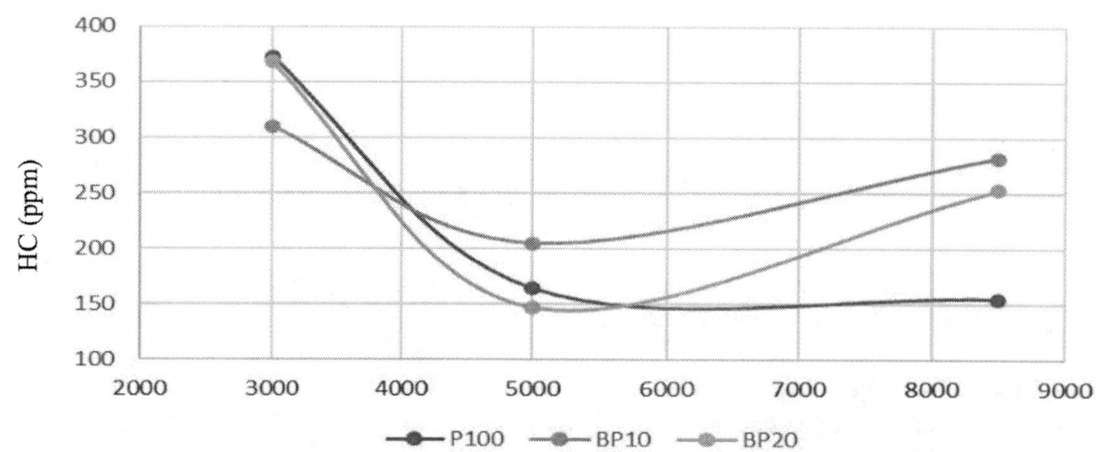

Figure 5. Comparison of exhaust emissions (HC) in fuel mixtures P100, BP10, BP20.

From the test results shown in figure 5, the highest HC is in the P100 mixture of 373 ppm with a rotation of 3000 rpm, while the lowest emission level is in the fuel mixture BP20 = 147 ppm. Starting from the average exhaust gas emission ratio (HC), compared to the HC government standard of 200-400 ppm, the BP20 fuel output does not meet the standard, but if the HC is reduced, the combustion will be better.

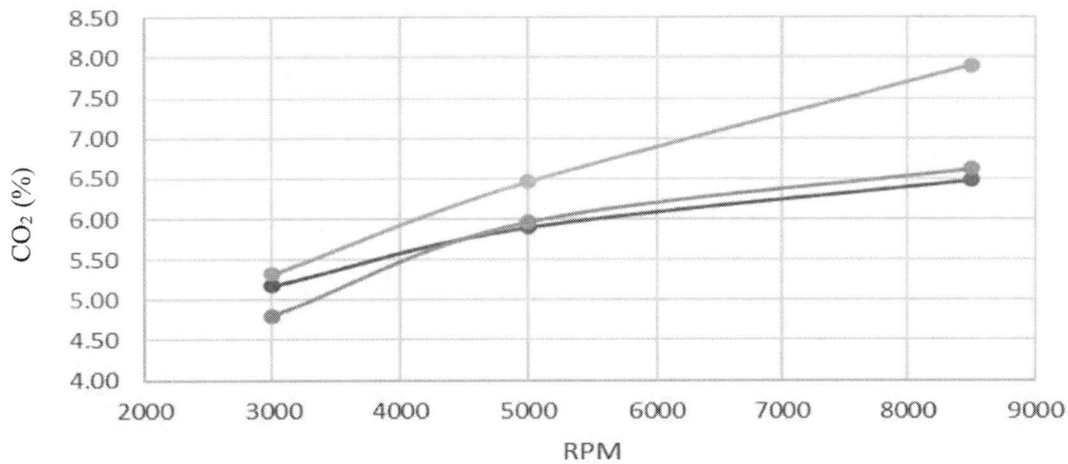

Figure 6. Comparison of exhaust emissions (CO_2) in fuel mixtures P100, BP10, BP20.

From tests carried out using a gas analyzer test tool, the results are shown in figure 6: CO_2 emissions at engine speed of 3,000 rpm with results P100 = 5.18%, BP10 = 4.80%, BP20 = 5.32% and at engine speed 5500 rpm emission level P100 = 590%, BP10 = 5.96%, BP20 = 6.46%, and at 8500 rpm the emission level is P100 = 6.48%, BP10 = 6.62%, BP20 = 7.90%. At 3000 rpm, the CO_2 in the BP10 mixture has the lowest levels compared to government standards with levels of 12-15%, so that the BP10 mixture does not meet the standards with a yield of 4.80%. If the CO_2 drops, the scale on the engine block is thick, it is advisable to check the engine.

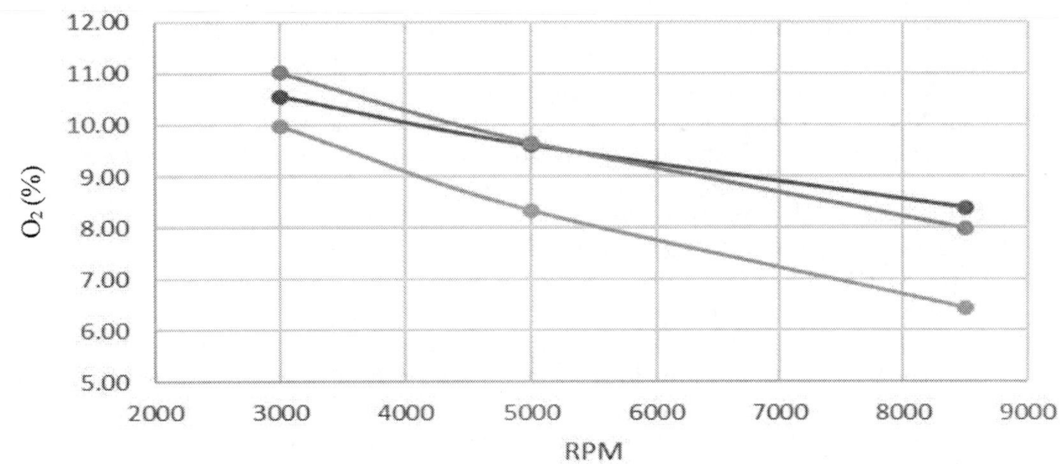

Figure 7. Comparison of exhaust emissions (O_2) in fuel mixtures P100, BP10, BP20.

The test results are shown in figure 7, the highest O_2 is in the BP10 fuel produced at 3000 rpm engine speed, namely 11.01%, while the lowest O_2 performance is in the BP20 fuel mixture with an emission level of 6.43% at 8500 rpm. Compared to the government standard level of 0.5-2%, the fuel yield of

BP20 has a grade of 6.43% which far exceeds the fuel standard. A high O2 level indicates a problem with the exhaust that may be leaking / clogging.

4. Conclusion

The effect of using a mixture of plastic fuel oil with Pertamax on engine power and torque performance was investigated. The highest torque was P100 fuel or 11.308 Nm pure plastic oil at a rotation of 4500 rpm. The highest power generated was observed in the mixture of BP20 or 80% Pertamax fuel and 20% plastic oil at 8.24 kw at 8250 rpm. The best exhaust emission for CO = 4.33% at 3000 rpm was with P100 fuel and the highest emission value obtained for gas O_2 = 6.43% at 8500 was with a BP20 fuel mixture. Of the three test results, good fuel was coded P100 (pure plastic oil) but because P100 is difficult to find, another option is to use plastic fuel code BP20.

Acknowledgments

Our gratitude goes to the Universitas Muhammadiyah Ponorogo, SMKN 1 Badegan Ponorogo, and PT. Pertamina Rayon Madiun who have helped so that this research could run smoothly. It also goes to the reviewers for their suggestions and input for the betterment of this article.

References

[1] Korlantas 2020 Perkembangan jumlah kendaraan bermotor Badan pusat statistik
[2] Jayanti N E, Hakam M and Santiasih I 2014 Emisi gas carbon monooksida (CO) dan hidrocarbon (HC) pada rekayasa jumlah blade turbo ventilator sepeda motor supra X 125 tahun 2006 *Rotasi* **16** 1
[3] Supriyatmojo B D S K, Putra W T and Winangun K 2018 Pengaruh penggunaan tiga metode injector cleaner terhadap emisi gas buang dan konsumsi bahan bakar pada honda vario injeksi 125 *Komputek* **2** 23
[4] Fadiah N 2013 Terhadap performa mesin dan emisi gas buang sepeda motor yamaha v-ixion *Abstrak* **02** 132–40
[5] Winangun K 2009 Uji emisi penggunaan bioetanol dari tetes tebu sebagai campuran premium dengan oktan booster pada sepeda motor yamaha vega ZR *J. Pendidik. Tek. Mesin* **19** 25–31
[6] Ferguson C R 2001 Internal combustion engines - applied thermosciences United States Copyright Act USA pp 253
[7] Wiranata Y D and Ansori A 2017 Campuran premium dan pertamax pada berbagai variasi pada sepeda motor new honda vario 110 FI *J. Univ. Negeri Surabaya* 183–90
[8] Ningrat A A W K, Kusuma I G B W, and Wayan I 2016 Pengaruh penggunaan bahan bakar pertalite terhadap akselerasi Mettek Sol **2(1)** pp. 59–67
[9] Mani M and Nagarajan G 2009 Influence of injection timing on performance, emission and combustion characteristics of a DI diesel engine running on waste plastic oil *Energy* **34** 1617-23
[10] Mani M, Nagarajan G and Sampath S 2011 Characterisation and effect of using waste plastic oil and diesel fuel blends in compression ignition engine *Energy* **36** 212–9
[11] Kareddula V K and Puli R K 2018 Influence of plastic oil with ethanol gasoline blending on multi cylinder spark ignition engine *Alexandria Eng. J.* **57** 2585–9
[12] Devaraj J, Robinson Y and Ganapathi P 2015 Experimental investigation of performance, emission and combustion characteristics of waste plastic pyrolysis oil blended with diethyl ether used as fuel for diesel engine *Energy* **85** 304–9
[13] Putra W, Bin Muhamad S, Bin Muhamad M and Bin Mohdzakaria M A 2019 Effectiveness teston hardness performance of plastic waste and sawdust composite *Int. J. Recent Technol. Eng.* **8** 273–80
[14] Suhendar A, Farid A and Fuhaid N 2016 Pengaruh pemakaian campuran bahan bakar premium dengan ethanol terhadap unjuk kerja motor bakar injeksi *Widya Tek.* **24(1)** pp 13-6
[15] Ilham W, Fadelan, Kuntang W 2019 Perencanaan alat pencacah rumput gajah dengan pisau lengkung kapasitas 110 kg/jam *Komputek* **3** p 22-32

IOP Conf. Series: Materials Science and Engineering **980** (2020) 012059 doi:10.1088/1757-899X/980/1/012059

[16] Heywood J B, 1988 *Internal Combustion Engine* (New York: Library of Congress Cataloging-iP.Publication Da) pp15-19
[17] Ismiyati, Marlita D and Saidah D 2014 Pencemaran udara akibat emisi gas buang kendaraan bermotor *J. Manaj. Transp. Logistik* **01** 241-8
[18] Tugaswati A T 2008 Emisi gas buang kendaraan bermotor dan dampaknya terhadap kesehatan *Komisi Penghapusan Bensin Bertimbel* **1** 1–11

IOP Conf. Series: Materials Science and Engineering **980** (2020) 012060 doi:10.1088/1757-899X/980/1/012060

The kinetic study of pyrolysis of lignite and polyethylene plastic bag waste using the thermogravimetric analysis

N Aulia, H Wijayanti*, and D R Wicakso

Department of Chemical Engineering, Lambung Mangkurat University, Jl. A. Yani Km. 35.5, South Kalimantan, Indonesia 70714

Email: hesti.wijayanti@ulm.ac.id

Abstract. Lignite or low rank coal has a great potential to be utilized as an alternative energy. In this study, the pyrolysis of lignite and polyethylene plastic bag waste with different compositions (0, 25, 50 and 75 wt.% of polyethylene) had been investigated using a thermogravimetric analyzer. The results showed that the decomposition temperature of lignite was in the range of 250-600^{0}C, while the decomposition temperature of lignite-polyethylene plastic bag waste mixtures was around 280-600^{0}C. The addition of polyethylene to the lignite during pyrolysis reaction contributed to the increasing mass decomposition from 32.16 wt.% (lignite only or 0 wt.% of polyethylene) to 72.70 wt.% (75 wt.% of polyethylene). Furthermore, the thermogravimetric data were used to perform the kinetic study. The obtained kinetic parameters (activation energy and pre-exponential factor) for both lignite and lignite-polyethylene plastic bag waste mixtures were in the range of 16.16-28.95 kJ/mol and 0.014-0.212 min^{-1}.

1. Introduction

Indonesia has considerable coal potential. Based on the data from Indonesia Investments [1], Indonesia was the fifth rank of the countries that has the largest coal reserves in the world. Indonesian coal deposit is 0.5% of all coal reserves in the world [2]. South Kalimantan is the second largest area that has coal resources in Indonesia with a total amount of 4.7 billion tons [3]. The coals from this area include the types of lignite and sub-bituminous coals. Lignite is a low grade coal (low rank coal) which is caused by the low energy density and its high moisture content so that it is less efficiently used in direct combustion to produce energy [5].

Plenty of research has been conducted to optimize the use of lignite as an efficient energy source such as by increasing the grade of lignite or by converting it into liquid fuels. There are many methods of converting coal such as gasification, hydrothermal liquefaction [6], or pyrolysis [7]. Pyrolysis is one of the best coal upgrading methods for a clean and efficient utilization of lignite. Lignite contains high adequate-oxygen content, so it is less profitable in the pyrolysis process. The addition of hydrogen in the pyrolysis can improve product quality. However, pure gas hydrogen is still expensive.

Plastic waste still becomes a big problem for the environment. Besides being unable to be decomposed by nature, waste plastic will also cause air pollution if burned. The plastic waste recycling processes cause a major problem in the environment. The plastic waste that is often found in Indonesia is polyethylene in the form of plastic bags. According to Baofeng et al. [6], elements of C and H contained in the plastic are very high, even higher than coal, and can interact during the pyrolysis process with lignite. It is hoped that plastic waste will help as the hydrogen "donor" during the lignite pyrolysis process.

The coal pyrolysis with polyethylene (PE) and polypropylene (PP) at various conditions demonstrates the enhancement of the liquid yields and the conversion of coals. The synergistic effects have been observed from the previous studies in the mixtures of PE and PP with low rank coal and lignite using hydrogen gas even though at low-set temperature [8]. All studies agree that the plastic contents have a contribution as the hydrogen supplier during coal pyrolysis[8].

In general, a thermogravimetric analysis (TGA) is used to study mass decomposition and kinetics over the pyrolysis of the solid materials including coal and plastic [11]. It plays a role in the determination of mass decomposition of the material linear to time and temperature. The kinetic parameters of this reaction are obtained by applying the Arrhenius law equation [11]. The knowledge of the kinetics of the thermal decomposition of the fuel feedstock is essential for further design of the pyrolysis systems and their conversion mechanisms.

In this research, the thermogravimetric analysis is conducted to investigate the mechanisms of pyrolysis of lignite/polyethylene plastic bag waste mixtures under the nitrogen atmosphere. By considering the overall processes as a one-step first order reaction, the kinetic model is determined to fit the thermogravimetric data.

2. Materials and methods

2.1 Materials

The raw materials utilized in this work included the low rank coal (lignite) from PT Adaro at Tabalong district and the polyethylene plastic bag waste obtained from PT Trijaya Abadi Banjarbaru, South Kalimantan. The particle size of lignite was less than 2 mm and the size of polyethylene plastic bag waste pelletized was less than 3 mm.

2.2 Pyrolysis

The pyrolysis reaction was conducted using a thermogravimetric analyzer Linseis STA PT 1600 with a heating rate of 10°C/min up to 600°C under 50 mL/min of nitrogen. The nitrogen flow was used to ensure an inert atmosphere condition. The mixture of lignite and pellet of PE were homogeneously mixed. Approximately 15 mg of samples and the PE pellets with the ratios of 0, 25, 50, and 75% from the total sample mixture were used for every run that were further labelled by PE, 25PE, 50PE, and 75PE.

2.3 Kinetic study

From the TGA results, the sample decomposition corresponding to the temperature of weight loss (TG) and the weight loss rate (DTG) was gained for the pyrolysis reaction [9, 11]. The weight loss rate (DTG) was calculated by the following equation 1

$$\frac{dW}{dt} = -\frac{1}{Wo}\left(\frac{dWt}{dt}\right) \tag{1}$$

The fraction of weight loss can be determined as the pyrolysis conversion x and can be computed by the following equation 2

$$x = \frac{Wo-Wt}{Wo-Wf} \tag{2}$$

where Wo is the initial mass of the sample; Wt is the mass at certain time t and Wf is the final mass that left at the end of pyrolysis [11].

By using the TGA data, the integral method was used to obtain the kinetic constants, such as the activation energy and the pre-exponential factor during pyrolysis of lignite and polyethylene plastic bag waste [10, 11]. Many previous studies have made an assumption that a solid material pyrolysis mechanism includes first order reaction, and the heating rate is considered constant during pyrolysis [9, 11]. Following the Arrhenius law equation, thus the calculation for the kinetics of lignite and polyethylene plastic bag waste pyrolysis reaction can be explained by the equation below:

$$\frac{dx}{dt} = Aexp\left(-\frac{E}{RT}\right)(1-x) \tag{3}$$

where A is pre-exponential factor, E is activation energy, T is temperature, t is time and x is obtained from Eq. (2).

Rewrite Eq. (3) at the heating rate H set at the constant value, H=dT/dt, and then integrating results:

$$ln\left[\frac{-ln(1-x)}{T^2}\right] = ln[\frac{AR}{HE}(1-\frac{2RT}{E})] - \frac{E}{RT} \tag{4}$$

The expression ln [AR/HE(1-2RT/E)] in Eq. (4) is essentially constant for the most values of E at the temperature range of the pyrolysis [9, 11]. Then, if the left side of Eq. (4) as the function of y is plotted versus 1/T as the x function, a straight line will be gained if it is assumed that the pyrolysis reaction as a first order reaction. The slope and intercept of the resulted line should be used to determine the activation energy (E) and the pre-exponential factor (A).

3. Results and discussion

3.1. The proximate analysis

The proximate analysis was conducted to determine the raw material (lignite and PE) condition prior to the pyrolysis reaction. This consisted of water content, ash content, volatile matter and fixed carbon analysis.

3.1.1 Water content analysis

The water content was analyzed using a crucible that had been dried in an oven at 105°C for an hour and cooled in a desiccator for 15 minutes. Then 0.5 gram of the sample (lignite and PE) was put into the crucible then heated at 105°C for 180 minutes and cooled in the closed crucible for 15 minutes in the desiccator and then weighed. Afterward, the crucible that contained the sample was heated again at 105°C for 60 minutes and cooled again for 15 minutes in a desiccator then weighed. Then the third step was repeated until the difference between the last and the previous water content results was 0.0003 grams.

3.1.2 Ash content analysis

The ash content was analyzed by calcining a crucible using a furnace with a temperature of 575°C for 120 minutes, then it was cooled in a desiccator for 15 minutes. Next, 0.5 grams of the sample from the previous water content analysis was heated at 105°C for 180 minutes until the weight was stable, then it was weighed. The next step was 2-stage calcination, which was done by calcining it at a temperature of 250°C (the heating rate was 10°C/min.) for 30 minutes, followed by the calcination from 250–575°C for 150 minutes. After that, the sample was cooled in a desiccator and weighed then calcined again at 575°C for 60 minutes. This step was repeated until the weight was stable.

3.1.3 Volatile Matter Analysis

This volatile matter analysis was carried out by calcining the crusible using a furnace at a temperature of 950°C for 30 minutes then cooled and weighed. Next, about 0.5 grams of the sample was put into the crucible and closed then weighed. The calcination was repeated at 950°C for 7 minutes then cooled and weighed again.

3.1.4 Fixed carbon analysis
The fixed carbon value was calculated from the values of water content, ash content and volatile matter by using the equation as follows:

$$\% \text{ Fixed Carbon} = 100 - (\%\text{Water Content} + \% \text{ Ash Content} + \%\text{Volatile Matter}) \quad (5)$$

The results of proximate analysis of the lignite and PE are presented in table 1.

Table 1. Proximate analysis for the samples of lignite and PE plastic bag waste.

Samples	Proximate Analysis			
	Water (%)	*Ash* (%)	*Volatile Matter* (%)	*Fixed Carbon* (%)
Lignite	17.10	6.51	59.15	17.24
PE	0.34	3.37	95.61	0.68

3.2. Thermal decomposition of the lignite and the lignite-PE plastic bag waste mixtures
The thermogravimetric analysis (TGA) plotted the weight decomposition of the samples due to the increasing pyrolysis temperature. The TGA results for the lignite and the lignite-PE plastic bag waste mixtures with various compositions are shown in figure 1.

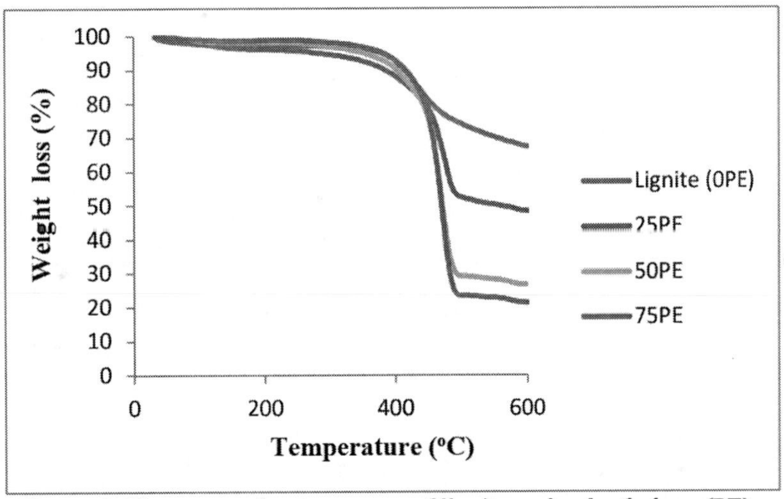

Figure 1. Thermogravimetry curves of lignite and polyethylene (PE) plastic waste pyrolysis.

Based on figure 1, it is obvious that the weight loss of all samples enlarge with the enhancement of temperature. In comparison, the curves of lignite and PE plastic bag waste have different trends in their thermal behavior possibly due to the difference in their structures [9].

In general, biomass pyrolysis consists of three main steps [9], so does lignite pyrolysis. The first step, which was almost the same for all samples (below 200°C), was water removal. Next, the second step that resulted in the main weight decomposition occurred at different temperature ranges. The lignite itself mainly decomposed at 250-600°C. The addition of PE slightly shifted the main decomposition temperature to 280-500°C. Zhou et al. [8] point out that the second step of lignite pyrolysis is called an active thermal decomposition. This can be the most complicated step because the chemical bonds of the materials start to decompose at high temperatures [9]. The final step is the continuous devolatilization step that occurred at over 600 and 500°C for the lignite and the lignite-PE mixtures, respectively. In the final pyrolysis reaction, the overall weight loss of lignite pyrolysis was 32.16 %, whereas the overall weight losses for co-pyrolysis of the lignite-PE plastic waste mixtures were 49.93%, 65.06%, and 72.20% due to the addition of 25PE, 50PE, and 75PE, respectively. The lower weight loss of the lignite compared to that of the mixtures possibly due to higher ash and fixed carbon contents [9, 11] as seen in table 1.

To conclude, the addition of PE plastic waste into the lignite pyrolysis through the thermogravimetry analysis (TGA) caused the final decomposition to occur with a lower temperature and the sample decomposition to increase.

3.3. Kinetic study

In order to determine the kinetic parameters of the pyrolysis reaction, namely activation energy and pre-exponential factor, equation 4 was applied. The linear model is established by plotting ln[-ln(1-x)/T^2] versus 1/T for pyrolysis of the lignite and that of the lignite-PE plastic waste mixtures as shown in figure 2.

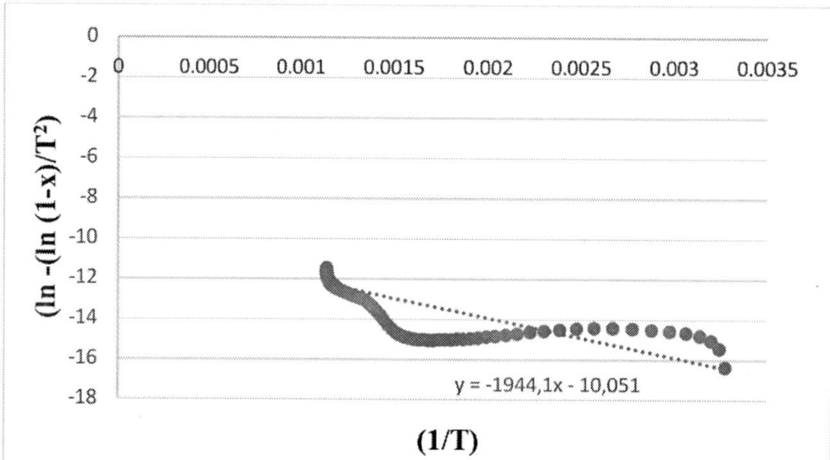

(a)

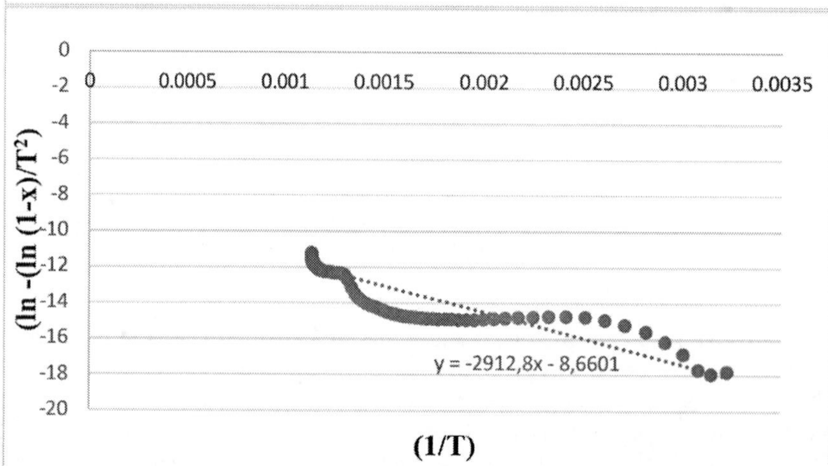

(b)

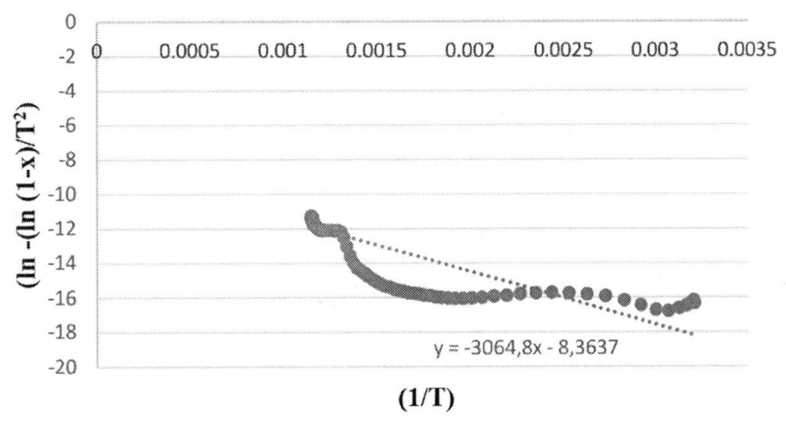

(c)

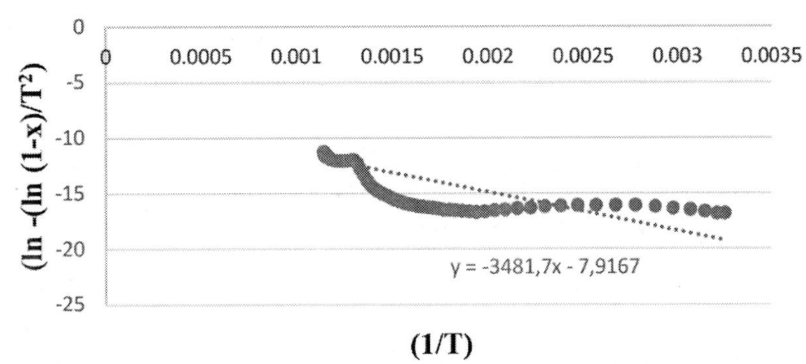

(d)

Figure 2. Plots of ln(-ln(1-x)/T2) vs 1/T of Lignite (0PE) (a), 25PE (b), 50PE (c), and 75PE (d) from the TGA data.

The results in figure 2 indicate that the data considerably fit the linear regression models with the R^2 values in the range of 0.69 to 0.86. Furthermore, the slope and intercept of each line can be used to compute the activation energy (E) and pre-exponential factor (A) as displayed in table 2.

Table 2. Kinetic parameters for pyrolysis of lignite and PE plastic bag waste.

Samples	E (kJ/mol)	A (min^{-1})	R^2
Lignite (0PE)	16.16	0.014	0.69
25PE	24.22	0.084	0.86
50PE	25.48	0.119	0.77
75PE	28.95	0.212	0.73

Data in table 2 confirm that the activation energy of lignite pyrolysis is 16.16 kJ/mol. During pyrolysis, the activation energy gets increasing to 24.22, 25.48, and 28.95 simultaneously with the increased PE compositions from 25, 50, to 75%, respectively. This result suggested that the energy

IOP Conf. Series: Materials Science and Engineering **980** (2020) 012060 doi:10.1088/1757-899X/980/1/012060

required to break the matrix of the lignite-PE plastic waste mixtures was higher than that of the lignite itself. It was because of the strong bonds in the matrix of the polymers in PE plastic waste. Hence, to destroy the bonds, it needed higher temperature and activation energy [9] than that of biomass, including lignite. Thus, the more PE used in the lignite pyrolysis, the more was the activation energy needed. The increase of activation energy as the PE compositions raised shows that the different pyrolysis reactivity is the result of different PE compositions.

4. Conclusion

This study investigated the thermal behaviors of lignite and polyethylene (PE) plastic bag waste using the thermogravimetric analyzer. The lignite started to decompose at a lower temperature than PE. The addition of PE plastic bag waste into the lignite caused the final decomposition to occur with a lower temperature while the sample decomposition got higher. The results from this kinetic study indicate that the first-order reactions of Arrhenius law equation satisfactorily fit for describing the pyrolysis mechanism of lignite and polyethylene (PE) plastic bag waste.

Acknowledgments

The authors wish to thank the Directorate of Research and Community Service, The Ministry of Education and Culture of Indonesia for the "Thesis Magister" Grant Year 2020 with the Project No:113.1.06/UN8.2/PG/2020.

References

[1] Indonesia Investments 2020 https://www.indonesia-investments.com/
[2] BP Statistical Review of World Energy June 2010 https://bakerinstitute.org/
[3] Rosyid F A and Adachi T 2016 *Forecasting on Indonesian Coal Production and Future Extraction Cost: A Tool for Formulating Policy on Coal Marketing* (Natural Resources:70 pp 677-696
[4] Speight J G *Handbook of Coal Analysis* (New Jersey: John Wiley & Sons, Inc.,) p 2005
[5] Shi Y, Li S and Hu H 2012 *Studies On Pyrolysis Charecteristic Of Lignite And Properties Of Its Pyrolysates* (Journal of Analytical and Applied Pyrolysis) pp 75-8
[6] Baofeng W, Yaru H and Jinjun Z 2015 *Sulfur Distribution during Hydrothermal Liquefaction of Lignite, Wheat Straw and Plastic Waste in Sub-Critical Water* (China Petroleum Processing and Petrochemical Technology) pp 24-30
[7] Chunmei Q, Min Z and Jianghong W, Puhai Y, Xu Y 2014 *Pyrolysis And Co-Pyrolysis Of Lignite And Plastic* (International Journal of Mining Science and Technology) pp 137-41
[8] Zhou L, Luo T and Huang *Q* 2019 *Co-Pyrolysis Characteristics And Kinetics Of Coal And Plastic Blends* (Energy Conversion) pp 705-10
[9] Çepeliogullar Ö and Pütün Ayse E 2013 *Thermal And Kinetic Behaviors Of Biomass And Plastic Waste* (Energy Conversion and Management) pp 263-70
[10] Liu, Q, Hu H, Zhou Q, Zhu S, Chen G 2003 *Effect Of Inorganic Matter On Reactivity And Kinetics Of Coal Pyrolysis* (Fuel 83) pp 713-8
[11] Cai J, Wang Y, Zhou L, Huang Q 2008 *Thermogravimetric Analysis And Kinetics Of Coal/Plastic Blends During Co-Pyrolysis In Nitrogen Atmosphere (*Fuel Processing Technology) pp 21-7

IOP Conf. Series: Materials Science and Engineering **980** (2020) 012061 doi:10.1088/1757-899X/980/1/012061

Increasing product quality of torrefied palm kernel shell batch model with internal surface area modification

K Karelius[1], M Dirgantara[2*], N Rumbang[3], N Kristian[2] and F Purwanto[1]
[1]Study Program of Chemistry, University of Palangka Raya, Palangka Raya, Indonesia
[2]Study Program of Physics, University of Palangka Raya, Palangka Raya, Indonesia
[3]Study Program of Agriculture Cultivation, University of Palangka Raya, Palangka Raya, Indonesia

Email: dirgantaramade@mipa.upr.ac.id

Abstract. As the largest palm oil-producing country in the world, Indonesia has abundant biomass potential from the production of crude palm oil (CPO). Palm oil production in 2018 in Indonesia is 34.94 million tons, where have by-products are empty bunches (23%), mesocarp fiber (12%), and palm kernel shell (5%). Palm kernel shells are a potential biomass that can upgrade the quality as a solid fuel through thermochemical processes. Torrefaction is one of the thermochemical processes where currently being used to increase the quality of biomass. The usual method used for the torrefaction process in the laboratory is the batch method. However, this method has a disadvantage when the capacity is improved make decreases the quality of the fuel produced. In this study, the quality of torrefied palm kernel shell from batch and batch torrefaction with internal surface area modification were compared. The torrefaction process was carried out at 275°C, with a residence time of 30 minutes. The results showed that the torrefaction of the modified batch method had a better heating value, and was proximate as a fuel.

1. Introduction

Biomass is one of the renewable energy that is potent and currently replaces fossil fuels in industries such as steam power plants [1]. Biomass is defined as biological material originating from life or living organisms with a carbon structure and a chemical mixture of organic matter containing hydrogen, nitrogen, oxygen, and a small number of atoms & other elements. However, in terms of utilization as energy, biomass is defined as all organic materials from plants, animals, products, or wastes from aquaculture industries such as agriculture, plantation, forestry, animal husbandry, and fisheries [2]. Biomass as an energy source is very important to maintaining the availability of energy can also help reduce the environmental impact of plantation or forestry waste [3,5].

As the most significant crude palm oil (CPO) production globally, Indonesia has a large amount of biomass as a by-product of this production. Indonesia is also committed to increasing palm oil products, which can be seen from the increased production of fresh fruit bunches (FFB) every year. In 2016, it amounted to 31.73 million tons to 45.86 million tons in 2019 [6]. By-products generated from each ton of FFB is 23 % empty bunches, 12% mesocarp fiber, and 5% palm kernel shells [7,8]. Compared to other solid waste from CPO production, the highest potential to be used as fuel by going through a thermochemical process to increase its quality is palm kernel shell.

Torrefaction is a thermochemical process to improve the quality of biomass as a fuel such as increasing heat value and carbon content [9,11]. There are various kinds of torrefaction methods, one of which is the most widely used batch method. The method's weakness is that the reaction time is directly proportional to the amount of biomass that is processed. This method causes the energy requirements of the torrefaction to be higher. Therefore, in this research, a modification of the biomass receptacle for increasing the surface area/heat contact of the biomass to streamline the torrefaction time.

2. Materials and methods

2.1. *Torrefaction process*

The Torrefaction process is shown in figure 1 [4] by modifying the torrefaction bait container into 2 (two) different models (model I and model II). Torrefaction model I had one container like the conventional batch method, while in model II, the container was modified into a rack to increase heat contact on the biomass surface. In figure 1, nitrogen gas was supplied at a rate of 2 Liters/minute for 10 minutes until the reactor in inert conditions. The torrefaction process was carried out at 275°C, with a residence time of 30 minutes [12].

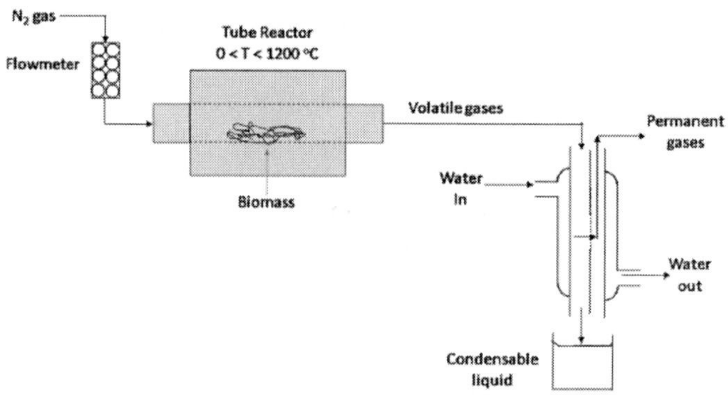

Figure 1. Torrefaction process [4].

Torrefaction process by the batch method used two container models intending to modify the surface area of biomass that was in contact with heat. The model I biomass container that was currently commonly used only consisted of one container. Model II was modified into a rack so that the heat contact to the biomass would be more efficient. The model, I and II containers, can be seen in figure 2.

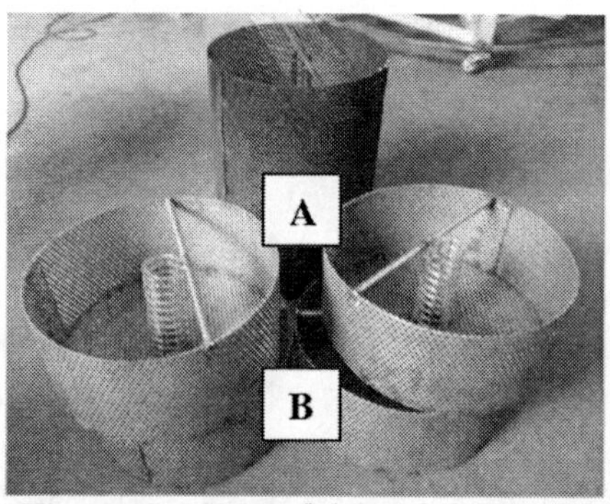

Figure 2. (A) Model I and (b) Model II

2.2. *Calorific value analysis*

Calorific value or higher heating value (HHV) was determined by following the ASTM D240 or EN 14918 procedure using a bomb calorimeter [4,13]. The samples measured were samples before treatment (control), after torrefaction model I dan II.

2.3. Proximate analysis
The proximate analysis to determine humidity, volatility, and ash content, in this study was tested following standard procedures of ASTM D3171, D3175, and D3174. The sample of this analysis controls was torrefaction model I and II [5, 13, 14].

3. Results and discussion
3.1. The visual products of torrefaction
The Torrefaction process in this study referred to Dirgantara et al. 2020, where the optimal conditions of the torrefaction process at 275 °C with a residence time of 30 minutes [12]. Visually, the results of the model I torrefaction looked less optimal where physically, it did not look much different from the control. Black color on the torrefied palm kernel shell was an indication that the carbonization process was successful during the torrefaction [4,15,16]. The torrefaction results with model II looked better than model I. The black color of the palm kernel shell was distributed evenly. The following results of torrefaction in models I and II were compared with controls (figure 3).

Figure 3. (A) Control, (b) Model and (C) Model II.

3.2. Calorific value
Calorific value is the main parameter that indicates the quality of biomass as a solid fuel[17,19]. More heat can be released by every gram of biomass when an increase in the calorific value [12]. The results of the analysis of the heating value are shown in figure 4. The heating value increased significantly after the torrefaction process, where the control of 4469.35 Cal/gram increased to 5072.48 Cal/gram (model I) and 5440.67 Cal/gram (model II). An increase in the carbon content to the palm shells during the torrefaction process caused an increase in HHV. This caused the elimination of volatile substances during the torrefaction process, such as water, acetic acid, and phenols in the palm shells [3]. During the process of torrefaction, thermal degradation of hemicellulose and lignin occurred in volatile substances. The formation of volatile substances, causing the ratio of O/C and H/C, would continue to decrease, thereby increasing the palm shell's carbon content [20]. This had led to an increase in the calorific value of the palm kernel shell resulting from torrefaction. Making the biomass container into a shelf (model II) in the torrefaction process, as shown in figure 2, made the heat more evenly distributed so that the thermochemical process on the palm kernel shell was better. With the even distribution of heat distributed causes more volatile compounds to be lost, so that the bound carbon value increases and had an impact on increasing the heating value.

3.3. Proximate analysis
Analysis Proximate used to determine the characteristics of the and quality of solid fuels concerning the use of solid fuels, namely to determine the relative amounts of water moist (moisture content), volatile substance (volatile matter), ash (ash content), and carbon moored (fixed carbon) [19,21]. The results of the proximate analysis are shown in figure 5. Moisture content in biomass affected the

calorific value. The higher the water content of solid fuels, the more significant the heating value needed in the combustion process, so that it required extra energy. Also, the high-water content would increase the risk of fire hazards when storing biomass for a long time. High humidity could potentially cause fermentation so that the temperature of the material increased, which could cause a fire if there was friction or was close to a heat source.

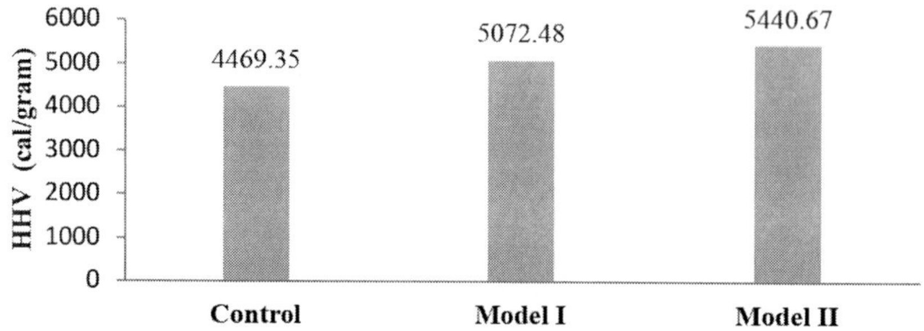

Figure 4. Calorific value of palm kernel shell control, model I and model II.

Moisture content in biomass affected the calorific value. The higher the water content of solid fuels, the more significant the heating value needed in the combustion process, so that it required extra energy [4,22]. Also, the high-water content would increase the risk of fire hazards when storing biomass for a long time. High humidity could potentially cause fermentation so that the temperature of the material increased, which could cause a fire if there was friction or was close to a heat source. The water content in model II was smaller than the model I; this condition indicated that model II was more efficient in reducing water content.

The high ash content in solid fuels resulted in lower total heat [21,23]. Low ash content was significant in the thermal conversion of biomass, increasing efficiency, mainly when the biomass contained potassium or halides such as chlorine. The ash content of the palm shell torrefaction product in Model I and II increased. The mass reduction that occurred during the process of torrefaction is not accompanied by degradation of the ash-forming inorganic component into the cause of this increase.

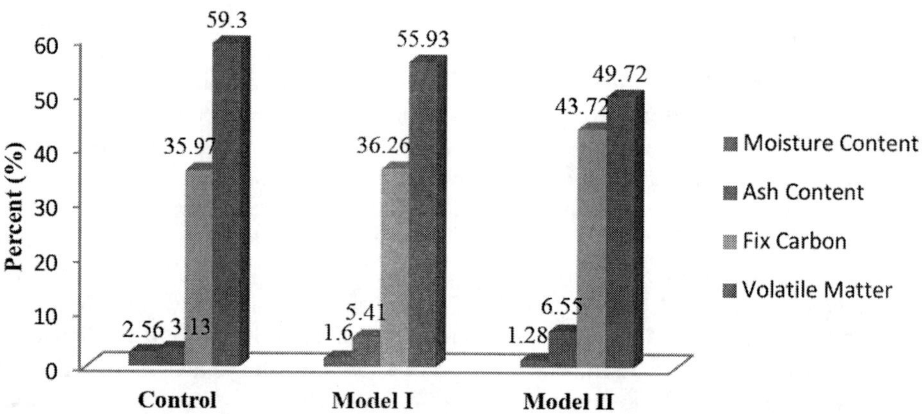

Figure 5. Proximate analysis torrefied palm kernel shell (control, model I and model II).

Volatile matter (VM) is the amount of substance lost when biomass is heated at a specific temperature and time. One of the factors influencing the rate of release of volatile compounds is the surface area of the biomass. The more surface area (torrefaction model II), the faster the rate of release of volatile compounds. The main chemical components that lost during the torrefaction process in biomass are hydrogen and oxygen [24]. Phenols and carbonyl are produced in the temperature range of 200 - 300 ° C from volatile compounds, especially oxygen. The reduction of hydrogen and oxygen content leads to an increase in the ratio of carbon to hydrogen and oxygen, which affects to increase in the calorific value products [25,26].

Fixed Carbon (FC) states the amount of carbon contained in material or biomass [4,12]. The Torrefaction process caused much volatile matter to be lost so that fixed carbon would continue to increase after torrefaction. Levels of fixed carbon were associated with volatile levels. The lower levels of volatiles in biomass caused higher levels of fixed carbon so that the heating value was also higher. Torrefaction products with model II had higher fixed carbon values than the model I torrefaction products. The large surface area of biomass in model II caused more volatile compounds that were lost affect edthe higher carbon content bound to the model II torrefaction product.

4. Conclusion

Modification of the biomass container in the torrefaction reactor (model II) batch method produced more optimal products than without modification (model I). This product quality can be seen from the higher heating value and lower volatile matter in model II. Thus, increasing the surface area/heat contact area of the biomass could streamline the batch method torrefaction process and produce better fuel quality.

Acknowledgments

This work was supported by Badan Pengelola Dana Perkebunan Kelapa Sawit (BPDPKS) who have funded through the Grant Palm Research Scheme.

References

[1] Perea-Moreno M-A, E. Samerón-Manzano and Perea-Moreno A J 2019 Biomass as Renewable Energy: Worldwide Research Trends *Sustainability* **11** 863

[2] Pranoto B, Pandin M, Fithri S R and Nasution S 2016 Peta Potensi Limbah Biomassa Pertanian dan Kehutanan Sebagai Basis Data Pengembangan Energi Terbarukan. *Ketenagalıstrıkan dan Energi Terbarukan* **12** 123–30

[3] Mafu L D, Neomagus H W J P, Everson R C, Carrier M, Strydom C A and Bunt J R 2016 Structural and chemical modifications of typical South African biomasses during torrefaction. *Bioresource Technology* **202** 192–97

[4] Mamvura T A, Pahla G and Muzenda E 2018 Torrefaction of waste biomass for application in energy production in South Africa. *South African Journal of Chemical Engineering* **25** 1–12

[5] Mohd Faizal H *et al.* 2018 Torrefaction of densified mesocarp fibre and palm kernel shell *Renewable Energy* **122** 419–28

[6] BPS Badan Pusat Statistik 2020 *https://www.bps.go.id/dynamictable/2015/09/04/839/produksi-tanaman-perkebunan-menurut-propinsi-dan-jenis-tanaman-indonesia-ribu-ton-2011-2019-.html* (accessed Jul. 14, 2020)

[7] Thaim T and Rasid R A 2016 Improvement Empty Fruit Bunch Properties through Torrefaction. *Australian Journal of Basic and Applied Sciences* **10** 114–21

[8] Saelor S, Kongjan P and Thong S O 2017 Biogas Production from Anaerobic Co-digestion of Palm Oil Mill Effluent and Empty Fruit Bunches *Energy Procedia* **138** 717–22

[9] Abdul Samad N A F, Jamin N A and Saleh S 2017 Torrefaction of Municipal Solid Waste in Malaysia *Energy Procedia* **138** 313–18

[10] Abdul Wahid F R A, Saleh S and Abdul Samad N A F 2017 Estimation of Higher Heating Value of Torrefied Palm Oil Wastes from Proximate Analysis. *Energy Procedia* **138** 307–12

[11] Alamsyah R, Siregar N C and Hasanah F 2017 Torrefaction study for energy upgrading on Indonesian biomass as low emission solid fuel. *IOP Conference Series: Earth and Environmental Science* **65** 012051

[12] Dirgantara M, Karelius, Cahyana B T, Suastika K G and Akbar A R 2020 Effect of Temperature and Residence Time Torrefaction Palm Kernel Shell On The Calorific Value and Energy Yield *Journal of Physics: Conference Series* **1428** 012010

[13] Nyoman Sukarta I and Sri Ayuni P 2016 Analisis Proksimat dan Nilai Kalor pada Pellet Biosolid yang Dikombinasikan Dengan Biomassa Limbah Bambu *JST (Jurnal Sains dan Teknologi)* **5** 8278

[14] Susanty W, Helwani Z and Zulfansyah 2018 Torrefaction of oil palm frond: The effect of process condition to calorific value and proximate analysis. *IOP Conference Series: Materials Science and Engineering* **345** 012016

[15] Bach Q-V and Skreiberg Ø 2016 Upgrading biomass fuels via wet torrefaction: A review and comparison with dry torrefactio. *Renewable and Sustainable Energy Reviews* **54** 665–77

[16] Asadullah M, Adi A M, Suhada N, Malek N H, Saringat M I and Azdarpour A 2014 Optimization of palm kernel shell torrefaction to produce energy densified bio-coal *Energy Conversion and Management* **88** 1086–93

[17] Erol M, Haykiri-Acma H and Küçükbayrak S 2010 Calorific value estimation of biomass from their proximate analyses data **35** 170–73

[18] Dirgantara M, Kristian N and Karelius 2019 Evaluasi Prediksi Nilai Higher Heating Value (HHV) Biomassa Berdasarkan Analisis Ultimate. *Jurnal Jejaring Matematika dan Sains* **1** 107–13

[19] Hosseinpour S, Aghbashlo M and Tabatabaei M 2018 Biomass higher heating value (HHV) modeling on the basis of proximate analysis using iterative network-based fuzzy partial least squares coupled with principle component analysis (PCA-INFPLS) *Fuel* **222** 1–10

[20] Karelius, Dirgantara M, Rumbang N, Suastika K G and Akbar A R M 2020 Torrefaction of palm kernel shell using COMB method and its physicochemical properties *Journal of Physics: Conference Series* **1422** 012005

[21] Sirisomboon P, Funke A and Posom J 2020 Improvement of proximate data and calorific value assessment of bamboo through near infrared wood chips acquisition *Renewable Energy* **147** 1921–31

[22] Pahla G, Mamvura T A, Ntuli F and Muzenda E 2017 Energy densification of animal waste lignocellulose biomass and raw biomass *South African Journal of Chemical Engineering* **24** 168–75

[23] Sukiran M A, Abnisa F, Wan W M A Daud, Abu Bakar N and Loh S K 2017 A review of torrefaction of oil palm solid wastes for biofuel production *Energy Conversion and Management* **149** 101–20

[24] Karelius, Dirgantara M, Rumbang N, Suastika K G, Prabawa I D G P and Ernawati L 2020 The Prediction of Optimal Torrefaction Condition Palm Kernel Shell based on Elemental Composition *International Conference on Industrial Technology* **2019** 46–51

[25] Nhuchhen D R and Abdul Salam P 2012 Estimation of higher heating value of biomass from proximate analysis: A new approach *Fuel* **99** 55–63

[26] Chen Y-C, Chen W-H, Lin B-J, Chang J-S and Ong H C 2017 Fuel Property Variation of Biomass Undergoing Torrefaction *Energy Procedia* **105** 108–12

Optimization of the hydrogenation and rafination process for cocoa butter substitute production using palm kernel oil in a small and medium scale industry

L Junaidi[1,2*], N Lestari[1*] and Y R Meutia[1]
[1]Center for Agro-Based Industry, Ministry of Industry, Jl. Ir. Juanda No 11 Bogor, West Java, Indonesia

Email: lukmanjunaidi@gmail.com, namilestari65@gmail.com

Abstract. The study on the optimization of hydrogenation and rafination process for cocoa butter substitute (CBS) production using palm kernel oil in small and medium scale industry was conducted. The CBS production process included: hydrogenation, cooling, bleaching, filtering, deodorizing, cooling, and packaging. The optimum condition of the hydrogenation process was obtained using hydrogen with 0.2% nickel catalyst, at 195-200°C, under 2-2.5 bar, using 6.12 hz agitation at 1700 rpm, for 12 hours process. The optimum condition of the rafination process was achieved by bleaching using 2 % bentonite at 80-95°C for 45 minutes and cooling it at 60°C, and deodorizing it at 240°C for 4 hours. The final product of CBS had similar quality characteristics as commercial cocoa butter and was well accepted by consumers. It contained no trans-fatty-acid, 0 nickel, low FFA, and a melting point of 37.1°C.

1. Introduction

Cocoa butter is usually used for the production of chocolates, coatings, and confectionery fillings. Cocoa butter has a unique characteristic that is solid at room temperature and melts quickly above 30-32°C [1,2]. This behavior produces a stable confectionery product that releases flavors in the mouth at body temperature without any undesirable waxy texture. Cocoa butter consists of three main triacylglycerols [1,3]. These triacylglycerols dominate the melting characteristic and solid fat content as a function of temperature and polymorphic transformations of chocolate, providing chocolate's textural and sensory properties [4,5].

Because cocoa butter is very expensive compared to other commercial vegetable fats and oils, many manufacturers are looking for alternatives to cocoa butter [2,6]. Cocoa butter alternatives consist of three categories: cocoa butter substitutes (CBS) produced from lauric fats, cocoa butter replacers (CBR) produced from the hydrogenation of vegetable oils, and cocoa butter equivalents (CBE) produced from polymorphic and non-lauric fats [2,7]. Among them, CBS is generally used in confectionery fillings such as compound chocolates and other confectionery products [8,9].

Crude palm kernel oil contains non-triacylglycerol components that must be partially or totally removed to become acceptable for human consumption. Crude oil is therefore submitted to several treatments whose objective is to remove the objectionable minor components with the least possible damage to the oil fraction and minimal losses of desirable constituents [10].

Today, palm oil is the largest source of vegetable oil needed by many industries in the world. Indonesia, as the largest producer of palm oil in the world, has an advantage in terms of resources compared to other palm oil-producing countries. Palm oil and its derivatives are widely used in the manufacturing of edible oils, margarine and softeners, ice cream, and dairy products [11].

The chocolate processing industry is one of the food industries which is quite promising. However, in its development, several obstacles must be faced by the chocolate processing industry, one of which is the supply of cocoa butter. As one of the main raw materials in the chocolate processing industry, cocoa butter is used as a raw material for chocolate products such as chocolate bars, chocolate spreads, chocolate coatings, and others. Cocoa fat used in chocolate processing affects the composition of the fat fraction in the chocolate formulation. The fat fraction in the chocolate formulation provides an important role in determining the texture, appearance, and handling of the process and product storage.

The fat fraction in chocolate products is mostly derived from cocoa fat and milk fat. Cocoa butter is the best fat for chocolate products. However, there are several limitations regarding the use of cocoa butter, including unstable supply, relatively high price between natural fats and oils, inadequate for use in hot climate areas, and varying qualities. Besides, the tempering process is needed for chocolate products that fully use cocoa butter in its formulation, because it will tend to experience blooming (the outer layer is white like moldy). Therefore, various attempts have been made to develop special fats as an alternative to the use of cocoa butter, one of which is CBS.

CBS has a high content of C 12 : 0 fatty acids (lauric acid). CBS laurate is generally produced from high-fat laurate which is generally produced from the tropics, such as coconut oil and palm kernel oil (PKO) produced from palm oil. The advantages of CBS include: having good oxidative stability (thus providing a longer shelf life), good eating quality, good quality release flavor, not giving a waxy sensation (no waxy after taste), and a very similar texture to chocolates. CBS can solidify quickly, providing good gloss quality and gloss retention, and most importantly CBS is available at a much cheaper price than the cocoa butter price. With all the advantages of CBS, the processing of palm kernel oil into CBS will increase along with the increasing market demand and the advance of oil processing engineering technology.

Hydrogenation is a process to change fat liquid oil into semi-plastic solid that can be used as margarine, special cooking oil, and other special products such as cocoa butter substitute. Hydrogenation is a saturation process to increase oxidative stability and melting points. In the hydrogenation process, hydrogen is reacted with the unsaturated double bonds contained in fatty acids. Oils have been hydrogenated since the 1930s to prolong their shelf life stability [12-14]. Depending on the level of the hydrogenation, oil can be modified into various oils with a variety of hardness. By controlling the PKO hydrogenation process (melting point around 27-28°C), hydrogenated products can be produced with various melting points, ranging from 32-41°C [14]. The chemical, physical and sensory properties of the final product strongly depend on the number of residual double bonds and on the contents of cis-trans isomers present in the mixture [14], which depend on various operating factors, including temperature, hydrogen pressure, catalyst, and circulation rate [16]. The effect of temperature on the hydrogenation has clearly been observed. Hydrogenation, like other chemical reactions, is accelerated by increasing the temperature [17]. Hydrogenated oil is more stable and does not become rancid as quickly as the unhydrogenated oil [18]. It has a higher melting point, and it is often used in frying and pastries for this reason. The basic hydrogenation process of edible oils means that the liquid oils change into semi-solid substances, which have desired melting characteristics and partially hydrogenated oils with increased stability [19].

The hydrogenation process is usually carried out in a three-phase semi-batch reactor where hydrogen gas is bubbled with pressure in hot vegetable fat in the presence of a catalyst [20]. The hydrogenation of edible oils is an important process in the fat industry because of its wide applications to produce margarine, shortenings, and frying oils [18,21].

Hydrogenated CBS generally still has free fatty acids, unpleasant odors, and high nickel residues. All these three constraints can be minimized by the process of rafination (refinement), generally used in physics including degumming, bleaching, and deodorization [22].

The research aims to develop the optimum hydrogenation and rafination process for the production of CBS from Refined Bleached Deodorized Palm Kernel Oil (RBDPKO) as a raw material for compound chocolate.

2. Materials and methods

2.1. Materials

Materials used in this study were RBDPKO (produced by Indonesian Palm Oil Research Institute), nickel, bentonite, and chemicals for analysis obtained from a chemical store in Bogor.

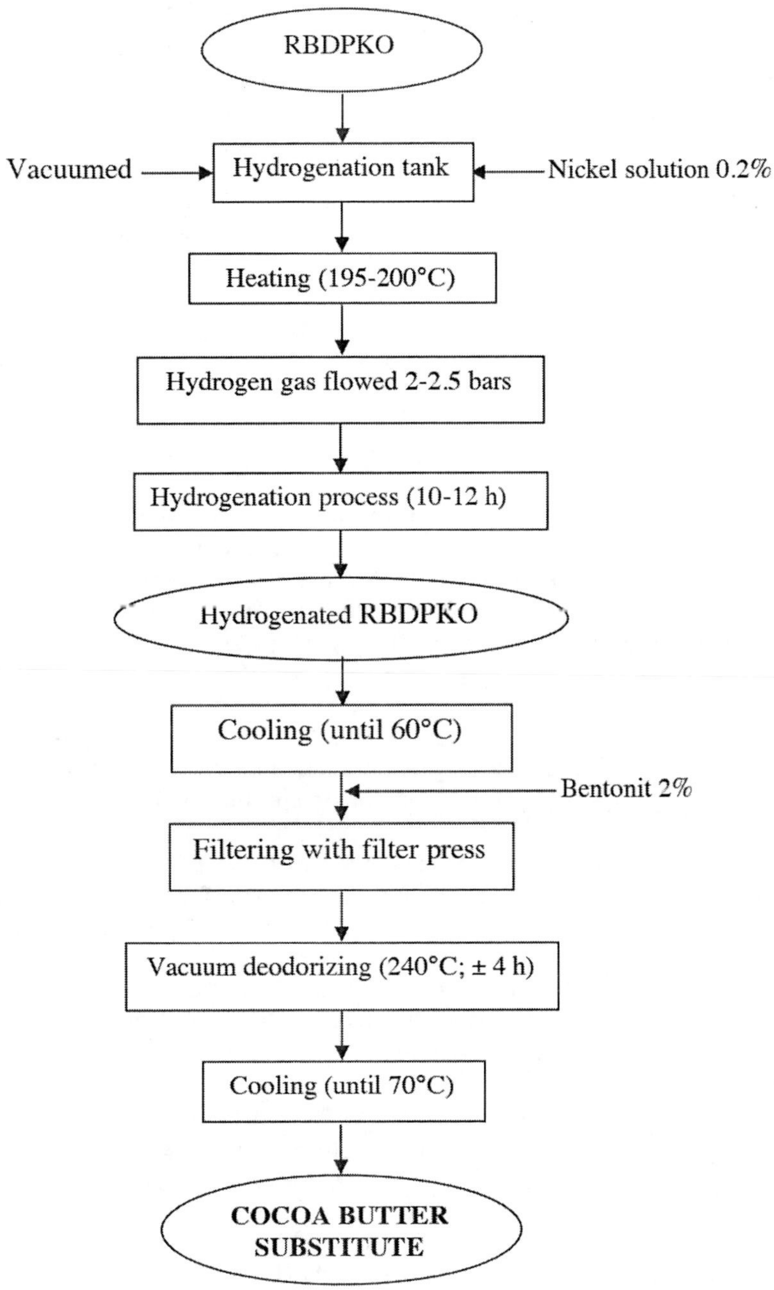

Figure 1. The process of CBS production.

IOP Conf. Series: Materials Science and Engineering **980** (2020) 012062 doi:10.1088/1757-899X/980/1/012062

2.2. Equipment
Pieces of equipment used in this study consisted of the pressurized hydrogenation reactor tank and the rafination equipment (bleaching tank, filter press, and deodorizer).

2.3. Methods
The research was carried out in several steps as follows:

2.3.1. Production of cocoa butter substitute using palm kernel oil through the hydrogenation and rafination process. Process production of CBS from RBDPKO raw material was carried out based on the flow process diagram as shown in figure 1. The process mainly consisted of 3 steps: hydrogenation, rafination, and deodorizing.

The variations of treatment in this study were: nickel concentration (0.1% and 0.2%), hydrogenation temperature (170°C and 200°C), and hydrogenation duration (10, 12, and 14 hours).

The treatments for rafination were: the use of 2% bentonite (bleaching earth) and bleaching process temperature of 80-95°C for 45 minutes. The bleaching process was followed by the filtering process and the deodorization process was carried out at 240°C for 4 hours.

2.3.2. Analysis of RBDPKO and CBS. The analysis conducted included: moisture content [23], impurities [24], melting point [25], iodine number [23], free fatty acid [23], solid fat content [26], fatty acid composition [27], trans fatty acid [27] and nickel content [28].

2.3.3. Techno-economic calculation. The techno-economic calculation was conducted for evaluating the feasibility of CBS. A simple techno-economic calculation including IRR, pay-back period (PBP) and break event point (BEP).

3. Results and discussion

3.1. Raw materials analysis
An evaluation of RBDPKO quality characteristics was required before conducted CBS processing. The analysis result of RBDPKO quality characteristics is presented in table 1.

The test result was compared to the Indonesian National Standard (SNI) for Refined Bleached Deodorized Palm Kernel Oil: SNI 01-0023-1998 [29]. Based on the data presented in table 1, the raw material used met the SNI requirements.

3.2. Hydrogenation process
The main step for producing CBS from RBDPKO is the hydrogenation process. The hydrogenation process was carried out in a pressurized hydrogenation tank 100 kg/batch capacity, with a maximum stirring speed of 1700 rpm, 2.5 bar pressure, the temperature at 190-200°C, with specific processing time.

The hydrogenation equipment, designed for small and medium scale industry, consisted of a hydrogenation tank with dimensions of 40 cm diameter and 80 cm length, made of 9 mm thick SS 304 pipe material, 4000-watt heating element, 1 HP gear motor, and 1 HP inverter. The equipment accessories consisted of a manometer, pressure gauge, safety valve, thermocouple, digital thermometer, hydrogen gas cylinder, gas regulator, glass and indicator lamp, pulley, and draining valve. The hydrogenation tank was coated with SS plate, covered with glass wool, and equipped with a control panel. The design and manufacture of the hydrogenation reactor tank are shown in figure 2.

The CBS manufacturing processes commenced with the hydrogenation process. This study was conducted to find out the optimum conditions of the hydrogenation process related to factors such as nickel concentration, pressure, temperature, and time of the hydrogenation process and the optimum condition of the refining process. The raw material used was RBDPKO. The hydrogenation process was carried out using hydrogen gas and nickel catalyst (0.1% and 0.2% concentrations), at the temperature of 170-175°C and 195-200°C and hydrogenation duration (10, 12, and 14 hours) until a CBS product that met company (commercial) standards was obtained.

IOP Conf. Series: Materials Science and Engineering **980** (2020) 012062 doi:10.1088/1757-899X/980/1/012062

Table 1. Quality characteristics of RBDPKO.

Parameter	Result	SNI Requirements
Moisture (%)	0.07	≤ 0.08
Impurities (%)	0	≤ 0.08
Free Fatty Acid (Oleic) (%)	1.33	≤ 0.1
Iodine Number, Wijs gram Iod/100 grams	14.8	14 - 19
Trans Fat (%)	0	--
Melting Point (^{0}C)	26.5	≤ 30
Nickel (mg/kg)	0	--
Fatty Acid Composition (%):		
C_8 : 0 caprilic	3.42	--
C_{10} : 0 capric	3.35	--
C_{12} : 0 lauric	48.54	42 - 52
C_{14} : 0 miristic	15.95	12 - 18
C_{16} : 0 palmitic	8.09	6 - 12
C_{18} : 0 stearic	2.44	1 - 3.5
C_{18} : 1 oleic	15.09	14 - 23
C_{18} : 2 linoleic	2.66	1 - 5
C_{18} : 3 linolenic	0	--

Figure 2. Hydrogenation equipment for the small and medium scale industry.

Based on the data collected, it could be concluded that the hydrogenation process reached its optimum conditions at the temperature of 195-200°C, the pressure of 2.5 bar, 12 hours of processing time, with 6.12 hz stirring at 1700 rpm.

3.3. The rafination process

The hydrogenated RBDPKO produced still had an undesirable odor, a rather dark color, and potentially contained high free fatty acid and nickel residues. To overcome this problem, it was necessary to conduct the rafination process, following the hydrogenation process. The rafination process could improve color, remove odors, reduce free fatty acid, and nickel content.

Generally, the rafination process was carried out in a physical manner which included degumming, filtering, bleaching, and deodorization. Filtering was carried out to remove nickel, followed by the bleaching process in the bleaching tank. The bleaching process was conducted by adding 2% bentonite at a temperature of 80-95°C for 45 minutes. The bleaching process was followed by the filtering process using a filter press to remove bentonite and other impurities. Finally, the deodorization process was carried out using a deodorizer at 240°C for 4 hours process.

The design and manufacture of the rafination equipment in this research consisted of a bleaching tank, filter press, and deodorizer. The rafination equipment can be seen in figures 3a and 3b.

The optimum rafination process conducted consisted of filtering process by the filter press, bleaching using 2% bentonite at the temperature of 85-90°C for 45 minutes, cooling at 60°C, deodorized in a deodorizing tank at the temperature of 240°C for 4 hours and cooling at 70°C. The study showed that the rafination process conducted was able to reduce nickel content to zero, give clearer color, and remove the distinctive smell of the hydrogenation.

3.4. Analysis of product quality characteristics

The result of the analysis of RBDPKO, commercial CBS, and CBS resulted from this research can be seen in tables 2 and 3, and the solid fat content can be seen in table 4.

Figure 3a. The rafination equipment for small and medium scale industry (bleaching and deodorizing).

Figure 3b. The rafination equipment for small and medium scale industry (filter press).

Table 2. Quality characteristics of RBDPKO and CBS PKO commercial.

Parameter	RBD PKO	CBS PKO Comercial
Moisture (%)	0.07	0.05
Impurities (%)	0	0.01
Free Fatty Acid (Oleic) (%)	1.33	0.06
Iodine Number, Wijs gram Iod/100 grams	14.8	1.12
Trans Fat (%)	0	< 0.005
Melting Point (^{0}C)	26.5	36.9
Nickel (mg/kg)	0	< 0.11
Fatty Acid Composition (%):		
C_8 : 0 caprilic	3.42	0.98
C_{10} : 0 capric	3.35	2.94
C_{12} : 0 lauric	48.54	54.5
C_{14} : 0 miristic	15.95	22.6
C_{16} : 0 palmitic	8.09	8.30
C_{18} : 0 stearic	2.44	9.95
C_{18} : 1 oleic	15.09	0.13
C_{18} : 2 linoleic	2.66	0.05
C_{18} : 3 linolenic	0	0.08

IOP Conf. Series: Materials Science and Engineering **980** (2020) 012062 doi:10.1088/1757-899X/980/1/012062

The key performance indicators of the hydrogenation reaction in producing CBS could be evaluated on several factors, including a decrease in the number of iodines, an increase in melting point, changes in the content of stearic acid, oleic acid, and linoleic acid, and physical appearances such as texture, color, and aroma [22].

The twelve hours hydrogenation process could produce a CBS palm kernel oil similar to the CBS palm kernel oil commercial. The study showed that the longer hydrogenation process would produce a higher melting point. In the 10-hour hydrogenation process, it could be observed that the melting point approached the expected melting point which was around 27.9°C. The melting point reached the expected melting point at 37.1°C using a 12-hour hydrogenation process. This value is relatively close to the melting point of commercial products ranging from 34.0-34.6% as shown in table 2.

Table 3. Quality characteristics of CBS PKO produced from this research.

Parameter	CBS PKO (10 hours process)	CBS PKO (12 hours process)	CBS PKO (14 hours process)
Moisture (%)	0.21	0.13	0.05
Impurities (%)	0.01	0.02	0.01
Free Fatty Acid (Oleic) (%)	0.56	0.57	0.06
Iodine Number, Wijs gram Iod/100 grams	37.14	0.54	1.12
Trans Fat (%)	8.5	0	2.5
Melting Point (^{0}C)	27.9	37.1	34.0
Nickel (mg/kg)	0.47	0	0.07
Fatty Acid Composition (%)			
C_8 : 0 caprylic	3.8	3.9	4.0
C_{10} : 0 capric	3.2	3.2	3.7
C_{12} : 0 lauric	48.6	50.0	55.9
C_{14} : 0 miristic	15.9	16.4	18.5
C_{16} : 0 palmitic	8.8	12.3	11.2
C_{18} : 0 stearic	17.5	13.0	13.3
C_{18} : 1 oleic	3.4	0.9	0.5
C_{18} : 2 linoleic	-	0.1	0.2
C_{18} : 3 linolenic	-	0.1	-

Based on the test results shown in table 3, it could be concluded that for the 12 hours hydrogenation process time, the CBS Iod number decreased from its original value, 14.8 grams Iod/100 grams to 0.54 grams Iod/100 grams. This result showed that the hydrogenation process was running well. In terms of texture and color, the resulting CBS was also similar to commercial CBS.

The performance of the hydrogenation process could also be evaluated from the change in stearic acid (C 18 : 0) content. The stearic acid content in RBDPKO originally 2.44%, increased to 13.0-17.5% after the hydrogenation process. The oleic acid (C 18 : 1) content which was originally around 15.09%, reduced to 0.9-3.4%. This was caused by the process of double bond saturation. The completion of the hydrogenation process can also be seen from the change in the texture of the product, which was originally liquid RBDPKO to become semi-solid CBS RBDPKO.

Table 4. The solid fat content of CBS RBDPKO.

Temperature (^{0}C)	Sort of Sample (hours process)	SFC (%)
20	10	61.24
20	12	77.75
20	14	76.62
30	10	3.21
30	12	27.51
30	14	21.18
35	10	nd
35	12	9.26
35	14	5.42
40	10	nd
40	12	0.87
40	14	0.96

The value of solid fat content (SFC) of CBS is very important for the process of producing processed chocolate products. To produce good quality chocolate products, the SFC value of CBS at temperatures of 35°C and 40°C must be of low value. The results of this research show that the value of SFC produced by a 12-hours hydrogenation process time ranged from 5.42-9.46% (temperature 35°C) and ranged from 0.87-0.96% (temperature 40°C). Generally the confectionary fats (chocolate) should exhibit approximately 63% SFC at 20°C, 40% at 25°C, and 0% at 37°C [30].

3.5. *Techno-economic calculation*

The application of a technology or diversification of a product must be followed by a techno-economic calculation or feasibility study. The feasibility parameters used in the techno-economic evaluation of the production of CBS were an internal rate of return (IRR), pay-back period (PBP), and break event point (BEP). The techno-economic calculation was based on a discount rate of 16%. Based on the calculations, the CBS production from RBDPKO in terms of techno-economic analysis was feasible to be developed on a small and medium scale industry. The techno-economic calculation was based on the production capacity of 70 liters of RBDPKO/day. The result of the techno-economic calculation of CBS production from RBDPKO was: IRR 22.98%, PBP for 4.35 years, and BEP at Rp 116,000,000.

4. Conclusion

The optimum condition of the hydrogenation process was obtained using hydrogen with 0,2% nickel catalyst, at 195-200 ^{0}C, under 2-2.5 bar, using 6.12 hz agitation at 1700 rpm, for 12 hours process. The optimum condition of the rafination process was obtained by bleaching using 2% bentonite at 80-95°C for 45 minutes and cooling at 60°C, and deodorizing at 240°C for 4 hours. The final product of CBS had similar quality characteristics as commercial cocoa butter and as well accepted by consumers. It contained no trans fatty acid, 0 nickel, low FFA, and a melting point of 37.1°C. Based on the simple techno-economic calculation, producing CBS using RBDPKO is feasible with the value of IRR 22.98%, PBP 4.35 years, and BEP Rp 116,000,000.

Acknowledgments
Thanks to the Center for Agro-based Industry, Ministry of Industry of the Republic of Indonesia for providing research funding and laboratory facilities for conducting this research, and PT Bina Karya Prima, PPKS, and PT Tama Cokelat Indonesia for facilitating and supporting this research.

References
[1] Gunstone F 2011 *Vegetable Oils in Food Technology: Composition, Properties and Uses* (New York: Wiley-CRC Press) pp 291–343
[2] Biswas N, Cheow Y L, Tan C P, Kanagaratnam S and Siow L F 2017 Cocoa butter substitute (CBS) produced from palm mid-fraction/palm kernel oil/palm stearin for confectionery fillings *J. AOCS* **94(2)** 235-45
[3] Bootello M A, Hartel R W, Garces R, Martínez-Force E and Salas J.J 2012 Evaluation of high oleic-high stearic sunflower hard stearins for cocoa butter equivalent formulation *Food Chem.* **134(3)** 1409-17
[4] Afoakwa E O 2010 *Chocolate Production and Consumption Patterns* In: E O Afoakwa (ed) *Chocolate science and technology* (West Sussex : Wiley & Blackwell) pp 1–10
[5] Biswas N, Cheow Y L, Tan C P and Siow L F 2016 Blending of palm mid-fraction, refined bleached deodorized palm kernel oil or palm stearin for cocoa butter alternative *J. AOCS* **93(10)** 1415-27
[6] Wei Y, Siewers V and Nielsen J 2017 Cocoa butter-like lipid production ability of non-oleaginous and oleaginous yeasts under nitrogen-limited culture conditions *Appl. Microbiol. Biotechnol* **101(9)** 3577-85
[7] Cisse V and Yemiscioglu F 2019 Cacao Butter and Alternatives Production *Cukurova Tarım ve Gıda Bilimleri Dergisi* **34(1)** 51-60
[8] Hassim N A M and Dian L N H M 2017 Usage of palm oil, palm kernel oil and their fractions as confectionery fats *J. Oil Palm Res.* **29(3)** 301-10
[9] Hussain N, Agus B A P, Rahim S N F A and Halim HA 2018 Comparison of quality characteristics between compound and pure milk chocolate *MOJ Food Proc. Technol.* **6(3)** 292-6
[10] Gibon V 2012 *Palm Oil and Palm Kernel Oil Refining and Fractionation Technology*. In *Palm Oil* (AOCS Press - Elsevier) pp 329-75
[11] Rios R V, Pessanha M D F, Almeida PFD, Viana C L and Lannes S C D S 2014 Application of fats in some food products *Food Sci. and Technol.* **34(1)** 3-15
[12] Alsobaai A M, Al Shaibani A M, Moustafa T and Derhem A 2012 Effect of hydrogenation temperature on the palm mid-fraction fatty acids composition and conversion *J. of King Saud University-Eng. Sci.* **24(1)** 45-51
[13] Xiao H 2007 *Low Trans Fatty Acid Containing Hydrogenated Edible Oils* (M.D Thesis, Miami University, Oxford Ohio USA)
[14] Hariyadi P 2009 High grade specialty fats dari sawit *Jurnal Info Sawit* **(special edition)** 41-4
[15] Ramirez E, Recasens F, Fernandez M and Larrayoz M A 2004 Sunflower oil hydrogenation on Pd/C in SC propane in a continuous recycle reactor *Am. Ins. Chem. Eng.* **50** 1-7
[16] Schmidt A and Schomaker R 2007 Partial hydrogenation of sunflower oil in a membrane reactor. *J. Molecul. Cat. A: Chem.* **271** 192–9
[17] Fernandez M B, Tonetto G M, Crapiste G H and Damiani D E 2007 Revisiting the hydrogenation of sunflower oil over a Ni-catalyst *J. Food Eng.* **82** 199-208
[18] Murzin D Y and Simakova I L 2008 Kinetic aspect of stereoselectivity in hydrogenation of fatty acids *J. Molecular Cat.: Chem.* **286** 156–61
[19] Kadhum A A H and Shamma M N 2017 Edible lipids modification processes: A review *Crit. Rev. Food Sci. Nutr.* **57(1)** 48-58
[20] Stankovic M, Krstic J, Gabrovska M, Radonjic V, Nikolova D, Loncarevic, D, and Jovanovic D 2017 Supported nickel-based catalysts for partial hydrogenation of edible oils. In *New*

IOP Conf. Series: Materials Science and Engineering **980** (2020) 012062 doi:10.1088/1757-899X/980/1/012062

Advances in Hydrogenation Processes: Fundamentals and Applications (Croatia-In Tech) 131

[21] Mba O I, Dumont M J and Ngadi M 2015 Palm oil: processing, characterization and utilization in the food industry–a review *Food Biosci.* **10** 26-41

[22] Siahaan D and Hasibuan H A 2012 Optimasi hidrogenasi minyak inti sawit skala 100 kg/batch dan rafinasi cocoa butter substitute yang dihasilkan *Proc. Insinas* Jakarta 37-42.

[23] Badan Standardisasi Nasional 1998 *SNI 01-3555-1998: Cara Uji Minyak dan Lemak.* (Jakarta- Badan Standardisasi Nasional)

[24] Badan Standardisasi Nasional 1992 *SNI 01-2891-1992: Cara Uji Makanan dan Minuman.* (Jakarta - Badan Standardisasi Nasional)

[25] Badan Standardisasi Nasional 2009 *SNI 3748-2009: Lemak Kakao.* (Jakarta - Badan Standardisasi Nasional)

[26] American Oil Chemists' Society 1999 *AOCS Official Method Cd 16b-93: Solid Fat Content (SFC) by Low-Resolution Nuclear Magnetic Resonance - The direct method 1999* (Urbana IL USA - AOCS)

[27] American Oil Chemists Society 2011 *Official Methods and Recommended Practices of the American Oil Chemist' Society* 4th ed. (Champaign ILUSA - AOCS)

[28] Badan Standardisasi Nasional 1998 *SNI 01-2896-1998: Cara Uji Cemaran Logam dalam Makanan* (Jakarta- Badan Standardisasi Nasional)

[29] Badan Standardisasi Nasional 1998 *SNI 01-0023-1998: Refined Bleached Deodorized Palm Kernel Oil (RBD Palm Kernel Oil)* (Jakarta- Badan Standardisasi Nasional)

[30] Biswas N, Cheow YL, Tan CP, Siow LF 2018 Physicochemical properties of enzymatically produced palm-oil-based cocoa butter substitute (CBS) with cocoa butter mixture *Eur. J. Lipid Sci. Technol.* **120(3)** 1-9

IOP Conf. Series: Materials Science and Engineering **980** (2020) 012063 doi:10.1088/1757-899X/980/1/012063

Preparation of the $NiCl_2$ intercalated on the bentonite as a catalyst in the cracking process of biodiesel

I W Sutapa[*], S Amalia and Rosmawaty

Department of Chemistry, Faculty Mathematics and Natural Science, University of Pattimura Jl. Ir. Putuhena Poka 97233, Ambon-Maluku- Indonesia

Email: wayansutapa@fmipa.unpatti.ac.id

Abstract. Biogasoline is a benign energy product that can be produced through the process of breaking down biodiesel. The purpose of this study is to produce biogasoline from cooking oil waste through the cracking process using $NiCl_2$ intercalated as a catalyst. Initially, waste from cooking oil was converted into biodiesel through a process of esterification and transesterification. After biodiesel with carbon chain lengths of 16-20 was acquired, the cracking process to produce biogasoline that had carbon chain lengths of 4-12 with the aid of a catalyst was conducted. The bentonite intercalated $NiCl_2$ catalyst was calcined at a temperature of 600°C for 2 hours, then characterized by XRD and SEM. The results of the diffractogram with XRD showed a peak around 2θ: 44.67° which was the peak of the Ni crystalline. The cracking process of biodiesel was carried out at a temperature of 450°C and 500°C then characterized using GC-MS. The chromatograms generated at a temperature of 450°C showed the presence of three compounds belonging to the faction of biogasoline they were methyl 6-hydroxycaproate ($C_7H_{14}O_3$), 4-hydroxymethyl-1,3-dioxolane ($C_4H_8O_3$), and methyl nonylate ($C_{10}H_{20}O_2$).

1. Introduction

A report from the Congressional Research Services (CRS) in 2003 to the Energy Commission in the United States Congress, stated that the level of fossil fuel use continues as it is today without increasing production efficiency, the discovery of new reserves, and the transition to alternative renewable energy sources. If this persists, then it is estimated that reserves of fossil fuel energy sources, especially petroleum, will only be sufficient for the next 30-50 years [1–5]. As a result of the increasing use of petroleum for fuel consumption and the depletion of petroleum reserves, it becomes a trigger to look for alternative and renewable energy. Biodiesel is one of the solutions to these problems [6,7]. Alternative energy sources such as hydrogen and biodiesel have caught the attention of researchers around the world. Biodiesel is an alternative fuel produced from renewable sources [8–11].

To date, the biodiesel produced from edible oil has a higher price compared to non-edible oil because it competes with consumption by the people. Therefore, now many biodiesel-based non-edible oils are being developed. Besides, biodiesel raw materials produced from non-edible oil are usually cheaper than conventional diesel fuel from petroleum. One source of non-edible oil is wasted cooking oil. In addition to increasing the added value of used cooking oil, the use of used cooking oil as a basis for making biodiesel is an excellent alternative way to handle waste [12,13].

Biodiesel production is generally carried out through a transesterifications reaction using a homogeneous catalyst (acid or base) [14–16]. Homogeneous catalysts are catalysts that have the same phase as reactants and products and have the advantage; they do not require high temperatures and pressures in the reaction. Homogeneous catalysts that are widely used in transesterification reactions are potassium hydroxide (KOH) and sodium hydroxide (NaOH) catalysts [3,17,18]. Biodiesel produced in the transesterification reaction of vegetable oils and animal fats with homogeneous catalysts generally has a hydrocarbon chain consisting of 16-20 carbon atoms [19,20].

The long biodiesel carbon chain can be converted into gasoline (gasoline) with a shorter chain length, namely C_4-C_{12}, through the cracking process. In petroleum and chemical geology, cracking is the process by which complex organic molecules such as hydrocarbons are broken down into simple units by cutting carbon-carbon bonds in the precursors. Simply put, cracking is the process of breaking down long-chain hydrocarbons into shorter ones. The result of cracking from biodiesel called biogasoline has several advantages over gasoline derived from fossils. It is environmentally friendly and can be reproduced in a short time. One method of cracking is breaking using a catalyst [21,22].

Several researchers have carried out the cracking reaction of cooking oil waste using zeolite/nickel catalysts. The natural zeolite was modified through the process: demineralization, dealumination, and activation, then followed by the impregnation of natural zeolite with $NiCl_2$. The determination of the optimal ratio is based on wasted cooking oil cracking products with a catalyst/sample ratio. GC-MS analysis showed C_9-C_{15} fraction hydrocarbon compounds dominated the cracking product composition of each catalyst/sample. The optimal conversion was confirmed by catalyst/sample cracking product 10:40 w/w with the composition of pentadecane compound ($C_{15}H_{32}$) of 44.60% and nonane compound (C_9H_{20}) of 4.47% [23–27].

Most of the content of bentonite is montmorillonite, with a structure that has a similarity to zeolite with great potential to be used as a catalyst for cracking. The bentonite is stable at high temperatures if the chemical elements contained in bentonite must be minimized by inserting ions, molecules, or compounds into their structures called intercalation. Ions, molecules, or compounds to be intercalated should have a commercial value that is cheap, easy to obtain, and of good quality. In our research goal, the cracking catalyst produced can be used on a large scale. Finally, the biogasoline that is produced can be consumed by all levels of society. Based on the background, this study investigated the catalyst $NiCl_2$ intercalated in bentonite and its applications as a catalyst in the biodiesel cracking process from cooking oil waste.

2. Materials and methods

2.1 Materials

The materials used in this study were cooking oil waste, methanol (Merck), ethanol (Merck), H_2SO_4 (Merck), KOH (Merck), anhydrous Na_2SO_4, commercial bentonite, $NiCl_2$, CTAB (Cetyl Trimethyl Ammonium Bromide) (Merck), phenolphthalein indicators, Whatman filter-40, and distilled water.

2.2 Bentonite preparation

A total of 250 g of bentonite was sieved using a 200 mesh sieve. Subsequently, bentonite was dispersed into distilled water and stirred using a magnetic stirrer for 3 hours. The suspension obtained was filtered using filter paper. The filtering results were dried in an oven at 110°C for 2 hours. The dried sample was crushed and sieved with a size of 200 mesh and then analyzed with XRD and SEM.

2.3 Waste cooking oil preparation

A total of 500 mL of used cooking oil was heated at 120 °C to evaporate water in used cooking oil. After heating, the oil was filtered with Whatman 40 filter paper to separate larger sized solid impurities.

2.4 Intercalation process

Fifty five gram of bentonite was dispersed into 1250 mL of distilled water and stirred for 3 hours to make a bentonite suspension. After 3 hours of bentonite suspension, 10 g of quaternary salt of Cetyl

Trimethyl Ammonium Bromide (CTAB) was added and stirred for 24 hours. Next, 11 gram of $NiCl_2$ catalyst was added to the suspension, and the stirring process was continued for another 24 hours. Then the suspension was filtered, and the sediment was dried in an oven at 110°C. After drying, the modified bentonite was crushed and sieved to 200 mesh size and then calcined at 600°C for 2 hours to obtain intercalated bentonite $NiCl_2$. XRD and SEM then analyzed the intercalated bentonite.

2.5 Esterification process
A total of 50 gram of sample was put into a three-neck flask, esterified with free fatty acids with methanol (comparison of oil with methanol 1: 3), added with 1 M H_2SO_4 catalyst as much as 1.25% of the total weight of oil and methanol, then refluxed at 60-65°C for 2 hours. The reaction mixture was cooled and formed in two layers, methanol and methyl ester (top layer), triglycerides (bottom layer), and then separated using a separating funnel.

2.6 Transesterification process
The separated triglyceride was then transesterified with methanol (oil and methanol ratio 1: 6), and a 1% KOH catalyst was added by the weight of the mixture of oil and methanol, refluxed at 60°C for 2 hours. The reaction mixture was cooled down, and two layers formed, methyl ester (biodiesel) in the upper layer and glycerol in the lower layer. Then the mixture was separated by a separating funnel, and then the methyl ester was evaporated. The methyl ester layer was washed with distilled water repeatedly using a separating funnel until the water layer was transparent. The final step was adding anhydrous Na_2SO_4 to bind the residual water then filtered with Whatman 40 filter paper. The resulting biodiesel product was then characterized using GC-MS.

2.7 Cracking process
Cracking of biodiesel from used cooking oil was carried out using a cracking reactor. A total of 50 mL of sample was inserted into the reactor, followed by the inclusion of 3 g of the catalyst $NiCl_2$ intercalated into the upper furnace. The cracking process was carried out with variations in the temperature in the lower furnace of 450°C and 500°C, and a constant temperature in the upper furnace of 350°C. Each heat was carried out for 60 minutes. Then the product produced was analyzed using GC-MS.

3. Results and discussion

3.1 Bentonite preparation
Bentonite preparation began with the washing process of bentonite. This process was carried out by dispersing bentonite into distilled water. Bentonite was sieved using a 200 mesh sieve. Sifted bentonite was then dispersed into distilled water and stirred using a magnetic stirrer for 3 hours (figure 1). This dispersing process aimed to remove impurities and was expected to increase the surface area of bentonite. The resulting suspension was filtered using filter paper. The filtering results were dried in an oven at 110°C for 2 hours to remove water molecules found in the bentonite pore. The dried sample was crushed and sieved with a 200 mesh sieve. This process aimed to reduce the size of the bentonite so that it increased the surface area of the bentonite. This process produced clean bentonite that was ready to use. Furthermore, it was analyzed using XRD and SEM on clean (pure) bentonite samples.

3.2 Intercalation process
The bentonite intercalation process was initiated by dispersing bentonite into the water. Bentonite with a wide surface can absorb water so that hydrates are formed between layers on the bentonite, which causes the bentonite to appear to expand. Bentonite has a layered structure with the ability to swell (swelling) and has cations that can be exchanged. The presence of water molecules allows cations such as Ca^{2+}, Mg^{2+}, Na^{+}, and K^{+} to be transferred because the water in the interlayer space can keep the

layers separate, thereby reducing the electric field between the surface of bentonite and interchangeable ions [28,29].

The intercalation process was continued by adding a cationic surfactant to the bentonite suspension. The cationic surfactant used was Cetyl Trimethyl Ammonium Bromide (CTAB) quaternary ammonium salt, and the mixture was stirred for 24 hours using a magnetic stirrer. The addition of CTAB organic molecules as quaternary ammonium served as a template. The CTAB surfactant, which included a cationic surfactant, would lead to pore formation. Templates functioned as agents that directed the bentonite crystal structure. In this case, the template was a surfactant cation functioned as a cation to neutralize the anionic framework ($[SiO_4]^{4-}$ or $[AlO_4]^{5-}$) [30–32]. This cationic compound would replace the clay surface cations and change the clay surface to be very hydrophobic. When the aluminate and silicate solutions were mixed with the surfactants, the surfactants would form micelles to reduce surface tension, hydrophobic groups would gather, and hydrophilic surfactant heads would move away from their hydrophobic groups so that a cylindrical circle was formed.

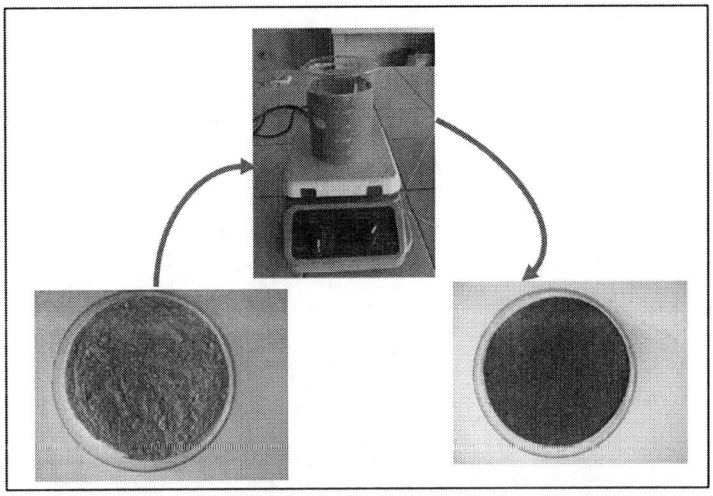

Figure 1. The cleaning process to produce the clean bentonite.

Electrostatically, the hydrophilic group (surfactant head) would interact with the primary building unit of the bentonite. The primary building unit of a bentonite structure was a tetrahedral consisting of a central Si or Al atom surrounded by four oxygen atoms, namely ($[SiO_4]^{4-}$ or $[AlO_4]^{5-}$) [28]. After 24 hours, $NiCl_2$ was inserted into the mixture of bentonite and CTAB. Ni^{2+} ions would enter through hydrophobic interactions and act as catalysts. Bentonite added with CTAB surfactant formed a complex compound with a quaternary ammonium salt, which became intercalated and then bound to Ni^{2+} ions. This mixture experienced structural compaction through a heating process at a temperature of 110°C for three days and produced a cream-colored solid. The solid was crushed and sifted using a 200 mesh sieve. This sifting aimed to increase the distance between intercalated bentonite layers, which in itself also increased the surface area of the bentonite. The next step was calcination to remove surfactants left behind in the intercalated bentonite pores after the heating process. Calcination was carried out at 600°C for 2 hours, which was expected to be able to decompose CTAB so that it left bentonite pores. XRD and SEM characterized the bentonite product.

3.3 Characterization using x-ray diffraction (XRD)

Characterization with XRD in this study aimed to provide information about the crystallization of a particular mineral. The results obtained can also be used to determine the magnitude of the shift in distance between layers of silicate clay. If intercalation occurred, we would see a change in the distance between the layers marked by variations in bass spacing. The pure bentonite diffractogram can be seen in figure 2.

Figure 2 shows that the pure bentonite used in this study contained montmorillonite selected at an angle of 2θ: 6.05° (d: 14.57 Å); 12.12° (d: 7.29 Å); 20.32° (d: 4.36 Å); 62.16° (d: 1.49 Å) and 73.89° (d: 1.28 Å). Other peaks that appeared at an angle of 2θ: 35.72° (d: 2.51 Å) and 55.52° (d: 1.65 Å) were the peaks resulted from quartz, while the feldspar peaks appeared at an angle of 2θ: 22.02° (d: 4.03 Å). The XRD diffractogram results show the bentonite used could provide enough cavities to continue the $NiCl_2$ intercalation process. This condition was characterized by a reasonably high montmorillonite content in the bentonite sample used. Montmorillonite is a material consisting of silica and alumina arranged octahedrally and tetrahedrally.

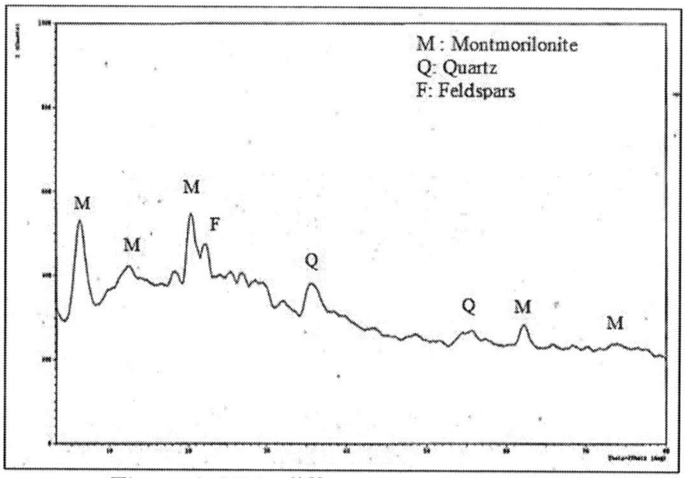

Figure 2. XRD diffractogram of bentonite.

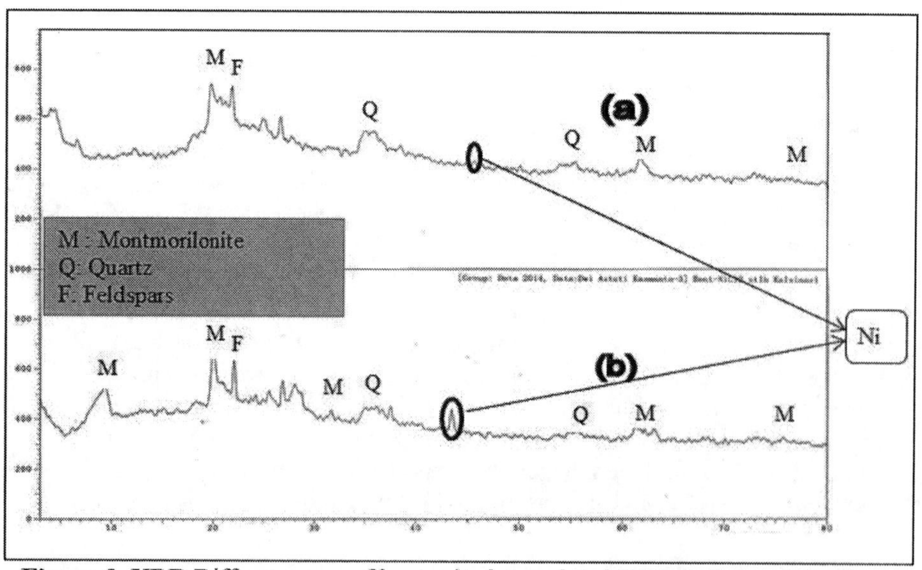

Figure 3. XRD Diffractogram of bentonite intercalated $NiCl_2$ before calcination (a), and bentonite intercalated $NiCl_2$ after calcination (b).

Figure 3 shows the crystallization of bentonite intercalated $NiCl_2$ before calcination. It can be seen that the intensity of peak sharpness was compared to the peak of pure bentonite diffractogram. Two new diffraction peaks appeared as Ni diffraction peaks, namely at a value of 2θ at 44.96° with a basal spacing of (d: 2.01 Å). The new peak appeared as a Ni peak, which was intercalated in the bentonite structure. Figure 3 shows the peaks with low intensity on the diffractogram before calcination was reduced. It indicates that the CTAB surfactant decomposed. In this diffractogram, a new peak appeared with a good sharpness intensity at a value of 2θ at 44.67° with a basal spacing (d: 2.03 Å). This peak shows the Ni crystalline phase, which experienced a shift in the 2θ number.

3.4 Characterization using Scanning Electron Microscope (SEM)

The results of micrographs with SEM at a magnification of 15,000x are shown in figure 4. In figure 4 (a) it can be seen that the porosity of pure bentonite before intercalation was relatively large compared to the porosity in intercalated bentonite $NiCl_2$ in figure 4 (b). It can be explained by the appearance of a black pore basin that was indicated as part of the pore of the bentonite. In contrast, the micrograph results in figure 4(b) show fewer pore basins compared to figure 4(a).

(a) (b) (c)

Figure 4. The bentonite micrographs (a). Pure bentonite; (b) $NiCl_2$ intercalated before calcination; (c) $NiCl_2$ intercalated after calcination.

This is presumably because, at the time of intercalation, a portion of the bentonite pore was covered by the surfactant and catalyst being embedded. As a result of the partial closure of pores in intercalated bentonite, the N_2 gas that passed through the pore was getting smaller so that the data from the gas absorption analysis showed a reduced surface area value. Figure 4(c) shows the porosity of $NiCl_2$ intercalated after calcination, showing a wider pore surface because the surfactant had decomposed in the calcination process.

3.5 Esterification of cooking oil waste

The esterification process was carried out by reacting methanol with cooking oil waste (ratio of oil and methanol 1: 3) and by adding 1 M H_2SO_4 catalyst as much as 1.25% of the total weight of oil and methanol. The mixture was refluxed for 2 hours at 60-65°C. The results obtained from the esterification reaction was a mixture of methyl ester and methanol in the upper layer and triglycerides in the lower layer (figure 5). The esterification reaction occurred from a carboxylic acid with alcohol with the help of an acid catalyst to form esters and water. The problem that arose with the use of acid catalysts was the formation of water in the mixture so that the addition of excess alcohol was carried out because the water formed would dissolve in methanol and not inhibit the reaction process. The acid catalyst used in this case was sulfuric acid because it relatively quickly converted free fatty acids into alkyl esters [33].

3.6 Transesterification of waste cooking oil triglycerides

The transesterification of used cooking oil triglycerides was carried out after the esterification reaction. The transesterification reaction was carried out by reacting used cooking oil triglycerides and methanol in a 1:6 mole ratio and adding KOH catalyst as much as 1% of the total weight of oil and methanol, for 2 hours at a temperature of 60-65°C. This condition was based on the optimization process that was conducted in another study. Next was the evaporation process of methyl ester (biodiesel) to remove the remaining methanol.

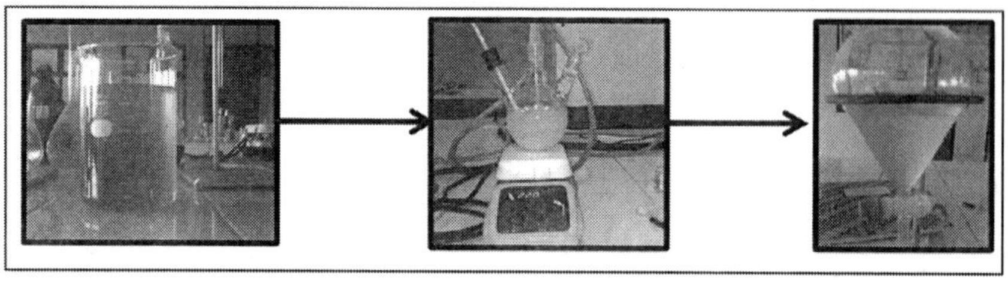

Figure 5. Esterification of the waste cooking oil route.

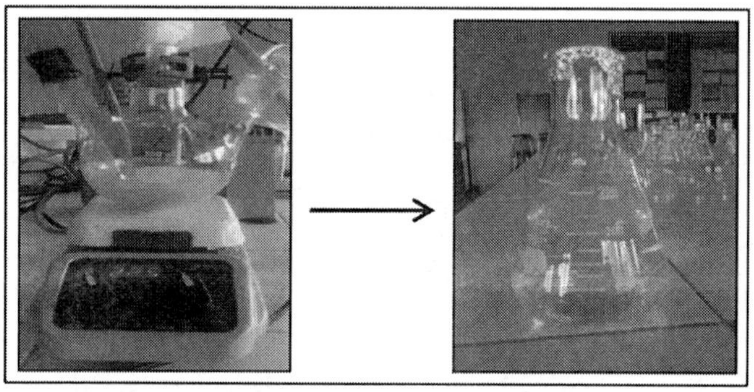

Figure 6. Transesterification of the waste cooking oil route.

The method of washing methyl esters (biodiesel) was done by using distilled water until the bottom layer was transparent. The washing product consisted of two layers, namely the top layer in the form of methyl ester (biodiesel) and the bottom layer was distilled water and the remaining glycerol (figure 6). The methyl ester (biodiesel) layer was above because the specific gravity of the water was higher than the methyl ester (biodiesel). The separated methyl esters (biodiesel) were then given anhydrous Na_2SO_4, which served to bind the remnants of water that was still contained in methyl esters (biodiesel). It was then filtered using Whatman 40 paper and characterized using GC-MS to determine the compounds contained in the methyl esters.

3.7 Biodiesel characterization using GC-MS

GC-MS analysis was performed on methyl esters (biodiesel) resulting from the transesterification reaction with a KOH catalyst. Through GC-MS, it was found that the type of methyl esters contained in methyl esters (biodiesel) from used cooking oil was produced from the transesterification reaction. The number of peaks indicates the number of transesterified methyl esters, each of which was completely separated, resulting in peaks with different retention times. The top that came out first was the ester with a short carbon chain. The column (stationary phase) used is non-polar, while in general, the ester was polar. Short-chain esters were polar from long-chain esters. The methyl ester with a longer chain would be held in the column while the short-chain ester would escape with the mobile phase-out of the column. The content of methyl esters resulting from GC-MS analysis on used cooking oil methyl esters can be seen in figure 7 and table 1.

IOP Conf. Series: Materials Science and Engineering **980** (2020) 012063 doi:10.1088/1757-899X/980/1/012063

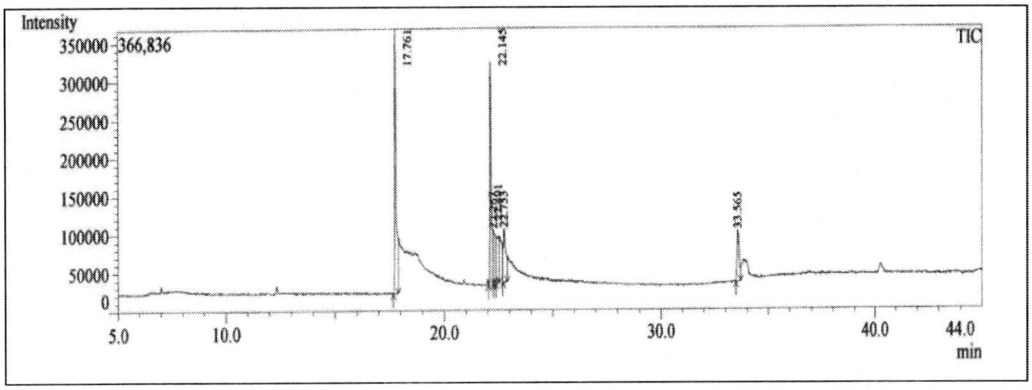

Figure 7. Methyl ester chromatogram of biodiesel from cooking oil waste.

Table 1. Methyl ester content of biodiesel from waste cooking oil.

No.	Compounds	Retention time (minute)	% Areas
1	Methyl palmitate	17.76	32.40
2	Methyl oleate	22.14	25.88
3	Methyl stearate	22.75	11.38

In figure 3, the chromatogram GC-MS test results showed six peaks, but only three peaks were detected as fatty acid methyl esters (table 1). The methyl esters were methyl palmitate ($C_{17}H_{34}O_2$) with the largest area of 32.40% with a retention time of 17.76 minutes (m/z = 270), methyl oleate ($C_{19}H_{36}O_2$) with a percentage area of 25.88% and a retention time of 22.14 minutes (m/z = 296), Methyl stearate ($C_{19}H_{38}O_2$) with a percentage area of 11.38% and a retention time of 22.75 minutes and (m/z = 298).

3.8 Biodiesel cracking process

Catalyst cracking is a way to break down complex hydrocarbons into simpler molecules so that they can improve the quality and also reduce the amount of residue produced. Cracking biodiesel from used cooking oil in isothermal conditions was carried out using a cracking reactor. A total of 50 mL of sample was put into the sample tube, followed by the inclusion of three gram of the intercalated catalyst $NiCl_2$ into the catalyst tube. The temperature of the upper furnace was fixed at 350°C. The cracking process was carried out with temperature variations in the lower furnaces of 450°C and 500°C. This temperature selection was based on the optimum conditions of trial and error tests carried out on the reactor.

3.9 Results of cracking process using GC-MS

GC-MS chromatogram analysis provided information on the number of different component compounds between the two temperature variations. Chromatogram results can be seen in figure 8. From the results of the chromatogram at 450°C, there were 11 peaks. The carbon chain breaking in biodiesel resulted in shorter carbon chain compounds. Among these 11 peaks, there were three peaks of compounds belonging to the gasoline fraction, they were thought to be methyl 6-hydroxacapaproic ($C_7H_{14}O_3$), 4-hydroxymethyl-1,3-dioxolane ($C_4H_8O_3$), and methyl nonylate ($C_{10}H_{20}O_2$).

Figure 9 shows that the carbon chain was cut from biodiesel (C_{16-20}) to biogasoline (C_{4-12}). The five short-chain compounds were successfully tracked well due to the acidity of the bentonite catalyst which had been intercalated with $NiCl_2$. The more active the acidic side of the catalyst was used, the more product was produced. The following chromatogram results of cracking at 500°C can be seen in figure 9. The results of the GC-MS chromatogram at 500°C produced seven peak compounds. These compounds underwent carbon chain cuts that were not too far from the biodiesel carbon chain (C_{16-20}).

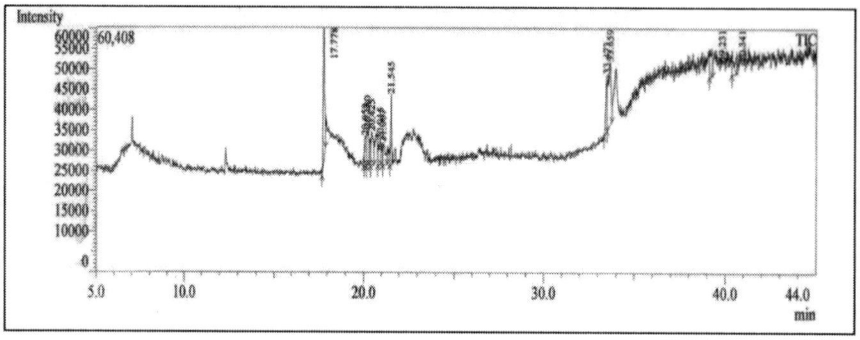

Figure 8. Biodiesel cracking chromatogram at 450 °C.

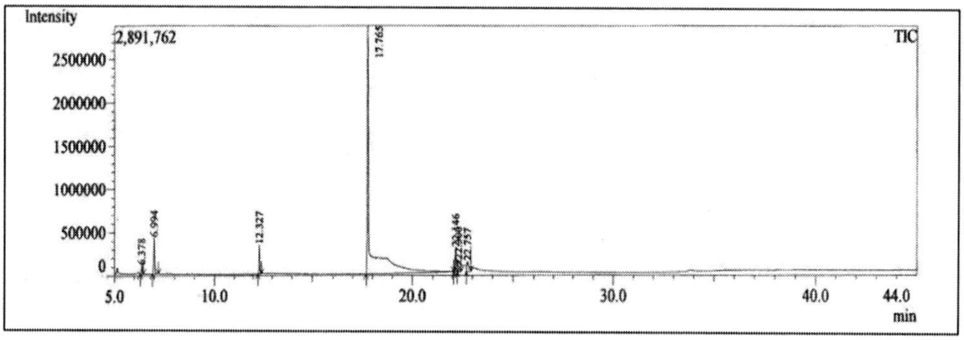

Figure 9. Biodiesel cracking chromatogram at 500 °C.

Three compounds underwent carbon chain-breaking, they were compounds dodecane ($C_{12}H_{26}$), methyl dodecanoic ($C_{13}H_{26}O_2$), and methyl tridecanoic ($C_{14}H_{28}O_2$). The only dodecane belonged to the gasoline fraction, while methyl dodecanoic and methyl tridecanoic belonged to the kerosene fraction which had a carbon chain length of 13-15 (C_{13-15}). Product conversions dominated by large molecular weight compounds occurred because of the polymerization and re-polymerization reactions. The polymerization reaction could occur between free radicals and carbon molecules that had double bonds to produce products with a higher number of C atoms. In contrast, the re-polymerization reaction was influenced by the length of cracking reaction time and the abundance of products produced. It would cause the catalyst to be deactivated by an abundant product so that it could cover the active site on the catalyst [24]. The GC-MS results obtained showed similarities in the products produced using the cracking process in other studies [27].

4. Conclusion

The biodiesel product contents resulted from cooking oil waste were methyl palmitate ($C_{17}H_{34}O_2$), methyl oleate ($C_{19}H_{36}O_2$), and methyl stearate ($C_{19}H_{38}O_2$). Based on the diffractogram of the intercalated bentonite $NiCl_2$ before the calcination stage, a new peak appeared as a Ni peak, namely at a value of 2θ 44.96°. After the calcination process, it was seen that the Ni peak shifted from 2θ to 44.66°. The results of the morphological study of intercalated bentonite $NiCl_2$ using SEM showed the $NiCl_2$ catalyst had covered a portion of the surface of the bentonite pore. Based on the results of GC-MS analysis on the cracking process at 450°C, it was found that there was a cutting of the carbon chain from biodiesel (C_{16-20}) to biogasoline (C_{4-12}). The compounds were thought to include methyl 6-hydroxy caproic ($C_7H_{14}O_3$), 4-hydroxymethyl-1,3-dioxolane ($C_4H_8O_3$), and methyl nonylate

($C_{10}H_{20}O_2$). The results of GC-MS chromatograms at 500 °C found that three compounds underwent carbon chain cutting Those compounds were suspected as dodecane ($C_{12}H_{26}$), methyl dodecanoic ($C_{13}H_{26}O_2$), and methyl tridecanoic ($C_{14}H_{28}O_2$).

References

[1] Centi G and Perathoner S 2020 Chemistry and energy beyond fossil fuels. A perspective view on the role of syngas from waste sources *Catalysis Today* **342** 4–12
[2] Sutapa I W, Bandjar A, Latupeirissa J and Toesik M M A F 2020 Development of Utilization of Tuna Bone Waste as a Catalyst for Biodiesel Production Using Padina sp. Seaweed and Palm Oil *J. Phys.: Conf. Ser.* **1463** 012010
[3] Darnoko D and Cheryan M 2000 Continuous production of palm methyl esters *J Amer Oil Chem Soc* **77** 1269–72
[4] Haas M J 2005 Improving the economics of biodiesel production through the use of low value lipids as feedstocks: vegetable oil soapstock *Fuel Processing Technology* **86** 1087–96
[5] Wahab A W, Karim A, Nafie N L, Satrimafitrah P, Triana and Sutapa I W 2019 Production of the nanoparticles using leaf of Muntingia calabura L. as bioreductor and potential as a blood sugar nanosensor *J. Phys.: Conf. Ser.* **1242** 012004
[6] Sutapa I W, Rosmawaty and Ropa H 2019 Non-edible oil of Cerbera manghas L. seed from Seram Island - Maluku as Oil Source of Biodiesel Production *J. Phys.: Conf. Ser.* **1341** 032001
[7] Xingang Z and Pingkuo L 2013 Substitution among energy sources: An empirical analysis on biomass energy for fossil fuel of China *Renewable and Sustainable Energy Reviews* **18** 194–202
[8] Ogunkunle O and Ahmed N A 2019 A review of global current scenario of biodiesel adoption and combustion in vehicular diesel engines *Energy Reports* **5** 1560–79
[9] Singh D, Sharma D, Soni S L, Sharma S and Kumari D 2019 Chemical compositions, properties, and standards for different generation biodiesels: A review *Fuel* **253** 60–71
[10] Willauer H D, Hardy D R, Baldwin J W, DiMascio F, Williams F W, Bradley M J and Hoheisel R 2018 Economic comparisons of littoral production of low carbon fuel from non-fossil energy sources and seawater *Journal of Cleaner Production* **170** 1473–83
[11] Soekamto N H, Liong S, Fauziah S, Wahid I, Firdaus, Taba P and Ahmad F 2018 Dengue antiviral activity of polar extract fromMelochia umbellata(Houtt) Stapf var. Visenia *J. Phys.: Conf. Ser.* **979** 012017
[12] Kadir Yesilyurt M and Cesur C 2020 Biodiesel synthesis from Styrax officinalis L. seed oil as a novel and potential non-edible feedstock: A parametric optimization study through the Taguchi technique *Fuel* **265** 117025
[13] Silveira E G, Barcelos L F T, Perez V H, Justo O R, Ramirez L C, Rêgo Filho L de M and de Castro M P P 2019 Biodiesel production from non-edible forage turnip oil by extruded catalyst *Industrial Crops and Products* **139** 111503
[14] Danish M, kale P, Ahmad T, Ayoub M, Geremew B and Adeloju S 2020 Conversion of flaxseed oil into biodiesel using KOH catalyst: Optimization and characterization dataset *Data in Brief* **29** 105225
[15] Agarwal M, Chauhan G, Chaurasia S P and Singh K 2012 Study of catalytic behavior of KOH as homogeneous and heterogeneous catalyst for biodiesel production *Journal of the Taiwan Institute of Chemical Engineers* **43** 89–94
[16] Abdullah, Rahmawati Sianipar R N, Ariyani D and Nata I F 2017 Conversion of palm oil sludge to biodiesel using alum and KOH as catalysts *Sustainable Environment Research* **27** 291–5
[17] Rashtizadeh E, Farzaneh F and Ghandi M 2010 A comparative study of KOH loaded on double aluminosilicate layers, microporous and mesoporous materials as catalyst for biodiesel production via transesterification of soybean oil *Fuel* **89** 3393–8
[18] Sutapa I W, Wahab A W, Taba P and Nafie N L 2018 Dislocation, crystallite size distribution and

lattice strain of magnesium oxide nanoparticles *J. Phys.: Conf. Ser.* **979** 012021
[19] Leung D Y C and Guo Y 2006 Transesterification of neat and used frying oil: Optimization for biodiesel production *Fuel Processing Technology* **87** 883–90
[20] Simsek S and Uslu S 2020 Comparative evaluation of the influence of waste vegetable oil and waste animal oil-based biodiesel on diesel engine performance and emissions *Fuel* **280** 118613
[21] Cheng J, Zhang Z, Zhang X, Fan Z, Liu J and Zhou J 2019 Continuous hydroprocessing of microalgae biodiesel to jet fuel range hydrocarbons promoted by Ni/hierarchical mesoporous Y zeolite catalyst *International Journal of Hydrogen Energy* **44** 11765–73
[22] Cheng J, Zhang Z, Zhang X, Liu J, Zhou J and Cen K 2019 Sulfonated mesoporous Y zeolite with nickel to catalyze hydrocracking of microalgae biodiesel into jet fuel range hydrocarbons *International Journal of Hydrogen Energy* **44** 1650–8
[23] Shirasaki Y, Nasu H, Hashimoto T and Ishihara A 2019 Effects of types of zeolite and oxide and preparation methods on dehydrocyclization-cracking of soybean oil using hierarchical zeolite-oxide composite-supported Pt/NiMo sulfided catalysts *Fuel Processing Technology* **194** 106109
[24] Silalahi I H, Sayekti E and Sianipar A 2013 Optimization of cooking oil cracking reaction using Zeolite/Nickel catalyst *J. Kimia Khatulistiwa* **2** 6
[25] Roddu A K, Wahab A W, Ahmad A, Taba P and Sutapa I W 2020 Theoretical analysis properties of gold nanoparticles resulted by bioreduction process *J. Phys.: Conf. Ser.* **1463** 012008
[26] Harimu L, Matsjeh S, Siswanta D, Sentosa S J and Sutapa I W 2019 Synthesis of poly (ethyl eugenyl oxyacetate) as carrier for separation of heavy metal ions Fe(III), Cr(III), Cu(II), Ni(II), Co(II), and Pb(II) using liquid-liquid extraction method *J. Phys.: Conf. Ser.* **1341** 032003
[27] Ishihara A, Kawaraya D, Sonthisawate T, Kimura K, Hashimoto T and Nasu H 2015 Catalytic cracking of soybean oil by hierarchical zeolite containing mesoporous silica-aluminas using a Curie point pyrolyzer *Journal of Molecular Catalysis A: Chemical* **396** 310–8
[28] Pandey S 2017 A comprehensive review on recent developments in bentonite-based materials used as adsorbents for wastewater treatment *Journal of Molecular Liquids* **241** 1091–113
[29] Said M I, Abustam E, Wahab A W, Taba P, Gani A and Wahid A M 2019 Effect of ethanol used in a degreasing process on Bali cattle bones on the physicochemical properties of extracted collagen *Bulgarian Journal of Agricultural Science* **5** 418–23
[30] Lawal I A, Klink M, Ndungu P and Moodley B 2019 Brief bibliometric analysis of "ionic liquid" applications and its review as a substitute for common adsorbent modifier for the adsorption of organic pollutants *Environmental Research* **175** 34–51
[31] Harimu L, Haeruddin, Baari M J and Sutapa I W 2020 Effectiveness of reducing cyanide levels in the dioscorea hispida dennst bulbs through soaking in seawater and interaction with ash scrub *J. Phys.: Conf. Ser.* **1463** 012013
[32] Misriyani, Wahab A W, Gunlazuardi J, Taba P and Shiomori K 2015 Synthesis and characterization of TiO_2 nanotube films for a photo-electrochemical corrosion prevention of stainless steel under uv light exposure *International Journal of Applied Chemistry* **11** 443–53
[33] Gerpen J V 2005 Biodiesel processing and production *Fuel Processing Technology* **86** 1097–107

IOP Conf. Series: Materials Science and Engineering **980** (2020) 012064 doi:10.1088/1757-899X/980/1/012064

The potentials of biochar from agricultural waste as a carrier material of biofertilizer for swamplands

E Maftuah[1], M Saleh[1] and E Pratiwi[2*]

[1]Indonesia Swampland Agriculture Research Institute (ISARI), Jl. Kebun Karet, Loktabat, Banjarbaru, South Kalimatan

[2]Soil Research Institute, Jl. Tentara Pelajar No 12, Bogor

Email: eni_balittra@yahoo.com

Abstract. Biochar has pores suitable for microbial habitat, and it contains carbon which can be used as an energy source by microbes. The research aims to determine the potential of some biochar as a carrier of biological fertilizer for swamplands. The treatment given was the type of biochar (rice husk, coconut shell, and palm empty fruit bunches) and the microbial type ie decomposer, Phosphate Solubilizing Bacteria (PSB), N Fixing Bacteria, and a consortium microbial. The design used was a Factorial Complete Randomized Design, 3 replications. Biochar analysis included organic C, total N, pH, CEC, SiO_2, ash content and water content. Calculation of microbial populations was done at 2, 6 and 10 weeks after inoculation. Biochar rice husks, oil palm empty fruit bunches and coconut shells were very effective as decomposer carrier material. The highest of population of N-fixing bacteria and PSB at 6 weeks after inoculation was in the biochar rice husk reached 0.33 x 10^6 cfu/g for N-fixing bacteria and 54.8 x 10^6 cfu/g for P solvent bacterial. Rice husk biochar can be used as a carrier material both for a single biofertilizer, as well as a consortium biofertilizer that consists of decomposer fungi, P solubilizing bacteria and N-fixing bacteria.

1. Introduction

Swampland has great potentials to be used as a productive agricultural area. Swampland in Indonesia covers 34.12 million Ha [1]. However, swamps have several problems among others because of high soil acidity and relatively low availability of nutrients in the soil [2,3]. Optimizing the swamp lands requires improvements in water, soil, land and plant management. One of the tipping points to increase swampland productivity is needed the support the availability of fertilizers both organic and inorganic.

The continuous use of inorganic fertilizers not only requires high cost, but also negatively impact soil quality. Increased awareness and understanding of the negative impacts of the use of inorganic fertilizers encourage the development of alternative technologies or fertilizer products that are more environmentally friendly [4,5]. Biofertilizer is an alternative type of fertilizer which, when used in accordance with soil conditions and its intended purpose, is able to increase soil fertility and plant productivity, efficient use of inorganic fertilizers, and reduce environmental impacts.

In general, microorganisms in biological fertilizers include N-fixing, Phosphate Solubilizing Bacteria, organic decomposers, and producing siderophores and growth hormones such as IAA, gibberellin, cytokinin and ethylene [6,7]. Shinha [8] suggested that biological fertilizers provide benefits for plant growth and increase yields through its role of fixing N, dissolving P and K, releasing plant growth regulators, and producing antibiotics, and decomposing organic matter.

One of the quality of biological fertilizer is determined by the carrier material. Biochar can be used as an effective carrier material for microbes that play an important role in the soil so that it can play a role in the manufacture of biological fertilizers [9]. Biochar has an ideal pore size for soil microbial growth [9,10]. The presence of biochar in the soil can be beneficial for bacteria, fungi, and other soil microbes. Chen [11] reported that biochar microbes are protected from external conditions, such as pH, toxic materials and competition with other microbes. Fungi can sporulate in biochar micro pores because in these pores the competition that occurs with other saprophytes is quite low [12]. Biochar can increase the diversity of microorganisms in the soil [13]. These advantages make biochar as usable as a soil enhancer and carrier agent [14,15,16]. This research aims to study the potentials of various types of biochar as a biological fertilizer carrier for swamps.

2. Materials and methods

2.1. Materials

The research was conducted in the laboratory of the Indonesian Swampland Research Institute from February to October 2015. Equipment for laboratory activities used namely: laminar flow (ESCO), autoclave (All American), incubator, analytical balance, petri dish, erlenmeyer, baker, measuring cup, glass spatula and stainless steel, etc. The materials used include decomposer isolates, Phosphate Solubilizing Bacteria (PSB) and N fixing bacteria, rice husk biochar, coconut shell biochar, oil palm empty fruit bunch biochar, and chemicals (merck), such as: $Ca(PO_4)_2$, KCl, $MgSO_47H_2O$, $MgSO_47H_2O$, $(NH_4)_2SO_4$, glucose, yeast, K_2HPO_4, NaCl, $CaCl_2$, Na_2MoO_4 and aquadest.

2.2. Methods

The treatment given was the type of biochar (rice husk, coconut shell, and palm empty fruit bunches)and the type of microbial decomposer, Phosphate Solubilizing Bacteria (PSB) and N fixing bacteria, and a consortium microbial i.e decomposer, PSB, N fixing bacteria. The design used was a Factorial Complete Randomized Design, three replications.

Biochar was prepared by pyrolysis at a temperature about 400°C using a retort kiln. The decomposer microbial used was *Trichoderma* sp, Phosphate Solubilizing Bacteria microbial was *Bacillus sp* and N-fastening microbial was *Azotobacter*. The microbes used had been tested by inter-isolate suitability. This test was carried out by inoculating 3-4 isolates namely microbial decomposers, Phosphate Solubilizing Bacteria, N fastening in one Petridish at room temperature (± 30 °C). At five days after inoculation, the growth of each isolate was observed, whether inhibition occurred or not. Synergic growing microbes were selected for the test formulations of these carriers.

The carrier material (biochar) was dried (± 14% moisture content) and then mashed to a size of 2 mm. About 100 grams of each carrier was put in a plastic bag, plus water to get about 30% water content and then sterilized. In each of these carriers the inoculated decomposer (*Trichoderma* sp), Phosphate Solubilizing Bacteria (*Bacillus sp*) and N fixing bacteria (*Azotobacter*), and the combination of the three microbes. The combination of 12 types of treatment, and each treatment was repeated 3 times.

The analysis of the biochar raw materials carried out included lignin, cellulose, hemicellulose, organic C and total N. Biochar analysis included organic C content, total N, pH, CEC, SiO_2, ash content and water content. The media used for fungi using PDA media, P solvent bacteria using Pikovskaya media and N-fixing bacteria using Nitrogen Free Bacteria (NFB) media. Calculation of microbial populations (Phosphate Solubilizing Bacteria, N-fixing bacteria and decomposer fungi) was done every week at 2, 6 and 10 weeks after inoculation.

3. Results and discussions

3.1. Biochar characteristics

The characteristics of the biochar raw materials used i.e rice husks, empty palm fruit bunches and coconut shells are presented in table 1. Organic materials used as biochar sources are of various quality. Coconut shell had the highest C/N ratio, and the lowest in rice husk. The content of lignin in rice husk and coconut shell wass almost the same, namely rice husk 32.69% while coconut shell 31.87% and lowest oil palm empty fruit bunch was 25.60%. Cellulose ranged from 43.46 to 49.63%, while hemicellulose was 19.65-23.16% (table 1).

Table 1. Cellulose content, hemicellulose and C/N ratio of several organic materials from biochar sources.

Type of material	Content (%)					C/N
	Lignin	Cellulose	Hemicellulose	C	N	
Rice husk	32,69	43,46	23,16	49,07	0,84	58,42
Empty palm fruit bunch	25,60	49,63	20,69	55,49	0,70	79,27
Coconut shells	31,87	46,36	19,65	55,52	0,42	132.19

The characteristics of biochar produced from the pyrolysis process are presented in table 2. The content of organic C varied, the highest in oil palm empty fruit bunches and the lowest in rice husks. The highest SiO_2 content of rice husk reached 34.83% and the lowest was at coconut shell 4.04%. The results of the biochar chemical analysis showed that rice husk biochar had a lower pH so it tended to be more suitable for microbes. Microbes can be affected by biochar both directly and indirectly through physicochemical changes. The pH values play a key role in microbial abundance [17]. Slightly alkaline or neutral conditions are preferred bacterial and fungal growth compared to weak acid conditions [18,19]. In general, biochar has a pH value that tends to be alkaline. There was a negative correlation between microbial enzyme activity and biochar pH value [20]. The characteristics of biochar produced from the pyrolysis process are presented in table 2.

Table 2. Characteristics of biochar used in research.

Type of biochar	C (%)	pH	CEC Cmol$^{(+)}$ kg^{-1}	SiO_2	Ash content	N	P	K	Ca	Mg	Fe
				%							
Rice husk	23.40	8.99	37.38	34.83	44.35	0.73	0.48	0.54	0.21	0.18	0.2
Empty palm fruit	42.33	9.39	9.93	4.90	27.09	0.99	0.49	8.65	0.43	0.67	0.5
Coconut shell	29.69	9.61	9.61	4.04	48.96	1.28	0.52	2.96	0.29	4.43	0.2

Fungal and bacterial populations are also related to pore spaces in biochar. Fungi and bacteria live and multiply in macropores. Microbial cells have a size of 0.5 – 5 μm, consisting mainly of bacteria, fungi, actinomycetes and algae in pores of 2 μm – 20 μm [21]. The existence and size, as well as the distribution of pores in biochar, provide suitable habitats for many microorganisms by protecting them from predators and drought, providing carbon for energy and minerals for nutrition [12,22]. Biochar pore structure, high surface area, ability to absorb dissolved organic matter, gases and inorganic nutrients tend to provide excellent habitat for microbes to move, grow and multiply, especially for bacteria, actinomycetes and arbuscular mycorrhizal fungi [23].

Biochar can be a microbial carrier because biochar is a relatively moist substance, the surface properties of biochar and slowly releasing minerals, the rate of decomposition is much slower than organic matter [23]. Soil microbial populations can be affected by the quality and quantity of biochar added. Biochar can function as a substrate of microorganisms depending on the composition of pyrolysis residual compounds [24, 25], however biochar can also be toxic to plants [26]. Biochar can act as a place provider, as well as a food source for microorganisms. Microorganisms that live on the surface of biochar are able to decompose enzymes needed for metabolism. Substrate C and inorganic nutrients absorbed on the surface of biochar and/or ash content become a food source for microorganisms.

3.2. Fungi population

The fungus used in this research was Trichoderma which acted as a decomposer. The results of the isolation study using PDA media at 2, 6 and 10 weeks as presented in table 3.

Table 3. The average number of fungi populations at 2, 6 and 10 weeks after inoculation on PDA media.

Treatments	Population mean x 10^4 (CFU/g)		
	2 WAI	6 WAI	10 WAI
Rice husk biochar + *Trichoderma*	Very abundant	Very abundant	Very abundant
Rice husk biochar + *Bacillus sp*	0	0	0
Rice husk biochar + *Azotobacter*	0	0	0
Rice husk biochar + microbial consortium	1,33	Very abundant	Very abundant
Coconut shell biochar + *Trichoderma*	Very abundant	Very abundant	Very abundant
Coconut shell biochar + *Bacillus sp*	0	0	0
Coconut shell biochar + *Azotobacter*	0	0	0
Coconut shell biochar + microbial consortium	Very abundant	Very abundant	Very abundant
Empty palm fruit bunches biochar + *Trichoderma*	Very abundant	Very abundant	Very abundant
Empty palm fruit bunches biochar + *Bacillus sp*	0	0	0
Empty palm fruit bunches biochar + *Azotobacter*	0	0	0
Empty palm fruit bunches biochar + microbial consortium	Very abundant	Very abundant	Very abundant

WAI= week after inoculation

All three types of biochar could be used as a fungi carrier media. It can be seen that the population of fungi in all the biochar types with the inoculation of *Trichoderma* until the 10th week was very high. Ogawa [27] suggested that biochar can be a habitat that is not suitable for saprophytic fungi, but not for mycorrhizal fungi. Biochar can be used as a carrier agent for arbuscular mycorrhizae because mycorrhiza can use biochar as a habitat for growth and a source of nutrition [28]. However, this will greatly depend on the nature of the biochar, as well as on the number of labile organic molecules present in biochar, because biochar can function as a source of C and energy for soil microbes. Biochar can function as a source of C and energy for soil microbes. Laborda [29] showed that fungi (*Trichoderma* and *Penicillium* spp) could contribute to coal depolymerization (hard coal, sub-bituminous coal and lignite) through the production of the phenoloxidase enzyme. Research by Ascough [30] which carried out research in the laboratory showed that by using SEM tools can clearly show the growth of fungi on the surface of the biochar and inside (interior) biochar for three months.

3.3.Bacteria population

The bacterial population in the pikovskaya media was used to identify P solubilizing bacteria as presented in table 4. The highest P-solubilizing bacterial population was in rice husk biochar media. Several studies have shown that the application of biochar can increase population and bacterial activity. Biochar increased the abundance of *Geobacter*, *Anaeromyxobacter* and *Clostridium* [31]

Provision of biochar can increase the population and activity of *Bacillus* in the soil, as well as increase P uptake, root and canopy growth and the combination of biochar and *Bacillus* is a prospect for sustainable agriculture [32]. Biochar rice husks can increase the population of the bacterium *Bacillus mucilaginosus* [33]. There was a positive correlation between the P content in biochar and the population of P solubilizing bacteria. The correlation of the effect of P content on biochar was in accordance with the regression equation $Y = -14844x^2 + 1888x - 3454$ $R^2 = 0.625$ [34].

Biochar could also be used as a carrier for N-fixing bacteria. In this study, Azotobacter bacteria were used to determine the effectiveness of biochar as a fungal-carrying agent. The results showed that only rice husk biochar could be used as this Azotobacter carrier agent (table 5).

Table 4. Average bacterial population numbers at 2, 6 and 10 weeks after inoculation on Pikovskaya media.

Treatments	Population mean x 10^6(CFU/g)		
	2 WAI	6 WAI	10 WAI
Rice husk biochar + *Trichoderma*	9.67	6.00	5.33
Rice husk biochar + *Bacillus sp*	37.33	54.67	28.30
Rice husk biochar + *Azotobacter*	39.00	18.67	14.00
Rice husk biochar + microbial consortium	41.00	18.00	10.00
Coconut shell biochar + *Trichoderma*	4.50	6.00	7.67
Coconut shell biochar + *Bacillus sp*	24.67	6.67	4.67
Coconut shell biochar + *Azotobacter*	5.00	19.00	11.00
Coconut shell biochar + microbial consortium	5.00	8.67	0.33
Empty palm fruit bunches biochar + *Trichoderma*	0.67	15.00	20.33
Empty palm fruit bunches biochar + *Bacillus sp*	16.00	5.00	5.33
Empty palm fruit bunches biochar + *Azotobacter*	0.67	6.33	4.67
Empty palm fruit bunches biochar + microbial consortium	6.33	0	0

WAI= week after inoculation

Table 5. The average bacterial population at 2, 6 and 10 weeks after inoculation on NFB media.

Treatments	Population mean x 10^6(CFU/g)		
	2 WAI	6 WAI	10 WAI
Rice husk biochar + *Trichoderma*	0	0	0
Rice husk biochar + *Bacillus sp*	0	0	0
Rice husk biochar + *Azotobacter*	7.67	0.33	0.33
Rice husk biochar + microbial consortium	5.50	0.33	0
Coconut shell biochar + *Trichoderma*	0	0	0
Coconut shell biochar + *Bacillus sp*	0	0	0
Coconut shell biochar + *Azotobacter*	0	0	0
Coconut shell biochar + microbial consortium	5.30	0	0
Empty palm fruit bunches biochar + *Trichoderma*	0	0	0
Empty palm fruit bunches biochar + *Bacillus sp*	0	0	0
Empty palm fruit bunches biochar + *Azotobacter*	0	0	0
Empty palm fruit bunches biochar + microbial consortium	0	0	0

WAI= week after inoculation

Azotobacter in addition to fixing N that is in the soil can also produce phytohormones such as cytokinins [35] and gibberellins, IAA [36] and exopolysaccharide production [35] which can increase root proliferation and plant growth Azotobacter sp. able to increase growth through nitrogen supply, supply of growth regulators, and make soil conditions more favorable for plant growth [37].

Biochar could increase the activity of N fixation by the N fixing bacterial. Güereña [38] reported that fix N in the common bean crop (*Phaseolus vulgaris*) was enhanced by biochar rice husk biochar, biochar bagase, corn stover biochar, corncob biochar, *Eucalyptus* biochar, *Delonix* biochar, and biochar tea. Application of tea biochar showed high nodule formation in ordinary pea plants.

Biofertilizers consisting of microbial consortia ie decomposers, P solubilizing bacteria and N-fixing bacteria were more effective using rice husk carriers than coconut shells and palm empty fruit bunches (tables 3, 4 and 5). The microbial population in the rice husk biochar media for *Trichoderma*, *Bacillus* and *Azotobacter* was the highest compared to the other biochar, and although at week 10 *Azotobacter* was not seen in all carriers. *Azotobacter* population in rice husk biochar was higher than in biochar cancorb and bamboo [39].

4. Conclusion

One of the quality of biochar is determined by biochar raw material. Biochar rice husks, oil palm empty fruit bunches and coconut shells are very effective as Trichoderma carrier material. However, rice husk biochar is better than coconut shell biochar and palm empty fruit bunches for Bacillus and Azotobacter carrier material. Rice husk biochar can be used as a carrier material both for a single biofertilizer, as well as a consortium biofertilizer that consists of decomposer fungi, P solubilizing bacteria, and N-fixing bacteria.

References

[1] Ritung S, *et al* W 2015 *Sumberdaya Lahan Pertanian Indonesia: Luas, Penyebaran, dan Potensi Ketersediaan* (Bogor: IAARD Press) p 98

[2] Haryono, Noor M, Syahbuddin H, and Sarwani M 2013 *Lahan Rawa: Penelitian dan Pengembangan* (Jakarta: IAARD Press) p 103

[3] Sulaiman A A, Subagyono K, Alihamsyah T, Noor M, Hermanto, Muharam A, Subiksa IGM, and Suwastika IW 2018 *Membangkitkan Lahan Rawa, Membangun Lumbung Pangan Indonesia* (Jakarta: IAARD Press) p 153

[4] Mukhlis and Fauziati N 2010 Substitusi pupuk NPK anorganik dengan pupuk organic Tithoganik pada tanaman kedelai di lahan pasang surut *Jurnal Agroscientie* **17(1)**15-9

[5] Mukhlis, Indrayati L 2012 *Pros. Sem. Nas. Hasil Penelitian Tanaman Aneka Kacang dan Umbi Tahun* (Malang: Pusat Litbang Tanaman Pangan) pp 244-9

[6] Suriadikarta D A and Simanungkalit R D M 2006 in Simanungkalit, R D M *et al. Pupuk Organik dan Pupuk Hayati* eds (Bogor: BB Litbang Sumberdaya Lahan Pertanian BPPP) pp 1-10

[7] Wolf D C and Wagner G H 2005 Principle and application of soil microbiology 2nd. *In* Silvia DM *et al.* (eds) *Carbon Transformation and Soil Organic Matter Formation.* (Upper Saddle River : NewJersey Pearson Education) pp 285-331

[8] Sinha R K, Valani D, Chauhan K, and Agarwal S 2014 Embarking on a second green revolution for sustainable agriculture by vermiculture biotechnology using earthworms: reviving the dreams of Sir Charles Darwin *Int J Agric Health Saf.* **1** 50–64

[9] Santi L P and Goenadi D H 2012 Pemanfaatan bio-char sebagai pembawa mikroba untuk pemantap agregat tanah Ultisol dari Taman Bogo-Lampung *Menara Perkebunan* **78(2)** 52-60

[10] Vanek S J, Thies J, Wang B, Hanley K and Lehmann J 2016 Pore-size and water activity effects on survival of rhizobium tropici in biochar inoculant carriers *J Microb Biochem Technol* **8** 296-306

[11] Chen WF, Zhang WM and Meng J 2013 Advances and prospects in research of biochar utilization in agriculture. *Sci Agric Sin* **46(16)** 3324–3333

[12] Saito M and Marumoto T 2002 Inoculation with arbuscular mycorrhizal fungi: The status quo in Japan and the future prospects *Plant and Soil* **244** 273–79

[13] Li-li H, Zhe-ke Z and Hui-min Y 2016 Effects on soil quality of biochar and straw amendment in conjunction with chemical fertilizers *J Integrative Agricult.* **15(0)** 60345–7

[14] Kheiry M V, Hafezi A M and Hesaraki S 2013 Bone Regeneration Using Nanotechnology–Calcium Silicate NanoComposites. UCT *J Res Sci Eng and Tech* **1(1)** 1-3

[15] Hale L, Luth M and Crowley D 2015 Biochar characteristics relate to its utility as an alternative soil inoculum carrier to peat and vermiculite *Soil Biology and Biochem* **81** 228-35

[16] Egamberdieva, Reckling M and Wirth S 2017 Biochar-based Bradyrhizobium inoculum improves growth of lupin (*Lupinus angustifolius* L.) under drought stress *Euro J Soil Biology* **78** 3842

[17] Deng Q, Hui D F, Wang J M, Iwuozo S, Yu C L, Jima T, Smart D, Reddy C and Dennis S 2015 Corn yield and soil nitrous oxide emission under different fertilizer and soil management: a three-year field experiment in Middle Tennessee *PLoS ONE* **10(4)** 14
[18] Marstorp H, Guan X and Gong P 2000 Relationship between dsDNA, chloroform labile C and ergosterol in soils of different organic matter contents and pH *Soil Biology & Biochem* **32** 879–82
[19] Rousk J, Brookes P C and Baath E 2009 Contrasting soil pH effects on fungal and bacterial growth suggests functional redundancy in carbon mineralization *Appl Environ Microb* **75** 1589–96
[20] Huang D, Liu L, Zeng G, Xu P, Huang C, Deng L, Wang R and Wan J 2017 The effects of rice straw biochar on indigenous microbial community and enzymes activity in heavy metal-contaminated sediment *Chemosphere* **174** 545-53
[21] Lal R 2006 *Encyclopedia of Soil Science*, Second Edition, II. (Boca Raton FL: Taylor & Francis)
[22] Warnock D D, Lehmann J, Kuyper T W and Rillig M C 2007 Mycorrhizal responses to biochar in soil-concepts and mechanisms *Plant and Soil* **300** 9-20
[23] Thies J E and Rillig M 2009 Characteristics of biochar: biological properties. In: Lehmann J, Joseph S (Eds.) *Biochar for Environmental Management: Science and Technology* (London: Earthscan) pp 85-105
[24] Ogawa M 1994 Symbiosis of people and nature in the tropics *Farming Japan* **28** 10–34
[25] Steiner C, Das K C, Garcia M, Förster B and Zech W 2008 Charcoal and smoke extract stimulate the soil microbial community in a highly weathered xanthic ferralsol. *Pedobiologia* **51(5-6)** 359-366
[26] McClellan T, Deenik J, Uehara G and Antal 2007 *Int. Ann. Meetings* (Louisiana: ASA-CSSA-SSA) http://ac-s.confex.com/crops/2007am/ techprogram/P35834. HTM
[27] Ogawa M and Yamabe Y 1986 Effects of charcoal on VA mycorrhizae and nodule formation of soybeans *Bulletin of the Green*. 108–133
[28] Hammer E C, Balogh-Brunstad Z, Jakobsen I, Olsson P A, Stipp S L S and Rillig M C 2014 A mycorrhizal fungus grows on biochar and captures phosphorus from its surfaces *Soil Biology & Biochem*. **77** pp 252-360
[29] Laborda F, Monistrol IF, Luna N and Fernandez M 1999 Processes of liquefaction/solubilization of Spanish coals by microorganisms *Applied Microb. and Biotech*. **52** pp 49–56
[30] Ascough P L, Bird M I, Meredith W, Wood R E, Snape C E, Brock F, Higham T F G, Large D, Apperley D, 2010 Hydropyrolysis: implications for radiocarbon pre-treatment and characterization of black carbon *Radiocarbon* **52** pp 1336–50
[31] Qiaoa J, Li X, and Li F 2018 Roles of different active metal-reducing bacteria in arsenic release from arsenic-contaminated paddy soil amended with biochar *J. Hazardous Materials*. **344** p958-67
[32] Saxena J, Rana G, and Pandey M 2013 Impact of addition of biochar along with Bacillus sp. on growth and yield of French beans *Scientia Horticulturae*. **162** pp 351-56
[33] Yang F, Zhou Y, Liu W, Tang W, Meng J, Chen W and Li X 2019 Strain-Specific Effects of Biochar and Its Water-Soluble Compounds on Bacterial Growth *Appl. Sci*. **9** 3209
[34] Balittra 2015 *Laporan hasil penelitian* (Banjarbaru: Balai Penelitian Pertanian Lahan Rawa)
[35] Hindersah R, Fitriatin B N and Setiawati M R 2003. *Proc. Int. Conf. Env. and Urban Management* 491-98
[36] Lenin G and Jayanthi M 2012 Efficiency of plant growth promoting rhizobacteria (PGPR) on enhancement of growth, yield and nutrient content of *Catharanthus roseus*. *Int J Res Pure Appl Microbiol* **2** 37–42
[37] Sembiring Y SV 2013 Kajian penggunaan mikroorganisme tanah untuk meningkatkan efisiensi pemupukan pada tanaman karet *J. Warta Perkaretan* **32(1)** 7 –15

[38] Güereña D T, Lehmann J, Enders E T, Karanja N and Neufeldt H 2015 Partitioning the contributions of biochar properties to enhanced biological nitrogen fixation in common bean *(Phaseolus vulgaris)* *Biol Fertil Soils* **51(4)** 479–91
[39] Yang X, Zhang S, Ju M and Liu L 2019 Preparation and modification of biochar materials and their application in soil remediation *Appl Sci* **9577** 1365

The utilization of agricultural waste as biochar for optimizing swampland: a review

A Susilawati*, E Maftuah and A Fahmi
Indonesian Swampland Agriculture Research Institute (ISARI), Indonesia
Jl. Kebun Karet- Loktabat- Banjarbaru- South Kalimatan- Indonesia

Email: ani.nbl@gmail.com

Abstract. The utilization of agricultural waste as biochar is one way to manage agricultural waste to encourage swamps optimization. There is quite a lot of agricultural waste in Indonesia estimated to reach 10.7 t y^{-1}, including rice-husks, cocoa pods, coconut-shells, oil palm shells, and corncobs. One of the management of agricultural waste is to make biochar from various sources of raw materials. This paper is intended to provide information about agricultural waste as biochar in agriculture to optimize swampland in Indonesia. The quality of biochar depends on raw materials, incinerator, combustion temperature, and the length of combustion. The application of biochar on agricultural land serves as a soil enhancer that can improve its chemical properties and soil physical properties. Improvements in the quality of the soil chemical and physical properties impact the availability of nutrients and water through the ability of biochar to retain nutrients and water. In the end, the addition of biochar has implications for increasing the productivity of food crops. In the future, it is expected that with the application of biochar, swamplands can increase productivity.

1. Introduction

Swampland is one of Indonesia's most extensive agroecology and has considerable potential for agricultural development now and in the future, but its utilization has not been carried out optimally. Tidal swamp land consists of mineral soil (sulfuric acid) reaching 7.55 million ha and peat reaching 1.37 million ha, while lebak swamp land consists of mineral land reaching 11.64 million ha and peat reaching 13.56 million ha [1]. However, swamps are generally marginal and infertile so proper management is needed. Swamp management requires specific technologies including improving soil conditions through amelioration. Farmers in swampland often use ameliorant materials such as lime, manure, and ash [2]. Proper management of swamps will make a real contribution to Indonesia's food security. However, along with the declining quality of the soil and the environment, efforts are needed to manage swampland wisely (wise use) based on sustainability and safe for the environment.

When this has begun to develop globally, biochar/charcoal from agricultural waste as an alternative soil enhancer. Biochar is produced through a process of thermal decomposition of organic matter under conditions of limited oxygen. Similar to charcoal produced from natural combustion, but different because of its function as an amendment to the ground [3,4]. Biochar can be produced from various biomass sources such as plant residues, plant waste or animal waste. The characteristics of biochar depend on the raw material and pyrolysis conditions such as temperature, time duration and oxygen supply [4]. Biochar has demonstrated the ability to improve soil chemical properties such as pH [5,6,7,8], nutrient availability [9], nutrient and water retention [10,11,12] and cation exchange capacity [10,6,7,13], reducing N fertilizer loss [14] and reducing nutrient leaching [15,16]. Also, on soil physical properties: increasing water holding capacity and available water [6], reducing soil density and strength [6] and can change the structure and abundance of soil biological communities [17]. Biochar can last a long time in the soil or has a relatively long effect, or is relatively resistant to microorganism attack, so the decomposition process is slow. Biochar is also able to reduce the washing of pesticides and nutrients and ultimately impact the quality of the environment. Biochar did not disturb the carbon-nitrogen balance, but it can hold and makes the water and nutrients more available to plants. Biochar structures with micropores can increase water holding capacity [18,19,20], and reduce nitrogen run off [21,22].

Biochar can be used to manufacture biological fertilizers because it is an effective carrier for microbes needed in the soil [18]. In the biochar micropore, the fungus can sporulate because the competition is relatively low with other saprophytes [23]. The diversity of microorganisms in the soil can be improved by the application of biochar [8]. The role of biochar in the soil can be a habitat for soil microbes, but it is not consumed by mycorrhizae in a short time so that the applied biochar can stay in the soil for a long time.

Biochar production is agricultural waste management that can be developed considering the abundant raw material sources. In the future, it is hoped that the optimization of swamps can be done by utilizing agricultural waste available in situ. This paper aims to provide information about the potential use of agricultural waste as biochar in agriculture, significantly to optimize swampland in Indonesia and describe the role of biochar in improving the quality of swamps, namely chemical properties, physical properties, soil biology, and land productivity.

2. Potential of agricultural waste biochar raw materials

Organic materials derived from living bodies, both animals and plants, are widely used as raw materials for agriculture, animal husbandry, construction, and so on, eventually producing waste products. This biomass waste will become a problem if it is not appropriately utilized, and eventually it will become waste that is not useful and can pollute the environment. The direct utilization of biomass waste is considered inefficient and needs to be converted into biochar first [24].

Biochar is a new term used to describe porous charcoal (finely pulverized charcoal) made from various biomass. Charcoal result of incomplete combustion with limited oxygen or no oxygen-called biochar [25,26,27]. Basic ingredients can be sourced from agricultural, livestock, forestry and household waste [28] which is abundant agricultural waste such as rice husk, rice field 1 ha with an average yield of 7 t^{-1} rice will yield about 1.54 tons of agricultural waste rice hulls for every harvest season [27].

The potential of biochar raw materials in Indonesia is relatively abundant in agricultural residual waste, especially those that are difficult to decompose or with high C / N ratios such as wood residues, coconut shells, rice husks, cocoa pods, corncobs, sufficiently available. Thahir [29] informed that the rice milling process could produce 16.3-28.0% of the husk every time. The coconut shell's potential is also enormous considering the coconut plant area reaches 3.79 million hectares with copra equivalent production of 2.94 million t^{-1} [30], where the proportion of shells from one coconut is around 15-19%. The area of cocoa plants is around 1.85 million ha [30] with the production of dried fruit skins around 830 thousand tons/year [31] and the proportion of corn cobs 21% of the weight of dried cobs [32]. So far, agricultural waste has not been used optimally, only limited to being used as a renewable energy source [33] and animal feed.

Results analysis of Sarwani [34] informs that nationally, the potential of agricultural biomass that can be converted into biochar is estimated to be around 10. 7 million tons, producing 3.1 million tons of biochar. The highest potential comes from rice husk which reaches 6. 8 million t y^{-1} and will produce biochar1. 77 million t y^{-1} or around 56. 48% of the total national biochar potential (table 1). The high potential of biomass to be used as biochar is highly dependent on availability and competition with other uses. Coconut shell biomass is widely used for other purposes, namely energy sources, whole corn cobs can be used for animal feed. Other sources that can be utilized are the rest of the twigs, cassava stems, oil palm empty fruit bunches, mangrove stems. In principle, the source of biochar raw material is agricultural waste.

Table 1. Estimated agricultural waste and it is potential as a biochar raw material.

Agricultural biomass	Total ($t\ y^{-1}$)	Assuming the proportion of biomass that can be converted (%)	Potential of biomass that can be converted into biochar ($t\ y^{-1}$)	Biochar / biomass ratio	Biochar potential ($t\ y^{-1}$)
Rice husk	13 612 343	50	6806172	0.26	1769605
Coconut shell	539. 644	50	269822	0.25	67456
Oil palm shell	6400 000	30	1920000	0.5	960000
Cocoa pods	1. 208553	50	604277	0.33	199411
Corn cobs	3652372	30	1095712	0.13	142443
Total	25412912		10695982		3138914

Source: Sarwani (34)

In swamps, the raw materials for making biochar with high levels of abundance include rice husks, corn straw, rice straw, kalakai, karamunting, galam, bamboo, palm oil cake, palm leaves, palm fronds, oil palm bunches, coconut shells and purun tikus. The characterization and analysis of cellulose content, hemicellulose, and C / N ratio of each biochar material are presented in table 2. The highest lignin levels are found in raw materials kalakai (35.76%) and the lowest in corn straw (17.03%). High lignin content makes it difficult for organic matter to decompose [35]. The highest cellulose content was found in organic matter from oil palm cake (34.45%) and the lowest was in organic matter from corn straw (17.03%), in contrast to hemicellulose content. Hemicellulose is an amorphous polymer associated with cellulose and lignin, easy to undergo depolymerization, hydrolysis by acids, are alkaline, and easily dissolve in water. Hemicellulose has a bond with lignin stronger than a bond with cellulose and easily binds to water. Hemicellulose levels differ in the type of needle-leaf wood and broad-leaf wood [36].

Table 2. Cellulose content, hemicellulose, and C/N ratio of some organic biochar sources.

Type of material	Content (%)					C/N
	Lignin	Cellulose	Hemicellulose	C	N	
Karamunting	27.7	50.8	20.5	52.6	1.1	46.9
Kalakai	35.7	42.6	21.1	53.8	1.1	48.1
Corn straw	17.0	34.0	37.6	54.8	1.1	48.9
Palm leaf	30.0	42.9	26.3	50.4	0.9	51.5
Rice husks	32.6	43.4	23.1	49.0	0.8	58.4
Purun tikus	26.3	43.8	19.1	50.6	0.8	60.3
Rice straw	24.4	40.4	30.7	51.1	0.8	60.9
Oil palm bunches	25.6	49.6	20.6	55.4	0.7	79.2
Bamboo	24.8	50.1	24.5	56.4	0.5	112.0
Palm midrib	25.4	50.1	24.3	55.9	0.4	124.9
Coconut shell	31.8	46.3	19.6	55.5	0.4	132.1
Oil palm cake	34.4	51.5	14.0	50.5	0.3	138.9
Galam	25.6	51.0	23.3	57.5	0.3	171.2

Sources: Maftuah dan Nursyamsi [37]

One way to increase the effectiveness and efficiency of biomass waste is by carbonization using pyrolysis technology. A pyrolysis is a complex event, where organic compounds in biomass are decomposed by heating without oxygen. Only the volatile matter is released, while the carbon remains in it, this process occurs at temperatures above 300 ° C for 4 - 7 hours [38].

The chemical and physical properties of biochar determine the quality of biochar (table 3), and these properties are determined by the type of raw material (softwood, hardwood, rice husk, etc.) and carbonization method (The type of combustion tool, temperature), and the form of biochar (solid, powder, activated carbon) [39,40]. The characteristics of biochar are influenced by the base material itself and have different effects on soil and plants [27].

Table 3. Characteristics of biochar physical-chemical properties.

Type of biochar	Pyrolysis temperature (°C)	C	pH	CEC Cmol /kg	SiO_2	Ash content	N	P	K	Ca	Mg	Fe
					%							
Rice husk[1)]	300-400	23.40	8.99	37.38	34.83	44.35	0.73	0.48	0.54	0.21	0.18	0.26
Palm bunches[1)]	350-450	42.33	9.39	9.93	4.90	27.09	0.99	0.49	8.65	0.43	0.67	0.53
Bamboo[1)]	350-450	50.03	9.30	9.30	3.21	11.26	1.04	0.45	3.18	0.16	0.13	0.16
Palm midrib[1)]	300-450	40.04	9.74	9.74	6.30	31.17	1.01	0.39	1.53	0.53	0.96	0.15
Coconut shell[1)]	250-350	29.69	9.61	9.61	4.04	48.96	1.28	0.52	2.96	0.29	4.43	0.29
Palm cake [1)]	350-500	23.73	8.30	8.30	3.90	59.32	0.87	0.44	0.72	0.09	0.30	0.04
Galam stem[1)]		45.06	9.04	9.04	2.13	23.15	0.54	0.42	1.14	0.43	0.05	0.14
Cocoa pods [2)]	250-350	33.04	10.8	-	-	-	0.83	0.33	11.25	-	-	-
Oil palm shells [2)]	250-350	49.18	8.2	-	-	-	1.61	0.25	0.04	-	-	-
Branch legume tree [3)]	-	18.11	9.4	7.05	-	-	0.58	0.1	1.11	-	-	-
Palm shells [4)]	500	25.62	-	4.58	-	-	1.32	0.07	0.08	-	-	-

Sources: [1)] Maftuah and Sosiawan [41]; [2)] Nurida [42]; [3)] Dariah [43]; [4)] Santi dan Goenadi [44]

Interest in biochar as a soil reformer has recently overgrown. In biochar production biochar, pyrolysis slow to have multiple benefits that include waste management, renewable energy production, climate change mitigation and adaptation, as well as agricultural productivity [45].

3. The Role of Biochar in Optimizing Swampland

Swamplands are lands throughout the year or during a long time in a year, always saturated with water or flooded [46]. Swamp is included in suboptimal land because it has low productivity due to internal and external factors [47]. Utilization of sub-optimal land, including swamps, will be the foundation of future expectations, but requires technological innovation to overcome obstacles following the land's characteristics and typology.

Tidal swamps are swamps that are affected by tides. Based on soil classification [48], tidal swamps are characterized by aquatic conditions (saturated with water) and have sulfidic material (iron sulfide) or pyrite, generally reacting with extreme acidity (pH <4) so it is often called acid sulphate soil [46]. This land generally has a low level of fertility and productivity so that agricultural development requires technological input.

The freshwater swamp is a swampland that is not influenced by tides but is inundated by river overflows for at least three months with a pool of at least 50 cm [46]. The chemical nature of the soil in the swamp swamplands is very dependent on the type of soil. Mineral soil (river sediment) has a clay texture and pH of 4.5 to 6.5. Every year the lebak land gets silt from the upper area (upstream area), so that the soil fertility is classified as moderate. Generally, N, P, and K values are low to moderate, but Ca and Mg content and CEC are generally moderate to high. The high or low nutrient content is influenced by the amount of nutrient contribution from the upstream area that enters through water run-off. This land is suitable for agricultural business. The problem only lies in the dynamics of water levels that are difficult to predict.

Peatlands consist of mostly organic matter which is weathered, with ash content equal to or less than 35%, peat depth equal to or more than 50 cm, and organic carbon content according to its weight of at least 12% [49]. Peatland development for agriculture faces biophysical land, socioeconomic and environmental constraints. Biophysical land conditions that often arise are subsidence, dry back, acidification, nutrient deficiency, low load-carrying capacity, and high porosity.

The utilization of biochar in swamps as ameliorant has not been done much. Biochar is a carbon-rich solid material formed through the combustion process of organic or biomass materials without or with little oxygen (pyrolysis) at 250-500 °C. Unlike organic matter, biochar is stable for thousands of years when biochar is mixed into the soil and can absorb carbon [25,50,51].

Biochar is a pyrolysis residue in charcoal that contains high carbon and is useful for agriculture especially, for improving physical, chemical, and biological soil properties. The addition of biochar can improve soil fertility and restore degraded soil quality [52,10]. Biochar can improve the soil by increasing pH, retaining water, retaining nutrients, increasing biota activity, and reducing pollution [11]. The availability of P, soil pH, K, and Ca-dd increases with biochar application biochar [6].

Biochar can increase plant productivity directly or indirectly. The influence of biochar directly through the donation of nutrients released, while indirectly through the improvement of soil buffering capacity in holding nutrients [11,53,12], increasing soil pH [8,5,7], and the increase in CEC, improvement of soil physical properties and their effects on microbial function and population. Biochar can reduce nutrient loss through leaching, to improve fertilizer efficiency [15,16]. The provision of biochar can reduce N fertilizer losses [14,21,22,54]. This also applies to P nutrients that are not retained by ordinary organic matter. Biochar is more able to survive in the soil than other organic materials [25]. The biochar function works, absorbs and binds water, and supplies calcium and magnesium elements to plants.

A good combination of air and water for soil microbial growth is in biochar. Useful carrier material for the microbes also exists in biochar, which can make biological fertilizer [18], biochar micropores, fungi can sporulate because low competition occurs with other saprophytes [23]. Nugraheni [57] added that biochar could increase soil fertility and provide suitable habitat for soil microbes that can even hold water and nutrients to be more available to plants.

Masulili [6] applied biochar to acid sulphate soils. It reduced Al-dd, dissolved Fe, increase porosity, pH, P, CEC, Ca-dd, and K-dd. The combination treatment of biochar and chicken manure can increase the yield of Inpara-3 rice. Improvement of soil chemical properties through biochar and other ameliorant administration also impacts increasing rice yields in acid sulphate soils [6].

According to Ferizal [56] the provision of biochar as a soil enhancer derived from the combustion of agricultural waste with limited oxygen has excellent ability as a material change in soil because organic C still survives in carbon black. Biological charcoal produced from this combustion will produce activated carbon, which contains minerals such as calcium or magnesium and inorganic carbon, biochar is widely used as ameliorant material to improve soil quality, especially marginal soils [57]. So that biochar can be a substitute for growing media in plants. The biochar structure, which has a micropore can increase water's holding capacity [18,19,20].

Soil chemical properties sub-optimal land can be improved by applying biochar, including acid sulphate tidal swamps and peatlands. Biochar can increase soil pH, total N, available P, and K-dd in an acid dry land, dry climate dry land, tidal swamps, and peatlands (table 4).

Table 4. The influence of biochar on soil chemical properties in swampland.

Treatment	pH H_2O	N total %	P tsd ppm	K-dd me/100 g	KTK me/100 g	Type of land	Reference
Control	4.84	0.11	26.55	0.37	6.29	Tidal swampland	Barus and Santri [58]
Rice husk biochar, 5 t / ha	5.13	0.13	27.87	0.43	7.60		
Rice husk biochar, 19 t / ha	5.22	0.14	30.15	0.48	8.34		
Control	4.18	0.93	120.19	0.53	-	peatlands	Maftuah and Nursyamsi [59]
Galam stem biochar 8 t / ha	4.24	0.95	156.58	0.59	-		
Rice husk biochar 8 t / ha	4.19	0.95	262.14	0.96	-		
Palm midrib biochar 8 t / ha	4.20	0.95	263.09	0.95	-		
Coconut shell biochar 8 t / ha	4.29	1.26	175.59	0.94	-		

Note: -; data not available

The increase in crop production occurs due to the application of biochar in swamps (table 5). Application of rice husk biochar increases rice yield in tidal land reaching 4.35 - 5.36 t / ha [58]. On peatlands, there was an increase in corn yield due to the administration of rice husk biochar to 59% [60]. Biochar also provides increased soybean crop production in sub-optimal land. Giving rice husk biochar 10 t / ha increases the dry weight of soybean seeds to reach 2.5-2.9 t/ha [61].

Table 5. Effect of biochar on swampland productivity.

Treatment	Corn dry pipil (t/ha)	Rice dry grain (t/ha)	Increased (%)	Type of land	Reference
Without biochar	-	5.25		Tidal swampland (acid sulphate soil)	Barus [62]
Rice husk Biochar, 2 t / ha	-	5.36	2		
Without biochar	3.59	-		Peatland	Simatupang [60]
Rice husk Biochar, 7.5 t / ha	5.71	-	59		
Coconut shell Biochar, 7.5 t / ha	4.29	-	19.5		

4. Conclusion

One effort to manage agricultural waste is by making biochar. Biochar utilization opportunities for agricultural land are tremendous, both in terms of the availability of raw materials and their functions. Biochar application can be a solution in the use and optimization of swampland. Biochar in swampland plays a role in improving soil properties both, physical, chemical, and biological so that land productivity can increase. Plant productivity also increases in line with the improvement in land quality. The effectiveness of biochar in improving soil properties is influenced by biochar type, biochar manufacturing process, dosage, application method, biochar size, and soil type. The use of biochar is an alternative effort to improve soil fertility and improve the environment in a sustainable and easy and inexpensive way.

Reference

[1] Nawangsih A A, Purwanto H, dan Agung W 1999 *Budidaya Cabai Hot Beauty* Cetakan kedelapan (Jakarta: Penebar Swadaya)

[2] Ritung S, et al 2015 *Sumberdaya lahan pertanian Indonesia: Luas, Penyebaran, dan Potensi Ketersediaan* (Jakarta: IAARD Press) p 97

[3] Maftu'ah E 2012 *Ameliorasi Lahan Gambut Terdegradasi dan Pengaruhnya terhadap Produksi Tanaman Jagung Manis* [Disertasi] Universitas Gadjah Mada Yogyakarta.

[4] Lehmann J and Joseph S 2009 *Biochar for Environmental Management.* 1st ed (London: Earthscan) p 416

[5] Sohi S, Lopez-Capel E, Krull E and Bol R 2009 *Biochar, climate change and soil: a review to guide future research CSIRO Land and Water Science Report* 05/09 February 2009

[6] Topoliantz S, Ponge J F and Ballof S 2007 Manioc peel and charcoal: a potential organic amendment for sustainable soil fertility in the tropics *Biol Fert Soils* **41** 15–21

[7] Masulili A, Utomo W H and Syechfani M S 2010 Rice husk biochar for rice based cropping system in acid soil: The characteristics of rice husk biochar and its influence on the properties of acid sulphate soils and rice growth in West Kalimantan, Indonesia *J. Agric. Sci.* **2 (1)** 39-47

[8] Yuan J H, Quan W R K and Wang R H 2011 Comparation of ameliorating effect on an acidic ultisol between four crop straw and their biochars *J. soil and Sediment* **11(5)** 741-50

[9] Li-li He, Zhe-ke Z and Hui-min Y 2016 Effects on soil quality of biochar and straw amendment in conjunction with chemical fertilizers *J. Integrative Agric.* **15(0)** 60345–7

[10] Chan K Y, Van Zwieten L, Meszaros I, Downie A and Joseph S 2008 Using poultry litter biochars as soil amendments *Aust J Soil Res.* **46** 437–44

[11] Glaser B, Lehmann J and Zech W 2002 Ameliorating physical and chemical properties of highly weathered soils in the tropics with charcoal: A review *Biol Fertil Soils* **35** 219-30
[12] Laird D A, Fleming P, Davis D D, Horton R, Wang B Q and Karlen D L 2010 Impact of *biochar* amandement on quality of typical Midwestern Agricultural Soil *Geoderma* **158 (3-4)** 443-9
[13] Zhang T, Fang C, Li P, Jiang R and Nie H 2013 Application of biochar for phosphate adsorption and recovery from wastewater *Adv. Materials Res.* **(750-752)** p 1389–92
[14] Van Zwieten L, Kimber S, Morris S, Chan KY, Downie A, Rust J, Joseph S and Cowie A 2010 Effect of biochar from slow pyrolysis of papermill waste on agronomic performance and soil fertility *Plant and Soil* **327** 235-46
[15] Steiner C, Teixeira WG, Lehmann J, Nehls T, de Macêdo J L V, Blum W E H and Zech W 2007 Long term effects of manure, charcoal and mineral fertilization on crop production and fertility on a highly weathered Central Amazonian upland soil *Plant soil* **291**: 275-90
[16] Novak J M, Busscher W J, Laird D L, Ahmedna M, Watts D W and Niandou M A S 2009 Impact of biochar amendment on fertility of a southeastern coastal plain soil *Soil Sci* **174** 105–12
[17] Yao Y, Gao B, Zhang M, Inyang M and Zimmerman A R 2012 Effect of biochar amendment on sorption and leaching of nitrate, ammonium, and phosphate in a sandy soil Chemosphere **89** 1467–71
[18] Pietikainen J, Kiikkila O and Fritze H 2000 Charcoal as a habitat for microbes and its effect on the microbial community of the underlying humus *Oikos* **89** 231–42
[19] Santi L P and Goenadi D H 2010 Pemanfaatan bio-char sebagai pembawa mikroba untuk pemantap agregat tanah Ultisol dari Taman Bogo-Lampung *Menara Perkebunan* **78(2)** 11-22
[20] Sutono S, Nurida N L 2012 Kemampuan biochar memegang air pada tanah bertekstur pasir *Buana Sains* **12 (1)** 45–52
[21] Yu O K, Brian Y R and Sam S 2013 Impact of biochar on the water holding capacity of loamy sand soil. *Int. J. Energy and Env. Eng.* **4** 44
[22] Clough T J, Condron L M, Kammann C, Müller C 2013 A Review of Biochar and Soil Nitrogen Dynamics. *Agronomy* **3** 275–93
[23] Knowles O A, Robinson B H, Contangelo A, Clucas L 2011 Biochar for the mitigation of nitrate leaching from soil amended with biosolids *Science of the Total Environment* **409** 3206-10
[24] Saito M, Marumoto T 2002 Inoculation with arbuscular mycorrhizal fungi: the status quo in Japan and the future prospects *Plant and Soil* **244** 273-9
[25] Bridgwater A 2003 Renewable fuels and chemicals by thermal processing of biomass *Chem. Eng. J.* **91** 87-102
[26] Lehmann J 2007 Bio-energy in the black *Front. Ecol. Environ.* **5(7)** 381–7
[27] Liang B J, Lenham D, Solomon S, Sohi J E, Thies J O, Skjemstad F J, Luizao M H, Engelhard, E G, Neves and Wirick 2008 Stability of Biomass drived Black Carbon in Soil *Geochimika et Cosmochimica* A. **72** 6069-6078.
[28] Gani A 2009 Arang Hayati Biochar Sebagai Komponen Perbaikan Produktivitas Lahan *Iptek Tanaman Pangan* **4(1)** 33-48.
[29] Shenbagavalli S and Mahimairaja S 2012 Production and characterization of biochar from different biological wastes *Int. J. Plant Animal and Environ. Sci.* **2 (1)** 197-201
[30] Thahir R Rachmat R and Suismono 2008 Pengembangan Agroindustri Padi *Dalam* Suyamto dkk (*Ed*) *Padi: Inovasi Teknologi dan Ketahanan Pangan* (Subang: Balai Besar Penelitian Padi) p 34 -76.
[31] Badan Pusat Statistik 2013 *Statistik Indonesia* (Jakarta: BPS) p 668
[32] Direktorat Jenderal Bina Produksi 2007 *Statistik Perkebunan Indonesia* (Jakarta: Dirjen Perkebunan Deptan)

[33] Nurida NL, A Rachman dan Sutono 2012 Potensi pembenah tanah biochar dalam pemulihan sifat tanah terdegradasi dan peningkatan hasil jagung pada Typic Kanhapludults lampung *Jurnal Penelitian Ilmu-Ilmu Kelaman* **12 (1)** 69-74
[34] Okimori Y, Ogawa M and Takahashi F 2003 Potential of CO_2 reduction by carbonizing biomass waste from industrial tree plantation in South Sumatra, Indonesia *Mitigation and Adaption Strategies for Global Change* **8** 261-80
[35] Sarwani M, Nurida N and Agus F 2013 Greenhouse emissions and land use issues related to the use of bioenergy in Indonesia *Jurnal Penelitian dan Pengembangan Pertanian* **32(2)** 56-66
[36] Pratikno H 2002 *Studi Pemanfaatan Berbagai Biomasa Flora untuk Peningkatan Ketersediaan P dan Bahan Organik Tanah pada Tanah Berkapur di DAS Brantas Hulu Malang Selatan* [Tesis] Universitas Brawijaya, Malang
[37] Achmadi S 1990 *Kimia Kayu* (Bogor: Institut Pertanian Bogor)
[38] Maftuah E and Nursyamsi D 2015 *Prosiding Sem. Nas. Masy. Biodiversitas Indonesia* (Yogyakarta: ISSN 2407-8050) **1(4)** 776-81
[39] Demirbas A 2005 Pyrolysis of Ground Wood In Irregular Heating Rate Conditions *J. Analyt. and Appl. Pyrolysis* **72** 243-8
[40] Ogawa, M 2006 Carbon sequestration by carbonization of biomass and forestation: three case studies mitigation and adaptation strategies for global change *Springer* **11(2)** p 421-36
[41] Amonette J and Joseph S 2009 Characteristics of Biochar: Micro-Chemical Properties In: *Biochar for Environmental Management: Science and Technology*, Lehmann, J. and S. Joseph (ed). (London: Earthscan-UK) p 33–52
[42] Maftuah E and Sosiawan H 2018 Biochar: ameliorant material that environmentally friendly for swampland *Proc. Int. Workshop and Seminar Innov. Environ.-Friendly Agricultural Tech. Supp. Sustainable Food Self-Sufficiency* (Surakarta: ISBN 978-602-344-252-2)
[43] Nurida N L, Dariah A and Rachman A 2009 Kualitas limbah pertanian sebagai bahan baku pembenah berupa biochar untuk rehabilitasi lahan *Pros. Sem. Nas. dan Dialog Sumberdaya Lahan Pertanian* (Bogor: BPPP_Deptan) 209-15
[44] Dariah A, Nurida N L and Sutono 2013 The effect of biochar on soil quality and maize production in upland in dry climate region *Proc. 11th Int. Conf. The East and Southeast Asia Fderation of Soil Sci. Societies* (Bogor: Society of Soil Science)
[45] Santi L P and Goenadi D H 2012 Pemanfaatan biochar cangkang sawit sebagai pembawa mikroba pemantap agregat *J. Penelitian Ilmu-Ilmu Kelaman* **12 (1)** 7-14
[46] Chan K Y, Van Zwieten L, Meszaros I, Downie A and Joseph S 2007 Agronomic value of greenwaste biochar as a soil amandment *Australian J. Soil Res.* **45** 629-34.
[47] Subagyo H *2006 Lahan Rawa Lebak in Karakteristik dan Pengelolaan Lahan Rawa*, (Bogor: Balai Besar Litbang Sumberdaya Lahan Pertanian) p 99–116
[48] Mulyani A and Sarwani M 2013 Karakteristik dan potensi lahan suboptimal untuk pengembangan pertanian di Indonesia *Jurnal Sumberdaya Lahan* **2** 47–56
[49] Soil Survey Staff 1999 *Soil Taxonomy A Basic System for Making and Interpreting Soil Surveys* 2nd Ed. (USA: USDA-NRCS Agric Handb.) 436
[50] Soil Survey Staff 2011 *Soil Taxonomy a Basic System of Soil Classification for Making and Interpreting Soil Surveys* 11th Ed. (United States: Depart.Agriculture Washington DC) p 754
[51] Renner R 2007 Rethinking biochar environ *Sci. Technol.* **41** 5932–33
[52] Fraser B 2010 High-tech charcoal fights climate change *Environ. Sci. Technol.* 548.
[53] Atkinson C J, Fitzgerald J D and Hipps N A 2010 Potential mechanisms for achieving agricultural benefits from biochar application to temperate soils: A review *Plant Soil* **337** 1–18
[54] Schulz H and Glaser B 2012 Effects of biochar compared to organic and inorganic fertilizers on soil quality and plant growth in a greenhouse experiment *J Plant Nutr Soil Sci* **175** 410–22
[55] Gao S and DeLuca T H 2016 Influence of biochar on soil nutrient transformations, nutrient leaching, and crop yield *Adv Plants Agric Res* **4(5)** 1–16

[56] Nugraheni S R, Prasetya A, and Sihana 2013 Processing Biochar from Solid Waste of Arenga Pinnata Flour Industry *J. Tek. Kim.* **11(1)** 31-6

[57] Ferizal M and Basri A B 2011 *Arang Hayati (Biochar) Sebagai Pembenah Tanah* (Aceh: BPTP) p 2

[58] Rondon M J, Renner, Lehmann R, Ramírez J, and Hurtado M 2007 Biological nitrogen fixation by common beans (*Phaseolus vulgaris* L.) increases with biochar additions *Bio. and Fert. in Soils* **63(6)** 699-708

[59] Barus J and Santri N 2016 *Pros. Sem. Nas. Asosiasi Biochar Indonesia* (Pontianak: ISBN) 978-602-72935-2-6

[60] Maftuah E, Nursyamsi D 2019 Effect of Biochar on Peat Soil Fertility and NPK Uptake by Corn *AGRIVITA J. Agricult. Sci.* **41(1)** 64–73

[61] Simatupang R S, Maftu'ah E and Subagio H 2017 Pengaruh pemberian formula bahan amelioran terhadap pertumbuhan dan hasil jagung di lahan gambut In Fahmuddin A.(Eds) *Pros. Kongr. Nas. Perkumpulan Masy. Gambut Indonesia (HGI)* VII ed. (Bogor: BBSDLP ISBN) 978-602-459-049-9

[62] Endriani and Kurniawan A 2018 Konservasi tanah dan karbon melalui pemanfaatan biochar pada pertanaman kedelai *J. Ilmiah Ilmu Terapan Univ. Jambi* **2 (2)** 93–106

[63] Barus J 2016 Utilization of crops residues as compost and biochar for improving soil physical properties and upland rice productivity *J. Degraded Min Lands Manag* **3(4)** 631–7

IOP Conf. Series: Materials Science and Engineering **980** (2020) 012066 doi:10.1088/1757-899X/980/1/012066

Alternate wetting and drying system (AWD) combined with farmyard manure to increase rice yield and reduce methane emission and water use

N Al Viandari*, T A Adriany and A Pramono
Indonesian Agricultural Environment Research Institute. Jalan Raya Jakenan-Jaken Km 5 Pati Central Java Indonesia

Email: nourmaviandari@gmail.com

Abstract. Rice cultivation in flooded conditions is one of the methane emission sources. The objective of this study was to determine the effect of water management and farmyard manure application on grain yield, CH4 emissions, and water use efficiency. This study was conducted at the Indonesian Agricultural Environment Research Institute, Pati Central Java from November 2019 to March 2020. The treatments consisted of two factors, a) water management, namely 1) continuous flooding, at 5 cm (CF) and 2) Alternate Wetting-Drying, at 15 cm (AWD) system; and b) farmyard manure namely 1) farmyard manure application 2) without farmyard manure. The observed parameters were rainfall and air temperature, CH4 emissions, water level and milled dry grain. The measurement of CH4 gas emissions was conducted using a closed chamber method. The result of this study indicated that AWD management could increase grain yield by 3.3-8.5% and also reduce CH4 emissions by 12-14% compared to continuous flooding. AWD combined with organic fertilizer application produced the highest grain 7.27 tons ha-1 and emitted the lowest CH4. The AWD management also saved 32-38% of water compared to the continuous flooding.

1. Introduction

Global food demand is increasing annually. The ways to suffice the demand and avoid food scarcity by keeping food security and ensuring food availability while climate change significantly impact natural resources. The land management strategies are needed to increase the rice yield and reduce the impact of climate change since the agricultural sector contributes to emitting greenhouse gases through methane emissions in rice cultivation. Rice cultivation emits methane, especially in flooding conditions.

Plant and soil ebullition are factors that cause a large number of CH_4 released from rice fields into the atmosphere. 35%-62% of methane is emitted through soil ebullition [1]. Methane is the result of anaerobic microbial decomposition of organic matter by methanogenic bacteria through methylotrophic, hydrogenotrophic, and acetoclastic routes [2]. Moreover, the aerenchyma vessels found in well-developed leaves, stems, and roots also cause gas exchange in anaerobic soil conditions [3]. In the long run and if not surmounted, it will cause global warming and impact climate change.

One of the obvious impacts of climate change is weather chaos that causes drought. In several areas, they have shorter periods of rain and less rainfall even in the coming rainy season so that drought has caused water scarcity whereas water is one of the primary needs for rice irrigation which is the largest water consumer in the agriculture sector [4]. An alternative way to reduce greenhouse gases in rice cultivation, especially CH_4, is the use of water management. Despite being a primary need, not all phases of rice growth require abundant water. At certain times, plants need a lot of water especially during the stages of seedling and panicle filling. In addition, the soil also needs to be desiccated to create aerobic conditions so that the roots can respire and microorganism's activity can increase soil fertility [5].

To regulate drying and wetting irrigation in rice cultivation, there is a developed and applied technology such as Alternate and Wetting Drying (AWD) system. AWD is a water saving management technique in irrigated rice cultivation that was initiated by the International Rice Research Institute (IRRI). At the beginning of the water management, the land is flooded and the water level will decrease naturally for several days (up to 10 days) or reach a certain level below the land surface [6]. Water conservation by the AWD technique in rice production can reduce CH_4 emissions [7]. Water use in producing rice can be cut down by as much as 25% using AWD. However, this water conservation not only reduces CH_4 emission by 11-97% but also decreases the yield by 13% - 33% [8]. Therefore, there need to be great techniques to reduce the impact of lowering yield. Reducing water consumption can be applied in the rice field without affecting the yield.

Kato [9] suggested that there are many factors that cause yield reduction; some of them are climate and soil properties (physical and chemical soil properties). The factors also include drainage frequency, duration of flooding, and water shortage situation. These conditions affect crop growth. Therefore, by adding organic matter, the plant can grow well without losing yield. It is also claimed that increasing organic matter contributes directly and affects soil properties. Organic matter not only helps stabilize soil aggregates and pores by binding to organic matter and soil organisms, but also very well and thoroughly affects the increasing the capacity of soil moisture mixed with mineral soils, especially in the top layer of soil with more organic material content and more water can be stored. The objective of this study was to determine the effect of water management and organic matter addition on grain yield, CH_4 emissions, and water use efficiency.

2. Materials and methods

2.1. *Time and location of the study*

This study was conducted at the Jaken Experimental Station, Indonesian Agricultural Environment Research Institute, Pati Central Java from November 2019 to March 2020 in the rainfed condition. The rainfall conditions are presented in figure 1. Soil Classification was silt loam, Aeric Endoaquepts. Endoaquepts soil type has the characteristics of a topsoil with a pH of 5.7, 34% sand texture, 56% dust, 10% clay, organic C 0.18%, N total 0.05%, C ratio / N 3.39, total P 108.3 ppm, total K 319.1 ppm, CEC 8.69 cmol kg-1, Micro element Fe 0.06%, Ca 0.51% and Mg 0.04% [10].

2.2. *Research design*

The treatments consisted of two factors. A) water management: 1) continuous flooding, at 5 cm (Cf) and 2) Alternate Wetting-Drying, at a 15 cm (Aw) System. B) providing organic matters treatment: 1) by giving farmyard manure (Or) 2) without giving farmyard manure (No). The daily water level measurement was done at 4:00-5:00 pm using piezometers which were made from PVC pipe and installed on each plot. The four treatments were laid out in Completely Randomized Block Design with four replications. The blocks of the experiment were 6x6 m. Plastic sheets set around 40 cm soil depth was used in each block to prevent water permeation.

A rice variety, Inpari 32, was planted at 20x20 cm. Farmyard manure was given about 3-ton ha^{-1} and applied during or after tillage and also 120 kg of N, 60 kg of P_2O5, and 90 kg of K_2O were applied in each block. Pests, diseases, and weeds were appropriately controlled.

2.3. Measurements

The observed parameters were rice yield, methane flux (CH_4), and surface water level. The closed chamber method was applied to measure CH_4 fluxes. The chamber made of plexiglass had 50: 50:100 of length, width, height. A 20 ml syringes were used to take the gas with 0, 10, 20, 30 minutes intervals. Gas samples were taken from 7 am to 9 am and stored in 10 ml vacuumed vials. Gas sample concentrations were analyzed by Gas Chromatography (GC) type Variant 450-GC (Varian Inc., CA, USA) equipped with a flame ionization detector (FID) for CH_4 analysis. Calculating methane emissions was done using the trapezoidal integration method formula [11], as seen in equation 1:

$$E = \frac{dc}{dt} x \frac{Vch}{Ach} x \frac{mW}{mv} x \frac{273{,}2}{273{,}2 + T} \tag{1}$$

E is flux (mgm^{-2} hari-^{1}), dc/dt is difference in CH_4 concentration per time (ppm minute^{-1}), Vch is Box volume (m^3), Ach is Box Area (m^2), mW is Molecul weight (g), mV is Molecule volume (22.4 l) and T is Average temperature (°C).

Rice yield (14% moisture content) was measured from 6 m^2 yield sampling in each block. AWD system was applied based on the daily measurement of the water surface level. The flow rate in the pipe was multiplied by the time of irrigation used to measure irrigation water usage (ton ha^{-1}). That measurement was applied to reflood the field and keep the water surface level of 15 cm.

3. Results and discussions

3.1. Grain yield

The highest average of rice yield was obtained by Aw-Or treatment of 7.27 ton ha^{-1} during one growing season. Each treatment caused significantly different yields (table 1). The lowest rice yield was obtained by continuous flooding and without providing farmyard manure. AWD system and farmyard manure were factors that significantly affected the gained yield.

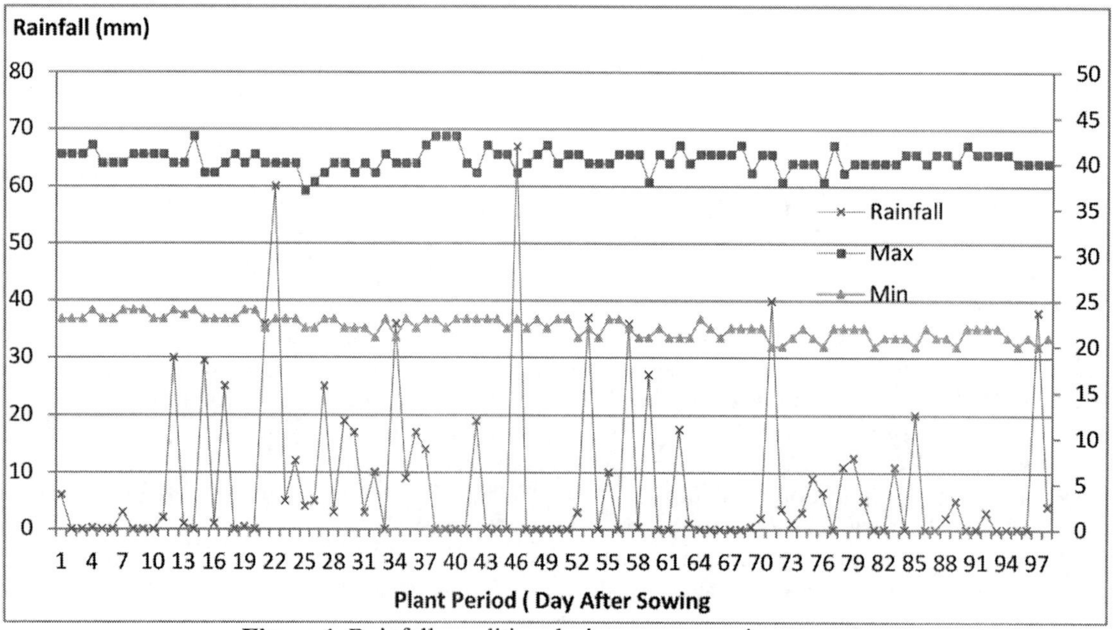

Figure 1. Rainfall condition during one a growing season.

IOP Conf. Series: Materials Science and Engineering **980** (2020) 012066 doi:10.1088/1757-899X/980/1/012066

Table 1. Rice yield in various treatments for water management combined with providing organic matter.

Treatment	Dry Milled Grain (ton ha^{-1})
Cf-Or	7.03^{ab}
Cf-No	6.42^{a}
Aw-Or	7.27^{b}
Aw-No	6.97^{ab}

Note: The numbers that followed by the same letter in the same column show no significant difference in the Tukey test of 5%. - Data source: primary research

Water management and providing inputs in the soil were important elements that influenced plant growth and also rice yield. Providing farmyard manure both in continuous flooding and AWD system led to a higher yield than without farmyard manure treatment. Providing organic matter such as farmyard manure affected soil permeability and aggregates [12]. Farmyard manure also increased physical properties and also nutrient uptake which gave positive impacts on plant growth, yield, and also yield component [13]. Organic matter contains various nutrients notably micronutrients that definitely affect the total of the tillers in each plant [14]. The number of productive tillers affects rice productivity. Hasanuzzaman [15] informed that increasing tillers in rice was influenced by the application of manure and inorganic fertilizer. In accordance with the number of tillers per square meter, the availability of nitrogen plays an important role in cell division. Satyanarayana [16] reported that providing farmyard manure can improve the N, P, and K uptake by 20, 12, and 9%. Farmyard manure treatment contributes to an increase in nutrients availability. Another study also reported that providing organic material in the form of manure can increase crop yield [17].

AWD-Or treatment gained a higher yield than the Cf-Or treatment while in both of the treatments farmyard manure was applied. This result indicated that the AWD system could save water irrigation without sacrificing the yield. The timing of the AWD system was an important key to avoid yield reduction. AWD system timing should be aligned with the plant growth stage. Carrijo [18] reported that applying AWD does not impact the yield both on the vegetative or reproductive stage.

3.2 Methane emission

The CH_4 flux measurements showed that all treatments had a similar scheme. The CH_4 flux increased with plant growth and reached a peak of 62 DAS (figure 2). At that age, plants entered the panicle growth stage. CH_4 flux decreased again when entering the ripening phase. In the maximum tillering and panicle growth stage the highest CH_4 fluxes were recorded [19]. It was related to the increasing availability of methanogenic bacteria that was detected in root exudates. The panicle initiation stage showed the highest flux since it had a correlation with the higher root vigor of biomass.

The reduction in CH_4 emissions occurred in the AW-Or treatment. The Aw-No and Aw-Or treatments reduced CH_4 emissions by 12% and 14% compared to conventional farming. The highest average of CH4 emissions was found in the Cf-No treatment of 386.5 mg CH_4 m^{-2} d^{-1}, while in the Aw-No and Aw-Or treatments were 331.9 mg CH_4 m^{-2} d^{-1} and 317.9 mg CH_4 m^{-2} d^{-1} (figure 3).

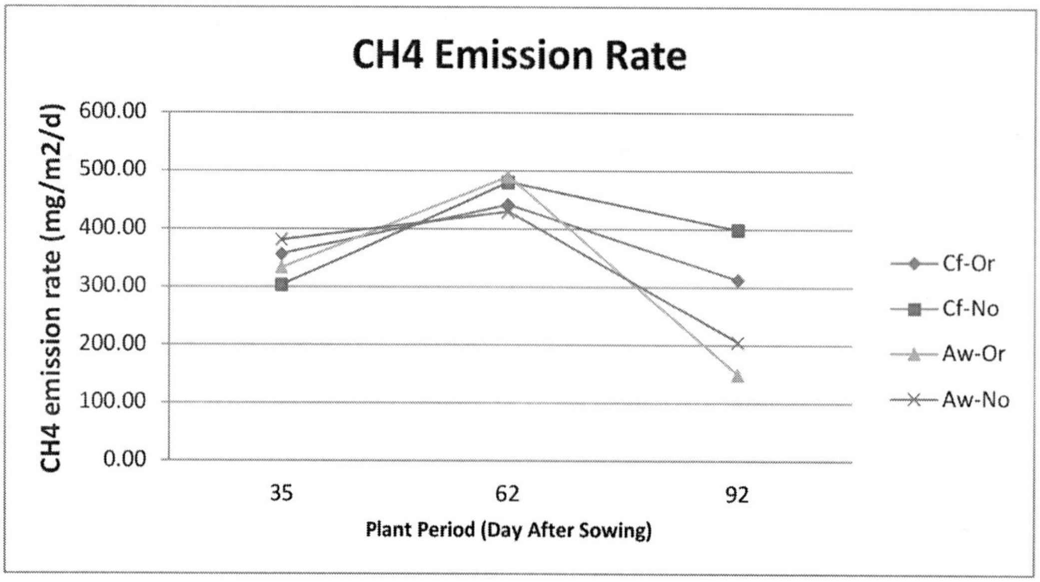

Figure 2. CH_4 Emission Rate in various treatments for water management combined with providing farmyard manure.

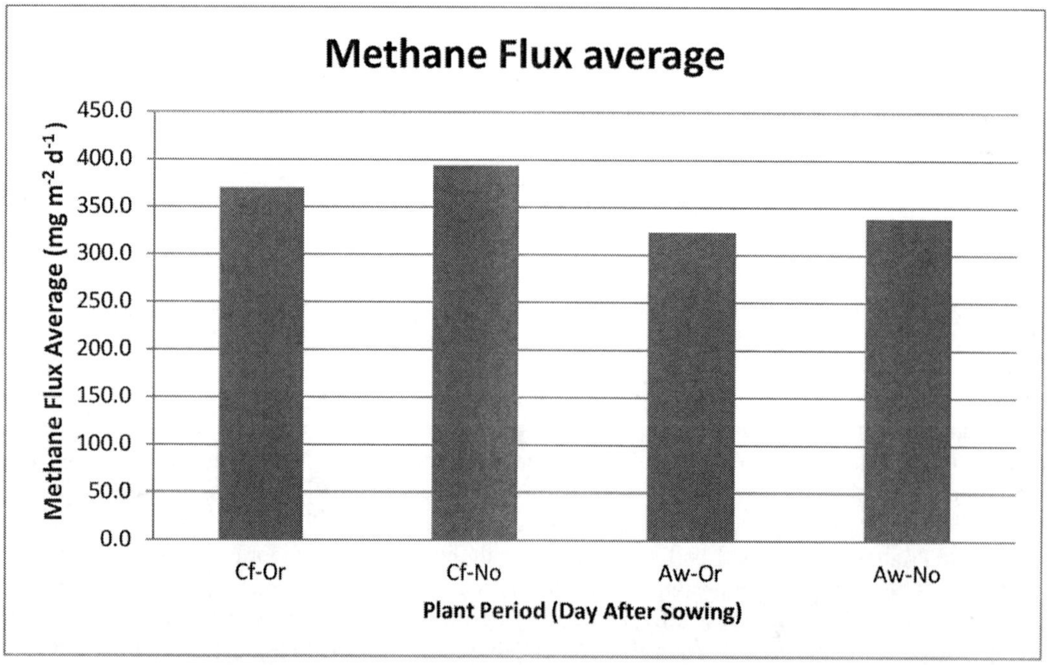

Figure 3. Flux average in various treatments for water management combined with farmyard manure.

This indicated that irrigation water management in paddy fields greatly affected CH_4 emissions. The amount of archaeal and bacterial communities was influenced by desiccation and reflooding [20]. The CH_4 formation in the soil was a microbiological process which was controlled by several factors, one of them was potential redox. The flooded condition would reduce the redox potential and provide a suitable environment for methanogen growth [21]. Oxidative conditions during the dry period in the active tillers and before the primordia phase relatively caused low CH_4 emission compared to the reductive conditions [22].

3.3. Water use

In the AWD treatment, there were wet and dry cycles during plant growth. The water level in the AWD treatment would decrease if there was no rain to a certain level (figure 4). The amount of water that could be saved during one growing season was 32% and 38% (table 2), respectively in the Aw-No cm and Aw-Or treatments. Similar results were found by other researchers who stated that more water input is almost 38% lower than inundated treatment [23] Several studies also summarized [24] that AWD could reduce percolation, drainage, and evaporation by 15-20% without sacrificing the yield.

The total amount of water used by Cf treatment for the plant growth period was much higher than the AWD treatment (table 2). The AWD treatment was more efficient to save water and indicated the actual water requirement on the field. These effects indicated that AWD was very beneficial because it could reduce the need for irrigation water and save more water than continuous flooding. Bouman [25] uttered that AWD systems are considered more effective in irrigation agriculture where farmers use water pumps to support irrigation so that in addition to increasing water efficiency, they can also reduce other inputs in the form of fuel costs for water pumps [26]. In addition, water-saving technology in rice cultivation is also needed to resolve the water scarcity and drought problem as a consequence of the climate change effect. AWD techniques can be applied to irrigate land when water is really needed by plant roots to grow and develop better. Sriphirom [27] reported that the AWD system gives benefits in the scarcity of water areas where there is no rainfall water during the tilling stage so the farmer can save water to irrigate the main stage of rice growth and avoid the yield reduction since the water scarcity.

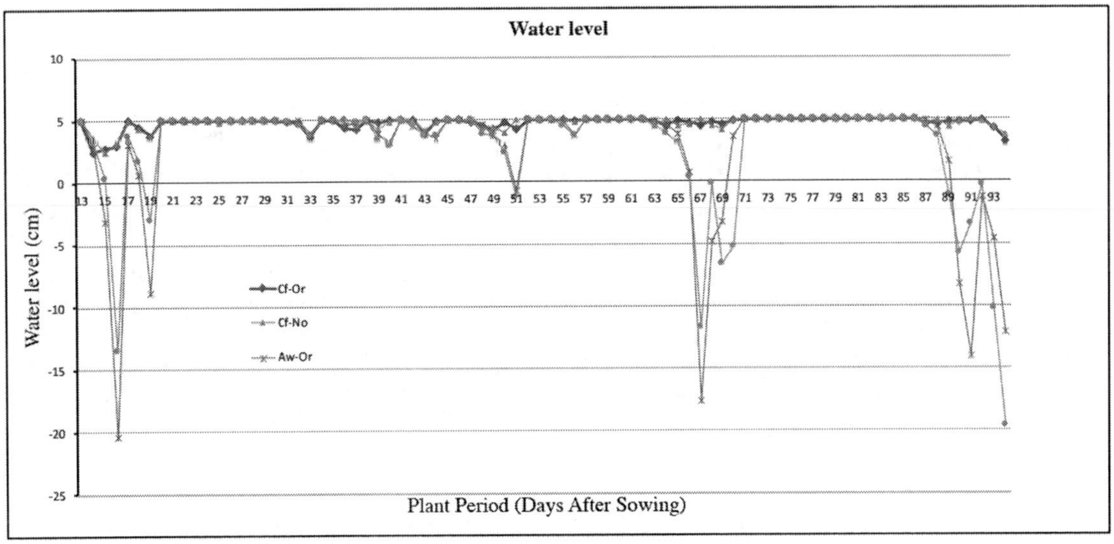

Figure 4. Water levels in various water managements.

Table 2. Water added in various treatments.

Parameter	Cf-Or	Cf-No	Aw-Or	Aw-No
Added water (t ha^{-1})	2144.32	2366.15	1450.34	1469.60
% water efficiency			0.32	0.38

4. Conclusion

The AWD treatment saved irrigation water and reduced greenhouse gas emissions by 12% and 14% and increased the rice yield in paddy fields by 3.3-8.5%. The treatment of Aw-Or reduced CH_4 emissions by 32-38% compared to flooded treatment. The highest grain yield was obtained by Aw-Or treatment which was 7.27 tons ha^{-1} and was significantly different from others. The Aw-Or treatment was a safe technique for reducing GHG emissions from rice fields.

Acknowledgments

We thank the researchers and technical staff of MIRSA team and Laboratory assistant of Green House Gas Research Group, and also to our beloved office, IAERI, for allowing us to use the facilities. The Ministry of Agriculture, Forestry, and Fisheries of Japan has supported this research via an international research project "Technology Development for Circulatory Food Production Systems Responsive to Climate Change: Development of Mitigation option for greenhouse gas emissions from agricultural lands in Asia (MIRSA-3).

References

[1] Wassmann, R, Neue H U, Alberto M C et al 1996 Fluxes and pools of methane in wetland rice soils with varying organic inputs *Environ. Monit. Assess* **42** pp. 163–73
[2] Yao H, Conrad R, Wassmann, R, and Neue H U 1999 Effect of soil characteristics on sequential reduction and methane production in sixteen rice paddy soils from China, the Philippines, and Italy. *Biogeochemistry* **47** 269–95
[3] Jayadeva, H, Setty T K, Prabhakara, Gowda R C, Devendra R, Malikarjun G B, Bandi A G 2009 Methane emission as influenced by different crop establishment techniques and organic manures *Agricultural Science Digest* **29** 4 241–45
[4] Thakur A, Uphoff N, Antony E 2010 An assessment of physiological effects of system of rice intensification (SRI) practices compared with recommended rice cultivation practices in India *Exp. Agric.* 46 77–98
[5] Fageria N 2007 Yield physiology of rice *J. Plant Nutr* **30** 843-79
[6] Richards, M and Sander B O 2014 Alternate wetting and drying in irrigated rice: Implementation guidance for policymakers and investors (Philipines)
[7] FAO 2017 *Climate Smart Agriculture: Policies, Practice and Financing for Food Security, Adaptation and Mitigation* (Rome: Electronic Publishing Policy and Support Branch Communication Division)
[8] Lagomarsino A, Agnelli A E, Linquist B, Adviento-Borbe M A, Agnelli A, Gavina G, et al 2016 Alternate wetting and drying of rice reduced CH4 emissions but triggered N_2O peaks in a clayey soil of Central Italy *Pedosphere* **26** 533–48
[9] Kato Y, Midori O, and Keisuke K 2009 field crops research yield potential and water use efficiency of aerobic rice (*Oryza Sativa* L.) in Japan *Field Crops Res.* **113** 328–34
[10] Setyanto, P, Pramono A, Adriany T A, and Susilawati H L 2018 Alternate wetting and drying reduces methane emission from a rice paddy in Central Java, Indonesia without yield loss *Soil Sci* 1–8
[11] Minamikawa K, Tokida T, Sudo S, Padre A, and Yagi K 2015 *Guidelines for Measuring CH4 and N2O Emissions from Rice Paddies by a Manually Operated Closed Chamber Method* (Japan: National Institute for Agro-Environmental Sciences)
[12] Singh Y, Singh, B, Khera T S, Meelu O P 1994 Integrated management of green manure, farmyard manure, and nitrogen fertilizer in a rice-wheat rotation in northeastern india. *Arid Soil Res. Rehab.* **8** 199–205
[13] Mondal S S and Chettri M 1998 Integrated nutrient management for sustaining productivity and fertility under rice (*oryza sativa*)-based cropping system. indian *J. Agr. Sci* 68 337–40
[14] Belefant-Miller H 2007 Poultry litter induces tillering in rice *J. Sustain. Agric* **31** 151-60

[15] Hasanuzzaman M, Ahamed K U, Rahmatullah N M, and Akhter N 2010 Plant growth characters and productivity of wetland rice (*oryza sativa* l.) as affected by application of different manures *Emir J Food Agric* **22** 46–58

[16] Satyanarayana V, Prasad P V V, Murthy V R K, and Boote K J 2002 Influence of integrated use of farmyard manure and inorganic fertilizers on yield and yield components of irrigated lowland rice *J. Plant Nutr* **25** 2081–90

[17] Sudarsono, Arif W, Melati M, And Aziz S A 2014 Growth and yield of organic rice with cow manure application *Agrivita* **36** 19–25

[18] Carrijo D R, Lundy M E, and Linquist B A 2017 Rice yields and water use under alternate wetting and drying irrigation: A meta-analysis *Field Crops Res.* **203** 173-80

[19] Gogoi N, and Baruah K K 2005 Methane emission characteristics and its relations with plant and soil parameters under irrigated rice ecosystem of northeast india *Chemosphere* **59** 1677–84

[20] Reim A, Hernández M, Klose M, Chidthaisong A, Yuttitham M, and Conrad R 2017 Response of methanogenic microbial communities to desiccation stress in flooded and rain-fed paddy soil from thailand *Frontiers in microbiology* **8** 785

[21] Watanabe T, Wang G, Katsutoshi T, Ohashi Y, Kimura M, Asakawa S 2010 Vertical changes in bacterial and archaeal communities with soil depth in Japanese paddy fields *Soil Sci* **56**

[22] Pramono A and Adriany T A, Setyanto P 2018 *Soil Quality for Food Security and Healthy* (Thailand)

[23] Rejesus, R M, Palis F G, Rodriguez D G P, Lampayan R M, Boumand B A M 2010 Impact of the alternate wetting and drying (AWD) water-saving irrigation technique : evidence from rice producers in the philippines impact of the alternate wetting and drying (awd) water-saving irrigation technique : evidence from rice producers in the philippines *Food Policy* **36**

[24] Bouman, B A M. Humphrey, E, Tuong, T P, and Barker, R 2007 Rice and water *Adv. Agron.* **92** 187-237

[25] Roderick, E T 2009. The impact of integrated pest management information dissemination methods on insecticide use and efficiency: evidence from rice producers in south *vietnam Review of Agric. Econ.* **314** 814-33

[26] Lampayan, R M, Rejesus, R M, Singleton, G R, and Bouman, B A 2015 Adoption and economics of alternate wetting and drying water management for irrigated lowland rice *Field Crops Res.* **170** 95–108

[27] Sriphirom P., Chidthaisong A., Towprayoon S., 2018, Rice Cultivation to Cope with Drought Situation by Alternate Wet and Dry (AWD) Water Management System: Case Study of Ratchaburi Province, Thailand, Chemical Engineering Transactions, **63**, 139-44

IOP Conf. Series: Materials Science and Engineering **980** (2020) 012067 doi:10.1088/1757-899X/980/1/012067

Water management and rice husk biochar application to solve acid sulfate soil problems to promote rice yield and reduce greenhouse gas emission

W A Yusuf* and Mukhlis
Indonesian Swampland Agricultural Research Institute, Banjarbaru 70712, Indonesia
Jalan Kebun Karet Loktabat, Banjarbaru

Email: annisa_balittra@yahoo.com

Abstract. The use of biochar for soil fertility improvement is gaining popularity due to its potential to improve soil quality and increase crop yield in swampland. Water management plays a key role in controlling various dynamic processes such as acidity production. The water management system in acid sulfate soil should have a proper leaching and flushing capacity of the oxidation product. When these soils are used for rice, we found that the most important constraints were (1) Acidity (which includes the combined effects of pH, Al toxicity, and P deficiency) and (2) Fe stress (which is due to the combined effects of Fe toxicity and deficiencies of other divalent cations such as calcium/Ca). The experiment was carried out to investigate the effects of water management and rice husk biochar and compost applied in combination (0,1% of each w/w, thus 0,1% compost + 1% rice husk biochar) on growth, the yield of rice and methane emission on acid sulfate soil. The Provision of 50% compost + 50% Biochar rice husk with washing water management every two weeks (P2) in acid sulfate soil could reduce methane emissions by up to 53.16% with the highest dried unhusked rice yield of 3.86 t/ha.

1. Introduction

The problem of acid sulfate soil originates from the presence of pyrite (FeS_2) in the soil layer. If this pyrite layer undergoes oxidation, Fe will be released to become Fe^{3+} and S will be oxidized to sulfate which causes the soil pH to decrease to become unusually acidic (pH <3.5), so that plants cannot grow. The high acidity in the soil due to the oxidation of pyrite will have an impact on increasing the solubility of toxic elements such as Fe and Al accompanied by available P and low base saturation and other nutrient deficiencies. The utilization of acid sulfate soil for lowland rice with a minimum thickness slows the pyrite oxidation process. Pyrite oxidation can be suppressed by maintaining the soil environment in reductive conditions. Continuous inundation on the soil in tidal lands results in: (1) an increase in the concentration of ferrous iron (Fe^{2+}), (2) a decrease in the value of soil redox potential (Eh) due to the reduction of $NO3^-$ and Fe^{3+} which results in the accumulation of NH_4^+ and Fe^{2+}, so that the soil condition becomes very reductive and (3) increase in soil pH due to the release of OH^- and H^+ consumption followed by a decrease in the activity of H^+.

One of the efforts to overcome the fertility problems of acid sulfate soils is by adding ameliorants such as biochar [1]. Panhwar [2] reported that the soils need to be improved with some soil amendments like the application of basalt, ground magnesium limestone, and organic materials (biofertilizer) that can increase the soil pH, improve soil nutrients and reduce the Al and Fe toxicity. As a soil amendment agent, biochar has unique physical and chemical characteristics. Biochar can improve soil physical and chemical properties, reduce nutrient leaching, optimize soil microbial population structure and richness, and biochar can promote crop growth, yield, and quality [3]. Plants grown in biochar-amended soil has greater root mass and root length densities, as well as total and individual root lengths for secondary and tertiary roots, compared with plants grown in unamended soil[4]. Biochar in acid Sulfate soils has great potential to reduce the content of toxic elements and their mobility in the soil due to their high surface area and mostly negative charge for most types of biochar and increase soil pH. The results of a study by Zhang [5] showed that the combination of organic fertilizer and biochar had improved soil nutrient status and soil fertility. In addition, biochar has strong adsorption functions, which can effectively adsorb toxic elements [6]. The amendment of rice husk biochar to acid sulfate soils from Malaysia was found to improve soil physicochemical properties and increase the yield of corn and rice. The rice husk biochar produced via a gasification process in a local rice mill possesses physicochemical properties that are able to neutralize the acidic nature of the soils [7]. Qayyum [8] reported that the application of biochar combined compost reduced absorption of P and increased the bioavailability of P. The experiment was carried out to investigate the effects of water management and rice husk biochar and compost applied in combination (0,1% of each w/w, thus 0,1% compost + 1% rice husk biochar) on growth, the yield of rice and methane emission on acid sulfate soil.

2. Materials and methods

2.1. Time and location

This experiment was carried out from July to October 2019 at the Balittra Experimental Garden (ISARI), Barito Kuala Regency, South Kalimantan Province, Indonesia (-5 ° 10'15 " S, 114 ° 36'12 " E), with an annual average temperature of 26-27°C. According to the "Schmit and Ferguson" classification, Barito Kuala is a Type B rainfall, which has 1-2 dry months a year. According to[9], the soil in the study location was included in the Typic Sulfaquent subgroup, seen from the soil pH value of 3.81, but after being oxidized the pH decreased to <3.5. The pyrite oxidation process produced Fe^{2+} and SO_4^{2-} and H_2 which resulted in a decrease in pH. According to Dent [10] that a pH value <2.5 or 3 after given H_2O_2 indicates strong acidity of sulfate. In addition, the decomposition of aluminum silicate minerals will free aluminum ions. These ions can be absorbed by soil colloids, and when hydrolyzed will donate H^+ ions, so that the soil becomes acidic due to the increased contribution of H^+ ions.

2.2. Research design

The design used in this study was a split-plot design. The main plot was water management, namely: (1) inundation, (2) leaching every 1 week (P1), (3) leaching every 2 weeks (P2). Subplots were the use of ameliorants, namely: (1) Control (without ameliorant), (2) 50% compost + 50% Biochar rice husk

2.3. Gas calculation

Gas was collected in the field using a syringe, then measured using Gas Chromatography (GC). Examples of gas in the field were taken from the closed chamber (Paralonfiber) using a syringe and then measured CH_4, also periodically at 30, 60, and 90 days afterward. The calculation of CH_4 flux in each treatment was done using equation 1.

$$E = \frac{Bm}{Vm} x \frac{\delta Csp}{\delta t} x \frac{V}{A} x \frac{273.2}{T + 273.2} \tag{1}$$

Where E is the emission of CH_4 (mg/m^2/day), V is the volume of the chamber (m^3), A is the area of the chamber (m^2), T is the temperature in chamber (°C), δCsp/δt is the rate of change in the concentration of CH_4 gas (ppm/minute), Bm is CH_4 gas molecular weight under standard conditions, and Vm is the gas volume at standard temperature and pressure is 22.41 liters at 23°K.

3. Results and discussion

3.1. Soil pH and Fe concentration

The characterization results showed that the study area had a high Fe^{2+} content of 319.58 mg.kg^{-1} and sulfate of 1097.97 mg.kg-1 and Aluminum of 15 cmol (+), kg^{-1}. The high Fe^{2+} and SO_4^{2-} content was a result of the oxidation of pyrite which was released into the soil solution. This pattern of Fe^{2+} content is the same as that reported by Anda [11]. They reported that in acid sulfate soils in Kalimantan the total Fe content ranged between 0.5 and 1.3% in the oxidation layer which was lower than that in the reduction layer, namely 1.0-2.3% with the pyrite content in the oxidation layer ranging from 0.09 - 0.32% lower than in the layer, reduction which ranges from 0.17-1.91%. The high concentration of Fe^{2+} in the soil was accompanied by low exchangeable K[12]. This might be related to P and Zn deficiency and H2S toxicity [2].

The results of preliminary soil analysis showed that the initial soil pH ranged from 3.92 to 4.80 with ferrous iron solubility ranging from 119.43 to 809.53 ppm (figure 1). One of the high content of Fe^{2+} was the result of intensive oxidation of pyrite and was released into the soil solution

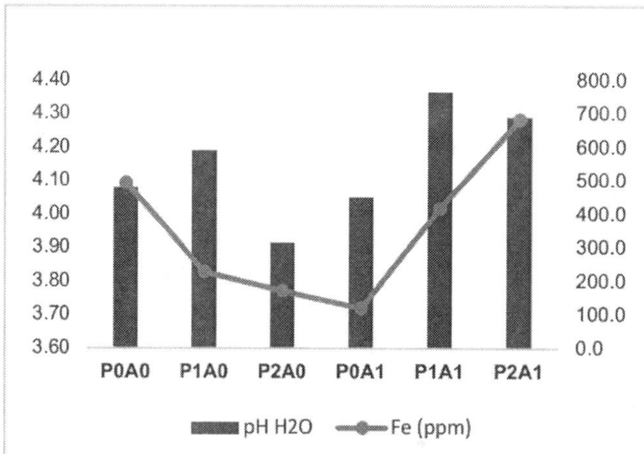

Figure 1. pH soil and Fe concentration in tidal swampland with water management and application of ameliorant.

3.2. Methane emissions

Methane emissions in acid Sulfate soils are influenced by, among other things, the type of ameliorant material given as indicated by the research data. Provision of ameliorant 50% compost + 50% rice husk biochar could reduce methane emissions released from rice cultivation in acid sulfate soil by 53.16% compared to without ameliorant (figure 2). Agricultural practices such as water systems also affect the amount of methane emissions released from paddy fields because they are related to the dominant inundation process during the growth period, thus affecting the activity of roots and processes that take place such as photosynthetic and respiration processes. The results showed that the water system with two weeks of washing (P2) reduced CH_4 emissions by 48.03% and washing every one week (P1) reduced methane emissions by 42.10% compared to without washing (P0) (figure 3).

This was caused by continuous flooding in the process of rice cultivation making soil conditions anaerobic, so the rate of degradation of organic matter was low.

Figure 2. Performance of rice in tidal swampland with water management and application ameliorant

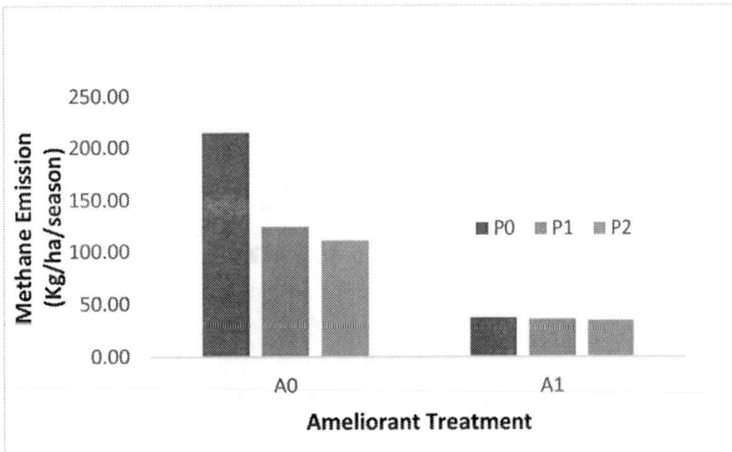

Figure 3. Methane emissions during one growing season in tidal swampland with water management and application ameliorant.

3.3. The yield of rice (ton/ha)

The ameliorant treatment tested in the field was able to approach the yield potential of the swamp rice used (Inpara 3) (figure 4). The highest yield of dry unhulled rice was seen in the treatment of 50% compost + 50% Biochar rice husk combined with washing every two weeks (P2), namely 3.86 t/ha, while the treatment with the lowest GKP yield was 2.42 t/ha, seen in treatment A0 (without ameliorant) with water management treatment without washing (P0). This is related to the higher availability of N, P, K nutrients. Increased availability of N, P, K nutrients in both types of acid Sulfate soils, both natural and intensive, was related to the provision of compost and N, P, K fertilization[1].

Washing every two weeks was the best treatment in water management as well as giving ameliorant 50% compost + 50% Biochar of rice husk in acid Sulfate land because P2 treatment (two weeks of washing) had the highest plant with a height reaching 110.4 cm and increased GKP yield by 20%. This was related to the quality of water used in the washing process because P2 treatment (washing every two weeks) used better quality water (large tide) than P1 treatment (washing every one week) using water at large and small tide. Water quality at high tide is shown in figure 5. The correlation between pH and DHL values in large and small tide showed a very significant negative correlation with r values, respectively, r = -0.846 ** (p = 0.004) and r = -0.521 * (p = 0.027). It appears that there was an insignificant positive relationship between growth and methane emissions as indicated by the value of R2 = 0.429 (figure 6).

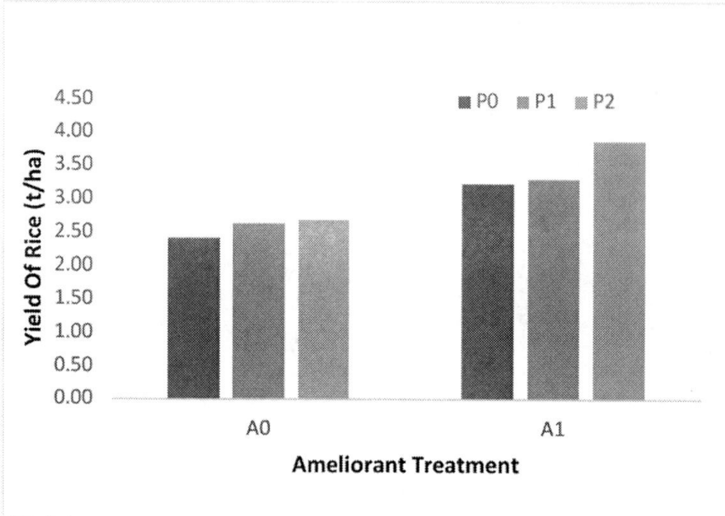

Figure 4. The yield of rice with water management and application of ameliorant in tidal swampland.

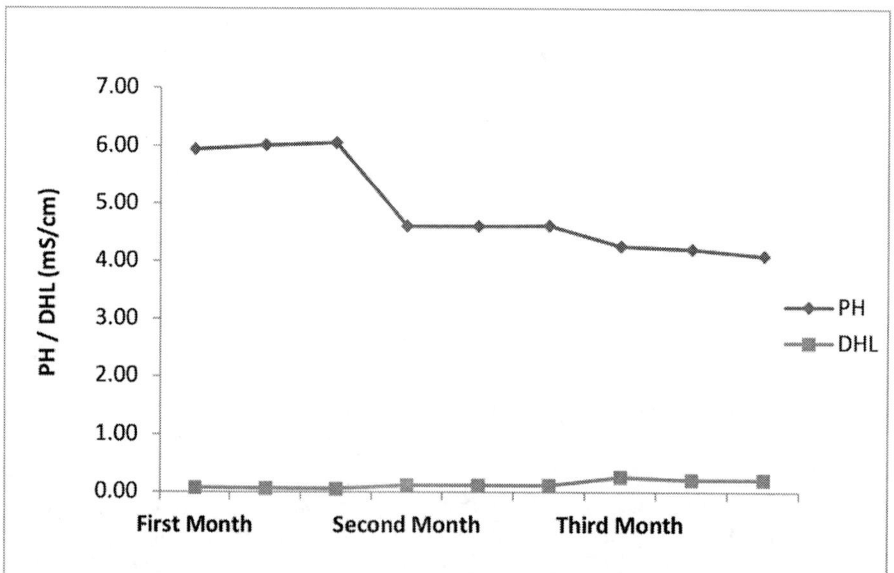

Figure 5. Tidal water quality in tertiary channels.

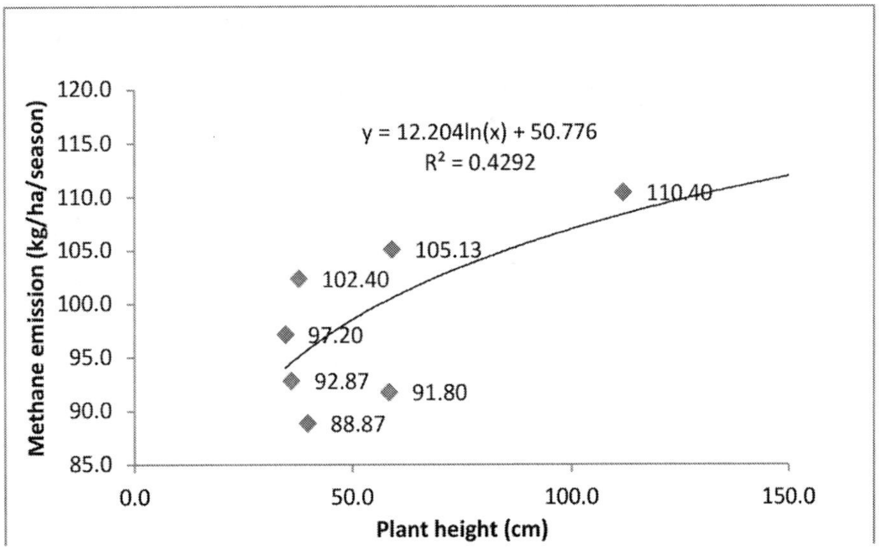

Figure 6. The relationship between plant growth and methane emissions.

4. Conclusion

The Provision of 50% compost + 50% Biochar rice husk with washing water management every two weeks (P2) in acid sulfate soil could reduce methane emissions by up to 53.16% with the highest dried unhusked rice yield of 3.86 t/ha.

Reference

[1]Annisa W and Nursyamsi D 2016 Pengaruh amelioran, pupuk dan sistem pengelolaan tanah sulfat masam terhadap hasil padi dan emisi metana *Jurnal Tanah dan Iklim* **40(2)** 135 – 45

[2]Panhwar Q A, Naher U A, Radziah O, Shamshuddin J and Razi I M O H D 2014 Bio-Fertilizer, ground magnesium limestone and basalt applications may improve chemical properties of Malaysian acid sulfate soils and rice growth *Pedosphere* **24 (6)** 827-35

[3]Lehmann J, Rillig M C, Thies J, Masiello C A, Hockaday W C and Crowley D 2011 Soil biology & biochemistry biochar effects on soil biota : A review. *Soil Biology and Biochem.* **43(9)** 1812-36

[4]Mitchell K, French E, Beckerman J, Iyer-Pascuzzi A, Volenec J and Gibson K 2018 Biochar alters the root systems of large crabgrass *Hort.Science* **53(3)** 354–9

[5]Zhang J, He Z, Tian H, Zhu G and Peng X 2007 Identification of aluminium-responsive genes in rice cultivars with different aluminium sensitivities *J. Experimental Botany* **58 (8)** 2269–78

[6]Gu Y et al 2017 Application of biochar reduces *Ralstonia solanacearum* infection via effects on pathogen chemotaxis, swarming motility, and root exudate adsorption. *Plant and Soil* **415** p 269–81

[7]Ajwa H A and Tabatabai M A 1993. Comparison of some methods for determination of sulfate in soils *Comm. in Soil Sci. and Plant Analysis*, **24:15-16** 1817-32

[8]Qayyum M F, Ashraf I, Abid, M and Steffens D 2015 Effect of biochar, lime, and compost application on phosphorus adsorption in a Ferralsol *J. Plant Nutrition and Soil Sci.* **178(4)**

[9]Soil Survey Staff 2014 *Keys to Soil Taxonomy*, 12th ed. (Washington-DC: USDA-Natural Resources Conservation Service)

[10]Dent D L and Pons L J 1995 A world perspective on acid Sulfate soils *Geoderma* **67** 263-76

[11]Anda M, Siswanto A B and Subandiono R E 2009 Properties of organic and acid sulfate soils and water of a "reclaimed" tidal backswamp in Central Kalimantan, Indonesia. *Geoderma* **149 (2009)** 54–65

[12]Ottow J C G, Benckiser G, Watanabe I 1982 Iron toxicity of rice as a multiple nutritional soil stress *Proc. Int. symposium on distribution, characteristics and utilization of problem soils* (Ibaraki: Tropical Agriculture Research Center) p 167–79

IOP Conf. Series: Materials Science and Engineering **980** (2020) 012068 doi:10.1088/1757-899X/980/1/012068

Effectiveness of compost and vermicompost from market organic waste to improve soil chemical properties

Syarifinnur[1]*, Y Nuraini[2], B Prasetya[2] and E Handayanto[2]
[1] Postgraduate Program, Faculty of Agriculture, Brawijaya University, Jl. Veteran, Malang 65145, Indonesia
[2] Department of Soil Science, Faculty of Agriculture, Brawijaya University, Jl. Veteran, Malang 65145, Indonesia

Email: syarifinnur@gmail.com

Abstract. The purpose of this study is to evaluate the results of compost and vermicompost derived from market organic waste and its ability to improve soil chemical properties. Seven treatments consisted of treatments without organic matter (control), compost consisted of three treatments (2.5, 5, and 10 tons/ha), and vermicompost consisted of three treatments (2.5, 5, and 10 tons/ha). The treatments were arranged using a Completely Randomized Design with three replications. Each treatment was mixed with soil and incubated for eight weeks. After eight weeks the parameters observed consisted of total N, total P, available P, total K, Organic-C, and pH. The results showed a change in soil chemical content. The highest increase in soil chemical content was shown in the treatment of vermicompost. Application vermicompost 10 tons/ha had a significant influence on changes in soil pH, organic C, total P, available P, total K, and Compost 10 tons/ha increased total N. Vermicompost application of 10 tons/ha could increase total P (215.99%), available P (465.34%), total K (37.78%), pH (12.38%) and organic-C (40.34 %). Compost 10 tons/ha increases total N by (54%). Market organic waste effectively used as vermicompost and compost that could improve the soil's chemical properties.

1. Introduction

The availability of nutrients in the soil is a critical factor in agricultural cultivation. Land degradation due to improper land management today has decreased soil properties that support plant growth [1]. This is also exacerbated by the excessive application of inorganic fertilizers resulting in reduced living things or microorganisms that exist in the soil, such as fungi, bacteria, earthworms, and others [2]. If this is not overcome by changing agricultural practices in the field over time, it will decrease the productivity of agricultural products because the soil is no longer able to provide suitable conditions for plants to grow well [3]. One way to increase soil fertility is by adding organic material. The application of organic matter to the soil is one of the methods used to improve soil quality to enhance the soil's physical, chemical, and biological properties [4]. In the cultivation of organic agriculture, organic material is used as the main essential ingredient to provide soil nutrition, not to feed plants. For this reason, the success of crop cultivation is primarily determined by the type of organic material supplied to realize sustainable agriculture [5].

One source of organic matter that can be utilized is organic matter from market waste [6]. In general, waste is one of the problems that exist in each region, and its quantity always increases from time to time. Proper waste management is one way to maintain sustainability not to be polluted and create a healthy environment [7]. Waste management that is generally done today has not solved the problem of waste as a whole, and even it has resulted in water, air, and soil pollution, especially at the final waste disposal site. For this reason, efforts are needed so that the waste can be reused for the benefit of humans, especially in agriculture.

The use of market waste as an alternative source of organic material has great potential. Organic matter availability from market waste can be converted as soil ameliorant always increase from time to time, and the source of organic matter available at any time. The method of converting organic matter is essential to producing good quality organic matter for applied to the soil [8]. The method that can be used is composting and vermicomposting methods [9]. Both of these methods have their respective advantages to produce organic matter that can improve soil conditions.

Compost and vermicompost's direct effect has been well known for its positive impact on increasing chemical, physical, biological soil content, and enhancing plant growth [10]. Vermicompost and compost have some differences in the content of specific ingredients. The process of phosphatase and β-glucosidase vermicomposting are higher than the composting process. Based on the high levels of phosphatase, this is why vermicompost becomes one of the potential organic materials to be able to increase the presence of P content in the soil [11].

The process of composting municipal waste using earthworms showed higher values of the elements N, P, K, Ca, Mg, and Na compared to composting without worms. Increased nutrient content that is higher comes from the enzymatic activity in the earthworm's body so that it affects the process of mobilization, and mineralization of these elements is higher when compared to the treatment without worms [12]. Vermicompost application also shows an increase in the number of growth hormones present in the ground, such as gibberellin, cytokinins, auxins, and humic acids, increasing plant growth [13]. This study aims to determine the effect of compost and vermicompost organic market waste on improving soil chemical properties.

2. Materials and methods

2. 1. Soil, compost and vermicompost product materials

The soil used in this was taken from agricultural land in Telaga Waru Village, Labuapi District, West Lombok, West Nusa Tenggara (8°37'32.6"S, 116°07'04.9"E). Soil characteristics topsoil (0-20 cm) used include clay texture class (Sand 42%, Clay 14%, slit 44%), moisture content 46.75%, pH (H_2O) 5.68, organic C 1.17%, Total N 0.15%, Total P 23.74 mg/100g, Available P 2.03 ppm, Total K 19.67 mg/100g, exchangeable cation of K, Na, Ca, Mg respectively 1.14, 0.63, 3.83 and 1.82 cmol/kg, CEC 15.45 cmol/kg and base saturation of 58,03%. The vermicompost's chemical composition was as follows: total N 1,77%, total P 0.43%, total K 2,20%, pH 7.30, and organic-C 19.38%. The compost's chemical composition was as follows: total N 1,95%, total P 0.29%, total K 4,96%, pH 7.6, and organic C 25.69%.

2. 2. Experimental incubation design

The incubation experiment in the laboratory was following the methods used by [14]. The experimental treatment was combinations of three doses of vermicompost (2.5, 5, 10 tons/ha), three doses of compost (2.5, 5, 10 tons/ha), and a control (0 ton/ha). Seven treatments were arranged in a completely randomized design with three replication. Each treatment combination of vermicompost and compost was mixed with 300 g of soil (diameter < 2 mm, dried oven), and water added to 80% of field capacity. Each treatment is covered with aluminum and given a hole to reduce the evaporation process. All incubation pots were placed at room temperature and periodically weighed. In the eighth week, all vermicompost and compost treatment were analyzed for the soil's chemical properties. At eight weeks, the observation was made for total N (Kjeldahl), total P (HCl 25%), available P (Bray I), total K (HCl 25%), pH (H_2O) and organic C (Walkley-Black). Methods of analysis, according to [15]. Data were subjected for analysis of variance (ANOVA) for total N, total P, available P, total K, pH, and organic C, followed by DMRT (Duncan's Multiple Range Test). The probability levels used for statistical significance were $P < 0.05$ for the tests. Statistical analysis of the data was carried out with SPSS 18.0 software program

3. Result and discussion

3. 1. Total nitrate content (N)

Applications compost and vermicompost increased the total N content of the soil. The highest increase in total N was the addition of 10 tons/ha of compost (figure 1). The increase in total N content from 0.13% to 0.20% or increased by 54% of the initial soil content. The increase in total N content was due to the rise in nitrogen mineralization, and the total N of compost was higher than vermicompost.

Several studies have shown mixed results related to the N mineralization process in soils. Research shows that compost has an average N value of total soil of 221.2 mg/kg higher than vermicompost of 196 mg/kg [16]. While another research shows the opposite, the total N soil in vermicompost application has a higher value of 138 mg/kg compared with compost application of 134 mg/kg [17]. The difference in soil nitrogen content is due to differences in material composition due to the treatment given, affecting the final result of the incubation process [18].

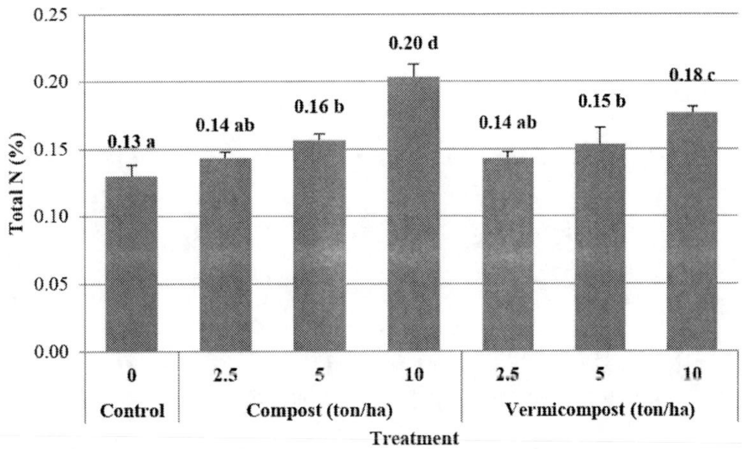

Figure 1. The effect of giving vermicompost and compost on the total soil N content (%).

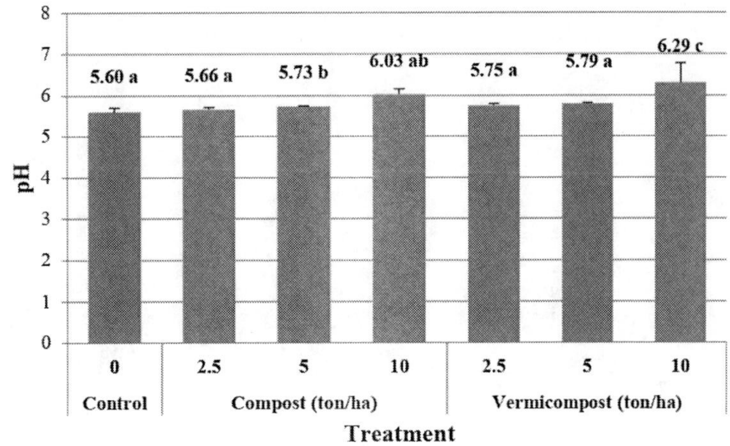

Figure 2. The effect of giving vermicompost and compost on the soil pH content.

3.2. pH

The application of compost and vermicompost increased soil pH compared to initial soil (figure 2). The increase in pH was directly proportional to the amount of organic matter added. Vermicompost 10 tons/ha gave the highest pH increase from 5.60 to 6.29 or an increase of 12.38%.

Some research showed that giving vermicompost has been able to improve soil pH conditions towards neutral. pH changes that occur due to the decomposition of organic matter to produce organic acids such as humic acid and phosphoric acid derived from earthworm decomposition to reduce the concentration of hydrogen ions (H +). Which causes the soil to become more acidic and increase hydrogen ions (OH-) [19].

3. 3. Total and available phosphorus content (P)

The application of compost and vermicompost significantly influenced the total content of P and available P soils. Application 10 ton/ha vermicompost provided the highest total and P available than compost and other doses (figure 3 and figure 4). The vermicompost application could increase the initial P content from 19.67 ppm to 58.33 or increase of 215.99%, and available P content had increased from 2.60 ppm to 13.96 ppm increased by 465.34%.

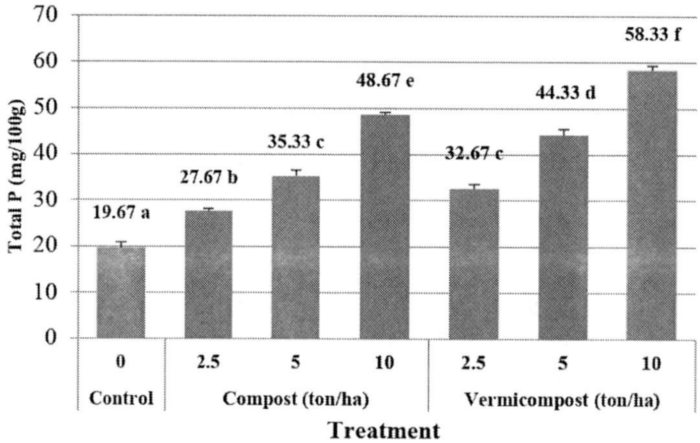

Figure 3. The effect of giving vermicompost and compost on the soil total phosphorus content (mg/100g).

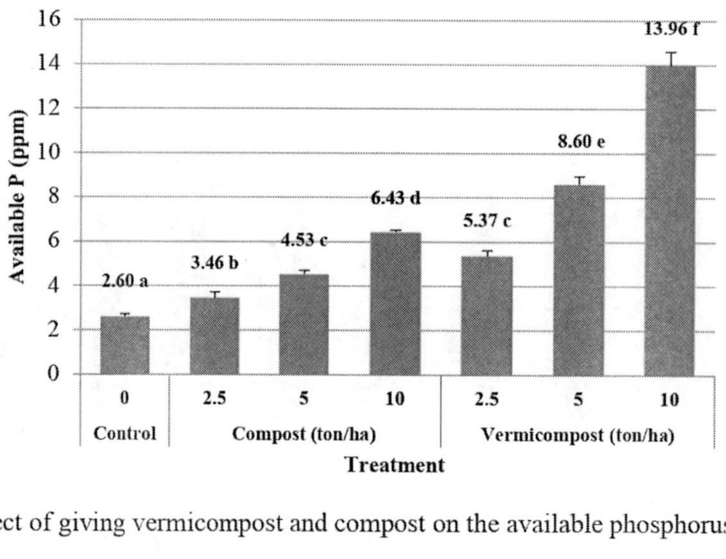

Figure 4. The effect of giving vermicompost and compost on the available phosphorus content (ppm).

The increase in the total P content and available P is due to the decomposition process and mineralization of organic matter that produces organic acids that help release P, so the availability of soil P is increasing [20]. The increase in soil P content is also due to the P assimilation mediated by microorganisms derived from the digestion of earthworms, increasing the presence of soil P content [21].

Some research reports the application of vermicompost has increased the value of available soil P content by 15.45 ppm higher than compost by 13.82 ppm and application of vermicompost is able to increase the presence of available P from 38 ppm to 81 ppm for dose 10 tons/ha [22]. Another study using vermicompost doses of 20, 30, and 40 tons/ha can increase the available P, 81, 90, and 109 ppm, respectively, from the initial soil sample. Based on this, vermicompost is a technology that can provide more P elements compared to other organic materials to improve soil properties [23] [11].

3. 4. Total potassium content (K)

Application of compost and vermicompost increased total K content compared to original K soil content. The highest increase of total K (91,41%) occurred in vermicompost treatment, from 17.67 to 37.00 (figure 5). The increase in soil K content that occurred was in line with the dose of organic matter used.

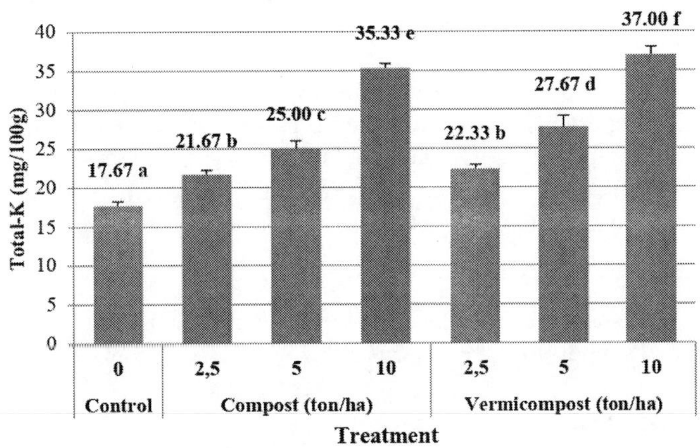

Figure 5. The effect of giving vermicompost and compost on the potassium (K) content (mg/100g).

An increase in the total K element in the soil is caused by changes in potassium distribution between the forms that can be exchanged and cannot be exchanged in the soil [24]. This change in total K content is due to the digestive system of earthworms can process organic waste material [25]. Besides that, it produces high levels of interchangeable potassium concentrations because earthworms microbial activity can increase organic matter degradation and release K to enhance K in the soil. [26]

3. 5. Organic C

Based on the analysis result, the application of vermicompost and compost had a significant effect on the increase in soil organic-C compared to initial soil organic C (control) (figure 6). The increase in soil organic C that occurred was in line with the dose of organic matter used. In the vermicompost application, 10 tons/ha gave the highest value compared to other treatments. The vermicompost dose of 10 tons/ha increased the organic C content by 40.34%, while the compost application increased the organic C content by 27.84%.

A similar research report vermicompost application has increased the organic C content from 0.95% to 1.30% or increased to 36.84%, while compost can increase organic C by 1.27 or increase by 33.88%. The increase in soil organic C from vermicompost applications, because of the mineralization process to the soil assisted by microorganisms from earthworm that ingested organic matter [27].

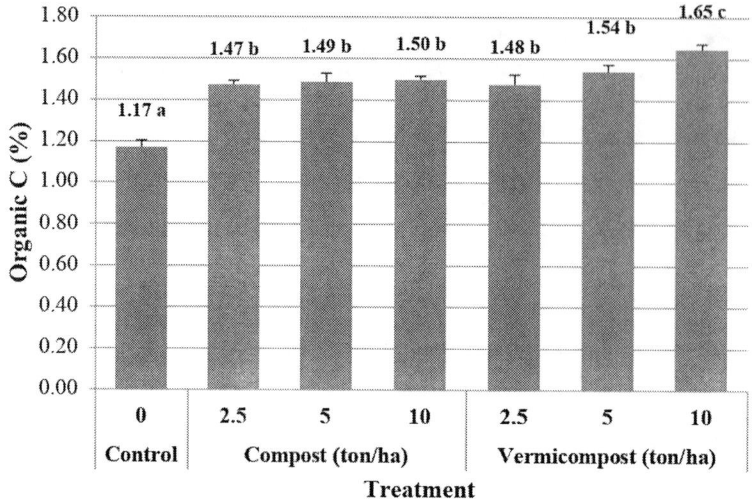

Figure 6. The effect of giving vermicompost and compost on the soil organic-C content (%).

4. Conclusion

Application vermicompost 10 tons/ha had a significant influence on changes in total-P, available-P, total-K pH, C-organic, and Compost 10 tons/ha increases total-N. Vermicompost application of 10 tons/ha could increase total P (215.99%), available P (465.34%), total K (91.41%), pH (12.38%) and organic C (40.34 %). Compost 10 tons/ha increased total N by (54%). Therefore, the application of vermicompost and compost could improve the soil chemical properties and be applied.

Acknowledgments
The authors thank the Lembaga Pengelola Dana Pendidikan (LPDP), Ministry of Finance, Republic of Indonesia, for financial support to carry out this study.

References

[1] Lenka N K, Jaiswal S P, Thakur J K, Lenka S, Mandal A, Dwivedi A K, Lakaria B L, Biswass A K, Shukla A K and Yashona D S 2017 *Curr. Sci.* **112** p 2434–9
[2] Savci S 2012 *APCBEE Procedia* **1** p 287–92
[3] Cowie A L, Orr B J, Castillo V M, Chasek P, Crossman N D, Erlewein A, Louwagie G, Maron M, Metternicht G, Minelli S, Tengberg A E, Walter S and Welton S 2018 *Environ. Sci. Policy* **79** 25–35
[4] Merante P, Dibari C, Ferrise R, Sánchez B, Iglesias A, Lesschen J P, Kuikman P, Yeluripati J, Smith P and Bindi M 2017 *Soil Tillage Res.* **165** 95–106
[5] Pommier T, Merroune A, Bettarel Y, Got P, Janeau J L, Jouquet P, Thu T D, Toan T D and Rochelle-Newall E 2014 *FEMS Microbiol. Ecol.* **90** 622–32
[6] Lupton S 2017 *J. Rural Stud.* **55** 83–99
[7] Rakkini V M and Kumar S 2017 *J. Civ. Eng. Environ. Sci.* **3** 13–7
[8] Villar I, Alves D, Pérez-díaz D and Mato S 2016 *Waste Manag.* **48** 409–17
[9] Fornes F, Mendoza-Hernández D, García-de-la-Fuente R, Abad M and Belda R M 2012 *Bioresour. Technol.* **118** 296–305
[10] Ngo P T, Rumpel C, Dignac M F, Billou D, Duc T T and Jouquet P 2011 *Ecol. Eng.* **37** 269–76
[11] Lazcano C, Gómez-Brandón M and Domínguez J 2008 *Chemosphere* **72** 1013–9
[12] Soobhany N, Mohee R and Garg V K 2015 *J. Environ. Chem. Eng.* **3** 2931–42
[13] Huang K, Li F, Wei Y, Fu X and Chen X 2014 *Bioresour. Technol.* **170** 45–52

[14] Handayanto E, Cadish G and Giller K E 1994 *Plant Soil* **160** 238–47
[15] Balittanah 2009 *Petunjuk Teknis: Analisis Kimia Tanah, Tanaman, Air dan Pupuk* (Bogor : Balai Penelitian Tanah) p 7-28
[16] Goswami L, Nath A, Sutradhar S, Sundar S, Kalamdhad A, Vellingiri K and Kim K 2017 *J. Environ. Manage.* **200** 243–52
[17] Asrin A F N, Hanom S A K and Ossain S H A K H 2019 J. Biol.Sci **28** 121–9
[18] Paradelo R, Moldes A B and Barral M T 2011 *Waste Manag. Res.* **29** 1177–84
[19] Manyuchi M, Chitambwe T, Phiri A, Muredzi P and Kanhukamwe Q 2013 *Int. J. Chem. and Env. Eng. Effect* **2**
[20] Fox T R, Comerford N B and McFee W W 1990 *Soil Sci. Soc. Am. J.* **54** 1763–7
[21] Najjari F and Ghasemi S 2018 *Sci. Hortic.* **232** 250–5
[22] Uz I and Tavali I E 2014 *Sci. World J.* **2014** 1 - 11
[23] Atiyeh R M, Subler S, Edwards C A, Bachman G, Metzger J D and Shuster W 2000 *Pedobiologia* (Jena) **44** 579–90
[24] Zhi-wei S, Tao S, Wen-jing D and Jing W 2019 *J. Environ. Manage.* **243** 269-272
[25] Parthasarathi K, Balamurugan M, Prashija K, Jayanthi L and Ameer Basha S 2016 *Int. J. Recycl. Org. Waste Agric.* **5** 65–86
[26] Ananthavalli R, Ramadas V, John Paul J A, Karunai Selvi B and Karmegam N 2019 *Bioresour. Technol.* **275** 394-401
[27] Pathma J and Sakthivel N 2012 *Springerplus* **1** 1–19

IOP Conf. Series: Materials Science and Engineering **980** (2020) 012069 doi:10.1088/1757-899X/980/1/012069

The utilization of agricultural waste for peatland management; in case chili cultivation

A Fahmi* and A Susilawati
Indonesian Swampland Agricultural Research Institute

Email: fahmi.nbl@gmail.com

Abstract. Peatland are known fragile and having a low bearing capacity for plant growth and production. One of the technologies that could be applied to improve peatland management is agricultural waste utilization. Agricultural waste such as ash and biochar from rice husk could be used as an ameliorant to improve the peatlands fertility. The objective of the experiment is to study peatland properties, carbon dioxide emission and yield of chili due to the application of ameliorant from agricultural waste. The experiment was conducted on the peatland that located around the farmer's settlement. The treatment of experiment was set up with three treatments i.e.; rice husk biochar (RHB), rice husk ash (RHA) and without ameliorant. The results showed that the application of agricultural waste improved some peatland chemical properties, reduced carbon dioxide gas emission, and increased the yield of chili.

1. Introduction

About 3% of peatlands in worldwide are found in Indonesia [1]. Most of them are located in tidal swamp areas and associated with alluvial soil which is utilized by farmers for rice field. A part of peatland has converted to the social and economic infrastructure. Peatland that located around the farmer's settlement has long been utilized by farmers as agricultural land, especially for horticulture commodity such as chili and tomatoes. However, peatlands are known having fragile properties and low bearing capacity for plant growth and production. Low nutrient levels and high acidity become the main problem in optimizing of peatlands to improve farmers' welfare.

Peatland is a nonrenewable natural resource, mismanagement of peatland causes serious land degradation. Drying land, clearing land with frequent burning by farmers causes peat oxidation and increases the danger of peat fires, which leads to peat loss and increases greenhouse gas emissions, and increases the risk of global warming. Instead, technology is needed in managing peatlands for agriculture that is environmentally friendly and economical.

Rice is a staple food for Indonesian, at least 15-20% of rice husk (RH) as agricultural waste is produced from each grain yield production [2], it means that 800 kg of husk could be obtained for each in 4 t ha^{-1} of grain yield. The utilization of raw RH is generally not recommended as soil ameliorant since it contains high carbon (C) : nitrogen (N) ratio of about 85:1 and is rich in silica and lignin which makes it difficult to be degraded [3]. Keeping in mind the high risk of the environment, as well as the welfare of farmers, a strategy should be adopted for an effective utilization of RH for agricultural management on peatland. However, the presence of high lignin content makes the RH less vulnerable to microbial attack. Pre-treatments are required for effective utilization RH as agricultural waste. A simple technology for utilization of RH is converted of RH to be rice husk biochar (RHB) or rice husk ash (RHA). The research was conducted to study the changes in peatland soil chemical properties, carbon dioxide emission, and yield of chili due to the application of ameliorant from agricultural waste.

2. Methods

The experiment was conducted on moderate peatland that located around the farmer's settlement, Kalampangan Village, Central Kalimantan. The experiment consisted of one treatment factor with six replications. The factor was soil amelioration with RH as raw material, which consisting of RHA, RHB and without amelioration (control) (table 1).

Table 1. Rate of ameliorant and base fertilizer were applied in the experiment.

Treatment code	Ameliorant (t ha^{-1})				Base fertilizer (kg ha^{-1})		
	RHA	RHB	Lime	Cattle manure	Urea	SP 36	KCl
RHA	16.5	0	2	15	125	125	125
RHB	0	15	2	15	125	125	125
Control	0	0	2	15	125	125	125

Soil parameters such as soil pH, soil C, soil electrical conductivity (EC), phosphorus (P), potassium (K), calcium (Ca), magnesium (Mg) and cation exchange capacity (CEC) in soil were analyzed in a laboratory based on Balai Penelitian Tanah [4]. All of the parameters were observed at the end of the experiment.

3. Results and discussion

Ameliorant application increased peat soil pH, higher neutralizing power of ameliorant shown by RHA (figure 1). Previously Masulili et al [5] also reported that RHB and RHA increased soil pH. The higher pH of RHA is possibly related to the presence of carbonates formed during the heating process [6]. According to Islabão et al [7] and Okon et al [8] application RHA improve soil pH of acid soil. This condition related to high base cations contained in RHA [7] as reflected in EC value (figure 3). According to Demeyer et al [9] while organic matter combustion is mineralized and basic cations are converted to their oxides, which can be hydrated and carbonated by exposure to atmospheric conditions after burning. Potassium and Na oxides in ash are mainly responsible for the neutralizing power of acidity [10].

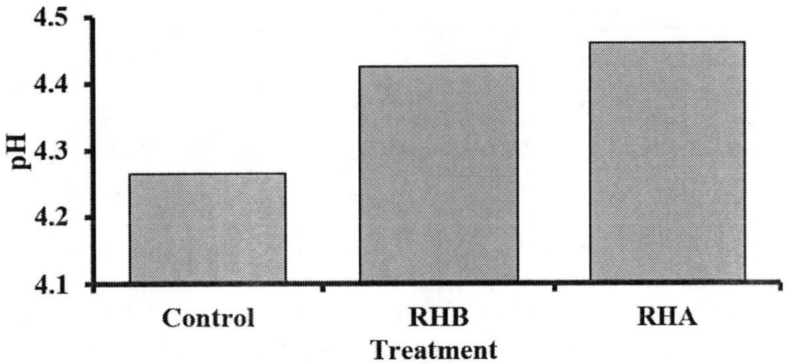

Figure 1. Peat soil pH due to soil amelioration with RHB and RHA.

Total C in peat soil determined with lost on ignition method. Ameliorant application decreased carbon in peat soil (figure 2), it was related to increasing mineral content in peat soil due to RHB and RHA application, as reflected in EC value (figure 3). Although EC did not provide a direct measurement of specific ions or salt compounds, it has been known that EC is correlated to concentrations of sodium (Na), ammonium (NH_4), K, Ca, Mg, iron (Fe), aluminium (Al), copper (Cu), zinc (Zn) and manganese (Mn). At least ameliorant such as RHB containing 50-70% mineral (as ash content) [5;11] and RHA contain 20-38% mineral (as ash content) [12; 13].

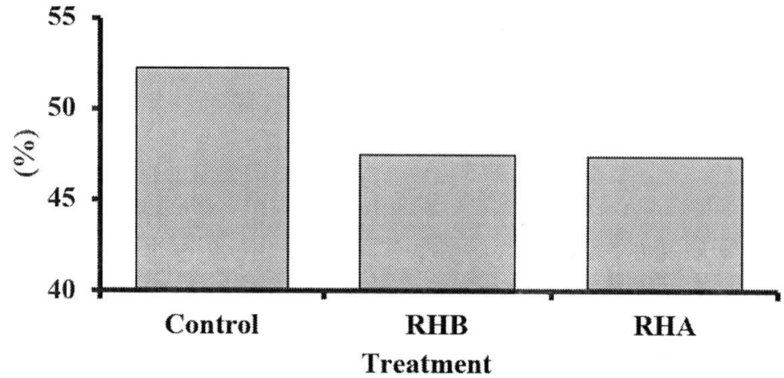

Figure 2. Soil C content due to soil amelioration with RHB and RHA.

Total C in peat soil determined with lost on ignition method. Ameliorant application decreased carbon in peat soil (figure 2), it was related to increasing mineral content in peat soil due to RHB and RHA application, as reflected in EC value (figure 3). Although EC did not provide a direct measurement of specific ions or salt compounds, it has been known that EC is correlated to concentrations of sodium (Na), ammonium (NH_4), K, Ca, Mg, iron (Fe), aluminium (Al), copper (Cu), zinc (Zn) and manganese (Mn). At least ameliorant such as RHB containing 50-70% mineral (as ash content) [5;11] and RHA contain 20-38% mineral (as ash content) [12; 13].

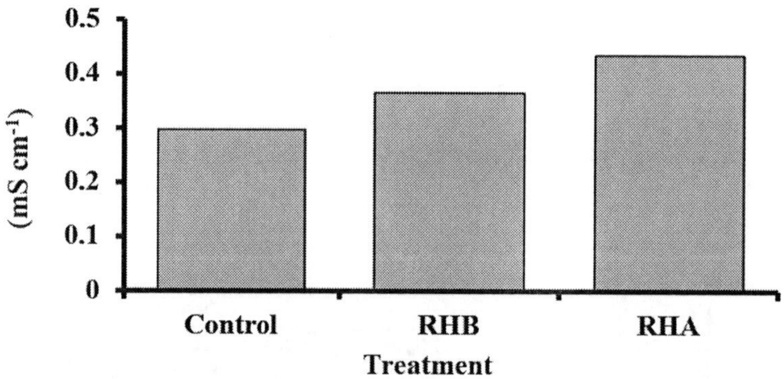

Figure 3. The value of EC due to soil amelioration with RHB and RHA.

Phosphorus and exchangeable K in peat soil increased due to RHB and RHA application (figures 4A and 4B). Similar results were reported by Oladele et al [14] and Okon et al [8]. Increasing concentrations of exchangeable P and K in peat soils are related to the P and K contained in the applied RHB and RHA, chemical analysis in the laboratory shows that RHB [6] and RHA [8] contain P and K which can be utilized by plants. Based on ameliorant types, a higher concentration of P and K in peat soil was shown by RHA application. Phosphorus and K contained in RHA was higher than RHB (table 2), this is in contrary to Masulili et al [5] reported that RHA application caused lower P in acid soil than RHB application. This condition may relate to the difference in P and K contained in RHA and RHB that used in both studies, in which the recent study showed P and K contain in RHA was higher than RHB while the opposite condition is shown from Masulili et al [5].

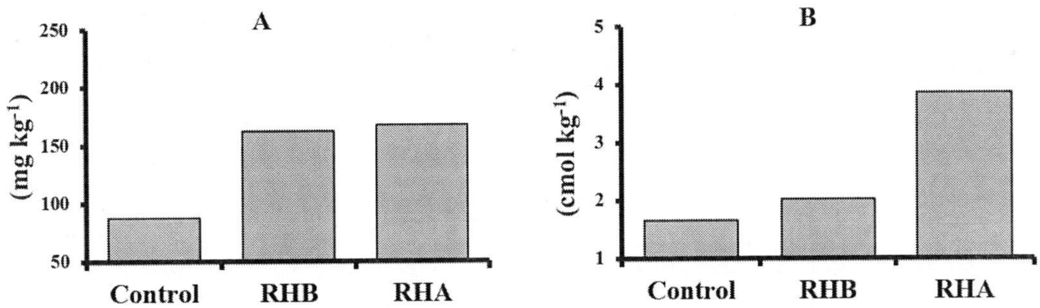

Figure 4. The concentration of exchangeable P (A) and K (B) of peat soil due to soil amelioration with RHB and RHA.

Table 2. Characterictic of ameliorant used in experiment.

Ameliorant	P (%)	K (%)
Cattle manure	1,902	3,459
RHA	0,297	1,310
RHB	0.048	1.107

The opposite of exchangeable P and K concentration in peat soil were shown by exchangeable Ca and Mg concentration, their concentration in peat soil was decreased due to RHB and RHA application (figures 5A and 5B). Previously, Wacal et al [15] also reported a similar condition, they stated that biochar act as a sink rather than a source for exchangeable Ca and Mg. Besides, according to Wacal et al [15] decreased in exchangeable Mg concentration in soil was related to competition between Mg and K in soil solution or adsorption site, RHA application increased the concentration of exchangeable K in peat soil (figure 4). The recent report is similar with Oyama and Hayashi [16] who found that RHA application increase exchangeable K and decreased exchangeable Mg.

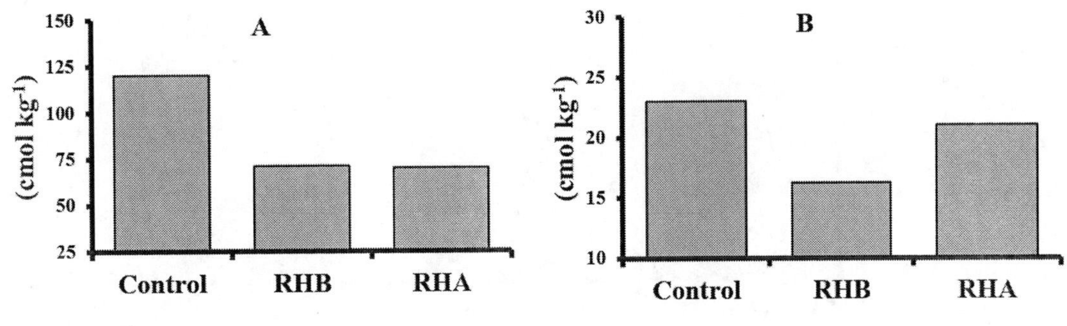

Figure 5. The concentration of exchangeable Ca (A) and Mg (B) of peat soil due to soil amelioration with RHB and RHA.

Peat soil has large CEC, ranges 52-160 cmol kg^{-1} [17], Gruba and Mulder [18] stated large CEC of soils relate with large organic C content. Organic matter has been well established as a source of CEC sites in the solid phase of soils [19]. The recent study showed that CEC of peat soil was 176 cmol kg^{-1}, RHB and RHA application decreased CEC of peat soil (figure 6). This is contrary to Masulili et al [5] and Okon et al [8], they reported that RHB and RHA application increased CEC of acid soil. These difference might be explained with soil as research material that used in each study, in which Masulili et al [5] used mineral soil with low CEC as material in their study, whereas recent study used peat soil with large CEC as material. In mineral soil, RHB and RHA application increased organic material content of soil and decreased the proportion of mineral content of soil. While in the peat soil, RHB and RHA application increased mineral content of soil and furthermore decreased the proportion of organic material content of soil.

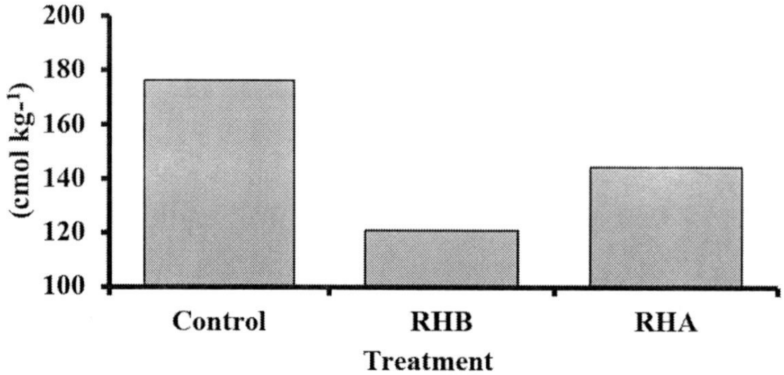

Figure 6. Soil CEC value due to soil amelioration with RHB and RHA.

As stated in the methodology, all of the experiment units receive 25 t ha^{-1} of chicken manure. Commonly, the application of organic fertilizer significantly enhanced CO_2 production, Johansen et al [20] reported that CO_2 emissions increased after cattle slurry application, this fact attributed to the supply of high amounts of easily available C and N, which stimulated microbial activity. The application RHB and RHA decreased CO_2 emission (figure 7). Previously, a similar result was reported by Sarkhot et al [21] that soil amendment with biochar resulted in a reduction in cumulative CO_2 flux, and RHB amendment had significantly reduced CO_2 emissions [22]. RHB and RHA have a high potential to sequester atmospheric CO_2 into more stable soil organic carbon as stated by Zimmerman et al [23]. These results suggest that RHB and RHA applications contribute to mitigating global warming.

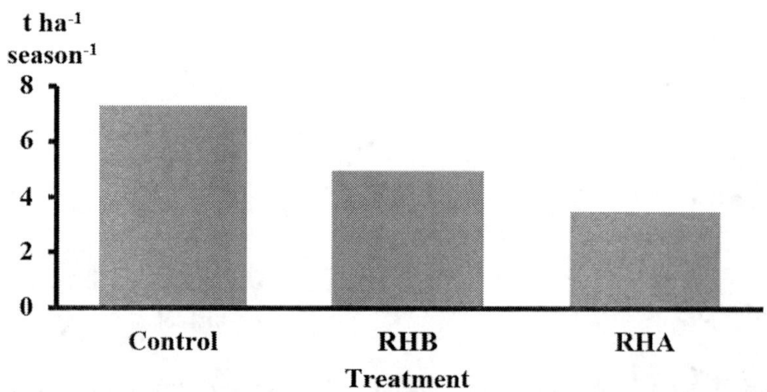

Figure 7. The CO_2 emission value due to soil amelioration with RHB and RHA.

Application of RHB and RHA increased chili yield, the highest yield (17.6 t ha^{-1}) was obtained in the RHA application while RHB application obtained 13.86 t ha^{-1} (figure 8). This result is in line with the report by Priyadharshini and Seran [24] who obtain the highest cowpea result due to applied RHA and chemical fertilizer. The highest result that obtained from RHA application was related to increasing soil quality mainly exchangeable K as shown in figures 1, 4 and 5. According to Gupta and Sengar [25] significant yield of tomato were increased by increasing application rates of potassium.

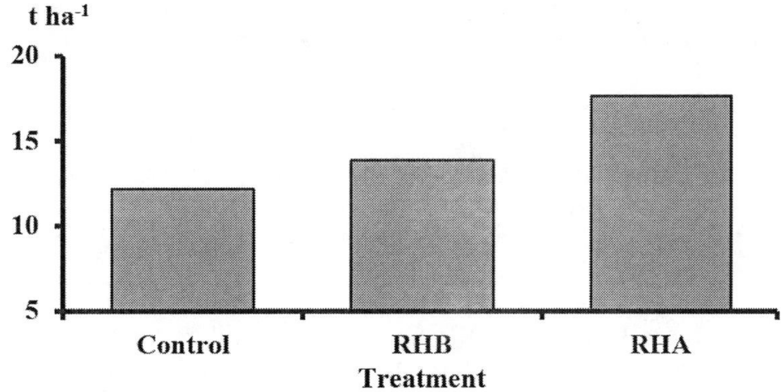

Figure 8. The yield of chili due to soil amelioration with RHB and RHA.

4. Conclusion

The utilization of agricultural waste as ameliorant such as RHB and RHA improved peat soil chemical properties, RHB and RHA application increased peat soil pH, P concentration, and exchangeable K in peat soil, increased chili yield soil, and decreased CO_2 emission. We argue that expanding the use of RHB and RHA in agricultural lands would be important for sequestering atmospheric CO_2 and mitigating climate change.

References

[1] Limpens J, Berendse F, Blodau C, Canadell J G, Freeman C, Holden J, Roulet N, Rydin H and Schaepman–Strub G 2008 Peatlands and the carbon cycle from local processes to global implications – a synthesis *Biogeo* **5** 1475–91

[2] LISD (Layanan Informasi Desa) 2019 Visited 2 of July 2020 https://8villages.com/full/petani/article/id/5cb83662d324d09579ad80f9

[3] Thiyageshwari S, Gayathri P, Krishnamoorthy R, Anandham R and Paul D 2018 Exploration of rice husk compost as an alternate organic manure to enhance the productivity of blackgram in typic haplustalf and typic rhodustalf *Int. J. Environ. Res. Pub. Health* **15(2)** 358

[4] Balai Penelitian Tanah 2005 *Analisis Kimia Tanah, Tanaman, Air dan Pupuk* Badan Penelitian dan Pengembangan Pertanian Departemen Pertanian (Bogor) p 136 (*in Indonesian*).

[5] Masulili A, Utomo W H and Syechfani M S 2010 Rice husk biochar for rice based cropping system in acid soil 1 The characteristics of rice husk biochar and its influence on the properties of acid sulfate soils and rice growth in west kalimantan Indonesia *J. Agric. Sci.* **2(1)** 39-47

[6] Singh C, Tiwari S, Gupta V K and Singh J S 2018 The effect of rice husk biochar on soil nutrient status, microbial biomass and paddy productivity of nutrient poor agriculture soils *Catena* **171** 485-493

[7] Islabão G O, Vahl L C, Timm L C, Paul D L and Kath A H 2014 Rice husk ash as corrective of soil acidity *R. Bras. Ci. Solo* **38** 934-41

[8] Okon P B, Amalu U C and Ogeh J S 2005 Effect of rice husk ash and phosphorus on some properties of acid sands and yield of okra *Commun. in Soil Sci. and Plant Anal.* **36** 833-45

[9] Demeyer A, Voundi N and Verloo M G 2001 Characteristics of wood ash and influence on soil properties and nutrient uptake an overview *Bioresour. Technol.* **73(3)** 287-95

[10] Ulery A L, Graham R C and Amrhein C 1993 Wood-ash composition and soil pH following intense burning *Soil Sci.* **156** 358-364

[11] Singh R, Srivastava P, Singh P, Sharma A K, Singh H, and Raghubansh A S 2019 Impact of rice-husk ash on the soil biophysical and agronomic parameters of wheat crop under a dry tropical ecosystem *Ecol. Indic.* **105** 505-15
[12] Jongpradist J, W Homtragoon, R Sukkarak, W Kongkitkul, and P Jamsawang 2018 Efficiency of rie husk ash as cementitious material in high-strength cement-admixed clay pornkasem *Adv. in Civil Eng.* doi.org/10.1155/2018/8346319.
[13] Menya E, Olupot P W, Storz H, Lubwama M, and Kiros Y 2018 Characterization and alkaline pretreatment of rice husk varieties in Uganda for potential utilization as precursors in the production of activated carbon and other value-added products *Waste Manag.* **8** 104-16
[14] Oladele S O, Adeyemo A J and Awodun M A 2019 Influence of rice husk biochar and inorganic fertilizer on soil nutrients availability and rain-fed rice yield in two contrasting soils *Geoderma* **336** 1-11
[15] Wacal C, Ogata N, Basalirwa D, Handa T, Sasagawa D, Acidri R, Ishigaki T, Kato M, Yamamoto S and Nishihara E 2019 Growth, seed yield, mineral nutrients and soil properties of sesame (*Seseamum indicum* L.) as influenced by biochar addition on upland field converted from paddy pp 229-40 In Biochart as Soil Amendment, Impact on Soil Properties and Suatainable Resources Management ed J M DeLaRosa Agronomy
[16] Koyama S and Hayashi H 2017 Rice yield and soil carbon dynamics over three years of applying rice hust charcoal to an andosol paddy field *Plant prod. Sci.* **20** 176-82
[17] Fahmi A 2012 Saling Tindak Tanah Gambut dan Substratum Bahan Sulfidik Serta Pengaruhnya Terhadap Sifat Kimia Tanah *Disertasi* Universiatas Gadjah Mada p 212
[18] Gruba P and J Mulder 2015 Tree species affect cation exchange capacity (CEC) and cation binding properties of organic matter in acid forest soils *Sci. Tot. Environ.* **511** 655–62
[19] Tan K H 2014 *Humic Matter in the Soil and the Environment; Principles and controversies* (Florida: CRC Press) 439 p.
[20] Johansen A, Carter M S, Jensen E S, Hauggard-Nielsen H and Ambus P 2013 Effects of digestate from anaerobically digested cattle slurry and plant materials on soil microbial community and emission of CO2 and N2O *Appl. Soil Ecol.* **63** 36-44
[21] Sarkhot D V, Berhe A A, and Ghezzehei T A 2012 Impact of biochar enriched with dairy manure eluent on carbon and nitrogen dynamics *J. Environ. qual.* **41** 1107-14
[22] Nguyen, D H, Scheer C, Rowlings D W and Grace P 2016 Rice husk biochar and crop residue amendment in subtropical cropping soils effect on biomass production, nitrogen use efficiency and greenhouse gas emissions *Biol. Fert. soils* **52** 261-70
[23] Zimmerman A R 2010 Abiotic and microbial oxidation of laboratory produced black carbon *Environ. Sci. & Tech.* **44** 1295–301
[24] Priyadharshini J and Seran T H 2009 Paddy husk ash as a source of potassium for growth and yield of cowpea (*Vigna unguiculata* L.) *J. Agric. Sci.* **4** 67-76
[25] Gupta C R and Sengar S S 2000 Response of tomato (Lycopersicon esculentum Mill.) to nitrogen and potassium fertilization in acidic soil of Bastar *Vegetable Sci.* **27** 94-5

IOP Conf. Series: Materials Science and Engineering **980** (2020) 012070 doi:10.1088/1757-899X/980/1/012070

The potential of anaerobic-constructed wetland system for wastewater treatment of rice straw pulping

Y Setiawan* and H Hardiani
Centre for Pulp and Paper-Ministry of Industry, Jl. Raya Dayeuh Kolot No. 132 Bandung-40258, Indonesia

Email: yusupsetiawan1960@gmail.com

Abstract. Wastewater of cooking rice straw in the manufacture of pulp contains high levels of pollutants that must be treated first before being discharged into the environment. The purpose of this research is to study the potential of anaerobic and constructed wetland systems in treating pulp-making wastewater from rice straws. Anaerobic reactor with a volume of 30 L was filled with 5 L domestic sludge as sludge seed. Wastewater with COD of 762 - 17,900 mg/L, BOD_5 of 354 – 8,000 mg/L, and TSS of 123 - 350 mg/L was fed into the anaerobic reactor. The organic loading rate was varied from 0.15 to 2.13 kg COD/m^3.day, during the residence of wastewater for 5-8 days. Anaerobic reactor effluent was further processed in a constructed wetland reactor that varies in height from a mixture of sandy soil of 10-30 cm. The hydraulic residence time of wastewater in the constructed wetland reactor was 6.7 - 11.6 days. Influent and effluent of anaerobic and constructed wetland reactors were analysed for COD, BOD_5, TSS, and pH parameters. The result indicated that wastewater treatment with anaerobic system could reduce COD of 63-72% and BOD_5 of 65-75%, and the constructed wetland system could reduce COD and BOD_5 of 96.5% and 96.1%, respectively.

1. Introduction

Rice straw is an agricultural waste which is available in quite a lot in Indonesia. Rice straw can be used as a raw material for making pulp which is a raw material for making special paper which has a high export value. Pulp making with rice straw raw materials is carried out by small scale industries which are generally located in villages close to the source of raw materials. The small-scale pulp and paper industry generally have old machines, consume a lot of water and energy so that it produces a lot of waste water. A serious environmental problem faced by the agro-based industry is the release of black liquor containing chemicals which are highly polluting the environment in the form of lignin and cooking chemicals that cannot be recovered through chemical recovery unit before being discharged into receiving water bodies. The small-scale pulp and paper industry generally have no chemical recovery plant that functions to recover cooking chemicals that are still contained in black liquor because it is not economical [1,2].

Wastewater containing lignin and cooking chemicals contain a high concentration of pollutants which is difficult to treat in wastewater treatment plants because lignin is not easily decomposed [1,3]. When this wastewater is discharged directly into the environment, it will pollute the receiving water. Therefore, it needs to be treated before being discharged into the environment. Since the pollutants contained in this wastewater are quite high, the treatment of wastewater requires several stages of treatment [4]. Wastewater treatment systems, which are used in the pulp and paper industry, are biological treatment processes such as activated sludge systems, oxidation ponds, activated sludge combined systems and oxidation ponds, and combined systems of anaerobic ponds and oxidation ponds [2]. One alternative wastewater treatment that is able to treat wastewater with high pollutants is an anaerobic biological treatment system. A wastewater treatment with anaerobic system is basically a system where wastewater flows into a reactor containing anaerobic sludge. During the wastewater flow in the reactor, pollutants are degraded by anaerobic microorganisms into simpler compounds. The mechanism of the decomposition process of organic substances by anaerobic microorganisms through the stages of the process of hydrolysis, acidogenesis, acetogenesis, and methanogenesis producing the final result of CH_4 (50-70%) and CO_2 [4]. Wastewater treatment with conventional anaerobic systems is usually in the form of open ponds with organic loads ranging from 0.04-0.30 kg COD/m^3.day and the hydraulic residence time of wastewater in ponds ranges from 30-50 days [5]. The disadvantage of this anaerobic ponds is the problem of odour and loss of methane gas as energy into the atmosphere [6,7].

Wastewater treatment with activated sludge aerobic and anaerobic systems using plastic media such as polypropylene (PP), polystyrene (PS), and polyethylene (PE) as the media for attaching microorganisms has been applied to industrial wastewater treatment [8]. The organic loading rate of this system could reach 3-4 times of the organic loading rate of activated sludge conventional system, 0.2-0.6 kg BOD/kg MLVSS. day with producing similar effluent quality [9]. This wastewater treatment requires quite high construction and operational costs. In addition, the effluent from the initial stage of wastewater treatment has mostly not met the quality standard due to the high pollutant concentration requiring polishing treatment [4].

The constructed wetland has a high level of biological activity that can convert pollutants contained in wastewater into non-toxic compounds. Constructed wetland is used for secondary and tertiary treatment in some cases [10,11]. Constructed wetlands can use local materials and local labour, which is a major advantage in developing countries such as Indonesia [12,13]. The mechanism of reducing pollutant levels that occur in constructed wetlands involves physical treatment such as sedimentation and filtration, chemical treatment such as adsorption and precipitation, and biological treatment such as bacterial metabolism, plant metabolism and plant adsorption. For this reason, it is necessary to investigate the feasibility of applying the constructed wetland subsurface as a viable technology for industrial wastewater treatment. Some studies show that vegetation used in constructed wetland bulrush species is not only able to remove organic and inorganic substances in wastewater but also is able to eliminate the heavy metal content [13–15]. In constructed wetland vegetation has a role in providing oxygen in the root zone, being the attachment of bacteria, and absorbing pollutants in wastewater [12,16].

This paper presents the results of research on wastewater treatment systems for making pulp from rice straw with anaerobic and constructed wetland systems. The purpose of this paper is to look at the potential application of anaerobic and constructed wetland treatment systems as an appropriate wastewater treatment technology for rice straw-based pulp and paper industries.

2. Materials and methods

2.1 Materials

Wastewater used in the experiment was mixed wastewater from cooking rice straw with soda process and pulp washing. The raw material for rice straw was taken from a pulp and paper industry located in West Java. Anaerobic reactor was made of plastic material that has a volume of 30 L. Anaerobic reactor was equipped with a feed channel, biogas channel, and treated wastewater channel. Recycled polypropylene (PP) plastic flakes of mineral water glass with a length of 2 cm and a width of 2 cm with a volume of about 9 L was filled into anaerobic reactors.

Seed sludge from anaerobic ponds treating domestic wastewater in the amount of 5 L with a total solid concentration of 1.8% was used as anaerobic reactor seed sludge. Three wetland reactors were made of plastic materials, each of which has an effective volume of 96.8 L, 124 L and 154 L. At the bottom of each the wetland reactor a pile of gravel about 2-5 cm in diameter was installed as high as about 10 cm. The surface of the gravel was covered by palm fibers about 1 cm high and the palm fiber surface was covered with sandy soil with varying height of about 10 cm (WL1), 20 cm (WL2) and 30 cm (WL3). Water plants used were bulrush (*Scirpus longii*) and water hyacinth.

2.2 Methods

Wastewater was fed continuously into an anaerobic reactor that already contains seed sludge and recycled polypropylene (PP) plastic flakes with wastewater flow rate adjusted to the hydraulic residence time of wastewater in the anaerobic reactor of 5-8 days. Flowing of wastewater was carried out in 2 stages, namely drainage of dilute wastewater as an adaptation of anaerobic seed sludge and drainage of concentrated wastewater. During the experiment, the pH of the wastewater entering the anaerobic reactor was maintained at a pH of 6.5-7.5. In the initial stage of the experiment or an adaptation stage of anaerobic seed sludge, dilute wastewater with a COD concentration of 762-1,800 mg / L was flowed into the anaerobic reactor for 54 days with the flow rate of 4.2 L/day so that the hydraulic retention time (HRT) of 5 days. Starting from the 62nd day to the 99th day, wastewater containing COD of 8.000-17.900 mg/L was flowed into the anaerobic reactor with the flow rate of 2.6 L/d so that the wastewater was adjusted to the hydraulic residence time of the wastewater in the anaerobic reactor of 8 days.

The anaerobic reactor effluent was collected in the storage tank which was further treated in the wetland. Each wetland reactor that has been planted with bulrush (*Scirpus longii*) and water hyacinth was first cleaned from the impurities using tap water until clean, before feeding an anaerobic reactor effluent. Then the wetland reactor was fed by an effluent anaerobic reactor containing COD of around 5,000 mg/L and BOD_5 of around 2,200 mg/L in the flow rate of 14.4 L/day. After 12 days the number of aquatic plants is calculated to see the rate of plant growth. The dimensions and operating conditions of the experimental constructed wetland reactor are as shown in table 1. Schematic diagram of wastewater treatment of rice straw pulping with anaerobic reactors-constructed wetlands system is shown in figure 1.

Table 1. The dimensions and operating conditions of the experimental constructed wetland reactor.

Dimensions:	WL1	WL2	WL3
- Length (m)	1.10	1.10	1.10
- Width (m)	0.40	0.40	0.40
- Height (m)	0.26	0.36	0.46
- Total volume (m^3)	0.094	0.124	0.154
- Water depth (m)	0.10	0.10	0.10
- Substrate depth (m)	0.16	0.26	0.36
- Plants name	bulrush (*Scirpus longii*) and water hyacinth		
- No. of rhizome/m^2	22		
Operation conditions:			
- OLR (g BOD_5/m^2.day)	13.9	8.9	8.9
- HRT (day)	6.72	8.67	10.61

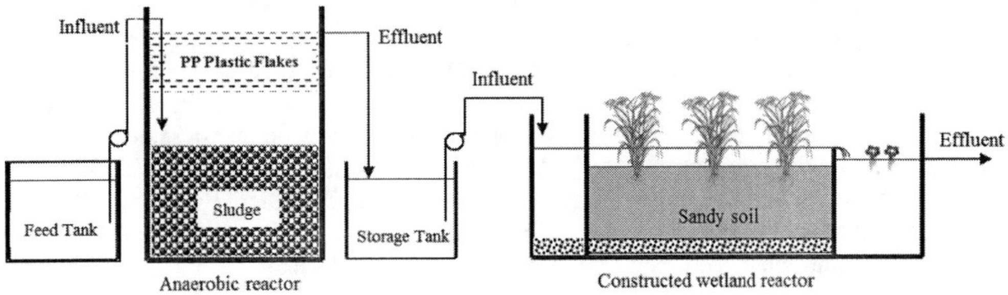

Figure 1. Schematic diagram of wastewater treatment of rice straw pulping with anaerobic reactors-constructed wetlands system.

IOP Conf. Series: Materials Science and Engineering **980** (2020) 012070 doi:10.1088/1757-899X/980/1/012070

Influent and effluent of anaerobic reactors and constructed wetlands which include COD, BOD, TSS, and pH parameters were analyzed using standard methods for the examination of water and wastewater [17].

3. Results and discussions

3.1. Characteristic of wastewater

Characteristics of pulp-making wastewater from rice straw has a neutral pH, with a relatively low total suspended solid (TSS) content. However, it contains high level of organic compounds as indicated by the parameters of COD and BOD_5 as in table 2. This wastewater contains lignin which is not easily decomposed by and anaerobic microorganism and the effluent from an anaerobic biological treatment system has mostly not met the quality standard due to the high pollutant concentration requiring polishing treatment [1,3,4].

Table 2. Characteristics of pulp-making wastewater from rice straw.

No.	Parameter	Unit	Concentration
1.	pH	-	7.0-7.1
2.	TSS	mg/L	123-350
3.	COD	mg/L	762-17,900
4.	BOD_5	mg/L	354-8,005

3.2. Performance of anaerobic reactor

At the beginning of the experiment (1st to 54th day), dilute wastewater was flowed into the anaerobic reactor aims to provide an opportunity for the anaerobic seed sludge to adapt to wastewater. This diluted wastewater run for about 2 months. Dilute wastewater with a concentration of COD around 762-1800 mg / L and BOD_5 around 354-845 mg / L are flowed into the anaerobic reactor at a rate of 4.2 L / day so that the hydraulic residence time (HRT) of wastewater in the anaerobic reactor is about 5 days. The organic loading rate during the adaptation process was between 0.1510-0.3566 kg COD/m^3.day as shown in figure 2. Under these conditions, the anaerobic seed sludge used was of good quality and the amount is optimum enough so that they can quickly adapt to this wastewater. This is indicated by the removal of pollutant parameters, especially COD and BOD_5 whose removal values tend to increase with the length of time of the experiment. As shown in figure 3, removal of COD and BOD_5 which on the 11th day showed a high removal value of 66% for COD and 68% for BOD_5. After that until the 54th day, COD and BOD_5 removal increased to 80-85% and 83-86%, respectively. This shows that the anaerobic reactor is in steady state.

On the 55th day until the end of the 99th day of the experiment, the anaerobic reactor was flowed with concentrated wastewater at a flow rate of 2.6 L / day with a COD concentration of 8,000-17,900 mg/L and a BOD_5 of 3,680-8,005 mg/L. In this condition the organic loading rate is around 0.9811-2.1953 kg COD/m^3.day and hydraulic residence time (HRT) of 8 days as shown in figure 2 and figure 3. This higher organic loading rate causes anaerobic sludge in the anaerobic reactor to a slight shock. This can be seen from the removal of COD and BOD_5 decreasing to 52.5% and 59.1% respectively. Even though the hydraulic residence time (HRT) has been extended to 8 days. But after that the anaerobic sludge in the anaerobic reactor can adapt again to the organic loading rate received which is indicated by the increasing return of COD and BOD_5 removal reaching 72.1% and 74.7%, respectively. At the end of the experiment, concentrations of COD and BOD_5 of 5,000 mg/L and BOD_5 of 2,200 mg/L could be achieved. Such concentrations of COD and BOD_5 are still above the quality standard requirements and require polishing treatment [1,3,4,16].

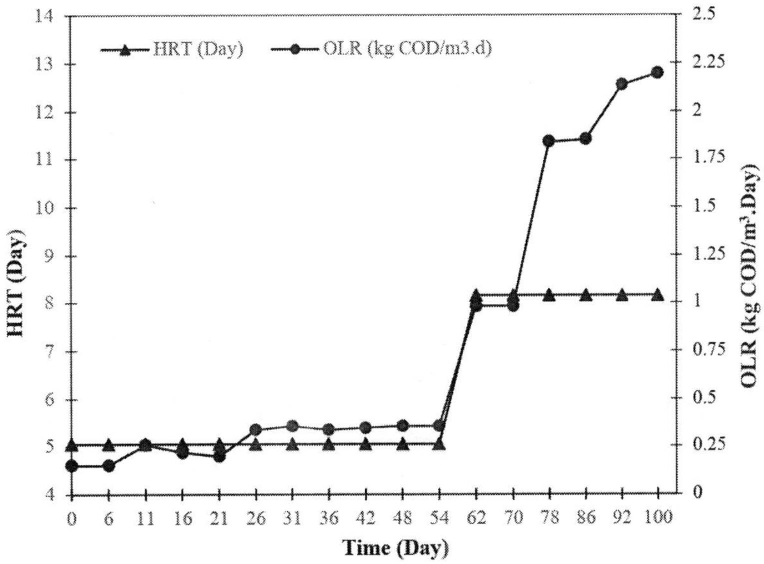

Figure 2. Hydraulic residence time (HRT) and organic loading (OLR) of anaerobic reactor.

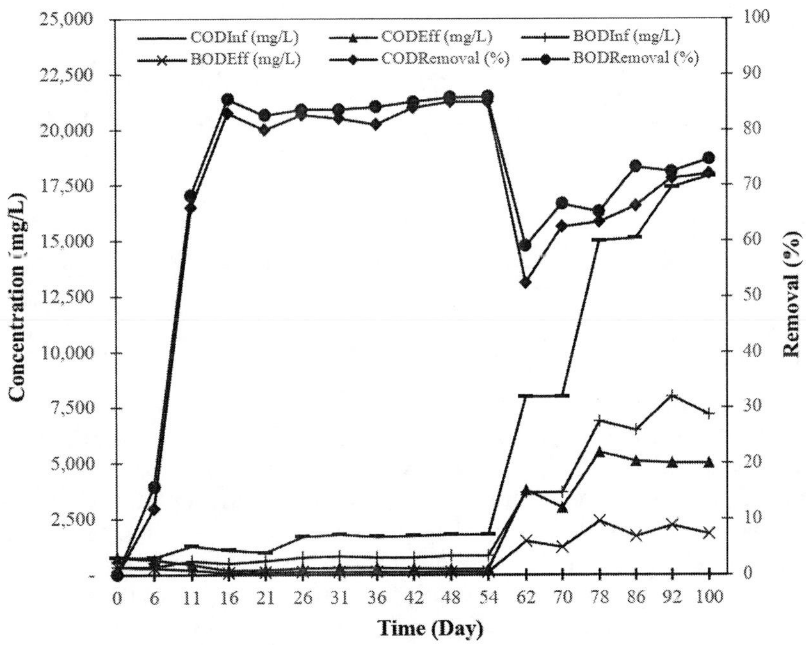

Figure 3. Influent, effluent and removal of anaerobic reactor.

When compared with conventional anaerobic treatment systems, open ponds, which have organic loading rates ranging from 0.04-0.30 kg COD/m^3.day and the hydraulic residence time (HRT) of 30-50 days [5], anaerobic reactors equipped with recycled polypropylene (PP) plastic flakes of mineral water glass in the amount of 30% of the reactor volume can increase organic loading rate and can reduce HRT by around 4 times. This can be caused by the number of microorganisms attached on the surface of recycled polypropylene (PP) plastic flakes. However, the number and types of microorganisms attached to recycled polypropylene (PP) plastic flakes need further study.

Table 3. The plants growth of bulrush (*Scirpus longii*).

Wet land reactor	Number of plant stems and growth rate		
	1st day	12th day	Growth rate (stems/day)
WL1	10	42	2.67
WL2	10	37	2.25
WL3	10	36	2.17

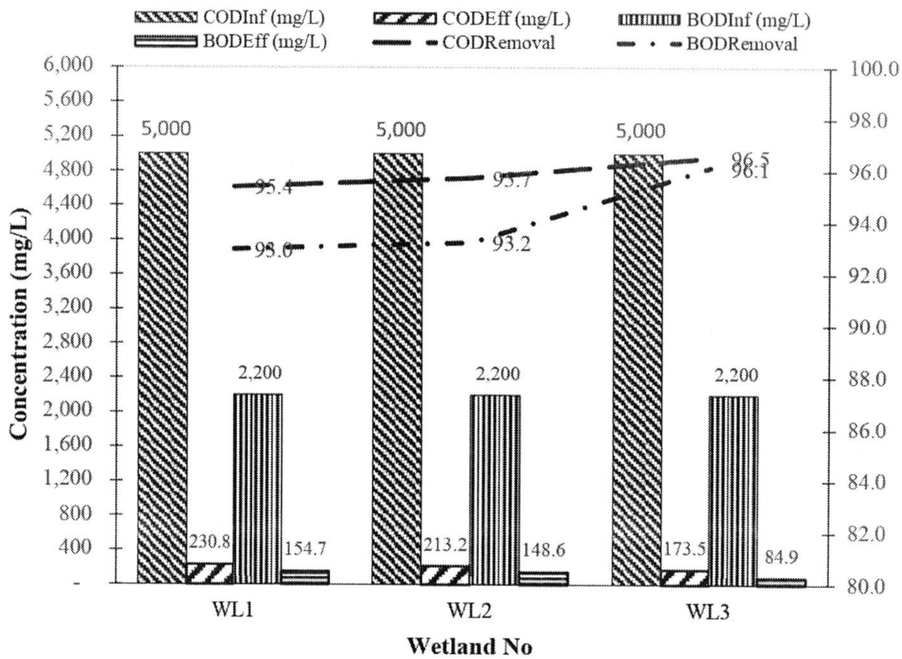

Figure 4. Influent, effluent and removal of constructed wetland.

3.3. Performance of constructed wetland reactor

From observations, bulrush (*Scirpus longii*) can grow in wetlands flowed by wastewater. Number of plant stems and growth rate is shown in table 3. The growth rate of bulrush (*Scirpus longii*) is 2.17-2.67 stems/day and increases with the length of planting time.

In the experimental conditions as shown in table 1 and the growth rate of bulrush (*Scirpus longii*) as shown in table 3, each constructed wetland produced effluent quality with the highest pollutant removals values as shown in figure 4. The highest removal of COD and BOD_5 were obtained by WL3 with COD and BOD_5 removal of 97.6% and 96.1%, respectively. The COD and BOD_5 concentrations were 173.5 mg/L and 84.9 mg/L, respectively. The performance of WL3 is slightly better compared to WL1 and WL2, which is likely to be caused by higher substrate in the form of higher sandy soil in WL3. The height of this substrate affects the number of microorganisms that grow both in the substrate itself and in the roots of aquatic plants [15,18]. In addition, WL3 has a longer HRT (10.61 days) and OLR (8.9 g BOD/m^2.day) which is lower than that of WL1 and WL2. The amount of OLR (8.9 g BOD/m^2.day) from WL3 is slightly larger than Vymazal (2010) recommendation of 6 g BOD/m^2.day for the treatment of various types of wastewater [10].

Overall, wastewater treatment with anaerobic-constructed wetland system can remove COD of 99% and BOD_5 of 96.1% with the effluent containing COD and BOD_5 of 173.5 mg/L and 84.9 mg/L, respectively. The ability to reduce the high concentration of wastewater pollutants, anaerobic-constructed wetland system has the potential to be applied as an appropriate wastewater treatment technology in the rice straw-based pulp and paper industry.

4. Conclusions

Anaerobic-constructed wetland treatment system could remove organic pollutants of rice straw pulping wastewater up to 99% for COD and 96.1% for BOD_5. This high ability in reducing of organic pollutants of rice straw pulping wastewater indicates that anaerobic-constructed wetland system has the potential to be applied as an appropriate wastewater treatment technology in the rice straw-based pulp and paper industry.

Acknowledgments
The authors gratefully acknowledge the fellow analysts, technical personnel, and related parties at the center for pulp and paper for all the assistance so that this research was accomplished.

References

[1] Doma H S and Abou-Elela S I 2003 Treatment of black liquor derived from non-woody feedstocks *Trans. Ecol. Environ.* **63** 453–62
[2] Singh D 2017 Treatment of pulp & paper industry wastewater *Int J Sci Eng Res* **8** 48–52
[3] Issa R M, Abou-sekkina M M, Bastawisy A E M, Khedr A M and El-helece W A 2014 New treatment of the black liquor produced from pulping of rice straw *Elixir ChemPhys* **33** 2276–8
[4] Hubbe M A, Metts J R, Hermosilla D, Blanco M A, Yerushalmi L, Haghighat F, Lindholm-lehto P, Khodaparast Z, Kamali M and Elliott A 2016 Wastewater Treatment and Reclamation : A Review of Pulp and Paper Industry Practices and Opportunities *BioResources* **11** 7953–8091
[5] Tchobanoglous G, Burton F L and Stensel H D 2003 *Wastewater Engineering, Treatment and Reuse* ed Metcalf & Eddy Inc. (McGraw-Hill Companies, Inc)
[6] Lier J B Van, Mahmoud N and Zeeman G 2008 Anaerobic Wastewater Treatment *Biological Wastewater Treatment: Principles Modelling and Design* ed M Henze, M C M van Loosdrecht, G A Ekama and D Brdjanovic (London, UK: IWA Publishing) pp 401–2
[7] Sari F P 2014 Enhanced biogas production from rice straw with various pretreatment : a review *Waste Technol.* **2** 17–25
[8] Tawfik A, Badr N, Taleb E A and El-Senousy W 2012 Sewage treatment in an up-flow anaerobic sponge reactor followed by moving bed biofilm reactor based on polyurethane carrier material *Desalin. water Treat.* **37** 350–8
[9] Chai S, Guo J, Chai Y, Cai J and Gao L 2014 Desalination and Water Treatment Anaerobic treatment of winery wastewater in moving bed biofilm reactors *Desalin. Water Treat.* **52** 1841–9
[10] Vymazal J 2010 Constructed Wetlands for Wastewater Treatment *Water* **2** 530–49
[11] Mustafa A 2013 Constructed Wetland for Wastewater Treatment and Reuse : A Case Study of Developing Country *Int. J. Environ. Sci. Dev.* **4** 20–4
[12] Try E, Sembiring J and Muntalif B S 2011 Optimization of leachate treatment efficiency by using constructed wetland *J. Tek. Lingkung.* **17** 1–10
[13] Khazaleh M and Gopalan B 2019 Constructed Wetland for Wastewater Treatment Constructed Wetland for Wastewater Treatment *J. Mod. Sci. Technol.* **6** 78–86
[14] Šereš M, Hnátková T, Vymazal J and Vaněk T 2017 Removal efficiency of constructed wetland for treatment of agricultural wastewaters *Chem. J. Mold.* **12** 45–52
[15] Skrzypiec K and Gajewska M H 2017 The use of constructed wetlands for the treatment of industrial wastewater *J. Water L. Dev.* **34** 233–40
[16] Zeb B S, Mahmood Q, Jadoon S, Pervez A, Irshad M, Bilal M and Bhatti Z A 2013 Combined Industrial Wastewater Treatment in Anaerobic Bioreactor Posttreated in Constructed Wetland *Biomed Res. Int.* **2013** 1–9
[17] APHA 1999 *Standard Methods for the Examination of Water and Wastewater* (Washington,

IOP Conf. Series: Materials Science and Engineering **980** (2020) 012070 doi:10.1088/1757-899X/980/1/012070

DC: American Public Health Association, American Water Works Association, Water Pollution Control Federation)
[18] Polprasert C 2007 *Organic Waste Recycling Technology and Management* (London: IWA Publishing)

IOP Conf. Series: Materials Science and Engineering **980** (2020) 012071 doi:10.1088/1757-899X/980/1/012071

Utilization of lightweight brick waste as soils stabilizing agent

Y F Arifin[1, 2]*and A S Kusworo[1]
[1] Civil Engineering Study Program, University of Lambung Mangkurat, Banjarbaru, Indonesia
[2] Wetland Based Material Study Center, University of Lambung Mangkurat, Banjarbaru, Indonesia

Email: y.arifin@ulm.ac.id

Abstract. The utilization of lightweight bricks as building walls, including houses, is getting higher. Beside its many advantages, lightweight brick also has disadvantages. One of them is that the remaining pieces cannot be used for its primary function and become waste. In the field, this waste is used as a material to stabilize the soil. However, there are no tests that examine its effectiveness as a stabilizing agent, the percentage of its composition, and what soil can be stabilized. This research focuses on laboratory tests on the use of lightweight brick waste for soil stabilization. The types of soil used in this study were lateritic, organic and expansive soil. The soils and waste were mixed at percentages of 2%, 4%, 6%, 8%, and 10% by weight. Some tests, such as the Atterberg limit, Standard Proctor compaction, and unconfined compression tests were carried out to determine its effect on the soil. The results showed that lightweight brick waste reduced the liquid limit, plasticity index, and increased soils shear strength. Different types of soil produced different levels of influence. Moisture content in the mixtures was also found to affect the results of soil stabilization.

1. Introduction

Lightweight brick has been widely used in the world since the 1980s, including Indonesia. Due to its high demand, one lightweight brick manufacturing company in Indonesia can produce one million cubic meters per year.

There are two types of lightweight bricks, namely Cellular Lightweight Concrete (CLC) and Autoclaved Aerated Concrete (AAC). The type commonly used today is the AAC. Its main ingredients are quartz sand, cement, lime, a little gypsum, water, and aluminum paste. The CLC type is rarely used because it is made of a material containing fly ash, which is still considered hazardous waste.

Lightweight brick has advantages over other conventional bricks. Aside from being lightweight, it attributes include low water absorption and high strength that qualifies it as a concrete brick walls [1]. However, the disadvantage of using lightweight bricks is that the remaining pieces cannot be used anymore. This is because the installation requires a flat surface and a thin distance between bricks. A survey in one real estate in Banjar District found that an average of $1m^3$ waste was produced from every $20m^3$ of lightweight brick used.

In Dian Anugerah Regency's real estate in Banjar Regency, the lightweight brick waste was mixed with soil embankment. A field dynamic cone penetrometer (DCP) [2] test was carried out on the soil with and without lightweight bricks. This resulted in the California Bearing Ratio (CBR) of 8.1% and 4.3%, respectively. The addition of the waste has succeeded in increasing the soil's bearing capacity almost two times. However, there is no study about the optimal percentage of lightweight brick

IOP Conf. Series: Materials Science and Engineering **980** (2020) 012071 doi:10.1088/1757-899X/980/1/012071

addition to achieving maximum strength and soil water content suitable for the mixture. A research should be carried out to examine the percentage of lightweight bricks in addition to soil.

Many methods have been carried out by researchers to increase the carrying capacity of the soil. This is carried out by mixing it with other materials both with natural ingredients [3–7] and chemistry. These materials include cement [8–10], lime, and fly ash [11,12], and mixtures both (i.e., cement and palm Kernel shell ash)[13]. Some of these materials are residual or waste material from the industry.

Industrial wastes that have been examined as soil stabilization agents include rubber tires [14], [15], tiles [16], and marble [17]. Waste tires and marble are only used as fillers and they require other material agents as additives such as cement or lime. Tile waste can be used as a stabilizing material. It is used as a filler and binder that react with the soil chemically [16]. No research has been carried out on the use of lightweight brick waste as a soil stabilization agent. This research focuses on the use of lightweight brick waste to increase soil shear strength.

2. Materials and Methods

2.1 *Materials used*

2.1.1 Soils. The three types of soil used in this study were lateritic, organic, and expansive soil. Lateritic and organic soils were taken from Banjarmasin and Banjarbaru, respectively. The expansive soil was a mixture of low plasticity clay and commercial bentonite sold as an industrial product. The basic soil properties used were summarized in table 1. Before adding to the stabilization agent, the chemical compound of the soil used was tested using X-ray Fluorescence (XRF) analyzer, and the results are summarized in figure 1.

2.1.2 Light brick waste. Light brick waste used is an AAC type. In this research, the term lightweight brick was abbreviated as LWB for simplification. The LWB had a dry volumetric weight of 650kg/m^3. The chemical composition of LWB was also shown in figure 1. The LWB was crushed and sieved using sieves No. 4 for compaction test, No. 20 for unconfined compression test (UCT), and No. 40 for Atterberg limits before it was used.

Table 1. Basic properties of soils used.

Properties		Soil		
		Lateritic	Organic	expansive
Specific Gravity		2.67	2.61	2.56
Water content	%	20.89	53.26	7.00
Gravel (> 2 mm)	%	7.96	0.31	2.50
Coarse sand (0.6-2.0 mm)	%	9.06	4.90	1.01
Medium sand (0.2-0.6 mm)	%	8.31	4.48	1.22
Fine sand (0.05-0.2 mm)	%	12.47	3.73	4.20
Silt and Clay (0.002-0.05)	%	25.01	35.28	41.64
Clay (<0.002mm)	%	37.19	51.31	49.44
Liquid limit	%	47.80	58.80	80.03
Plastic limit	%	23.98	36.61	25.00
Plasticity Index	%	23.82	22.19	55.03

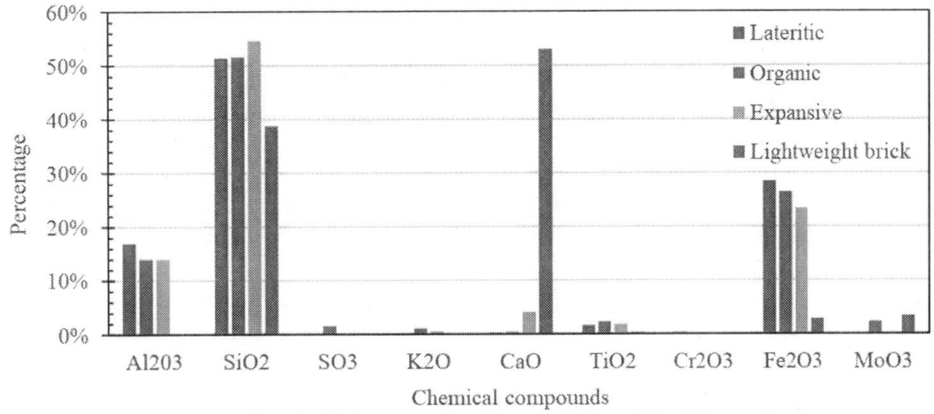

Figure 1. Chemical compounds of material used in this study.

2.2. *Methods and procedures*

2.2.1 Samples preparation. Soils and *LWB* waste samples were mixed by percentages of 2%, 4%, 6%, 8%, and 10% *LWB* content based on the dry weight, after considering the moisture content of the soil. The sample's initial density and moisture content were determined based on the results of the standard Proctor compaction test (ASTM D 698-07) [18]. The standard proctor compaction curves of lateritic and organic clay were shown in figure 2. The maximum dry densities of lateritic and organic clay were 1.7t/m^3 and 1.44t/m^3 respectively. Furthermore, the optimum moisture content of lateritic and organic clay were 20% and 30% respectively. Considering field application, the percentage of 0.95% maximum dry density was used i.e., 1.6t/m^3 and 1.37 t/m^3 for lateritic soil and organic clay respectively. The moisture content of the mixture was also determined by the compaction result (i.e., the moisture content of 15%, 20%, 25% for lateritic soils, and 25%, 30%, and 35% for organic clay). The moisture content and density of expansive soil were assumed to be the same as lateritic soils because of the limited available bentonite. The samples were statically compacted to achieve the density mentioned above.

2.2.2 Physical properties and shear strength tests. The Atterberg limits test [19] was performed to investigate the effect of stabilizing agents on the soil's basic properties. Shear strength compacted samples were determined using the unconfined compression test (UCT) [20]. X-ray diffraction test was also conducted to investigate the effect of the stabilizing agent on the mineralogy of samples.

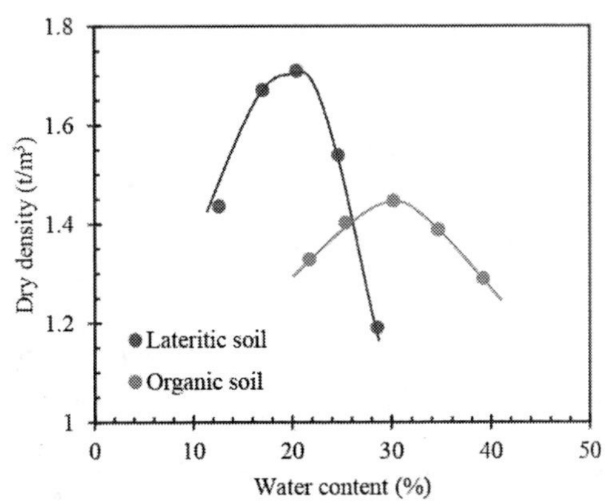

Figure 2. Standard Proctor compaction curves of lateritic and organic soils.

3. Results and Discussion

Figure 3 shows the Atterberg limits test results which include the liquid limit (LL) and plasticity index (PI). Plastic limit (PL) was represented in a plasticity index (i.e., PI = LL-PL). Consistently, each soil's data was represented by different colors, i.e., blue for lateritic, yellow for organic and green for expansive soils. Moreover, the initial data of the soil was presented in black.

Figure 3(a) shows the effect of *LWB* addition to the soil liquid limit. As shown in the figure, the initial LL samples were 80.03%, 58.80%, and 47.80% respectively. Expansive, organic, and lateritic soils were reduced to 74.8%, 54.2%, and 43.1% at 10% *LWB* content. LL samples were reduced by an average of 4%. It can be concluded that LL samples are affected by lightweight brick waste because its value decreases by an increasing percentage of *LWB*.

Figure 3(b) shows a graph of the relationship between the percentage of lightweight brick waste with PI. The initial PI of expansive, organic, and lateritic samples were 55.03%, 22.19%, and 23.82%, respectively, decreased to 54.2%, 17.36%, and 16.36%. Although the LL reduction was little, increasing the percentage of *LWB* also gave rise to PL, as shown by a significant decrease in PI.

Figure 3(c) shows the plasticity chart which was used to determine the soil classification by USCS method (ASTM D2487)[21]. This curve also clarifies the type of soil used in this study, namely clay with high plasticity (CH), organic soil with high plasticity (OH), and clay with low plasticity (CL), respectively, for expansive, organic, and lateritic soils. The addition of *LWB* results in the movement of data towards different soil types, are represented by arrows. This occurred in the lateritic sample that was turned into silt soil with low plasticity (ML). Meanwhile, the other soils turned to silt soils with high plasticity and organic with low plasticity, respectively, for expansive and organic soils. Changes in clay type into silt soils with larger grain sizes indicate the aggregation of clay granules due to the addition of *LWB*.

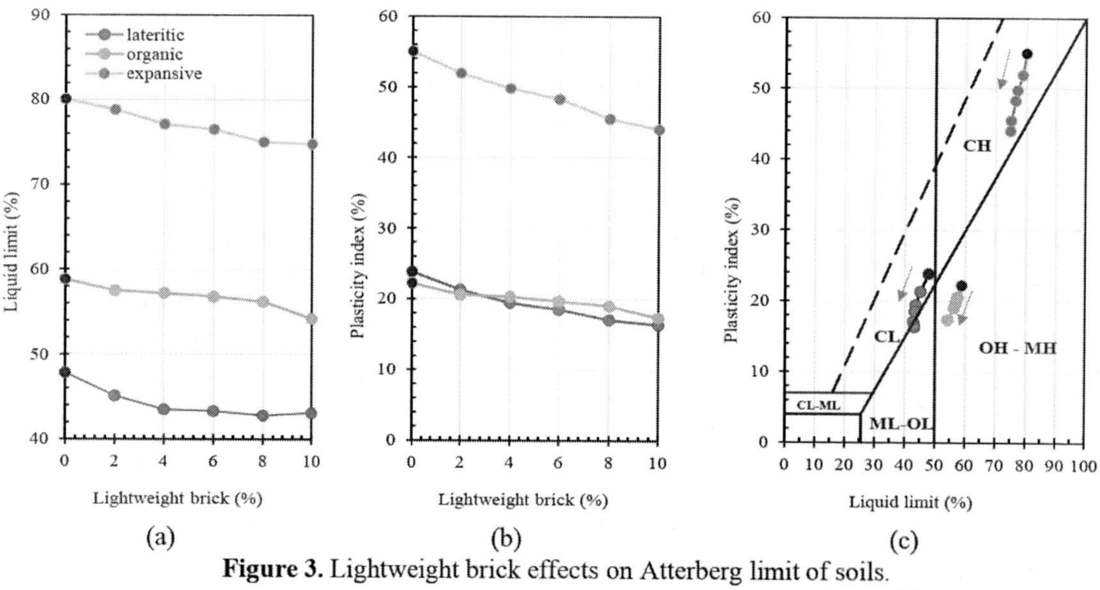

Figure 3. Lightweight brick effects on Atterberg limit of soils.

Figure 4 shows typical UCT results obtained in this study. The figure shows the compressive strength versus strain curves of the materials tested at different moisture contents and represented by 4% *LWB*. As shown in figures 4(a) and 4(b), the samples tested at the optimum moisture content (as shown in figure 2 produced the highest maximum compressive strength (q_u). Conversely, a sample which was compacted at higher moisture content than the optimum moisture content produced the lowest qu. These results show that the samples were still characterized as clay soils. The compaction at optimum moisture content resulted to maximum density followed by other beneficial behaviors, including maximum shear strength.

Figure 4(c) shows the curve of compressive strength versus strain relationships for samples of expansive soil mixed with 4% *LWB*. The highest maximum compressive strength was found at the highest moisture content, which was at w = 25%. This happened because the assumptions of the moisture content used for this sample are smaller than its optimum moisture content. The optimum moisture content for expansive soils was generally higher. The addition of bentonite increased soil plasticity and optimum soil moisture content [22]. However, these results provided valuable information where the addition of *LWB* to the soil was carried out. It was implemented at optimum moisture content or smaller to obtain favorable effects. The result shown in figure 5 is the relationship between moisture content and undrained cohesion (c_u) of samples. Undrained cohesion is a parameter of soil shear strength where $c_u = ½q_u$. As shown in figures 5(a) and 5(b), the peak of c_u occurs in a

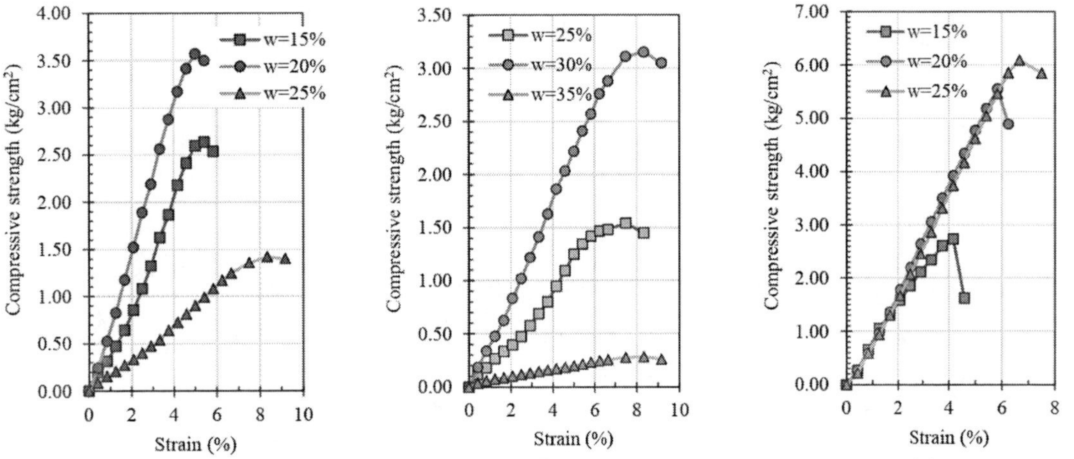

(a) (b) (c)

Figure 4. Compressive strength sample versus strain of soils with 4% lightweight brick content (a) lateritic soil, (b) organic soil, and (c) expansive soil.

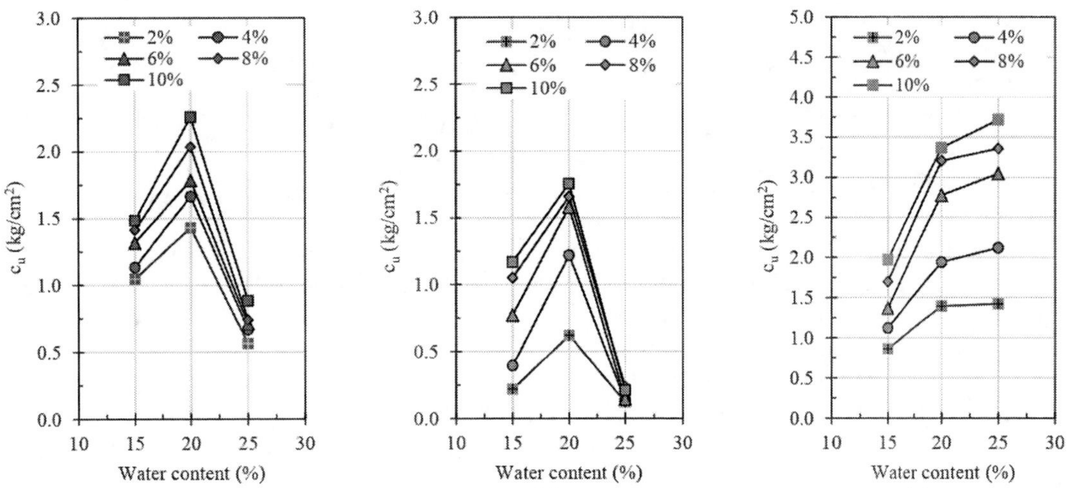

(a) (b) (c)

Figure 5. Cohesion undrained of compacted soils-LWB mixtures as a function of water content (a) lateritic soil, (b) organic soil, and (c) expansive soil.

compacted sample at the optimum moisture content for lateritic and organic soils. Furthermore, for expansive soils compacted at moisture content less than the optimum moisture content, the c_u increased by the rise of moisture content.

Figure 6 shows the undrained cohesion of the sample as a function of mild brick waste content. Figure 6(a) shows the c_u value of the sample compacted at the lowest moisture content (or at dry of optimum for lateritic and organic soils). For lateritic soils, cu increased from 1.23 kg/cm^2 to 2.26 kg/cm^2 or increased 1.8 times at 10% LWB percentage. Similar results were also obtained in the organic-*LWB* clay mixture where c_u increased from 0.24kg/cm^{2}to 1.76kg/cm^2 or increased 7.5 times at the addition of 10% *LWB*. c_u increases by the rise in the percentage of lightweight brick waste. The highest increment is seen in figure 6(b) for samples compacted at the optimum moisture content (i.e., lateritic and organic soils). As for expansive soils, the value of c_u increased from 0.926kg/cm^2 to 3.37kg/cm^2 or 3.6 times at 10% *LWB* content.

Figure 6(c) shows the c_u sample compacted at higher moisture content than the optimum moisture content (also called the wet of optimum) for lateritic and organic samples. As shown in the figure, c_u lateritic samples 0.55kg/cm^2 at 2% *LWB* increased to 0.88kg/cm^2 or 1.6 times the initial c_u. For organic soils, c_u increases from 0.12kg/cm^2 to 0.21kg/cm^2 or 1.7 times at 2% and 10% *LWB* content. These values are smaller than c_u samples without *LWB* compacted at the optimum moisture content (figure 6(b)). Different results were seen in expansive soils compacted at the dry of optimum. c_u increased significantly from 1.62kg/cm^2 at 2% *LWB* to 3.72kg/cm^2 at 10% *LWB*. This result confirmed the previous results where the best result of *LWB*-soil mixtures was compacted at the optimum moisture content or less.

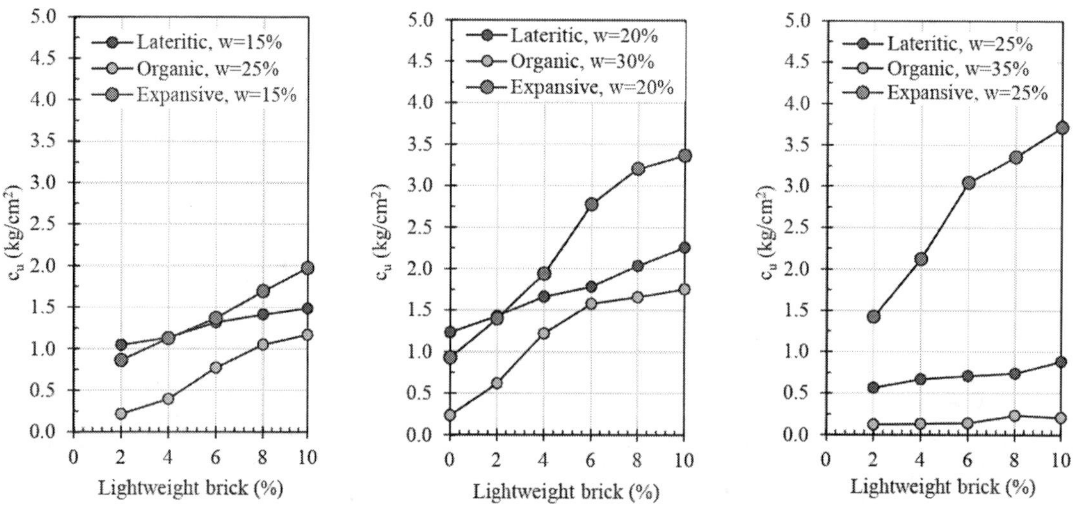

(a) (b) (c)

Figure 6. Cohesion undrained samples as a function of LWB content.

Figure 7 presents the results of X-ray diffraction analysis conducted on soil samples. This was mixed with 2% and 10% *LWB* at optimum moisture content for lateritic soils organic clays. As shown in figures 7(a) and 7(b), lateritic soils and organic clays contained dominant kaolinite and quartz. Furthermore, *LWB* contained dominant quartz and calcite (CaCO3). The presence of calcite was confirmed by comparing it with the XRD pattern reported by [23]. From figure 7(a), an increase in the intensity of quartz is recognized at 2θ between 30°-40° followed by the loss of the calcite peak at intervals of 25°-30°. This behavior showed an increase in ionic bonds with Si producing SiO. Figure 7(a) also revealed the emergence of low-intensity calcite peaks in lateritic soils added by 10% *LWB* due to the remaining non-bound calcites.

The XRD results of organic clay and *LWB* mixtures are shown in figure 7(b). There was little change in the intensity and peaks in both samples with 2% and 10% *LWB*. Change only takes place in an increase in quartz intensity. These results confirmed that the ionic bond was not stable. Therefore, the excess water in the sample resulted in a slight increase in the sample's shear strength. This was obtained for lateritic and organic samples compacted at the high moisture content (figure 6(c)).

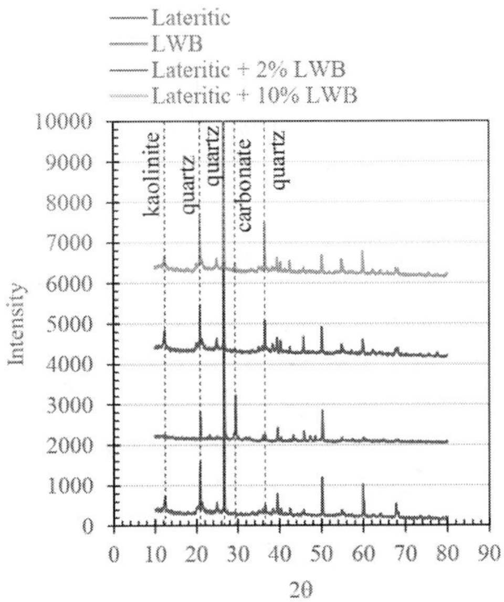

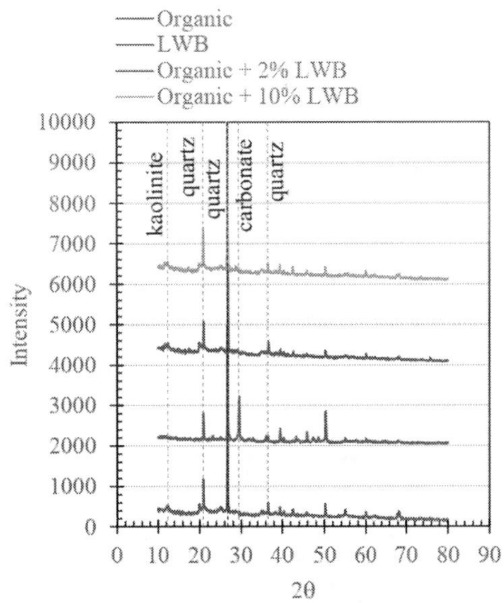

(a) (b)

Gambar 7. X-ray diffraction result of lateritic and organic soils-LWB mixtures.

4. Conclusions

Stabilization of lateritic soils, organic clays, and expansive soils using lightweight brick waste has been presented and discussed. LL and PI samples were affected by *LWB* because the values decreased by an increase in the percentage of waste. The rise in the percentage of *LWB* waste also increased PL shown by significantly reducing PI values. The tendency for soil to change occurred by an increase in the *LWB* content. The data evolution in the plasticity chart indicated the result.

Increasing the *LWB* percentage also raised the shear strength of the soil, as designated by an increase in undrained soil cohesion. Initial moisture content affected the shear strength of the *LWB*-mixed soil sample. Compactions at optimum moisture content or smaller produced the best effect of *LWB*. Conversely, compactions at higher moisture content than the optimum produced less shear strength than the soil's strength without a stabilizing agent. The XRD results confirmed the presence of ionic bonds by reducing the calcite peak and increasing the intensity of the quartz peak of soils-*LWB* mixtures.

References

[1] SNI03-0349 1989 Bata beton untuk pasangan dinding Badan Stand. Nas.
[2] ASTM D6951 2018 Standard Test Method for Use of the Dynamic Cone Penetrometer in Shallow Pavement Applications ASTM Int. West Conshohocken PA
[3] Ashour T and Wu W 2010 The influence of natural reinforcement fibers on erosion properties of earth plaster materials for straw bale buildings J. Build. Apprais **5** 329–40
[4] Anggraini V, Asadi A, Huat B B K and Nahazanan H 2015 Performance of Chemically Treated Natural Fibres and Lime in Soft Soil for the Utilisation as Pile-Supported Earth Platform Int. J. Geosynth. Gr. Eng. **1** 1–14

[5] Brahmachary T K, Ahsan M K and Rokonuzzaman M 2019 Impact of rice husk ash (RHA) and nylon fiber on the bearing capacity of organic soil SN Appl. Sci. **1** 1–13
[6] Arifin Y F, Misnawati and Normelani E 2019 The use of natural fiber from oil palm empty fruit bunches for soft soil stabilization IOP Conf. Ser. Mater. Sci. Eng. **669** 2–10
[7] Arifin Y F, Misnawati and Ridha M 2020 Laboratory Compaction Method of Soft Clay and Natural Plant Fiber/Shell Mixtures IOP Conf. Ser. Earth Environ. Sci. **499**
[8] Shooshpasha I and Shirvani R A 2015 Effect of cement stabilization on geotechnical properties of sandy soils Geomech. Eng. **8** 17–31
[9] Biswal D R and Sahoo U C 2018 Strength and Stiffness Studies of Cement Stabilized Granular Lateritic Soil Soil Testing, Soil Stability and Ground Improvement **2** 320–36
[10] Amadi A A and Osu A S 2016 Effect of curing time on strength development in black cotton soil – Quarry fines composite stabilized with cement kiln dust (CKD) J. King Saud Univ. - Eng. Sci., **30** 305–12
[11] Sharma N K, Swain S K and Sahoo U C 2012 Stabilization of a Clayey Soil with Fly Ash and Lime: A Micro Level Investigation Geotech. Geol. Eng. **30** 1197–205
[12] Choobbasti A J, Ghodrat H, Vahdatirad M J, Firouzian S, Barari A, Torabi M and Bagherian A 2010 Influence of using rice husk ash in soil stabilization method with lime Front. Earth Sci. China b 471–80
[13] Arifin Y F and Rahman G 2019 Stabilization of Soft Soil with Cement and Palm Kernel Shell Ash Admixture MATEC Web Conf. **280** 04011
[14] Nivedya M K and Veeraragavan A 2017 Sand–Tire Chip Mixtures for Sustainable Geoengineering Applications Sustain. Issues Civ. Eng. ed G L Sivakumar Babu et al (Springer) 35–50
[15] Mashiri M S, Vinod J S, Sheikh M N and Tsang H H 2015 Shear strength and dilatancy behaviour of sand-tyre chip mixtures Soils Found. **55** 517–28
[16] Al-Bared M A M, Marto A and Latifi N 2018 Utilization of Recycled Tiles and Tyres in Stabilization of Soils and Production of Construction Materials – A State-of-the-Art Review KSCE J. Civ. Eng. **22** 3860–74
[17] Bhavsar S N, Joshi H B, Shrof P K and Ankit P 2014 Impact of Marble Powder on Engineering Properties of Black Cotton Soil Impact of Marble Powder on Engineering Properties of Black Cotton Soil Int. J. Sci. Res. Dev. **2** 136
[18] ASTM D698 2007 Standard Test Methods for Laboratory Compaction Characteristics of Soil Using Standard Effort ASTM Int 1-15
[19] ASTM D4318 2010 Standard Test Methods for Liquid Limit, Plastic Limit, and Plasticity Index of Soils ASTM Int 1–14
[20] ASTM D2166 2013 Standard Test Method for Unconfined Compressive Strength of Cohesive Soil ASTM Int. 1–7
[21] ASTM D2487 2006 Standard Practice for Classification of Soils for Engineering Purposes (Unified Soil Classification System) ASTM Int. West Conshohocken, PA 1–5
[22] Arifin Y F and Sambelum 2019 Bentonite Enhanced Soil as an Alternative Landfill Liner in Rikut Jawu, South Barito IOP Conf. Ser. Earth Environ. Sci. **239**
[23] Galván-Ruiz M, Hernández J, Baños L, Noriega-Montes J and Rodríguez-García M E 2009 Characterization of Calcium carbonate, calcium oxide, and calcium hydroxide as starting point to the improvement of lime for their use in construction J. Mater. Civ. Eng. **21** 694–98

IOP Conf. Series: Materials Science and Engineering **980** (2020) 012072 doi:10.1088/1757-899X/980/1/012072

Overcoming constraint of tidal swampland with water management with one-way flow system to increase of rice growth

W A Yusuf[1]* and E Setiawati[2]
[1]Indonesian Swampland Agricultural Research Institute, Banjarbaru 70712, Indonesia
[2]Institute of Research and Standardization of Industry, Ministry of Industry, Jl. Panglima Batur Barat No.2, Banjarbaru South Kalimantan, Indonesia

Email: annisa_balittra@yahoo.com

Abstract. The paper addresses the issue whether water management is the key and the only practical way to reclaim and ameliorate a large area of tidal swampland. It reviews recent developments in water management practices with special emphasis on leaching acid sulphate soils are found throughout the length and breadth of the earth. Success in water management is the key for the reconciliation between agricultural yields and environmental sustainability. Water management of tidal swampland, which is dominated by acid sulphate soils, is a serious problem and challenges to reconcile agricultural practices and sustainable quality of ecosystem. Mismanagement of water dynamics may bring about catastrophic consequences instead of friendly environment. The local knowledge/wisdom in managing water supply and drainage to sustain rice yields with a two-way flow system of channels that allows a supply of fresh water during spring tide and drainage during tide has been introduced. The traditional system for reclaiming the tidal swampland by digging a drainage channel called *handil*. The experiment was carried out to investigate the effects of water management on soil chemistry of tidal swampland. Water management with a one-way flow system reduced of soil acidity, toxic elements and increased rice growth in tidal swampland.

1. Introduction

Lowland areas, especially swampland, have a high potential to be used as a strategic choice for development OF agricultural production area to cope with food security [1]. Acid sulphate soil is one type of soil found in tidal swampland area and classified as marginal land because the soil has unprofitable properties with a low fertility rate. The presence of sulfidic material which contains a lot of pyrite is the specific characteristic of acid sulphate soil. This pyrite has unique properties and depends on the state of the water. Pyrite in acid sulphate land has become a serious obstacle in the development of tidal swampland for rice cultivation. The pyrite formed is stable under such reduction condition unless to oxidized was produced of sulphuric acid [2-4]. The oxidation of the pyrite occurs in tidal swampland as a result of artificial drainage or natural phenomena such as falling water levels, e.g. droughts [5]. Water management is the key to success in the development of tidal swampland to provide water supply as well as maintain reductive conditions. Lack of water or excess of water will create a disaster. Crops that experience water deficit will result in a decrease of product quality and quantity. Excessive water will create the decrease of harvesting yield and diseases outbreak. Extreme condition of excessive water will results in soil leaching, erosion and flood [6].

The concept of water management is based on hydrology in tidal swampland i.e the amplitude of inundation (Type A, B, C dan D). The water management system applied by farmers (at tertiary and quarter level) on tidal swampland is generally two-way flow. Wignyosukarto [7] reports that A two-way flow system that allows the supply of fresh water during high tide to overcome the process of washing land in the tidal swamp and is called a *tabat* system that can control the water level in the field, based on local knowledge. The one-way flow system of water management in swampland, which supplies water from the secondary system to the field (tertiary system). This system is mainly designed to leach out the acid accumulating in the water.

The local farmers have developed the traditional technology system for reclaiming the tidal swampland by digging a drainage channel called a *handil*. The word *handil* comes from the Dutch word 'aandeel', meaning a share or part of the work and have several functions: (1) Draining excess water and acidic water from agricultural land, (2) Providing fresh irrigation water during tidal waves. This irrigation is limited to the topography and distance of the river to the land, (3) Communication channel for transportation of people and goods [8]. The water management in tidal swampland should meet at least two objectives i.e: (1) to supplying enough water for leaching out pyrite oxidation products and diluting acid water, (2) to drainage of excess water and to keeping the potential acid sulphate soil under reduced conditions in order to avoid excessive oxidation. The experiment was carried out to investigate the effects of water management on soil chemistry of tidal swampland.

2. Materials and methods

The research was carried out in the Indonesian Swampland Agricultural Research Institute greenhouse by mimicking field conditions using a completely randomized factorial design. The treatment consisted of two factors, namely: the first factor of the washing system includes: (1) one-way flow system (water was not returned), (2) two-way flow system (water was returned). The second factor was amelioration, namely: (1) without ameliorant, (2) ameliorant with dolomit, (3) ameliorant with calcite. NPK fertilizer was given at 100% of the recommended dosage determined based on the results of soil testing using DSS software for rice in swamps. Fertilizer N (Urea) was given 2 times, at the age of 7 days after planting (HST) 1/3 part, then 2/3 part based on BWD (Leaf Color Chart). Whereas P (SP 36) and K (KCl) fertilizers wre given at the beginning of planting. The rice variety used was the Inpara 2 rice variety. The nursery was carried out outside the plantation business area to save time and seedlings aged between 20-25 days in the nursery.

Observation of water and soil quality which was carried out periodically includes: Water pH and electrical conductivity (EC) were conducted at 2 mst, 4 mst, 6 mst and 8 mst. As for growth, plant height was also observed. Experimental data analysis was performed using analysis of variance to determine the average treatment. If the effect was significantly different (F count> F table), then proceed with Duncan's multiple range test (DMRT) to find out which treatment was significantly different. This analysis used SAS 9.10 software for Windows.

3. Results and discussion

3.1. *Dynamic of water acidity (pH)*

The increase in soil pH is due to the liberation of OH^- and H^+ consumption followed by a decrease in the activity of H^+ ions [9]. In addition, the increase in soil pH is determined by the ratio of H+ consumption / electron consumption. In figure 1, it can be seen that at the beginning of inundation the soil pH continued to increase until it reached the peak (observation 6 mst) and again decreased at 8 mst observation. The lowest acidity (the high of pH value) at 8 wap was shown in one-way flow system with ameliorant dolomit reached 5.28. The decrease in pH after reaching the peak at 8 mst observation was due to precipitation of Fe^{2+} with carbonates and sulfides which would contribute to H^+ ions so that soil pH decreased. The decrease in pH at 2 WAP was due to dissociation of H_2CO_3 ions to ion H^+ and ion HCO_3 at a pH ranging from 4-5 that would emit H^+ ions [10]. Ion H^+ would increase the concentration of H^+ ions in soil solution, so that the pH decreased.

IOP Conf. Series: Materials Science and Engineering **980** (2020) 012072 doi:10.1088/1757-899X/980/1/012072

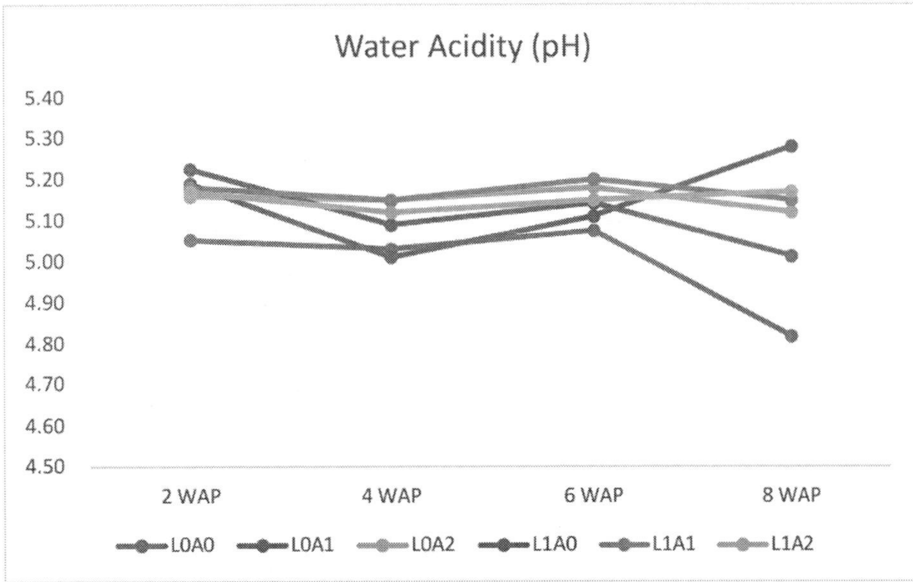

Note: L0=one way flow system; L1=two way flow system; Ao=without ameliorant; A1=ameliorant with dolomite; A2=ameliorant with calcite

Figure 1. Dynamic of water acidity for the two water management treatments and the three ameliorant systems.

3.2. Dynamics of salinity (EC)

The result showed that water management with one-way flow system gave lower water salinity compared to water management with two-way flow system (figure 2). Inundation of paddy soil initially would reduce the soil DHL to a stable value. It is seen in the figure 2 that DHL land continued to decline until the 8 WAP observation period. The decline in the value of DHL was due to the precipitation of Fe^{3+} as $Fe_3(OH)_8$ and FeS as well as loss of CO_2 and conversion of $RCOO^-$ to CH_4. This high salinity resulted in inhibition of absorption of water and nutrients by plants. This is also a result of the inundation process. Rice was tolerant to salinity with values ranging from 4 - 10 dSm^{-1} [11], but for flooded soils the highest DHL values were between 2 - 4 dS m^{-1} [12]. The high salinity caused plant poisoning mainly by Na^+ and Cl^- ions.

3.3. Water potential redox

Water potential redox for the two water management treatments and the three ameliorant systems was shown in figure 3. Redox potential value determines the soil conditions. At a redox potential value> 300 mV soils under oxidative conditions, then the value of Eh 100 - 300 mV soils in slightly reduced conditions, whereas the reduction soil conditions range from 100 - (-100) mV, and potential values> -100 mV indicate soils in very conditions reduced. Oxidants (electron acceptors) and reductants (donor electrons) limit the oxidation reduction process in stagnant soils. If the acceptor electrons are more than the reductant, the redox potential tends to increase (positive), but if the reductant (donor electron) is more than the oxidant, the redox potential tends to decrease (negative) [9]. The highest redox potencial value was shown in one-way water treatment and lime ameliorant treatment which reached +315.15 mV which showed oxidative conditions.

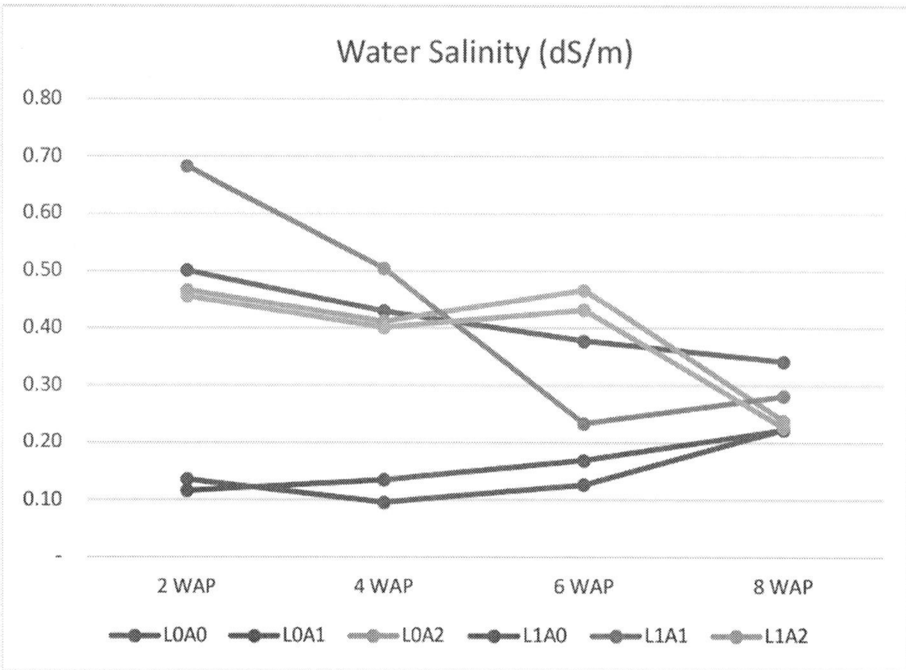

Note: L0=one way flow system; L1=two way flow system; Ao=without ameliorant; A1=ameliorant with dolomite; A2=ameliorant with calcite

Figure 2. Dynamics salinity for the two water management treatments and the three ameliorant systems.

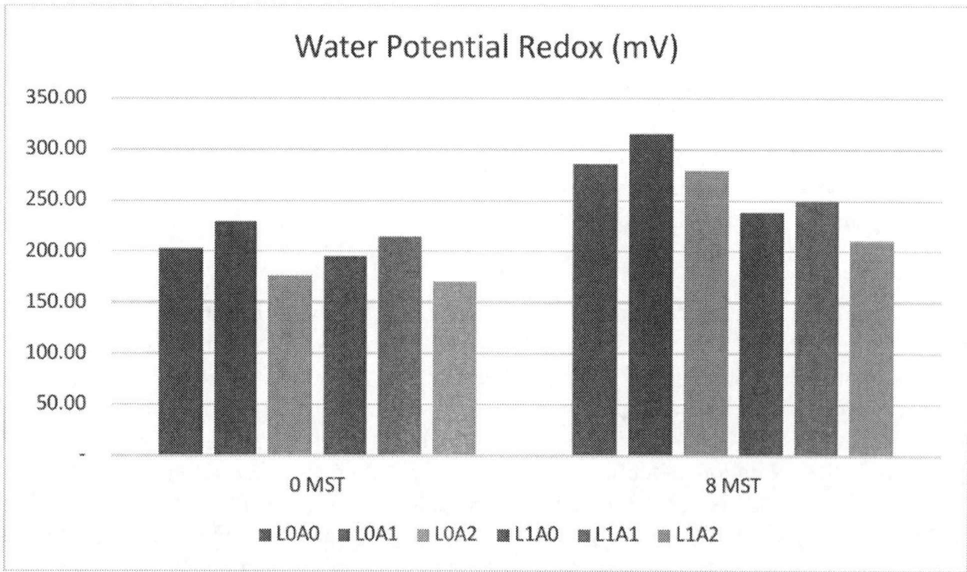

Note: L0=one way flow system; L1=two way flow system; Ao=without ameliorant; A1=ameliorant with dolomite; A2=ameliorant with calcite

Figure 3. Water potential redox for the two water management treatments and the three ameliorant systems.

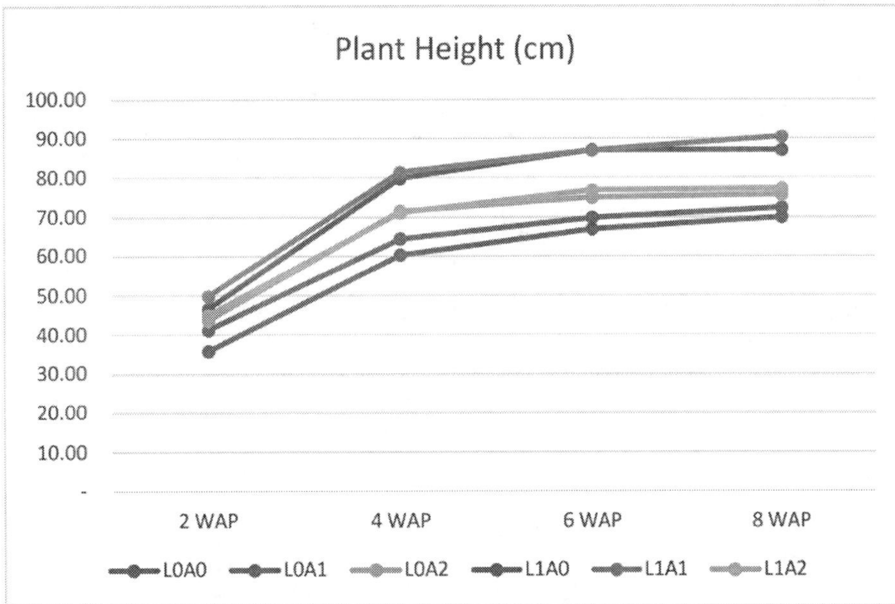

Note: L0=one way flow system; L1=two way flow system; Ao=without ameliorant; A1=ameliorant with dolomite; A2=ameliorant with calcite

Figure 4. Plant height of rice for the two water management treatments and the three ameliorant systems.

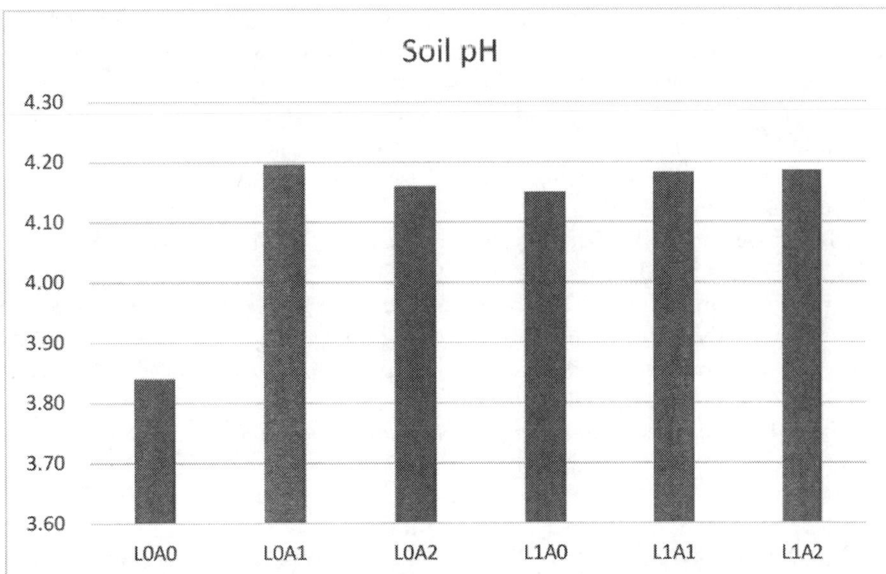

Note: L0=one way flow system; L1=two way flow system; Ao=without ameliorant; A1=ameliorant with dolomite; A2=ameliorant with calcite

Figure 5. Soil acidity for the two water management treatments and the three ameliorant systems in the 8 WAP.

3.4. Plant height of rice

The result showed that effect of water management treatments with two-way flow system with ameliorant dolomit gave higher plant height of rice compared to other treatment (figure 4). And the lowest plant height was shown in the water management with one-way flow system without ameliorant reached 72.33 cm. This is related to the increase in soil pH value reached 4.18 (figure 5). Calcification applications using agricultural lime in acid soils can neutralize soil pH, increase the concentration of alkalinity and total hardness, increase the availability of carbon for photosynthesis, and create a pH buffer system for waters [13].

4. Conclusion

The success of the use of tidal swamps to become agricultural land is determined by the appropriate water management system. Water management treatments with one-way flow system without ameliorant gave higher plant height of rice reached 90.50 cm. The highest plant height in tidal swampland was shown in water management with one-way flow system with ameliorant dolomite.

References

[1] Alihamsyah T 2004 Potential and land reform to increase rice production swamp Rice and rice economy of Indonesia (Jakarta: Department of research and agricultural development)

[2] Lin C, O'Brien K, Lancaster G, Sullivan L A and McConchie D 2000 An improved analytical procedure for determination of total actual acidity (TAA) in acid sulfate soils *Science of the Total Environment* **262** 57-61

[3] Ward N J, Sullivan L A, Bush R T and Lin C 2002 Assessment of peroxide oxidation for acid sulfate soil analysis. 1. Reduced inorganic sulfur *Australian Journal of Soil Research* **40** 433-42

[4] Ahern C R, McElnea A E, Sullivan L A 2004 *Acid sulfate soils laboratory methods guidelines* (Queensland: Queensland Department of Natural Resources, Mines and Energy, Indooroopilly) pp i-12.4

[5] Reid R J and Butcher C S 2011 Positive and negative impacts of plants on acid production in exposed acid sulphate soils *Plant and Soil* **349** 183-90

[6] Saleh E 2013 Water management of swampland as adaptation toward the climate change in South Sumatra. *Procedding International Seminar On Climate Change and Food Security*, Oct 24-25 2013 Palembang, Indonesia. Volume: I (Palembang: Sriwijaya University) pp 198

[7] Wignyosukarto B S 2013 Leaching and flushing of acidity in the reclamation of acid sulphate soil, Kalimantan, Indonesia. *Irrigation and Drainage* **62** 75-81

[8] Sulaiman A A, Sulaeman Y and Minasny B 2019 A framework for the development of wetland for agricultural use in Indonesia. *Resources* **8** 34

[9] Reddy K R and Delaune R D 2008 *The Bigeochemistry of wetland: Science and application* (New York: CRC Press) pp 800

[10] Chesworth W, Cortizas A M and Garcia-Rodeja E 2006 The redex-pH approach to the geochemistry of the earth's land surface, with application to peatlands *Peatlands: Evolution and Records of Environmental and Climate Changes* ch 8, ed I P Martini, A M Cortizas and W Chesworth (Amsterdam: Elsevier) pp 175-95

[11] Noor M 2004 *Lahan Rawa: Sifat dan Pengelolaan Tanah Bermasalah Sulfat Masam* (Jakarta : PT. Raja Grafindo Persada) pp 241

[12] Ponnamperuma F N 1977 Behavior of minor elements in paddy soils *IRRI Research Paper Series* (Manila: The International Rice Research Institute) p 15

[13] Boyd C E, Wood C W and Thunjai T 2002 Aquaculture pond bottom soil quality management. *Pond Dinamics/Aquaculture Collaborate Research Support Program* (Oregon: Oregon State University) p 41

IOP Conf. Series: Materials Science and Engineering **980** (2020) 012073 doi:10.1088/1757-899X/980/1/012073

Activated carbon from Nypa *(Nypa fruticans)* leaves applied for the Fe and Mn removal

I Syauqiah[1,2]*, M Elma[1,2,3], D P Mailani[1] and N Pratiwi[1]

[1]Department of Chemical Engineering, Faculty of Engineering, Lambung Mangkurat University, Banjarbaru 70714, Indonesia
[2]Wetland-Based Materials Research Center, Lambung Mangkurat University, Banjarbaru 70714, Indonesia
[3]Material and Membrane Research Group (M^2ReG), Lambung Mangkurat University, Jl. A. Yani KM 36, Banjarbaru, South Kalimantan 70714, Indonesia

Email: isna_tk@ulm.ac.id

Abstract. Nypa palm (*Nypa fruticans*) is a type of palm-shaped mangrove. A nypa leaf contains 28.9% cellulose so that it is possible to produce an activated carbon applied for heavy metal removal in the solution. Thus, this study aimed to remove Fe and Mn in the $FeCl_3$ and $MnCl_2$ solution using the activated carbon produced from nypa leaves. In the experimental procedure, 5 grams activated carbon from nypa leaves (the particles that could pass through 40, 80, and 120 mesh size and retained on the sizes of 50, 100, and 150 mesh) were added and undergone in a batch system (for 60 and 120 minutes). The concentrations of Fe and Mn in the solution were investigated using the Atomic Absorption Spectrophotometer (AAS). The results showed that there were significant decreases in the Fe and Mn concentrations at the 60-minutes contact time with 150 mesh particle size (the absorption capacity values of Fe and Mn were 59.96 and 96.94 wt.%, respectively). This was due to the bigger surface area of the adsorbent and the longer contact time resulting in better interaction between the activated carbon from nypa leaves to remove Fe and Mn concentrations in the solution.

1. Introduction

Nypa palm *(Nypa fruticans)* is a palm species with special characteristics, and it grows throughout pond estuaries in brackish water. Nypa leaves as a part of the nypa palm (*Nypa fruticans*) contain 28.9% cellulose[1]. Pore structure in the cellulose is able to absorb heavy metals[2]. Some factors that may affect adsorption capacity are contact time, surface area, solubility of adsorbate, molecular size, temperature and degree of acidity (pH)[3]. Ali et al.[4] found that nypa leave adsorbent successfully adsorbed the metal chromium (Cr VI) up to 94.41 wt.% removal.

Indonesia is one of the countries which has very large land areas. Based on the results of an updated compilation of land map evaluation, inherited land data (legacy soil data), and soil survey until 2011, peatland area in Indonesia is around 14.9 million ha[5]. Additionally, the regulation of the Minister of Health of the Republic of Indonesia Number 492/MENKES/ PER/IV/2010 states that the maximum permitted levels of Fe and Mn concentrations in the raw water are 0.3 and 0.1 $mg.L^{-1}$, respectively. Water sources in peat or swamp areas are generally shallow with special characteristics, such as having dark brown to black (145 - 850 PtCo) color and high organic content (138 - 1560 $mg.L^{-1}$ $KMnO_4$) as well as containing iron (Fe) ≤ 25 $mg.L^{-1}$, manganese (Mn) with levels ≤ 1 $mg.L^{-1}$ and sour acids (pH 3.7-5.3)[6, 7]. Fe and Mn concentrations that exceed the permitted levels may cause some health problems. Meanwhile, based on quality standard parameters, peat water that exceeds the maximum permitted level is for clean water.

One method that can be used to reduce Fe and Mn contents in the water is through an adsorption method. Adsorbents with natural ingredients have several advantages for overcoming dangerous and toxic metals in the environment, including relatively low cost, easy to obtain, renewable and environmentally friendly[8]. Therefore, this study aims is to investigate the activated carbon fabricated from nypa leaves as the Fe and Mn removal in the solution.

2. Materials and methods

2.1 Adsorbent preparation
Nypa leaves used in this study were collected from the coastal area in Kotabaru, South Kalimantan-Indonesia. Nypa leaves were washed with water for several times to remove impurities then dried under the sun. The leaves were cut into small pieces to dry easily. Next, the leaves were calcined in the furnace in the air (300°C for 1 hour). The carbon formed was crushed using a grinder and sieved with different sizes (the particles that passed through 40, 80 and 120 mesh and retained in 50, 100 and 150 mesh size). The appearance of carbon formed with various sizes can be seen in figure 1.

(a) (b) (c)

Figure 1. Carbon Powder Based on Different Size
(a) 50 mesh; (b) 100 mesh; (c) 150 mesh

2.2 Proximate analysis
A proximate test equipment was employed to measure the water content, ash content and crude fibre[9].

2.3 Activation process
The adsorbent was activated by dissolving the powder into 0.1 M citric acid solution ($C_6H_8O_7$) and stirred for 2 hours. The adsorbent was then filtered and dried in an oven at 100°C for 2 hours. The dried adsorbent was washed using distilled water to neutralize the pH then dried again in an oven at 80°C for 2 hours. Finally, the nypa leaf activated carbon was ready to be contacted with the artificial metal solution[9].

2.4 Adsorption process (the effect of particle size and contact time)
100 ml of each $FeCl_3$ and $MnCl_2$ solution of 0.5 $mg.L^{-1}$ were put into different Erlenmeyer flasks. Then 5 grams of nypa leaf activated carbon with different sizes (the particles that retained in 50, 100, and 150 mesh) were poured into the Erlenmeyer flasks which had been filled up with the artificial metal solution. Afterward, the mixed solution was stirred using a shaker with a speed of 100 rpm. The

stirring process was carried out for 30 and 60 minutes. Last, the solution was filtered, and the filtrate was analyzed using Atomic Absorption Spectrophotometer (AAS) to determine the Fe and Mn contents in the artificial solution[9].

3. Results and discussion

3.1 Characterization

3.1.1 Proximate analysis

The proximate analysis was performed to determine the moisture content, ash content and crude fiber content in the carbonized samples of nypa leaf powder. The results of the proximate analysis are presented in table 1.

Table 1. Characteristics of nypa leaves after carbonization.

Characteristics	Contents (%)
Moisture	3.66
Ash	10.08
Rough fibre	18.26

Based on table 1, it is recorded that the water content of nypa leaf carbon powder is 3.66%. Indeed, before the carbonization, the samples were dried, so that the water content in the samples was quite low. The low water content in the samples caused the pores of the adsorbent to become bigger and made the surface area wider[10]. Because the samples were made of natural ingredients, it did not only contain carbon but also some other minerals. Mostly the mineral disappeared during the carbonization and activation process, while the rest remained in the samples. The remaining mineral was assumed to be ash[11]. As seen in Table 1, the value of ash content in nypa leaf carbon powder in this study is 10.08 wt.%. The values of water content and ash content in nypa leaf carbon powder qualify the Active Charcoal Quality Standards based on the 1995 SNI (Indonesian Standard), where the maximum water content and ash content in the active carbon powder are 15 and 10 wt.%, respectively. Furthermore, fiber is a constituent component of the cell walls in plants, which usually includes cellulose, hemicellulose and lignin. The crude fiber content in the nypa leaf carbon powder in this study was 18.26 wt.%. As a comparison, the results of previous research on agar solid waste revealed that the crude fiber content of agar solid waste was 38.05 wt%. This value was higher than that of the nypa leaf carbon powder in this present research. The high content of crude fiber was caused by the fact that most of the waste was fiber concentrate which no longer could be extracted. Besides, the crude fiber level also indicated the carbon in the samples. In other words, the higher the level of crude fiber, the higher was the carbon contained in the samples[12]. The higher carbon content also caused the surface area to widen [13-23]. The presence of nitrogen functional groups in amine functional groups on the carbon surface will increase the adsorption capacity.

3.2 Batch adsorption

3.2.1 Fe adsorption using Nypa leaf activated citric acid ($C_6H_8O_7$) 0.1 M

The Fe adsorption process using nypa leaves (*Nypa fruticans*) activated citric acid ($C_6H_8O_7$) 0.1 M was carried out in the artificial $FeCl_3$ solution (with an initial concentration of 0.5 mg.L^{-1}). This solution would be contacted against the adsorbent with different contact time (30 and 60 minutes) and different sizes of adsorbent (the particles that retained on 50, 100, and 150 mesh). The effect of contact time and adsorbent size on Fe concentration are presented in table 2.

Table 2. Effect of contact time and adsorbent size on Fe concentration (mg/L).

Time (minute)	Adsorbent Size (mesh) 50	100	150
30	0.410	0.260	0.257
60	0.318	0.243	0.200

The result above indicated that the concentration of Fe decreased when the contact time was longer. The smallest value of Fe concentration was at 60 minutes with 150 mesh adsorbent size, which was equal to 0.2 mg.L^{-1}. The longer contact time created the diffusion process and the attachment of Fe^{3+} ions to the surface of the adsorbent was better that subsequently caused the Fe concentration to decrease [24]. Moreover, the smaller the particle size of the adsorbent, the bigger was the surface area of the adsorbent that caused a smaller Fe concentration. The correlation between the contact time and particle size of the adsorbent on Fe adsorption can be seen in figure 2.

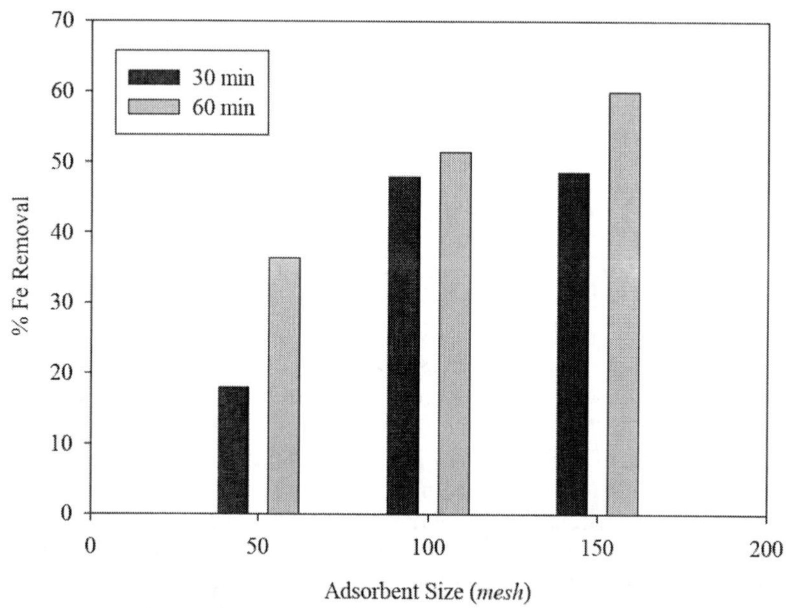

Figure 2. Adsorption curve of fe at different contact time and adsorbent particle size.

The trend of the adsorption showed that the removal percentage increased along with the increase of contact time. This was due to the longer contact time that resulted in better interaction between nypa leaf adsorbent with Fe^{3+} ions, thus more Fe^{3+} ions were adsorbed by the nypa leaf adsorbent through the Fe^{3+} ion exchange reaction with H^+ ions in the activated nypa leaf adsorbent. The maximum removal percentage reduction of Fe at the 60-minutes contact time with a particle size of 150 mesh was 59.96 wt.%. This is in accordance with a theory which states the longer contact time used, the greater decrease in Fe level because the process of the adsorbate absorption is more optimal [25]. Furthermore, the higher the size of the carbon mesh, the greater was the capacity of adsorption because the particle size became smaller so that the surface area of the activated carbon was greater, which made more active parts to absorb [26]. As a comparison with the Granular Activated Carbon (GAC) adsorbent, the maximum removal percentage of Fe with the same dose and contact time in Fe solution was 99.5% [27].

3.2.2 Mn adsorption using Nypa leaf activated citric acid ($C_6H_8O_7$) 0.1 M

The Mn adsorption process using nypa (*Nypa fruticans*) leaf activated citric acid (C6H8O7) 0.1 M was carried out in $MnCl_2$ solution with an initial concentration of 0.5 mg.L^{-1}. This solution would be contacted against the adsorbent with different contact time (30 and 60 minutes) and different size of the adsorbent (the particles that retained on 50, 100 and 150 mesh). The effect of contact time and adsorbent size on Mn concentration are presented in table 3.

Table 3. The effect of contact time and adsorbent size on Mn.

Time (minute)	Adsorbent Size (mesh) 50	100	150
30	0.110	0.098	0.037
60	0.119	0.076	0.015

Table 3 shows that the concentration of Mn decreases when the time was longer. The smallest Mn concentration was at 60 minutes with 150 mesh adsorbent size, which was 0.15 mg.L^{-1}. Similar to the Fe adsorption, the absorption of Mn^{2+} ions also occurred better along with the longer contact time and the smallest size of the adsorbent. The correlation between the contact time and particle size of the adsorbent on Mn adsorption can be seen in figure 3.

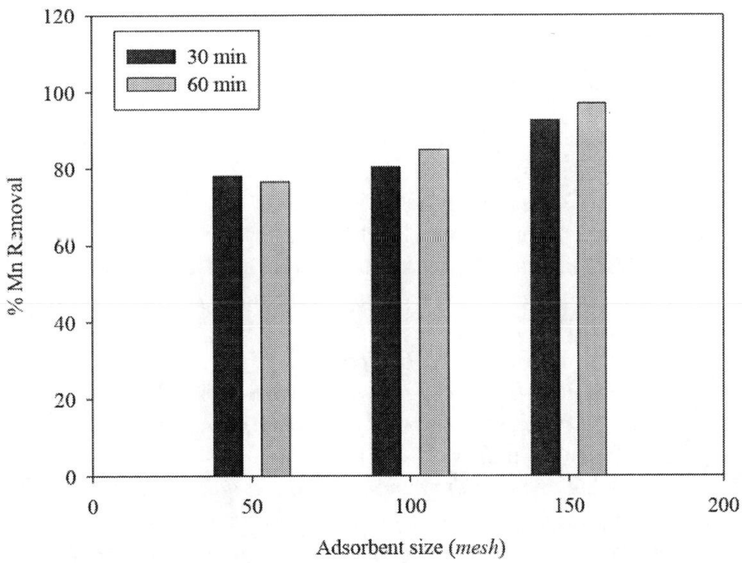

Figure 3. Adsorption curve of Mn at different contact time and adsorbent particle size

The longer contact time resulted in greater interaction between the nypa leaf adsorbent and the Mn^{2+} ions, thus more Mn^{2+} ions are absorbed by the nypa leaf adsorbent through the Mn^{2+} ion exchange reaction with the H^+ ion in the nypa leaf adsorbent that had been activated. From figure 2, it can be seen that an increase in absorption of Mn metal over time. However, at 50 mesh particle size, higher adsorption occurred at 30 minutes and decreased at 60 minutes. This condition was caused by the small surface area of the adsorbent. It made the adsorbent was no longer able to absorb because the Mn^{2+} ions had accumulated or became very dense in a faster time. Thus, Mn^{2+} ions which had been bound to the adsorbent would return to the solution or so-called desorption[28]. In this study, the maximum removal of Mn at 60 minutes contact time with a particle size of 150 mesh was 96.94 wt.%.

However, in the range of 30 to 60 minutes, the difference in the reduction of Mn levels was relatively small. It means that 30 minutes is the optimum time for nypa leaf adsorbent to absorb Mn in the solution. As a comparison with the granular activated carbon (GAC) adsorbent, the maximum removal percentage of Mn with the same dose and contact time in Fe solution was 97.4% [27].

4. Conclusion

In conclusion, based on the results of the proximate analysis test on the nypa leaf powder, the nypa leaves are sufficient to be used as an adsorbent referring to the characteristics of the Active Charcoal Standard based on SNI 1995 (Indonesian Standard). The longer contact time and the bigger surface area of the adsorbent resulted in more interactions between the active carbon of nypa leaves and Fe^{3+} and Mn^{2+} ions. Therefore, more Fe^{3+} and Mn^{2+} ions were absorbed by the nypa leaf adsorbent. It occurred through the interaction between Fe^{3+} and Mn^{2+} ions and H^{+} ions on the activated nypa leaf adsorbent.

Acknowledgments

The authors would like to express their deep gratitude for the facilities provided by the Process Laboratory in the Department of Chemical Engineering, Faculty of Engineering of Lambung Mangkurat University and Banjarbaru Research and Standardization Center, South Kalimantan, Indonesia, for the help of implementing proximate analysis. Isna Syauqiah acknowledges the financial support research grant from the Lecture Mandatory Research Grant of Lambung Mangkurat University 2019-2020. Last, Muthia Elma is thankful for the Lecture Mandatory Research Grant of Lambung Mangkurat University 2020.

References

[1] Tamunaidu P and Saka S 2011 *Chemical characterization of various parts of nipa palm (Nypa fruticans)* Industrial Crops and Products **34(3)** 1423-28

[2] Handayani A W 2010 *Penggunaan Selulosa Daun Nanas sebagai Adsorben Logam Berat Cd(II)*, in *Kimia* (Surakarta: Universitas Sebelas Maret)

[3] Hasyim U H 2017 *Pengaruh konsentrasi HCl dan massa adsorben dalam pengolahan limbah pelumas bekas dengan kajian keseimbangan adsorpsi bentoint terhadap logam Fe* Jurnal Integrasi Proses **6** 191-6

[4] Ali Y, M M W R et al. 2016 *Nypa fruticans As A Potential Low Cost Adsorbent To Up Take Heavy Metals From Industrial Wastewater* Ijaber **2** 1359-71

[5] Ritung S et al. 2011 *Peta Lahan Gambut Indonesia, skala 1:250.000*, B.B.P.d.P.S.L. Pertanian, Editor Kementerian Pertanian

[6] Nainggolan P 2008 *Efektivitas Penurunan Kadar Fe dan Mn Sumur Gali dengan Menggunakan Sistem Upflow Berdasarkan Jenis dan Ketebalan Media Saringan dari Dusun I Kikik Kecamatan Hamparan Perak Tahun 2007* in *Fakultas Kesehatan Masyarakat* Universitas Sumatera Utara (Medan: Universitas Sumatera Utara)

[7] Mahmud et al. 2020 *Effect of Two Stages Adsorption as Pre-Treatment of Natural Organic Matter Removal in Ultrafiltration Process for Peat Water Treatment.* Materials Science Forum **988** 114-21

[8] Diantariani N P, Sudiarta I W and Elantiani N K 2008 *Proses Biosorpsi dan Desorpsi Ion Cr(Vi) Pada Biosorben Rumput Laut Eucheuma spinosum.* Jurnal Kimia **2**

[9] Ristianingsih Y et al. 2019 *Biosorben Daun Ketapang dengan Pengaktifan Asam Sitrat sebagai Pereduksi Merkuri.* Jurnal Teknologi Agro-Industri **6(2)** 92-9

[10] Safariyanti S J, Rahmalia W and Shofiyani A 2018 *Sintesis dan Karakterisasi Karbon aktif dari Tempurung Buah Nipah (Nypa fruticans) Menggunakan Aktivator Asam Klorida.* Jurnal Kimia Khatulistiwa **7**

[11] Suhendarwati L, Suharto B and Susanawati L D 2014 *Pengaruh Konsentrasi Larutan Kalium Hidroksida pada Abu Dasar ampas Tebu Teraktivasi.* Jurnal Sumber Daya Alam & Lingkungan **1**

[12] Faujiah F 2012 *Pemanfaatan Karbon Aktif Dari Limbah Padat Industri Agar-Agar Sebagai Adsorben Logam Berat Dan Bahan Organik Dari Limbah Industri Tekstil* in *Departemen Teknologi Hasil Perairan* Institut Pertanian Bogor: Bogor

[13] Rahman S K et al. 2020 *Functionalization of hybrid organosilica based membranes for water desalination – Preparation using Ethyl Silicate 40 and P123.* Materials Today: Proceedings

[14] Maimunawaro et al. 2020 *Deconvolution of carbon silica templated thin film using ES40 and P123 via rapid thermal processing method.* Materials Today: Proceedings

[15] Lestari R A et al. 2020 *Organo Silica Membranes for Wetland Saline Water Desalination: Effect of membranes calcination temperatures.* E3S Web Conf. **148** p 07006

[16] Elma M et al. 2020 *Coagulation as pretreatment for membrane-based wetland saline water desalination.* Asia-Pacific Journal of Chemical Engineering p e2461

[17] Elma M et al. 2020 *The Performance of Membranes Interlayer-Free Silica-Pectin Templated for Seawater Desalination via Pervaporation Operated at High Temperature of Feed Solution.* Materials Science Forum **981** p 349-55

[18] Syauqiyah I et al. 2019 *Interlayer-free Silica-carbon Template Membranes from Pectin and P123 for Water Desalination.* MATEC Web Conf. **280** p 03017

[19] Rampun E L A et al. 2019 *Interlayer-free silica–pectin membrane for sea-water desalination.* Membrane Technology **2019**(12) p 5-9

[20] Rahma A et al. 2019 *Removal of natural organic matter for wetland saline water desalination by coagulation-pervaporation.* Jurnal Kimia Sains dan Aplikasi **22**(3) p 85-92

[21] Pratiwi A E et al. 2019 *Deconvolution of pectin carbonised template silica thin-film: synthesis and characterisation.* Membrane Technology **2019**(9) p 5-8

[22] Mujiyanti D R, Elma M and Amalia M 2019 *Interlayer-free Glucose Carbonised Template Silica Membranes for Brine Water Desalination.* MATEC Web Conf. **280** p 03010

[23] Elma M et al. 2019 *Fabrication of Interlayer-free P123 Caronised Template Silica Membranes for Water Desalination: Conventional Versus Rapid Thermal Processing (CTP vs RTP) Techniques.* in *IOP Conference Series: Materials Science and Engineering* IOP Publishing

[24] Nor F, Sunarto W and Prasetya A T 2014 *Sintesis Biomassa Bulu Ayam Teraktivasi NaoH /Na2SO3 Aplikasinya Penurun Kadar Tembaga Limbah Elektroplating.* Indo. J. Chem. Sci. **3**

[25] Syauqiah I, Amalia M and Kartini H A 2011 *Analisis Variasi Waktu Dan Kecepatan Pengaduk Pada Proses Adsorpsi Limbah Logam Berat Dengan Arang Aktif.* Info Teknik **12**

[26] Sukir, *Pembuatan dan Karakterisasi Karbon Aktif dari Sekam Padi*, in *Kimia*2008, Institut Teknologi Bandung: Bandung.

[27] Goher M E et al. 2015 *Removal of aluminum, iron and manganese ions from industrial wastes using granular activated carbon and Amberlite IR-120H.* The Egyptian Journal of Aquatic Research **41(2)** 155-64

[28] Zaini H, Abubakar S and Zaimahwati 2017 *Penyisihan Mangan (Ii) Dalam Air Limbah Dengan Metode Kolom Menggunakan Ampas Tebu Sebagai Adsorben.* Proceeding Seminar Nasional Politeknik Negeri Lhokseumawe

The effect of addition of bacterium *Pseudomonas aeruginosa* on biodegradation of methyl orange dye by brown-rot fungus *Gloeophyllum trabeum*

A S Purnomo*, F D Rahmadini, R Nawfa, and S R Putra
Department of Chemistry, Faculty of Science and Data Analytics, Institut Teknologi Sepuluh Nopember, ITS Sukolilo Campus, Surabaya, 60111, Indonesia

Email: adi_setyo@chem.its.ac.id; adi.spurnomo@yahoo.com

Abstract. The methyl orange (MO), one of common textile dyes from azo groups, has negative impact in human life and the environment. Therefore, many attempts have been devoted to find the most effective method for MO degradation. Brown-rot fungus *Gloeophyllum trabeum* has identified as the biodegradable agent of MO, but its efficiency is still low, and it requires a long incubation time. In this work, the biodegradable performance of brown-rot fungus *Gloeophyllum trabeum* was investigated for MO degradation in the presence of bacterium *Pseudomonas aeruginosa* with various volumes (2-10 mL, 1 mL = 5.05 x 10^{12} Colony Forming Unit (CFU)). The addition of 10 mL of bacteria into *G. trabeum* culture showed the maximum degradation of 88.67% in potato dextrose broth (PDB) medium for the 7-day incubation. The identified metabolites were 4-[((4-dimethyliminio) cyclohexa-2,5-dien-1-ylidene) hydrazinyl] phenolate ($C_{14}H_{15}N_3O$, compound 1), 4-[(4-iminio-cyclohexa-2,5-dien-1-ylidene) hydrazinyl] benzenesulfonate ($C_{12}H_{10}N_3O_3S$, compound 2), 4-[(hydroxy-4-iminio-cyclohexa-2,5-dien-1-ylidene) hydrazinyl] benzenesulfonate ($C_{12}H_{10}N_3O_4S$, compound 3), 4-[((4-dimethyliminio)-hydroxy-cyclohexa-2,5-dien-1-ylidene) hydrazinyl] methoxy benzenesulfonate ($C_{15}H_{16}N_3O_5S$, compound 4), and 4-[((4-dimethyliminio)-dihydroxy-cyclohexa-2,5-dien-1-ylidene) hydrazinyl] dimethoxy benzenesulfonate ($C_{16}H_{18}N_3O_7S$, compound 5). Based on the identification of metabolic products, the mixed cultures transformed MO via three pathways: (1) desulfonylation, (2) demethylation, and (3) hydroxylation. These results indicate that *P. aeruginosa* can enhance MO biodecolorization by *G. trabeum*.

1. Introduction

The development and progress of the textile industry in addition to providing many benefits to the community also has a negative impact on the environment because this industry always produces textile dye liquid waste in the production process. A large number of dyes in the textile industry cannot fully be absorbed into the colored fiber fabric so that it will be released as waste [1]. Wastewater that contains dyes is very dangerous because most dyes are difficult to biodegrade (non-biodegradable), resistant, and toxic. Thus, if dyes are released into the aquatic environment without prior treatment, it can cause serious pollution to water sources which result in damage to the ecosystem environment and threats to human health [2]. Methyl Orange (MO) is a type of azo dyes that is found in textile industry waste and is a toxic compound and harms the environment [3]. To prevent the risk posed by such pollution, efforts are needed to reduce the entry of dyes into the aquatic environment. A variety of methods have been developed for handling dyestuff liquid waste from the textile industry [4]. Bioremediation is considered a quite cheap and efficient method for handling dyes waste [5-10]. Brown-rot fungi are one of the microorganisms that can degrade some pollutants including dyes [11-14] by involving the Fenton reaction [15-18].

Other research has been carried out to degrade MO dye by utilizing the brown-rot fungus *Gloeophyllum trabeum* which can produce radical hydroxides through the Fenton reaction [13]. Purnomo et al. [13] reported that *G. trabeum* was able to degrade MO in liquid PDB media by 47.53% for 14 days of incubation. The yield of MO degradation by *G. trabeum* was still relatively low and requires a long incubation time, hence a culture modification is needed to improve its ability.

Some studies suggest that the use of mixed cultures of fungi and bacteria can increase the ability of culture degradation [11, 14, 19-23]. One type of bacterium that can degrade some organic pollutants is *Pseudomonas aeruginosa* because it can produce degrading enzymes such as azoreductase [23-27]. *P. aeruginosa* had been reported to be able to enhance DDT degradation in mixed culture with some wood-rot fungi [28, 29]. Therefore, in this study, the effect of the addition of *P. aeruginosa* on the ability of biodegradation of MO by *G. trabeum* was investigated as an innovation in the effort to deal with the problem of textile dye waste.

2. Materials and methods

2.1. Chemicals

A number of chemicals needed are as follows. Methyl Orange (MO) was purchased from SAP Chemicals, potato dextrose agar (PDA), nutrient agar (NA), and nutrient broth (NB) were purchased from Merck, Germany, while potato dextrose broth (PDB) was purchased from Himedia, India.

2.2 Culture condition

2.2.1 Regeneration of G. trabeum. The brown-rot fungus *G. trabeum* NBRC 6509 species was used as the first degradation agent. The fungal mycelium was taken with a sterile ose needle and inoculated into a petri dish containing the sterile PDA medium then incubated at 30±0.5 °C for 7 days until the mycelium covered the entire surface of the PDA medium [30].

2.2.2 *Regeneration* of *P. aeruginosa*. The bacterium *P. aeruginosa* NBRC 3080 was used as a second degradation agent. One bacterial colony of *P. aeruginosa* was taken with a sterile ose needle and inoculated into a petri dish containing the sterile NA medium then incubated at 37±0.5 °C for 24 hours [31, 32].

2.2.3 *Preparation of Liquid Culture of G. trabeum*. One petri dish containing *G. trabeum* in the PDA medium (2.2.1) was homogenized by using a sterile blender that contained 25 mL of sterile distilled water. One milliliter of homogenate was inoculated into 8 mL of sterile PDB medium in 100 mL of Erlenmeyer and then pre-incubated at 30±0.5 °C for 7 days under a static condition [33].

2.2.4 *Preparation of liquid culture* of *P. aeruginosa*. One colony of *P. aeruginosa* (2.2.2) was inoculated into 10 mL of sterile NB medium. The culture was incubated at 37±0.5 °C for 24 hours with shaking at a speed of 180 rpm as a starter. After 24 hours, one milliliter of *P. aeruginosa* bacterium starter was inoculated into 500 mL of sterile NB medium and then pre-incubated at 37±0.5 °C under a shaking condition with a shaker at a speed of 180 rpm until it reached a maximum OD of the stationary phase of *P. aeruginosa* bacterium [34, 35].

2.3 Biodegradation of MO

G. trabeum culture (biotic control), after pre-incubation (2.2.3), was added with 1 mL of 2000 mg/L MO (final concentration 100 mg/L), followed by the addition of 10 mL of PDB medium (the final volume was 20 mL) and then incubated for 7 days at 30 °C. The bio-decolorization of MO by the mixed cultures was performed by mixing the pre-incubated *G. trabeum* culture (2.2.3) and the pre-incubated *P. aeruginosa* culture (2.2.4) as much as 2, 4, 6, 8, and 10 mL (1 mL = 5.05 × 10^{12} CFU), followed by the addition of MB (the final concentration was 100 mg/L) and PDB to the total volume of 20 mL. As an abiotic control, 1 mL of 2000 mg/L MO (the final concentration was 100 mg/L) was added into 19 mL of PDB medium (the final volume was 20 mL) without an addition of fungal and/or bacterial cultures. All cultures and abiotic control were incubated under a static condition for 7 days at 30±0.5 °C. After 7 days, the cultures were separated by using a centrifuge at 3000 rpm for 15 minutes. The supernatant was taken and filtered by using Whatman filter paper. The supernatant absorbance was measured with a UV-Vis spectrophotometer. The remaining supernatant was stored in a refrigerator for the characterization of metabolites. The calculation of the percentage of MO dye decolorization was according to equation 1 at a wavelength of 465 nm [14].

$$\%\ \text{decolorization} = \frac{Ac - At}{Ac} \times 100\% \tag{1}$$

where Ac is absorbance abiotic control, and At is absorbance treatment.

2.4 Product metabolite identification

The identification of metabolites was performed by using the LC-TOF/MS (Impact II, Bruker). The source of ionization was the electrospray ionization (ESI) with a mass range of 50-500. The elution method used was the gradient method with a flow rate of 0.2 mL/min in the first 3 minutes and the next 7 minutes using a flow rate of 0.4 mL/min. The mobile phase was methanol and water with a ratio of 99:1 in the initial 3 minutes and 61:39 for the remaining 7 minutes. The column was the Acclaim TM RSLC 120 C18 with a size of 2.1 x 100 mm with a column temperature of 33 °C [12-14].

3. Results and discussions

3.1. The effect of addition of P. aeruginosa on biodegradation of MO by G. trabeum

In this study, the addition of *P. aeruginosa* on the biodegradation of MO by brown-rot fungus *G. trabeum* was investigated. The MO degradation by *G. trabeum* fungus was also carried out as the biotic control. The absorbance profile of MO degradation is shown in figure 1. The maximum absorbance of MO for abiotic control is detected at a wavelength of 465 nm. However, in the biotic control and the mixed cultures, the absorbance of peaks decrease which indicates decolorization. Besides, the other peaks appears at the wavelength of 300-390 nm indicating the peaks of metabolites. Since the MO peak appears in the wavelength range of 400-600 nm which is the wavelength region of visible light, the metabolites might be colorless [3]. The absorbance of biotic control of *G. trabeum* is lower than that of the abiotic control, indicating *G. trabeum* decolorized MO. Besides, the biotic control peak is lower than the mixed cultures with the addition of 2, 4, and 6 mL of bacteria. This suggests that the mixed cultures still adapt to the addition of MO. The MO degradation by the mixed cultures of *G. trabeum* and *P. aeruginosa* occurs maximally in the additions of 8 and 10 mL of bacteria as evidenced by the lower peaks at 465 nm compared with the other peaks.

The ability of mixed cultures of *G. trabeum* and *P. aeruginosa* to degrade MO was determined quantitatively by measuring the percent of decolorization (table 1). The percent of MO decolorization by the mixed cultures increased along with the increase in the amount of added bacteria. The more the bacteria added, the greater were the biosurfactants as well as dye degradation enzymes produced by *P. aeruginosa*. Thus, it caused an increase in the MO decolorization. Based on table 1, the degradation of MO by mixed culture occurs optimally in the addition of 10 mL of bacteria (1 mL = 5.05 x 10^{12} CFU) approximately 88.67%. However, the percent of decolorization at the additions of 8 mL to 10 mL of bacteria is not significantly different due to the competition of bacteria for surviving that might occur rather than to decolorize MO. Some toxic metabolites might be produced during the stationary phase under an abundant population of bacteria [22].

IOP Conf. Series: Materials Science and Engineering **980** (2020) 012074 doi:10.1088/1757-899X/980/1/012074

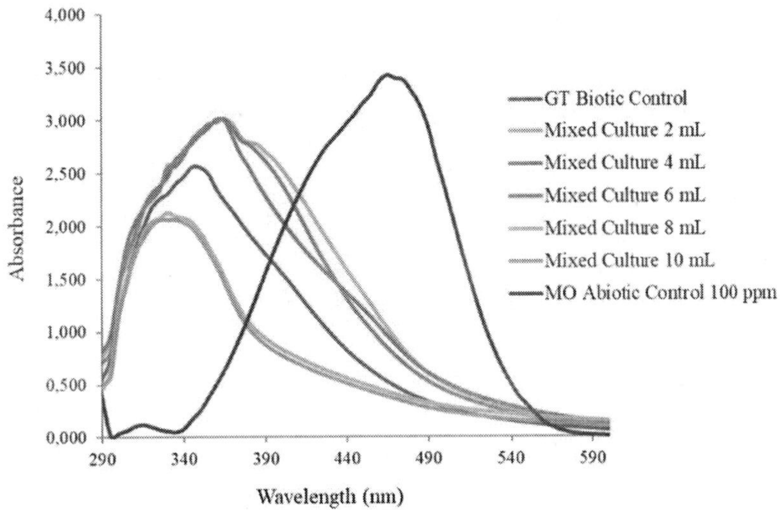

Figure 1. Profile of absorbance of MO degradation.

Table 1. % Decolorization of MO.

Amount of bacteria addition (mL)	decolorization (%)
G. trabeum (biotic control)	84.67 ± 0.14a
2	69.92 ± 0.02b
4	71.96 ± 0.04b
6	75.32 ± 0.01b
8	87.53 ± 0.02a
10	88.67 ± 0.02a

The significant differences occur in the mixed cultures with the additions of 6 mL and 8 mL of bacteria, from 75.32% to 87.53%. It suggests that each degradation agent can adapt to each other when living in the same medium in degrading MO. The ability of *G. trabeum* to degrade MO is related to its ability of producing oxalic acid and H_2O_2 which play a role in the Fenton reaction [36]. In addition to involving the Fenton reaction, *G. trabeum* also produces several extracellular enzymes to degrade wood and some pollutants, such as endoglucases and xylanases that are known produced by *G. trabeum* in the very large quantities [37]. The ability of bacterium *P. aeruginosa* in degrading MO is also related to its ability to produce a degrading enzyme, i.e. enzyme azoreductase, lignin peroxidase, DCIP reductase, and tyrosinase [38, 39]. Apart from producing degrading enzymes, Scheibenbogen et al. [40] reported that *P. aeruginosa* was able to produce biosurfactants of rhamnolipid types which could also be applied in increased degradation of hydrocarbon pollutants.

3.2 Metabolites identification

The profile of chromatogram of MO metabolites is shown in figure 2. It can be seen that the treatment MO peak is lower compared with the control MO peak, identified at the retention time of 7.29 minutes. Besides the MO peak, in the treatment chromatogram, there are 5 new peaks appear that do not yet exist in the control chromatogram. They are identified as the metabolites produced during the process of MO degradation by the mixed cultures of *G. trabeum* and *P. aeruginosa*. Identification of metabolites is determined based on the similarity between MS spectrum and time retention from database (table 2). As seen in table 2, the peak at the retention time of 1.29 minutes has m/z 241 which is identified as 4-[((4-

dimethyliminio)-cyclohexa-2,5-dien-1-ylidene) hydrazinyl] phenolate ($C_{14}H_{15}N_3O$, compound 1). The peak at the retention time of 2.55 minutes has m/z 276 identified as 4-[(4-iminio-cyclohexa-2,5-dien-1-ylidene) hydrazinyl] benzenesulfonate ($C_{12}H_{10}N_3O_3S$, compound 2). Meanwhile, the peak at the retention time of 2.72 minutes has m/z 292 identified as 4-[(hydroxy-4-iminio-cyclohexa-2,5-dien-1-ylidene) hydrazinyl] benzenesulfonate ($C_{12}H_{10}N_3O_4S$, compound 3), and the peak at retention time of 4.38 mins has m/z 350 identified as 4-[((4-dimethyliminio)-hydroxy-cyclohexa-2,5-dien-1-ylidene) hydrazinyl] methoxy benzenesulfonate ($C_{15}H_{16}N_3O_5S$, compound 4). The peak at the retention time of 5.54 minutes has m/z 396 identified as 4-[((4-dimethyliminio)-dihydroxy-cyclohexa-2,5-dien-1-ylidene) hydrazinyl] dimethoxy-benzenesulfonate ($C_{16}H_{18}N_3O_7S$, compound 5). Based on this identification of metabolites, the MO degradation pathways by the mixed cultures of *G. trabeum* and *P. aeruginosa* were proposed in this study (figure 3).

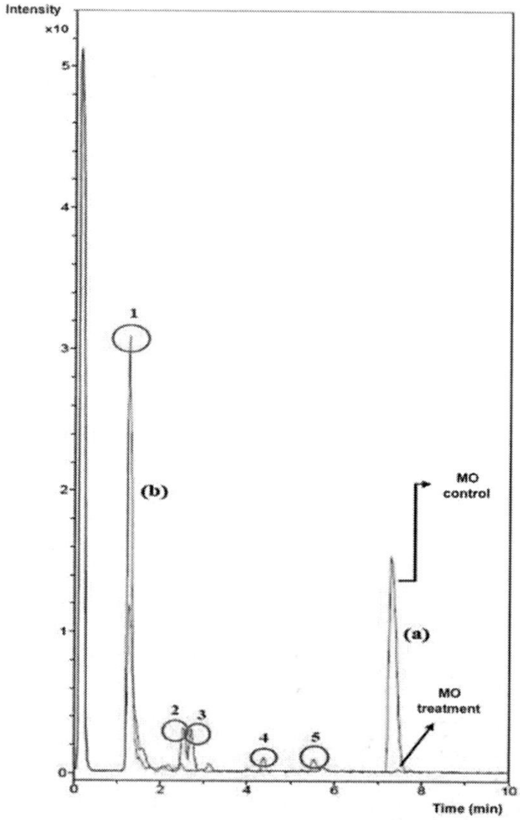

Figure 2. Chromatogram profiles of MO degradation by the mixed cultures of *G. trabeum* and *P. aeruginosa*. Purple chromatogram: control (a); green chromatogram: treatment (b).

There were three MO degradation pathways by the mixed cultures of *G. trabeum* and *P. aeruginosa* proposed. The first pathway: the ionized MO was transformed into compound 1 by desulfonylation. The second pathway: the ionized MO was transformed into compound 2 by demethylation, followed by hydroxylation to compound 3. Last, the third pathway: the ionized MO was transformed into compound 4 by hydroxylation and methylation, followed by hydroxylation and methylation again into compound 5. Hydroxylation occurred due to the presence of radical hydroxides that might be produced from the Fenton reaction by *G. trabeum* that attacked MO [20]. Previously, *G. trabeum* transformed MO via three pathways: (1) demethylation, followed by hydroxylation reactions; (2) hydroxylation, followed by demethylation; and (3) desulfonylation [13].

Table 2. Metabolite products of MO degradation by the mixed cultures.

RT (min)	Mr	Name	Molecular formula	Structure
1.29	241	4 – [((4-dimethyliminio)-cyclohexa-2,5-diene-1-ylidene) hydrazinyl] phenolate	$C_{14}H_{15}N_3O$	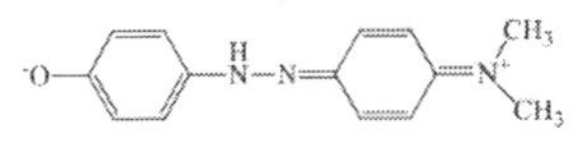
2.55	276	4 – [(4-iminio-cyclohexa-2,5-diene-1-ylidene) hydrazinyl] benzenesulfonate	$C_{12}H_{10}N_3O_3S$	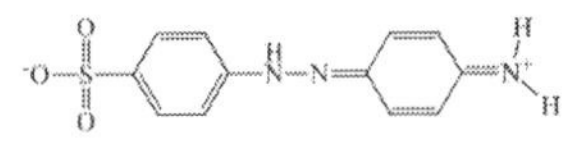
2.72	292	4 – [(hydroxy-4-iminio-cyclohexa-2,5-diene-1-ylidene) hydrazinyl] benzenesulfonate	$C_{12}H_{10}N_3O_4S$	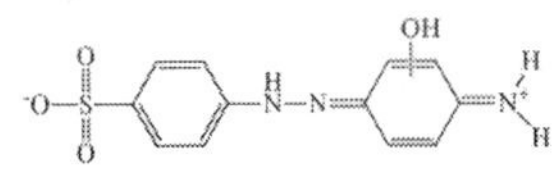
4.38	350	4 – [((4-dimethyliminio) -hydroxy-cyclohexa-2,5-diene-1-ylidene) hydrazinyl] -metoxy-benzensulfonate	$C_{15}H_{16}N_3O_5S$	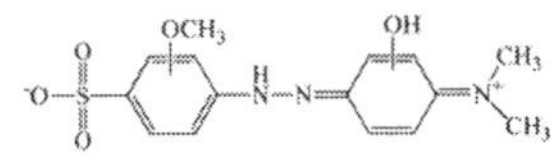
5.54	396	4 – [((4-dimethyliminio)-dihidroxy-cyclohexa-2,5-diene-1-ylidene) hydrazinyl] -dimethoxy-benzensulfonate	$C_{16}H_{18}N_3O_7S$	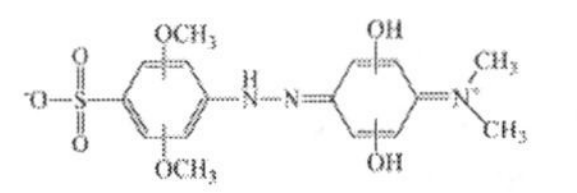

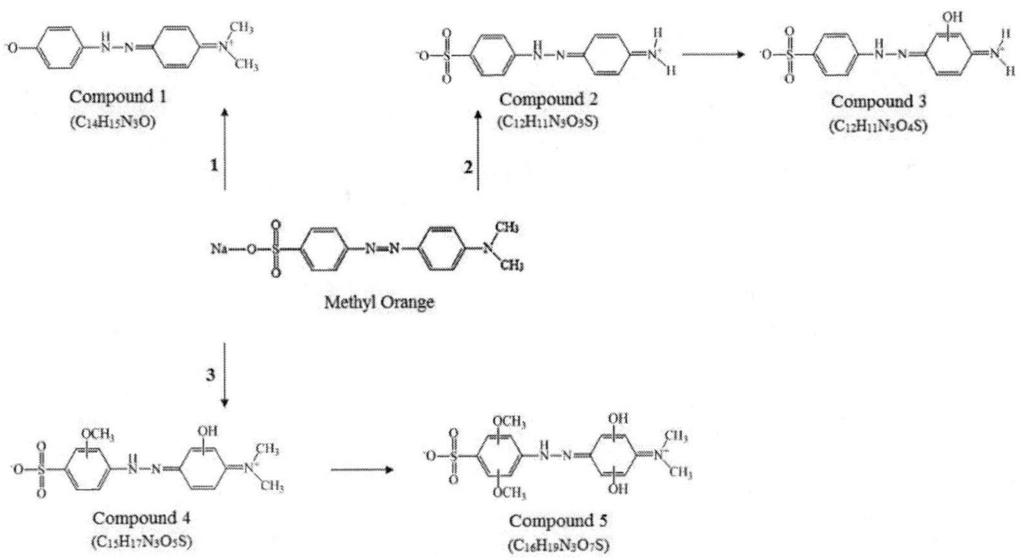

Figure 3. Proposed MO degradation pathways by mixed cultures of *G. trabeum* and *P. aeruginosa*.

4. Conclusion

The addition of bacterium *P. aeruginosa* increased the ability of brown-rot fungus *G. trabeum* to decolorize MO. The highest MO decolorization was obtained in the addition of 10 mL (1 mL = 5.05 x 10^{12} CFU) of *P. aeruginosa* with decolorization percentage by 88.67% in the PDB liquid medium for the 7-day incubation. The mixed cultures of *G. trabeum* and *P. aeruginosa* transformed MO into 4-[((4-dimethyliminio)-cyclohexa-2,5-dien-1-ylidene) hydrazinyl] phenolate ($C_{14}H_{15}N_3O$, compound 1), 4-[(4-iminio-cyclohexa-2,5-dien-1-ylidene) hydrazinyl] benzenesulfonate ($C_{12}H_{10}N_3O_3S$, compound 2), 4-[(hydroxy-4-iminio-cyclohexa-2,5-dien-1-ylidene) hydrazinyl] benzenesulfonate ($C_{12}H_{10}N_3O_4S$, compound 3); 4-[((4-dimethyliminio)-hydroxy-cyclohexa-2,5-dien-1-ylidene) hydrazinyl] methoxy benzenesulfonate ($C_{15}H_{16}N_3O_5S$, compound 4); and 4-[((4-dimethyliminio)-dihydroxy-cyclohexa-2,5-dien-1-ylidene) hydrazinyl] dimethoxy-benzenesulfonate ($C_{16}H_{18}N_3O_7S$, compound 5). Based on the identification of metabolic products, the mixed cultures transformed MO via three pathways: (1) desulfonylation; (2) demethylation; and (3) hydroxylation. This research finally indicate that bacterium *P. aeruginosa* can enhance the decolorization of the MO dye by brown-rot fungus *G. trabeum*.

Acknowledgments
This study was funded by Deputy for Research and Development Strengthening, Indonesian Ministry of Research and Technology/National Research and Innovation Agency, under the Basic Research Scheme Number: 3/E1/KP.PTNBH/2020 and Amendment Contract Number: 3/AMD/E1/KP.PTNBH/2020.

References

[1] Selvam K, et al. 2003 *Bioresour Technol* **88** 115
[2] Hamdaoui O and Chiha M 2006 *Acta Chim Slov* **54**, 407
[3] Alamddine L and El jamal MM 2009 *J Chem Technol Metall* **44** 127
[4] Moreira MT, et al. 2000 *Biotechnol Lett* **22** 1499
[5] Purnomo AS, et al. 2010 *Chemosphere* **80** 619
[6] Purnomo AS, et al. 2010 *Int Biodeterior Biodegrad* **64** 397
[7] Purnomo AS, et al. 2013 *Int Biodeterior Biodegrad* **82** 40
[8] Purnomo AS, et al. 2014 *Environ Sci Pollut Res* **21** 11305
[9] Purnomo AS 2017 1st ed. Springer Nature, Switzerland
[10] Purnomo AS, et al. 2017 *Curr Microbiol* **74** 320
[11] Rizqi HD and Purnomo AS 2017 *World J Microbiol Biotechnol* **33** 92
[12] Purnomo AS, et al. 2020 *AIP Conf Proc* **2237** 020002
[13] Purnomo AS, et al. 2019 *Intl J Environ Sci Tech* **16** 7555
[14] Purnomo AS and Mawaddah MO 2020 *Biodiversitas* **21(5)** 2297
[15] Contreas D, et al. 2005 *Int Biodeterior Biodegrad* **57** 63
[16] Purnomo AS, et al. 2010 *Int Biodeterior Biodegrad* **64** 560
[17] Purnomo AS, et al. 2011 *Int Biodeterior Biodegrad* **65** 691
[18] Purnomo AS, et al. 2011 *Int Biodeterior Biodegrad* **65** 921
[19] Sariwati A, et al. 2017 *Curr Microbiol* **74** 1068
[20] Sariwati A and Purnomo AS 2018 *Indon J Chem* **18** 75
[21] Setyo PA, et al. 2018 *Res J Chem Environ* **22** 151
[22] Grizca BE and Setyo PA 2018 *J Physics: Conf Ser* **1095** 102015
[23] Purnomo AS, et al. 2020 *Heliyon*, **6(6)**, e04027
[24] Asad S, et al. 2007 *Bioresour Technol* **98** 2082
[25] Nawfa R, et al. 2019 *Asian J Chem* **31(10)** 2367
[26] Chen KC, et al. 2003 *Biotechnology* **101** 57
[27] Moosvi S, et al. 2005 *World J Microbiol Biotechnol* **21** 667
[28] Boelan EG and Purnomo AS 2019 *Hayati J Biosci* **26** 90
[29] Purnomo AS, et al. 2017 *J Microbiol Biotechnol* **27(7)** 1306

[30] Purnomo AS, et al. 2008 *J Biosci Bioeng* **105** 614
[31] Purnomo AS, et al. 2020 *J Idn Chem Soc* **3(1)** 53
[32] Purnomo AS, et al. 2019 *Mal J Fund Appl Sci* **15(3)** 377
[33] Purnomo AS, et al. 2019 *J Microbiol Biotechnol* **29** 1424
[34] Wahyuni S, et al. 2016 *Asian J Chem* **28** 2731
[35] Wahyuni S, et al. 2017 *Asian J Chem* **29** 1119
[36] Hatakka A 1994 *FEMS Microbiol Rev* **13**, 125
[37] Maier RM, et al. 2009 *Environ Microbiol second Edition*. UK: Academic Press Elsevier
[38] Sarayu K and Sandhya S 2010 *Appl Biochem Biotech* **160(4)** 1241
[39] Jadhav JP, et al. 2010 *Biodegradation* **21(3)** 453
[40] Scheibenbogen K, et al. 1994 *J Chem Technol Biotechnol* **59** 53

IOP Conf. Series: Materials Science and Engineering **980** (2020) 012075 doi:10.1088/1757-899X/980/1/012075

Degradation of linear alkylbenzene sulfonate (LAS) using TiO_2-chitosan composite as a photocatalyst

N A Rizky, U Irawati* and T Rohman

Department of Chemistry, Faculty of Mathematics and Natural Sciences, Lambung Mangkurat University Banjarbaru 70714, Indonesia

Email: uirawati@ulm.ac.id

Abstract. Linear Alkylbenzene Sulfonate (LAS) is an anionic surfactant that is widely used as an active ingredient in synthetic detergent formulation. Being a component in domestic waste, LAS can potentially be harmful to the environment. In this research, the degradation of LAS using TiO_2-chitosan composite as a photocatalyst was investigated. The optimum dose of TiO_2-chitosan to be used was determined, and the effect of pH towards the degradation effectiveness was also studied. The results showed that a dose of 1 g/100 mL composite at pH 6 gave an optimum result for the LAS degradation. In this optimum condition, the decrease in LAS concentration was up to 95.20% from its initial concentration.

1. Introduction

Linear Alkylbenzene Sulfonate (LAS) is one of the most common anionic surfactants that can be found as a component in commercial detergents. The use of LAS particularly increased after ABS (alkyl benzene sulfonates), another type of anionic surfactant, was banned due to its nature for being practically non-biodegradable [1]. Despite LAS being biodegradable in the environment, detergent wastewater that contains LAS still need to be properly treated before being released to the environment. This is because the degradation of LAS will not take place effectively in an anaerobic condition [2, 3]. Besides, high turbidity will also prevent the penetration of sunlight that slows down the degradation process [4]. In places with a high population density, domestic waste often contains a high load of detergent wastewater, and it tends to have very high turbidity [5]. Under this condition, the rate of LAS degradation in the domestic wastewater would be very low which might lead to a buildup of LAS concentration in the water to a toxic level. It has been reported that LAS concentration between 2-10 mg/L in the water can potentially be harmful to the aquatic organisms living in the polluted water [6].

Dealing with the negative impact of LAS in the environment, an effective treatment to remove LAS from the water is necessary. The use of photocatalyst, such as titanium oxide (TiO_2) to degrade LAS is an interesting option as it accelerates the degradation of LAS into non-harmful compound. Indeed, titanium dioxide has been widely used as a photocatalyst for various purposes due to its superior properties, such as high chemical stability and low toxicity [7]. Nevertheless, its application to treat wastewater still has some drawbacks. Because it is in the form of a fine powder, the recovery from the treated wastewater will be a difficult process. Moreover, adding a high dose of TiO_2 powder will cause an increase in turbidity and reduce the intensity of penetrating light needed for the photodegradation process of the targeted waste. To overcome these difficulties, TiO_2 is often immobilized into other materials as its solid support [8].

In this research, a composite material was prepared by immobilizing TiO_2 onto chitosan beads (CS). The TiO_2/CS composite was then utilized to remove LAS from the solution. The effect of added dose and pH was investigated to find an optimum condition for LAS removal using this composite. As a comparison, LAS removal by photodegradation using TiO_2 powder and by adsorption using chitosan beads were also examined.

2. Materials and methods

2.1. Materials and instrumentation

Standard glasswares (*Pyrex*) were used in this research. The samples were illuminated using two 30W UV-lamps (*Phillips*) in a closed cabinet. The inner sides of the cabinet were covered with black plastic sheets to minimize light penetration from the outside. The chitosan used was food-grade chitosan (100 mesh, MW = 50,000-80,000 Da). Other chemicals used were TiO_2 powder, NaOH, linear alkylbenzene sulphonate (LAS) powder, H_2SO_4, methylene blue solution, EDTA, $CHCl_3$, HCl, $NaH_2PO_4.H_2O$, CH_3COOH, and epichlorohydrin. All chemicals were reagent-grade from commercial sources (Merck) and used as received. UV-Vis analysis was done using *Genesys* 10S UV-VIS Spectrophotometer.

2.2. Preparation of TiO_2/CS composites

Chitosan beads used in the composites were prepared based on the method suggested by Zhao, Xiao, Zhang, Su, & Tan [9]. Two grams of chitosan was dissolved in 100 mL 2% (w/w) aqueous acetic acid solution. After a 45-minutes of stirring, 0.4 grams of TiO_2 was added, followed by adding 1 mL of epichlorohydrin into the solution as a crosslinking agent. The mixture was continuously stirred for 5 hours. To form the desired beads, the chitosan solution was dripped into a 0.25 M NaOH solution using a dropping pipette. The formed beads were let to stay in the solution overnight, and then the beads were thoroughly washed using distilled water. The composites were put into a beaker containing 0.15% (w/v) EDTA solution and gently stirred for 4 hours, and then the composites were washed using distilled water. Afterwards, the composites were put into a 0.05 M NaOH solution, stirred gently for two hours before rinsed with distilled water. The resulting TiO_2/CS composites were filtered out and dried in the oven at 60°C. Some cross-linked chitosan beads (CS) were also prepared through the same procedure without the addition of TiO_2 in the process. To verify the immobilization of TiO_2 onto the chitosan beads in the composites, both the TiO_2/CS composites and CS were analysed using FTIR Spectrophotometer.

2.3. Determination of the optimum dose

One hundred millilitres of LAS with a 100 ppm concentration was placed in a beaker glass. The desired amounts of TiO_2/CS composites were added into the solution. In this study, three different doses of composites were used, i.e. 0.5, 1, and 2 grams of composites for 100 mL of LAS solution, and then the mixture was irradiated using the UV lamps in the closed cabinet for 5 hours. After the irradiation, the mixture was then centrifuged to separate the liquid phase from the composites. The liquid phase was withdrawn while the remained LAS concentration in it was analysed. The LAS concentration was analysed by MBAS methods, as suggested by the Standard Methods for the Examination of Water and Wastewater on anionic surfactant analysis. The same procedure was repeated using TiO_2 powder and CS with the same dose variations. To study the LAS removal through adsorption by CS, a set of experiments using CS where the LAS solution and CS were kept in a dark closed cabinet without UV irradiation was also conducted. This was to minimize the photodegradation process of LAS, so that the adsorption by CS became the dominant process taking place in the LAS removal.

2.4. Determination of optimum pH
One hundred millilitres of LAS with a 100-ppm concentration was placed in a beaker glass. The pH of the LAS solution was adjusted by adding 1 N NaOH or 1 N HCl, dropwise. The range of pH being studied in the research was from pH 2 to 10. After the desired pH was reached, TiO_2/CS composites were added into the solution. The amount of the composites added was fixed at the optimum dose that was determined based on the previous step. The same procedure of samples irradiation using UV lights and LAS analyses was repeated. Similar to the determination of the optimum dose, the determination of optimum pH was also conducted using TiO_2 powder and CS for both treatments with UV lamps and in the dark cabinet.

3. Results and discussion

3.1. FTIR analysis
To verify that TiO_2 had been successfully immobilized onto the chitosan beads, the functional groups of the TiO_2/CS and CS were analysed using FTIR spectrophotometer and compared to each other, along with the FTIR spectrum of TiO_2 powder. The FTIR spectra of TiO_2/CS, CS and TiO_2 powder are shown in figure 1.

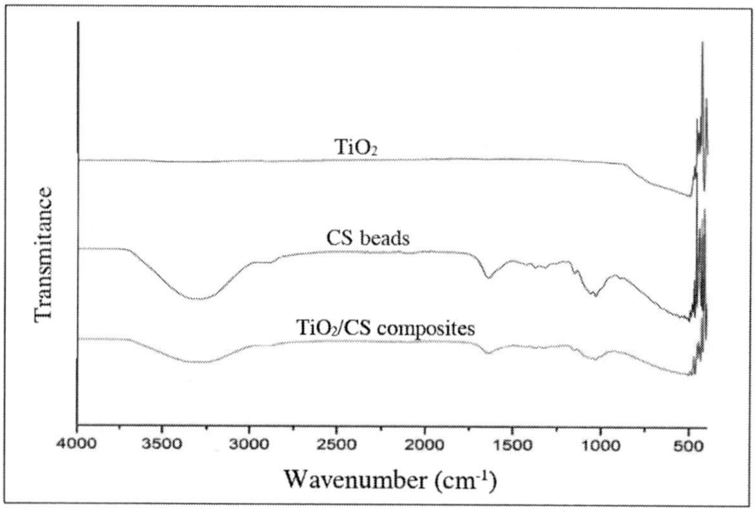

Figure 1. FTIR spectra of TiO_2, CS beads and TiO_2/CS composites.

From the figure, TiO_2/CS and CS appeared to have the same pattern. The peaks were around 3290 and 1650 cm^{-1}, and the characteristic of peaks for chitosan could be observed on the FTIR spectra of both TiO_2/CS composites and CS beads. However, TiO_2/CS spectrum showed peaks at wavenumbers that were less than 500 cm^{-1} which were not present in the CS spectrum. These peaks came from TiO_2, an inorganic compound that did not give peaks at wavenumbers higher than 500 cm^{-1}, as shown in the spectrum of TiO_2 in figure 1. This confirmed that TiO_2 powder added in the composite preparation had been successfully immobilized into the structure of the chitosan beads. The TiO_2 might be attached on the chitosan beads through hydrogen bonding, as implied from a decrease of intensity at the peak around 3400 cm^{-1}.

3.2. Optimum dose
To determine an optimum dose in each treatment, various amounts of composite, CS beads, and TiO_2 powder were added into 100 mL LAS solution, where the initial LAS concentration was 100 ppm. The results are displayed in figure 2 that shows the decrease of LAS concentration in the solution as a function of the dose being added to the solutions.

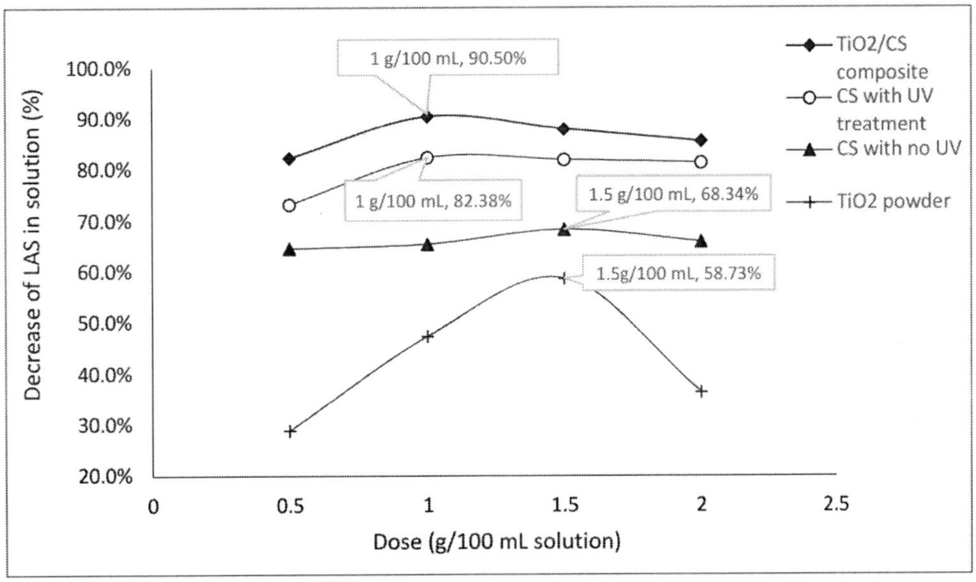

Figure 2. The decrease in LAS concentration for each treatment with various doses.

From the figure, it can easily be observed that TiO_2/CS showed the best result in decreasing the LAS concentration from the solution. One gram of TiO_2/CS composites in a 100 mL sample was enough to remove 90% LAS from the solution. A higher dose did not increase the removal percentage significantly, implying that a 1 g composite/100 mL sample was enough to have an optimum result. It is interesting to note that the LAS removal appeared to be more effective through adsorption instead of the photodegradation. This was indicated by a better removal percentage when the LAS solution was treated with CS, even when the mixture was kept in a dark cabinet with a minimum exposure to light. The optimum result obtained was a 58.73% removal of LAS when LAS solution was treated by TiO_2 powder which was reached when using 1.5 grams of TiO_2 powder for a 100 mL solution. When CS was used and the solution was kept in a dark cabinet, the LAS removal was up to 68.34% from the initial concentration. This result shows a considerably lower value than a study conducted by Kurniyati, Sumarni, & Latifah [10]. They reported that chitosan bead was able to reduce the concentration of ABS, another type of anionic surfactant, up to around 90% from its initial concentration. However, their result which showed a higher removal was probably because their study was not set up in a condition where the UV light was minimized and, thus, enabled a surfactant removal through the photodegradation process.

Nevertheless, the presence of UV light is also able to remove LAS from the solution as it facilitates the photodegradation of LAS. This was indicated by the increase in LAS removal when the solution was treated with CS beads and irradiated with UV lights. It can be deduced that the LAS removal occurred not only through adsorption by the CS beads, but also through the photodegradation process by the UV light. As shown in figure 2, not only the LAS removal increased up to 82.38%, but this optimum result also came with a lower dose of beads. The addition of TiO_2 into the chitosan beads enhanced the LAS removal by catalysing the photodegradation of LAS. This result corresponds to the report by Zainal, Hui, Hussen, & Abdullah [11] who pointed out that removal of contaminants using a TiO_2-chitosan composites would occur through adsorption and photodegradation. Meanwhile, it has to be noted that in other treatments, a higher dose than the optimum one no longer increased the LAS removal when it came to the treatment with TiO_2. An excessive dose leads to a quite significant decrease in its efficiency to remove LAS. Since TiO_2 powder is practically insoluble in water, a high load of TiO_2 in water would make the particles being suspended in it then blocked UV lights from penetrating the solution. As highlighted by Kansal, Ali, & Kapoor [12], a high load of TiO_2 will reduce the effectiveness of photodegradation process as the UV light is scattered by the TiO_2 particles.

3.3. Optimum pH

It has been reported that pH often plays a key role in waste removal, including in treatment using photocatalysts. This is mainly because pH will determine the compound species present in the system and, thus, it will influence the interaction between said species and the photocatalysts [13]. In this study, after the optimum dose had been determined, the study was repeated by using the optimum dose for each treatment but at various pH. The result is presented in figure 3, where the removal of LAS is shown as a function of the solution pH.

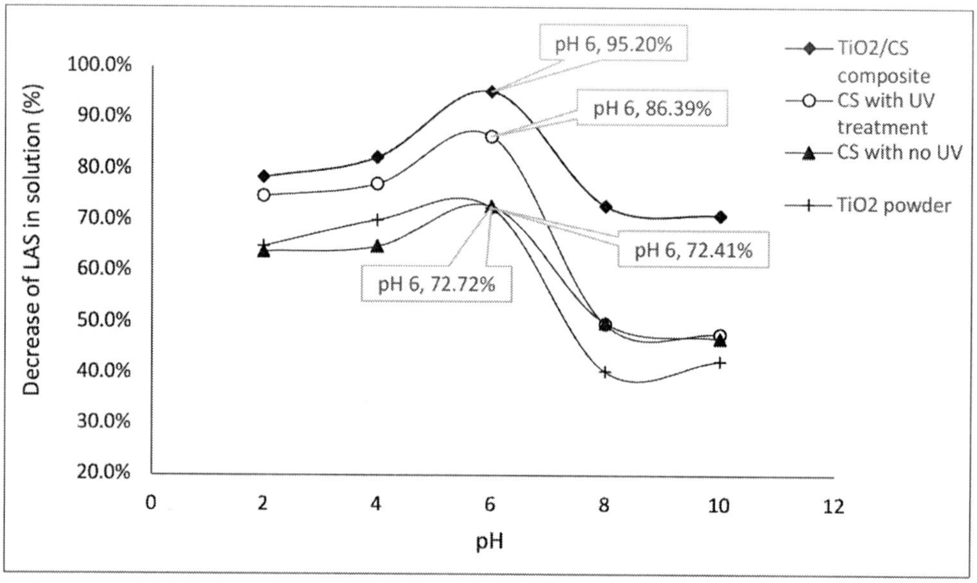

Figure 3. The decrease in LAS concentration at different pH.

Similar to the previous results, the highest percentage of LAS removal from the solution was achieved when LAS was treated using TiO_2/CS composites. All of the treatments gave optimum results at pH 6. A higher pH led to a performance drop for all of the treatments quite significantly. The effect of pH can be explained by how CS and TiO_2 become positively charged in an acidic pH. Due to the presence of its amine groups, CS will become a polycationic in an acidic pH as the amine groups are protonated, making them available to have electrostatic interaction with LAS [14]. The surface of TiO_2 also becomes negatively charged in an acidic pH, facilitating the adsorption of LAS onto it [15]. However, as indicated from a lower LAS removal in a lower pH, it appeared that when the pH was too low, there would be a competition between the positive charges of CS and the TiO_2 surface with the excessive proton to interact with LAS molecules. Hence, a slightly acidic pH (pH $\approx$6) is a favourable condition to have an optimum result. When pH is alkaline, the amine groups in CS are less protonated, and makes it less effective to adsorb LAS molecules. The same thing occurs for the TiO_2 case. In alkaline pH, the surface of TiO_2 will be negatively charged which leads to repulsion forces between it and the LAS molecules. It is also worth to note that the attachment of LAS to the surface of TiO_2 seemed to play an important role in the photocatalytic degradation of LAS. In figure 2 where the pH of the solution was not adjusted (pH $\approx$7.5), the LAS removal was much lower than when the pH was adjusted to be slightly acidic (pH $\approx$6). At the optimum pH, the LAS removal by TiO_2 powder through photodegradation process was up to 72.41% from its initial concentration. Under this condition, the LAS removal through the photodegradation process was even as effective as adsorption by CS conducted in a dark closed cabinet (72.72%). The same pattern can also be observed in a treatment using TiO_2/CS composite where a slight adjustment of pH can enhance the LAS removal effectiveness.

4. Conclusion

From this study, it can be concluded that TiO_2/CS composite is a potential material to be used in removing LAS from solution. The composite gave a better result than when the LAS solution was treated with TiO_2 powder and CS beads, as the removal took place through adsorption and photocatalytic degradation. The removal of LAS in this study was highly influenced by pH, and the optimum results were achieved at pH 6, or a slightly acidic pH. A dose of 1-gram composite per 100 mL solution at pH 6 was enough to remove LAS up to 95% from its initial concentration.

References

[1] Scheibel JJ 2004 The evolution of anionic surfactant technology to meet the requirements of the laundry detergent industry *J Surfactants Deterg* **7(4)** 319-28
[2] Cowan-Ellsberry C, Belanger S, Dorn P, Dyer S, McAvoy D, Sanderson H, Versteeg D, Ferrer D, Stanton K 2014 Environmental safety of the use of major surfactant classes in North America *Crit Rev Env Sci Tec* **44(17)** 1893-993
[3] Scott MJ and Jones MN 2000 The biodegradation of surfactants in the environment *Biochim Biophys Acta* **1508(1)** 235-51
[4] Oyama T, Aoshima A, Horikoshi S, Hidaka H, Zhao J and Serpone N 2004 Solar photocatalysis, photodegradation of a commercial detergent in aqueous TiO_2 dispersions under sunlight irradiation *Sol Energy* **77(5)** 525-32
[5] Abd El-Gawad HS 2014 Aquatic environmental monitoring and removal efficiency of detergents *Water Science* **28(1)** 51-64
[6] Steber J 2007 The ecotoxicity of cleaning product ingredients in *Handbook for Cleaning/ Decontamination of Surfaces* ed Johansson I and Somasundaran P (Amsterdam: Elsevier) pp 721-746.
[7] Daghrir R, Drogui P and Robert D 2013 Modified TiO_2 for environmental photocatalytic applications: a review *Ind Eng Chem Res* **52(10)** 3581-99
[8] Khataee AR, Fathinia M, Aber S and Zarei M 2010 Optimization of photocatalytic treatment of dye solution on supported TiO_2 nanoparticles by central composite design: intermediates identification *J Hazard Mater* **181(1-3)** 886-97
[9] Zhao X, Xiao G, Zhang X, Su H and Tan T 2012 The effect of Ni^{2+} and Cu^{2+} on the photocatalytic degradation of dyes by the chitosan–Tio_2 complex *Appl Biochem Biotech* **168(1)** 183-97
[10] Kurniyati R, Sumarni W and Latifah L 2015 Pengaruh chitosan beads dan chitosan beads ikat silang asam sitrat sebagai penurun kadar fosfat dan ABS (Alkyl Benzene Sulfonate) pada limbah laundry," *Indo J Chem Sci* **4(1)** 36-41
[11] Zainal Z, Hui LK, Hussein MZ and Abdullah AH 2009 Characterization of TiO_2–chitosan/glass photocatalyst for the removal of a monoazo dye via photodegradation–adsorption process *J Hazard Mater* 164(1) 138-45
[12] Kansal SK, Ali AH, Kapoor S 2010 Photocatalytic decolorization of biebrich scarlet dye in aqueous phase using different nanophotocatalysts *Desalination* **259(1-3)** 147-55
[13] Akpan UG and Hameed BH 2009 Parameters affecting the photocatalytic degradation of dyes using TiO_2-based photocatalysts: a review *J Hazard Mater* **170(2-3)** 520-29
[14] Ngah WW, Teong LC and Hanafiah MM 2011 Adsorption of dyes and heavy metal ions by chitosan composites: A review *Carbohyd polym* **83(4)** 1446-56
[15] Gaya UI and Abdullah AH 2008 Heterogeneous photocatalytic degradation of organic

IOP Conf. Series: Materials Science and Engineering **980** (2020) 012075 doi:10.1088/1757-899X/980/1/012075

contaminants over titanium dioxide: a review of fundamentals, progress and problems *J Photoch Photobio C* **9(1)** 1-12

IOP Conf. Series: Materials Science and Engineering **980** (2020) 012076 doi:10.1088/1757-899X/980/1/012076

Water quality assessment and a study of current palm oil mill effluent (POME) treatment by ponding system method

J Jumadi*, A Kamari and S T S Wong
Department of Chemistry, Faculty of Science and Mathematics, Universiti Pendidikan Sultan Idris, 35900, Tanjung Malim, Perak, Malaysia.

E-mail: julianajumadi@gmail.com

Abstract. Palm oil is one of the main drivers of Malaysian economy that becomes major contributor to the Malaysian Gross Domestic Product (GDP) in the agriculture sector. The demand for palm oil-based product is expected to increase proportionally to the growing population, thus palm oil plantations have been growing rapidly to meet the global demand. Hence, to ensure sustainable management of palm oil production, avoid a significant effect to environmental pollution, the Roundtable on Sustainable Palm Oil (RSPO) and Malaysian Sustainable Palm Oil (MSPO) have been established to minimize environmental issues related to the palm oil industry in Malaysia. This paper provides the physicochemical characteristics and treatment techniques currently available for treating palm oil mill effluent (POME) nowadays. In addition, the correspondence between biological oxygen demand (BOD) and suspended solid (SS) of raw POME samples with the rainfall distribution pattern were also evaluated. The sampling took place at the cooling pond of a palm oil mill throughout 2019 and had been analyzed using water quality standard methods. The results indicated that the total monthly rainfall significantly influenced the concentration of BOD and SS parameters, of which in dry season (February) the highest BOD (36,200 mg/L) and lowest SS (10,522 mg/L) were recorded.

1. Introduction

In early 1870, the British government was introduced the palm oil tree (*Elaeis guineensis)* natively from West Africa to Malaysia for landscape decorative purposes. Afterwards, the palm oil tree was developed as an industrial crop to reduce dependency on rubber and tin. Historically, commercial palm oil cultivation was first encountered at Tennamaran Estate, Selangor in 1917 and the cultivation had increased rapidly in Malaysia started in early 1960 [1,2]. This is according to the introduction of government's agricultural diversification program for planting palm oil. Thus, the palm oil planted area was expanded remarkably from 54 674 hectare in 1960 to more than 5 million hectares in 2019 [3].

E. guineensis is a monoecious species as male and female reproductive systems exist together on the same plant that can grow up to approximately 10-30 metres tall with an average life-span around 25-30 years. The ideal growing condition for palm oil was in rainy tropical lowland regions that fall within 10 degrees north or south of the Earth's equator, where large scale palm crop can be found particularly at Malaysia and Indonesia [4]. It takes 36 months to start bearing the fresh fruit bunches (FFBs) that can be harvested all year round. Every single tree can produce averagely 10 tonnes of FFBs per hectare weighing between 10 and 15 kg with up to 3000 dark purple fruitlet per bunch. The fleshy fruit of each oil palm is virtually spherical or elongated in shape that contains 49% oil at mesocarp and about 50% endosperm (figure 1) [5].

Palm fruits generate two different types of oils: (i) palm kernel oil from the seeds/kernel and (ii) crude palm oil from the mesocarp. The palm oil able to extracted 4.0 tonnes of crude palm oil, 0.5 tonnes of palm kernel oil, and 0.6 tonnes of palm kernel cake per hectare [6]. Palm oil is one of the highest yield oil crops, which required lesser land that other oil-producing crop. It is estimated that cultivation average yields of palm oil, soybean, rapeseed, and sunflower, are able to produce about 18.5, 0.38, 0.70 and 0.45 tonnes of oil per hectare per year, respectively [7,8]. The extremely versatile characteristics of oil palm such as semi-solid at ambient temperature, longer shelf-life, stable at high heat, resistant to oxidation, odourless and colourless become the most useful and consumed vegetable oil worldwide. Nearly 50% of the packed products can be found in grocery stores, anything about food to personal care products are mostly produced from palm oil [9]. Palm oil is entirely free of genetically modified organism (GMO), cholesterol, and trans-fatty acid [10].

Table 1. The global production of palm oil in 2019 [11].

Country	Production (million metric tonnes)	%
Indonesia	41.50	58
Malaysia	20.80	29
Thailand	2.90	4
Colombia	1.63	2
Nigeria	1.02	1
Others	3.17	6

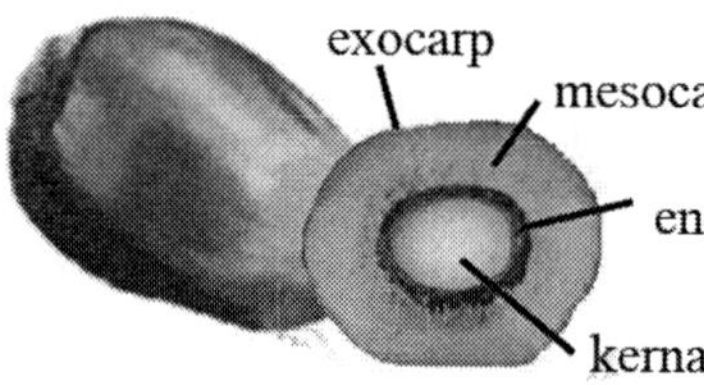

Figure 1. Structure of an oil palm fruit [5].

According to the data reported by the United States Department of Agriculture in 2020, Indonesia and Malaysia covered over 87% of world supply of palm oil including with other 49 countries (table 1) [11]. The palm oil industry in Malaysia has grown rapidly. The total production of crude palm oil in 2018 and 2019 was reported as 19,515,496 and 19,858,367 tonnes [6], respectively and it has a consequential contribution to economic development and growth in Malaysia. Therefore, this paper will summarize in detail the characteristic of POME and the biological reactions involved in the ponding system, along with the factor that could affect its efficiency. Besides, the observation of water quality in the palm oil mill pond treatment, and studies about the relationship between POME parameter (BOD and SS) and rainfall distribution will also be evaluated.

2. A study of Malaysian palm oil industry

2.1. Palm oil mill effluent (POME)

POME is a liquid waste produced in the extraction process of palm oil. It has been estimated that more than 0.75 m^3 of POME will be normally produced for each tonne of FFB production. In 2019, Malaysia produced roughly 19.85 million tonnes of crude palm oil (CPO), expected around 1.5 m^3 of water is needed in producing one tonne of CPO and generated close to 14.89 million m^3 of POME annually [12,13]. Due to enormous water consumption throughout the palm oil extraction process, thus half amount of it will end up as liquid effluent. They are five major separation stages in palm oil extraction process, namely: (i) sterilisation (created 36% of total POME), (ii) bunch stripping, (iii) digestion and pressing, (iv) oil clarification and purification (60% of POME) and (v) nut cracking (4% of POME) [14].

Generally, the liquid waste from palm oil extraction is non-toxic because there is no chemical addition in the extraction process, and it makes it as the best food source of nutrients for microorganisms [15]. The raw or untreated POME has a high amount of nutrient and organic matter. The high total number of organic matter content is due to the existence of different kinds of sugars such as arabinose, xylose, glucose, galactose and mannose, which act as carbon sources to the microbial community [16]. In Malaysia, there are 436 water pollution sources in 2013 contributed by palm oil mills industry as stated in the annual Malaysia Environment Quality Report.

2.2. Characterisation of POME

The characteristics of raw POME are listed in table 2. Generally, the physical of raw or untreated POME is dark-brown colour with a colloidal mixture consisting of 95 to 96% water, 0.6 to 0.7% oil and 4 to 5% total solids including 2 to 4% suspended solid particularly fragments from the palm fruit [17]. It is also acidic because of the organic and free fatty acid production in the fermentation process, it is non-toxic, has a distinctive odor, and is discharged at a temperature between 80-90°C [18].

Fresh POME also possesses a high number of total solids, biochemical oxygen demand, chemical oxygen demand, oil, and grease. The characteristics of POME may vary based on days, factories, batches, processing techniques, the quality or type of fruit, and the chopping season [19].

Table 2. Physicochemical characteristics of untreated POME [20,21,22].

Parameters	Units	Range
Temperature	°C	80 - 90
pH	-	4 - 5
Biochemical Oxygen Demand (BOD)	mg/L	25,000 - 65,714
Chemical Oxygen Demand (COD)	mg/L	44,300 - 102,696
Total Solid (TS)	mg/L	40,500 - 72,058
Total Suspended Solids (TSS)	mg/L	12,800 - 18,000
Total Volatile Solids (TVS)	mg/L	34,000 - 49,300
Oil and Grease (O&G)	mg/L	145 - 9,341
Ammonical-Nitrogen (NH_3-N)	mg/L	35 - 103
Total Kjeldahl nitrogen (TKN)	mg/L	750 - 770

Moreover, POME is known as an effluent that contains high composition and concentration of carbohydrate, protein, nitrogenous compound, lipid, and minerals that can be utilized as a food source by aquatic organisms [23]. Also, it comprises various essential nutrient elements namely nitrogen, phosphorous, calcium, magnesium, and potassium in POME that can be reused or returned to the plantation in the form of fertilizer [24]. Besides all the nutrients, POME usually contains heavy metals such as zinc (Zn) and iron (Fe), which are toxic and non-biodegradable which can affect aquatic living organisms and cause adverse effects to humans health [25]. Table 3 shows the element content in raw POME.

Table 3. Elements content in raw POME [20,26].

Element	Concentration (mg/L)	Element	Concentration (mg/L)
Potassium	3713.99	Strontium	3.14
Magnesium	467.58	Cobalt	1.80
Calcium	275.16	Copper	1.82
Phosphorous	180	Iron	1.42
Sodium	50.21	Gallium	1.37
Bismuth	7.61	Manganese	1.92
Aluminum	3.31	Zinc	1.60
Rubidium	3.87	Barium	1.63

2.3. Regulation control of effluent discharge

Table 4 shows the parameters limit for watercourse discharge of liquid effluent from three world's largest production of palm oil industry. It is clear that regulation on the limit for watercourse discharge of effluent is necessary in order to protect the watercourse and coastal ecosystems. Generally, the government of a country will appoint an organization to monitor the quality of effluent and put a punishment to those who do not comply with the regulation. Even though it has been under a very strict regulation implemented by the government for each palm oil exporter country, in many cases, the palm oil factories are failed to comply with the standard discharge limit. This issue can be related to the difference in POME properties and the overflow of effluent during heavy rain to a nearby river [27].

Table 4. Parameters limit for watercourse discharge of effluent from oil palm industry in Malaysia, Indonesia and Thailand [28,29,30].

Parameters	Units	Limit Discharge		
		Malaysia	Indonesia	Thailand
BOD	mg/L	100	250	100
COD	mg/L	*	500	1000
Total Solid	mg/L	*	N/A	N/A
Suspended Solid	mg/L	400	300	150
Oil and Grease	mg/L	50	30	25
Ammoniacal nitrogen	mg/L	150	20	N/A
Total Nitrogen	mg/L	200	N/A	50
pH	-	5 - 9	6 – 9	5 - 9
Temperature	°C	45	N/A	40

Note, * No discharge standard after 1984, N/A Not Available

2.4. Current POME treatment: ponding system

Ponding system is the widely used treatment method for POME in Malaysia with more than 85% of mills have adopted the treatment system since it is inexpensive, has a low operational maintenance, has a simplicity and ease of handling [31]. Basically, biological treatment processes of the ponding system can be divided into anaerobic, facultative, aerobic and aerated [32]. The energy demand consumption for ponding systems is low due to the absence of mechanical mixing and rarely operation control or monitoring. According to Zainal *et al.* [31], ponding system has been found to show a productively reduced in the concentration of pollutants up to 100-1725 mg/L for COD, 100-610 mg/L for BOD and 100-200 mg/L for ammoniacal nitrogen.

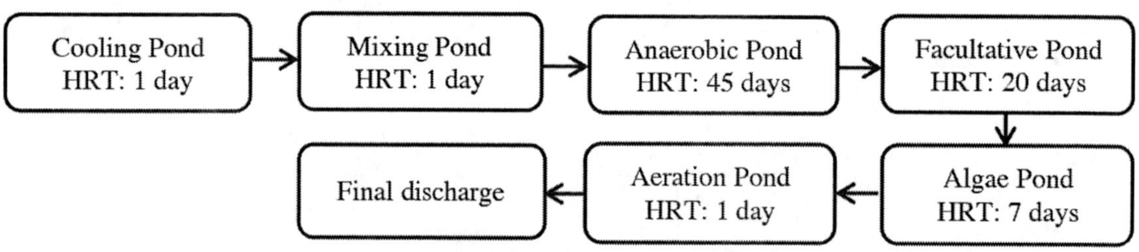

Figure 2. Layout plan for POME waste water treatment plant.

Formation of scum and the accumulation of solid sludge at the bottom of the system pond is considering one shortcoming of the treatment. When scum and sludge mix together for a long time it will lower the effectiveness of the system by decreasing the volumetric proportion and required long hydraulic retention time for degradation. Besides, the microbial in charge for waste breakdown are kept in suspension and bubbles rise to surface of water [33]. Other disadvantages of ponds system are requiring large land area, control and monitoring systems are difficult because of the pond size. Despite of these weaknesses, this system offers an easy technology, lower construction and maintenance cost, is easy to manage and operate as the technology required is relatively uncomplicated, therefore skilled worker is not required. The layout plan for POME waste water treatment plant are summarized in figure 2.

2.4.1 *Anaerobic pond.* Anaerobic digestion is one of the most useful biological treatments for POME treatment which gives very significant BOD reduction up to 90%, COD and O&G. However, it has less tolerance to nitrogen, phosphorus and pathogen number. In anaerobic pond treatment, microorganism is used to breakdown all organic matters (carbohydrates, protein, oil, and fat) in the effluent without the presence of oxygen or other oxidizing chemicals to produce methane and carbon dioxide. Anaerobic ponds are usually 7-19 foot depth or as deep as possible without reaching groundwater to minimize the presence of oxygen and maximizing heat retention. At such depths, it will result in lowering oxygen transfer and improving the performance of the microorganism to digest the organic matter. There are three steps of biological conversion of organic matter: (i) hydrolysis (organic polymers such as polysaccharides, protein, and lipid are broken down to monomer by hydrolytic bacteria), (ii) acidogenesis (fermentation forming simpler organic compound volatile fatty acid) and (iii) methanogenesis (volatile fatty acid into methane and carbon dioxide) [34]. Anaerobic processes also produce renewable energy source in the form of biogas methane, and other by–product such as hydrogen sulfide, ammonia, carbon dioxide, a variety of amine, and volatile acid compound [35]. The illustration of the anaerobic pond is shown in figure 3.

The advantage of anaerobic ponds impacts on the high removal of organic load, the tolerance to high TSS and O&G level effluent, low operating input, production of valuable by-product methane gas, production of less sludge than other technologies [17]. It has been estimated that up to 28 m^3 of biogas (70% methane, 30% carbon dioxide and a trace amount of hydrogen sulfide) was produced for 1 m^3 of POME which equivalent to more than 168 kWh of electricity generating capacity (1 m^3 = 6kWh) [45]. Besides the entire aforementioned advantages anaerobic pond also has weaknesses: it has difficulty to collect and utilize the methane gas since the pond is normally designed in the open-air system [18]. Other disadvantages are effluent needs further treatment, long retention time, low process stability, and low start-up period [17]. Since anaerobic pond also produces hydrogen sulphide, several of amine and volatile acid compound that can cause an obnoxious and offensive odor. At the anaerobic zone, methane gas has also been produced which can cause 21 times higher heat-trapping ability of greenhouse gas than carbon dioxide [36].

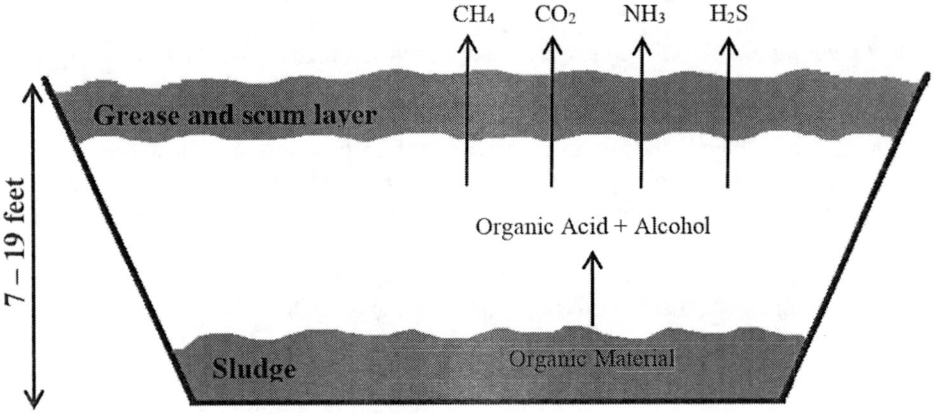

Figure 3. Illustration of Anaerobic Pond [40].

2.4.2 Facultative pond. In the facultative ponding system, it has three distinct layer combinations of aerobic, facultative and anaerobic layers. At the first layer (top zone), algae use carbon dioxide, ammonia, phosphates and sunlight for the encouragement of photosynthesis and metabolic processes, thus ultimately releasing oxygen. Meanwhile, at the second layer (middle zone) the condition is rather in between aerobic and anaerobic operation. With the absence of minimal volume of oxygen, the facultative bacteria tend to grow more rapidly and oxidise the organics in middle zone. At third layer (bottom zone) with the total absence of oxygen, the anaerobic bacteria come in contact with the settled solids and organic matter in the wastewater and consume it as food which will then release hydrogen sulphide, carbon dioxide and methane [36].

Figure 4 shows the illustration of facultative pond. Facultative organisms can perform with or without dissolved oxygen at range 8-9 foot depth. The advantages of facultative ponds are moderately effective in removing organic load, the design of the pond is simple and cheap to construct, easy to operate, low odor when operated properly, no energy required, and little operational input needed [37]. Several disadvantages of facultative ponds are settled sludges and inert materials require periodic removal, produce unpleasant odors when aerobic blanket disappears or when effluent overloaded, no significant nutrient or pathogen removal and may produce some methane.

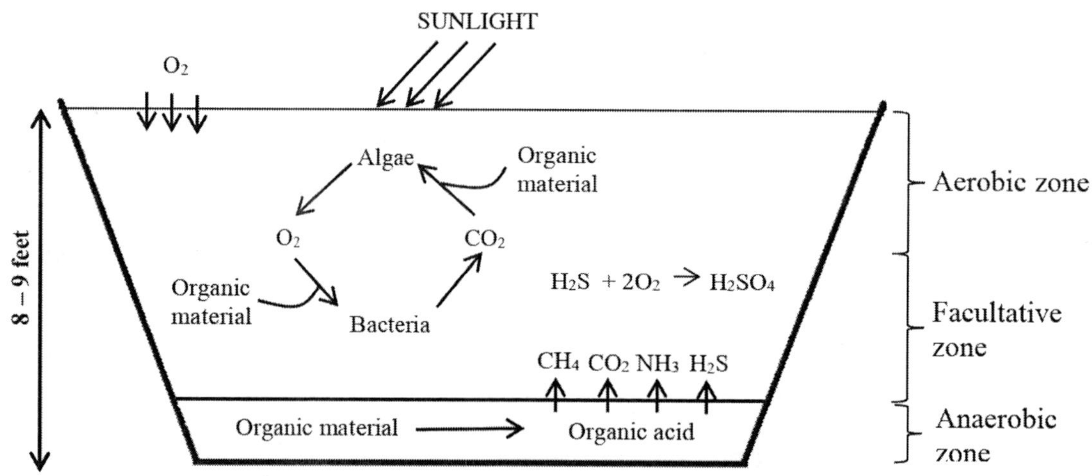

Figure 4. Illustration of Facultative Pond [40].

2.4.3 Aerobic/algae pond. Aerobic ponds are also known as algae ponds where oxygen is present throughout the entire depth of the pond all the time in order to promote the photosynthesis process of the algae or microorganism. Photosynthesis is a biological process that occur naturally in which algae use carbon dioxide along with the presence of sunlight to produce oxygen and carbohydrates. In wastewater treatment ponds, oxygen released from the photosynthesis process is used by the bacteria to stabilize the suspended organic material in wastewater. Bacteria use the oxygen to break down organic matter into simpler compounds, and releasing carbon dioxide for the use of algae. Breaking down organic material will reduce BOD level. At night, the photosynthesis activity by the algal cell will reduce and the algae use dissolved oxygen and discharge CO_2 in the process of respiration that lower the pH and cause to the longer hydraulic retention time.

To ensure the light able to penetrate fully around the pond and support the photosynthesis process so the deep of the ponds usually 6-7 feet. Therefore, sunlight is an important factor in aerobic pond. Aerobic bacteria and algae will stabilize the organic wastes for affluent and emit dissolved oxygen as a by-product of photosynthesis. The illustration of aerobic pond is shown in figure 5.

IOP Conf. Series: Materials Science and Engineering **980** (2020) 012076 doi:10.1088/1757-899X/980/1/012076

The advantages of aerobic ponds are the resistance to organic and hydraulic shock loads, no continuous sludge handling, fewer mechanical problems, low operational, and energy cost. Meanwhile, the disadvantages of the aerobic ponds are it requires a large land area, climatic condition, vegetation (algae), labour intensive, and costly sludge removal [38].

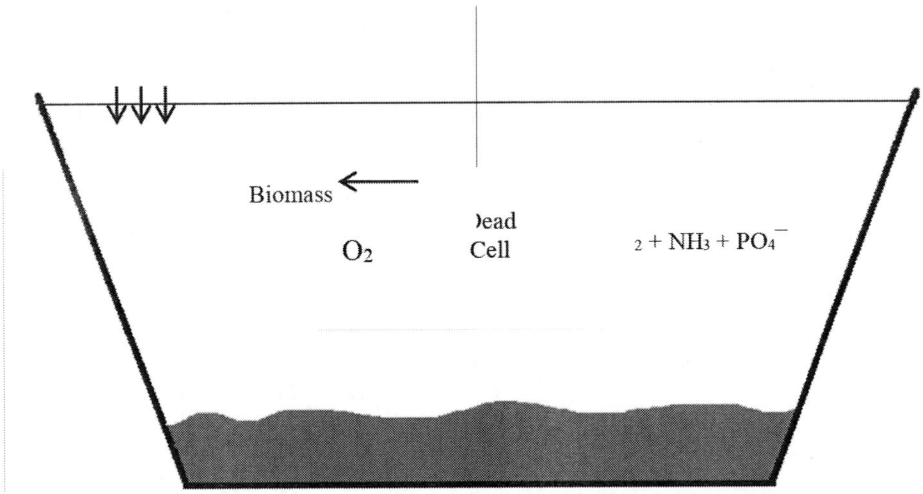

Figure 5. Illustration of Aerobic Pond [40].

2.4.4 Aerated pond. Aeration pond (6-20 feet depth) is a process that adding oxygen provided by aerator to increase the dissolved oxygen levels in the pond thus maximize the aerobic microorganism activity in the consumption of pollutant in a reasonable time. The second function of the aerated pond was to keep microbial biomass in suspension with the continuously stir thus avoiding a build-up of sludge and also ensure that the entire pond in contact with the oxygen. Mixing will also exhaust the carbon dioxide from the system, as a result, the store of carbon dioxide is depleted and algal growth becomes limited. The microbes biodegrade on the organic pollutants and create an accumulate of bacteria floc on the bottom of the tank or lagoon which can easily settle out which called as activated sludge.

The advantages of aerobic ponds are capable to adapt high organic and hydraulic loads, very efficient for BOD and pathogens reduction, less production of offensive odours with the right design and maintenance, less land required, and the treated water can be recycled and reused when a secondary maturation/settling pond was conducted followed with the aerated pond. Meanwhile, the disadvantages are it requires high energy consumption to turn on the aerator system, high capital and operating costs depending on the price of land and electricity, skilled operation workers for operation and maintenance, professional design and construction management, sludge and possibly effluent might require further treatment [39].

3. Materials and methods

The water quality assessment was conducted to observe and collect the background information of the POME in ponding system and how the total average rainfall was affected by its parameter. The POME from cooling pond was chosen for the monitoring study that was obtained from the palm oil mill located at Trolak, Perak, Malaysia. Basically, a total of 12 monthly samplings had been collected every month started from January to December 2019. Since the fresh POME had a temperature ranging between 75 - 90° C, it was allowed to cool around 40 - 45°C before being collected into polyethylene bottles. The sample was kept at low temperature 4°C in airtight bottle to prevent biodegradation due to the microbial activity. The physicochemical parameter (BOD and SS) of POME were measured in situ by using recommended method suggested by APHA Standard Method Examination of Water and Wastewater. The BOD and SS parameter was determined by using APHA method 5210 and 2540, respectively.

4. Results and discussion

4.1. Water quality monitoring

Figure 6 showed the assessment of the water quality trends of POME at cooling pond though over the year of 2019. The use of the monitoring was to determine the parameters trend of the POME and to study the relationships between climate changes and the parameters of the POME. According to the data released by the Malaysian Meteorological Department, the rainfall pattern in Perak showed the lesser rainfall distribution from January to February, June to July and December than the rest of the month at below 120 mm of average rainfall. While April and August to November recorded highest average rainfall above 300 mm.

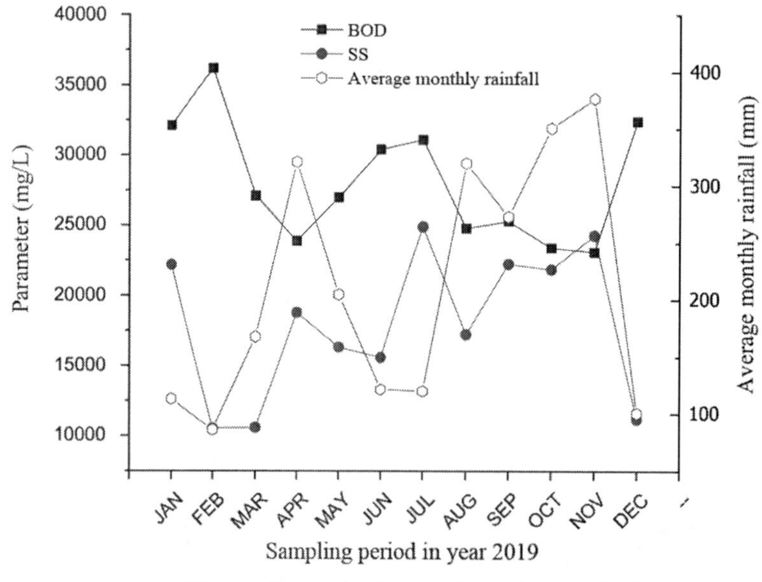

Figure 6. Monitoring studies of raw POME.

From figure 6 it clearly showed that the rainfall pattern and trends significantly affected the concentration of BOD and SS in the POME. As we can see, in February which received the lowest amount of rainfall with monthly average rainfall approximately 85.96 mm, gave the highest reading of BOD at 36,200 mg/L and the lowest reading of SS at 10,522 mg/L. However, at highest monthly average rainfall (November) shows the lowest BOD and highest SS at 23,100 mg/L and 24,302 mg/L, respectively. The turbidity and TSS reading might be increased at heavy rainfall season due to pond banks erosion, and an increase in flow rates during heavy rainfall would keep suspended solid instead of settle at the bottom of pond [43]. Meanwhile, the decrease of BOD concentration in rainfall season was caused by the lower content of organic carbon in suspended solids and a lower biodegradability of organic carbon [44]. Thus, the result revealed that the relationship of total rainfall distribution would change the trends of BOD and SS concentration.

4.2. Factors affecting ponding treatment
There are several conditions that might affect ponding treatment biological activity, primarily climatic condition, vegetation and seasonal changes.

4.2.1 Climatic condition. Sunlight, temperature and wind were the three primary climatic conditions that might affect the biological activity of the pond treatment. First, sunlight was the key factor for photosynthesis performance and influenced the oxygen production in a pond. The depth of pond was also important to determine the light penetrates through the pond which encouraged the algae to grow, and the production of oxygen. Second, temperature also affected the microorganisms growth rate and activity. As temperatures increased during summer, the bacteria and algae activity also increased. While the microorganisms activity was the lowest during the winter. At extremely cold temperature, Ice and snow would cover the surface pond that limits the sunlight penetration, as well as in raining season. Third, the wind produced natural mixing to the pond which transfers the oxygen throughout the pond and its surface. The effectiveness of the biological treatment in the pond would be enhanced when sufficient mixing occurred between oxygen, algae, bacteria and organic matter. In addition, the mixing operator was also to control the growing of vegetation on the pond surface [40].

4.2.2 Vegetation. Overabundant duckweed (*lemnaceae*) would create a floating mat on the surface of the pond which restricting UV penetration that inhibited algae growth. This would lowering the dissolved oxygen levels and BOD treatment since the duckweed prevented the photosynthesis process of algae in the pond. For a good reason, a little quantity of duckweed could be advantageous to control algae thrive and very effective at adsorbing nutrients. However, it just took it into its cell wall without rib the nutrient and would realize the nutrient back into the water once it died [41].

There were many different forms of algae growing in wastewater treatment ponds, they were commonly green and blue-green algae. Green algae would be predominating when pond conditions and treatment yield the best result which gave the green color to the ponds. While blue-green algae were filamentous that showed bad pond treatment condition and often formed unsightly and odorous mats. The pond content blue–green algae will have high organic loading, low dissolved oxygen (DO), warm water temperature conditions, and low in nutrients [42].

4.2.3 Seasonal changes. Treatment efficiency decrease in the raining season were reduced in sunlight penetrated period and limits the quantity of photosynthesis process, which reduced the dissolved oxygen in the pond. Raining season would also decline the bacterial activity than decreasing the treatment efficiency and the entire pond might go anaerobic. While in the summer period the entire pond would obtain maximum oxygen levels for algae photosynthesis activity. Warm water temperatures could increase bacteria action to produce the best environment for efficient ponding treatment [40].

5. Conclusions and recommendations
As the palm oil industry continues to expand to fulfil the global demand, the generation of POME is also expected to increase significantly. However, there are several pose of a challenges and issue related to the POME, in term to ensure that the final discharge effluent obeys with the standards limit set up by the Department of Environment Malaysia. Undoubtedly, the conventional ponding system was still the most favoured POME treatment method used in most palm oil mill in Malaysia, although it required a high retention time and large treatment areas. This was due to the capability of ponding system to reduce the BOD and degrade the contaminant at low operational cost and maintenance for enormous amount of POME under various climate change scenarios. However, further studies should be conducted to improve the rate of degradation to reducing the hydraulic retention time, by introducing new methods or enhancing the existing treatment method. Beside that, anaerobic digestion of ponding system has generated enormous amount of methane gas that can cause emission of greenhouse gases (GHG) which contribute to global warming. Therefore, installing the biogas capture

system in pond is necessary, and in the future, the methane emission would have the potential to be turned into renewable energy. From the monitoring of water quality, the analysis of BOD and SS parameter of POME showed that, in general the average monthly rainfall significantly influenced the concentration of the parameters. The trends found in this study indicated that the reading of BOD was the highest and lower in SS at dry season, and vice versa in rainy season.

Acknowledgement

The authors would like to express their sincere gratitude to Department of Chemistry, Faculty of Science and Mathematics, Universiti Pendidikan Sultan Idris for providing the instrumentation services for this research.

References

[1] Ooi Z X, Ismail H, Bakar A A and Teoh Y P 2014 *BioResources* 9 7926-40
[2] Awalludin M F, Sulaiman O, Hashim R and Nadhari W N A W 2015 *Renew. Sust. Energ. Rev.* **50** 1469-84
[3] Malaysian Palm Oil Council 2020 The oil palm tree
[4] Woittiez L S, van Wijk M T, Slingerland M, van Noordwijk M and Giller K E 2017 *Eur. J. Agron.* **83** 57-5
[5] Kwasi P 2002 Small-scale palm oil processing in Africa FAO *Agricultural Services Bulletin* 148
[6] Malaysian Palm Oil Board 2020 Malaysian Palm Oil Industry
[7] Murphy D J 2014 *J. Oil Palm Res.* **26** 1-24
[8] Basiron Y 2007 *Eur. J. Lipid Sci. Tech.* **109** 289-95
[9] World Wildlife Fund 2020 8 Things to know about palm oil
[10] Dian N L H M, Hamid R A, Kanagaratnam S, Isa W R A, Hassim N A M, Ismail N H, Omar Z and Sahri M M 2017 *J. Oil Palm Res.* **29** 487-511
[11] United States Department of Agriculture 2020 World Agricultural Production
[12] Kumaran P, Hephzibah D, Sivasankari R, Saifuddin N and Sahamsuddin A 2016 *Renew. Sust. Energ. Rev.* **56** 929-40
[13] Wu T Y, Mohammad A W, Jahim J M and Anuar N 2010 *J. Environ. Manage.* **91** 1467-90
[14] Liew W L, Kassim M A, Muda K, Loh S K and Affam A C 2015 *J. Environ. Manage.* **149** 222-35
[15] Madaki Y S and Seng L 2013 *Int. J. Sci. Environ. Technol.* **2** 1138-22
[16] Ng Y S, Lim C R and Chan D J C 2016 *J. Environ. Chem. Eng.* **4** 4890-6
[17] Bala J D, Lalung J and Ismail N 2014 *Int. J. Sci. Res. Publ.* **4** 1-24
[18] Loh S K, Lai M E, Ngatiman M, Lim W S, Choo Y M, Zhang Z and Salimon J 2013 *J. Oil Palm Res.* **25** 273-81
[19] Poh P E, Yong W J and Chong M F 2010 *Ind. Eng. Chem. Res.* **49** 11732-40
[20] Ma A N 2000 *Palm Oil Dev.* **30** 1-10
[21] Chin M J, Poh P E, Tey B T, Chan E S and Chin K L 2013 *Renew. Sust. Energ. Rev.* **26** 717-26
[22] Bala J D, Lalung J, Al-Gheethi A A S, Kaizar H and Ismail N 2018 *Water Conserv. Sci. Eng.* **3** 139-56
[23] Islam M A, Yousuf A, Karim A, Pirozzi D, Khan M R and Wahid Z A 2018 *J. Clean. Prod.* **172** 1779-87
[24] Muhrizal S, Shamshuddin J, Fauziah I and Husni M A H 2006 *Geoderma* **131** 110-22
[25] Jawad R J, Ismail M H S and Siajam S I 2018 *IIUM Eng. J.* **19** 1
[26] Tan S P, Kong H F, Bashir M J K, Lo P K, Ho C D and Ng C A 2017 *Bioresour. Technol.* **245** 915-24
[27] Shahrifun N S A, Ab'lah N N, Hussain H, Aris A, Omar Q and Ahmad N 2015 *Malaysian J. Civ. Eng.* **27** 144-51
[28] Department of Environment Malaysia 2020 Environmental Quality (Prescribed Premises) (Crude Palm Oil) Regulation 1977

IOP Conf. Series: Materials Science and Engineering **980** (2020) 012076 doi:10.1088/1757-899X/980/1/012076

[29] Badan Pemeriksa Keuangan Republik Indonesia 2020 Peraturan Pemerintah (PP) No. 82 Tahun 2001 Pengelolaan Kualitas Air Dan Pengendalian Pencemaran Air
[30] Department of Industrial Work Thailand 1997 Environmental Management Guideline for the Palm Oil Industry
[31] Zainal N H, Jalani N F, Mamat R and Astimar A A 2017 *J. Oil Palm Res.* **29** 528-40
[32] Chan Y J and Chong M F 2019 Palm Oil Mill Effluent (POME) Treatment—Current Technologies, Biogas Capture and Challenges *Green Technologies for the Oil Palm Industry* (Singapore: Springer) pp 71-92
[33] Hassan M A, Yacob S, Shirai Y and Hung Y-T 2004 *Handbook of Industrial and Hazardous Wastes Treatment* (New York, USA: CRC Press) pp 776-96
[34] Adela B N, Muzzammil N, Loh S K and Choo Y M 2014 *Asian J. Microbiol. Biotechnol. Environ. Sci.* **16** 225-31
[35] Wang J, Mahmood Q, Qiu J -P, Li Y -S, Chang Y -S, Chi L -N and Li X -D 2015 *BioMed Res. Int.* **2015** 617861
[36] Rupani P F, Singh R P, Ibrahim M H and Esa N 2010 *World Appl. Sci. J.* **11** 70-81
[37] United States Environmental Protection Agency 2002 Wastewater Technology Fact Sheet: Facultative Lagoons, US Environmental Protection Agency.
[38] Abdurahman N H, Rosli Y M and Azhari N H 2013 The Performance Evaluation of Anaerobic Methods for Palm Oil Mill Effluent (POME) Treatment: A Review *International Perspectives on Water Quality Management and Pollutant Control*
[39] United States Environmental Protection Agency 2000 Wastewater Technology Fact Sheet Aerated, Partial Mix Lagoons, US Environmental Protection Agency
[40] Wisconsin Department of Natural Resources 2015 Ponds, Lagoons, and Natural Systems Study Guide
[41] Muradov N, Taha M, Miranda A F, Kadali K, Gujar A, Rochfort S, Stevenson T, Ball A S and Mouradov A 2014 *Biotechnol. Biofuels* **7** 30
[42] Sen B, Alp M T, Sonmez F, Kocer M A T and Canpolat O 2013 Relationship of Algae to Water Pollution and Waste Water Treatment *Water Treatment*
[43] Fondriest Environmental, Inc. 2014 Turbidity, Total Suspended Solids and Water Clarity. Fundamentals of Environmental Measurements
[44] Servais P, Seidl M and Mouchel J M 1999 *Water Environ. Res.* **71** 408-17
[45] Association Québécoise de la Production d'Énergie Renouvelable 2020 How much energy is there in biogas?

Comparing the effectiveness of chitosan and conventional coagulants for coal wastewater treatment

A Maria[1], E Mayasari[1], U Irawati[1]* and Zulfikurrahman[2]
[1]Chemistry Department, Faculty of Mathematics and Natural Sciences, Lambung Mangkurat University Banjarbaru 70714, INDONESIA
[2]Civil Maintenance, CPBL Support Departement, Kelanis Site, PT Adaro Indonesia, Barito Selatan, Kalimantan Tengah

Email: uirawati@ulm.ac.id

Abstract. In this research, the potential of chitosan to be used as a coagulant to treat coal wastewater was investigated, in comparison to a conventional coagulant, i.e: $Al_2(SO_4)_3$ or aluminum sulfate, and Poly Aluminium Chloride (PAC). The parameters being studied were turbidity, pH, TDS, and TSS. The result of this research showed that chitosan worked as a more efficient coagulant to treat coal wastewater compared to alum and PAC, in terms of the needed dose of application. The optimum dose of chitosan was 20 mg/L that gave a 100% decrease in turbidity and TSS. On the other hand, a dose of 120 mg/L of alum was needed to have an optimum result, where the turbidity and TSS were decreased up to 100%. When PAC was used as a coagulant, the optimum dose was 120 ppm that reduced turbidity and TSS to 99.50% and 99.58%, respectively. Coagulation by chitosan, alum, and PAC were all influenced by pH, where the optimum pH for all three coagulants was within a range of neutral pH.

1. Introduction

One of the characteristics of coal wastewater released from coal processing activities is its high turbidity due to a high level of TSS. Before being released to the environment, coal industries often treat this wastewater by adding coagulants into it to reduce the turbidity. Aluminum-based coagulants, which include aluminum salts and polyaluminium chloride (PAC) are some of the most extensively-used coagulants. However, the use of aluminum-based coagulants comes with some drawbacks. Aluminum salts often yield sludge with a high level of metal concentration coming from the coagulant residual [1]. There is also a possibility of an increase in the concentration of dissolved aluminum ions in the treated water. It has been suggested that a high level of aluminum might be one of the causes of Alzheimer's disease [2]. Due to the potential negative impact on the use of conventional aluminum-based coagulants, an alternative coagulant is needed to have a more environmentally friendly treatment process on coal wastewater. In this paper, the potential of chitosan as a natural coagulant is presented. Comparison of the dose needed to have an optimum result in decreasing turbidity for between chitosan and conventional coagulants, i.e. PAC and $Al_2(SO_4)_3$ (aluminum sulfate) was investigated. The effect of pH on the effectiveness of each coagulant in decreasing turbidity and TSS level was also studied.

IOP Conf. Series: Materials Science and Engineering **980** (2020) 012077 doi:10.1088/1757-899X/980/1/012077

2. Materials and methods

2.1. Materials

Coal wastewater samples were collected from the Kelanis site, one of the coal processing locations in South Barito, Central Kalimantan. Chemicals used were food-grade chitosan (100 mesh, MW = 50,000-80,000 Da); NaOH, HCl, CH_3COOH, $Al_2(SO_4)_3$, and PAC powder. All chemicals were reagent-grade from commercial sources (*Merck*) and used as received. To prepare a 2% (w/v) chitosan solution, 20 grams of chitosan was dissolved in 1 L CH_3COOH 2%. The solution was then labeled as (C-Chit). The other coagulants, i.e $Al_2(SO_4)_3$ and PAC were also prepared as 2% (w/v) solutions by dissolving them in distilled water. Each of the conventional coagulant solution was labeled as C-Al and C-PAC, respectively.

2.2. Experimental setup

To study the coagulation process, experiments were conducted by using a standard jar-test apparatus with six 1L-beakers. To determine the optimum dose, each beaker was filled with coal wastewater sample, and a specified amount of chitosan coagulant solution was added into it. The amount of chitosan coagulant solution added varied in each beaker. The mixture in each beaker was mixed rapidly with a 100-rpm speed for one minute, followed by a slow mixing at a 45-rpm speed for 20 minutes. The mixture was then let to settle for a 30-minute of sedimentation time. The supernatants were then withdrawn from the liquid phase for the analysis. The parameters being analyzed were turbidity, TSS level, and pH. All of the analyses were conducted in compliance with Indonesian National Standards (SNI, *Standar Nasional Indonesia*) for analysis of water and wastewater quality. The standard jar-test apparatus was also used to investigate the effect of pH on the coagulation efficiency. Coal wastewater samples were added into the beakers, and then the pH in each of the beaker was adjusted by adding 1.0 N NaOH and or 1.0 N HCl to have a set of samples with various pH, ranged from 4 to 9. Into each beaker, a specified amount of chitosan coagulant was added. The amount added was by the dose giving the most optimum result in decreasing turbidity, as obtained in the previous step. The samples were then rapidly mixed with a 100-rpm speed for one minute, followed by a flocculation process where the mixing speed was set at 45 rpm for 20 minutes, and a sedimentation time of 30 minutes. Supernatants were then withdrawn from the liquid phase. The pH, turbidity and TSS level from the supernatants were then analyzed. The same procedures to determine the optimum dose along with investigation on the effect of pH were repeated by using Al2(SO4)3 and PAC as coagulants.

3. Results and discussion

3.1. Optimum dose

In coagulation, the dose is one of the most crucial factors affecting the effectiveness of the coagulation process. While a dose that is too low would not give an optimum result, using an excessive amount of coagulant can lead to deflocculation [3]. The optimum dose can be different from one type of coagulant to another, and in many cases, the optimum dose also depends on the initial turbidity of the water being treated. Figure 1 shows the comparison between the decrease in turbidity of the coal wastewater samples when chitosan, PAC, and $Al_2(SO_4)_3$ were used as coagulants within the same dose range.

The result showed that while a 20 mg/L dose of chitosan coagulant was enough to have a 100% decrease in turbidity, samples treated by PAC and $Al_2(SO_4)_3$ practically did not show a decrease in turbidity nor TSS level. Pictures presented in figures 2-4 showed the clear difference of turbidity reduction when each of the coagulant being studied was used within the same dose range. Based on this result, the dose range of both PAC and $Al_2(SO_4)_3$was then increased until optimum results were obtained. The dose range that gave optimum results for each coagulant being investigated, in reducing turbidity and TSS levels is presented in figures 5 and 6.

IOP Conf. Series: Materials Science and Engineering **980** (2020) 012077 doi:10.1088/1757-899X/980/1/012077

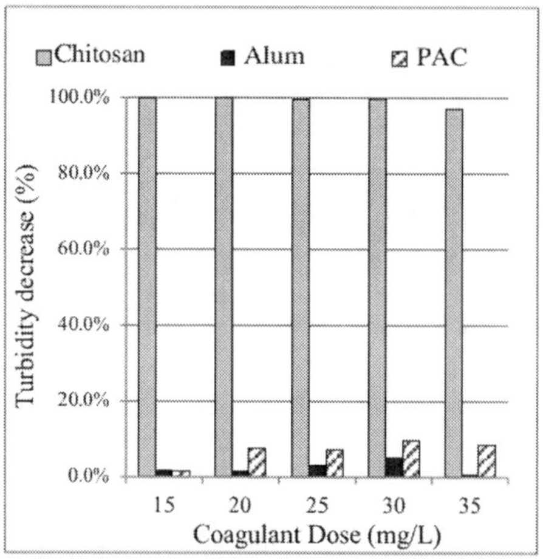

Figure 1. Turbidity decrease of coal wastewater treated by each type of coagulants.

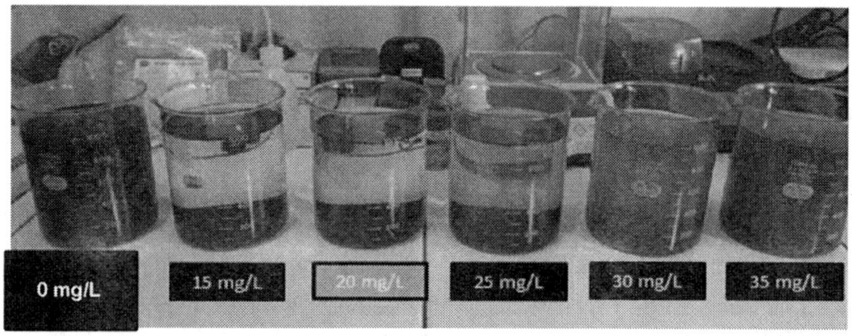

Figure 2. Coal wastewater samples after treated with chitosan as coagulant with a dose range of 0-35 mg/L.

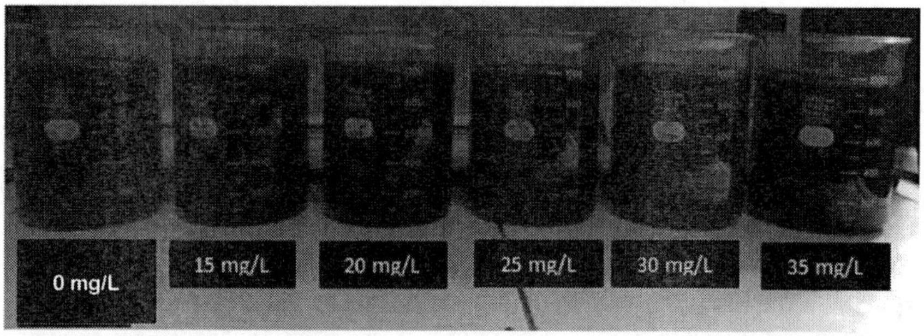

Figure 3. Coal wastewater samples after treated with $Al_2(SO_4)_3$ as coagulant with a dose range of 0-35 mg/L.

The results clearly showed that chitosan needs a much lower dose to give satisfying results, compared to both $Al_2(SO_4)_3$ and PAC. While chitosan needed a dose of 20 mg/L to have an optimum result, the optimum dose for $Al_2(SO_4)_3$ and PAC were 120 mg/L and 105 mg/L, respectively. The difference in the dose needed is likely due to the mechanisms of coagulation taking place. As coagulant-flocculant, chitosan can work through two mechanisms, i.e: charge neutralization and bridging mechanism [4, 5]. Chitosan can destabilize the colloidal particles in the suspension as chitosan is a biopolymer with polycationic properties. This is due to the presence of its amine groups that become protonated in an acidic medium. The long-chain structure of chitosan enabling it to form a bridge between the destabilized colloidal particles forms flocs that would then precipitate. In samples with a high initial TSS level, this mechanism of coagulation-flocculation seemed to work more effectively than the charge neutralization mechanism as the main mechanism of coagulation-flocculation by $Al_2(SO_4)_3$. In treatment using $Al_2(SO_4)_3$, hydrolyzed species were formed once $Al_2(SO_4)_3$ comes in contact with water that will be adsorbed by the colloidal particles [6]. However, a high dose of $Al_2(SO_4)_3$ will be needed to form flocs that are big enough to precipitate through a sweep-floc mechanism [7]. As a polymer, PAC can also work as coagulant-flocculant through the bridging mechanism. However, this mechanism can only take place in a high dose of PAC. If PAC is used in a low dose, the flocculation-coagulation can only occur through the charge-neutralization mechanism. The same pattern when chitosan was evaluated for its performance compared to conventional coagulants was observed by Ahmad *et al* [8]. They reported that chitosan needed a much lower dose (as low as 0.5 g/L) to decrease turbidity in palm oil mill effluent (POME) compared to $Al_2(SO_4)_3$ and PAC, which needed a dose of 8.0 and 6.0 g/l, respectively.

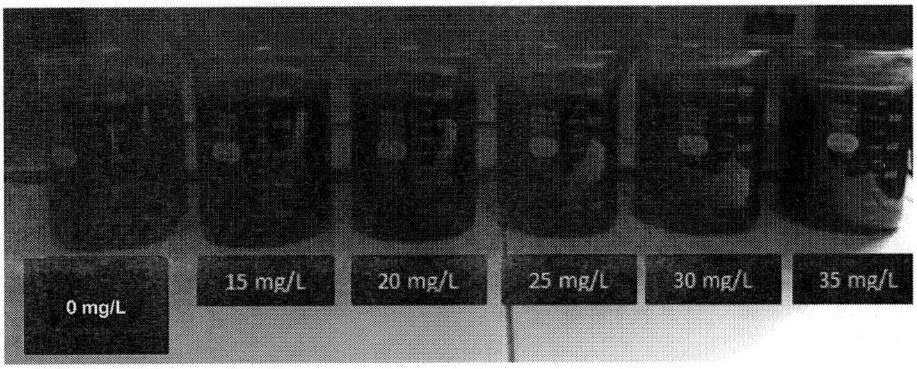

Figure 4. Coal wastewater samples after treated with PAC as coagulant with a dose range of 0-35 mg/L

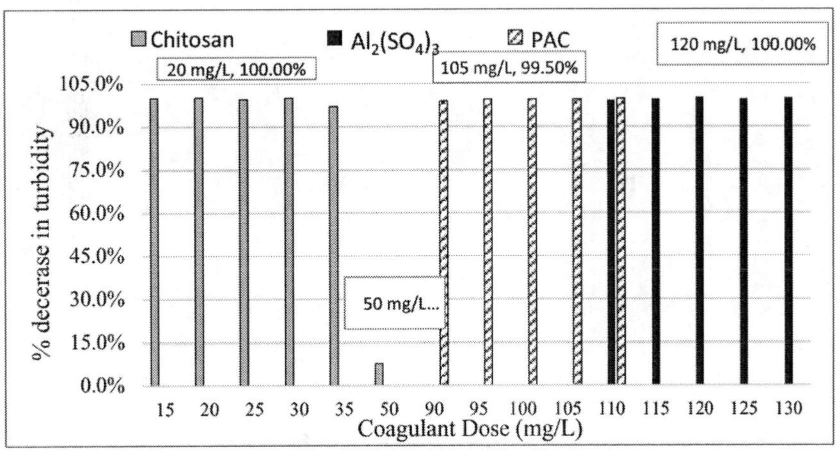

Figure 5. Optimum dose range for each type of coagulants in decreasing turbidity.

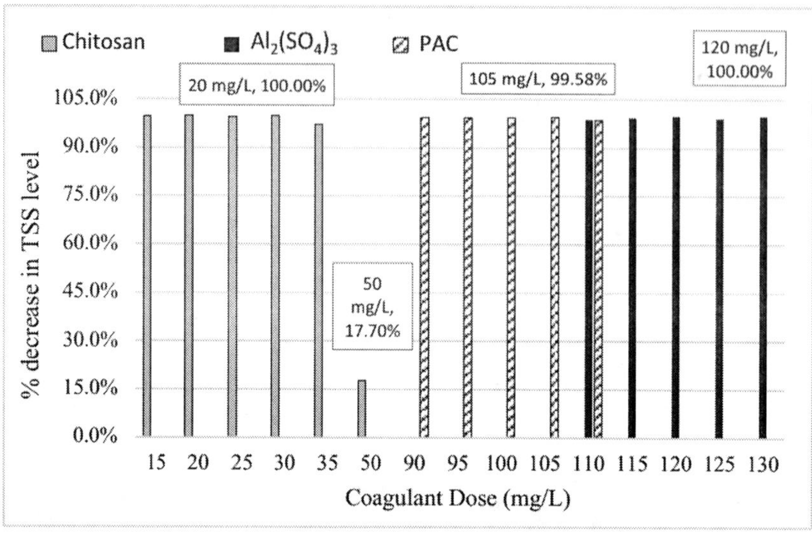

Figure 6. The optimum dose range for each type of coagulants in decreasing TSS level.

It is interesting to be noted that the performance of chitosan appeared to be very sensitive to overdosing. When the dose was increased to 50 mg/L, the decrease in turbidity and TSS value reduced significantly, indicating the occurrence of deflocculation. This is probably because of the nature of chitosan for being polycationic lead to an excess of positive charges in the system when the concentration of chitosan is above its optimum dose. As highlighted by Guibal *et al* [9], this excess of positive charge might cause the particles to be re-stabilized and breaking the formed flocs.

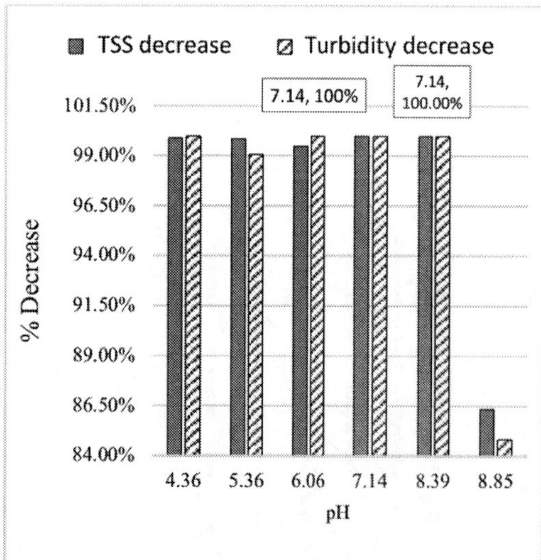

Figure 7. TSS level and turbidity decrease in treatment using chitosan at different pH.

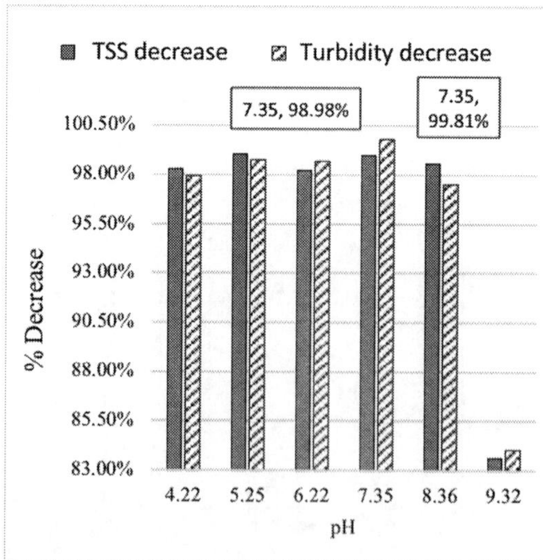

Figure 8. TSS level and turbidity decrease in treatment using $Al_2(SO_4)_3$ at different pH.

3.2. Optimum pH in coagulation

Another important factor in a coagulation process that has been well documented is pH [10]. In coagulation, pH will determine the presence of the protonated form of the coagulants. The figures below present how pH affected the performance of each coagulant in treating coal wastewater. Figures 7, 8, and 9 each show the decrease in turbidity and TSS levels when coal wastewater was treated using chitosan, $Al_2(SO_4)_3$, and PAC, respectively.

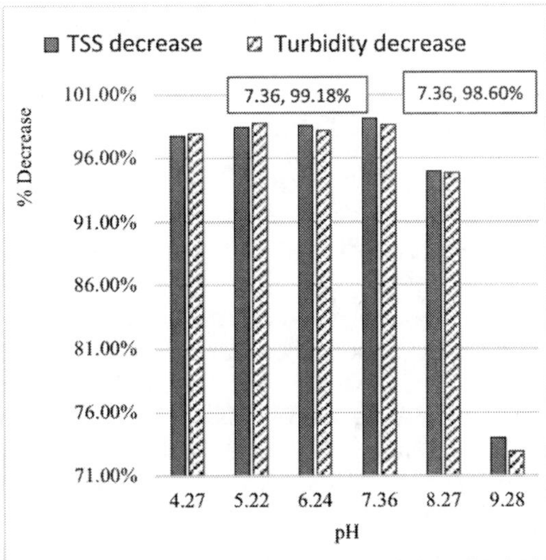

Figure 9. TSS level and turbidity decrease in treatment using PAC at different pH

From all of those figures, it can be observed that all of the coagulants work in a pH range where the system was acidic to a slightly alkaline condition. Once the pH of the system went to a more alkaline condition, the performance of all coagulants dropped significantly, as shown by a steep decline in the decrease in turbidity and TSS level. For all the coagulants, an optimum result was obtained when the pH is within a neutral range. This result is in accordance with some other studies that also showed a preferable condition for optimum coagulation was within a neutral pH range [9,11,12]. Particularly for chitosan, an acidic condition will ensure that the amine groups are protonated, enabling efficient charge neutralization in the coagulation process [4,13,14].

4. Conclusion

The result of this study has demonstrated that chitosan is a strong potential to be used as a natural coagulant as an alternative to conventional coagulant in treating coal wastewater. Chitosan worked more efficiently as it needed a much lower dose than $Al_2(SO_4)_3$ and PAC to give the result in decreasing turbidity and TSS levels in the coal wastewater samples. However, more research is still needed to explore the feasibility of using chitosan on a much larger scale than the laboratory scale that has been conducted in this study.

Acknowledgments

We would like to thank PT Adaro Indonesia who had generously allowed us to conduct this research in their facility.

References

[1] Lee KE, Morad N, Teng TT and Poh BT 2012 Development, characterization and the application of hybrid materials in coagulation/flocculation of wastewater: A review *Chem Eng J* **203** 370-86

[2] Matilainen A, Vepsäläinen M and Sillanpää M 2010 Natural organic matter removal by coagulation during drinking water treatment: A review *Adv Colloid Interfac* **159(2)** 189-97

[3] Wei H, Gao B, Ren J, Li A and Yang H 2018 Coagulation/flocculation in dewatering of sludge a review," *Water Res* **143** 608-31

[4] Bhalkaran S and Wilson LD 2016 Investigation of self-assembly processes for chitosan-based coagulant-flocculant systems: a mini-review *Int J Mol Sci* **17(10)** 1662

[5] Renault F, Sancey B, Badot PM and Crini G 2009 Chitosan for coagulation/flocculation processes–an eco-friendly approach *Eur Polym J* **45(5)** pp 1337-48

[6] Mahmudabadi TZ, Ebrahimi AA, Eslami H, Mokhtari M, Salmani MH, Ghaneian MT, Mohamadzadeh M and Pakdaman M 2018 Optimization and economic evaluation of modified coagulation–flocculation process for enhanced treatment of ceramic-tile industry wastewater *AMB Express* **8(1)** 1-12

[7] Hu CY, Lo SL, Chang CL, Chen FL, Wu YD and Ma JL 2013 Treatment of highly turbid water using chitosan and aluminum salts *Sep Purif Technol* **104** 322-26

[8] Ahmad AL, Sumathi S and Hameed BH 2006 Coagulation of residue oil and suspended solid in palm oil mill effluent by chitosan, alum and PAC *Chem Eng J* **118(1-2)** 99-105

[9] Guibal E, Van Vooren M, Dempsey BA and Roussy J 2006 A review of the use of chitosan for the removal of particulate and dissolved contaminants *Separ Sci Technol* **41(11)** 2487-514

[10] Naceradska J, Pivokonska L and Pivokonsky M 2019 On the importance of pH value in coagulation," *J Water Supply Res T* **68(3)** 222-30

[11] Pan JR, Huang C, Chen S and Chung YC 1999 Evaluation of a modified chitosan biopolymer for coagulation of colloidal particles *Colloid Surface A* **147(3)** 359-64

[12] Saritha V, Srinivas N and Vuppala NS 2017 Analysis and optimization of coagulation and flocculation process *Applied Water Science* **7(1)** 451-60

[13] Escudero-Oñate C, Martínez-Francés E 2018 A Review of Chitosan-Based Materials for the Removal of Organic Pollution from Water and Bioaugmentation *Chitin-Chitosan-Myriad Functionalities in Science and Technology* (London:IntechOpen)

[14] Fabris R, Chow CW, Drikas M 2010 Evaluation of chitosan as a natural coagulant for drinking water treatment," *Water Sci Technol* **61(8)** 2119-28

IOP Conf. Series: Materials Science and Engineering **980** (2020) 012078 doi:10.1088/1757-899X/980/1/012078

Rapid start-up of expanded granular sludge bed (EGSB) reactor using granulated anaerobic bacteria in pharmaceutical wastewater treatment: pilot scale

R A Malik*, H Vistanty, A Mukimin and N Zen
Center of Industrial Pollution Prevention Technology, Indonesia Ministry of Industry, Jln. Ki mangunsarkoro No 5, Semarang, Center Java, 50136

Email: rizalmalik72@gmail.com

Abstract. The start-up process is one of the most critical processes in wastewater treatment. The performance of a biological agent (bacteria) and/or the system can be predicted when the start-up process is carried out. The purpose of the start-up process is an adaptation of biological agent, and also as a granulation process for bacteria to form granule. The aim of this research was to study the effect of granular sludge in enhancing start-up time. The research was done using a 7 m tall stainless-steel reactor with 1,2 m x 1,2 m cubic shape-based and seeded with granulated anaerobic bacteria. The Reactor was operated continuously with configuration as follows: 4 h of HRT with up-flow velocity (V_{up}) of 1.75 m/h and recirculation ratio of 1:2. The main parameters observed were pH, alkalinity, and COD. The result showed that the effluent pH increased from 5,1 (inlet) to 6,9 on the first day of start-up and achieved pH 8,2 on the 4th day, the alkalinity also showed an increase from 219,05 $CaCO_3$ mg/l to 470,3 CaCO3 mg/l. and the reduction of COD was 41,1% on the 4th day. It can be concluded that seeding the reactor with granulated anaerobic bacteria had successfully accelerated the start-up process.

1. Introduction

Pharmaceutical industries are some of the most important industries to support human life, yet the activity of the industry would give a negative impact on the environment and even more on human health if the wastewater is not treated well. Most pharmaceutical industries producing antibiotics will have antibiotics in their wastewater. The presence of antibiotics in wastewater will disrupt the biological wastewater treatment and lead to inefficient wastewater treatment processes. While the chemical treatment is not suitable for treating pharmaceuticals wastewater [1], there are numbers of technologies that could be used to treating pharmaceuticals wastewater such as photocatalytic [2], mixed aerobic-physical technology (membrane filtration/reverse osmosis) [3], ozone and peroxide oxidation [4]. Recently, anaerobic technologies have started to be a promising option for treating pharmaceuticals wastewater due to their high degradation efficiency and lower operational cost.

Anaerobic digestion for treating industrial wastewater has been developed for more than 30 years, Pol and Lettinga [5] have proposed anaerobic digestion as a promising technology for treating industrial wastewater. By now, most industries have been using or at least trying to use this technology for treating their wastewater. The reason that industries prefer to use this technology is due to high effectiveness, minimum maintenance, and operational cost, no chemical needed, and also less to none dangerous sludge production compared to other conventional treatments such as aerobic and physical-chemistry treatments [6]. A lot of research about anaerobic technology has been done to achieve higher treatment efficiency and a more versatile system. Expanded granular sludge bed (EGSB) is the latest anaerobic technology that has been introduced to industries, which is an improvement from the famous up-flow anaerobic sludge blanket (UASB) technology. EGSB was first introduced in the mid-1990s, the major difference between EGSB and UASB is higher up-flow velocity and it can handle a wider variety of the wastewater characteristic such as low to high strength organic loading rate (OLR) or even wastewater that contains recalcitrant substances [7].

In the application of advanced anaerobic treatment such as UASB and EGSB, granules formation is extremely important to ensure the effectiveness of the anaerobic degradation process. Anaerobic granules are important because they do not only support active biofilm but also provide settleability to allow maximum contact between anaerobic microorganism and wastewater. Even though the granulation process is essential to UASB and EGSB system, yet it is one of the major concerns in the start-up process. Anaerobic granulation is known as having a long startup process, it depends on the source of bacterial seed, desired operational condition, and wastewater characteristic. The granulation time varies from 30 [8] to 120 days [9] but is mostly done for more than 30 days.

The speed of the granulation process depends on many aspects such as bacterial morphology, substrate similarity, surface charge compatibility, and ability to produce specific polymers to form flock or granules [7], to achieve functional granules it would require a variety of naturally symbiotic microorganism consortia and it would be a time-consuming process. Previous works done by Liu [10] show a significant difference in the number of the microbial community in seed sludge and granulated sludge after 58 days granulation process. In general, the time required for the startup process is not only affected by the sludge source but also by feed wastewater properties and sludge retention time (SRT) [11]. In addition to microorganism dynamic during the startup process, the conventional startup method uses gradual dilution wastewater as feed, starting from 10% to 90%, this method provides a high chance of startup process but takes more time. During the startup and operational processes, pH, alkalinity, and COD reduction are used as main parameters to control and monitor the stability or health of the anaerobic system.

However, the method to speed up the start-up process remains unknown. Thus this work is aimed at figuring out the use of granulated anaerobic sludge in enhancing the start-up process time in pilot-scale EGSB reactor for treating pharmaceuticals wastewater.

2. Materials and methods

2.1. Materials

2.1.1. Anaerobic bacteria inocula. Inocula used in this research was granulated anaerobic sludge obtained from The Research and technology laboratory, Center of Industrial Pollution Prevention Technology. The granular sludge has 3-7 mm in diameter.

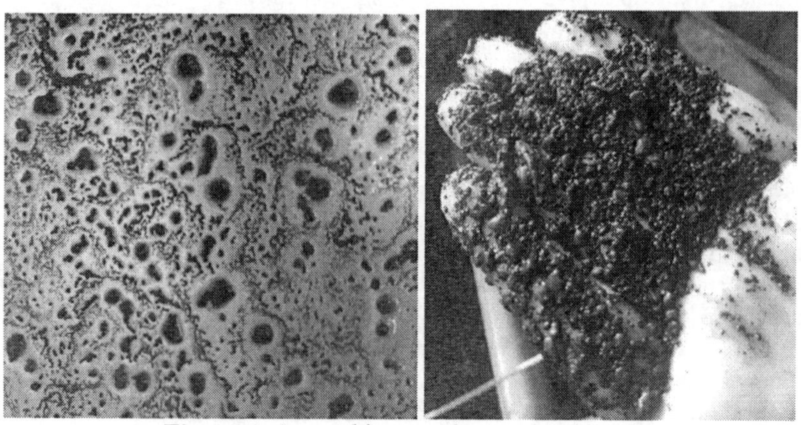

Figure 1. Anaerobic granules used as inocula.

2.1.2. Reactor configuration. The EGSB reactor comprised a cubic shape column of 7 m in total height (6,5m working height) and 1,2 m x 1,2 m cubic shape base made from 4mm thick stainless steel 306L with a total working volume of 9,36 m^3. Inlet and circulation flow were supplied by a digital pump (DAB water technology), nutrition and phosphate buffer was supplemented using a dosing pump.

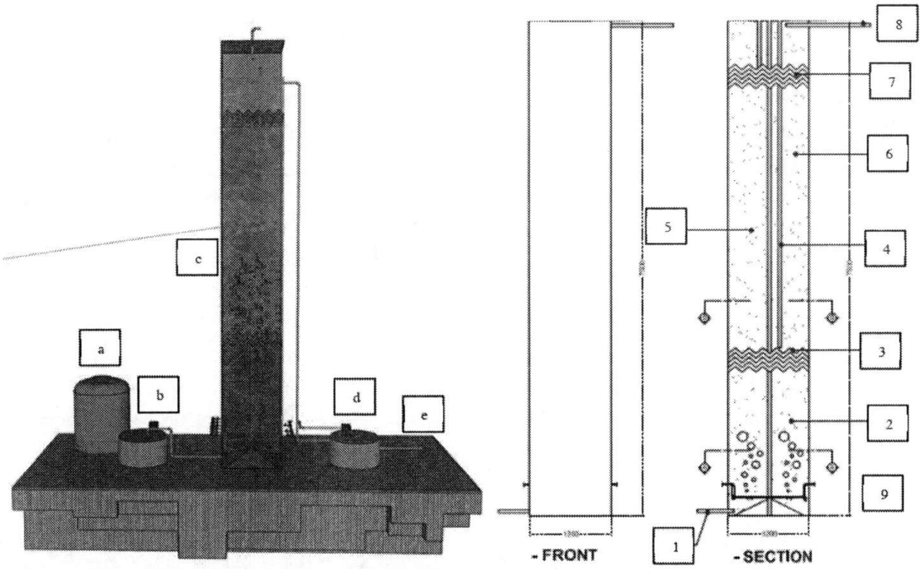

Figure 2. 3D modelling of EGSB reactor units, (a) Wastewater storage tank; (b) Equalization tank; (c) EGSB reactor; (d) Circulation and settling tank; (e) Outlet, and schematic diagrams of reactors flow, (1) Inlet pipe; (2) Settled Granular sludge; (3) Baffle separator/gas capture; (4) gas pipe; (5) Splattered granular sludge; (6) Processed wastewater; (7) Separator/ settled baffle; (8) Reactor outlet; (9) Circulation pipe.

2.1.3. Nutrient. To maintain sludge performance and to prevent shock loading during startup process, supplementary nutrition was required. 5 litres of nutrient used in this research was modified methanogenium media (DSMZ 141) consisting of (g/l) : KCl (Merck) 0,34; $MgCl_2.6H_2O$ (Merck) 4; $MgSO_4.7H_2O$ (Merck) 3,45; NH_4Cl (Merck) 0,25; $CaCl.2H_2O$ (Merck) 0,14; K_2HPO_4 (Merck) 0,14; NaCl (Merck) 18; Yeast extract (Oxoid) 2; $NaHCO_3$ (Merck) 5, and 2 litres of stock solution of trace elements consisting of (g/l) : $MgSO_4.7H_2O$ (Merck) 1,5; MnSO4.H2O (merck) 3; NaCl (Merck) 0,5; $FeSO_4.7H_2O$ (Merck) 1; $CoSO_4.7H_2O$ (Merck) 0,18; $CaCl_2.2H_2O$ (Merck) 0,1; $ZnSO_4.7H_2O$ (Merck) 0,18; $CuSO_4.5H_2O$ (Merck) 0,01; H_3BO_3 (Merck) 0,01; $NaMoO_4.2H_2O$ (Merck) 0,01; NiCl $_2.6H_2O$ (Merck) 0,03. Additional 3 litres stock buffer phosphate containing (g/l) $NaHCO_3$ 3,6 and KH_2PO_4 6,8 were also used to help maintain the system pH. Before used in the experiment, all of the nutrients above were stored at 4°C refrigerator to prevent decaying or contamination which could lead to the decrease in its quality.

2.2. Methods

2.2.1 Granular sludge seeding. The volume of granular sludge seeded into the reactor was about 28,5% of the total reactor volume, this was the recommended amount of anaerobic sludge inocula [7]. Inoculated sludge was then diluted with tap water until the reactor was filled. Inoculated sludge was then left in the reactor for about 24 hours to settle. On the day after, tap water was added again until the water flowing through the reactor outlet, this was done to wash out debris and other unwanted materials brought from the sludge. Before the startup experiment, nutrition was supplied using a dosing pump to maintain the sludge growth during the startup process.

2.2.2 Wastewater source and experimental configuration. The raw pharmaceutical wastewater was collected from the industry wastewater tank and then pumped into the experiment wastewater storage tank. The startup process was carried out by using diluted pharmaceuticals industry wastewater (50% wastewater: 50% tap water), on the start-up process wastewater maximum loading rate must not exceed 50% [7]. Dilution was carried out in an equalization tank and stirred using an overhead mixer. Experimental configuration was done by diluting wastewater and tap water in an equalization tank, then the influent was pumped into the reactor through an inlet pipe in the base of the reactor, thus make the granular sludge on the base pushed vertically along with wastewater inlet and circulated wastewater. Influent and circulation pump were configured to achieve a vertical up-flow speed (V_{up}) 1,75 m/hr with inlet: circulation ratio was 1:2 and HRT was kept at 4 hours throughout the startup duration. The reactor was operated continuously for 4 days. Before the experiment began, 5 liters of nutrient was diluted until the final volume was 20 liters, 2 liters of trace elements were diluted until the final volume was 10 liters and 3 liters of buffer phosphate were diluted until 6 liters, all of the materials were then stored in a separate container before injected to the reactor using dosing pump through the inlet channel. Before the experimental startup process, half of the supplementary nutrition was gradually supplemented into the reactor and then recirculated for 24 hours, this was done to maximize the contact between nutrition and anaerobic granules. At the initial startup process, diluted wastewater and supplementary nutrition were pumped into the reactor and the startup process began. Inlet and outlet samples were collected using a clean plastic bottle sample every 4 hours and then the samples were brought to the laboratory to analyze their pH, alkalinity, and COD content.

2.2.3 Analysis. Chemical parameters bicarbonate alkalinity and COD were measured according to APHA 2320 and APHA 5220 D respectively while pH was measured using Krisbow KW06-744 pH and temperature tester according to SNI 6869.11:2009. The data from the tested parameters were then collected and evaluated as per day result, this was due to the characteristic of inlet wastewater that would change every 24 hours depends on the industrial production process.

3. Results and discussion

3.1 pH changes during startup

The acidic condition value that varied from 5,1 to 5,9 (figure 3) in the inlet occurred due to accidental fermentation of sugar and starch content in the wastewater during storage in the industrial wastewater tank and also the mixture of wastewater from antibiotic production facilities which produced very acidic wastewater could lead to the inlet pH becomes acidic[12]. A high level of sugar and starch content in pharmaceuticals wastewater might occur due to the syrup and capsule or tablet production process. The lowest outlet pH value occurred on the first day of the experiment which was 6,9, this happened because the anaerobic granules were still in the adaptation process to the wastewater. On the other hand, on the first day, the circulation of the effluent did not occur yet meanwhile, the main aspect of the EGSB reactor was circulation/recirculation. Circulation was critical to prevent pH drop in the anaerobic system [13]. The effluent pH increased gradually day by day, from 7,3 on the second day to 8,2 on the third and fourth days. Thus gradual increment happened due to the circulation that started and the granules that started to adapt to the system condition. Maintaining a pH above 6,6 in the anaerobic system is very important since the success of the anaerobic system relies on complete degradation processes from fermentation-acidogenesis-acetogenesis and methanogenesis. However, the drop in pH in the system would inhibit methanogens bacteria to grow and utilize acetate to produce methane [14]. Alkalinity in the system would also affect the system pH. Buffer phosphate supplementation also helped the system to maintain pH on the reactor. The works done by Kisphergher [15] show that the fermentation process without bicarbonate supplementation could lead to a pH drop from 6,5 to 5,5.

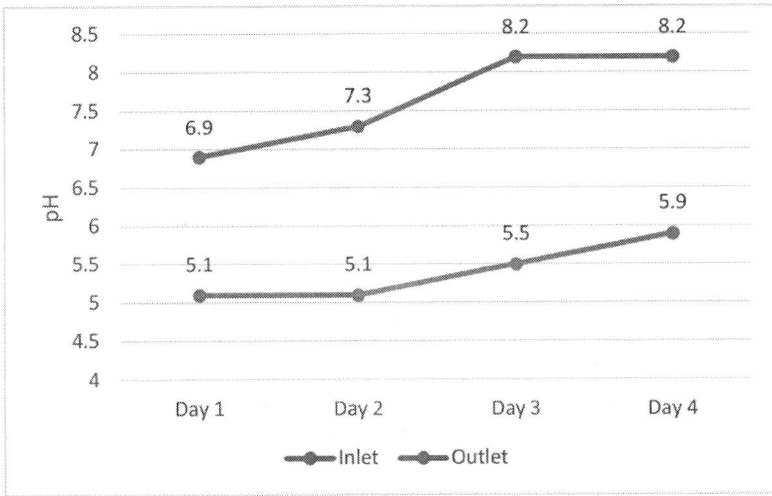

Figure 3. pH fluctuation on inlet and outlet during the experiment.

3.2 Alkalinity

Alkalinity is an important parameter to support methanogenesis in the anaerobic system. Inlet alkalinity varied from 219,05 mg/l to 277,63 mg/l (figure 4), the fluctuation of inlet characteristic happened due to the different quantity or different products produced by the industry. Overall, alkalinity in our EGSB reactor effluent increased from 470,3 on the first day to 519,1 and 499,83 mg/l on the third and fourth days respectively. Theoretically, the proposed amount of alkalinity to enhance bio methanation in the anaerobic process is about 2500 mg/l, it is needed to supply the system for the necessary co-substrate and to prevent pH drop below 6,5 [16]. The alkalinity of CaCO3 is also important to support the anaerobic system if there is excess production of volatile acids and other acids from the fermentation process and acidogenesis or acetogenesis. $CaCO_3$ alkalinity was not only provided from the addition of buffer phosphate but also the regeneration of VFA (volatile fatty acid) alkalinity during acetate to methane conversion [17].

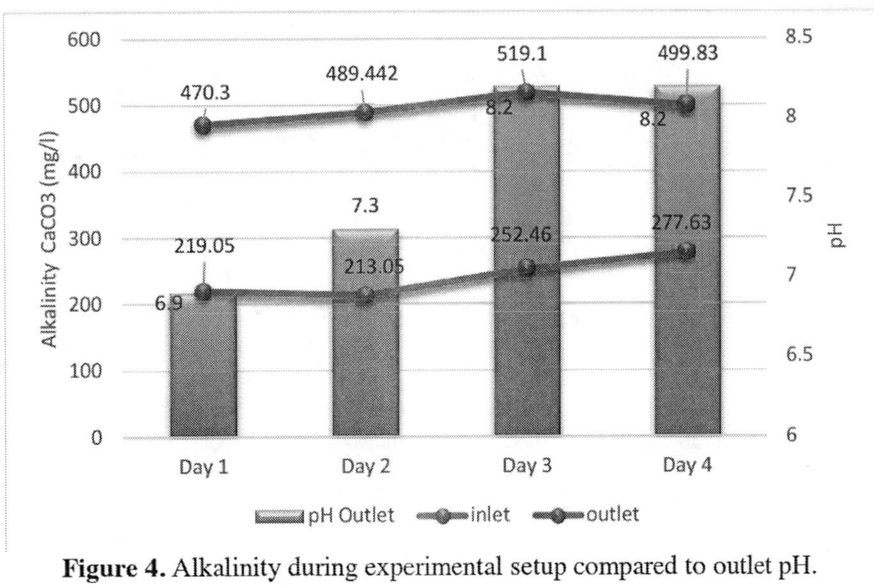

Figure 4. Alkalinity during experimental setup compared to outlet pH.

Figure 4 shows that during the startup process, pH and alkalinity started to elevate on the second day. This phenomenon proves that the supplementary nutrition and phosphate buffer had a significant role in preventing bacterial community to collapse when introduced to wastewater. On the third day of the experiment, the trend of alkalinity outlet was relatively stagnant even though inlet alkalinity was fluctuating, and buffer phosphate supplementation stopped. This showed that the anaerobic system configuration was capable to maintain system stability during the startup process.

3.3 COD reduction during startup

COD is an essential chemical parameter in the wastewater treatment plant as the success of wastewater treatment is determined by COD reduction. Figure 5 shows the diluted wastewater used in this research was relatively stable varied from 1198,68 mg/l at the lowest and 1284,85 at the highest. The lowest COD reduction percentage was on the first day of the startup process which was 33,75% due to system shock loading. Shock loading on the first day of the startup is already understood by many researchers, thus made the percentage of COD reduction was low. Nutrition and phosphate buffer supplementation on the first day of startup helped the system to deal with shock loading. The data (figure 5) shows that after surpassing the first day, anaerobic granules recovered thus lead to the increasing of COD reduction percentage to 38,34%, 38,5%, and 40,14% in the second, third, and fourth days respectively. Thus, it is proven that with the current COD loading rate, the EGSB system supplemented with anaerobic granular sludge could easily adapt to the wastewater characteristic and successfully surpass the startup process in a short time.

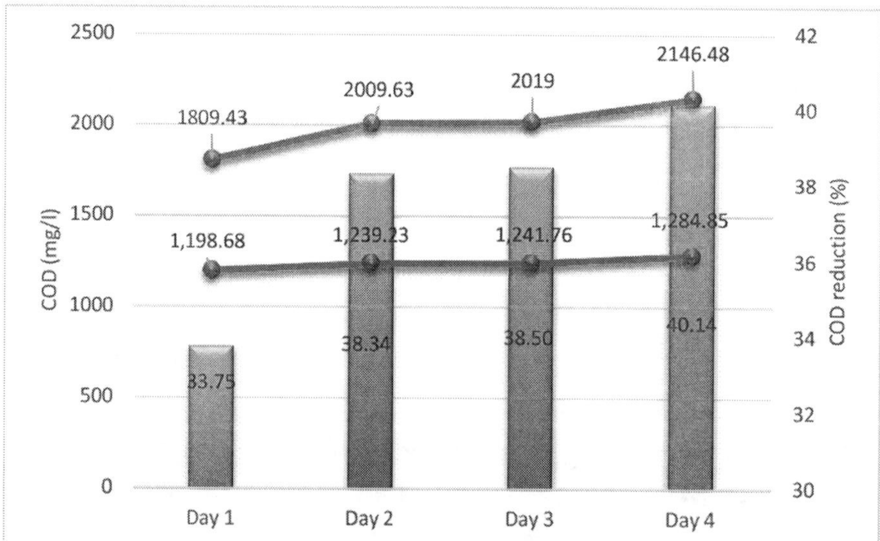

Figure 5. COD reduction during experimental startup.

COD reduction in the anaerobic system or start-up period was determined by the reactor setup such as HRT, COD inlet, OLR, and wastewater: diluter mixtures. Another start-up configuration done by Liu [10] shows a 70% of soluble COD reduction in 21 days startup using the EGSB reactor. However, the result was obtained by using longer retention time (12 hours) lower V_{up} (0,1m/h) and lower OLR, and longer startup time. This setup could be useful when the source of the seed was not a granulated sludge and the targeted wastewater was low-strength wastewater. Another study done by Zhang [18] found an accelerated EGSB reactor start-up that was done in 34 days using stored annamox (anaerobic ammonium oxidation bacteria) sludge for treating wastewater with high nitrogen content. The other configuration of the anaerobic startup process using the UASB reactor for treating the pharmaceutical wastewater has successfully gained 80-85% COD reduction during 120 days of startup time [19].

IOP Conf. Series: Materials Science and Engineering **980** (2020) 012078 doi:10.1088/1757-899X/980/1/012078

4. Conclusion

The result of this work concluded that the start-up process using the EGSB system using directly 50:50 wastewater to diluter (tap water) ratio and supplemented with granulated sludge has successfully accelerated the start-up time to less than a week. Supplementing proper nutrition, trace elements, and buffer phosphate at the initial startup process has been proved to prevent seed to collapse due to shock loading stress.

Acknowledgments

This study was financially supported by DIPA BBTPPI, Indonesia Ministry of Industry, and analytical work was supported by Instrumental Laboratory, water and wastewater laboratory BBTPPI.

References

[1]Mayahabate S P,Gupta S K, and Joshi S G 1988 Biological Treatment of Pharmaceutical wastewater in India differs because of various types of medicines like allopathic, ayurvedic and homeopathic. Several studies of different types of wastewater treatment which are medicine *Water. Air. Soil Pollut.* **38** 189–97

[2]Kanakaraju D, Glass B D, and Oelgemöller M 2014 Titanium dioxide photocatalysis for pharmaceutical wastewater treatment *Environ. Chem. Lett.* **12** 27–47

[3]Deegan M, Shaik B, Nolan K, Urell K, Oelgemoller M, Tobin J and Morrisey A 2011 Treatment options for wastewater effluents from pharmaceutical companies *Int. J. Environ. Sci. Technol.* **8** 649–66

[4]Balcioğlu I A and Ötker M 2003 Treatment of pharmaceutical wastewater containing antibiotics by O3 and O3/H2O2 processes *Chemosphere.* **50** 85–95

[5]Hulshoff Pol L and Lettinga G 1986 New technologies for anaerobic wastewater treatment *Water Sci. Technol.* **18** 41–53

[6]Bernard O, Hadj-Sadok Z, Dochain D, Genovesi A, and Steyer J P 2001 Dynamical model development and parameter identification for an anaerobic wastewater treatment process *Biotechnol. Bioeng.* **75** 424–38

[7]Lettinga G, Field J, Van Lier J, Zeeman G, and Hulshoff Pol L W 1997 Advanced anaerobic wastewater treatment in the near future *Water Sci. Technol.* **35** 5–12

[8]Boušková A, Dohányos M, Schmidt J E, and Angelidaki I 2005 Strategies for changing temperature from mesophilic to thermophilic conditions in anaerobic CSTR reactors treating sewage sludge *Water Res.* **39** 1481–88

[9]González J S, Rivera A, Borja R, and Sánchez E 1998 Influence of organic volumetric loading rate, nutrient balance and alkalinity: COD ratio on the anaerobic sludge granulation of an UASB reactor treating sugar cane molasses *Int. Biodeterior. Biodegrad.* **41** 127–31

[10] Liu Y, Hiuang L, Dong G, Liu G, Wu X, Wang C, Liu X, Wang L 2017 Enhanced granulation and methane recovery at low load by downflow sludge circulation in anaerobic treatment of domestic wastewater *Bioresour. Technol.* **249** 851–7

[11] Tian Z, Zhang Y, Li Y, Chi Y, and Yang M 2014 Rapid establishment of thermophilic anaerobic microbial community during the one-step startup of thermophilic anaerobic digestion from a mesophilic digester *Water Res.* **69** 9–19

[12] Saleem M 2007 Pharmaceutical wastewater treatment : A physicochemical study *J. Res.* **18** 125–34

[13] Vistanty H and Malik R A 2019 Enhanced Performance of Multi-Stage Anaerobic Digestion of Tofu Wastewater: Role of Recirculation *J. Ris. Teknol. Pencegah. Pencemaran Ind.* **10** 29–37

[14] Van Lier J B, Mahmoud N, and Zeeman Z 2008 *Anaerobic Wastewater Treatment.*, Biological wastewater treatment;princiles modelling an design, ed Henze M, van Loosdrecht M C M, Ekama B D, Brdjanovic D IWA Publishing (UK) 401-42

[15] Kispergher E M, D'Aquino C A, da Costa Junior L C, de Mello T C, Weinschutz R, and Mathias A L 2017 Effect of organic load and alkalinity on dairy wastewater biomethanation *Eng.*

Agric. **37** 820–7

[16] Tauseef S M, Abbasi T, and Abbasi S A 2013 Energy recovery from wastewaters with high-rate anaerobic digesters *Renew. Sustain. Energy Rev.* **19** 704–41

[17] Işik M and Sponza D T 2005 Effects of alkalinity and co-substrate on the performance of an upflow anaerobic sludge blanket (UASB) reactor through decolorization of Congo Red azo dye *Bioresour. Technol.* **96** 633–43

[18] Zhang W, Zhang Y, Li L, Zhang X, and Jin Y 2014 Fast start-up of expanded granular sludge bed (EGSB) reactor using stored Anammox sludge*Water Sci. Technol.* **69** 1469–74

[19] Rodríguez-Martinez J, Garza-Garcia Y, Aguilera-Carbo A, Martínez-Amador S Y, and Sosa-Santillan G J 2005 Influence of nitrate and sulfate on the anaerobic treatment of pharmaceutical wasterwater *Eng. Life Sci.* **5** 568–73

IOP Conf. Series: Materials Science and Engineering **980** (2020) 012079 doi:10.1088/1757-899X/980/1/012079

Comparison of ferrate (FeO_4^{2-}) and ultrasonic waves ability as *Coliform* antibacterial in Kahayan River water Central Kalimantan

N Kurniawati[1]*, S Sunariyati[2], K Karelius[3], L Hakim[1], W Krestina[2], and D A P Wardani[3]

[1] Departement of Physics, Faculty of Mathematic and Natural Science, Palangka Raya University, Palangka Raya, Indonesia
[2] Departement of Biology, Faculty of Mathematic and Natural Science, Palangka Raya University, Palangka Raya, Indonesia
[3] Departement of Chemestry, Faculty of Mathematic and Natural Science, Palangka Raya University, Palangka Raya, Indonesia

Email: nenykurniawati@mipa.upr.ac.id

Abstract. This study aims to compare ferrate (FeO_4^{2-}) and exposure of ultrasonic waves as an antibacterial coliform in the Kahayan River water in Central Kalimantan. Sampling in this study used the Grab Sampling method with the Kemmerer Sampler tool. Ferrate ($FeO_4$2-) was made by reacting to a solution of $Fe(NO)_3$ with NaOCl in alkaline conditions. Application of Ferrate (Fe(VI)) done by varying the concentration (0.03125; 0.0625; 0.125; 0.25; 0.5 mmol/L) to get the best concentration, and follow by varying times (0, 5, 10, 15, 20, 25, 30 minutes) at optimum concentration. Ultrasonic wave exposure was carried out with varying frequencies to get the optimum frequency and followed by variations in the exposure time at the optimum frequency. The coliform test was carried out by the MPN method with stages of detection tests, affirmation tests, and colony counting. The MPN analysis showed that ferrate could kill coliform bacteria at a concentration of 0.0625 mmol/L during the 5 minutes stirring time with 100% efficiency. Ultrasonic wave exposure could kill coliform bacteria optimally at a 40 kHz frequency during 3 hours of exposure time with 96% efficiency These results showed that ferrate was more effective in killing coliform bacteria compared with ultrasonic wave exposure.

1. Introduction

Water is essential for human life. The source of water used is not always clean. For drinking water purposes, water must be treated to reduce odor, taste, clarity, or eliminate diseases caused by pathogens [1]. More than 1 billion people cannot get clean water sources. Many people suffer from illness and even death due to water contamination [2]. Un-disinfected water triggers the development of *Escherichia coli* and causes *gastroenteritis* in Walkerton, Ontario (2000), *cryptosporidiosis* in Milwaukee, Wisconsin (1993), and *cholera* in Peru (1991) [3,4].

The Indonesian Central Bureau of Statistics reports that 75% of Indonesia's rivers are polluted by pollutants, just as the Kahayan river water in 2008 was heavily polluted [5]. *Coliform* bacteria contaminate the Kahayan River. The biological indicator indicates that water is trash, or fecal contamination is *coliform* bacteria in the water [6].

One way to deal with coliform bacteria contamination is by utilizing exposure to ultrasonic waves. Ultrasonic waves reproduce sound waves with frequencies above 20 kHz. Ultrasonic waves propagate in a diffuse pattern, while the angle of spread of the waves is inversely proportional to the working frequency. Ultrasonic waves can be generated by using a piezoelectric transducer or a ferroelectric transducer. This transducer will produce high-frequency sound waves [7].

In 1960, the discovery that the ship's sonar emitted sound waves could cause fish to die. Since then, various researches have been developed on the ultrasonic potential to kill bacteria [8]. Acoustic cavitation generated by ultrasonic waves can inactivate bacteria through several physicals, mechanical and chemical effects. When collapsed, cavitation bubbles generate enough energy to weaken or irritate through the following processes:

- The induction of cavitation produces a force as a result of resonance on the surface of bacterial cells. This force creates pressure and pressure gradient due to the collapse of gas bubbles entering or near the bacterial cell wall. Within a certain time, this mechanical fatigue causes damage to the bacterial cells. It also depends on the frequency.
- A shear force formed, which is induced by microstreaming in the bacterial cell.
- Chemical groaning occurs due to the formation of radicals (H· and OH·) during cavitation in water media, on bacterial cell walls, and weakens the cell walls to disintegration.
- The end product of this water's sonochemical degradation is hydrogen peroxide (H_2O_2), which is a potent bactericide [9].

A study on ultrasonic waves to kill bacteria was carried out by Zhou et al. in 2009 to kill *E.coli* bacteria on spinach leaves [10]. Besides, there is the use of ultrasonic waves to reduce the amount of *Salmonella typhimurium* in milk [11], study the effect of ultrasonic waves on beef broth [12], and control microbes in water [13–15].

Research on the sterilization of Kahayan River water was previously carried out by Suastika et al. in 2015 [7]. Based on this research, it was reported that the optimum frequency of deactivating *E. coli* in water samples was 27.6 kHz with an efficiency of 37.8%.

The accepted treatment method to protect public health apart from ultrasonic treatment is filtration and disinfection [2]. The filtration method is considered a weakness in removing viruses or spores that are small in size. Filtration may not be able to 99% reduce contamination. A disinfectant is a chemical treatment that is adopted to remove harmful and safe microorganisms. Some of the disinfectants that have been used include hydrogen peroxide, UV light, permanganate, or a combination of chloramine [16,17]. However, there are several obstacles to applying this disinfectant because it can cause side effects [18–21].

Ferrate [FeO_4^{2-}] is an environmentally friendly treatment that can be used in drinking water treatment. Several studies have used ferrates in the field of water treatment technology and the wastewater industry. The nature of ferrate can kill various viruses and bacteria in water and wastewater treatment processes. The potassium ferrate was reported to reduce 50% color, 30% COD, and kill 10% bacteria in the water than aluminum sulfate and ferric sulfate [22]. There is still limited research on the experimental of ferrate(VI) as a disinfectant focused only on drinking water is available in our acknowledgment.

Iron, which has an oxidation number of +2 and +3, can be strongly oxidized in the environment, iron with a higher oxidation number such as + 4, + 5, and +6, too [23]. Ferrate, which is iron with an oxidation state of +6, is of concern because it can be synthesized in an environmentally friendly manner [24,25]. Several papers published the Fe(VI) reaction mechanism (especially the FeO_4^{2-} oxidation). For example, ferrate can oxidize lactam antibiotics, and ferrate is very effective in reducing lactam antibacterial activity. Fe(VI) as an inorganic compound, is more stable in the pH range 9-10 but reacts slowly in water because it forms ferrate and oxygen ions[26–28], as in equation 1.

$$2FeO_4 + 5H_2O \rightarrow 2Fe + 3/2O_2 + 10OH^- \quad (1)$$

Fe (VI) is unstable at pH below nine, and ferrate ion will quickly auto-decompose within seconds to minutes. This reaction follows second order kinetics. The auto-decomposition rate of Fe (VI) increases with increasing pH, experiencing protonation from the more reactive FeO_4^{2-} [29]. The potential advantage of using Fe (VI) in the water treatment process is that it does not produce harmful toxicological side effects [30].

Disinfectants designed to kill harmful organisms (e.g., bacteria and protozoa) and oxidizers degrade various contaminants. Fe (VI) is useful in deactivating (*escherichia coli*, *sphaerotilus*, *bacillus subtilis*, *bacillus licheniformis* RB1-1B, *mycobacterium frederiksbergense* M8-6, *mycobacterium setense* M9-4) [31–33], viruses (f2 coliphage, Q -coliphage and Bacteriophage MS2) [32,34,35], algae (Microcystin-LR) [36] and other pathogens [37] in synthetic buffer solutions and secondary effluents. Besides, even Fe(VI) in low doses can inactivate 99.9% of various microorganisms, both on a laboratory scale or in early experiments [38]. Many researchers have used Fe(VI) to remove fecal contamination from various world regions with *coliform* mortality rates above 99.9% [39,40].

Fe (VI) is a powerful disinfectant and oxidant useful for treating drinking water in the U.S. for more than a decade[41]. In many ways, it can be considerably affordable, a sample alternative to ozonation, and thus it is well suited for use in small systems. Like ozone and other disinfectants such as chlorine dioxide and chloramines, Fe (VI) is a highly selective oxidant, but it does not lead to the formation of any know hazardous by-products (brominated or chlorinated DBPs), bromate or chlorite). Ferrate is considered a ''green'' chemical for water treatment, because it does not produce any known toxic by-products [40]. In this paper, the synthesis of ferrate (Fe(VI)) was analyzed using the wet synthesis method as a disinfectant in the Kahayan river water treatment.

2. Materials and methods

This study's method is divided into three stages: water sampling and *coliform* bacteria examination, sonication using ultrasonic, and oxidation using ferrate (Fe (VI)).

2.1. Water sampling and coliform bacteria examination

Kahayan River water sampling was carried out by varying several sampling points by taking into account the river depth, river width, and water discharge. Kahayan River has a depth of 7 m, a width of 500 m, the most massive water discharge is 2,716 m^3/s, and the smallest water discharge is 26.30 m^3/s. Kahayan River water sampling was carried out in January-April when the water discharge reached its most massive water discharge. Based on these data, the Kahayan River water sampling carried out at 6 points, namely 3 points from the width of the Kahayan River at the 125 m point; 250 m; 375 m and 2 points from the depth of the Kahayan River at a point of 1.4 m; 5.6 m [42,43].

Coliform bacteria examination uses a quantitative examination, with the following stages:

a. Estimation Test. The estimation test is a specific test for the detection of Coliform bacteria. This is a preliminary test of *coliform* bacteria's presence or absence based on the formation of acids and gases caused by lactose fermentation by Coli bacteria. The method used was the three tubes series *Most Probable Number* (MPN) method. The principle of the MPN method is to calculate the closest estimate through the estimator test [44].
b. Confirmative Test. The positive results in the prediction test for raw water were followed by a reinforcement test using the MPN series three methods. The media used was *Brilliant Green Lactose Bile* (BGLB). This medium is a selective medium for coliform, so it can inhibit other bacteria's growth and increase growth for coliform. The observation of river water samples in the MPN tube tested with three series of tubes using BGLB media where the air bubbles formed in the durham tubes after 24 hours in the form of carbon dioxide gas produced due to lactose fermentation in BGLB media by coliform bacteria.
c. Complimentary Test. This test is a final analysis of a water sample to detect the presence of coliform. The medium used was *Eosin Methylene Blue* (EMB) [45].

d. The calculation of the efficiency of reducing the number of bacterial colonies was then carried out to see the percentage of reduction in bacterial colonies using equation 2:

$$\% \text{ Efficiency} = \frac{MPN_{ktr} - MPN_{per}}{MPN_{ktr}} \times 100 \qquad (2)$$

where% efficiency is the percentage reduction in bacteria, MPN_{ktr} is the control MPN index, and MPN_{per} is the treatment's MPN index [43].

2.2. Oxidation using ferrate (Fe (VI)).

2.2.1 Synthesis of ferrates (Fe (VI)) and the Kahayan River water treatment. Ferrate was synthesized in the following way:

1. Manufacture of 5,25% NaOCl solution
2. NaOCl solution with a 5,25% concentration was made by diluting 43,75 mL of 12% NaOCl from the bleach solution to 100 mL.
3. Making $FeCl_3$ 0,3 M solution
4. The solution of $FeCl_3$ with a concentration of 0,3 M was prepared by dissolving 4,058 grams of FeCl3· 6H2O with aquabides to 50 ml.
5. Synthesis of ferrates from clothing bleach solutions
6. Ferrate synthesis was carried out by reacting 40 mL of clothes whitening solution containing 5,25% NaOCl with 12 grams of NaOH. The solution is stirred until NaOH is completely dissolved, then added 1 mL $FeCl_3$. The solution is stirred again until the solution turns purple, then closed and allowed to stand for a day. Furthermore, it was analyzed by UV-Vis spectrophotometer to determine the maximum wavelength of ferrate.

2.2.2 The Kahayan River water treatment using ferrate as disinfectant. Solution containing ferrate ions. This solution was used as a disinfectant in the Kahayan river water by adding a volume of ferrate solution to the Kahayan river water sample. First, the optimum dosage of effective ferrate ions was determined as a disinfectant for *coliform* bacteria that lived in the Kahayan river water under natural environmental conditions, by adding a solution of ferrate ions to 250 mL of Kahayan river water with a concentration of 0,5; 1,5; 2,5; and 3,5 mmol/L [46], with stirring for 30 minutes [1]. Then, each sample was planted into MPN media for 24 hours to identify the remaining *coliform* bacteria after treatment.

Then, the optimization of the contact time of ferrate ions with *coliform* in the Kahayan river water was carried out. The concentration used in this optimization referred to [1] which is 0,03125 mmol/L with stirring variations at time 0; 5; 10; 15; 20; 25 and 30 minutes. Furthermore, each sample was planted into MPN media for 24 hours to identify *coliform* bacteria left after treatment.

2.3. Sonication using ultrasonic

At this stage, the Kahayan River water samples treated with ultrasonic waves. The stages of sonication are as follows:

1. The water sample was placed in a beaker, and at the top of the glass, an ultrasonic speaker was placed. To prevent other waves' attenuation during the ultrasonic exposure process, the sample glass and speaker were placed in a glass container filled with sterile water.
2. Generating ultrasonic waves using a piezo tweeter speaker with a frequency of 20 kHz. The activation of the speaker operation used an ultrasonic wave generator (signal generator).
3. Place the piezo speaker on the top of the container containing the Kahayan River water sample for 2 hours.
4. The process repeated for 20 kHz, 30 kHz, 40 kHz, 50 kHz, and 60 kHz [42].
5. Based on the optimum frequency data above, processes 1 and 2 repeated with time variations of 1, 2, 3, 4, and 5 hours [43].

3. Results and discussion

3.1. Oxidation using ferrate (Fe (VI)).

3.1.1. Synthesis of ferrate. Ferrate synthesis was carried out by dissolving NaOH solids with a clothes whitening solution containing 5,25% NaOCl. Furthermore, the solution Fe(III) was added later with the iron species intention to be oxidized by NaOCl in an alkaline atmosphere. In the oxidation reaction of Fe(III) to Fe(VI), $FeCl_3$ was added drop by drop. This reaction produced a dark purple solution which showed that Fe (III) has been oxidized to Fe (VI). The following reaction shows ferrate formation reaction in equation 3.

$$2Fe_3+ + 3ClO- +10OH- \rightarrow 2FeO_4^{(2-)} + 3Cl- + 5H_2O \quad (3)$$

This solution is allowed to stand for a day with the aim that the fermentation produced is stable. The stability of ferrate was shown by the color of the solution formed. This solution was then analyzed by UV-Vis spectroscopy to determine the maximum wavelength of ferrate in the visible wave. The measurement results obtained that the maximum wavelength was 510 nm, as reported by Karelius and Asi (2017) showed in figure 1 [47].

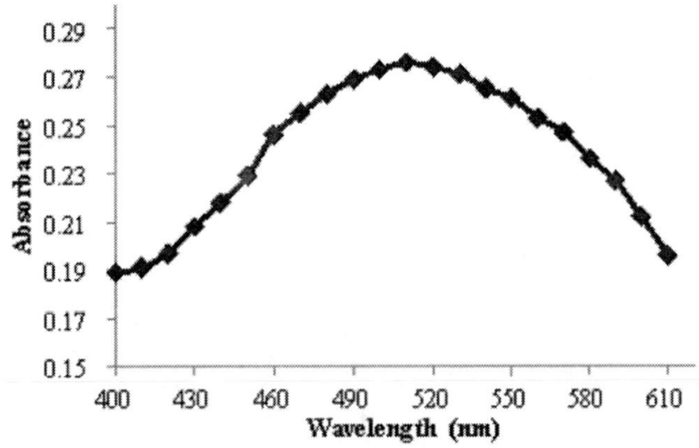

Figure 1. Maximum wavelength of ferrate.

3.1.2. Quantification of Fe(VI). Electrochemical, volumetric, and various spectrophotometric methods are some of the methods used for the quantitative determination of Fe (VI), *[48,49]*. Of these various methods, the UV-Vis spectral method offers simple steps, fast experimental work, and accurate results *[49]*, to determine the concentration and purity of Fe (VI) using this method. The determination of the concentration and purity of Fe (VI) is as follows,

- Determination of the concentration of Fe (VI) in water by absorbance of UV-Vis based on Beer's law;
- Fe (VI) ion in water appeared dark purple with a maximum absorbance peak at 510 nm;
- The molar absorbance coefficient (ε) was determined to be 1070 $M^{-1}cm^{-1}$;
- Calculating the concentration of Fe (VI) using the following equation 4 [48],

$$\varepsilon = \frac{A}{BC} \quad (4)$$

where the notation "*ε*" equals 1070 $M^{-1}cm^{-1}$ (molar absorbance coefficient), "*A*" is UV-Vis absorbance by sample, "*B*" represents Fe (VI) concentration (M) in the examined sample, and "*C*" was the light path (cm) of the quartz cell used in this study. The concentration of ferrate ions in the solution was 365,85 mmol/L.

3.1.3 Ferrate ions as disinfectants. The initial condition of the Kahayan river water in pH of 5.39 has a total suspended solid (TSS) of 849 mg/L, turbidity of 420 mg/L, and iron, sodium and chlorine total content respectively of 0.229; 1.67; 3.30 mg/L. The presence of coliform in the Kahayan river was carried out using the MPN method and BGLB media. This is the community that uses the Kahayan River for various activities including being used as a latrine so that fecal *coliform* bacteria can ensure pollution.

This study used ferrate ions as a disinfectant for *coliform* bacteria that lives in the Kahayan river water, a source of raw water for PDAM of Palangka Raya. The first step taken was to determine the optimum dose of effective ferrate ions as a disinfectant for *coliform* bacteria that live in the Kahayan river water under natural environmental conditions, by adding ferrates to 250 mL of Kahayan river water at a concentration of 0.5; 1.5; 2.5 and 3.5 mmol/L[46], stirring for 30 minutes [1]. The results of the treatment can be seen in the following figure 2.

Figure 3 was based on the data generated from observations of effective ferrate as a disinfectant. Observation of river water samples in MPN tubes treated with ferrates' addition at a concentration of 0,5; 1,5; 2,5 and 3,5 mmol/L with stirring for 30 minutes did not show a positive test for the presence of *coliform* bacteria in 4 treatments with the addition of ferrate. No carbon dioxide gas was formed in the Durham tube with BGLB media after 24 hours (figure 3).

The use of Fe (VI) in several water sources in several cities in the world is known to kill almost 99,9% of the total *coliforms* in water sources. The effect of significant Fe(VI) as a disinfectant on effective microorganisms against environmentally safe *coliforms* is not mutagenic or carcinogenic [50].

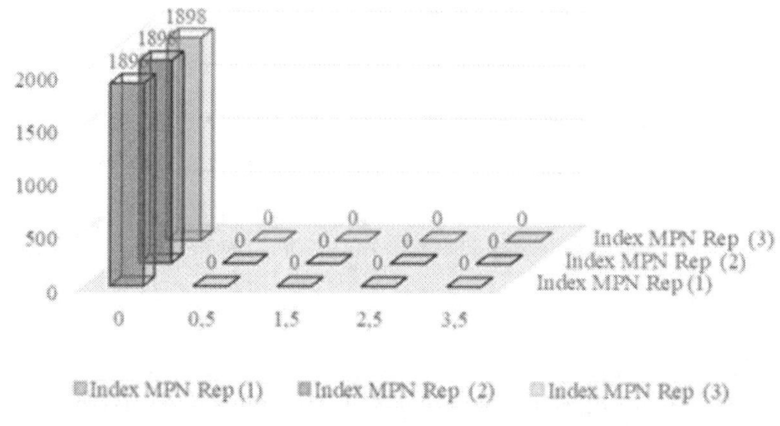

Figure 2. Calculation of ferrate treatment as disinfectant.

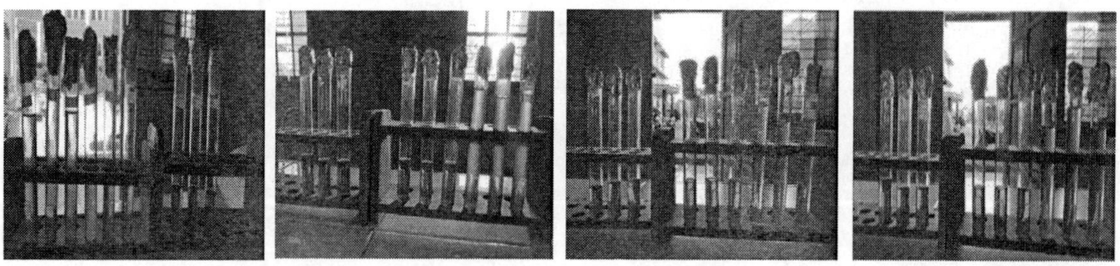

(a) (b) (c) (d)

Figure 3. The MPN tube of river water sample using ferrate after 24 hours. Concentration of (a) 0,5; (b) 1,5; (c) 2,5; (d) 3,5 mmol/L.

Figure 2, the calculation of *coliform* bacteria in several variations of ferrate concentration after 24 hours calculation of the number of *coliform* bacteria showed that ferrate was very effective as a disinfectant for *coliform* bacteria in the Kahayan river water. Ferrate ions with the lowest concentration of 0,5 mmol/L could function as a disinfectant for *coliform* bacteria in the Kahayan river water with 100% inactivation efficiency. Based on these results it is deemed necessary to re-optimize the concentration of ferrate ions added to the Kahayan river water as a disinfectant. The concentration used refers to Nguema [1] which is 0.03125 mmol/L with variations in the stirring time varying 0; 5; 10; 15; 20; 25 and 30 minutes.

The results of the treatment can be seen in figure 4. Stirring time at 0.03125 mmol/L ferrate ion concentration was found to affect the efficiency of inactivation. Figure 5 shows that at the time of stirring 0 minutes (without stirring) it turns out that ferrate ions at a concentration of 0.03125 mmol/L still gave positive test results on *coliform* bacteria with an efficiency of 88.97%. While from the stirring time 5 minutes onwards, the test did not give positive results for *coliform* bacteria. These results indicated that at the stirring time of only 5 minutes ferrate ions with a concentration of 0.03125 mmol/L were able to inactivate *coliform* bacteria in the Kahayan river water with 100% efficiency.

The final goal of this research is to find out the concentration of ferrate which can inactivate *coliform* bacteria in the Kahayan river water optimally under natural environmental conditions, which can later be applied to water treatment processes in PDAM Palangka Raya city by minimizing mechanical processes such as stirring. One possible way to achieve this is to find the concentration of ferrate that is able to inactivate *coliform* bacteria in the Kahayan river water optimally even without stirring.

Following figure 5 are the results of the ferrate treatment at a concentration of 0.03125; 0.0625; 0.125; 0.25; 0.5 mmol/L without stirring. The results of the addition of ferrate ions at various concentrations in the Kahayan river water containing *coliform* bacteria show that at the concentration of ferrate ions 0,03125; 0,0625; 0,125 mmol/L, still gave positive test results on *coliform* bacteria, with each inactivation efficiency of 76,92%, 85,725 and 99,42%. Ferrate ions with these concentrations had not inactivated *coliform* bacteria in the Kahayan river water to 100%. In contrast, the ferrate ion concentration of 0,25 and 0,5 mmol/L without stirring did not give positive test results on *coliform* bacteria. These results indicated that at a concentration of 0,25 mmol/L the ferrate ion could inactivate *coliform* bacteria in the Kahayan river water without stirring with 100% efficiency.

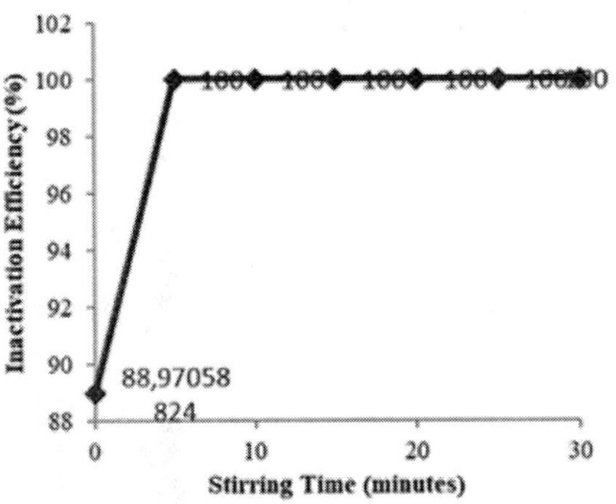

Figure 4. The inactivation efficiency of *coliform* bacteria in the variation of stirring.

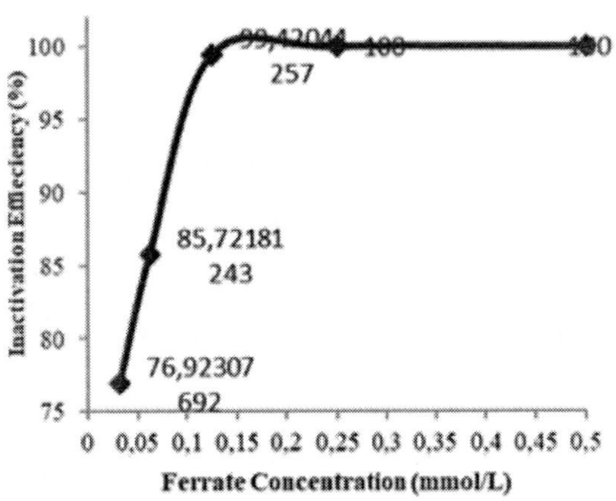

Figure 5. The inactivation efficiency of coliform bacteria in the variation of ferrate concentration.

3.2. Sonication using ultrasonic

The first stage in this research was to determine the optimum frequency of exposure to ultrasonic waves. Water samples of 250 mL were exposed to ultrasonic waves for 2 hours [12], with variations of 20 kHz, 30 kHz, 40 kHz, 50 kHz, and 60 kHz. The water sample was placed in a beaker, and at the top of the glass, an ultrasonic speaker was placed. To prevent other waves' attenuation during the ultrasonic exposure process, the sample glass and speaker were placed in a glass container filled with sterile water. Then the bacteria count was checked after 2 hours of exposure.

The raw data obtained from the examination was the MPN index, from this data, it was used to calculate the inactivation efficiency of coliform bacteria, the higher the efficiency value indicated the high level of coliform bacteria inactivation. Figure 6 shows the results of the examination of coliform bacteria in water samples with variations in the frequency of exposure to ultrasonic waves. The highest efficiency to reduce the number of bacteria was at a frequency of 40 kHz which reached a value of 85.72%, which means that after 2 hours of treatment there were 85.72% of bacteria lost from water samples, this shows that ultrasonic waves could be used as anti-bacterial in river water Kahayan with sufficient frequency 40 kHz.

Based on the results of efficiency of coliform colony reduction with frequency, water samples were exposed to ultrasonic waves with a frequency of 40 kHz, with variations in an exposure time of 1 hour, 2 hours, 3 hours, 4 hours, and 5 hours. After that, the water sample removed, and the number of coliform colonies examined.

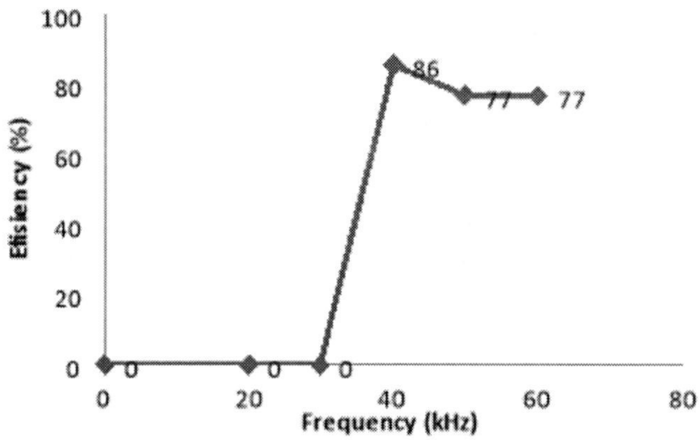

Figure 6. Graph of the efficiency of coliform colony reduction with frequency.

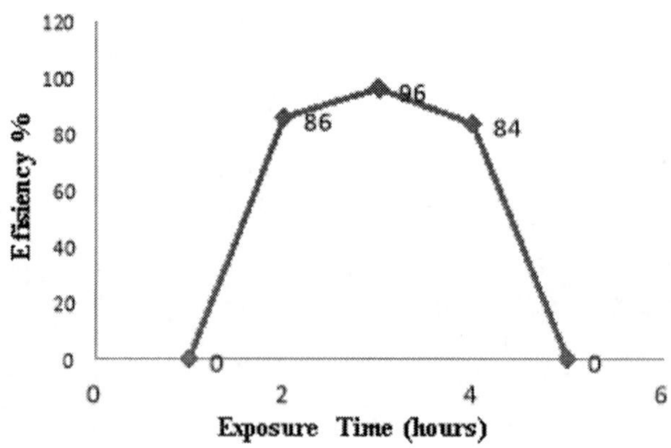

Figure 7. Effect of exposure time on the efficiency of the number of bacterial colonies.

The results of the coliform examination after incubation for 24 hours presented in figure 7. It shows that by comparing the number of bacterial colonies in control, the reduction in the number of *coliforms* was at least when exposed to 3 hours, while at 5 hours, there was no change in the number of bacterial colonies. From the graph we can see, the highest efficiency for reducing the number of bacteria at 96% exposure time. This data shows that after 3 hours of treatment, 96% of bacteria had lost from water samples, this shows that ultrasonic waves with a 40 kHz frequency could be used as anti-bacteria in Kahayan River water with an optimal exposure time of 3 hours. However, these results show that ferrate was more effective in killing coliform bacteria than ultrasonic wave exposure.

4. Conclusion

Fe (VI) is an effective disinfectant for the treatment of The Kahayan river water. Laboratory research the application of Ferrate (Fe (VI)) in the MPN test results showed that at the concentration of Ferrat (Fe (VI)) 0,03125 mmol/L positive results were only seen at 0 minutes contact time. The inactivation efficiency of Ferrate (Fe (VI)) in this treatment was 88.97%. The variation of concentration treatment showed at the 0 minutes contact time the MPN test were positive at concentration of 0.03125; 0.0625; and 0.125 mmol/L and negative at concentration of 0.25; 0.5 mmol/L. This study's result shows that the concentration of 0.25 mmol/L Ferrate (Fe (VI)) could function as a disinfectant for *coliform* bacteria in Kahayan river with 100% inactivation efficiency.

After the ultrasonic wave exposure treatment with a 40 kHz frequency by varying the exposure time, the sufficient time to reduce the number of coliform colonies was 3 hours, with 96% efficiency. Based on this study's result, it is concluded that ultrasonic waves can also be used as coliform anti-bacteria, but these results show that ferrate is more effective in killing coliform bacteria compared with ultrasonic wave exposure.

Acknowledgments

The research team would like to thank the Ministry of Research and Technology/the National Research and Innovation Agency of the Republic of Indonesia for funding this research through the 2020 Higher Education Operational Assistance Grand.

References

[1] Nguema P F and Jun M 2016 Application of Ferrate (VI) as Disinfectant in Drinking Water Treatment Processes: A Review 11

[2] Rose J B 2002 Viewpoint: Water Quality Security *Environ. Sci. Technol.* **36** 246A-50A

[3] Richardson S D, Simmons J E and Rice G 2002 Peer reviewed: disinfection by products: the next generation

[4] Hunter P R 2003 Drinking water and diarrhoeal disease due to Escherichia coli *J. Water Health* **1** 65–72

[5] Badan Pusat Statistik 2017 Status Kualitas Air Sungai, 2007 - 2016

[6] Hujjatusnaini N 2009 Kajian Tentang Kualitas Mikrobiologi Berdasarkan Nilai MPN Coliform, Coliform fekal, dan Jumlah Koloni Bakteri Escherihia coli, Kualitas Fisik dan Kimia Air Minum Isi Ulang Di Kota Palangkaraya Sebagai Bahan Penunjang Praktikum Mikrobiologi. (Tesis) *Disertasi dan Tesis Program Pascasarj. UM* **0**

[7] Suastika K G, Martani N S and Hartanto T J 2015 Rancang Bangun Sistem Pembangkit Gelombang Ultrasonik Sebagai Metode Alternatif Menurunkan Jumlah Bakteri E. Coli Pada Proses Penjernihan Air *Pros. Simp. Fis. Nas. XXVII Univ. Udayana Bali* **1** 546–52

[8] Gao S, Lewis G D, Ashokkumar M and Hemar Y 2014 Inactivation of microorganisms by low-frequency high-power ultrasound: 1. Effect of growth phase and capsule properties of the bacteria *Ultrason. Sonochem.* **21** 446–53

[9] Joyce E, Phull S S, Lorimer J P and Mason T J 2003 The development and evaluation of ultrasound for the treatment of bacterial suspensions. A study of frequency, power and sonication time on cultured Bacillus species *Ultrason. Sonochem.* **10** 315–8
[10] Zhou B, Feng H and Luo Y 2009 Ultrasound Enhanced Sanitizer Efficacy in Reduction of Escherichia coli O157 : H7 Population on Spinach Leaves *J. Food Sci.* **74** M308–13
[11] Juraga E, Šalamon B S, Herceg Z and Jambrak A R 2011 Application of High Intensity Ultrasound Treatment on Enterobacteriae Count in Milk *Mljekarstvo J. Dairy Prod. Process. Improv.* **61** 11
[12] Puspasari N, Surtono A and Warsito W 2014 Efek Frekuensi Gelombang Ultrasonik Terhadap Mikroba Pada Air Kaldu Daging Sapi *J. Teori Dan Apl. Fis.* **2** 171–7
[13] Broekman S, Pohlmann O, Beardwood E S and de Meulenaer E C 2010 Ultrasonic treatment for microbiological control of water systems *Ultrason. Sonochem.* **17** 1041–8
[14] Al-Juboori R A, Aravinthan V and Yusaf T 2015 Impact of pulsed ultrasound on bacteria reduction of natural waters *Ultrason. Sonochem.* **27** 137–47
[15] Gómez-López V M, Gogate P R and Gogate P R 2017 Reconditioning of Vegetable Wash Water by Physical Methods *Trends Food Saf. Prot.*
[16] Jiang J-Q and Lloyd B 2002 Progress in the development and use of ferrate (VI) salt as an oxidant and coagulant for water and wastewater treatment *Water Res.* **36** 1397–408
[17] Jiang J Q 2007 Research progress in the use of ferrate (VI) for the environmental remediation *J. Hazard. Mater.* **146** 617–23
[18] Rook J J 1974 Formation of haroforms during chlorination of natural waters *Water Treat Exam* **23** 234–43
[19] Haag W R and Hoigne J 1983 Ozonation of bromide-containing waters: kinetics of formation of hypobromous acid and bromate *Environ. Sci. Technol.* **17** 261–7
[20] Gombos E, Barkács K, Felföldi T, Vértes C, Makó M, Palkó G and Záray G 2013 Removal of organic matters in wastewater treatment by ferrate (VI)-technology *Microchem. J.* **107** 115–20
[21] Kumar M A, Rohitha C N, Kulkarni S J and Narender N 2010 Bromination of aromatic compounds using ammonium bromide and Oxone® *Synthesis* **2010** 1629–32
[22] Li C, Li X Z and Graham N 2005 A study of the preparation and reactivity of potassium ferrate *Chemosphere* **61** 537–43
[23] Jeannot C, Malaman B, Gerardin R and Oulladiaf B 2002 Synthesis, crystal and magnetic structures of the sodium ferrate (IV) Na4FeO4 studied by neutron diffraction and Mössbauer techniques *J. Solid State Chem.* **165** 266–77
[24] Licht S, Ghosh S, Naschitz V, Halperin N and Halperin L 2001 Fe(VI) Catalyzed Manganese Redox Chemistry: Permanganate and Super-Iron Alkaline Batteries *J. Phys. Chem. B* **105** 11933–6
[25] Sharma V K, Mishra S K and Nesnas N 2006 Oxidation of Sulfonamide Antimicrobials by Ferrate(VI) [FeVIO42-] *Environ. Sci. Technol.* **40** 7222–7
[26] Karlesa A, De Vera G A D, Dodd M C, Park J, Espino M P B and Lee Y 2014 Ferrate(VI) Oxidation of β-Lactam Antibiotics: Reaction Kinetics, Antibacterial Activity Changes, and Transformation Products *Environ. Sci. Technol.* **48** 10380–9
[27] Lee D G and Gai H 1993 Kinetics and mechanism of the oxidation of alcohols by ferrate ion *Can. J. Chem.* **71** 1394–400
[28] Cho M, Lee Y, Choi W, Chung H and Yoon J 2006 Study on Fe(VI) species as a disinfectant: Quantitative evaluation and modeling for inactivating Escherichia coli *Water Res.* **40** 3580–6
[29] Rush J D, Zhao Z and Bielski B H J 1996 Reaction of ferrate (VI)/ferrate (V) with hydrogen peroxide and superoxide anion-a stopped-flow and premix pulse radiolysis study *Free Radic. Res.* **24** 187–98
[30] de Luca S J, Idle C N and Chao A C 1996 Quality improvement of biosolids by ferrate (VI) oxidation of offensive odour compounds *Water Sci. Technol.* **33** 119–30

[31] Makky E A, Park G-S, Choi I-W, Cho S-I and Kim H 2011 Comparison of Fe (VI)(FeO42-) and ozone in inactivating Bacillus subtilis spores *Chemosphere* **83** 1228–33
[32] Jiang J-Q, Wang S and Panagoulopoulos A 2007 The role of potassium ferrate (VI) in the inactivation of Escherichia coli and in the reduction of COD for water remediation *Desalination* **210** 266–73
[33] Gombos E, Felföldi T, Barkács K, Vértes C, Vajna B and Záray G 2012 Ferrate treatment for inactivation of bacterial community in municipal secondary effluent *Bioresour. Technol.* **107** 116–21
[34] Kazama F 1994 Inactivation of coliphage Qβ by potassium ferrate *FEMS Microbiol. Lett.* **118** 345–9
[35] Karaatli T 1998 *Disinfection of surface waters by ferrate* (MS thesis, Middle East Technical University, Turkey)
[36] Kanari N, Ostrosi E, Ninane L, Neveux N and Evrard O 2005 Synthesizing alkali ferrates using a waste as a raw material *JOM* **57** 39–42
[37] Licht S, Naschitz V and Ghosh S 2002 Silver mediation of Fe (VI) charge transfer: Activation of the K2FeO4 super-iron cathode *J. Phys. Chem. B* **106** 5947–55
[38] Jiang J-Q, Stanford C and Alsheyab M 2009 The online generation and application of ferrate (VI) for sewage treatment—A pilot scale trial *Sep. Purif. Technol.* **68** 227–31
[39] Ling F, Wang J-G, Liu Q-F, Li M, Ye L-T and Gong X-N 2010 Prevention of Ichthyophthirius multifiliis infestation in goldfish (Carassius auratus) by potassium ferrate (VI) treatment *Vet. Parasitol.* **168** 212–6
[40] Lee Y and von Gunten U 2010 Oxidative transformation of micropollutants during municipal wastewater treatment: Comparison of kinetic aspects of selective (chlorine, chlorine dioxide, ferrateVI, and ozone) and non-selective oxidants (hydroxyl radical) *Water Res.* **44** 555–66
[41] Lee Y, Um I and Yoon J 2003 Arsenic (III) oxidation by iron (VI)(ferrate) and subsequent removal of arsenic (V) by iron (III) coagulation *Environ. Sci. Technol.* **37** 5750–6
[42] Hakim L, Sunariyati S, Karelius, Kurniawati N, Krestina W and Wardani D A P 2020 Ultrasonic wave as anti-bacteria of coliform in the Kahayan River, Palangka Raya, Indonesia *J. Phys. Conf. Ser.* **1517** 012054
[43] Neny Kurniawati K, Luqman Hakim D A P W and Dwi Tyas Setiawan F P 2020 Pengaruh Variasi Waktu Paparan Gelombang Ultrasonik dalam Mengurangi Jumlah Bakteri coliform pada Sampel Air Sungai Kahayan *Risal. Fis.* **4** 9–13
[44] Putri A M and Kurnia P 2018 Identifikasi Keberadaan Bakteri Coliform dan Total Mikroba dalam Es Dung-Dung di Sekitar Kampus Universitas Muhammadiyah Surakarta [Identification of Coliform Bacteria and The Total Mikrobes in Dung-Dung Ice around Universitas Muhammadiyah Surakarta Campus] *Media Gizi Indones.* **13** 41–8
[45] Pradhika I 2019 Hubungan MPN dan CFU *Lab. Mikrobiol. Standar*
[46] Zahari A, Tazi A and Blaghen M 2015 Study of the antibacterial activity of the potassium permanganoferrate K3FexMnyO8 *Int. J. Innov. Sci. Res.* **13** 523–9
[47] Karelius K and Asi N B 2017 Sintesis Ferrat (FeO42-) dari Fe (NO3) 2 dan NaOCl sebagai Pendegradasi Methylene Blue *J. Sains Dan Terap. Kim.* **10** 1–7
[48] Licht S, Naschitz V, Halperin L, Halperin N, Lin L, Chen J, Ghosh S and Liu B 2001 Analysis of ferrate (VI) compounds and super-iron Fe (VI) battery cathodes: FTIR, ICP, titrimetric, XRD, UV/VIS, and electrochemical characterization *J. Power Sources* **101** 167–76
[49] Luo Z, Strouse M, Jiang J-Q and Sharma V K 2011 Methodologies for the analytical determination of ferrate (VI): a review *J. Environ. Sci. Health Part A* **46** 453–60
[50] Sharma V K, Kazama F, Jiangyong H and Ray A K 2005 Ferrates (iron (VI) and iron (V)): environmentally friendly oxidants and disinfectants *J. Water Health* **3** 45–58

IOP Conf. Series: Materials Science and Engineering **980** (2020) 012080 doi:10.1088/1757-899X/980/1/012080

Low concentration lead ion adsorption determination performance using activated carbon from bambu betung (*Dendrocalamus asper*)

A Zakaria[1]*, N Yuliani[3], A Oktaviani[4] and Fachrurrazie[2]
[1)]Industrial Waste Treatment Departement, [2)] Analytical Chemistry, Politeknik AKA Bogor, Jl. Pangeran Sogiri No.283, Bogor, 16154, Indonesia
[3)] Departement of Biology, [4)]Departement of Chemistry, Faculty of Mathematics and Natural Science Universitas Nusa Bangsa, Gg. Walet Kukupu No.02/06, Bogor, 16166, Indonesia

Email: ahmad-zakaria@kemenperin.go.id

Abstract. Adsorption is one alternative that is used to reduce lead (Pb) at low concentrations. Activated charcoal from *Dendrocalamus asper* (Schult) known as local name "Bambu Betung" has better performance in low concentration Pb adsorption at the liquid waste. The activation by added Na_2CO_3 was carried out for 24 hours and followed by neutralization of the filtrate until it reaches pH 7. Determination of optimum adsorption using adsorbent weight, adsorbate pH and contact time was treated concurrently including the design of surface method response statistics central composite design (RSM CCD). An optimal condition conducted to obtain these adsorption Langmuir and Freundlich isothermal mechanism. CCD processing was managed to obtained optimum conditions of pH, weight, and time. The results of the isothermal equation were in accordance with the Langmuir model because the absorption model has given an r value of 0.998 through the Langmuir equation y = 0.047x + 0.009. Overall, the optimum adsorption capacity of activated charcoal from *D.asper* (Schult) was performed at 21.28 mg per gram adsorbent. Based on the characteristics of betung bamboo activated charcoal test using SEM-EDX it was shown that there was no contamination of lead ion, this is the same as activated charcoal in general.

1. Introduction

The industry has one of the adversarial consequences as the primary contributor to pollution through water and atmosphere. Industrial wastewater frequently carries toxic heavy metal particles such as Hg, Cd, Pb, Cu, Zn, and Ni [1]. Heavy metals that penetrate currents will pollute the waterbody. During expanding over contaminating streams, heavy metals will further sink to deposits that have a dwelling time of up to thousands of years and heavy metals will also accumulate in living bodies through several ways, specifically throughout the respiratory tract, food channels, and through the skin [2].

Lead (Pb) is one of the heavy metals generally enclosed in industrial waste which is usually originating from batteries, ceramics, and paint discharged into the waterbody. Several methods to reduce or even eliminate heavy metals have been carried out with physical, chemical processes, which include precipitation, coagulation, and ion exchange. Although the method mentioned earlier is still relatively expensive.

An adsorption process is an effective purification and separation technique used in the industry because it is considered more economical in water and waste treatment [3]. Adsorption is a technique often used to reduce heavy metal ions in wastewater [4]. One of the adsorbents that commonly used is activated charcoal. Activated charcoal is a material consisting of high absorptive porous carbon that has experienced a reaction with chemicals after carbonization to improve its absorption [5]. The main requirement in making activated charcoal is to contain carbon elements [6]. The activated charcoal could also produce from biomass such as bamboo betung (*Dendrocalamus asper*) [5].

2. Materials and methods

2.1 Materials and equipment

The materials used in this study included betung bamboo charcoal; 1.5% Na_2CO_3; manufactured Pb waste solution; $Pb(NO_3)_2$ 1000 ppm from Merck; Pb (II) 200 ppm; NaOH 0.1 N; HCl 0.1 N; and distilled water.

The instruments and tools used in this study were: Shimadzu AA-4000 atomic absorption spectrophotometer, Agilent pH meter, mortar, sieve, analytical balance, shaker, 100 mL Erlenmeyer flask, Whatman 42 filter paper, 50 mL volumetric pipette, measuring flask, a 500 ml glass cup, bulb and, stirring rod.

2.2 Methods

The study consisted of four stages, namely testing the characteristics of activated charcoal, sample preparation, preliminary research, isothermal adsorption. In the initial process, the activated charcoal was crushed using a sifted mortar to obtain relatively uniform and fine particles. After that, preliminary research was carried out to determine the ideal range for adsorption optimization testing. The determination of isothermal adsorption used the Langmuir and Freundlich equations to get the optimum adsorption capacity [7].

The Langmuir and Freundlich equation is presented as equations 1 and 2.

The Langmuir equation:

$$\frac{Ce}{Qe}=\frac{1}{ab}+\frac{1}{a}Ce \tag{1}$$

Freundlich equation:

$$\log Qe=\log K+\frac{1}{n}\log Ce \tag{2}$$

Where Ce is concentration of adsorbate in solution (mg / L), Qe is concentration of adsorbate absorbed per program of adsorbent (mg / g), a is optimum adsorption capacity (mg / g) in the Langmuir equation, b is empirical constant in press. Langmuir, K is optimum adsorption capacity (mg / g) in the Freundlich equation, and n is empirical constant in the Freundlich equation.

2.2.1. Sample preparation. The sieve method was used to equalize the particle size, activated charcoal a then rocked at the mesh number 80. The study consisted of four stages, namely testing the characteristics of activated charcoal, sample preparation, preliminary research, isothermal adsorption.

2.2.2. Pb ion measurement. Pb Ion Measurement with Atomic Absorption Spectrophotometer. Atomic Absorption Spectrophotometer was under the following conditions:

Cathode Lamp: Pb Wavelength	: 283.3 nm
Burning Height	: 7.0 mm
Gas	: Acetylene-Air
Acetylene Flow Rate	: 2.0 L / min
Airflow Rate	: 15.0 L / min

The measurement started by reading the low concentration standard series to high concentration reading for the sample solution. The measurement of the standard series, it obtained a linear equation between the adsorption of cathode lamp rays to the concentration of the analyte.

2.2.3. Characterization betung bamboo active charcoal. The characterization of betung bamboo active charcoal was by Scanning Electron Microscopes-Energy Dispersive X-ray Spectroscopy (SEM-EDX). The adsorbent tip of the spatula was applied to the specimen cylinder holder and then inserted into the chamber on the SEM-EDX device. The Sample chamber was condition into a vacuum state. Electron fired onto the surface of the sample. The beam firing process could observe on a computer screen, where the outside of the sample shot by the beam scanned over the entire area of the observation area.

3. Results and discussion

The method regarding obtaining betung bamboo activated charcoal consisted of individual processes, specifically dehydration, carbonation, also chemical activation methods. Through the dehydration, the method intended to reduce the water content carried in raw forms. Linked to the carbonation process, this goal of carbonation proceeded to separate oxygen and hydrogen.

The introduction to the carbonation process occurred through the enlargement of carbon pores, directly through the time of carbonation. The absorption persisted negative impact because the pores of the were blocked over by tar. Therefore, this continued underpinned with the activation process. Activation hold a treatment of charcoal that endeavored to increase the pore by improving hydrocarbon bonds or oxidizing surface molecules. The charcoal turned both physical and chemical properties and the comprising area spread and influenced the adsorption power.

The activation method was carried out for 24 hours, then the neutralization process was carried out until the filtrate had a pH of 7. Penetration aimed to remove Na2O which turned into NaOH when reacted with water molecules. Na2O must be extracted in this process because the pressure did not cover the pores of activated charcoal so that the adsorption process was not optimal. The optimum value based on the analysis of Pb metal ion adsorption in previous studies was pH 7, with a bio-adsorbent weight of 50 mg, and a contact time of 127.10 minutes.

Table 1. Isotherm adsorption determination of bamboo betong activated carbon.

C	Ce	Qe	Langmuir		Freundlich	
			cc	ce/qe	log ce	log qe
0	0	0	0	0	0	0
10	0,828571	9,17	0,8286	0,0903	-0,0817	0,96844
20	2,8571	17,14	2,8571	0,1667	0,45593	1,23408
30	6,3857	23,61	6,3857	0,2704	0,80521	1,37317
50	28,4714	21,53	28,4714	1,3225	1,45441	1,33302
60	39,5857	20,41	39,5857	1,9391	1,59754	1,30993

3.1. Results of determination of adsorption isotherms

The data experiment Pb ion isothermal adsorption by activated betung bamboothat following 2 equation models, Langmuir and Freundlich equations. The layers that occurred on the surface of the adsorbent against the adsorbate result based on the correlation coefficient and optimum adsorption capacity. The results achieved by the study lead to developing the Langmuir model because this gave an r-value of 0.998 with equation $y = 0.047x + 0.009$. The Freundlich equation eventually received r of 0.790 with equation $y = 0.186x + 1.084$. The Langmuir equation would be implemented to the adsorption of Pb^{2+} metal ions by betung bamboo activated charcoal. Based on that we proposed the Pb^{2+} metal ions built a single layer on the surface of betung bamboo activated charcoal. Meanwhile, from the Langmuir equation, the optimum capacity of activated charcoalwas 21.28 mg / g of activated charcoal. Isothermal adsorption curves are presented in table 1.

3.2. Characteristics of bamboo betung active charcoal

Following activating betung bamboo, when measurement characteristics using SEM that supported to recognize the structure and contour of smaller-scale surfaces and EDX that could detect elements in the surface body. The activated charcoal pore hold a diameter between 6.031 μm to 21.44 μm CO_3 and is discussed in the pore charcoal besides comprises impurities.bActivated charcoal manages to be clean figure 1b, the diameter of activated charcoal ranges from 1.117 μm to 6.701 μm.

The pore size means thought to affect the adsorption power of charcoal. If the pore size of the adsorbent gets less, the ability of the adsorption increases, assuming that the adsorbed component can enter its porous cavity. The greater amount of adsorbent would provide a greater surface area for the adsorbate to be desorbed. In addition, the more amount of adsorbent would also give a greater chance of contact with adsorbate molecules (Sembodo, 2006). EDX test results showed the highest carbon content is inactivated charcoal is 90.03% compared to activated charcoal which is 78.28%.

This happened because before the charcoal was activated figure 1a. The element carbon in charcoal binds with hydrogen and oxygen, can form hydroxy phenols, carboxylic acids, and others. During the process of activation and further calcification of the elements, hydrogen and oxygen will evaporate so that the carbon content of the activated charcoal increases [5].

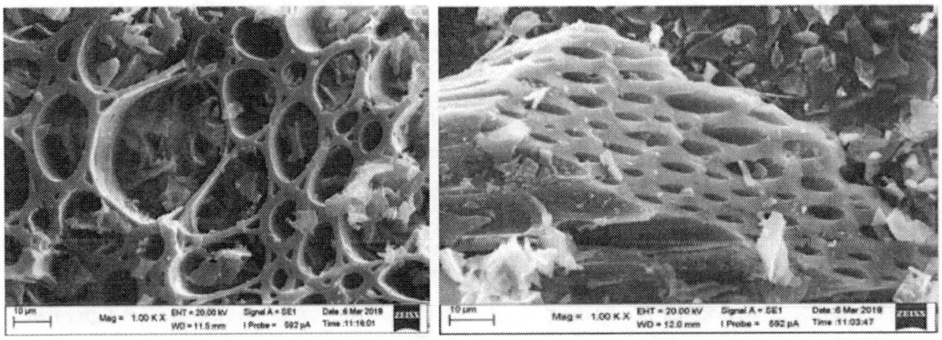

a b

Figure 1. Betong bamboo activated charcoal SEM test results a) untreated, b) treated

Table 2. EDX betung bamboo test results.

Elements	Density Activated Carbon Untreated (%)	Density Activated Carbon Treated (%)
C	78,28	90,03
O	10,23	7,01
Mg	0,17	0,32
Si	1,87	0,21
F	0,44	0,23
Cl	0,39	0,36
K	4,14	0,33
Cu	1,06	1,03
In	3,42	0,47

EDX test results for betung bamboo charcoal activated or not activated are presented in table 2. EDX test results showed the highest carbon content was in activated charcoal. This happened because before the charcoal was activated. The carbon element in charcoal binds with hydrogen and oxygen, can form hydroxy phenols, carboxylic acids, and others.The data also showed there is no contamination of Lead element from activated charcoal in term of making process. During the process of activation and further calcification of the elements, hydrogen and oxygen will evaporate so that the carbon content of the activated charcoal increases [5].

4. Conclusion

Based on the characteristics of betung bamboo activated charcoal test using SEM EDX, method shows that the surface shape of betung bamboo activated charcoal has clean pores in compared to un-activated charcoal, this is the same as activated charcoal in general. Optimum determination of isothermal adsorption of Pb metal ions by the appropriate betung bamboo activated charcoal is the Langmuir equation with a correlation (r) of 0.998 and the optimum capacity of activated bamboo betung charcoal is 21.28 mg/g.

Acknowledgments

I would like to express my special thanks of gratitude Politeknik AKA Bogor as well as our director who gave me the golden opportunity to do this wonderful project on the topic.

References

[1] Sanusi HS, Syamsu S, and Sarjirun S. 1985. *Kandungan dan Distribusi Logam Berat Pada Berbagai Komoditi Ikan Laut yang di slurkan lewat TPI Pasar (Ikan Jakarta)* (Bogor: FPIK IPB)

[2] Darmono. 2001. *Lingkungan Hidup dan Pencemaran Hubungannya dengan Toksikologi Senyawa Logam* (Jakarta: Universitas Indonesia)

[3] Al-Asher, Banat S F, Al-Omari R and Duvnjak Z 2000 Prediction of BinarySorption Isotherm for The Sorption of Heavy Metal by Pine bark Using Single Isoterm Data *Chemosphore* **41** 659-65

[4] Selvi K, Pattabhi S and Kardivelu K 2001 Removal of Cr (VI) from Aqueous Solution by Adsoption Onto Activated Carbon *Bioresour Technol* **80**

[5] Putri, Eka E, Anita S, Bali S 2017 *Potensi Arang Aktif Bambu Betung(Dendrocalamus asper) Sebagai Adsorben Ion* Mn^{2+} *dan* NO_3^- (Bengkalis: Dalam Air Sumur Bor Buruk Balu)

[6] Rijali A, Usman M, and Zulkarnain 2015 Pembuatan dan Karakterisasi Karbon Aktif dari Bambu Betung dengan Aktivasi Menggunakan Activating AgentH_2O *Jurnal FMIPA* **2** 102

[7] ZakariaA, Eti R, Irmanida S. dan Yustinus P 2012 *Adsorpsi Cu(II) menggunakan Zeolit Sintetis* dari Abu Terbang Batubara. Proc.Pertemuan Ilmiah IPTEK BAHAN2012 (PT BIN-BATAN) p191

The implementation of green industry standard at batik industry to develop eco-friendly batik

L Indrayani* and M Triwiswara
Center For Craft and Batik, Ministry of Industry, Indonesia

Email: indrayanililin@gmail.com

Abstract. Since UNESCO has established batik as Indonesia's cultural heritage, batik has become increasingly recognized internationally. However, many parties are concerned about environmental problems due to the waste from the batik production process. Therefore, the batik industry must change its thinking about the concept of environmentally friendly industries by applying green industry principles. This paper will explain the application of green industry principles that can be fulfilled at every stage of the batik production process. The technical and management requirements are covered by the Green Industry Standard (GIS) for the batik industry which is stipulated by the Minister of Industry Regulation Number 39 of 2019. GIS includes the efficiency of input materials, namely raw and auxiliary materials as well as the selection of alternative materials that are more environmentally friendly, energy-efficient and minimize water consumption and pay attention to the principles of water conservation, good waste treatment, and limitation of air emissions due to the use of technology and low-carbon energy options. It is expected that with the application of green industry principles, the batik industry will not only experience an increase in the economic sector but become an environmentally industry.

1. Introduction

Batik has received the International recognition by UNESCO as a non-material Indonesian cultural heritage (Masterpieces of Intangible Heritage of Humanity) on September 2009. This is confirmed by the government through Presidential Decree No. 33 Year 2009 by setting October 2 as national batik day. This recognition is followed by the emergence of many new batik industrial centers. The batik industry sector has a strategic role in development, especially to foster employment absorption and its contribution in fostering the growth of the national creative industry. Along with the improvement of the batik industry, the environmental problems are also increasing. One of UNESCO's assessment criteria which is stated as good is that batik is considered to have an eco-friendly value or green value. The batik industry is demanded to continue to improve in order to improve its performance standards and performance in all fields, including in terms of handling the potential for environmental pollution due to the emergence of waste from the batik industry.

Environmental problems in the batik industry usually start from the production process which resulted in a waste of raw materials, water, energy, and the disposal of waste that will burden the environment. One of the dominant environmental problems today is liquid waste from the process of making batik. It is estimated that the use of water in the process of making batik on average is approximately 25 - 50 m^2 per meter of batik. Data from the Ministry of Industry in 2017 shows that batik production in Indonesia averages 500 million meters per year, meaning 25 million m^3 of water per year. Water supply for the batik industry per year is equivalent to supplying fresh water needs for 2,500 house holds [1]. This can be seen from the excessive use of synthetic dyes and acid base, and use of recycling batik wax is not optimal. This is an example of an activity in the batik industry which not only results in waste but can also result in a high burden on its liquid waste so that it has the potential to pollute the environment.

Therefore, it absolutely needs a strategic effort that is a set of policies and regulations about carrying out efforts to prevent pollution and damage to the environment through efforts to select environmentally friendly raw materials, increase efficient use of resources (raw materials, energy and water) at each stage of production, renewal the use or improvement of low carbon production technology, the selection of effective and efficient process types, the design of environmentally friendly products and the minimization of waste, bearing in mind that it is a challenge to implement a comprehensive form of environmental management efforts in the batik industry that is predominantly small and medium enterprise, not only oriented towards improving the economic sector but also more concerned with environmental sustainability.

To overcome the above problems, a strategic effort is needed in the form of a set of policies and regulations concerning efforts to prevent pollution and environmental damage. These efforts are carried out at every stage of production starting from the selection of environmentally friendly raw materials, increasing the efficiency of resource use (raw materials, energy, and water), renewing the use or increasing of low-carbon production technology, selecting effective and efficient types of processes, product design environmentally friendly until the end, namely waste handling. This is a challenge in itself for the batik industry which is dominated by small and medium enterprises, which are not only oriented towards improving the economic sector but are also more concerned with environmental sustainability.

Green industry is one of the objectives of industrial operations as stipulated in article 3 of Law No. 3 Year 2014 concerning Industry [3]. Green industry is an industry that in its production process prioritizes efforts to efficiency and effectiveness in the use of resources in a sustainable manner so as to be able to harmonize industrial development with the preservation of environmental functions and can provide benefits to the community. To encourage the realization of the green industry, the Ministry of Industry facilitates green industry certification services to industries that have contributed to the country's economy, benefits society and participating in preserving environmental functions through the efficient use of resources and the application of environmentally friendly production processes. The certification facilitation aims to motivate the industry to further increase efforts towards a green industry. Certainly, the industries that obtain green industry certification must meet the terms and conditions set out in Green Industry Standards (GIS). The Ministry of Industry is developing GIS for various industrial groups. GIS for the batik industry is stipulated by Regulation of the Minister of Industry Number 39 Year 2019 [4].

2. Methodology

The research method is a qualitative descriptive approach that includes technical requirements and management requirements. The technical management requirements include the selection of raw materials, auxiliary materials, energy, water, production processes, products, packaging, waste and greenhouse gas emissions. While in the aspects of policy and organizational management, strategic planning, implementation and monitoring, management review, corporate social responsibility, and employment. The research was carried out by conducting a literature study and study of Minister of Industry Regulation Number 39 of 2019 concerning GIS for the batik industry [4].

3. Results and Discussions

According to Indonesia Nasional Standard or SNI 0239 : 2014, the definition of batik is a handicraft as a result of using hot wax or batik wax as a color barrier with the main tool of batik wax sticks in the form of writing or stamp canting forming a specific motif that has meaning. The batik industry is an industry with the Indonesian Business Field Standard Classification No. 13134 which includes batik

IOP Conf. Series: Materials Science and Engineering **980** (2020) 012081 doi:10.1088/1757-899X/980/1/012081

wax, whether done by written, stamp, or a combination of stamp and written. The production process in the batik industry consists of several steps starting from designing motifs, making patterns, batik process, coloring (dyeing), washing and rinsing, drying to producing finished products of batik.

The application of GIS in the batik industry is a challenge as the majority of batik industries are small and medium enterprises. The environmental problems that often arise in the batik industry are mainly due to the production process often resulting in waste of material and energy as well as due to the disposal of waste which will burden the environment and the higher cost of waste treatment. The batik industry annually produces the highest levels of CO_2 emissions when compared to other small and medium industrial sectors that still depend on the industry for fossil fuels such as kerosene. A large number of batik industry also still uses batik wax, chemicals in the form of dyes and excessive acid and base solutions which all have a negative impact on the environment and society.

Excessive use of water by the batik industry also results in scarcity of fresh water, while the limitation of fresh water is a major problem for many other regions. On one hand, the batik industry requires large amounts of water as raw material, but disposes of it into liquid waste in the final stages of the process. Overall, the main source of batik industry liquid waste comes from processes related to the dyeing process. In addition to the high content of dyes, the batik industry liquid waste also contains synthetic materials which have the potential to contain heavy metals that are difficult to dissolve or difficult to decompose so they can be categorized as hazardous wastes that cause environmental pollution. Therefore GIS needs to be implemented in the batik industry. The advantage of the application is not merely to increase efficiency, save production costs, and fulfill the market for environmentally friendly products, but more than that, namely the fulfillment of laws and regulations, conservation of raw materials and energy, and efforts to preserve the environment.

Green industry standard technical requirements for the batik industry:

3.1. Raw Materials

3.1.1 Sources of raw materials. Raw materials for the batik industry consist of fabrics originating from natural and synthetic processes in the form of cotton, silk, woven, etc. Fabric as the raw material can be obtained through the purchase of origin of textile and non-mechine weaving equipment raw materials, both from internal sources (local / domestic) and external (import through agents.

3.1.2. The ratio of products to the use of raw materials. One indicator of the efficiency of resource use is seen from the amount of optimization of raw material consumption. This value is determined by the ratio of the finished product to the use of raw materials. The limit value of the ratio of products to raw materials desired in GIS for the batik industry is 98% (ninety eight percent), the ratio value of products to raw material usage is the ratio between batik cloth products produced in m^2 using raw materials such as cloth, silk and others. others in m^2.

In the batik industry, often in one production process, it must have a reject product, not small in quantity, due to an error in the production process. With a product to raw material ratio of 98%, it has a different meaning, namely the maximum number of rejected products that are allowed at 2%. Therefore, batik industry players need to record and analyze the causes of product failure (reject product) in order to take corrective action. These corrective actions are useful to minimize the number of reject products.

3.2. Auxiliary materials

3.2.1. Types of auxiliary materials. The auxiliary materials in the batik industry are batik wax, natural dyes, synthetic dyes, fixators, and other auxiliary materials such as fixators, catalysts, soda ash, water glass, etc. One way to reduce the negative impact on the environment and human health is to limit the content of harmful dyes used in these auxiliaries.

3.2.2. Auxiliary material specifications

3.2.2.1 Batik wax. In GIS, the criteria for recycle of batik wax of the former batik industry are required to process the used night and reuse for the production process up to 95% (nine five percent) of the total batik wax used by the production process. The calculation for batik wax recycle used equation 1.

$$\textit{Recycle wax} = (\textit{number of wax recoverable per year per wax usage per year}) \times 100\% \quad (1)$$

This showed that the rest of batik wax that was allowed to be carried in liquid waste in WWTP is only 5% (five percent). Therefore batik industrialists need to use batik wax as efficiently as possible and carry out the stages of handling batik wax in each process starting from the stages of batik, *pelorodan* (wax removal) until the processing of liquid waste on every batik fabric products both written batik, stamp or combination in order to meet the above numbers. Some efforts conducted to save the batik wax can be done in the following ways including recording the identification of the use of the batik wax for batik process and perform a recovery system (recovery) of batik wax in batik wax removal process using the wax trap method, the *kowen* system. Based on the data from the Clean Batik Initiative (CBI) [2], it was stated that *kowen* system could capture evening products as much as 20% (twenty percent) of the batik wax used in the production process. Restrictions on the use of wax on each product and the batik process on GIS for the batik industry are presented in the data in table 1.

Table 1. The value of wax use limitations on each product and the type of dyeing process[5,6].

No	Wax Criteria	Limitation on the number of wax (g/m^2) cloth both sides
1	Written batik single dyeing process	276
2	Written batik with double dyeing process	316
3	Written batik with triple dyeing process	336
4	Written batik with double wax removal with *tembokan* motif	476
5	Written batik with triple wax removal without *tembokan* motif	356
6	Stamp batik with single dyeing proses	320
7	Stamp and combination batik with double dyeing process	360
8	Stamp and combination batik with triple dyeing process	380
9	Stamp and combination batik with double wax removal with *tembokan* motif	520
10	Stamp and combination batik with twice wax removal without *tembokan* motif	400

3.2.2.2 Dyes In the batik dyeing process there are 2 (two) processes, namely coloring using synthetic and natural dyes. Synthetic dyes is a dye derived from the reaction of chemicals, while natural dyes is derived from natural ingredients which are generally derived from plants (roots, stems, leaves, skin, flowers, etc.). Currently batik dyeing using natural dyes began to be developed as a substitute for synthetics dyes, considering that Indonesia is a country that has the largest biodiversity in the world, then changes in the batik sector will have a very wide impact on the Indonesian economy. Economic improvement is caused by an increase in economic value natural resources from the use of local plants become natural dyes resources. This is one of the real efforts to develop environmentally friendly batik for the sustainable batik industry.

In the GIS for the batik industry, the use of a quantity of dyes also has a maximum limit value for both natural dyes and synthetic dyes. That is for the efficient use of dyes in the production process without reducing the color quality of batik fabric products and minimizing the quantity and quality of

IOP Conf. Series: Materials Science and Engineering **980** (2020) 012081 doi:10.1088/1757-899X/980/1/012081

batik industry liquid waste. Table 2 presents the maximum value limitation for the use of natural dyes and synthetic dyes as well as supporting materials which are the criteria in GIS for the batik industry.

Table 2. Maximum quantitative limitation for utilization of natural and synthetic dyes and auxiliary for batik industry [5,6].

No	Type of Dyes	Dyes and Auxiliry materials	Maximum Quantitafif Limition
1	Natural Dyes	All color item	200 g/m^2 both sides cloth
2	Synthetic Dyes Naphtol (Dark Color)	Naftol Dyes	3,6g/m^2 both sides cloth
		Diazonium Salt	10,8 g/m^2 both sides cloth
		Coustic	1,8 g/m^2 both sides cloth
3	Synthetic Dyes Naphtol (Light Color)	Dyes	1,8 g/m^2 both sides cloth
		Diazotized base or salt	5,4 g/m^2 both sides cloth
		Coustic	0,9 g/m^2 both sides cloth
4	Synthetic Dyes Indigosol (Dark Color)	Indigosol Dyes	2,4 g/m^2 both sides cloth
		Nitrit	4,8 g/m^2 both sides cloth (double dyeing)
		HCl	6 cc/m^2 both sides cloth
5	Syntetic Dyes Indigosol (Light Color)	Indigosol Dyes	1,2 g/m^2 both sides cloth
		Nitrit	2,4 g/m^2 both sides cloth (double dyeing)
		HCl	3 cc/m^2 both sides cloth
6	Syntetic Dyes Indigosol (Dark Color)	Indigosol Dyes	16 g/m^2 both sides cloth
		Coustic	2 g/m^2
		Ash Soda	4 g/m^2
		Water Glass	0,4 kg/m^2
		Water	1,6 kg/m^2
7	Syntetic Dyes Indigosol (Light Color))	Indigosol Dyes	8 g/m^2
		Coustic	1 g/m^2
		Ash Soda	2 g/m^2
		Water Glass	0.2 kg/m^2
		Water	0.8 kg/m2

3.3. Energy consumption

The energy in the batik industry is generally used for the process of producing batik using both canting, tasting, and wax removal. But most of the energy is used in the wax removal process. In the batik industry, the use of energy does not only use electricity but it is also possible to use other types of energy such as gas, kerosene, and even firewood. The amount of energy consumption reduction in the batik industry is calculated from the amount of savings obtained by implementing an energy saving program. To quantify the large decrease in energy consumption it is assumed that there will be a reduction in energy and emissions based on the type of energy implemented at a certain period. Table 3 presents the data on the types and quantities of maximum energy used at each stage of the industrial production process according to the criteria in the GIS for the batik industry.

3.4. Water consumption

3.4.1. Use of process water. The use of water in the batik industry plays a very important role considering that in every stage of the batik production process uses water. Efforts to save water by using water as efficiently as possible are beneficial in maintaining the sustainability of resources and conserving water. In the green industry standard, it is necessary to record industrial water use (source and amount of water demand) for the production and utility processes, as well as real production data for the last 1 (one) year. Table 4 presents the data on the use of process water for batik cloth with the coloring process natural and synthetic dyes that meet the GIS criteria.

Table 3. Data on the type and quantity of maximum energy used at each stage of the industrial production process [5,6].

No	Type Energy Consumtion	Maximum Quantity /m² cloth	Note
1	LPG (liquid petroleum gasses) or kerosene	3.7 MJ	Batik process , stamp process and wax removal
2	Biomass	2.3 MJ	Wax removal
3	Electricity	0.2 Kwh	Batik process, lighting, water pump,etc

Table 4. The batik fabric water use data with natural and sintetic dyeing process [5,6].

No	Water consumtion	Maximum limitation
1	Natural dyes	10 L/m² cloth
2	Synthetic Dyes	50 L/m² cloth

3.4.2. Process water recycling ratio. The efficient use of water can be interpreted by using less water to produce the same amount of product. In addition, the efficiency of water use is also indicated by the criteria for the ratio of water reuse. Calculation of the minimum process water recycling ratio can be calculated using the equation 2

$$D_A = (R_A / T_A) X 100\% \quad (2)$$

Where, D_A is water recycling (%), R_A is the amount of water returned to the production process in the last 1 (one) year (m^3) period, and T_A is the amount of water used for the production process in the last 1 (one) year (m^3) period.

Table 5 presents data on the water reuse ratio for the batik industry with the natural and synthetic dyes processes that meet the GIS criteria. From table 5 it showed that the batik industrialist must make efforts to use water recycling not only for water efficiency and saving but for water conservation efforts. One of the efforts to use water reuse is to separate concentrated liquid waste from dilute liquid waste and to dispose of the flushing wastewater. The second and third rinsing wastewater or so on does not need to be directly discharged to the WWTP but can be reused for other processes, such as washing equipment.

Table 5. The minimum limit value is the use of the water recycling ratio for the batik industry with the natural and sintetic dyeing process [5,6].

No	Reuse water ratio	Minimum limitation
1	Natural Dyes	75%
2	Synthetic Dyes	30 %

3.5. Products

To maintain the quality of the products produced by the batik industry, batik products must meet certain quality standards. This is to protect the comfort and health of consumers and reduce the negative impact on the environment. Table 6 presents the quality standards of batik products for the batik industry with types of batik products that meet GIS criteria.

Table 6. Quality standar batik product for batik industry [5,6].

No	Batik Product	Product Quality Standard
1	Written Batik	INS 8302:2016: Written Batik – Cloth-Identification, Quality requirements and test methods
2	Stamp Batik	INS 8303:2016 : Stamp Batik – Cloth- Identification, Quality requirements and test methods
3	Combination Batik	INS 8304:2016 : Combinaition Batik – Cloth-Identification, Quality requirements and test methods

3.6. Packaging

The reduction of negative impacts on the environment and health is done by limiting the use of hazardous substances in packaging materials. Packaging can be made of plastic or paper according to the size, quality, and amount of batik cloth. This package is intended to improve the appearance of batik cloth that is packaged and protect the goods from the surrounding influences. In the GIS for the batik industry, plastic is only used as a secondary packaging, with the minimum amount possible, as a protector. Paper is used as primary packaging.

3.7. Waste

3.7.1. Means of liquid waste management. Most of the batik industry in Indonesia is a small-scale industry, perhaps even a home industry that involves the local community as a buffer for the creative economy. Efforts to manage waste in a significant quantity are unlikely to be completed by small-scale batik industry. For example, in the manufacture of Wastewater Treatment Plant (WWTP) and its operation, it is impossible for small-scale batik industrialist to independently do it. Therefore, the government needs to arrange for the facilitation of assistance for the manufacture and operation of small scale WWTP for the batik industry. Several approaches to using the communal WWTP for the batik industry are located in an industrial center area. It is necessary to procure portable liquid waste processing equipment that can alternately process small-scale batik industries. Likewise, liquid waste processing technology is needed, which is an appropriate technology for the medium-scale batik industry, which has a higher production capacity, which must manage its batik waste independently.

3.7.2. Fulfillment of liquid waste parameters to environmental quality standards in accordance with statutory provisions. Based on the statutory provisions, each region that has a batik industrial center has the authority to compile regional regulations that apply to its respective regions. So far, the regulation regarding batik waste follows the provisions or regulations related to the textile industry regulations. Therefore, the green industry standard for the batik industry is allowed to dispose of its waste on condition that it meets the environmental quality standards set by the regulations in each region. Laboratory test results on samples of batik industrial waste must go through a nationally accredited laboratory.

The Ministry of Environment Decree No. 5 of 2014 concerns that the quality standards for wastewater does not specifically regulate the quality standards for the batik industry, but regulations regarding the limits of wastewater that can be discharged into the environment can refer to the quality standards for the textile industry[3]. Therefore, almost all regulations related to regional batik wastewater also follow the Ministry of Environment Decree No. 5 of 2014 concerning the quality standards for wastewater determined by each region in accordance with their authority[3].

3.7.3. Means of exhaust and air emissions management. The batik industry emits flue gas emissions but the quantity is not as large as other large manufacturing industries but the batik industry is still obliged to comply with the technical requirements, namely supporting requirements in relation to compliance with ambient emission quality standards, and noise for example adequate conditions of room air ventilation and technical requirements the other. Examination of the existence and operational conditions of the means of managing exhaust emissions and air need to be considered properly.

3.7.4. Facilities for hazardous waste management. In the process of making batik, it always involves acids and bases as raw materials, namely in the coloring process and also as a supporting material, namely the initial mordanting process and the process of strengthening the color (fixation). According to Article 5 paragraph 3 and paragraph 4, Government Regulation Number 101 of 2014 concerning management of hazardous waste, expired acid-base conditions used in the process of making batik are hazardous waste with category 1, namely waste that has explosive characteristics, flammable, reactive, infectious, and / or corrosive [3]. In the appendix section Government Regulation No. 101 of 2014 explains that batik waste is waste with waste code number 22, namely for the textile industry, including bleaching and dyeing of fibers and textile materials, making fireproof, coating, coating, or absorbing clothing, with batik business process with wax is done by writing, stamp or combination with waste in the form of sludge from WWTP which is included in category 2 [3].

The legal consequences of Article 11 and Article 12 of Government Regulation Number 101 of 2014 concerning hazardous waste management stipulate that batik waste that categorizes acids and bases in batik waste is category 1 and the mud (sludge) of WWTP batik category 2 is that batik industrialist must follow the provisions of the treatment, transportation, and storage of hazardous waste. This is certainly not a simple thing that can be adhered to by the batik industry. Batik industrialists need to make certain efforts to comply with the regulation, one of them is by having a hazardous waste Temporary Storage (TS) completed with a hazardous waste management permit.

3.7.5. Means of solid waste management. The implementation of waste management includes waste reduction and handling. The batik industry is required to make waste reduction and waste management. Waste management includes sorting, collecting, transporting, processing and final processing of waste.

3.8. O2 emission levels
Emissions for the batik industry are generally generated from sources derived from the use of energy in each batik-making process, either directly or indirectly. Energy usage calculations can be used to verify carbon emission calculations for the batik industry. The maximum carbon emission limit according to GIS for the batik industry is 1.58 kg CO_2 equivalent / m^2 of the finished products.

Green Industry Standard Management Requirements for the Batik Industry

3.9. Policy and organization
3.9.1. Green industry policy. The batik industry must have a written policy on the application of a green industry. The principle of application is accompanied by a policy document on the application of the principle of green industry, which at least contains the target of savings or efficiency in the use of raw materials, energy, water, CO_2 emission reduction and waste (hazardous waste and non hazardous waste) reduction in a period of 1 (one) year, which is set by the top leadership (owner of the batik industry).

3.9.2. Organization. The existence of implementing the implementation of the green industry has an organizational structure with clear duties and obligations and is determined by the owner of the batik industry. This can be demonstrated through increased training or human resources capacity enhancement of human resources capacity on green industry principles.

IOP Conf. Series: Materials Science and Engineering **980** (2020) 012081 doi:10.1088/1757-899X/980/1/012081

3.9.3. Socialization. Policies on the application of the green industry must be disseminated to all employees and all people involved in the batik industry.

3.10. Planning

The batik industry must establish policies for implementing green industry principles that can be measured through established technical and management requirements. The policy is written in the program to achieve the goals and objectives of the application of green industry principles. The program is also a joint policy that will be carried out from the top leadership to the smallest part of the batik industry.

3.11. Implementation and Monitoring

3.11.1. Implementation of the program. The program is carried out in the form of activities according to the schedule and reported regularly to the policies of the top leaders (owners) of the batik industry. Implementation of the program, at least includes efficiency in the use of raw materials, energy use, water use, GHG emission reduction, and waste reduction (hazardous and non-hazardous).

3.11.2. Program monitoring. The owner of the batik industry is obliged to monitor the implementation of the program to determine the success rate of implementing green industry principles in his company. The monitoring program is carried out regularly and is complemented by a report from the program at least once a year.

3.12. Implementation of management review

3.12.1. Implementation. The batik industry conducts management review. The report on the results of management conducts management reviews on a periodic basis in the last 1 (one) year period.

3.12.2. Company consistency. The batik industry consistently uses reports before and against the results reports after the follow-up act of monitoring, continued requirements or audit results, to take corrective actions.

3.13. Cooperate responsibility response

The batik industry must have a social concern for the sustainable CSR (Cooperate Responsibility Response) program. One example of a possible form of CSR is the partnership program. The concept of the partnership program can be done one of them by placing the private sector as a component of good governance, then its existence is expected to contribute to development activities. For example, the large-scale batik industry produces batik cloth and clothing products and those that produce a variety of fabric products, including batik, have relatively not yet carried out CSR programs optimally in accordance with the standard CSR program criteria and existing regulations. This CSR program does not only apply to the large scale batik industry but can also be carried out by other large-scale industries which are located close to the centers of the small-scale batik industry so that it is expected to be able to make a large contribution both in the form of funds and technology in environmental management, especially in management batik wastewater at small industrial centers.

3.14. Employment

In the batik industry there are often minor work accidents. Therefore, the provision of labor facilities under physical provisions, reporting of regulations and their implementation. Providing facilities at least includes: training of workers (Law No. 13 of 2003) health checks (Ministry of Manpower and Transmigration Decree No. 2 of 1980) monitoring the workplace environment (Ministry of Manpower and Transmigration Decree No. 5 Year 2018) provision of first aid kits (Ministry of Manpower and Transmigration Decree No. 15 Year 2008) the provision of personal protective equipment (Ministry of Manpower and Transmigration Decree No. 8 Year 2010)

IOP Conf. Series: Materials Science and Engineering **980** (2020) 012081 doi:10.1088/1757-899X/980/1/012081

4. Conclusion

The implementation of green industry principles following the Minister of Industry Decree Number 39 Year 2019 concerning Green Industry Standards (GIS) for the batik industry, it is hoped that the batik industry can make the batik industry, which in its production process puts forward efforts to utilize resources efficiently and effectively in a sustainable manner so that it can able to harmonize industrial development with the preservation of environmental functions and can provide benefits to the community and make batik as an eco-friendly product.

Acknowledgments

I would like to thank Center For Craft and Batik and Centre For Green Industry, Ministry of Industry that their support from organizations and providing opportunities for this research.

References

[1] Balai Besar Kerajinan dan Batik. 2010. Produksi bersih pada industri batik. Yogyakarta
[2] EKONID, (2011), Clean Batik Initiative (CBI), Jakarta
[3] The Minister of Industry. (2014). Green Industry Concept and Implementation. Jakarta.
[4] The Minister of Industry Decree No. 39 Year 2019 concerning Green Industry Standards (GIS) for batik industry.
[5] GTZ/P3U. 2013. Good house keeping. Bonn
[6] The Minister of Industry Republik Indonesia. 2019. Green industry award assessment guidelines. Jakarta

IOP Conf. Series: Materials Science and Engineering **980** (2020) 012082 doi:10.1088/1757-899X/980/1/012082

The potential of food waste as bioplastic material to promote environmental sustainability: A review

M O Ramadhan* and M N Handayani
Study Program of Agro-industry Technology Education, Faculty of Technology and Vocational Education, Universitas Pendidikan Indonesia, Jalan Dr. Setiabudhi No 207 Bandung 40154, West Java, Indonesia

Email: ramadhanoka@upi.edu

Abstract. Food waste is a challenge for sustainable development since it can increase greenhouse gas emissions and other environmental issues. Food processing and manufacturing are known to raise food waste. Meanwhile, plastic waste is a big problem in environmental pollution. This study aims to explore the potential of food waste as bioplastic material as an environmentally friendly alternative packaging. The method used was an effective literature review. The results showed that many food wastes have the potential to be developed as bioplastics if they contain biopolymers such as starch, cellulose, or other biopolymers. Some of the food wastes are obtained from the food processing industry such as sludge waste, cassava peel, banana peel, pineapple peel, durian seed, jackfruit seed, avocado seed, and chicken feather. The Development of bioplastics from food waste has a double benefit that can solve two problems indirectly, namely reducing plastic waste and food waste at the same time, thereby promoting environmental sustainability.

1. Introduction

Environmental sustainability has been a real challenge for a living being due to the rising of population, urbanization, and standard of living [1]. It also becomes a major concern in the Sustainable Development Goals (SDGs) since it declared as the three of the world's biggest challenge: responsible consumption and production (SDGs Number 12), attention to climate action (SDGs Number 13), and life below water (SDGs Number 14) [2].

For years, plastic waste has become one of the biggest problems due to their excessive use, its difficulty in decomposition, also its huge of mass cause a lot of negative impact to landfill and water pollution [3]. Especially for single-use plastic commonly used for plastic packaging including objects that can only be used once before they are thrown away, generated nearly 50% of the plastic waste in 2015 [4]. The most possible solution is to substitute synthetic polymeric materials by biodegradable materials [3]. After usage, it will be degraded by microorganisms and minimalize the adverse ecological impact [5].

On the other side, food waste is known as one of the materials that lead to a huge environmental impact. In 2008, food waste contributes to global anthropogenic greenhouse gas (GHG) emissions about 19%-29%, releasing more than 16,000 megatons of carbon dioxide equivalent ($MtCO_2e$) to the atmosphere, and other environmental issues [6], Food waste has been proved by several researchers that it can be developed into environmentally friendly biopolymers [7].

The purpose of this study is to summarize any kind of food waste that proved can be developed into bioplastic material with potential applications in food packaging to promote environmental sustainability.

2. Food waste

The Food and Agriculture Organization (FAO) defined food waste as follows: "Food loss is defined as 'the decrease in quantity or quality of food.' Food waste is part of food loss and refers to discarding or alternative (nonfood) use of food that is safe and nutritious for human consumption along the entire food supply chain, from primary production to end household consumer-level" [8].

Food waste has a different definition in the food supply chain (FSC), even though it is most easily determined at the retail and consumer stages, lead to 'food' for human consumption [9]. Food loss is qualitative, such as a decrease in nutritional value, and unwanted changes in taste, color, texture, or quantitatively measured by volume or weight reduction [10]. The source of food waste comes from various stages in the food supply chain. Start from the beginning of FSC including production (damage during harvesting, late time of harvesting, etc.), distribution (out-grade, contamination, damages during transport, etc.), also the consumption of food (left-over food, poor cooked, served too much, etc.) [10].

Food waste is generally classified into two categories-avoidable and unavoidable. Avoidable food waste is the waste that could be prevented (such as through better planning), whereas unavoidable food waste is food waste that cannot be sold or eaten [11]. Examples of avoidable food waste are food that is thrown away, left-over food, and over-purchasing, while the unavoidable one such as bones, shell, feather, and more.

The current report shows that there are 1.3 billion tons of food wasted globally every year [12]. Food waste also contributes to greenhouse gas (GHG) emissions during the final disposal process at the final disposal site (uncontrolled release of methane) and during activities related to food production, processing, manufacturing, transportation, storage, and distribution. Other environmental impacts are value reduction and pollution of natural resources such as soil, water, and biological life behind it. These tons of food waste not only affect the environment but also affect the economic side. The economic impact is the waste of the costs to solve the chain reaction problems [13].

3. Plastic waste

For decades, the world dependence on petroleum-based synthetic polymers or commonly known by the public as plastics has become the one most appealing material used for a wide variety of applications. More than 400 million tons of plastic are produced every year in the world [14]. This record is proportional to their characteristics of the material due to low price, ease of production, strong endurance, low density, and its aesthetic look.

In 2015, 36% of the world's plastic production is designed as plastic packaging with single-use/disposable purposes. These include, among other items, food packaging, grocery bags, cutlery, bottles, straws, cups, and more. It also contributes to 47% of plastic waste generated globally. Therefore, plastics currently the most common and persistent pollutants in the world. In the end, plastics are recycled into other things, burned by incineration, landfilled, dumped, or littered in the environment. If current consumption patterns do not change and the development of environmentally friendly alternative materials is not carried out, there will be around 12 billion tons of plastics wasted in landfills and the environment by 2050. Being part of the natural ecosystem bringing billions and long-lasting negative effects on the environment both life on land and below the water [14].

There are a lot of plastic forms, such as plastic bags, that have been found to poison the respiratory passages and stomachs of various species in the ocean. Animals such as dolphins, turtle mistakenly see a plastic bag as their food. There is an emerging fact that toxic chemicals were absorbed during the transfer of the manufacturing process from inserting plastic into animal tissue, finally entering the human's food chain as well [15]. Although the prevention of plastic waste is a priority, reducing the

IOP Conf. Series: Materials Science and Engineering **980** (2020) 012082 doi:10.1088/1757-899X/980/1/012082

use of plastic is certainly not easy. Therefore researchers, policies, and industries are currently trying to develop new alternative materials that are biodegradable and environmentally friendly.

4. Bioplastic

Biodegradable polymers are known as bioplastic made from polymers derived from biological sources [16]. These polymers can be degraded environmentally by microorganisms and water in compost piles [17]. Bioplastic can be categorized as petroleum-based biodegradable polymers (fossil-based), bioplastics from mixed source (bio-petroleum), and renewable resource-based polymers (naturally from plants and animals) [18].

There are so many advantages for the environment due to the use of bioplastic such as potentially lower the carbon footprint and GHG emissions, lower energy cost in manufacturing, reduction of permanent litter, and much safer to the environment [19]. Bioplastics also have advantages in the characteristics of the material such as much greater water vapor permeability than standard plastic, less oily feel, good printability, softer, and more tactile.

The method of bioplastic making may be different for each material used, the bioplastic characteristics produced, and various product configurations [20]. According to previous research, figure 1 summarized the complex process for each material, each process has a variety of methods, ingredients, and compositions used.

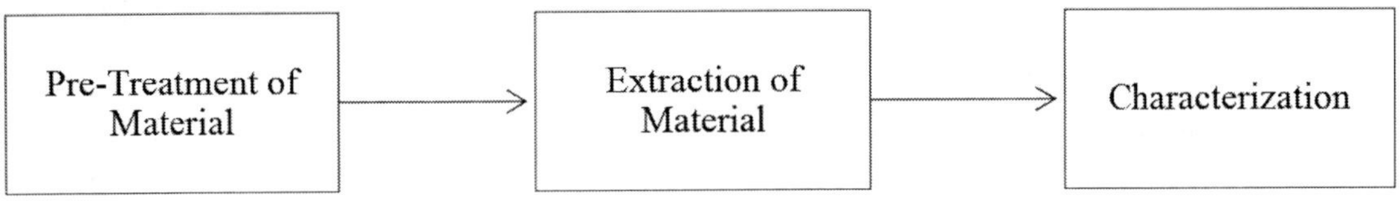

Pre-Treatment including processes such as grinding, drying, hydrolyzing the material. Not all parts of the waste are used in the bioplastic manufacturing process, such as extracting only its starch and cellulose. And the most important part is characterizing materials such as adding plasticizer agent, odor controlling agent, and biological material [20].

5. Potential of food waste as bioplastic material

5.1. Sludge waste from food industry

The sludge that is freshly activated from wastewater treatment in a food processing industry can be used to make biodegradable plastic. Active sludge contains various types of microorganisms that can be used to produce PHB (Poly-b-hydroxybutyrate), produced by various bacteria as microbial polyester and stored in cells in the form of granules. Some bacterial species which naturally produce PHB are *Ralstonia eutrophes, Alcaligenes, Pseudomonas, Bacillus, Rhodococcus, Staphylococcus and Micrococcus* [43]. This material can be suitable for the synthesis of environmentally friendly plastics material [21]. PHB production costs are very high, so the use of activated sludge is expected to be more efficient.

PHB production from activated sludge is carried out as follows: fresh activated sludge is collected from food processing wastewater and enriched in synthetic media as a carbon source using acetic acid. After the enrichment process, the bacterial strain was isolated on a nutrient agar plate by spreading mud. Based on the characteristics of the colony five different strains of bacteria were observed. Filtering the isolated strains for PHB production. At the end of the process, the culture is smeared on a glass slide, stained with black Sudan B, and examined under a process of contrast microscope for PHB beads. All five strains were determined to accumulate PHB granules [22]. The production and recovery process of PHB from activated sludge can reduce sludge waste that requires further treatment and at the same time increase the efficiency of PHB production.

5.2. Cassava peel

Because of its starch content, cassava peel is an agricultural waste that can be used as a bioplastic material. Starch is universal, renewable, and easily obtained so that it becomes potential material for bioplastic manufacturing [23]. The study of making bioplastics from cassava peel starch has been carried out by combining the materials of starch and chitosan as a plasticizer by using sorbitol. The results showed that the best mechanical properties for bioplastics with a tensile strength value of 1.37 MPa was obtained at the addition of sorbitol 30% with a ratio of starch: chitosan (7:3). Young modulus value of 3.7 MPa was obtained at the addition of sorbitol 30% and a starch: chitosan ratio of 8: 2, and an elongation value of 26.55% was obtained with 50% sorbitol addition, and starch: chitosan ratio of 9:1 [24].

Making bioplastics from cassava peel begins with the extraction of starch from cassava peel. Furthermore, making cassava peel starch solution 10% (w/v), while chitosan was dissolved in glacial acetic acid with a concentration of 20% (w/v). Cassava peel starch was mixed with chitosan at a certain ratio and stirred at gelatinization temperature (80°C) for 25 minutes. Then mixed homogeneously with sorbitol at a certain concentration. Then, print the homogeneous solution on a plate with a thickness of 2.5 mm and dried at 60°C for 5 hours to form a bioplastic sheet [24].

Comparison with other research has shown that the addition of microcrystalline cellulose (MCC) can increase the tensile strength value up to 9.12 MPa by using only 20% sorbitol. In the contrary to tensile strength, a decrease in elongation at break was reported. The improvement in bioplastics with reinforcing MCC could be attributed to the strong hydrogen bond between hydroxyl groups of the interface of both MCC fillers and starch matrix. This formation is also influenced by the percolation mechanism [44].

5.3. Banana peel

Banana peels as agricultural processing waste can be used in making bioplastics because they contain cellulose, starch, pectin, and other polymers. Cellulose is modified to obtain thermoplastic materials through acetylation (cellulose acetate) [25]. The use of pectin in making bioplastics functions as an emulsifier that increases intermolecular bonds in the film [26].

The utilization of banana peels has been done by adding citric acid to prevent browning so that pectin produced by banana peels is brighter. Pectin of Kepok banana peel extracted with citric acid at 90°C for 3 hours, has a water content of 11.55%, ash content of 3.060%, methoxyl content of 3.966%. Bioplastics produced from pectin Kepok banana peel has a film thickness ranging from 0.00311 – 0.00387 cm, permeability 35.30 – 63.63%, with traction strength of 2.6286 – 10.5620 MPa, extension value 16.66 – 58.33%. The best bioplastics in the study resulted from the addition of 5 grams of pectin [26].

Types of banana peels that have good characteristics to be used as bioplastics are yellow banana peels compared to green bananas and red bananas. although the difference between the three was very slight. However, the transparency of visual appearance and ease of production is found in green bananas due to their higher starch content [45].

5.4. Pineapple peel

Pineapple peel is a waste that comes from household consumption as well as the pineapple processing industry. The main constituent of the peel is extracted as cellulose through refluxing with acid or alkaline solutions [27]. Cellulose is a natural polymer, consisting of glucose units with a homogeneous chain structure. Cellulose can be converted to carboxy methylcellulose (CMC) through the etherification process.

Pure cellulose potentially can be made from pineapple by refluxing pineapple skin powder with 0.5 M HCl and 1 M NaOH solution at 90°C for 1 hour and 2 hours. Then, the solid material bleached with calcium hypochlorite. Pure cellulose was immersed in a mixture of isopropyl alcohol and NaOH for 12 hours. After that, it was reacted with chloroacetic acid at 55°C for about 6 hours. The optimum conditions for carboxymethylation are 5 g cellulose, 13.0 g chloroacetic acid and 40% w / v 50 mL NaOH. The optimized product has a DS of 0.50 and can be used as an element in bioplastics [28].

5.5. Durian seed

Durian seeds are waste from food processing as well as part of durian fruit that is not consumed because it feels slimy and itchy on the tongue. Even so, the seeds have nutrients including protein, carbohydrates, fats, calcium, and phosphorus [29]. Carbohydrates in the form of starch in durian seeds have the potential to be used as material for making bioplastics. However, starch-based bioplastics have some disadvantages such as low mechanical strength and less resistance to water [30].

A starch-based bioplastic study has been carried out that utilize starch from durian seeds with the addition of chitosan as a filler, and plasticizer by using sorbitol. Durian seed starch used in the study had a water content of 12.73%, an/ ash content of 0.51%, the starch content of 76.65%, amylose content of 22.34%, amylopectin content of 54.32%, the protein content of 11.61%, and fat content of 0.61%. The result showed bioplastics had optimal mechanical properties at a heating temperature of 70°C with a composition between durian seed starch and chitosan of 7:3 grams and the addition of 20 grams of sorbitol [31].

5.6. Jackfruit seed

Jackfruit seed, which constitutes 8-15% of the jackfruit, is potential food wastes due to its high starch content [32]. It can be used for bioplastics production as raw material. A study on making bioplastics from jackfruit seed starch has been carried out. The starch had a moisture content of 6.04%, amylose content of 16.39%, ash content of 1.08%, starch content of 70.22%, amylopectin content of 53.83%, protein content of 4.68%, fat content of 0.54 %. The bioplastic making was a combination of starch, chitosan, and sorbitol as a plasticizer. The best bioplastic was obtained by the ratio of starch: chitosan (w / w) = 8: 2 and sorbitol concentration of 25% with a tensile strength of 13,524 MPa [33].

Meanwhile, other studies developed bioplastic from jackfruit seeds using glycerol as a plasticizer. The starch concentrations used ranged from 2-6% w/w and glycerol from 20 to 60 g / 100 g starch. The study states that jackfruit seed starch can be used as raw material for bioplastic, with the character of low opacity, moderate permeability of water vapor, and relatively high mechanical stability, by gelatinized starch dispersions process using glycerol [33].

5.7. Avocado seed

Avocado seed, which represents 10-22% from the total weight of avocado [34], is potential food wastes due its to high starch content [35]. This starch content is a raw material for making bioplastic, as same as jackfruit seed. A bioplastic development study from avocado seed starch has been carried out by adding chitosan and glycerol. Avocado seed starch used had a moisture content of 1.087%, an/ ash content of 1.007%, the starch content of 67.6950%, amylose content of 32.4739%, amylopectin content of 35.3212%, the protein content of 10.440%, the fat content of 1.860%, peak viscosity is 3847 cP by setting the gelatinization temperature at 85.17°C. The best bioplastic conditions of avocado seeds were obtained at 90 ° C with a ratio of starch: chitosan (w / w) = 7: 3 and glycerol 0.2 ml / g with a tensile strength of 5.096 MPa, breaking extension 14.016%, Modulus Young 36.359 MPa. Bioplastics with chitosan as fillers and glycerol as plasticizers have a smooth and soft fracture surface and few cavities [36].

5.8. Chicken feather

The fur is the by-products of animal sources besides feet, head, bones, and viscera [.37]. Usually, the hair is discarded or burned which can be an environmental problem [38]. Though fur is one of the potential resources to make biodegradable materials. Feathers contain 90% of a protein known as keratin which is a worldwide by-product of the poultry industry.

A few researchers conducted studies to develop potential ingredients from fur and keratin in making bioplastic [39,40,41]. There is a study of bioplastic synthesis using keratin from chicken feathers. Keratin solution can be extracted from chicken feathers, then produce plastic films by mixing it with glycerol (2%). The mixture is stirred at 60°C for 5 hours with constant magnetic stirring. The mixture was then poured into an aluminum weighing vessel and dried in an oven at 60°C for a day. The resulting bioplastics had good mechanical and thermal properties and was proven to be biodegradable [42].

6. Conclusion

Bioplastic can be made from polymers derived from biological sources, one of them is food waste. It comes from the food processing industry or household consumption such as sludge waste, cassava peel, banana peel, pineapple peel, durian seed, jackfruit seed, avocado seed, and chicken feather. The development of bioplastic from food waste has a double benefit that can solve two problems, indirectly reducing plastic waste and food waste at the same time, thereby promoting environmental sustainability.

References

[1] AlQattan N, Acheampong M, Jaward F M, Ertem F C, Vijayakumar N and Bello T 2018 Reviewing the potential of Waste-to-Energy (WTE) technologies for sustainable development goal (SDG) numbers seven and eleven *Renew. Energy Focus* **27** 97–110

[2] GA U 2015 Transforming our world: the 2030 agenda for sustainable development. (New York: Division for Sustainable Development Goals)

[3] Gonçalves de Moura I, Vasconcelos de Sá A, Lemos Machado Abreu A S and Alves Machado, A V 2017 Bioplastics from agro-wastes for food packaging applications *Food Packag*. 223-63

[4] Giacovelli C 2018 *Single-use plastics: a roadmap for sustainability* (UN Environment Program Report)

[5] Mohanty A K, Misra M and Hinrichsen G 2000 Biofibres, biodegradable polymers and biocomposites: An overview *Macro. Mater. and Engineer.* **276(1)** 1–24

[6] Vermeulen S J, Campbell B M, and Ingram J S I 2012 Climate change and food systems *Ann. Review of Env. and Resour.* **37(1)** 195–222

[7] Shah A A, Hasan F, Hameed A and Ahmed S 2008 Biological degradation of plastics: a comprehensive review. *Biotech. Advan.* **26(3)** 246–65

[8] Footprint F F W 2014 *Full-Cost Accounting Final Report*

[9] Parfitt J, Barthel M and MacNaughton S 2010 Food waste within food supply chains: quantification and potential for change to 2050 *Philo. Trans. of the Royal Soc. B: Bio. Scien.* **365(1554)** 3065–81.

[10] Buzby J C and Hyman J 2012 Total and per capita value of food loss in the united states *Food Policy* **37(5)** 561-70

[11] Bagherzadeh M, Inamura M and Jeong H 2014 *Food Waste Along the Food Chain* (OECD Publishing) 71

[12] Kojima R and Ishikawa M 2013 *Prevention and Recycling of Food Wastes in Japan: Policies and Achievements* (Japan: Kobe University)

[13] Lipinski B, Hanson C, Lomax J, Kitinoja L, Waite R and Searchinger T 2013 Reducing food loss and waste *World Resour. Instit. Work. Paper* **1** 1-40.

[14] Geyer R, Jambeck J R and Law K L 2017 Production, use, and fate of all plastics ever made *Scien. Advan.* **3(7)** e1700782

[15] Jambeck J R, Geyer R, Wilcox C, Siegler T R, Perryman M, Andrady A and Law K L 2015 Plastic waste inputs from land into the ocean *Science* **347(6223)** 768-71

[16] Chen Y J 2014 Bioplastics and their role in achieving global sustainability *J. of Chem. and Pharm. Research* **6(1)** 226–31

[17] Zheng Y, Yanful E K and Bassi A S 2005 A review of plastic waste biodegradation *Critic. Rev.*

IOP Conf. Series: Materials Science and Engineering **980** (2020) 012082 doi:10.1088/1757-899X/980/1/012082

in Biotech. **25(4)** 243–50
[18] Reddy M M, Misra M and Mohanty A K 2012 Bio-based materials in the new bio-economy *Chem. Engin. Prog.* **108(5)** 37-42
[19] Pilla S 2011 *Handbook of Bioplastics and Biocomposites Engineering Applications* (John Wiley & Sons) 81
[20] Hagemann R and D'Amico D 2009 *Bio-plastic Composite Material, Method of Making Same, and Method of Using Same* (US 2009/0110654 A1)
[21] Fang L, Wenqing L, Darin R and Tingyue G 1998 Production of poly-b-hydroxybutyrate on molasses by recombinant Escherichia coli *Biotech. Letters* **20 (4)** 345–8
[22] Kumar M S, Mudliar S N, Reddy K M K and Chakrabarti T 2004 Production of biodegradable plastics from activated sludge generated from a food processing industrial wastewater treatment plant *Biores. Tech.* **95(3)** 327-30
[23] Ma X, Chang P R, Yang J and Yu J 2009 Preparation and properties of glycerol plasticized - pea starch/zinc oxide bio nano composit *Carbo. Poly.* **75** 472-8
[24] Fathanah U, Lubis M R, Rosnelly C M and Moulana R 2013 Making and characterizing bioplastic from cassava (*manihot utilissima*) peel starch with sorbitol as plasticizer *Proc. the 7th International Conf. of Chemical Engineering on Science and Appl*
[25] Rana G K, Singh Y, Mishra S P and Rahangdale H K 2018 Potential use of banana and its by-products: a review. *Int. J. Curr. Microbiol. Appl. Sci.* **7(6)** 1827-32
[26] Chodijah S, Husaini A and Zaman M 2019 Extraction of pectin from banana peels (*musa paradiasica* fomatypica) for biodegradable plastic films *JPhCS* **1167(1)** 012061
[27] Xu W, Reddy N and Yang Y 2009 Extraction characterization and potential applications of cellulose in corn kernels and distillers' dried grains with solubles (DDGS) *Carbohydr. Polym.* **76** 521-27
[28] Chumee J and Khemmakama P 2014 Carboxymethyl cellulose from pineapple peel: useful green bioplastic *advan. Mater. Resear.* **979** 366-9
[29] Jufri M, Dewi R, Ridwan A and Firli 2006 Studi kemampuan pati biji durian sebagai bahan pengikat dalam tablet ketoprofen secara granulasi basah *M. Ilmu Kefarm. Depart. Farm. FMIPA* **3(2)**
[30] Ban W 2006 Influence of natural biomaterials on the elastic properties of starch-derived films: an optimization study *J. of Appl. Poly. Science* **15** 30-8
[31] Ginting M H S, Kristiani M, Amelia Y and Hasibuan R 2016 The effect of chitosan, sorbitol, and heating temperature bioplastic solution on mechanical properties of bioplastic from durian seed starch (Durio zibehinus) *Int. J. of Engin. Res. and Appli.* **6** 33-8
[32] Mukprasirt A and Sajjaanantakul K 2004 Physico-chemical properties of flour and starch from jackfruit seeds (artocarpus heterophyllus lam.) compared with modified starches. *Int. J. of Food Scien. & Tech.* **39(3)** 271-6
[33] Lubis M, Harahap M B, Manullang A, Ginting M H S and Sartika M 2017 Utilization starch of jackfruit seed (*Artocarpus heterophyllus*) as raw material for bioplastics manufacturing using sorbitol as plasticizer and chitosan as filler *JPhCS* **801(1)** 012014
[34] Gómez-López V M 1999 Characterization of avocado (*persea americana* mill.) varieties of low oil content *J. of Agri. and Food Chem.* **47(7)** 2707-10
[35] Kahn V 1987 Characterization of starch isolated from avocado seeds *J. of Food Scien.* **52(6)** 1646-8
[36] Ginting M H S, Tarigan F R and Singgih A M 2015 Effect of gelatinization temperature and chitosan on mechanical properties of bioplastics from avocado seed starch (*Persea Americana* Mill) with plasticizer glycerol *The Int. J. Of Engin. And Scien.* **4** 36-43
[37] Zhu G Y *et al* 2010 Hydrolysis technology and kinetics of poultry waste to produce amino acids in subcritical water *J. of Analy. and App. Pyrol.* **88(2)** 187-91
[38] Jin E, Reddy N, Zhu Z and Yang Y 2011 Graft polymerization of native chicken feathers for thermoplastic applications *J. of Agri. and Food Chem.* **59(5)** 1729-38

IOP Conf. Series: Materials Science and Engineering **980** (2020) 012082 doi:10.1088/1757-899X/980/1/012082

[39] Tanabe T, Okitsu N, Tachibana A and Yamauchi K 2002 Preparation and characterization of keratin–chitosan composite film *Biomaterials* **23(3)** 817-25

[40] Ullah A, Vasanthan T, Bressler D, Elias A L and Wu J 2011 Bioplastics from feather quill *Biomacromolecul* **12(10)** 3826-32

[41] Moore G R P, Martelli S M, Gandolfo C, do Amaral Sobral P J and Laurindo J B 2006 Influence of the glycerol concentration on some physical properties of feather keratin films *Food Hydro.* **20(7)** 975-82

[42] Ramakrishnan N, Sharma S, Gupta A and Alashwal B Y 2018 Keratin based bioplastic film from chicken feathers and its characterization *Int. J. of Bio. Macromol.* **111** 352-8

[43] Ansari S and Fatma T 2014 Polyhydroxybutyrate-a biodegradable plastic and its various formulations *Int. J. of Innov. Res. in Scien. Engin. and Tech.* **3(2)** 9598-602

[44] Maulida S M and Tarigan P 2016 Production of starch based bioplastic from cassava peel reinforced with microcrystalline cellulose avicel PH101 using sorbitol as plasticizer *Int. J. Phys. Conf. Ser* **710** 12012

[45] Singh R and Dakshinamoorthy A N D *Comparative Study of Bioplastic Sheets from Different Varieties of Banana Peel [Online]* Available at: <http://reports.ias.ac.in/report/20428/comparative-study-of-bioplastic-sheets-from-different-varieties-of-banana-peel> [Accessed 23 October 2020]

IOP Conf. Series: Materials Science and Engineering **980** (2020) 012083 doi:10.1088/1757-899X/980/1/012083

A review: The use of mangrove for biomonitoring on aquatic environment

R Wilda, A M Hamdan* and R Rahmi
Department of Environmental Engineering, Faculty of Science and Technology,UIN Ar-Raniry Banda Aceh, Indonesia, 23111

Email: abd.mujahid.hamdan@gmail.com

Abstract. Mangroves have been investigated widely as plants that can absorb and accumulate heavy metals in their tissues. Due to that ability, mangroves have been used to reduce heavy metals in the aquatic environment. Furthermore, mangroves have been used for biomonitoring of heavy metals pollution. The use of mangroves for biomonitoring on aquatic environments has been considered as a cheap, rapid, and sufficient method. The mangrove is an organism that has the ability to absorb the contaminants of heavy metal and to function as fine pollutants trap. However, to develop this technique for wide use, research and investigation are still needed. This paper is aimed to describe the future direction of mangrove studies of using mangroves as a biomonitoring agent. Based on the review, *Rhizophopora mucronata* and *Avicennia marina* are mangrove species that have a promising ability to be used for biomonitoring in the aquatic environment.

1. Introduction

Industrialization and urbanization have increased the anthropogenic contribution of heavy metals in the biosphere [1-4]. Domestic waste is a major source that contributes to the pollution of the environment [5]. The effect of anthropogenic has altered metal mobility and diversity [6-7]. Human activities in industrial sector, such as mining industry, as well as domestic activities, have a significant contribution to the increasing of heavy metals both in the air and the soil [8-9]. According to [10] the aquatic environments of Indonesia are suspected in certain areas to be contaminated by organic contaminants such as oil, detergent, and organic materials in the form of domestic sewage from the auto industry. In addition, heavy metals are one type of contaminants that attracts attention globally. The heaviest metals contaminant encountered in the aquatic environtment [11,12,13]. Sources of heavy metal pollutants can be from mining operations [14] printing, electronic industry, waste [15] traffic activities, dust emissions, and agricultural activities [16]. The pollutions in the water not only can cause a decline in water quality, but also affect public and environmental health and environment health [17,18]. Therefore, the monitoring of heavy metals is necessary to provide a base controlling and maintaining environmental continuity.

Biomonitoring is an environmental monitoring technique that uses certain organisms or bioindicators that can provide information on the changes and quality of the environment [19-23]. Biomonitoring can be an excellent technique for monitoring surface water pollution such us rivers and ocean pollution [24]. Biomonitoring has been used to analyze various types of environments, both environments that change due to anthropogenic activities such as industry and housing, as well as environmental changes due to the influx of pollutants from natural sources [25,26].

In practice, the indicators used in biomonitoring can be animals, microorganisms, humans, and plants. One of the plants that are extensively used in biomonitoring is mangrove plants. Mangrove plants are plants that are commonly found in Indonesia such as in Sumatra, Java, Papua, Sulawesi, and Kalimantan. Mangroves are a community of plants that live in highly saline environments that are affected by tides [27]. The ecosystems of mangroves are intertidal ecosystem where are located between the marine and terrestrial environments in tropical and subtropical areas [28]. Mangrove forest ecosystems provide a variety of ecological benefits, including protection against flooding, prevention of coastline erosion, buffering salinity, and abundance of biodiversity [28].

Ecosystems of mangrove have the ability to tolerate heavy metal in their environment [29,30]. A number of studies have indicated that the plants are able to limit the mobility of contaminants in estuary environments [31]. One of the ecological functions of this plant is to resist toxic material [32,13] and to absorb heavy metals from the environment [33]. Therefore, mangroves have been used as bioremediation agents for heavy metals in the aquatic environment [34,35]. Because of its ability to accumulate heavy metals, the presence of heavy metals in their leaves can be a proxy of heavy metals abundance in the aquatic environment. The heavy metals that can be absorbed by mangroves including Cr, As, Al, Cd, Cu, Mn, Fe, Mo, Pb, Ni and Zn [36,37]. This paper aimed to review the mangroves' abilities as a heavy metal indicator in the aquatic environment.

2. Accumulated heavy metals in the mangroves folliage

The mangrove species differ in their ability to tolerate and absorb heavy metals from the environment into their folliage. The difference is caused by the dissimilarity of the root system of each species [33]. In certain conditions, mangroves will be stressed and they cannot grow properly [38]. For example, mangrove species such as *Excoecaria agallocha* have a higher toxicity in the absorption of Pb in the roots than in the leaf tissue [39]. Heavy metals such as Cu, Zn, Cd, and Hg generally indicate high bioconcentration factors at the root, while the concentration factor for the leaf is usually much lower than the other parts [40,38]. Meanwhile, the type of *Avicennia marina* in translating Pb to leaves is lower than to the roots [41].

Table 1 shows several reports that use mangrove plants to reduce heavy metal content in water. Although different species, mangrove plants have a function as pollutant traps in absorbing various heavy metal elements in their environment. *Avicennia marina* and *Rhizophora mucronata* are mangrove spesies that have an excellent capability to absorb heavy metals such as Hg, Mn, Zn, Cr, Cu, Cd and Pb [40-43]. *A. marina* is a mangrove that is mostly found in coastal areas. The root of *R. mucronata* is hanging and has oval-shaped leaves [44]. According to [45], the roots folliage are the higher tissues in accumulation of heavy metals, because the roots are in direct contact to the sediments. Meanwhile, the absorption ability of heavy metals in mangrove species is differentiated by their root system [33]. The root surface of *Rhizophora* is broader than the root surface of *Avicennia*, so the ability of *Rhizophora* tends to be greater than *Avicennia* [42].

3. Mechanism of mangroves in accumulating heavy metals

Heavy metals from industrial waste are one of the most destructive contaminants for aquatic ecosystems. Heavy metals introduced into the water will be precipitated to the sediment. The sedimentation process occurs due to the unbiodegradable character of heavy metals, then it will affect the life of aquatic biota such as shellfish, shrimp, and crabs [50]. Furthermore, they affect the food chain system. The heavy metals that transport into the aquatic environment are generally ionic. The heavy metals in the form of compounds such as oxides, sulfides, hydroxides, and carbonate compounds are dissolved easily in the water [37]. A high concentration of heavy metals both essential and non-essential can affect plant growth, even they can threaten plant growth [51]. Therefore, the plants have mechanisms to minimize metal toxicity [18,52-57].

Table 1. Mangroves species that have been used in biomonitoring of heavy metals in the aquatic environment.

Species	Metals	Concentration in mangrove plants Root (ppm)	Leaf (ppm)	Stem (ppm)	Fruit (ppm)	References
Avicennia marina	Cu	23.674	16.567	21.674		Handayani, 2006 [42]
	Cd	15.303	16.567			
	Zn	21.143				
Rhizphora mucronata	Cu	24.431				
	Cd	21.342				
	Zn	19.546				
Avicennia marina	Cu	7,17	7,76			Awaliya, 2018 [46]
	Pb	2	2,30			
Avicennia marina	Pb	2,19	3,54	5,89	1,71	Arisandy, 2012 [47]
Avicennia marina	Pb	0,0912				Wulandari, 2018 [33]
Rhizophora mucronata	Pb	0,0916				
Avicennia marina	Pb	105,60	110,81	66,70		Heriyanto, 2011[48]
	Cu	15,28	8,03	10,39		
	Hg	0,002	0,026	0,059		
Avicennia marina	Hg	13,83				Ali, 2012 [49]
	Pb	5,21				
Rhizophora mucronata	Hg	13,82	110,81	66,70		
	Pb	5,19	8,03	10,39		
Sonneratai alba	Pb	4,15	3,74			Khairuddin, 2018 [35]
	Cd	0,19	0,24			
Rhizophora Apicullata	Pb	1,85	3,21			
	Cd	0,18	0,41			

The presence of heavy metals has an impact on the aquatic plants, including mangroves plant. The heavy metals in the mangrove tissues are entered by the mechanism of transportation of heavy metals from the environment into plant organs [58]. The heavy metals that enter into the mangroves tissues are accumulated. The heavy metals are in the form of anion and cation that are through the root systems [59,56]. The absorption occurs through the root epidermis [43]. The process of root system in absorbing is called rhizofiltration. This is a process in which plant roots absorb and precipitate heavy metals [60,37]. According to Rohmawati [61] the accumulation of heavy metals into plant roots by molecular transport in the root membrane, then form a complex metal transport to the xylem. Furthermore, absorption occurs in the two processes, including (i) absorption of ions directly into a meristem cell, and (ii) absorption of ions in the leaves. The mechanism of heavy metal accumulation in mangroves can be seen in figure 1.

Figure 1 is the absorption process of heavy metals in mangroves. The mechanism of absorbing heavy metals depends on the root surface and nutrients [62]. The heavy metal transportation and absorption through plasma membrane and tissue of secondary carrier such as a channel protein or H^+ carrier protein, where membrane negative potential encourages cation uptake through secondary carrier [33]. Then, the cations are released into the xylem which is assisted by a protein-carrying membrane. Phytocelatin and metallothionein are transport tissues that have a fundamental role in translocating of heavy metals [33]. Phytocelatin is a group of proteins that contain amino acids such as cysteine, glycine, and glutamic acid. These proteins will induce plants if the plant is stressed by heavy metals [63,64]. Phytocrystalline binds the heavy metal ions, and then transport them to vacuoles. While metallothionein tissue space stored for excess heavy metal ions, then it also transports protein for excess heavy metals from cells to other cells. Factors that influence the transportation of organic substances in plants are: 1) pressure from the roots that push water upwards, 2) leaf transpiration, and 3) capillarity of xylem [65].

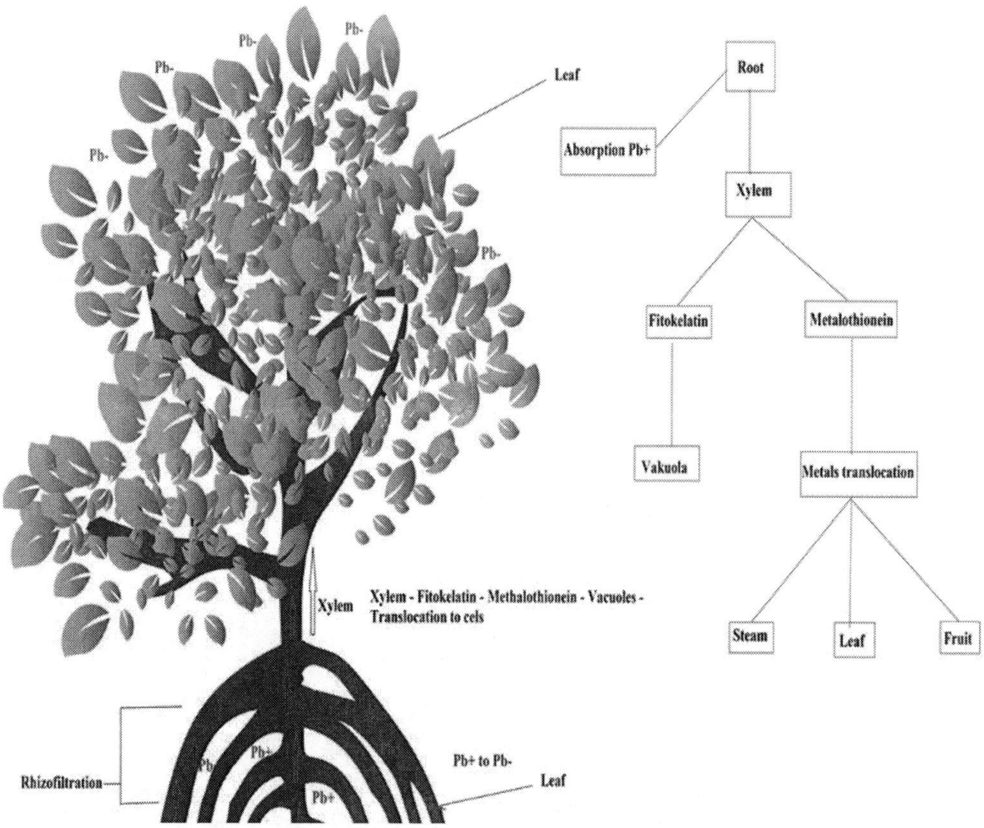

Figure 1. Mechanisms of heavy metals accumulations of mangroves plants.

The ability of the accumulation of heavy metals in mangrove plants can be determined by using bio concentration factor (BCF) and translocation factor (TF) [66,40]. BCF is used to determine the ability of plants to accumulate heavy metals in roots or leaves to the heavy metal concentrations within sediments [33]. The values > 1000 indicates high ability of acculation, the values of ≥ 250 are medium ability and the values of BCF <250 are low ability. The translocation factor can be used to measure the amount of heavy metal transferred from one organ to another [2, 67-50].

4. Potential use of mangroves

The mangrove forest ecosystem is one of the most prolific and unique [70]. Mangrove vegetation that grows in coastal waters is part of the coastal ecosystem which has the highest level of productivity compared to other coastal ecosystems. The presence of mangrove ecosystems in coastal waters became vital because mangrove vegetation has the ability to accumulate heavy metals and to reduce the concentration of pollutants in the water [71]. Mangrove are mainly in tropical and subtropical zones [72]. Indonesia has the largest mangrove forest in the world. The Indonesian mangroves forest covers more than 50% of Asia's and nearly 25% of the world's mangrove forests [73]. Indonesia has long coastline, ± 95,181 km where a portion of the coastal area is covered with mangrove forests [74]. The Indonesia mangrove forests are amazingly varied because of the remarkable diversity of physiological conditions on the coast of Indonesia [75,76].

In Indonesia, the mangrove ecosystem is one of the resources endangered in the coastal zone. Anthropogenic is a major contributor to the degradation of mangrove forests, such as agriculture that uses chemicals, industry, and mining [77]. In the last three decades nearly 50% of the total degradation rate of mangrove forests in Indonesia had been lost. About 6.7 million ha is left to be around 3.2 million ha [78]. Logging is one of the causes for the decreasing of mangrove population. Excessive logging has a significant effect on species diversity and natural wealth. Mangrove populations that are converted into functions such as land use as fish ponds, which then influence the concentration of heavy metals in mangrove sediments [79]. The use of mangrove plants in Indonesia is usually used for risk reduction to catastrophic events such as abrasion and a drop of the land as a result of increasing population around the coastal that could result in the rob flood. The root system of mangroves reduces soil erosion and helps to stabilize the nearby coastal landscape. Another role of the mangrove ecosystem could be developed as a medium to neutralize the heavy metals in its surroundings. The followings are the roles of mangrove forests that are very important for the environment and humans: (1) to protect coastal areas from distractions and to provide habitat for different animal species, (2) as pollinators and carbon storage [80] (3) as a barrier to waves and protect from coastal erosion, (4) as an upbringing as a place to feed biota and spawn areas for various kinds of aquatic biota, (5) as strategic places for people who like fishing and as place for ecotourism (6) economically mangrove plants can be used as construction materials, firewood, roofs, docks, traditional medicine and handicrafts [81] mangroves are plants that have ability as hyperacumulator.

Moreover, another role of the mangrove as a beach buffer against natural disasters such as tsunami of hurricane waves and extreme waves [82]. Mangroves do not only play an important role in ensuring the sustainability of coastal ecosystems, but also in providing important socio-economic benefits for communities around the coast [83]. Awareness of the use of mangrove waters is very influential on the state of the environment is polluted, and mangrove ecosystems that have various functions such as the function of ecological, social and economic [84]. Therefore, a monitoring of metal pollutants is necessary to provide a basis for controlling pollution, and biomonitoring is one way to tackle the problem of pollution since it reflects metal biocapability [85].

5. Conclusion

The use of mangroves for biomonitoring of heavy metals is very appropriate. Based on the review, *Rhizophopora mucronata* and *Avecennia marina* are mangrove species that have a promising ability to be used for biomonitoring in aquatic environment. The mangrove is an organism that has the ability to absorb the contaminants of heavy metal and to function as fine pollutants trap. The mangroves also play a major role in the prevention of disasters.

References

[1] Lasat M M 2002 Phytoextraction of toxic metals *Journal of Envi Quality,* **31(1)** 109

[2] Nagajyoti P C, Lee K D and Sreekanth T V M 2010 Heavy metals, occurrence and toxicity for plants: a review *Environt. Chem, letters* **8(3)** 199-216

[3] Wang S L, Xu X-R, Sun Y X, Liu J L and Li H B 2013 Heavy metal pollution in coastal areas of South China: A review Marine Pollution Bulletin **76(1-2)** 7–15 doi:10.1016/j.marpolbul.2013.08.025

[4] Harnani B R D 2017 Kemampuan *Avicennia marina* dan *Avicennia alba* untuk menurunkan konsentrasi tembaga (Cu) di muara sungai Wonorejo surabaya, (*Doctoral dissertation, Institut Teknologi Sepuluh Nopember*)
[5] Vijay R, Khobragade P and Mohapatra P K 2011 Assessment of groundwater quality in Puri City, India: an impact of anthropogenic activities *Environt. monitoring and assessment*, **177(1-4)** 409-18
[6] He B Li R Chai M and Qiu G 2013 Threat of heavy metal contamination in eight mangrove plants from the futian mangrove forest, China *Environt Geoc and Health* **36(3)** 467–76
[7] Dewi P K, Hastuti E D and Budihastuti R 2018 Kemampuan akumulasi logam berat tembaga (Cu) pada akar mangrove jenis avicennia marina (forsk.) dan rhizophora mucronata (lamk.) di lahan tambak *J. Akademika, Biologi* **7(4)** 14-9
[8] Lisa N 2013 Skripsi profil penyebaran logam berat di sekitar TPA pakusari Jember
[9] Irhamni I 2017 Serapan logam berat esensial dan non esensial pada air lindi TPA kota Banda Aceh dalam mewujudkan pembangunan berkelanjutan *J. Serambi, Engineering* **2(1)**
[10] Makkasau A M, Sjahrul N, Jalaluddin and I Raya 2011 Teknik fitoremediasi fitoplankton suatu alternatif pemulihan lingkungan laut yang tercemar ion logam Cd2+ dan Cr6+ Pendidikan Guru Sekolah Dasar **7(2)** 155-68
[11] Budiastuti P M, Raharjo and N A Y Dewanti 2016 Analisis pencemaran logam berat timbal di perairan babon kecamatan genuk semarang *J. Kesehatan Masyarakat (e-Journal),* **4 (5)** 119-25
[12] Sun Z, Chen J, Wang X and Lv C 2016 Heavy metal accumulation in native plants at a metallurgy waste site in rural areas of Northern China *Ecological. Engineering*, **86** 60-8
[13] Fadillah S 2017 Cemaran logam berat timbal (Pb) pada daging ikan cendro (tylosurus crocodilus) di pesisir krueng raya kabupaten aceh besar (contamination of heavy metal lead (Pb) in flesh of needlefish (tylosurus crocodilus) at Krueng Raya Aceh Besar) *J. Ilmiah Mahasiswa Veteriner*, **1(3)** 391-7
[14] Herawati N, Suzuki S, Hayashi K, Rivai IF and Koyoma H 2000 Cadmium, copper and zinc levels in rice and soil pf japan , Indonesia and China by soil type *Bull environ contam toxicol* **64 (1)** 33-9
[15] Rainbow P S and Luoma SN 2011 Metal tocity, uptake and bioaccumulation inaquatic invertebrates-modelling zin incrustaceans Aquat. Toxicol. **105 (3-4)** 455-65
[16] Caregnato F F, Koller C E, MacFarlane G R and Moreira J C 2008 The glutathione antioxidant system as a biomarker suite for the assessment of heavy metal exposure and effect in the grey mangrove, Avicennia marina (Forsk.) Vierh *Marine Pollution Bulletin*, **56(6)** 1119-27
[17] Rondonuwu S B 2014 Fitoremediasi limbah merkuri menggunakan tanaman dan sistem reaktor *J. Ilmiah Sains*, **14(1)** 52-9
[18] Jupriyati Ruri,Nirwani Soenardjo and Chrisna Adhi Suryono 2013 Akumulasi logam berat timbal (Pb) dan pengaruhnya terhadap histologi akar mangrove avicennia marina (forssk). Vierh. Di perairan Mangunharjo Semarang *J. Of Marine Research*, **3 (1)** 61-8
[19] Day J 2000 Biomonitoring: Appropriate technology for the 21st Century. 1 st WARFSA, **1(2)** 1-8
[20] Cairns J 2005 The crucial link between natural systems and society Mankind quarterly, **45(3)**, 289-308
[21] Komarawidjaja Wage and Titiresmi 2006 Teknik biomonitoring-sebagai alternatif "tool" pemantauan kualitas lingkungan perairan. Jakarta Teknologi Lingkungan, 144-47

[22] Aina L C S D E R and Kaswinarni F 2016 Biomonitoring pencemaran sungai Silugonggo kecamatan juwana berdasarkan kandungan logam berat (Pb) pada ikan lundu. Bioma *J. Ilmiah biologi*, **5(2)**

[23] Riswandi A, Mahmudi M, Kurniawan A and Salamah L N 2019 Biofilm application as biomonitoring agent in heavy metals Pb2+ and Cr6+ in ngimboh coastal, ujung pangkah, Gresik *J. Of wetlands environmental management,* **7(2)** 134-9

[24] Dewi E S, Ni'mah K and Kaswinarni F 2019 The content of heavy metal lead (Pb) on baung fish (Hemibagrus nemurus) as biomonitoring pollution of Wulan River of Demak Regency *J. Of Physics: Conference Series,* **1217(1) p 012128**

[25] Owoade O K, Olise F S, Obioh I B, Olaniyi H B, Ferrero L and Bolzacchini E 2009. EDXRF elemental assay of airborne particulates: a case study of an iron and steel industry, Lagos, Nigeria *Scientific Research and Essay,* **4(11)** 1342–47

[26] Taiwo A M, Beddows D C S, Calzolai G, Harrison R M, Lucarelli F, Nava S and Vecchi R 2014 Receptor modelling of airborne particulate matter in the vicinity of a major steelworks site *Science of the total environment*, **490** 488-500

[27] Panjaitan G C 2009 Akumulasi logam berat tembaga (Cu) dan timbal (Pb) pada pohon avicennia marina di hutan mangrove. *Skripsi.* Jurusan budaya hutan. Universitas Sumatera Utara. Medan.

[28] Lewis M, Pryor R and Wilking L 2011 Fate and effects of anthropogenic chemicals in mangrove ecosystems: A review *Environt. Pollution*, **159** 2328–46

[29] Nouri H, Borujeni S C, Nirola R, Hassanli A, Beecham S, Alaghmand S and Mulcahy D 2017 Application of green remediation on soil salinity treatment: a review on halophytoremediation *Processsafety and environmental protection*, **107** 94-107

[30] Liang Y, Sun W, Zhu YG and Christie P 2007 Mechanisms of siliconmediated alleviation of abiotic stresses in higher plants: a review *Environ. Pollut*, **147(2)** 422–8

[31] Chaudhuri P, Nath B and Birch G 2014 Accumulation of trace metals in grey mangrove Avicennia marina fine nutritive roots: The role of rhizosphere processes *Marine Pollution Bulletin*, **79(1-2)** 284–92

[32] Setiawan H 2013 Akumulasi dan distribusi logam berat pada begetasi mangrove di pesisir Sulawesi selatan *J. Ilmu kehutanan*, **7(1)** 12-24

[33] Wulandari T, Hastuti R B and Hastuti E D 2018 Kemampuan akumulasi timbal (Pb) pada akar mangrove jenis avicennia marina (forsk.) Dan rhizophora mucronata (lamk.) Di lahan Tambak Mangunharjo Semarang *J. Akademika Biologi*, **7(1)** 89-96

[34] Nath B, Birch G and Chaudhuri P 2014 Assessment of sediment quality in Avicennia marina-dominated embayments of Sydney estuary: the potential use of pneumatophores (aerial roots) as a bio-indicator of trace metal contamination *Science of the total environment*, **472** 1010-22

[35] Khairuddin M Y and Syukur A 2019 Analisis kandungan logam berat pada tumbuhan mangrove *J. Biologi Tropis*, **18(1)** 69-79

[36] Almahasheer H, Serrano O, Duarte C M and Irigoien X 2018 Remobilization of heavy metals by mangrove leaves *Frontiers in Marine Science*, **5** 484

[37] Utami R, Rismawati W and Sapanli K 2018 Pemanfaatan mangrove untuk mengurangi logam berat di perairan In *seminar nasional hari air sedunia*, **1(1)** 141-53

[38] Yan Z, Sun X, Xu Y, Zhang Q and Li X 2017 Accumulation and tolerance of mangroves to heavy metals: a review *Current Pollution Reports*, **3(4)** 302–17

[39] Yan Z and Tam N F Y 2013 Differences in lead tolerance between Kandelia obovata and Acanthus ilicifolius seedlings under varying treatment times *Aquatic toxicology*, **126** 154-62
[40] MacFarlane G R, Pulkownik and M D Burchett 2007 Accumulation and distribution of heavy metals in grey mangrove, avicennia marina (forsk) vierh: biological indication potential *Environt. Pollution*, **123** 139-51
[41] MacFarlane G R and M D Burchett 2001 Photosynthetic pigments and peroxides activity as indicators of heavy metal stress in the grey mangrove avicennia marina (forsk.) Veirh *Marine Pollution Bulletin*, **42** 233-40
[42] Handayani T 2006 Bioakumulasi logam berat dalam mangrove rhizopora mucronata dan avicennia marina di muara angke Jakarta *J. Teknologi Lingkungan*, **7(3)**
[43] Subiandono E, Bismark M and Heriyanto N M 2013 Kemampuan avicennia marina (forsk.) Vierh. Dan rhizophora apiculata bl. Dalam penyerapan polutan logam berat *J. Penelitian Hutan dan Konservasi Alam*, **10(1)** 93-102
[44] Duke N C 2007 Rhizophora apiculata, R. Mucronata, R. Stylosa, R. annamalai, R.Lamarckii (indo– west pacific stilt mangrove) *Permanent Agriculture Resources*, **2 (1)**
[45] Kusumastuti W 2009 Evaluasi lahan basah bervegetasi mangrove dalam mengurangi pencemaran lingkungan: studi kasus di desa kepetingan kabupaten sidoarjo. Thesis (tidak dipublikasikan) Universitas Diponegoro
[46] Awaliyah H F, Yona D and Pratiwi D C 2018 Akumulasi logam berat (pb dan cu) pada akar dan daun mangrove avicennia marina di sungai lamong, jawa timur *DEPIK J. Ilmu-Ilmu Perairan, Pesisir dan Perikanan*, **7(3)** 187-97
[47] Arisandy K R, Herawati E Y and Suprayitno E 2012 Akumulasi logam berat timbal (Pb) dan gambaran histologi pada jaringan avicennia marina (forsk.) Vierh di perairan pantai jawa timur *J. Penelitian Perikanan*, **1(1)** 15-25
[48] Heriyanto N M 2017 Kandungan logam berat pada tumbuhan tanah, air ikan dan udang di hutan mangrove *J. Penelitian hutan tanaman*, **8(4)** 197-205
[49] Ali Munawar and Rina 2012 Kemampuan tanaman mangrove untuk menyerap logam berat merkuri (hg) dan timbal (pb). Universitas pembangunan nasional "veteran" jawa timur *J. Ilmiah Teknik Lingkungan*, **2(2)**
[50] Yalcin G, Narin I and Soylak M 2008 Multivariate analysis of heavy metal contents of sediments from gumusler creek, nigde, turkey *Environt. Geology*, **54** 1155-63
[51] Hall J L 2002 Cellular mechanisms for heavy metal detoxification and tolerance *J. Of experimental botany*, **53(366)** 1-11
[52] Lindsey HD, James MM and Hector MG 2004 An assessment of metal contamination in mangrove sediments and leaves from punta mala bay, pacific panama *Marine Pollution Bulletin*, **50** 547-52
[53] Dalvi A A and Bhalerao S A 2013 Response of plants towards heavy metal toxicity: an overview of avoidance, tolerance and uptake mechanism *Ann Plant Sci*, *2*(**9**) 362-8
[54] Siahaan M T A, Ambariyanto and Yulianto B 2013 Pengaruh pemberian timbal (Pb) dengan konsentrasi berbeda terhadap klorofil, kandungan timbal pada akar dan daun, serta struktur histologi jaringan akar anakan mangrove Rhizophora sp. Mucronata *J. Of Marine Research*, **2(2)** 111-9

[55] Hamzah F and Pancawati Y 2013 Fitoremidiasi logam berat dengan menggunakan mangrove (phytoremiditation of heavy metals using mangroves) *ILMU KELAUTAN. Indonesian J. Of Marine Sciences*, **18(4)** 203-12
[56] Sruthi P, Shackira A M and Puthur J T 2017 Heavy metal detoxification mechanisms in halophytes: an overview *Wetlands ecology and management*, **25(2)** 129-48
[57] Mulyadi E, Laksmono R and Aprianti D 2017 Fungsi mangrove sebagai pengendali pencemar logam berat *J. Ilmiah teknik lingkungan*, **1** (Edisi Khusus): 24- 35
[58] Kurniawan A and Mustikasari D 2019 Review: Mekanisme akumulasi logam berat di ekosistem pascatambang timah *J. ilmu lingkungan*, **17(3)** 408-15
[59] Machado W, Gueiros B B, Lisboa-Filho S D and Lacerda L D 2005 Trace metals in mangrove seedlings: role of iron plaque formation *Wetlands Ecology and Management*, **13(2)** 199-206
[60] Irawanto R 2014 Konsentrasi logam berat (Pb dan Cd) pada bagian tumbuhan akuatik Coix lacryma-jobi (Jali). In *Seminar Nasional Konservasi dan Pemanfaatan Sumber Daya Alam* 2015. Sebelas Maret University
[61] Rohmawati 2007 Uji daya akumulasi tumbuhan *avicennia marina* terhadap logam berat (Cu, Cd Dan Hg) di pantai kenjeran surabaya. Skripsi tidak diterbitkan. Malang: Jurusan biologi fakultas sains dan teknologi universitas islam negeri malang
[62] Hidayanti N 2013 Mekanisme Fisiologis Tumbuhan Hiperakumulator Logam Berat= Heavy Metal Hyperaccumulator Plant Physiology Mechanism *J. Teknologi Lingkungan*, **14(2)** 75-82
[63] Prasad M B K, Ramanathan A L, Shrivastav S K and Saxena R 2006 Metal fractionation studies in surfacial and core sediments in the Achankovil river basin in India *Environmental monitoring and assessment*, **121(1-3)** 77-102
[64] Priyanto B and Prayitno J 2006 Fitoremediasi sebagai sebuah teknologi pemulihan pencemaran, khususnya logam
[65] Rungkut W, Timur S J and Febriana E 2017 Kandungan logam berat timbal (Pb) pada akar dan daun mangrove avicennia marina (forsk.) Di kawasan mangrove
[66] Zhang W, Cai Y, Tu C and Ma L Q 2002 Arsenic speciation and distribution in an arsenic hyperaccumulating plant *Science of The Total Environment*, **300(1-3)** 167
[67] Suteja Y and Dirgayusa I G N P 2018 Bioaccumulation and translocation of chromium on crabs and mangroves in Mati River estuary, Bali, Indonesia. *Aquaculture, Aquarium, Conservation and Legislation*, **11(2)** 469-75
[68] Takarina N D and Pin T G 2017 Bioconcentration factor (BCF) and translocation factor (TF) of heavy metals in mangrove trees of Blanakan fish farm *Makara Journal of Science*, **21(2)** 4
[69] Abohassan R A 2013 Heavy metal pollution in Avicennia marina mangrove systems on the Red Sea coast of Saudi Arabia *J. Of King Abdulaziz University: Metrology, Environment and Arid Land Agricultural Sciences*, **142(579)** 1-38
[70] Kaewtubtim, P., Meeinkuirt, W., Seepom, S and Pichtel, J 2016. Heavy metal phytoremediation potential of plant species in a mangrove ecosystem in Pattani Bay, Thailand. *Applied Ecology and Environmental Research*, **14(1)** 367-82
[71] Hamzah F and Setiawan A 2010 Akumulasi Logam Berat Pb, Cu, dan Zn di Hutan Mangrove Muara Angke, Jakarta Utara *J. Ilmu dan Teknologi Kelautan Tropis*, **2(2)** 41-52
[72] Moreira I T A, Oliveira O M C, Triguis J A, Queiroz A F S, Ferreira S L C, Martin C M S, Silva A C M and Falcao B A 2013 Phytoremediation in mangrove sediments impacted by persistent

total petroleum hydrocarbons (TPH's) using Avicennia schaueriana. *Marine Pollution Bulletin,* **67 (1-2)** 130-6

[73] Onrizal O 2010 Perubahan tutupan hutan mangrove di pantai timur sumatera utara periode 1977-2006 *J. Biologi Indonesia*, **6(2)** 163-72

[74] Kusmana C 2014 Distribution and current status of mangrove forests in Indonesia. In *Mangrove ecosystems of Asia* (pp. 37-60). Springer, New York, NY

[75] Eddy S, Iskandar I, Ridho M R and Mulyana A 2017 Dampak aktivitas antropogenik terhadap degradasi hutan mangrove di Indonesia

[76] Polidoro BA, Carpenter KE, Collins L, Duke NC, Ellison AM, Ellison JC, Farnsworth EJ, Fernando ES, Kathiresan K, Koedam NE, Livingstone SR, Miyagi T, Moore GE, Nam VN, Ong JE, Primavera JH, Salmo SG, Sanciangco JC, Sukardjo S, Wang Y and Yong JWH 2010 The loss of species: mangrove extinction risks and geographic areas of global concern. PLoS One 5(4): e10095.**doi:10.1371**/*journal.pone*.0010095

[77] Ilman M, Wibisono I T C and Suryadiputra I N M 2011 State of the Art Information on Mangrove Ecosystems in Indonesia. Wetlands International-Indonesia Programme. Bogor

[78] Fitri R Y and Anwar K 2014 Kebijakan pemerintah terhadap pelestarian hutan mangrove di kecamatan tebing tinggi kabupaten bengkalis. Jom FISIP, **1(2)** 1-15

[79] Ngole-Jeme V M, Fonge B A, Tabot P T and Mumbang C 2016. Impact of logging activities in a tropical mangrove on ecosystem diversity and sediment heavy metal concentrations *J. Of Coastal Conservation*, **20 (3)** 245-55

[80] Purnobasuki H 2012 Pemanfaatan hutan mangrove sebagai penyimpan karbon. *Buletin PSL Universitas Surabaya*, **28(3-5)** 1-6

[81] Saprudin and Halidah 2012 Potensi dan nilai manfaat jasa lingkungan hutan mangrove di kabupaten sinjai sulawesi selatan *Jurnal. Penelitian Hutan dan Konservasi Alam* , **9(3)** 213-9

[82] Amma P K G and Bhaskaran P K 2020 Role of mangroves in wind-wave climate modeling. A review. *Of Coastal Conservation*, **24(2) doi:10.1007**/s11852

[83] Suratman M N 2008 Carbon sequestration potential of mangroves in southeast asia. *Managing Forest Ecosystems*, 297–315 **doi:10.1007**/978-1

[84] Rizal A, Sahidin A and Herawati H 2018 Economic value estimation of mangrove ecosystems in Indonesia. *Biodiversity International Journal*, *2*(**1)** 98-100

[85] Tien C J and Chen C S 2013 Patterns of metal accumulation by natural river biofilms during their growth and seasonal succession. *Archives of environmental contamination and toxicology*, **64(4)** 605-16

Biodegradation of buried crude oil in soil microcosm by fungal co-culture

D H Y Yanto[1]* and A Hidayat[2]

[1]Research Center for Biomaterials, Indonesian Institute of Sciences (LIPI). Jl. Raya Bogor Km. 46 Cibinong Bogor 16911, Indonesia.
[2]Forest Research and Development Center, Research, Development and Innovation Agency, Ministry of Environmental and Forestry. Jl. Gunung Batu No.5, Bogor 16610, Indonesia

Email: dede@biomaterial.lipi.go.id

Abstract. Crude oil pollution in soil is a major environmental problem over the world. After long-term contamination, crude oil can be buried in the soil and leaving the toxic and persistence compounds. Treatment of buried crude oil in soil is a major challenge. In this study, buried crude oil in soil (20 cm-depth) was treated by the co-culture of three fungal strains (*Pestalotiopsis* sp.NG007/*Polyporus* sp. S133/*Trametes hirsuta* D7 with composition 1/1/1 and pre-grown in wood meal) placed on the top or bottom layer of the soil. Biodegradation behavior of crude oils were analyzed at different depth interval (0–5 cm, 5–15 cm, and 15–20 cm) with constant oxygen diffusion. The results showed that both position of fungal co-culture could degrade crude oil 63 – 92% after 30 days. Placing the co-culture in the bottom layer position of the soil demonstrated more effective degradation for all the depth interval than that of the top layer position. During the biodegradation, the co-culture produced oxidative enzymes such as catechol 1,2-dioxygenase, catechol 2,3-dioxygenase, laccase, manganese peroxidase, and lignin peroxidase with significant activities. This study offers an important strategy to remediate crude oil contaminated soil, specifically buried crude oil.

1. Introduction

Exploitation of crude oil continues to grow and plays a significant role in the modern industry. Some protocols and regulations have governed the use of crude oil, but unconsciously crude oil becomes a source of contaminants in the environment and ecosystems around the world [1,2]. The average global crude oil spread is around 9.1 million tons per year and is predicted will continue to grow [3]. Contamination of crude oil can occur in aquatic and terrestrial environments. Oil spills cases in Indonesia were recorded in 1999, about 4000 barrels of oils had contaminated sea in Cilacap - Central Java; in 2009 occurred in Manyar, Gresik - East Java; in 2014 occurred in Indramayu, West Java (along 7 km of coastline was polluted by crude oil) and 200 barrels of crude oil contaminated the swamp river in Kab. Siak – Riau.

Crude oils that contaminated the aquatic and terrestrial environments exhibit unique characteristics. In order to achieve an appropriate target in the bioremediation of crude oil, understanding these characteristics is important. Since the crude oil contaminated soil, spilled oil is generally slow-moving or static. The oil collects in depressions or against natural and man-made barriers, usually, the size of the affected area is small and it is easy to define the location and amount of surface oil, only light oils spread to form a thin layer; often considerable pooling of oil weathering slows considerably after approximately 24 hours [4]. Besides, soil characterization, enzymes involved during biodegradation, and metabolites are major factors in the bioremediation of crude oils [5].

Degradation of crude oil in contaminated soil has been reported widely. Strain *Polyporus* sp. S133 degrades by 26% crude oil 15000 ppm in soil [6]. *Bacillus* sp. degrade crude oil (80-89%) from the initial concentration of 5000 ppm for 5 days [7]. *Etreptomyces* sp. ERI-CPDA-1 has the ability to extract diesel oil by 98% from an initial concentration of 1000 ppm for 7 days [8]. *Dietzia cinnamea*, *Hoyosella altamirensis* and *Vibrio alginolyticus* bacteria isolated from the Gulf of Kuwait are able to degrade crude oil by 39% (initial concentration of 1000 ppm) for 14 days [9]. However, research on degradation of buried crude oil in soil is still limited. After long-term contamination, crude oil can be buried in the soil and leaving toxic and persistence compounds. Treatment of buried crude oil in the soil is a major challenge. Therefore, degradation of buried crude oil in soil (20 cm-depth) is important to be investigated. This study uses the co-culture of three fungal strains placed on the top or bottom layer of the soil to degrade crude oil. Biodegradation behavior was analyzed at different depth intervals of the soil.

2. Materials and methods

Crude oil was obtained from Taiyo Petroleum Co., Ltd. (Japan). Agar, malt extract, glucose, silica gel C-200, polypeptone, and other solvents were purchased from Wako Pure Chemical Industries, Ltd. (Japan). *Pestalotiopsis* sp. NG007 and *Polyporus* sp. S133 were previously reported to have the ability to degrade crude oil [10–12]. *T. hirsuta* D7, a basidiomycete, was newly isolated from a fungal body that grew on rotten wood near the River Pakning in the peat swamp forest Bengkalis, Riau, Indonesia [13]. All three fungi were cultured on malt extract agar (MEA) medium containing 20 g L^{-1} malt extract, 20 g L^{-1} glucose, 1 g L^{-1} polypeptone, and 20 g L^{-1} agar.

2.1. *Pre-culture in liquid medium*

Pestalotiopsis sp. NG007, *Polyporus* sp. S133, and *T. hirsuta* D7 were individually pre-cultured in 20 mL malt extract (ME) liquid medium containing 20 g L^{-1} ME, 20 g L^{-1} glucose, and 1 g L^{-1} polypeptone. The pH of the medium was adjusted to 4.5. Strains were pre-cultured under dark conditions at 25°C for 7 d. Each culture was homogenized at 5000 rpm for 10 min before being applied to soil artificially contaminated with crude oils.

2.2. *Pre-grown in wood meal*

The wood meal was obtained from Daigo Mokuzai Co., Ltd. (Osaka, Japan). Several 5-mm disks of an actively growing strain: NG007, S133, or D7 in MEA medium as described in section 2.2, were transferred to 3-h autoclaved wood meal (100 g) combined with 10% (w/w) glucose, 15% (w/w) shitake nutrient (a kind of nutrient for fungal growth purchased from Showa Sangyo Co. Ltd., Tokyo, Japan), and 60% (v/w) distilled water, and then incubated for approximately one month. Pre-grown cultures were maintained in a cool room at 4°C prior to use.

2.3. *Penetration test of individual fungus*

Fungi NG007, S133, and D7 were individually placed on the top or bottom layer of soil (containing 10% w/w glucose and 60% v/w distilled Water). The ability of the fungus to penetrate the soil with 20 cm depth was investigated every day for 17 days.

2.4. *Experimental design*

Soil consisting of 10% w/w glucose and 60% v/w distilled water (non-slurry soil) was artificially contaminated by crude oil to a final concentration of 1000 mg kg^{-1} soil. The artificially contaminated soil was mixed until it was homogenous then was placed in cylinder flask 2 L with 20 cm-depth. Pre-grown fungi NG007, S133, and D7 were mixed and added to soil with composition 1/1/1 and final total concentration of 10% w/w of the artificially contaminated soil. In this experiment, the fungal co-culture was placed on the top or bottom layer of soil. During incubation, oxygen diffusion was applied with constant flowrate. The soil was divided into 3 depth intervals (0–5 cm), (5–15 cm), and (15–20 cm), respectively from the top layer of soil. Degradation of crude oil was investigated after 15 and 30-day incubation in each interval. Abiotic control was performed with no addition of fungal co-culture into soil mixture containing 1000 mg kg^{-1} crude oil.

2.5. *Chemical analysis of remaining PHCs*
Description of the biodegradation analysis of crude oils and their fractions (aliphatic, aromatic, resin, and asphaltene) was provided in our previous study [10]. The crude oil that remained in each depth interval of the soil mixture was consecutively extracted (taken about 30 gram from each depth interval) using 100 ml *n*-hexane, 100 ml dichloromethane, and 100 ml chloroform. After evaporation, the total biodegradation of crude oil was analyzed using the gravimetric method. *n*-Hexane was used to separate asphaltene from the PHCs. The n-hexane-insoluble fraction that remained in the filter paper after filtration mostly consisted of asphaltene. The weight of asphaltene was analyzed by the gravimetric method. n-Hexane-soluble fractions were further separated to obtain aliphatic, aromatic, and resin fractions using purification on a silica gel (C-200) column eluted with 100 ml of n-hexane, 100 ml of toluene, and 200 ml of chloroform: methanol (1:1 (v/v)), respectively. The aliphatic and aromatic fractions were subjected to gas chromatography (GC-FID Shimadzu 2014) with a TC-5 capillary column (30 m, id x 0.25 mm x 0.25 μm). The carrier gas, helium, was delivered at a constant flow rate of 1.5 ml min^{-1} with a column pressure of 100 kPa and an interface temperature of 280°C. The column temperature was started at 60°C, increased at 10°C min^{-1} to 280°C, and then maintained at 280°C for 10 min. The injection volume was 2 μl and the injector temperature was maintained at 280°C. Aliphatic and aromatic biodegradation was calculated in the post-run analysis with a calculated minimum area of 100 using the following equation:

$$\boldsymbol{B} = \frac{\sum \boldsymbol{A0} - \sum \boldsymbol{A1}}{\sum \boldsymbol{A0}} \times \mathbf{100\%} \qquad (1)$$

where *B* (%) is the biodegradation of the aliphatics or aromatics; $\sum \boldsymbol{A0}$ is the total area of the aliphatic or aromatic fraction from the abiotic control; $\sum \boldsymbol{A1}$ is the total area of the aliphatic or aromatic fraction after treatments.

2.6. *Enzymatic assay*
The enzymatic activities of catechol 1,2-dioxygenase (C12O), catechol 2,3-dioxygenase (C23O), laccase, manganese peroxidase (MnP), and lignin peroxidase (LiP) were assayed using a UV-Vis spectrophotometer (Shimadzu UV-1600, Japan). C12O activity was assayed by the method described by Nakazawa and Atsushi [14]. C23O activity was assayed as described by Nozaki [15]. Laccase activity was measured as described by Leonowicz and Grzywnowics [16]. MnP activity was determined based on the oxidation of 2,6-dimethoxyphenol (2,6-DMP) by the MnP system to form a quinine dimer [17]. LiP activity was determined by the oxidation of veratryl alcohol to form veratryl aldehyde at 310 nm [18]. All experiments were performed at 25°C. Activities were expressed as units per gram wet soil matter (U g^{-1}), in which one enzyme unit (U) was defined as that forming 1.0 μmol of the product per minute under the assay conditions.

3. Results and discussion

3.1. *Composition of crude oil*
Crude oil used in this study was composed 49.3% aliphatic, 34.0% aromatic, 8% resin, and 8.67% asphaltene with the API gravity number 17.9 [10]. The aliphatic fraction contains C12 to C30 with the

majority of middle carbon numbers and the aromatic fraction contains several polyaromatic hydrocarbons (PAHs) such as anthracene (ANT) and phenanthrene (PHE) [12]. Based on the API gravity number, the crude oil sample was categorized as heavy crude oil.

3.2. Penetration of fungal growth

Penetration of individual fungal culture NG007, S133 and D7, respectively were investigated in 20 cm-depth of soil contaminated with crude oil. Each culture was placed in the top or bottom layer of the soil. As shown in figures 1.a and 1.b, all the fungi placed in the bottom layer grow faster than that of in the top layer. The penetration of mycelial fungi in the soil was influenced by the type of fungi, moisture content of the soil and type and percentage of bulking agents [19]. This study found that placement of the fungi also affected the ability of fungi to penetrate the soil. For individual fungal strain, NG007 demonstrated a faster growth compared with strain D7 and S133, respectively. Strain NG007 was only able to reach a depth of downward penetration as far as 14.7 cm after 17 days of incubation when it was placed in the top position. However, when it was placed in the bottom position, strain NG007 was able to cover the entire upper layer in 5 days. These phenomena were also found in strain D7 and S133 which was able to grow until the upper layer in 13 and 15 days respectively when they were placed in bottom position figure 1.b. For fungal co-culture, the growing penetration rate detected in the soil was lower than that of all the single strain. The fungal co-culture covered all the soil within 15 and 25 days when they were placed in the top or bottom position, respectively. This might be caused by the time-need interaction of all the fungal strains in the co-culture at the initial stage of growth.

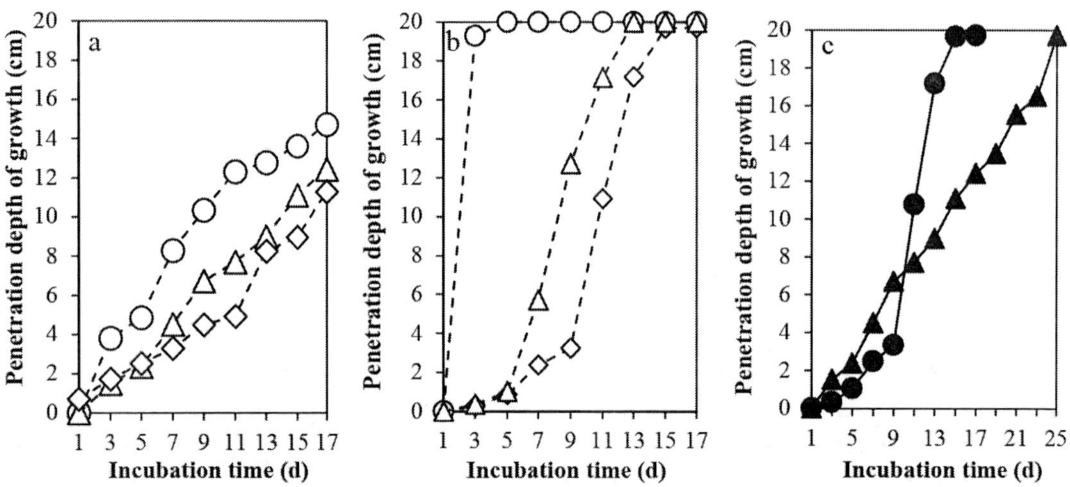

Figure 1. Penetration depth of individual fungal growth placing in the top layer (a) or bottom layer (b), and mixture of fungal co-culture growth (c). Symbol in Figure 1.a and 1.b (*Pestalotiopsis* sp. NG007 (○), *T. hirsuta* D7 (△), and *Polyporus* sp. S133 (◇)). Symbol in Figure 1.c (top layer (●) and bottom layer (▲))

3.3. Biodegradation of crude oil

Long-term contamination by crude oil allows the components of crude oil to settle in the soil for a long time. Some fractions like aliphatic and aromatic may move to the bottom layer easily than the higher fractions (resin and asphaltene). This will lead to the increasing concentration of aliphatic and aromatic fractions in the bottom layer. Aliphatic and aromatic are generally more susceptible to biodegradation than resin and asphaltene. Therefore, the degradation results for TPH were generally lower from top to bottom (figure 2.a). In addition, placing the fungal co-culture in the bottom position made the biodegradation enhanced in all the layer because of the higher activities of enzyme produced.

The study found that the ability of fungi to degrade crude oil was influenced by the distance of the crude oil sample from the fungi and the growth ability of the fungi to penetrate the soil. The efficiency of degradation of crude oil in a distance 20 cm from the initial fungi decreased 9.8 and 26.8% when the fungi were placed in the bottom and top of the soil, respectively. Besides, because of the oxygen-dependent growth, some fungi preferred to grow in an upward direction than to penetrate down the soil.

The fungal co-culture degraded 63–92% crude oil in the soil for 30 days. All the fractions in crude oil could be degraded by the fungal co-culture with a degradation rate 42–90% for aliphatic fraction, 86–98% for aromatic fraction, 85–100% of asphaltene fraction, and 55–93% of resin fraction, respectively. Strain NG007, D7, and S133 had been reported to be able to degrade PAHs and crude oils [10]. A combination of all these fungi has been reported had a beneficial effect in total petroleum hydrocarbons (TPHs) degradation. This study supports previous results that the fungal co-culture is also able to degrade buried crude oil in the soil to a depth of 20 cm.

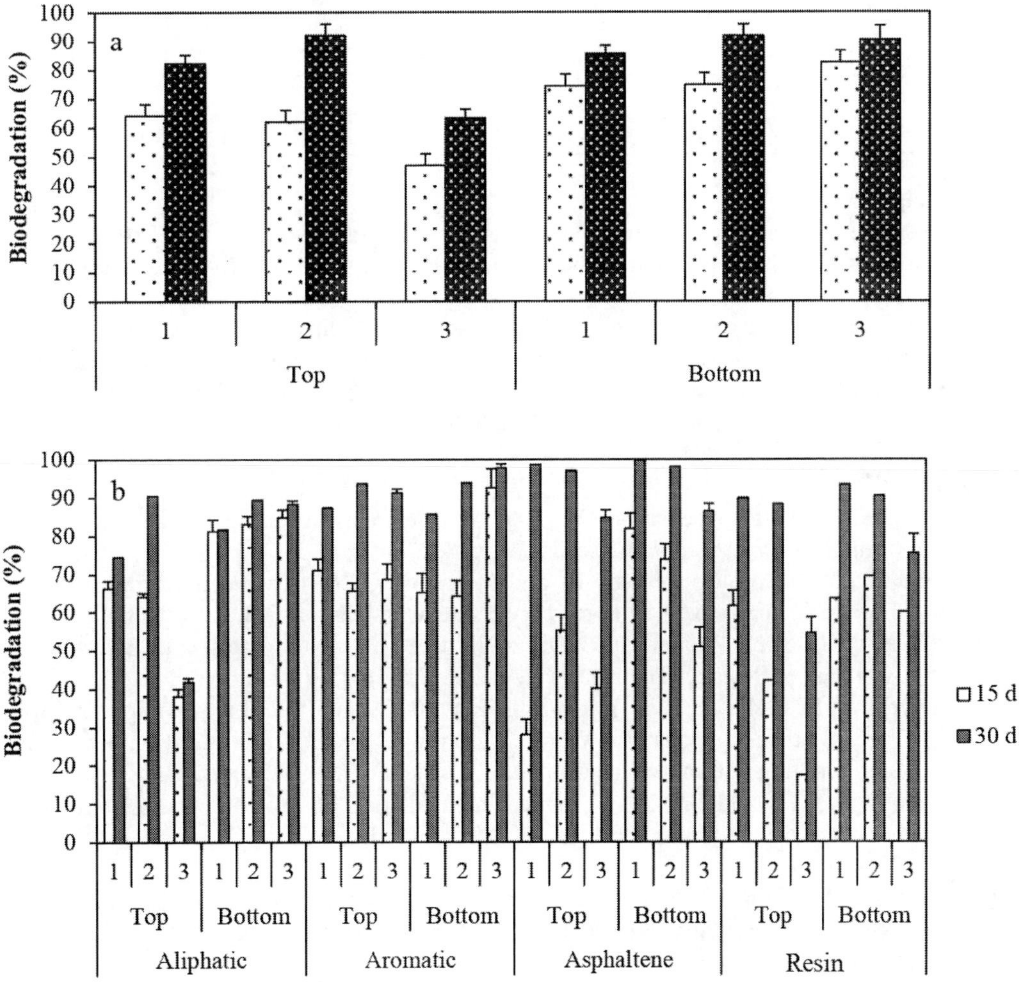

Figure 2. TPH degradation (a) and fractions degradation (b) of crude oil by mix culture of NG007/D7/S133 1/1/1 in soil for 15 and 30 days at different interval of depth 0–5 cm (1), 5–15 cm (2), and 15–20 cm (3).

3.4. Enzyme activities

Crude oil is a complex mixture that contains aliphatic, aromatic, resin and asphaltene compounds. The co-culture NG007/S133/D7 1/1/1 had been reported able to degrade crude oil by the combination of enzymes produced by the co-culture. In this study, during the incubation of the co-culture in soil buried crude oil, several oxidative enzyme such as C12O, C23O, lac, MnP and LiP were detected (table 1). Dioxygenase enzymes such C12O and C23O had been reported initiating the degradation of aliphatic fractions in crude oil. Also, ligninolytic enzymes such as lac, MnP and LiP may contribute to the acceleration of crude oil biodegradation by attacking the aromatic, resin and asphaltene fractions. Placement of the initial position of the co-culture affected the degradation and enzyme production in different depth interval of soil. Generally, the activities decreased when the distance of soil from the initial place of co-culture increased. For example, when the co-culture was placed in the top position, activities of enzymes generally decrease from top to bottom layer.

Table 1. Enzyme activities of fungal co-culture in soil-buried crude oil.

Depth (cm)	Enzyme activities (U g^{-1})									
	Bottom					Top				
	C12O	C23O	Lac	MnP	LiP	C12O	C23O	Lac	MnP	LiP
1 (0 - 5)	135	20	103	31	30	248	49	1185	225	26
2 (5 - 15)	120	29	163	14	13	73	29	1411	92	15
3 (15 - 20)	145	49	1924	129	40	103	33	123	17	30

4. Conclusions

Both positions of fungal co-culture degraded the crude oil 63–92% after 30 days. The co-culture placed in the bottom layer position of the soil demonstrated more effective degradation for all the soil depth interval than the top layer position. During the biodegradation of the crude oil, the co-culture produced several oxidative enzymes such as catechol 1,2-dioxygenase, catechol 2,3-dioxygenase, laccase, manganese peroxidase, and lignin peroxidase with significant activities indicating the important role of these enzymes during the degradation. This study offers an important strategy to remediate crude oil contaminated soil, specifically for soil buried crude oil.

References

[1] Adekunle AA and Adebambo OA 2007 Petroleum hydrocarbon utilization by fungi isolated from *Detarium Senegalense* seeds *J. Am. Sci.* **3** 69–76
[2] Ojumu TV, Bello OO, Sonibare JA and Solomon BO 2004 Evaluation of microbial systems for bioremediation of petroleum refinery effluents in Nigeria. Afr. J. Biotechnol. 4 31–5
[3] National Research Council 2003 Oil in the Sea III: Inputs, Fates, and Effects. The National Academy of Science, The National Academies Press, Washington, D.C.
[4] Owens EH, Taylor E and Parker-Hall HA 2007 Spill Site Investigation in Environmental Forensic Investigations. In: Wang, Z., Stout, S.A. (Eds). Oil spill environment forensics, fingerprinting and source identification. Academic Press, Elsevier, United State of America.
[5] Li Q, Liu J & Gadd GM 2020 Fungal bioremediation of soil co-contaminated with petroleum hydrocarbons and toxic metals Appl. Microbiol. Biotechnol. https://doi.org/10.1007/s00253-020-10854-y
[6] Ayu KR, Hadibarata T, Toyama T, Tanaka Y and Mori K 2011 Bioremediation of Crude Oil by White Rot Fungi Polyporus sp. S133 J. Microbiol. Biotechno., 21 995–1000
[7] Margesin R and Schinner F 2001 Biodegradation and bioremediation of hydrocarbons in extreme environments Appl. Microbiol. Biotechnol. 56 650–63
[8] Balachandran C, Duraipandiyan V, Balakrishna K and Ignacimuthu S 2012 Petroleum and polycyclic aromatic hydrocarbons (PAHs) degradation and naphthalene metabolism in

Streptomyces sp. (ERI-CPDA-1) isolated from oil contaminated soil Bioresour. Technol. 112 83–90

[9] Al-Awadhi H, Dashti N, Kansour M, Sorkhoh N and Radwan S 2012 Hydrocarbon-utilizing bacteria associated with biofouling materials from offshore waters of the Arabian Gulf Int. Biodeterior. Biodegrad.. 69 10–6

[10] Yanto DHY and Tachibana S 2013 Biodegradation of petroleum hydrocarbons by a newly isolated Pestalotiopsis sp. NG007 Int. Biodeterior. Biodegrad 85 438–50

[11] Yanto DHY, Hidayat A and Tachibana S 2017 Periodical biostimulation with nutrient addition and bioaugmentation using mixed fungal cultures to maintain enzymatic oxidation during extended bioremediation of oily soil microcosms Int. Biodeterior. Biodegrad. 116 112–23

[12] Yanto DHY and Tachibana S 2014 Potential of fungal co-culturing for accelerated biodegradation of petroleum hydrocarbons in soil J. Hazard. Mater. 278 454–63

[13] Hidayat A and Yanto DHY 2018 Biodegradation and metabolic pathway of phenanthrene by a new tropical fungus, Trametes hirsuta D7 J. Environ. Chem. Eng. 6 2454–60

[14] Nakazawa T and Atsushi N 1970 Pyrocatechase (Pseduomonas). In: Colowick, S.P., Kaplan, N.O. (Eds.), Methods in Enzymology, vol. 17. Academic Press, New York. pp. 518–522. Part 1

[15] Nozaki M 1970 Metapyrocatechase (Pseudomonas). In: Colowick, S.P., Kaplan, N.O (Eds.), Methods in Enzymology, vol.17. Academic Press, New York, pp. 522–25

[16] Leonowicz A and Grzywnowicz K 1981 Quantitative estimation of laccase forms in some white rot fungi using syringaldazine as a substrate Enzyme Microbial Technol. 3 55–8

[17] Wariishi H, Valli K and Gold MH 1992 Manganese (II) oxidation by manganese peroxidase from the basidiomycete Phanerochaete chrysosporium-kinetic mechanism and role of chelators J. Biol. Chem. 267 23688–95

[18] Tien M and Kirk TK 1984 Lignin-degrading enzyme from Phanerochaete chrysosporium: Purification, characterization, and catalytic properties of a unique H2O2-requiring oxygenase Proc. Natl. Acad. Sci. 81 2280–4

[19] Meysami P, Baheri H 2003 Pre-screening of fungi and bulking agents for contaminated soil bioremediation Adv. Environ. Res. 7 881–7

AUTHOR INDEX